# The Weibull Distribution

## A Handbook

Horst Rinne

Justus-Liebig-University
Giessen, Germany

CRC Press
Taylor & Francis Group
Boca Raton London New York

CRC Press is an imprint of the
Taylor & Francis Group an **informa** business

A CHAPMAN & HALL BOOK

Chapman & Hall/CRC
Taylor & Francis Group
6000 Broken Sound Parkway NW, Suite 300
Boca Raton, FL 33487-2742

International Standard Book Number-13: 978-1-4200-8743-7 (Hardcover)

### Library of Congress Cataloging-in-Publication Data

Rinne, Horst.
   The Weibull distribution : a handbook / Horst Rinne.
       p. cm.
   Includes bibliographical references and index.
   ISBN 978-1-4200-8743-7 (alk. paper)
   1. Weibull distribution--Handbooks, manuals, etc. I. Title.

QA273.6.R56 2008
519.2'4--dc22                                                                    2008033436

**Visit the Taylor & Francis Web site at**
**http://www.taylorandfrancis.com**

**and the CRC Press Web site at**
**http://www.crcpress.com**

# Contents

# Preface

More than fifty years have passed since WALODDI WEIBULL presented "his" distribution to the international scientific community (see WEIBULL (1951)), but the origins extend back to the 1920s. The 1951 publication marked the start of triumphant progress of this distribution in statistical theory as well as in applied statistics. Hundreds of authors around the globe have contributed to its development and the number of publications has far exceeded expectations. I think the time has come to compile the findings that are distributed in dozens of scientific journals and hundreds of research papers. My thanks go to all those who in the past have worked on the WEIBULL distribution. This handbook could not have been written without these men and women.

The idea of writing a monograph on the WEIBULL distribution first came to me in the late 1960s when —at the Technical University of Berlin (West) — I was preparing my Habilitation's thesis on strategies of maintenance and replacement. Since that time I have carefully compiled what has been written on this distribution, and I hope that no major contribution has escaped my attention. I had done some research of my own, e.g., on the discovery of the WEIBULL distribution and on capability indices for WEIBULL distributed characteristics, and I had given lectures on WEIBULL analysis to students and to practitioners, but only by the end of my university career did I have the leisure to complete the work on this text.

This book is intended to be a general desk reference for all those people who have to model statistical data coming from various fields, but mainly from the life sciences and the engineering sciences. The reader should have a basic knowledge of calculus and probability theory as well as statistical theory, but all the subjects treated are first presented in a general setting before application to the WEIBULL distribution. The book is self-explanatory and self-contained. Each section starts with a detailed "suggested reading" giving a path to follow for any reader wishing to learn more. I also have included numerous illustrative examples and exercises that expand special topics. The text includes more than 90 tables and more than 100 figures; as an old Chinese proverb says: "A picture says more than thousand words." I have refrained from showing what parts of a WEIBULL analysis might be done by statistical software packages and from reproducing any program code. Instead in Chapter 7 I give hints about software dealing with WEIBULL applications. In preparing this text I have applied MATHEMATICA (developed by WOLFRAM RESEARCH, Inc.) to do algebraic calculations and formula manipulation, and GAUSS (developed by APTECH SYSTEMS, Inc.) to do numerical calculations.

The book is divided into three parts, but most important are the first two parts. Part I is based in the fields of probability theory and gives a careful and thorough mathematical description of the WEIBULL distribution and all its features. Chapter 1 starts with the fascinating history of its discovery and also reports on several physical explanations for this distribution. The WEIBULL distribution is mainly used to model lifetime and duration data, so I present in Chapter 2 those six functions that primarily serve to describe lifetime as a random variable, e.g., the failure density, the failure and reliability functions, the hazard rate and its integral and the mean residual life function. I show how these functions are affected by the three parameters (location, scale and shape parameters) of the WEIBULL. I further show how the moments and percentiles depend on these parameters. Chapter 3 shows how the WEIBULL fits into different systems of statistical distributions, how it is related to other familiar distributions and what kinds of modifications exist. Chapter 4 is devoted to WEIBULL processes and WEIBULL renewal theory, topics that are important in the context of reliability growth. Chapter 5 on order statistics is of utmost importance because lifetime data almost naturally arrive in ascending order. Chapter 6 reports on characterizations of the WEIBULL and deals with the question of finding assumptions that uniquely determine this distribution.

Part II, the core of this book, reports on WEIBULL analysis. The introductory Chapter 7 provides a survey of WEIBULL applications in tabular form and provides some statistical software packages and consulting corporations doing WEIBULL analysis. The inference process heavily depends on how the data have been sampled, so Chapter 8 reports on collecting lifetime data, with an emphasis on techniques such as censoring and acceleration by practicing stress to shorten test duration. Chapters 9 through 15 are devoted to the estimation of the three WEIBULL parameters using classical as well as BAYESIAN and further approaches, comprising graphical, linear maximum likelihood and miscellaneous techniques. The maximum likelihood method, presented in Chapter 11, is the most versatile and considers all types of censored data. I have also written a chapter on parameter estimation under accelerated life testing (Chapter 16) and on parameter estimation for mixed WEIBULL models (Chapter 17). Inference of WEIBULL processes is dealt with in Chapter 18. Knowledge of certain percentiles (median life or the upper and lower quartiles) and of the reliability to survive a given age is very important for all practitioners as is the prediction of future random quantities such as the time to the next failure or the number of failures within a future time span. Chapters 19 and 20 are devoted to these topics. WEIBULL parameter testing is presented in Chapter 21. Chapter 22 provides different types of goodness-of-fit tests and methods to discriminate between WEIBULL and other distributions and to select the better of several WEIBULL distributions.

Part III contains what is compulsory for a scientific book: lists of abbreviations and notations, author and subject indexes, a detailed bibliography and a table of the gamma, digamma, and trigamma functions that are often used in WEIBULL analysis.

It is difficult to write a book of this scope without the help and input of many people. PAUL A. SAMUELSON, the second winner of the NOBEL prize in economics in 1970, once said that each scientist stands upon the shoulders of his predecessors. First and foremost I am indebted to all those who in the past have worked on the WEIBULL distribution and increased knowledge of this distribution. Their contributions are listed in the bibliography.

Thanks are offered to my former secretary, INGE BOJARA, and one of my former students, STEFFEN RAU, for converting parts of my manuscript into LATEX files. I further acknowledge with thanks the help of my former assistant, DOROTHEA REIMER, who was responsible for the layout and the creation of the LATEX style file.

Special thanks are due to my niece, MONICA WOOD, and my grandnephew, SIMON WOOD, who are both citizens of the U.K. and native English speakers. Their help in polishing my "German" English is gratefully acknowledged.

I would like to acknowledge PAUL A. TOBIAS, who read an early version of my text and whose comments prompted me to incorporate some forgotten topics. BOB STERN of the TAYLOR & FRANCIS GROUP was a great help in bringing this text to publication. I thank the AMERICAN SOCIETY FOR QUALITY CONTROL, the AMERICAN STATISTICAL ASSOCIATION, IEEE, and MARCEL DEKKER, Inc., for their kind permission to reprint tables originally published in their journals and books.

I intended to finish this text soon after my retirement from office in 2005, but, unfortunately, my work was interrupted twice by severe illness. Without the surgical skill of Dr. PETER ROTH and Dr. PAUL VOGT of the Giessen University Clinics this book would not have been completed at all. Last but not least I want to express my deepest appreciation to my wife BRIGITTE who always encouraged me to continue with my research and to overcome the setbacks to my health.

Dr. HORST RINNE, Professor Emeritus of Statistics and Econometrics
Department of Economics and Management Science
Justus–Liebig–University, Giessen, Germany

# List of Figures

# List of Tables

# I

# Genesis, theory and description

# 1 History and meaning of the WEIBULL distribution

For more than half a century the WEIBULL distribution has attracted the attention of statisticians working on theory and methods as well as in various fields of applied statistics. Hundreds or even thousands of papers have been written on this distribution (see the bibliography in Part III) and the research is ongoing.[1] Together with the normal, exponential, $\chi^2-$, $t-$ and $F-$distributions the WEIBULL distribution is — without any doubt — the most popular model in modern statistics. It is of utmost interest to theory–orientated statisticians because of its great number of special features and to practitioners because of its ability to fit to data from various fields, ranging from life data to weather data or observations made in economics and business administration, in hydrology, in biology or in the engineering sciences (see Chapter 7).

The aim of this first section is to trace the development of the distribution from its early beginnings in the 1920s until 1951, when it was presented to a greater public by WEIBULL. We will also comment on the physical meaning and some interpretations of the WEIBULL distribution. An enumeration and an explanation of how and where it has been applied successfully is delayed until the introductory section of Part II.

## 1.1 Genesis of the WEIBULL distribution[2]

Quite often a scientific method, a procedure, a theorem or a formula does not bear the name of its true discoverer or its original author. A famous example is the GAUSS distribution, better known under the neutral name "normal distribution." The earliest published derivation [as a limiting form of the binomial probability distribution $\Pr(X = k) = \binom{n}{k} P^k (1 - P)^{n-k}$ for $P$ fixed and $n \to \infty$] seems to be that of ABRAHAM DE MOIVRE (1667 – 1754) in a pamphlet dated November 12, 1733 and written in Latin. An English translation of this pamphlet with some additions can be found in the second edition (1738) of his famous book entitled *The Doctrine of Chance; or a Method of Calculating the Probability of Events in Play*, dedicated to NEWTON. In 1774 PIERRE SIMON DE LAPLACE (1749 – 1827) obtained the normal distribution, as an approximation to the hypergeometric distribution and four years later he advocated the tabulation of the probability integral $\Phi(u) = \int_{-\infty}^{u} \left( \sigma \sqrt{2 \pi} \right)^{-1} \exp(-x^2/2) \, \mathrm{d}x$. Finally, in "Theoria Motus Corporum Coelesticum" ($\hateq$ Theory of motion of celestial bodies) from 1809 and in "Bestimmung der Genauigkeit von Beobachtungen" ($\hateq$ Determining the accuracy of observations) from 1816 CARL FRIEDRICH GAUSS (1777 – 1855) derived the distribution as a kind of error

---

[1]  Asking for references on the keyword "WEIBULL distribution" on the Internet, e.g., using *Google*, retrieves 119,000 entries.

[2]  Suggested reading for this section: HALLINAN (1993), RINNE (1995).

law governing the size of errors made when measuring astronomical distances.

The history and discovery of what is now known as the WEIBULL distribution is even more exciting than that of the normal distribution, which lasted 60 to 70 years. The key events in the derivation of the WEIBULL distribution took place between 1922 and 1943. This process is so astonishing because there were three groups of persons working independently with very different aims, thus forming a chain of three links. WALODDI WEIBULL (1887 – 1979) is the last link of this chain. The distribution bears his name for good reasons because it was he who propagated this distribution internationally and interdisciplinarily. Two of the approaches leading to this distribution are located in engineering practice and are more or less heuristic whereas the third approach, the oldest and purely scientific one, is based on theoretical reasoning and is located in research on statistical methods and probability theory.

### 1.1.1   Origins in science[3]

The WEIBULL distribution with **density function** (DF)[4]

$$f(x \mid a, b, c) = \frac{c}{b}\left(\frac{x-a}{b}\right)^{c-1} \exp\left\{-\left(\frac{x-a}{b}\right)^{c}\right\}; \quad x \geq a; \; a \in \mathbb{R}; \; b, c \in \mathbb{R}^{+}; \quad (1.1a)$$

**cumulative distribution function** (CDF)

$$F(x \mid a, b, c) := \int_{a}^{x} f(u \mid a, b, c)\,\mathrm{d}u = 1 - \exp\left\{-\left(\frac{x-a}{b}\right)^{c}\right\}; \quad (1.1b)$$

and **hazard rate** (HR)

$$h(x \mid a, b, c) := \frac{f(x \mid a, b, c)}{1 - F(x \mid a, b, c)} = \frac{c}{b}\left(\frac{x-a}{b}\right)^{c-1} \quad (1.1c)$$

is a member of the family of extreme value distributions. These distributions are the limit distributions of the smallest or the greatest value, respectively, in a sample with sample size $n \to \infty$.

For a finite sample of size $n$ with each sample variable $X_i$ ($i = 1, 2, \ldots, n$) being independently and identically distributed with DF $f(x)$ and CDF $F(x)$, we define the two statistics

$$Y_n := \min_{1 \leq i \leq n} \{X_i\} \quad (1.2a)$$

and

$$Z_n := \max_{1 \leq i \leq n} \{X_i\}. \quad (1.2b)$$

---

[3]  Suggested reading for this section: EPSTEIN (1960), GUMBEL (1954, 1958), LEADBETTER (1974), THEILER/ TÖVISSI (1976).

[4]  The functions given by (1.1a) to (1.1c) and their parameters will be discussed thoroughly in Sections 2.2 to 2.4.

The distribution of $Y_n$, the **sample minimum**, is found as follows:

$$\Pr(Y_n > y) \quad = \quad \Pr(X_i > y \ \forall \, i)$$

$$= \quad \Pr(X_1 > y, \ X_2 > y, \ \ldots, \ X_n > y)$$

$$= \quad \left[1 - F(y)\right]^n.$$

Therefore,

$$F_{Y_n}(y) := \Pr(Y_n \le y) = 1 - \left[1 - F(y)\right]^n \tag{1.3a}$$

is the CDF and

$$f_{Y_n}(y) = n \, f(y) \left[1 - F(y)\right]^{n-1} \tag{1.3b}$$

is the corresponding DF.

The CDF of $Z_n$, the **sample maximum**, is obtained as follows:

$$F_{Z_n}(z) := \Pr(Z_n \le z) \quad = \quad \Pr(X_i \le z \ \forall \, i)$$

$$= \quad \Pr(X_1 \le z, \ X_2 \le z, \ \ldots, \ X_n \le z)$$

$$= \quad \left[F(z)\right]^n. \tag{1.4a}$$

The corresponding DF is

$$f_{Z_n}(z) = n \, f(z) \left[F(z)\right]^{n-1}. \tag{1.4b}$$

The limiting distribution for $n \to \infty$ of both $Y_n$ and $Z_n$ is degenerated. So it is quite natural to investigate under which circumstances there will exist a non–trivial limiting extreme value distribution and what it looks like. The search for an answer started in the 1920s. The opening paper from 1922 is a fundamental contribution of LADISLAUS VON BORTKIEWICZ,[5] 1868–1931, on the range and the mean range in samples from the normal distribution as a function of the sample size.

He drew attention to the fact that the largest normal values are new variates having distributions of their own. He thus deserves credit for having clearly stated the problem. In the following year, 1923, RICHARD VON MISES, 1883–1953, introduced the fundamental notion of the expected largest value without using this term. This was the first step toward a knowledge of the asymptotic distribution for normal observations.

The first study of the largest value for other than the normal distribution was by E.L. DODD (1923). He was the first to calculate the median of the largest value. The next progress was by L.H.C. TIPPETT (1925) who calculated the numerical values of the probabilities for the largest normal value and the mean range.

The first paper based on the concept of a class of initial distributions was by MAURICE FRÉCHET, 1878–1973. He was also the first to obtain an asymptotic distribution of the largest value. FRÉCHET's paper of 1927, published in a remote Polish journal, never gained the recognition it merited. Because R.A. FISHER, 1890–1963, and L.H.C. TIPPETT published a paper in 1928 that is now referred to in all works on extreme values.

---

[5] Biographical and bibliographical notes on most of the personalities mentioned in this section can be found on the Internet under http://www-history.mcs.st-and.ac.uk/history/.

They found again, independently of FRÉCHET, his asymptotic distribution and constructed two other asymptotes. The key passage on the limiting distribution of the sample maximum reads as follows (Fisher/Tippett, 1928, p. 180):

> Since the extreme member of a sample $m \cdot n$ may be regarded as the extreme member of a sample $n$ of the extreme members of samples of $m$, and since, if a limiting form exists, both of these distributions will tend to the limiting form as $m$ is increased indefinitely, it follows that the limiting distribution must be such that the extreme member of a sample of $n$ from such a distribution has itself a similar distribution.
>
> If $P$ is the probability of an observation being less than $x$, the probability that the greatest of a sample of $n$ is less than $x$ is $P^n$, consequently in the limiting distribution we have the functional equation
>
> $$P^n(x) = P(a_n x + b_n),$$
>
> the solutions of this functional equation will give all the possible forms.

FISHER and TIPPETT as well as FRÉCHET started from a stability postulate, which means the following: Assume we have $m$ samples each of size $n$. From each sample the largest value is taken, and the maximum of the $m$ samples of size $n$ is, at the same time, the maximum in a sample of size $m \cdot n$. Therefore, FISHER and TIPPETT say, the distribution of the largest value in a sample of size $m \cdot n$ should be the same as the distribution of the largest value in a sample of size $n$, except for a linear transformation. This postulate is written in the above cited functional equation. FISHER and TIPPETT then continue to show that the solutions can be members of three classes only, writing (Fisher/Tippett, 1928, p. 182–183):

> The only possible limiting curves are therefore:
>
> $$I. \quad \mathrm{d}P = e^{-x-e^{-x}}\,\mathrm{d}x, \dots$$
> $$II. \quad \mathrm{d}P = \frac{k}{x^{k+1}}\, e^{-x^{-k}}\,\mathrm{d}x, \dots$$
> $$III. \quad \mathrm{d}P = k\,(-x)^{k-1}\, e^{-(-x)^k}\,\mathrm{d}x, \dots$$

This is exactly the order of enumeration which later led to the names of these classes: extreme value distribution (for the maximum) of type I, type II and type III.

Because of

$$\min_i\{X_i\} = -\left(\max_i\{-X_i\}\right), \tag{1.5}$$

the distributions of the asymptotically smallest and largest sample values are linked as follows, $Y$ being the minimum variable and $Z$ the maximum variable:

$$\Pr(Y \le t) = \Pr(Z \ge -t) = 1 - \Pr(Z \le -t), \tag{1.6a}$$

$$F_Y(t) = 1 - F_Z(-t), \tag{1.6b}$$

$$f_Y(t) = f_Z(-t). \tag{1.6c}$$

The $X_i$ are supposed to be continuous. Taking the notation of FISHER and TIPPETT, we thus arrive at the following pairs of extreme value distributions, giving their DFs.

Table 1/1:　Comparison of the extreme value densities

| | Maximum | Minimum | |
|---|---|---|---|
| Type I | $f(x) = \exp(-x - e^{-x}), \ x \in \mathbb{R}$ | $f(x) = \exp(x - e^x), \ x \in \mathbb{R}$ | (1.7a/b) |
| Type II | $f(x) = \dfrac{k}{x^{k+1}} \exp(-x^{-k}), \ x \geq 0$ | $f(x) = \dfrac{k}{(-x)^{k+1}} \exp\left[-(-x)^{(-k)}\right],$ $x \leq 0$ | (1.8a/b) |
| Type III | $f(x) = k\,(-x)^{k-1} \exp[-(-x)^k],$ $x \leq 0$ | $f(x) = k\,x^{k-1} \exp(-x^k), \ x \geq 0$ | (1.9a/b) |

The left–hand (right–hand) part of Fig. 1/1 shows the densities of the largest (smallest) sample value, where $k = 2$ has been used. The graph of the maximum–density is the reflection of the type–equivalent minimum–density, reflected about the ordinate axis.

Figure 1/1: Densities of the extreme value distributions

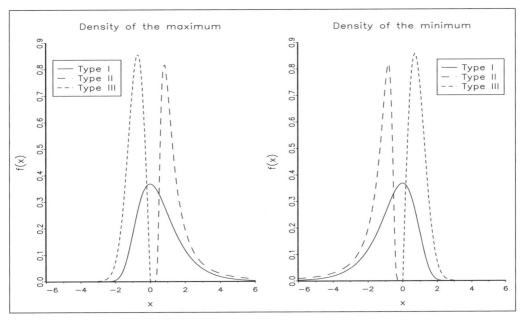

Comparing (1.9b) to (1.1a) when $a = 0$, $b = 1$ and $c = k$, one easily recognizes that the WEIBULL distribution is the extreme value distribution of type III for the sample minimum.[6] In their paper of 1928 FISHER and TIPPETT did not write down the formulas pertaining to the sample minimum because considering (1.6c), these formulas seemed to

---

[6] The derivation of the extreme value distribution rests on the assumption of independent sample values. GALAMBOS (1981) investigates what type of dependence might also lead to a WEIBULL distribution.

be obvious to them.[7] Furthermore, they did not think it would be necessary to general-ize the distributions by introducing a location parameter $a$ as well as a scale parameter $b$. The main part of their paper deals with the type–I–maximum distribution being the most general one of these distributions. Let $X_2$ have a type–II–maximum distribution, then $X = \ln\left(X_2^k\right) = k\,\ln(X_2)$ has the type–I–maximum distribution (1.7a). If $X_3$ has a type–III–maximum distribution, then $X = -\ln\left[\left(-X_3\right)^k\right] = -k\,\ln\left(-X_3\right)$ will also have the type–I–maximum distribution with DF (1.7a). That is why the type–I distribution is referred to as *the* extreme value distribution.[8] FISHER and TIPPETT further show that a type–I distribution appears if the sample comes from a normal distribution.

The general proof of what kind of parent distribution will lead to what type of extreme value distribution was given in 1943 by B.V. GNEDENKO, 1912–1995, based on a paper by R. VON MISES (1936). The resulting type of extreme value distribution depends on the behavior of the sampled distribution on the relevant side, i.e., on the left–hand (right-hand) side for the distribution of the minimum (maximum). The main results are:

- Type I will come up if the sampled distribution is unlimited toward the relevant side and is of **exponential type** on that side, meaning that $F(x)$, the CDF of the sampled distribution, is increasing toward unity with $x \to \infty$ (decreasing toward zero with $x \to -\infty$) at least as quickly as an exponential function. Prototypes are the exponential, normal and $\chi^2$–distributions.[9]

- Type II will come up if the sampled distribution has a range which is unlimited from below (for the minimum type) or unlimited from above (for the maximum type) and if its CDF is of **CAUCHY–type**. This means, that for some positive $k$ and $A$:

$$\lim_{x \to \infty} x^k \left[1 - F(x)\right] = A \text{ in case of a maximum or}$$

$$\lim_{x \to -\infty} (-x)^k F(x) = A \text{ in case of a minimum.}$$

The convergence of $F(x)$ or $\left[1 - F(x)\right]$ is slower than exponential. The prototype is the CAUCHY distribution itself.

- Type III will come up if the sampled distribution has a range which is bounded from above (for the maximum type) or bounded from below (for the minimum type), the bound being $x_0$.[10] Besides, $F(x)$ must behave like

$$\beta\,(x_0 - x)^\alpha \text{ for some } \alpha,\ \beta > 0 \text{ as } x \to x_0^- \text{ (case of a maximum) and like}$$

$$\beta\,(x - x_0)^\alpha \text{ for some } \alpha,\ \beta > 0 \text{ as } x \to x_0^+ \text{ (case of a minimum).}$$

---

[7] The title of their paper clearly tells us that they had in mind the sample minimum too.

[8] Another name is **doubly exponential distribution**.

[9] REVFEIM (1984) gives an interesting physical model — located in meteorology — which leads to the type–I distribution.

[10] We will revert to this approach in Sect. 1.2.1.

A prototype is the uniform distribution over some interval $[A, B]$.

We finally mention that a type–II distribution is called **FRÉCHET distribution**. A type–I distribution which has been studied extensively and applied frequently by E.J. GUMBEL, 1891 – 1966, see GUMBEL (1954, 1958), is called **GUMBEL distribution**.[11]

## 1.1.2  Origins in practice

In the 1930s we can locate two approaches in the engineering sciences that led to the WEIBULL distribution. These approaches are independent among themselves and independent of the mathematical/statistical papers on extreme values.

### 1.1.2.1  Grinding of material[12]– ROSIN, RAMMLER and SPERLING

An important branch in material sciences is the grinding and pulverization of solid material. There are a lot of technical applications for which the distribution of ground material according to the diameter of the "seed," normally in the $\mu m$-band, is of high interest. Examples are sand of quartz, cement, gypsum, coal or pulverized metals. The reaction time, the yield or the quality of a (chemical) process using this ground material depends on the size of the seed and its surface. Engineers have developed several methods to determine size distributions by experiment.

A very intuitive method is sieving. A given quantity $M$ of the material is poured on a set of sieves with standardized width of mesh $x_1 < x_2 < \ldots < x_t$. Then, the remainder $R_1$, $R_2$, $\ldots$, $R_t$ on each sieve is measured as a proportion of $M$. In most cases the relationship between $x$, the width of mesh ($\widehat{=}$ diameter of the seed), and the relative remainder $R$ can be described reasonably well by the following **formula**, first given by **ROSIN/RAMMLER/SPERLING** (1933):

$$R(x) = \exp\left[ - (x/b)^c \right]. \tag{1.10}$$

(1.10) is nothing but the complement of the WEIBULL cumulative distribution function in (1.1b) with $a = 0$. (1.10) has two parameters. $b$ is called the **finesse of grinding**, i.e., that width of mesh associated with a remainder $R = \exp(-1) \approx 0.3679$. Formally, $b$ is a scale parameter.[13] Parameter $c$ has no dimension and is a measure of the **uniformity of grinding**.

How did ROSIN, RAMMLER and SPERLING arrive at their formula? — RAMMLER (1937, p. 163) reports as follows:

---

[11] KOGELSCHATZ (1993) gives a biography of EMIL JULIUS GUMBEL. Gumbel's photo can be found under www.ub.heidelberg.de.

[12] Suggested reading for this section: STANGE (1953a,b).

[13] In another context when $x$ stands for lifetime, $b$ is termed **characteristic life** (see Sect. 2.2.1).

| Original German text | English translation |
|---|---|
| Wahrscheinlichkeitsbetrachtungen, die wir zunächst anstellten, lieferten für die Gewichtsverteilungskurve die Form $$y_G = f(x) = \frac{dD}{dx} = -\frac{dR}{dx} = a\,x^2\,e^{-b\,x^2}, \quad (4)$$ | We started from probabilistic reasoning[14] which led to the weight distribution of the form $$y_G = f(x) = \frac{dD}{dx} = -\frac{dR}{dx} = a\,x^2\,e^{-b\,x^2}, \quad (4)$$ |
| worin $a$ und $b$ Verteilungsfestwerte sind, ..., $y_G$ gibt bestimmungsgemäß den Gewichtsanteil aller Körner von der Korngröße $x$ an. Die praktische Nachprüfung zeigte, daß diese Gleichung keine allgemeine Gültigkeit beanspruchen konnte, sondern, daß die Exponenten nicht untereinander gleich und nicht gleich 2 sein konnten, während der allgemeine Aufbau offenbar zutraf. | $a$ and $b$ being fixed parameter values,[15] ..., according to the rules $y_G$ is the weight proportion of all seeds having size $x$. Practical verification showed that the formula could not claim general validity but that the exponents could not be equal to one another and could not be equal to 2 whereas the general structure of the formula seemed to be correct. |
| Auf verschiedenen Wegen vorgehend fanden schließlich Sperling einerseits, Rosin und ich andererseits annähernd gleichzeitig die gesuchte Gleichung der Korngrößenzusammensetzung. Sperling ging anschauungsgemäß von der häufigsten Verteilungsform ... aus und legte ihr die Beziehung $$f(x) = a\,x^m\,e^{-b\,x^n} \quad (5)$$ | Finally, working on different routes Sperling on the one hand and Rosin and I on the other hand found the equation of the composition of the seeds nearly at the same time.[16] Sperling — by visual perception — started from the most frequent form of a distribution[17] ... taking the equation $$f(x) = a\,x^m\,e^{-b\,x^n}. \quad (5)$$ |
| bei. Er führte dann einen der beiden Exponenten auf den anderen zurück, indem er $$m = n - 1 \quad (6)$$ | Then, he reduced one of the exponents to the other exponent by equating $$m = n - 1. \quad (6)$$ |
| setzte, womit die Gleichung in geschlossener Form integrierbar wird $$D = F(x) = \frac{a}{n\,b}\left(1 - e^{-b\,x^n}\right). \quad (7)$$ | This leads to an equation to be integrated in closed form $$D = F(x) = \frac{a}{n\,b}\left(1 - e^{-b\,x^n}\right). \quad (7)$$ |

| Original German text | English translation |
|---|---|
| Für $x_{max} = s$ wird $D = 100$, woraus man weiter erhält | $x_{max} = s$ leads to $D = 100$ and further to |
| $$a = \frac{100\,n\,b}{1 - e^{-b\,s^n}} \cdot \qquad (8)$$ | $$a = \frac{100\,n\,b}{1 - e^{-b\,s^n}} \cdot \qquad (8)$$ |
| Hierin kann $e^{-b\,s^n} \approx 0$ gesetzt werden. Dann ergibt sich | In (8) one can set $e^{-b\,s^n} \approx 0$. So the result is |
| $$D = F(x) = 100\left(1 - e^{-b\,x^n}\right), \qquad (9)$$ | $$D = F(x) = 100\left(1 - e^{-b\,x^n}\right), \qquad (9)$$ |
| $$R = 100 - F(x) = 100\,e^{-b\,x^n}, \qquad (10)$$ | $$R = 100 - F(x) = 100\,e^{-b\,x^n}, \qquad (10)$$ |
| $$f(x) = 100\,n\,b\,x^{n-1}\,e^{-b\,x^n}. \qquad (11)$$ | $$f(x) = 100\,n\,b\,x^{n-1}\,e^{-b\,x^n}. \qquad (11)$$ |
| Die Ableitung ist also spekulativer Natur, indem von einer der Hauptformen der Kornverteilungskurve erfahrungsgemäß angepaßten Gleichung ausgegangen wird … | The derivation is speculative in essence because it starts from an equation which is fitted to the main form of the distribution curve by experience … |

RAMMLER continues to describe the approach made by him and ROSIN giving the same result. Their heuristic and graphic–oriented approach is based on grinding experiments and their observation that the passage through the mesh — expressed as a function of the finesse of the sieve — behaved like a general parabola. After several algebraic manipulations the equation for $R$ is the same as (10). Further, RAMMLER (1937) reports on a paper by J.G. BENNETT (1936) based on their paper of 1933. The decisive part that slightly connects the WEIBULL distribution, heuristically discovered as the law governing the size distribution of ground material, and the WEIBULL distribution, theoretically derived as the limit distribution of the smallest sample value by FISHER and TIPPETT, reads as follows (RAMMLER, 1937, p. 164):

---

[14] The paper does not contain any details about this kind of reasoning. (Remark of the author)

[15] $D$ is the passage through the mesh and is thus the complement to $R$. (Remark of the author)

[16] At this point RAMMLER cites the paper from 1933 of ROSIN, RAMMLER and SPERLING. (Remark of the author)

[17] This is a density which is skewed to the right and has a mode not on the border of the domain. (Remark of the author)

| Original German text | English translation |
|---|---|
| Bennett hat nun versucht, durch physikalische und wahrscheinlichkeitstheoretische Betrachtungsweise des Zerkleinerungsvorgangs das praktische Ergebnis theoretisch zu unterbauen. Er geht dabei von der bekannten Überlegung aus, daß, da die Geschwindigkeit der Druckfortpflanzung sehr groß, verglichen mit der elastischen Deformation ist, der Bruch an Schwächestellen ohne Rücksicht auf ihre Lage, eintreten wird. | Bennett now tried to support the practical findings by physical and probabilistic reasoning. He starts from the well–known consideration that the breaking takes place at **weak points** (emphasis added) irrespective of their position, because the velocity of the propagation of pressure is very high compared to the elastic deformation. |

The weakness of material and its breaking are explicitly dealt with in the third and last approach undertaken by WEIBULL in 1939.

### 1.1.2.2 Strength of material[18]— WEIBULL

In 1939 the Swedish engineer WALODDI WEIBULL[19] published two reports on the strength of material in a series edited by the Royal Swedish Institute for Engineering Research. In "A Statistical Theory of the Strength of Material" (1939a) the distribution function is introduced, empirically based on observations experimentally obtained from tensile, bending and torsional tests on rods made of stearic acid and plaster–of–Paris. In "The Phenomenon of Rupture in Solids" (1939b) one can find the moments of this distribution along with graphical and tabular aids for estimation of the parameters. As with ROSIN/RAMMLER/SPERLINGS's approach WEIBULL's reasoning at the decisive part of his paper is empirical and heuristic and lacks theoretical argumentation. The relevant text in WEIBULL (1939a, p. 7) reads as follows:

> In a general manner, the distribution curve $S_1$ (Here is a typographical error: $S_1$ should be $S_\ell$.) for any length $\ell$ may be computed from the distribution curve $S_1$ for the unit of length according to the formula
> $$1 - S_\ell = (1 - S_1)^\ell \text{ or}$$
> $$\log(1 - S_1) = \ell \cdot \log(1 - S_1). \tag{4}$$
> (The left–hand side of (4) should be $S_\ell$.) In this case the volume $V$ of the stressed system is proportional to the length $\ell$, and if $S_0$ is the distribution curve for that length of rod which corresponds to the unit of volume, we have
> $$\log(1 - S) = V \cdot \log(1 - S_0). \tag{5}$$

---

[18] Suggested reading for this section: HALLINAN (1993), HELLER (1985), WEIBULL (1939a,b, 1949, 1951, 1952, 1961, 1967a,b), WEIBULL/ODQUIST (1956).

[19] Biographical notes may be found in the excursus "The life and œuvre of WALODDI WEIBULL" further down.

If we now put $B = -\log(1 - S)$ and call $B$ the "risk of rupture," we find that $B$ is proportional to the volume and to $\log(1 - S_0)$ which is a function of the tensile strength $\sigma$ alone .... In respect of such materials (WEIBULL thinks of isotropic materials) we find that the risk of rupture $\mathrm{d}B$ for a small volume element $\mathrm{d}v$ is determined by the equation

$$\mathrm{d}B = -\log(1 - S_0)\,\mathrm{d}v. \tag{6}$$

As has been mentioned in the above, $\log(1 - S_0)$ is a function of $\sigma$ only and is negative because $1 - S_0 < 1$. Hence we have

$$\mathrm{d}B = -n(\sigma)\,\mathrm{d}v. \tag{7}$$

If the distribution of stresses in the body is arbitrary, the risk of rupture is

$$B = \int n(\sigma)\,\mathrm{d}v \tag{8}$$

and the probability of rupture

$$S \equiv 1 - e^{-B} = 1 - e^{-\int n(\sigma)\,\mathrm{d}v}. \tag{9}$$

Notice the distinction made by WEIBULL between $B$, called "risk of rupture," and $S$ termed "probability of rupture." Nowadays, in life testing $B$ is known as the **cumulative hazard rate** (CHR), i.e., the integral of (1.1c) leading to $H(x) = [(x - a)/b]^c$, and (9) gives the well–known relation (see Table 2/1) $F(x) = 1 - \exp\left[-H(x)\right]$. Some pages later the text reads (WEIBULL, 1939a, p. 27/28):

The system of coordinates which is best suited to the computation of distribution curves follows from, and is motivated by, equation (9) according to which the probability of rupture is $S = 1 - e^{-\int n(\sigma)\,\mathrm{d}v}$. For a uniform distribution of stresses throughout the volume we have $S = 1 - e^{-V \cdot n(\sigma)}$ and, consequently

$$\log\log\frac{1}{1 - S} = \log n(\sigma) + \log V. \tag{76}$$

Hence it follows that if $\log\log\dfrac{1}{1 - S}$ is plotted as an ordinate and an arbitrary function $f(\sigma)$ as an abscissa in a system of rectangular coordinates, then a variation of the volume $V$ of the test specimen will only imply a parallel displacement, but no deformation of the distribution function. This circumstance, of course, facilitates the study, especially if the material function assumes the form

$$n(\sigma) = \left(\frac{\sigma - \sigma_u}{\sigma_0}\right)^m \tag{77}$$

and if we take $f(\sigma) = \log(\sigma - \sigma_u)$ as abscissa, since in that case we obtain

$$\log\log\frac{1}{1 - S} = m\,\log(\sigma - \sigma_u) - m\,\log\sigma_0 + V \tag{78}$$

[(78) contains another typographical error. It should read $\log V$ instead of $V$.] and the distribution will be linear.

If one sets $V = 1$ and re–transforms (78) one reaches — written with other symbols — the WEIBULL distribution function (1.1b) given at the beginning of Sect. 1.1.1.[20] WEIBULL demonstrated the empirical relevance of (78) by a great number of strength experiments with very different kinds of materials, e.g., porcelain, concrete, cotton, iron and aluminium. WEIBULL even describes how to estimate $\sigma_u$ from the graph mentioned along with (78). This method is "trial and error" and consists in varying the value of $\sigma_u$ until the curved cloud of points turns into a linearly ordered cloud of points, see the text commenting on Fig. 2/8.

After World War II WEIBULL in his publications from 1949, 1951, 1952, 1959 and 1961 presented "his" distribution to the international scientific community. The key paper is WEIBULL (1951). Unfortunately, when reproducing his law of distribution some typographical errors crept into his papers. Additionally, he used the same symbol with differing meanings in different papers.[21] So in the beginning of the WEIBULL analysis there was some confusion among the scientists. The NATO–sponsored monograph WEIBULL (1961) marks the beginning of a paradigmatic change in life testing. Up to the end of the 1950s lifetime in the engineering sciences was nearly always modeled by the exponential distribution, see EPSTEIN (1958), which then was gradually substituted by the more flexible WEIBULL distribution.

### Excursus:   The life and œuvre of WALODDI WEIBULL[22]

ERNST HJALMAR WALDODDI WEIBULL was born on June 18, 1887. In the 17th century his family immigrated into Sweden from Schleswig–Holstein, at that time closely connected to Denmark, but today a part of Germany.

He became a midshipmen in the Royal Swedish Coast Guard in 1904 and was promoted to sub-lieutenant in 1907, to Captain in 1916 and to Major in 1940. By then he had finished the military schools and simultaneously taken courses of the Royal Institute of Technology at Stockholm University, finally graduating as Fil. Lic. in 1924. He left active military service in 1917 and acted in German and Swedish industries as an inventor (ball and roller bearings, electric hammers) and as a consulting engineer, for example, with SAAB, an enterprize engaged in producing automobiles, trucks and arms (combat aircrafts).

He published his first scientific paper on the propagation of explosive waves in 1914. Taking part in expeditions to marine destinations all over the world on a research ship he could use his newly developed technique of explosive charges to determine the type of ocean bottom sediments and their thickness. This method is used today in offshore oil explorations.

WEIBULL became a full professor at the Royal Institute of Technology in 1924, and was awarded the degree of Ph.D.h.c. at the University of Uppsala in 1932. In 1941 a donation from the Swedish arms factory BOFORS gave him a personal research professorship in Technical Physics at the Royal

---

[20]  WEIBULL (1939b, p. 16/17) arrives at (1.1b) by starting from $S = 1 - \exp\left\{ -V \left( \dfrac{\sigma - \sigma_u}{\sigma_0} \right)^m \right\}$ and by introducing $\sigma_V := \sigma_0 \, V^{-1/m}$.

[21]  Details pertaining to these observations can be found in Sect. 2.2.4.

[22]  The greater part of this excursus rests upon R.A. HELLER (1985). Photos of WEIBULL can be found under www.york.ac.uk and www.bobabernethy.com.

Institute of Technology, Stockholm.

In 1953 WEIBULL retired from the Royal Institute of Technology and became professor emeritus. For most people retirement is the end of a professional career, but not for WEIBULL. His activities just started in this time. He became a consultant to the Fatigue Branch of the U.S. Air Force Material Laboratory. For 14 years he conducted research and wrote many papers and technical reports which provide valuable information and data on material properties and on the analysis of probability distributions. He conducted work on turbine fatigue and studied new methods of estimating the parameters of his distribution, WEIBULL (1967a,b). His work on the planning and interpretation of fatigue data is monumental and resulted in his book from 1961. In 1963, at the invitation of the late Professor ALFRED FREUDENTHAL, he became Visiting Professor at Columbia University's Institute for the Study of Fatigue and Reliability. Here he met E.J. GUMBEL.

In the course of his long and productive career, W. WEIBULL received many honors:
- the honorary doctorate from the University of Uppsala in 1932,
- the Polholm Medal in 1940,
- the ASME medal (American Society of Mechanical Engineers)[23] in 1972,
- the Great Gold Medal from the Royal Swedish Academy of Engineering Sciences in 1978, personally presented to him by the King of Sweden.

We do not hesitate to state that some people blamed WEIBULL for having been strongly engaged in the arms industry and military research. But as is widely known, inventions and knowledge gained in the military sphere turn out to be of value in everyday life, sooner or later.

W. WEIBULL did not only leave behind an enormous scientific œuvre but a great family too. He was proud of his nine children and numerous grand- and great–grandchildren. He worked to the last day of his remarkable life which completed on October 12, 1979, in Annecy, France.

## 1.2 Physical meanings and interpretations of the WEIBULL distribution[24]

There always exists at least one physical model behind each statistical distribution. This physical model helps to interpret the distribution and aids to identify and to select an appropriate distribution for a given set of empirical data. With respect to the WEIBULL distribution one can find several physical models that will be presented in the following sections.[25]

### 1.2.1 The model of the weakest link

Perhaps the oldest physical model leading to a WEIBULL distribution is connected with its extreme value origin. Given a physical system consisting of $n$ identical units or items

---

[23] The other recipient of the ASME medal in the same year was astronaut NEIL ARMSTRONG, the first man on the moon, who probably did not know that his successful voyage was partly due to the pioneering work of W. WEIBULL.

[24] Suggested reading for this section: EPSTEIN (1960), LEMOINE/WENOCUR (1985), MALIK (1975), REVFEIM (1984), STAUFFER (1979).

[25] There are early papers which deny the possibility of giving a physical meaning or interpretation to the WEIBULL distribution, e.g., GORSKI (1968).

connected in series, i.e. the system operates as long as all $n$ units operate, and it fails with the first failure of one of these units. An example of such a **series system** is a chain which is as strong as its weakest link or, in other words, the life of the chain ends with the shortest lifetime of a link.

Let $X_i$ $(i = 1, \dots, n)$ be the random lifetimes of the serially linked units and let the $X_i$ be independently and identically distributed with CDF $F(x)$. $Y_n$ denotes the lifetime of the series system which is given by

$$Y_n = \min_{1 \leq i \leq n} \{X_i\}.$$

The CDF of $Y_n$ has already been derived in (1.3a):

$$F_{Y_n}(y) := \Pr(Y_n \leq y) = 1 - \left[1 - F(y)\right]^n.$$

In general, it is unpleasant to work with this formula as it involves powers of $1 - F(y)$. One can avoid this for large $n$ by using a technique given in CRAMÉR (1971, p. 371). We define the random variable $U_n$ as

$$U_n := n \, F(Y_n). \tag{1.11a}$$

For any fixed $u$ in $[0, n]$ we have

$$
\begin{aligned}
\Pr(U_n \leq u) &= \Pr\left[n \, F(Y_n) \leq u\right] \\
&= \Pr\left[Y_n \leq F^{-1}\left(\frac{u}{n}\right)\right].
\end{aligned}
\tag{1.11b}
$$

Substituting into (1.3a) one easily derives the following distribution function of $U_n$:

$$G_n(u) = 1 - \left(1 - \frac{u}{n}\right)^n. \tag{1.11c}$$

As $n \to \infty$, the sequence of random variables $U_n$ converges in distribution to a random variable $U$ because the sequence of CDFs $G_n(u)$ converges for any $u$ to the CDF

$$G(u) := \lim_{n \to \infty} G_n(u) = 1 - e^{-u}, \; u \geq 0, \tag{1.11d}$$

with corresponding DF

$$g(u) := \lim_{n \to \infty} g_n(u) = e^{-u}, \; u \geq 0. \tag{1.11e}$$

It is clear from (1.11a) that the sequence of random variables $Y_n$ converges in distribution to a random variable $Y$ where

$$Y = F^{-1}\left(\frac{U}{n}\right). \tag{1.11f}$$

Hence, if we can compute $F^{-1}(\cdot)$, then we can determine the distribution of $U$.

We will now give three examples with different underlying distributions $F(x)$, all leading to a WEIBULL distribution for $Y_n$.

Example 1: $X_i$ $(i = 1, 2, \dots, n)$ is **uniformly distributed** in $[a, b]$.

The CDF of each $X_i$ is

$$F(x) = \begin{cases} 0 & \text{for} & x < a, \\ \dfrac{x-a}{b-a} & \text{for} & a \le x \le b, \\ 1 & \text{for} & x > b. \end{cases} \qquad (1.12\text{a})$$

(1.11a) becomes

$$U_n = n \frac{Y_n - a}{b-a}, \qquad (1.12\text{b})$$

so

$$Y_n \sim a + \frac{b-a}{n} U, \qquad (1.12\text{c})$$

where $\sim$ reads "is distributed as." We can state that for large $n$ the variate $Y_n$ is exponentially distributed with[26]

$$G_n(y) = 1 - \exp\left(-n \frac{y-a}{b-a}\right) \quad \text{for} \ a \le y \le b \qquad (1.12\text{d})$$

and

$$g_n(y) = \frac{n\,y}{b-a} \exp\left(-n \frac{y-a}{b-a}\right) \quad \text{for} \ a \le y \le b. \qquad ■ \\ (1.12\text{e})$$

Example 2: $X_i$ $(i = 1, 2, \ldots, n)$ is **exponentially distributed**, i.e., for each $X_i$ we have

$$F(x) = \begin{cases} 0 & \text{for} & x < a, \\ 1 - \exp\left(-\dfrac{x-a}{b}\right) & \text{for} & x \ge a. \end{cases} \qquad (1.13\text{a})$$

Here, (1.11a) turns into

$$U_n = n \left(1 - \exp\left[-\frac{Y_n - a}{b}\right]\right). \qquad (1.13\text{b})$$

Hence,

$$Y_n \sim a + b \ln\left[\frac{1}{1 - \dfrac{U}{n}}\right] = a + b \left[\frac{U}{n} + \left(\frac{U}{n}\right)^2 + \ldots\right]. \qquad (1.13\text{c})$$

Neglecting terms $(U/n)^k$ which are $\text{O}\left(\dfrac{1}{n}\right)$ for powers greater than one we have

$$Y_n \sim a + b \frac{U}{n}. \qquad (1.13\text{d})$$

Thus $Y_n$ is, for large $n$, exponentially distributed with CDF

$$G_n(y) = \begin{cases} 0 & \text{for} & y < a, \\ 1 - \exp\left[-n \dfrac{Y_n - a}{b}\right] & \text{for} & y \ge a. \end{cases} \qquad ■ \\ (1.13\text{e})$$

---

[26] The exponential distribution is a special case $(c = 1)$ of the WEIBULL distribution.

Example 3: $X_i$ $(i = 1, 2, \ldots, n)$ are supposed to be iid where the common distribution is the **power distribution** with

$$F(x) = \begin{cases} 0 & \text{for} & x < a, \\ \left(\dfrac{x-a}{b}\right)^c & \text{for} & a \leq x \leq a+b, \\ 1 & \text{for} & x > a+b, \end{cases} \tag{1.14a}$$

and

$$f(x) = \begin{cases} \dfrac{c}{b}\left(\dfrac{x-a}{b}\right)^{c-1} & \text{for} & a \leq x \leq a+b, \\ 0 & \text{elsewhere.} \end{cases} \tag{1.14b}$$

It follows from (1.11a) that

$$U_n = n\left(\frac{Y_n - a}{b}\right)^c. \tag{1.14c}$$

Hence,

$$Y_n \sim a + b\left(\frac{U}{n}\right)^{1/c}. \tag{1.14d}$$

So, for large $n$, $Y_n$ has the CDF

$$G_n(y) = 1 - \exp\left[-n\left(\frac{y-a}{b}\right)^c\right] \quad \text{for } a \leq y \leq a+b \tag{1.14e}$$

and the DF

$$g_n(y) = n\,c\left(\frac{y-a}{b}\right)^{c-1}\exp\left[-n\left(\frac{y-a}{b}\right)^c\right] \quad \text{for } a \leq y \leq a+b. \quad \blacksquare \tag{1.14f}$$

Limiting distributions of the kind obtained in the preceding three examples are called type–III asymptotic distributions of the smallest values. These distributions arise when two conditions are met.

- The range, over which the underlying density is defined, is bounded from below, i.e., $F(x) = 0$ for $x \leq x_0$ for some finite $x_0$.

- $F(x)$ behaves like $\beta\,(x - x_0)^\alpha$, for some $\alpha$, $\beta > 0$ as $x \to x_0^+$.

Under these conditions, $Y_n = \min_{1 \leq i \leq n}\{X_i\}$ is asymptotically distributed like the random variable $x_0 + (U/n\,\beta)^{1/\alpha}$. The associated CDF is

$$G_n(y) = \begin{cases} 0 & \text{for} & y < x_0, \\ 1 - \exp\left[-n\,\beta\,(y - x_0)^\alpha\right] & \text{for} & y \geq x_0. \end{cases} \tag{1.15}$$

### 1.2.2 Two models of data degradation leading to WEIBULL distributed failures

In some reliability studies, it is possible to measure physical degradation as a function of time, e.g., the wear of a disc break or a broken block. In other applications actual physical degradation cannot be observed directly, but measures of product performance degradation, e.g., the output of a device, may be available. Both kinds of data are generally referred to as **degradation data** and may be available continuously or at specific points in time where measurements are taken.

Most failures can be traced to an underlying degradation process. In some applications there may be more than one degradation variable or more than one underlying degradation process. Using only a single degradation variable the failure would occur when this variable has reached a certain critical level. MEEKER/ESCOBAR (1998, Chapter 13) give many examples and models of degradation leading to failure. Generally, it is not easy to derive the failure distribution, e.g., the DF or the CDF, from the degradation path model. These authors give an example leading to a lognormal failure distribution. We will give two examples of degradation processes which will result in the WEIBULL failure distribution.

A promising approach to derive a failure time distribution that accurately reflects the dynamic dependency of system failure and decay on the state of the system is as follows: System state or wear and tear is modeled by an appropriately chosen random process — for example, a diffusion process — and the occurrences of fatal shocks are modeled by a POISSON process whose rate function is state dependent. The system is said to fail when either wear and tear accumulates beyond an acceptable or safe level or a fatal shock occurs.

The **shot–noise model**[27] supposes that the system is subjected to "shots" or jolts according to a POISSON process. A jolt may consist of an internal component malfunctioning or an external "blow" to the system. Jolts induce stress on the system when they occur. However, if the system survives the jolt, it may then recover to some extent. For example, the mortality rate of persons who have suffered a heart attack declines with the elapsed time since the trauma. In this case, the heart actually repairs itself to a certain degree. The shot–noise model is both easily interpretable and analytically tractable.

The system wear and tear is modeled by a **BROWNIAN motion** with positive drift. The system fails whenever the wear and tear reaches a certain critical threshold. Under this modeling assumption, the time to system failure corresponds to the first passage time of the BROWNIAN motion to the critical level, and this first passage time has an **inverse GAUSSIAN distribution**, which is extremely tractable from the viewpoint of statistical analysis.

Based on the two foregoing models, an appropriate conceptual framework for reliability modeling is the following: Suppose that a certain component in a physical system begins operating with a given strength or a given operational age (e.g., extent of wear–and–tear or stress), denoted by $x$, that can be measured in physical units. Suppose that, as time goes on, component wear–and–tear or stress builds up (loss of strength with increasing age), perhaps in a random way. (The concept of wear–and–tear buildup is dual to that of

---

[27] A reference for this model and its background is COX/ISHAM (1980).

declining strength.) For instance, consider the thickness of tread on a tire. The tread wears down with use. In some cases this wear may be offset, but only in part, by maintenance and repair. Such considerations suggest modeling component strength (or susceptibility to failure) by a stochastic process $X = \{X(t),\ t \geq 0\}$ with starting state corresponding to the initial level of strength (or initial operational age) $x$. This process $X$ should tend to drift downward (decrease) with time as wear builds up; if $X$ is the operational age or wear–and–tear process, then it should tend to drift upward. The component may fail when either wear alone has reduced strength below some safe level or at the occurrence of some hazardous event (e.g., an external shock such as the tire's abrupt encounter with a sharp portion of road surface) severe enough to overcome current strength. We denote by $\tau$ the time of passage of the $X$ process to the critical level. It seems reasonable that the rate of fatal shocks should be modeled as a decreasing function of component strength or an increasing function of component wear–and–tear. We denote by $k(x)$ the POISSON killing rate associated with state $x$ and by $T$ the time to failure of the component. With the above conventions and modeling assumptions in place, we can express the probability of surviving beyond time $t$, starting with strength or operational age $x$, as follows:

$$P_x(T > t) = \mathrm{E}_x\left\{ \exp\left( -\int_0^t k\big[X(s)\big]\,\mathrm{d}s \right) I_{\{\tau > t\}} \right\}. \qquad (1.16a)$$

Suppose that component strength evolves in accordance with a **diffusion process** $X = \{X(t), t \geq 0\}$ having drift parameter $\mu(x)$ and diffusion coefficient $\sigma^2(x)$ in state $x > 0$. Then $\mu(x)$ can be interpreted as the rate at which wear and tear builds up in state $x$. (Alternatively, if component wear builds up according to $X$, then $\mu(x)$ can be interpreted as the rate at which strength declines in state $x$.) If $T$ is the time to failure of the component, we assume that

$$\Pr\Big(T \leq h \,|\, T > s,\ X(s) = x\Big) = k(x)\,h + \mathrm{o}(h)\ \forall\ s,$$

so that $k(x)$ can be interpreted as the POISSON rate of occurrence of a traumatic shock of magnitude sufficient to overcome, or "kill," a component of strength $x$.

Now let $w(x,t)$ be the probability that a component of strength $x$ survives beyond time $t$, that is

$$w(x,t) = P_x(T > t). \qquad (1.16b)$$

Indeed, $w(x,t)$ coincides with the right–hand side of (1.16a) if $\tau$ is the first passage time of the diffusion $X$ to the critical level. It follows from the backward differential equation for the KAC functional of the diffusion process $X$ that $w(x,t)$, satisfies

$$\frac{\partial w(x,t)}{\partial t} = -k(x)\,w(x,t) + \mu(x)\,\frac{\partial w(x,t)}{\partial t} + \frac{\sigma^2(x)}{2}\,\frac{\partial^2 w(x,t)}{\partial x^2}. \qquad (1.16c)$$

The initial condition for this differential equation is determined by the critical strength or wear threshold $\Delta$, allowing $0 \leq \Delta \leq +\infty$. If the diffusion process represents strength, then

$$w(x,t) = \begin{cases} 1 & \text{if } x > \Delta, \\ 0 & \text{otherwise.} \end{cases}$$

If it represents wear or stress, then

$$w(x,t) = \begin{cases} 1 & \text{if } x < \Delta, \\ 0 & \text{otherwise.} \end{cases}$$

The solution of (1.16c) is not always easy to find. Depending on the infinitesimal parameters $\mu(x)$ and $\sigma^2(x)$, on the killing rate function $k(x)$ and on the failure threshold $\Delta$, the solution sometimes may be given in explicit form and in other cases we have to use a computational algorithm. Different examples are given by LEMOINE/WENOCUR (1985).

We will present a subclass of the model described leading to some well–known lifetime distributions, the WEIBULL distribution being one of them. In this subclass the buildup of wear (or loss of strength) is assumed to be deterministic, and thus $\sigma^2(x) = 0$. Under this assumption, the system state process $X = \{X(t),\ t \geq 0\}$ satisfies the following equation

$$X(t) = x + \int_0^t \mu\big[X(s)\big]\,\mathrm{d}s. \tag{1.17a}$$

$X$ may also be expressed by the differential equation

$$\mathrm{d}X(t) = \mu\big[X(s)\big]\,\mathrm{d}t \tag{1.17b}$$

with initial condition $X(0) = x$. The probability of system survival through time $t$ is then given by

$$\Pr(T > t) = \exp\left\{-\int_0^t k\big[X(s)\big]\,\mathrm{d}s\right\} I_{\{\tau > t\}}. \tag{1.18a}$$

(1.18a) can be interpreted in terms of standard reliability theory. To simplify the notation, suppose for now that $\Delta = +\infty$ and $x = 0$ and write $\Pr(T > t)$ in place of $w(x,t)$. First, we need to recall the hazard rate function $h(t)$, which is usually defined by

$$h(t) = \frac{-\mathrm{d}\Pr(T > t)/\mathrm{d}t}{\Pr(T > t)}, \quad \text{see Table 2/1 and (2.4).}$$

Integrating $h(\cdot)$ over $[0, t]$ and assuming $\Pr(T = 0) = 0$ show that

$$\int_0^t h(s)\,\mathrm{d}s = -\ln\big[\Pr(T > t)\big].$$

Consequently,

$$\exp\left[-\int_0^t h(s)\,\mathrm{d}s\right] = \Pr(T > t). \tag{1.18b}$$

Comparing (1.18a) and (1.18b), we see that

$$h(t) = k\big[X(t)\big]. \tag{1.18c}$$

Therefore, the failure rate at time $t$ is equal to the killing rate that corresponds to the "system state" at time $t$.

Moreover, equations (1.17b) and (1.18c) enable us to reinterpret reliability distributions in terms of a killing rate function and a system evolution process. We will only investigate the WEIBULL distribution (1.1a–c) with $a = 0$:

$$\Pr(T > t) = \exp\left[-\left(\frac{t}{b}\right)^c\right], \quad t \geq 0. \tag{1.19a}$$

Then

$$k\big[X(t)\big] = h(t) = \frac{\mathrm{d}(t/b)^c}{\mathrm{d}t} = \frac{c}{b}\left(\frac{t}{b}\right)^{c-1}. \tag{1.19b}$$

Suppose that $c \neq 1$. Taking $k(x) = x^{c-1}$ gives

$$X(t) = t\left[c\left(\frac{1}{b}\right)^c\right]^{1/(c-1)}. \tag{1.19c}$$

Setting $\delta := \left[c\left(\frac{1}{b}\right)^c\right]^{1/(c-1)}$ will simplify (1.19c) to

$$X(t) = \delta\, t. \tag{1.19d}$$

This implies that the wear rate function is constant, i.e.

$$\mu(x) = \delta. \tag{1.19e}$$

Thus a component's lifetime has a WEIBULL distribution when the wear rate is constant and independent of the state and the killing rate is a non–trivial power function of the state. When the failure threshold $\Delta$ is finite, we will get a truncated WEIBULL distribution.

If $c = 1$, i.e., the WEIBULL distribution turns into an exponential distribution, then $k\big[X(t)\big] = (1/b)^c \;\forall t$. Thus, either the killing rate function is constant or the wear rate is zero or both. So, the exponential distribution as a failure model is a strong assumption.

### 1.2.3   The hazard rate approach

Engineers are interested in studying "wearing styles" and these may be expressed by a suitable chosen hazard rate $h(t)$. Setting $h(t) = \beta t$, $\beta > 0$, $t \geq 0$ would mean that the hazard rate varies only linearly, so we make it more general to cover the non–linear case by choosing a power function:

$$h(t) = \beta\, t^\gamma,\ \gamma > 0. \tag{1.20a}$$

Now,

$$h(t) = \frac{f(t)}{R(t)},\ \ \text{see (2.4)},$$

and

$$R(t) = \exp\left(-\frac{\beta}{\gamma+1}\, t^{\gamma+1}\right),\ \text{see Table 2/1}. \tag{1.20b}$$

Define

$$\eta := \gamma + 1\ \text{ and }\ \theta := \frac{\beta}{\gamma+1}\,;$$

then

$$
\begin{aligned}
R(t) &= \exp(-\theta\, t^\eta)\\
f(t) &= \eta\, \theta\, t^{\eta-1}\, \exp\!\left(\theta\, t^\eta\right)
\end{aligned}
\tag{1.20c}
$$

is the WEIBULL distribution in (1.1a–b) written with another set of parameters. Thus, WEIBULL's parameter $\eta$ and $\theta$ do have a physical meaning when looked at in terms of $\beta$ and $\gamma$, the parameters of the basic hazard rate model.

Now we modify the hazard rate of (1.20a) to

$$h(t) = \beta\big[(p\,t)^{\gamma_1} + (\overline{p}\,t)^{\gamma_2}\big], \tag{1.21a}$$

where $p$ is the fraction $(0 \le p \le 1)$ of working time and $\gamma_1,\ \gamma_1 > 1$ is the corresponding intensity of wear. $\overline{p} := 1 - p$ is the portion of time when the device is not busy but still ages with intensity $\gamma_2,\ 1 \le \gamma_2 < \gamma_1$. For this type of intermittently working device, we can find $f(t)$ and $R(t)$ from $h(t)$ of (1.21a). After substituting:

$$
\left.
\begin{aligned}
\eta_i &:= \gamma_i + 1\\
\theta_i &:= \gamma_i\, \beta/(\gamma_i + 1)
\end{aligned}
\right\} i = 1, 2,
$$

we have

$$f(t) = \Big[\eta_1\, \theta_1\, (p\,t)^{\eta_1-1} + \eta_2\, \theta_2\, (\overline{p}\,t)^{\eta_2-1}\Big]\, \exp\!\left(-\theta_1\, p^{\eta_1-1}\, t^{\eta_1}\right)\, \exp\!\left(-\theta_2\, \overline{p}^{\,\eta_2-1}\, t^{\eta_2}\right). \tag{1.21b}$$

We now look at the two limiting cases. When the device is working constantly $(p = 1)$, the general density (1.21b) turns into

$$f(t) = \eta_1\, \theta_1\, t^{\eta_1-1}\, \exp\!\left(-\theta_1\, t^{\eta_1}\right)$$

with $\eta_1$ as the new intensity of wear. Let $p = 0$ (the device is not working at all but is still subject to aging), then

$$f(t) = \eta_2\,\theta_2\,t^{\eta_2-1}\,\exp\!\big(-\theta_2\,t^{\eta_2}\big)$$

with $\eta_2$ as the new intensity of aging.

Finally, we modify (1.21a) to

$$h(t) = \beta\big[(p\,t)^\gamma + \overline{p}\big]. \tag{1.22a}$$

Physical interpretation of (1.22a) is that the device wears when it works; otherwise it has a constant hazard rate. (1.22a) generates, after proper substitution,

$$f(t) = \big[\theta_2\,\overline{p} + \eta_1\,\theta_1\,(p\,t)^{\eta_1-1}\big]\,\exp\!\big(-\overline{p}\,\theta_2\,t\big)\,\exp\!\big(-\theta_1\,p^{\eta_1-1}\,t^{\eta_1}\big). \tag{1.22b}$$

### 1.2.4   The broken-stick model

Another physical interpretation of the WEIBULL distribution not related to failure is given by STAUFFER (1979) in the context of forestry, where it gives good fit to the tree–diameter frequency distribution, see GREEN et al. (1993). This approach shows that the WEIBULL distribution may be used as a size distribution too.

The derivation starts with the so–called broken stick model which was proposed by MACARTHUR (1957, 1960) for the non–overlapping allocation of a critical resource among competing species. He suggested that a limiting resource should be expected to be shared among competing species in amounts proportional to the expected sizes of the pieces of a randomly broken stick of unit length. If there are $k$ pieces (= species) arranged in decreasing order, the expected size of the $i$–th piece is given by

$$\mathrm{E}(X_i) = \frac{1}{k}\sum_{j=i}^{k}\frac{1}{j}. \tag{1.23}$$

Assuming the abundances, i.e., the number of individuals, of the species are proportional to these resource allocations, the broken–stick model yields a ranked–abundance list representing the ranking of the species in order of decreasing abundance or cardinality. If $k$ is large, ranked–abundance lists can be converted into species–abundance curves by letting the abundance $X$ be a continuous random variable with DF $f(x)$ and CDF $F(x)$. Then,

$$\begin{aligned}
F(x) &= \Pr(X \le x) \\
     &= 1 - \Pr(X > x) \\
     &= 1 - \frac{i}{k},
\end{aligned} \tag{1.24a}$$

where $i$ is the number of the species corresponding to $x$ on the ranked–abundance list.

Equating the expected value $E(X_i)$ with $x_i$,

$$x_i = \frac{1}{k} \sum_{j=i}^{k} \frac{1}{j} \tag{1.24b}$$

$$\approx \frac{1}{k} \int_i^k \frac{1}{t} \, dt, \quad \text{for } k \text{ being large,}$$

$$= \frac{1}{k} \ln\left(\frac{k}{i}\right),$$

so that

$$\frac{i}{k} \approx e^{-k x_i}.$$

Hence,

$$F(x) = 1 - e^{-k x}$$

and

$$f(x) = k \, e^{-k x}.$$

This is the exponential distribution, a special case of the WEIBULL distribution (1.1a,b) where $a = 0$, $c = 1$ and $b = 1/k$.

In the context of forestry the analogues to species and abundances are the trees and the diameter of the trees. It will now be shown how the broken–stick model can be generalized to produce a ranked–diameter allocation list yielding a WEIBULL tree–diameter curve.

We look at the WEIBULL distribution in a slightly reparameterized version with $a = 0$:

$$f(y) = \frac{\gamma}{\beta} y^{\gamma-1} \exp\left(-y^\gamma/\beta\right), \tag{1.25a}$$

$$F(y) = 1 - \exp\left(-y^\gamma/\beta\right). \tag{1.25b}$$

Comparing (1.25a,b) with (1.1a,b) we have the following parameter relations:

$$\gamma = c; \quad \beta = b^c.$$

Now we reverse the procedure which led to the derivation of the exponential–abundance curve from the broken–stick model, applying it here to the WEIBULL distribution

$$F(y) = 1 - \Pr(Y > y)$$

$$= 1 - \frac{i}{k}$$

from the ranked–abundance list. Hence,

$$\frac{i}{k} = \exp\left(-y_i^\gamma/\beta\right).$$

Solving for $y_i$ we find

$$
\begin{aligned}
y_i &= \left[\beta \ln(k/i)\right]^{1/\gamma} \\
&= \left[\beta \int_i^k \frac{1}{t}\, dt\right]^{1/\gamma} \\
&\approx \left[\beta \sum_{j=i}^k \frac{1}{j}\right]^{1/\gamma}.
\end{aligned}
$$

Note that $y_i$ can be expressed in terms of the broken–stick model (1.24b):

$$
\begin{aligned}
y_i &= (\beta k)^{1/\gamma}\left[\frac{1}{k}\sum_{j=i}^k \frac{1}{j}\right]^{1/\gamma} \\
&= (\beta k)^{1/\gamma}\, x_i^{1/\gamma}.
\end{aligned}
$$

This generalization of the broken–stick model produces a WEIBULL tree–diameter curve. So $\beta$ and $\gamma$ effect the perturbation of the broken–stick abundances.

STAUFFER (1979) gives the following interpretations (with proofs) of the parameters $\beta$ and $\gamma$.

- The reciprocal of $\gamma$ acts as an index of non–randomness for the diameters (or abundances):
    - when $1/\gamma = 1$, the diameters are random as described by the original broken–stick model;
    - when $1/\gamma < 1$, the diameters become more regular or uniform;
    - when $1/\gamma > 1$, the diameters become more aggregated or dissimilar.

  Thus, $\gamma$ is related to the numerical differences between the abundances or diameters.

- The parameter $\beta$ is inversely proportional to the total frequency of allocations $k$. It can be shown that
  $$
  \beta = const./k,
  $$
  where $const.$ depends on $\gamma$ and that
    - when $\gamma = 1$, $const. = 1$, so that $\beta = 1/k$;
    - when $\gamma < 1$, $const. > 1$ and $\beta > 1/k$;
    - when $\gamma > 1$, $const. < 1$ and $\beta < 1/k$.

# 2 Definition and properties of the WEIBULL distribution

This chapter gives a complete description of the WEIBULL distribution in the sense of probability theory. Several functions representing a WEIBULL random variable will be discussed in their dependence on the three parameters attached to this distribution. Moments, moment ratios, generating functions, percentiles and other characteristic features are analyzed too.

## 2.1 Functions describing lifetime as a random variable

The time span, for which a natural unit is alive or for which a technical unit is functioning, generally is not predetermined or fixed; instead it has to be regarded as a random variable $X$ called **lifetime**. Generally this variable is continuous and non–negative.

There exist several functions which completely specify the distribution of a random variable. In the context of lifetime, six mathematically equivalent functions have evolved:

- the failure density,
- the failure distribution,
- the reliability function,
- the hazard rate,
- the cumulative hazard rate, and
- the mean residual life function.

Each of these functions completely describes the distribution of a random lifetime, and any one of them unequivocally determines the other five as may be seen from Tab. 2/1.

These six formulas are not the only possible ways to represent a random variable $X$. Other representations include the moment generating function $\mathrm{E}[\exp(t\,X)]$, the characteristic function $\mathrm{E}[\exp(i\,t\,X)]$, the MELLIN transform $\mathrm{E}(X^t)$, the density quantile function $f[F^{-1}(P)]$, and the total time on test transform $\int_0^{F^{-1}(P)} R(x)\,\mathrm{d}x$ for $0 \leq P \leq 1$, where $F^{-1}(P)$ is the inverse cumulative distribution function. Some of these functions will be discussed along with their application to the WEIBULL distribution in subsequent sections.

1. The density function (abbreviated by DF) $f(x)$ satisfying

$$f(x) \geq 0 \ \forall \, x \ \text{ and } \int\limits_{-\infty}^{+\infty} f(x)\,\mathrm{d}x = 1$$

is called **failure density** and gives the chance of a unit to fail or to die at the age of $x$. The **unconditional probability** of a newly born or newly produced unit to die or to fail around the age of $x$ is given by

$$\Pr(x - \Delta x/2 < X \le x + \Delta x/2) \approx f(x)\,\Delta x; \quad \Delta x \text{ small.} \qquad (2.1a)$$

The unconditional probability to reach an age between times $x_\ell$ and $x_u$, $x_\ell < x_u$, is given by

$$\Pr(x_\ell < X \le x_u) = \int_{x_\ell}^{x_u} f(x)\,\mathrm{d}x. \qquad (2.1b)$$

Normally a failure density is skewed to the right.

2. The cumulative distribution function (CDF) is called **life distribution** or **failure distribution**:

$$F(x) := \Pr(X \le x) = \int_0^x f(z)\,\mathrm{d}z. \qquad (2.2)$$

$F(x)$ gives the probability of failing up to time $x$ or having a life span of at most $x$. $F(x)$ is a non–increasing function of $x$ satisfying

$$\lim_{x \to -\infty} F(x) = 0 \quad \text{and} \quad \lim_{x \to +\infty} F(x) = 1.$$

3. The complementary cumulative distribution function (CCDF) is called **survival (survivor) function** or **reliability function**:

$$R(x) := \Pr(X > x) = \int_x^\infty f(z)\,\mathrm{d}z. \qquad (2.3)$$

$R(x)$ is the probability of surviving an age of $x$. $R(x)$ is a non–decreasing function of $x$ satisfying

$$\lim_{x \to -\infty} R(x) = 1 \quad \text{and} \quad \lim_{x \to +\infty} R(x) = 0.$$

4. The **hazard rate** or **instantaneous failure rate** (HR; also known as **rate function** or **intensity function**) is defined as

$$h(x) := \lim_{\Delta x \to 0} \frac{\Pr(x < X \le x + \Delta x \mid X > x)}{\Delta x} = \frac{f(x)}{R(x)} \qquad (2.4)$$

and satisfies

$$h(x) \ge 0 \;\forall\, x \quad \text{and} \quad \int_0^\infty h(x)\,\mathrm{d}x = \infty.$$

$h(x)\,\Delta x$ ($\Delta x$ small) is the approximate probability of an $x$–survivor to fail immediately after having reached the age of $x$. In contrast to (2.1a) this is a **conditional**

**probability**; the condition "$X > x$" means survivorship of $x$. The reciprocal of the HR is known as **MILL's ratio**. In reliability work $h(x)$ is popular because it has the intuitive interpretation as the amount of risk associated with an item which has survived to time $x$. In demography and actuarial science $h(x)$ is known as the **force to mortality**.[1] In lifetables $h(x)$ is approximated by the probability that a person of age $x$ will die within the following year. Some notions of aging refer to the HR:

$$h'(x) = \frac{\mathrm{d}h(x)}{\mathrm{d}x} \begin{cases} > 0 \ \forall\, x \ \text{ means } \textbf{positive aging}; \\ = 0 \ \forall\, x \ \text{ means } \textbf{no aging}; \\ > 0 \ \forall\, x \ \text{ means } \textbf{negative aging}. \end{cases}$$

5. The **cumulative hazard rate** $H(x)$, CHR for short, is given by

$$H(x) := \int_0^x h(z)\,\mathrm{d}z. \tag{2.5}$$

CHR is a non–decreasing function of $x$ satisfying $H(0) = 0$ and $\lim_{x\to\infty} H(x) = \infty$. $H(x)$ is not normalized to the unit interval $[0,1]$. Because $H(x) = -\ln[R(x)]$, $H(x)$ has an exponential distribution with a mean of one. Thus $H^{-1}[-\ln(1-Y)]$ generates a random number for Monte Carlo simulation when $Y$ is uniformly distributed on $[0,1]$. CHR parallels the renewal function from renewal theory, see Sect. 4.4.

6. The function

$$\mu(x) := \mathrm{E}(X - x \mid X > x) \quad = \quad \frac{\int_x^\infty z\,f(z)\,\mathrm{d}z}{R(x)} - x$$

$$= \quad \frac{\int_x^\infty R(z)\,\mathrm{d}z}{R(x)}, \text{ if } \lim_{x\to\infty} x^2 f(x) = 0 \tag{2.6}$$

is the **mean residual life** (MRL) of an $x$–survivor or the additional expected life of an item aged $x$. It satisfies the following three conditions given by SWARTZ (1973):

$$\mu(x) \geq 0; \quad \frac{\mathrm{d}\mu(x)}{\mathrm{d}x} \geq -1; \quad \int_0^\infty \frac{1}{\mu(x)}\,\mathrm{d}x = \infty.$$

---

[1] If $X$ is discrete, the hazard rate is defined as

$$h(x_k) := \frac{\Pr(X = x_k)}{\sum_{j=k}^\infty \Pr(X = x_j)}, \quad k \in \mathbb{Z}.$$

The **probable life** or **total expected life** of an item having survived to $x$ is given by $x + \mu(x)$. Especially,

$$\mu(0) = \mathrm{E}(X) = \mu = \int\limits_0^\infty R(z)\,\mathrm{d}z \qquad (2.7)$$

is the mean age of a new item.

Table 2/1 shows the relationships among the six functions defined above.

## 2.2  Failure density

The distribution, originally introduced by WEIBULL in 1939, depends on three parameters. The corresponding DF will be discussed in Sect. 2.2.1, whereas the densities having two parameters or one parameter only will be presented in Sect. 2.2.2. The properties of the reduced WEIBULL density, a special one–parameter version to which all other densities may be linked easily, are analyzed thoroughly in Sect. 2.2.3.

### 2.2.1  Three-parameter density

A random variable $X$ has a three–parameter WEIBULL distribution with parameters $a$, $b$ and $c$ if its DF is given by

$$f_X(x \mid a,b,c) = \frac{c}{b}\left(\frac{x-a}{b}\right)^{c-1} \exp\left\{-\left(\frac{x-a}{b}\right)^c\right\}, \quad x \geq a. \qquad (2.8)$$

This is the most general form of the classical WEIBULL distribution; newer forms with more than three parameters are presented in Sect. 3.3.7. The fact that the distribution of $X$ is given by (2.8) is noted as

$$X \sim We(a,b,c)$$

for short.

The first parameter $a$ is defined on $\mathbb{R}$, i.e., $-\infty < a < +\infty$, and is measured in the same unit as the realization $x$ of $X$, normally a unit of time (*sec., min., h., day, month*). In the context of $X$ being a lifetime $a$ is called **delay**, **guarantee time**, **minimum life**, **safe life**, **shelf age**, more generally it is termed **origin** or **threshold**. So the domain of support for $f_X(x \mid a.b, c)$ is $x \geq a$. For $x$ being a duration which normally cannot be negative and for $a$ being the minimum duration, the domain of $a$ will not be $\mathbb{R}$ but the smaller interval $[0, \infty)$.

From a statistical point of view $a$ is a **location parameter**. Changing $a$ when the other parameters are held constant will result in a parallel movement of the density curve over the abscissa (see Fig. 2/1, where $F_X(x \mid a, 1, 2)$ is depicted for $a = 0$, $0.5$ and $1$). Enlarging (reducing) $a$ causes a movement of the density to the right (to the left) so that $a$ is called **shift parameter** or **translation parameter** too. Formally the shifting is expressed as

$$f_X(x \mid a,b,c) = f_X(x - \delta \mid a + \delta, b, c) \qquad (2.9a)$$

with the special case

$$f_X(x \mid a,b,c) = f_X(x - a \mid 0, b, c). \qquad (2.9b)$$

Table 2/1:   Relations among the six functions describing stochastic lifetime

| to \ from | $f(x)$ | $F(x)$ | $R(x)$ | $h(x)$ | $H(x)$ | $\mu(x)$ |
|---|---|---|---|---|---|---|
| $f(x)$ | — | $\displaystyle\int_0^x f(z)\,dz$ | $\displaystyle\int_x^\infty f(z)\,dz$ | $\dfrac{f(x)}{\displaystyle\int_x^\infty f(z)\,dz}$ | $-\ln\left\{\displaystyle\int_x^\infty f(z)\,dz\right\}$ | $\dfrac{\displaystyle\int_0^\infty z\,f(x+z)\,dz}{\displaystyle\int_x^\infty f(z)\,dz}$ |
| $F(x)$ | $F'(x)$ | — | $1-F(x)$ | $\dfrac{F'(x)}{1-F(x)}$ | $-\ln\{1-F(x)\}$ | $\dfrac{\displaystyle\int_x^\infty [1-F(z)]\,dz}{1-F(x)}$ |
| $R(x)$ | $-R'(x)$ | $1-R(x)$ | — | $\dfrac{-R'(x)}{R(x)}$ | $-\ln[R(x)]$ | $\dfrac{\displaystyle\int_x^\infty R(z)\,dz}{R(x)}$ |
| $h(x)$ | $h(x)\exp\left\{-\displaystyle\int_0^x h(z)\,dz\right\}$ | $1-\exp\left\{-\displaystyle\int_0^x h(z)\,dz\right\}$ | $\exp\left\{-\displaystyle\int_0^x h(z)\,dz\right\}$ | — | $\displaystyle\int_0^x h(z)\,dz$ | $\dfrac{\displaystyle\int_x^\infty \exp\left\{-\int_0^z h(v)\,dv\right\}dz}{\exp\left\{-\displaystyle\int_0^x h(z)\,dz\right\}}$ |
| $H(x)$ | $-\dfrac{d\{\exp[-H(x)]\}}{dx}$ | $1-\exp\{-H(x)\}$ | $\exp\{-H(x)\}$ | $H'(x)$ | — | $\dfrac{\displaystyle\int_x^\infty \exp\{-H(z)\}\,dz}{\exp\{-H(x)\}}$ |
| $\mu(x)$ | $\dfrac{1+\mu'(x)}{\mu^2(x)}\times\mu(0)\times{}$ $\times\exp\left\{-\displaystyle\int_0^x \dfrac{1}{\mu(z)}\,dz\right\}$ | $1-\dfrac{\mu(0)}{\mu(x)}\times{}$ $\times\exp\left\{-\displaystyle\int_0^x \dfrac{1}{\mu(z)}\,dz\right\}$ | $\dfrac{\mu(0)}{\mu(x)}\times{}$ $\times\exp\left\{-\displaystyle\int_0^x \dfrac{1}{\mu(z)}\,dz\right\}$ | $\dfrac{1}{\mu(x)}\{1+\mu'(x)\}$ | $\ln\left\{\dfrac{\mu(x)}{\mu(0)}\right\}+{}$ $+\displaystyle\int_0^x \dfrac{1}{\mu(z)}\,dz$ | — |

Figure 2/1: WEIBULL densities with differing values of the location parameter

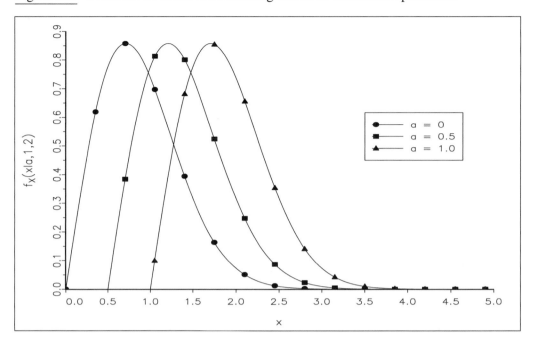

The second parameter $b$ in (2.8) has the domain $(0, \infty)$ and is measured in the same unit as $x$. When the random variable $X$ is a lifetime, $b$ is called **characteristic life** for it is typical that the CDF of all Weibull variates with the same $a$ and $b$ but varying $c$ will intersect at the point with coordinates $x = a + b$ and $F_X(a + b \,|\, a, b, c) \approx 0.6321$ (cf. Sect. 2.3) i.e., the chance of surviving the minimum life $a$ by $b$ units of time is roughly 63.21%. From the statistical point of view $b$ is a **scale parameter**. Changing $b$ while $a$ and $c$ are held constant will alter the density at $x$ in the direction of the ordinate (see Fig. 2/2 where $F_X(x \,|\, 0, b, 2)$ is depicted for $b = 0.5$, 1 and 2). Enlarging $b$ will cause a compression or reduction of the density and reducing $b$ will magnify or stretch it while the scale on the abscissa goes into the opposite direction. This means that a growing (shrinking) $b$ will cause the variation of $X$ to become greater (smaller). Formally the scaling function of $b$ can be seen in

$$f_X(x \,|\, a, b, c) = \delta \, f_X([x - a]\,\delta + a \,|\, a, b\,\delta, c), \quad \delta > 0, \tag{2.10a}$$

with the special case

$$f_X(x \,|\, 0, b, c) = \frac{1}{b} \, f_X\left(\frac{x}{b} \,\Big|\, 0, 1, c\right). \tag{2.10b}$$

Taken together the parameters $a$ and $b$ cause a linear transformation of a lifetime $x$ into a number $u$ without any dimension:

$$x \xrightarrow{a,b} \frac{x - a}{b} =: u.$$

Starting with $f_X(x \,|\, 0, 1, c)$ the combination of (2.9a) and (2.10a) results in the well–known relation between the DF of a random variable and its linear transformation:

$$f_X(x \,|\, a, b, c) = \frac{1}{b} \, f_X\left(\frac{x - a}{b} \,\bigg|\, 0, 1, c\right). \tag{2.11}$$

Figure 2/2: WEIBULL densities with differing values of the scale parameter

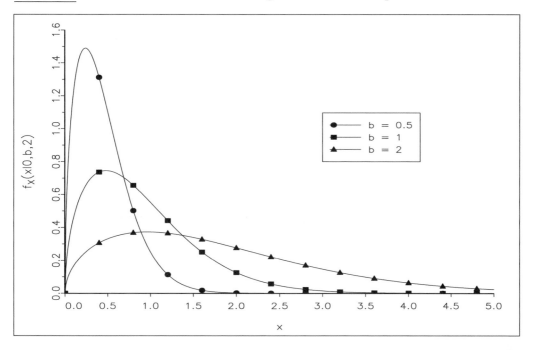

The third parameter in (2.8) has the domain $(0, \infty)$ and bears no dimension. It is called **WEIBULL–slope**, because it gives the slope of the CDF when graphed on WEIBULL–probability–paper (see Sect. 2.3) or the slope of the CHR graphed on WEIBULL–hazard–paper, (see Sect. 2.5). From the statistical point of view $c$ is a **form parameter** or **shape parameter**. Varying $c$ causes the graph of the DF to change its form which can be seen in Fig. 2/3. There $f_X(x \mid 0, 1, c)$ is depicted for $c = 0.5$, 1, 2, 3.6 and 6.5. The shape parameter is responsible for the appearance of a WEIBULL density. When $c < 1$, the exponential part of the density (2.8) dominates, and the curve is J–shaped. When $c > 1$, the effect of the polynomial part of the density becomes more pronounced, and the density curve becomes skewed unimodal. As each WEIBULL density can be derived from this special version by means of (2.11), it will be sufficient to study the behavior of this one–parameter case depending on $c$ only.

Looking at (2.8) one recognizes that the three parameters $a$, $b$ and $c$ perform a joined linear and power transformation of $x$:

$$x \xrightarrow{a,b,c} \left( \frac{x-a}{b} \right)^c = u^c.$$

The relation between an arbitrary three–parameter WEIBULL density and the special density with $a = 0$, $b = 1$ and $c = 1$ is given by

$$f_X(x \mid a, b, c) = \frac{c}{b} \left( \frac{x-a}{b} \right)^{c-1} f_X \left( \left[ \frac{x-a}{b} \right]^c \,\Big|\, 0, 1, 1 \right). \tag{2.12}$$

Finally it is shown that (2.8) really is a DF, meaning that the integral of (2.8) over $[a, \infty)$ is normalized to one:

$$
\int_a^\infty f_X(x \mid a, b, c)\,\mathrm{d}x \;=\; \int_a^\infty \frac{c}{b}\left(\frac{x-a}{b}\right)^{c-1}\exp\left\{-\left(\frac{x-a}{b}\right)^c\right\}\,\mathrm{d}x
$$

$$
=\; \int_0^\infty c\,u^{c-1}\exp\{-u^c\}\,\mathrm{d}u = \left[-\exp\{-u^c\}\right]_0^\infty
$$

$$
=\; \lim_{u\to\infty}\left(-\exp\{-u^c\}\right) + \exp\{-0^c\} = 1.
$$

The result is achieved by substitution:

$$
u = \frac{x-a}{b} \quad\text{and}\quad \mathrm{d}u = \frac{1}{b}\,\mathrm{d}x.
$$

Figure 2/3: WEIBULL densities with differing values of the shape parameter

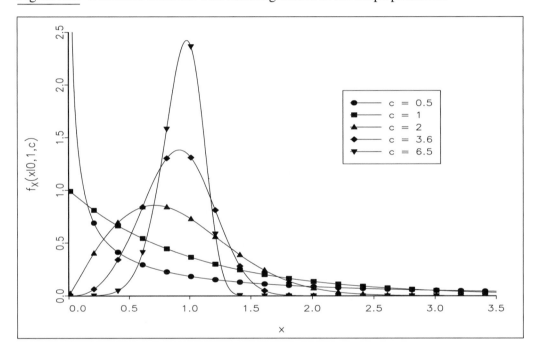

### 2.2.2  Two- and one-parameter densities

Neutralizing the parameters $a$, $b$ and $c$ by setting $a = 0$ and/or $b = 1$ and/or $c = 1$ results in three two–parameter distributions and three one–parameter distributions. Not all of them are of equal importance in application.

The densities of the two–parameter versions are

$$
f_X(x \mid 0, b, c) \;=\; \frac{c}{b}\left(\frac{x}{b}\right)^{c-1}\exp\left\{-\left(\frac{x}{b}\right)^c\right\}, \tag{2.13}
$$

$$
f_X(x \mid a, 1, c) \;=\; c\,(x-a)^{c-1}\exp\left\{-(x-a)^c\right\}, \tag{2.14}
$$

$$
f_X(x \mid a, b, 1) \;=\; \frac{1}{b}\exp\left\{-\left(\frac{x-a}{b}\right)\right\}. \tag{2.15}
$$

Density curves belonging to (2.13) – (2.15) can be seen in Figures 2/1 to 2/3.

(2.13) is by far the most often found two–parameter WEIBULL distribution, called the **scale–shape–version**. The reason is that items normally start to fail after the age of $x = 0$. If $a \neq 0$ but known, one can work with (2.9b) so that the shifted variate $X - a$ can be modeled by (2.13). Density (2.14), the **location–shape–version**, is unscaled, but shifted only. Finally (2.15), the **shift–scale–version** is nothing but the generalized exponential distribution, (see Sect. 3.2.1).

The densities of the one–parameter distributions are

$$f_X(x \,|\, 0, 1, c) \;=\; c\,x^{c-1}\,\exp(-x^c), \tag{2.16}$$

$$f_X(x \,|\, 0, b, 1) \;=\; \frac{1}{b}\,\exp\left(-\frac{x}{b}\right), \tag{2.17}$$

$$f_X(x \,|\, a, 1, 1) \;=\; \exp\{-(x - a)\}. \tag{2.18}$$

Corresponding curves can be found in Figures 2/1 to 2/3 also.

The functions (2.17) and (2.18) are nothing but special cases of the generalized exponential distribution given by (2.15), either only scaled or only shifted. So $c = 1$ always leads to an exponential distribution. (2.16), the shape–version, is called **reduced** or **normalized WEIBULL density**. It will be analyzed in Sect. 2.2.3.

The reduced or normalized form within a distribution family, having, among other parameters, a location parameter and a scale parameter, results by setting the location parameter equal to zero and the scale parameter equal to one. Each member of such a family emerges from the reduced form by a simple linear transformation. If $U$, the **reduced variable**, has the density $f_U(u)$, then

$$X = a + b\,U$$

has the density

$$f_X(x) = \begin{cases} \dfrac{1}{b}\, f_U\left(\dfrac{x - a}{b}\right) & \text{for } b > 0, \\[2ex] -\dfrac{1}{b}\, f_U\left(\dfrac{x - a}{b}\right) & \text{for } b < 0. \end{cases}$$

In most cases it will be sufficient to derive results pertaining to the reduced form from which the results for the general location–scale–form may be found by re–transformation. In the following sections this idea is applied to the WEIBULL distribution.

It is to be noticed that, in general, **standardizing** a variate $X$ to

$$Y := \frac{X - \mathrm{E}(X)}{\sqrt{\mathrm{Var}(X)}}$$

giving $\mathrm{E}(Y) = 0$ and $\mathrm{Var}(Y) = 1$ does not result in the reduced form unless $\mathrm{E}(X) = a$ and $\sqrt{\mathrm{Var}(X)} = b$, as is the case with the normal distribution.

### 2.2.3 Analysis of the reduced WEIBULL density[2]

In this section the shape of the reduced density

$$f_U(u \mid c) = c\, u^{c-1} \exp(-u^c); \quad c > 0, \; u \geq 0 \tag{2.19}$$

in its dependence on $c$ will be investigated, especially the curvature (convex $\hat{=}$ concave upward) or concave ($\hat{=}$ concave downward)), the mode and the points of inflection.

The behavior of $f_U(u \mid c)$ with $u \to 0$ or $u \to \infty$ is a follows:

$$\lim_{u \to 0} f_U(u \mid c) \;=\; \begin{cases} \infty & \text{for} \;\; 0 < c < 1, \\ 1 & \text{for} \;\; c = 1, \\ 0 & \text{for} \;\; c > 1, \end{cases}$$

$$\lim_{u \to \infty} f_U(u \mid c) \;=\; 0 \; \forall\, c > 0.$$

The first and second derivatives of (2.19) with respect to $u$ are

$$f_U'(u \mid c) \;=\; c\, u^{c-2} \left(c - 1 - c\, u^c\right) \exp(-u^c), \tag{2.20a}$$

$$f_U''(u \mid c) \;=\; c\, u^{c-3}\left\{2 + c\left[c - 3 - 3(c-1)u^c + c\, u^{2c}\right]\right\} \exp(-u^c). \tag{2.20b}$$

The possible **extreme values** of $f_U(u \mid c)$ are given by the roots $u^*$ of $f_U'(u \mid c) = 0$:

$$u^* = \left(\frac{c-1}{c}\right)^{1/c}, \tag{2.21a}$$

where the density is

$$f_U(u^* \mid c) = c\left(\frac{c-1}{c\, e}\right)^{(c-1)/c}. \tag{2.21b}$$

The possible **points of inflection** result from $f_U''(u \mid c) = 0$, a quadratic equation in $v = c\, u^c$ with roots

$$u^{**} = \left[\frac{3(c-1) \pm \sqrt{(5c-1)(c-1)}}{2c}\right]^{1/c}, \tag{2.22a}$$

where the density is

$$f_U(u^{**} \mid c) = c\left[\frac{3(c-1) \pm \sqrt{(5c-1)(c-1)}}{2c}\right]^{(c-1)/c} \exp\left\{-\frac{3(c-1) \pm \sqrt{(5c-1)(c-1)}}{2c}\right\}. \tag{2.22b}$$

When studying the behavior of $f_U(u \mid c)$, it is appropriate to make a distinction of six cases regarding $c$.

---

[2] Suggested readings for this section: FAREWELL/PRENTICE (1977), HAHN/GODFREY/RENZI (1960), LEHMAN (1963), PLAIT (1962).

First case: $0 < c < 1$

Since $u$ must be positive, one can see from (2.21a) that there is no extreme value $u^*$. From
(2.22a) it is to be noticed that $u^{**}$ is negative for $c \geq 1/5$ and imaginary if $1/5 < c < 1$,
i.e. there is no inflection point as well. The relations

$$\lim_{u \to 0} f_U(u \mid c) = \infty, \qquad\qquad \lim_{u \to \infty} f_U(u \mid c) = 0,$$

$$f'_U(u \mid c) < 0 \quad \text{for} \quad u > 0, \qquad f''_U(u \mid c) > 0 \quad \text{for} \quad u > 0$$

mean that the density comes down from positive infinity at $u = 0$, decreases monotonically
and is convex. An example for this case is $f_X(x \mid 0, 1, 0.5)$ in Fig. 2/3.

Second case: $c = 1$

Here (2.19) is the density of the reduced exponential distribution having a left–sided maximum $u^* = 0$ with $f_U(0 \mid 1) = 1$. As in the first case the density is decreasing, monotone
and convex (see $f_X(x \mid 0, 1, 1)$ in Fig. 2/3).

Third case: $1 < c \leq 2$

Under this circumstance (2.21a) has an admissible solution[3] $u^* \in \left(0, \sqrt{0.5}\,\right]$ for which
(2.21b) is negative, i.e., a genuine maximum or mode will exist. The ordinate value
$f_U(u^* \mid c)$ grows as $u^*$ moves from $0^+$ to $\sqrt{0.5}$ with $c$ going from $1^+$ to 2. Concerning
inflection points there will be only one which is to the right of $u = \sqrt{0.5}$ as is seen from
(2.22a) when the positive sign is used.[4] Because of

$$f_U(0 \mid c) = f_U(\infty \mid c) = 0 \quad \text{for} \quad c > 1,$$

the density is not monotone, it rises to a mode and then falls. The density is concave over
$(0, u^{**})$, where $f''_U(u \mid c) < 0$, and convex over $(u^{**}, +\infty)$, where $f''_U(u \mid c) > 0$. An
example of this case is $f_X(x \mid 0, 1, 2)$ in Fig. 2/3.

Fourth case: $c > 2$

Studying (2.21a) shows again a critical point $u^* < 1$ which after substituting into (2.20b)
gives $f''_U(u^* \mid c) < 0$; thus, there is a mode. But (2.22a) now has two positive values, one
to the left and the other one to the right of the mode. The left (right) point of inflection
$u^{**}_\ell$ ($u^{**}_r$) results from (2.22a) when the negative (positive) sign is used. As $f''_U(u \mid c) < 0$
over $(u^{**}_\ell, u^{**}_r)$, the density is convex whereas it is concave to the left of $u^{**}_\ell$ and to the right
of $u^{**}_r$. The graphs of $f_X(x \mid 0, 1, 3.6)$ and $f_X(x \mid 0, 1, 6.5)$ in Fig. 2/3 depict the behavior
in this case.

Fifth case: $c \to 0$

The first derivative (2.20a) tells that the density has a great negative slope in the near vicinity to the right of $u = 0$. The density becomes still steeper with $c \to 0$. For $u \gg 0$ the
density goes to zero as does its slope. So for $c \to 0$ the density takes the form of the letter
L and moves toward the axes of the coordinate system (see Fig. 2/4). In the limit $c = 0$ the
density concentrates at $u = 0$, meaning that all items of such a population have any life at
all.

---

[3] LEHMAN (1963, p. 34) is wrong when he states that $u^* \in (0, 1)$.

[4] The negative sign leads to a real, but non–positive value of $u^{**}$ which does not count.

Figure 2/4:  Reduced WEIBULL densities with $c \to 0$

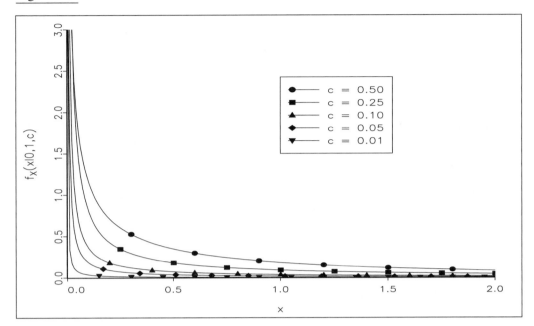

<u>Sixth case:</u>  $c \to \infty$

It can be seen from (2.21a) that the mode $u^*$ moves very slowly to the right and approaches $u^*_{\max} = 1$ as $c \to \infty$. The density at $u^*$ gradually becomes $c/e$ which in turn grows without limits with $c$. Thus the ordinate at the mode becomes infinite (see Fig. 2/5). The left–hand inflection point, which comes into existence for $c \geq 2$ and is always to the left of $u^*$, moves to the right according to

$$u^{**}_\ell \approx \left[ \left( 3 - \sqrt{5} \right) \Big/ 2 \right]^{1/c} \approx 0.3820^{1/c}, \quad c \gg 2.$$

In the limit $(c \to \infty)$ the left–hand inflection point and the mode coincide.

The right–hand inflection point follows an interesting path (see Fig. 2/6). After coming into existence, i.e., for $c > 1$, it is always to the right of the mode $u^*$, e.g., having the values

$$u^{**}_r = 0 \text{ for } c \to 1^+; \quad u^{**}_r = 1 \text{ for } c = \sqrt{2};$$

$$u^{**}_r = \sqrt{3/2} \approx 1.2247 \text{ for } c = 2; \quad u^{**}_r = 1 \text{ for } c \to \infty.$$

So as $c$ grows, $u^{**}_r$ moves to the right for a while and then returns approaching 1 which is the limit of the mode $u^*$ for $c \to \infty$. To find the value of $c$ which provides the rightmost inflection point, (2.22a) has to be differentiated with respect to $c$ and to be equated to 0. For that purpose one defines

$$g(c) := \left[ 3\,(c-1) + \sqrt{(5\,c-1)\,(c-1)} \right] \Big/ (2\,c)$$

so that

$$k(c) := g(c)^{1/c} = u^{**}_r.$$

Figure 2/5: Reduced WEIBULL densities with $c \to \infty$

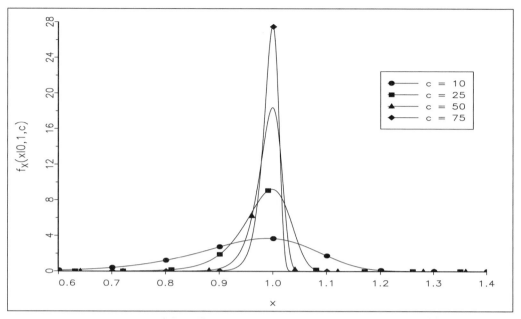

It is advisable to determine $\mathrm{d}k(c)/\mathrm{d}c = \mathrm{d}u_r^{**}/\mathrm{d}c$ using the formula of logarithmic derivation leading to

$$k'(c) = k(c) \left\{ \frac{g'(c)}{g(c)} \frac{1}{c} - \frac{\ln[g(c)]}{c^2} \right\}.$$

Setting $k'(c)$ equal to zero makes $k(c)$ maximal. $k'(c) = 0$ leads to

$$c\,g'(c) = g(c)\,\ln[g(c)] \tag{2.23}$$

because $c > 0$ and $k(c) > 0$. The numerical solution of (2.23) is $c \approx 2.4625$, so $u_r^{**} \approx 1.2451$ is the farthest right inflection point with density $f_U(1.2451 \,|\, 2.4625) \approx 0.3218$.

In the limit $(c \to \infty)$ the density concentrates at $u = 1$ meaning that all units of the corresponding population live until 1 and then perish simultaneously.

Fig. 2/6 summarizes all findings on the previous pages by showing in the left picture the simultaneous behavior of mode and inflection points as a function of $c$. The corresponding picture on the right traces the movement of the densities at these points.

The right–hand picture shows a crossing of the functions, depicting the densities at $u_\ell^{**}$ and $u_r^{**}$. This happens for $c \approx 3.447798$ giving $u_\ell^{**} \approx 0.57750732$ and $u_r^{**} \approx 1.21898219$ and density $f \approx 0.77351174$. For $c < 3.447798$ the density at $u_\ell^{**}$ is smaller than at $u_r^{**}$ leading to a positively skewed density ($\hat{=}$ skewed to the right). The opposite happens for $c > 3.447798$ leading to a negatively skewed density ($\hat{=}$ skewed to the left). From densities of equal height at the left and right inflection points one might conclude that for this special value of $c$ the density must be symmetric. But this is not true for two reasons:

Figure 2/6: Movement of mode and inflection points and their corresponding densities

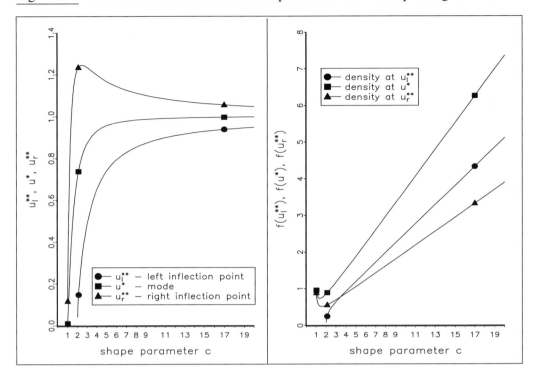

- Left and right inflection points with equal densities and equal distance to the mode are necessary but not sufficient for symmetry.

- Here, the inflection points having equal densities do not have equal distances to the mode $u^* = 0.905423260$. The left–hand inflection point has distance $0.32791628$ while the right–hand inflection point is $0.31355850$ away from the mode. By the way, irrespective of $c$ the left–hand inflection always has a greater distance to the mode than the right–hand inflection point.

So for $c \approx 3.447798$ the WEIBULL density is approximately symmetric. Further inquiry into the **skewness** will be done in Sect. 2.8.1, using the positions of mode and median; in Sect. 2.9.2 with regard to mean, median and mode; and finally in Sect. 2.9.4, using moment ratios.

The results pertaining to the general WEIBULL variable $X = a + b\,U$, $a \neq 0$ and/or $b \neq 1$, are

Mode:

$$x^* = a + b \left( \frac{c-1}{c} \right)^{1/c}, \quad c \geq 1. \tag{2.24}$$

Inflection points:

$$x_r^{**} = a + b \left[ \frac{3\,(c-1) + \sqrt{(5\,c-1)\,(c-1)}}{2\,c} \right]^{1/c}, \quad c > 1; \qquad (2.25a)$$

$$x_\ell^* = a + b \left[ \frac{3\,(c-1) - \sqrt{(5\,c-1)\,(c-1)}}{2\,c} \right]^{1/c}, \quad c > 2. \qquad (2.25b)$$

<u>$c \to 0$</u>: The density concentrates at $x = a$.

<u>$c \to \infty$</u>: The density concentrates at $x = a + b$.

### 2.2.4  Differing notations[5]

A look at the numerous papers on the WEIBULL distribution reveals a lot of differing notations, and readers working with several papers may easily become confused. Especially the scale parameter and the shape parameter are involved. This sections only enumerates the presentations of the DF, but all other functions and characteristic features are affected too. The reader might rewrite the latter if needed using the parameter relations given here. Primarily, there are two reasons for having differing notations:

- Some forms are easier to handle in typesetting.
- Other forms facilitate the construction of estimators and test functions or the mathematical manipulation of the distribution.

WEIBULL himself used differing notations of his distribution. The presentation (2.8), which will be used throughout this book, is in accordance with WEIBULL's first papers (1939a,b), albeit he used other symbols:

- $\sigma$ for the variable $x$,[6]
- $\sigma_u$ for the shift parameter $a$,
- $\sigma_o$ for the scale parameter $b$, and
- $m$ for the shape parameter $c$.

Following World War II, WEIBULL (1949) extended the application of his distribution from strength phenomena to cyclic fatigue phenomena. His new presentation used the slightly modified form

$$f_X(x \,|\, a, b_1, c) = c\,b_1\,(x-a)^{c-1} \exp\{-b_1\,(x-a)^c\}, \qquad (2.26a)$$

where the original letters have been changed to have a closer resemblance to (2.8). The scale parameter has been "brought upstairs" into the numerator and includes the effect of the shape parameter $c$. Parameter $b_1$ is a combined scale–shape factor

$$b_1 = \frac{1}{b^c} \quad \text{or} \quad b = b_1^{-1/c}. \qquad (2.26b)$$

---

[5]  Suggested reading for this section: HALLINAN (1993).

[6]  The symbol $\sigma$ is commonly used in engineering to represent stress and should not be confused with the same symbol for a population standard deviation used in statistics.

WEIBULL then introduced his distribution to the U.S. in a frequently–cited article (WEIBULL, 1951) with unfortunate typographical errors. Those errors have spawned many significant articles premised on an accidental and unintended function which correctly reads

$$f_X(x \mid a, b_2, c) = c \, \frac{(x-a)^{(c-1)}}{b_2} \, \exp\left\{ -\frac{(x-a)^c}{b_2} \right\}. \tag{2.27a}$$

Here $b_2$ is the reciprocal of $b_1$:

$$b_2 = \frac{1}{b_1} = b^c \text{ or } b = b_2^{1/c}. \tag{2.27b}$$

In a follow–up discussion published in the journals, some readers noted that although the printed formula looked like (2.27a) the graphs used to illustrate WEIBULL's article from 1951 appeared to be actually represented by (2.8). WEIBULL (1952) acknowledged that positioning of the parenthesis in the 1951 article had been an "awkward misprint," and that equation (2.8) was indeed intended to be the form used.

In the spirit of simplification and concerning the ease of computation some authors presented yet another form of the exponential in which the scale parameter of (2.8) is placed in the numerator to give

$$f_X(x \mid a, b_3, c) = c \, b_3 \left[ b_3 \, (x-a) \right]^{c-1} \exp\left\{ - \left[ b_3 \, (x-a) \right]^c \right\}. \tag{2.28a}$$

Here $b_3$ is the reciprocal of the genuine scale parameter $b$:

$$b_3 = \frac{1}{b} \text{ or } b = \frac{1}{b_3}. \tag{2.28b}$$

In his monumental *Fatigue Testing and Analysis of Results*, published in 1961 on behalf of NATO, WEIBULL introduced yet another formulation:

$$f_X(x \mid a, b, c_1) = c_1^{-1} \, b^{-1/c_1} \, (x-a)^{(1/c_1)-1} \, \exp\left\{ - \left( \frac{x-a}{b} \right)^{1/c_1} \right\} \tag{2.29a}$$

in which the shape parameter has changed to $c_1$, the reciprocal of $c$:

$$c_1 = \frac{1}{c} \text{ or } c = \frac{1}{c_1}. \tag{2.29b}$$

This reformulation is motivated by working with the ln–transformation $Y := \ln X$ of the WEIBULL distributed random variable $X$ in the inference process (see Part II). $Y$ has a type–I extreme–value distribution of the minimum (see Sections 3.2.2 and 3.3.4) and the scale parameter of that distribution equals the reciprocal of $c$.

Equations (2.8) and (2.26a) to (2.29a) are equivalent formulations for the WEIBULL distribution, each representing the same idea of using a combined linear and power transformation of a random variable which then has an exponential distribution. This idea is legitimately traceable to (and named for) its proponent. Each of the foregoing five formulations and still some more are in use, but there is a trend in the literature on the WEIBULL distribution toward the simple looking form (2.8), where each parameter has a function of its own and is not mingled with another parameter.

Finally, some other formulas are presented. Combining the reciprocal scale parameter $b_3$ und the reciprocal shape parameter $c_1$ into one formula gives a nicer looking expression than (2.29a):

$$f_X(x \mid a, b_3, c_1) = \frac{b_3}{c_1} \left[ b_3 \left( x - a \right) \right]^{(1/c_1)-1} \exp\left\{ -\left[ b_3 \left( x - a \right) \right]^{1/c_1} \right\}. \tag{2.30}$$

Another still nicer looking form than (2.29a) combines $c_1$ and the scale–shape factor $b_1 = 1/b^c = 1/b^{1/c_1} = b^{-1/c_1}$ to give

$$f_X(x \mid a, b_1, c_1) = \frac{b_1}{c_1} \left( x - a \right)^{(1/c_1)-1} \exp\left\{ -b_1 \left( x - a \right)^{1/c_1} \right\}. \tag{2.31}$$

Sometimes one can find a rather strange looking formula where even the scale parameter has undergone a shifting:

$$f_X(x \mid a, b_4, c) = \frac{c}{b_4 - a} \left( \frac{x - a}{b_4 - a} \right)^{c-1} \exp\left\{ -\left( \frac{x - a}{b_4 - a} \right)^{c} \right\} \tag{2.32a}$$

with

$$b_4 = b + a \quad \text{or} \quad b = b_4 - a. \tag{2.32b}$$

After having introduced moments (in Sect. 2.9) and percentiles (in Sect. 2.8) still other formulas of the WEIBULL distribution will be given where moments and percentiles as functional parameters are used instead of the function parameters $a$, $b$ and $c$.

## 2.3 Failure distribution (CDF) and reliability function (CCDF)

Integrating (2.19) with respect to $v$ within the limits 0 and $u$ gives the CDF of the reduced WEIBULL distribution

$$
\begin{aligned}
F_U(u \mid c) &= \Pr(U \leq u) \\
&= \int_0^u c\, v^{c-1} \exp(-v^c) \, dv \\
&= 1 - \exp(-u^c). 
\end{aligned} \tag{2.33a}
$$

The CDF of the three–parameter version is

$$F_X(x \mid a, b, c) = 1 - \exp\left\{ -\left( \frac{x - a}{b} \right)^{c} \right\}, \tag{2.33b}$$

so

$$F_X(x \mid a, b, c) = F_U\left( \frac{x - a}{b} \,\middle|\, c \right). \tag{2.33c}$$

Fig. 2/7 shows in the left part several graphs of (2.33a) and in the right part some graphs belonging to (2.33b).

Figure 2/7:  WEIBULL failure distribution with differing parameter values

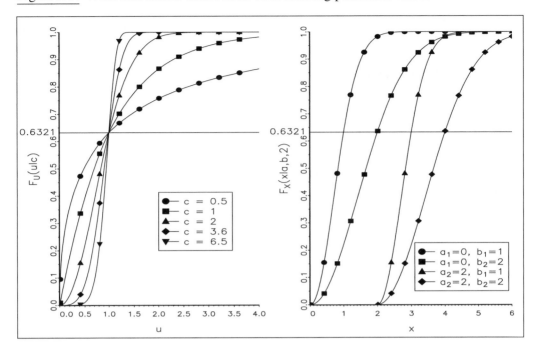

Looking at Fig. 2/7, especially at the left side, reveals two interesting features.

1. All curves belonging to the reduced WEIBULL distribution intersect in one point with coordinates $u = 1$ and $F_U(1 \,|\, c) = 1 - \exp(-1) \approx 0.6321$. For the three–parameter version on the right-hand side, one finds

$$F_X(a + b \,|\, a, b, c) \;=\; 1 - \exp\left\{-\left(\frac{a + b - a}{b}\right)^c\right\}$$

$$=\; 1 - \exp(-1) \approx 0.6321.$$

So there is a probability of approximately 63.21% that a WEIBULL distributed life-time is at most $b$ time units greater than its minimum $a$. The name "characteristic life" given to $b$ may be explained by the fact that with respect to (2.13), which is the most frequently found version in practice, the parameter value $b$ can be read off on the abscissa underneath $F = 0.6321$. Therefore it is characteristic to all WEIBULL distribution functions with $a = 0$ and the same value of $b$ to have one common point $(0.6321, b)$ in the $F$–$x$–plain. In case of the three–parameter distribution this point is shifted horizontally by the amount of $a$: $(0.6321, a + b)$ (see the graphs with $a_2 = 2$ on the right-hand side in Fig. 2/7).

2. The distribution functions have differing curvature that is predetermined by the be-havior of the corresponding DF:

   - $0 < c \le 1$ — $F$ has no point of inflection and is concave.
   - $c > 1$ — $F$ is convex as long as the density increases, i.e., until $x \ge x^*$ ($\,\widehat{=}\,$ mode).

     $F$ has an inflection point at $x^*$, and to the right of $x^*$ it is concave.

The reliability function (CCDF) is the complementary function to $F$:

$$R_U(u\,|\,c) = 1 - F_U(u\,|\,c) = \exp(-u^c) \qquad (2.34a)$$

and

$$R_X(x\,|\,a,b,c) = 1 - F_X(x\,|\,a,b,c) = \exp\left\{-\left(\frac{x-a}{b}\right)^c\right\}. \qquad (2.34b)$$

The characteristic point now has coordinates $(0.3679, a+b)$. The curvature is opposite to that of $F$.

The CDF of a WEIBULL random variable can be transformed to a straight line. Starting with (2.33b) and writing $F$ instead of $F_X(x\,|\,a,b,c)$, one has

$$1 - F = \exp\left\{-\left(\frac{x-a}{b}\right)^c\right\}.$$

First taking the reciprocal

$$\frac{1}{1-F} = \exp\left\{\left(\frac{x-a}{b}\right)^c\right\},$$

then the natural logarithm

$$\ln\left(\frac{1}{1-F}\right) = \left(\frac{x-a}{b}\right)^c,$$

and finally the decimal or common logarithm

$$\log\ln\left(\frac{1}{1-F}\right) = c\,\log\left(\frac{x-a}{b}\right)$$

$$\underbrace{\log\left[-\ln(1-F)\right]}_{=:\,\widetilde{y}} = c\,\underbrace{\log(x-a)}_{=:\,\widetilde{x}}\ \underbrace{-c\,\log b}_{=:\,\widetilde{b}}, \qquad (2.35a)$$

results in a linear equation

$$\widetilde{y} = c\,\widetilde{x} + \widetilde{b} \qquad (2.35b)$$

with slope $c$ and intercept $\widetilde{b} = -c\,\log b$. The values to be plotted on the abscissa are $\widetilde{x} = \log(x-a)$, the decimal logarithms of the re–shifted lifetime. The ordinate values are the double logarithms of $(1-F)^{-1}$.

Paper with axes scaled in the just described manner is termed **WEIBULL–probability–paper**. It is a useful tool when estimating the parameters of this distribution (see Chapter 9) and when testing whether a sample comes from a WEIBULL population (see Sect. 22.1). For an easy finding of the (estimated) $b$– and $c$–values the paper often is supplemented to give a nomogram. Fig. 2/8 shows an example of the WEIBULL–probability–paper with some CDF–graphs. A straight line appears only when $a$ is known so that $F_X(x\,|\,a,b,c)$ is plotted over the matching $x-a$. In Fig. 2/8 the parameters are $a=5$ with $b=50$ and

$c = 3.5$. If instead $F_X(x \,|\, 5, 50, 3.5)$ is plotted over $x - 0$ (respectively, over $x - 13.5$), i.e., $a$ has erroneously been chosen too small (respectively, too large) then the graph results in a curve which is concave (respectively, convex).

Figure 2/8: Several population CDFs on WEIBULL probability paper

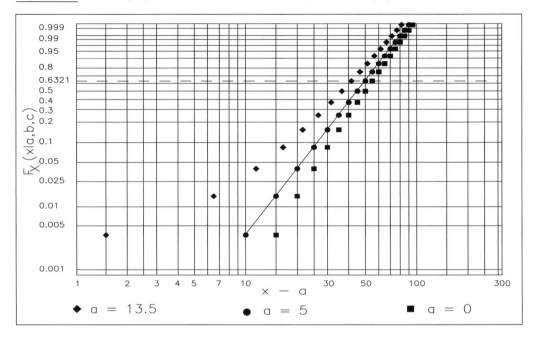

## 2.4   Hazard rate (HR)[7]

The general formula (2.4) for an HR becomes

$$h_U(u \,|\, c) \;=\; \frac{f_U(u \,|\, c)}{R_U(u \,|\, c)} = \frac{c\,u^{c-1}\,\exp(-u^c)}{\exp(-u^c)}$$

$$=\; c\,u^{c-1} \tag{2.36a}$$

in the case of a reduced WEIBULL distribution and

$$h_X(x \,|\, a, b, c) \;=\; \frac{f_X(x \,|\, a, b, c)}{R_X(x \,|\, a, b, c)} = \frac{\dfrac{c}{b}\left(\dfrac{x-a}{b}\right)^{c-1}\exp\left\{-\left(\dfrac{x-a}{b}\right)^c\right\}}{\exp\left\{-\left(\dfrac{x-a}{b}\right)^c\right\}}$$

$$=\; \frac{c}{b}\left(\frac{x-a}{b}\right)^{c-1} \tag{2.36b}$$

otherwise. The relation between these hazard rates is

$$h_X(x \,|\, a, b, c) = \frac{1}{b}\,h_U\left(\frac{x-a}{b}\,\bigg|\, c\right). \tag{2.36c}$$

---

[7]   Suggested reading for this and the following sections: BARLOW/MARSHALL/PROSCHAN (1963), KAO (1959), LEHMAN (1963), MUKHERJEE/ROY (1987).

Crucial to the behavior of a WEIBULL HR and also to the aging process is the shape parameter, so it suffices to study (2.36a). Fig. 2/9 shows the course of (2.36a) for several values of $c$. There is no common point of intersection.

Figure 2/9: Hazard rates of several reduced WEIBULL distributions

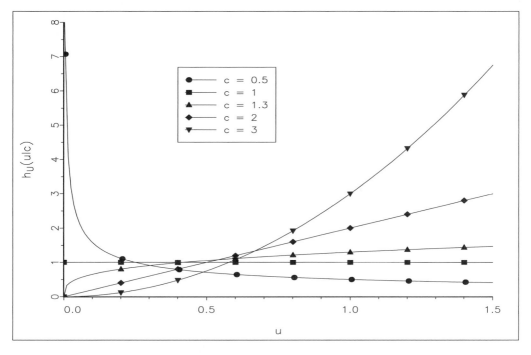

The HR in (2.36a) is a power function, and for the purpose of its discussion several cases as in Sect. 2.2.3 have to be considered. The knowledge of the first and second derivatives of (2.36a) with respect to $u$ is useful:

$$h_U'(u \mid c) \quad = \quad \frac{\mathrm{d}h_U(u \mid c)}{\mathrm{d}u} = c\,(c-1)\,u^{c-2}, \tag{2.37a}$$

$$h_U''(u \mid c) \quad = \quad \frac{\mathrm{d}^2 h_U(u \mid c)}{\mathrm{d}u^2} = c\,(c-1)\,(c-2)u^{c-3}. \tag{2.37b}$$

First case: $0 < c < 1$
This case is characterized by

$$\lim_{u \to 0} h_U(u \mid c) = \infty;$$

$$\lim_{u \to \infty} h_U(u \mid c) = 0;$$

$$h_U'(u \mid c) < 0 \ \text{ for } \ u > 0;$$

$$h_U''(u \mid c) > 0 \ \text{ for } \ u > 0.$$

Thus the hazard rate comes down from positive infinity at $u = 0$, is monotonically decreasing and convex (see $h_U(u \mid 0.5)$ in Fig. 2/9). With $0 < c < 1$ the WEIBULL distribution reveals a delayed and negative aging process. The probability of a $u$–survivor to fail immediately after $u$ is smaller the higher the age reached. During the mission of the items a screening takes place in the sense that weak items with hidden defects are sorted out at young age. So in the course of time only robust and "healthy" items remain in the population or as LEHMAN (1963, p. 36) writes: "So once the obstacles of early youth have been hurdled, life can continue almost indefinitely." Distributions having $h' < 0$ are called early–failure–distributions because the majority of failures in such a population takes place at young age. Lifetime distributions with this property have been found in practice, e.g., with semiconductors and other electronic parts. "Certain types of business enterprizes follow this pattern." (LEHMAN, 1963, p. 37). PROSCHAN (1963) gives an explanation for this type of aging which he observed for the air conditioning system of Boeing–airplanes.

Second case: $c = 1$
This case leads to the exponential distribution which, as is generally known, has a constant and age–independent HR:

$$h_U(u \,|\, 1) = 1 \ \forall \, u \geq 0.$$

So items of a population with $c = 1$ will not undergo any aging.

Third case: $1 < c < 2$
Because of

$$\lim_{u \to 0} h_U(u \,|\, c) = 0;$$

$$\lim_{u \to \infty} h_U(u \,|\, c) = \infty;$$

$$h'_U(u \,|\, c) > 0 \ \text{ for } \ u \geq 0;$$

$$h''_U(u \,|\, c) < 0 \ \text{ for } \ u \geq 0;$$

the HR is starting in the origin, increases monotonically, but is concave (see $h_U(u \,|\, 1.3)$ in Fig. 2/9). Members of such a population have the property of positive aging: the greater the age reached the higher the probability of an immediate failure.

Fourth case: $c = 2$
The HR is now given by

$$h_U(u \,|\, 2) = 2\,u \ \forall \, u \geq 0,$$

i.e. there is a linear (because of $h'' = 0$) and positive aging (because of $h' > 0$). This version of a WEIBULL distribution is also called **linear hazard rate distribution** or **RAYLEIGH distribution**.

Fifth case: $c > 2$
Because of

$$h'_U(u \,|\, c) > 0 \ \text{ and } \ h''_U(u \,|\, c) > 0,$$

the aging is positive and accelerated. A large value of $c$ indicates high dependability, or very few early deaths, then all failures can be anticipated almost simultaneously at or near $u = c$, respectively, at $x = b + c$ in the general case.

Summarizing the results above one can ascertain that depending on $c$, there are three types of aging:

- $0 < c < 1$ — negative aging,
- $c = 1$ — no aging,
- $c > 1$ — positive aging.

The HR being monotone in either case the WEIBULL distribution does not allow to model the bathtub–form or U–form of the HR which is often found in practice. A bathtub HR may be realized, within another family of distributions, for example, either using a HJORTH distribution (HJORTH, 1980), or with either a mixed or a composite WEIBULL distribution (see KAO (1959) or Sect. 3.3.6.3 and Sect. 3.3.6.4). All properties of distributions with monotone HR, as, for example, stated by BARLOW/MARSHALL/PROSCHAN (1963), apply to the WEIBULL distribution. Some of these properties will be discussed in the context of aging criteria in Sect. 2.7.

With respect to equipment or machinery having a monotonically increasing HR, a policy of preventative maintenance seems to be advisable. The withdrawal age of a still working piece of equipment might be determined by some chosen value $h$ of the HR, i.e., a preventative replacement takes place if

$$h_X(x_h \,|\, a, b, c) = \frac{c}{b} \left( \frac{x_h - a}{b} \right)^{c-1} = h, \tag{2.38a}$$

i.e. when the risk of failure within the next moment has reached a given critical value $h$. From (2.38a) the replacement life or retired life follows as

$$x_h = a + \left( h \, b^c / c \right)^{1/(c-1)}. \tag{2.38b}$$

Analysis of $h_X(x \,|\, a, b, c)$ has so far been done mostly in the time domain, i.e., in terms of its behavior with respect to time or age. It is quite logical to treat $h_X(X \,|\, a, b, c)$ as a transform of $X$ or as a random function with an easily deducible probability distribution and thus to analyze the failure rate in the frequency "domain." This **hazard rate transform** of the WEIBULL variable has been studied by MUKHERJEE/ROY (1987). Incidentally, some characterizations of the WEIBULL law have been found, and failure time and failure rate have been stochastically compared.

## 2.5 Cumulative hazard rate (CHR)

For some purposes, especially in testing, the cumulated or integrated hazard rate

$$H_U(u \,|\, c) \;=\; \int_0^u h_U(v \,|\, c)\,\mathrm{d}v = \int_0^u c\, v^{c-1}\,\mathrm{d}v$$

$$=\; u^c, \tag{2.39a}$$

respectively

$$H_X(x \mid a, b, c) = \left(\frac{x - a}{b}\right)^c = H_U\left(\frac{x - a}{b} \,\middle|\, c\right), \tag{2.39b}$$

is needed. CHR increases

- monotonically, but with falling rate for $0 < c < 1$;

- linearly (or with constant rate) for $c = 1$;

- monotonically, but with growing rate for $c > 1$.

Because CHR easily stretches itself over several powers of ten, the ordinate in Fig. 2/10 has a logarithmic scale. The CHR–graphs of all WEIBULL variables with the same value of $a$ and $b$ intersect in one point which has the coordinates $H = 1$ and $x = a + b$.

Figure 2/10: Cumulative hazard rate for a reduced WEIBULL variable

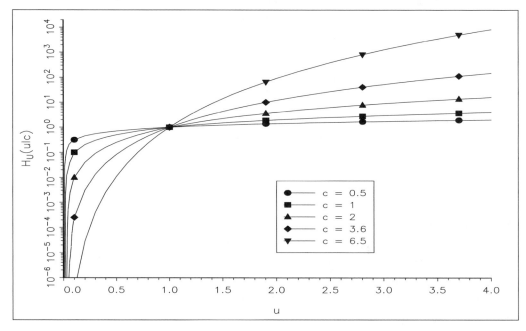

Taking the decimal logarithm of (2.39a,b) results in a linear function:

$$\log H_U(u|c) \;=\; c \log u \tag{2.40a}$$

$$\log H_X(x|a, b, c) \;=\; c \log(x - a) - c \log b. \tag{2.40b}$$

So the **WEIBULL–hazard–paper** is a paper with both axes being scaled logarithmicly, the vertical axis for the log–cumulated–hazard and the horizontal axis for the logarithm of the re–shifted lifetime $x - a$. This paper serves the same purposes as the WEIBULL–probability–paper of Sect. 2.3, with the latter being more popular.

Figure 2/11: Several population CDFs on WEIBULL hazard paper

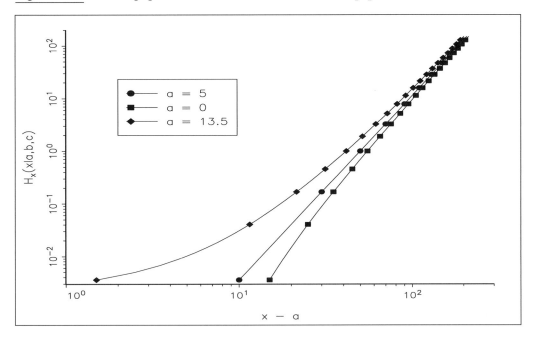

The parameter values in Fig. 2/11 have been chosen as in Fig. 2/8. A straight line shows up in the hazard–paper only if $a$ has been correctly chosen, otherwise a concave curve ($a$ too small) or a convex curve ($a$ too large) will result.

## 2.6 Mean residual life function (MRL)[8]

Before going into the details of the WEIBULL–MRL we have to introduce and to comment upon the underlying variate called **residual life**:

$$Y \mid x := X - x \mid X \geq x. \tag{2.41a}$$

$Y \mid x$ is a conditional variate, the condition being that an item has survived $x$ units of time. Thus, $Y \mid x$ is the additional or further lifetime of an $x$–survivor. The distribution of $Y \mid x$ is found by left–truncating the distribution of the original variate $X$ at $x$ and shifting the origin to this point of truncation. A variate which is closely related to (2.41a) and which must not be confounded with $X - x \mid X \geq x$ is the variate $X \mid X \geq x$, termed the **conditional age**. The latter is the age at failure of an item which has survived at least $x$ units of time. $X \mid X \geq x$ is a truncated variate only,[9] but it is not shifted.

---

[8]  Suggested reading for this section:  AROIAN (1965), KAO (1959), LEEMIS (1986), RAJA RAO/TALWALKER (1989), SWARTZ (1973), TANG/LU/CHEW (1999).

[9]  Truncation of distributions will be intensively dealt with in Sect. 3.3.5.

The following formulas show some functions describing $Y \mid x$. First, the general expression is given, followed by the special case of the WEIBULL distribution. The subscript $X$ of a function symbol indicates that this function belongs to the original, unconditioned variate $X$.

Density function

$$
\begin{aligned}
f(y \mid x) &= \frac{f_X(y+x)}{R_X(x)}, \quad y \geq 0 \\
&= \frac{c}{b}\left(\frac{y+x-a}{b}\right)^{c-1} \exp\left\{\left(\frac{x-a}{b}\right)^c - \left(\frac{y+x-a}{b}\right)^c\right\} \quad (2.41\text{b})
\end{aligned}
$$

Cumulative distribution function

$$
\begin{aligned}
F(y \mid x) &= \frac{F_X(x+x) - F_X(x)}{R_X(x)}, \quad y \geq 0 \\
&= 1 - \exp\left\{\left(\frac{x-a}{b}\right)^c - \left(\frac{y+x-a}{b}\right)^c\right\} \quad (2.41\text{c})
\end{aligned}
$$

Complementary cumulative distribution function

$$
\begin{aligned}
R(y \mid x) &= \frac{R_X(y+x)}{R_X(x)}, \quad y \geq 0 \\
&= \exp\left\{\left(\frac{x-a}{b}\right)^c - \left(\frac{y+x-a}{b}\right)^c\right\} \quad (2.41\text{d})
\end{aligned}
$$

Hazard rate

$$
\begin{aligned}
h(y \mid x) &= \frac{f(y \mid x))}{R(y \mid x)} = \frac{f_X(y+x)}{R_X(y+x)} = h_X(y+x), \quad y \geq 0 \\
&= \frac{c}{b}\left(\frac{y+x-a}{b}\right)^{c-1} \quad (2.41\text{e})
\end{aligned}
$$

The hazard rate of the residual life follows the same course as that of the complete life after shifting.

Cumulative hazard rate

$$
\begin{aligned}
H(y \mid x) &= \int_0^y h(z \mid x)\,dz = \int_0^y h_X(z + x)\,dz \\
&= \int_x^{x+y} h_X(v)\,dv = H_X(y + x) - H_X(x) \quad v = x + z \\
&= \left(\frac{y + x - a}{b}\right)^c - \left(\frac{x - a}{b}\right)^c
\end{aligned}
\tag{2.41f}
$$

Percentile function[10]

$$
\begin{aligned}
y_P = F^{-1}(P \mid x) &= F_X^{-1}\left(P + F_X(x)[1 - P]\right) - x, \quad 0 < P < 1 \\
&= a + \sqrt[c]{(x - a)^c - b^c \ln(1 - P)} - x
\end{aligned}
\tag{2.41g}
$$

Moments about zero[11]

$$
\begin{aligned}
E(Y^r \mid x) &= \int_0^\infty y^r f(y \mid x)\,dy = \int_0^\infty y^r \frac{f_X(y + x)}{R_X(x)}\,dy \\
&= r \int_0^\infty y^{r-1} R(y \mid x)\,dy \\
&= \frac{r}{R_X(x)} \int_0^\infty y^{r-1} R_X(y + x)\,dy
\end{aligned}
\tag{2.41h}
$$

The MRL–function is the special case of (2.41h) for $r = 1$, denoted by

$$
\mu(x) := E(Y \mid X \geq x),
\tag{2.41i}
$$

providing the interpretation of $\mu(x)$ as the average remaining lifetime of an item that has survived to at least time $x$. When $X$ has a DF $f_X(x)$ and a finite mean $\mu$, then

$$
\mu(x) = \frac{1}{R_X(x)} \int_x^\infty z\, f_X(z)\,dz - x
\tag{2.41j}
$$

---

[10] See Sect. 2.8 for more details on this concept.

[11] See Sect. 2.9.1 for more details on this concept.

or

$$\mu(x) = \frac{1}{R_X(x)} \int\limits_x^\infty R_X(z)\,\mathrm{d}z. \qquad (2.41\text{k})$$

Whereas (2.41j) is useful for the computation of $\mu(x)$, (2.41k) is useful for obtaining relationships between $\mu(x)$ and the CDF $F_X(x) = 1 - R_X(x)$.

SWARTZ (1973, p. 108) gives the following <u>theorem</u>:

If $\mu(x)$ is the MRL of a random variable $X$ with survival function $R_X(x)$ and with finite mean $\mu = \mathrm{E}(X)$, then

1.

$$\mu(x) \geq 0; \qquad (2.42\text{a})$$

2.

$$\mu(0) = \mu; \qquad (2.42\text{b})$$

3. if $R_X(x)$ is absolutely continuous, $\mu'(x)$ will exist and

$$\mu'(x) \geq -1; \qquad (2.42\text{c})$$

4.

$$\int\limits_0^\infty \frac{1}{\mu(x)}\,\mathrm{d}x \quad \text{diverges}; \qquad (2.42\text{d})$$

5.

$$R_X(x) = \frac{\mu(0)}{\mu(x)} \exp\left\{ -\int\limits_0^x \frac{1}{\mu(z)}\,\mathrm{d}z \right\}. \qquad (2.42\text{e})$$

The proofs are straightforward and are omitted.

Starting from (2.41j) and using the relationship

$$R_X(x) = \exp\left\{ -H_X(x) \right\}$$

from Table 2/1, another representation

$$\mu(x) = \exp\left\{ H_X(x) \right\} \int\limits_0^x \exp\left\{ -H_X(z) \right\}\,\mathrm{d}z$$

follows which is useful in finding the derivative of $\mu(x)$ with respect to $x$:

$$\mu'(x) = \frac{\mathrm{d}\mu(x)}{\mathrm{d}x} = \mu(x)\,h_X(x) - 1. \qquad (2.43)$$

As $\mu(x) \geq 0$ and $h_X(x) \geq 0$, the result (2.42c) is proven. A consequence of (2.43) is that

- MRL is increasing, if $\mu(x)\,h_X(x) > 1$;

- MRL is constant, if $\mu(x)\, h_X(x) = 1$;

- MRL is decreasing, if $\mu(x)\, h_X(x) < 1$.

A last general result on the MRL–function of a random variable $X$ defined over $a \leq x < \infty$, $a \in \mathbb{R}$, is

$$\mu(x) = \frac{1}{R_X(a)}\left[a - x + \int_a^\infty R_X(z)\,\mathrm{d}z\right] = \mu - x \quad \text{for } x \leq a, \qquad (2.44a)$$

$$\mu(x) = \frac{1}{R_X(x)}\int_x^\infty R_X(z)\,\mathrm{d}z = \frac{1}{R_X(x)}\int_0^\infty R_X(x+v)\,\mathrm{d}v \quad \text{for } x > a. \qquad (2.44b)$$

(2.44b) follows with $z = x + v$, whereas (2.44a) is a consequence of

$$\mu_r' = \mathrm{E}(X^r) = \int_{-\infty}^{+\infty} x^r\, f_X(x)\,\mathrm{d}x = a^r + r\int_a^{+\infty} x^{r-1}\, R_X(x)\,\mathrm{d}x, \qquad (2.45a)$$

if $\lim_{x\to\infty} x^{r+1} f(x) = 0$ for any $r < \infty$.

Proof of (2.45a):

Rewrite the last term of the above expression for $\mu_r'$ as follows:

$$\int_a^\infty [1 - F_X(x)]\, r\, x^{r-1}\,\mathrm{d}x.$$

Denote $y = 1 - F_X(x)$ and $\mathrm{d}v = r\, x^{r-1}\,\mathrm{d}x$ and integrate this expression by parts. The result is

$$x^r\, [1 - F_X(x)]\Big|_a^\infty + \mu_r'.$$

The upper limit of the first term vanishes if $\lim_{x\to\infty} x^{r-1} f(x) = 0$ for any finite $r$ and $F_X(a) = 0$ by definition of CDF. Hence, (2.45a) is proven. ∎

Corollary to (2.45a):

If the distribution is defined over the range $0 \leq x < \infty$, then

$$\mu_r' = r\int_0^{+\infty} x^{r-1}\, R_X(x)\,\mathrm{d}x, \qquad (2.45b)$$

with the special case

$$\mu_1' = \mu = \int\limits_0^{+\infty} R_X(x)\, dx. \tag{2.45c}$$

Evaluation of the general MRL–function (2.41j) with respect to the WEIBULL distribution gives[12]

$$\mu(x) = B(x) - (x - a), \tag{2.46a}$$

where

$$B(x) := \exp\left\{\left(\frac{x-a}{b}\right)^c\right\} \int\limits_x^\infty (z-a)\, \frac{c}{b}\left(\frac{z-a}{b}\right)^{c-1} \exp\left\{-\left(\frac{z-a}{b}\right)^c\right\}\, dz \tag{2.46b}$$

is the $a$–**reduced total expected life** of an $x$–survivor.[13] Let

$$y = \left(\frac{x-a}{b}\right)^c \quad \text{resp.} \quad v = \left(\frac{z-a}{b}\right)^c$$

so that

$$dz = \frac{b}{c}\left(\frac{z-a}{b}\right)^{-(c-1)},$$

and (2.46b) becomes

$$B(x) = b\,\exp\left\{\left(\frac{x-a}{b}\right)^c\right\} \int\limits_y^\infty e^{-v}\, v^{1/c}\, dv. \tag{2.46c}$$

This definite integral may be either evaluated as an incomplete gamma function or reduced to a $\chi^2$–distribution or a type–III–function of the PEARSONIAN distribution system. The first two courses shall be pursued in the sequel. But before doing that, the special case $c = 1$, the exponential distribution, shall be considered. Then, (2.46c) leads to

$$B(x) = b\left(1 + \frac{x-a}{b}\right) = b + x - a,$$

and (2.46a) gives

$$\mu(x) = b.$$

This result is to be expected because, due to the lacking memory of this distribution, the mean $E(X - a)$ as well as $\mu(x)$ are both equal to $b$.

---

[12] RAJA RAO/TALWALKER (1989) derived lower and upper bounds for the WEIBULL–MRL–function.

[13] Working with $z - a$ instead of $z$ under the integral facilitates the manipulation of (2.46b), but it requires a compensating correction, i.e., the subtraction of $x - a$ instead of $x$ in (2.46a).

Evaluation of (2.46c) by the **complementary incomplete gamma function**[14]

$$\Gamma(k \,|\, u) := \int_u^\infty e^{-v} \, v^{k-1} \, \mathrm{d}v \tag{2.47a}$$

gives

$$B(x) = b \exp\left\{\left(\frac{x-a}{b}\right)^c\right\} \Gamma\left(1 + \frac{1}{c} \,\Big|\, \left[\frac{x-a}{b}\right]^c\right)$$

$$= \frac{b}{c} \exp\left\{\left(\frac{x-a}{b}\right)^c\right\} \Gamma\left(\frac{1}{c} \,\Big|\, \left[\frac{x-a}{b}\right]^c\right). \tag{2.47b}$$

Calculating in this way is not normally supported by statistical software because routines to evaluate (2.47a) are rarely included.

Evaluation of (2.46c) by the $\chi^2$**–distribution** should not be a problem with any software. The CDF of a $\chi^2$–distribution with $\nu$ degrees of freedom is given by

$$F_{\chi^2}(z \,|\, \nu) = \frac{1}{2^{\nu/2} \, \Gamma\left(\frac{\nu}{2}\right)} \int_0^z e^{-u/2} \, u^{(\nu/2)-1} \, \mathrm{d}u \tag{2.48a}$$

and the CCDF by

$$R_{\chi^2}(z \,|\, \nu) = 1 - F_{\chi^2}(z \,|\, \nu) = \frac{1}{2^{\nu/2} \, \Gamma\left(\frac{\nu}{2}\right)} \int_z^\infty e^{-u/2} \, u^{(\nu/2)-1} \, \mathrm{d}u. \tag{2.48b}$$

Now let $v = u/2$ in (2.46c), which then becomes

$$B(x) = b \exp\left\{\left(\frac{x-a}{b}\right)^c\right\} \int_{2y}^\infty e^{-u/2} \, (u/2)^{1/c} \, \mathrm{d}u/2$$

$$= b \exp\left\{\left(\frac{x-a}{b}\right)^c\right\} 2^{-(c+1)/c} \int_{2y}^\infty e^{-u/2} \, (u/2)^{1/c} \, \mathrm{d}u. \tag{2.49a}$$

From comparison of (2.49a) with (2.48b) the following representation of $B(x)$ is evident:

$$B(x) = b \exp\left\{\left(\frac{x-a}{b}\right)^c\right\} \Gamma\left(1 + \frac{1}{c}\right) R_{\chi^2}\left[2\left(\frac{x-a}{b}\right)^c \,\Big|\, 2\left(1 + \frac{1}{c}\right)\right]$$

$$= \frac{b}{c} \exp\left\{\left(\frac{x-a}{b}\right)^c\right\} \Gamma\left(\frac{1}{c}\right) \left\{1 - F_{\chi^2}\left[2\left(\frac{x-a}{b}\right)^c \,\Big|\, 2\left(1 + \frac{1}{c}\right)\right]\right\} \tag{2.49b}$$

---

[14] The complete gamma function (EULER integral of the second kind) is given by

$$\Gamma(k) = \int_0^\infty e^{-v} \, v^{k-1} \, \mathrm{d}v.$$

For more details on the gamma function and its relatives see the excursus in Sect. 2.9.1.

Fig. 2/12 gives the graphs of MRL belonging to the reduced WEIBULL distribution having $c = 0.5$, 1, 2 and 3.5. MRL is a monotone function and

- increasing for $0 < c < 1$ (The hazard rate is decreasing.);
- constant for $c = 1$ (The hazard rate is constant.);
- decreasing for $c > 2$ (The hazard rate is increasing.).

The above mentioned relationships between MRL and HR will be further discussed in the following section, dealing with aging criteria.

Figure 2/12: MRL function of several reduced WEIBULL distributions

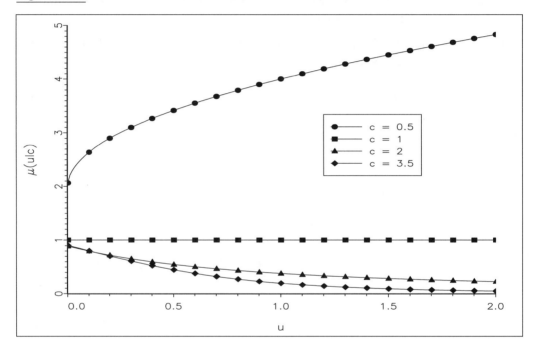

## 2.7   Aging criteria[15]

In the context of statistical lifetime analysis, aging does not mean that a unit becomes older in the sense of time. Rather it is a notion pertaining to residual life. Aging is thus the phenomenon that a chronological older unit has a shorter residual life in some statistical sense than a newer or chronological younger unit. Lifetime distributions are mostly characterized with respect to aging by the behavior of

- their HR $h(x)$ or
- their MRL $\mu(x)$.

---

[15]   Suggested reading for this section: BARLOW/MARSHALL/PROSCHAN (1963), BARLOW/PROSCHAN (1975), BRYSON/SIDDIQUI (1969), LEEMIS (1986), MAKINO (1984), SHAKED/LI (1997).

But other functions may be used too, as will be demonstrated soon. The aging criteria presented below are linked as shown in Fig. 2/13.

Figure 2/13: Chains of implications for several aging criteria

$$
\begin{array}{c}
\text{IAF} \\
\Updownarrow \\
\text{PF}_2 \implies \text{IHR} \quad \nearrow \quad \text{IHRA} \implies \text{NBU} \quad \searrow \quad \text{NBUE} \implies \text{HNBUE} \\
\Updownarrow \quad \searrow \quad \quad \text{DMRL} \quad \nearrow \\
\text{IIHR}
\end{array}
$$

$$
\begin{array}{c}
\text{DAF} \\
\Updownarrow \\
\text{DHR} \quad \nearrow \quad \text{DHRA} \implies \text{NWU} \quad \searrow \quad \text{NWUE} \implies \text{HNWUE} \\
\Updownarrow \quad \searrow \quad \quad \text{IMRL} \quad \nearrow \\
\text{DIHR}
\end{array}
$$

The upper chain refers to positive aging. A proof of the upper chain starting with IHR can be found in BRYSON/SIDDIQUI (1969). By an analogous argumentation the lower chain, referring to negative or inverse aging, may be proved too.

The following presentation of aging criteria starts by considering the class of distributions with a monotone hazard rate. The WEIBULL family is a member of that class.

Definitions:

1. A distribution function $F(x)$ is called an **IHR–distribution** (**i**ncreasing **h**azard **r**ate) or **IFR–distribution** (**i**ncreasing **f**ailure **r**ate) if its HR $h(x)$ is non–decreasing in $x$:

$$h(x_2) \geq h(x_1) \ \forall \, x_2 > x_1.$$

2. A distribution function $F(x)$ is called a **DHR–distribution** (**d**ecreasing **h**azard **r**ate) or **DFR–distribution** (**d**ecreasing **f**ailure **r**ate) if its HR $h(x)$ is non–increasing in $x$:

$$h(x_2) \leq h(x_1) \ \forall \, x_2 > x_1. \qquad \blacksquare$$

The WEIBULL distribution is

- IHR for $c \geq 1$,
- DHR for $0 \leq c \leq 1$.

The exponential distribution with $c = 1$ is the only continuous distribution which is IHR as well as DHR.[16]

Remarks:

1. The classification introduced above is also applicable to distributions which because of a non–existing DF function have no HR. Since the CCDF of the **residual life** $Y \mid x := X - x \mid X \geq x$ can be written as

$$R(y \mid x) = \frac{R(x + y)}{R(x)}$$

$$= \frac{\exp\left\{ -\int_0^{x+y} h(v) \, dv \right\}}{\exp\left\{ -\int_0^{x} h(v) \, dv \right\}} \quad \text{(see Tab. 2/1)}$$

$$= \exp\left\{ -\int_x^{x+y} h(v) \, dv \right\}$$

it follows that the conditional survival probability $R(y \mid x)$ is an increasing (decreasing) function of $x$, the conditioning age, if and only if the HR $h(x)$ is increasing (decreasing). So the following definition for distributions with non–existing HR is obvious: $F(x)$ is an IHR (DHR) distribution if $R(y \mid x)$ with arbitrary but fixed $y \geq 0$ is monotonically increasing (decreasing) in $x$, $0 < x < \infty$.

2. The IHR (DHR) property may also be characterized by the residual life $Y \mid x$ using the notions of "stochastically larger (smaller)":

Definition: A random variable $X_1$ with distribution function $F_1(x)$ is called **stochastically larger (smaller)** than a random variable $X_2$ with distribution function $F_2(x)$, in symbols:

$$X_1 \succeq X_2 \quad \text{resp.} \quad X_1 \preceq X_2,$$

if

$$F_1(x) \leq F_2(x) \ \forall \, x \quad \text{or} \quad \Pr(X_1 > x) \geq \Pr(X_2 > x) \ \forall \, x$$

resp.

$$F_1(x) \geq F_2(x) \ \forall \, x \quad \text{or} \quad \Pr(X_1 > x) \leq \Pr(X_2 > x) \ \forall \, x. \quad \blacksquare$$

---

[16] The **geometric distribution** with probability function

$$\Pr(X = j) = P \, (1 - P)^j; \quad j = 0, 1, 2, \ldots; \quad 0 < P < 1$$

is the only discrete distribution which is IHR as well as DHR.

Obviously

$$Y \,|\, x_1 \succeq Y \,|\, x_2 \ \forall \, x_1 \le x_2 \quad \text{resp.} \quad Y \,|\, x_1 \preceq Y \,|\, x_2 \ \forall \, x_1 \le x_2$$

if, and only if, the underlying distribution function $F(x)$ is IHR (DHR). Therefore, under the aspect of the behavior of the corresponding residual life the IHR (DHR) property is a reasonable mathematical description of aging (anti–aging).

<u>Theorem 1:</u> $F(x)$ is IHR (DHR) if and only if $\ln[R(x)]$ is concave (convex).

<u>Proof of theorem 1:</u> As $H(x) = -\ln[R(x)]$, the conditional survival function $R(y \,|\, x)$ is given by $R(y \,|\, x) = \exp\{-[H(x+y) - H(x)]\}$. So $F(x)$ is IHR (DHR) if and only if the difference $H(x+y) - H(x)$ is increasing (decreasing) in $x$ for fixed $y$.   ■

The logarithm of the general WEIBULL survival function is

$$\ln[R_X(x \,|\, a, b, c)] = -\left(\frac{x-a}{b}\right)^c.$$

Its second derivative with respect to $x$ is given by

$$\frac{\mathrm{d}^2\{\ln[R_X(x \,|\, a, b, c)]\}}{\mathrm{d}x^2} = -\frac{c}{b}\frac{c-1}{b}\left(\frac{x-a}{b}\right)^{c-2},$$

so

$$\frac{\mathrm{d}^2\{\ln[R_X(x \,|\, a, b, c)]\}}{\mathrm{d}x^2} \begin{cases} < \ 0 & \text{for} \ \ 0 < c < 1 \ \text{(concave)}, \\ = \ 0 & \text{for} \ \ c = 1, \\ > \ 0 & \text{for} \ \ c > 1 \ \text{(convex)}. \end{cases}$$

<u>Theorem 2:</u> IHR distributions have finite moments of any order.

<u>Remark:</u> For DHR distributions theorem 2 is not always true.

<u>Theorem 3:</u> For an IHR distribution the following inequality holds

$$\mathrm{E}(X^r) \le r! \left[\mathrm{E}(X)\right]^r; \ \ r = 1, 2, 3, \dots; \tag{2.50a}$$

whereas for a DHR distribution with existing moments we have

$$\mathrm{E}(X^r) \ge r! \left[\mathrm{E}(X)\right]^r; \ \ r = 1, 2, 3, \dots \quad ■ \tag{2.50b}$$

<u>Corollary to theorem 3:</u> For $r = 2$, (2.50a) gives

$$\mathrm{E}(X^2) \le 2\,\mu^2.$$

Combining this inequality with the representation $\sigma^2 = \mathrm{E}(X^2) - \mu^2$ for the variance results in

$$\sigma^2 = \mathrm{E}(X^2) - \mu^2 \le \mu^2,$$

so that CV, the **coefficient of variation**, of an IHR distribution is appraised as

$$CV = \frac{\sigma}{\mu} \leq 1.$$

Applying this reasoning to a DHR distribution using (2.50b) results in the appraisal

$$CV = \frac{\sigma}{\mu} \geq 1. \qquad\qquad ∎$$

CV for the WEIBULL distribution will be discussed in Sect. 2.9.3.

Instead of looking at the HR given by

$$h(x) = \frac{f(x)}{1 - F(x)}, \qquad\qquad (2.51a)$$

a generalization of the HR–classes of distributions is found by introducing a quotient of the form

$$h^*(x \,|\, y) := \frac{f(x)}{F(x + y) - F(x)}, \qquad\qquad (2.51b)$$

i.e. the number 1 in the denominator of (2.51a) is replaced by $F(x + y) < 1$.

Definition: A DF $f(x)$ is called **PÓLYA density of order 2**, $PF_2$ for short, If for all $y > 0$ the quotient (2.51b) is monotonically increasing in $x$.

Another equivalent definition says that for $PF_2$–distributions, the density is log–concave. Log–concavity of the density is the strongest aging property (see the upper part of Fig. 2/13).

Theorem 4: A distribution belongs to the IHR–class if its density is $PF_2$.

The inversion of theorem 4 is not necessarily true so that an IHR–distribution must not belong to the $PF_2$–class. Fig. 2/14 shows the graph of several logarithmized WEIBULL densities in the reduced case ($a = 0;\ b = 1$), which are concave for $c > 1$.

Figure 2/14: Logarithmized WEIBULL densities

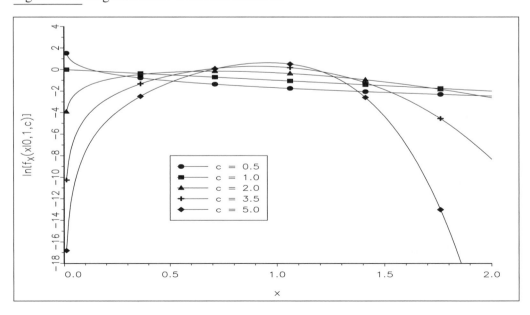

By introducing the **hazard rate average**

$$
\left.
\begin{aligned}
\overline{h(x)} \;\; &:= \;\; \frac{1}{x} \int_0^x h(v)\,\mathrm{d}v \\[2mm]
&= \;\; \frac{H(x)}{x} \\[2mm]
&= \;\; -\frac{\ln[R(x)]}{x}
\end{aligned}
\right\},
\tag{2.52a}
$$

the assumption of monotonicity of the HR may be weakened if instead the monotonicity of the hazard rate average is requested.

<u>Definitions:</u>

1. A distribution function $F(x)$ is called an **IHRA–distribution** (increasing hazard rate average) or **IFRA–distribution** (increasing failure rate average) if its hazard rate average is non–decreasing in $x$:

$$
\overline{h(x_2)} \geq \overline{h(x_1)} \;\; \forall \, x_2 > x_1.
$$

2. A distribution function $F(x)$ is called a **DHRA–distribution** (decreasing hazard rate average) or **DFRA–distribution** (decreasing failare rate average) if its hazard rate average is non–increasing in $x$:

$$
\overline{h(x_2)} \leq \overline{h(x_1)} \;\; \forall \, x_2 > x_1.
$$

For the general WEIBULL distribution having an arbitrary location parameter $a \in \mathbb{R}$ (2.52a) has to be modified slightly to give

$$
\left.
\begin{aligned}
\overline{h(x-a)} \;\; &:= \;\; \frac{1}{x-a} \int_a^x h_X(v \,|\, a,b,c)\,\mathrm{d}v \\[2mm]
&= \;\; \frac{H_X(x \,|\, a,b,c)}{x-a} \\[2mm]
&= \;\; \frac{\ln[R_X(x \,|\, a,b,c)]}{x-a}\,.
\end{aligned}
\right\}
\tag{2.52b}
$$

Using (2.39b) the WEIBULL hazard rate average is given by

$$
\overline{h(x-a)} = \left(\frac{x-a}{b}\right)^c \Big/ (x-a) = \frac{1}{b}\left(\frac{x-a}{b}\right)^{c-1}, \quad x > a,
\tag{2.52c}
$$

so that it is increasing for $c \geq 1$ and decreasing for $0 < c \leq 1$. Generally the following theorem 5 holds.

<u>Theorem 5:</u> If $F(x)$ is an IHR- (a DHR-) distribution then $F(x)$ is IHRA (DHRA) too.

---

**Excursus:   Mean hazard rate**

The hazard rate average defined in (2.52a,b) is a simple or unweighed average of the HR over the interval $[0,x]$ resp. $[a,x]$. MAKINO (1984) introduced the **mean hazard rate** as the density–weighted average of the HR over its complete domain of support. So the mean hazard rate is nothing else but the expectation of the HR:

$$
\begin{aligned}
\mathrm{E}\big[h(X)\big] \quad &:= \quad \int\limits_0^\infty h(x)\, f(x)\, \mathrm{d}x \\[2mm]
&= \quad \int\limits_0^\infty \frac{[R'(x)]^2}{R(x)}\, \mathrm{d}x.
\end{aligned}
\left.\vphantom{\int_0^\infty}\right\}
\tag{2.53a}
$$

For a WEIBULL distribution this general formula turns into

$$
\mathrm{E}\big[h(X\,|\,a,b,c)\big] = \Big(\frac{c}{b}\Big)^2 \int\limits_a^\infty \Big(\frac{x-a}{b}\Big)^{2\,(c-1)} \exp\Big\{-\Big(\frac{x-a}{b}\Big)^c\Big\}\, \mathrm{d}x,
$$

which after some manipulation is

$$
\mathrm{E}\big[h(X\,|\,0,b,c)\big] = \frac{c}{b} \int\limits_0^\infty w^{1-1/c}\, e^{-w}\, \mathrm{d}w.
\tag{2.53b}
$$

The integral in (2.53b) does not exist for $0 < c \le 0.5$, and for $c > 0.5$ it is easily transformed into the complete gamma function so that

$$
\begin{aligned}
\mathrm{E}\big[h(X\,|\,0,b,c)\big] \quad &= \quad \frac{c}{b} \int\limits_0^\infty w^{(2\,c-1)/c-1}\, e^{-w}\, \mathrm{d}w \\[2mm]
&= \quad \frac{c}{b}\, \Gamma\Big(\frac{2\,c-1}{c}\Big),\ c > 0.5.
\end{aligned}
\tag{2.53c}
$$

Fig. 2/15 shows the course of $\mathrm{E}\big[h(X\,|\,0,1,c)\big]$ which is nearly linear for greater values of $c$, say $c > 4$.

Figure 2/15: Mean hazard rate of the WEIBULL distribution for $b = 1$

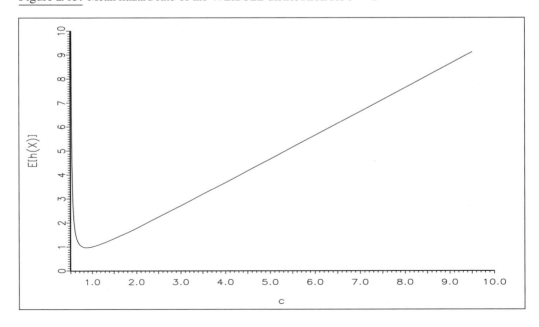

MAKINO (1984) used the mean hazard rate to approximate the WEIBULL distribution by a normal distribution, equating their mean hazard rates. The solution is $c \approx 3.43927$.

---

The probability of an $x$–survivor living another $y$ units of time is

$$R(y \mid x) := \Pr(Y > y \mid X \geq x) = \frac{R(x + y)}{R(x)},$$

whereas the probability of a new item living more than $y$ units of time is

$$\Pr(Y > y \mid X \geq 0) = R(y).$$

Definitions:

- A distribution function $F(x)$ is called **NBU** (**n**ew **b**etter than **u**sed) if

$$R(y) \geq \frac{R(x + y)}{R(x)} \quad \text{for } x > 0, \ y > 0.$$

- A distribution function $F(x)$ is called **NWU** (**n**ew **w**orse than **u**sed) if

$$R(y) \leq \frac{R(x + y)}{R(x)} \quad \text{for } x > 0, \ y > 0. \qquad ■$$

It is to be noticed that $F(x)$ is NBU (NWU) if and only if $Y \mid x$ is stochastically smaller (larger) than $X$:

$$Y|x \preceq X \ \text{ resp. } \ Y|x \succeq X.$$

Theorem 6:

a) $F(x)$ is NBU if and only if $H(\cdot)$ is **superadditive**, i.e.,

$$H(x + y) \geq H(x) + H(y) \text{ for } x, \ y > 0.$$

b) $F(x)$ is NWU if and only if $H(\cdot)$ is **subadditive**, i.e.,

$$H(x + y) \leq H(x) + H(y) \text{ for } x, \ y > 0. \qquad ■$$

According to (2.41b) the mean residual life of an $x$–survivor is

$$\mathrm{E}(Y \mid x) = \mu(x) = \frac{1}{R(x)} \int\limits_{x}^{\infty} R(z) \, \mathrm{d}z,$$

whereas the mean life of a new item is

$$\mathrm{E}(Y \mid x = 0) = \mu(0) = \int\limits_{0}^{\infty} R(z) \, \mathrm{d}z.$$

Definitions:

- A distribution function $F(x)$ is called **NBUE** (**n**ew **b**etter than **u**sed in **e**xpectation) if

$$\int\limits_x^\infty R(z)\,dz \leq \mu(0)\,R(x)\ \text{ for }\ x > 0.$$

- A distribution function $F(x)$ is called **NWUE** (**n**ew **w**orse than **u**sed in **e**xpectation) if

$$\int\limits_x^\infty R(z)\,dz \geq \mu(0)\,R(x)\ \text{ for }\ x > 0.\qquad\blacksquare$$

<u>Theorem 7:</u> A non–improvable appraisal for a NBUE–distribution is given by

$$F(x) \leq \frac{x}{\mu(0)}\ \text{ for }\ x < \mu(0).$$

<u>Theorem 8:</u> If $F(x)$ is NBU (NWU) then $F(x)$ is NBUE (NWUE) too. (The reversal is not always true.)

Definitions: A distribution function $F(x)$ is called **HNBUE** (**h**armonic **n**ew **b**etter than **u**sed in **e**xpectation) if

$$\int\limits_x^\infty R(z)\,dz \leq \mu(0)\,\exp\left[-x/\mu(0)\right]\ \text{ for }\ x > 0 \qquad (2.54a)$$

and **HNWUE** (**h**armonic **n**ew **w**orse than **u**sed in **e**xpectation) if

$$\int\limits_x^\infty R(z)\,dz \geq \mu(0)\,\exp\left[-x/\mu(0)\right]\ \text{ for }\ x > 0.\ \blacksquare \qquad (2.54b)$$

The term HNBUE resp. HNWUE is to be explained as follows. Starting from the mean residual life

$$\mu(x) = \frac{1}{R(x)}\int\limits_x^\infty R(z)\,dz,$$

(2.54b) may be written as

$$\frac{1}{\dfrac{1}{x}\int\limits_0^x \dfrac{1}{\mu(z)}\,dz} \leq \mu(0).$$

This inequality says that the harmonic mean of the expected residual life $\mu(z)$ in $[0, x]$ is at most the harmonic mean of $\mu(0)$.

BRYSON/SIDDIQUI (1969) introduced another aging criterion which they called **specific aging factor** and which is defined as

$$A(x,y) = \frac{R(x)\,R(y)}{R(x+y)}\;;\;\; x,\,y \geq 0. \tag{2.55}$$

Notice the interchangeability of the arguments $x$ and $y$ and the relationship to NBU und NWU. If a distribution is NBU (NWU), its specific aging factor is $A(x,y) \geq 1$ resp. $A(x,y) \leq 1$.

Definitions:

A distribution function $F(x)$ is called **IAF** (**i**ncreasing **a**ging **f**actor) if

$$A(x_2,y) \geq A(x_1,y)\;\forall\,y \geq 0,\; x_2 \geq x_1 \geq 0,$$

and **DAF** (**d**ecreasing **a**ging **f**actor) if

$$A(x_2,y) \leq A(x_1,y)\;\forall\,y \geq 0,\; x_2 \geq x_1 \geq 0. \qquad\blacksquare$$

A generalization of the hazard rate average $\overline{h(x)}$, defined above, is the **interval hazard rate average**:

$$\overline{h(x,y)} \;\; := \;\; \frac{1}{y}\int\limits_{x}^{x+y} h(v)\,\mathrm{d}v$$

$$= \;\; \frac{H(x+y) - H(x)}{y}. \tag{2.56}$$

$\overline{h(y)}$ and $\overline{h(0,y)}$ are related as follows:

$$\overline{h(y)} = \overline{h(0,y)}.$$

Definitions:

A distribution function $F(x)$ is called **IIHR** (**i**ncreasing **i**nterval **h**azard **r**ate average) if

$$\overline{h(x_2,y)} \geq \overline{h(x_1,y)}\;;\forall\,y \geq 0,\; x_2 \geq x_1 \geq 0,$$

and **DIHR** (**d**ecreasing **i**nterval **h**azard **r**ate average) if

$$\overline{h(x_2,y)} \leq \overline{h(x_1,y)}\;\forall\,y \geq 0,\; x_2 \geq x_1 \geq 0. \qquad\blacksquare$$

Finally, if the mean residual life $\mu(x)$ is monotone in $x$, it may be used to characterize the aging process too.

Definitions:

A distribution function $F(x)$ is called an **IMRL**– (a **DIHR**–)distribution if its mean residual life is non–decreasing in $x$:

$$\mu(x_2) \geq \mu(x_1)\;\forall\,x_2 > x_1 \quad \text{(IMRL)};$$

$$\mu(x_2) \leq \mu(x_1)\;\forall\,x_2 > x_1 \quad \text{(DMRL)}. \qquad\blacksquare$$

## 2.8   Percentiles and random number generation[17]

For a continuous random variable $X$, the percentile of order $P$ is that realization $x_P$ of $X$ that satisfies

$$F(x_P) = P, \ 0 < P < 1.$$

Because of $F(x) = \Pr(X \le x)$ this means:

$$\Pr(X \le x_P) = P.$$

So $x_P$ is that value of $X$, e.g., that special life, reached by $P\,100\%$ of the population. An individual percentile is an indicator of location, a difference of two percentiles $x_{P_2} - x_{P_1}$, $P_2 > P_1$, is a measure of variation, and suitable combinations of more than two percentiles may be used to indicate skewness and kurtosis. The **percentile function** $F^{-1}(P) = x_P$ is a means for generating random numbers of $X$ with distribution function $F(x)$.

### 2.8.1   Percentiles

Percentiles of a WEIBULL distribution are easily found because its CDF can be inverted in closed form. Starting from (2.33a), the CDF of the reduced WEIBULL variable, this inversion is

$$
\begin{aligned}
P &= F_U(u_P \,|\, c) \\[4pt]
&= 1 - \exp(-u_P^c) \\
\Longrightarrow \ln(1 - P) &= -u_P^c \\
u_P^c &= -\ln(1 - P) \\
u_P &= \left[ -\ln(1 - P) \right]^{1/c} = \left( \ln\left[ \frac{1}{1 - P} \right] \right)^{1/c}.
\end{aligned}
\tag{2.57a}
$$

The general WEIBULL percentiles are given by

$$
\begin{aligned}
x_P &= a + b\, u_P \\[4pt]
&= a + b \left[ -\ln(1 - P) \right]^{1/c} = a + b \left( \ln\left[ \frac{1}{1 - P} \right] \right)^{1/c}.
\end{aligned}
\tag{2.57b}
$$

Fig. 2/16 shows the course of the reduced WEIBULL percentile–function (2.57a) for several values of $c$. There is a common point of intersection with coordinates $P \approx 0.6321$, $u_{0.6321} = 1$.

---

[17] Suggested reading for this section: ISHIOKA (1990), JOHNSON (1968), MONTANARI/CAVALLINI/ TOMMASINI/CACCIARI/CONTIN (1995), RAMBERG/TADIKAMALLA (1974), TADIKAMALLA/ SCHRIBER (1977).

Some special percentiles are

$$u_{0.5} = (\ln 2)^{1/c} \approx 0.69315^{1/c} \qquad - \textbf{median}, \qquad (2.58\text{a})$$

$$u_{0.25} = \left[\ln (4/3)\right]^{1/c} \approx 0.28768^{1/c} \qquad - \text{lower quartile}, \qquad (2.58\text{b})$$

$$u_{0.75} = \left[\ln 4\right]^{1/c} \approx 1.38629^{1/c} \qquad - \textbf{upper quartile}, \qquad (2.58\text{c})$$

$$u_{0.6321} \approx 1 \qquad\qquad\qquad - \textbf{characteristic life}, \qquad (2.58\text{d})$$

$$u_{0.1} = \left[\ln (10/9)\right]^{1/c} \approx 0.10536^{1/c} \qquad - \text{first decile.}[18] \qquad (2.58\text{e})$$

In reliability engineering

$$x_Q^* = x_{1-P}, \quad Q := 1 - P$$

is called **reliable life**, i.e., the reliable life of order $Q$ is the percentile of order $1 - P$. $x_Q^*$ has a survival probability of $Q$.

Figure 2/16: Percentile function for the reduced WEIBULL distribution

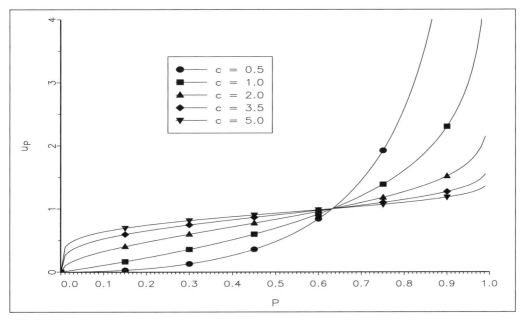

A necessary, but not sufficient, condition for a DF function to be symmetric is the **coincidence** of its **mode** and its **median**. Given $a$ and $b$, the coincidence depends on $c$. The condition is

$$(\ln 2)^{1/c} = \left(\ln \left[\frac{c-1}{c}\right]\right)^{1/c}$$

$$\implies c = \frac{1}{1 - \ln 2} \approx 3.258891, \qquad (2.59\text{a})$$

---

[18] The first decile plays an import role in assessing the life of ball–bearings and roller–bearings, see DOURGNON/REYROLLE (1966) or MCCOOL (1966, 1974a).

so that the realization which is being mode as well as median is

$$x^* = x_{0.5} \approx a + b \, (\ln 2)^{1/3.258891} \approx a + 0.893628 \, b. \tag{2.59b}$$

For $c < 3.258891$ $(c > 3.258891)$ the mode is to the left (to the right) of the median indicating positive (negative) skewness. Further inquiry into the skewness will be done in Sections 2.9.2 and 2.9.4.

Because of (2.57b) one has the opportunity to simply substitute one of the function parameters $a$ or $b$ by a percentile. Substituting $a$, the distribution function (2.33b) turns into

$$F_X(x \mid x_P, b, c) = 1 - \exp\left\{ -\left[ \frac{x - x_P}{b} + \left( \ln\left[ \frac{1}{1-P} \right] \right)^{1/c} \right]^c \right\}, \tag{2.60a}$$

with regard to $b$ the result is

$$F_X(x \mid a, x_P, c) = 1 - \exp\left\{ -\left( \ln\left[ \frac{1}{1-P} \right] \right) \left( \frac{x - a}{x_P - a} \right)^c \right\}. \tag{2.60b}$$

$P \approx 0.6321$ leads to the simple looking expressions

$$F_X(x \mid x_{0.6321}, b, c) = 1 - \exp\left\{ -\left[ \frac{x - x_{0.6321}}{b} + 1 \right]^c \right\}$$

$$F_X(x \mid a, x_{0.6321}, c) = 1 - \exp\left\{ -\left( \frac{x - a}{x_{0.6321} - a} \right)^c \right\}.$$

### 2.8.2  WEIBULL random numbers

Because the CDF $F_X(x \mid a, b, c)$ of a three–parameter WEIBULL variable is of a closed explicit form, it is easy to generate pseudo–random numbers through the probability integral transformation. Let $Z$ be the random variable uniformly distributed in $(0, 1)$ and, specifically, setting

$$Z = F_X(X \mid a, b, c) = 1 - \exp\left\{ -\left( \frac{X - a}{b} \right)^c \right\}$$

and finally inverting this transformation, we obtain

$$X = a + b \left[ -\ln(1 - Z) \right]^{1/c} = a + b \left( \ln\left[ \frac{1}{1-Z} \right] \right)^{1/c}. \tag{2.61}$$

Thus, after generating a pseudo–random observation from the uniform $(0, 1)$–population, the required WEIBULL observation of $X$ can be generated from (2.61) for specified values of the parameters $a$, $b$ and $c$.

Another way to simulate a WEIBULL observation is to make use of any efficient exponential simulation algorithm. Realizing that $[(X - a)/b]^c$ is distributed as reduced exponential, we can simulate the required WEIBULL observation of $X$ by

$$X = a + b \, W^{1/c}, \tag{2.62}$$

where $W$ denotes the reduced exponential pseudo–random observation already simulated.

ISHIOKA (1990) has discussed the generation of WEIBULL random numbers using the ratio of uniform pseudo–random observations. MONTANARI et al. (1995) have compared several approaches to generate WEIBULL observations. RAMBERG/TADIKAMALLA (1974) have demonstrated how gamma variates can be generated by using WEIBULL sampling mechanisms.

## 2.9 Moments, cumulants and their generating functions

Moments and cumulants are the expected values of certain functions of a random variable. They serve to numerically describe the variable with respect to given characteristics, as e.g., location, variation, skewness and kurtosis. Sect. 2.9.1 gives the definitions of several types of moments, shows their interrelations and how they may be developed by using so–called generating functions. The following Sections 2.9.2 to 2.9.4 go into the details of measuring location, variation, skewness and kurtosis by means of moments.

### 2.9.1 General formulas[19]

The **moments about zero** (alternatively **uncorrected moments** or **raw moments**) play a key role for all other kinds of moments because the latter can be — more or less easily — expressed by zero–moments. Further, for the WEIBULL family it suffices to develop the moments of the reduced variable $U$ because those of the general variable $X = a + b\,U$ are easily derived from the $U$–moments.

The expected value of $X^r$ is termed the $r$–th moment about zero of the random variable $X$:

$$\mu_r'(X) := \mathrm{E}(X^r). \tag{2.63a}$$

Generally, $r$ is any real number, but for the most part $r$ is taken as a non–negative integer. With regard to the reduced WEIBULL variable we get

$$
\begin{aligned}
\mu_r'(U) = \mathrm{E}(U^r) \;&=\; \int_0^\infty u^r\, f_U(u\,|\,c)\,\mathrm{d}u \\[2mm]
&=\; \int_0^\infty u^r\, c\,u^{c-1}\, \exp\{-u^c\}\,\mathrm{d}u \qquad (2.63b) \\[2mm]
&=\; \int_0^\infty v^{r/c}\, e^{-v}\,\mathrm{d}v. \qquad (2.63c)
\end{aligned}
$$

The transition from (2.63b) to (2.63c) rests upon the substitution

$$v = u^c,$$

---

[19] Suggested reading for Sections 2.9.1 – 2.9.3: FLEHINGER/LEWIS (1959), JOHNSON/KOTZ/KEMP (1992), Kao (1959), KÜBLER (1979), LEHMAN (1963), MCEWEN/PARRESOL (1991), RAJA RAO/TALWALKER (1989).

so that

$$\mathrm{d}v = c\,u^{c-1}\,\mathrm{d}u \quad \text{resp.} \quad \mathrm{d}u = \frac{1}{c}\,u^{1-c}\,\mathrm{d}v.$$

(2.63c) is nothing but the well–known gamma function (see the following excursus), so we have

$$\mu'_r(U) \quad = \quad \Gamma\left(\frac{r}{c}+1\right) \tag{2.63d}$$

$$=: \quad \Gamma_r, \text{ for short in the sequel.} \tag{2.63e}$$

The mean or expected value of the reduced WEIBULL variable is thus

$$\mathrm{E}(U) = \mu_U = \Gamma_1. \tag{2.63f}$$

It will be discussed in Sect. 2.9.2.

## Excursus:   Gamma function and its relatives[20]

### A.  Complete gamma function

There exist three equivalent definitions of the complete gamma function $\Gamma(z)$, gamma function for short.

Definition 1, due to LEONHARD EULER (1707 – 1783):

$$\Gamma(z) := \int\limits_0^\infty t^{z-1}\,e^{-t}\,\mathrm{d}t, \; z > 0$$

Definition 2, due to CARL FRIEDRICH GAUSS (1777 – 1855):

$$\Gamma(z) := \lim_{n\to\infty} \frac{n!\,n^z}{z\,(z+1)\ldots(z+n)}, \; z \neq 0, -1, -2, \ldots$$

Definition 3, due to KARL WEIERSTRASS (1815 – 1897):

$$\frac{1}{\Gamma(z)} := z\,e^{\gamma\,z} \prod_{n=1}^\infty \left[\left(1+\frac{z}{n}\right)e^{-z/n}\right],$$

where

$$\gamma = \lim_{n\to\infty} \left\{1 + \frac{1}{2} + \frac{1}{3} + \ldots + \frac{1}{n} - \ln\,n\right\} \approx 0.577215665$$

is EULER's constant.

Integrating by parts in definition 1 gives the following **recurrence formula** for $\Gamma(z)$,

$$\Gamma(z+1) = z\,\Gamma(z),$$

enabling a definition of $\Gamma(z)$ over the entire real line,[21] except where $z$ is zero or a negative integer, as

$$\Gamma(z) = \begin{cases} \int\limits_0^\infty t^{z-1}\,e^{-t}\,\mathrm{d}t, & z > 0 \\[2ex] z^{-1}\,\Gamma(z+1), & z < 0, z \neq -1, -2, \ldots \end{cases}$$

---

[20] Suggested reading: ABRAMOWITZ/STEGUN (1965)

[21] The gamma function is also defined for a complex variable $t$ and for a complex parameter $z$, provided that the respective real part is positive.

The general recurrence formula is

$$\Gamma(n+z) = \Gamma(z) \prod_{i=1}^{n} (z+i-1).$$

From definition 1 we have

$$\Gamma(1) = 1 = 0!$$

so that using the recurrence formula

$$\Gamma(n+1) = n!, \quad n = 1, 2, 3, \ldots$$

holds. From definition 3 it can be shown that

$$\Gamma\left(\frac{1}{2}\right) = \sqrt{\pi} \approx 1.77245,$$

and furthermore,

$$\Gamma\left(n+\frac{1}{2}\right) = \frac{1 \cdot 3 \cdot 5 \cdot \ldots \cdot (2n-1)}{2^n} \sqrt{\pi}, \; n = 1, 2, 3, \ldots$$

The table in Part III gives values of $\Gamma(z)$ for $1 \leq z \leq 2$.

There exist a lot of approximations for $\Gamma(z)$. The following two formulas have been obtained using some form of **STIRLINGS's expansion** (JAMES STIRLING, 1692 – 1770) for the gamma function:

1) $\Gamma(z) \approx \sqrt{2\pi}\, z^{z-1/2}\, e^{-z}\, \exp\left\{\dfrac{1}{12\,z} - \dfrac{1}{360\,z^3} + \dfrac{1}{1260\,z^5} - \cdots\right\},$

2) $\Gamma(z) \approx \sqrt{2\pi}(z-1)^{z-1/2} e^{-z-1} \exp\left\{\dfrac{1}{12(z-1)} - \dfrac{1}{360(z-1)^3} + \dfrac{1}{1260(z-1)^5} - \dfrac{1}{1680(z-1)^7} + \cdots\right\}.$

The remainder terms in both formula are each less in absolute value than the first term that is neglected, and they have the same sign.

Two **polynomial approximations** are

1) $\Gamma(1+x) = 1 + \displaystyle\sum_{i=1}^{5} a_i\, x^i + \varepsilon(x)$ for $0 \leq x \leq 1,$ where $|\varepsilon(x)| \leq 5 \cdot 10^{-5},$

with

| | | | | |
|---|---|---|---|---|
| $a_1$ | $=$ | $-0.5748646$ | $a_4 = $ | $0.4245549$ |
| $a_2$ | $=$ | $0.9512363$ | $a_5 = $ | $-0.1010678$ |
| $a_3$ | $=$ | $-0.6998588$ | | |

2) $\Gamma(1+x) = 1 + \sum_{j=1}^{8} b_j x^j + \varepsilon(x)$ for $0 \le x \le 1$, where $|\varepsilon(x)| \le 3 \cdot 10^{-7}$,

with

$$b_1 = -0.577191652 \qquad\qquad b_5 = -0.756704078$$

$$b_2 = 0.988205891 \qquad\qquad b_6 = 0.482199394$$

$$b_3 = -0.897056937 \qquad\qquad b_7 = -0.193527818$$

$$b_4 = 0.918206857 \qquad\qquad b_8 = 0.035868343$$

### B. Polygamma functions

The first derivative of $\ln \Gamma(z)$ is called the **digamma function** or **psi function**:

$$\psi(z) := \frac{\mathrm{d}}{\mathrm{d}z}\left[\ln \Gamma(z)\right] = \frac{\Gamma'(z)}{\Gamma(z)} = \frac{\mathrm{d}\Gamma(z)/\mathrm{d}z}{\Gamma(z)}$$

leading to

$$\Gamma'(z) = \psi(z)\,\Gamma(z).$$

Similarly

$$\psi'(z) := \frac{\mathrm{d}}{\mathrm{d}z}\left[\psi(z)\right] = \frac{\mathrm{d}^2}{\mathrm{d}z^2}\left[\ln \Gamma(z)\right]$$

is called **trigamma function** which leads to

$$\frac{\mathrm{d}^2\Gamma(z)}{\mathrm{d}z^2} = \Gamma(z)\left[\psi^2(z) + \psi'(z)\right].$$

Generally,

$$\psi^{(n)}(z) := \frac{\mathrm{d}^n}{\mathrm{d}z^n}\left[\psi(z)\right] = \frac{\mathrm{d}^{n+1}}{\mathrm{d}z^{n+1}}\left[\ln \Gamma(z)\right]; \; n = 1, 2, 3, \ldots$$

is called $(n+2)$–**gamma function** or **polygamma function**. $\psi''$, $\psi^{(3)}$, $\psi^{(4)}$ are the tetra-, penta- and hexagamma functions, respectively.

The recurrence formula $\Gamma(z+1) = z\,\Gamma(z)$ yields the following recurrence formulas for the psi function:

$$\psi(z+1) = \psi(z) + z^{-1}$$

$$\psi(z+n) = \psi(z) + \sum_{i=1}^{n}(z+i-1)^{-1}; \; n = 1, 2, 3, \ldots,$$

and generally for polygamma functions

$$\psi^{(n)}(z+1) = \psi^{(n)}(z) + (-1)^n\, n!\, z^{-n-1}.$$

Further we have

$$
\begin{aligned}
\psi(z) &= \lim_{n \to \infty} \left[ \ln n - \sum_{i=0}^{n} (z+i)^{-1} \right] \\
&= -\gamma - \frac{1}{z} + \sum_{i=1}^{\infty} \frac{z}{i\,(z+i)}, \ z \neq -1, -2, \ldots \\
&= -\gamma + (z-1) \sum_{i=0}^{\infty} \left[ (i+1)\,(i+z) \right]^{-1}.
\end{aligned}
$$

Particular values of $\psi(z)$ are

$$
\psi(1) = -\gamma, \quad \psi(0.5) = -\gamma - 2\ln 2 \approx -1.963510.
$$

The table in Part III gives values of $\psi(z)$ and $\psi'(z)$ for $1 \leq z \leq 2$. Fig. 2/17 shows the courses of $\Gamma(z)$ and $\psi(z)$.

The reflection formula is

$$
\psi(1-z) = \psi(z) + \pi \cot(z\,\pi).
$$

Asymptotic expansions are

$$
\psi(z) = \ln z - \frac{1}{2\,z} - \frac{1}{12\,z^2} + \frac{1}{120\,z^4} - \frac{1}{252\,z^6} + \cdots
$$

Figure 2/17: Gamma function and psi function

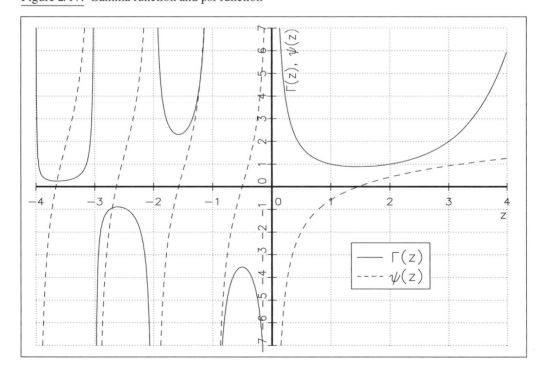

(leading to $\psi(z) \approx \ln(z - 0.5)$, provided $z \geq 2$) and

$$\psi'(z) = \frac{1}{z} + \frac{1}{2\,z^2} + \frac{1}{6\,z^3} - \frac{1}{30\,z^5} + \frac{1}{42\,z^7} - \frac{1}{30\,z^9} + \cdots$$

$$\psi''(z) = -\frac{1}{z^2} - \frac{1}{z^3} - \frac{1}{2\,z^4} + \frac{1}{6\,z^6} - \frac{1}{6\,z^8} + \frac{3}{10\,z^{10}} - \frac{5}{6\,z^{15}} + \cdots$$

## C. Incomplete gamma function

The incomplete gamma function is defined as

$$\gamma(z \mid u) := \int_0^u t^{z-1}\, e^{-t}\, \mathrm{d}t.$$

Another notation is

$$\Gamma_u(z) = \gamma(z \mid u).$$

The complement is

$$\Gamma(z \mid u) := \int_u^\infty t^{z-1}\, e^{-t}\, \mathrm{d}t,$$

i.e.,

$$\gamma(z \mid u) + \Gamma(z \mid u) = \Gamma(z).$$

An infinite series formula for $\gamma(z \mid u)$ is

$$\gamma(z \mid u) = \sum_{i=0}^{\infty} \frac{(-1)^i}{i!}\, \frac{u^{z+i}}{z+i}.$$

The following recurrence formulas hold

$$\gamma(z+1 \mid u) = z\,\gamma(z \mid u) - u^z\, e^{-u},$$
$$\Gamma(z+1 \mid u) = z\,\Gamma(z \mid u) - u^z\, e^{-u}.$$

The **incomplete gamma function ratio**

$$\frac{\gamma(z \mid u)}{\Gamma(z)} = \frac{\Gamma_u(z)}{\Gamma(z)}$$

is related to the distribution function $F(x \mid \lambda, c)$ of a gamma–distributed random variable $X$ with density

$$f(x \mid \lambda, c) = \begin{cases} \dfrac{\lambda\,(\lambda\,x)^{c-1}\,\exp(-\lambda\,x)}{\Gamma(c)} & \text{for} \quad x > 0;\ \lambda,\ c \in \mathbb{R}^+ \\ 0 & \text{for} \quad x \quad \text{else} \end{cases}$$

as

$$F(x \mid \lambda, c) = \begin{cases} 0 & \text{for} \quad x < 0 \\[2mm] \dfrac{\gamma(c \mid \lambda x)}{\Gamma(c)} & \text{for} \quad x \ge 0. \end{cases}$$

KARL PEARSON (1857 – 1936) has tabulated[22] an especially parameterized version of the incomplete gamma function ratio:

$$I(u, p) := \frac{\Gamma_{u\sqrt{p+1}}(p+1)}{\Gamma(p+1)}.$$

The moments about zero of a general WEIBULL variable are linked to those of the reduced WEIBULL variable. Inserting $X = a + b\,U$ into (2.63a) and using (2.63d), we get

$$\begin{aligned} \mu_r'(X) = \mathrm{E}(X^r) &= \mathrm{E}\left[(a + b\,U)^r\right] \\[2mm] &= \sum_{j=0}^{r} \binom{r}{j} a^j\, b^{r-j}\, \mathrm{E}(U^{r-j}) \\[2mm] &= \sum_{j=0}^{r} \binom{r}{j} a^j\, b^{r-j}\, \Gamma\!\left(\frac{r-j}{c} + 1\right). \end{aligned} \tag{2.64a}$$

With the abbreviation given in (2.63e) the first four raw moments are

$$\mu_1'(X) = \mathrm{E}(X) \quad = \quad a + b\,\Gamma_1 =: \mu_X, \quad \textbf{mean of } X, \tag{2.64b}$$

$$\mu_2'(X) = \mathrm{E}(X^2) \quad = \quad a^2 + 2\,a\,b\,\Gamma_1 + b^2\,\Gamma_2, \tag{2.64c}$$

$$\mu_3'(X) = \mathrm{E}(X^3) \quad = \quad a^3 + 3\,a^2\,b\,\Gamma_1 + 3\,a\,b^2\,\Gamma_2 + b^3\,\Gamma_3, \tag{2.64d}$$

$$\mu_4'(X) = \mathrm{E}(X^4) \quad = \quad a^4 + 4\,a^3\,b\,\Gamma_1 + 6\,a^2\,b^2\,\Gamma_2 + 4\,a\,b^3\,\Gamma_3 + b^4\,\Gamma_4. \tag{2.64e}$$

The **raw moment generating function** $M_X(t)$ of a random variable $X$, if it exists (i.e., is finite), is the expectation

$$M_X(t) := \mathrm{E}\!\left(e^{t\,X}\right). \tag{2.65a}$$

When $M_X(t)$ exist for some interval $|t| < T$, where $T > 0$, then $\mu_r'(X)$ is the coefficient of $t^r/r!$ in the TAYLOR–expansion of $M_X(t)$:

$$M_X(t) = \mathrm{E}\!\left(e^{t\,X}\right) = 1 + \sum_{r \ge 1} \frac{t^r}{r!}\, \mu_r'(X), \tag{2.65b}$$

or stated otherwise

$$\mu_r'(X) = \left.\frac{\mathrm{d}^r M_X(t)}{\mathrm{d}t^r}\right|_{t=0}. \tag{2.65c}$$

---

[22] K. PEARSON (1922): Tables of the incomplete $\Gamma$–Function; H.M. Stationery Office, London.

With regard to the reduced WEIBULL variable we get

$$M_U(t) = \mathrm{E}(e^{tU}) = \int_0^\infty c\, u^{c-1}\, \exp(t\, u - u^c)\, \mathrm{d}u. \tag{2.65d}$$

There exists no closed form for the integral in (2.65d). In general, all other generating functions, when applied to the WEIBULL distribution, also have no closed form. Combining (2.63d) with (2.65b), $M_U(t)$ may be expressed as

$$M_U(t) = 1 + \sum_{r \geq 1} \frac{t^r}{r!}\, \Gamma\!\left(\frac{r}{c} + 1\right). \tag{2.65e}$$

The $r$–th moment of $X$ about a constant $a$ is $\mathrm{E}\big[(X - a)^r\big]$. When $a = \mu = \mathrm{E}(X)$, we get the $r$–th **moment about the mean** (also called **central moment** or **corrected moment**):

$$\mu_r(X) := \mathrm{E}\big[(X - \mu)^r\big]. \tag{2.66a}$$

The raw moments and central moments are linked by either

$$\mu_r(X) = \sum_{j=0}^{r} \binom{r}{j} (-1)^j\, \mu^j\, \mu'_{r-j} \tag{2.66b}$$

or

$$\mu'_r(X) = \sum_{j=0}^{r} \binom{r}{j} \mu^j\, \mu_{r-j}. \tag{2.66c}$$

So the central moments of the reduced WEIBULL variable $U$ are

$$\mu_r(U) = \mathrm{E}\big[(U - \Gamma_1)^r\big] = \sum_{j=0}^{r} \binom{r}{j} (-1)^j\, \Gamma_1^j\, \Gamma_{r-j}, \tag{2.67a}$$

and especially for $r = 1,\ 2,\ 3,\ 4$:

$$\mu_1(U) = 0, \tag{2.67b}$$

$$\mu_2(U) = \Gamma_2 - \Gamma_1^2 =: \mathrm{Var}(U) =: \sigma_U^2,\ \textbf{variance of } U, \tag{2.67c}$$

$$\mu_3(U) = \Gamma_3 - 3\,\Gamma_2\,\Gamma_1 + 2\,\Gamma_1^3, \tag{2.67d}$$

$$\mu_4(U) = \Gamma_4 - 4\,\Gamma_3\,\Gamma_1 + 6\,\Gamma_2\,\Gamma_1^2 - 3\,\Gamma_1^4. \tag{2.67e}$$

For the general WEIBULL variable $X$, we get

$$\begin{aligned}
\mu_r(X) &= \mathrm{E}\left[\big\{(a + b\,U) - (a + b\,\Gamma_1)\big\}^r\right] \\
&= b^r\, \mathrm{E}\big[(U - \Gamma_1)^r\big] \\
&= b^r\, \mu_r(U).
\end{aligned} \tag{2.67f}$$

Especially,

$$\mu_2(X) = b^2 \left(\Gamma_2 - \Gamma_1^2\right) =: \text{Var}(X) := \sigma_X^2 \qquad (2.67\text{g})$$

is the variance of $X$.

The **central moment generating function** of a random variable $X$ is the expectation

$$Z_X(t) \; := \; \text{E}\!\left[e^{(X-\mu)\,t}\right]$$

$$= \; e^{-t\,\mu}\, M_X(t), \qquad (2.68\text{a})$$

if it exists. $\text{E}\!\left[(X-\mu)^r\right] = \mu_r$ is linked to $Z_X(t)$ by

$$Z_X(t) = 1 + \sum_{r \geq 1} \frac{t^r}{r!}\,\mu_r. \qquad (2.68\text{b})$$

There exist several other types of moments which are in close relation to the raw moments or the central moments, respectively. Generating functions for these moments may be found too.

1. **Standardized moments:**

$$\mu_r^* := \text{E}\!\left[\left(\frac{X-\mu}{\sigma}\right)^r\right]; \; \mu = \text{E}(X); \; \sigma = \sqrt{\text{Var}(X)} = \sqrt{\mu_2(X)} \qquad (2.69\text{a})$$

with

$$\mu_r^* := \frac{\mu_r}{\sigma^r}. \qquad (2.69\text{b})$$

2. **Ascending factorial raw moments:**

$$\mu'_{[r]} := \text{E}\!\left(X^{[r]}\right), \; X^{[r]} := X\,(X+1)\,(X+2)\ldots(X+r-1), \qquad (2.70\text{a})$$

with

$$\mu'_{[1]} \; = \; \mu'_1; \; \mu'_{[2]} = \mu'_2 + \mu; \; \mu'_{[3]} = \mu'_3 + 3\,\mu'_2 + 2\,\mu;$$

$$\mu'_{[4]} \; = \; \mu'_4 + 6\,\mu'_3 + 11\,\mu'_2 - 6\,\mu$$

and generating function

$$NF_X(t) := \text{E}\!\left[(1-t)^{-X}\right], \qquad (2.70\text{b})$$

where

$$\mu'_{[r]} = \left.\frac{d^r\, NF_X(t)}{dt^r}\right|_{t=0}.$$

3. **Descending factorial raw moments**

$$\mu'_{(r)} := \text{E}\!\left[X^{(r)}\right], \; X^{(r)} := X\,(X-1)\,(X-2)\ldots(X-r+1), \qquad (2.71\text{a})$$

with

$$\mu'_{(1)} \; = \; \mu'_1; \; \mu'_{(2)} = \mu'_2 - \mu; \; \mu'_{(3)} = \mu'_3 - 3\,\mu'_2 + 2\,\mu;$$

$$\mu'_{(4)} \; = \; \mu'_4 - 6\,\mu'_3 + 11\,\mu'_2 - 6\,\mu$$

and generating function

$$Nf_X(t) := \mathrm{E}\big[(1+t)^X\big], \tag{2.71b}$$

where

$$\mu'_{(r)} = \frac{\mathrm{d}^r Nf_X(t)}{\mathrm{d}t^r}\bigg|_{t=0}.$$

### 4. Ascending factorial central moments

$$\mu_{[r]} := \mathrm{E}\Big[(X-\mu)^{[r]}\Big],\ X-\mu)^{[r]} := (X-\mu)\,(X-\mu+1)\,(X-\mu+2)\ldots(X-\mu+r-1), \tag{2.72a}$$

with

$$\mu_{[1]} = 0;\quad \mu_{[2]} = \mu_2;\quad \mu_{[3]} = \mu_3 + 3\,\mu_2;$$
$$\mu_{[4]} = \mu_4 + 6\,\mu'_3 + 11\,\mu'_2.$$

The corresponding generating function is

$$ZF_X(t) := (1-t)^\mu\,\mathrm{E}\big[(1-t)^X\big] \tag{2.72b}$$

where

$$\mu_{[r]} = \frac{\mathrm{d}^r ZF_X(t)}{\mathrm{d}t^r}\bigg|_{t=0}.$$

### 5. Descending factorial central moments

$$\mu_{(r)} := \mathrm{E}\big[(X-\mu)^{(r)}\big],\ (X-\mu)^{(r)} := (X-\mu)\,(X-\mu-1)\,(X-\mu-2)\ldots(X-\mu-r+1), \tag{2.73a}$$

with

$$\mu_{(1)} = 0;\quad \mu_{(2)} = \mu_2;\quad \mu_{(3)} = \mu_3 - 3\,\mu_2;$$
$$\mu_{(4)} = \mu_4 - 6\,\mu_3 + 11\,\mu_2.$$

The generating function is

$$Zf_X(t) := (1+t)^{-\mu}\,\mathrm{E}\big[(1+t)^X\big] \tag{2.73b}$$

leading to

$$\mu_{(r)} = \frac{\mathrm{d}^r Zf_X(t)}{\mathrm{d}t^r}\bigg|_{t=0}.$$

### Absolute raw moments

$$\nu'_r := \mathrm{E}[\,|X|^r\,] \tag{2.74a}$$

of odd order $r$ for a WEIBULL distribution must be evaluated numerically when $a < 0$. For even order $r$ we have

$$\nu'_r = \mu'_r.$$

**Absolute central moments**

$$\nu_r := \mathrm{E}[\,|X - \mu|^r\,] \qquad (2.74b)$$

have to be evaluated numerically for odd order $r$ when $X$ has a WEIBULL distribution, but

$$\nu_r = \mu_r \ \text{ for } \ r = 2\,k; \ k = 1, 2, \dots$$

Whereas the raw moment generating function $M_X(t) = \mathrm{E}(e^{tX})$ may not exist for each random variable, the **characteristic function**

$$\varphi_X(t) := \mathrm{E}(e^{i\,t\,X}), \ \ i := \sqrt{-1} \qquad (2.75a)$$

always exists and is uniquely determined by the CDF of $X$ when $X$ is continuous. It satisfies

$$\varphi_X(0) \ = \ 1, \qquad (2.75b)$$

$$|\varphi_X(t)| \ \leq \ 1, \qquad (2.75c)$$

$$\varphi_X(-t) \ = \ \overline{\varphi_X(t)}, \qquad (2.75d)$$

where the overline denotes the complex conjugate. $\varphi_X(t)$ has properties similar to those of $M_X(t)$, e.g.,

$$\mu_r' = i^r\,\varphi_X^{(r)}(0), \qquad (2.75e)$$

where

$$\varphi_X^{(r)}(0) := \left. \frac{\mathrm{d}^r \varphi_X(t)}{\mathrm{d}t^r} \right|_{t=0}.$$

The characteristic function uniquely determines the DF $f_X(x)$ as

$$f_X(x) = \frac{1}{2\,\pi} \int_{-\infty}^{\infty} e^{-i\,t\,x}\,\varphi_X(t)\,\mathrm{d}t. \qquad (2.75f)$$

If $X_1$ and $X_2$ are independent random variables with characteristic functions $\varphi_1(t)$ and $\varphi_2(t)$, respectively, then the characteristic function of the sum $X_1 + X_2$ is

$$\varphi_{X_1+X_2}(t) = \varphi_1(t)\,\varphi_2(t),$$

and that of the difference $X_1 - X_2$ is

$$\varphi_{X_1-X_2}(t) = \varphi_1(t)\,\varphi_2(-t).$$

A concept closely related to the notion of a raw moment is that of a **cumulant**. The cumulant of order $r$ is denoted by $\kappa_r$. Cumulants and raw moments are defined by the following identity:[23]

$$\exp\{\kappa_1\,t + \kappa_2\,t^2/2! + \kappa_3\,t^3/3! + \dots\} = 1 + \mu_1'\,t + \mu_2'\,t^2/2! + \mu_3'\,t^3/3! \dots \qquad (2.76a)$$

---

[23] Some authors use $i\,t$ (with $i = \sqrt{-1}$) instead of $t$.

Evaluation of (2.76a) leads to

$$\begin{aligned}
\mu_1' &= \mu = \kappa_1 = \mathrm{E}(X), \\
\mu_2' &= \kappa_2 + \kappa_1^2, \\
\mu_3' &= \kappa_3 + 3\,\kappa_2\,\kappa_1 + \kappa_1^3, \\
\mu_4' &= \kappa_4 + 4\,\kappa_3\,\kappa_1 + 3\,\kappa_2^2 + 6\,\kappa_2\,\kappa_1^2 + \kappa_1^4,
\end{aligned}$$

generally

$$\mu_r' = \sum_{j=1}^{r} \binom{r-1}{j-1} \mu_{r-j}'\,\kappa_j. \qquad (2.76b)$$

For the reverse relation we have

$$\begin{aligned}
\kappa_1 &= \mu_1' = \mu, \\
\kappa_2 &= \mu_2' - \mu = \mu_2 \ \text{(variance)}, \\
\kappa_3 &= \mu_3' - 3\,\mu_2'\,\mu + 2\,\mu^3 = \mu_3 \ \text{(third central moment)}, \\
\kappa_4 &= \mu_4' - 4\,\mu_3' - 3\,(\mu_2')^2 + 12\,\mu_2'\,\mu^2 - 6\,\mu^4 = \mu_4 - 3\,\mu_2^2.
\end{aligned}$$

(2.76a) may be stated by using generating functions. The natural logarithm of the raw moment generating function $M_X(t)$, defined in (2.65a), is the **cumulant generating function** $K_X(t)$:

$$K_X(t) = \ln M_X(t) = \sum_{r \geq 1} \frac{t^r}{r!}\,\kappa_r. \qquad (2.77a)$$

(Note the missing term belonging to $r = 0$!) Because of

$$M_{X+a}(t) = \mathrm{E}\big[e^{t\,(X+a)}\big] = e^{t\,a}\,M_X(t),$$

we have

$$K_{X+a}(t) = t\,a + K_X(t). \qquad (2.77b)$$

So, for $r \geq 2$ the coefficients of $t^r/r!$ in the TAYLOR–expansion of $K_{X+a}(t)$ and $K_X(t)$ are identical, and the cumulants for $r \geq 2$ are not affected by the addition of a constant:

$$\left.\begin{aligned}
\kappa_r(X + a) &= \kappa_r(X), \ r \geq 2, \\
\kappa_1(X + a) &= a + \kappa_1(X).
\end{aligned}\right\} \qquad (2.77c)$$

For this reason the cumulants have also been called **semivariants** or **half–invariants**. Putting $a = -\mu$ shows that, for $r \geq 2$, the cumulants $\kappa_r$ are functions of the central moments $\mu_r$. In fact

$$\kappa_2 = \mu_2 = \sigma^2,$$

$$\kappa_3 = \mu_3,$$

$$\kappa_4 = \mu_4 - 3\mu_2^2,$$

$$\kappa_5 = \mu_5 - 10\mu_3\mu_2.$$

Let $X_1, X_2, \ldots, X_n$ be independent random variables and $X = \sum_{j=1}^{n} X_j$. Provided the relevant functions exist, we have[24]

$$K_X(t) = \sum_{j=1}^{n} K_{X_j}(t). \tag{2.78a}$$

It follows from (2.78a):

$$\kappa_r\left(\sum_{j=1}^{n} X_j\right) = \sum_{j=1}^{n} \kappa_r(X_j) \ \forall \ r, \tag{2.78b}$$

so the cumulant of a sum equals the sum of the cumulants, finally explaining the name of this statistical concept.

Yet another concept is that of the **information generating function**, defined as

$$T_X(t) := \mathrm{E}\big[\{f(X)\}^t\big], \tag{2.79a}$$

from which the **entropy** $I(X)$ of the distribution results as

$$I(X) = -\frac{\mathrm{d}T_X(t)}{\mathrm{d}t}\bigg|_{t=1}. \tag{2.79b}$$

With respect to the reduced WEIBULL distribution, (2.79a) leads to

$$T_U(t) = c^t \int_0^{\infty} u^{t(c-1)} e^{-tu^c} \,\mathrm{d}u. \tag{2.80a}$$

Substituting $t\,u^c = v$ (taking $t > 0$ and noting that $c > 0$), we obtain

$$T_U(t) = \frac{c^{t-1}}{t^{[1+t(c-1)]/c}} \Gamma\left[\frac{1+t(c-1)}{c}\right], \tag{2.80b}$$

so that the **WEIBULL entropy** is

$$I_c(U) = -T'_U(1) = \frac{c-1}{c}\gamma - \ln c + 1, \tag{2.80c}$$

---

[24] The corresponding proposition for the raw moment generating function is

$$M_X(t) = \prod_{j=1}^{n} M_{X_j}(t).$$

$\gamma$ being EULER's constant ($\gamma \approx 0.57722$). Fig. 2/18 shows $I_c(U)$ as a function of $c$. The entropy takes its maximum $(\gamma - 1) - \ln \gamma + 1 \approx 1.12676$ for $c = \gamma$. The limiting behavior is $\lim\limits_{c \to 0} I_c(U) = \lim\limits_{c \to \infty} I_c(U) = -\infty$.

**Figure 2/18:** Entropy of the reduced WEIBULL distribution

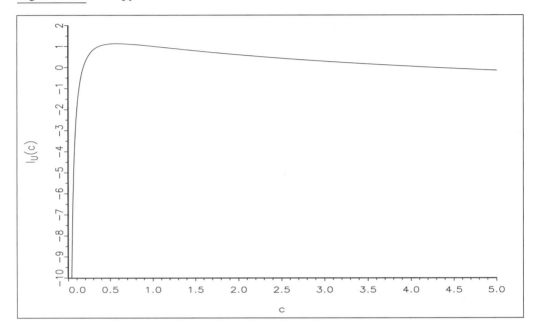

To describe a random variable with respect to its form and to see similarities between the distributions of different variables, **ratios of moments** and **of cumulants** have been introduced. There exist two types of moment ratios built upon the central moments:

1. $\alpha$–**coefficients of order** $r$

$$\alpha_r := \frac{\mu_r}{\mu_2^{r/2}}; \ r = 2, 3, \ldots \tag{2.81a}$$

The $\alpha$–coefficients are identical to the reduced moments, see (2.69a,b):

$$\alpha_r = \mu_r^*.$$

2. $\beta$–**coefficients of order** $r$

$$\beta_r := \begin{cases} \beta_{2k+1} & := \dfrac{\mu_3 \, \mu_{2k+3}}{\mu_2^{k+3}}; & k = 0, 1, 2, \ldots \\[2ex] \beta_{2k} & := \dfrac{\mu_{2k+2}}{\mu_2^{k+1}}; & k = 1, 2, \ldots \end{cases} \tag{2.81b}$$

The ratios of cumulants are known as $\gamma$–coefficients and are defined as

$$\gamma_r := \frac{\kappa_{r+2}}{\kappa_2^{r/2+1}} ; \quad r = 1, 2, \ldots \tag{2.81c}$$

Each of the three types of ratios is invariant with respect to a linear transformation $a + b\,X$, $b > 0$, of the random variable $X$. For $b < 0$ the sign of the ratios changes when $r$ is of odd order.

The application of $\alpha_3 = \mu_3/\mu_2^{3/2}$ as a measure of skewness and of $\alpha_4 = \mu_4/\mu_2^2$ as a measure of kurtosis to the WEIBULL distribution will be discussed in Sect. 2.9.4. These two most frequently used ratios are linked to the $\beta$– and $\gamma$–coefficients as follows:

$$\alpha_3 \quad := \quad \frac{\mu_3}{\mu_2^{3/2}} = \text{sign}(\mu_3)\,\sqrt{\beta_1} = \text{sign}(\mu_3)\sqrt{\frac{\mu_3^2}{\mu_2^3}} \tag{2.82a}$$

$$= \quad \gamma_1 = \frac{\kappa_3}{\kappa_2^{3/2}} , \tag{2.82b}$$

$$\alpha_4 \quad = \quad \frac{\mu_4}{\mu_2^2} = \beta_2 \tag{2.83a}$$

$$= \quad \gamma_2 + 3 = \frac{\kappa_4}{\kappa_2^2} + 3. \tag{2.83b}$$

### 2.9.2   Mean and its relation to mode and median

The mean or expected value of the reduced WEIBULL variable $U$ is given (see (2.63f)) by

$$\text{E}(U) = \mu_U = \Gamma\left(1 + \frac{1}{c}\right) =: \Gamma_1.$$

Fig. 2/19 shows how $\text{E}(U)$ depends on $c$ in a non–monotone manner. The course of $\text{E}(U)$ is determined by the behavior of $\Gamma(x)$, $x > 0$ (see Fig. 2.17). $\Gamma(x)$ comes down from positive infinity at $x = 0$ and falls with convex curvature to $x_0 \approx 1.4616321450$, where

$$\min_x \Gamma(x) = \Gamma(x_0) \approx 0.8856031944.$$

Thereafter $\Gamma(x)$ grows with convex curvature to positive infinity for $x \to \infty$. Setting

$$x = 1 + \frac{1}{c},$$

this behavior of the gamma function means to $\text{E}(U)$:

- $\displaystyle \lim_{c \to 0} \Gamma\left(1 + \frac{1}{c}\right) \quad = \quad \infty,$

- $\displaystyle \lim_{c \to \infty} \Gamma\left(1 + \frac{1}{c}\right) \quad = \quad 1,$

- $\displaystyle \min_c \Gamma\left(1 + \frac{1}{c}\right) \quad = \quad \Gamma\left(1 + \frac{1}{c_0}\right) \approx 0.8856031944$ with $c_0 = (x_0 - 1)^{-1} \approx 2.16623.$

A small value of $c$ implies a great mean, but — see Sect. 2.9.3 — a great variance too. The high value of the mean, occurring with early–time–failure distributions ($c < 1$), is thus essentially determined by the long lifetime of the few robust items in the population.

Figure 2/19: Mean, median and mode of the reduced WEIBULL distribution

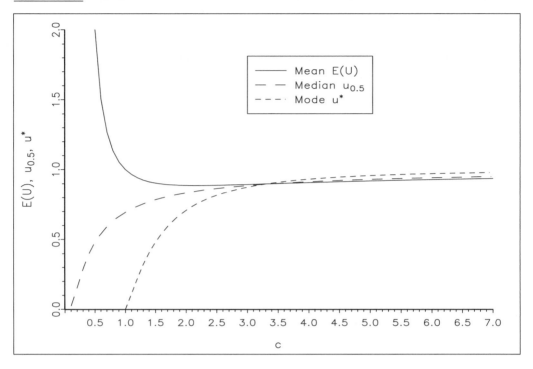

Table 2/2 shows how the mean $E(U) = \Gamma(1 + 1/c)$, the median $u_{0.5} = (\ln 2)^{1/c}$ and the mode $u^* = \left[(c - 1)/c\right]^{1/c}$ vary with $c$ in the vicinity of $c = 3.3$. There exists no value of $c$ resulting in $E(U) = u_{0.5} = u^*$. Instead, we have three different values of $c$ where two of these parameters coincide:

- $u^* = u_{0.5} \approx 0.89362816$ for $c \approx 3.25889135$,

- $u^* = E(U) \approx 0.89718563$ for $c \approx 3.31246917$,

- $E(U) = u_{0.5} \approx 0.89892230$ for $c \approx 3.43954065$.

Looking at the size–relation of these three location parameters one can make some statements on the skewness of the WEIBULL distribution depending on $c$:

- $u^* < u_{0.5} < E(U)$ for $c < 3.25889135$

    indicates a positively skewed density with a relatively long right–hand tail;

- $u^* > u_{0.5} > E(U)$ for $c > 3.43954065$

Table 2/2:  Mean, median and mode of the reduced WEIBULL distribution for
$3.20 \leq c \leq 3.50$

| $c$ | $E(U)$ | $u_{0,5}$ | $u^*$ |
|---|---|---|---|
| 3.200000 | 0.8956537 | 0.8917805 | 0.8895036 |
| 3.210000 | 0.8957894 | 0.8920987 | 0.8902205 |
| 3.220000 | 0.8959252 | 0.8924151 | 0.8909305 |
| 3.230000 | 0.8960612 | 0.8927296 | 0.8916337 |
| 3.240000 | 0.8961973 | 0.8930424 | 0.8923303 |
| 3.250000 | 0.8963334 | 0.8933532 | 0.8930202 |
| 3.258891 | 0.8964546 | 0.8936282 | 0.8936282 |
| 3.260000 | 0.8964697 | 0.8936623 | 0.8937036 |
| 3.270000 | 0.8966060 | 0.8939696 | 0.8943806 |
| 3.280000 | 0.8967424 | 0.8942752 | 0.8950512 |
| 3.290000 | 0.8968788 | 0.8945790 | 0.8957155 |
| 3.300000 | 0.8970153 | 0.8948810 | 0.8963736 |
| 3.310000 | 0.8971519 | 0.8951813 | 0.8970256 |
| 3.312469 | 0.8971856 | 0.8952552 | 0.8971856 |
| 3.320000 | 0.8972885 | 0.8954799 | 0.8976715 |
| 3.330000 | 0.8974251 | 0.8957769 | 0.8983114 |
| 3.340000 | 0.8975618 | 0.8960721 | 0.8989455 |
| 3.350000 | 0.8976985 | 0.8963657 | 0.8995737 |
| 3.260000 | 0.8964697 | 0.8936623 | 0.8937036 |
| 3.370000 | 0.8979719 | 0.8969479 | 0.9008128 |
| 3.380000 | 0.8981086 | 0.8972365 | 0.9014239 |
| 3.390000 | 0.8982453 | 0.8975236 | 0.9020294 |
| 3.400000 | 0.8983820 | 0.8978090 | 0.9026295 |
| 3.410000 | 0.8985187 | 0.8980929 | 0.9032241 |
| 3.420000 | 0.8986553 | 0.8983752 | 0.9038133 |
| 3.430000 | 0.8987920 | 0.8986559 | 0.9043972 |
| 3.439541 | 0.8989223 | 0.8989223 | 0.9049495 |
| 3.440000 | 0.8989286 | 0.8989351 | 0.9049759 |
| 3.450000 | 0.8990651 | 0.8992127 | 0.9055494 |
| 3.460000 | 0.8992016 | 0.8994889 | 0.9061178 |
| 3.470000 | 0.8993381 | 0.8997635 | 0.9066812 |
| 3.480000 | 0.8994745 | 0.9000366 | 0.9072395 |
| 3.490000 | 0.8996109 | 0.9003083 | 0.9077929 |
| 3.500000 | 0.8997472 | 0.9005785 | 0.9083415 |

indicates a negatively skewed density with a relatively long left–hand tail;

- $u^* \approx u_{0.5} \approx E(U)$ for $3.25889135 \leq c \leq 3.43954065$

indicates a roughly symmetric density.

Further details on skewness will be presented in Sect. 2.9.4.

Looking at Fig. 2.19 one can see that $E(U)$ may be approximated by 1 for $c > 0.58$ with a maximum error of 12.5%. A more accurate approximating formula is

$$E(U) \approx 1 - \frac{\gamma}{c} + \left( \frac{\pi^2}{6} + \gamma \right) \bigg/ (2\,c^2) \approx 1 - \frac{0.57722}{c} + \frac{0.98905}{c^2}, \qquad (2.84)$$

which is based on a TAYLOR–expansion of $\Gamma(1 + \frac{1}{c})$ around $c = 1$ neglecting terms of order greater than two.

The mean of the general WEIBULL variable $X$ is

$$E(X) = a + b\,\Gamma\left( 1 + \frac{1}{c} \right) = a + b\,\Gamma_1, \qquad (2.85)$$

where higher values of $a$ and/or $b$ lead to a high value of $E(X)$. Generally,

$$E(X) \; < \; a + b \;\; \text{for} \;\; c > 1,$$
$$E(X) \; = \; a + b \;\; \text{for} \;\; c = 1,$$
$$E(X) \; > \; a + b \;\; \text{for} \;\; c < 1.$$

From (2.85) we have the opportunity to substitute one of the parameters $a$, $b$ or $c$ in the DF function by $\mu_X = E(X)$. Because $c$ can only be isolated from (2.85) by numerical methods and — in the case of $\Gamma_1 < 1$ (see Fig. 2/18) — not yet unequivocally, $a$ and $b$ are the only candidates for this process. Substituting $a$ in (2.8) gives

$$f_X(x \,|\, \mu_X, b, c) = \frac{c}{b} \left( \frac{x - \mu_X}{b} + \Gamma_1 \right)^{c-1} \exp\left\{ - \left( \frac{x - \mu_X}{b} + \Gamma_1 \right)^c \right\}, \qquad (2.86a)$$

and substituting $b$ leads to

$$f_X(x \,|\, a, \mu_X, c) = \frac{c\,\Gamma_1}{\mu_X - a} \left( \frac{x - a}{\mu_X - a}\,\Gamma_1 \right)^{c-1} \exp\left\{ - \left( \frac{x - a}{\mu_X - a}\,\Gamma_1 \right)^c \right\}. \qquad (2.86b)$$

When $a = 0$, the last formula turns into the simple expression

$$f_X(x \,|\, 0, \mu_X, c) = c \left( \frac{\Gamma_1}{\mu_X} \right)^c x^{c-1} \exp\left\{ - \left( \frac{x}{\mu_X}\,\Gamma_1 \right)^c \right\}. \qquad (2.86c)$$

Another possibility of substituting $b$ is to use the standard deviation $\sigma_X$ (see Sect. 2.9.3). Writing the WEIBULL CDF with $a = 0$ by once substituting $b$ by $\mu_X$ and another time by $\sigma_X$ gives the opportunity to construct a nomogram to find $F_X(x \,|\, 0, b, c)$ as was done by KOTELNIKOV (1964).

From (2.85) we have

$$b = \frac{E(X) - a}{\Gamma\left(1 + \dfrac{1}{c}\right)} . \tag{2.87a}$$

When $a = 0$ the simple relation

$$b = \frac{E(X)}{\Gamma\left(1 + \dfrac{1}{c}\right)} \tag{2.87b}$$

tells us that there are numerous pairs of $b$ and $c$ leading to the same mean. Fig. 2/20 shows for four values (1, 2, 3 and 4) of $E(X)$ the $b$–$c$ combinations with equal mean, and Fig. 2/21 depicts four densities with $E(X) = 1$.

Figure 2/20: Combinations of $b$ and $c$ leading to the same mean

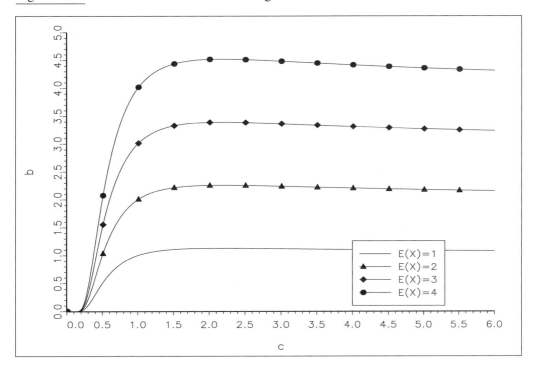

### 2.9.3 Variance, standard deviation and coefficient of variation

The **variance** of the reduced WEIBULL distribution is

$$\mathrm{Var}(U) = \sigma_U^2 = \Gamma\left(1 + \frac{2}{c}\right) - \Gamma\left(1 + \frac{1}{c}\right)^2 = \Gamma_2 - \Gamma_1^2, \tag{2.88a}$$

Figure 2/21: Four densities with $E(X) = 1$ and different $b$–$c$ combinations

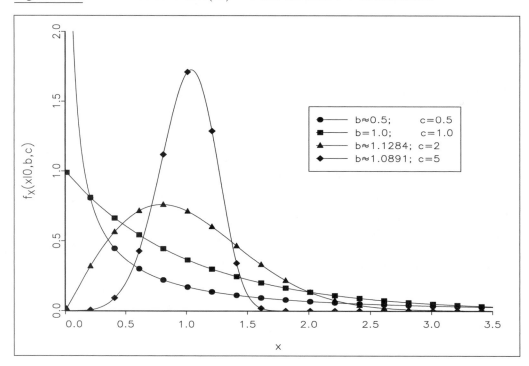

from which we derive the **standard deviation**

$$\sigma_U = \left(\Gamma_2 - \Gamma_1^2\right)^{1/2} \tag{2.88b}$$

and the **coefficient of variation**

$$\text{CV}(U) = \frac{\sigma_U}{\mu_U} = \frac{\left(\Gamma_2 - \Gamma_1^2\right)^{1/2}}{\Gamma_1}. \tag{2.88c}$$

Fig. 2/22 shows how these three measures of dispersion depend on $c$. All three are convexly decreasing functions of $c$, coming down from positive infinity at $c = 0$ and converging to zero with $c \to \infty$. For larger values of $c$ the standard deviation and coefficient of variation coincide because $\lim\limits_{c\to\infty} \Gamma(1 + 1/c) = 1$. Small $c$ means large dispersion and large $c$ means small dispersion. An approximating formula for $\text{Var}(U)$ is

$$\text{Var}(U) \approx \frac{\pi^2}{6\,c^2} \approx 1.64493\,c^{-2}. \tag{2.89}$$

Similar to (2.84) this approximation results from a TAYLOR–development of the gamma function around $c = 1$, terminating after the quadratic term and neglecting all powers of $c$ smaller than $-2$.

Figure 2/22: Variance, standard deviation and coefficient of variation of the reduced WEIBULL distribution

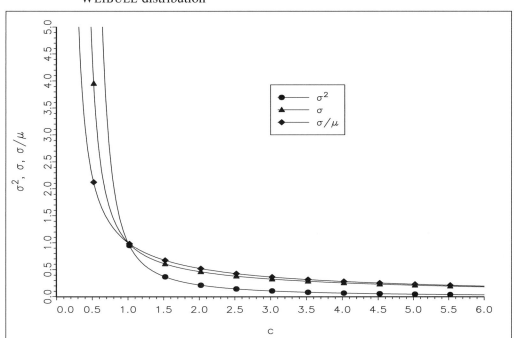

For a general WEIBULL distribution the three measures of dispersion are

$$\text{Var}(X) \;=\; \sigma_X^2 \;=\; b^2\,\text{Var}(U) \;=\; b^2\big[\Gamma_2 - \Gamma_1^2\big], \tag{2.90a}$$

$$\sigma_X \;=\; b\,\sigma_U \;=\; b\big[\Gamma_2 - \Gamma_1^2\big]^{1/2}, \tag{2.90b}$$

$$\text{CV}(X) \;=\; \frac{\sigma_X}{\mu_X} \;=\; \frac{b\big[\Gamma_2 - \Gamma_1^2\big]^{1/2}}{a + b\,\Gamma_1}\,. \tag{2.90c}$$

Variance and standard deviation now depend on $b$ too and are increasing with $b$. The coefficient of variation is depending on all three parameters. Using (2.90b), the scale parameter $b$ appearing in the density formula (2.8) may be substituted by $\sigma_X$:

$$f_X(x\,|\,a, \sigma_X, c) = \frac{c}{\sigma_X^c}\big[\Gamma_2 - \Gamma_1^2\big]^{c/2} (x - a)^{c-1} \exp\left\{-\left(\frac{x - a}{\sigma_X}\right)^c \big[\Gamma_2 - \Gamma_1^2\big]^{c/2}\right\}.$$
$$\tag{2.91}$$

### 2.9.4 Skewness and kurtosis[25]

The most popular way to measure the skewness and the kurtosis of a DF rests upon ratios of moments. But there exist other approaches, mainly based on percentiles, which will be presented too.

---

[25] Suggested reading for this section: BENJAMINI/KRIEGER (1999), COHEN (1973), DUBEY (1967a), GROENEVELD (1986, 1998), GURKER (1995), HINKLEY (1975), LEHMAN (1963), MACGILLIVRAY (1992), ROUSU (1973), SCHMID/TREDE (2003), VAN ZWET (1979).

In Sect. 2.9.1 we have given three types of ratios based on moments; see (2.81a–c). Skewness is measured by

$$\alpha_3 \;=\; \frac{\mu_3}{\mu_2^{3/2}}\,, \tag{2.92a}$$

$$=\; \operatorname{sign}(\mu_3)\,\sqrt{\beta_1}, \tag{2.92b}$$

$$=\; \gamma_1. \tag{2.92c}$$

With respect to a WEIBULL distribution this measure turns into

$$\alpha_3 = \frac{\Gamma_3 - 3\,\Gamma_2\,\Gamma_1 + 2\,\Gamma_1^3}{\left(\Gamma_2 - \Gamma_1^2\right)^{3/2}}\,;\;\; \Gamma_i := \Gamma(1 + i/c). \tag{2.93}$$

Fig. 2/23 shows how $\alpha_3$ depends on $c$.

Figure 2/23: Moment ratios for a WEIBULL distribution

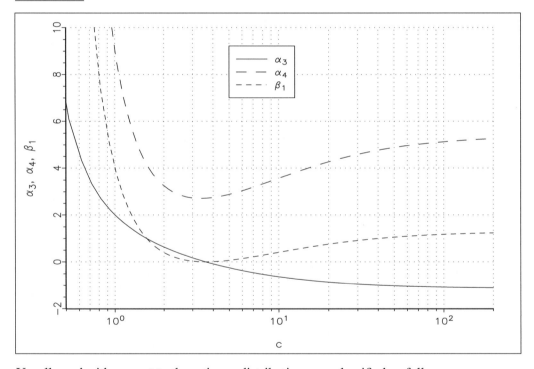

Usually and with respect to the ratio $\alpha_3$ distributions are classified as follows:

- The density is called **symmetric** if $\alpha_3 = 0$. For $c_0 \approx 3.60$ — more accurately: $c_0 = 3.6023494257197$ as calculated by COHEN (1973) — we have $\alpha_3 = 0$. But with this value of $c$ the WEIBULL distribution is not perfectly symmetric. For instance, mean, median and mode do not coincide when $c = c_0$:

$$\mathrm{E}(X) \;=\; a + 0.90114\,b$$

$$x_{0.5} \;=\; a + 0.90326\,b$$

$$x^* \;=\; a + 0.91369\,b,$$

and the left and right inflection points are not in equal distance to the mode:

$$x_\ell^{**} = a + 0.60137\,b$$

$$x_r^{**} = a + 1.21337\,b.$$

Other means to measure skewness as to be found in Tab. 2/3 will show divergent statements on the value of $c$ resulting into symmetry of a WEIBULL density.

- The density is called **positively skewed** (skewed to the right or steep on the left–hand side) showing a relatively long right–hand tail if $\alpha_3 > 0$. This happens for $c < c_0$.

- The density is called **negatively skewed** (skewed to the left or steep on the right–hand side) showing a relatively long left–hand tail if $\alpha_3 < 0$. This occurs for $c > c_0$.

Note, that $\alpha_3$ is a large positive value for small $c$, decreases to zero at about $c_0 \approx 3.60$ and then becomes moderately negative. It is not correct to say — as can be found in some papers, e.g., LEHMAN (1963) — that $\alpha_3$ converges to zero with $c \to \infty$. To study the asymptotic behavior of $\alpha_3$, we look at

$$\beta_1 = \frac{\left(\Gamma_3 - 3\,\Gamma_2\,\Gamma_1 + 2\,\Gamma_1^3\right)^2}{\left(\Gamma_2 - \Gamma_1^2\right)^3}\;;\;\; \Gamma_i := \Gamma(1 + i/c).$$

The limit of $\beta_1$ for $c \to \infty$ is an indeterminate expression of type $0/0$. After having applied L' HOSPITAL's rule six times we get

$$\lim_{c \to \infty} \beta_1 = \frac{20\left[\psi''(1)\right]^2}{90\left[2\,\psi'(1)\right]^3} \approx 1.298567591. \tag{2.94a}$$

Because of (2.92b) the result for $\alpha_3$ is

$$\lim_{c \to \infty} \alpha_3 \approx -1.13955. \tag{2.94b}$$

**Excursus:  Measuring skewness**

The measurement of skewness has attracted the attention of statisticians for over one hundred years. In 1896 GALTON introduced the percentile oriented measure[26]

$$ga := \frac{x_{0.8} - x_{0.5}}{x_{0.5} - x_{0.2}}\;;\;\; 0 \le ga < \infty. \tag{2.95}$$

$ga = 1$ indicates symmetry, and $ga < 1$ is interpreted as skewness to the left. In 1901 BOWLEY suggested

$$bo := \frac{(x_{0.75} - x_{0.5}) - (x_{0.5} - x_{0.25})}{x_{0.75} - x_{0.25}}\;:\;\; -1 \le bo \le +1. \tag{2.96}$$

---

[26] Percentile orientation of an index means that it is less affected by the tail behavior or — dealing with sample data — by outliers.

For a symmetric distribution we have $bo = 0$, whereas $bo \approx +1$ ($bo \approx -1$) indicates strong right (left) skewness. If $x_{0.8}$ and $x_{0.2}$ in $ga$ are replaced by $x_{0.75}$ and $x_{0.25}$, respectively, then

$$ga = \frac{1 + bo}{1 - bo}$$

is a monotonic increasing function of $bo$.

A natural generalization of BOWLEY's coefficient is the **skewness function**, see MACGILLIVRAY (1992):

$$\gamma_\alpha(F) := \frac{\left[F^{-1}(1 - \alpha) - x_{0.5}\right] - \left[x_{0.5} - F^{-1}(\alpha)\right]}{F^{-1}(1 - \alpha) - F^{-1}(\alpha)}; \quad 0 < \alpha < 0.5; \qquad (2.97)$$

where $F^{-1}(1 - \alpha)$ is the percentile of order $1 - \alpha$ belonging to the CDF $F(x)$. We have

- $|\gamma_\alpha(F)| \leq 1$,

- $\gamma_\alpha(F) = 0$ for a symmetric density,

- $\gamma_\alpha(F) \approx 1$ for extreme right skewness,

- $\gamma_\alpha(F) \approx -1$ for extreme left skewness.

The **tilt factor** introduced by HINKLEY (1975) starts from $\gamma_\alpha(F)$ and is defined as

$$\tau_\alpha(F) := \frac{1 + \gamma_\alpha(F)}{1 - \gamma_\alpha(F)} = \frac{F^{-1}(1 - \alpha) - x_{0.5}}{x_{0.5} - F^{-1}(\alpha)}; \quad 0 < \alpha < 0.5. \qquad (2.98)$$

$\tau_\alpha(F)$ is a generalization of GALTON's coefficient.

It seems reasonable to formulate some requirements to be satisfied by a general skewness parameter $\gamma$ for a continuous random variable $X$. The following three requirements are given by VAN ZWET (1979):

1. $\gamma(a + b\,X) = \gamma(X)$ for any $b > 0$ and $-\infty < a < \infty$,

2. $\gamma(-X) = -\gamma(X)$,

3. If $X$ and $Y$ are two random variables for which $X <_c Y$ (i.e., $X$ $c$–precedes $Y$ in the sense of VAN ZWET, see below), then $\gamma(X) \leq \gamma(Y)$.

The first requirement assures that a skewness measure be location and scale invariant. The second requirement assures that changing the sign of a random variable changes the sign of the skewness, but not its magnitude. The $c$–precedence means that if $X$ and $Y$ have CDFs $F$ and $G$, respectively, then $X <_c Y$ if and only if $G^{-1}(F(x))$ is convex. In this case there is good reason to conclude that $Y$ is at least as skew to the right as $X$.

With respect to the (reduced) WEIBULL CDF

$$F_U(u \,|\, c) = 1 - \exp\left[-u^c\right]$$

assume that there are two parameter values of $c$ with $c_2 < c_1$. It is easy to show that $F_2^{-1}\big(F_1(u)\big) = u^{c_1/c_2}$ which is convex for $c_2 < c_1$. Hence as $c$ increases, the skewness decreases.

$\alpha_3$ defined in (2.93) satisfies the aforementioned three requirements as does the measure

$$gr := \frac{\mathrm{E}(X) - x_{0.5}}{\mathrm{E}\big(|X - x_{0.5}|\big)} \tag{2.99}$$

suggested by GROENEVELD (1986) (see Tab. 2/3). But all skewness measures resting upon the mean–median–mode inequality, as, e.g.,

$$\frac{\mathrm{E}(X) - x^*}{\sigma} \quad \text{or} \quad \frac{\mathrm{E}(X) - x_{0.5}}{\sigma},$$

do not satisfy the third requirement.

Fig. 2/24 shows how BOWLEY's and GALTON's measures of skewness vary with $c$. We have

- $ga = 1$ for $c \approx 3.3047279$ and $\lim\limits_{c \to \infty} ga = 0.743231$,

- $bo = 0$ for $c \approx 3.2883129$ and $\lim\limits_{c \to \infty} bo = -0.118433$.

Figure 2/24: BOWLEY's and GALTON's measures of skewness for the WEIBULL distribution

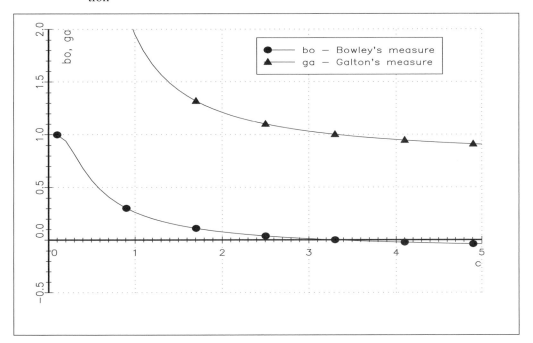

Tab. 2/3 summarizes how the order of mean, median and mode changes with $c$ and what is the sign of $\alpha_3$ and $gr$ in the respective interval for $c$.

Table 2/3:   Skewness properties of a WEIBULL density

| $c$ | Relationship of mean $\mu$, median $x_{0.5}$ and mode $x^*$ | Sign of $\alpha_3$ | Sign of $gr$ |
|:---:|:---:|:---:|:---:|
| $c < 3.2589$ | $x^* < x_{0.5} < \mu$ | $+$ | $+$ |
| $3.2589 < c < 3.3125$ | $x_{0.5} < x^* < \mu$ | $+$ | $+$ |
| $3.3125 < c < 3.4395$ | $x_{0.5} < \mu < x^*$ | $+$ | $+$ |
| $3.4395 < c < 3.6023$ | $\mu < x_{0.5} < x^*$ | $+$ | $-$ |
| $c > 3.6023$ | $\mu < x_{0.5} < x^*$ | $-$ | $-$ |

The **kurtosis** of a DF is measured by

$$\alpha_4 = \frac{\mu_4}{\mu_2^2}, \tag{2.100a}$$

$$= \beta_2, \tag{2.100b}$$

$$= \gamma_2 + 3. \tag{2.100c}$$

The reference value for $\alpha_4$ is 3, the value taken by $\alpha_3$ when $X$ is normally distributed. A distribution having

- $\alpha_4 = 3$ is called mesokurtic,

- and platykurtic for $\alpha_4 < 3$,

- and leptokurtic for $\alpha_4 > 3$ .

$\alpha_4 > 3$ means that there is an excess of values in the neighborhood of the mean as well as far away from it, with a depletion of the flanks of the curve representing the distribution. $\alpha_4 < 3$ indicates a flatter top than the normal distribution, e.g., occurring with a rectangular distribution.

With respect to the WEIBULL distribution we have

$$\alpha_4 = \frac{\Gamma_4 - 4\,\Gamma_3\,\Gamma_1 + 6\,\Gamma_2\,\Gamma_1^2 - 3\,\Gamma_1^4}{\left(\Gamma_2 - \Gamma_1^2\right)^2}, \quad \Gamma_i := \Gamma(1 + i/c). \tag{2.101}$$

Fig. 2/23 shows how $\alpha_4$ varies with $c$. $\alpha_4$ is decreasing until it reaches a minimum of $\alpha_4 \approx 2.710513$ at $c \approx 3.360128$. Thereafter, $\alpha_4$ is increasing, but has a finite limit:[27] $\lim_{c\to\infty} \alpha_4 = 5.4$. When $c \approx 2.25$ and $c \approx 5.75$, we have $\alpha_4 = 3$, i.e., the kurtosis of a normal distribution. The WEIBULL distribution has less kurtosis than a normal distribution for $2.25 < c < 5.75$, outside that interval the kurtosis is higher than 3.

---

[27] The limit of $\alpha_4$ for $c \to \infty$ is an indeterminate expression of type 0/0. After having applied L' HOSPI-TAL's rule four times, we get

$$\lim_{c\to\infty} \alpha_4 = \frac{72\left[\psi'(1)\right]^2 + 24\,\psi'''(1)}{6\left[2\,\psi'(1)\right]^2} = 5.4.$$

A limit of $\infty$ as reported by LEHMAN (1963) is wrong.

Finally, we mention a percentile–oriented measure for kurtosis $(L)$ given in SCHMID/TREDE (2003), defined as the product of a measure of tail

$$T := \frac{x_{0.975} - x_{0.025}}{x_{0.875} - x_{0.125}} \tag{2.102a}$$

and a measure of peakedness

$$P := \frac{x_{0.875} - x_{0.125}}{x_{0.75} - x_{0.25}}, \tag{2.102b}$$

resulting in

$$L := T\,P = \frac{x_{0.975} - x_{0.025}}{x_{0.75} - x_{0.25}}. \tag{2.102c}$$

When the random variable is normally distributed, we have $L = 2.9058$. This value is reached by a WEIBULL variable for $c \approx 6.85$.

Summarizing the findings on measuring skewness and kurtosis of a WEIBULL distribution one can state: There exist values of the shape parameter $c$ for which the shapes of a WEIBULL distribution and a normal distribution are almost identical, i.e., for $3.35 \lesssim c \lesssim 3.60$. DUBEY (1967a) has studied the quality of correspondence of these two distributions. His results will be commented upon in Sect. 3.2.4.

When $\beta_2 = \alpha_4$ is plotted as a function of $\beta_1 = \alpha_1^3$ in the so–called PEARSON–diagram in Sect. 3.1.1, we will see which other distribution families closely resemble the WEIBULL distribution.

# 3 Related distributions

This chapter explores which other distributions the WEIBULL distribution is related to and in what manner. We first discuss (Section 3.1) how the WEIBULL distribution fits into some of the well–known systems or families of distributions. These findings will lead to some familiar distributions that either incorporate the WEIBULL distribution as a special case or are related to it in some way (Section 3.2). Then (in Sections 3.3.1 to 3.3.10) we present a great variety of distribution models that have been derived from the WEIBULL distribution in one way or the other. As in Chapter 2, the presentation in this chapter is on the theoretical or probabilistic level.

## 3.1 Systems of distributions and the WEIBULL distribution[1]

Distributions may be classified into families or systems such that the members of a family

- have the same special properties and/or
- have been constructed according to a common design and/or
- share the same structure.

Such families have been designed to provide approximations to as wide a variety of observed or empirical distributions as possible.

### 3.1.1 PEARSON system

The oldest system of distributions was developed by KARL PEARSON around 1895. Its introduction was a significant development for two reasons. Firstly, the system yielded simple mathematical representations — involving a small number of parameters — for histogram data in many applications. Secondly, it provided a theoretical framework for various families of sampling distributions discovered subsequently by PEARSON and others. PEARSON took as his starting point the skewed binomial and hypergeometric distributions, which he smoothed in an attempt to construct skewed continuous density functions. He noted that the probabilities $P_r$ for the hypergeometric distribution satisfy the difference equation

$$P_r - P_{r-1} = \frac{(r-a)\,P_r}{b_0 + b_1\,r + b_2\,r^2}$$

for values of $r$ inside the range. A limiting argument suggests a comparable differential equation for the probability density function

$$f'(x) = \frac{\mathrm{d}f(x)}{\mathrm{d}x} = \frac{(x-a)\,f(x)}{b_0 + b_1\,x + b_2\,x^2}\,. \tag{3.1}$$

---

[1] Suggested reading for this section: BARNDORFF–NIELSEN (1978), ELDERTON/JOHNSON (1969), JOHNSON/KOTZ/BALAKRISHNAN (1994, Chapter 4), ORD (1972), PEARSON/HARTLEY (1972).

The solutions $f(x)$ are the density functions of the **PEARSON system**.

The various types or families of curves within the PEARSON system correspond to distinct forms of solutions to (3.1). There are three main distributions in the system, designated types I, IV and VI by PEARSON, which are generated by the roots of the quadratic in the denominator of (3.1):

- type I with DF

$$f(x) = (1 + x)^{m_1} (1 - x)^{m_2}, \quad -1 \leq x \leq 1,$$

results when the two roots are real with opposite signs (The **beta distribution of the first kind** is of type I.);

- type IV with DF

$$f(x) = (1 + x^2)^{-m} \exp\{-\nu \tan^1(x)\}, \quad -\infty < x < \infty,$$

results when the two roots are complex;

- type VI with DF

$$f(x) = x^{m_2} (1 + x)^{-m_1}, \quad 0 \leq x < \infty,$$

results when the two roots are real with the same sign. (The **F–** or **beta distribution of the second kind** is of type VI.)

Ten more "transition" types follow as special cases.

A key feature of the PEARSON system is that the first four moments (when they exist) may be expressed explicitly in terms of the four parameters $(a, b_0, b_1$ and $b_2)$ of (3.1). In turn, the two moments ratios,

$$\beta_1 = \frac{\mu_3^2}{\mu_2^3} \quad \text{(skewness)},$$

$$\beta_2 = \frac{\mu_4}{\mu_2^2} \quad \text{(kurtosis)},$$

provide a complete taxonomy of the system that can be depicted in a so–called **moment–ratio diagram** with $\beta_1$ on the abscissa and $\beta_2$ on the ordinate. Fig. 3/1 shows a detail of such a diagram emphasizing that area where we find the WEIBULL distribution.

- The limit for all distributions is given by

$$\beta_2 - \beta_1 = 1,$$

or stated otherwise: $\beta_2 \leq 1 + \beta_1$.

- The line for type III (**gamma distribution**) is given by

$$\beta_2 = 3 + 1.5 \beta_1.$$

It separates the regions of the type–I and type–VI distributions.

- The line for type V, separating the regions of type–VI and type–IV distributions, is given by

$$\beta_1 (\beta_2 + 3)^2 = 4 (4 \beta_2 - 3 \beta_1) (2 \beta_2 - 3 \beta_1 - 6)$$

Figure 3/1:  Moment ratio diagram for the PEARSON system showing the WEIBULL distribution

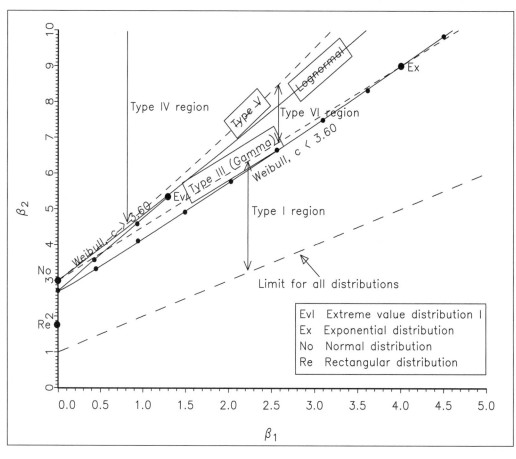

or — solved for $\beta_2$ — by

$$\beta_2 = \frac{3\left(-16 - 13\,\beta_1 - 2\,\sqrt{(4 + \beta_1)^3}\,\right)}{\beta_1 - 32}\,,\quad 0 \le \beta_1 < 32.$$

We have also marked by dots the positions of four other special distributions:

- **uniform** or **rectangular distribution** ($\beta_1 = 0;\ \beta_2 = 1.8$),
- **normal distribution** ($\beta_1 = 0;\ \beta_2 = 3$),
- **exponential distribution** ($\beta_1 = 4;\ \beta_2 = 9$) at the crossing of the gamma–line and the lower branch of the WEIBULL–distribution–line,
- **type–I extreme value distribution** ($\beta_1 \approx 1.2986;\ \beta_2 = 5.4$) at the end of the upper branch of the WEIBULL–distribution–line,

and by a line corresponding to the **lognormal distribution** which completely falls into the type–VI region.

When $\beta_2$ is plotted over $\beta_1$ for a WEIBULL distribution we get a parametric function depending on the shape parameter $c$. This function has a vertex at $(\beta_1, \beta_2) \approx (0, 2.72)$

corresponding to $c \approx 3.6023$ with a finite upper branch (valid for $c > 3.6023$) ending at $(\beta_1, \beta_2) \approx (1.2928, 5.4)$ and an infinite lower branch (valid for $c < 3.6023$). One easily sees that the WEIBULL distribution does not belong to only one family of the PEARSON system. For $c < 3.6023$ the WEIBULL distribution lies mainly in the type–I region and extends approximately parallel to the type–III (gamma) line until the two lines intersect at $(\beta_1, \beta_2) = (4.0, 9.0)$ corresponding to the exponential distribution (= WEIBULL distribution with $c = 1$). The WEIBULL line for $c > 3.6023$ originates in the type–I region and extends approximately parallel to the type–V line. It crosses the type–III line into the type–VI region at a point with $\beta_1 \approx 0.5$ and then moves toward the lognormal line, ending on that line at a point which marks the type–I extreme value distribution. Hence, the WEIBULL distribution with $c > 3.6023$ will closely resemble the PEARSON type–VI distribution when $\beta_1 \geq 0.5$, and for $\beta_1$ approximately greater than 1.0 it will closely resemble the lognormal distribution.

### 3.1.2   BURR system[2]

The **BURR system** (see BURR (1942)) fits cumulative distribution functions, rather than density functions, to frequency data, thus avoiding the problems of numerical integration which are encountered when probabilities or percentiles are evaluated from PEARSON curves. A CDF $y := F(x)$ in the BURR system has to satisfy the differential equation:

$$\frac{\mathrm{d}y}{\mathrm{d}x} = y\,(1 - y)\,g(x, y), \quad y := F(x), \tag{3.2}$$

an analogue to the differential equation (3.1) that generates the PEARSON system. The function $g(x, y)$ must be positive for $0 \leq y \leq 1$ and $x$ in the support of $F(x)$. Different choices of $g(x, y)$ generate various solutions $F(x)$. These can be classified by their functional forms, each of which gives rise to a family of CDFs within the BURR system. BURR listed twelve such families.

With respect to the WEIBULL distribution the BURR type–XII distribution is of special interest. Its CDF is

$$F(x) = 1 - \frac{1}{(1 + x^c)^k}\,; \quad x, c, k > 0, \tag{3.3a}$$

with DF

$$f(x) = k\,c\,x^{c-1}\,(1 + x^c)^{-k-1} \tag{3.3b}$$

and moments about the origin given by

$$\mathrm{E}\big(X^r\big) = \mu_r' = k\,\Gamma\Big(k - \frac{r}{c}\Big)\,\Gamma\Big(\frac{r}{c} + 1\Big) \Big/ \Gamma(k+1) \ \text{ for } ck > r. \tag{3.3c}$$

It is required that $c\,k > 4$ for the fourth moment, and thus $\beta_2$, to exist. This family gives rise to a useful range of value of skewness, $\alpha_3 = \pm\sqrt{\beta_1}$, and kurtosis, $\alpha_4 = \beta_2$. It may be generalized by introducing a location parameter and a scale parameter.

---

[2]   Suggested reading for this section: RODRIGUEZ (1977), TADIKAMALLA (1980a).

Whereas the third and fourth moment combinations of the PEARSON families do not over-lap[3] we have an overlapping when looking at the BURR families. For depicting the type–XII family we use another type of moment–ratio diagram with $\alpha_3 = \pm\sqrt{\beta_1}$ as abscissa, thus showing positive as well as negative skewness, and furthermore it is upside down. Thus the upper bound in Fig. 3/2 is referred to as "lower bound" and conversely in the following text. The parametric equations for $\sqrt{\beta_1}$ and $\beta_2$ are

$$\sqrt{\beta_1} = \frac{\Gamma^2(k)\,\lambda_3 - 3\,\Gamma(k)\,\lambda_2\,\lambda_1 + 2\,\lambda_1^3}{\left[\Gamma(k)\,\lambda_2 - \lambda_1^2\right]^{3/2}} \tag{3.3d}$$

$$\beta_2 = \frac{\Gamma^3(k)\,\lambda_4 - 4\,\Gamma^2(k)\,\lambda_3\,\lambda_1 + 6\,\Gamma(k)\,\lambda_2\,\lambda_1^2 - 3\,\lambda_1^4}{\left[\Gamma(k)\,\lambda_2 - \lambda_1^2\right]^{3/2}} \tag{3.3e}$$

where

$$\lambda_j := \Gamma\left(\frac{j}{c} + 1\right)\,\Gamma\left(k - \frac{j}{c}\right)\ ;\ j = 1, 2, 3, 4.$$

Figure 3/2: Moment ratio diagram for the BURR type-XII family and the WEIBULL distribution

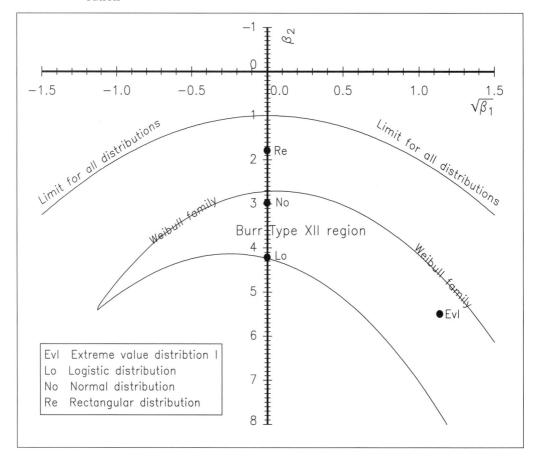

RODRIGUEZ states: "The type–XII BURR distributions occupy a region shaped like the prow of an ancient Roman galley. As $k$ varies from $4/c$ to $+\infty$ the $c$–constant curve moves in a counter–clockwise direction toward the tip of the 'prow'. For $c < 3.6$ the $c$–constant lines terminate at end–points with positive $\sqrt{\beta_1}$. For $c > 3.6$ the $c$–constant lines terminate at end–points with negative $\sqrt{\beta_1}$." These end–points form the **lower bound** of the type–XII region. RODRIGUEZ (1977) gives the following parametric equations of the end–points:

$$\lim_{k \to \infty} \sqrt{\beta_1} = \frac{\Gamma_3 - 3\,\Gamma_2\,\Gamma_1 + 2\,\Gamma_1^3}{\left(\Gamma_2 - \Gamma_1^2\right)^{3/2}}, \tag{3.4a}$$

$$\lim_{k \to \infty} \beta_2 = \frac{\Gamma_4 - 4\,\Gamma_3\,\Gamma_1 + 6\,\Gamma_2\,\Gamma_1 - 3\,\Gamma_1^4}{\left(\Gamma_2 - \Gamma_1^2\right)^2}, \tag{3.4b}$$

where

$$\Gamma_i := \Gamma\left(1 + \frac{i}{c}\right).$$

This lower bound is identical to the WEIBULL curve in the $\left(\sqrt{\beta_1}, \beta_2\right)$–plane; compare (3.4a,b) to (2.93) and (2.101). The identification of the lower bound with the WEIBULL family can also be explained as follows, starting from (3.3a):

$$
\begin{aligned}
\Pr\left[X \le \left(\frac{1}{k}\right)^{1/c} y\right] &= 1 - \left(1 + \frac{y^c}{k}\right)^{-k} \\
&= 1 - \exp\left\{-k \ln\left(1 + \frac{y^c}{k}\right)\right\} \\
&= 1 - \exp\left\{-k\left[\frac{y^c}{k} - \frac{1}{2}\left(\frac{y^c}{k}\right)^2 + \cdots\right]\right\} \\
\Rightarrow\ &1 - \exp\left(-y^c\right) \text{ for } k \to \infty.
\end{aligned}
$$

Hence, the WEIBULL family is the limiting form ($k \to \infty$) of the BURR type–XII family.

We finally mention the **upper bound** for the BURR type–XII region. In the negative $\sqrt{\beta_1}$ half–plane the bound is given for $c = \infty$ stretching from the point $\left(\sqrt{\beta_1}, \beta_2\right) \approx (-1.14, 5.4)$ to $\left(\sqrt{\beta_1}, \beta_2\right) = (0.4, 2)$ which is associated with the **logistic distribution**. In the positive $\sqrt{\beta_1}$ half–plain this bound corresponds to BURR type–XII distributions for which $k = 1$ and $c > 4$ has been proved by RODRIGUEZ (1977).

### 3.1.3 JOHNSON system

The transformation of a variate to normality is the basis of the JOHNSON system. By analogy with the PEARSON system, it would be convenient if a simple transformation could be found such that, for any possible pair of values $\sqrt{\beta_1}, \beta_2$, there will be just one member of the corresponding family of distributions. No such single simple transformation is available, but JOHNSON (1949) has found sets of three such transformations that, when combined, do provide one distribution corresponding to each pair of values $\sqrt{\beta_1}$ and $\beta_2$.

The advantage with JOHNSON's transformation is that, when inverted and applied to a normally distributed variable, it yields three families of density curves with a high degree of shape flexibility.

Let $T$ be a standardized normal variate, i.e., having $\mathrm{E}(T) = 0$ and $\mathrm{Var}(T) = 1$, then the system is defined by

$$T = \gamma + \delta \, g(Y).  \tag{3.5a}$$

Taking

$$
\begin{aligned}
g(Y) &= \ln\left[Y + \sqrt{(1 + Y^2)}\right] \\
&= \sinh^{-1} Y
\end{aligned}
\tag{3.5b}
$$

leads to the $S_U$ **system with unbounded range**: $-\infty < Y < \infty$.

$$g(Y) = \ln Y  \tag{3.5c}$$

leads to the **lognormal family** $S_L$ with $Y > 0$. The third **system** $S_B$ **with bounded range** $0 < Y < 1$ rests upon the transformation

$$g(Y) = \ln\left(\frac{Y}{1 - Y}\right).  \tag{3.5d}$$

The variate $Y$ is linearly related to a variate $X$ that we wish to approximate in distribution:

$$Y = \frac{X - \xi}{\lambda}.  \tag{3.5e}$$

The density of $Y$ in the $S_U$ system is given by

$$f(y) = \frac{\delta}{\sqrt{2\pi}} \, \frac{1}{\sqrt{1 + y^2}} \, \exp\left\{-\frac{\left[\gamma + \delta \, \sinh^{-1} y\right]^2}{2}\right\}, \quad y \in \mathbb{R},  \tag{3.6a}$$

with

$$\sqrt{\beta_1} = -\left[\frac{1}{2} \omega \, (\omega - 1)\right]^{1/2} A^{-3/2} \left[\omega \, (\omega + 2) \sinh(3\,B) + 3 \sinh B\right]  \tag{3.6b}$$

$$\beta_2 = \left[a_2 \cosh(4\,B) + a_1 \cosh(2\,B) + a_0\right] \Big/ (2\,B^2),  \tag{3.6c}$$

where

$$\omega := \exp(1/\delta^2), \quad B := \gamma/\delta, \quad A := \omega \cosh(2\,B) + 1,$$

$$a_2 := \omega^2 \left(\omega^4 + 2\,\omega^3 + 3\,\omega^2 - 3\right), \quad a_1 := 4\,\omega^2 \, (\omega + 2), \quad a_0 := 3\,(2\,\omega + 1).$$

In the $S_L$ system the density is that of a lognormal variate:

$$f(y) = \frac{\delta}{\sqrt{2\pi}} \, \frac{1}{y} \, \exp\left\{-\frac{\left[\gamma + \delta \, \ln y\right]^2}{2}\right\}, \quad y > 0,  \tag{3.7a}$$

with

$$\sqrt{\beta_1} = (\omega - 1)(\omega + 2)^2 \tag{3.7b}$$

$$\beta_2 = \omega^4 + 2\omega^3 + 3\omega^2 - 3. \tag{3.7c}$$

The density in the $S_B$ system is given by

$$f(y) = \frac{\delta}{\sqrt{2\pi}} \frac{1}{y(1-y)} \exp\left\{ -\frac{(\gamma + \delta \ln[y/(1-y)])^2}{2} \right\}, \quad 0 < y < 1, \tag{3.8}$$

and there exist no general and explicit formulas for the moments.

Looking at a moment–ratio diagram (see Fig. 3/1) one sees that the system $S_B$ holds for the region bounded by $\beta_1 = 0$, the bottom line $\beta_2 - \beta_1 - 1 = 0$ and the lognormal curve, given by $S_L$. $S_U$ holds for the corresponding region above $S_L$. In relation to the PEARSON system, $S_B$ overlaps types I, II, III and part of VI; similarly $S_U$ overlaps types IV, V, VII and part of VI. As for shapes of the density, $S_U$ is unimodal, but $S_B$ may be bimodal. The WEIBULL curve is completely located in the $S_L$ region, thus telling that the WEIBULL distribution is a member of the JOHNSON $S_L$ system.

### 3.1.4  Miscellaneous

There exists a great number of further families. Some of them are based on transformations like the JOHNSON system, e.g., the TUKEY's lambda distributions for a variate $X$ and starting from the reduced uniform variable $Y$ with DF

$$f(y) = 1 \quad \text{for} \quad 0 < y < 1$$

and defined by

$$X := \begin{cases} \dfrac{Y^\lambda - (1 - Y)^\lambda}{\lambda} & \text{for} \quad \lambda \neq 0, \\[2ex] \ln\left(\dfrac{Y}{1 - Y}\right) & \text{for} \quad \lambda = 0. \end{cases}$$

Other families are based on expansions e.g., the GRAM–CHARLIER series, the EDGE-WORTH series and the CORNISH–FISHER expansions.

Last but not least, we have many families consisting of two members only. This is a dichotomous classification, where one class has a special property and the other class is missing that property. With respect to the WEIBULL distribution we will present and investigate

- location–scale distributions,
- stable distributions,
- ID distributions and
- the exponential family of distributions.

A variate $X$ belongs to the **location–scale family** if its CDF $F_X(\cdot)$ may be written as

$$F_X(x \mid a, b) = F_Y\left(\frac{x - a}{b}\right) \tag{3.9}$$

and the CDF $F_Y(\cdot)$ does not depend on any parameter. $a$ and $b$ are called location parameter and scale parameter, respectively. $Y := (X - a)/b$ is termed the **reduced variable** and $F_Y(y)$ is the reduced CDF with $a = 0$ and $b = 1$. The three–parameter WEIBULL distribution given by (2.8) evidently does not belong to this family, unless $c = 1$; i.e., we have an exponential distribution (see Sect. 3.2.1). Sometimes a suitable transformation $Z := g(X)$ has a location–scale distribution. This is the case when $X$ has a two–parameter WEIBULL distribution with $a = 0$:

$$F(x) = 1 - \exp\left\{ - \left( \frac{x}{b} \right)^c \right\}. \tag{3.10a}$$

The log–transformed variate

$$Z := \ln X$$

has a type–I extreme value distribution of the minimum (see Sect. 3.2.2) with CDF

$$F(z) = 1 - \exp\left\{ - \exp\left( \frac{z - a^*}{b^*} \right) \right\}, \tag{3.10b}$$

where

$$a^* = \ln b \quad \text{and} \quad b^* = 1/c.$$

This transformation will be of importance when making inference on the parameters $b$ and $c$ (see Chapters 9 ff).

Let $X, X_1, X_2, \ldots$ be independent identically distributed variates. The distribution of $X$ is **stable in the broad sense** if it is not concentrated at one point and if for each $n \in \mathbb{N}$ there exist constants $a_n > 0$ and $b_n \in \mathbb{R}$ such that $(X_1 + X_2 + \ldots + X_n)/a_n - b_n$ has the same distribution as $X$. If the above holds with $b_n = 0$ for all $n$ then the distribution is **stable in the strict sense**.[4] For every stable distribution we have $a_n = n^{1/\alpha}$ for some characteristic exponent $\alpha$ with $0 < \alpha \leq 2$. The family of GAUSSIAN distributions is the unique family of distributions that are stable with $\alpha = 2$. CAUCHY distributions are stable with $\alpha = 1$. The WEIBULL distribution is not stable, neither in the broad sense nor in the strict sense.

A random variable $X$ is called **infinitely divisible distributed** ($\hat{=}$ IDD) if for each $n \in \mathbb{N}$, independent identically distributed ($\hat{=}$ iid) random variables $X_{n,1}, X_{n,2}, \ldots, X_{n,n}$ exist such that

$$X \overset{\mathrm{d}}{=} X_{n,1} + X_{n,2} + \ldots + X_{n,n},$$

where $\overset{\mathrm{d}}{=}$ denotes "equality in distribution." Equivalently, denoting the CDFs of $X$ and $X_{n,1}$ by $F(\cdot)$ and $F_n(\cdot)$, respectively, one has

$$F(\cdot) = F_n(\cdot) * F_n(\cdot) * \ldots * F_n(\cdot) =: F_n^*(\cdot)$$

---

[4] Stable distributions are mainly used to model certain random economic phenomena, especially in finance, having distributions with "fat tails," thus indicating the possibility of an infinite variance.

where the operator $*$ denotes "convolution." All stable distributions are IDD. The WEIBULL distribution is not a member of the IDD family.

A concept related to IDD is **reproductivity through summation**, meaning that the CDF of a sum of $n$ iid random variables $X_i$ $(i = 1, \ldots, n)$ belongs to the same distribution family as each of the terms in the sum. The WEIBULL distribution is not reproductive through summation, but it is **reproductive through formation of the minimum** of $n$ iid WEIBULL random variables.

<u>Theorem:</u> Let $X_i \overset{iid}{\sim} We(a, b, c)$ for $i = 1, \ldots, n$ and $Y = \min(X_1, \ldots, X_n)$, then
$Y \sim We(a, b\, n^{-1/c}, c)$.

<u>Proof:</u> Inserting $F(x)$, given in (2.33b), into the general formula (1.3a)

$$F_Y(y) = 1 - \left[1 - F_X(y)\right]^n$$

gives

$$
\begin{aligned}
F_Y(y) &= 1 - \left[\exp\left\{-\left(\frac{y-a}{b}\right)^c\right\}\right]^n \\
&= 1 - \exp\left\{-n\left(\frac{y-a}{b}\right)^c\right\} \\
&= 1 - \exp\left\{-\left(\frac{y-a}{b\, n^{-1/c}}\right)^c\right\} \Rightarrow We\left(a, b\, n^{-1/c}, c\right).
\end{aligned}
$$

Thus the scale parameter changes from $b$ to $b/\sqrt[c]{n}$ and the location and form parameters remain unchanged. ∎

The **exponential family** of continuous distributions is characterized by having a density function of the form

$$f(x \mid \theta) = A(\theta)\, B(x)\, \exp\left\{Q(\theta) \circ T(x)\right\}, \tag{3.11a}$$

where $\theta$ is a parameter (both $x$ and $\theta$ may, of course, be multidimensional), $Q(\theta)$ and $T(x)$ are vectors of common dimension, $m$, and the operator $\circ$ denotes the "inner product," i.e.,

$$Q(\theta) \circ T(x) = \sum_{j=1}^{m} Q_j(\theta)\, T_j(x). \tag{3.11b}$$

The exponential family plays an important role in estimating and testing the parameters $\theta_1, \ldots, \theta_m$ because we have the following.

<u>Theorem:</u> Let $X_1, \ldots, X_n$ be a random sample of size $n$ from the density $f(x \mid \theta_1, \ldots, \theta_m)$ of an exponential family and let

$$S_j = \sum_{i=1}^{n} T_j(X_i).$$

Then $S_1, \ldots, S_m$ are jointly **sufficient** and **complete** for $\theta_1, \ldots, \theta_m$ when $n > m$. ∎

Unfortunately the three–parameter WEIBULL distribution is not exponential; thus, the principle of sufficient statistics is not helpful in reducing WEIBULL sample data. Sometimes a distribution family is not exponential in its entirety, but certain subfamilies obtained by fixing one or more of its parameters are exponential. When the location parameter $a$ and the form parameter $c$ are both known and fixed then the WEIBULL distribution is exponential with respect to the scale parameter.

## 3.2    WEIBULL distributions and other familiar distributions

Statistical distributions exist in a great number as may be seen from the comprehensive documentation of JOHNSON/KOTZ (1992, 1994, 1995) et al. consisting of five volumes in its second edition. The relationships between the distributions are as manifold as are the family relationships between the European dynasties. One may easily overlook a relationship and so we apologize should we have forgotten any distributional relationship, involving the WEIBULL distribution.

### 3.2.1    WEIBULL and exponential distributions

We start with perhaps the most simple relationship between the WEIBULL distribution and any other distribution, namely the exponential distribution. Looking to the origins of the WEIBULL distribution in practice (see Section 1.1.2) we recognize that WEIBULL as well as ROSIN, RAMMLER and SPERLING (1933) did nothing but adding a third parameter to the two–parameter exponential distribution (= generalized exponential distribution) with DF

$$f(y \mid a, b) = \frac{1}{b} \exp\left[-\frac{y - a}{b}\right]; \ \ y \geq a, \ a \in \mathbb{R}, \ b > 0. \tag{3.12a}$$

Now, let

$$Y = \left(\frac{X - a}{b}\right)^c, \tag{3.12b}$$

and we have the reduced exponential distribution ($a = 0$, $b = 1$) with density

$$f(y) = e^{-y}, \ y > 0, \tag{3.12c}$$

then $X$ has a WEIBULL distribution with DF

$$f(x \mid a, b, c) = \frac{c}{b}\left(\frac{x - a}{b}\right)^{c-1} \exp\left[-\left(\frac{x - a}{b}\right)^c\right]; \ \ x \geq a, \ a \in \mathbb{R}, \ b, \ c > 0. \tag{3.12d}$$

The transformation (3.12b) is referred to as the **power–law transformation**. Comparing (3.12a) with (3.12d) one recognizes that the exponential distribution is a special case (with $c = 1$) of the WEIBULL distribution.

### 3.2.2    WEIBULL and extreme value distributions

In Sect. 1.1.1 we reported on the discovery of the three types of extreme value distributions. Introducing a location parameter $a$ ($a \in \mathbb{R}$) and a scale parameter $b$ ($b > 0$) into the functions of Table 1/1, we arrive at the following general formulas for extreme value distributions:

**Type–I–maximum distribution:**[5]   $Y \sim Ev_I(a,b)$

$$f_I^M(y) = \frac{1}{b} \exp\left\{ -\frac{y-a}{b} - \exp\left[ -\frac{y-a}{b} \right] \right\}, \quad y \in \mathbb{R}; \qquad (3.13a)$$

$$F_I^M(y) = \exp\left\{ -\exp\left[ -\frac{y-a}{b} \right] \right\}; \qquad (3.13b)$$

**Type–II–maximum distribution:**   $Y \sim Ev_{II}(a,b,c)$

$$f_{II}^M(y) = \frac{c}{b} \left( \frac{y-a}{b} \right)^{-c-1} \exp\left\{ -\left( \frac{y-a}{b} \right)^{-c} \right\}, \quad y \geq a, \, c > 0; \quad (3.14a)$$

$$F_{II}^M(y) = \left\{ \begin{array}{ll} 0 & \text{for } y < a \\ \exp\left\{ -\left( \frac{y-a}{b} \right)^{-c} \right\} & \text{for } y \geq a \end{array} \right\}; \qquad (3.14b)$$

**Type–III–maximum distribution:**   $Y \sim Ev_{III}(a,b,c)$

$$f_{III}^M(y) = \frac{c}{b} \left( \frac{a-y}{b} \right)^{c-1} \exp\left\{ -\left( \frac{a-y}{b} \right)^{c} \right\}, \quad y < a, \, c > 0; \quad (3.15a)$$

$$F_{III}^M(y) = \left\{ \begin{array}{ll} \exp\left\{ -\left( \frac{a-y}{b} \right)^{c} \right\} & \text{for } y < a \\ 1 & \text{for } y \geq a \end{array} \right\}. \qquad (3.15b)$$

The corresponding distributions of

$$X - a = -(Y - a) \qquad (3.16)$$

are those of the minimum and are given by the following:

**Type–I–minimum distribution:**   $X \sim Ev_i(a,b)$  or  $X \sim Lw(a,b)$

$$f_I^m(x) = \frac{1}{b} \exp\left\{ \frac{x-a}{b} - \exp\left[ \frac{x-a}{b} \right] \right\}, \quad x \in \mathbb{R}; \qquad (3.17a)$$

$$F_I^m(x) = 1 - \exp\left\{ -\exp\left[ \frac{x-a}{b} \right] \right\}; \qquad (3.17b)$$

---

[5]  On account of its functional form this distribution is also sometimes called **doubly exponential distribution**.

**Type–II–minimum distribution**:   $X \sim Ev_{ii}(a, b, c)$

$$f_{II}^m(x) \;=\; \frac{c}{b} \left(\frac{a-x}{b}\right)^{-c-1} \exp\left\{-\left(\frac{a-x}{b}\right)^{-c}\right\}, \; x < a, \; c > 0; \quad (3.18a)$$

$$F_{II}^m(x) \;=\; \left\{ \begin{array}{ll} 1 - \exp\left\{-\left(\dfrac{a-x}{b}\right)^{-c}\right\} & \text{for} \quad x < a \\[2ex] 1 & \text{for} \quad x \geq a \end{array} \right\}; \quad (3.18b)$$

**Type–III–minimum distribution** or WEIBULL distribution:
$X \sim Ev_{iii}(a, b, c)$   or   $X \sim We(a, b, c)$

$$f_{III}^m(x) \;=\; \frac{c}{b} \left(\frac{x-a}{b}\right)^{c-1} \exp\left\{-\left(\frac{x-a}{b}\right)^{c}\right\}, \; x \geq a, \; c > 0; \quad (3.19a)$$

$$F_{III}^M(x) \;=\; \left\{ \begin{array}{ll} 0 & \text{for} \quad x < a \\[2ex] 1 - \exp\left\{-\left(\dfrac{x-a}{b}\right)^{c}\right\} & \text{for} \quad x \geq a \end{array} \right\}. \quad (3.19b)$$

The transformation given in (3.16) means a reflection about a vertical axis at $y = x = a$, so the type–III–maximum distribution (3.15a/b) is the **reflected WEIBULL distribution**, which will be analyzed in Sect. 3.3.2.

The type–II and type–III distributions can be transformed to a type–I distribution using a suitable logarithmic transformation. Starting from the WEIBULL distribution (3.19a,b), we set

$$Z = \ln(X - a), \quad X \geq a, \quad (3.20)$$

and thus transform (3.19a,b) to

$$f(z) \;=\; c \exp\left\{c\,(z - \ln b) - \exp\left[c\,(z - \ln b)\right]\right\} \quad (3.21a)$$

$$F(z) \;=\; 1 - \exp\left\{-\exp\left[c\,(z - \ln b)\right]\right\}. \quad (3.21b)$$

This is a type–I–minimum distribution which has a location parameter $a* = \ln b$ and a scale parameter $b^* = 1/c$. That is why a type–I–minimum distribution is called a **Log–WEIBULL distribution**;[6] it will be analyzed in Sect. 3.3.4.

We finally mention a third transformation of a WEIBULL variate, which leads to another member of the extreme value class. We make the following reciprocal transformation to a WEIBULL distributed variate $X$:

$$Z = \frac{b^2}{X - a}. \quad (3.22)$$

---

[6] We have an analogue relationship between the normal and the lognormal distributions.

Applying the well–known rules of finding the DF and CDF of a transformed variable, we arrive at

$$f(z) \;=\; \frac{c}{b} \left(\frac{z}{b}\right)^{-c-1} \exp\left\{-\left(\frac{z}{b}\right)^{-c}\right\}, \tag{3.23a}$$

$$F(z) \;=\; \exp\left\{-\left(\frac{z}{b}\right)^{-c}\right\}. \tag{3.23b}$$

Comparing (3.23a,b) with (3.14a,b) we see that $Z$ has a type–II–maximum distribution with a zero location parameter. Thus a type–II–maximum distribution may be called an **inverse WEIBULL distribution**, it will be analyzed in Sect. 3.3.3.

### 3.2.3 WEIBULL and gamma distributions[7]

In the last section we have seen several relatives of the WEIBULL distributions originating in some kind of transformation of the WEIBULL variate. Here, we will present a class of distributions, the gamma family, which includes the WEIBULL distribution as a special case for distinct parameter values.

The **reduced form** of a gamma distribution with only **one parameter** has the DF:

$$f(x \mid d) = \frac{x^{d-1} \exp(-x)}{\Gamma(d)}; \;\; x \geq 0, \, d > 0. \tag{3.24a}$$

$d$ is a shape parameter. $d = 1$ results in the reduced exponential distribution. Introducing a scale parameter $b$ into (3.24a) gives the **two–parameter gamma distribution**:

$$f(x \mid b, d) = \frac{x^{d-1} \exp(-x/b)}{b^d \, \Gamma(d)}; \;\; x \geq 0; \, b, d > 0. \tag{3.24b}$$

The **three–parameter gamma distribution** has an additional location parameter $a$:

$$f(x \mid a, b, d) = \frac{(x-a)^{d-1} \exp\left(-\dfrac{x-a}{b}\right)}{b^d \, \Gamma(d)}; \;\; x \geq a, \, a \in \mathbb{R}, \, b, d > 0. \tag{3.24c}$$

STACY (1965) introduced a second shape parameter $c$ $(c > 0)$ into the two–parameter gamma distribution (3.24b). When we introduce this second shape parameter into the three–parameter version (3.24c), we arrive at the **four–parameter gamma distribution**, also named **generalized gamma distribution**,[8]

$$f(x \mid a, b, c, d) = \frac{c \, (x-a)^{c \, d-1}}{b^{c \, d} \, \Gamma(d)} \exp\left\{-\left(\frac{x-a}{b}\right)^c\right\}; \;\; x \geq a; \, a \in \mathbb{R}; \, b, \, c, \, d > 0. \tag{3.24d}$$

---

[7] Suggested reading for this section: HAGER/BAIN/ANTLE (1971), PARR/WEBSTER (1965), STACY (1962), STACY/MIHRAM (1965).

[8] A further generalization, introduced by STACY/MIHRAN (1965), allows $c < 0$, whereby the factor $c$ in $c \, (x-a)^{c \, d-1}$ has to be changed against $|c|$.

The DF (3.24d) contains a number of familiar distributions when $b$, $c$ and/or $d$ are given special values (see Tab. 3/1).

Table 3/1:   Special cases of the generalized gamma distribution

| $f(x \mid a, b, c, d)$ | Name of the distribution |
|---|---|
| $f(x \mid 0, 1, 1, 1)$ | reduced exponential distribution |
| $f(x \mid a, b, 1, 1)$ | two–parameter exponential distribution |
| $f(x \mid a, b, 1, \nu)$ | ERLANG distribution, $\nu \in \mathbb{N}$ |
| $f(x \mid 0, 1, c, 1)$ | reduced WEIBULL distribution |
| $f(x \mid a, b, c, 1)$ | three–parameter WEIBULL distribution |
| $f\left(x \mid 0, 2, 1, \dfrac{\nu}{2}\right)$ | $\chi^2$–distribution with $\nu$ degrees of freedom, $\nu \in \mathbb{N}$ |
| $f\left(x \mid 0, \sqrt{2}, 2, \dfrac{\nu}{2}\right)$ | $\chi$–distribution with $\nu$ degrees of freedom, $\nu \in \mathbb{N}$ |
| $f\left(x \mid 0, \sqrt{2}, 2, \dfrac{1}{2}\right)$ | half–normal distribution |
| $f(x \mid 0, \sqrt{2}, 2, 1)$ | circular normal distribution |
| $f(x \mid a, b, 2, 1)$ | RAYLEIGH distribution |

LIEBSCHER (1967) compares the two–parameter gamma distribution and the lognormal distribution to the WEIBULL distribution with $a = 0$ and gives conditions on the parameters of these distributions resulting in a stochastic ordering.

### 3.2.4   WEIBULL and normal distributions[9]

The relationships mentioned so far are exact whereas the relation of the WEIBULL distribution to the normal distribution only holds approximately. The quality of approximation depends on the criteria which will be chosen in equating these distributions. In Sect. 2.9.4 we have seen that there exist values of the shape parameter $c$ leading to a skewness of zero and a kurtosis of three which are typical for the normal distribution. Depending on how skewness is measured we have different values of $c$ giving a value of zero for the measure of skewness chosen:

- $c \approx 3.60235$ for $\alpha_3 = 0$,

- $c \approx 3.43954$ for $\mu = x_{0.5}$, i.e., for mean = median,

- $c \approx 3.31247$ for $\mu = x^*$, i.e., for mean = mode,

- $c \approx 3.25889$ for $x^* = x_{0.5}$, i.e., for mode = median.

---

[9] Suggested for this section: DUBEY (1967a), MAKINO (1984).

Regarding the kurtosis $\alpha_4$, we have two values of $c$ ($c \approx 2.25200$ and $c \approx 5.77278$) giving $\alpha_4 = 3$.

MAKINO (1984) suggests to base the approximation on the mean hazard rate $\mathrm{E}\big[(h(X)\big]$; see (2.53a). The standardized normal and WEIBULL distributions have the same mean hazard rate $\mathrm{E}\big[h(X)\big] = 0.90486$ when $c \approx 3.43927$, which is nearly the value of the shape parameter such that the mean is equal to the median.

In the sequel we will look at the closeness of the CDF of the standard normal distribution

$$\Phi(\tau) = \int\limits_{-\infty}^{\tau} \frac{1}{\sqrt{2\pi}}\, e^{-t^2/2}\, \mathrm{d}t \tag{3.25a}$$

to the CDF of the standardized WEIBULL variate

$$T = \frac{X - \mu}{\sigma} = \frac{X - (a + b\Gamma_1)}{b\sqrt{\Gamma_2 - \Gamma_1^2}}, \quad \Gamma_i := \Gamma\left(1 + \frac{i}{c}\right),$$

given by

$$\begin{aligned}
F_W(\tau) &= \Pr\left(\frac{X - \mu}{\sigma} \le \tau\right) = \Pr\big(X \le \mu + \sigma\tau\big) \\[2mm]
&= 1 - \exp\left[-\left(\frac{\mu + \sigma\tau - a}{b}\right)^c\right] \\[2mm]
&= 1 - \exp\left[-\left(\Gamma_1 + \tau\sqrt{\Gamma_2 - \Gamma_1^2}\right)^c\right],
\end{aligned} \tag{3.25b}$$

which is only dependent on the shape parameter $c$. In order to achieve $F_W(\tau) \ge 0$, the expression $\Gamma_1 + \tau\sqrt{\Gamma_2 - \Gamma_1^2}$ has to be non–negative, i.e.,

$$\tau \ge -\frac{\Gamma_1}{\sqrt{\Gamma_2 - \Gamma_1^2}}. \tag{3.25c}$$

The right-hand side of (3.25c) is the reciprocal of the coefficient of variation (2.88c) when $a = 0$ und $b = 1$.

We want to exploit (3.25b) in comparison with (3.25a) for the six values of $c$ given in Tab. 3/2.

Table 3/2: Values of $c$, $\Gamma_1$, $\sqrt{\Gamma_2 - \Gamma_1^2}$ and $\Gamma_1/\sqrt{\Gamma_2 - \Gamma_1^2}$

| $c$ | $\Gamma_1$ | $\sqrt{\Gamma_2 - \Gamma_1^2}$ | $\Gamma_1/\sqrt{\Gamma_2 - \Gamma_1^2}$ | Remark |
|---|---|---|---|---|
| 2.25200 | 0.88574 | 0.41619 | 2.12819 | $\alpha_4 \approx 0$ |
| 3.25889 | 0.89645 | 0.30249 | 2.96356 | $x^* \approx x_{0.5}$ |
| 3.31247 | 0.89719 | 0.29834 | 3.00730 | $\mu \approx x^*$ |
| 3.43954 | 0.89892 | 0.28897 | 3.11081 | $\mu \approx x_{0.5}$ |
| 3.60235 | 0.90114 | 0.27787 | 3.24306 | $\alpha_3 \approx 0$ |
| 5.77278 | 0.92573 | 0.18587 | 4.98046 | $\alpha_4 \approx 0$ |

Using Tab. 3/2 the six WEIBULL CDFs, which will be compared with $\Phi(\tau)$ in Tab. 3/3, are given by

$$F_W^{(1)}(\tau) = 1 - \exp\left[-(0.88574 + 0.41619\,\tau)^{2.25200}\right],$$

$$F_W^{(2)}(\tau) = 1 - \exp\left[-(0.89645 + 0.30249\,\tau)^{3.25889}\right],$$

$$F_W^{(3)}(\tau) = 1 - \exp\left[-(0.89719 + 0.29834\,\tau)^{3.31247}\right],$$

$$F_W^{(4)}(\tau) = 1 - \exp\left[-(0.89892 + 0.28897\,\tau)^{3.43954}\right],$$

$$F_W^{(5)}(\tau) = 1 - \exp\left[-(0.90114 + 0.27787\,\tau)^{3.60235}\right],$$

$$F_W^{(6)}(\tau) = 1 - \exp\left[-(0.92573 + 0.18587\,\tau)^{5.77278}\right].$$

Tab. 3/3 also gives the differences $\Delta^{(i)}(\tau) = F_W^{(i)}(\tau) - \Phi(\tau)$ for $i = 1,\,2,\,\ldots,\,6$ and $\tau = -3.0\,(0.1)\,3.0$. A look at Tab. 3/3 reveals the following facts:

- It is not advisable to base a WEIBULL approximation of the normal distribution on the equivalence of the kurtosis because this leads to the maximum errors of all six cases: $\left|\Delta^{(1)}(\tau)\right| = 0.0355$ for $c \approx 2.52200$ and $\left|\Delta^{(6)}(\tau)\right| = 0.0274$ for $c \approx 5.77278$.

- With respect to corresponding skewness (= zero) of both distributions, the errors are much smaller.

- The four cases of corresponding skewness have different performances.

  - We find that $c = 3.60235$ (attached to $\alpha_3 = 0$) yields the smallest maximum absolute difference $\left(\left|\Delta^{(5)}(\tau)\right| = 0.0079\right)$ followed by $\max_\tau\left|\Delta^{(4)}(\tau)\right| = 0.0088$ for $c = 3.43945$, $\max_\tau\left|\Delta^{(3)}(\tau)\right| = 0.0099$ for $c = 3.31247$ and $\max_\tau\left|\Delta^{(2)}(\tau)\right| = 0.0105$ for $c = 3.25889$.

  - None of the four values of $c$ with zero–skewness is uniformly better than any other. $c \approx 3.60235$ leads to the smallest absolute error for $-3.0 \leq \tau \leq -1.8$, $-1.0 \leq \tau \leq -0.1$ and $1.0 \leq \tau \leq 1.6$; $c \approx 3.25889$ is best for $-1.5 \leq \tau \leq -1.1$, $0.2 \leq \tau \leq 0.9$ and $2.1 \leq \tau \leq 3.0$; $c \approx 3.31247$ is best for $\tau = -1,6$, $\tau = 0.1$ and $\tau = 1.9$; and, finally, $c \approx 3.43954$ gives the smallest absolute error for $\tau = -1.7$, $\tau = 0.0$ and $1.7 \leq \tau \leq 1.8$.

  - Generally, we may say, probabilities associated with the lower (upper) tail of a normal distribution can be approximated satisfactorily by using a WEIBULL distribution with the shape parameter $c \approx 3.60235$ $(c \approx 3.25889)$.

Another relationship between the WEIBULL and normal distributions, based upon the $\chi^2$–distribution, will be discussed in connection with the FAUCHON et al. (1976) extensions in Sections 3.3.7.1 and 3.3.7.2.

### 3.2.5  WEIBULL **and further distributions**

In the context of the generalized gamma distribution we mentioned its special cases. One of them is the $\chi$–**distribution** with $\nu$ degrees of freedom which has the DF:

$$f(x \mid \nu) = \frac{2}{\sqrt{2}\,\Gamma\!\left(\frac{\nu}{2}\right)} \left(\frac{x}{\sqrt{2}}\right)^{\nu-1} \exp\left\{-\left(\frac{x}{\sqrt{2}}\right)^2\right\}, \quad x \geq 0,\ \nu \in \mathbb{N}. \qquad (3.26a)$$

(3.26a) gives a WEIBULL distribution with $a = 0$, $b = \sqrt{2}$ and $c = 2$ for $\nu = 2$. Substituting the special scale factor $\sqrt{2}$ by a general scale parameter $b > 0$ leads to the following WEIBULL density:

$$f(x \mid 0, b, 2) = \frac{2}{b} \left(\frac{x}{b}\right) \exp\left\{-\left(\frac{x}{b}\right)^2\right\}, \qquad (3.26b)$$

which is easily recognized as the density of a **RAYLEIGH distribution**.

<u>Table 3/3:</u>  Values of $\Phi(\tau)$, $F_W^{(i)}(\tau)$ and $\Delta^{(i)}(\tau)$ for $i = 1, 2, \ldots, 6$

| $\tau$ | $\Phi(\tau)$ | $F_W^{(1)}(\tau)$ | $\Delta^{(1)}(\tau)$ | $F_W^{(2)}(\tau)$ | $\Delta^{(2)}(\tau)$ | $F_W^{(3)}(\tau)$ | $\Delta^{(3)}(\tau)$ | $F_W^{(4)}(\tau)$ | $\Delta^{(4)}(\tau)$ | $F_W^{(5)}(\tau)$ | $\Delta^{(5)}(\tau)$ | $F_W^{(6)}(\tau)$ | $\Delta^{(6)}(\tau)$ |
|---|---|---|---|---|---|---|---|---|---|---|---|---|---|
| $-3.0$ | .0013 | – | – | – | – | .0000 | $-.0013$ | .0000 | $-.0013$ | .0001 | $-.0013$ | .0031 | .0018 |
| $-2.9$ | .0019 | – | – | .0000 | $-.0019$ | .0000 | $-.0019$ | .0001 | $-.0018$ | .0002 | $-.0017$ | .0041 | .0023 |
| $-2.8$ | .0026 | – | – | .0001 | $-.0025$ | .0001 | $-.0025$ | .0003 | $-.0023$ | .0005 | $-.0020$ | .0054 | .0029 |
| $-2.7$ | .0035 | – | – | .0003 | $-.0032$ | .0004 | $-.0031$ | .0007 | $-.0028$ | .0011 | $-.0024$ | .0070 | .0036 |
| $-2.6$ | .0047 | – | – | .0008 | $-.0039$ | .0009 | $-.0037$ | .0014 | $-.0033$ | .0020 | $-.0026$ | .0090 | .0043 |
| $-2.5$ | .0062 | – | – | .0017 | $-.0046$ | .0019 | $-.0043$ | .0026 | $-.0036$ | .0034 | $-.0028$ | .0114 | .0052 |
| $-2.4$ | .0082 | – | – | .0031 | $-.0051$ | .0035 | $-.0047$ | .0043 | $-.0039$ | .0053 | $-.0028$ | .0143 | .0061 |
| $-2.3$ | .0107 | – | – | .0053 | $-.0054$ | .0058 | $-.0050$ | .0068 | $-.0040$ | .0080 | $-.0027$ | .0178 | .0070 |
| $-2.2$ | .0139 | – | – | .0084 | $-.0055$ | .0089 | $-.0050$ | .0101 | $-.0038$ | .0115 | $-.0024$ | .0219 | .0080 |
| $-2.1$ | .0179 | .0000 | $-.0178$ | .0125 | $-.0054$ | .0131 | $-.0048$ | .0144 | $-.0035$ | .0159 | $-.0019$ | .0268 | .0089 |
| $-2.0$ | .0228 | .0014 | $-.0214$ | .0178 | $-.0049$ | .0185 | $-.0043$ | .0199 | $-.0029$ | .0215 | $-.0013$ | .0325 | .0098 |
| $-1.9$ | .0287 | .0050 | $-.0237$ | .0245 | $-.0042$ | .0252 | $-.0035$ | .0266 | $-.0021$ | .0283 | $-.0004$ | .0392 | .0105 |
| $-1.8$ | .0359 | .0112 | $-.0247$ | .0327 | $-.0032$ | .0334 | $-.0025$ | .0348 | $-.0011$ | .0365 | .0006 | .0470 | .0110 |
| $-1.7$ | .0446 | .0204 | $-.0242$ | .0426 | $-.0020$ | .0432 | $-.0013$ | .0446 | .0001 | .0462 | .0017 | .0559 | .0113 |
| $-1.6$ | .0548 | .0325 | $-.0223$ | .0543 | $-.0005$ | .0549 | .0001 | .0562 | .0014 | .0576 | .0028 | .0661 | .0113 |
| $-1.5$ | .0668 | .0476 | $-.0192$ | .0679 | .0010 | .0684 | .0016 | .0695 | .0027 | .0708 | .0040 | .0777 | .0109 |
| $-1.4$ | .0808 | .0657 | $-.0150$ | .0835 | .0027 | .0839 | .0031 | .0848 | .0041 | .0858 | .0051 | .0909 | .0101 |

<u>Table 3/3:</u> Values of $\Phi(\tau)$, $F_W^{(i)}(\tau)$ and $\Delta^{(i)}(\tau)$ for $i = 1, 2, \ldots, 6$ (Continuation)

| $\tau$ | $\Phi(\tau)$ | $F_W^{(1)}(\tau)$ | $\Delta^{(1)}(\tau)$ | $F_W^{(2)}(\tau)$ | $\Delta^{(2)}(\tau)$ | $F_W^{(3)}(\tau)$ | $\Delta^{(3)}(\tau)$ | $F_W^{(4)}(\tau)$ | $\Delta^{(4)}(\tau)$ | $F_W^{(5)}(\tau)$ | $\Delta^{(5)}(\tau)$ | $F_W^{(6)}(\tau)$ | $\Delta^{(6)}($ |
|---|---|---|---|---|---|---|---|---|---|---|---|---|---|
| −1.3 | .0968 | .0868 | −.0100 | .1012 | .0044 | .1015 | .0047 | .1022 | .0054 | .1029 | .0061 | .1057 | .0085 |
| −1.2 | .1151 | .1108 | −.0043 | .1210 | .0060 | .1212 | .0062 | .1216 | .0065 | .1220 | .0069 | .1223 | .0072 |
| −1.1 | .1357 | .1374 | .0018 | .1431 | .0074 | .1431 | .0075 | .1432 | .0075 | .1432 | .0075 | .1407 | .0050 |
| −1.0 | .1587 | .1666 | .0079 | .1673 | .0087 | .1672 | .0085 | .1669 | .0082 | .1665 | .0078 | .1611 | .0024 |
| −0.9 | .1841 | .1980 | .0139 | .1937 | .0096 | .1934 | .0093 | .1927 | .0087 | .1919 | .0079 | .1835 | −.000 |
| −0.8 | .2119 | .2314 | .0195 | .2221 | .0103 | .2217 | .0098 | .2207 | .0088 | .2194 | .0076 | .2079 | −.003 |
| −0.7 | .2420 | .2665 | .0245 | .2525 | .0105 | .2519 | .0099 | .2506 | .0086 | .2489 | .0070 | .2345 | −.007 |
| −0.6 | .2743 | .3030 | .0287 | .2847 | .0104 | .2840 | .0097 | .2823 | .0080 | .2803 | .0060 | .2631 | −.011 |
| −0.5 | .3085 | .3405 | .0320 | .3185 | .0100 | .3177 | .0091 | .3157 | .0072 | .3134 | .0049 | .2937 | −.014 |
| −0.4 | .3446 | .3788 | .0342 | .3538 | .0092 | .3528 | .0082 | .3506 | .0060 | .3480 | .0035 | .3263 | −.018 |
| −0.3 | .3821 | .4175 | .0354 | .3902 | .0081 | .3891 | .0071 | .3868 | .0047 | .3840 | .0019 | .3608 | −.021 |
| −0.2 | .4207 | .4563 | .0355 | .4275 | .0067 | .4264 | .0057 | .4239 | .0032 | .4210 | .0003 | .3968 | −.023 |
| −0.1 | .4602 | .4948 | .0346 | .4654 | .0052 | .4643 | .0041 | .4618 | .0016 | .4588 | −.0014 | .4343 | −.025 |
| 0.0 | .5000 | .5328 | .0328 | .5036 | .0036 | .5025 | .0025 | .5000 | −.0000 | .4971 | −.0029 | .4730 | −.027 |
| 0.1 | .5398 | .5699 | .0301 | .5417 | .0019 | .5407 | .0009 | .5383 | −.0015 | .5355 | −.0043 | .5125 | −.027 |
| 0.2 | .5793 | .6060 | .0268 | .5795 | .0003 | .5786 | −.0007 | .5763 | −.0029 | .5737 | −.0055 | .5524 | −.026 |
| 0.3 | .6179 | .6409 | .0229 | .6167 | −.0012 | .6158 | −.0021 | .6138 | −.0041 | .6114 | −.0065 | .5925 | −.025 |
| 0.4 | .6554 | .6742 | .0188 | .6528 | −.0026 | .6521 | −.0033 | .6503 | −.0051 | .6483 | −.0071 | .6323 | −.023 |
| 0.5 | .6915 | .7059 | .0144 | .6878 | −.0037 | .6871 | −.0043 | .6857 | −.0058 | .6840 | −.0074 | .6713 | −.020 |
| 0.6 | .7257 | .7358 | .0101 | .7212 | −.0046 | .7207 | −.0051 | .7196 | −.0062 | .7183 | −.0075 | .7092 | −.016 |
| 0.7 | .7580 | .7639 | .0059 | .7528 | −.0052 | .7525 | −.0056 | .7517 | −.0063 | .7508 | −.0072 | .7455 | −.012 |
| 0.8 | .7881 | .7901 | .0020 | .7826 | −.0056 | .7824 | −.0058 | .7819 | −.0062 | .7815 | −.0067 | .7799 | −.008 |
| 0.9 | .8159 | .8143 | −.0016 | .8102 | −.0057 | .8102 | −.0057 | .8101 | −.0059 | .8100 | −.0059 | .8119 | −.004 |
| 1.0 | .8413 | .8366 | −.0047 | .8357 | −.0056 | .8358 | −.0055 | .8360 | −.0053 | .8363 | −.0050 | .8415 | .0001 |
| 1.1 | .8643 | .8570 | −.0074 | .8590 | −.0053 | .8592 | −.0051 | .8597 | −.0047 | .8603 | −.0040 | .8683 | .0039 |
| 1.2 | .8849 | .8754 | −.0095 | .8800 | −.0049 | .8803 | −.0046 | .8810 | −.0039 | .8819 | −.0030 | .8921 | .0072 |
| 1.3 | .9032 | .8921 | −.0111 | .8989 | −.0043 | .8992 | −.0040 | .9001 | −.0031 | .9012 | −.0020 | .9131 | .0099 |

Table 3/3: Values of $\Phi(\tau)$, $F_W^{(i)}(\tau)$ and $\Delta^{(i)}(\tau)$ for $i = 1, 2, \ldots, 6$ (Continuation)

| $\tau$ | $\Phi(\tau)$ | $F_W^{(1)}(\tau)$ | $\Delta^{(1)}(\tau)$ | $F_W^{(2)}(\tau)$ | $\Delta^{(2)}(\tau)$ | $F_W^{(3)}(\tau)$ | $\Delta^{(3)}(\tau)$ | $F_W^{(4)}(\tau)$ | $\Delta^{(4)}(\tau)$ | $F_W^{(5)}(\tau)$ | $\Delta^{(5)}(\tau)$ | $F_W^{(6)}(\tau)$ | $\Delta^{(6)}(\tau)$ |
|---|---|---|---|---|---|---|---|---|---|---|---|---|---|
| 1.4 | .9192 | .9070 | −.0122 | .9155 | −.0037 | .9159 | −.0033 | .9169 | −.0023 | .9182 | −.0010 | .9312 | .0120 |
| 1.5 | .9332 | .9203 | −.0129 | .9301 | −.0031 | .9306 | −.0026 | .9317 | −.0015 | .9330 | −.0002 | .9465 | .0133 |
| 1.6 | .9452 | .9321 | −.0131 | .9427 | −.0025 | .9432 | −.0020 | .9444 | −.0008 | .9458 | .0006 | .9592 | .0140 |
| 1.7 | .9554 | .9424 | −.0130 | .9535 | −.0019 | .9540 | −.0014 | .9552 | −.0003 | .9566 | .0012 | .9695 | .0140 |
| 1.8 | .9641 | .9515 | −.0126 | .9627 | −.0014 | .9632 | −.0009 | .9643 | .0002 | .9657 | .0016 | .9777 | .0136 |
| 1.9 | .9713 | .9593 | −.0120 | .9704 | −.0009 | .9708 | −.0004 | .9719 | .0006 | .9732 | .0019 | .9840 | .0127 |
| 2.0 | .9772 | .9661 | −.0112 | .9767 | −.0005 | .9772 | −.0001 | .9781 | .0009 | .9793 | .0021 | .9889 | .0116 |
| 2.1 | .9821 | .9719 | −.0103 | .9819 | −.0002 | .9823 | .0002 | .9832 | .0011 | .9843 | .0021 | .9924 | .0103 |
| 2.2 | .9861 | .9768 | −.0093 | .9861 | .0000 | .9865 | .0004 | .9873 | .0012 | .9882 | .0021 | .9950 | .0089 |
| 2.3 | .9893 | .9810 | −.0083 | .9895 | .0002 | .9898 | .0005 | .9905 | .0012 | .9913 | .0020 | .9968 | .0075 |
| 2.4 | .9918 | .9845 | −.0073 | .9921 | .0003 | .9924 | .0006 | .9929 | .0011 | .9936 | .0018 | .9980 | .0062 |
| 2.5 | .9938 | .9874 | −.0064 | .9941 | .0004 | .9944 | .0006 | .9949 | .0011 | .9954 | .0016 | .9988 | .0050 |
| 2.6 | .9953 | .9899 | −.0055 | .9957 | .0004 | .9959 | .0006 | .9963 | .0010 | .9968 | .0014 | .9993 | .0039 |
| 2.7 | .9965 | .9919 | −.0046 | .9969 | .0004 | .9971 | .0005 | .9974 | .0008 | .9977 | .0012 | .9996 | .0031 |
| 2.8 | .9974 | .9935 | −.0039 | .9978 | .0004 | .9979 | .0005 | .9982 | .0007 | .9985 | .0010 | .9998 | .0023 |
| 2.9 | .9981 | .9949 | −.0033 | .9985 | .0003 | .9985 | .0004 | .9987 | .0006 | .9990 | .0008 | .9999 | .0017 |
| 3.0 | .9987 | .9960 | −.0027 | .9989 | .0003 | .9990 | .0003 | .9991 | .0005 | .9993 | .0007 | .9999 | .0013 |

## Excursus: RAYLEIGH distribution

This distribution was introduced by J.W. STRUTT (Lord RAYLEIGH) (1842 – 1919) in a problem of acoustics. Let $X_1$, $X_2$, $\ldots$, $X_n$ be an iid sample of size $n$ from a normal distribution with $E(X_i) = 0 \ \forall \ i$ and $\mathrm{Var}(X_i) = \sigma^2 \ \forall \ i$. The density function of

$$Y = \sqrt{\sum_{i=1}^{n} X_i^2},$$

that is, the distance from the origin to a point $(X_1, \ldots, X_n)$ in the $n$–dimensional EUCLIDEAN space, is

$$f(y) = \frac{2}{(2\sigma^2)^{n/2} \Gamma(n/2)} \, y^{n-1} \exp\left\{ -\frac{y^2}{2\sigma^2} \right\}; \quad y > 0, \ \sigma > 0. \tag{3.27}$$

With $n = \nu$ and $\sigma = 1$, equation (3.27) is the $\chi$–distribution (3.26a), and with $n = 2$ and $\sigma = b$, we have (3.26b).

---

The hazard rate belonging to the special WEIBULL distribution (3.26b) is

$$h(x \mid 0, b, 2) = \frac{2}{b}\frac{x}{b} = \frac{2}{b^2}\,x. \tag{3.28}$$

This is a linear hazard rate. Thus, the WEIBULL distribution with $c = 2$ is a member of the class of **polynomial hazard rate distributions**, the polynomial being of degree one.

Physical processes that involve nucleation and growth or relaxation phenomena have been analyzed by GITTUS (1967) to arrive at the class of distribution functions that should characterize the terminal state of replicates. He starts from the following differential equation for the CDF:

$$\frac{\mathrm{d}F(x)}{\mathrm{d}x} = K\,x^m\,\left[1 - F(x)\right]^n; \quad K > 0,\; m > -1,\; n \geq 1. \tag{3.29a}$$

For $n = 1$ the solution is

$$F(x)_{m,1} = 1 - \exp\left\{-K\,\frac{x^{m+1}}{m+1}\right\}. \tag{3.29b}$$

(3.29b) is a WEIBULL distribution with $a = 0$, $b = \left[(m+1)/K\right]^{1/(m+1)}$ and $c = m + 1$. For $n > 1$ the solution is

$$F(x)_{m,n} = 1 - \left\{\frac{(n-1)\,K}{m+1}\,x^{m+1} + 1\right\}^{1/(1-n)}. \tag{3.29c}$$

The effects of $m$, $n$ and $K$ on the form of the CDF, given by (3.29c), are as follows:

- Increasing the value of $m$ increases the steepness of the central part of the curve.

- Increasing the value of $K$ makes the curve more upright.

- Increasing the value of $n$ reduces its height.

The normal and the logistic distributions are very similar, both being symmetric, but the logistic distribution has more kurtosis ($\alpha_4 = 4.2$). It seems obvious to approximate the **logistic distribution** by a symmetric version of the WEIBULL distribution. Comparing the CDF of the reduced logistic distribution

$$F(\tau) = \frac{1}{1 - \exp\left(-\pi\tau/\sqrt{3}\right)}$$

with the CDF of a reduced WEIBULL distribution in the symmetric cases $F_W^{(2)}(\tau)$ to $F_W^{(5)}(\tau)$ in Tab. 3/3 gives worse results than the approximation to the normal distribution. Fig. 3/3 shows the best fitting WEIBULL CDF ($c \approx 3.60235$ giving $\alpha_3 = 0$) in comparison with the logistic and normal CDFs.

Figure 3/3: CDFs of the standardized logistic, normal and WEIBULL distributions

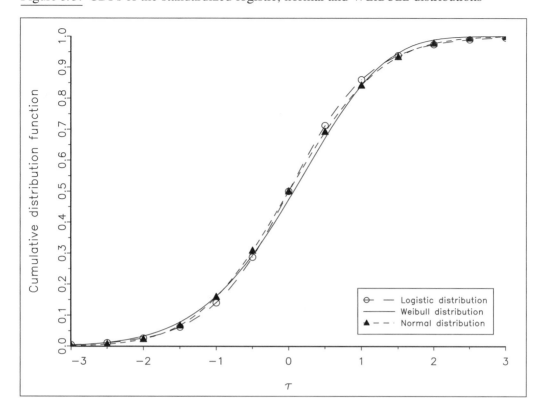

## 3.3 Modifications of the WEIBULL distribution[10]

A great number of distributions have been developed that take the classical WEIBULL distribution (2.8) as their point of departure. Some of these modifications have been made in response to questions from practice while other generalizations have their origin in pure science. We hope that the grouping of the WEIBULL offshoots in the following ten subsections is neither overlapping nor incomplete.

### 3.3.1 Discrete WEIBULL distribution[11]

A main area of application for the WEIBULL distribution is lifetime research and reliability theory. In these fields we often encounter failure data measured as discrete variables such as number of work loads, blows, runs, cycles, shocks or revolutions. Sometimes a device is inspected once an hour, a week or a month whether it is still working or has failed in the meantime, thus leading to a lifetime counted in natural units of time. In this context, the **geometric** and the **negative binomial distributions** are known to be discrete alternatives for the exponential and gamma distributions, respectively. We are interested, from the

---

[10] Suggested reading for this section: MURTHY/XIE/JIANG (2004).

[11] Parameter estimation for discrete WEIBULL distributions is treated by ALI KHAN/KHALIQUE/ABOUAMMOH (1989).

viewpoints of both theory and practice, what discrete distribution might correspond to the WEIBULL distribution. The answer to this question is not unique, depending on what characteristic of the continuous WEIBULL distribution is to be preserved.

The **type–I discrete WEIBULL distribution**, introduced by NAKAGAWA/OSAKI (1975), retains the form of the continuous CDF. The **type–II discrete WEIBULL distribution**, suggested by STEIN/DATTERO (1984), retains the form of the continuous hazard rate. It is impossible to find a discrete WEIBULL distribution that mimics both the CDF and the HR of the continuous version in the sense that its CDF and its HR agree with those of the continuous WEIBULL distribution for integral values of the variate.

We will first present these two types and compare them to each other and to the continuous version. Finally, we mention another approach, that of PADGETT/SPURRIER (1985) and SALVIA (1996). This approach does not start from the continuous WEIBULL distribution but tries to generalize the notions of hazard rate and mean residual life to the discrete case. But first of all we have to comment on the functions that generally describe a discrete lifetime variable.

**Excursus: Functions for a discrete lifetime variable**

We define and redefine the following concepts:
- probability of failure within the $k$–th unit of time (**probability mass function**)

$$P_k := \Pr(X = k); \quad k = 0, 1, 2, \ldots; \tag{3.30a}$$

- failure distribution CDF

$$F(k) := \Pr(X \leq k) = \sum_{i=0}^{k} P_i; \quad k = 0, 1, 2, \ldots \tag{3.30b}$$

with

$$F(\infty) = 1, \quad F(-1) := 0 \text{ and } P_k = F(k) - F(k-1); \tag{3.30c}$$

- survival function CCDF

$$\begin{aligned} R(k) &:= \Pr(X > k) = 1 - F(k) \\ &= \sum_{i=k+1}^{\infty} P_i; \quad k = 0, 1, 2, \ldots \end{aligned} \tag{3.30d}$$

with

$$R(\infty) = 0, \quad R(-1) := 1 \text{ and } P_k = R(k-1) - R(k). \tag{3.30e}$$

The most frequently used and wide–spread definition of the hazard rate in the discrete case, see footnote 1 in Chapter 2, is

$$\left. \begin{aligned} h_k &:= \Pr(X = k \mid X \geq k); \quad k = 0, 1, 2, \ldots, \\ &= \frac{\Pr(X = k)}{\Pr(X \geq k)} \\ &= \frac{F(k) - F(k-1)}{1 - F(k-1)} \\ &= \frac{R(k-1) - R(k)}{R(k-1)} \end{aligned} \right\} \tag{3.31a}$$

with corresponding cumulative hazard rate

$$H(k) := \sum_{i=0}^{k} h_i. \tag{3.31b}$$

We notice the following properties of (3.31a):[12]

- $h_k$ is a conditional probability, thus

$$0 \le h_k \le 1. \tag{3.32a}$$

- These conditional probabilities and the unconditional probabilities $P_k$ are linked as follows:

$$\left.\begin{array}{rcl} P_0 & = & h_0 \\ P_k & = & h_k \, (1 - h_{k-1}) \cdot \ldots \cdot (1 - h_0); \ \ k \ge 1. \end{array}\right\} \tag{3.32b}$$

- The survival function $R(k)$ and the hazard rate $h_k$ are linked as

$$R(k) = (1 - h_0) \, (1 - h_1) \cdot \ldots \cdot (1 - h_k); \ \ h = 0, 1, 2, \ldots \tag{3.32c}$$

- The mean $E(X)$, if it exists, is given by

$$E(X) = \sum_{i=1}^{k} R(k) = \sum_{k=1}^{\infty} \prod_{j=0}^{k} (1 - h_j). \tag{3.32d}$$

The relationships between $h(x)$, $H(x)$ on the one side and $F(x)$, $R(x)$ on the other side in the continuous case, which are to be found in Tab. 2/1, do not hold with $h_k$ and $H(k)$ defined above, especially

$$R(k) \ne \exp\left[ - H(k) \right] = \exp\left[ - \sum_{i=0}^{k} h_i \right];$$

instead we have (3.32c). For this reason ROY/GUPTA (1992) have proposed an alternative discrete hazard rate function:

$$\lambda_k := \ln\left( \frac{R(k-1)}{R(k)} \right); \ \ k = 0, 1, 2, \ldots \tag{3.33a}$$

With the corresponding cumulative function

$$\left.\begin{array}{rcl} \Lambda(k) & := & \sum_{i=0}^{k} \lambda_i \\ & = & \ln R(-1) - \ln R(k) \\ & = & - \ln R(k), \ \text{ see (3.30e)} \end{array}\right\}, \tag{3.33b}$$

we arrive at

$$R(k) = \exp\left[ - \Lambda(k) \right]. \tag{3.33c}$$

We will term $\lambda_k$ the **pseudo–hazard function** so as to differentiate it from the hazard rate $h_k$.

---

[12] For discrete hazard functions, see SALVIA/BOLLINGER (1982).

The **type–I discrete WEIBULL distribution** introduced by NAKAGAWA/OSAKI (1975) mimics the CDF of the continuous WEIBULL distribution. They consider a probability mass function $P_k^I$ $(k = 0, 1, 2, \ldots)$ indirectly defined by

$$\Pr(X \geq k) = \sum_{j=k}^{\infty} P_j^I = q^{k^\beta}; \quad k = 0, 1, 2, \ldots; \ 0 < q < 1 \text{ and } \beta > 0. \qquad (3.34a)$$

The probability mass function follows as

$$P_k^I = q^{k^\beta} - q^{(k+1)^\beta}; \quad k = 0, 1, 2, \ldots; \qquad (3.34b)$$

and the hazard rate according to (3.31a) as

$$h_k^I = 1 - q^{(k+1)^\beta - k^\beta}; \quad k = 0, 1, 2, \ldots, \qquad (3.34c)$$

The hazard rate

- has the constant value $1 - q$ for $\beta = 1$,

- is decreasing for $0 < \beta < 1$ and

- is increasing for $\beta > 1$.

So, the parameter $\beta$ plays the same role as $c$ in the continuous case. The CDF is given by

$$F^I(k) = 1 - q^{(k+1)^\beta}; \quad k = 0, 1, 2, \ldots; \qquad (3.34d)$$

and the CCDF by

$$R^I(k) = q^{(k+1)^\beta}; \quad k = 0, 1, 2, \ldots \qquad (3.34e)$$

Thus, the pseudo–hazard function follows as

$$\lambda_k^I = \left[ k^\beta - (k+1)^\beta \right] \ln q; \quad k = 0, 1, 2, \ldots \qquad (3.34f)$$

and its behavior in response to $\beta$ is the same as that of $h_k^I$.

Fig. 3/4 shows the hazard rate (3.34c) in the upper part and the pseudo–hazard function (3.34f) in the lower part for $q = 0.9$ and $\beta = 0.5, \ 1.0, \ 1.5$. $h_k^I$ and $\lambda_k^I$ are not equal to each other but $\lambda_k^I(q, \beta) > h_k^I(q, \beta)$, the difference being the greater the smaller $q$ and/or the greater $\beta$. $\lambda_k^I$ increases linearly for $\beta = 2$, which is similar to the continuous case, whereas $h_k^I$ is increasing but is concave for $\beta = 2$.

Figure 3/4: Hazard rate and pseudo–hazard function for the type-I discrete WEIBULL distribution

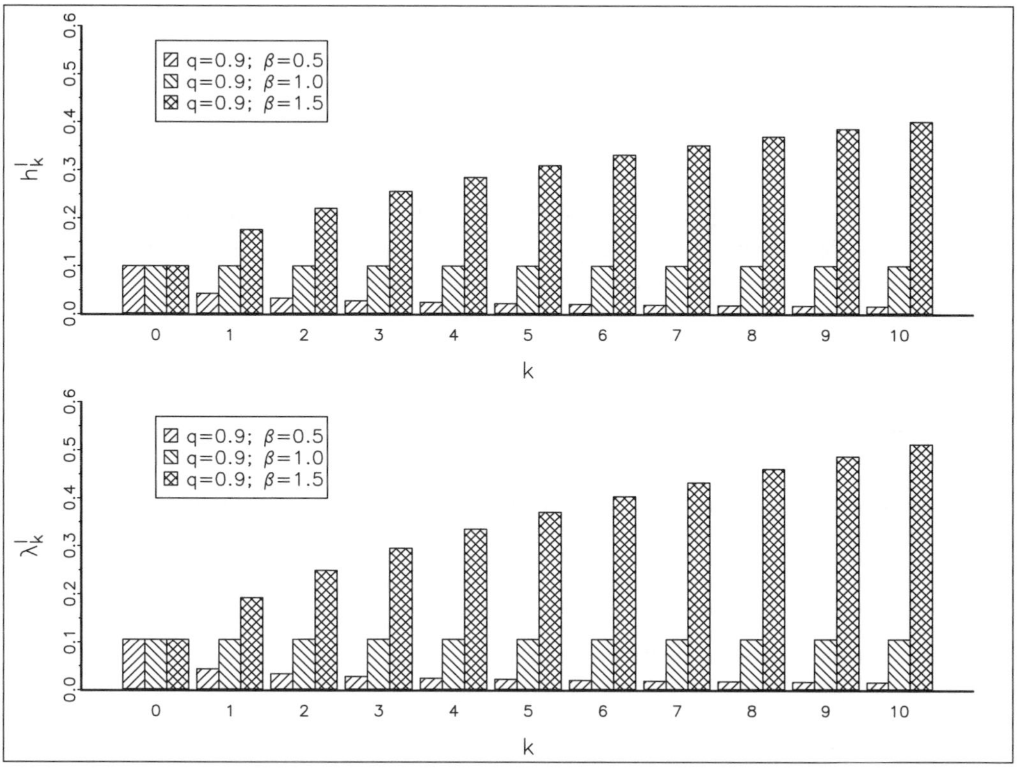

Compared with the CDF of the two–parameter continuous WEIBULL distribution

$$F(x \mid 0, b, c) = 1 - \exp\left\{ - \left( \frac{x}{b} \right)^c \right\}, \quad x \geq 0, \tag{3.35}$$

we see that (3.34d) and (3.35) have the same double exponential form and they coincide for all $x = k + 1 \ (k = 0, 1, 2, \ldots)$ if

$$\beta = c \ \text{ and } \ q = \exp(-1/b^c) = \exp(-b_1),$$

where $b_1 = 1/b^c$ is the combined scale–shape factor of (2.26b).

<u>Remark:</u> Suppose that a discrete variate $Y$ has a geometric distribution, i.e.,

$$\Pr(Y = k) = p \, q^{k-1}; \ \ k = 1, 2, \ldots; \ p + q = 1 \ \text{ and } \ 0 < p < 1;$$

and

$$\Pr(Y \geq k) = q^k.$$

Then, the transformed variate $X = Y^{1/\beta}$, $\beta > 0$, will have

$$\Pr(X \geq k) = \Pr(Y \geq k^\beta) = q^{k^\beta},$$

and hence $X$ has the discrete WEIBULL distribution introduced above. When $\beta = 1$, the discrete WEIBULL distribution reduces to the geometric distribution. This transformation is the counterpart to the power–law relationship linking the exponential and the continuous WEIBULL distributions.

The moments of the type–I discrete WEIBULL distribution

$$\mathrm{E}(X^r) = \sum_{k=1}^{\infty} k^r \left( q^{k^\beta} - q^{(k+1)^\beta} \right)$$

have no closed–form analytical expressions; they have to be evaluated numerically. ALI KHAN et al. (1989) give the following inequality for the means $\mu_d$ and $\mu_c$ of the discrete and continuous distributions, respectively, when $q = \exp(-1/b^c)$:

$$\mu_d - 1 < \mu_c < \mu_d.$$

These authors and KULASEKERA (1994) show how to estimate the two parameters $\beta$ and $q$ of (3.34b).

The **type–II discrete WEIBULL distribution** introduced by STEIN/DATTERO (1984) mimics the HR

$$h(x \,|\, 0, b, c) = \frac{c}{b} \left( \frac{x}{b} \right)^{c-1} \tag{3.36}$$

of the continuous WEIBULL distribution by

$$h_k^{II} = \left\{ \begin{array}{ll} \alpha\, k^{\beta-1} & \text{for} \quad k = 1, 2, \ldots, m \\ 0 & \text{for} \quad k = 0 \text{ or } k > m \end{array} \right\}, \quad \alpha > 0,\ \beta > 0. \tag{3.37a}$$

$m$ is a truncation value, given by

$$m = \left\{ \begin{array}{ll} \text{int}\left[ \alpha^{-1/(\beta-1)} \right] & \text{if} \quad \beta > 1 \\ \infty & \text{if} \quad \beta \leq 1 \end{array} \right\}, \tag{3.37b}$$

which is necessary to ensure $h_k^{II} \leq 1$; see (3.32a).

(3.36) and (3.37a) coincide at $x = k$ for

$$\beta = c \quad \text{and} \quad \alpha = c/b^c,$$

i.e., $\alpha$ is a combined scale–shape factor. The probability mass function generated by (3.37a) is

$$
P_k^{II} = \left\{ \begin{array}{l} h_1^{II} \\ h_k^{II} \left( 1 - h_{k-1}^{II} \right) \ldots \left( 1 - h_1^{II} \right) \quad \text{for} \quad k \geq 2, \end{array} \right.
$$

$$
= \alpha\, k^{\beta-1} \prod_{j=1}^{k-1} (1 - \alpha\, j^{\beta-1}); \quad k = 1, 2, \ldots, m. \tag{3.37c}
$$

The corresponding survival or reliability function is

$$R^{II}(k) = \prod_{j=1}^{k}\left(1 - h_k^{II}\right) = \prod_{j=1}^{k}\left(1 - \alpha\,j^{\beta-1}\right);\quad k = 1, 2, \ldots, m. \tag{3.37d}$$

(3.37d) in combination with (3.33b,c) gives the pseudo–hazard function

$$\lambda_k^{II} = -\ln\left(1 - \alpha\,k^{\beta-1}\right). \tag{3.37e}$$

For $\beta = 1$ the type–II discrete WEIBULL distribution reduces to a geometric distribution as does the type–I distribution.

The discrete distribution suggested by PADGETT/SPURRIER (1985) and SALVIA (1996) is not similar in functional form to any of the functions describing a continuous WEIBULL distribution. The only item connecting this distribution to the continuous WEIBULL model is the fact that its hazard rate may be constant, increasing or decreasing depending on only one parameter. The discrete hazard rate of this model is

$$h_k^{III} = 1 - \exp\left[-d\,(k+1)^{\beta}\right];\quad k = 0, 1, 2, \ldots;\ d > 0,\ \beta \in \mathbb{R}, \tag{3.38a}$$

where $h_k^{III}$ is

- constant with $1 - \exp(-d)$ for $\beta = 0$,
- increasing for $\beta > 0$,
- decreasing for $\beta < 0$.

The probability mass function is

$$P_k^{III} = \exp\left[-d\,(k+1)^{\beta}\right]\prod_{j=1}^{k}\exp\left(-d\,j^{\beta}\right);\quad k = 0, 1, 2, \ldots; \tag{3.38b}$$

and the survival function is

$$R^{III}(k) = \exp\left[-d\sum_{j=1}^{k+1} j^{\beta}\right];\quad k = 0, 1, 2, \ldots \tag{3.38c}$$

For suitable values of $d > 0$ the parameter $\beta$ in the range $-1 \le \beta \le 1$ is sufficient to describe many hazard rates of discrete distributions. The pseudo–hazard function corresponding to (3.38a) is

$$\lambda_k^{III} = d\,(k+1)^{\beta};\quad k = 0, 1, 2, \ldots; \tag{3.38d}$$

which is similar to (3.37a).

### 3.3.2 Reflected and double WEIBULL distributions

The modifications of this and next two sections consist of some kind of transformation of a continuous WEIBULL variate. The **reflected WEIBULL distribution**, introduced by COHEN (1973), originates in the following linear transformation of a classical WEIBULL variate $X$:

$$Y - a = -(X - a) = a - X.$$

This leads to a reflection of the classical WEIBULL DF about a vertical axis at $x = a$ resulting in

$$f_R(y \mid a, b, c) = \frac{c}{b} \left(\frac{a-y}{b}\right)^{c-1} \exp\left\{-\left(\frac{a-y}{b}\right)^c\right\}; \quad y < a; \; b, \, c > 0; \qquad (3.39\text{a})$$

$$F_R(y \mid a, b, c) = \left\{ \begin{array}{ll} \exp\left\{-\left(\frac{a-y}{b}\right)^c\right\} & \text{for} \quad y < a, \\ 1 & \text{for} \quad y \ge a. \end{array} \right\} \qquad (3.39\text{b})$$

This distribution has been recognized as the type–III maximum distribution in Sect. 3.2.2. The corresponding hazard rate is

$$h_R(y \mid a, b, c) = \frac{c}{b} \left(\frac{a-y}{b}\right)^{c-1} \frac{\exp\left\{-\left(\frac{a-y}{b}\right)^c\right\}}{1 - \exp\left\{-\left(\frac{a-y}{b}\right)^c\right\}}, \qquad (3.39\text{c})$$

which — independent of $c$ — is increasing, and goes to $\infty$ with $y$ to $a^-$.

Some parameters of the reflected WEIBULL distribution are

$$\text{E}(Y) = a - b\,\Gamma_1, \qquad (3.39\text{d})$$

$$\text{Var}(Y) = \text{Var}(X) = b^2 \left(\Gamma_2 - \Gamma_1^2\right), \qquad (3.39\text{e})$$

$$\alpha_3(Y) = -\alpha_3(X), \qquad (3.39\text{f})$$

$$\alpha_4(Y) = \alpha_4(X), \qquad (3.39\text{g})$$

$$y_{0.5} = a - b\,(\ln 2)^{1/c}, \qquad (3.39\text{h})$$

$$y^* = a - b\,(1 - 1/c)^{1/c} \qquad (3.39\text{i})$$

Figure 3/5: Hazard rate of the reflected WEIBULL distribution ($a = 0, \; b = 1$)

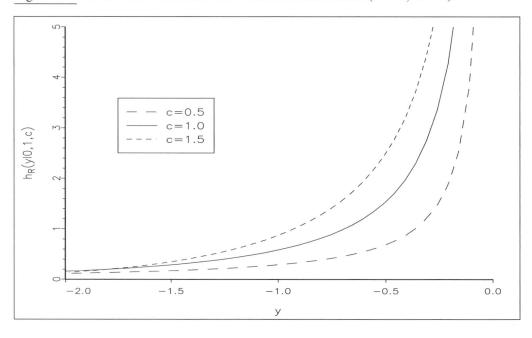

By changing the sign of the data sampled from a reflected WEIBULL distribution, it can be viewed as data from the classical WEIBULL model. Thus the parameters may be estimated by the methods discussed in Chapters 9 ff.

Combining the classical and the reflected WEIBULL models into one distribution results in the **double WEIBULL distribution** with DF

$$f_D(y \mid a, b, c) = \frac{c}{2\,b} \left| \frac{a-y}{b} \right|^{c-1} \exp\left\{ -\left| \frac{a-y}{b} \right|^c \right\}; \quad y,\, a \in \mathbb{R};\ b,\, c > 0; \qquad (3.40a)$$

and CDF

$$F_D(y \mid a, b, c) = \begin{cases} 0.5 \, \exp\left\{ -\left( \dfrac{a-y}{b} \right)^c \right\} & \text{for} \quad y \le a, \\[3mm] 1 - 0.5 \, \exp\left\{ -\left( \dfrac{y-a}{b} \right)^c \right\} & \text{for} \quad y \ge a. \end{cases} \qquad (3.40b)$$

The density function is symmetric about a vertical line in $y = a$ (see Fig. 3/6 for $a = 0$) and the CDF is symmetric about the point ($y = a$, $F_D = 0.5$). When $c = 1$, we have the **double exponential distribution** or **LAPLACE distribution**.

In its general three–parameter version the double WEIBULL distribution is not easy to analyze. Thus, BALAKRISHNAN/KOCHERLAKOTA (1985), who introduced this model, concentrated on the reduced form with $a = 0$ and $b = 1$. Then, (3.40a,b) turn into

$$f_D(y \mid 0, 1, c) = \frac{c}{2} \, |y|^{c-1} \, \exp\{ -|y|^c \} \qquad (3.41a)$$

and

$$F_D(y \mid 0, 1, c) = \begin{cases} 0.5 \, \exp\left[ -(-y)^c \right] & \text{for} \quad y \le 0, \\[2mm] 1 - 0.5 \, \exp\left[ -y^c \right] & \text{for} \quad y \ge 0. \end{cases} \qquad (3.41b)$$

(3.41a) is depicted in the upper part of Fig. 3/6 for several values of $c$. Notice that the distribution is bimodal for $c > 1$ and unimodal for $c = 1$ and has an improper mode for $0 < c < 1$. The lower part of Fig. 3/6 shows the hazard rate

$$h_D(y \mid 0, 1, c) = \begin{cases} \dfrac{c\,(-y)^{c-1} \, \exp\left\{ -(-y)^c \right\}}{2 - \exp\{ -(-y)^c \}} & \text{for} \quad y \le 0, \\[4mm] \dfrac{c\,y^{c-1} \, \exp\{ -y^c \}}{\exp\{ -y^c \}} & \text{for} \quad y \ge 0, \end{cases} \qquad (3.41c)$$

which — except for $c = 1$ — is far from being monotone; it is asymmetric in any case.

Figure 3/6: Density function and hazard rate of the double WEIBULL distribution ($a = 0$, $b = 1$)

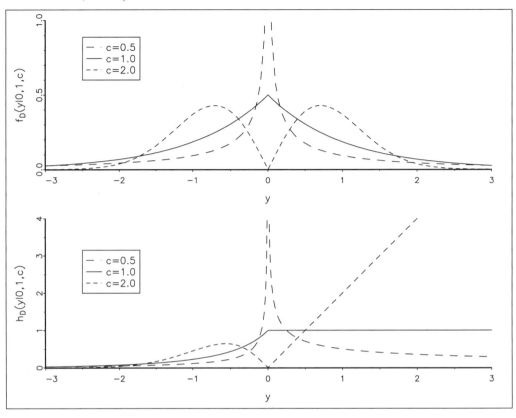

The moments of this reduced form of the double WEIBULL distribution are given by

$$E(Y^r) = \left\{ \begin{array}{ll} 0 & \text{for } r \text{ odd,} \\ \Gamma\left(1 + \dfrac{r}{c}\right) & \text{for } r \text{ even.} \end{array} \right\} \qquad (3.41\text{d})$$

Thus, the variance follows as

$$\text{Var}(Y) = \Gamma\left(1 + \frac{2}{c}\right) \qquad (3.41\text{e})$$

and the kurtosis as

$$\alpha_4(Y) = \Gamma\left(1 + \frac{4}{c}\right) \Big/ \Gamma^2\left(1 + \frac{2}{c}\right). \qquad (3.41\text{f})$$

The absolute moments of $Y$ are easily found to be

$$E\big[|Y|^r\big] = \Gamma\left(1 + \frac{r}{c}\right), \quad r = 0, 1, 2, \ldots \qquad (3.41\text{g})$$

Parameter estimation for this distribution — mainly based on order statistics — is considered by BALAKRISHNAN/KOCHERLAKOTA (1985), RAO/NARASIMHAM (1989) and RAO/RAO/NARASIMHAM (1991).

### 3.3.3   Inverse WEIBULL distribution[13]

In Sect. 3.2.2. we have shown that — when $X \sim We(a, b, c)$, i.e., $X$ has a classical WEIBULL distribution — the transformed variable

$$Y = \frac{b^2}{X - a}$$

has the DF

$$f_I(y \mid b, c) = \frac{c}{b} \left(\frac{y}{b}\right)^{-c-1} \exp\left\{-\left(\frac{y}{b}\right)^{-c}\right\}; \quad y \geq 0; \; b, c > 0 \qquad (3.42a)$$

and the CDF

$$F_I(y \mid b, c) = \exp\left\{-\left(\frac{y}{b}\right)^{-c}\right\}. \qquad (3.42b)$$

This distribution is known as **inverse WEIBULL distribution**.[14]   Other names for this distribution are **complementary WEIBULL distribution** (DRAPELLA, 1993), **reciprocal WEIBULL distribution** (MUDHOLKAR/KOLLIA, 1994) and **reverse WEIBULL distribution** (MURTHY et al., 2004, p. 23).The distribution has been introduced by KELLER et al. (1982) as a suitable model to describe degradation phenomena of mechanical components (pistons, crankshafts) of diesel engines.

The density function generally exhibits a long right tail (compared with that of the commonly used distributions, see the upper part of Fig. 3/7, showing the densities of the classical and the inverse WEIBULL distributions for the same set of parameter values). In contrast to the classical WEIBULL density the inverse WEIBULL density always has a mode $y^*$ in the interior of the support, given by

$$y^* = b \left(\frac{c}{1 + c}\right)^{1/c}. \qquad (3.42c)$$

and it is always positively skewed.

Inverting (3.42b) leads to the following percentile function

$$y_P = F_I^{-1}(P) = b \left(-\ln P\right)^{1/c}. \qquad (3.42d)$$

The $r$–th moment about zero is given by

$$\begin{aligned}
\mathrm{E}(Y^r) &= \int_0^\infty y^r \left(\frac{c}{b}\right) \left(\frac{y}{b}\right)^{-c-1} \exp\left[-\left(\frac{y}{b}\right)^{-c}\right] \mathrm{d}y \\
&= b^r \, \Gamma\left(1 - \frac{r}{c}\right) \quad \text{for } r < c \text{ only.} \qquad (3.42e)
\end{aligned}$$

The above integral is not finite for $r \geq c$. As such, when $c \leq 2$, the variance is not finite. This is a consequence of the long, fat right tail.

---

[13]   Suggested reading for this section: CALABRIA/PULCINI (1989, 1990, 1994), DRAPELLA (1993), ERTO (1989), ERTO/RAPONE (1984), JIANG/MURTHY/JI (2001), MUDHOLKAR/KOLLIA (1994).

[14]   The distribution is identical to the type–II maximum distribution.

The hazard rate of the inverse WEIBULL distribution

$$h_I(y \mid b,c) = \frac{\left(\frac{c}{b}\right)\left(\frac{y}{b}\right)^{-c-1}\exp\left\{-\left(\frac{y}{b}\right)^{-c}\right\}}{1-\exp\left\{-\left(\frac{y}{b}\right)^{-c}\right\}} \tag{3.42f}$$

is an upside down bathtub (see the lower part of Fig. 3/7) and has a behavior similar to that of the lognormal and inverse GAUSSIAN distributions:

$$\lim_{y\to 0} h_I(y \mid b,c) = \lim_{y\to\infty} h_I(y \mid b,c) = 0$$

with a maximum at $y_h^*$, which is the solution of

$$\frac{\left(\frac{b}{y}\right)^c}{1-\exp\left[-\left(\frac{b}{y}\right)^c\right]} = \frac{c+1}{c}.$$

Figure 3/7: Density function and hazard rate of the inverse WEIBULL distribution

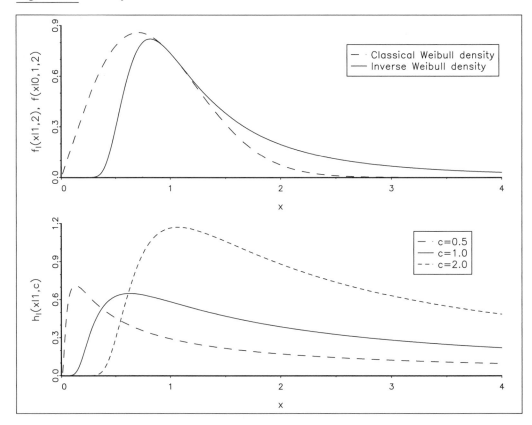

CALABRIA/PULCINI (1989, 1994) have applied the ML–method to estimate the two parameters. They also studied a BAYESIAN approach of predicting the ordered lifetimes in a future sample from an inverse WEIBULL distribution under type–I and type–II censoring.

### 3.3.4  Log-WEIBULL distribution[15]

In Sect. 3.2.2 we have introduced the following transformation of $X \sim We(a, b, c)$ :

$$Y = \ln(X - a); \ \ X \geq a.$$

$Y$ is called a **Log–WEIBULL variable**. Starting from

$$F_X(t \mid a, b, c) = \text{Pr}(X \leq t) = 1 - \exp\left\{-\left(\frac{t - a}{b}\right)^c\right\},$$

we have

$$
\begin{aligned}
\text{Pr}(Y \leq t) &= \text{Pr}\left[\ln(X - a) \leq t\right] \\
&= \text{Pr}\left[X - a \leq e^t\right] = \text{Pr}\left[X \leq a + e^t\right] \\
&= 1 - \exp\left\{-\left(e^t/b\right)^c\right\} \\
&= 1 - \exp\left\{-\exp\left[c\,(t - \ln b)\right]\right\}.
\end{aligned}
\tag{3.43a}
$$

(3.43a) is nothing but $F_L(t \mid a^*, b^*)$, the CDF of a type–I–minimum distribution (see (3.17b)) with location parameter

$$a^* := \ln b \tag{3.43b}$$

and scale parameter

$$b^* := 1/c. \tag{3.43c}$$

Provided, $a$ of the original WEIBULL variate is known the log–transformation results in a distribution of the location–scale–family which is easy to deal with. So one approach to estimate the parameters $b$ and $c$ rests upon a preceding log–transformation of the observed WEIBULL data.

DF and HR belonging to (3.43a) are

$$f_L(t \mid a^*, b^*) = \frac{1}{b^*}\exp\left\{\frac{t - a^*}{b^*} - \exp\left[\frac{t - a^*}{b^*}\right]\right\} \tag{3.43d}$$

and

$$h_L(t \mid a^*, b^*) = \frac{1}{b^*}\exp\left(\frac{t - a^*}{b^*}\right). \tag{3.43e}$$

(3.43d) has been graphed in Fig. 1/1 for $a^* = 0$ and $b^* = 1$. The hazard rate is an increasing function of $t$. The percentile function is

$$t_P = a^* + b^*\ln\left[-\ln(1 - P)\right]; \ \ 0 < P < 1. \tag{3.43f}$$

In order to derive the moments of a Log–WEIBULL distribution, we introduce the reduced variable

$$Z = c\,(Y - \ln b) = \frac{Y - a^*}{b^*}$$

---

[15]  Suggested reading for this section: KOTZ/NADARAJAH (2000), LIEBLEIN/ZELEN (1956), WHITE (1969).

with

$$F_L(z \mid 0, 1) \;=\; 1 - \exp\!\big(-e^z\big), \; z \in \mathbb{R}, \tag{3.44a}$$

$$f_L(z \mid 0, 1) \;=\; \exp\!\big(z - e^z\big). \tag{3.44b}$$

The corresponding raw moment generating function is

$$
\begin{aligned}
M_Z(\theta) \;&=\; \mathrm{E}\!\big(-e^{\theta Z}\big) \\[2mm]
&=\; \int_{-\infty}^{\infty} e^{\theta z}\, \exp\!\big(z - e^{-z}\big)\,\mathrm{d}z, \\[2mm]
&=\; \int_{0}^{\infty} u^{\theta}\, e^{-u}\,\mathrm{d}u, \text{ setting } u = e^z, \\[2mm]
&=\; \Gamma(1 + \theta). \tag{3.44c}
\end{aligned}
$$

Using (2.65c) we easily find

$$
\begin{aligned}
\mathrm{E}(Z) \;&=\; \left.\frac{\mathrm{d}\Gamma(1+\theta)}{\mathrm{d}\theta}\right|_{\theta=0} = \Gamma'(1) \\[2mm]
&=\; \psi(1)\,\Gamma(1) \\[2mm]
&=\; -\gamma \approx -0.577216; \tag{3.44d}
\end{aligned}
$$

$$
\begin{aligned}
\mathrm{E}\big(Z^2\big) \;&=\; \left.\frac{\mathrm{d}^2\Gamma(1+\theta)}{\mathrm{d}\theta^2}\right|_{\theta=0} = \Gamma''(1) \\[2mm]
&=\; \Gamma(1)\,\big[\psi^2(1) + \psi'(1)\big] \\[2mm]
&=\; \gamma^2 + \pi^2/6 \approx 1.97811; \tag{3.44e}
\end{aligned}
$$

$$
\begin{aligned}
\mathrm{E}\big(Z^3\big) \;&=\; \left.\frac{\mathrm{d}^3\Gamma(1+\theta)}{\mathrm{d}\theta^3}\right|_{\theta=0} = \Gamma'''(1) \\[2mm]
&=\; \Gamma(1)\,\big[\psi^3(1) + 3\,\psi(1)\,\psi'(1) + \psi''(1)\big] \\[2mm]
&=\; -\gamma^3 - \frac{\gamma\,\pi^2}{2} + \psi''(1) \approx -5.44487; \tag{3.44f}
\end{aligned}
$$

$$
\begin{aligned}
\mathrm{E}\big(Z^4\big) \;&=\; \left.\frac{\mathrm{d}^4\Gamma(1+\theta)}{\mathrm{d}\theta^4}\right|_{\theta=0} = \Gamma^{(4)}(1) \\[2mm]
&=\; \Gamma(1)\,\Big\{\psi^4(1) + 6\,\psi^2(1)\,\psi'(1) + 3\,\big[\psi'(1)\big]^2 + 4\,\psi(1)\,\psi''(1) + \psi'''(1)\Big\} \\[2mm]
&=\; \gamma^4 + \gamma^2\,\pi^2 + \frac{3\,\pi^4}{20} - 4\,\gamma\,\psi''(1) \approx 23.5615. \tag{3.44g}
\end{aligned}
$$

(3.44c–d) lead to the following results for a general Log–WEIBULL variable:

$$\mathrm{E}(Y) = a^* + b^*(-\gamma) \approx a^* - 0.577216\, b^*; \tag{3.45a}$$

$$\mathrm{Var}(Y) = (b^*)^2\, \pi^2/6 \approx 1.644934\,(b^*)^2; \tag{3.45b}$$

$$\alpha_3 \approx -1.13955; \tag{3.45c}$$

$$\alpha_4 \approx 5.4. \tag{3.45d}$$

So the Log–WEIBULL density is negatively skewed with a mode $y^* = a^*$ and leptokurtic.

Continuing with the reduced Log–WEIBULL variable, we can state the following property of the first order statistic (sample minimum) in a sample of size $n$

$$Z_{1:n} = \min_{1 \le i \le n}(Z_i); \quad Z_i \text{ iid};$$

$$\begin{aligned}
\Pr\bigl(Z_{1:n} \le t\bigr) &= 1 - \bigl[1 - F_L(t\,|\,0,1)\bigr]^n \\
&= 1 - \exp[-n\,e^t] \\
&= F_L\bigl[t + \ln(n)\bigr],
\end{aligned} \tag{3.46a}$$

which implies that $Z_{1:n}$ has the same distribution as $Z - \ln(n)$, and hence

$$\mathrm{E}\bigl(Z_{1:n}\bigr) = -\gamma - \ln(n) \tag{3.46b}$$

$$\mathrm{Var}\bigl(Z_{1:n}\bigr) = \mathrm{Var}(Z) = \pi^2/6. \tag{3.46c}$$

We will revert to order statistics in greater detail in Sect. 5.

Parameter estimation of the Log–WEIBULL distribution can be viewed as a special case of the estimation for extreme value distributions. DEKKERS et al. (1989) and CHRISTOPEIT (1994) have investigated estimation by the method of moments. More information about statistical inference is provided in the monograph by KOTZ/NADARAJAH (2000). The first, but still noteworthy case study applying the Log–WEIBULL distribution as the lifetime distribution for ball bearing was done by LIEBLEIN/ZELEN (1956).

We finally note that the relationship between WEIBULL and Log–WEIBULL variables is the reverse of that between normal and log–normal variables. That is, if $\log X$ is normal then $X$ is called a log-normal variable, while if $\exp(Y)$ has a WEIBULL distribution then we say that $Y$ is a Log–WEIBULL variable. Perhaps a more appropriate name than Log–WEIBULL variable could have been chosen, but on the other hand one might also argue that the name "log-normal" is misapplied.

### 3.3.5 Truncated WEIBULL distributions[16]

In common language the two words *truncation* and *censoring* are synonyms, but in statistical science they have different meanings. We join COHEN (1991, p. 1) when we say

---

[16] Suggest reading for this section: AROIAN (1965), CROWDER (1990), GROSS (1971), HWANG (1996), MARTINEZ/QUINTANA (1991), MCEWEN/PARRESOL (1991), MITTAL/DAHIYA (1989), SHALABY (1993), SHALABY/AL–YOUSSEF (1992), SUGIURA/GOMI (1985), WINGO (1988, 1989, 1998).

**truncation** is a notion **related to populations** whereas **censoring** is **related to samples**. A truncated population is one where — according to the size of the variable — some part of the original population has been removed thus restricting the support to a smaller range. A truncated distribution is nothing but a conditional distribution,[17] the condition being that the variate is observable in a restricted range only. The most familiar types of truncation, occurring when the variate is a lifetime or a duration, are

- **left–truncation** In such a lower truncated distribution the smaller realizations below some left truncation point $t_\ell$ have been omitted. This will happen when all items of an original distribution are submitted to some kind of screening, e.g., realized by a burn–in lasting $t_\ell$ units of time.

- **right–truncation** An upper truncated distribution is missing the greater realizations above some right truncation point $t_r$. This might happen when the items of a population are scheduled to be in use for a maximum time of $t_r$ only.

- **double–truncation** A doubly truncated distribution only comprises the "middle" portion of all items.

**Censored samples** are those in which sample specimens with measurements that lie in some restricted areas of the sample space may be identified and thus counted, but are not otherwise measured. So the censored sample units are not omitted or forgotten but they are known and are presented by their number, but not by the exact values of their characteristic.

In a truncated population a certain portion of units is eliminated without replacement thus the reduced size of the population, which would be less than 1 or 100%, has to be rescaled to arrive at an integrated DF amounting to 1; see (3.48a). Censored samples are extensively used in life–testing. We will present the commonly used types of censoring in Sect. 8.3 before turning to estimation and testing because the inferential approach depends on whether and how the sample has been censored.

Starting with a three–parameter WEIBULL distribution, the general truncated WEIBULL model is given by the following CDF

$$
\begin{aligned}
F_{DT}(x \mid a,b,c,t_\ell,t_r) &= \frac{F(x \mid a,b,c) - F(t_\ell \mid a,b,c)}{F(t_r \mid a,b,c) - F(t_\ell \mid a,b,c)}; \quad a \le t_\ell \le x \le t_r < \infty, \\[2mm]
&= \frac{\exp\left\{-\left(\dfrac{t_\ell - a}{b}\right)^c\right\} - \exp\left\{-\left(\dfrac{x - a}{b}\right)^c\right\}}{\exp\left\{-\left(\dfrac{t_\ell - a}{b}\right)^c\right\} - \exp\left\{-\left(\dfrac{t_r - a}{b}\right)^c\right\}} \\[2mm]
&= \frac{1 - \exp\left\{\left(\dfrac{t_\ell - a}{b}\right)^c - \left(\dfrac{x - a}{b}\right)^c\right\}}{1 - \exp\left\{\left(\dfrac{t_\ell - a}{b}\right)^c - \left(\dfrac{t_r - a}{b}\right)^c\right\}} \, . \quad (3.47a)
\end{aligned}
$$

This is also referred to as the **doubly truncated WEIBULL distribution**. We notice the following special cases of (3.47a):

---

[17] See Sect. 2.6.

- $(t_\ell > a, \ t_r = \infty)$ gives the **left truncated WEIBULL distribution** with CDF

$$F_{LT}(x \,|\, a, b, c, t_\ell, \infty) = 1 - \exp\left\{ \left( \frac{t_\ell - a}{b} \right)^c - \left( \frac{x - a}{b} \right)^c \right\}. \qquad (3.47b)$$

- $(t_\ell = a, a < t_r < \infty)$ gives the **right truncated WEIBULL distribution** with CDF

$$F_{RT}(x \,|\, a, b, c, a, t_r) = \frac{1 - \exp\left\{ -\left( \dfrac{x - a}{b} \right)^c \right\}}{1 - \exp\left\{ -\left( \dfrac{t_r - a}{b} \right)^c \right\}}. \qquad (3.47c)$$

- $(t_\ell = a; t_r = \infty)$ leads to the original, non–truncated WEIBULL distribution.

In the sequel we will give results pertaining to the doubly truncated WEIBULL distribution. Special results for the left and right truncated distributions will be given only when they do not follow from the doubly truncated result in a trivial way.

The DF belonging to (3.47a) is

$$
\begin{aligned}
f_{DT}(x \,|\, a, b, c, t_\ell, t_r) &= \frac{f(x \,|\, a, b, c)}{F(t_r \,|\, a, b, c) - F(t_\ell \,|\, a, b, c)}; \quad a \le t_\ell \le x \le t_r < \infty; \\[2ex]
&= \frac{\dfrac{c}{b} \left( \dfrac{x - a}{b} \right)^{c-1} \exp\left\{ -\left( \dfrac{x - a}{b} \right)^c \right\}}{\exp\left\{ -\left( \dfrac{t_\ell - a}{b} \right)^c \right\} - \exp\left\{ -\left( \dfrac{t_r - a}{b} \right)^c \right\}} \\[2ex]
&= \frac{\dfrac{c}{b} \left( \dfrac{x - a}{b} \right)^{c-1} \exp\left\{ \left( \dfrac{t_\ell - a}{b} \right)^c - \left( \dfrac{x - a}{b} \right)^c \right\}}{1 - \exp\left\{ \left( \dfrac{t_\ell - a}{b} \right)^c - \left( \dfrac{t_r - a}{b} \right)^c \right\}}. \quad (3.48a)
\end{aligned}
$$

The shape of $f_{DT}(x \,|\, a, b, c, t_\ell, t_r)$ is determined by the shape of $f(x \,|\, a, b, c)$ over the interval $t_\ell \le x \le t_r$ (see Fig. 3/8) with the following results:

- $(c < 1)$ or $(c > 1$ and $t_\ell > x^*) \implies f_{DT}(\cdot)$ is decreasing,
- $c > 1$ and $t_r < x^* \implies f_{DT}(\cdot)$ is increasing,
- $c > 1$ and $t_\ell < x^* < t_r \implies f_{DT}(\cdot)$ is unimodal,

where $x^* = a + b \left[ (c - 1)/c \right]^{1/c}$ is the mode of the non–truncated distribution.

The hazard rate belonging to (3.47a) is

$$
\begin{aligned}
h_{DT}(x \,|\, a, b, c, t_\ell, t_r) &= \frac{f_{DT}(x \,|\, a, b, c)}{1 - F_{DT}(x \,|\, a, b, c, t_\ell, t_r)}; \quad a \le t_\ell \le x \le t_r < \infty; \\[2ex]
&= \frac{f(x \,|\, a, b, c)}{F(t_r \,|\, a, b, c) - F(x \,|\, a, b, c)} \\[2ex]
&= \frac{f(x \,|\, a, b, c)}{1 - F(x \,|\, a, b, c)} \, \frac{1 - F(x \,|\, a, b, c)}{F(t_r \,|\, a, b, c) - F(x \,|\, a, b, c)}. \quad (3.48b)
\end{aligned}
$$

The first factor on the right–hand side of (3.48b) is the hazard rate (2.36b) of the non–truncated distribution. The second factor is a monotonically increasing function, approaching $\infty$ as $x \to t_r$. So $h_{DT}(\cdot)$ can be either decreasing, increasing or bathtub–shaped. The bathtub shape will come up in the doubly and right truncated cases when $0 < c < 1$.

Figure 3/8: Densities of truncated and non-truncated WEIBULL distributions

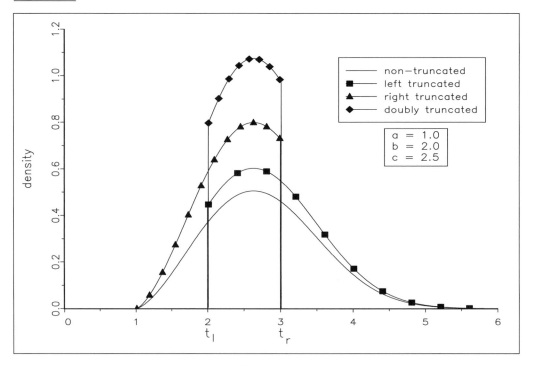

The percentile function belonging to (3.47a) is

$$F_{DT}^{-1}(P) = a + b \left\{ \ln \left[ \frac{1}{1 - F(t_\ell \mid a,b,c) - P\left[F(t_r \mid a,b,c) - F(t_\ell \mid a,b,c)\right]} \right] \right\}^{1/c} . \tag{3.48c}$$

MCEWEN/PARRESOL (1991) give the following formula for the raw moments of the doubly truncated WEIBULL distribution:

$$E\left(X_{DT}^r\right) = \frac{\exp\left\{\left(\dfrac{t_\ell - a}{b}\right)^c\right\}}{1 - \exp\left\{\left(-\dfrac{t_r - a}{b}\right)^c\right\}} \sum_{j=0}^{r} \binom{r}{j} b^{r-j} a^j \times$$

$$\left\{ \gamma\left[\frac{r-j}{c} + 1 \left| \left(-\frac{t_r - a}{b}\right)^c\right] - \gamma\left[\frac{r-j}{c} + 1 \left| \left(\frac{t_\ell - a}{b}\right)^c\right] \right\}, \tag{3.48d}$$

where $\gamma[\cdot \mid \cdot]$ is the incomplete gamma function (see the excursus in Sect. 2.9.1). So, the moments can be evaluated only numerically. Looking at the reduced WEIBULL variate $U$, i.e., $a = 0$ and $b = 1$, (3.48d) becomes a little bit simpler:

$$\mathrm{E}\big(U_{DT}^r\big) = \frac{\exp\{t_\ell^c\}}{1 - \exp\{-t_r^c\}} \left\{ \gamma\left[\frac{r}{c} + 1 \,\Big|\, t_r^c\right] - \gamma\left[\frac{r}{c} + 1 \,\Big|\, t_\ell^c\right] \right\}. \tag{3.48e}$$

In the left truncated case (3.48d,e) turn into[18]

$$\mathrm{E}\big(X_{LT}^r\big) = \exp\left\{\left(\frac{t_\ell - a}{b}\right)^c\right\} \sum_{j=0}^{r} \binom{r}{j} b^{r-j}\, a^j \times$$

$$\left\{ \Gamma\left(\frac{r-j}{c} + 1\right) - \gamma\left[\frac{r-j}{c} + 1 \,\Big|\, \left(\frac{t_\ell - a}{b}\right)^c\right] \right\} \tag{3.49a}$$

and

$$\mathrm{E}\big(U_{LT}^r\big) = \exp\{t_\ell^c\} \left\{ \Gamma\left(\frac{r}{c} + 1\right) - \gamma\left[\frac{r}{c} + 1 \,\Big|\, t_\ell^c\right] \right\}. \tag{3.49b}$$

When the distribution is right truncated we have

$$\mathrm{E}\big(X_{RT}^r\big) = \frac{1}{1 - \exp\left\{-\left(\frac{t_r - a}{b}\right)^c\right\}} \sum_{j=0}^{r} \binom{r}{j} b^{r-j}\, a^j\, \gamma\left[\frac{r-j}{c} + 1 \,\Big|\, \left(\frac{t_r - a}{b}\right)^c\right]$$

$$\tag{3.50a}$$

and

$$\mathrm{E}\big(U_{RT}^r\big) = \frac{1}{1 - \exp\{-t_r^c\}} \gamma\left[\frac{r}{c} + 1 \,\Big|\, t_r^c\right]. \tag{3.50b}$$

SUGIURA/GOMI (1985) give a moment–ratio diagram showing the right truncated WEIBULL distribution in comparison with the doubly truncated normal distribution.

Testing hypotheses on the parameters of a truncated WEIBULL distribution is treated by CROWDER (1990) and MARTINEZ/QUINTANA (1991). Parameter estimation is discussed by MITTAL/DAHIYA (1989) and WINGO (1988, 1989, 1998) applying the maximum likelihood approach and by SHALABY (1993) and SHALABY/AL–YOUSSEF (1992) using a BAYES approach. See CHARERNKAVANICH/COHEN (1984) for estimating the parameters as well as the unknown left point of truncation by maximum likelihood. HWANG (1996) gives an interesting application of the truncated WEIBULL model in reliability prediction of missile systems.

### 3.3.6 Models including two or more distributions

This section presents models which result when several variates or several distributions are combined in one way or another whereby at least one of these variates or distributions has to be WEIBULL. We will first (Sect. 3.3.6.1) study the distribution of the sum of WEIBULL variates. Then (Sect. 3.3.6.2) we turn to the lifetime distribution of a system made up

---

[18] AROIAN (1965) gives an evaluation based on the $\chi^2$–distribution.

by linking two or more randomly failing components. Sections 3.3.6.3 to 3.3.6.5 deal with composing and with discrete as well as continuous mixing; topics which have been explored intensively in theory and practice.

### 3.3.6.1 WEIBULL folding

The distribution of the **sum of** two or more **variates** is found by convoluting or folding their distributions. We will only report on the sum of iid WEIBULL variates. Such a sum is of considerable importance in renewal theory (Sect. 4.4) and for WEIBULL processes (Sect. 4.3).

First, we will summarize some general results on a sum

$$Y = \sum_{i=1}^{n} X_i$$

of $n$ iid variates. The DF and CDF of $Y$ are given by the following recursions:

$$f_n(y) = \int_R f(x) f_{n-1}(y - x) \, dx; \quad n = 2, 3, \ldots; \tag{3.51a}$$

$$F_n(y) = \int_R f(x) F_{n-1}(y - x) \, dx; \quad n = 2, 3, \ldots; \tag{3.51b}$$

with

$$f_1(\cdot) = f(\cdot) \quad \text{and} \quad F_1(\cdot) = F(\cdot).$$

$R$, the area of integration in (3.51a,b), contains all arguments for which the DFs and CDFs in the integrand are defined. The raw moments of $Y$ are given by

$$
\begin{aligned}
\mathrm{E}(Y^r) &= \mathrm{E}\big[(X_1 + X_2 + \ldots + X_n)^r\big] \\
&= \sum \frac{r!}{r_1! \, r_2! \ldots r_n!} \mathrm{E}\big(X_1^{r_1}\big) \mathrm{E}\big(X_2^{r_2}\big) \ldots \mathrm{E}\big(X_n^{r_n}\big).
\end{aligned}
\tag{3.51c}
$$

The summation in (3.51c) runs over all $n$–tuples $(r_1, r_2, \ldots, r_n)$ with $\sum_{i=1}^{n} r_i = r$ and $0 \le r_i \le r$. The number of summands is $\binom{r+n-1}{r}$. In particular we have

$$\mathrm{E}(Y) = \sum_{i=1}^{n} \mathrm{E}(X_i). \tag{3.51d}$$

The variance of $Y$ is

$$\mathrm{Var}(Y) = \sum_{i=1}^{n} \mathrm{Var}(X_i). \tag{3.51e}$$

Assuming

$$X_i \sim We(a, b, c); \quad i = 1, 2, \ldots, n,$$

(3.51a,b) turn into

$$f_n(y) = \int\limits_{na}^{y} \frac{c}{b} \left(\frac{x-a}{b}\right)^{c-1} \exp\left\{-\left(\frac{x-a}{b}\right)^c\right\} f_{n-1}(y-x)\, dx; \quad n = 2,3,\ldots; \quad (3.52a)$$

$$F_n(y) = \int\limits_{na}^{y} \frac{c}{b} \left(\frac{x-a}{b}\right)^{c-1} \exp\left\{-\left(\frac{x-a}{b}\right)^c\right\} F_{n-1}(y-x)\, dx; \quad n = 2,3,\ldots. \quad (3.52b)$$

Whereas the evaluation of (3.51d,e) for WEIBULL variates is quite simple there generally do not exist closed form expressions for the integrals[19] on the right–hand sides of (3.52a,b), not even in the case $n = 2$.

An exception is the case $c = 1$ for all $n \in \mathbb{N}$, i.e. the case of folding identically and independently exponentially distributed variates. The $n$–fold convolution is a gamma distribution, more precisely an ERLANG distribution as $n$ is a positive integer. For $n = 2$ and $c = 1$ we have

$$f_2(y \,|\, a, b, 1) = \frac{y - 2a}{b^2} \exp\left[-\left(\frac{y - 2a}{b}\right)\right], \quad y \geq 2a,$$

$$F_2(y \,|\, a, b, 1) = 1 - \exp\left[-\frac{y - 2a}{b}\right] \left(\frac{y - 2a + b}{b}\right), \quad y \geq 2a.$$

For $n \in \mathbb{N}$, $c = 1$ and $a = 0$, we have the handsome formulas

$$f_n(y \,|\, 0, b, 1) = \frac{y^{n-1}}{b^n \, (n-1)!} \exp\left(-\frac{y}{b}\right),$$

$$F_n(y \,|\, 0, b, 1) = 1 - \exp\left(-\frac{y}{b}\right) \sum_{i=0}^{n-1} \frac{y^i}{b^i \, i!}.$$

Turning to a genuine WEIBULL distribution, i.e., $c \neq 1$, we will only evaluate (3.52a,b) for $n = 2$ and $a = 0$ when $c = 2$ or $c = 3$.

$\boxed{c = 2}$

$$f_2(y \,|\, 0, b, 2) = \frac{\exp\left[-\left(\frac{y}{b}\right)^2\right]}{2\, b^3} \left\{2\, b\, y - \sqrt{2\pi}\, \exp\left[\frac{1}{2}\left(\frac{y}{b}\right)^2\right](b^2 - y^2)\, \text{erf}\left(\frac{y}{b\sqrt{2}}\right)\right\}$$

$$F_2(y \,|\, 0, b, 2) = 1 - \exp\left[-\left(\frac{y}{b}\right)^2\right] - \frac{\exp\left[-\frac{1}{2}\left(\frac{y}{b}\right)^2\right]}{b} \sqrt{\frac{\pi}{2}}\, \text{erf}\left(\frac{y}{b\sqrt{2}}\right)$$

---

[19] In Sect. 3.1.4 we have already mentioned that the WEIBULL distribution is not reproductive through summation.

$\boxed{c=3}$

$$f_2(y\,|\,0,b,3) \;=\; \frac{\exp\!\left[-\left(\frac{y}{b}\right)^3\right]}{16\,b^{9/2}\,y^{5/2}}\left\{6\,(b\,y)^{3/2}\,(y^3-2\,b^3)+\sqrt{3\,\pi}\,\exp\!\left[\frac{3}{4}\left(\frac{y}{b}\right)^3\right]\times\right.$$

$$\left.\left[4\,b^6-4\,(b\,y)^3+3\,y^6\right]\operatorname{erf}\!\left[\frac{\sqrt{3}}{2}\left(\frac{y}{b}\right)^{3/2}\right]\right\}$$

$$F_2(y\,|\,0,b,3) \;=\; 1-\frac{\exp\!\left[-\left(\frac{y}{b}\right)^3\right]}{2}-\frac{\exp\!\left[-\frac{1}{4}\left(\frac{y}{b}\right)^3\right]}{4\,(b\,y)^{3/2}}\sqrt{\frac{\pi}{3}}\,(2\,b^3+3\,y^3)\operatorname{erf}\!\left[\frac{\sqrt{3}}{2}\left(\frac{y}{b}\right)^{3/2}\right]$$

$\operatorname{erf}(\cdot)$ is the **GAUSSIAN error function**:

$$\operatorname{erf}(\tau)=\frac{2}{\sqrt{\pi}}\int_0^\tau e^{-t^2}\,\mathrm{d}t, \tag{3.53a}$$

which is related to $\Phi(z)$, the CDF of the standardized normal variate by

$$\Phi(\tau)=\frac{1}{2}\left[1+\operatorname{erf}(\tau/\sqrt{2})\right] \quad \text{for} \;\; \tau\geq 0. \tag{3.53b}$$

Figure 3/9: Densities of a two-fold convolution of reduced WEIBULL variates

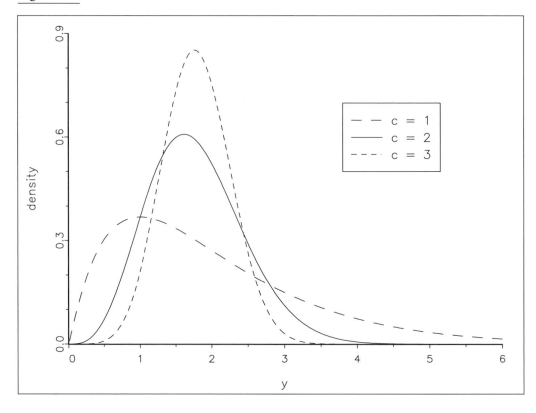

#### 3.3.6.2 WEIBULL models for parallel and series systems

Two simple systems in reliability theory are the **series system** and the **parallel system**, each consisting of $n$ components that will fail at random. For component $i$ $(1 \le i \le n)$ let $F_i(t) = \Pr(X_i \le t)$ be the probability of failure up to time $t$, then $R_i(t) = 1 - F_i(t) = \Pr(X_i > t)$ is the probability of its surviving of time $t$.

A series system is working as long as all its $n$ components are working, so that the series survival probability of time $t$ is given by

$$
\begin{aligned}
R_S(t) &= \Pr(X_1 > t,\ X_2 > t,\ \ldots,\ X_n > t) \\
&= \prod_{i=1}^{n} R_i(t).
\end{aligned} \tag{3.54a}
$$

(3.54a) holds only for independently failing components. $R_S(t)$ is a decreasing function of $n$ with $R_S(t) \le R_i(t)\ \forall\ i$. So, a series system is more reliable the less components are combined into it. The series system's failure distribution follows as

$$
F_S(t) = 1 - \prod_{i=1}^{n} R_i(t) = 1 - \prod_{i=1}^{n} \left[ 1 - F_i(t) \right] \tag{3.54b}
$$

with DF

$$
\begin{aligned}
f_S(t) &= \sum_{i=1}^{n} \left\{ \prod_{j=1, j \ne i}^{n} \left[ 1 - F_j(t) \right] \right\} f_i(t), \\
&= R_S(t) \sum_{i=1}^{n} \frac{f_i(t)}{R_i(t)}.
\end{aligned} \tag{3.54c}
$$

Thus, the HR of a series system simply is

$$
\begin{aligned}
h_S(t) &= \frac{f_S(t)}{R_S(t)} = \sum_{i=1}^{n} \frac{f_i(t)}{R_i(t)} \\
&= \sum_{i=1}^{n} h_i(t),
\end{aligned} \tag{3.54d}
$$

i.e., the sum of each component's hazard rate.

A parallel system is working as long as at least one of its $n$ components is working and fails when all its components have failed. The failure distribution CDF is thus

$$
\begin{aligned}
F_P(t) &= \Pr(X_1 \le t,\ X_2 \le t,\ \ldots,\ X_n \le t) \\
&= \prod_{i=1}^{n} F_i(t)
\end{aligned} \tag{3.55a}
$$

for independently failing units. $F_P(t)$ is a decreasing function of $n$ with $F_P(t) \le F_i(t) \ \forall \ i$. Thus, a parallel system is more reliable the more components it has. The system's reliability function or survival function is

$$R_P(t) = 1 - \prod_{i=1}^{n} F_i(t) = 1 - \prod_{i=1}^{n} \left[ 1 - R_i(t) \right] \tag{3.55b}$$

and its DF is

$$
\begin{aligned}
f_P(t) &= \sum_{i=1}^{n} \left\{ \prod_{j=1, j \ne i}^{n} F_j(t) \right\} f_i(t) \\
&= F_P(t) \sum_{i=1}^{n} \frac{f_i(t)}{F_i(t)} .
\end{aligned}
\tag{3.55c}
$$

The HR follows as

$$
\begin{aligned}
h_P(t) &= \frac{f_P(t)}{R_P(t)} = \frac{F_P(t)}{1 - F_P(t)} \sum_{i=1}^{n} \frac{f_i(t)}{F_i(t)} \\
&= \frac{F_P(t)}{R_P(t)} \sum_{i=1}^{n} \frac{f_i(t)}{R_i(t)} \frac{R_i(t)}{F_i(t)} \\
&= \frac{F_P(t)}{R_P(t)} \sum_{i=1}^{n} h_i(t) \frac{R_i(t)}{F_i(t)} ,
\end{aligned}
\tag{3.55d}
$$

i.e., HR is a weighted sum of the individual hazard rates, where the weights $R_i(t)/F_i(t)$ are decreasing with $t$. This sum is multiplied by the factor $F_P(t)/R_P(t)$ which is increasing with $t$.

Before evaluating (3.54a–d) and (3.55a–d) using WEIBULL distributions, we should mention that the two preceding systems are borderline cases of the more general "**k–out–of–n–system**", which is working as long as at least $k$ of its $n$ components have not failed. Thus, an series system is a "$n$–out–of–$n$" system and a parallel system is a "1–out–of–$n$" system. The reliability function of a "$k$–out–of–$n$" system is given by

$$R_{k,n}(t) = \sum_{m=k}^{n} \left[ \sum_{Z(m,n)} R_{i_1}(t) \cdot \ldots \cdot R_{i_m}(t) \cdot F_{i_{m+1}}(t) \cdot \ldots \cdot F_{i_n}(t) \right] . \tag{3.56}$$

$Z(m,n)$ is the set of all partitions of $n$ units into two classes such that one class comprises $m$ non–failed units and the other class $n-m$ failed units. An inner sum gives the probability that exactly $m$ of the $n$ units have not failed. The outer sum rums over all $m \ge k$.

We first evaluate the series system, also known as **competing risk model** or **multi–risk model**, with WEIBULL distributed lifetimes of its components.[20] Assuming $X_i \overset{\text{iid}}{\sim}$

---

[20] MURTHY/XIE/JIANG (2004) study multi–risk models with inverse WEIBULL distributions and hybrid models for which one component has a WEIBULL distributed lifetime and the remaining components are non–WEIBULL.

$We(a, b, c)$; $i = 1, 2, \ldots, n$; (3.54a) turns into

$$R_S(t) = \exp\left\{-n\left(\frac{t-a}{b}\right)^c\right\} = \exp\left\{-\left(\frac{t-a}{b\,n^{-1/c}}\right)^c\right\}. \tag{3.57a}$$

(3.57a) is easily recognized as the CCDF of a variate $T \sim We(a, b\,n^{-1/c}, c)$, the minimum of $n$ iid WEIBULL variates (see Sect. 3.1.4). Assuming $X_i \overset{\text{iid}}{\sim} We(a, b_i, c)$; $i = 1, 2, \ldots, n$; i.e., the scale factor is varying with $i$ and the series system's lifetime $T$ is still WEIBULL with

$$R_S(t) = \exp\left\{-(t-a)^c \sum_{i=1}^n \frac{1}{b_i^c}\right\}. \tag{3.57b}$$

Now, the scale parameter of $T$ is $\left(\sum_{i=1}^n b_i^{-c}\right)^{-1/c}$. $T$ has no WEIBULL distribution when the shape parameters and/or the location parameters of the components' lifetimes are differing.

In the general case, i.e., $X_i \sim We(a_i, b_i, c_i)$ and independent, assuming $c_1 < c_2 < \ldots < c_n$ we can state some results on the hazard rate $h_S(t)$ of (3.54d).

- The asymptotic properties are
    ◇ $h_S(t) \approx h_1(t)$ for $t \to 0$,
    ◇ $h_S(t) \approx h_n(t)$ for $t \to \infty$.

  For small $t$ the hazard rate of the series system is nearly the same as that for the component with the smallest shape parameter. For large $t$, it is approximately the hazard rate for the component with the largest shape parameter.

- $h_S(t)$ can have only one of three possible shapes:
    ◇ $h_S(t)$ is decreasing when $c_1 < c_2 < \ldots < c_n < 1$.
    ◇ $h_S(t)$ is increasing when $1 < c_1 < c_2 < \ldots < c_n$.
    ◇ $h_S(t)$ is bathtub–shaped when $c_1 < \ldots < c_{j-1} < 1 < c_j < \ldots < c_n$.

Parameter estimation in this general case is extensively discussed by MURTHY/XIE/JIANG (2004, pp. 186–189) and by DAVISON/LOUZADA–NETO (2000), who call this distribution a **poly-WEIBULL model**.

A parallel system with WEIBULL distributed lifetimes of its components is sometimes called a **multiplicative WEIBULL model**, see (3.55a). Assuming $X_i \overset{\text{iid}}{\sim} We(a, b, c)$; $i = 1, 2, \ldots, n$, (3.55a) turns into

$$F_P(t) = \left[1 - \exp\left\{-\left(\frac{t-a}{b}\right)^c\right\}\right]^n. \tag{3.58a}$$

(3.58a) is a special case of the **exponentiated WEIBULL distribution**, introduced by MUDHOLKAR/SRIVASTAVA (1993) and analyzed in subsequent papers by MUDHOLKAR et al. (1995), MUDHOLKAR/HUTSON (1996) and JIANG/MURTHY (1999). Whereas in (3.58a) the parameter $n$ is a positive integer, it is substituted by $\nu \in \mathbb{R}$ to arrive at the exponentiated WEIBULL distribution.

The DF belonging to (3.58a) is

$$
\begin{aligned}
f_P(t) &= n\, f_X(t) \left[ F_X(t) \right]^{n-1} \\
&= n \frac{c}{b} \left( \frac{x-a}{b} \right)^{c-1} \exp\left\{ -\left( \frac{x-a}{b} \right)^c \right\} \left[ 1 - \exp\left\{ -\left( \frac{x-a}{b} \right)^c \right\} \right]^{n-1} \quad (3.58b)
\end{aligned}
$$

$f_P(t)$ is (see the upper part of Fig. 3/10)

- monotonically decreasing for $n\,c \le 1$ with $f_P(0) = \infty$ for $n\,c < 1$ and $f_P(0) = 1/b$ for $n\,c = 1$,

- unimodal for $n\,c > 1$ with — according to MUDHOLKAR/HUTSON (1996) — a mode approximated by

$$
t^* = a + b \left\{ \frac{1}{2} \left[ \frac{\sqrt{c\,(c - 8n + 2\,cn + 9\,cn^2)}}{cn} - 1 - \frac{1}{n} \right] \right\}^n . \qquad (3.58c)
$$

It is noteworthy that the effect of the product $c\,n$ on the shape of the DF for the exponentiated WEIBULL distribution is the same as that of $c$ in the simple WEIBULL distribution.

The HR belonging to (3.58a) is

$$
\begin{aligned}
h_P(t) &= \frac{f_P(t)}{1 - F_P(t)} \\
&= n \frac{c}{b} \left( \frac{x-a}{b} \right)^{c-1} \frac{\exp\left\{ -\left( \frac{x-a}{b} \right)^c \right\} \left[ 1 - \exp\left\{ -\left( \frac{x-a}{b} \right)^c \right\} \right]^{n-1}}{1 - \left[ 1 - \exp\left\{ -\left( \frac{x-a}{b} \right)^c \right\} \right]^n} \quad (3.58d)
\end{aligned}
$$

For small $t$ we have

$$
h_P(t) \approx \left( \frac{n\,c}{b} \right) \left( \frac{t-a}{b} \right)^{n\,c-1} ,
$$

i.e., the system's hazard rate can be approximated by the hazard rate of a WEIBULL distribution with shape parameter $n\,c$ and scale parameter $b$. For large $t$ we have

$$
h_P(t) \approx \left( \frac{c}{b} \right) \left( \frac{t-a}{b} \right)^{c-1} ,
$$

i.e., the approximating hazard rate is that of the underlying WEIBULL distribution. The shape of the HR is as follows (also see the lower part of Fig. 3/10):

- monotonically decreasing for $c \le 1$ and $n\,c \le 1$,

- monotonically increasing for $c \ge 1$ and $n\,c \ge 1$,

- unimodal for $c < 1$ and $n\,c > 1$,

- bathtub–shaped for $c > 1$ and $\nu\,c < 1$, $\nu \in \mathbb{R}^+$.

Figure 3/10: Densities and hazard rates of multiplicative WEIBULL models $(a = 0,\ b = 1)$

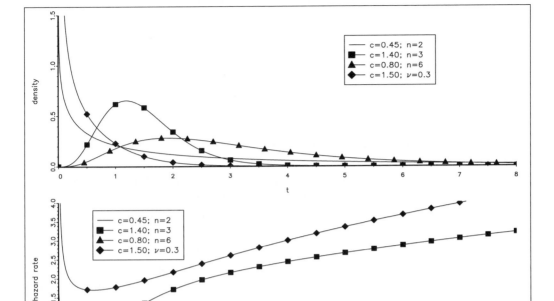

The raw moments belonging to (3.58a), but with $a = 0$, are

$$\mu'_r = n\,b^r\,\Gamma\left(1 + \frac{r}{c}\right) \sum_{j=0}^{n-1}(-1)^j \binom{n-1}{j} \frac{1}{(j+1)^{1+r/c}}.$$ (3.58e)

The percentile function is simply

$$F_P^{-1}(P) = a + b\left\{-\ln\left[1 - P^{1/n}\right]\right\}^{1/c},\ 0 < P < 1.$$ (3.58f)

General results on a multiplicative WEIBULL model with components having differing parameter values are not easy to find. When the components all have a two–parameter WEIBULL distribution $(a_i = 0\ \forall\ i)$ and assuming $c_1 \le c_2 \le \ldots \le c_n$ and $b_i \ge b_j$ for $i < j$ if $c_i = c_j$, MURTHY/XIE/JIANG (2004, p. 198/9) state the following results:

$$F_P(t) \approx F_0(t)\ \text{for small}\ t$$ (3.59a)

with $F_0(t)$ as the two–parameter WEIBULL CDF having

$$c_0 = \sum_{i=1}^{n} c_i \ \text{and} \ b_0 = \sum_{i=1}^{n} b_i^{c_i/c_0},$$

$$F_P(t) \approx (1 - k) + k\,F_1(t)\ \text{for large}\ t,$$ (3.59b)

where $k$ is the number of components with distribution identical to $F_1(t)$ having $c_1$ and $b_1$. The DF can be decreasing or may have $k$ modes. The HR can never have a bathtub shape, but has four possible shapes (decreasing, increasing, $k$ modal followed by increasing or decreasing followed by $k$ modal). The asymptotes of the HR are

$$
h_P(t) \approx \left\{ \begin{array}{ll} h_0(t) = \dfrac{c_0}{b_0} \left( \dfrac{t}{c_0} \right)^{c_0-1} & \text{for} \quad t \to 0 \\[4mm] h_1(t) = \dfrac{c_1}{b_1} \left( \dfrac{t}{b_1} \right)^{c_1-1} & \text{for} \quad t \to \infty. \end{array} \right\} \tag{3.59c}
$$

### 3.3.6.3 Composite WEIBULL distributions

A **composite distribution**, also known as **sectional model** or **piecewise model**, is constructed by knotting together pieces of two or more distributions in such a manner that the resulting distribution suffices some smoothness criteria. The task is somewhat similar to that of constructing a spline function. In reliability theory and life testing the composite WEIBULL distribution, first introduced by KAO (1959) to model the lifetime distribution of electronic tubes, is primarily used to arrive at a non–monotone hazard rate which is not encountered in only one WEIBULL distribution. The bathtub shape, regularly found in life–tables as the behavior of the conditional probability of death at age $x$, is of interest, see ELANDT–JOHNSON/JOHNSON (1980, Chapter 7.5). We will present the WEIBULL composite distribution having $n \geq 2$ WEIBULL subpopulations. Of course, it is possible to join WEIBULL and other distributions into a so–called hybrid composite model.

An $n$–fold composite WEIBULL CDF is defined as

$$
F_{nc}(x) = F_i(x \,|\, a_i, b_i, c_i) \ \text{ for } \ \tau_{i-1} \leq x \leq \tau_i; \ i = 1, 2, \dots, n, \tag{3.60a}
$$

where $F_i(x \,|\, a_i, b_i, c_i) = 1 - \exp\left\{ - \left( \dfrac{x - a_i}{b_i} \right)^{c_i} \right\}$ is the $i$–th component in CDF form. The quantities $\tau_i$ are the knots or points of component partition. Because the end points are $\tau_0 = a_1$ and $\tau_n = \infty$, for a $n$–fold composite model there are only $n - 1$ partition parameters. The $n$–fold composite WEIBULL DF is

$$
f_{nc}(x) = f_i(x \,|\, a_i, b_i, c_i) \tag{3.60b}
$$

$$
= \frac{c_i}{b_i} \left( \frac{x - a_i}{b_i} \right)^{c_i-1} \exp\left\{ - \left( \frac{x - a_i}{b_i} \right)^{c_i} \right\} \text{ for } \tau_{i-1} \leq x \leq \tau_i; \ i = 1, 2, \dots, n; \tag{3.60c}
$$

with corresponding HR

$$
h_{nc}(x) = \frac{f_i(x \,|\, a_i, b_i, c_i)}{1 - F_i(x \,|\, a_i, b_i, c_i)} = h_i(x \,|\, a_i, b_i, c_i) \text{ for } \tau_{i-1} \leq x \leq \tau_i; \ i = 1, 2, \dots, n.
$$
$$
\tag{3.60d}
$$

In the interval $[\tau_{i-1}; \tau_i]$ the portion

$$
P_i = F_i(\tau_i \,|\, a_i, b_i, c_i) - F_{i-1}(\tau_{i-1} \,|\, a_{i-1}, b_{i-1}, c_{i-1}); \ i = 1, \dots, n, \tag{3.60e}
$$

with

$$F_0(\cdot) = 0 \ \text{ and } \ F_n(\cdot) = 1,$$

is governed by the $i$–th WEIBULL distribution.

We first look at a composite WEIBULL model requiring only that the CDFs join continuously:

$$F_i(\tau_i \,|\, a_i, b_i, c_i) = F_{i+1}(\tau_i \,|\, a_{i-1}, b_{i+1}, c_{i+1}); \ i = 1, \ldots, n. \tag{3.61a}$$

Fulfillment of (3.61a) does not guarantee that the DFs and the HRs join without making a jump (see Fig. 3/11). At those endpoints, where two CDFs have to be joined, seven parameters are involved: three parameters for each distribution and the partition parameter $\tau_i$. Because of the restriction (3.61a) six of the seven parameters are independent. The most common way to construct (3.60a) is to fix the six parameters $\tau_i$, $a_i$, $b_i$, $c_i$, $b_{i+1}$, $c_{i+1}$ ($i = 1, 2, \ldots, n-1$) and to shift the right–hand CDF with index $i+1$, i.e., to determine $a_{i+1}$ so as to fulfill (3.61a). Thus, the resulting location parameter $a_{i+1}$ is given by

$$a_{i+1} = \tau_i - b_{i+1} \left( \frac{\tau_i - a_i}{b_i} \right)^{c_i/c_{i+1}} ; \ i = 1, 2, \ldots, n-1. \tag{3.61b}$$

Fig. 3/11 shows the CDF, DF and HR together with the WEIBULL probability plot for a three–fold composite model with the following parameter set:

$$a_1 = 0; \ b_1 = 1; \ c_1 = 0.5; \ \tau_1 = 0.127217 \implies P_1 = 0.3;$$

$$a_2 = -0.586133; \ b_2 = 2; \ c_2 = 1; \ \tau_2 = 1.819333 \implies P_2 = 0.4;$$

$$a_3 = 1.270987; \ b_3 = 0.5; \ c_3 = 2 \implies P_3 = 0.3.$$

When we want to have a composite WEIBULL model with both the CDFs and the DFs joining continuously, we have to introduce a second continuity condition along with (3.61a):

$$f_i(\tau_i \,|\, a_i, b_i, c_i) = f_{i+1}(\tau_i \,|\, a_{i+1}, b_{i+1}, c_{i+1}); \ i = 1, 2, \ldots, n-1. \tag{3.62a}$$

(3.61a) and (3.62a) guarantee a continuous hazard rate, too. In order to satisfy (3.61a) and (3.62a) the seven parameters coming up at knot $i$ have to fulfill the following two equations:

$$a_i = \tau_i - \left( \frac{c_{i+1}}{c_i} \frac{b_i^{c_i/c_{i+1}}}{b_{i+1}} \right)^{c_{i+1}/(c_i - c_{i+1})} , \tag{3.62b}$$

$$a_{i+1} = \tau_i - b_{i+1} \left( \frac{\tau_i - a_i}{b_i} \right)^{c_i/c_{i+1}} , \tag{3.62c}$$

so five out of the seven parameters are independent.

Figure 3/11: Composite WEIBULL distribution with continuously joining CDFs

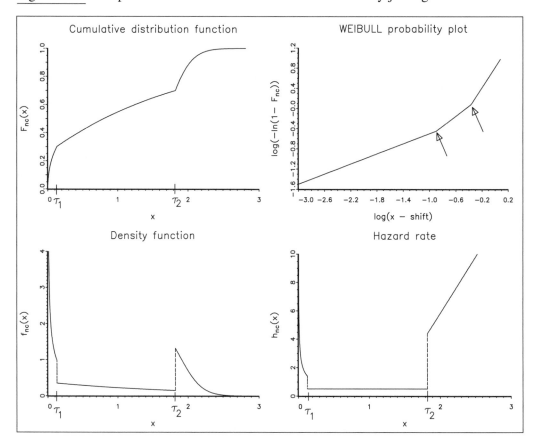

A common procedure to construct a composite WEIBULL distribution supposes that at least the knots $\tau_i$ and the shape parameters $c_i$ are fixed. Then, the algorithm is as follows:

1. Given $\tau_1$, $b_1$, $c_1$, $b_2$, $c_2$, the location parameters $a_1$ and $a_2$ of the first two distributions are determined using (3.62b,c).

2. For the following knots ($i = 2, \ldots, n - 1$) the parameters $\tau_i, a_i$ (from the preceding step), $b_i$, $c_i$ and $c_{i+1}$ are known. $b_{i+1}$, the scale parameter of distribution $i + 1$, is calculated as

$$b_{i+1} = \frac{c_{i+1}}{c_i} b_i^{c_i/c_{i+1}} \left(\tau_i - a_i\right)^{-(c_i - c_{i+1})/c_{i+1}}. \qquad (3.62d)$$

Then $a_{i+1}$ follows from (3.62c).

Fig. 3/12 gives an example for a three–fold composite and completely continuous WEIBULL model. The predetermined parameters are

$$\tau_1 = 1; \; b_1 = 1; \; c_1 = 0.5; \; b_2 = 2; \; c_2 = 1.0; \; \tau_2 = 2.0; \; c_2 = 3.$$

The algorithm yields

$$a_1 = 0; \; a_2 = -1; \; b_3 \approx 7.862224; \; a_3 \approx -7.$$

The portions $P_i$ of the three–sectional distributions are

$$P_1 \approx 0.6321; \; P_2 \approx 0.1448; \; P_3 \approx 0.2231.$$

Figure 3/12: Composite WEIBULL distribution with continuously joining CDFs, DFs and HRs

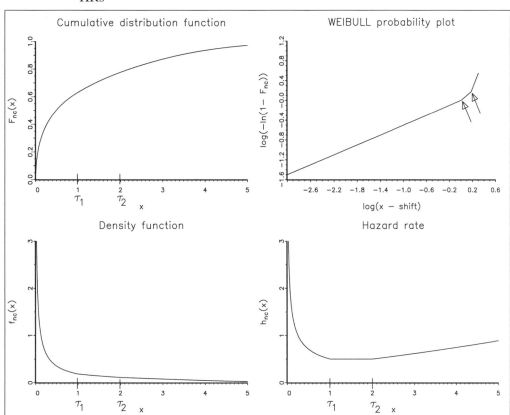

Moments of composite WEIBULL distribution have to be determined by numerical integration. Percentiles are easily calculated because (3.60a) can be inverted in closed form.

There are mainly two approaches to estimate the parameters of a composite WEIBULL distribution. The graphical approach rests either upon the WEIBULL probability–plot, see KAO (1959) or MURTHY/XIE/JIANG (2004), or on the WEIBULL hazard–plot, see ELANDT–JOHNSON/JOHNSON (1980), and adheres to the kinks (Fig. 3/11 and 3/12) that will also come up in the empirical plots. The numerical approach consists in a maximum likelihood estimation; see AROIAN/ROBINSON (1966) and COLVERT/BOARDMAN (1976).

### 3.3.6.4 Mixed WEIBULL distributions[21]

The mixing of several WEIBULL distributions is another means to arrive at a non–monotone hazard rate, but this will never have a bathtub–shape as will be shown further down. A

---

[21] Suggested reading for this section: Since the first paper on mixed WEIBULL models by KAO (1959), the literature on this topic has grown at an increasing pace. MURTHY/XIE/JIANG (2004) give an up–to–date overview of this relative to the classical WEIBULL distribution.

**discrete** or **finite mixture**[22] is a linear combination of two or more WEIBULL distributions that — contrary to the composite model of the preceding section — have no restricted support.[23] A simple explanation of discretely mixed distributions is the following: The population under consideration is made up of $n \geq 2$ subpopulations contributing the portion $\omega_i$, $i = 1, 2, \ldots, n; 0 < \omega_i < 1; \sum_{i=1}^{n} \omega_i = 1$; to the entire population. For example, a lot of items delivered by a factory has been produced on $n$ production lines which are not working perfectly identically so that some continuous quality characteristic $X$ of the items has DF $F_i(x)$ when produced on line $i$.

Before turning to WEIBULL mixtures we summarize some general results on mixed distributions:

**Mixed DF**

$$f_m(x) = \sum_{i=1}^{n} \omega_i \, f_i(x); \quad 0 < \omega_i < 1; \quad \sum_{i=1}^{n} \omega_i = 1; \tag{3.63a}$$

**Mixed CDF und CCDF**

$$F_m(x) = \sum_{i=1}^{n} \omega_i \, F_i(x); \tag{3.63b}$$

$$R_m x) = \sum_{i=1}^{n} \omega_i \, R_i(x); \tag{3.63c}$$

**Mixed HR**

$$h_m(x) = \frac{f_m(x)}{R_m(x)}$$

$$= \sum_{i=1}^{n} g_i(x) \, h_i(x), \tag{3.63d}$$

where $h_i(x) = f_i(x)/R_i(x)$ is the hazard rate of subpopulation $i$ and the weights $g_i(x)$ are given by

$$g_i(x) = \frac{\omega_i \, R_i(x)}{R_m(x)}, \quad \sum_{i=1}^{n} g_i(x) = 1 \tag{3.63e}$$

and are thus varying with $x$;

**Mixed raw moments**

$$\mathrm{E}(X^r) = \sum_{i=1}^{n} \omega_i \, \mathrm{E}(X_i^r), \text{ where } \mathrm{E}(X_i^r) = \int x_i^r \, f_i(x) \, \mathrm{d}x. \tag{3.63f}$$

---

[22] A **continuous** or **infinite mixture**, where the weighting function is a DF for one of the WEIBULL parameters, will be presented in Sect. 3.3.6.5. In this case the WEIBULL parameter is a random variable.

[23] We will not present **hybrid mixtures** where some of the mixed distributions do not belong to the WEIBULL family. Some papers on hybrid mixtures are AL–HUSSAINI/ABD–EL–HAKIM (1989, 1990, 1992) on a mixture of an inverse GAUSSIAN and a two–parameter WEIBULL distribution, CHANG (1998) on a mixture of WEIBULL and GOMPERTZ distributions to model the death rate in human life–tables, LANDES (1993) on a mixture of normal and WEIBULL distributions and MAJESKE/HERRIN (1995) on a twofold mixture of WEIBULL and uniform distributions for predicting automobile warranty claims.

From (3.63f) we have two special results:

**Mean of the mixed distribution**

$$\mathrm{E}(X) = \sum_{i=1}^{n} \omega_i \, \mathrm{E}(X_i);$$ (3.63g)

**Variance of the mixed distribution**

$$
\begin{aligned}
\mathrm{Var}(X) &= \mathrm{E}\big\{[X - \mathrm{E}(X)]^2\big\} = \mathrm{E}(X^2) - [\mathrm{E}(X)]^2 \\
&= \sum_{i=1}^{n} \omega_i \, \mathrm{E}(X_i^2) - \left[\sum_{i=1}^{n} \omega_i \, \mathrm{E}(X_i)\right]^2 \\
&= \sum_{i=1}^{n} \omega_i \left[\mathrm{Var}(X_i) + \mathrm{E}(X_i)\right]^2 - \left[\sum_{i=1}^{n} \omega_i \, \mathrm{E}(X_i)\right]^2 \\
&= \sum_{i=1}^{n} \omega_i \mathrm{Var}(X_i) + \sum_{i=1}^{n} \omega_i \left[\mathrm{E}(X_i) - \mathrm{E}(X)\right]^2.
\end{aligned}
$$ (3.63h)

Thus, the variance of the mixed distribution is the sum of two non–negative components. The first component on the right–hand side of (3.63h) is the mean of the $n$ variances. It is termed the **internal variance** or **within–variance**. The second component is the **external variance** or **between–variance**, showing how the means of the individual distributions vary around their common mean $\mathrm{E}(X)$.

Inserting the formulas describing the functions and moments of $n$ WEIBULL distributions into (3.63a–h) gives the special results for a mixed WEIBULL model. Besides parameter estimation (see farther down) the behavior of the density and the hazard rate of a mixed WEIBULL distribution have attracted the attention of statisticians.

We start by first developing some approximations to the CDF of a mixed WEIBULL model where the $n$ subpopulations all have a two–parameter WEIBULL distribution[24] and assuming in the sequel, without loss of generality, that $c_i \le c_j$, $i < j$, and $b_i > b_j$, when $c_i = c_j$. Denoting

$$y_i = \left(\frac{x}{b_i}\right)^{c_i}$$

we have

$$\lim_{x \to 0} \left(\frac{y_i}{y_1}\right) = \left\{ \begin{array}{ll} 0 & \text{if} \quad c_i > c_1 \\[2mm] \left(\dfrac{b_1}{b_i}\right)^{c_1} & \text{if} \quad c_i = c_1 \end{array} \right\}$$ (3.64a)

and

$$\lim_{x \to \infty} \left(\frac{y_i}{y_1}\right) = \left\{ \begin{array}{ll} \infty & \text{if} \quad c_i > c_1 \\[2mm] \left(\dfrac{b_1}{b_i}\right)^{c_1} > 1 & \text{if} \quad c_i = c_1 \end{array} \right\}.$$ (3.64b)

---

[24] The results for a three–parameter WEIBULL distribution are similar because the three–parameter distribution reduces to a two–parameter distribution under a shifting of the time scale.

From (3.64a,b) JIANG/MURTHY (1995) derive the following results from which the behavior of the hazard and density functions may be deduced:

1. For small $x$, i.e., very close to zero,

$$F_m(x) = \sum_{i=1}^{n} \omega_i \, F_i(x \,|\, 0, b_i, c_i)$$

can be approximated by

$$F_m(x) \approx g \, F_1(x \,|\, 0, b_1, c_1), \qquad\qquad\qquad (3.65a)$$

where

$$g = \sum_{j=1}^{m} \omega_j \left( \frac{b_1}{b_j} \right)^{c_1} \qquad\qquad\qquad (3.65b)$$

and $m$ is the number of the subpopulations with the common shape parameter $c_1$. When $m = 1$, then $g = \omega_1$.

2. For large $x$, $F_m(x)$ can be approximated by

$$F_m(x) \approx 1 - \omega_1 \left[ 1 - F_1(x \,|\, 0, b_1, c_1) \right]. \qquad\qquad (3.65c)$$

From (3.65a,b) the **density of the mixed WEIBULL model** can be approximated by

$$f_m(x) \approx g \, f_1(x \,|\, 0, b_1, c_1) \ \text{ for small } \ x \qquad\qquad (3.66a)$$

and

$$f_m(x) \approx \omega_1 \, f_1(x \,|\, 0, b_1, c_1) \ \text{ for large } \ x, \qquad\qquad (3.66b)$$

implying that $f_m(x)$ is increasing (decreasing) for small $x$ if $c_1 > 1$ ($c_1 < 1$). The shape of $f_m(x)$ depends on the model parameters, and the possible shapes are

- decreasing followed by $k - 1$ modes ($k = 1, 2, \ldots, n - 1$),
- $k$–modal ($k = 1, 2, \ldots, n$).

Of special interest is the two–fold mixture. The possible shapes in this case are

- decreasing,
- unimodal,
- decreasing followed by unimodal,
- bimodal.

Although the two–fold mixture model has five parameters ($b_1$, $b_2$, $c_1$, $c_2$ and $\omega_1$), as $\omega_2 = 1 - \omega_1$, the DF–shape is only a function of the two shape parameters, the ratio of the two scale parameters and the mixing parameter. JIANG/MURTHY (1998) give a parametric characterization of the density function in this four–dimensional parameter space.

From (3.66a,c) the **hazard rate of the mixed WEIBULL model** can be approximated by

$$h_m(x) \approx g \, h_1(x \,|\, 0, b_1, c_1) \text{ for small } x \qquad\qquad (3.67a)$$

and

$$h_m(x) \approx h_1(x \,|\, 0, b_1, c_1) \text{ for large } x. \qquad\qquad (3.67b)$$

Thus, $h_m(x)$ is increasing (decreasing) for small $x$ if $c_1 > 1$ $(c_1 < 1)$. The shape of $h_m(x)$ depends on the model parameters leading to the following possible shapes; see GUPTA/GUPTA (1996):

- decreasing,
- increasing,
- decreasing followed by $k$ modes $(k = 1, 2, \ldots, n - 1)$,
- $k$ modal followed by increasing $(k = 1, 2, \ldots, n)$.

A special case of a decreasing mixed HR has already been given by PROSCHAN (1963).[25] He proved that a mixture of exponential distributions, being WEIBULL distribution with $c_i = 1$; $i = 1, 2, \ldots, n$, which have differing scale parameters $b_i$, will lead to a mixed model with decreasing hazard rate.

We now turn to the hazard rate of a two–fold WEIBULL mixture, which has been studied in some detail. This hazard rate follows from (3.63d,e) as

$$h_m(x) = \frac{\omega_1 R_1(x \,|\, 0, b_1, c_1)}{R_m(x)} h_1(x \,|\, 0, b_1, c_1) + \frac{(1 - \omega_1) R_2(x \,|\, 0, b_2, c_2)}{R_m(x)} h_2(x \,|\, 0, b_2, c_2).$$

(3.68a)

Assuming $c_1 \leq c_2$ JIANG/MURTHY (1998) show

$$h_m(x) \to h_1(x \,|\, 0, b_1, c_1) \text{ for large } x,$$ 

(3.68b)

$$h_m(x) \to g \, h_1(x \,|\, 0, b_1, c_1) \text{ for small } x,$$ 

(3.68c)

with

$$g = \omega_1 \text{ for } c_1 < c_2 \text{ and } g = \omega_1 + (1 - \omega_1) \left(\frac{b_1}{b_2}\right)^{c_1} \text{ for } c_1 = c_2.$$

(3.68b,c) imply that for small and large $x$ the shape of $h_m(x)$ is similar to that of $h_1(x \,|\, .)$. When $c_1 < 1$ $(c_1 > 1)$, then $h_1(x|.)$ is decreasing (increasing) for all $x$. As a result, $h_m(x)$ is decreasing (increasing) for small and large $x$, so $h_m(x)$ cannot have a bathtub shape.[26] The possible shapes of (3.68a) are

- decreasing,
- increasing,
- uni–modal followed by increasing,
- decreasing followed by uni–modal,
- bi–modal followed by increasing.

The conditions on the parameters $b_1$, $b_2$, $c_1$, $c_2$ and $\omega$, leading to these shapes, are given in JIANG/MURTHY (1998). Fig. 3/13 shows a hazard rate (right–hand side) along with its corresponding density function (left–hand side) for the five possible hazard shapes.

---

[25] A paper with newer results on mixtures of exponential distributions is JEWELL (1982).

[26] One can even show that $n$–fold mixtures of WEIBULL distributions $(n > 2)$ can never have a bathtub shape; thus falsifying the assertion of KROHN (1969) that a three–fold mixture of WEIBULL distributions having $c_1 < 1$, $c_2 = 1$ and $c_3 > 1$ leads to a bathtub shape.

Figure 3/13: Hazard rate and density function of some two-fold WEIBULL mixtures

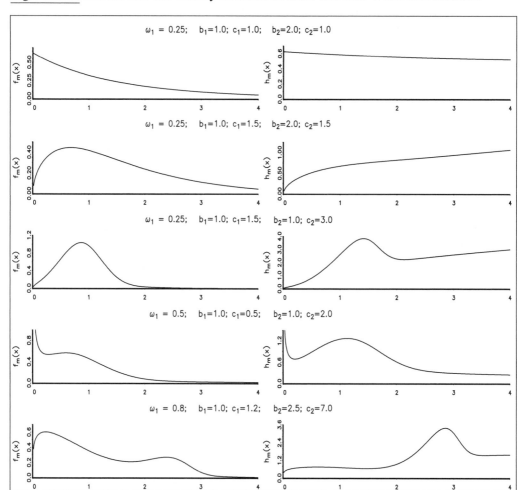

The literature on **parameter estimation** for the mixture model is vast. So, we will cite only some key papers. **Graphical methods** based the WEIBULL probability plot seem to be dominant. The first paper based on this approach is KAO (1959) on a two–fold mixture. His approach has been applied by several authors introducing minor modifications. A newer paper circumventing some of the drawbacks of older approaches is JIANG/MURTHY (1995). The graphical method may lead to satisfactory results for a two–fold mixture in well–separated cases, i.e., $b_1 \gg b_2$ or $b_1 \ll b_2$. Some papers on **analytical methods** of parameter estimation are

- RIDER (1961) and FALLS (1970) using the method of moments,
- ASHOUR (1987a) and JIANG/KECECIOGLU (1992b) applying the maximum likelihood approach,
- WOODWARD/GUNST (1987) taking a minimum–distance estimator,
- CHENG/FU (1982) choosing a weighted least–squares estimator,
- ASHOUR (1987b) and AHMAD et al. (1997) considering a BAYES approach.

### 3.3.6.5   Compound WEIBULL distributions[27]

The weights $\omega_i$, used in mixing $n$ WEIBULL densities $f_i(x \mid a_i, b_i, c_i)$, may be interpreted as the probabilities of $n$ sets of values for the parameter vector $(a, b, c)$. Generally, each parameter of a parametric distribution may be regarded as a continuous variate, so that the probabilities $\omega_i$ have to be substituted against a density function. The resulting continuous mixture is termed **compound distribution**. In the sequel we will discuss only the case where one of the parameters, denoted by $\Theta$, is random. Let $\theta$ be a realization of $\Theta$. The density of $X$, given $\theta$, is conditional and denoted by $f_X(x \mid \theta)$. In the context of compounding it is termed **parental distribution**. The density of $\Theta$, denoted by $f_\Theta(\theta)$, is termed **compounding distribution** or **prior distribution**.[28]

The **joint density** of the variates $X$ and $\Theta$ is given by

$$f(x, \theta) = f_X(x \mid \theta)\, f_\Theta(\theta), \tag{3.69a}$$

from which the **marginal density** of $X$ follows as

$$f(x) = \int f(x, \theta)\, \mathrm{d}\theta = \int f_X(x, \theta) f_\Theta(\theta)\, \mathrm{d}\theta. \tag{3.69b}$$

This is the DF of the compound distribution. In statistics and probability theory a special notation is used to express compounding. For example, let the parental distribution be normal with random mean $\Theta$, $\theta$ being a realization of $\Theta$, and fixed variance $\mathrm{Var}(X \mid \theta) = \sigma_*^2$, i.e., $X \mid \theta \sim No(\theta, \sigma_*^2)$. Supposing a prior distribution, which is also normal, $\Theta \sim No(\xi, \sigma_{**}^2)$, compounding in this case is denoted by

$$No(\Theta, \sigma_*^2) \bigwedge_{\Theta} No(\xi, \sigma_{**}^2).$$

$\bigwedge$ is the **compounding operator**. We shortly mention that in this example compounding is reproductive,

$$No(\Theta, \sigma_*^2) \bigwedge_{\Theta} No(\xi, \sigma_{**}^2) = No(\xi, \sigma_*^2 + \sigma_{**}^2).$$

Compound WEIBULL distributions have been intensively discussed with respect to a **random scale factor**. This scale factor is the reciprocal of the combined scale–shape parameter introduced by W. WEIBULL; see (2.26a,b):

$$\beta := 1/b^c = b^{-c}. \tag{3.70a}$$

Thus, the parental WEIBULL DF reads as

$$f_X(x \mid \beta) = c\,\beta\,(x - a)^{c-1}\,\exp\{-\beta\,(x - a)^c\}. \tag{3.70b}$$

The results pertaining to this conditional WEIBULL distribution which have been reported by DUBEY (1968) und HARRIS/SINGPURWALLA (1968) are special cases of a more general **theorem of ELANDT–JOHNSON (1976)**:

---

[27] Suggested reading for this section: DUBEY (1968), ELANDT–JOHNSON (1976), HARRIS/SINGPUWALLA (1968, 1969).

[28] The term *prior distribution* is used in BAYES inference, see Sect. 14, which is related to compounding.

Let

$$F_X(x, \boldsymbol{\alpha} \mid \lambda) = \begin{cases} 0 & \text{for} \quad x \le x_0 \\ 1 - \exp\{-\lambda\, u(x, \boldsymbol{\alpha})\} & \text{for} \quad x > x_0 \end{cases} \tag{3.71a}$$

be the CDF of the parental distribution, which is of **exponential type** with $\lambda$ as a "special" parameter and $\boldsymbol{\alpha}$ as a vector of "ordinary" parameters. $u(x, \boldsymbol{\alpha})$ is an increasing function of $x$ with

$$u(x, \boldsymbol{\alpha}) \longrightarrow 0 \quad \text{as} \quad x \to x_0,$$
$$u(x, \boldsymbol{\alpha}) \longrightarrow \infty \quad \text{as} \quad x \to \infty.$$

(The parental WEIBULL distribution (3.70b) is of this type with $\lambda = \beta$ and $u(x, \boldsymbol{\alpha}) = (x-a)^c$.) Further, let the compounding distribution have a raw moment generating function

$$\mathrm{E}_\Lambda(e^{t\,\lambda}) =: M_\Lambda(t).$$

Then, the CDF of the compound distribution is given by

$$F(x) = \begin{cases} 0 & \text{for} \quad x \le x_0 \\ 1 - M_\Lambda\big[-u(x, \boldsymbol{\alpha})\big] & \text{for} \quad x > x_0, \end{cases} \tag{3.71b}$$

provided that $M_\Lambda\big[-u(x, \boldsymbol{\alpha})\big]$ exists for $x > x_0$. ∎

We will apply this theorem to a parental WEIBULL distribution of type (3.70b) where the stochastic scale factor, denoted by $B$, has either a uniform (= rectangular) prior or a gamma prior. These prior distributions are the most popular ones in BAYES inference.

**First case:** $We(a, \sqrt[c]{1/B}, c) \underset{B}{\wedge} Re(\xi, \delta)$

The uniform or rectangular distribution over $[\xi, \xi + \delta]$ has DF

$$f_B(y) = \begin{cases} 1/\delta & \text{for} \quad \xi \le y \le \xi + \delta \\ 0 & \text{for} \quad \xi \text{ elsewhere,} \end{cases} \tag{3.72a}$$

denoted by

$$B \sim Re(\xi, \delta).$$

The corresponding raw moment generating function is

$$M_B(t) = \frac{\exp(\xi\, t)}{\delta\, t} \big[\exp(\delta\, t) - 1\big]. \tag{3.72b}$$

Putting

$$t = -u(x, a, c) = -(x - a)^c$$

into (3.71b) yields the following CDF of the compound WEIBULL distribution (or **WEIBULL–uniform distribution**):

$$F(x \mid a, \delta, \xi, c) = 1 - \frac{\exp\big\{-\xi\,(x-a)^c\big\} - \exp\big\{-(\xi+\delta)\,(x-a)^c\big\}}{\delta\,(x-a)^c} \tag{3.72c}$$

with corresponding DF

$$
\left.
\begin{aligned}
&f(x\,|\,a,\delta,\xi,c) = \frac{c}{\delta} \times \\
&\frac{\exp\{-(\xi+\delta)(x-a)^c\}\left[\exp\{\delta\,(x-a)^c\}\left(1+\xi\,(x-a)^c\right)-(\xi+\delta)(x-a)^{c}-1\right]}{(x-a)^{c+1}}
\end{aligned}
\right\}
$$
(3.72d)

With some integration and manipulation we find

$$
\mathrm{E}(X) = a + \frac{c}{\delta\,(c-1)}\,\Gamma\!\left(1+\frac{1}{c}\right)\left[(\xi+\delta)^{1-1/c} - \xi^{1-1/c}\right],
$$
(3.72e)

$$
\left.
\begin{aligned}
\mathrm{Var}(X) \;=\;& \frac{c}{\delta\,(c-1)}\,\Gamma\!\left(1+\frac{2}{c}\right)\left[(\xi+\delta)^{1-1/c} - \xi^{1-1/c}\right] - \\
& \frac{c^2}{\delta^2\,(c-1)^2}\,\Gamma^2\!\left(1+\frac{1}{c}\right)\left[(\xi+\delta)^{1-1/c} - \xi^{1-1/c}\right]^2.
\end{aligned}
\right\}
$$
(3.72f)

When the prior density (3.72a) degenerates ($\delta \to 0$), the formulas (3.72c–f) show the familiar results of the common three–parameter distribution for $b = \xi^{-1/c}$.

**Second case**: $We(a, \sqrt[c]{1/B}, c) \underset{B}{\wedge} Ga(\xi,\delta,\gamma)$

The three–parameter gamma prior $Ga(\xi,\delta,\gamma)$ over $[\xi, \infty)$ has DF

$$
f_B(y) = \frac{\delta^\gamma\,(y-\xi)^{\gamma-1}}{\Gamma(\gamma)}\,\exp\{-\delta\,(y-\xi)\}
$$
(3.73a)

with raw moment generating function

$$
M_B(t) = \frac{\delta^\gamma\,\exp\{\xi\,t\}}{(\delta-t)^\gamma}.
$$
(3.73b)

Taking

$$
t = -u(x,a,c) = -(x-a)^c
$$

in (3.73b) turns (3.71b) into the following CDF of this compound WEIBULL distribution (or **WEIBULL–gamma distribution**):

$$
F(x\,|\,a,\delta,\xi,\gamma,c) = 1 - \frac{\delta^\gamma\,\exp\{-\xi\,(x-a)^c\}}{\left[\delta+(x-a)^c\right]^\gamma}
$$
(3.73c)

with DF

$$
f(x\,|\,a,\delta,\xi,\gamma,c) = \frac{c\,\delta^\gamma\,(x-a)^{c-1}\,\{\gamma+\xi\,[\delta+(x-a)^c]\}\,\exp\{-\xi\,(x-a)^c\}}{\left[\delta+(x-a)^c\right]^{\gamma+1}}.
$$
(3.73d)

In the special case $\xi = 0$, there exist closed form expressions for the mean and variance:

$$
\mathrm{E}(X\,|\,\xi=0) = a + \frac{\delta^{1/c}\,\Gamma\!\left(\gamma-\frac{1}{c}\right)\Gamma\!\left(\frac{1}{c}+1\right)}{\Gamma(\gamma)},
$$
(3.73e)

$$\text{Var}(X \mid \xi = 0) = \frac{\delta^{2/c}}{\Gamma(\gamma)} \left\{ \Gamma\left(\gamma - \frac{2}{c}\right) \Gamma\left(\frac{2}{c} + 1\right) - \frac{1}{\Gamma(\gamma)} \Gamma^2\left(\frac{1}{c} + 1\right) \right\}. \qquad (3.73f)$$

We further mention that in the special case $\xi = 0$ and $\delta = 1$, (3.73c,d) turn into the BURR type–XII distribution; see (3.3a,b). Finally, we refer to HARRIS/SINGPURWALLA (1969) when parameters have to be estimated for each of the two compound WEIBULL distributions.

### 3.3.7 WEIBULL distributions with additional parameters

Soon after the WEIBULL distribution had been presented to a broader readership by WEIBULL in 1951, statisticians began to re–parameterize this distribution. By introducing additional parameters, those authors aimed at more flexibility to fit the model to given datasets. But, the more parameters that are involved and have to be estimated, the greater will be the risk to not properly identify these parameters from the data.

In the preceding sections, we have already encountered several WEIBULL distributions with more than the traditional location, scale and shape parameters, namely:

- the doubly truncated WEIBULL distribution (3.47a) with five parameters,
- the left and the right truncated WEIBULL distributions (3.47b,c) with four parameters each,
- the WEIBULL competing risk model (3.57a,b) with four parameters,
- the multiplicative WEIBULL model (3.58a,b) and its generalization, the exponentiated WEIBULL model, each having four parameters.

In this section we will present further WEIBULL distributions, enlarged with respect to the parameter set.

#### 3.3.7.1 Four-parameter distributions

##### MARSHALL/OLKIN extension

In their paper from 1997, MARSHALL/OLKIN proposed a new and rather general method to introduce an additional parameter into a given distribution having $F(x)$ and $R(x)$ as CDF and CCDF, respectively. The CCDF of the new distribution is given by

$$G(x) = \frac{\alpha R(x)}{1 - (1 - \alpha) R(x)} = \frac{\alpha R(x)}{F(x) + \alpha R(x)}, \quad \alpha > 0. \qquad (3.74)$$

Substituting $F(x)$ and $R(x)$ by the CDF and CCDF of the three–parameter WEIBULL distribution, we arrive at the following CCDF of the **extended WEIBULL distribution**:

$$R(x \mid a, b, c, \alpha) = \frac{\alpha \exp\left\{ -\left(\frac{x - a}{b}\right)^c \right\}}{1 - (1 - \alpha) \exp\left\{ -\left(\frac{x - a}{b}\right)^c \right\}}. \qquad (3.75a)$$

The corresponding CDF is

$$F(x \mid a, b, c, \alpha) = \frac{1 - \exp\left\{-\left(\frac{x-a}{b}\right)^c\right\}}{1 - (1-\alpha) \exp\left\{-\left(\frac{x-a}{b}\right)^c\right\}} \tag{3.75b}$$

with DF

$$f(x \mid a, b, c, \alpha) = \frac{\dfrac{\alpha\, c}{b} \left(\dfrac{x-a}{b}\right)^{c-1} \exp\left\{-\left(\dfrac{x-a}{b}\right)^c\right\}}{\left[1 - (1-\alpha) \exp\left\{-\left(\dfrac{x-a}{b}\right)^c\right\}\right]^2} \tag{3.75c}$$

and HR

$$h(x \mid a, b, c, \alpha) = \frac{\dfrac{c}{b} \left(\dfrac{x-a}{b}\right)^{c-1}}{1 - (1-\alpha) \exp\left\{-\left(\dfrac{x-a}{b}\right)^c\right\}} . \tag{3.75d}$$

The hazard rate given by (3.75d) has the following behavior for growing $x$.

- for $\alpha \geq 1$ and $c \geq 1$ it is increasing,
- for $\alpha \leq 1$ and $c \leq 1$ it is decreasing,
- for $c > 1$ it is initially increasing and eventually increasing, but there may be one interval where it is decreasing,
- for $c < 1$ it is initially decreasing and eventually decreasing, but there may be one interval where it is increasing,

When $|1 - \alpha| \leq 1$, moments of this distribution can be given in closed form. Particularly we have

$$\mathrm{E}\left[(X - a)^r\right] = \frac{r}{b\,c} \sum_{j=0}^{\infty} \frac{(1-\alpha)^j}{(1+j)^{r/c}} \, \Gamma\left(\frac{r}{c}\right). \tag{3.75e}$$

MARSHALL/OLKIN proved that the distribution family generated by (3.74) possesses what they call **geometric–extreme stability**: If $X_i$, $i \leq N$, is a sequence of iid variates with CCDF (3.74) and $N$ is geometrically distributed, i.e. $\mathrm{Pr}(N = n) = P(1 - P)^n$; $n = 1, 2, \ldots$; then the minimum (maximum) of $X_i$ also has a distribution in the same family.

**LAI et al. extension**

LAI et al. (2003) proposed the following extension of the WEIBULL model,[29] assuming $a = 0$ and using the combined scale–shape parameter[30] $b_1 = 1/b^c$:

$$F(x \mid 0, b_1, c, \alpha) = 1 - \exp\{-b_1\, x^c\, e^{\alpha x}\}; \quad b_1,\, c > 0; \quad \alpha \geq 0. \tag{3.76a}$$

---

[29] In a recent article NADARAJAH/KOTZ (2005) note that this extension has been made some years earlier by GURVICH et al. (1997); see also LAI et al. (2005).

[30] For parameter estimation under type–II censoring, see NG (2005).

The corresponding DF, HR and CHR are

$$f(x \mid 0, b_1, c, \alpha) = b_1 (c + \alpha x) x^{c-1} \exp\{\alpha x - b_1 x^c e^{\alpha x}\}, \qquad (3.76b)$$

$$h(x \mid 0, b_1, c, \alpha) = b_1 (c + \alpha x) x^{c-1} \exp\{\alpha x\}, \qquad (3.76c)$$

$$H(x \mid 0, b_1, c, \alpha) = b_1 x^c \exp\{\alpha x\}. \qquad (3.76d)$$

We have the following relationships with other distributions:

- $\alpha = 0$ gives the general WEIBULL distribution.

- $c = 0$ results in the type–I extreme value distribution for the minimum, also known as a log–gamma distribution.

- (3.76a) is a limiting case of the beta–integrated model with CHR

$$H(x) = b_1 x^c (1 - g x)^k; \;\; 0 < x < g^{-1};$$

and $g > 0$, $k < 0$. Set $g = 1/\nu$ and $k = \alpha \nu$. With $\nu \to \infty$ we have $(1 - x/\nu)^{-\alpha \nu} \to \exp\{\alpha x\}$, and $H(x)$ is the same as (3.76d).

Moments of this extended WEIBULL model have to be evaluated by numeric integration. Percentiles $x_P$ of order $P$ follow by solving

$$x_P = \left[ -\frac{\ln(1 - P)}{b_1} \exp\{-\alpha x_P\} \right]^{1/c}. \qquad (3.76e)$$

The behavior of the hazard rate (3.76c) is of special interest. Its shape solely depends on $c$, and the two other parameters $\alpha$ and $b_1$ have no effect. With respect to $c$ two cases have to be distinguished (see Fig. 3/14):

- $c \geq 1$

  HR is increasing. $\left[ h(x \mid \cdot) \to \infty \text{ as } x \to \infty, \text{ but } h(0 \mid \cdot) = 0 \text{ for } c > 1 \text{ and } h(0 \mid \cdot) = b_1 c \text{ for } c = 1. \right]$

- $0 < c < 1$

  $h(x \mid \cdot)$ is initially decreasing and then increasing, i.e. it has a bathtub shape with $\lim_{x \to 0} h(x \mid \cdot) = \lim_{x \to \infty} h(x \mid \cdot) = \infty$. The minimum of $h(x \mid \cdot)$ is reached at

$$x_{h,\min} = \frac{\sqrt{c} - c}{\alpha}, \qquad (3.76f)$$

  which is increasing when $\alpha$ decreases.

We finally mention that LAI et al. (2003) show how to estimate the model parameters by the ML–method.

Figure 3/14: Hazard rate of the LAI et al. extension of the WEIBULL model

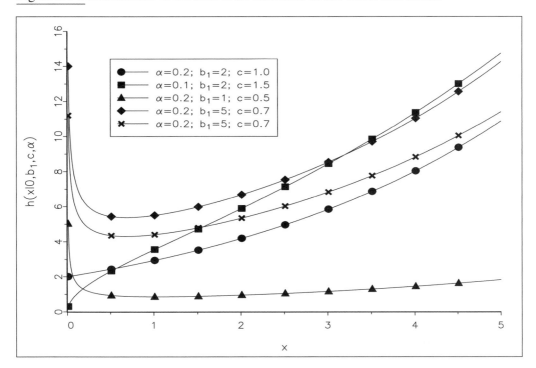

**Extensions by modifying the percentile function**

The percentile function of the general three–parameter WEIBULL distribution is given by (see (2.57b)):

$$x_P = a + b\left[-\ln(1-P)\right]^{1/c}. \tag{3.77}$$

Modifying this function and introducing a fourth parameter $\nu$ and re–inverting to $P = F(x \mid \ldots)$ is another way to arrive at extended WEIBULL models. We will discuss two approaches along this line.

MUDHOLKAR et al. (1996) transformed (3.77) into

$$x_P = a + b\left[\frac{1-(1-P)^\lambda}{\lambda}\right]^{1/c}, \quad \lambda \in \mathbb{R}. \tag{3.78a}$$

The limit of the ratio $[1-(1-P)^\lambda]/\lambda$ in (3.78a) for $\lambda \to 0$ is $-\ln(1-P)$, which is similar to the definition of the **BOX–COX–transformation**, used to stabilize the variance of a time series. So, for $\lambda = 0$ we have the general WEIBULL distribution. Solving for $P = F(x \mid \cdot)$ gives

$$F(x \mid a,b,c,\lambda) = 1 - \left[1 - \lambda\left(\frac{x-a}{b}\right)^c\right]^{1/\lambda}. \tag{3.78b}$$

The support of (3.78b) depends not only on $a$ in the sense that $x \geq a$, but also on some further parameters:

- $\lambda \le 0$ gives the support $(a, \infty)$.

- $\lambda > 0$ gives the support $(a, b/\lambda^{1/c})$.

The DF and HR belonging to (3.78b) are

$$f(x \mid a, b, c, \lambda) = \frac{c}{b} \left( \frac{x-a}{b} \right)^{c-1} \left[ 1 - \lambda \left( \frac{x-a}{b} \right)^c \right]^{-1+1/\lambda}, \tag{3.78c}$$

$$h(x \mid a, b, c, \lambda) = \frac{c}{b} \left( \frac{x-a}{b} \right)^{c-1} \left[ 1 - \lambda \left( \frac{x-a}{b} \right)^c \right]^{-1}. \tag{3.78d}$$

The behavior of (3.78d) is as follows:

- $c = 1$ and $\lambda = 0 \ \Rightarrow \ h(x \mid \cdot) = 1/b = \text{constant}$,

- $c < 1$ and $\lambda > 0 \ \Rightarrow \ h(x \mid \cdot)$ is bathtub–shaped,

- $c \le 1$ and $\lambda \le 0 \ \Rightarrow \ h(x \mid \cdot)$ is decreasing,

- $c > 1$ and $\lambda < 0 \ \Rightarrow \ h(x \mid \cdot)$ is inverse bathtub–shaped,

- $c \ge 1$ and $\lambda \ge 0 \ \Rightarrow \ h(x \mid \cdot)$ is increasing

(see Fig. 3/15).

Figure 3/15: Hazard rate of the MUDHOLKAR et al. (1996) extension of the WEIBULL model

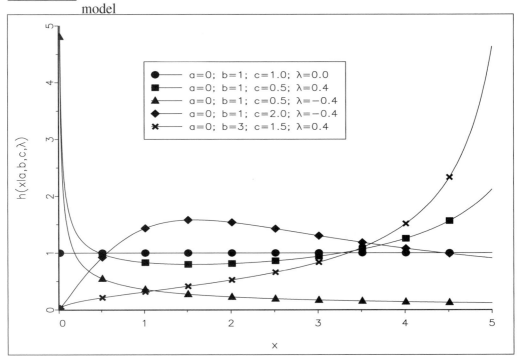

The $r$–th moment of $(X - a)$ according to MUDHOLKAR/KOLLIA (1994) is

$$
E\left[(X-a)^r\right] = \begin{cases} b^r \dfrac{\Gamma\left(\dfrac{1}{\lambda}\right)\Gamma\left(\dfrac{r}{c}+1\right)}{\Gamma\left(\dfrac{1}{\lambda}+\dfrac{r}{c}+1\right)\lambda^{r/c+1}} & \text{for } \lambda > 0, \\[6mm] b^r \dfrac{\Gamma\left(-\dfrac{r}{c}-\dfrac{1}{\lambda}\right)\Gamma\left(\dfrac{r}{c}+1\right)}{\Gamma\left(1-\dfrac{1}{\lambda}\right)(-\lambda)^{r/c+1}} & \text{for } \lambda < 0. \end{cases} \tag{3.78e}
$$

Hence, if $\lambda > 0$ and $c > 0$, moments of all orders exist. For $\lambda < 0$ the $r$–th moment exists if $\lambda/c \leq -r^{-1}$, and in case of $1/c < 0$ — see the next paragraph — if $1/c \geq -r^{-1}$. We should mention that MUDHOLKAR et al. (1996) show how to estimate the parameters by ML.

MUDHOLKAR/KOLLIA (1994) have studied a more generalized percentile function than (3.78a), circumventing its discontinuity at $\lambda = 0$. Instead, they proposed the following percentile function, assuming $a = 0$ and $b = 1$:

$$
x_P = c\left[\frac{1-(1-P)^\lambda}{\lambda}\right]^{1/c} - c, \quad \lambda \in \mathbb{R}. \tag{3.79a}
$$

Here, $c$ may be negative, too, so that the inverse WEIBULL distribution (Sect. 3.3.3) is included in this model. Solving for $P = F(x\,|\,\cdot)$ leads to

$$
F_G(x\,|\,0,1,c,\lambda) = 1 - \left[1-\lambda\left(1+\frac{x}{c}\right)^c\right]^{1/\lambda} \tag{3.79b}
$$

with DF and HR

$$
f_G(x\,|\,0,1,c,\lambda) = \left[1-\lambda\left(1+\frac{x}{c}\right)^c\right]^{\frac{1}{\lambda}-1}\left(1+\frac{x}{c}\right)^{c-1}, \tag{3.79c}
$$

$$
h_G(x\,|\,0,1,c,\lambda) = \left(1+\frac{x}{c}\right)^{c-1}\left[1-\lambda\left(1+\frac{x}{c}\right)^c\right]^{-1}. \tag{3.79d}
$$

The support of these last three functions depends on the parameters $c$ and $\lambda$ as follows:

- $c < 0$
  * If $\lambda < 0$, we have $x \in (-\infty, -c)$.
  * If $\lambda > 0$, we have $x \in (-\infty, c/\lambda^{1/c} - c)$.
- $c > 0$
  * If $\lambda < 0$, we have $x \in (-c, \infty)$.
  * If $\lambda > 0$, we have $x \in (-c, c/\lambda^{1/c} - c)$.

The raw moments of $X$, distributed according to (3.79b), can be expressed by the moments given in (3.78e). (3.79a) includes the following special cases:

- $\lambda = 0,\ c > 0 \Rightarrow$ reduced WEIBULL distribution,
- $\lambda = 0,\ c < 0 \Rightarrow$ inverse reduced WEIBULL distribution,

- $\lambda = 1$, $c = 1$ $\Rightarrow$ uniform distribution over $[0, 1]$,
- $\lambda = 0$, $c = 1$ $\Rightarrow$ exponential distribution,
- $\lambda = -1$, $c = \infty$ $\Rightarrow$ logistic distribution,

The limiting cases ($c \to \infty$, $\lambda \to 0$) are interpreted according to L' HOPITAL's rule.

## XIE et al. extension

XIE et al. (2002) proposed a distribution that we have enlarged to a four–parameter model by introducing a location parameter $a$. The CDF is

$$F(x) = 1 - \exp\left\{\lambda\alpha\left(1 - \exp\left[\left(\frac{x-a}{\alpha}\right)^\beta\right]\right)\right\}, \quad x \geq a, \tag{3.80a}$$

with $\lambda, \alpha, \beta > 0$ and $a \in \mathbb{R}$. The corresponding DF and HR are

$$f(x) = \lambda\beta\left(\frac{x-a}{\alpha}\right)^{\beta-1} \exp\left\{\left(\frac{x-a}{\alpha}\right)^\beta\right\}\exp\left\{\lambda\alpha\left(1 - \exp\left[\left(\frac{x-a}{\alpha}\right)^\beta\right]\right)\right\}, \tag{3.80b}$$

$$h(x) = \lambda\beta\left(\frac{x-a}{\alpha}\right)^{\beta-1} \exp\left\{\left(\frac{x-a}{\alpha}\right)^\beta\right\}. \tag{3.80c}$$

The behavior of HR is determined by the shape parameter $\beta$:

- $\beta \geq 1$ $\Rightarrow$ $h(x)$ is increasing from $h(a) = 0$, if $\beta > 1$, or from $h(a) = \lambda$, if $\beta = 1$.
- $0 < \beta < 1$ $\Rightarrow$ $h(x)$ is decreasing from $x < x^*$ and decreasing for $x > x^*$, where

$$x^* = a + \alpha\left(1/\beta - 1\right)^{1/\beta},$$

  i.e., $h(x)$ has a bathtub shape.

We find the following relations to other distributions:

- For $\alpha = 1$ we have the original version of this distribution introduced by CHEN (2000).

- When $\alpha$ is large, then $1 - \exp\left\{\left(\frac{x-a}{\alpha}\right)^\beta\right\} \approx -\exp\left(\frac{x-a}{\alpha}\right)^\beta$, so that (3.80a) can be approximated by a WEIBULL distribution.

- When $\lambda\alpha = 1$, this model is related to the exponential power model of SMITH/BAIN (1975).

## KIES extension

KIES (1958) introduced the following four–parameter version of a WEIBULL distribution that has a finite interval as support:

$$F(x) = 1 - \exp\left\{-\lambda\left(\frac{x-a}{b-x}\right)^\beta\right\}, \quad 0 \leq a \leq x \leq b < \infty, \tag{3.81a}$$

with $\lambda, \beta > 0$. The corresponding DF and HR are

$$f(x) \;=\; \dfrac{\lambda\,\beta\,(b-a)\,\exp\!\left\{-\lambda\left(\dfrac{x-a}{b-x}\right)^{\beta}\right\}(x-a)^{\beta-1}}{(b-x)^{\beta+1}}\,, \tag{3.81b}$$

$$h(x) \;=\; \dfrac{\lambda\,\beta\,(b-a)\,(x-a)^{\beta-1}}{(b-x)^{\beta+1}}\,. \tag{3.81c}$$

$h(x)$ is increasing for $\beta \geq 1$, and it has a bathtub shape for $0 < \beta < 1$. The approach of KIES represents another way to doubly truncate the WEIBULL distribution (see Sect. 3.3.5). We will revert to (3.81a) in Sect. 3.3.7.2, where this model is extended by a fifth parameter.

**FAUCHON et al. extension**

In Sect. 3.2.3 we have presented several extensions of the gamma distribution. One of them is the four–parameter gamma distribution, originally mentioned by HARTER (1967). Its density is given by

$$f(x\,|\,a,b,c,d) = \dfrac{c}{b\,\Gamma(d)}\left(\dfrac{x-a}{b}\right)^{c\,d-1}\exp\!\left\{-\left(\dfrac{x-a}{b}\right)^{c}\right\}; \; x \geq a. \tag{3.82a}$$

(3.82a) contains a lot of other distributions as special cases for special values of the parameters (see Tab. 3/1). For $d = 1$ we have the three–parameter WEIBULL distribution, $c = 1$ leads to gamma distributions, and $a = 0$, $b = 2$, $c = 1$, $d = \nu/2$ gives a $\chi^2$–distribution with $\nu$ degrees of freedom. As the $\chi^2$–distribution is related to a normal distribution in the following way

$$X = \dfrac{1}{\sigma^2}\sum_{j=1}^{\nu} Y_j^2 \; \text{ and } \; Y_j \overset{\text{iid}}{\sim} No(0,\sigma^2) \;\Longrightarrow\; X \sim \chi^2(\nu),$$

and as both, the $\chi^2$– and the WEIBULL distributions, are linked via the general gamma distribution, we can find a bridge connecting the WEIBULL and the normal distributions. This relationship, which was explored by FAUCHON et al. (1976), gives an interesting interpretation of the second form parameter $d$ in (3.82a).

First, we will show how to combine iid normal variates to get a three–parameter WEIBULL distribution:

1. $Y_1,\ Y_2 \overset{\text{iid}}{\sim} No(0,\sigma^2) \;\Longrightarrow\; (Y_1^2 + Y_2^2) \sim \sigma^2\,\chi^2(2)$,

2. $\sqrt{Y_1^2 + Y_2^2} \sim \sigma\,\chi(2) = We\big(0,\sigma\,\sqrt{2},2\big)$ (see Tab. 3/1).

More generally,

$$Y_1,\ Y_2 \overset{\text{iid}}{\sim} No(0,\sigma^2) \;\Longrightarrow\; a + \sqrt[c]{Y_1^2 + Y_2^2} \sim We\Big(a,\big(2\,\sigma^2\big)^{1/c},c\Big).$$

So, when we add $\nu = 2$ squared normal variates which are centered and of equal variance, the second form parameter $d$ in (3.82a) always is unity. If, instead of $\nu = 2$, we add $\nu = 2\,k$ such squared normal variates, the second form parameter will be $d = k$:

$$Y_j \overset{\text{iid}}{\sim} No(0,\sigma^2),\ j = 1,2,\ldots,2\,k \;\Longrightarrow\; a + \sqrt[c]{\sum_{j=1}^{2k} Y_j^2} \sim We\Big(a,\big(2\,\sigma^2\big)^{1/c},c,k\Big).$$

We mention some moments belonging to (3.82a):

$$\mathrm{E}(X^r) = \frac{b^r}{\Gamma(d)} \sum_{i=0}^{r} \binom{r}{i} \left(\frac{a}{b}\right)^{r-i} \Gamma\left(d + \frac{i}{c}\right), \tag{3.82b}$$

$$\mathrm{E}(X) = a + b\, \frac{\Gamma\left(d + \dfrac{1}{c}\right)}{\Gamma(d)}, \tag{3.82c}$$

$$\mathrm{Var}(X) = \left(\frac{b}{\Gamma(d)}\right)^2 \left[\Gamma(d)\,\Gamma\left(d + \frac{2}{c}\right)\,\Gamma^2\left(d + \frac{1}{c}\right)\right]. \tag{3.82d}$$

The hazard rate behaves as follows: When

- $(1 - d\,c)/[c\,(c - 1)] > 1$, then it is
  * bathtub–shaped for $c > 1$,
  * upside–down bathtub–shaped for $0 < c < 1$.
- Otherwise it is
  * constant for $c = 1$,
  * increasing for $c > 1$,
  * decreasing for $c < 1$.

A five–parameter extension of (3.82a) given by FAUCHON et al. (1976) will be presented in the following section.

### 3.3.7.2 Five-parameter distributions

**PHANI et al. extension**

PHANI (1987) has introduced a fifth parameter into (3.81a) in order to attain more flexibility in fitting the distribution to given datasets. Instead of only one exponent $\beta$ in (3.81a), this distribution has different exponents $\beta_1$ und $\beta_2$ for the nominator and denominator in the exp–term:

$$F(x) = 1 - \exp\left\{-\lambda \frac{(x - a)^{\beta_1}}{(b - x)^{\beta_2}}\right\}, \quad 0 \le a \le x \le b < \infty, \tag{3.83a}$$

$$f(x) = \frac{\lambda\,(x - a)^{\beta_1 - 1}\left[(b\,\beta_1 - a\,\beta_2) + (\beta_2 - \beta_1)\,x\right]}{(b - x)^{\beta_2 + 1}} \exp\left\{-\lambda \frac{(x - a)^{\beta_1}}{(b - x)^{\beta_2}}\right\} \tag{3.83b}$$

$$h(x) = \frac{\lambda\,(x - a)^{\beta_1 - 1}\left[(b\,\beta_1 - a\,\beta_2) + (\beta_2 - \beta_1)\,x\right]}{(b - x)^{\beta_2 + 1}}. \tag{3.83c}$$

The hazard rate has a bathtub shape when $0 < \beta_1 < 1$ and $0 < \beta_2 < 1$. All other $\beta_1$–$\beta_2$–combinations lead to an increasing hazard rate.

**FAUCHON et al. extension**

The FAUCHON et al. (1976) extension in the preceding section rests upon the generalized gamma distribution and its relation to the $\chi^2$–distribution, the latter being the sum of squared and centered normal variates. If, instead, the normal variates have differing means unequal to zero but are still homoscedastic, we know that:

$$Y_j \sim No(\mu_j, \sigma^2); \; j = 1, 2, \ldots, 2k \text{ and independent} \implies Y = \sum_{j=1}^{2k} Y_j^2 \sim \sigma^2 \chi^2(2k, \lambda),$$

i.e., we have a non–central $\chi^2$–distribution:

$$f(y) = \eta^k \, y^{k-1} \, \exp\big\{ -(\lambda + \eta \, y)\big\} \sum_{j=0}^{\infty} \frac{(\eta \, \lambda \, y)^j}{j! \, \Gamma(j+k)}$$

with non–centrality parameter

$$\lambda = \frac{1}{2\,\sigma^2} \sum_{j=1}^{2k} \mu_j^2$$

and scale factor

$$\eta = 2\,\sigma^2.$$

So, if we start with $Y_j = No(\mu_j, \sigma^2)$, we get

$$X = a + \sqrt[c]{\sum_{j=1}^{2\,k} Y_j^2} \sim We\big(a, \eta^{-1/c}, c, k, \lambda\big)$$

with a CDF given by

$$F(x) = e^{-\lambda} \, \frac{c}{2\,k} \sum_{j=0}^{\infty} \frac{\lambda^j}{j! \, \Gamma(j+k)} \int_{a}^{(x-a)^c} \left(\frac{t-a}{b}\right)^{c\,(k+j)-1} \exp\left\{-\left(\frac{t-a}{b}\right)^c\right\} \mathrm{d}t$$

$$\tag{3.84a}$$

and corresponding DF

$$f(x) = e^{-\lambda} \, \frac{c}{b} \left(\frac{x-a}{b}\right)^{c\,k-1} \exp\left\{-\left(\frac{x-a}{b}\right)^c\right\} \sum_{j=0}^{\infty} \frac{\lambda^j}{j! \, \Gamma(j+k)} \left(\frac{x-a}{b}\right)^{c\,k}.$$

$$\tag{3.84b}$$

The new, fifth, parameter in this WEIBULL distribution is the non–centrality parameter $\lambda$. The raw moments of this distribution can be expressed by the **KUMMER–function**, but have to be evaluated numerically:

$$\mathrm{E}\big(X^r\big) = b^r \, e^{-\lambda} \, \frac{\Gamma\left(k + \dfrac{r}{c}\right)}{\Gamma(k)} \, K\left(k + \frac{r}{c}, k, \lambda\right),$$

$$\tag{3.84c}$$

where $K(\cdot)$ is the KUMMER–function (E.E. KUMMER, 1810 – 1893):

$$K(\alpha, \beta, z) := \sum_{j=0}^{\infty} \frac{(\alpha)_j \, z^j}{(\beta)_j \, j!}$$

with

$$\begin{aligned}
(\alpha)_j &:= \alpha\,(\alpha+1)\ldots(\alpha+j-1), \; (\alpha)_0 := 1; \\
(\beta)_j &:= \beta\,(\beta+1)\ldots(\beta+j-1), \; (\beta)_0 := 1.
\end{aligned}$$

### 3.3.8 WEIBULL distributions with varying parameters

Mixed and compound WEIBULL distributions are models with varying parameters, but that variation is at random and cannot be attributed to some other distinct variable. In this section we will present some approaches where the WEIBULL parameters are functionally dependent on one or more other variables. In Sect. 3.3.8.1 the explanatory variable is time itself; in Sect. 3.3.8.2 the WEIBULL parameters will be made dependent on measurable variables other than time, called **covariates**.

#### 3.3.8.1 Time-dependent parameters

The scale parameter $b$ and/or the shape parameter $c$ have been made time–varying in some generalizations of the WEIBULL distribution where the latter is the model for a stochastic duration. We first present an approach of ZUO et al. (1999) with both, $b$ and $c$, dependent on time.

Let $Y(t)$ be the **degradation** of a device at time $t$. $Y(t)$ is assumed to be a random variable with CDF

$$F(y \mid t) = \Pr\big[Y(t) \leq y\big],$$

which at each point of time is a two–parameter WEIBULL distribution. Of course, $Y(t)$ is a non–decreasing function of $t$. The device will not fail as long as the degradation has not surpassed some given critical value $D$. Thus, the reliability function (CCDF) of a device is given by

$$R(t) = \Pr(T \geq t) = \Pr\big[Y(t) \leq D\big].$$

ZUO et al. suppose the following WEIBULL CHR with respect to $t$:

$$H(t) = \left(\frac{D}{b(t)}\right)^{c(t)}, \tag{3.85a}$$

where $b(t)$ and $c(t)$ are specified as

$$b(t) = \alpha\, t^\beta \exp(\gamma\, t), \tag{3.85b}$$

$$c(t) = \alpha\left(1 + \frac{1}{t}\right)^\beta \exp(\gamma/t). \tag{3.85c}$$

The parameters $\alpha$, $\beta$ and $\gamma$ have to be such that $H(t)$ goes to infinity as $t \to \infty$. Starting with (3.85a) and using Tab. 2/1 we arrive at

$$R(t) = \exp\left\{-\left[\frac{D}{b(t)}\right]^{c(t)}\right\}, \tag{3.85d}$$

$$f(t) = -\frac{\mathrm{d}\exp\left\{-\left[\dfrac{D}{b(t)}\right]^{c(t)}\right\}}{\mathrm{d}t}$$

$$= \exp\left\{-\left[\frac{D}{b(t)}\right]^{c(t)}\right\}\frac{\mathrm{d}\left[D/b(t)\right]^{c(t)}}{\mathrm{d}t}. \tag{3.85e}$$

SRIVASTAVA (1974) studies a WEIBULL model with a scale parameter which periodically changes between $b_1$ and $b_2$, attributed to two different conditions of usage of the device. Introducing $\tau_i(t)$; $i = 1, 2$; as the cumulative times, starting at $t = 0$, spent in phase $i$, the hazard rate $h(t)$ is

$$h(t) = \frac{c}{b_1} \left[ \frac{\tau_1(t)}{b_1} \right]^{c-1} + \frac{c}{b_2} \left[ \frac{\tau_2(t)}{b_2} \right]^{c-1}. \tag{3.86a}$$

This approach is easily recognized to be a special case of a two–component series system or a competing risk model with continuously joining hazard rates; see (3.54d). The difference between (3.86a) and (3.54d) is that we have two subpopulations with independent variables $\tau_1(t)$ and $\tau_2(t)$ instead of only one variable $t$. Some other functions describing the time to failure of this model are

$$H(t) = \left[ \frac{\tau_1(t)}{b_1} \right]^{c-1} + \left[ \frac{\tau_2(t)}{b_2} \right]^{c-1}, \tag{3.86b}$$

$$R(t) = \exp \left\{ - \left[ \frac{\tau_1(t)}{b_1} \right]^{c-1} - \left[ \frac{\tau_2(t)}{b_2} \right]^{c-1} \right\}, \tag{3.86c}$$

$$f(t) = \left\{ \frac{c}{b_1} \left[ \frac{\tau_1(t)}{b_1} \right]^{c-1} + \frac{c}{b_2} \left[ \frac{\tau_2(t)}{b_2} \right]^{c-1} \right\} \exp \left\{ - \left[ \frac{\tau_1(t)}{b_1} \right]^{c} - \left[ \frac{\tau_2(t)}{b_2} \right]^{c} \right\}. \tag{3.86d}$$

ZACKS (1984) presents a WEIBULL model where the shape parameter is given by

$$c \left\{ \begin{array}{ll} = & \text{for} \;\; 0 \le t \le \tau \\[2mm] > & \text{for} \;\; t > \tau \end{array} \right\} \tag{3.87a}$$

with the following hazard rate

$$h(t) = \left\{ \begin{array}{ll} \dfrac{1}{b} & \text{for} \;\; 0 \le t \le \tau \\[4mm] \dfrac{1}{b} + \dfrac{c}{b} \left( \dfrac{t - \tau}{b} \right)^{c-1} & \text{for} \;\; t > \tau. \end{array} \right\} \tag{3.87b}$$

ZACKS calls this an **exponential–WEIBULL distribution**. In the first phase of lifetime up to the change point $\tau$ the device has a constant hazard rate that afterwards is super–positioned by a WEIBULL hazard rate. This model is nothing but a special two–fold composite or sectional model (see Sect. 3.3.6.3). Some other functions describing the time to failure of this model are

$$H(t) = \left\{ \begin{array}{ll} \dfrac{t}{b} & \text{for} \;\; 0 \le t \le \tau \\[4mm] \dfrac{\tau}{b} + \left( \dfrac{t - \tau}{b} \right)^{c} & \text{for} \;\; t > \tau, \end{array} \right\} \tag{3.87c}$$

$$F(t) = 1 - \exp \left\{ \frac{t}{b} - \left[ \left( \frac{t - \tau}{b} \right)_+ \right]^{c} \right\}, \; t \ge 0, \tag{3.87d}$$

$$f(t) = \frac{1}{b} e^{-t/b} \left\{ 1 + c \left[ \left( \frac{t - \tau}{b} \right)_+ \right]^{c-1} \right\} \exp \left\{ - \left[ \left( \frac{t - \tau}{b} \right)_+ \right]^{c} \right\}, \; t \ge 0, \tag{3.87e}$$

where $(y)_+ := \max(0, y)$. ZACKS gives the following formula for raw moments

$$E(X^r) = b^r \, \mu'_r(c, \tau/b) \tag{3.87f}$$

with

$$\mu'_r(c, \tau) = r! \left[ 1 - e^{-1/b} \sum_{j=0}^{r} \frac{1}{b^j \, j!} \right] + e^r \sum_{j=0}^{r} \binom{r}{j} \tau^{r-j} M_j(c) \tag{3.87g}$$

and

$$M_j(c) = \begin{cases} 1 & \text{for} \quad j = 0 \\ j \displaystyle\int_0^\infty x^{j-1} \exp\{-(x + x^c)\} \, dx & \text{for} \quad j \geq 1. \end{cases}$$

### 3.3.8.2 Models with covariates[31]

When the WEIBULL distribution, or any other distribution, is used to model lifetime or some other duration variable, it is often opportune to introduce one or more other variables that explain or influence the spell length. Two such approaches will be discussed

- The scale parameter $b$ is influenced by the supplementary variable(s). This gives the **accelerated life model**.
- The hazard rate $h(t)$ is directly dependent on the covariate(s). The prototype of this class is the **proportional hazard model**.

Farther down we will show that the WEIBULL accelerated life model is identical to the WEIBULL proportional hazard model.

In reliability applications a supplementary variable $s$ represents the **stress** on the item such that its lifetime is some function of $s$. The stress may be electrical, mechanical, thermal etc. In life testing of technical items the application of some kind of stress is a means to shorten the time to failure (see Sect.16), thus explaining the term **accelerated life testing**. The general form of the WEIBULL accelerated life model with one stress variable $s$ is

$$F(t \mid s) = 1 - \exp\left\{ -\left[ \frac{t}{b \, \beta(s)} \right]^c \right\}, \quad t \geq 0. \tag{3.88}$$

The function $\beta(s)$ is the so–called **acceleration factor** which —in the parametrization above —must be a decreasing function of $s$, leading to a higher probability of failure up to time $t$ the higher the stress exercised on an item. Three types of $\beta(s)$ are extensively used:

- The **ARRHENIUS–equation**[32]

$$\beta(s) = \exp(\alpha_0 + \alpha_1/s) \tag{3.89a}$$

gives the **ARRHENIUS–WEIBULL distribution**.

---

[31]  Suggested reading for this section: LAWLESS (1982), NELSON (1990).

[32]  S.A. ARRHENIUS (1859 – 1927) was a Swedish physico–chemist who — in 1903 — received the NO-BEL prize in chemistry for his theory of electrolytical dissociation. (3.89a) in its original version describes the dependency of temperature on the speed of reaction of some kind of material.

- The **EYRING–equation** (H. EYRING, 1901 – 1981) is an alternative to (3.89a):

$$\beta(s) = \alpha_0 \, s^{\alpha_1} \, \exp(\alpha_2 \, s) \tag{3.89b}$$

and leads to the **EYRING–WEIBULL distribution**.

- The equation

$$\beta(s) = \alpha_0/s^{\gamma_1} \tag{3.89c}$$

gives the **power WEIBULL distribution**.

An acceleration factor which depends on a vector $s$ of $m$ stress variables $s_i$ is given by

$$\beta(s) = \exp\left\{ \sum_{i=1}^{m} \gamma_i \, s_i \right\} = \exp\left( \gamma' \, s \right). \tag{3.89d}$$

The hazard rate corresponding to (3.88) is

$$h(t \,|\, s) = \frac{c}{b \, \beta(s)} \left( \frac{t}{b \, \beta(s)} \right)^{c-1}. \tag{3.90a}$$

Thus, the ratio of the hazard rates for two items with covariates values $s_1$ and $s_2$ is

$$\frac{h(t \,|\, s_1)}{h(t \,|\, s_2)} = \left[ \frac{\beta(s_2)}{\beta(s_1)} \right]^c, \tag{3.90b}$$

which is independent of $t$. Looking at the CDF of (3.88) for two different stress levels $s_1$ and $s_2$, we can state

$$F(t \,|\, s_1) = F\left( \frac{\beta(s_2)}{\beta(s_1)} \, t \,\Big|\, s_2 \right). \tag{3.90c}$$

So, if an item with stress $s_1$ has a lifetime $T_1$, then an item with stress $s_2$ has a lifetime $T_2$ given by

$$T_2 = \frac{\beta(s_2)}{\beta(s_1)} \, T_1. \tag{3.90d}$$

Because $\beta(s)$ is supposed to be decreasing with $s$, we will have $T_2 < T_1$ for $s_2 > s_1$.

The moments belonging to (3.88) are the same as those of a common two–parameter WEIBULL distribution with the scale parameter $b$ replaced by $b \, \beta(s)$.

A proportional hazard model is characterized by

$$h(t \,|\, s) = g(s) \, h_0(t), \tag{3.91a}$$

where $h_0(t)$ is called the **baseline hazard rate**. After its introduction by COX (1972) the proportional hazard model has been extensively applied in the social and economic sciences to model the sojourn time of an individual in a given state, e.g., in unemployment or in a given job. The argument of the scalar–valued function $g(\cdot)$ may be a vector too. The most popular proportional hazard model takes the WEIBULL hazard rate as its baseline hazard rate, turning (3.91a) into

$$h(t \mid s) = g(s) \frac{c}{b} \left(\frac{t}{b}\right)^{c-1} \tag{3.91b}$$

with CDF

$$F(t \mid s) = 1 - \exp\left\{-g(s)\left(\frac{t}{b}\right)^{c}\right\}, \quad t \geq 0. \tag{3.91c}$$

When

$$\frac{1}{g(s)} = [\beta(s)]^{c},$$

the WEIBULL proportional hazard model and the WEIBULL accelerated life model coincide.

It is also instructive to view (3.88) through the distribution of $Y = \ln T$, when the acceleration factor — in most general form — is $\beta(s)$, i.e., it depends on a vector of stress variables. The distribution of $Y$ is of extreme value form (see (3.43a–d)) and has DF

$$f(y \mid s) = \frac{1}{b^{*}} \exp\left\{\frac{y - a^{*}(s)}{b^{*}} - \exp\left[\frac{y - a^{*}(s)}{b^{*}}\right]\right\}, \quad y \in \mathbb{R}, \tag{3.92a}$$

with

$$b^{*} = \frac{1}{c} \quad \text{and} \quad a^{*}(s) = \ln\left[b\,\beta(s)\right].$$

Written another way, (3.92a) is the DF of

$$Y = a^{*}(s) + b^{*} Z, \tag{3.92b}$$

where $Z$ has a reduced extreme value distribution with DF $\exp\{z - e^{z}\}$. (3.92b) is a **location–scale regression model** with error $Z$. The constancy of $c$ in (3.88) corresponds to the constancy of $b^{*}$ in (3.92b), so $\ln T$ has a constant variance. A variety of functional forms for $\beta(s)$ or $a^{*}(s)$ is often employed together with either (3.88) or (3.92a). The most popular one is perhaps the log–linear form for which

$$a^{*}(s) = \gamma_{0} + \gamma' s \tag{3.92c}$$

with $\gamma_{0} = \ln b$ and $\beta(s) = \gamma' s = \sum_{i=1}^{m} \gamma_{i} s_{i}$. This linear regression can be estimated by a multitude of methods, see LAWLESS (1982, pp. 299ff.).

We finally mention a generalization of the accelerated life model where the stress changes with time; i.e., stress is growing with time. This approach is applied in life testing to shorten the testing period. NELSON (1990) shows how to estimate the parameters for a **step–stress model** where the stress is raised at discrete time intervals.

### 3.3.9 Multidimensional WEIBULL models[33]

In Sect. 3.3.6.2 we have studied the lifetime distribution of two special types of a multi–component system, i.e., the parallel and the series systems, under the rather restrictive assumption of independently failing components. When the failure times of the components are dependent, we need a multivariate distribution that is not simply the product of its marginal distributions. There are mainly two approaches to introduce multivariate extensions of the one–dimensional WEIBULL distribution:

- One simple approach is to transform the bivariate and multivariate exponential distribution through power transformation. As the extension of the univariate exponential distribution to a bi- or multivariate exponential distribution is not unique, there exist several extensions all having marginal exponentials; see KOTZ/BALAKRISHNAN/JOHNSON (2000, Chap. 47). A similar approach involves the transformation of some multivariate extreme value distribution.

- A very different approach is to specify the dependence between two or more univariate WEIBULL variates so that the emerging bi- or multivariate WEIBULL distribution has WEIBULL marginals.

We start (Sect. 3.3.9.1) by presenting — in detail — several bivariate WEIBULL models and finish by giving an overview (Sect. 3.3.9.2) on some multivariate WEIBULL models.

#### 3.3.9.1 Bivariate WEIBULL distributions

We will first present some bivariate WEIBULL distributions, **BWD** for short, that are obtained from a bivariate exponential distribution, **BED** for short, by power transformation. A genuine BED has both marginal distributions as exponential. As KOTZ/BALAKRISHNAN/JOHNSON (2000) show, many of such BEDs exist. The BED models of FREUND (1961) and MARSHALL/OLKIN (1967) have received the most attention in describing the statistical dependence of component's life in two–component systems. These two systems rest upon a clearly defined physical model that is both simple and realistic. So, it is appropriate to study possible WEIBULL extensions to these two BEDs.

#### BWDs based on FREUND's BED

FREUND (1961) proposed the following failure mechanism of a two–component system: Initially the two components have constant failure rates, $\lambda_1$ and $\lambda_2$, with independent DFs when both are in operation:

$$f_i(x_i \mid \lambda_i) = \lambda_i \exp\{-\lambda_i x_i\}; \;\; x_i \geq 0, \; \lambda_i > 0; \; i = 1, 2. \tag{3.93a}$$

But the lifetimes $X_1$ and $X_2$ are dependent because a failure of either component does not result in a replacement, but changes the parameter of the life distribution of the other component to $\lambda_i^*$, mostly to $\lambda_i^* > \lambda_i$ as the non–failed component has a higher workload.

---

[33]  Suggested reading for this section: CROWDER (1989), HANAGAL (1996), HOUGAARD (1986), KOTZ/BALAKRISHNAN/JOHNSON (2000), LEE (1979), LU (1989, 1990, 1992a,b), LU/BHATTACHARYYA (1990), MARSHALL/OLKIN (1967), PATRA/DEY (1999), ROY/MUKHERJEE (1988), SARKAR (1987), SPURRIER/WEIER (1981) and TARAMUTO/WADA (2001).

There is no other dependence. The time to first failure is exponentially distributed with parameter

$$\lambda = \lambda_1 + \lambda_2.$$

The probability that component $i$ is the first to fail is $\lambda_i/\lambda$, whenever the first failure occurs. The distribution of the time from first failure to failure of the other component is thus a mixture of two exponential distributions with parameters $\lambda_1^*$ or $\lambda_2^*$ in proportions $\lambda_2/\lambda$ and $\lambda_1/\lambda$, respectively. Finally, the joint DF of $X_1$ and $X_2$ is

$$f_{1,2}^{FR}(x_1,x_2) = \begin{cases} f_1(x_1\,|\,\lambda_1)\,R_2(x_1\,|\,\lambda_2)\,f_2(x_2-x_1\,|\,\lambda_2^*) & \text{for } 0 \le x_1 < x_2 \\ f_2(x_2\,|\,\lambda_2)\,R_1(x_2\,|\,\lambda_2)\,f_1(x_1-x_2\,|\,\lambda_1^*) & \text{for } 0 \le x_2 < x_1 \end{cases} \quad (3.93\mathrm{b})$$

$$= \begin{cases} \lambda_1\,\lambda_2^*\,\exp\{-\lambda_2^*\,x_2 - (\lambda_1+\lambda_2-\lambda_2^*)\,x_1\} & \text{for } 0 \le x_1 < x_2 \\ \lambda_2\,\lambda_1^*\,\exp\{-\lambda_1^*\,x_1 - (\lambda_1+\lambda_2-\lambda_1^*)\,x_2\} & \text{for } 0 \le x_2 < x_1, \end{cases} \quad (3.93\mathrm{c})$$

where

$$R_i(x_i\,|\,\lambda_i) = \exp\{-\lambda_i\,x_i\}.$$

$f_{1,2}^{FR}(x_1,x_2)$ is not continuous at $x = x_1 = x_2$, unless $\lambda_1\,\lambda_2^* = \lambda_2\,\lambda_1^*$. The joint survival function CCDF is

$$R_{1,2}^{Fr}(x_1,x_2) = \begin{cases} \dfrac{1}{\lambda_1+\lambda_2-\lambda_2^*}\Big[\lambda_1\,\exp\{-(\lambda_1+\lambda_2-\lambda_2^*)\,x_1 - \lambda_2^*\,x_2\} + \\ \qquad (\lambda_2-\lambda_2^*)\,\exp\{-(\lambda_1+\lambda_2)\,x_2\}\Big] \quad \text{for } 0 \le x_1 < x_2 \\[2mm] \dfrac{1}{\lambda_1+\lambda_2-\lambda_2^*}\Big[\lambda_2\,\exp\{-(\lambda_1+\lambda_2-\lambda_1^*)x_2 - \lambda_1^*x_1\} + \\ \qquad (\lambda_1-\lambda_1^*)\,\exp\{-(\lambda_1+\lambda_2)\,x_1\}\Big] \quad \text{for } 0 \le x_2 < x_1. \end{cases}$$
$$(3.93\mathrm{d})$$

Provided $\lambda_1 + \lambda_2 \ne \lambda_i^*$, the marginal DF of $X_i$ is

$$f_i(x_i) = \frac{1}{\lambda_1+\lambda_2-\lambda_i^*}\begin{Bmatrix} (\lambda_i-\lambda_i^*)\,(\lambda_1+\lambda_2)\,\exp\{-(\lambda_1+\lambda_2)\,x_i\} + \\ \lambda_i^*\,\lambda_{3-i}\,\exp\{-\lambda_i^*\,x_i\} \end{Bmatrix}; \quad x_i \ge 0.$$
$$(3.93\mathrm{e})$$

(3.93e) is not exponential but rather a mixture of two exponential distributions for $\lambda_i > \lambda_i^*$, otherwise (3.93e) is a weighted average. So (3.93c,d) do not represent a genuine BED.

We now substitute the exponential distributions in (3.93a) by WEIBULL distributions in the following parametrization:

$$f_i(y_i\,|\,\lambda_i,c_i) = c_i\,\lambda_i\,(\lambda_i\,y_i)^{c_i-1}\,\exp\{-(\lambda_i\,y_i)^{c_i}\}; \quad y_i \ge 0;\ \lambda_i,c_i > 0;\ i = 1,2. \quad (3.94\mathrm{a})$$

When one component fails, the remaining lifetime of the other component is still WEIBULL, but with possibly changed parameters $\lambda_i^*$ and $c_i^*$. Time is reset on the surviving component. The shape parameter $c_i^*$ might be equal to $c_i$, but the scale parameter $\lambda_i^*$ is not equal to $\lambda_i$ generally, which makes the hazard rate of the non–failed component change.

As time is reset on the surviving component, it has a shifted WEIBULL distribution[34] and (3.93b) turns into the following BWD:

$$f_{1,2}(y_1, y_2) = \begin{cases} f_1(y_1 \mid \lambda_1, c_1) \exp\{-(\lambda_2 \, y_1)^{c_2}\} \times \\ c_2^* \lambda_2^* [\lambda_2^* (y_2 - y_1)]^{c_2^*-1} \exp\{-[\lambda_2^* (y_2 - y_1)]^{c_2^*}\} \text{ for } 0 \le y_1 < y_2 \\ f_2(y_2 \mid \lambda_2, c_2) \exp\{-(\lambda_1 \, y_2)^{c_1}\} \times \\ c_1^* \lambda_1^* [\lambda_1^* (y_1 - y_2)]^{c_1^*-1} \exp\{-[\lambda_1^* (y_1 - y_2)]^{c_1^*}\} \text{ for } 0 \le y_2 < y_1. \end{cases}$$
(3.94b)

(3.94b) yields (3.93c) for the special case of all shape parameters equal to one.

LU (1989) shows that the above WEIBULL extension to FREUND's BED is not equal to a bivariate WEIBULL model obtained by using a direct power transformation of the marginals of FREUND's BED resulting in

$$f_{1,2}^*(y_1, y_2) = \begin{cases} f_1(y_1 \mid \lambda_1, c_1) \exp\{-[(\lambda_2 - \lambda_2^*) \, y_1]^{c_1}\} \times \\ f_2(y_2 \mid \lambda_2^*, c_2) \text{ for } 0 \le y_1^{c_1} < y_2^{c_2} \\ f_2(y_2 \mid \lambda_2, c_2) \exp\{-[(\lambda_1 - \lambda_1^*) \, y_2]^{c_2}\} \times \\ f_1(y_1 \mid \lambda_1^*, c_1) \text{ for } 0 \le y_2^{c_2} < y_1^{c_1}. \end{cases}$$
(3.95)

SPURRIER/WEIER (1981) modified the FREUND idea of a hazard rate change to derive another bivariate WEIBULL model. Let $U$ be the time to the first failure and $W$ the time from the first to the second failure. So the time of system failure is $V = U + W$. The time $U$ until the first failure is distributed as the minimum of two iid WEIBULL variates with shape parameter $c$ and scale factor $\lambda$, so

$$f_U(u) = 2 \, c \, \lambda \, (\lambda \, u)^{c-1} \exp\{-2 \, (\lambda \, u)^c\}, \quad u \ge 0$$
(3.96a)

(see Sect. 3.1.4). Upon the first failure the remaining component changes its scale factor to $\theta \, \lambda, \theta > 0$. Then we have the following conditional density of $W \mid u$:

$$f_{W|u}(w) = \frac{c \, \lambda \, \theta \, [\lambda \, (\theta \, w + u)]^{c-1} \exp\{-[\lambda \, (\theta \, w + u)]^c\}}{\exp\{-(\lambda \, u)^c\}}, \quad w \ge 0.$$
(3.96b)

The joint DF of $U$ and $W$ is

$$\begin{aligned} f(u, w) = f_U(u) \, f_{W \mid u}(w) \;&=\; 2 \, c^2 \, \theta \, \lambda^{2c} \, u^{c-1} \, (\theta \, w + u) \times \\ &\quad \exp\{-[\lambda \, (\theta \, w + u)]^c - (\lambda \, u)^c\} \,; \quad u, w \ge 0. \end{aligned}$$
(3.96c)

The DF of the time to system failure $V = U + W$ does not in general have a closed form and has to be determined by numerical integration of (3.96c) over the set $\{(u, w) \, : \, u + w = v\}$.

---

[34] The shift or location parameter is equal to the time of the first failure of a component.

Three important special cases of (3.96c) are

**Case I:** $c = 1$; $\theta = 1$ $\implies$ The lifetimes of the two components are iid exponential variates.

**Case II:** $c = 1$; $\implies$ The model reduces to a reparametrization of the FREUND model.

For cases I and II we have :

$$f(v) = \begin{cases} 2\theta\lambda(2-\theta)^{-1}\exp\{-\theta\lambda v\}\left[1-\exp\{-(2-\theta\lambda v)\}\right]; & v \geq 0 \text{ and } \theta \neq 2, \\ 4\lambda^2 v\exp\{-2\lambda v\}; & v \geq 0 \text{ and } \theta = 2. \end{cases}$$

**Case III:** $\theta = 1$; $\implies$ The lifetimes of the two components are iid WEIBULL variates.

In this case we get

$$f(v) = 2c\lambda^c v^{c-1}\left[\exp\{-(\lambda v)^c\}\exp\{-2(\lambda v)^c\}\right], \quad v \geq 0.$$

We finally mention that LU (1989) has generalized the model of SPURRIER/WEIER by introducing different shape parameters for the two components and by also allowing a change of the shape parameter upon the first failure.

### BWDs based on MARSHALL/OLKIN's BED

The physical model behind the MARSHALL/OLKIN (1967) BED differs from that of FREUND's BED insofar as there is a third failure mechanism that hits both components simultaneously. To be more precise, their model is as follows: The components of a two–component system fail after receiving a shock which is always fatal. The occurrence of shocks is governed by three independent POISSON processes $N_i(t;\lambda_i)$; $i = 1,2,3$. By $N(t;\lambda) = \{N(t), t \geq 0; \lambda\}$, we mean a POISSON process (see Sect. 4.2) with parameter $\lambda$, giving the mean number of events per unit of time. Events in the processes $N_1(t;\lambda_1)$ and $N_2(t;\lambda_2)$ are shocks to components 1 and 2, respectively, and events in the process $N_3(t;\lambda_3)$ are shocks to both components. The joint survival function of $X_1$ and $X_2$, the lifetimes of the two components, is:

$$\begin{aligned} R_{1,2}^{MO}(x_1,x_2) &= \Pr(X_1 > x_1, X_2 > x_2) \\ &= \Pr\{N_1(x_1;\lambda_1) = 0, N_2(x_2;\lambda_2) = 0, N_3(\max[x_1,x_2];\lambda_3) = 0\} \\ &= \exp\{-\lambda_1 x_1 - \lambda_2 x_2 - \lambda_3 \max[x_1,x_2]\}. \end{aligned} \quad (3.97a)$$

$$= \begin{cases} \exp\{-\lambda_1 x_1 - (\lambda_2 + \lambda_3)x_2\} & \text{for } 0 \leq x_1 \leq x_2 \\ \exp\{-(\lambda_1 + \lambda_3)x_1 - \lambda_2 x_2\} & \text{for } 0 \leq x_2 \leq x_1. \end{cases} \quad (3.97b)$$

The marginal distributions are genuine one–dimensional exponential distributions:

$$R_i(x_i) = \exp\{-(\lambda_i + \lambda_3)x_i\}; \quad i = 1,2. \quad (3.97c)$$

The probability that a failure on component $i$ occurs first is

$$\Pr(X_i < X_j) = \frac{\lambda_i}{\lambda_1 + \lambda_2 + \lambda_3}; \quad i, j = 1, 2; \; i \neq j;$$

and we have a positive probability that both components fail simultaneously:[35]

$$\Pr(X_1 = X_2) = \frac{\lambda_3}{\lambda_1 + \lambda_2 + \lambda_3}.$$

The latter probability is responsible for the **singularity of the distribution** along the line $x_1 = x_2$. MARSHALL/OLKIN (1967) show that the joint survival function (3.97a,b) can be written as a mixture of an absolutely continuous survival function

$$R_c(x_1, x_2) = \left\{ \begin{array}{l} \dfrac{\lambda_1 + \lambda_2 + \lambda_3}{\lambda_1 + \lambda_2} \exp\big\{ - \lambda_1 x_1 - \lambda_2 x_2 - \lambda_3 \max[x_1, x_2]\big\} - \\[3mm] \dfrac{\lambda_3}{\lambda_1 + \lambda_2} \exp\big\{ - (\lambda_1 + \lambda_2 + \lambda_3) \max[x_1, x_2]\big\} \end{array} \right\} \tag{3.97d}$$

and a singular survival function

$$R_s(x_1, x_2) = \exp\big\{ - (\lambda_1 + \lambda_2 + \lambda_3) \max[x_1, x_2]\big\} \tag{3.97e}$$

in the form

$$R_{1,2}^{MO}(x_1, x_2) = \frac{\lambda_1 + \lambda_2}{\lambda_1 + \lambda_2 + \lambda_3} R_c(x_1, x_2) + \frac{\lambda_3}{\lambda_1 + \lambda_2 + \lambda_3} R_s(x_1, x_2). \tag{3.97f}$$

The joint DF of the MARSHALL/OLKIN model is

$$f_{1,2}^{MO}(x_1, x_2) = \left\{ \begin{array}{ll} \lambda_1 (\lambda_2 + \lambda_3) R_{1,2}^{MO}(x_1, x_2) & \text{for} \quad 0 \leq x_1 < x_2 \\[2mm] \lambda_2 (\lambda_1 + \lambda_3) R_{1,2}^{MO}(x_1, x_2) & \text{for} \quad 0 \leq x_2 < x_1 \\[2mm] \lambda_3 R_{1,2}^{MO}(x, x) & \text{for} \quad 0 \leq x_1 = x_2 = x. \end{array} \right\} \tag{3.97g}$$

A first WEIBULL extension to this model has been suggested by MARSHALL/OLKIN themselves, applying the power–law transformations $X_i = Y_i^{c_i}$; $i = 1, 2$; thus changing (3.97a,b) into

$$R_{1,2}(y_1, y_2) = \exp\big\{ - \lambda_1 y_1^{c_1} - \lambda_2 y_2^{c_2} - \lambda_3 \max\big[y_1^{c_1}, y_2^{c_2}\big]\big\} \tag{3.98a}$$

$$= \left\{ \begin{array}{ll} \exp\big\{ - \lambda_1 y_1^{c_1} - (\lambda_2 + \lambda_3) y_2^{c_2}\big\} & \text{for} \quad 0 \leq y_1^{c_1} \leq y_2^{c_2} \\[2mm] \exp\big\{ - (\lambda_1 + \lambda_3) y_1^{c_1} - \lambda_2 y_2^{c_2}\big\} & \text{for} \quad 0 \leq y_2^{c_2} \leq y_1^{c_1}. \end{array} \right\} \tag{3.98b}$$

---

[35] The correlation coefficient between $X_1$ and $X_2$ is $\varrho(X_1, X_2) = \Pr(X_1 = X_2) > 0$.

The areas in the $y_1$–$y_2$–plane, where the two branches of (3.98b) are valid, may be expressed by $y_2 \geq y_1^{c_1/c_2}$ and $y_2 \leq y_1^{c_1/c_2}$.

Set $y_1$ or $y_2$ to 0 in (3.98a,b) to get the marginal survival functions:

$$R_i(y_i) = \exp\{-(\lambda_i + \lambda_3)\,y_i^{c_i}\}, \quad i = 1, 2. \tag{3.98c}$$

These are genuine univariate WEIBULL distributions. Whereas the joint survival function is absolutely continuous (see Fig. 3/16), the joint DF is not. It is given by

$$f_{1,2}(y_1, y_2) = \begin{cases} \lambda_1(\lambda_2+\lambda_3)c_1 y_1^{c_1-1} c_2 y_2^{c_2-1} \exp\{-\lambda_1 y_1^{c_1} - (\lambda_2+\lambda_3)y_2^{c_2}\}; & 0 \leq y_1^{c_1} < y_2^{c_2} \\ \lambda_2(\lambda_1+\lambda_3)c_1 y_1^{c_1-1} c_2 y_2^{c_2-1} \exp\{-(\lambda_1+\lambda_3)y_1^{c_1} - \lambda_2 y_2^{c_2}\}; & 0 \leq y_2^{c_2} < y_1^{c_1} \\ \lambda_3 c_1 y_1^{c_1-1} \exp\{-(\lambda_1+\lambda_2+\lambda_3)y_1^{c_1}\}; & 0 \leq y_1^{c_1} = y_2^{c_2}. \end{cases} \tag{3.98d}$$

The third branch of (3.98d) represents the discrete part of the joint DF and it lies on the curve $y_2 = y_1^{c_1/c_2}$ in the $y_1$–$y_2$–plane. This locus is a straight line through the origin for $c_1 = c_2$. The discontinuity is clearly seen in Fig. 3/17, where we have marked some points of discontinuity on the surface grid. The discontinuity is best seen in the contour plot.

Figure 3/16: Joint survival function of a BWD of MARSHALL/OLKIN type
(surface plot and contour plot)

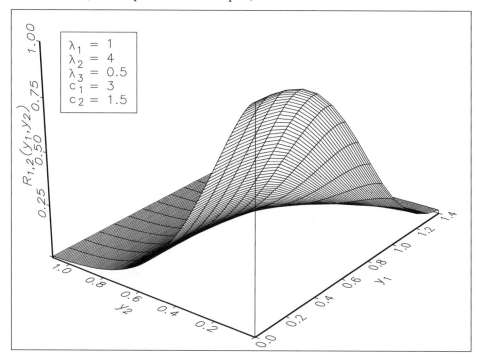

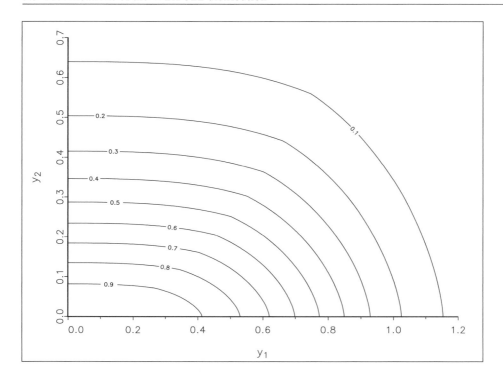

Figure 3/17: Joint density function of a BWD of MARSHALL/OLKIN type
(surface plot and contour plot)

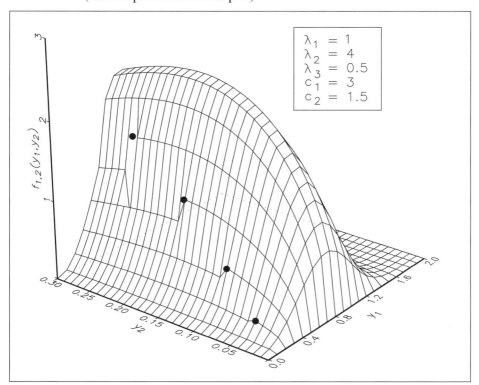

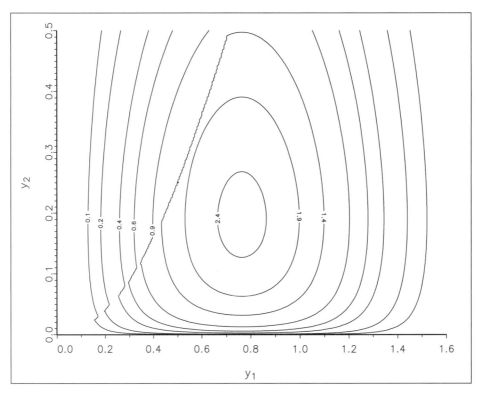

LU (1989) has proposed a second WEIBULL extension to MARSHALL/OLKIN's BED by generalizing their fatal shock model, which is given by three homogeneous POISSON processes, to non–homogeneous POISSON processes $N_i(t; \lambda_i, c_i)$ with power-law intensities[36] $\lambda_i c_i (\lambda_i t)^{c_i-1}$; $i = 1, 2, 3$. As in the original MARSHALL/OLKIN model's the processes $N_1$ and $N_2$ cause failure of the component 1 and 2, respectively, while events in the process $N_3$ are shocks under which both components will fail simultaneously. The joint survival function for the lifetimes $Y_1, Y_2$ of the two components are

$$R_{1,2}(y_1, y_2) = \Pr(Y_1 > y_1, Y_2 > y_2)$$

$$= \Pr\left\{ N_1(y_1; \lambda_1; c_1) = 0, \; N_2(y_2; \lambda_2, c_2) = 0, \; N_3\left( \max[y_1, y_2]; \lambda_3, c_3 \right) \right\}$$

$$= \exp\left\{ - \lambda_1 y_1^{c_1} - \lambda_2 y_2^{c_2} - \lambda_3 \left( \max[y_1, y_2] \right)^{c_3} \right\}. \tag{3.99a}$$

$$= \left\{ \begin{array}{ll} \exp\left\{ - \lambda_1 y_1^{c_1} - \lambda_2 y_2^{c_2} - \lambda_3 y_2^{c_3} \right\} & \text{for} \quad 0 \leq y_1 \leq y_2 \\ \exp\left\{ - \lambda_1 y_1^{c_1} - \lambda_2 y_2^{c_2} - \lambda_3 y_1^{c_3} \right\} & \text{for} \quad 0 \leq y_2 \leq y_1. \end{array} \right\} \tag{3.99b}$$

(3.99a/b) differ from (3.98a,b) in two respects:

- The LU extension has one more parameter $c_3$.

- The admissible regions for the two branches are different from those of the first extension, and the borderline is given by $y_1 = y_2$.

---

[36] This intensity function looks like a WEIBULL hazard rate. Its relation to a WEIBULL process will be discussed in Sect. 4.3.

(3.99a,b) have the marginals

$$R_i(y_i) = \exp\big\{ - \lambda_3\, y_i^{c_3} - \lambda_i\, y_i^{c_i} \big\}; \quad i = 1, 2; \tag{3.99c}$$

which, in general, are not WEIBULL. Of course, this extension has no absolutely joint DF as well.

We mention the following special cases of (3.99a,b):

- When $c_1 = c_2 = c_3 = c$, (3.99a,b) coincide with the original MARSHALL/OLKIN extension, where the marginals and the minimum $T = \min[Y_1, Y_2]$; i.e., the lifetime of the series system, are all WEIBULL variates with

$$
\begin{aligned}
R(y_1) &= \exp\big\{ - (\lambda_1 + \lambda_3)\, y_1^c \big\}, \\
R(y_2) &= \exp\big\{ - (\lambda_2 + \lambda_3)\, y_2^c \big\}, \\
R(t) &= \exp\big\{ - (\lambda_1 + \lambda_2 + \lambda_3)\, t^c \big\}.
\end{aligned}
$$

- The univariate minimum–type distribution, introduced by FRIEDMAN/GERTSBAKH (1980) has the survival function

$$R(x) = \exp\big\{ - \lambda^*\, x - \lambda^{**}\, x^c \big\}.$$

It belongs to a series system with two statistically dependent components having exponential and WEIBULL as lifetime distributions. This model is appropriate where system failure occurs with constant and/or increasing intensity. With $c_1 = c_2 = 1$ and $c_3 = c$, (3.99a) turns into

$$R(y_1, y_2) = \exp\big\{ - (\lambda_1\, y_1 - \lambda_2\, y_2 - \lambda_3\, \big( \max[y_1, y_2] \big)^c \big\},$$

which can be regarded as a bivariate extension of the minimum–type distribution because it has marginals of the univariate minimum–type.

- The univariate linear hazard rate distribution with $h(x) = a + b\,x$ has the survival function

$$R(x) = \exp\big\{ - a\,x - 0.5\,b\,x^2 \big\}.$$

This is a special case $(c = 2)$ of the univariate minimum–type distribution. Taking $c_1 = c_2 = 2$ and $c_3 = 1$ in (3.99a), we get the following bivariate extension of the linear hazard rate distribution:

$$R(y_1, y_2) = \exp\big\{ - \lambda_1\, y_1^2 - \lambda_2\, y_2^2 - \lambda_3\, \max[y_1, y_2] \big\},$$

which has marginals of the form of the linear hazard rate distribution. With $c_1 = c_2 = 1$ and $c_3 = 2$, we would get the same bivariate model.

### BWD's based on LU/BHATTACHARYYA (1990)

LU/BHATTACHARYYA (1990) have proposed several new constructions of bivariate WEIBULL models. A first family of models rests upon the general form

$$R_{1,2}(y_1, y_2) = \exp\left\{ - \left( \frac{y_1}{b_1} \right)^{c_1} - \left( \frac{y_2}{b_2} \right)^{c_2} - \delta\, \psi(y_1, y_2) \right\}. \tag{3.100}$$

Different forms of the function $\psi(y_1, y_2)$ together with the factor $\delta$ yield special members of this family:

- $\delta = 0$ and any $\psi(y_1, y_2)$ lead to a BWD with independent variables $Y_1$ and $Y_2$.

- $\delta = \lambda_3$ and $\psi(y_1, y_2) = \max\left(y_1^{c_1}, y_2^{c_2}\right)$ together with $\lambda_1 = b_1^{-c_1}$, $\lambda_2 = b_2^{-c_2}$ gives the MARSHALL/OLKIN extension (3.98a).

- LU/BHATTACHARYYA proposed the two following functions:

$$\psi(y_1, y_2) = \left\{ \left(\frac{y_1}{b_1}\right)^{c_1/m} + \left(\frac{y_2}{b_2}\right)^{c_2/m} \right\}^m,$$

$$\psi(y_1, y_2) = \left[1 - \exp\left\{-\left(\frac{y_1}{b_1}\right)^{c_1}\right\}\right]\left[1 - \exp\left\{-\left(\frac{y_2}{b_2}\right)^{c_2}\right\}\right].$$

- A BWD to be found in LEE (1979):

$$R_{1,2}(y_1, y_2) = \exp\left\{ -\lambda_1\, d_1^c\, y_1^c - \lambda_2\, d_2^c\, y_2^c - \lambda_3\, \max\left[d_1^c\, y_1^c, d_2^c\, y_2^c\right] \right\}$$

is of type (3.100) for $c_1 = c_2 = c$.

All BWD models presented up to here have a singular component. A BWD being absolutely continuous is given by LEE (1979):

$$R_{1,2}(y_1, y_2) = \exp\left\{ -(\lambda_1\, y_1^c + \lambda_1\, y_2^c)^\gamma \right\}; \quad y_1, y_2 \geq 0; \qquad (3.101a)$$

with $\lambda_1, \lambda_2 > 0$, $c > 0$ and $0 < \gamma \leq 1$. The corresponding DF is

$$f_{1,2}(y_1, y_2) = \gamma^2\, c^2\, (\lambda_1\, \lambda_2)^\gamma\, (y_1\, y_2)^{c-1}\left[(y_1\, y_2)^c\right]^{\gamma-1} \times$$
$$\exp\left\{-(\lambda_1\, y_1^c + \lambda_2\, y_2^c)^\gamma\right\}. \qquad (3.101b)$$

The surface plot and the contour plot of (3.101b) in Fig. 3/18 show the continuity of this BWD. The marginals of (3.101a) are all WEIBULL:

$$\left. \begin{aligned} R_1(y_1) &= \exp\{-\lambda_1^\gamma\, y_1^{c\gamma}\}, \\ R_2(y_2) &= \exp\{-\lambda_2^\gamma\, y_2^{c\gamma}\}. \end{aligned} \right\} \qquad (3.101c)$$

The minimum of $Y_1$ and $Y_2$ is also WEIBULL with shape parameter $c\gamma$.

Figure 3/18: Joint density function of the absolutely continuous BWD given by LEE (surface plot and contour plot)

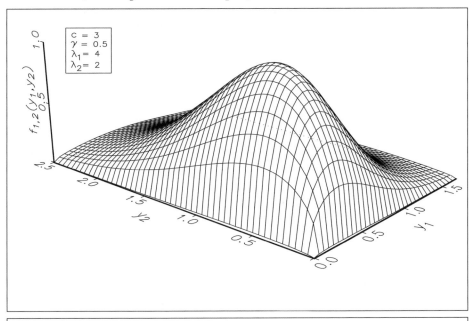

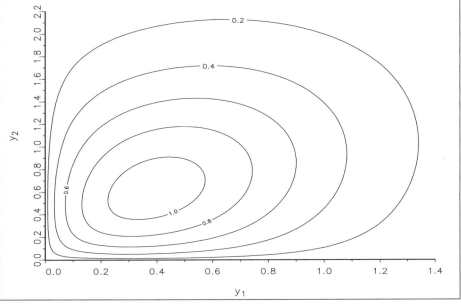

### 3.3.9.2   Multivariate WEIBULL distributions

MARSHALL/OLKIN (1967) have proposed the following multivariate exponential distribution (MVE) for a vector $\boldsymbol{x}$ of $p$ variates giving the joint survival function

$$
\begin{aligned}
R(\boldsymbol{x}) \;=\;& \Pr(X_1 > x_1, X_2 > x_2, \ldots, X_p > x_p) \\
\;=\;& \exp\Bigg\{ -\sum_{i=1}^{p} \lambda_i\, x_i - \sum_{i<j} \lambda_{ij}\, \max[x_i, x_j] - \\
& \sum_{i<j<k} \lambda_{ijk}\, \max[x_i, x_j, x_k] - \ldots - \lambda_{12\ldots p}\, \max[x_2, x_2, \ldots, x_p] \Bigg\}.
\end{aligned}
$$

$$(3.102a)$$

To obtain a more compact notation for (3.102a), let $S$ denote the set of vectors $(s_1, s_2, \ldots, s_p)$ where each $s_j = 0$ or $1$ but $(s_1, s_2, \ldots, s_p) \neq (0, 0, \ldots, 0)$. For any vector $\boldsymbol{s} \in S$, $\max(s_i\, x_i)$ is the maximum of the $x_i$'s for which $s_i = 1$. Thus,

$$
R(\boldsymbol{x}) = \exp\Bigg\{ -\sum_{\boldsymbol{s} \in S} \lambda_{\boldsymbol{s}}\, \max(s_i\, x_i) \Bigg\}. \tag{3.102b}
$$

The zero–one–variable $s_j$ in $\boldsymbol{s}$ indicates which component(s) of the $p$–component system is (are) simultaneously hit by a POISSON process with shock rate $\lambda_{\boldsymbol{s}}$. For example, for $p = 3$ we get

$$
\begin{aligned}
R(x_1, x_2, x_3) \;=\;& \exp\big\{ -\lambda_{100}\, x_1 - \lambda_{010}\, x_2 - \lambda_{001}\, x_3 \\
& -\lambda_{110}\, \max[x_1, x_2] - \lambda_{101}\, \max[x_1, x_2] - \lambda_{011}\, \max[x_2, x_3] \\
& -\lambda_{111}\, \max[x_1, x_2, x_3] \big\}.
\end{aligned}
$$

The $k$–dimensional marginals ($k = 1, 2, \ldots, p - 1$) are all exponential. But the MVE of dimension $k \geq 2$ is not absolutely continuous because a singular part is present. At least one of the hyperplanes $x_i = x_j$ ($i \neq j$); $x_i = x_j = x_k$ ($i, j, k$ distinct), etc., has a positive probability.

Introducing the power transformations $X_i = Y_i^{c_i}$; $i = 1, \ldots, p$; gives a first multivariate WEIBULL distribution — MWD for short — with joint survival function

$$
R(\boldsymbol{y}) = \exp\Bigg\{ -\sum_{\boldsymbol{s} \in S} \lambda_{\boldsymbol{s}}\, \max\left(s_i\, y_i^{c_i}\right), \quad \boldsymbol{y} \geq \boldsymbol{o}. \tag{3.103a}
$$

This MWD is not absolutely continuous but has WEIBULL marginals of all dimensions $k = 1, 2, \ldots, p$. HANAGAL (1996) discusses the following special case of (3.103a):

$$
R(\boldsymbol{y}) = \exp\Bigg\{ -\sum_{i=1}^{p} \lambda_i\, y_i^{c} - \lambda_0 \left( \max[y_1, y_2, \ldots, y_p] \right)^{c} \Bigg\}. \tag{3.103b}
$$

For $\lambda_0 = 0$ the joint distribution is the product of $p$ independent one–dimensional WEIBULL distributions. The one–dimensional marginals of (3.103b) are

$$R(y_i) = \exp\{-(\lambda_i + \lambda_0)\,y_i^c\}. \qquad (3.103c)$$

The distribution of $T = \min(Y_1, \ldots, Y_n)$, i.e., of the system's lifetime, is

$$R(t) = \exp\left\{-\left(\sum_{i=0}^{p} \lambda_i\right) t^c\right\}. \qquad (3.103d)$$

A MWD which is absolutely continuous and which has WEIBULL marginals of all dimensions has been proposed by HOUGAARD (1986) and later on by ROY/MUKHERJEE (1988). The joint survival function is

$$R(\boldsymbol{y}) = \exp\left\{\left[-\sum_{i=1}^{p} \lambda_i\,y_i^c\right]^{\nu}\right\}; \quad \lambda_i > 0; \; c, \nu > 0; \; y_i \geq 0. \qquad (3.104)$$

$T = \min\left(\dfrac{\lambda_1}{a_1}\,Y_1, \ldots, \dfrac{\lambda_p}{a_p}\,Y_p\right)$ has a one–dimensional WEIBULL distribution with shape parameter $c\,\nu$ when $a_i \geq 0$ and such that $\left(\sum_{i=2}^{p} a_i^c\right)^{1/c} = 1$.

CROWDER (1989) has extended (3.104) in two aspects:

- The exponent $c$ is made specific for variate $Y_i$.
- He introduced another parameter $\kappa > 0$.

Thus, the MWD of CROWDER has the joint survival distribution

$$R(\boldsymbol{y}) = \exp\left\{\kappa^{\nu} - \left[\kappa + \sum_{i=1}^{p} \lambda_i\,y_i^{c_i}\right]^{\nu}\right\}. \qquad (3.105)$$

$\kappa = 0$ gives (3.104). The marginal distributions for $Y_i$ are each WEIBULL when $\kappa = 0$ or $\nu = 1$. For general $\kappa$ and $\nu$ the marginal distributions are such that $\left(\kappa + \lambda_i\,Y_i^{c_i}\right)^{\nu} - \kappa^{\nu}$ is exponential with unit mean.

PATRA/DEY (1999) have constructed a class of MWD in which each component has a mixture of WEIBULL distributions. Specifically, by taking

$$Y_i = \sum_{j=1}^{m} a_{ij}\,Y_{ij} \text{ with } Y_{ij} \sim We(0, \lambda_{ij}, c_{ij}); \;\; i = 1, 2, \ldots, p, \qquad (3.106a)$$

where $(a_{i1}, \ldots, a_{im})$ is a vector of mixing probabilities ($a_{ij} \geq 0 \,\forall\, i, j$ and $\sum_{j=1}^{m} a_{ij} = 1$) and by further taking an exponentially distributed variate $Z$ with DF

$$f(z) = \lambda_0\,\exp\{-\lambda_0\,z\}, \qquad (3.106b)$$

where the $Y_i$ and $Z$ are all independent, they consider the multivariate distribution of

$$X_i = \min(Y_i, Z); \;\; i = 1, \ldots, p.$$

The joint survival function of $(X_1, \ldots, X_p)$ is

$$
\begin{aligned}
R(\boldsymbol{x}) &= \prod_{i=1}^{p} \Pr(X_i > x_i) \left[\Pr(Z > x_0)\right]^{1/p} \\
&= \prod_{i=1}^{p} \sum_{j=1}^{m} a_{ij} \exp\left\{-\left(\lambda_{ij}\, x_i^{c_{ij}} + \frac{\lambda\, x_0}{r}\right)\right\},
\end{aligned}
\tag{3.106c}
$$

where $x_0 = \max(x_1, \ldots, x_p) > 0$. The special case $m = p = 2$ is studied by PATRA/DEY in some detail.

### 3.3.10  Miscellaneous

In this last section of Chapter 3 we will list some relatives to the WEIBULL distribution which do not fit into one of the classes or models mentioned above.

The model proposed by VODA (1989) — called **pseudo–WEIBULL distribution** — has the DF

$$
g(y \mid b, c) = \frac{c}{b\,\Gamma\!\left(1 + \frac{1}{c}\right)} \left(\frac{y}{c}\right)^c \exp\left\{-\left(\frac{y}{c}\right)^c\right\}; \quad b, c, y > 0.
\tag{3.107a}
$$

(3.107a) is easily recognized to be

$$
g(y \mid b, c) = \frac{y}{\mu^*}\, f(y \mid 0, b, c),
\tag{3.107b}
$$

where $\mu^*$ is the mean and $f(y \mid 0, b, c)$ is the DF (2.8) of a conventional WEIBULL distribution. The CDF belonging to (3.107a) is given by

$$
\begin{aligned}
G(y \mid b, c) &= \int_0^y g(y \mid b, c)\, \mathrm{d}u \\
&= \frac{\Gamma_{\sqrt{b^c y}}\left(1 + \frac{1}{c}\right)}{\Gamma\!\left(1 + \frac{1}{c}\right)}
\end{aligned}
\tag{3.107c}
$$

and — because of its dependence on the incomplete gamma function $\Gamma.(\cdot)$ (see the excursus on the gamma function in Sect. 2.9.1) — has to be evaluated by numerical integration. The moments are given by

$$
\mathrm{E}(Y^r) = \frac{b^{r/c}\,\Gamma\!\left(1 + \frac{r+1}{c}\right)}{\Gamma\!\left(1 + \frac{1}{c}\right)}.
\tag{3.107d}
$$

We have

$$
\mathrm{E}(Y) = \frac{b^{1/c}\,\Gamma(1 + 2/c)}{\Gamma(1 + 1/c)},
\tag{3.107e}
$$

$$
\mathrm{Var}(Y) = b^{2/c} \left\{\frac{\Gamma(1 + 3/c)}{\Gamma(1 + 1/c)} - \frac{\Gamma^2(1 + 2/c)}{\Gamma^2(1 + 1/c)}\right\}.
\tag{3.107f}
$$

(3.107a) gives a gamma density — see (3.24) — with $b$ and $d = 2$:

$$
f(x \mid b, 2) = \frac{x}{b^2} \exp(-x/b).
$$

But the pseudo–WEIBULL distribution is a generalization neither of the gamma distribution nor of the WEIBULL distribution.

FREIMER et al. (1989) developed what they call an **extended WEIBULL distribution** which rests upon a generalization of the percentile function $x_P = a + b \left[ -\ln(1 - P) \right]^{1/c}$ of the conventional three–parameter WEIBULL distribution:[37]

$$
u_P = \begin{cases} d \left\{ \left[ -\ln(1 - P) \right]^{1/d} - 1 \right\} & \text{for} \quad d \neq 0 \\ \ln \left[ -\ln(1 - P) \right] & \text{for} \quad d = 0. \end{cases} \tag{3.108a}
$$

Inverting, i.e., solving for $P = F(u)$, gives the CDF

$$
F(u \mid d) = \begin{cases} 1 - \exp \left\{ -\left( 1 + \dfrac{u}{d} \right)^d \right\} & \text{for} \quad d \neq 0 \\ 1 - \exp \left\{ -e^u \right\} & \text{for} \quad d = 0. \end{cases} \tag{3.108b}
$$

The support of $F(u \mid d)$ depends on $d$ as follows:
- $d < 0$ gives $-\infty < u < -d$,
- $d > 0$ gives $-d < u < \infty$,
- $d = 0$ gives $-\infty < u < \infty$.

SRIVASTAVA (1989) calls his special version a **generalized WEIBULL distribution**. It has some resemblance to a WEIBULL model with time–depending parameters (see Sect. 3.3.8.1). The general CDF

$$
F(x) = 1 - e^{-\Psi(x)} \tag{3.109}
$$

with non–decreasing function $\Psi(x)$ where $\Psi(0) = 0$ and $\Psi(\infty) = \infty$. The special case

$$
\Psi(x) = \left( \frac{x - a}{b} \right)^c; \quad a \in \mathbb{R};\ b, c > 0
$$

yields the conventional three–parameter WEIBULL distribution.

In Fig. 2/8 we have presented the WEIBULL-probability–paper. On this paper the following transformation of the conventional three–parameter WEIBULL CDF

$$
\ln \left\{ -\ln \left[ 1 - F(y \mid a, b, c) \right] \right\} = -c \ln b + c \ln(x - a) \tag{3.110}
$$

gives a straight line. The SLYMEN/LACHENBRUCH (1984) model rests upon a generalization of the WEIBULL transformation (3.110):

$$
\ln \left\{ -\ln \left[ 1 - F(x) \right] \right\} = \alpha + \beta \, w(x), \tag{3.111a}
$$

where $w(x)$ is a monotonically increasing function depending on one or more parameters satisfying

$$
\lim_{x \to 0} w(x) = -\infty \quad \text{and} \quad \lim_{x \to \infty} w(x) = \infty.
$$

---

[37] See also the extensions given in (3.78a) – (3.79b).

They discussed the special function

$$w(x) = \frac{x^\nu - x^{-\nu}}{2\,\nu}; \quad x, \nu \geq 0. \tag{3.111b}$$

Re–transformation of (3.111a) using (3.111b) gives

$$F(x \mid \alpha, \beta, \nu) = 1 - \exp\left\{-\exp\left[\alpha + \frac{\beta\left(x^\nu - x^{-\nu}\right)}{2\,\nu}\right]\right\}, \quad x \geq 0. \tag{3.111c}$$

The corresponding HR is

$$h(x \mid \alpha, \beta, \nu) = \frac{\beta}{2}\left(x^\nu + x^{-\nu-1}\right)\exp\left\{\alpha + \frac{\beta\left(x^\nu - x^{-\nu}\right)}{2\,\nu}\right\}, \tag{3.111d}$$

which is monotonically increasing when

$$\beta > \frac{2\left[(\nu+1)\,x^{-\nu-2} - (\nu-1)\,x^{\nu+2}\right]}{(x^{\nu-1} + x^{-\nu-1})^2};$$

otherwise it may start with a decreasing part but finally (for $x$ great) it will increase.

# 4  WEIBULL processes and WEIBULL renewal theory

This chapter contains the description of two types of stochastic processes, both involving a WEIBULL distribution.[1] These processes are motivated by the statistical enquiry into repairable systems. When upon the failure of a system it is repaired in such a manner that its reliability is just like that of a brand new system, the stochastic behavior of such a system is modeled by a **WEIBULL renewal process**, for the time to failure being WEIBULL (see Sect. 4.4). ASCHER/FEINGOLD (1984) recommended the phrase "same–as–new" to name a renewal process. If the system's reliability does not change after a repair, i.e. the repaired system is in the same condition after the repair as just before the failure, then the appropriate model is a **non–homogeneous POISSON process**, NHPP for short. The **WEIBULL process** (see Sect. 4.3), is such a NHPP with an intensity function which is of WEIBULL hazard rate type. The term "minimal repair" has also been used to describe the effect of a failure and subsequent repair on a system modeled by an NHPP. ASCHER/FEINGOLD recommend to use the phrase "same–as–old" to characterize this kind of model.

For a better understanding and classification of the processes mentioned above, we first give a concise overview on stochastic processes (Sect. 4.1). This is followed in Sect. 4.2 by a short report on POISSON processes, especially on the **homogeneous POISSON process**, HPP for short. Results for the HPP may easily be reformulated so that they pertain to an NHPP.

## 4.1  Stochastic[2] processes — An overview

According to the most general definition, a stochastic[3] process is any family of random variables. Synonyms are "chance process" or "random process." More precisely, the following **definition** holds

Let $[\Omega, \mathcal{A}, P]$ be a probability space and $\mathbb{T} \in \mathbb{R}$ an arbitrary, but non–random index set. A function $X$ that maps $\Omega \times \mathbb{T}$ into $\mathbb{R}$, i.e., $X : \Omega \times \mathbb{T} \to \mathbb{R}$, is called a one–dimensional, real–valued **stochastic process**. The set of realization of $X$, $\{X(t, \omega) \,|\, \omega \in \Omega, \ t \in \mathbb{T}\}$ is termed the **state space**, whereas $\mathbb{T}$ is called the **parameter space**.  ■

Two rather simple classifications adhere to these last two spaces. We have

- a **point process** for a countable state space and

---

[1]  The inference of these processes, i.e., of the WEIBULL process, will be presented in Chapter 18.

[2]  Suggested reading for this section: BARTLETT (1962), DOOB (1953), FELLER (1966), KARLIN (1969), PARZEN (1962), ROSS (1980).

[3]  The Greek word $\sigma\tau o\chi\alpha\sigma\tau\iota\kappa o\varsigma$ means "to guess."

- a **diffusion process** (real valued process) for a state space being $\mathbb{R}$ or any compact subset of $\mathbb{R}$.

With respect to the parameter space $\mathbb{T}$, we have

- a **discrete parameter process**, sometimes called a **stochastic series** or a **stochastic chain** for $\mathbb{T}$ being denumerable and

- a **continuous parameter process** or **continuous process** for short when $\mathbb{T}$ is non–countable.

We get

- a fixed number $X(t^*, \omega^*)$ for a fixed $t^*$ and a fixed $\omega^*$,

- a random variable $X(t^*, \omega)$ for a fixed $t^*$, i.e., in the cross section, with a distribution function generally depending on $t^*$,

- a sequence path, a trajectory or a realization $X(t, \omega^*)$ of the process for a fixed $\omega^*$ in the longitudinal section,

- the whole process for $t$ and $\omega$ both varying.

A stochastic process is thus interpretable in two ways, either as a collection, an ensemble of functions $(X_t)_{t \in \mathbb{T}}$, from which one is chosen at random by realizing a special $\omega^*$, or as — in the case of $\mathbb{T}$ being denumerable — a series of random variables $X_1, X_2, \ldots, X_t, \ldots$.

In most applications the parameter $t$ represents time, but it may also be the altitude, the latitude, the longitude or the coordinates on a plane or in a solid.[4]

The distinguishing features of a stochastic process $\{X_t\}$ are the relationships among the random variables $X_t$, $t \in \mathbb{T}$. These relationships are specified by giving the joint distribution function of every finite family $X_{t_1}, \ldots, X_{t_n}$ of variables of the process. The joint distribution can often be specified in terms of other distributions associated with the process. A stochastic process may be considered as well defined once its state space, index parameter and family of joint distributions are prescribed. However, in dealing with continuous parameter processes certain difficulties arise, which will not be discussed here.

We now describe some of the classical types of stochastic processes characterized by different dependence relationships among $X_t$. Unless otherwise stated, we take $\mathbb{T} = \mathbb{R}$.

### Processes with independent increments

If the random variables

$$X_{t_2} - X_{t_1}, \ X_{t_3} - X_{t_2}, \ \ldots, \ X_{t_n} - X_{t_{n-1}}$$

are independent for all choices of $t_1, \ldots, t_n$ satisfying

$$t_1 < t_2 < \ldots < t_n,$$

then we say $\{X_t\}$ is a process with **independent increments**. If the index set contains a smallest index $t_0$, it is also assumed that

$$X_{t_0}, \ X_{t_1} - X_{t_0}, \ X_{t_2} - X_{t_1}, \ \ldots, \ X_{t_n} - X_{t_{n-1}}$$

---

[4] When $t$ is a vector we rather speak of a **random field**.

are independent. If the index set is discrete, $\mathbb{T} = \{0, 1, 2, \ldots\}$, then a process with independent increments reduces to a sequence of random variables $Z_0 = X_0$, $Z_i = X_i - X_{i-1}$ ($i = 1, 2, 3, \ldots$), in the sense that knowing the individual distributions of $Z_0$, $Z_1$, $\ldots$ enables one to determine the joint distribution of any finite set of the $X_i$. In fact,

$$X_i = Z_0 + Z_1 + \ldots + Z_i; \quad i = 0, 1, 2, \ldots$$

## Martingales

Let $\{X_t\}$ be a real–valued process with discrete or continuous parameter set. We say $\{X_t\}$ is a **martingale**[5] if, for any $t_1 < t_2 < \ldots < t_{n-1}$

$$E(X_{t_{n+1}} \mid X_{t_1} = x_1, \ldots, x_{t_n} = x_n) = x_n$$

for all values of $x_1, \ldots, x_n$. It is easily verified that the process $X_n = Z_1 + \ldots + Z_n$; $n = 1, 2, \ldots$; is a discrete time martingale if the $Z_i$ are independent with means equal to zero. Similarly, if $X_t$, $0 \le t < \infty$ has independent increments, whose means are zero, then $\{X_t\}$ is a continuous time martingale.

## MARKOV processes

A MARKOV process, named after A.A. MARKOV (1856 – 1922), is a process with the property that, given the value of $X_t$, the values of $X_s$, $s > t$, do not depend on the values of $X_u$, $u < t$; i.e., the probability of any particular future behavior of the process, when its present state is known exactly, is not altered by additional knowledge concerning its past behavior. In formal terms a process is said to be **Markovian** if

$$\Pr\big(a < X_t \le b \mid X_{t_1} = x_1, X_{t_2} = x_2, \ldots, X_{t_n} = x_n\big) = \Pr\big(a < X_t \le b \mid X_{t_n} = x_n\big), \tag{4.1a}$$

whenever $t_1 < t_2 < \ldots < t_n < t$. Let $A$ be an interval on the real line. The function

$$P(x, s; t, A) = \Pr(X_t \in A \mid X_s = x), \quad t > s, \tag{4.1b}$$

is called the **transition probability function**. It is basic to the study of the structure of MARKOV processes. (4.1b) may be expressed as follows:

$$\Pr(a < X_t \le b \mid X_{t_1} = x_1, X_{t_2} = x_2, \ldots, X_{t_n} = x_n) = \Pr(x_n, t_n; t, A), \tag{4.1c}$$

when $A = \{\xi \mid a < \xi \le b\}$. The probability distribution of $(X_{t_1}, X_{t_2}, \ldots, X_{t_n})$ can be computed in terms of (4.1b) and the initial distribution function of $X_{t_1}$. A **MARKOV chain** has a discrete index set and a countable state space.

## Stationary processes

A stationary process $\{X_t\}$ for $t \in \mathbb{T}$, when $\mathbb{T}$ could be one of the sets $(-\infty, \infty)$, $[0, \infty)$, $\{\ldots, -1, 0, 1, \ldots\}$ or $\{1, 2, \ldots\}$, is said to be **strictly stationary** if the joint distribution function of the families of random variables $(X_{t_1+h}, X_{t_2+h}, \ldots, X_{t_n+h})$ and

---

[5] The term *martingale* comes from the Provençal name of the French community Martigues. It has a long history in a gambling context, where originally it meant a system for recouping losses by doubling the stake after each loss. In modern statistics the concept still has to do with gambling and means a "fair" game.

$(X_{t_1}, X_{t_2}, \ldots, X_{t_n})$ are the same for all $h > 0$ and arbitrary selections $t_1, t_2, \ldots, t_n$ of $\mathbb{T}$. The distribution functions are invariant with respect to a translation of all time coordinates. This condition asserts that in essence the process is in probabilistic equivalence and that the particular times at which we examine the process are of no relevance.

A stochastic process $\{X_t\}$ is said to be **wide sense stationary** or **covariance stationary** if it possesses finite second moments and if

$$\mathrm{Cov}(X_t, X_{t+h}) = \mathrm{E}(X_t\, X_{t+h}) - \mathrm{E}(X_t)\,\mathrm{E}(X_{t+h})$$

depends only on $h$ for all $t \in \mathbb{T}$.

**Remark:**   A MARKOV process is said to have stationary transition probabilities if $P(x, s;\ t, A)$ defined in (4.1b) is a function only of $t - s$. Remember that $P(x, s;\ t, A)$ is a conditional probability, given the present state. Therefore, there is no reason to expect that a MARKOV process with stationary transition probabilities is a stationary process, and this is indeed the case.

Besides the four main types discussed above, there are other well–known stochastic processes characterized by special and additional properties. Here, only a brief presentation of point processes will be given because the POISSON process, the WEIBULL process and the WEIBULL renewal process belong to this class.

A **point process** is obtained when one considers a sequence of events occurring in continuous time, individual events, e.g., failures of a system, being distinguished only by their position in time. Let $T_n;\ n = 0, 1, 2, \ldots;$ be a sequence of positive random variables such that

$$T_0 = 0,\ T_n < T_{n+1}\ \text{and}\ T_n \longrightarrow T_\infty \leq \infty,$$

$T_n$ representing the instant of occurrence of the $n$–th event. The total number of events in the interval from 0 to $t$ is a random variable $N_t$ that may be written in the form

$$N_t = \sum_{n=0}^{\infty} I_{\{T_N \leq t\}}, \tag{4.2a}$$

where $I_A$ is the indicator of set $A$. The family $N = (N_t,\ 0 \leq t \leq \infty)$ is a **counting process** and

$$N_t = n\ \text{if}\ T_n \leq t < T_{n+1}. \tag{4.2b}$$

Note that $N$ is an increasing process and its trajectories are right–continuous step functions with upward jumps of magnitude 1. Each of the random sequences $\{T_n\}$ and $\{N_t\}$ is known as a point process. Typical examples include the POISSON processes and the renewal processes. The processes $\{T_n\}$ and $\{N_t\}$ are linked as follows:

$$N_t = \max\{n \,|\, T_n \leq t\},\ N_t = 0\ \text{for}\ t < T_1, \tag{4.3a}$$

$$T_{N_t} \leq t < T_{N_{t+1}}, \tag{4.3b}$$

$$N_t < n \iff T_n > t, \tag{4.3c}$$

$$N_t \leq n \iff T_{n+1} > t. \tag{4.3d}$$

From (4.3c,d) we immediately get

$$\Pr(N_t < n) = \Pr(T_n > t), \tag{4.3e}$$

$$\Pr(N_t \le n) = \Pr(T_{n+1} > t). \tag{4.3f}$$

(4.3e,f) lead to

$$
\begin{aligned}
\Pr(N_t = n) &= \Pr(N_t \le n) - \Pr(N_t < n) \\
&= \Pr(T_{n+1} > t) - \Pr(T_n > t).
\end{aligned}
\tag{4.3g}
$$

The **intensity** of a counting process is a family of random variables $\lambda(t)$ such that relative to a history $(\mathcal{F}_s)$

$$\mathrm{E}\big(N_t - N_s \mid \mathcal{F}_s\big) = \mathrm{E}\left( \int_s^t \lambda(u)\,\mathrm{d}u \mid \mathcal{F}_s \right), \ s \le t, \tag{4.4a}$$

which implies that the random variables

$$M_t = N_t - \int_0^t \lambda(u)\,\mathrm{d}u, \tag{4.4b}$$

form a martingale relative to $(\mathcal{F}_t)$.

## 4.2 POISSON processes[6]

The most popular counting process $\{N_t\}$ is the **homogenous POISSON process**, POISSON process for short, named after S.D. POISSON (1781 – 1840). This process rests upon the following assumptions:

1. The increments $N_{t_0} - N_{t_1}, \ldots, N_{t_n} - N_{t_n-1}$ for $t_0 < t_1 < \ldots < t_n$ are mutually independent; thus, the POISSON process is a process with independent increments.

2. The discrete random variable

$$X_h := N_{t_0+h} - N_{t_0},$$

i.e., the increment in $(t_0, t_0 + h]$, depends only on the interval length $h$ and neither on $t_0$ nor on the value of $N_{t_0}$. The increments in intervals of equal length are thus stationary.

3. The probability of at least one event happening in a time period of length $h$ is essentially proportional to $h$:

$$p(h) := \Pr(X_h \ge 1) = \lambda h + o(h); \ \ h \to 0, \ \lambda > 0. \tag{4.5a}$$

The proportional factor $\lambda$ is the **intensity** of the occurrence of events.[7]

---

[6] Suggested reading for this section is the same as that for Sect. 4.1.

[7] $f(x) = o(x)$, $x \to 0$, is the usual symbolic way of writing the relation $\lim_{x \to 0} f(x)/x = 0$.

4. The probability of two or more events happening in time $h$ is $o(h)$.

The last assumption is tantamount to excluding the possibility of the simultaneous occurring of two or more events.

Let $P_m(t)$ denote the probability that exactly $m$ events occur in an interval of length $t$, i.e.

$$P_m(t) := \Pr(X_t = m); \quad m = 0, 1, 2, \ldots.$$

The fourth assumption from above can be stated in the form

$$\sum_{m=2}^{\infty} P_m(h) = o(h)$$

and clearly

$$p(h) = P_1(h) + P_2(h) + \ldots \tag{4.5b}$$

Because of the assumption of independence (first assumption), we have

$$\begin{aligned} P_0(t+h) &= P_0(t) \cdot P_0(h) \\ &= P_0(t) \left[1 - p(h)\right] \text{ (see (4.5b))} \end{aligned} \tag{4.5c}$$

and therefore

$$\frac{P_0(t+h) - P_0(t)}{h} = -P_0(t) \frac{p(h)}{h} . \tag{4.5d}$$

On the basis of the third assumption we know that $p(h)/h \to \lambda$ for $h \to 0$. Therefore, the probability $P_0(t)$ that the event has not happened during $(0, t]$ satisfies the differential equation

$$P_0(t) = -\lambda P_0(t), \tag{4.5e}$$

whose well–known solution is

$$P_0(t) = c e^{-\lambda t}.$$

The constant $c$ is determined by the initial condition $P_0(0) = 1$, which implies $c = 1$. Thus the complete solution to (4.5e) is

$$P_0 = e^{-\lambda t}. \tag{4.5f}$$

The following excursus shows that for every integer $m$ the probability is given by the **POISSON probability function** $\left(X \sim Po(\lambda)\right)$

$$\Pr(X_t = m) = \frac{(\lambda t)^m}{m!} e^{-\lambda t}; \quad m = 0, 1, 2, \ldots \tag{4.5g}$$

**Excursus:   Construction of the POISSON distribution**

For every $m$, $m \geq 1$, it is easy to see that

$$P_m(t+h) = P_m(t) P_0(h) + P_{m-1}(t) P_1(h) + \sum_{i=2}^{m} P_{m-i}(t) P_i(h). \tag{4.6a}$$

By definition $P_0(h) = 1 - p(h)$. The fourth assumption implies that

$$P_1(h) = p(h) + o(h) \ \text{ and } \ \sum_{i=2}^{m} P_{m-i}(t)\, P_i(h) \leq \sum_{i=2}^{m} P_i(h) = o(h), \tag{4.6b}$$

since obviously $P_i(t) \leq 1$. Therefore, with the aid of (4.6b) we rearrange (4.6a) into the form

$$\begin{aligned} P_m(t+h) - P_m(t) &= P_m(t)\left[P_0(h) - 1\right] + P_{m-1}(t)\, P_1(h) + \sum_{i=2}^{m} P_{m-i}(t)\, P_i(h) \\ &= -P_m(t)\, p(h) + P_{m-1}(t)\, P_1(h) + \sum_{i=2}^{m} P_{m-i}(t)\, P_i(h) \\ &= -P_m(t)\, \lambda h + P_{m-1}(t)\, \lambda h + o(h). \end{aligned} \tag{4.6c}$$

(4.6c) gives

$$\frac{P_m(t+h) - P_m(t)}{h} \to -\lambda\, P_m(t) + \lambda\, P_{m-1}(t) \ \text{ as } \ h \to 0$$

leading to the following combined difference–differential–equations

$$P'_m(t) = -\lambda\, P_m(t) + \lambda\, P_{m-1}(t); \ m = 1, 2, \ldots, \tag{4.6d}$$

subject to the initial conditions

$$P_m(0) = 0; \ m = 1, 2, \ldots$$

In order to solve (4.6d) we introduce the auxiliary functions

$$Q_m(t) = P_m(t)\, e^{\lambda t}; \ m = 0, 1, 2, \ldots \tag{4.6e}$$

Substituting (4.6e) into (4.6d) gives

$$Q'_m(t) = \lambda\, Q_{m-1}(t); \ m = 0, 1, 2, \ldots, \tag{4.6f}$$

where $Q_0(t) = 1$ and the initial conditions are $Q_m(0) = 0$ for $m = 1, 2, \ldots$ Solving (4.6f) recursively yields

$$\begin{aligned} Q'_1(t) &= \lambda & \Longrightarrow \quad Q_1(t) &= \lambda t + c & \Longrightarrow \quad Q_1(t) &= \lambda t \\[4pt] Q'_2(t) &= \lambda^2 t & \Longrightarrow \quad Q_2(t) &= \frac{\lambda^2 t^2}{2} + c & \Longrightarrow \quad Q_2(t) &= \frac{\lambda^2 t^2}{2} \\[4pt] Q'_3(t) &= \frac{\lambda^3 t^2}{2} & \Longrightarrow \quad Q_3(t) &= \frac{\lambda^3 t^3}{3 \cdot 2} + c & \Longrightarrow \quad Q_3(t) &= \frac{\lambda^3 t^3}{3!} \\[4pt] &\ \vdots & \vdots \qquad &\ \vdots & \vdots \qquad &\ \vdots \\[4pt] Q'_m(t) &= \frac{\lambda^m t^{m-1}}{(m-1)!} & \Longrightarrow \quad Q_m(t) &= \frac{\lambda^m t^m}{m!} + c & \Longrightarrow \quad Q_m(t) &= \frac{\lambda^m t^m}{m!}. \end{aligned}$$

Resolving (4.6e) gives the POISSON probability function

$$P_m(t) = \frac{(\lambda t)^m}{m!}\, e^{-\lambda t}; \ m = 0, 1, 2, \ldots$$

The mean number of occurrences in time $t$ is thus the expectation of the POISSON distribution:

$$E(X_t) = \sum_{m=0}^{\infty} m \frac{(\lambda t)^m}{m!} e^{-\lambda t} = \lambda t, \tag{4.7a}$$

also termed **mean value function**

$$\Lambda(t) := \lambda t. \tag{4.7b}$$

The **intensity function** or **arrival rate**, giving the mean rate of occurrences for $t \to \infty$, is

$$\lambda(t) := \frac{d\Lambda(t)}{dt} = \lambda, \tag{4.7c}$$

which is constant with respect to time. We further have

$$
\begin{aligned}
E(N_t - N_s) &= \int_s^t \lambda(u)\,du \\
&= \Lambda(t) - \Lambda(s) \\
&= \lambda\,(t - s),\ s < t. 
\end{aligned}
\tag{4.7d}
$$

As the events "no occurrence in $(0, t]$" and "$T_1$, the time to the first occurrence, is greater than $t$" are equivalent, (4.5f) also gives

$$\Pr(T_1 > t) = e^{-\lambda t}. \tag{4.8a}$$

The second assumption of the POISSON process states that $X_h$ does not depend on $t_0$, so the point $t_0$ may be taken as the moment of any occurrence and the **inter–arrival time** $Y$ between any two successive events is distributed as $T_1$ in (4.8a), the latter being the CCDF of the **exponential distribution**, $Ex(\lambda)$ for short. With respect to the inter–arrival time $Y$ of the POISSON process we have the following results:

$$
\begin{aligned}
\Pr(Y \le t) &= F_Y(t) = 1 - e^{-\lambda t}, \tag{4.8b} \\
f_Y(t) &= \lambda e^{-\lambda t}, \tag{4.8c} \\
E(Y) &= \lambda^{-1}. \tag{4.8d}
\end{aligned}
$$

$T_n$, the **time of the $n$–th event** for a POISSON process, is thus the sum of $n$ independent, identically and exponentially distributed variables:

$$T_n = \sum_{i=1}^{n} Y_i,\ Y_i \stackrel{iid}{\sim} Ex(\lambda). \tag{4.9a}$$

As is well known, this sum has a **gamma distribution**[8] with DF

$$f_n(t) = \frac{\lambda (\lambda t)^{n-1} \exp\{-\lambda t\}}{\Gamma(n)}; \quad t > 0, \; n = 1, 2, \dots \tag{4.9b}$$

For integer $n$ the CDF can be computed as

$$F_n(t) = 1 - \exp\{-\lambda t\} \sum_{i=0}^{n-1} \frac{(\lambda t)^i}{i!}; \quad t > 0; \; n = 1, 2, \dots \tag{4.9c}$$

The mean time to the $n$–th event is

$$E(T_n) = n/\lambda; \quad n = 1, 2, \dots \tag{4.9d}$$

A fundamental property of the POISSON process is **conditional uniformity**: For each $t$, given that $N_t = n$ and regardless of the rate $\lambda$, the conditional distribution of the arrival times $T_1, \dots, T_n$ is that of order statistics[9] $X_{1:n}, \dots, X_{n:n}$ engendered by independent random variables $X_1, \dots, X_n$, each uniformly distributed on $(0, t]$. From this property many computational relationships can be deduced.

For a **non–homogeneous Poisson process** (NHPP) on $\mathbb{R}^+$, the first assumption of the HPP, the independent increments property, is retained, but the arrival rate (intensity function) $\lambda(t)$ now is a function of $t$. Whereas for a HPP

$$\lim_{h \to 0} \Pr\big[N_{t+h} - N_t = 1 \,|\, N_s \text{ for } s \le t\big] \big/ h = \lambda \tag{4.10a}$$

and

$$\lim_{h \to 0} \Pr\big[N_{t+h} - N_t \ge 2 \,|\, N_s \text{ for } s \le t\big] \big/ h = 0 \tag{4.10b}$$

hold, the arrival–rate function of an NHPP is given by

$$\lim_{h \to 0} \Pr\big[N_{t+h} - N_t = 1 \,|\, N_s \text{ for } s \le t\big] \big/ h = \lambda(t) \tag{4.11a}$$

instead of (4.10a) with (4.10b) remaining as is. For an NHPP the random variable $N_t - N_s$, $t > s$, has a POISSON distribution with parameter (= mean)

$$E(N_t - N_s) = \int_s^t \lambda(u) \, du = \Lambda(t) - \Lambda(s), \tag{4.11b}$$

i.e.,

$$\Pr(N_t - N_s = m) = \frac{[\Lambda(t) - \Lambda(s)]^m}{m!} e^{-[\Lambda(t) - \Lambda(s)]}; \quad m = 0, 1, 2, \dots \tag{4.11c}$$

---

[8] A gamma distribution with integer value $n$ is also termed ERLANG distribution, named after the Danish engineer A.K. ERLANG (1878 – 1929).

[9] For order statistics see Chapter 5.

The NHPP is appropriate for situations in which the independent property holds but that of stationary increments fails. Such a situation may be encountered with a repairable system involving changes in its reliability as it ages. For example, when a complex system is in the development stage, early prototypes will often contain design flaws. During the early testing phase, design changes are made to correct such problems. If the development program is succeeding, one would expect a tendency toward longer times between failures. When this occurs, such systems are said to be undergoing **reliability growth**; see DUANE (1964) or CROW (1974). On the other hand, if a deteriorating system is given only **minimal repairs** when it fails, one would expect a tendency toward shorter times between failures as the system ages. If the intensity function $\lambda(t)$ is increasing, the times between failures tend to be shorten, and if it is decreasing, they tend to be longer.

We will denote the mean function giving the expected number of events of an NHPP in $(0, t]$ by

$$\Lambda(t) = \mathrm{E}(N_t) = \int_0^t \lambda(u)\, \mathrm{d}u. \tag{4.12}$$

$\Lambda(t)$ is

- a strictly increasing continuous function,
- defined for non–negative $t$ and
- satisfying $\Lambda(0) = 0$ and $\Lambda(\infty) = \infty$.

Some common examples of a mean function are

1. $\Lambda(t) = \lambda\,t, \;\; \lambda > 0$, for a HPP;

2. $\Lambda(t) = \left(\dfrac{t}{b}\right)^c ; \;\; b, c > 0$, for a WEIBULL process (see Sect. 4.3);

3. $\Lambda(t) = a\,\ln(1 + b\,t); \;\; a, b > 0$;

4. $\Lambda(t) = -\ln\left\{ \Phi\big[ -p\,\ln(\lambda t)\big]\right\}$, where $\Phi[.]$ is the standard GAUSSIAN CDF and $p > 0$;

5. $\Lambda(t) = a\left[\exp(b\,t) - 1\right]$, which is related to the GUMBEL–extreme–value density.

Let $T_1, T_2, \ldots$ be the waiting times or arrival times of an NHPP, and let

$$Y_i = T_i - T_{i-1}; \;\; T_0 = 0; \;\; i = 1, 2, \ldots$$

be the $i$–th inter–arrival time. Some special relations exist between the distribution function $G_1(t)$ of $Y_1 \equiv T_1$ and $\Lambda(t)$, which characterize the law of an NHPP; see PARZEN (1962, p. 138):

$$G_1(t) \;=\; 1 - \exp\{-\Lambda(t)\}, \;\; t \geq 0, \tag{4.13a}$$

$$\lambda(t) \;=\; \frac{\mathrm{d}\Lambda(t)}{\mathrm{d}t} = \frac{G_1'(t)}{1 - G_1(t)}, \;\; t \geq 0. \tag{4.13b}$$

The cumulative distribution of $Y_i$, given that the $(i-1)$–th event occurred at time $t_{i-1}$, is

$$G_i\big(t\,|\,t_{i-1}\big) = \frac{G_1\big(t + t_{i-1}\big) - G_1\big(t_{i-1}\big)}{1 - G_1\big(t_{i-1}\big)}\; ;\quad i = 2,3,\ldots \qquad (4.13c)$$

Hence, in an NHPP the times between successive events are not independent and not identically distributed.

## 4.3   WEIBULL processes[10]

A special NHPP is the **Weibull process**. In the literature this model has been called by many different names:

- **power law process** because of the form of its mean–value function (4.14a), see RIGDON/BASU (1989);
- **RASCH–WEIBULL process**, see MØLLER (1976);
- **WEIBULL intensity function**, see CROW (1974);
- **WEIBULL–POISSON process**, see BAIN/ENGELHARDT (1991a).

The WEIBULL process has the mean–value function

$$\Lambda(t) = \left(\frac{t}{b}\right)^c\; ;\quad t \geq 0;\; b, c > 0, \qquad (4.14a)$$

which is strictly monotonically increasing, and the intensity function

$$\lambda(t) = \frac{\mathrm{d}\Lambda(t)}{\mathrm{d}t} = \frac{c}{b}\left(\frac{t}{b}\right)^{c-1}. \qquad (4.14b)$$

The notions of intensity function and hazard rate, both having the same formula, should not be confused with one another. The latter is a relative value of failure for non–repairable systems, whereas the former is an absolute rate of failure for repairable systems. We further want to stress that what is WEIBULL distributed in a WEIBULL process is $T_1$, the time to the first occurrence or first failure, whereas $T_2, T_3, \ldots$ and the inter–arrival times $Y_i = T_i - T_{i-1}$ for $i \geq 2$ are not WEIBULL.

Before listing the probability formulas of some random variables associated with the WEIBULL process, we give a **characterization**; see MØLLER (1976).

A necessary and sufficient condition for a NHPP with strictly monotonically increasing mean–value function $\Lambda(t)$ to be a WEIBULL process is:
For every $n \geq 2$ the stochastic vector

$$(T_1/T_n,\ T_2/T_n,\ \ldots,\ T_{n-1}/T_n)$$

is stochastically independent of $T_n$. ∎

**Proof:** An NHPP can be transformed to an HPP by changing the time scale with the aid of the mean–value function $z := \Lambda(t)$ because the stochastic process $\{M(z);\ z \geq 0\}$, defined

---

[10]   Suggested reading for this section: BAIN/ENGELHARDT (1991a), CROW (1974), LEE/BELL/MASON (1988), MØLLER (1976), MURTHY/XIE/JIANG (2004), PARZEN (1962), RIGDON/BASU (1962).

by $M(z) = N[\Lambda^{-1}(z)]$, is an HPP with intensity 1. The relation between the arrival–time process $\{U_n; n = 1, 2, \ldots\}$ corresponding to $\{M(z); z \geq 0\}$ and the arrival–time process $\{T_n; n = 1, 2, \ldots\}$ from the NHPP is given by $U_n = \Lambda(T_n)$. Therefore, for all $n \geq 2$, the conditional distribution of $(\Lambda(T_1), \ldots, \Lambda(T_{n-1}))$, given that $T_n = t_n$, will be the same as the distribution of $n - 1$ ordered, independent observations from a rectangular distribution over the interval $(0, \Lambda(T_n))$, as shown by PARZEN (1962, p. 143). An immediate result is that the conditional distribution of $(T_1/T_n, \ldots, T_{n-1}/T_n)$, given $T_n = t_n$, is the same as the distribution of $n - 1$ ordered, independent observations from the distribution

$$F(x) = \frac{\Lambda(x\, t_n)}{\Lambda(t_n)} \quad \text{for} \ \ x \in (0, 1]. \tag{4.15}$$

Thus, $(T_1/T_n, \ldots, T_{n-1}/T_n)$ is stochastically independent of $T_n$ precisely in the case where the function $\Lambda(x\, t_n)/\Lambda(t_n)$ is independent of $t_n$ or, in other words, where $\Lambda(.)$ satisfies the functional equation $\Lambda(x\, t)\, \Lambda(1) = \Lambda(x)\, \Lambda(t)$. The positive, continuous, non–decreasing solution of this functional equation, under the condition that $\Lambda(t)$ need only be defined for the set of positive real numbers, is given by $\Lambda(t) = (t/b)^c$, where $b$ and $c$ are positive constants. ∎

We now turn to some random variables which come up in a WEIBULL process.

1. **Number of events $N_t$**

In any interval $(s, t]$ we have

$$\Pr(N_t - N_s = m) = \frac{\left(\dfrac{t^c - s^c}{b^c}\right)^m}{m!} \exp\left\{-\frac{t^c - s^c}{b^c}\right\}; \quad m = 0, 1, 2, \ldots; \tag{4.16a}$$

$$\mathrm{E}(N_t - N_s) = \frac{t^c - s^c}{b^c}. \tag{4.16b}$$

Especially for the interval $(0, t]$ we get

$$\Pr(N_t = m) = \frac{(t/b)^{c\, m}}{m!} \exp\left\{-\left(\frac{t}{b}\right)^c\right\}, \tag{4.16c}$$

$$\mathrm{E}(N_t) = \left(\frac{t}{b}\right)^c. \tag{4.16d}$$

From (4.16d) we have

$$\mathrm{E}\left[\frac{N_t}{t}\right] = \frac{t^{c-1}}{b^c}. \tag{4.16e}$$

On taking the logarithm of both sides, we get the following linear equation:

$$y = (c - 1)\, x - c \ln b, \tag{4.16f}$$

where

$$y = \ln\left\{\mathrm{E}\left[\frac{N_t}{t}\right]\right\} \quad \text{and} \ \ x = \ln t.$$

(4.16f) is the basis for a graphical approach to estimate the parameters $b$ and $c$ from $k$ observations $(y_i = \ln\{n_i/t_i\}, \; x_i = \ln t_i), \; i = 1, 2, \ldots, k$, where $n_i$ is the realization of $N_{t_i}$; see DUANE (1964) and Sect. 18.4.

If at given points in time $t_1, \ldots, t_r$ the corresponding numbers of events are $N_{t_1}, \ldots, N_{t_r}$, the joint distribution of successive differences of the latter variables is given by

$$\Pr\left[N_{t_1} = n_1, N_{t_2} - N_{t_1} = n_2, \ldots, N_{t_r} - N_{t_{r-1}} = n_r\right]$$

$$= \prod_{i=1}^{r} \left(\frac{t_i^c - t_{i-1}^c}{b^c}\right)^{n_i} \exp\left\{-\frac{t_i^c - t_{i-1}^c}{b^c}\right\} \bigg/ n_i!$$

$$= \binom{n}{n_1 \ldots n_r} \frac{\exp\left\{-\left(\frac{t_r}{b}\right)^c\right\}}{n! \, b^{cn}} \prod_{i=1}^{r} \left(t_i^c - t_{i-1}^c\right)^{n_i}, \qquad (4.16g)$$

where $t_0 = 0$ and $n = \sum_{i=1}^{r} n_i$.

2. **Arrival time $T_n$**

Since the transformed arrival times

$$V_i = \left(\frac{T_i}{b}\right)^c; \quad i = 1, 2, \ldots, n \qquad (4.17a)$$

have the same distributions as the first $n$ arrival times from an HPP with intensity one, see (3.12b,c), it follows that $T_n$ conforms to a **generalized gamma distribution** with density

$$f_n(t) = \frac{1}{(n-1)!} \frac{c}{b} \left(\frac{t}{b}\right)^{nc-1} \exp\left\{-\left(\frac{t}{b}\right)^c\right\}. \qquad (4.17b)$$

The corresponding CDF is

$$F_n(t) = \gamma\left(n \left| \left(\frac{t}{b}\right)^c\right.\right) \bigg/ \Gamma(n), \qquad (4.17c)$$

i.e., an incomplete gamma function ratio. We get

- the common two–parameter WEIBULL distribution for $n = 1$ and
- the gamma distribution for $c = 1$.

The mean of $T_n$ is

$$E(T_n) = \frac{b \, \Gamma\left(n + \frac{1}{c}\right)}{\Gamma(n)}, \qquad (4.17d)$$

which reduces to $n \, b$ for $c = 1$, see (4.9d), and to $b \, \Gamma(1 + 1/c)$ for $n = 1$, see (2.85).

3. **Conditional arrival times**

The arrival times $T_1, T_2, \ldots, T_{n-1}$, conditioned on $T_n = t_n$, are distributed as $n - 1$ order statistics from a distribution with CDF

$$
F(t) = \left\{
\begin{array}{lll}
0 & \text{for} & t \leq 0, \\[2mm]
(t/t_n)^c & \text{for} & 0 < t \leq t_n, \\[2mm]
1 & \text{for} & t > t_n.
\end{array}
\right\} \tag{4.18}
$$

4. **Inter–arrival time $Y_i$**

Using (4.13a–c) we find the following CDF of the inter–arrival times $Y_i = T_i - T_{i-1}$; $i = 2, 3, \ldots$, given that $(i - 1)$–th event occurred at time $t_{i-1}$:

$$
G_i(y \,|\, t_{i-1}) = 1 - \exp\left\{ -\left(\frac{y + t_{i-1}}{b}\right)^c + \left(\frac{t_{i-1}}{b}\right)^c \right\}; \quad i = 2, 3, \ldots \tag{4.19}
$$

This is a left–truncated WEIBULL distribution, the truncation point being $t_{i-1}$; see (3.47b). As $T_i = T_{i+1} + Y_i$, (4.19) also is the CDF of the conditional arrival time of the $i$–th event, given the arrival time $t_{i-1}$ of the preceding event.

We finally mention two generalizations of the WEIBULL process:

1. The intensity function given in (4.14b) may be enlarged by incorporating one or more covariates; see (3.91a–c).

2. When only every $k$–th event of the WEIBULL process is recorded as an event we get the so–called **modulated WEIBULL process**; see MURTHY et al. (2004). The joint density function for the recorded times $t_1, t_2, \ldots, t_n$ of the first $n$ events is

$$
f(t_1, \ldots, t_n) = \frac{\exp\left\{ -\left(\frac{t_n}{b}\right)^c \right\}}{\left[\Gamma(k)\right]^n} \left(\frac{c}{b}\right)^n \prod_{i=1}^{n} \left(\frac{t_i}{b}\right)^{c-1} \left[\left(\frac{t_i}{b}\right)^c - \left(\frac{t_{i-1}}{b}\right)^c\right]^{k-1}; \quad k = 1, 2, \ldots \tag{4.20}
$$

## 4.4  WEIBULL renewal processes

In this section we first give some general results on renewal theory (Sect. 4.4.1). This is followed by a thorough analysis of the ordinary WEIBULL renewal process (Sect. 4.4.2), ordinary WRP for short.

### 4.4.1  Renewal processes[11]

First, we present, in a rather informal way, the most popular types of renewal processes (RP for short) encountered in theory and practice. Then, we describe those random variables that are associated with every renewal process and that express different aspects of its performance.

---

[11]  Suggested reading for this section: BARLOW/PROSCHAN (1965, 1975), BAXTER/SCHEUER/ BLISCHKE/McCONALOGUE (1981, 1982), FELLER (1966, Vol. II), GNEDENKO/BELJAJEW/ SOLOWJEW (1968), ROSS (1970), SMITH (1958).

An **ordinary renewal process**, also known as **normal** or **simple renewal process**, is a sequence of iid non–negative random variables $Y_1, Y_2, \ldots$ These variates may be the lifetimes of a system or of a system's components which are renewed instantaneously upon failure. Renewal may be a replacement either by a completely new item or by a renewing repair. Unless stated otherwise the replacement or the repair are of negligible duration. An ordinary RP is characterized by the fact that we start the process in $t = 0$ with a new item so that $Y_1$, the time to the first failure, is a non–truncated lifetime. Then, $Y_1, Y_2, \ldots$ all have the same distribution function $G(y) = \Pr(Y_i \leq y)$, called the **underlying distribution**. This is the decisive difference to the counting processes presented in the preceding sections where — with the only possible exception of the HPP — the inter–arrival times $Y_i$ have differing distributions.

The next two types of RPs are characterized by the fact that $Y_i$ is an incomplete lifetime. A **delayed** or **modified renewal process** starts at $t = 0$ with an item that has been on duty for either a known or an unknown span of time, called **backward recurrence time**. So, $Y_1$ is the **remaining** or **residual lifetime**, also called **forward recurrence time** of the item working at $t = 0$. The distribution $G_1(y)$ of $Y_1$ is thus different from the underlying distribution $G(y)$, valid for $Y_i$, $i \geq 2$. A **stationary renewal process** is a special delayed process that has been started a long time before $t = 0$. So, the residual lifetime $Y_1$ of a stationary RP has the limit distribution of the forward recurrence time.

A **superposed renewal process** consists of $k \geq 2$ RPs that are running parallel and independently, an example being a weaving mill, where $k$ weaving machines of the same type are working. In general, the renewal distances of a superposed process are not iid variates. An exception is the superposition of $k$ HPPs resulting in a superposed process which is again of the POISSON type.

An **alternating renewal process** is characterized to be in one of two states at any time. For example, a machine may either be running (= state 1) or be down and under repair (= state 2), and these two states alternate. The sojourn times $Y_i$ and $Z_i$ in state 1 and state 2 have differing distribution functions $G_Y(y)$ and $G_Z(z)$. In $t = 0$ the process starts in state 1 with probability $P$ and in state 2 with probability $1 - P$.

The feature of a **cumulative renewal process** is that some kind of payment $W_i$ is attached to the $i$–th renewal, e.g., the cost of replacement. $W_i$ is a variate which may either be dependent or be independent of $Y_i$. The most interesting feature for a cumulative renewal process is the total payment accrued in some interval $(0, t]$.

There are four random variables associated to a RP (see Fig. 4/1) which are of special interest:

1. $T_n$, the time elapse from $t = 0$ up to and including the moment of the $n$–th renewal, $n = 1, 2, \ldots$ (The item installed in $t = 0$ is no renewal.),

2. $N_t$, the number of renewals in $(0, t]$,

3. $B_t$, the age of an item at any point $t$ (This the backward recurrence time to the preceding renewal.),

4. $F_t$, the residual or remaining lifetime of an item at any point $t$ on the time axis. (This is the forward recurrence time to the following renewal.)

Figure 4/1: Random variables of a renewal process

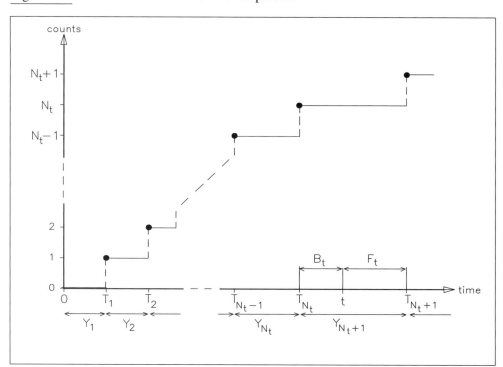

**Time to the $n$–th renewal**

If, upon a failure, the renewal takes place immediately and is completed in a time span of negligible length, the waiting time $T_n$ until the $n$–th renewal is given by

$$T_n = \sum_{i=1}^{n} Y_i, \ \ n \geq 1. \tag{4.21a}$$

$\{T_n\}$ is stochastic process. For a given $n$, $T_n$ is a continuous variable with CDF

$$F_n(t) = \Pr(T_n \leq t) = \Pr(Y_1 + \ldots + Y_n \leq t). \tag{4.21b}$$

$F_n(t)$ may be expressed by the so–called **convolution formula** or **convolution integral**. In the case of a normal RP we have

$$F_n(t) = G^{(n)}(t). \tag{4.21c}$$

$G^{(n)}(t)$ is the $n$–fold convolution of the underlying distribution $G(\cdot)$ with itself and is achieved recursively:

$$G^{(n)}(t) = \int_{0}^{t} G^{(n-1)}(t - u) \, \mathrm{d}G(u); \ \ n = 2, 3, \ldots, \tag{4.21d}$$

with $G^{(1)}(\cdot) = G(\cdot)$. For a delayed RP we have to use $G^{(1)}(\cdot) = G_1(y)$, the CDF of the forward recurrence time. The corresponding density functions are

$$
f_n(t) = g^{(n)}(t) = \left\{ \begin{array}{ll} \int_0^t g^{(n-1)}(t-u)\, g(u)\, \mathrm{d}u & - \quad \text{normal renewal process} \\ \int_0^t g^{(n-1)}(t-u)\, g_1(u)\, \mathrm{d}u & - \quad \text{delayed renewal process.} \end{array} \right\}
$$

(4.21e)

The mean and variance of $T_n$ are simply:

$$
\mathrm{E}(T_n) = \mathrm{E}(Y_1) + (n-1)\,\mathrm{E}(Y), \tag{4.21f}
$$

$$
\mathrm{Var}(T_n) = \mathrm{Var}(Y_1) + (n-1)\,\mathrm{Var}(Y), \tag{4.21g}
$$

where — for a delayed RP — $\mathrm{E}(Y_1)$ and $\mathrm{Var}(Y_1)$ are different from $\mathrm{E}(Y)$ and $\mathrm{Var}(Y)$.

There are only a few underlying distributions $G(\cdot)$ that lead to a simple analytical expression for (4.21c):

- Exponential distribution: $G(t) = 1 - \exp\{-\lambda t\}$; $g(t) = \lambda\,\exp\{-\lambda t\}$

$$
F_n(t) = 1 - \sum_{i=0}^{n-1} \frac{(\lambda t)^i}{i!}\,\exp\{-\lambda t\} \tag{4.22a}
$$

$$
f_n(t) = \lambda\,\frac{(\lambda t)^{n-1}}{(n-1)!}\,\exp\{-\lambda t\} \tag{4.22b}
$$

- ERLANG distribution:

$$
G(t) = 1 - \sum_{j=0}^{c-1} \frac{(\lambda t)^j}{j!}\,\exp\{-\lambda t\};
$$

$$
g(t) = \lambda\,\frac{(\lambda t)^{c-1}}{(c-1)!}\,\exp\{-\lambda t\}
$$

$$
F_n(t) = 1 - \sum_{i=0}^{n\,c-1} \frac{(\lambda t)^i}{i!}\,\exp\{-\lambda t\} \tag{4.23a}
$$

$$
f_n(t) = \lambda\,\frac{(\lambda t)^{n\,c-1}}{(n\,c-1)!}\,\exp\{-\lambda t\} \tag{4.23b}
$$

- Normal distribution: $G(t) = \Phi\left(\dfrac{t-\mu}{\sigma}\right)$;

$$
g(t) = \varphi\left(\frac{t-\mu}{\sigma}\right) = \frac{1}{\sigma\sqrt{2\pi}}\,\exp\left\{-\frac{(t-\mu)^2}{2\sigma^2}\right\}
$$

$$
F_n(t) = \Phi\left(\frac{t-n\,\mu}{\sigma\sqrt{n}}\right) = \int_{-\infty}^{(t-n\,\mu)/(\sigma\sqrt{n})} \frac{1}{\sqrt{2\pi}}\,\exp\left\{-\frac{u^2}{2}\right\}\,\mathrm{d}u \tag{4.24a}
$$

$$
f_n(t) = \frac{1}{\sigma\sqrt{n}}\,\varphi\left(\frac{t-n\,\mu}{\sigma\sqrt{n}}\right) = \frac{1}{\sigma\sqrt{2\pi n}}\,\exp\left\{-\frac{(t-n\,\mu)^2}{2n\,\sigma^2}\right\} \tag{4.24b}
$$

When $n$ is large, one can use a central limit theorem and approximate $F_n(t)$ of a normal RP:

$$F_n(t) \approx \Phi\left(\frac{t - n\,\mathrm{E}(Y)}{\sqrt{n\,\mathrm{Var}(Y)}}\right), \quad n \text{ large.} \tag{4.25}$$

As

$$\max_{1 \le i \le n} Y_i \le \sum_{i=1}^{n} Y_i$$

and regarding a normal RP, we have the following upper bound

$$F_n(t) = \Pr(T_n \le t) \le \Pr\left(\max_{1 \le i \le n} Y_i \le t\right) = [G(t)]^n. \tag{4.26}$$

**Number of renewals in $(0, t]$**

The most important variable in any RP is $N_t$, the number of renewals in $(0, t]$. Note that

$$\begin{aligned} \Pr(N_t \ge k) &= \Pr(T_k \le t) \\ &= F_k(t), \end{aligned} \tag{4.27a}$$

resulting into

$$\begin{aligned} \Pr(N_t = k) &= \Pr(N_t \ge k) - \Pr(N_t \ge k + 1) \\ &= \Pr(T_k \le t) - \Pr(T_{k+1} \le t) \\ &= F_k(t) - F_{k+1}(t); \quad k = 0, 1, \ldots \end{aligned} \tag{4.27b}$$

Then we get

$$\begin{aligned} \Pr(N_t = 0) &= \Pr(T_1 > t) \\ &= 1 - G_1(t). \end{aligned}$$

It is easy to show that $N_t$ has finite moments of all orders:

$$\begin{aligned} \mathrm{E}(N_t^r) &= \sum_{k=0}^{\infty} k^r \left[F_k(t) - F_{k+1}(t)\right] \\ &= \sum_{k=1}^{\infty} \left[k^r - (k-1)^r\right] F_k(t). \end{aligned} \tag{4.27c}$$

The **renewal function**

$$M(t) := \sum_{k=1}^{\infty} F_k(t) \tag{4.27d}$$

giving the mean number of renewals in $(0, t]$ is of special interest. Its derivation with respect to $t$ is known as the **renewal density**:

$$m(t) := \frac{\mathrm{d}M(t)}{\mathrm{d}t} = \sum_{k=1}^{\infty} f_k(t). \tag{4.27e}$$

We obtain an **integral equation for the renewal function** as follows. From (4.21c,d) we have

$$F_{k+1}(t) = G^{(k+1)}(t) = \int_0^t G^{(k)}(t-u)\,\mathrm{d}G(u); \quad k = 0, 1, 2, \ldots, \tag{4.28a}$$

where $F_1(t) = G(t)$ for a normal RP. Replacing $F_{k+1}(t)$ in (4.27d) by (4.28a) leads to

$$M(t) = G(t) + \sum_{k=1}^\infty \int_0^t G^{(k)}(t-u)\,\mathrm{d}G(u) \tag{4.28b}$$

or

$$M(t) = G(t) + \int_0^t M(t-u)\,\mathrm{d}G(u). \tag{4.28c}$$

(4.28c) is known as the **fundamental renewal equation**. For a delayed RP, (4.28c) turns into

$$M_d(t) = G_1(t) + \int_0^t M_d(t-u)\,\mathrm{d}G(u). \tag{4.28d}$$

The renewal density may also be expressed by an integral equation. Upon differentiating (4.28c,d) with respect to $t$, we get

$$m(t) = g(t) + \int_0^t m(t-u)\,g(u)\,\mathrm{d}u, \tag{4.28e}$$

$$m_d(t) = g_1(t) + \int_0^t m_d(t-u)\,g(u)\,\mathrm{d}u. \tag{4.28f}$$

The following excursus shows how to determine higher moments of $N_t$.

---

**Excursus:  Higher moments of $N_t$**

Integral equations for higher moments of $N_t$ may be derived by considering the **binomial moments**

$$M_k(t) := \mathrm{E}\binom{N_t}{k} = \mathrm{E}\left(\frac{N_t\,[N_t-1]\ldots[N_t-k+1]}{1\cdot 2\cdot\ldots\cdot k}\right). \tag{4.29a}$$

We then have

$$\begin{aligned} M_k(t) &= \sum_{j=0}^\infty \binom{j}{k}\left[G^{(j)}(t) - G^{(j+1)}(t)\right] \\ &= \sum_{j=k}^\infty \binom{j-1}{k-1} G^{(j)}(t). \end{aligned} \tag{4.29b}$$

The convolution of $M_k(t)$ and $M(t)$ is

$$
\begin{aligned}
\int_0^t M_k(t-u)\,\mathrm{d}M(u) &= \sum_{j=k}^\infty \sum_{i=1}^\infty \binom{j-1}{k-1} \int_0^t G^{(n)}(t-u)\,\mathrm{d}G(u) \\
&= \sum_{j=k}^\infty \sum_{i=1}^\infty \binom{j-1}{k-1} G^{(j+i)}(t) \\
&= \sum_{p=k+1}^\infty \sum_{q=0}^{p-k-1} \binom{q+k-1}{k-1} G^{(p)}(t); \quad p=i+j,\ q=j-k \\
&= \sum_{p=k+1}^\infty \binom{p-1}{k} G^{(p)}(t) \\
&= M_{k+1}(t).
\end{aligned}
\tag{4.29c}
$$

The moments of $N_t$ may be derived from the $M_k(t)$ by the relationship

$$
\mathrm{E}\big(N_t^k\big) = \sum_{j=1}^k \mathfrak{S}_{(k,j)}\, M_j(t)\, j!,
\tag{4.30a}
$$

where $\mathfrak{S}_{(k,j)}$ are **STIRLING numbers of the second kind**[12] defined by

$$
\mathfrak{S}_{(k,j)} = \frac{1}{j!} \sum_{i=0}^j (-1)^{j-i} \binom{j}{i} i^k
$$

or recursively by

$$
\begin{aligned}
\mathfrak{S}_{(k,1)} &= 1\ \forall\, k \ \text{and}\ \mathfrak{S}_{(k,j)} = 1\ \forall\, k=j;\ k=1,2,\ldots; \\
\mathfrak{S}_{(k,j)} &= j\,\mathfrak{S}_{(k-1,j)} + \mathfrak{S}_{(k-1,j-1)},\ 2\le j\le k, \\
\mathfrak{S}_{(k,j)} &= 0,\ j>k.
\end{aligned}
$$

In particular we get

$$
\begin{aligned}
\mathrm{E}\big(N_t\big) &= M_1(t) = M(t), & (4.30b) \\
\mathrm{E}\big(N_t^2\big) &= M_1(t) + 2\,M_2(t), & (4.30c) \\
\mathrm{E}\big(N_t^3\big) &= M_1(t) + 6\,M_2(t) + 6\,M_3(t), & (4.30d) \\
\mathrm{E}\big(N_t^4\big) &= M_1(t) + 14\,M_2(t) + 36\,M_3(t) + 24\,M_4(t). & (4.30e)
\end{aligned}
$$

Thus, the variance of $N_t$ is

$$
\begin{aligned}
\mathrm{Var}\big(N_t\big) &= \mathrm{E}\big(N_t^2\big) - \big[\mathrm{E}\big(N_t\big)\big]^2 \\
&= 2\,M_2(t) + M(t)\big[1 - M(t)\big].
\end{aligned}
\tag{4.30f}
$$

---

[12] $\mathfrak{S}_{(k,j)}$ is the number of ways of partitioning a set of $k$ elements into $j$ non–empty subsets.

There are only a few underlying distributions which allow a direct evaluation of (4.28c) or (4.28d)

- Exponential distribution; see (4.22a,b)

$$\Pr\big(N_t = k\big) \;=\; F_k(t) - F_{k+1}(t) = \frac{(\lambda\,t)^k}{k!}\, e^{-\lambda\,t} \tag{4.31a}$$

$$M(t) \;=\; \sum_{k=0}^{\infty} k\,\frac{(\lambda\,t)^k}{k!}\, e^{-\lambda\,t} = \lambda\,t \tag{4.31b}$$

$$m(t) \;=\; \lambda \tag{4.31c}$$

- ERLANG distribution; see (4.23a,b)

$$\Pr\big(N_t = k\big) \;=\; e^{-\lambda\,t} \sum_{j=k\,c}^{(k+1)\,c-1} \frac{(\lambda\,t)^j}{j!} \tag{4.32a}$$

$$M(t) \;=\; e^{-\lambda\,t} \sum_{k=1}^{\infty} \sum_{j=k\,c}^{\infty} \frac{(\lambda\,t)^j}{j!}$$

$$\;=\; \frac{1}{c}\left(\lambda\,t + \sum_{j=1}^{c-1} \frac{b^j}{1-b^j}\Big[1 - \exp\big\{-\lambda\,t\,\big(1 - b^j\big)\big\}\Big]\right) \tag{4.32b}$$

where $b := \exp\{2\,\pi\,i\,c\} = \cos(2\,\pi/c) + i\,\sin(2\,\pi/c);\ \ i = \sqrt{-1}.$

- Normal distribution; see (4.24a,b)

$$\Pr\big(N_t = k\big) \;=\; \Phi\!\left(\frac{t - k\,\mu}{\sigma\,\sqrt{k}}\right) - \Phi\!\left(\frac{t - (k+1)\,\mu}{\sigma\,\sqrt{k+1}}\right) \tag{4.33a}$$

$$M(t) \;=\; \sum_{k=1}^{\infty} \Phi\!\left(\frac{t - k\,\mu}{\sigma\,\sqrt{k}}\right) \tag{4.33b}$$

$$m(t) \;=\; \sum_{k=1}^{\infty} \frac{1}{\sigma\,\sqrt{k}}\,\varphi\!\left(\frac{t - k\,\mu}{\sigma\,\sqrt{k}}\right) \tag{4.33c}$$

One possibility to compute values of the renewal function consists of applying the **LAPLACE–STIELTJES transform**

$$M^*(s) := \int_0^{\infty} e^{-s\,t}\,\mathrm{d}M(t). \tag{4.34a}$$

Taking transform in (4.28c), we obtain

$$M^*(s) = G^*(s) + M^*(s)\,G^*(s), \tag{4.34b}$$

so that

$$M^*(s) = \frac{G^*(s)}{1 - G^*(s)}, \tag{4.34c}$$

or equivalently

$$G^*(s) = \frac{M^*(s)}{1 + M^*(s)}. \tag{4.34d}$$

Thus, we see that $M(t)$ is determined by $G(t)$, and conversely, $G(t)$ is determined $M(t)$. If there is an easy way to back–transform (4.34c), one has a solution of (4.28c).

Numerous methods for computing the renewal function $M(t)$ have been proposed. A comprehensive review of these approaches is given by CHAUDURY (1995). Sect. 4.4.2.2 shows how to proceed when $G(t)$ is WEIBULL.

There exists a great number of formulas giving bounds on $M(t)$, e.g.:

$$M(t) \approx G(t), \quad \text{if } G(t) \ll 1, \tag{4.35a}$$

$$G(t) \leq M(t) \leq \frac{G(t)}{1 - G(t)}, \tag{4.35b}$$

$$M(t) \geq \frac{t}{\mu} - 1, \quad \mu \text{ being the mean of } G(.), \tag{4.35c}$$

$$M(t) \leq (\geq) \frac{t}{\mu}, \quad \text{if } G(.) \text{ is NBUE (NWUE)}, \tag{4.35d}$$

$$\frac{t}{\mu} - 1 \leq M(t) \leq \frac{t}{\mu}, \quad \text{if } G(.) \text{ is NBUE.} \tag{4.35e}$$

With respect to the asymptotic behavior of $N_t$, its moments and its distribution, we mention the following results:

$$\frac{N_t}{t} \xrightarrow{\text{a.s.}} \frac{1}{\mu} \quad \text{as } t \to \infty, \tag{4.36a}$$

(a.s. stands for "almost surely")

$$\lim_{t \to \infty} \frac{M(t)}{t} = \frac{1}{\mu}, \tag{4.36b}$$

$$\lim_{t \to \infty} m(t) = \frac{1}{\mu}, \quad \text{if } \lim_{t \to \infty} g(t) = 0, \tag{4.36c}$$

$$\lim_{t \to \infty} \big[ M(t + \varepsilon) - M(t) \big] = \frac{\varepsilon}{\mu} \quad \textbf{(BLACKWELL's theorem).} \tag{4.36d}$$

SMITH (1959) has shown that under rather weak conditions on $G(t)$ there exist constants $a_n$ and $b_n$ such that the $n$–th cumulant[13] of $N_t$ is given by

$$\kappa_n(t) = a_n t + b_n + o(1) \quad \text{as } t \to \infty. \tag{4.37a}$$

---

[13] See (2.76) – (2.78) for details on cumulants.

In particular we have

$$\kappa_1(t) \;=\; M(t) \;=\; \mathrm{E}(N_t) \;\approx\; \frac{t}{\mu_1'} + \left(\frac{\mu_2'}{2\,(\mu_1')^2} - 1\right), \tag{4.37b}$$

$$\kappa_2(t) \;=\; \mathrm{Var}(N_t) \;\approx\; \frac{\mu_2' - (\mu_1')^2}{(\mu_1')^3}\, t + \left(\frac{5\,(\mu_2')^2}{4\,(\mu_1')^4} - \frac{2\,\mu_3'}{3\,(\mu_1')^3} - \frac{\mu_2'}{2\,(\mu_1')^2}\right), \tag{4.37c}$$

where

$$\mu_k' = \int\limits_0^\infty t^k\, \mathrm{d}G(t).$$

Using a central limit theorem for $N_t$, we finally state

$$N_t \stackrel{\text{approx}}{\sim} No\big(\kappa_1(t), \kappa_2(t)\big). \tag{4.37d}$$

**Recurrence times $B_t$ and $F_t$**

The **backward recurrence time** at any point $t$ is

$$B_t = t - T_{N_t}, \tag{4.38a}$$

whereas the **forward recurrence time** at $t$ is

$$F_t = T_{N_t+1} - t. \tag{4.38b}$$

The sum

$$L_t = B_t + F_t = T_{N_t+1} - T_{N_t} \tag{4.38c}$$

is the **total life length** of the item in service at time $t$. $\{B_t;\ t \geq 0\}$, $\{F_t;\ t \geq 0\}$ are homogenous MARKOV processes and statistically equivalent to the processes $\{Y_i;\ i = 1, 2, \ldots\}$, $\{T_n;\ n = 1, 2, \ldots\}$ and $\{N_t;\ t \geq 0\}$ in the sense that, given the realization of one of these processes, the realization of any other process is to be determined uniquely.

Assuming that the underlying distribution $G(.)$ is **non–lattice**,[14] we have the following results on the distributions of $B_t$, $F_t$ and $L_t$ for finite $t$ and for $B$, $F$ and $L$ for $t \to \infty$ (stationary case); see BLUMENTHAL (1967). The density $f_{L_t}(x)$ of $L_t$ can be obtained by noting that the probability that $L_t$ lies between $(x, x + \Delta x)$ equals the probability that a renewal occurs at time $\tau$ in the interval $(t - x, t)$ and that the item installed at that time has life between $x$ and $x + \Delta x$. If $t < x$, $L_t$ in the interval $(x, x + \Delta x)$ can also occur if the

---

[14] A discrete distribution is a **lattice** (or periodic) distribution if its discontinuity points are of the form

$$a + k\,d; k = 1, 2, \ldots,$$

when $a$ and $d$ $(d > 0)$ are constants.

first item fails between $(x, x + \Delta x)$. Thus, since

$$\Pr\left[\text{renewal in } (\tau,\ \tau + \mathrm{d}\tau)\right] \;=\; m(\tau)\,\mathrm{d}\tau, \tag{4.39a}$$

$$f_{L_t}(x) \;=\; g(x)\left[M(t) - M(t-x)\right] + \delta(x-t)\left[1 + M(t-x)\right], \qquad x \geq 0, \tag{4.39b}$$

where

$$\delta(u) = \begin{cases} 1, & \text{if } u \geq 0, \\ 0 & \text{otherwise.} \end{cases}$$

Using the facts that

$$\lim_{t \to \infty} \left[M(t) - M(t-x)\right] = \frac{x}{\mu} \quad \text{(Blackwell's theorem)}$$

and

$$\lim_{t \to \infty} \delta(x - t) = 0,$$

we find the asymptotic or stationary density of $L_t$ to be

$$f_L(x) = \lim_{t \to \infty} f_{L_t}(x) = \frac{x}{\mu}\, g(x). \tag{4.39c}$$

Also, for the moments of $L_t$, we see that

$$\lim_{t \to \infty} \int_0^\infty x^r f_{L_t}(x)\,\mathrm{d}x = \frac{1}{\mu} \int_0^\infty x^{r+1}\, g(x)\,\mathrm{d}x; \quad r = 1, 2, \ldots, \tag{4.39d}$$

whenever the expectation on the right exists.

Similar arguments give

- the density of $B_t$

$$f_{B_t}(x) = \left\{ \begin{array}{ccc} m(t - x)\left[1 - G(x)\right] & \text{for} & x < t \\[2mm] 0 & \text{for} & x \geq t \end{array} \right\}, \tag{4.40a}$$

   which puts mass $1 - G(t)$ at $x = t$, and

- the density of $F_t$

$$f_{F_t}(x) = g(t + x) + \int_0^t m(t - u)\, g(x + u)\,\mathrm{d}u, \quad x \geq 0. \tag{4.40b}$$

The joint density of $B_t$ and $F_t$ is

$$f_{B_t, F_t}(x, y) = m(t - x)\, g(x + y), \quad x \leq t,\ y \geq 0. \tag{4.40c}$$

In the limit we have

$$f_B(x) = \lim_{t \to \infty} f_{B_t}(x) = f_F(x) = \lim_{t \to \infty} f_{F_t}(x) = \frac{1 - G(x)}{\mu} \tag{4.41a}$$

$$f_{B,F}(x, y) = \lim_{t \to \infty} f_{B_t, F_t}(x, y) = \frac{g(x + y)}{\mu}; \quad x,\, y \geq 0, \tag{4.41b}$$

since $m(t - x)$ approaches $1/\mu$. The random variables $B_t$ and $F_t$ have the same limiting distribution as $t \to \infty$ but are not independent. Mean, variance and covariance of the limiting distribution are[15]

$$\mathrm{E}(B) = \mathrm{E}(F) = \frac{1}{2} \left( \mu + \frac{\sigma^2}{\mu} \right) > \frac{\mu}{2}, \tag{4.41c}$$

$$\mathrm{Var}(B) = \mathrm{Var}(F) = \frac{\mu'_3}{3\,\mu} - \left( \frac{\mu'_2}{2\,\mu} \right)^2, \tag{4.41d}$$

$$\mathrm{Cov(B,F)} = \frac{\mu'_3}{6\,\mu} - \left( \frac{\mu'_2}{2\,\mu} \right)^2, \tag{4.41e}$$

where $\mu$ and $\sigma^2$ are the mean and variance of the underlying distribution and $\mu'_r$ is $r$–th moment about the origin. Mean and variance of the limiting distribution of $L$ follow as

$$\mathrm{E}(L) = 2\,\mathrm{E}(B) = 2\,\mathrm{E}(F) = \mu + \frac{\sigma^2}{\mu}, \tag{4.41f}$$

$$\mathrm{Var}(L) = 2 \left[ \mathrm{Var}(B) + \mathrm{Cov}(B, F) \right],$$

$$= \frac{\mu'_3}{\mu} - \frac{\mu'_2}{\mu^2}. \tag{4.41g}$$

### 4.4.2  Ordinary WEIBULL renewal process[16]

The underlying distribution for the ordinary WRP is the common reduced WEIBULL distribution[17] with DF

$$g(y) = c\,y^{c-1}\,\exp\{-y^c\} \tag{4.42a}$$

and CDF

$$G(y) = 1 - \exp\{-y^c\}, \tag{4.42b}$$

which are the same for all $Y_i$ in $T_n = \sum_{i=1}^n Y_i$, the time from $t = 0$ to the instant of the $n$–th renewal, $n = 1, 2, \ldots$. We will first discuss (Sect. 4.4.2.1) the distribution of $T_n$. Then, in Sect. 4.4.2.2, we turn to the number of renewals in $(0, t]$, particularly to the renewal function $M(t)$ and to the renewal density $m(t)$. In Sect. 4.4.2.3 we will make some comments on the forward and backward recurrence times $F_t$ and $B_t$.

---

[15]  COLEMAN (1981) gives the moments of $F_t$ for finite $t$.

[16]  Suggested reading for this section: BAXTER/SCHEUER/BLISCHKE/MCCONALOGUE (1981, 1982), CONSTANTINE/ROBINSON (1997), FROM (2001), GARG/KALAGNANAM (1998), LOMICKI (1966), SMITH/LEADBETTER (1963), WHITE (1964a).

[17]  The results achieved for this special version of the WEIBULL distribution are easily generalized to the two–parameter distribution with a scaling parameter $b \neq 1$ by appropriate re–scaling of the variable.

#### 4.4.2.1   Time to the $n$–th renewal

The DF [and CDF] of $T_n$ is the $n$–fold convolution of $g(y)$ [and $G(y)$] with itself:

$$f_n(t) = g^{(n)}(t) = \int_0^t g^{(n-1)}(t-y)\, g(y)\, \mathrm{d}y, \qquad (4.43\text{a})$$

$$F_n(t) = G^{(n)}(t) = \int_0^t G^{(n-1)}(t-y)\, G(y)\, \mathrm{d}y, \qquad (4.43\text{b})$$

where $g^{(1)}(y) = g(y)$ from (4.42a) and $G^{(1)}(y) = G(y)$ from (4.42b). As stated in Sect. 3.3.6.1, there generally do exist closed form expressions for the integrals on the right–hand sides of (4.43a,b). We will give two approximations to $F_n(t)$:

- a power series expansion of $t^c$, introduced by WHITE (1964a),
- an infinite series of appropriate POISSONIAN functions of $t^c$, introduced by LOMICKI (1966).

WHITE (1964a) finds the following infinite series expansion of $F_n(t)$:

$$F_1(t) \;=\; G(t) = \sum_{j=1}^{\infty} (-1)^{j+1}\, a_1(j)\, \frac{t^{cj}}{j!}, \qquad (4.44\text{a})$$

$$
\begin{aligned}
F_{n+1}(t) &= \int_0^t F_n(t)\, \mathrm{d}F_1(t) \\[2mm]
&= \sum_{j=n+1}^{\infty} (-1)^{j+n+1}\, a_{n+1}(j)\, \frac{t^{cj}}{j!},
\end{aligned}
\qquad (4.44\text{b})
$$

where

$$a_1(j) \;\equiv\; 1 \ \forall\, j \qquad (4.44\text{c})$$

$$a_{n+1}(j) \;=\; \sum_{i=n}^{j-1} a_n(i)\, \gamma(i,j), \qquad (4.44\text{d})$$

$$\gamma(i,y) \;=\; \frac{\Gamma(1+i\,c)\, \Gamma\big[c\,(j-i)+1\big]\, j!}{\Gamma(1+j\,c)\, i!\, (j-i)!}. \qquad (4.44\text{e})$$

(4.44d,e) show that the coefficients $a_{n+1}(j)$ in (4.44b) also depend on the WEIBULL shape parameter $c$ also.

LOMICKI (1966) gives the following series expansion of $F_n(t)$:

$$F_n(t) = \sum_{j=n}^{\infty} \alpha_n(j)\, D_j(t^c). \qquad (4.45\text{a})$$

The components of (4.45a) are defined as follows:

$$D_j(t^c) = \sum_{i=j}^{\infty} \frac{(t^c)^i}{i!}\, \exp\left(-t^c\right); \quad j = 1, 2, \ldots \qquad (4.45\text{b})$$

Thus, $D_j(t^c)$ is the CCDF of a POISSON distribution with parameter $\lambda \tau = 1 \cdot t^c$. $D_j(t^c)$ may be expressed by the CDF of a $\chi^2$–distribution with $\nu = 2\,j$ degrees of freedom:

$$D_j(t^c) = F_{\chi^2}\big(2\,t^c \,|\, \nu = 2\,j\big)$$

$$= \int_0^{2\,t^c} \frac{x^{(\nu/2)-1}\,e^{-x/2}}{2^{\nu/2}\,\Gamma(\nu/2)}\,\mathrm{d}x. \tag{4.45c}$$

The coefficients $\alpha_n(j)$ in (4.45a) are to be computed by the following algorithm:

1.

$$\gamma(r) = \frac{\Gamma(1+r\,c)}{\Gamma(1+r)}\,;\quad r =, 1, 2, \ldots \tag{4.45d}$$

2.

$$b_{n+1}(j) = \sum_{i=n}^{j=1} b_n(i)\,\gamma(j-i); \quad \left\{ \begin{array}{l} n = 0, 1, 2, \ldots; \\ j = n+1, n+2\ldots \end{array} \right\} \tag{4.45e}$$

$$b_0(j) = \gamma(j); \quad j = 0, 1, 2, \ldots \tag{4.45f}$$

3.

$$a_n^*(j) = \sum_{i=n}^{j} (-1)^{i+n} \binom{j}{i} \frac{b_n(i)}{\gamma(n)}, \quad \left\{ \begin{array}{l} n = 0, 1, 2, \ldots; \\ j = n, n+1, \ldots \end{array} \right\} \tag{4.45g}$$

4.

$$\alpha_n(n) = a_n^*(n) \tag{4.45h}$$

$$\alpha_n(j) = \sum_{i=n}^{j} a_n^*(j) - \sum_{i=n}^{j-1} a_i^*(j-1); \quad j > n \tag{4.45i}$$

For numerical calculations the backward recurrence formula

$$\alpha_n(j) = \alpha_{n+1}(j) + \big[a_n^*(j) - a_n^*(j-1)\big] \tag{4.45j}$$

can be useful in conjunction with (4.45h).

Fig. 4/2 shows the CDFs $F_n(t)$ for $n = 1,\ 2,\ 3,\ 4,\ 5$ each combined with $c = 0.5,\ 1,\ 1.5$ and 2. The calculation was done with WHITE's approach. The rate of convergence is very slow, especially when $t$ is great. Greater values of $t$ are needed when $n$ and/or $c$ are great.

Figure 4/2:  Cumulative distribution function of $T_n$

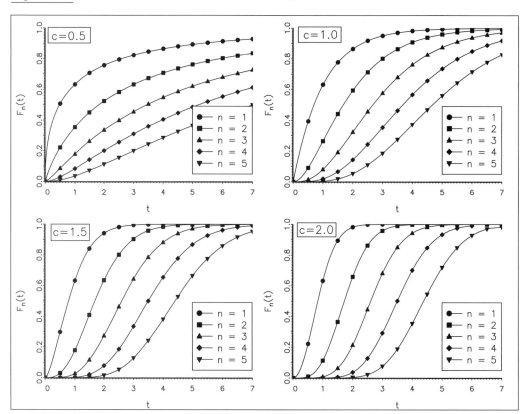

#### 4.4.2.2   Number of renewals $N_t$

The numerical evaluation of the renewal function $M(t)$, and the renewal intensity $m(t)$ when the underlying distribution $G(\cdot)$ is WEIBULL, has attracted the attention of several authors. Research in this area has adopted one of the following two approaches:

1. development of algorithms for explicitly computing the convolutions of the underlying distribution,

2. computing the renewal function by inverting its LAPLACE–STIELTJES transform.

We start our overview on the WEIBULL renewal function by citing a general approximation $m_a(t)$ to the renewal intensity $m(t)$, given by BARTHOLOMEW (1963):

$$m_a(t) = g(t) + \left[G(t)\right]^2 \bigg/ \int_0^t \left[1 - G(u)\right] \, \mathrm{d}u. \qquad (4.46\mathrm{a})$$

This approximation has the following properties:

- $\displaystyle \lim_{t \to \infty} m_a(t) = \lim_{t \to \infty} m(t) = \frac{1}{\mathrm{E}(Y)}$ ,

- $m_a(0) = m(0) = g(0)$,

- $m_a'(0) = m'(0)$  and  $m_a''(0) = m''(0)$,

- $m_a(0)$ is an upper bound for $m(t)$ for all $t$, if $g(t)/[1 - G(t)]$ is a non–decreasing function of $t$; i.e., $G(t)$ has a non–decreasing hazard rate.

Applying (4.46a) to the reduced WEIBULL distribution we get

$$m_a(t) = c\,t^{c-1}\,\exp\{-t^c\} + \frac{\left[1 - \exp\{-t^c\}\right]^2}{\Gamma\left(1 + \dfrac{1}{c}\right) - \dfrac{1}{c}\Gamma\left(\dfrac{1}{c}\Big| t^c\right)} \,. \tag{4.46b}$$

Fig. 4/3 shows (4.46b) for several values of the shape parameter $c$.

Figure 4/3: BARTHOLOMEW's approximation to the WEIBULL renewal intensity

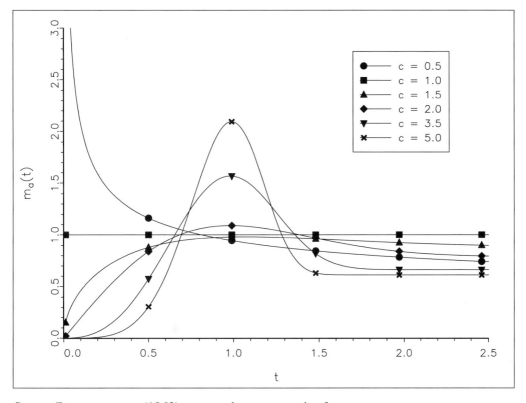

SMITH/LEADBETTER (1963) proposed a power series for

$$M(t) = G(t) + \int\limits_0^t M(t - u)\,\mathrm{d}G(u),$$

when $G(t) = 1 - \exp\{-t^c\}$, giving

$$M(t) = \sum_{k=1}^{\infty} \frac{(-1)^{k-1}\,A_k\,t^{c\,k}}{\Gamma(1 + k\,c)}, \tag{4.47a}$$

where

$$A_k = \frac{\Gamma(1 + kc)}{k!} - \sum_{j=1}^{k-1} \frac{\Gamma(1 + jc)}{j!} A_{k-j} \quad \text{and} \quad A_1 = \Gamma\left(1 + \frac{1}{c}\right). \tag{4.47b}$$

(4.47a) converges for all $t$ and all values of the shape parameter $c > 0$, but the series converges too slowly, however, for large values of $t$, especially for large $c > 1$. Convergence is rapid for all $t$ if $0 < c \leq 1$.

The two infinite series expansion for $F_n(t)$ given in the preceding section can be extended to approximate $M(t)$. The approach of WHITE (1966a), given in (4.44), results in the following series expansions of $M(t) = M_1(t)$ and higher binomial moments $M_n(t)$; see (4.29):

$$M(t) = \sum_{j=1}^{\infty}(-1)^{j+1} d_j \frac{t^{cj}}{j!}, \tag{4.48a}$$

where

$$d_1 = 1, \tag{4.48b}$$

$$d_j = 1 - \sum_{i=1}^{j-1} d_i\, \gamma(i,j), \tag{4.48c}$$

and $\gamma(i,j)$ given by (4.44e). With respect to higher binomial moments, we get

$$M_{n+1}(t) = \int_0^t M_n(t-u)\, \mathrm{d}M(u)$$

$$= \sum_{j=n+1}^{\infty}(-1)^{j+n+1} d_{n+1}(j)\frac{t^{cj}}{j!}, \tag{4.49a}$$

where

$$d_1(1) = d_j \tag{4.49b}$$

$$d_{n+1}(j) = \sum_{i=n}^{j-1} d_n(i)\, d_1(j-i)\, \gamma(i,j). \tag{4.49c}$$

The LOMICKI (1966) approach gives the following series expansion of $M(t)$:

$$M(t) = \sum_{j=1}^{\infty} c(j)\, D_j\!\left(t^c\right), \tag{4.50a}$$

where $D_j\!\left(t^c\right)$ is defined in (4.45b) and

$$c(j) = \sum_{n=1}^{j} \alpha_n(j), \quad j = 1,2,\ldots, \tag{4.50b}$$

where $\alpha_n(j)$ is given in (4.45i).

BAXTER et al. (1981, 1982) have extensively tabulated $\mathrm{E}(N_t) = M(t)$, $\mathrm{Var}(N_t)$ and $\int_0^t M(u)\,\mathrm{d}u$ for five underlying distribution functions, one of them being the reduced WEIBULL distribution. They used an algorithm that generates piecewise polynomial approximations to recursively defined convolutions $G^{(n)}(t)$ for $G \in C^2[0,\infty)$, assuming that $g = G'$ is bounded. The essence of the algorithm is the choice of a cubic–spline representation of $G^{(n)}(t)$ for each $n$, thus providing an accurate approximation that preserves both positivity and monotonicity.

CONSTANTINE/ROBINSON (1997) focused attention on the calculation of $M(t)$ when $c > 1$. They used a LAPLACE transform technique to form uniformly convergent series of damped exponential terms for $M(t)$:

$$M(t) = \frac{t}{\mu} + \frac{\sigma^2 - \mu^2}{2\,\mu^2} + \sum_{j=1}^{\infty} \frac{\exp\{t\,s_j\}}{-s_j\,g'(s_j)}, \tag{4.51a}$$

where $\mu$ and $\sigma^2$ are the mean and the variance of the reduced WEIBULL distribution:

$$\mu = \Gamma\left(1 + \frac{1}{c}\right); \quad \sigma^2 = \Gamma\left(1 + \frac{2}{c}\right) - \mu^2$$

and

$$
\begin{aligned}
g(s) &= c \int_0^{\infty} \exp\{-s\,u\}\,\exp\left\{-u^c\right\}u^{c-1}\,\mathrm{d}u \\
&= \int_0^{\infty} e^{-x}\,\exp\left\{-s\,x^{1/c}\right\}\mathrm{d}x,
\end{aligned}
\tag{4.51b}
$$

which is a special case of an integral known as FAXEN's integral. The $s_j$ are simple complex zeros of $g(s) = 1$, and $[-s_j\,g'(s_j)]^{-1}$ are the residues of the function

$$U(s) = \frac{g(s)}{s\,[1 - g(s)]}$$

at the poles $s = s_j$. To actually compute (4.51a), it is sometimes necessary to compute 500 or more $s_j$ values, especially for small $t$. The authors do give very good approximations to $s_j$, however. They recommend the SMITH/LEADBETTER (1963) series for small $t$, see (4.47), and their own exponential series for large $t$. They do not recommend use of the exponential series (4.51a) for $1 < c \leq 1.2$. The approach (4.51a) is much more difficult for the case $0 < c < 1$ because $g(s)$ has a branch–point at $s = 0$ and the $U(s)$ series does not exist.

FROM (2001) presents an approach which can be applied even if $0 < c < 1$. This approach is similar to that of GARG/KALAGNANAM (1998) approximating a modified rational function near the origin and switches to an asymptotic linear function for larger values of $t$. The FROM two–piece approximation with switch–over point $\mu + k\,\sigma$ given by

$$M(t) = \left\{\begin{array}{ll} \widehat{M}(t), & \text{if } 0 \leq t \leq \mu + k\,\sigma \\ \widetilde{M}(t), & \text{if } t > \mu + k\,\sigma \end{array}\right\}, \tag{4.52a}$$

where $k$ is a suitably chosen real number with $\mu + k\,\sigma \geq 0$,

$$\widehat{M}(t) = \frac{G(t)\left[1 + a_1\left(\dfrac{t}{\mu}\right) + a_2\left(\dfrac{t}{\mu}\right)^2 + \ldots + a_{n+1}\left(\dfrac{t}{\mu}\right)^{n+1}\right]}{1 + a_{n+2}\left(\dfrac{t}{\mu}\right) + a_{n+3}\left(\dfrac{t}{\mu}\right)^2 + \ldots + a_{2n+1}\left(\dfrac{t}{\mu}\right)^n} \tag{4.52b}$$

and

$$\widetilde{M}(t) = \frac{t}{\mu} + \frac{\sigma^2 - \mu^2}{2\,\mu^2}, \quad \text{cf. (4.37b)}. \tag{4.52c}$$

FROM shows how to determine the coefficients $a_1, a_2, \ldots, a_{2\,n+1}$ and the factor $k$ for $G(t)$ being WEIBULL or truncated normal.

We close this section with Fig. 4/4 showing $E(N_t)$ and $\text{Var}(N_t)$ for several values of $c$ and for $0 \leq t \leq 20$. The coordinates of the graphs were taken from the tables of BAXTER et al. (1981).

Figure 4/4:  Mean and variance of the number of WEIBULL renewals

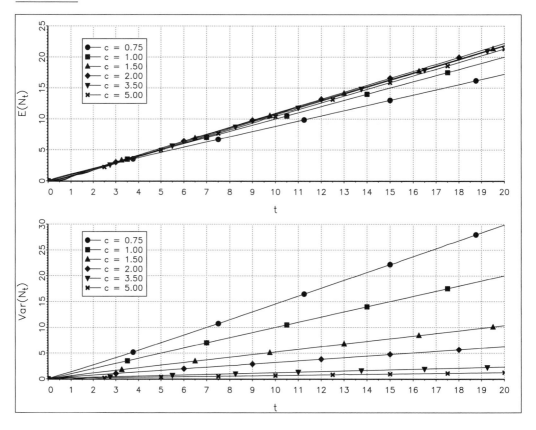

### 4.4.2.3  Forward and backward recurrence times

In (4.38a-c) we introduced the backward recurrence time $B_t$, the forward recurrence time $F_t$ and the total life length $L_t = B_t + F_t$ of an item in service at time $t$ and gave general formulas for the distributions and moments of $B_t$, $F_t$ and $L_t$. These formulas can only be evaluated numerically when the underlying distribution is WEIBULL and $t$ is finite. In the limiting case $(t \to \infty)$, however, we achieve rather simple closed–form expressions in the reduced WEIBULL case:

$$f_L(x) \;=\; \frac{x}{\mu}\, g(x), \quad \text{cf. (4.39c)},$$

$$=\; \frac{c}{\Gamma\!\left(1+\dfrac{1}{c}\right)}\, x^c \exp\{-x^c\}\,; \tag{4.53a}$$

$$f_B(x) \;=\; f_F(x) \;=\; \frac{1-G(x)}{\mu}\,, \quad \text{cf. (4.41a)},$$

$$=\; \frac{\exp\{-x^c\}}{\Gamma\!\left(1+\dfrac{1}{c}\right)}\,; \tag{4.53b}$$

$$f_{B,F}(x,y) \;=\; \frac{g(x+y)}{\mu}\,, \quad \text{cf. (4.41b)},$$

$$=\; \frac{c}{\Gamma\!\left(1+\dfrac{1}{c}\right)}\, (x+y)^{c-1} \exp\{-(x+y)^c\}\,; \tag{4.53c}$$

$$\mathrm{E}(B) \;=\; \mathrm{E}(F) \;=\; \frac{1}{2}\!\left(\mu + \frac{\sigma^2}{\mu}\right)\,, \quad \text{cf. (4.41c)},$$

$$=\; \frac{\Gamma\!\left(1+\dfrac{2}{c}\right)}{2\,\Gamma\!\left(1+\dfrac{1}{c}\right)}\,; \tag{4.53d}$$

$$\mathrm{Var}(B) \;=\; \mathrm{Var}(F) \;=\; \frac{\mu_3'}{3\,\mu} - \left(\frac{\mu_2'}{2\,\mu}\right)^2\,, \quad \text{cf. (4.41d)},$$

$$=\; \frac{\Gamma\!\left(1+\dfrac{3}{c}\right)}{3\,\Gamma\!\left(1+\dfrac{1}{c}\right)} - \left[\frac{\Gamma\!\left(1+\dfrac{2}{c}\right)}{2\,\Gamma\!\left(1+\dfrac{1}{c}\right)}\right]^2\,; \tag{4.53e}$$

$$\mathrm{Cov}(B,F) \;=\; \frac{\mu_3'}{6\,\mu} - \left(\frac{\mu_2'}{2\,\mu}\right)^2\,, \quad \text{cf. (4.41e)},$$

$$=\; \frac{\Gamma\!\left(1+\dfrac{3}{c}\right)}{6\,\Gamma\!\left(1+\dfrac{1}{c}\right)} - \left[\frac{\Gamma\!\left(1+\dfrac{2}{c}\right)}{2\,\Gamma\!\left(1+\dfrac{1}{c}\right)}\right]^2\,; \tag{4.53f}$$

$$\rho(B,F) = \frac{\mathrm{Cov}(B,F)}{\sqrt{\mathrm{Var}(B)\,\mathrm{Var}(F)}} = \frac{2\,\Gamma\!\left(1+\dfrac{1}{c}\right)\Gamma\!\left(1+\dfrac{3}{c}\right) - 3\,\Gamma^2\!\left(1+\dfrac{1}{c}\right)}{4\,\Gamma\!\left(1+\dfrac{1}{c}\right)\Gamma\!\left(1+\dfrac{3}{c}\right) - 3\,\Gamma^2\!\left(1+\dfrac{1}{c}\right)}\,. \tag{4.53g}$$

The correlation between $B$ and $F$ as indicated by (4.53g) behaves as follows:

$$\rho(B,F) \begin{cases} > 0 & \text{for} & 0 < c < 1, \\ = 0 & \text{for} & c = 1, \\ < 0 & \text{for} & c > 1. \end{cases}$$

$$\mathrm{E}(L) = \mu + \frac{\sigma^2}{\mu}, \quad \text{cf. (4.41f)},$$

$$= \frac{\Gamma\left(1 + \dfrac{2}{c}\right)}{\Gamma\left(1 + \dfrac{1}{c}\right)}; \tag{4.53h}$$

$$\mathrm{Var}(L) = \frac{\mu_3'}{\mu} - \frac{\mu_2'}{\mu^2}, \quad \text{cf. (4.41g)},$$

$$= \frac{\Gamma\left(1 + \dfrac{1}{c}\right)\Gamma\left(1 + \dfrac{3}{c}\right) - \Gamma\left(1 + \dfrac{2}{c}\right)}{\left[\Gamma\left(1 + \dfrac{1}{c}\right)\right]^2}. \tag{4.53i}$$

Fig. 4/5 shows the joint density (4.53c) for $c = 3$. The density is symmetric as is indicated by corresponding marginal densities (4.53b) and by equal means (4.53d) and variances (4.53e).

Figure 4/5: Joint density of the stationary Weibull forward and backward recurrence times

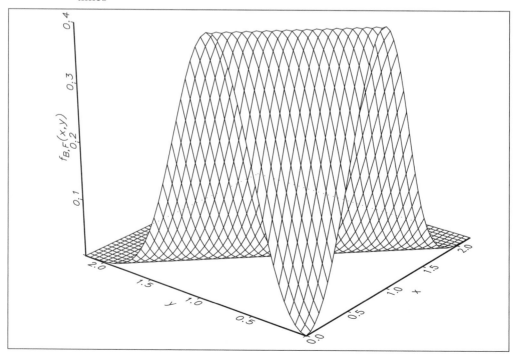

# 5 Order statistics and related variables

Order statistics and their functions play an important role in probability theory to characterize distributions (see Chapter 6) and in the inferential process to estimate parameters of a distribution and to test hypotheses on these parameters (see Chapters 9 to 22).

## 5.1 General definitions and basic formulas[1]

### 5.1.1 Distributions and moments of order statistics

Let $X_1, X_2, \ldots, X_n$ be iid with CDF $F(x)$. The variables $X_i$ being arranged in ascending order and written as

$$X_{1:n} \leq X_{2:n} \leq \ldots \leq X_{n:n}$$

are called **order statistics**. The CDF of $X_{r:n}$ is given by

$$
\begin{aligned}
F_{r:n}(x) &= \Pr\left(X_{r:n} \leq x\right) \\
&= \Pr(\text{at least } r \text{ of the } X_j \text{ are less than or equal to } x) \\
&= \sum_{i=r}^{n} \binom{n}{i} \left[F(x)\right]^i \left[1 - F(x)\right]^{n-i},
\end{aligned}
\tag{5.1a}
$$

because the term in the summand is the binomial probability that exactly $i$ of $X_1, \ldots, X_n$ are less than or equal to $x$. $F_{r:n}(x)$ can be written as the **incomplete beta function ratio** or **beta distribution function** $[X \sim Be(a,b)]$:

$$
\begin{aligned}
F_{r:n}(x) &= I_{F(x)}(r, n-r+1) \\
&= \frac{\int_0^{F(x)} u^{r-1} (1-u)^{n-r} \, \mathrm{d}u}{B(r, n-r+1)},
\end{aligned}
\tag{5.1b}
$$

where $B(r, n-r+1)$ is the **complete beta function**

$$
B(a,b) = \int_0^1 u^{a-1} (1-u)^{b-1} \, \mathrm{d}u = \frac{\Gamma(a)\,\Gamma(b)}{\Gamma(a+b)} .
\tag{5.1c}
$$

(5.1a,b) hold whether $X$ is discrete or continuous. In the sequel we will always assume that $X$ is continuous. Then, the random variables

$$
Z_r = F\left(X_{r:n}\right) - F\left(X_{r-1:n}\right); \quad r = 1, 2, \ldots, n+1,
\tag{5.1d}
$$

---

[1] Suggested reading for this section BALAKRISNAN/COHEN (1991), BALAKRISHNAN/RAO (1998), DAVID (1981), GALAMBOS (1978), KRISHNAIAH/RAO (1988), SARHAN/GREENBERG (1962).

with $F(X_{0:n}) = 0$ and $F(X_{n+1:n}) = 1$ are called **coverages**. Furthermore, when $f(x)$, the DF of $X$, exists, differentiation of (5.1b) with regard to (5.1c) gives the DF of $X_{r:n}$:

$$f_{r:n}(x) = \frac{n!}{(r-1)!\,(n-r)!} \left[F(x)\right]^{r-1} \left[1 - F(x)\right]^{n-r} f(x). \qquad (5.1e)$$

Some special values of $r$ are

- $r = 1$

  $X_{1:n}$ is the sample minimum with DF

$$f_{1:n}(x) = n\,f(x) \left[1 - F(x)\right]^{n-1} \qquad (5.2a)$$

  and CDF

$$F_{1:n}(x) = \sum_{i=1}^{n} \binom{n}{i} \left[F(x)\right]^{i} \left[1 - F(x)\right]^{n-i} = 1 - \left[1 - F(x)\right]^{n}. \qquad (5.2b)$$

- $r = n$

  $X_{n:n}$ is the sample maximum with DF

$$f_{n:n}(x) = n\,f(x) \left[F(x)\right]^{n-1} \qquad (5.3a)$$

  and CDF

$$F_{n:n}(x) = \left[F(x)\right]^{n}. \qquad (5.3b)$$

- $r = k + 1$ for $n = 2\,k + 1$

  $X_{k+1:2\,k+1}$ is the sample median.[2]

The **joint DF** of $X_{r:n}$ and $X_{s:n}$, $1 \le r < s \le n$, is

$$f_{r,s:n}(x,y) = \frac{n!}{(r-1)!\,(s-r-1)!\,(n-s)!} \left[F(x)\right]^{r-1} \left[F(y) - F(x)\right]^{s-r-1} \times$$

$$\left[1 - F(y)\right]^{n-s} f(x)\,f(y), \quad x < y. \qquad (5.4a)$$

Even if $X_1, \ldots, X_n$ are independent, order statistics are not independent random variables. The joint CDF of $X_{r:n}$ and $X_{s:n}$ may be obtained by integration of (5.4a) as well as by a direct argument valid also in the discrete case.

- For $x < y$ we have:

$$F_{r,s:n}(x,y) = \Pr(\text{at least } r\ X_i \le x \text{ and at least } s\ X_i \le y)$$

$$= \sum_{j=s}^{n} \sum_{k=r}^{j} \Pr(\text{exactly } k\ X_i \le x \text{ and exactly } j\ X_i \le y)$$

$$= \sum_{j=s}^{n} \sum_{k=r}^{j} \frac{n!}{k!\,(j-k)!\,(n-j)!} \left[F(x)\right]^{k} \left[F(y) - F(x)\right]^{j-k} \times$$

$$\left[1 - F(y)\right]^{n-j}, \quad x < y. \qquad (5.4b)$$

---

[2] The DF of the sample median for $n = 2\,k$ is given farther down in Sect. 5.1.2.

- For $x \geq y$ the inequality $X_{s:n} \leq y$ implies $X_{r:n} \leq x$, so that

$$F_{r,s:n}(x,y) = F_{s:n}(y), \quad x \geq y. \tag{5.4c}$$

Next we discuss the joint DF of $k$ order statistics $X_{r_1:n}, \ldots, X_{r_k:n}$, $1 \leq r_1 < \ldots < r_k \leq n$, $1 \leq k \leq n$, for $x_1 \leq x_2 \leq \ldots \leq x_k$:

$$f_{r_1,r_2,\ldots,r_k:n}(x_1, x_2, \ldots, x_n) = \frac{n!}{(r_1-1)!\,(r_2-r_1-1)!\ldots(n-r_k)!} \left[F(x_1)\right]^{r_1-1} \times$$
$$\left[F(x_2) - F(x_1)\right]^{r_2-r_1-1} \times \ldots \times \left[1 - F(x_k)\right]^{n-r_k} \times$$
$$f(x_1) \times f(x_2) \times \ldots \times f(x_k). \tag{5.5a}$$

The joint DF of the first $r$ order statistics $X_{1:n}, X_{2:n}, \ldots, X_{r:n}$ is

$$f_{1,2,\ldots,r:n}(x_1, x_2, \ldots, x_r) = \frac{n!}{(n-r)!} \left[1 - F(x_{r:n})\right]^{n-r} \prod_{i=1}^{r} f(x_i); \quad x_1 \leq x_2 \leq \ldots \leq x_r. \tag{5.5b}$$

The joint DF of all $n$ order statistics simply is

$$f_{1,2,\ldots,n:n}(x_1, x_2, \ldots, x_n) = n! \prod_{i=1}^{n} f(x_i); \quad x_1 \leq x_2 \leq \ldots \leq x_n. \tag{5.5c}$$

From the marginal DF of $X_{r:n}$ in (5.1e) and the joint DF of $X_{r:n}$ and $X_{s:n}$ in (5.4a), $r < s$, we obtain

- **the conditional DF** of $X_{s:n}$, given $X_{r:n} = x$:

$$f_{s:n}(y \mid X_{r:n} = x) = \frac{(n-r)!}{(s-r-1)!\,(n-s)!} \left\{ \frac{F(y) - F(x)}{1 - F(x)} \right\}^{s-r-1} \times$$
$$\left\{ \frac{1 - F(y)}{1 - F(x)} \right\}^{n-s} \left\{ \frac{f(y)}{1 - F(x)} \right\}, \quad x \leq y, \tag{5.6a}$$

(Since $[F(y) - F(x)]/[1 - F(x)]$ and $f(y)/[1 - F(x)]$ are the CDF and the DF, respectively, of the parent distribution truncated on the left at $X_{r:n} = x$, we can state:
The conditional distribution of $X_{s:n}$, given $X_{r:n} = x$, is the same as the distribution of the $(s-r)$–th order statistic in a sample of size $n - r$ from a population with distribution $F(.)$ truncated on the left at $x$.)

- the conditional DF of $X_{r:n}$, given $X_{s:n} = y$:

$$f_{r:n}(x \mid X_{s:n} = y) = \frac{(s-1)!}{(r-1)!\,(s-r-1)!} \left\{ \frac{F(x)}{F(y)} \right\}^{r-1} \left\{ \frac{F(y) - F(x)}{F(y)} \right\}^{s-r-1} \times$$
$$\frac{f(x)}{F(y)}; \quad x \leq y. \tag{5.6b}$$

(Since $F(x)/F(y)$ and $f(x)/F(y)$ are the CDF and DF of the parent distribution truncated on the right at $X_{s:n} = y$, respectively, we see that the conditional distribution of $X_{r:n}$, given $X_{s:n} = y$, is the same as the distribution of the $r$–th order statistic in a sample of size $s - 1$ from a population with distribution $F(.)$ truncated on the right at $y$.)

We will denote the **single moments** of order statistics, $E(X_{r:n}^k)$, by $\mu_{r:n}^{(k)}$, $1 \leq r \leq n$. They follow as

$$\mu_{r:n}^{(k)} = \int_{-\infty}^{\infty} x^k f_{r:n}(x) \, dx$$

$$= \frac{n!}{(r-1)!(n-r)!} \int_{-\infty}^{\infty} x^k [F(x)]^{r-1} [1 - F(x)]^{n-r} f(x) \, dx. \qquad (5.7a)$$

The most import case of (5.7a) is the mean, denoted by $\mu_{r:n}$:

$$\mu_{r:n} = E(X_{r:n}) = n \binom{n-1}{r-1} \int_{-\infty}^{\infty} x [F(x)]^{r-1} [1 - F(x)]^{n-r} \, dF(x). \qquad (5.7b)$$

Since $0 \leq F(x) \leq 1$, it follows that

$$|\mu_{r:n}| \leq n \binom{n-1}{r-1} \int_{-\infty}^{\infty} |x| \, dF(x), \qquad (5.7c)$$

showing that $\mu_{r:n}$ exists provided $E(X)$ exists, although the converse is not necessarily true.

An alternative formula for $\mu_{r:n}$ may be obtained by integration by parts in

$$\mu_{r:n} = \int_{-\infty}^{\infty} x \, dF_{r:n}(x).$$

To this end, note that for any CDF $F(x)$ the existence of $E(X)$ implies

$$\lim_{x \to -\infty} x \, F(x) = 0 \quad \text{and} \quad \lim_{x \to \infty} x \, [1 - F(x)] = 0,$$

so that we have

$$E(X) = \int_{-\infty}^{\infty} x \, dF(x)$$

$$= \int_{-\infty}^{0} x \, dF(x) - \int_{0}^{\infty} x \, d[1 - F(x)]$$

$$= \int_{0}^{\infty} [1 - F(x)] \, dx - \int_{-\infty}^{0} F(x) \, dx. \qquad (5.8a)$$

This general formula gives $\mu_{r:n} = \mathrm{E}(X_{r:n})$ if $F(x)$ is replaced by $F_{r:n}(x)$ :

$$\mu_{r:n} = \int_0^\infty \left[ 1 - F_{r:n}(x) \right] \mathrm{d}x - \int_{-\infty}^0 F_{r:n}(x)\,\mathrm{d}x. \tag{5.8b}$$

We may also write

$$\mu_{r:n} = \int_0^\infty \left[ 1 - F_{r:n}(x) - F_{r:n}(-x) \right] \mathrm{d}x, \tag{5.8c}$$

and when $f(x)$ is symmetric about $x = 0$, we have

$$\mu_{r:n} = \int_0^\infty \left[ F_{n-r+1}(x) - F_{r:n}(x) \right] \mathrm{d}x. \tag{5.8d}$$

**Product moments** of order statistics may be defined similarly:

$$
\begin{aligned}
\mu_{r,s:n}^{(k,\ell)} &= \mathrm{E}\left( X_{r:n}^k \, X_{s:n}^\ell \right) \\
&= \frac{n!}{(r-1)!\,(s-r-1)!\,(n-s)!} \iint\limits_{-\infty < x < y < \infty} x^k \, x^\ell \left[ F(x) \right]^{r-1} \times \\
& \quad \left[ F(y) - F(x) \right]^{s-r-1} \left[ 1 - F(y) \right]^{n-s} f(x)\,f(y)\,\mathrm{d}x\,\mathrm{d}y.
\end{aligned}
\tag{5.9a}
$$

The most important case derived from (5.9a) is the covariance of $X_{r:n}$ and $X_{s:n}$:

$$
\begin{aligned}
\mathrm{Cov}\left( X_{r:n}, X_{s:n} \right) &= \mathrm{E}\left( X_{r:n}\,X_{s:n} \right) - \mathrm{E}\left( X_{r:n} \right) \mathrm{E}\left( X_{s:n} \right) \\
&= \mu_{r,s:n} - \mu_{r:n}\,\mu_{s:n}.
\end{aligned}
\tag{5.9b}
$$

For the computation of moments of order statistics and for checking the results, we need some identities and recurrence relations. By using the basic identity

$$\left( \sum_{i=1}^n X_{i:n}^k \right)^\ell = \left( \sum_{i=1}^n X_i^k \right)^\ell, \tag{5.10a}$$

several **identities** for single and product moments of order statistics can be established which primarily serve the **purpose of checking**. By choosing $\ell = 1$ and taking expectations on both sides, we get the identity

$$\sum_{i=1}^n \mu_{i:n}^{(k)} = n\,\mathrm{E}\left( X^k \right) = n\,\mu_{1:1}^{(k)}. \tag{5.10b}$$

Similarly, by taking $k = 1$ and $\ell = 2$, we obtain

$$\sum_{i=1}^n X_{i:n}^2 + 2 \sum_{i=1}^{n-1} \sum_{j=i+1}^n X_{i:n}\,X_{j:n} = \sum_{i=1}^n X_i^2 + 2 \sum_{i=1}^{n-1} \sum_{j=i+1}^n X_i\,X_j. \tag{5.10c}$$

Now taking expectations on both sides leads to

$$\sum_{i=1}^{n} \mu_{i:n}^{(2)} + 2 \sum_{i=1}^{n-1} \sum_{j=i+1}^{n} \mu_{i,j:n} = n \, \mathrm{E}(X^2) + n \, (n-1) \big[\mathrm{E}(X)\big]^2, \qquad (5.10\mathrm{d})$$

which, when used together with (5.10b), yields an identity for product moments of order statistics

$$\sum_{i=1}^{n-1} \sum_{j=i+1}^{n} \mu_{i,j:n} = \binom{n}{2} \big[\mathrm{E}(X)\big]^2 = \binom{n}{2} \mu_{1:1}^2. \qquad (5.10\mathrm{e})$$

Starting from (5.10a), one can establish the triangle rule leading to a **recurrence relation for single moments** of order statistics:

$$r \, \mu_{r+1:n}^{(k)} + (n-r) \, \mu_{r:n}^{(k)} = n \, \mu_{r:n-1}^{(k)}. \qquad (5.11\mathrm{a})$$

For even values of $n$, say $n = 2\,m$, by setting $r = m$ and $k = 1$ in (5.11a), we immediately obtain the relation

$$\frac{1}{2}\Big[\mu_{m+1:2\,m} + \mu_{m:2\,m}\Big] = \mu_{m:2\,m-1} \qquad (5.11\mathrm{b})$$

telling that the expected value of the median in a sample of even size $2\,m$ is exactly equal to the expected value of the median in a sample of odd size $2\,m - 1$.

A similar **recurrence relation for the product moments** of order statistics is given by[3]

$$(r-1) \, \mu_{r,s:n}^{(k,\ell)} + (s-r) \, \mu_{r-1,s:n}^{(k,\ell)} + (n-s+1) \, \mu_{r-1,s-1:n}^{(k,\ell)} = n \, \mu_{r-1,s-1:n-1}^{(k,\ell)}. \qquad (5.11\mathrm{c})$$

We only mention two bounds for moments of order statistics; for more see ARNOLD/ BALAKRISHNAN (1989). For a continuous parent distribution with mean $\mu$ and variance $\sigma^2$ the following inequalities hold

$$\mu_{1:n} \ \geq \ \mu - \frac{(n-1)\,\sigma}{\sqrt{2\,n-1}}, \qquad (5.12\mathrm{a})$$

$$\mu_{n:n} \ \leq \ \mu + \frac{(n-1)\,\sigma}{\sqrt{2\,n-1}}. \qquad (5.12\mathrm{b})$$

### 5.1.2 Functions of order statistics

We will study some linear functions of order statistics and start with two special sums of two order statistics. The **median in a sample of even size** $n = 2\,m$ is defined as

$$\widetilde{X} := \frac{X_{m:2\,m} + X_{m+1:2\,m}}{2}. \qquad (5.13\mathrm{a})$$

Its DF $f_{\widetilde{X}}(y)$ may be derived from the joint DF of two order statistics (5.4a) by setting $n = 2\,m$, $r = m$ and $s = m+1$ and by using standard transformation methods. The mean

---

[3] More recurrence relatives may be found in ARNOLD/BALAKRISHNAN (1989).

of $\widetilde{X}$ has been commented upon in (5.11b). The sum $Z = X_{m:2\,m} + X_{m+1:2\,m}$ has DF

$$f_Z(z) = \frac{(2\,m)!}{2\,(m-1)!} \int\limits_{-\infty}^{+\infty} \{F(x)\,[1-F(z-x)]\}^{m-1}\,f(x)\,f(z-x)\,\mathrm{d}x, \qquad (5.13\mathrm{b})$$

so $\widetilde{X} = Z/2$ has DF

$$f_{\widetilde{X}}(y) = 2\,f_Z(2\,y) = \frac{(2\,m)!}{(m-1)!} \int\limits_{-\infty}^{\infty} \{F(x)\,[1-F(2\,y-x)]\}^{m-1}\,f(x)\,f(2\,y-x)\,\mathrm{d}x.$$

$$(5.13\mathrm{c})$$

The CDF of $\widetilde{X}$ is given by

$$F_{\widetilde{X}}(y) = \frac{2}{B(m,m)} \int\limits_{-\infty}^{y} [F(x)]^{m-1} \Big\{ [1-F(x)]^m - [1-F(2\,y-x)]^m \Big\}\,f(x)\,\mathrm{d}x. \quad (5.13\mathrm{d})$$

Another measure of central tendency in a sample – besides the median – is the **mid–range**:

$$M := \frac{X_{1:n} + X_{n:n}}{2} \qquad (5.14\mathrm{a})$$

with DF

$$f_M(y) = 2\,n\,(n-1) \int\limits_{-\infty}^{y} [F(2\,y-x) - F(x)]^{n-2}\,f(x)\,f(2\,y-x)\,\mathrm{d}x \qquad (5.14\mathrm{b})$$

and CDF

$$F_M(y) = n \int\limits_{-\infty}^{y} [F(2\,y-x) - F(x)]^{n-1}\,f(x)\,\mathrm{d}x. \qquad (5.14\mathrm{c})$$

The **difference** of two arbitrary order statistics

$$W_{rs} := X_{s:n} - X_{r:n}; \ \ 1 \le r < s \le n \qquad (5.15\mathrm{a})$$

has DF

$$f_{W_{rs}}(y) = \frac{n!}{(r-1)!\,(s-r-1)!\,(n-s)!} \int\limits_{-\infty}^{\infty} [F(x)]^{r-1}\,[F(y+x) - F(x)]^{s-r-1} \times$$

$$[1 - F(y+x)]^{n-s}\,f(x)\,f(y+x)\,\mathrm{d}x. \qquad (5.15\mathrm{b})$$

One special case of (5.15a) is the **spacing of order $r$**:

$$W_r := X_{r:n} - X_{r-1:n}; \ \ r = 1, 2, \ldots, n \ \text{ and } \ X_{0:n} \equiv 0 \qquad (5.16\mathrm{a})$$

with DF

$$f_{W_r}(y) = \frac{n!}{(r-2)!\,(n-r)!} \int\limits_{-\infty}^{\infty} [F(x)]^{r-2}\,[1 - F(y+x)]^{n-r}\,f(x)\,f(y+x)\,\mathrm{d}x \quad (5.16\mathrm{b})$$

and mean

$$E\big(W_{r+1}\big) = E\big(X_{r+1:n} - X_{r:n}\big) = \binom{n}{r} \int\limits_{-\infty}^{\infty} \big[F(x)\big]^r \big[1 - F(x)\big]^{n-r} \, dx; \;\; r = 1, 2, \ldots, n-1.$$

(5.16c)

The sum of the first $r$ spacings $W_i$ gives $X_{r:n}$:

$$X_{r:n} = \sum_{i=1}^{r} W_i; \;\; r = 1, 2, \ldots, n,$$

whereas the sum of the **normalized spacings** $n\,W_1$, $(n-1)\,W_2$, $\ldots$, $W_n$ gives the **total time on test** when $n$ items are put on test without replacement of failing items and the test lasts until the last failure (see Sect. 8.3.2.4).

Another special case of (5.15a) is the **range** $W$:

$$W := X_{n:n} - X_{1:n}$$

(5.17a)

with DF

$$f_W(y) = n\,(n-1) \int\limits_{-\infty}^{\infty} \big[F(x+y) - F(x)\big]^{n-2} f(x)\, f(x+y)\, dx,$$

(5.17b)

CDF

$$F_W(y) = n \int\limits_{-\infty}^{\infty} \big[F(x+y) - F(x)\big]^{n-1} f(x)\, dx$$

(5.17c)

and mean and variance

$$E(W) \;=\; E\big(X_{n:n}\big) - E\big(X_{1:n}\big)$$

(5.17d)

$$\mathrm{Var}(W) \;=\; \mathrm{Var}\big(X_{n:n}\big) - 2\,\mathrm{Cov}\big(X_{1:n}, X_{n:n}\big) + \mathrm{Var}\big(X_{1:n}\big).$$

(5.17e)

We now proceed to more **general linear functions of order statistics**:

$$L_n = \sum_{i=1}^{n} a_{in}\, X_{i:n}.$$

A major use of such functions arises in the estimation of location and scale parameters $a$ and $b$ for a location–scale distribution with a DF of the form

$$f_X(x \mid a, b) = \frac{1}{b}\, g\!\left(\frac{x-a}{b}\right); \, a \in \mathbb{R},\; b > 0,$$

where $g(\cdot)$ is parameter–free. Denoting the reduced variate by

$$U := \frac{X - a}{b}$$

and the moments of its order statistics by

$$\alpha_{r:n} \;:=\; E\big(U_{r:n}\big)$$

(5.18a)

$$\beta_{r,s:n} \;:=\; \mathrm{Cov}\big(U_{r:n}, U_{s:n}\big); \;\; r, s = 1, 2, \ldots, n,$$

(5.18b)

it follows that

$$U_{r:n} = \frac{X_{r:n} - a}{b},$$

so that

$$\mathrm{E}\big(X_{r:n}\big) \;=\; a + b\,\alpha_{r:n}, \tag{5.18c}$$

$$\mathrm{Cov}\big(X_{r:n}, X_{s:n}\big) \;=\; b^2\,\beta_{r,s:n}. \tag{5.18d}$$

Thus $\mathrm{E}\big(X_{r:n}\big)$ is linear in the parameters $a$ and $b$ with known coefficients, and $\mathrm{Cov}\big(X_{r:n}, X_{s:n}\big)$ is known apart from $b^2$. Therefore, the GAUSS–MARKOV least squares theorem may be applied, in a slightly generalized form since the covariance matrix is not diagonal. This gives the best linear unbiased estimators (BLUEs)

$$\widehat{a} \;=\; \sum_{i=1}^{n} a_{in}\, X_{i:n}, \tag{5.18e}$$

$$\widehat{b} \;=\; \sum_{i=1}^{n} b_{in}\, X_{i:n}, \tag{5.18f}$$

where the coefficients $a_{in}$ and $b_{in}$, which are functions of the $\alpha_{i:n}$ and $\beta_{i,j:n}$, can be evaluated once and for all, depending on the DF $g(.)$ of the reduced variate. This technique will be applied in Chapter 10.

### 5.1.3   Record times and record values[4]

We turn to the investigation of a specific sequence $L(m)$, $m \geq 1$, of random sample size and of the corresponding random maximum $Z_{L(m)}$. Let $X_1, X_2, \ldots$ be independent random variables with common continuous distribution function $F(x)$. Let $L(1) = 1$ and, for $m \geq 2$, let

$$L(m) = \min \big\{ j \,:\, j > L(m-1),\ X_j > X_{L(m-1)} \big\}. \tag{5.19}$$

The sequences $L(m)$ and $X_{L(m)}$ can be interpreted as follows. Consider an infinite sequence $X_1, X_2, \ldots$ of continuously iid random variables. Then let us go through the sequence $X_1, X_2, \ldots$ in their order of observation with the aim of picking out larger and larger terms. Obviously, the first largest is $X_1$. Then, for any $k$, if $Z_k = X_1$, we ignore $X_2, \ldots, X_k$, and we take that $X$ as the next one, i.e. $X_{L(2)}$, when for the first time, $Z_k > X_1$. We then continue the process. In other words, the investigation of $L(m)$ gives an insight into the position of those observations that change $Z_k$. The variates $X_{L(m)} = Z_{L(m)}$ thus form the increasing sequence $Z_1 < Z_{L(2)} < \cdots$,

---

[4]  This section is mainly based on the survey article "A record of records" by NEVZEROV/BALAKRISHNAN in BALAKRISHNAN/RAO (1998). DALLAS (1982) gives results on record values from the WEIBULL distribution.

**Definitions:** The sequence $L(m)$, $m \geq 1$, defined in (5.19) is called sequence of **upper record times**, record times for short. $m$ is the counter for the records and $L(m)$ is the random position of the $m$–th record in the series $X_1, X_2, \ldots$.

The corresponding $X$–value, i.e., $X_{L(m)} = Z_{L(m)}$, is called an **upper record**, record for short.

The sequence $\Delta(m) = L(m) - L(m-1)$, $m \geq 2$, is called the sequence of **inter–record times**.

The sequence $N(n)$, $n = 1, 2, \ldots$, denotes the **number of records** within the first $n$ variates of $X_1, X_2, \ldots$.

**Remark**: If we replace $>$ in (5.19) by $<$ we speak of **lower records**. In the sequel we will only deal with upper records, because the theory would be the same for lower records. In fact, we can obtain the lower records from the upper records by changing the sequence $X_1, X_2, \ldots$ to $-X_1, -X_2, \ldots$ or in the case when $X$'s are positive to $1/X_1, 1/X_2, \ldots$.[5]

$L(m)$ and $N(n)$ do not depend on the parent distribution $F(x)$. This is evident, because if $X_j \geq X_i$, then $F(X_j) \geq F(X_i)$ and the sequence $F(X_j)$ is a sequence of independent uniform variates. Hence, for arbitrary $F(x)$, $L(m)$ can be defined in (5.19) by the additional assumption that the variables $X_j$ are independent and uniformly distributed in $[0, 1]$.

We will first give the distributions of $L(m)$ and $N(n)$, respectively. To this end we introduce indicators $\xi_1, \xi_2, \ldots$ defined as follows:

$$\xi_i = \left\{ \begin{array}{ll} 1 & \text{if } X_i \text{ is a record,} \\ 0 & \text{otherwise,} \end{array} \right\} \tag{5.20a}$$

where $\Pr(\xi_i = 1) = 1/i$; $i = 1, 2, \ldots$. Then

$$N(n) = \sum_{i=1}^{n} \xi_i \tag{5.20b}$$

and

$$\Pr\left[L(m) \geq n\right] = \Pr\left[N(n) \leq m\right] = \Pr\left(\sum_{i=1}^{n} \xi_i \leq m\right).[6] \tag{5.20c}$$

A first result is the following **theorem**:

The sequence $L(1), L(2), \ldots$ is a MARKOV chain where

$$\Pr\left[L(m) = k \mid L(m-1) = \ell\right] = \frac{\ell}{k\,(k-1)}, \quad \text{if } k > \ell \geq m - 1 \geq 2, \tag{5.21a}$$

---

[5] For a discrete distribution one can introduce **weak records**. For it, we have to use the sign "$\geq$" in (5.19) instead of "$>$". In this case any repetition of a record value is a record too.

[6] The relation between $L(m)$ and $N(n)$ is the same as that between the number of renewals $N_t$ in $(0, t]$ and the time to the $n$–th renewal $T_n$; cf. (4.27a).

and

$$\Pr\left[L(m) > k \mid L(m-1) = \ell\right] = \frac{\ell}{k}, \quad \text{if } k \geq \ell. \qquad \blacksquare \qquad (5.21b)$$

The joint and marginal distributions of record times are as follows:

- if $1 < k_2 < k_3 < \ldots < k_m$, then:

$$\Pr\left[L(2) = k_2, \ldots, L(m) = k_m\right] = \frac{1}{k_m\,(k_m - 1)\,(k_{m-1} - 1) \cdot \ldots \cdot (k_2 - 1)}, \tag{5.22a}$$

- for any $m \geq 2$ and $k \geq m$:

$$\Pr\left[L(m) = k\right] = \sum_{1 < k_2 < \ldots < k_{m-1} < k} \frac{1}{k_m\,(k_m - 1) \cdot \ldots \cdot (k_2 - 1)}, \tag{5.22b}$$

- in particular for $m = 2$:

$$\Pr\left[L(2) = k\right] = \frac{1}{k\,(k-1)}, \quad k \geq 2. \tag{5.22c}$$

To connect the distributions of $L(m)$ and $N(n)$, we have to use (5.20c) and

$$\begin{aligned}
\Pr\left[L(m) = n\right] &= \Pr\left[N(n-1) = m-1,\ \xi_n = 1\right] \\
&= \frac{1}{n}\Pr\left[N(n-1) = m-1\right].
\end{aligned}$$

Let $\mathfrak{s}_{(r,j)}$, denote the **STIRLING number of the first kind**[7] defined by

$$x\,(x-1) \cdot \ldots \cdot (x - r + 1) = \sum_{j=0}^{r} \mathfrak{s}_{(r,j)}\,x^j \tag{5.23a}$$

or recursively by

$$\mathfrak{s}_{(r,j)} = \mathfrak{s}_{(r-1,j-1)} - (r-1)\,\mathfrak{s}_{(r-1,j)}; \quad r \geq j \geq 2 \tag{5.23b}$$

with

$$\mathfrak{s}_{(r,r)} = 1; \quad \mathfrak{s}_{(r,r-1)} = -\binom{r}{2}, \quad \mathfrak{s}_{(r,1)} = (-1)^{r-1}\,(r-1)!.$$

Then

$$\Pr\left[L(m) = k\right] = \frac{\left|\mathfrak{s}_{(k-1,m-1)}\right|}{k!}, \quad k \geq m \geq 2, \tag{5.24a}$$

and

$$\Pr\left[L(m) = k\right] \approx \frac{(\ln k)^{m-2}}{k^2\,(m-2)!} \quad \text{as } k \to \infty. \tag{5.24b}$$

---

[7] $(-1)^{r-j}\,\mathfrak{s}_{(r,j)}$ is the number of permutations of $r$ symbols which have exactly $j$ cycles.

We further have

$$\Pr\big[N(n) = k\big] \;=\; \frac{\big|\mathfrak{s}_{(k,n)}\big|}{n!}\,,\quad n \ge k \ge 1, \tag{5.25a}$$

$$\Pr\big[N(n) = k\big] \;\approx\; \frac{(\ln n)^{k-1}}{n\,(k-1)!}\,,\quad \text{as } n \to \infty. \tag{5.25b}$$

Fig. 5/1 shows (5.24a) for $m = 2,\ 3$ and $4$ in the upper part and (5.25a) for $n = 5,\ 10,\ 15$ in the lower part.

Figure 5/1: Probability distributions of $L(m)$ and $N(n)$

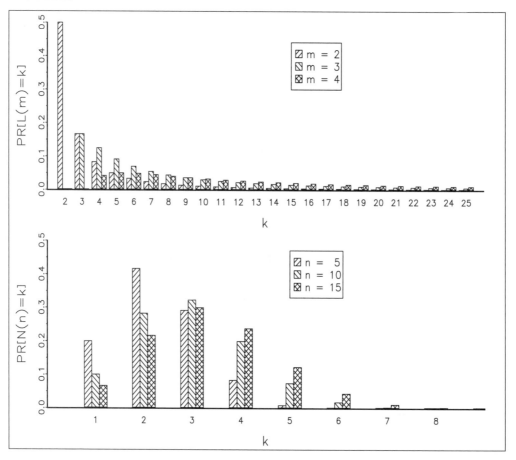

For the number of records $N(n)$ in the first $n$ sample units, we have the following moments

$$\mathrm{E}\big[N(n)\big] \;=\; \sum_{i=1}^{n} \frac{1}{i}\,, \tag{5.26a}$$

$$\mathrm{Var}\big[N(n)\big] \;=\; \sum_{i=1}^{n} \frac{1}{i} - \sum_{i=1}^{n} \frac{1}{i^2}\,, \tag{5.26b}$$

with $\mathrm{E}\big[N(n)\big]/\ln n \to 1$ and $\mathrm{Var}\big[N(n)\big] \to 1$ as $n \to \infty$.

With respect to the record times $L(m)$ we see that $\Pr[L(2) = k] = [k(k-1)]^{-1}$ for $k \geq 2$. This means that $\mathrm{E}[L(2)] = \infty$[8] and hence $\mathrm{E}[L(m)] = \infty$ for any $m \geq 2$. But

$$\mathrm{E}[\ln L(m)] = n - \gamma + O(m^2/2^m), \quad m \to \infty, \qquad (5.27a)$$

$$\mathrm{Var}[\ln L(m)] = n - \pi^2/6 + O(m^3/2^m), \quad m \to \infty, \qquad (5.27b)$$

$\gamma \approx 0.5772$ being EULER's constant. $N(n)$ and $\ln L(m)$ are asymptotically normal:

$$\Pr\left[\frac{N(n) - n}{\sqrt{\ln n}} \leq x\right] = \Phi(x), \quad n \to \infty, \qquad (5.28a)$$

$$\Pr\left[\frac{L(m) - m}{\sqrt{m}} \leq x\right] = \Phi(x), \quad m \to \infty, \qquad (5.28b)$$

$\Phi(x)$ being the CDF of the standard normal distribution.

With respect to the **inter–record times** $\Delta(m) = L(m) - L(m-1)$, $m \geq 2$, we state the following distributional results:

$$\Pr[\Delta(m) > k] = \int_0^\infty (1 - e^{-x})^k \frac{x^{m-2} e^{-x}}{(m-2)!} \, dx$$

$$= \int_0^1 u^k \frac{-[\ln(1-u)]^{m-2}}{(m-2)!} \, du$$

$$= \sum_{i=0}^k \binom{k}{i} (-1)^i (1+i)^{1-m}; \quad k = 0, 1, \ldots; m = 2, 3, \ldots. \quad (5.29a)$$

$$\Pr[\Delta(m) = k] = \int_0^\infty (1 - e^{-x})^{k-1} \frac{x^{m-2} e^{-2x}}{(m-2)!} \, dx$$

$$= \Pr[\Delta(m) > k - 1] - \Pr[\Delta(m) > k]; \quad k = 1, 2, \ldots. \quad (5.29b)$$

From (5.29a,b) it follows that $\mathrm{E}[\Delta(m)] = \infty$ for any $m = 2, 3, \ldots$, but the logarithmic moments can be approximated:

$$\mathrm{E}[\ln \Delta(m)] \approx m - 1 + \gamma. \qquad (5.29c)$$

We further have

$$\lim_{m \to \infty} \Pr\left[\frac{\ln \Delta(m) - m}{\sqrt{m}} \leq x\right] = \Phi(x). \qquad (5.29d)$$

---

[8] Notice the meaning of this statement. If a disaster has been recorded by $X_1$, then the value of $X_2$ bringing an even larger disaster has probability 0.5, but the expected waiting time to a larger disaster is infinity.

Fig. 5/2 shows (5.29b) for $m = 2$, 3 and 4.

Figure 5/2:  Probability function $\Pr\big[\Delta(m) = k\big]$ for $m = 2$, 3, 4

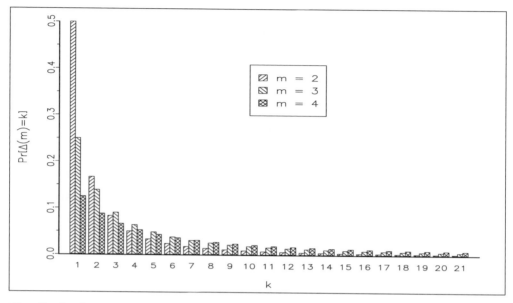

The distribution of the record values

$$X(m) := X_{L(m)}$$

depends on $F(x)$ and $f(x)$ of the parent distribution in the following way:

$$\Pr\big[X(m) \le x\big] = F_{X(m)}(x) = \frac{1}{(m-1)!} \int_{0}^{-\ln[1-F(x)]} u^{m-1}\, e^{-u}\, \mathrm{d}u; \quad m = 1, 2, \ldots$$

The DF of $X(m)$ is

(5.30a)

$$f_{X(m)} = \frac{f(x)}{(m-1)!} \big\{ -\ln[1 - F(x)] \big\}^{m-1}; \quad m = 1, 2, \ldots,$$   (5.30b)

and the joint DF of $X(\ell)$ and $X(m)$, $1 \le \ell < m$, is

$$f_{X(\ell), X(m)}(x, y) = \frac{1}{(\ell - 1)!\,(m - \ell - 1)!} \frac{f(x)}{1 - F(x)} \big\{ -\ln[1 - F(x)] \big\}^{\ell - 1} \times$$
$$f(y) \big\{ -\ln[1 - F(y)] + \ln[1 - F(x)] \big\}^{m - \ell - 1}, \quad x < y. \text{(5.30c)}$$

The MARKOV structure of record values implies

$$\Pr\big[X(m + 1) \ge x \,\big|\, X(m) = u\big] = \frac{\Pr(X \ge x)}{\Pr(X > u)}, \quad x > u.$$   (5.31a)

For any $m > 1$ and $n > 1$ the following **relationship between order statistics and record values** exists almost surely:

$$\Pr\big[X(m) > x \,\big|\, X(m-1) = u\big] = \Pr\big[X_{n:n} > x \,\big|\, X_{n-1:n} = u\big]. \tag{5.31b}$$

## 5.2   WEIBULL order statistics[9]

Let $X_{1:n}$ be the **first** or smallest **order statistic** in a sample of size $n$, where $X_i \sim We(a,b,c); i = 1,2,\dots,n$; i.e.:

$$f(x) \;=\; \frac{c}{b}\left(\frac{x-a}{b}\right)^{c-1} \exp\left\{-\left(\frac{x-a}{b}\right)^c\right\},$$

$$F(x) \;=\; 1 - \exp\left\{-\left(\frac{x-a}{b}\right)^c\right\}.$$

Then, using (5.2a,b), we have the familiar result — cf. Sect. 3.1.4 and (3.57a):

$$\begin{aligned}
f_{1:n}(x) &= n\,f(x)\left[1 - F(x)\right]^{n-1} \\[4pt]
&= \frac{n\,c}{b}\left(\frac{x-a}{b}\right)^{c-1}\exp\left\{-n\left(\frac{x-a}{b}\right)^c\right\} \\[4pt]
&= \frac{c}{b^*}\left(\frac{x-a}{b^*}\right)^{c-1}\exp\left\{-\left(\frac{x-a}{b^*}\right)^c\right\},
\end{aligned} \tag{5.32a}$$

where

$$b^* = b\,n^{-1/c} \tag{5.32b}$$

and

$$\begin{aligned}
F_{1:n}(x) &= 1 - \left[1 - F(x)\right]^n \\[4pt]
&= 1 - \exp\left\{-n\left(\frac{x-a}{b}\right)^c\right\} \\[4pt]
&= 1 - \exp\left\{-\left(\frac{x-a}{b^*}\right)^c\right\}.
\end{aligned} \tag{5.32c}$$

It is readily seen from (5.32a,c) that $X_{1:n}$ is also distributed as a WEIBULL variate:

$$X_{1:n} \sim We(a, b^*, c).$$

---

[9] Suggested reading for this section:   BALAKRISHNAN/COHEN (1991), BALAKRISHNAN/JOSHI (1981), DAVID/GROENEVELD (1982), FREIMER/MUDHOLKAR/LIN (1989), HARTER (1988), KAMPS (1991), KHAN/KHAN/PARVEZ (1984), KHAN/PARVEZ/YAQUB (1983), KHAN/YAQUB/PARVEZ (1983), LIEBLEIN (1955), LOCHNER/BASU/DIPONZIO (1974), MALIK/TRUDEL (1982), MOHIE–EL– DIN/MAHMOUD/ABO–YOUSSEF (1991), NIGM/EL–HAWARY (1996), PATEL (1975), PATEL/READ (1975).

In other words, the WEIBULL distribution is "closed under minima." For the raw moments of $X_{1:n}$ we have the familiar expression, see (2.64a):

$$E\left(X_{1:n}^r\right) = \sum_{j=0}^{r} \binom{r}{j} a^j \, (b^*)^{r-j} \, \Gamma\left(1 + \frac{r-j}{c}\right). \tag{5.32d}$$

In the sequel when discussing the $r$–th order statistic, $1 \leq r \leq n$, of a WEIBULL variate, we take the reduced WEIBULL variate $U \sim We(0, 1, c)$ with DF

$$f(u) = c \, u^{c-1} \, \exp\left\{-u^c\right\}$$

and CDF

$$F(u) = 1 - \exp\left\{-u^c\right\}.$$

The corresponding results for the three–parameter distribution $X \sim We(a, b, c)$ can be obtained by using the linear transformation

$$X_{r:n} = a + b \, U_{r:n}.$$

The DF of $U_{r:n}$, $1 \leq r \leq n$, is

$$f_{r:n}(u) = r \binom{n}{r} c \, u^{c-1} \, \exp\left\{-u^{(n-r+1)\,c}\right\} \, [1 - \exp(-u^c)]^{r-1}. \tag{5.33}$$

From (5.33) we obtain the **$k$–th raw moment of $U_{r:n}$** to be

$$
\begin{aligned}
E\left(U_{r:n}^k\right) &= r \binom{n}{r} \int_0^\infty u^k \, c \, u^{c-1} \, \exp\left\{-u^{(n-r+1)\,c}\right\} \, \left[1 - \exp\left(-u^c\right)\right]^{r-1} du \\
&= r \binom{n}{r} \sum_{i=0}^{r-1} (-1)^i \binom{r-1}{i} \int_0^\infty u^k \, c \, u^{c-1} \, \exp\left\{-(n-r+i+1) \, u^c\right\} du \\
&= r \binom{n}{r} \Gamma\left(1 + \frac{k}{c}\right) \sum_{i=0}^{r-1} \frac{(-1)^i \binom{r-1}{i}}{(n-r+i+1)^{1+(k/c)}}.
\end{aligned} \tag{5.34}
$$

(5.34) is due to LIEBLEIN (1955).

The computing of moments of order statistics heavily relies on recurrence relations. Besides the general recurrence relation given in (5.11a), there exist special relations for the WEIBULL distribution; see NIGM/EL–HAWARY (1996). We will present a special recurrence relation proposed by BALAKRISHNAN/JOSHI (1981). With respect to any arbitrary CDF $F(x)$, we have the following two recursions for the CDF $F_{r:n}(x)$ of $X_{r:n}$:

$$F_{r:n}(x) = F_{r-1:n}(x) - \binom{n}{r-1} [F(x)]^{r-1} \, [1 - F(x)]^{n-r+1} \tag{5.35a}$$

with $F_{0:n}(x) = 1 \; \forall \, x$, so that (5.35a) is true for $r = 1$ as well,

$$(r-1) \, F_{r:n}(x) + (n-r+1) \, F_{r-1:n}(x) = n \, F_{r-1:n-1}(x). \tag{5.35b}$$

On substituting $F_{r-1:n}(x)$ from (5.35b) into equation (5.35a), we get

$$F_{r:n}(x) = F_{r-1:n-1}(x) - \binom{n-1}{r-1} [F(x)]^{r-1} [1 - F(x)]^{n-r+1}. \qquad (5.35c)$$

Then using

$$E(U_{r:n}^k) = k \int_0^\infty x^{k-1} [1 - F_{r:n}(u)]\, du$$

and substituting for $F_{r:n}(u)$ from (5.32c), we get

$$E(U_{r:n}^k) = E(U_{r-1:n-1}^k) + k \binom{n-1}{r-1} \int_0^\infty x^{k-1} [1 - \exp\{-u^c\}]^{r-1} [\exp\{-u^c\}]^{n-r+1}\, du. \qquad (5.35d)$$

For $m \geq 1$, consider the integral

$$J_k(p,m) = \int_0^\infty x^{k-1} [1 - \exp\{-u^c\}]^m [\exp\{-u^c\}]^p\, du$$

$$= J_k(p, m-1) - J_k(p+1, m-1) \qquad (5.35e)$$

by writing $[1 - \exp\{-u^c\}]^m$ as $[1 - \exp\{-u^c\}]^{m-1} [1 - \exp\{-u^c\}]$ and splitting the integral in two. Also,

$$J_k(p,0) = \int_0^\infty x^{k-1} [\exp\{-u^c\}]^p\, du$$

$$= \frac{\Gamma(k/c)}{c\, p^{k/c}}. \qquad (5.35f)$$

Thus $J_k(p,0)$, $p \geq 1$, can be calculated by using the gamma function. The function $J_k(p,m)$ for $m \geq 1$ can now be obtained by using (5.35e) recursively. Finally, (5.35d) can be written as

$$E(U_{r:n}^k) = E(U_{r-1:n-1}^k) + k \binom{n-1}{r-1} J_k(n-r+1,\, r-1). \qquad (5.35g)$$

Starting with

$$E(U_{1:1}^k) = k\, J_k(1,0), \qquad (5.35h)$$

we obtain all the single raw moments of order statistics from a reduced WEIBULL distribution avoiding the usage of the gamma function to a great extent.

PATEL/READ (1975) give the following **bounds on the first moments** of $U_{r:n}$:

$$E(U_{r:n}) \leq \left\{ \sum_{i=1}^r (n-i+1)^{-1} \right\}^{1/c} \quad \text{for } c > 1 \qquad (5.36a)$$

$$E(U_{r:n}) \geq \left\{ \sum_{i=1}^r (n-i+1)^{-1} \right\}^{1/c} \quad \text{for } c < 1. \qquad (5.36b)$$

The joint DF of $U_{r:n}$ and $U_{s:n}$ $(1 \le r < s \le n)$ is

$$
\begin{aligned}
f_{r,s:n}(u,v) \;=\; & \frac{n!}{(r-1)!\,(s-r-1)!\,(n-s)!}\, c^2\,(u\,v)^{c-1}\,\exp\{-u^c\}\,\times \\
& \exp\{-(n-s+1)\,v^c\}\,\bigl[1-\exp\{-u^c\}\bigr]^{r-1} \\
& \bigl[\exp\{-u^c\}-\exp\{-v^c\}\bigr]^{s-r-1},\; 0 < u < v < \infty. \quad (5.37)
\end{aligned}
$$

From (5.37) we obtain the **product moment of $U_{r:n}$ and $U_{s:n}$** as [10]

$$
\begin{aligned}
\mathrm{E}\bigl(U_{r:n}\,U_{s:n}\bigr) \;=\; & \int_0^\infty \int_0^v u\,v\,f_{r,s:n}(u,v)\,\mathrm{d}u\,\mathrm{d}v \\[6pt]
=\; & \frac{n!\,c^2}{(r-1)!\,(s-r-1)!\,(n-s)!} \sum_{i=0}^{r-1}\sum_{j=0}^{s-r-1}(-1)^{s-r-1-j+i}\,\times \\
& \binom{r-1}{i}\binom{s-r-1}{j}\int_0^\infty\int_0^v \exp\{-(i+j+1)\,u^c\}\,\times \\
& \exp\{-(n-r-j)\,v^c\}\,(u\,v)^c\,\mathrm{d}u\,\mathrm{d}v \quad (5.38\mathrm{a}) \\[6pt]
=\; & \frac{n!}{(r-1)!\,(s-r-1)!\,(n-s)!} \sum_{i=0}^{r-1}\sum_{j=0}^{s-r-1}(-1)^{s-r-1-j+i}\,\times \\
& \binom{r-1}{i}\binom{s-r-1}{j}\phi_c(i+j+1,\,n-r-j). \quad (5.38\mathrm{b})
\end{aligned}
$$

This result is due to LIEBLEIN (1955) and $\phi_c(a,b)$ is LIEBLEIN's $\phi$–function defined by

$$
\phi_c(a,b) = c^2 \int_0^\infty \int_0^y \exp\{-a\,x^c - b\,y^c\}\,x^2\,y^2\,\mathrm{d}x\,\mathrm{d}y. \quad (5.38\mathrm{c})
$$

Through a differential equation approach, LIEBLEIN has derived an explicit algebraic formula for the $\phi$–function:

$$
\phi_c(a,b) = \frac{\Gamma^2\!\left(1+\dfrac{1}{c}\right)}{(a\,b)^{1+(1/c)}}\, I_{a/(a+b)}\!\left(1+\frac{1}{c},\,1+\frac{1}{c}\right) \quad \text{for } a \ge b, \quad (5.38\mathrm{d})
$$

where $I_p(c,d)$ is PEARSON's **incomplete beta function** defined as

$$
I_p(c,d) = \frac{\Gamma(c+d)}{\Gamma(c)\,\Gamma(d)} \int_0^p t^{c-1}\,(1-t)^{d-1}\,\mathrm{d}t,\; 0 < p \le 1.
$$

---

[10] Recurrence relations between product moments $\mathrm{E}\bigl(U_{r:n}^j\,U_{s:n}^k\bigr)$ are given by KHAN/PARVEZ/YAQUB (1983).

When $a < b$, $\phi_c(a, b)$ may be computed from the identity

$$\phi_c(a, b) + \phi_c(b, a) = \frac{\Gamma^2\left(1 + \dfrac{1}{c}\right)}{(a\,b)^{1+(1/c)}}. \tag{5.38e}$$

(5.38a) together with $\mathrm{E}(U_{r:n})$ and $\mathrm{E}(U_{s:n})$ leads to the **covariance**

$$\mathrm{Cov}(U_{r:n}, U_{s:n}) = \mathrm{E}(U_{r:n}\,U_{s:n}) - \mathrm{E}(U_{r:n})\,\mathrm{E}(U_{s:n}).$$

LOCHNER/BASU/DIPONZIO (1974) give the following **approximation to the covariance of order statistics** from the reduced WEIBULL distribution:

$$\mathrm{Cov}(U_{r:n}, U_{s:n}) \approx A(n, r, s) \cdot (a + b\,n + c\,r + d\,s), \tag{5.39}$$

where

$$
\begin{aligned}
A(n, r, s) &= \left[r/(n+2)\,(n+1-r)\right] c^{-2} \left[-\ln(1 - r/(n+1))\right]^{(1/c)-1} \times \\
&\qquad \left[-\ln(1 - s/(n+1))\right]^{1/c-1}
\end{aligned}
$$

with the coefficients $a$, $b$, $c$ and $d$ from Tab. 5/1.

Table 5/1: Coefficients to approximate the covariance of order statistics from the reduced WEIBULL distribution

| $c$ | $a$ | $b$ | $c$ | $d$ |
|-----|-----|-----|-----|-----|
| 1.5 | 1.3426 | −0.0270 | 0.0006 | 0.0212 |
| 2.0 | 1.2761 | −0.0214 | −0.0016 | 0.0190 |
| 2.5 | 1.2664 | −0.0186 | −0.0039 | 0.0171 |
| 3.0 | 1.2721 | −0.0169 | −0.0059 | 0.0155 |
| 3.5 | 1.2830 | −0.0157 | −0.0076 | 0.0142 |

There exists a number of tables giving the means, variances and covariances of order statistics from the reduced WEIBULL distribution:

- WEIBULL (1959) used LIEBLEIN's expression for $\mathrm{E}(U_{r:n})$ to tabulate the means, variances and covariances to five decimal places for $n = 1(1)15$, $r = 1(1)n$ and $c = 1/\alpha$ with $\alpha = 0.1(0.1)0.6(0.2)1.0$.

- WEIBULL (1967a) presented means, variances and covariances of all order statistics for $n = 5(5)20$ and $c^{-1} = 0.1(0.1)1.0$ to five decimal places.

- GOVINDARAJULU/JOSHI (1968), based on LIEBLEIN's results, tabulated the means, variances and covariances to five decimal places for $n = 1(1)10$ and for $c = 1.0(0.5)3.0(1.0)10$.

- MCELHONE/LARSEN (1969) tabulated $\mathrm{E}(U_{r:n})$ to six significant figures for $n = 1(1)25$ and $c = 1(1)10$, $\mathrm{Var}(U_{r:n})$ and $\mathrm{Cov}(U_{r:n}, U_{s:n})$ for $c = 1(1)5$ for the same set of $n$–values.

- HARTER (1970) tabulated $E(U_{r:n})$ to five decimal places for $n = 1(1)40$, $r = 1(1)n$ and $c = 0.5(0.5)4.0(1.0)8.0$.
- BALAKRISHNAN/CHAN (1993a) proposed tables for the means, variances and covariances of all order statistics for $n$ up to 20 and $c = 1/5$, $1/4$, $1/3$, $1/2$, $1.5(0.5)3$, $4(2)10$.

Table 5/2 presents — as a sample — means, variances and covariances of all order statistics for $n = 10$ and $c = 2$ in the upper part and $c = 3$ in the lower part.

Table 5/2: Means, variances and covariances of all order statistics for $n = 10$ and $c = 2$ and 3

| $r$ \ $s$ | $c = 2$ $\mathrm{Var}(U_{r:10})$ for $r = s$; $\mathrm{Cov}(U_{r:10}, U_{s:10})$ for $s > r$ | | | | | | | | | | $E(U_{r:10})$ |
|---|---|---|---|---|---|---|---|---|---|---|---|
| | 1 | 2 | 3 | 4 | 5 | 6 | 7 | 8 | 9 | 10 | |
| 1 | 0.02146 | 0.01523 | 0.01218 | 0.01025 | 0.00886 | 0.00776 | 0.00684 | 0.00601 | 0.00521 | 0.00430 | 0.28025 |
| 2 | — | 0.02462 | 0.01974 | 0.01664 | 0.01439 | 0.01262 | 0.01113 | 0.00979 | 0.00848 | 0.00700 | 0.43184 |
| 3 | — | — | 0.02692 | 0.02273 | 0.01969 | 0.01728 | 0.01525 | 0.01342 | 0.01163 | 0.00960 | 0.55605 |
| 4 | — | — | — | 0.02935 | 0.02546 | 0.02237 | 0.01976 | 0.01740 | 0.01509 | 0.01246 | 0.67054 |
| 5 | — | — | — | — | 0.03229 | 0.02841 | 0.02513 | 0.02214 | 0.01923 | 0.01589 | 0.78316 |
| 6 | — | — | — | — | — | 0.03615 | 0.03202 | 0.02825 | 0.02456 | 0.02032 | 0.89971 |
| 7 | — | — | — | — | — | — | 0.04168 | 0.03685 | 0.03208 | 0.02658 | 1.02662 |
| 8 | — | — | — | — | — | — | — | 0.05050 | 0.04409 | 0.03663 | 1.17408 |
| 9 | — | — | — | — | — | — | — | — | 0.06772 | 0.05656 | 1.36427 |
| 10 | — | — | — | — | — | — | — | — | — | 0.12092 | 1.67572 |

| $r$ \ $s$ | $c = 3$ $\mathrm{Var}(U_{r:10})$ for $r = s$; $\mathrm{Cov}(U_{r:10}, U_{s:10})$ for $s > r$ | | | | | | | | | | $E(U_{r:10}$ |
|---|---|---|---|---|---|---|---|---|---|---|---|
| | 1 | 2 | 3 | 4 | 5 | 6 | 7 | 8 | 9 | 10 | |
| 1 | 0.02269 | 0.01369 | 0.00998 | 0.00785 | 0.00643 | 0.00537 | 0.00453 | 0.00380 | 0.00314 | 0.00243 | 0.41448 |
| 2 | — | 0.01945 | 0.01422 | 0.01122 | 0.00920 | 0.00769 | 0.00649 | 0.00545 | 0.00450 | 0.00348 | 0.56264 |
| 3 | — | — | 0.01789 | 0.01414 | 0.01161 | 0.00971 | 0.00821 | 0.00690 | 0.00570 | 0.00441 | 0.66949 |
| 4 | — | — | — | 0.01717 | 0.01412 | 0.01183 | 0.01000 | 0.00842 | 0.00696 | 0.00539 | 0.76043 |
| 5 | — | — | — | — | 0.01700 | 0.01427 | 0.01208 | 0.01018 | 0.00842 | 0.00653 | 0.84459 |
| 6 | — | — | — | — | — | 0.01731 | 0.01468 | 0.01240 | 0.01027 | 0.00797 | 0.92730 |
| 7 | — | — | — | — | — | — | 0.01825 | 0.01544 | 0.01282 | 0.00997 | 1.01316 |
| 8 | — | — | — | — | — | — | — | 0.02016 | 0.01679 | 0.01310 | 1.10838 |
| 9 | — | — | — | — | — | — | — | — | 0.02435 | 0.01913 | 1.22512 |
| 10 | — | — | — | — | — | — | — | — | — | 0.03749 | 1.40417 |

Looking at

$$E(U^r) = \Gamma\left(1 + \frac{r}{s}\right)$$

one sees that **negative moments** of order $r > -c$ do exist for the reduced WEIBULL variate. KHAN/KHAN/PARVEZ (1984) discussed the negative moments of all WEIBULL order statistics of order greater than $-c$.

PATEL (1975) derived **bounds on moments of linear functions** of WEIBULL order statistics. The bounds are obtained by using JENSEN's inequality on the expected values of some functions of the CDF $F(\cdot)$ which are convex (concave) and strictly increasing. Defining

$$B_i := \sum_{j=n-i+1}^{n} \frac{1}{j}, \quad C_i := -\sum_{j=i}^{n} \frac{1}{j} \quad \text{and} \quad m := c^{-1}$$

and demanding $c \geq 1$ we have the following bounds for the **means of the midrange**

$$
\begin{align}
E\big[(U_{1:n} + U_{n:n})/2\big] &\leq \big[(B_1)^m + (B_n)^m\big]/2 \tag{5.40a}\\
E\big[(U_{1:n} + U_{n:n})/2\big] &\leq \big[(B_1 + B_2)/2\big]^m \tag{5.40b}\\
E\big[(U_{1:n} + U_{n:n})/2\big] &\geq \big[\{-\ln(1 - e^{C_1})\}^m + \{-\ln(1 - e^{C_n})\}^m\big]/2 \tag{5.40c}\\
E\big[(U_{1:n} + U_{n:n})/2\big] &\geq \big[\ln(1 - \exp\{(C_1 + C_2)/2\})\big]^m \tag{5.40d}
\end{align}
$$

and for the **means of the range**

$$
\begin{align}
E(U_{n:n} - U_{n:n}) &\leq (B_n - B_1)^m \tag{5.41a}\\
E(U_{n:n} - U_{1:n}) &\leq (B_n)^m - \{-\ln(1 - e^{C_1})\}^m \tag{5.41b}\\
E(U_{n:n} - U_{1:n}) &\geq \{-\ln(1 - e^{C_n})\}^m - (B_1)^m. \tag{5.41c}
\end{align}
$$

The intervals given by (5.40a) and (5.40c) are smaller than those given by (5.40b) and (5.40d). (5.41b) is nearer to the true value of $E(U_{n:n} - U_{1:n})$ than (5.41a). The paper of PATEL also gives bounds on the means of the $m$–th midrange $E\big[(U_{m:n} + U_{n-m+1:n})/2\big]$ and the $m$–th range $E(U_{n-m+1} - U_{m:n})$ and on the means of spacings.[11]

MALIK/TRUDEL (1982) derived the DF of the **quotient**

$$Z := U_{r:n}/U_{s:n}, \quad 1 \leq r < s \leq n,$$

**of two WEIBULL order statistics** using the MELLIN transform technique:

$$
\left.
\begin{aligned}
f_Z(z) &= \frac{n!}{(r-1)!\,(s-r-1)!\,(n-s)!} \sum_{i=0}^{s-r-1} \sum_{j=0}^{r-1} (-1)^{i+j} \binom{s-r-1}{i}\binom{r-1}{j} \times \\[2mm]
&\qquad \frac{c\,z^{c-1}}{\big[(n-s+i+1) + (s-r-i+j)\,z^c\big]^2}, \quad 0 < z \leq 1.
\end{aligned}
\right\}
\tag{5.42a}
$$

---

[11] DAVID/GROENEVELD (1982) discuss the expected length of spacings, whereas FREIMER et al. (1989) discuss the distribution of extreme spacings $U_{n:n} - U_{u-1:n}$.

Some special cases of (5.42a) are

- the ratio of the extreme order statistics $Z = U_{1:n}/U_{n:n}$:

$$f_z(Z) = n\,(n-1) \sum_{i=0}^{n-2} (-1)^i \binom{n-2}{i} \frac{c\,z^{c-1}}{\left[(i+1) + (n-i-1)\,z^c\right]^2},\ 0 < z \le 1,\ (5.42\text{b})$$

- the ratio of consecutive order statistics $Z = U_{r:n}/U_{r+1:n},\ r = 1,2,\dots,n-1$:

$$f_z(Z) = \frac{n!}{(r-1)!\,(n-r-1)!} \sum_{j=0}^{r-1} (-1)^j \binom{r-1}{j} \frac{c\,z^{c-1}}{\left[(n-r) + (j+1)\,z^c\right]^2},\ 0 < z \le 1.$$

$$(5.42\text{c})$$

If one is interested in the **simulation of order statistics** from a three–parameter WEIBULL distribution, i.e., $X_i \overset{\text{iid}}{\sim} We(a,b,c)$, NEWBY (1979) proposed a generator that promises to be faster than first generating $n$ iid variates $X_1, \dots, X_n$, see (2.61), and afterwards sorting them in ascending order. The sequence of order statistics is directly generated by

$$\left.\begin{aligned} h_0 &= 0, \\ h_{r+1} &= h_r - \ln(Z_r)/(n-r);\ r = 0,1,\dots,n-1; \\ X_{r+1:n} &= a + b\,\bigl(h_{r+1}\bigr)^{1/c}, \end{aligned}\right\} \qquad (5.43)$$

where $Z_r$ is uniformly distributed in $(0,1)$.

## 5.3 WEIBULL record values[12]

For reasons given in the preceding section we confine ourselves to the discussion of $U \sim We(0,1,c)$. Applying (5.30a,b) to the distribution of the reduced WEIBULL variate $U$, we arrive at the following CDF and DF of the $m$–th record value $U(m)$ in an iid sequence $U_1, U_2, \dots$:

$$\begin{aligned} F_{U(m)}(u) &= \frac{1}{\Gamma(m)} \int_0^{u^c} v^{m-1}\,e^{-v}\,\mathrm{d}v \\[2mm] &= 1 - \frac{\Gamma(m\,|\,u^c)}{\Gamma(m)};\ u > 0\ \text{and}\ m = 1,2,\dots; \qquad (5.44\text{a}) \end{aligned}$$

$$f_{U(m)}(u) = \frac{c\,u^{c\,m-1}}{\Gamma(m)}\,\exp\{-u^c\};\ u > 0\ \text{and}\ m = 1,2,\dots \qquad (5.44\text{b})$$

We see that (5.44a,b) are the CDF and DF of STACY's generalized gamma distribution, cf. (3.24d), where $a = 0$, $b = 1$ and $m$ is the second shape parameter besides $c$. From (5.44b) the $r$–th raw moment of $U(m)$ is obtained to be

$$\mathrm{E}\bigl[U(m)^r\bigr] = \frac{\Gamma\left(m + \dfrac{r}{c}\right)}{\Gamma(m)};\ r \ge 1,\ m = 1,2,\dots \qquad (5.44\text{c})$$

---

[12] Suggested reading for this section: DALLAS (1982), KAMPS (1991), PAWLAS/SZYNAL (2000a,b), SULTAN/BALAKRISHNAN (1999/2000) and HOINKES/PADGETT (1994) for the maximum likelihood estimation of the WEIBULL parameters based on observed record values.

In particular, we have

$$E[U(m)] = \frac{\Gamma\left(m + \frac{1}{c}\right)}{\Gamma(m)} \tag{5.44d}$$

and

$$\mathrm{Var}[U(m)] = \frac{\Gamma(m)\,\Gamma\left(m + \frac{2}{c}\right) - \Gamma^2\left(m + \frac{1}{c}\right)}{\Gamma^2(m)}. \tag{5.44e}$$

We get the following recurrence relation:

$$E[U(m+1)^r] = \left(1 + \frac{r}{m\,c}\right) E[U(m)^r]. \tag{5.44f}$$

The joint DF of $U(\ell)$ and $U(m)$, $1 \le \ell < m$, is

$$f_{U(\ell),U(m)}(u, v) = \frac{c^2}{\Gamma(\ell)\,\Gamma(m - \ell)}\,(u\,v)^{c-1}\,u^{c\,(\ell-1)}\left[v^c + u^c\right]^{c-1} \times$$
$$\exp\{-v^2\}; \quad 0 < u < v; \; \ell = 1, 2, \ldots, m - 1. \tag{5.45a}$$

From (5.45a) the product moment of order $1 + 1$ follows as

$$E[U(\ell)\,U(m)] = \frac{\Gamma\left(\ell + \frac{1}{c}\right)}{\Gamma(\ell)}\,\frac{\Gamma\left(m + \frac{2}{c}\right)}{\Gamma\left(m + \frac{1}{c}\right)}, \quad 1 \le \ell \le m - 1, \tag{5.45b}$$

so that the covariance of $U(\ell)$ and $U(m)$ is

$$\mathrm{Cov}[U(\ell)\,U(m)] = \frac{\Gamma\left(\ell + \frac{1}{c}\right)}{\Gamma(\ell)}\left[\frac{\Gamma\left(m + \frac{2}{c}\right)}{\Gamma\left(m + \frac{1}{c}\right)} - \frac{\Gamma\left(m + \frac{1}{c}\right)}{\Gamma(m)}\right], \quad 1 \le \ell \le m - 1. \tag{5.45c}$$

With respect to product moments of higher order than $1 + 1$, we can state the following recurrence relations:

$$E[U(\ell)^r\,U(\ell+1)^s] = \frac{\ell\,c}{\ell\,c + r}\,E[U(\ell+1)^r]; \quad \ell \ge 1; \; r, s = 1, 2, \ldots; \tag{5.45d}$$

$$E[U(\ell)^r\,U(m)^s] = \frac{\ell\,c}{\ell\,c + r}\,E[U(\ell+1)^r\,U(m)^s];$$
$$1 \le \ell \le m - 2; \; r, s = 1, 2, \ldots; \tag{5.45e}$$

$$E\big[U(\ell)^r\,U(\ell+2)^s\big] = \Big(1+\frac{s}{c}\Big)\,E\big[U(\ell)^r\,U(\ell+1)^s\big] \;-\ell\,\Big\{E\big[U(\ell+1)^r\,U(\ell+2)^s\big]$$

$$-\,E\big[U(\ell+1)^{r+s}\big]\Big\};\;\; \ell\ge 1;\; r,s=1,2,\ldots; \tag{5.45f}$$

$$E\big[U(\ell)^r\,U(m+1)^s\big] = \Big(1+\frac{s}{c\,(m-\ell)}\Big)E\big[U(\ell)^r\,U(m)^s\big] -\frac{\ell}{m-\ell}\Big\{E\big[U(\ell+1)^r\,U(m+1)^s\big]$$

$$-\,E\big[U(\ell+1)^r\,U(m)^s\big]\Big\};\;\; 1\le\ell\le m-2;\; r,s=1,2,\ldots. \tag{5.45g}$$

## 5.4   Log-WEIBULL order statistics[13]

Starting from the two–parameter WEIBULL distribution with scale parameter $b$ and shape parameter $c$, the logarithmic transformation converts this scale–and–shape–parameter distribution into a more tractable location–and–scale–parameter distribution; see Sect. 3.3.4. If $X \sim We(0,b,c)$ with CDF

$$F_X(x) = 1 - \exp\Big\{-\Big(\frac{x}{b}\Big)^c\Big\},$$

then

$$Y = \ln X$$

has a type–I–minimum distribution with location parameter

$$a^* = \ln b,\;\; a^* \in \mathbb{R},$$

and scale parameter

$$b^* = 1/c,\;\; b^* > 0;$$

i.e., the CDF and DF of $Y$ are

$$F_Y(y) = 1 - \exp\Big\{-\exp\Big[\frac{y-a^*}{b^*}\Big]\Big\},\;\; y\in\mathbb{R}, \tag{5.46a}$$

$$f_Y(y) = \frac{1}{b^*}\exp\Big\{\frac{y-a^*}{b^*} - \exp\Big(\frac{x-a^*}{b^*}\Big)\Big\},\;\; y\in\mathbb{R}. \tag{5.46b}$$

In the sequel we will only discuss the order statistics $Z_{r:n}$ of the reduced Log–WEIBULL variate $Z = (y-a^*)/b^*$ having

$$F_Z(z) = 1 - \exp\big\{-e^z\big\},\;\; z\in\mathbb{R}, \tag{5.47a}$$

$$f_Z(z) = \exp\big\{z - e^z\big\},\;\; z\in\mathbb{R}. \tag{5.47b}$$

---

[13]  Suggested reading for this section: HARTER (1988), LIEBLEIN (1953), WHITE (1964b, 1967b, 1969).

Applying (5.1e) to (5.47a,b) we get the following DF of $Z_{r:n}$:

$$f_{r:n}(z) = \frac{n!}{(r-1)!\,(n-r)!}\,\exp\{z - (n-r+1)\,e^z\}\,\left[1 - \exp\{-e^z\}\right]^{r-1}.\quad (5.47c)$$

For the first order statistic (5.47c) changes to

$$f_{1:n}(z) = n\,\exp\{z - n\,e^z\} = \exp\{(z + \ln n) - e^{z+\ln n}\} \qquad (5.48a)$$

with corresponding CDF

$$\begin{aligned}
F_{1:n}(z) &= 1 - \left[1 - F_Z(z)\right]^n \\
&= 1 - \exp\{-n\,e^z\} = 1 - \exp\{-e^{z+\ln n}\}. \qquad (5.48b)
\end{aligned}$$

We thus see that $Z_{1:n}$ again has a Log–WEIBULL distribution, but with an additional location parameter $(-\ln n)$; i.e., $Z_{1:n}$ has the same distribution as $Y = Z - \ln n$ and the Log–WEIBULL distribution is "closed under minima" like the ordinary WEIBULL distribution; see (5.32a–c). So the raw moments of $Z_{1:n}$ can easily be traced back to those of $Z$; see (3.44d–g):

$$E(Z_{1:n}) = E(Z) - \ln n = -\gamma - \ln n \approx -0.577215 - \ln n, \qquad (5.48c)$$

$$E(Z_{1:n}^2) = E\left[(Z - \ln n)^2\right] = \frac{\pi^2}{6} + (\gamma + \ln n)^2, \qquad (5.48d)$$

$$Var(Z_{1:n}) = Var(Z - \ln n) = Var(Z) = \frac{\pi^2}{6} \approx 1.97811. \qquad (5.48e)$$

Applying the recurrence relation (5.11a) we can express the raw moments of $Z_{r:n}$, $r > 1$, in terms of the raw moments of $Z_{1:n}$, e.g.:

$$E(Z_{2:n}) + (n-1)\,E(Z_{1:n}) = n\,E(Z_{1:n-1})$$

$$\begin{aligned}
\implies E(Z_{2:n}) &= -(n-1)\left[-\gamma - \ln n\right] + n\left[-\gamma - \ln(n-1)\right] \\
&= -\gamma + (n-1)\,\ln n - n\,\ln(n-1).
\end{aligned}$$

WHITE (1969) has proposed a method to compute $E(Z_{r:n}^k)$ directly without calculating $E(Z_{s:n}^k)$ for all $s < r$. For $k = 1$ the formula is

$$E(Z_{r:n}) = -\gamma - \sum_{j=0}^{r-1}(-1)^j\,\binom{n}{j}\,\Delta^j \ln(n-j), \qquad (5.49a)$$

where $\Delta^j$ is the $j$–th forward difference operator. For $k = 2$ WHITE gives

$$\left.\begin{aligned}
E(Z_{r:n}^2) &= \frac{\pi^2}{6} + \gamma^2 + 2\gamma \sum_{j=0}^{r-1}(-1)^j \binom{n}{j}\,\Delta^j \ln(n-j) + \\
&\quad \sum_{j=0}^{r-1}(-1)^j \binom{-n}{j}\,\Delta^j\left[\ln(n-j)\right]^2.
\end{aligned}\right\} \qquad (5.49b)$$

Combining (5.49a) and (5.49b) gives

$$\mathrm{Var}(Z_{r:n}) = \mathrm{E}(Z_{r:n}^2) - [\mathrm{E}(Z_{r:n})]^2$$

$$= \frac{\pi^2}{6} + \sum_{j=0}^{r-1}(-1)^j \binom{n}{j} \Delta^j [\ln(n-j)]^2 - \left\{\sum_{j=1}^{r-1}(-1)^j \binom{n}{j} \Delta^j \ln(n-j)\right\}^2 (5.49c)$$

The problem in applying these formulas for the numerical computation, for large $n$, is the buildup of rounding errors in the logarithms. So, the logarithm must be evaluated with a great number of decimal places.

For estimating the parameters $a^*$ and $b^*$ of the Log–WEIBULL distribution — or equivalently the parameters $b = e^{a^*}$ and $c = 1/b^*$ of the ordinary WEIBULL distribution — from order statistics (see (5.18a–f) and Sect. 10.2) we need the covariances of $Z_{r:n}$ and $Z_{s:n}$ which are built upon the cross moments $\mathrm{E}(Z_{r:n} Z_{s:n})$. The latter follow from the joint DF of $Z_{r:n}$ and $Z_{s:n}$:

$$f_{r,s:n}(u,v) = \frac{n!}{(r-1)!(s-r-1)!(n-s)!}[1-\exp\{-e^u\}]^{r-1}$$

$$[\exp\{-e^u\} - \exp\{-e^v\}]^{s-r-1} \times$$

$$\exp\{-(n-s+1)e^v - e^u + u + v\}; \ 1 \le r < s \le n, \ u < v, \ (5.50a)$$

as

$$\mathrm{E}(Z_{r:n} Z_{s:n}) = \iint\limits_{-\infty < u < v < \infty} u \, v \, f_{r,s:n}(u,v) \, \mathrm{d}u \, \mathrm{d}y. \tag{5.50b}$$

Making the change of variable

$$u = \ln y \quad \text{and} \quad v = \ln z,$$

we have

$$\mathrm{E}(Z_{r:n} Z_{s:n}) = C \iint\limits_{-\infty < y < z < \infty} \ln y \, \ln z \, [1 - e^{-y}]^{r-1} [e^{-y} - e^{-z}]^{s-r-1} \times \\ \exp\{(n-s+1)z\} e^{-y} \, \mathrm{d}z \, \mathrm{d}y, \right\}$$

$$(5.50c)$$

where

$$C = \frac{n!}{(r-1)!\,(s-r-1)!\,(n-s)!}.$$

Expanding the integrand in (5.50c) by the binomial theorem we get

$$\mathrm{E}(Z_{r:n} Z_{s:n}) = C \sum_{p=1}^{r-1} \sum_{q=1}^{r-s-1} (-1)^{p+q} \binom{r-1}{p}\binom{s-r-1}{q} \theta(p+s-r-q, n-s+q+1),$$

$$(5.50d)$$

where

$$\theta(\ell, m) = \int\limits_{0}^{\infty} \int\limits_{y}^{\infty} \ln y \, \ln z \, \exp\{-\ell \, y - m \, z\} \, dz \, dy. \qquad (5.51a)$$

WHITE (1964b) then uses results from LIEBLEIN (1954) to reduce the double integral in (5.51a) such that it is suitable to numerical calculations:

$$\theta(\ell, m) = \frac{\dfrac{\pi^2}{6} + [\gamma + \ln(\ell + m)]^2}{m\,(\ell + m)} + \frac{(\gamma + \ln \ell)\,\ln\left(\dfrac{m}{\ell + m}\right)}{\ell\,m} - \frac{L(1) - L\left(\dfrac{m}{\ell + m}\right)}{\ell\,m},$$

$$(5.51b)$$

where $L(x)$ is **EULER's dilogarithm function**

$$L(x) = - \int\limits_{0}^{x} \frac{\ln(1 - t)}{t} \, dt \qquad (5.51c)$$

with

$$L(1) = \pi^2/6 \quad \text{and} \quad L(x) = \sum_{k=1}^{\infty} \frac{x^k}{k^2}.$$

Table 5/3 is compiled from WHITE (1964b)[14] and gives $\mathrm{E}(Z_{r:10})$ together with $\mathrm{Cov}(Z_{r:10}, Z_{s:10}) = \mathrm{E}(Z_{r:10}\, Z_{s:10}) - \mathrm{E}(Z_{r:10})\,\mathrm{E}(Z_{s:10})$ for $1 \leq r \leq s \leq 10$, so that $\mathrm{Var}(Z_{r:10})$ is also included as $\mathrm{Cov}(Z_{r:10}, Z_{r:10})$.

Table 5/3: Means, variances and covariances of all order statistics of the reduced Log-WEIBULL variate for $n = 10$

| $\,^{s}$ $r$ | \multicolumn{10}{c}{$\mathrm{Var}(Z_{r:10})$ for $r = s$; $\mathrm{Cov}(Z_{r:10}, Z_{s:10})$ for $s > r$} | $\mathrm{E}(Z_{r:10})$ |
|---|---|---|---|---|---|---|---|---|---|---|---|
| | 1 | 2 | 3 | 4 | 5 | 6 | 7 | 8 | 9 | 10 | |
| 1 | 1.64493 | 0.61876 | 0.35919 | 0.24260 | 0.17615 | 0.13282 | 0.10185 | 0.07803 | 0.05824 | 0.03962 | −2.87980 |
| 2 | — | 0.64586 | 0.37650 | 0.25489 | 0.18536 | 0.13991 | 0.10738 | 0.08231 | 0.06148 | 0.04184 | −1.82620 |
| 3 | — | — | 0.39702 | 0.26954 | 0.19637 | 0.14842 | 0.11403 | 0.08749 | 0.06538 | 0.04453 | −1.26718 |
| 4 | — | — | — | 0.28739 | 0.20986 | 0.15888 | 0.12221 | 0.09387 | 0.07021 | 0.04786 | −0.86808 |
| 5 | — | — | — | — | 0.22686 | 0.17211 | 0.13261 | 0.10200 | 0.07639 | 0.05213 | −0.54361 |
| 6 | — | — | — | — | — | 0.18958 | 0.14641 | 0.11282 | 0.08463 | 0.05785 | −0.25745 |
| 7 | — | — | — | — | — | — | 0.16581 | 0.12812 | 0.09635 | 0.06603 | 0.01204 |
| 8 | — | — | — | — | — | — | — | 0.15191 | 0.11471 | 0.07893 | 0.28369 |
| 9 | — | — | — | — | — | — | — | — | 0.14879 | 0.10319 | 0.58456 |
| 10 | — | — | — | — | — | — | — | — | — | 0.17143 | 0.98987 |

---

[14] WHITE gives tables for means, variances and covariances of all order statistics for sample sizes from $n = 2$ to $n = 20$.

## 5.5 Order statistics and record values for several related WEIBULL distributions

We conclude this chapter with a short section on order statistics and related variables pertaining to the double, the truncated, the extended and the inverse WEIBULL distributions.

In Sect. 3.3.2 we have presented the **double WEIBULL distribution**; see (3.40a,b). The order statistics of the reduced double WEIBULL variate with DF

$$f_D(x \,|\, 0, 1, c) = \frac{c}{2} \, |x|^{c-1} \, \exp\{-|x|^c\}; \quad x \in \mathbb{R}, \, c > 0 \qquad (5.52a)$$

and CDF

$$F_D(x \,|\, 0, 1, c) = \left\{ \begin{array}{ll} 0.5 \, \exp\{-|x|^c\} & \text{for} \quad x \leq 0 \\[2mm] 1 - 0.5 \, \exp\{-x^c\} & \text{for} \quad x > 0 \end{array} \right\} \qquad (5.52b)$$

have been studied in some detail by BALAKRISHNAN/KOCHERLAKOTA (1985). (5.52a,b) in conjunction with (5.1e) give the following DF of $X_{r:n}$:

$$f_{r:n}(x) = \left\{ \begin{array}{ll} C_{r,n} \dfrac{c \, |x|^{c-1}}{2^r} \, \exp\{-r \, |x|^c\} \left[ 1 - \dfrac{1}{2} \exp\{-|x|^c\} \right]^{n-r} & \text{for} \quad x \leq 0 \\[4mm] C_{r,n} \dfrac{c \, x^{c-1}}{2^{n-r+1}} \, \exp\{-(n-r+1) \, x^c\} \left[ 1 - \dfrac{1}{2} \exp\{-x^c\} \right]^{r-1} & \text{for} \quad x > 0 \end{array} \right\} \qquad (5.53a)$$

with

$$C_{r,n} := \frac{n!}{(r-1)! \, (n-r)!} \, .$$

The $k$–th raw moment of $X_{r:n}$ found by BALAKRISHNAN/KOCHERLAKOTA is

$$\mathrm{E}\big(X_{r:n}^k\big) = C_{r,n} \, \Gamma\left(1 + \frac{k}{c}\right) \left\{ (-1)^k \sum_{i=0}^{n-r} (-1)^i \frac{\dbinom{n-r}{i}}{2^{r+i} \, (r+i)^{1+(k/c)}} + \right.$$

$$\left. \sum_{i=0}^{r-1} (-1)^i \frac{\dbinom{r-1}{i}}{2^{n-r+1+i} \, (n-r+1+i)^{1+(k/c)}} \, . \right\} \qquad (5.53b)$$

In computing the first moments $\mathrm{E}\big(X_{r:n}\big)$, one should observe the symmetry relation

$$\mathrm{E}\big(X_{r:n}\big) = -\mathrm{E}\big(X_{n-r+1:n}\big).$$

With respect to the mixed moment $\mathrm{E}\big(X_{r:n} \, X_{s:n}\big)$, $1 \leq r < s \leq n$, the authors derived the rather complicated expression

$$
\begin{aligned}
\mathrm{E}\big(X_{r:n}\,X_{s:n}\big) &= \\
&= C_{r,s,n}\Bigg\{ \sum_{i=0}^{n-s}\sum_{j=0}^{s-r-1}(-1)^{s-r+i-1-j}\binom{n-s}{i}\binom{s-r-1}{j}2^{-(s+i)}\,\phi_c(i+j+1,\,s-j-1) + \\
&\quad \sum_{i=0}^{r-1}\sum_{j=0}^{s-r-1}(-1)^{s-r+i-1-j}\binom{r-1}{i}\binom{s-r-1}{j}2^{-(n-r+i+1)}\,\phi_c(i+j+1,\,n-r-j) - \\
&\quad \sum_{i=0}^{s-r-1}\sum_{j=0}^{s-r-1-i}(-1)^{i+j}\binom{s-r-1}{i}\binom{s-r-1-i}{j}\frac{\Gamma^2\!\left(1+\dfrac{1}{c}\right)}{2^{n-s+r+i+j+1}}\,\frac{1}{\big[(r+i)(n-s+1+j)\big]^{1+(1/c)}}\Bigg\}
\end{aligned}
$$

$$(5.53c)$$

with

$$
C_{r,s,n} := \frac{n!}{(r-1)!\,(s-r-1)!\,(n-s)!}
$$

and $\phi_c(a,b)$ as LIEBLEIN's $\phi$–function defined in (5.38b,c).

In Sect. 3.3.5 we have introduced three types of **truncated WEIBULL distributions**. Order statistics of the reduced form of the doubly truncated WEIBULL distributions with DF

$$
f_{DT}(x\,|\,c,t_\ell,t_r) = \frac{c\,x^{c-1}\,\exp\{-x^c\}}{P-Q},\qquad 0\le t_\ell \le x \le t_r < \infty \tag{5.54a}
$$

and CDF

$$
F_{DT}(x\,|\,c,t_\ell,t_r) = \frac{\exp\{-t_\ell^c\}-\exp\{-x^c\}}{P-Q},\qquad 0\le t_\ell \le x \le t_r < \infty, \tag{5.54b}
$$

where

$$
P = 1-\exp\{-t_r^c\}\quad\text{and}\quad Q = 1-\exp\{-t_\ell^c\}
$$

have been studied by KHAN et al. (1983) and MOHIE–EL–DIN et al. (1991) with the special aim to establish recurrence relations for the raw moments of $X_{r:n}$ in a sample from (5.54a,b). Two general results for the raw moments of $X_{r:n}$ from any truncated distribution are

$$
\mathrm{E}\big(X_{1:n}^k\big) = t_\ell^k + k\int_{t_\ell}^{t_r} x^{k-1}\big[1-F_{DT}(x)\big]^n\,\mathrm{d}x;\quad n\ge 1;\ k=1,2,\dots; \tag{5.55a}
$$

$$
\begin{aligned}
\mathrm{E}\big(X_{r:n}^k\big) &= \mathrm{E}\big(X_{r-1:n-1}^k\big) + \\
&\binom{n-1}{r-1}k\int_{t_\ell}^{t_r} x^{k-1}\big[F_{DT}(x)\big]^{r-1}\big[1-F_{DT}(x)\big]^{n-r-1}\,\mathrm{d}x;\quad 2\le r\le n; n\ge 2; k=1,2,\dots,
\end{aligned}
$$

$$(5.55b)$$

where $F_{DT}(x)$ is the CDF of the interesting doubly truncated distribution and under the assumptions

$$\lim_{x \to t_\ell} \left\{ x^k \left[ F_{DT}(x) \right]^{r-1} \left[ 1 - F_{DT}(x) \right]^{n-r+1} \right\} = \lim_{x \to t_r} \left\{ x^k \left[ F_{DT}(x) \right]^{r-1} \left[ 1 - F_{DT}(x) \right]^{n-r+1} \right\}$$
$$= 0.$$

Now, let

$$Q^* = \frac{1 - Q}{P - Q} \text{ and } P^* = \frac{1 - P}{P - Q},$$

then $1 - F_{DT}(x \mid c, t_\ell, t_r)$ from (5.54b) may be written as

$$1 - F_{DT}(x \mid c, t_\ell, t_r) = -P^* + \frac{\exp\{-x^c\}}{P - Q}$$
$$= -P^* + \frac{1}{c} x^{1-c} f_{DT}(x \mid c, t_\ell, t_r). \qquad (5.56a)$$

Putting the value of $1 - F_{DT}(x \mid c, t_l, t_r)$ into the general formula (5.55a), we get for the moments of the first order statistic $X_{1:n}$:

$$E(X_{1:n}^k) = t_\ell^k + k \int_{t_\ell}^{t_r} x^{k-1} \left[ 1 - F_{DT}(x \mid c, t_\ell, t_r) \right]^{n-1} \left[ -P^* + \frac{1}{c} x^{1-c} \right] f_{DT}(x \mid c, t_\ell, t_r) \, dx$$

and after some manipulation

$$E(X_{1:n}^k) = Q^* t_\ell^k - P^* E(X_{1:n-1}^k) + \frac{k}{nc} E(X_{1:1}^{k-c}) \qquad (5.56b)$$

with the special relation for $n = 1$

$$E(X_{1:1}^k) = Q^* t_\ell^k - P^* t_r^k + \frac{k}{c} E(X_{1:1}^{k-c}). \qquad (5.56c)$$

With (5.56a) inserted into (5.55b) and using the recurrence relation (5.11a), the raw moments of $X_{r:n}$, $2 \le r \le n$, have the following recurrence:

$$E(X_{r:n}^k) = Q^* E(X_{r-1:n-1}^k) - P^* E(X_{r:n-1}^k) + \frac{k}{nc} E(X_{r:n}^{k-c}) \qquad (5.56d)$$

with the special relation for $r = n$

$$E(X_{n:n}^k) = Q^* E(X_{n-1:n-1}^k) - P^* t_r^k + \frac{k}{nc} E(X_{n:n}^{k-c}). \qquad (5.56e)$$

The **extended WEIBULL distribution**, introduced in Sect. 3.3.10, has been studied by FREIMER et al. (1989) with respect to the **extreme spacings**

$$W_n = X_{n:n} - X_{n-1:n} \text{ and } W_1 = X_{2:n} - X_{1:n}$$

in order to analyze the tail character. The authors defined a distribution to have a **medium right tail** when, as $n \to \infty$, $a_n X_{n:n} + b_n$ converges in distribution to $-\ln(Y)$, where $Y$ denotes the reduced exponential distribution with DF $f(y) = \exp\{-y\}$. The extended WEIBULL distribution with CDF given by (3.108b) has a right tail which is always medium, but the length of the left tail measured by $W_1$ always depends on the parameter $d$ of the extended WEIBULL distribution.

The **inverse WEIBULL distribution** of Sect. 3.3.3 has been studied with respect to its lower record values by PAWLAS/SZYNAL (2000a) and with respect to its lower generalized order statistics by PAWLAS/SZYNAL (2000b) in order to give characterizations of this distribution.

# 6 Characterizations[1]

Characterization problems deal with the question of finding assumptions which determine the distribution of a variate, at least to the extent that the distribution function belongs to a certain family of distributions. Characterization theorems are located on the borderline between probability theory and mathematical statistics, and utilize numerous classical tools of mathematical analysis, such as the theory of functions of complex variables, differential equations of different types, theory of series and, last but not least, the theory of functional equations. The latter approach with respect to the WEIBULL distribution is presented in Sect. 6.1.

The characterization may be based on nearly every function to be constructed for a variate. Sect. 6.2 gives several WEIBULL characterizations using conditional moments. WEIBULL characterization theorems based on different aspects of order statistics will be found in Sect. 6.3. Miscellaneous approaches using different arguments, e.g., record values, truncation, entropy, FISHER information, quantiles or the distribution of the random hazard rate, are presented in Sect. 6.4. Chapter 6 closes with characterization theorems for several related WEIBULL distributions.

Not all theorems characterizing a WEIBULL distribution are new. Some of them are generalizations of characterizations for the exponential distribution because a WEIBULL variate is a monotonic transformation of an exponential variate.

## 6.1  WEIBULL characterizations based on functional equations[2]

Let $X$ be an exponential variate with CDF

$$F(x \mid b) = 1 - \exp\{-x/b\}, \quad x \ge 0, \ b > 0. \tag{6.1a}$$

An important and useful characterization of $X$ is its **lack of memory** which can be stated as

$$\Pr(X > x + y \mid X > y) = \Pr(X > x) \ \forall \, x, y \ge 0; \tag{6.1b}$$

i.e., the conditional probability of surviving another $x$ units of time, given survival up to $y$, is the same as surviving $x$ units of time for a new item. The technique commonly employed

---

[1]  Suggested general reading on characterizations: GALAMBOS/KOTZ (1978) and KAGAN/LINNIK/RAO (1973), on characterizations by means of order statistics: BALAKRISHNAN/RAO (1998, Part IV), on characterizations of the WEIBULL distribution using more than one approach: JANARDAN/TANEJA (1979a,b) and ROY/MUKHERJEE (1986).

[2]  Suggested reading for this section: JANARDAN/TANEJA (1979b), MOOTHATHU (1990), ROY/ MUKHERJEE (1986), SHIMIZU/DAVIES (1981), WANG (1976).

in proving this characterization is the well–known **CAUCHY functional equation**

$$\phi(x+y) = \phi(x) + \phi(y) \ \forall \, x, y \geq 0 \tag{6.2a}$$

or equivalently

$$\psi(x+y) = \psi(x) \, \psi(y) \ \forall \, x, y \geq 0. \tag{6.2b}$$

It is well known that if $\phi(\cdot)$ and $\psi(\cdot)$ are continuous, then the solution for (6.2a) is

$$\phi(x) = A \, x \ \forall \, x \geq 0, \ \phi(1) = A, \tag{6.3a}$$

and the solution for (6.2b) is

$$\psi(x) = B^x \ \forall \, x \geq 0, \ \psi(1) = B. \tag{6.3b}$$

The assumption of continuity on $\phi(\cdot)$ and $\psi(\cdot)$ can be relaxed to measurability.

A generalization of the exponential distribution (6.1a) is the WEIBULL distribution with CDF

$$F(x \,|\, b, c) = 1 - \exp\left\{-\left(\frac{x}{b}\right)^c\right\}; \ x \geq 0; \ b, c > 0. \tag{6.4}$$

A characterization of the WEIBULL distribution in the spirit of the memoryless property of the exponential distribution (6.1b) is presented by the property

$$\Pr\left(X > \sqrt[c]{x^c + y^c} \,\middle|\, X > y\right) = \Pr(X > y) \ \forall \, x, y \geq 0. \tag{6.5}$$

To prove that (6.5) is a characterizing property of (6.4) we need to solve the functional equation:

$$\phi\left(\sqrt[c]{x^c + y^c}\right) = \phi(x) + \phi(y) \ \forall \, x, y \geq 0 \tag{6.6a}$$

or equivalently

$$\psi\left(\sqrt[c]{x^c + y^c}\right) = \psi(x) \, \psi(y) \ \forall \, x, y \geq 0, \tag{6.6b}$$

where $c \neq 0$ is fixed. The solution is given in Proposition 1 where we shall assume $\phi(\cdot)$ and $\psi(\cdot)$ are right– or left–continuous. The proof of Theorem 1 establishing the characterizing property of (6.5) for (6.4) needs only the right–continuity.

**Proposition 1**

Suppose $\phi(\cdot)$ and $\psi(\cdot)$ are real–valued right– or left–continuous functions on $\mathbb{R}^+ = \{x \,|\, x \geq 0\}$ and $c \neq 0$. Then $\phi(\cdot)$ and $\psi(\cdot)$ satisfy the functional equations (6.6a,b) for all $x, y \in \mathbb{R}^+$ if and only if

$$\phi(x) = Ax^c \ \forall \, x \in \mathbb{R}^+, \ \phi(1) = A, \tag{6.7a}$$

and

$$\psi(x) = B^{x^c} \ \forall \, x \in \mathbb{R}^+, \ \psi(1) = B. \tag{6.7b}$$

Hint: If $c < 0$, we shall take the domain of $\phi(\cdot)$ and $\psi(\cdot)$ to be $\mathbb{R}^+ \backslash \{0\}$. ∎

**Proof of Proposition 1**

The sufficient conditions are obvious. To show the necessary condition for $\phi(\cdot)$, letting $y = x\,(m-1)^{1/c};\ m = 2, 3 \ldots$ , we have

$$\phi\big(x\,m^{1/c}\big) = \phi(x) + \phi\big[x\,(m-1)^{1/c}\big] \text{ for } x \geq 0 \text{ and } m = 2, 3, \ldots \qquad (6.8a)$$

(6.8a) leads to
$$\phi\big(x\,m^{1/c}\big) = m\,\phi(x) \text{ for } x \geq 0 \text{ and } m = 2, 3, \ldots \qquad (6.8b)$$

Consequently

$$\phi\big[x\,(n/m)^{1/c}\big] = (n/m)\,\phi(x) \text{ for } x \geq 0 \text{ and } n, m = 2, 3, \ldots \qquad (6.8c)$$

Letting $x = 1$, we get

$$\phi\big(r^{1/c}\big) = r\,\phi(1) \text{ for all rational } r \in \mathbb{R}^+. \qquad (6.8d)$$

Now, suppose $\phi(\cdot)$ is right–continuous. Let $x \in \mathbb{R}^+$ and $\{r_n\}$ be a sequence of rational numbers such that $r_n \geq x$ and $\lim_{n\to\infty} r_n = x$, then

$$\phi\big(x^{1/c}\big) = \lim_{n\to\infty} \phi\big(r_n^{1/c}\big) = \lim_{n\to\infty} r_n\,\phi(1) = x\,\phi(1). \qquad \blacksquare$$
$$(6.8e)$$

Similarly, we can prove that (6.7b) is the solution of (6.6b).

**Theorem 1**

Let $c \neq 0$ and let $X$ be a non–degenerate variate with $\Pr(X \geq 0) = 1$. Then $X$ has the WEIBULL distribution (6.4) if and only if $X$ satisfies

$$\Pr\big(X > \sqrt[c]{x^c + y^c}\,\big|\,X > y\big) = \Pr(X > x) \ \forall\, x, y \geq 0, \qquad (6.9)$$

with $c > 0$. If $c < 0$ there is no variate possessing property (6.9). $\qquad \blacksquare$

**Proof of Theorem 1**

It can easily be verified that (6.9) is a necessary condition for $X$ to be WEIBULL. To prove that is also a sufficient condition, denote $R(x) = \Pr(X > x)$. Then condition (6.9) is equivalent to
$$R\big(\sqrt[c]{x^c + y^c}\big) = R(x)\,R(y) \ \forall\, x, y \geq 0. \qquad (6.10a)$$

It follows from the definition of $R(\cdot)$ that it is right–continuous. Therefore, from Proposition 1, the solution for $R(\cdot)$ is

$$R(x) = B^{x^c} \ \forall\, x \geq 0 \text{ and } B = R(1). \qquad (6.10b)$$

Since $R(\cdot)$ is non–increasing with $R(0) = 1$ and $\lim_{x\to\infty} R(x) = 0$, there is no solution of $c < 0$. If $c > 0$, the solution is

$$R(x) = \exp\left\{-\left(\frac{x}{b}\right)^c\right\} \text{ for } b > 0 \text{ and all } x \geq 0. \qquad \blacksquare$$
$$(6.10c)$$

We give two applications of Theorem 1 in reliability theory:

1. We rewrite (6.9) as

$$\Pr(X > x \mid X > y) = \Pr\left(X > \sqrt[c]{x^c + y^c}\right), \; x \geq y \geq 0. \tag{6.11}$$

Suppose that $X$, the lifetime of an item, has the WEIBULL distribution (6.4) and that the item has not failed up to $y > 0$. Then the conditional probability that it will be working at time $x \geq y$ can be found by calculating the unconditional probability $\Pr\left(X > \sqrt[c]{x^c + y^c}\right)$.

2. Now we rewrite (6.9) as

$$\Pr(X > x + y \mid X > y) = \Pr\left(X > \sqrt[c]{(x + x)^c - y^c}\right); \; x, y \geq 0. \tag{6.12}$$

Since, for each $x \geq 0$, the function $f(y) = (x + y)^c - y^c$ is decreasing (increasing) in $y$ if and only $c \leq 1$ ($c \geq 1$), the distribution of $X$ is a decreasing (increasing) hazard rate distribution if and only if $c \leq 1$ ($c \geq 1$).

ROY/MUKHERJEE (1986) used the multiplicative version (6.2b) of the CAUCHY functional equation to characterize the WEIBULL distribution via its hazard function

$$H(x) \;=\; \left(\frac{x}{b}\right)^c; \; x \geq 0; \; b, c > 0; \tag{6.13a}$$

$$=\; \lambda\, x^c \text{ with } \lambda = b^{-c}. \tag{6.13b}$$

From (6.13b) it is clear that for all $x > 0$, $y > 0$,

$$H(x\,y)\, H(1) = H(x)\, H(y), \;\; H(1) > 0. \tag{6.14}$$

If conversely (6.14) holds for all $x > 0$ and all $y > 0$, then $H(x) = \lambda x^c$ for some $c > 0$, $\lambda > 0$, as guaranteed by the multiplicative version of the CAUCHY functional equation. Thus we have for $H(1) > 0$ the following.

**Theorem 2**

$$H(x\,y)\, H(1) = H(x)\, H(y) \;\; \forall\, x, y > 0 \text{ iff } X \sim We(0, b, c). \qquad \blacksquare$$

The papers of SHIMIZU/DAVIES (1981) and MOOTHATHU (1990) also rely on a functional equation approach to prove several characterization theorems for the WEIBULL distribution. We give the theorems without proofs.

**Theorem 3** (SHIMIZU/DAVIES, 1981)

Let $N$ be an integer valued variate independent of the $X$'s in $\{X_1, X_2, \ldots, X_N\}$, which are iid, such that $\Pr(N \geq 2) = 1$ and that the distribution of $\ln N$ has a finite mean and is not concentrated on a lattice $\rho, 2\rho, \ldots,$ for any $\rho > 0$. If the random variable

$$Y = N^{1/c} \min\{X_1, X_2 \ldots, X_N\}$$

has the same distribution $F(\cdot)$ as $X_1$, then there exists a positive number $\lambda$ such that

$$F(x) = 1 - \exp\{-\lambda x^c\}. \qquad \blacksquare$$

We will encounter a similar theorem in Sect. 6.3, namely Theorem 11, which assumes a fixed sample size $n$ instead of a random sample size $N$.

**Theorem 4** (SHIMIZU/DAVIES, 1981)

Let $m \geq 2$ be a fixed positive number and let $Y_1, Y_2, \ldots, Y_m$ be a set of positive random variables independent of the $X$'s in $\{X_1, \ldots, X_m\}$ and satisfying

$$\Pr\left(\sum_{j=1}^{m} Y_j^c = 1\right) = 1$$

and

$$\Pr\left(\frac{\ln Y_i}{\ln Y_j} \text{ is irrational for some } i \text{ and } j\right) > 0.$$

If the random variable

$$Z = \min\left\{\frac{X_1}{Y_1}, \frac{X_2}{Y_2}, \ldots, \frac{X_m}{Y_m}\right\}$$

has the same distribution as $X_1$, then there exists a positive constant $\lambda$ such that the distribution function of $X_i$; $i = 1, \ldots, n$ is

$$F(x) = 1 - \exp\{-\lambda x^c\}. \qquad \blacksquare$$

For Theorem 5 we need some prerequisites.

1. For fixed integer $m \geq 2$ and fixed real number $c > 0$, consider the functions $d_i(\cdot, c)$; $i = 1, \ldots, m$ defined on $\mathbb{R}_+^{m-1}$ as follows, where $\boldsymbol{y} = (y_1, \ldots, y_{m-1})' \in \mathbb{R}_+^{m-1}$:

$$d_1(\boldsymbol{y}, c) = 1/(1 + y_1^c + \ldots, y_{m-1}^c)^{1/c},$$
$$d_j(\boldsymbol{y}, c) = y_{j-1} d_1(\boldsymbol{y}, c); \quad j = 2, \ldots, m.$$

2. Let $X_1, \ldots, X_m$ be iid variates and let $\boldsymbol{Y} = (Y_1, \ldots, Y_{m-1})'$ be a vector of $m - 1$ positive variates independent of the $X$'s.

3. Let

$$W = \min\left\{\frac{X_1}{d_1(\boldsymbol{Y}, c)}, \frac{X_2}{d_2(\boldsymbol{Y}, c)}, \ldots, \frac{X_m}{d_m(\boldsymbol{Y}, c)}\right\}.$$

**Theorem 5** (MOOTHATU, 1990)

$X_1$ is WEIBULL with $F(x) = 1 - \exp\{-\lambda x^c\}$ if and only if $W$ is independent of $\boldsymbol{Y}$. In either case $W$ has the same distribution as $X_1$. $\qquad \blacksquare$

## 6.2   WEIBULL characterizations based on conditional moments[3]

In the preceding section we started with a characterization of the exponential distribution, i.e., the lack of memory property (6.1b), which was generalized to a characterization of the WEIBULL distribution. Here, we will start in the same way by first giving a characterization of the exponential distribution by conditional moments. SHANBHAG (1970) proved that $X$ is exponentially distributed if and only if

$$\mathrm{E}(X \mid X > y) = y + \mathrm{E}(X) \ \forall \, y > 0. \tag{6.15a}$$

(6.15) might be seen as another way of expressing "lack of memory." A simple transformation of (6.15a) results in

$$\mathrm{E}(X - y \mid X > y) = \mathrm{E}(X) \tag{6.15b}$$

indicating that the mean of the residual lifetime $X - y$ after having survived up to $y$ is the same as the mean lifetime of a new item ($y = 0$).

A first generalization of (6.15a) so that the case of a WEIBULL distribution is also covered was given by HAMDAN (1972) with the following.

**Theorem 6** (HAMDAN, 1972)

An absolutely continuous variate $X$ has CDF

$$F(x) = \begin{cases} 1 - \exp\{-h(x)/h(b)\} & \text{for} \quad x \in [\alpha, \beta) \\ 0 & \text{for} \quad x \notin [\alpha, \beta), \end{cases} \tag{6.16a}$$

where the interval $[\alpha, \beta)$ is closed on the right, whenever $\beta$ is finite, $b$ is a positive constant and $h(\cdot)$ is a strictly increasing differentiable function from $[\alpha, \beta)$ onto $[0, \infty)$, if and only if

$$\mathrm{E}\big[h(X) \mid X > y\big] = h(y) + h(b) \ \text{ for } \ y \in [\alpha, \beta). \qquad \blacksquare \tag{6.16b}$$

**Proof of Theorem 6**

The necessity of (6.16b) can be verified directly. To prove the sufficiency of (6.16b) let $G(x)$ be the CDF of $X$. Then (6.16b) may be put in the form

$$\big[1 - G(y)\big]\big[h(y) + h(b)\big] \ = \ \int_{y}^{\beta} h(x)\,\mathrm{d}G(x)$$

$$= \ \mathrm{E}\big[h(X)\big] - \int_{\alpha}^{y} h(x)\,\mathrm{d}G(x). \tag{6.17a}$$

---

[3] Suggested reading for this section: HAMDAN (1972), OUYANG (1987), SHANBHAG (1970), TAL-WALKER (1977).

Integration by parts yields

$$\left[1 - G(y)\right]\left[h(y) + h(b)\right] = \mathrm{E}\left[h(X)\right] - h(y)\,G(y) + \int_{\alpha}^{y} h'(x)\,G(x)\,\mathrm{d}x. \tag{6.17b}$$

Differentiating (6.17b) with respect to $y$ gives

$$h(b)\,\frac{\mathrm{d}G(y)}{\mathrm{d}y} = h'(y)\left[1 - G(y)\right] \tag{6.17c}$$

and hence

$$G(y) = 1 - K\,\exp\left\{-h(y)/h(b)\right\}, \tag{6.17d}$$

where $K$ is a constant. Because of $G(\alpha) = 0$, we get $K = 1$, noting that $h(\alpha) = 0$.    ■

When $h(x) = x^c$, $x \in [0, \infty)$ and $c > 0$, (6.16b) characterizes the two–parameter WEIBULL distribution. Comparing (6.4) with (6.16a), the condition (6.16b) results in

$$\mathrm{E}\left(X^c \,\big|\, X > y\right) = y^c + b^c. \tag{6.18}$$

Obviously, SHANBHAG's (1970) characterization of the exponential distribution with $\mathrm{E}(X) = b$ and $F(x) = 1 - \exp\{x/b\}$ corresponds to $c = 1$.

Another special case pertaining to Theorem 6 may be derived with

$$h(x) = -\ln(1 - x), \quad x \in [0, 1],$$

and $b$ such that $h(b) = 1$, i.e., $b = 1 - e^{-1}$. Then the CDF in (6.16a) reduces to that of the uniform distribution in $[0, 1]$ with $F(x) = x$ for $0 \le x \le 1$. In fact, all CDFs for powers of $X$, e.g., $X^c$, in $[0, 1]$ are characterized by choosing $h(x) = -\ln(1 - x^c)$ and $b$ such that $h(b) = 1$, i.e. $b = (1 - e^{-1})^{1/c}$.

TALWALKER (1977) extended HAMDAN's characterization (1972) so that it covers not only the WEIBULL distribution and the exponential distribution as its special cases but also BURR's distribution, the beta distribution of the second kind and the PARETO distribution.

**Theorem 7** (TALWALKER, 1977)

An absolutely continuous variate $X$ has CDF

$$F(x) = \left\{ \begin{array}{ll} 1 - \left[\dfrac{h(x) + \dfrac{g(k)}{\psi(k) - 1}}{h(\alpha) + \dfrac{g(k)}{\psi(k) - 1}}\right]^{\psi(k)/[1 - \psi(k)]} & \text{for } x \in [\alpha, \beta) \\[4ex] 0 & \text{otherwise,} \end{array} \right\} \tag{6.19a}$$

where the interval $[\alpha, \beta)$ is closed on the right. Whenever $\beta$ is finite, $h(\cdot)$ is a real valued, continuous and differentiable function on $[\alpha, \beta)$ with $\mathrm{E}\left[h(X)\right] = k$ and $g(\cdot)$ and $\psi(\cdot)$ are finite, real valued functions of $k$ if and only if

$$\mathrm{E}\left[h(X) \,\big|\, X > y\right] = h(y)\,\psi(k) + g(k). \qquad\qquad ■ \tag{6.19b}$$

**Proof of Theorem 7**

Since it is easy to verify the necessity of the condition, we prove its sufficiency. Given that

$$E\big[h(x)\,|\,x \geq y\big] = h(y)\,\psi(k) + g(k),$$

we get

$$\int\limits_{y}^{\beta} h(x)\,\mathrm{d}F(x) = \big[1 - F(y)\big]\,\big[h(y)\,\psi(k) + g(k)\big]. \tag{6.20a}$$

Differentiating (6.20a) with respect to $y$ yields

$$-h(y)\,F'(y) = \psi(h)\,h'(y)\,\big[1 - F(y)\big] - \mathrm{d}F(y)\,\big[h(y)\,\psi(k) + g(k)\big]$$

$$\implies \quad F'(y)\,\big\{h(y)\,\big[\psi(k) - 1\big] + g(k)\big\} = \psi(k)\,h'(y)\,\big[1 - F(y)\big]$$

$$\implies \quad \frac{F'(y)}{1 - F(y)} = \psi(k)\,\frac{h'(y)}{h(y)\,\big[\psi(k) - 1\big] + g(k)}$$

$$\implies \quad \frac{-F'(y)}{1 - F(y)} = \frac{\psi(k)}{1 - \psi(k)}\,\frac{h'(y)}{h(y) + g(k)\big/\big[\psi(k) - 1\big]}$$

$$\implies F(y) = \begin{cases} 1 - \left[\dfrac{h(y) + \dfrac{g(k)}{\psi(k) - 1}}{h(\alpha) + \dfrac{g(k)}{\psi(k) - 1}}\right]^{\psi(k)/[1-\psi(k)]} & \text{for } y \in [\alpha, \beta) \\[4ex] 0 & \text{otherwise.} \end{cases} \quad\blacksquare$$

If we set

$$\begin{aligned} h(x) &= \exp\{-x^c\}, \ \ x \in [0, \infty) \ \text{ and } \ c > 0, \\ \psi(k) &= k \ \text{ with } \ 0 < k < 1, \\ g(k) &= 0, \end{aligned}$$

we get

$$F(x) = 1 - \big[\exp\{-x^c\}\big]^{k/(1-k)}.$$

Since $0 < k < 1$, we have $\lambda := k/(1-k) > 0$ so that

$$F(x) = 1 - \big[\exp\{-x^c\}\big]^{\lambda} = 1 - \exp\{-\lambda\,x^c\}$$

is the WEIBULL distribution with scale factor $\lambda = b^{-c}$.

OUYANG (1987) gave the following characterizing theorem that incorporates the theorems of SHANBHAG (6.15a) and of HAMDAN (6.18) as special cases.

**Theorem 8** (OUYANG, 1987)

Let $X$ be a variate with continuous CDF

$$
F(x) = \left\{
\begin{array}{ll}
0 & \text{for } x < \alpha \\[2mm]
1 - \exp\left\{ -\dfrac{1}{d}\left[ h(x) - h(\alpha) \right] \right\} & \text{for } x \in [\alpha, \beta) \\[2mm]
1 & \text{for } x \geq \beta
\end{array}
\right\}
\tag{6.21a}
$$

where $d$ is a nonzero constant, $h(x)$ is a real–valued continuous function defined on $[\alpha, \beta)$, possessing a continuous derivation on $(\alpha, \beta)$ with $\lim_{x \uparrow \beta} h(x) = \pm\infty$ and $\mathrm{E}\left[h(x)\right] = h(\alpha) + d$, if and only if

$$
\mathrm{E}\left[h(X) \mid X > y\right] = h(y) + d \ \forall \, y \in [\alpha, \beta).
\qquad\blacksquare
\tag{6.21b}
$$

**Proof of Theorem 8**

The part of necessity can be verified directly, so that the sufficiency has to been shown. We have

$$
\mathrm{E}\left[h(X) \mid X > y\right] = \frac{1}{1 - F(y)} \int_y^\beta h(x)\, \mathrm{d}F(x).
$$

Hence (6.21b) becomes

$$
\begin{aligned}
\left[1 - F(y)\right]\left[h(y) + d\right] &= \int_y^\beta h(x)\, \mathrm{d}F(x) \\[2mm]
&= \mathrm{E}\left[h(x)\right] + \int_\alpha^y h(x)\, \mathrm{d}\left[1 - F(x)\right].
\end{aligned}
\tag{6.22a}
$$

Integration by parts and using $F(\alpha) = 0$ gives

$$
\left[1 - F(y)\right]\left[h(y) + d\right] = \mathrm{E}\left[h(X)\right] + h(y)\left[1 - F(y)\right] - h(\alpha) - \int_\alpha^y h'(x)\left[1 - F(y)\right] \mathrm{d}x
\tag{6.22b}
$$

or equivalently

$$
d\left[1 - F(y)\right] = \mathrm{E}\left[h(X)\right] - h(\alpha) - \int_\alpha^y h'(x)\left[1 - F(y)\right] \mathrm{d}x.
\tag{6.22c}
$$

As $F(\cdot)$ and $h'(\cdot)$ are continuous by assumptions, the right–hand side of (6.22c) is differentiable. Consequently, so is the left–hand side. Now differentiation of (6.22c) with respect to $y$ yields

$$
d\left[1 - F(y)\right]' = -h'(y)\left[1 - F(y)\right]
$$

or equivalently

$$\frac{\left[1 - F(y)\right]'}{1 - F(y)} = -\frac{h'(y)}{d}. \tag{6.22d}$$

If (6.22d) is integrated from $\alpha$ to $x$, then the continuity of $F(x)$ gives

$$\ln\left[1 - F(y)\right] - \ln\left[1 - F(\alpha)\right] = \frac{1}{d}\left[h(x) - h(\alpha)\right]. \tag{6.22e}$$

Since $F(\alpha) = 0$, we can rewrite (6.22e) as

$$1 - F(x) = \exp\left\{-\frac{1}{d}\left[h(x) - h(\alpha)\right]\right\}$$

or

$$F(x) = 1 - \exp\left\{-\frac{1}{d}\left[h(x) - h(\alpha)\right]\right\}. \qquad \blacksquare$$

We remark

- If we take $h(x) = x$ and $d = \mathrm{E}(X)$, then Theorem 8 reduces to SHANBHAG's characterization of the exponential distribution.

- If $h(\cdot)$ is a strictly increasing function from $[\alpha, \beta)$ onto $[0, \infty)$ and $d = h(b)$, where $b$ is a positive constant, then Theorem 8 reduces to HAMDAN's characterization, cf. Theorem 6.

**Corollary to Theorem 8**

Let $X$ be a variate with continuous distribution function $F(x)$ and $F(x) < 1 \; \forall \, x \in [0, \infty)$. If

$$\mathrm{E}\left(X^c \,|\, X > y\right) = x^c + \frac{1}{\lambda}; \; \forall \, c > 0, \; \lambda > 0 \text{ and } \forall \, x \in [0, \infty),$$

then

$$F(x) = 1 - \exp\left\{-\lambda\, x^c\right\}, \quad x \in [0, \infty).$$

In particular, if $c = 1$, then $F(x) = 1 - \exp\{-\lambda\, x\}$ is the exponential distribution. $\quad\blacksquare$

**Proof of Corollary to Theorem 8**

Let $h(x) = x^c$, $x \in [0, \infty)$ and $d = 1/\lambda$; then it can be verified that $h(x)$ satisfies the assumptions of Theorem 8. Therefore, (6.21a) reduces to

$$F(x) = 1 - \exp\left\{-\lambda x^c\right\}, \quad \text{for } x \in [0, \infty).$$

Thus we have $X \sim We(0, \lambda^{-c}, c)$. $\quad\blacksquare$

OUYANG (1987) gave another characterization theorem which rests upon conditional moments where the condition is $X \leq y$, i.e., truncation on the right, instead of $X > y$ (truncation on the left).

**Theorem 9** (OUYANG, 1987)

Let $X$ be a variate with continuous distribution function $F(x)$, $F(x) > 0 \ \forall \ x \in (\alpha, \beta]$. Then

$$F(x) = \begin{cases} 0 & \text{for} \quad x \leq \alpha \\ \exp\left\{\dfrac{1}{d}\left[h(\beta) - [h(x)]\right]\right\} & \text{for} \quad x \in (\alpha, \beta] \\ 1 & \text{for} \quad x > \beta \end{cases} \qquad (6.23a)$$

where $d$ is a nonzero constant and $h(x)$ is a real–valued monotone function continuously differentiable on $(\alpha, \beta]$ with $\lim\limits_{x \downarrow \alpha} = \pm\infty$ and $\mathrm{E}\big[h(x)\big] = h(\beta) + d$, if and only if

$$\mathrm{E}\big[h(X) \,\big|\, X \leq y\big] = h(y) + d \ \forall \ x \in (\alpha, \beta]. \qquad (6.23b)$$

The interval $(\alpha, \beta]$ is open on the right whenever $\beta = +\infty$, and in this case $\lim\limits_{x \to \infty} h(x) = 0$. ∎

We omit the proof of Theorem 9 and give the following.

**Corollary to Theorem 9**

Let $X$ be a random variable with continuous distribution function $F(x)$, $F(x) > 0 \ \forall \ x \in (0, \infty)$. If $\mathrm{E}\big\{\ln\big[1 - \exp\big(-\lambda X^c\big)\big] \,\big|\, X \leq y\big\} = \ln\big[1 - \exp\big(-\lambda y^c\big)\big] - 1$ for $c > 0$, $\lambda > 0$ and for all $y \in (0, \infty)$, then $F(x) = 1 - \exp\big\{-\lambda x^c\big\}$. ∎

**Proof of Corollary to Theorem 9**

Let $h(x) = \ln\big[1 - \exp\big(-\lambda x^c\big)\big]$, $x \in (0, \infty)$ and $d = -1$, then $h(x)$ satisfies the assumption of Theorem 9. Therefore, (6.23a) reduces to

$$\begin{aligned} F(x) &= \exp\big\{\ln\big[1 - \exp\big(-\lambda x^c\big)\big]\big\} \\ &= 1 - \exp\big(-\lambda x^c\big), \ x \in (0, \infty). \end{aligned}$$

Thus we have $X \sim We(0, \lambda^{-1/c}, c)$. ∎

## 6.3   WEIBULL characterizations based on order statistics[4]

A lot of WEIBULL characterizations have been developed which rest upon order statistics in one way or the other. The oldest characterization in this class is that of DUBEY (1966e). Before citing his theorem we have to mention the following four formulas needed for the proof and which pertain to the minimum $Y := \min(X_1, \ldots, X_n)$ of $n$ iid variates $X_i$ each having $F(x)$ as its CDF and $f(x)$ as its DF:

$$F_n(y) = 1 - \big[1 - F(y)\big]^n \qquad (6.24a)$$

$$f_n(y) = n\, f(y) \big[1 - F(y)\big]^{n-1}. \qquad (6.24b)$$

---

[4] Suggested reading for this section: DUBEY (1966e), GALAMBOS (1975), JANARDAN/SCHAEFFER (1978), JANARDAN/TANEJA (1979a,b), KHAN/ALI (1987), KHAN/BEG (1987), RAO/SHANBHAG (1998).

From (6.24a,b) we recover the common distribution of each $X_i$:

$$F(x) \;=\; 1 - \big[1 - F_n(x)\big]^{1/n} \tag{6.24c}$$

$$f(x) \;=\; \frac{1}{n}\, f_n(x)\, \big[1 - F_n(x)\big]^{(1/n)-1}. \tag{6.24d}$$

**Theorem 10** (DUBEY, 1966c)

Let $X_1, \ldots, X_n$ be $n$ iid variates having a three parameter WEIBULL distribution with location parameter $a$, shape parameter $c$ and the combined scale–shape parameter[5] $\beta = b^c$ and let $Y$ the minimum of the $X's$. Then $Y$ obeys the WEIBULL law with the same parameters $a$ and $c$ but the combined scale–shape parameter $\beta/n$. Conversely, if $Y$ has a WEIBULL distribution with $\mu$ and $d$ as location parameter and scale parameter, respectively, and combined scale–shape parameter $\sigma$, then each $X_i$ obeys the WEIBULL law with the same parameters $\mu$ and $d$ for location and shape but $n\,\sigma$ as its combined scale–shape factor. ∎

**Proof of Theorem 10**

The DF of each $X_i$ in the parametrization given above is

$$f(x) = c\,\beta^{-1}\,(x-a)^{c-1}\,\exp\big\{-\beta^{-1}\,(x-a)^c\big\}; \;\; x \ge a \in \mathbb{R},\; \beta, c > 0,$$

and its CDF is

$$F(x) = 1 - \exp\big\{-\beta^{-1}\,(x-a)^c\big\}.$$

From (6.24b) the DF of $Y$ is found to be

$$f_n(y) = n\,c\,\beta^{-1}\,(x-a)^{c-1}\,\exp\big\{-n\,\beta^{-1}\,(x-a)^c\big\}.$$

Hence, $Y$ has the combined scale–shape parameter $\beta/n$. Now we shall prove the converse of Theorem 10. Here the DF of $Y$ is given by

$$f_n(y) = d\,\sigma^{-1}\,(y-\mu)^{d-1}\,\exp\big\{-\sigma^{-1}\,(y-\mu)^d\big\}$$

and its CDF by

$$F_n(y) = 1 - \exp\big\{-\sigma^{-1}\,(y-\mu)^d\big\}.$$

From (6.25d) we get the DF of each $X_i$ as

$$f(x) = (\sigma\,n)^{-1}\,d\,(x-\mu)^{d-1}\,\exp\big\{-(\sigma\,n)^{-1}\,(x-\mu)^d\big\},$$

where the combined scale–shape parameter is $\sigma\,n$. ∎

DUBEY's characterization is a special case or an application of a more general method for characterizing distributions and which is given by GALAMBOS (1975). This method, called the **method of limit laws**, can be summarized as follows: Let $X_{1:n} \le X_{2:n} \le \ldots \le X_{n:n}$ denote the order statistics of a sample of size $n$. Assume that a transformation $T_n\, X_{1:n}$ (or

---

[5] This parametrization of the WEIBULL distribution – see the proof of Theorem 10 – shows the effect of taking the minimum most clearly.

$T_n X_{n:n}$), for a single $n$, where $n$ is either fixed or is a random variable depending on a parameter $t$, reduces the distribution $G(x)$ of $T_n X_{1:n}$ to a function which does not depend on $n$. Let further $T_n$ be such that it admits a way to generate an infinite sequence $n(j)$ such that the distribution of $T_{n(j)} X_{1:n(j)}$ is also $G(x)$. Then, if the limit law of $T_n X_{1:n}$ exists and $G(x)$ is in the domain of attraction of a possible limit law $\Theta(x)$, then $G(x) = \Theta(x)$.

We can prove the following result using this method.

**Theorem 11**

Let $n^{1/c} X_{1:n}$ be distributed as $X$ for $n \geq 2$. Let further $F(x)$ of $X$ satisfy the property that $\lim_{x \downarrow 0} F(x)/x = \lambda > 0$, $\lambda$ finite, then $F(x)$ is given by the following WEIBULL CDF:

$$F(x) = 1 - \exp\{-\lambda x^c\}, \quad x > 0. \qquad \blacksquare \qquad (6.25a)$$

**Proof of Theorem 11**

In this case $T_n X_{1:n} = n^{1/c} X_{1:n}$ and $G(x) = F(x)$. Let $n(j) = n^{j/c}$, where $j$ is a positive integer. If for one $n \geq 2$, $G(x) = F(x)$, then for each $j$

$$T_{n(j)} X_{1:n(j)} = n(j) X_{1:n(j)}$$

also has $F(x)$ as its own CDF. Therefore, and for each $j \geq 1$,

$$F(x) = 1 - \left[1 - F\left(\frac{x}{n(j)}\right)\right]^{n(j)}, \quad F(0) = 0, \qquad (6.25b)$$

and by the asymptotic behavior of $F(x)$ for small $x$'s, it follows that as $j \to \infty$, the right–hand side of (6.25b) tends to (6.25a). $\qquad \blacksquare$

The following two theorems rest upon the distributional equivalence of order statistics or functions thereof.

**Theorem 12** (RAO/SHANBHAG, 1998)

Let $X_1, \ldots, X_n$, $n \geq 2$, be iid positive random variables and $a_1, \ldots, a_n$ positive real numbers not equal to 1, such that the smallest closed subgroup of $\mathbb{R}$ containing $\ln a_1, \ldots, \ln a_n$ equals $\mathbb{R}$ itself. Then for some $m \geq 1$,

$$\min\{a_1 X_1, \ldots, a_n X_n\} \quad \text{and} \quad X_{1:m}$$

have the same distribution if and only if the survivor function of $X_1$ is of the form

$$R(x) = \exp\{-\lambda_1 x^{\alpha_1} - \lambda_2 x^{\alpha_2}\}, \quad x \in \mathbb{R}^+, \qquad (6.26)$$

with $\lambda_1, \lambda_2 \geq 0$, $\lambda_1 + \lambda_2 > 0$ and $\alpha_r$; $r = 1, 2$; positive numbers such that $\sum_{i=1}^{n} a_i^{\alpha_r} = m$.
$\blacksquare$

**Corollary to Theorem 12**

If $\alpha_1 = \alpha_2$, the distribution corresponding to (6.26) is WEIBULL. $\qquad \blacksquare$

For the proof Theorem 12 and its corollary see RAO/SHANBHAG (1998).

**Theorem 13** (JANARDAN/TANEJA, 1979a)

Let $X$ be a variate having an absolutely continuous strictly increasing distribution function $F(x) = 0$ for $x \leq 0$ and $F(x) < 1$ for all $x > 0$. A necessary and sufficient condition for $X$ to have the WEIBULL distribution with $F(x) = 1 - \exp\{-\lambda x^c\}$ is that for any fixed $r$ and two distinct numbers $s_1$ and $s_2$ $(1 < r < s_1 < s_2 \leq n)$, the distribution of the statistics

$$V_i = \left(X_{s_1:n}^c - X_{r:n}^c\right)^{1/c}; \quad i = 1, 2$$

and

$$W_i = X_{s_i-n:n-r}$$

are identical for $i = 1, 2$ and $c > 0$. ∎

For a proof see the paper of JANARDAN/TANEJA (1979a).

The next two theorems are based upon the distributional independence of functions of order statistics.

**Theorem 14** (JANARDAN/SCHAEFFER, 1978)

Let $F(\cdot)$ be an absolutely continuous distribution function of $X$ with $F(0) = 0$ and DF

$$f(x) = c\,\lambda\,x^{c-1}\,\exp\{-\lambda x^c\}, \quad x \geq 0. \tag{6.27a}$$

Let $X_{i:n}$; $i = 1, \ldots, n$ be the order statistics of a random sample of size $n$. A necessary and sufficient condition of $X$ to have the DF (6.27a) is that the statistics

$$\left(X_{m+1:n}^c - X_{m:n}^c\right)^{1/c} \text{ and } X_{m:n}, \quad m \text{ fixed and } 1 \leq m \leq n-1 \tag{6.27b}$$

be independent. ∎

**Proof of Theorem 14**

Let $Y = X^c$ and let $g(y)$ be the DF of $Y$. Then $Y_{1:n} \leq Y_{2:n} \leq \ldots \leq Y_{n:n}$ are the order statistics from the above distribution. For fixed $m$, $1 \leq m \leq n-1$, SRIVASTAVA (1967) has shown that $Y_{m+1:n} - Y_{m:n}$ and $Y_{m:n}$ are independent iff $g(y) = \lambda \exp\{-\lambda y\}$; $\lambda > 0$, $y > 0$. Hence, $X_{m+1:n}^c - X_{m:n}^c$ and $X_{m:n}^c$ are independent iff $X$ has the DF given in (6.27a). ∎

**Theorem 15** (JANARDAN/TANEJA, 1979b)

The statistics $X_{1:n}$ and $D = \left(X_{2:n}^c - X_{1:n}^c\right)^{1/c}$ are stochastically independent if and only if the random variable $X$, from which $X_{1:n}$ and $X_{2:n}$ are the first two order statistics in a sample of size $n$, has the WEIBULL distribution with DF (6.27a). ∎

The proof of this theorem is done by showing that the conditional distribution of $D$, given $X_{1:n} = y$, is independent of $y$. For details see the paper of JANARDAN/TANEJA (1979b).

Finally, we have two characterizations of the WEIBULL law which are based on conditional moments of order statistics.

**Theorem 16** (KHAN/ALI, 1987)

Let $X$ be a continuous random variable with $F(x)$ as CDF, $F(0) = 0$, and $E(X^r) < \infty$ for $r > 0$. If $F(x) < 1$ for $x < \infty$, then $F(x) = 1 - \exp\{-\lambda x^c\}$; $\lambda, c > 0$ if and only if for $k < n$, $1 \leq i \leq n - k$

$$E\left[X_{k+i:n}^c \mid X_{k:n} = x\right] = x^c + \frac{1}{\lambda} \sum_{j=0}^{i-1} \frac{1}{n-k-j} \, . \qquad \blacksquare \qquad (6.28)$$

**Theorem 17** (KHAN/BEG, 1987)

Let $X$ be a continuous random variable with $F(x)$ as CDF such that $F(0) = 0$ and $F(x)$ has a second derivative on $(0, \infty)$ and its first derivative is not vanishing on $(0, \infty)$ so that in particular $F(x) < 1 \ \forall \, x \geq 0$. Let $c > 0$ and let $F(x)$ have a raw moment of order $2c$. If for some integer $k$, $0 < k < n$,

$$\text{Var}\left(X_{k+1:n}^c \mid X_{k:n} = x\right) = \quad \text{constant}$$

then $F(x) = 1 - \exp\{-\lambda x^c\}$ for $x \geq 0$, where $\lambda > 0$ is given by $\lambda^{-2} = c(n-k)^2$, and conversely. $\qquad \blacksquare$

For the proofs of Theorems 16 and 17 the reader is referred to the original papers.

## 6.4  Miscellaneous approaches of WEIBULL characterizations

We first give two theorems that rely on the independence of suitably chosen random variables. As early as 1968 the following theorem was stated and proven.

**Theorem 18** (SHESHADRI, 1968)

Let $X$ and $Y$ be two positive and independently distributed variables such that the quotient $V = X/Y$ has the DF

$$f(v) = \frac{\lambda v^{\lambda-1}}{(1+v^\lambda)^2} \, ; \quad v > 0, \ \lambda > 1. \qquad (6.29)$$

The random variables $X$ and $Y$ have the WEIBULL distribution with the same scale parameter if $X^\lambda + Y^\lambda$ is independent of $X/Y$. $\qquad \blacksquare$

CHAUDHURI/CHANDRA (1990) have given Theorem 19 characterizing the generalized gamma distribution (see Sect. 3.2.3), with DF

$$f(x) = \frac{c \, x^{cd-1}}{b^{cd} \, \Gamma(d)} \, \exp\left\{-\left(\frac{x}{b}\right)^c\right\}, \quad x > 0, \qquad (6.30a)$$

which incorporates the two–parameter WEIBULL distribution as the special case with $d = 1$.

**Theorem 19** (CHAUDHURI/CHANDRA, 1990)

A necessary and sufficient condition that $X_i;\ i = 1, 2, \ldots, n$ and $n \geq 2$ be independently distributed with DF (6.30a) is that

$$V_i = \frac{X_i^c}{\sum\limits_{j=0}^{i-1} X_j^c}\ ;\ \ j = 1, 2, \ldots, n \tag{6.30b}$$

be independently distributed with DF

$$f_i(v) = \frac{1}{B(k_i, d_i)}\ \frac{v}{d_i\,(1 + v)^{c_{i+1}}}\ ,\ \ v > 0, \tag{6.30c}$$

where $B(\cdot, \cdot)$ is the complete beta function and $k_i = \sum\limits_{j=0}^{i-1} c_i$.     ∎

The next two theorems are built upon truncated random variables. COOK (1978) used the following truncated variables:

$$Z_j(a_j) = \left\{ \begin{array}{ccc} X_j^c - a_j & \text{for} & x_j > a_j \\[2mm] 0 & \text{for} & x_j \leq a_j \end{array} \right\}$$

and supposed for some $s,\ 1 \leq s \leq n - 1$, that $\{Z_1, \ldots, Z_s\}$ and $\{Z_{s+1}, \ldots, Z_n\}$ are independent.

**Theorem 20** (COOK, 1978)

For all $a_j \geq 0$, $E\left\{ \prod\limits_{j=1}^{n} Z_j(a_j) \right\}$ depends on the parameters only through a function of $\sum_{j=1}^{n} a_j$ if and only if all $X_j$ are mutually independent and follow the WEIBULL distribution.     ∎

**Theorem 21** (JANARDAN/TANEJA 1979b)

A non–negative variate $X$ with finite expectation has a WEIBULL distribution if and only if, for some $\alpha > 0$,

$$\alpha\,E\{(X - s)^+\}\,E\{(X - t)^+\} = E\left\{ \left(X - \left[s^c + t^c\right]^{1/c}\right)^+ \right\} \tag{6.31}$$

for all $s$ and $t$ belonging to a dense subset of the positive half of the real line. Here $(X - u)^+ = \max(X - u, 0)$ denotes the positive part of $X - u$, where $u \geq 0$.     ∎

SCHOLZ (1990) gave a characterization of the three–parameter WEIBULL distribution in terms of relationships between particular triads of percentiles as defined by

$$C = \left\{ (u, v, w) :\ 0 < u < v < w < 1,\ \ln(1 - u)\ln(1 - w) = [\ln(1 - v)]^2 \right\}, \tag{6.32a}$$

i.e., $u, v, w$ are the orders of the percentiles

$$x_P = F^{-1}(P) = \inf\{x : F(x) \geq P\} \text{ for } 0 < P < 1. \qquad (6.32b)$$

For the three–parameter WEIBULL distribution with CDF,

$$F(x) = 1 - \exp\left\{-\left(\frac{x-a}{b}\right)^c\right\} \text{ for } x \geq a \in \mathbb{R}; \ b, c > 0; \qquad (6.32c)$$

it is known that for some fixed $t$, namely $t = a$, the following relation holds between its percentiles given by (2.57b):

$$x_u x_w - x_v^2 = t\left[x_u + x_w - 2x_v\right] \ \forall \ (u, v, w) \in C. \qquad (6.32d)$$

Theorem 22 states that this relationship characterizes the three–parameter WEIBULL distribution (6.32c).

**Theorem 22** (SCHOLZ, 1990)

A random variable $X$ with CDF $G(x)$ and percentiles $G^{-1}(P) = x_P$ satisfying the relationship (6.32d) is either degenerate or $X$ has the three–parameter WEIBULL distribution with $t = a$. ■

For the proof see SCHOLZ (1990).

There exist several characterization theorems for a WEIBULL distribution based on record values (see Sect. 5.1.3). KAMPS (1991) built upon upper record values $X_{L(m)}$ defined via the record times $L(m)$, $m \in \mathbb{N}_{\not=}$, as

$$L(m) = \min\left\{j : j > L(m-1), X_j > X_{L(m-1)}\right\}, \ m \geq 1, \ L(0) = 1. \qquad (6.33a)$$

Further, KAMPS used a specially parameterized WEIBULL distribution.

**Theorem 23** (KAMPS, 1991)

Let $\alpha > 0$, $p > 1$, $k \leq m$, $F^{-1}(0^+) \geq 0$ and $\mathrm{E}(X^{\alpha p + \varepsilon}) < \infty$. Then we have

$$\mathrm{E}\left(X_{L(m)}^\alpha\right) \leq \frac{(k!)^{1/p}}{m!} \Gamma\left(\frac{(m+1)p - (k+1)}{p-1}\right)^{1-1/p} \left[\mathrm{E}\left(X_{L(k)}^{\alpha p}\right)\right]^{1/p}. \qquad (6.33b)$$

In the case $k < m$, equality in (6.33b) holds iff $F(\cdot)$ is the following WEIBULL distribution

$$F(x) = 1 - \exp\left\{-\left(\frac{x}{c}\right)^{\alpha(p-1)/(m-k)}\right\}, \ x > 0, \ c > 0. \qquad ■$$

$$(6.33c)$$

PAWLAS/SZYNAL (2000a) built upon the **$k$–th upper record values** defined via the $k$–th upper record times $L_m^{(k)}$:

$$L_{m+1}^{(k)} = \min\left\{j > L_m^{(k)} : X_{j:j+k-1} > X_{L_m^{(k)}:L_{m+k-1}^{(k)}}\right\}; \ L_m^{(k)} = 1. \qquad (6.34a)$$

The sequence $\left\{Y_m^{(k)}, m \geq 1\right\}$, where $Y_m^{(k)} = X_{L_m^{(k)}}$, is called the sequence of $k$–th upper record values.

**Theorem 24** (PAWLAS/SZYNAL, 2000a)

Fix a positive integer $k \geq 1$ and let $r$ be a non–negative integer. A necessary and sufficient condition for $X$ to be distributed with DF

$$f(x) = \lambda\, c\, x^{c-1}\, \exp\{-\lambda\, x^c\}; \quad x \geq 0;\ c, \lambda > 0 \tag{6.34b}$$

is that

$$\mathrm{E}\left[\left(Y_m^{(k)}\right)^{r+c}\right] = \mathrm{E}\left[\left(Y_{m-1}^{(k)}\right)^{r+c}\right] + \frac{r+c}{k\,\lambda\,c}\,\mathrm{E}\left[\left(Y_m^{(k)}\right)^r\right] \quad \text{for}\ \ m = 1, 2, \ldots \quad\blacksquare \tag{6.34c}$$

ROY/MUKHERJEE (1986) gave several WEIBULL characterizations. One of them departs from the hazard rate

$$h(x) = \frac{f(x)}{1 - F(x)}\,,$$

but takes the argument to be a random variable $X$ so that $h(X)$ is a variate too. The characterization pertains to the distribution of $h(X)$.

**Theorem 25** (ROY/MUKHERJEE, 1986)

- $h(\cdot)$ is strictly increasing with $h(0) = 0$.
- $h(X)$ is WEIBULL with shape parameter $c' > 1$ and scale factor $\lambda'$, for the parametrization see (6.27a), iff $X$ is WEIBULL with shape parameter $c > 1$, $(1/c + 1/c') = 1$, and $\lambda' = \lambda\,(c\,\lambda')^c$. $\quad\blacksquare$

They further showed that the **FISHER–information** is minimized among all members of a class of DF's with

- $f(x)$ continuously differentiable on $(0, \infty)$,
- $x\, f(x) \to 0$ as $x \to 0^+$, $x^{1+c}\, f(x) \to 0$ as $x \to \infty$,
- $\int x^c\, f(x)\, dx = 1$ and $\int x^{2c}\, f(x)\, dx = 2$,

when $f(x)$ is the WEIBULL distribution with scale factor equal to 1 and $c$ as its shape parameter.[6] Another characterization theorem given by ROY/MUKHERJEE (1986) says that the **maximum entropy distribution** — under certain assumptions — is the WEIBULL distribution.

## 6.5   Characterizations of related WEIBULL distributions

In Sect. 3.3.6.5 we have presented several compound WEIBULL distributions, one of them being the **WEIBULL–gamma distribution** (see (3.73c,d)), where the compounding distribution is a gamma distribution for $B$ in the WEIBULL parametrization

$$f(x \mid B) = c\, B\, (x - a)^{c-1}\, \exp\{-B\,(x-a)^c\}.$$

Here, we take the parametrization

$$f(x \mid \beta) = c\,\beta^{-1}\,(x-a)^{c-1}\,\exp\{-\beta^{-1}\,(x-a)^c\} \tag{6.35a}$$

---

[6]  GERTSBAKH/KAGAN (1999) also characterized the WEIBULL distribution by properties of the FISHER–information but based on type–I censoring.

and the compounding distribution of $\beta$ is the following gamma distribution:

$$f(\beta) = \frac{\eta^\delta \exp\{-\eta \beta^{-1}\}}{\Gamma(\delta)\,\beta^{\delta+1}}\,; \quad \beta \geq 0;\; \eta, \delta > 0. \tag{6.35b}$$

Then the DF of this WEIBULL–gamma distribution is

$$f(x) = \frac{c\,\delta\,\eta^\delta\,(x-a)^{c-1}}{\left[\eta + (x-a)^c\right]^{\delta+1}}\,, \quad x \geq a. \tag{6.35c}$$

The characterization theorem of this compound WEIBULL distribution rests upon $Y = \min(X_1, \ldots, X_n)$, where the $X_i$ are iid with DF given by (6.35c), and is stated by DUBEY (1966e) as follows.

**Theorem 26** (DUBEY, 1966a)

Let $Y = \min(X_1, \ldots, X_n)$ of the compound WEIBULL distribution with parameters $a, c, \delta$ and $\eta$. Then $Y$ obeys the compound WEIBULL distribution with parameter $a, c, n\,\delta$ and $\eta$. Conversely, if $Y$ has a compound WEIBULL distribution with parameters $\mu, \lambda, \theta$ and $\sigma$, then each $X_i$ obeys the compound WEIBULL law with the parameters $\mu, \lambda, \theta/n$ and $\sigma$. ∎

The proof of Theorem 26 goes along the same line as that of Theorem 10 in Sect. 6.3.

EL–DIN et al. (1991) gave a characterization of the **truncated WEIBULL distribution** (see Sect. 3.3.5) by moments of order statistics. EL–ARISHY (1993) characterized a **mixture of two WEIBULL distributions** by conditional moments $E(X^c \mid X > y)$.

**Theorem 27** (EL–ARISHY, 1993)

$X$ follows a mixture of two WEIBULL distributions with common shape parameter $c$ and scale parameters $b_1$ and $b_2$, if and only if

$$E(X^c \mid X > y) = y^c + \frac{1}{b_1} + \frac{1}{b_2} - \frac{h(y)}{b_1\,b_2\,c\,y^{c-1}}\,, \tag{6.36}$$

where $h(\cdot)$ denotes the hazard rate.

The **inverse WEIBULL distribution** of Sect. 3.3.3 was characterized by PAWLAS/SZYNAL (2000b) using moments of the $k$–th record values $Y_m^{(k)}$ (for a definition see the introduction to Theorem 24). PAWLAS/SZYNAL (2000b) stated and proved the following.

**Theorem 28**

Fix a positive integer $k \geq 1$ and let $r$ be a positive integer. A necessary and sufficient condition for a random variable $X$ to have an inverse WEIBULL distribution with CDF

$$F(x) = \exp\left\{-\left(\frac{b}{x}\right)^c\right\}\,; \quad x > 0;\; c, b > 0; \tag{6.37a}$$

is that

$$E\left[\left(Y_m^{(k)}\right)^r\right] = \left(1 - \frac{r}{c\,(m-1)}\right) E\left[\left(Y_{m-1}^{(k)}\right)^r\right] \tag{6.37b}$$

for all positive integers $m$ such that $(m-1)\,c > r$. ∎

# II

# Applications and inference

# 7 WEIBULL applications and aids in doing WEIBULL analysis

The field of applications of the WEIBULL distribution and its relatives is vast and encompasses nearly all scientific disciplines. Using these distributions, data have been modeled which originate from such distinct areas as the biological, environmental, health, physical and social sciences. So we can give only a sample of problems that come from rather different areas and that have been solved successfully. In this section the presentation of these papers consists in only enumerating them by the topics treated. The ways in which these problems were solved in the papers, mostly by applying one or the other inferential procedure for estimating, testing, forecasting or simulation, will not be treated in this section but is postponed to Chapter 9 through 22.

The Tables 7/1 through 7/12 in Sect.7.1 are based on the screening of scientific journals. Monographs and books, especially on reliability theory or on life testing, contain many more applications, but they are not noticeably different from those listed in the following tables. Most recent examples — in journals which have appeared after 2003/4 — can be found on the Internet via a search engine such as *Google*. Besides the great number of papers and examples in monographs showing the successful application of the WEIBULL model, there are some critical papers that show possible perils of unguarded fitting of WEIBULL distributions, e.g., GITTUS (1967), GORSKI (1968), or MACKISACK/STILLMAN (1996).

Section 7.2 lists several statistical software packages capable of performing WEIBULL analysis as well as consulting corporations capable of performing reliability analysis based on the WEIBULL.

## 7.1 A survey of WEIBULL applications

Table 7/1:  WEIBULL applications in material science, engineering, physics and chemistry

| Author(s) | Topic(s) |
|---|---|
| ALMEIDA, J.B. (1999) | failure of coatings |
| BERRETTONI, J.N. (1964) | corrosion resistance of magnesium alloy plates, leakage failure of dry batteries, reliability of step motors and of solid tantalum capacitors |
| BOORLA, R. / ROTENBERGER, K. (1997) | flight load variation in helicopters |
| CACCIARI, M. et al. (1995) | partial discharge phenomena |

| Author(s) | Topic(s) |
|---|---|
| CONTIN, A. et al. (1994) | discharge inference |
| DURHAM, S.D. / PADGETT, W.J. (1997) | failure of carbon fiber composites |
| FANG, Z. et al. (1993) | particle size |
| FOK, S.L. et al. (2001) | fracture of brittle material |
| FRASER, R.P. / EISENKLAM, P. (1956) | size of droplets in spray |
| GALLACE, L. (1973/4) | power–hybrid burn–in |
| GOTTFRIED, P. / ROBERTS, H.R. (1963) | failure of semiconductors and capacitors |
| KAO, J.H.K. (1956) | failure of electron tubes |
| KESHRAN, K. et al. (1980) | fracture strength of glass |
| KOLAR–ANIC, L. et al. (1975) | kinetics of heterogeneous processes |
| KULKARNI, A. et al. (1973) | strength of fibers from coconut husks |
| KWON, Y.W. / BERNER, J. (1994) | damage in laminated composites |
| LIEBLEIN, J. / ZELEN, M. (1956) | ball bearing failures |
| LIERTZ, H. / OESTREICH, U.H.P. (1976) | reliability of optical wave guides for cables |
| MACKISACK, M.S. / STILLMAN, R.H. (1996) | life distribution of copper–chrome–arsenate treated power poles |
| MCCOOL, J.I. (1974a) | ball bearing failures |
| MU, F.C. et al. (2000) | dielectric breakdown voltage |
| NOSSIER, A. et al. (1980) | dielectric breakdown voltage |
| PADGETT, W.J. et al. (1995) | failure of carbon fiber composites |
| PERRY, J.N. (1962) | semiconductor burn–in |
| PHANI, K.K. (1987) | tensile strength of optical fibers |
| PLAIT, A. (1962) | time to relay failure |
| PROCASSINI, A.A. / ROMANO, A. (1961, 1962) | time to transistor failure |
| QUREISHI, F.S. / SHEIKH, A.K. (1997) | adhesive wear in metals |
| ROSIN, P. / RAMMLER, E. (1933) | fineness of coal |
| SCHREIBER, H.H. (1963) | failure of roller bearings |
| SHEIKH, A.K. et al. (1990) | pitting corrosion in pipelines |

| Author(s) | Topic(s) |
|---|---|
| SINGPURWALLA, N.D. (1971) | fatigue data on several materials |
| TSUMOTO, M. / IWATA, M. (1975) | impulse breakdown of polyethylene power cable |
| TSUMOTO, M. / OKIAI, R. (1974) | impulse breakdown of oil–filled cable |
| VIERTL, R. (1981) | lifetime of roller bearings |
| WANG, Y. et al. (1997) | dielectric breakdown voltage |
| WEIBULL, W. (1939a) | strength of rods of stearic acids and of plaster–of–Paris |
| WEIBULL, W. (1939b) | strength of steel |
| WEIBULL, W. (1951) | yield strength of steel, size distribution of fly ash, fiber strength of Indian cotton |
| WOLSTENHOLME, L.C. (1996) | tensile strength of carbon fiber |
| XU, S. / BARR, S. (1995) | fracture in concrete |
| YANG, L. et al. (1995) | latent failures of electronic products |
| YAZHOU, J. et al. (1995) | machining center failures |

Table 7/2: WEIBULL applications in meteorology and hydrology

| Author(s) | Topic(s) |
|---|---|
| APT, K.E. (1976) | atmospheric radioactivity data |
| BARROS, V.R. / ESTEVAN, E.A. (1983) | wind power from short wind records |
| BOES, D.C. (1989) | flood quantile estimation |
| CARLIN, J. / HASLETT, J. (1982) | wind power |
| CONRADSEN, K. et al. (1984) | wind speed distribution |
| DIXON, J.C. / SWIFT, R.H. (1984) | directional variations of wind speed |
| DUAN, J. et al. (1998) | precipitation in Pacific Northwest |
| HENDERSON, G. / WEBBER, N. (1978) | wave heights in the English Channel |
| HEO, J.H. et al. (2001) | flood frequency |
| JANDHYALA, V.K. et al. (1999) | trends in extreme temperatures |
| JIANG, H. et al. (1997) | raindrop size |
| LUN, I.Y.F. / LAM, J.C. (2000) | wind speed distribution |

| Author(s) | Topic(s) |
| --- | --- |
| NATHAN, R.J. / MCMAHON, T.A. (1990) | flood data |
| NESHYBA, S. (1980) | size of Antarctic icebergs |
| PAVIA, E.J. / O'BRIEN, J.J. (1986) | wind speed over the ocean |
| PETRUASKAS, A. / AAGARD, P.M. (1971) | design wave heights from historical storm data |
| REVFEIM, K.J.A. (1983) | volume of extreme rainfalls |
| SCHÜTTE, T. et al. (1987) | thunderstorm data |
| SEGURO, J.V. / LAMBERT, T.W. (2000) | wind speed distribution |
| SEKINE, M. et al. (1979) | weather clutter |
| SELKER, J.S. / HAITH, D.A. (1990) | precipitation distribution |
| TALKNER, P. / WEBER, R.O. (2000) | daily temperature fluctuations |
| TULLER, S.E. / BRETT, A.C. (1984) | wind speed data |
| VAN DER AUWERA, L. et al. (1980) | wind power distribution |
| VODA, V.G. (1978) | reliability aspects in meteorology |
| WILKS, D.S. (1989) | rainfall intensity |
| WONG, R.K.W. (1977) | hydrometeorological data |
| ZHANG, Y. (1982) | annual flood extremes |

Table 7/3:    WEIBULL applications in medicine, psychology and pharmacy

| Author(s) | Topic(s) |
| --- | --- |
| ABRAMS, K. et al. (1996) | cancer clinical trial data |
| ACHCAR, J.A. et al. (1985) | following–up data |
| ANDERSON, K.M. (1991) | coronary heart disease |
| BERRETTONI, J.N. (1964) | life expectancy of ethical drugs |
| BERRY, G.L. (1975) | carcinogenesis experimental data, human task performance times |
| CAMPOS, J.L. (1975) | cell survival, neoplastic survival |
| CHRISTENSEN, E.R. / CHEN, C.–Y. (1985) | multiple toxicity model |
| DASGUPTA, N. et al. (2000) | coronary artery disease |
| DEWANJI, A. et al. (1993) | tumori–genetic potency |

| Author(s) | Topic(s) |
|---|---|
| DYER, A.R. (1975) | relationship of systolic blood pressure, serum cholesterol and smoking to 14–year mortality |
| GOLDMAN, A.I. (1986) | survivorship and cure |
| HEIKKILÄ, H.J. (1999) | pharmacokinetic data |
| HOMAN, S.M. et al. (1987) | alcoholism relapse among veterans |
| IDA, M. (1980) | reaction time |
| MORRIS, C. / CHRISTIANSEN, C. (1995) | human survival times |
| OGDEN, J.E. (1978) | shelf aging of pharmaceuticals |
| PAGONIS, V. et al. (2001) | thermoluminescence glow |
| PETO, R. / LEE, P. (1973) | time to death following exposure to carcinogenesis |
| PIKE, M.C. (1962) | time to death following exposure to carcinogenesis |
| PORTIER, C.J. / DINSE, G.E. (1987) | tumor incidence in survival/sacrifice experiments |
| RAO, B.R. et al. (1991) | relapse rate of leukemia patients |
| SCHWENKE, J.R. (1987) | pharmacokinetic data |
| STRUTHERS, C.A. / FAREWELL,V.I. (1989) | time to AIDS data |
| WHITTENMORE, A. / ALTSCHULER, B. (1976) | lung cancer incidence in cigarette smokers |
| WILLIAMS, J.S. (1978) | survival analysis from experiments on patients |

Table 7/4:   WEIBULL applications in economics and business administration

| Author(s) | Topic(s) |
|---|---|
| BERRETTONI, J.N. (1964) | return goods after shipment, number of down times per shift |
| LOMAX, K.S. (1954) | business failures |
| MELNYK, M. / SCHWITTER, J.P. (1965) | life length of machinery |
| NAPIERLA, M. / POTRZ, J. (1976) | life length of complex production equipment |
| NEWBY, M.J. / WINTERTON, J. (1983) | duration of industrial stoppages |
| SILCOCK, H. (1954) | labor turnover |
| THORELLI, H.B. / HIMMELBAUER, W.G. (1967) | distribution of executive salaries |
| TSIONAS, E.G. (2000, 2002) | stock returns |
| UNGERER, A. (1980) | depreciation of the capital stock in an economy |

Table 7/5:   WEIBULL applications in quality control — acceptance sampling

| Author(s) | Topic(s) |
|---|---|
| BALASOORIYA, U. et al. (2000) | progressively censored sampling plans for variables |
| BALASOORIYA, U. / LOW, C.-K. (2004) | type–I progressively censored sampling plans for variables |
| BALOGH, A. / DUKATI, F. (1973) | sampling plans |
| FERTIG, K.W. / MANN, N.R.(1980) | sampling plans for variables |
| GOODE, H.P. / KAO, J.H.K. (1961a,b, 1962, 1963, 1964) | sampling plans for attributes |
| HARTER, H.L. / MOORE, A.H. (1976) | sequential sampling plans for attributes |
| HARTER, H.L. et al. (1985) | sequential sampling plans for reliability testing |
| HOSONO, J. et al. (1980) | sampling plans for variables |

| Author(s) | Topic(s) |
|---|---|
| JUN, C.–H. et al. (2006) | sampling plans for variables under sudden death testing |
| MCCOOL, J.I. (1966) | sampling plans for variables under type–II censoring |
| NG, H.K.T. et al. (2004) | progressive censoring plans |
| OFFICE OF THE ASSISTANT SECRETARY OF DEFENSE (1961, 1962, 1963, 1965) | sampling plans for attributes |
| SCHNEIDER, H. (1989) | sampling plans for variables |
| SOLAND, R.M. (1968b) | BAYESIAN sampling plans for attributes |
| SUN, D. / BERGER, J.O. (1994) | BAYESIAN sequential sampling plans |
| VAN WAGNER, F.R. (1966) | sequential sampling plans for variables |
| WU, J.W. et al. (2001) | failure–censored life tests |
| ZELEN, M. / DANNEMILLER, M.C. (1961) | robustness of life testing procedures |

Table 7/6:  WEIBULL applications in quality control — statistical process control

| Author(s) | Topic(s) |
|---|---|
| ERTO, P. / PALLOTTA, G. (2007) | SHEWHART type charts for percentiles |
| FRAHM, P. (1995) | $\overline{X}$–charts and block replacement |
| GHARE, P.M. (1981) | process control with SPRT |
| JOHNSON, N.L. (1966) | CUSUM control charts |
| KANJI, G.K. / ARIF, O.H. (2001) | median rankit control charts |
| MCWILLIAMS, T.P. (1989) | in–control–time of chart–controlled processes |
| MUKHERJEE, S.P. / SINGH, N.K. (1998) | process capability index |
| NELSON, P.R. (1979) | control charts for WEIBULL processes |
| PADGETT, W.J. / SPURRIER, J.D. (1990) | SHEWHART charts for percentiles |
| PARKHIDEH, B. / CASE, K.E. (1989) | dynamic $\overline{X}$–chart |
| RAHIM, M.A. (1998) | economic $X$–control charts |

| Author(s) | Topic(s) |
|---|---|
| RAMALHOTO, M.F. / MORAIS, M. (1998, 1999) | EWMA charts for the scale parameter |
| | SHEWHART charts for the scale parameter |
| RINNE, H. et al. (2004) | process capability index |
| ROY, S.D. / KAKOTY, S. (1997) | CUSUM control charts with WEIBULL in–control–time |
| ZHANG, L. / CHEN, G. (2004) | EWMA charts for the mean |

Table 7/7:   WEIBULL applications in maintenance and replacement

| Author(s) | Topic(s) |
|---|---|
| FRAHM, P. (1995) | block replacement and $\overline{X}$–charts |
| GEURTS, J.H.J. (1983) | replacement policies and preventative maintenance |
| GLASSER, G.J. (1967) | age replacement |
| LOVE, G.E. / GUO, R. (1996) | preventative maintenance |
| MUNFORD, A.G. / SHAHANI, A.K. (1973) | inspection policy |
| NAKAGAWA, T. (1978) | block replacement |
| NAKAGAWA, T. (1979) | inspection policy |
| NAKAGAWA, T. (1986) | preventative maintenance |
| NAKAGAWA, T. / YASUI, K. (1978) | block replacement |
| NAKAGAWA, T. / YASUI, K. (1981) | age replacement |
| SHAHANI, A.K. / NEWBOLD, S.B. (1972) | inspection policies |
| TADIKAMALLA, P.R. (1980b) | age replacement |
| WHITE, J.S. (1964a) | renewal analysis |
| ZACKS, S. / FENSKE, W.J. (1973) | inspection policy |

Table 7/8:   WEIBULL applications in inventory control

| Author(s) | Topic(s) |
|---|---|
| CHEN, J.M. / LIN, C.S. (2003) | replenishment |
| NIRUPAMA DEVI, K. et al. (2001) | perishable inventory models |
| TADIKAMALLA, P.R. (1978) | inventory lead time |
| WU, J.W. / LEE, W.C. (2003) | EOQ models for items with WEIBULL deterioration |

Table 7/9:   WEIBULL applications in warranty

| Author(s) | Topic(s) |
|---|---|
| KAR, T.R. / NACHLOS, J.A. (1997) | coordinated warranty and burn–in strategies |
| MAJESKE, K.D. / HERRIN, G.D. (1995) | automobile warranty |
| MANN, N.R. (1970c) | warranty bounds |
| MANN, N.R. (1970d) | warranty periods |
| MANN, N.R. / SAUNDERS, S.C. (1969) | evaluation of warranty assurance |
| OHL, H.L. (1974) | time to warranty claim |
| PARK, K.S. / YEE, S.R. (1984) | service costs for consumer product warranty |

Table 7/10: WEIBULL applications in biology and forestry

| Author(s) | Topic(s) |
|---|---|
| BAHLER, C. et al. (1989) | germination data of alfalfa |
| GREEN, E.J. et al. (1993) | tree diameter data |
| JOHNSON, R.A. / HASKELL, J.H. (1983) | lumber data |
| RAWLINGS, J.O. / CURE, W.W. (1985) | ozone effects in crop yield |
| RINK, G. et al. (1979) | germination data of sweet gum |
| STALLARD, N. / WHITEHEAD, A. (2000) | animal carciogenicity |
| WEIBULL, W. (1951) | length of Cyrtoideae, breadth of phasolus vulgaris beans |

Table 7/11: WEIBULL applications in geology, geography and astronomy

| Author(s) | Topic(s) |
|---|---|
| HAGIWARA, Y. (1974) | earthquake occurrences |
| HUGHES, D.W. (1978) | module mass distribution |
| HUILLET, T. / RAYNAUD, H.F. (1999) | earthquake data |
| RAO, P. / OLSON, R. (1974) | size of rock fragments |

Table 7/12: WEIBULL applications in miscellaneous fields

| Author(s) | Topic(s) |
|---|---|
| BERRY, G.L. (1981) | human performance descriptor |
| CHIN, A.C. et al. (1991) | traffic conflict in expressway merging |
| FREEMAN, D.H. et al. (1978) | fitting of synthetic life–tables |
| SHARIF, H. / ISLAM, M. (1980) | forecasting technological change |
| YAMADA, S. et al. (1993) | software reliability growth |

## 7.2 Aids in WEIBULL analysis

WEIBULL analysis as presented in the following chapters is mathematically complicated and cannot be done by using only a pencil, a sheet of paper and a simple pocket-calculator, instead computer aid is demanded. We know of no software package which has implemented all the methods presented in the sequel.

Those readers who have installed a mathematical–statistical **programming language** on their computer, e.g., *C++*, *FORTRAN*, *GAUSS*, *MATLAB*, *PASCAL* or *R*, and know how to write programs will have no difficulties in converting the formulas given here. We have been concerned about writing the formulas in an algorithmic way so that they might be easily programmed. Each programming language offers a wide variety of routines for optimization and simulation and other statistical techniques. Furthermore, each programming language has user groups of its own to exchange programs and experiences.

**Statistical software packages** such as *BMDP*, *SAS*, *SPSS* or *SYSTAT* offer subroutines for doing parts of WEIBULL analysis, e.g., parameter estimation from censored data by the maximum likelihood method or by a linear technique supported by hazard and/or probability plotting. For example, *SAS JMP* has an excellent survival analysis capability, incorporating both graphical plotting and maximum likelihood estimation and covering, in addition to WEIBULL, the exponential, lognormal and extreme value models.

There are **consultants and software houses** specializing in doing WEIBULL analysis. Upon entering "Weibull analysis and software" into *Google*, the reader will get about 40,600 answers. A corporation which is very often found among these answers is *ReliaSoft*. It has offices around the world. *Relia Soft* maintains two public Websites: `ReliaSoft.com` and `weibull.com` dedicated to meeting the needs of the reliability community. *ReliaSoft*'s *Weibull++* software has become the standard for reliability and life data analysis for thousands of companies worldwide. Another rather comprehensive software comes from *Relex*. The reader is recommended to browse through the different site maps offered by *Google* to see what is best suited to meet his or her special needs. But the reader should remember that software packages are cookbooks offering recipes. For their proper understanding, a handbook such as this one.

# 8 Collecting life data[1]

Most applications of the WEIBULL distribution pertain to life data for either technical or biological units. The inferential process for life data heavily depends on how these data have been compiled. So, before entering into the details of WEIBULL inference, it is advisable to look at the different ways to gather information on the length of life.

Life data may originate from either the field, i.e., from the reports of customers using these units (**field data**), or from the laboratory of the producer who tests a sample of units according to a given plan before selling (**lab data**). Sect. 8.1 compares these two types of data. Normally, lab data are superior to field data, so Chapters 9 ff. are mainly devoted to lab data.

In Sect. 8.2 we first enumerate and describe the parameters that determine a life testing plan and thus influence the inferential process. Sect. 8.3 is devoted to those types of plans, i.e., to those settings of the parameters, that are predominant in practice. The optimal parameter setting is rather complicated, but there is a paper of NG et al. (2004) treating this topic for progressively type–II censoring.

## 8.1 Field data versus laboratory data

A first option for obtaining information about the life length of a product is to collect data on the running times of the items as reported by their buyers. A second and competitive option is to run specially designed experiments (life tests) on a sample of items by either the producer himself or an entrusted testing laboratory. With respect to cost, to time and to statistics these two options show marked differences:

1. Presumably, the first option is the cheaper one. We do not have any cost for planning, running and evaluating an experiment. The buyer delivers the information almost free of charge. But, on the other hand, he does not receive any information about the reliability or the propensity to fail when buying the product. The latter fact is certainly not sales promoting.

2. With the first option we generally have to wait for a long time until a sufficient number of failures have been reported, especially when the product is very reliable, i.e., possesses longevity. Contrary to the second option, there does not exist the possibility to shorten the time to failure by operation under stress, i.e., by some type of accelerated life testing.

3. The data coming from the first option will probably be very heterogeneous as the conditions of running and maintaining the units in the field will differ from customer to customer. The data will be homogeneous when they originate from a well–defined

---

[1] Suggested reading for this chapter: BARLOW/PROSCHAN (1988), COHEN (1963, 1991), NELSON (1990), RINNE (1978).

and properly controlled life testing experiment conforming to those conditions under which the product is scheduled to work.

The above discussion clearly leads to the recommendation to build the inference of the life distribution $F(x)$ or of any parameters of $X$ on data from planned experiments. There are situations where field data are generated in the way as lab data. For example, we quite often have field data that are progressively or randomly censored.

## 8.2   Parameters of a life test plan

The entirety of rules and specifications according to which a life test has to be run, is called a **life test plan**. Among these prescriptions there are some of purely technical contents while other specifications are of relevance for the statistician who is to analyze the data. In the sequel we will present only the latter type of rules which will determine the estimating and testing formulas applied to the data. These statistically important parameters are the basis for establishing certain types of testing plans in Sect. 8.3.

A first and most relevant information for the statistician is the **number $n$ of specimens put on test in the beginning.** $n$ need not be equal to the **final** or **total sample size $n^*$** of tested items under the plan. $n^*$ depends on a second directive indicating whether **failing items** have **to be replaced** by new ones while the test is still under way. The coding of this instruction is done by the parameter $R$:

$$R = \left\{ \begin{array}{llll} 0 & \text{no replacement,} & n^* = n, \\ 1 & \text{replacement,} & n^* \geq n. \end{array} \right\} \tag{8.1a}$$

It is supposed that replacement takes place instantaneously or takes so little time with respect to the operating time that it may be neglected. So we always have $n$ specimens running when $R = 1$. Plans with $R = 1$ have a variable total sample size

$$n^* = n + A(D), \tag{8.1b}$$

where $A(D)$ is random and represents the accumulated failures (= replacements) in $(0, D]$, $D$ being the end of testing. It should be obvious that the replacement specimens and the first $n$ specimens have to come from the same population. As the quality of any inferential procedure depends directly on the total sample size $n^*$, plans with $R = 1$ will be better or at least as good as plans with $R = 0$. In plans with $R = 0$ the random moments of failure $X_{i:n}$; $i = 1, \ldots, n$, are lifetimes and arrive in naturally ascending order, i.e., they are order statistics; see the left part of Fig. 8/1. In plans with $R = 1$ the observable times of failure will be noted by $X_{i:n}^*$. $X_{1:n} = X_{1:n}^*$ always is a lifetime, but $X_{i:n}^*$, $i \geq 2$ are not necessarily lifetimes; they are time spans from the beginning of the test to the $i$–th failure in a superimposed renewal process consisting of $n$ simple renewal processes, see the right part of Fig. 8/1.

Figure 8/1: Example of an $(R = 0)$–plan and an $(R = 1)$–plan

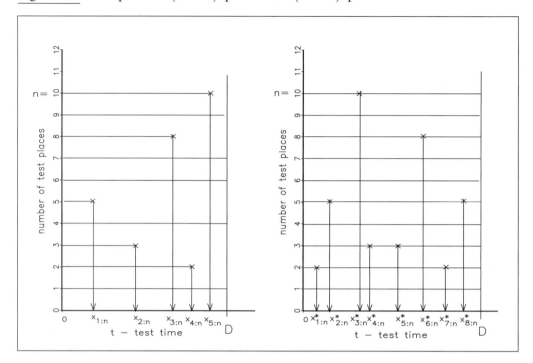

$(R = 1)$–plans are preferable in one or more of the following situations:

1. A specimen does not have a high value or price, at least relative to the other cost parameters of the plan.

2. There are only a few testing machines, i.e., the testing capacity is low and expensive, so that the machines should not be idle until the end of the test when there are early failures.

3. A greater number of failures and completed life spans are needed.

A third statistically relevant information of a life test plan concerns **the way the failure times are recorded**, either exactly or approximately. When there is a monitoring of all specimens on test, either by personnel or by some device, we will precisely register each failure and the life span is exact and free of any measurement error. Exact failure times cannot be observed when test units are inspected in time distances, generally periodically, e.g., once a day or a week. Periodic inspections are used, for example, because

- they are more convenient,
- they are less time consuming or less expensive,
- it is sometimes impossible to test for failures when the units are being used.

If units are inspected only once to see if the unit has failed (e.g., in a destructive stress test), the resulting data are called "quantal–response" data. If units undergo two or more inspections, the data are called "grouped" or "interval" data. Suppose that

- a sample of units are all put on test at the same time,
- all units are initially operating,
- inspections are at specified times.

Then, the resulting data will consist of the number of failures in each interval and the number of units that survived until the end of the test. Thus the failure times falling into the interval $(t_i, t_i + \Delta_i]$ have to be estimated. A reasonable guess would be $t_i + \Delta_i/2$, resulting in a maximum error of $\pm\Delta_i/2$. It is evident that inferences based on grouped data are less reliable than those using exact data, especially when the sample size is small and we have only a small number of failures per interval. Whether the interval width should be constant depends on the shape of the density function. If — for example — the density function is skewed to the right, the width should increase with lifetime. MEEKER (1986) gives guidelines for choosing statistically efficient inspection times and the approximate sample size that achieve a specified degree of precision for estimating a particular quantile of a two–parameter WEIBULL distribution. The parameter $G$ encodes the way data are recorded.

$$G = \left\{ \begin{array}{l} 0 \quad - \quad \text{no grouping,} \\ 1 \quad - \quad \text{grouping.} \end{array} \right\} \tag{8.2}$$

The case $G = 1$ has to be supplemented by information on how the grouping has to be done, e.g., equidistant.

Life testing is time consuming and will last a very long time when the specimens are highly reliable. So engineers and statisticians have developed life test plans with reduced running time. Engineers have proposed to shorten the time to failure by exercising some sort of **stress** on the test units so that they will fail earlier than under normal operating conditions. These approaches are called **accelerated life tests** (ALT). A critical point with ALT is the specification of the life–stress relationship. In this book we will only briefly describe ALT in conjunction with WEIBULL lifetime distributions (see Chapter 16). More details and a lot of examples on ALT are to be found in the monograph of NELSON (1990). The coding of a stress instruction is done by the parameter $S$:

$$S = \left\{ \begin{array}{l} 0 \quad - \quad \text{no stress,} \\ 1 \quad - \quad \text{stress.} \end{array} \right\} \tag{8.3}$$

The case $S = 1$ has to be supplemented by stating how ALT has to be done.

Below are some common types of acceleration of tests:

1. **High usage rate** — Two common ways of doing such a compressed time testing are

   - **Faster** — The product is run with a higher speed than normal, e.g., rolling bearings run at about three times their normal speed.

   - **Reduced off time** — Products which are off much of the time in actual use are run a greater fraction of time, e.g., washer or dryer, toaster or coffee maker.

2. **Specimen design** — Life of some products can be accelerated through the size, geometry and finish of specimens. Generally large specimens fail sooner than small ones.

3. **Stress loading** — Overstress testing consists of running a product at higher than normal levels of some accelerating stresses to shorten product life or to degrade some product performance faster. Typical accelerating stresses are temperature, pressure, voltage, mechanical load, thermal cycling, humidity, salt, sulphur, ozone, solar, radiation and vibration. In biological applications accelerated conditions arise when large doses of a chemical or radiological agent are given. The stress loading can be applied in various ways:

- **Constant stress** — Each specimen is run at a constant stress level until it fails or the test is terminated by censoring (see below).
- **Step stress** — In step–stress loading, a specimen is subjected to successively higher levels of stress. A specimen is first subjected to a specified constant stress for a specified length of time. If it does not fail, it is subjected to a higher stress level for a specified time. The stress on a specimen is thus increased step by step until it fails or the end of the test has come. Usually all specimens go through the same specified pattern of stress levels and test times, but it is also possible to apply different patterns to different specimens.
- **Progressive stress** — In progressive stress loading, a specimen undergoes a continuously increasing level of stress. The most popular case is a linearly increasing stress,[2] so that the stress at time $t$ is given by $s(t) = a\,t$. Different groups of specimens may undergo different progressive stress patters, i.e., which have different linear factors $a_1, \ldots, a_k$ in $s(t) = a_i\,t$.
- **Cyclic stress** — For example, insulation under alternating current voltage sees a sinusoidal stress. Also, many metal components repeatedly undergo a mechanical stress cycle.
- **Random stress** — Some products in use undergo randomly changing levels of stress, e.g., airplane structural components undergo wind buffeting.

Engineers reduce the testing time by shortening the individual lifetimes whereas statisticians limit the test duration $D$ by fixing it more or less directly via some sort of **censoring**. Of course, censoring may be combined with acceleration. Theory and practice have developed a great variety of censoring. A typology of test plans and how to specify $D$ will be presented in the following section. When lifetime data are incomplete this may be due either to the sampling plan itself or to the **unplanned withdrawal** of test units during the test. For example, in a medical experiment one or more of the subjects may leave town and be lost to follow–up or suffer an accident or lose their lives due to some illness other than that under study.

## 8.3 Types of life test plans

Summarizing the specifications enumerated so far we can briefly describe a life test plan by the **quintuple** $\{n,\ R,\ G,\ S,\ D\}$. The topic of Sect. 8.3.1 through 8.3.4 is how $D$ is specified by various types of censoring. The impact of censoring on the inferential process, especially on parameter estimation, will be discussed in Chapters 9 through 11, especially

---

[2] Such type of stress loading — often called **ramp–test** — in conjunction with a WEIBULL distribution for lifetime has been analyzed by BAI et al. (1992, 1997).

in Sect. 11.3 through Sect. 11.8 for ML estimation. Not every inferential approach is applicable to censored data. Estimation by the method of moments (Chapter 12) is not possible when data are censored, whereas the maximum likelihood approach (Chapter 11) is the most versatile method.

### 8.3.1   Introductory remarks

A data sample is said to be censored when, either by accident or design, the value of the variate under investigation is unobserved for some of the items in the sample. **Censoring** and **truncation** have to be distinguished.[3] Truncation is a concept related to the population and its distribution (see Sect. 3.3.5). Truncation of a distribution occurs when a range of possible variate values is either ignored or impossible to observe. Truncation thus modifies the distribution. Censoring modifies the selection of the random variables and is thus related to sampling. A censored observation is distinct from a missing observation in that the order of the censored observation relative to some of the uncensored observations is known and conveys information regarding the distribution being sampled.

We further mention that **censoring** may be either **informative** (= dependent) or **non–informative** (= independent); see BARLOW/PROSCHAN (1988), EMOTO/MATTHEWS (1990) or LAGAKOS (1979). Suppose unit lifetime, $X$, depends on an unknown vector of parameters, $\boldsymbol{\theta}$. A set of instruction determines when observation of a unit stops. This set is non–informative relative to $\boldsymbol{\theta}$ if it does not provide additional information about $\boldsymbol{\theta}$ other than that contained in the data. When $n$ items are put on test, and the test is stopped at the $r$–th observed failure ($\hat{=}$ type–II singly censoring), the stopping rule depends only on $r$ and is clearly independent of life distribution parameters since $r$ is fixed in advance. The same is true when we stop testing at time $T$ ($\hat{=}$ type–I singly censoring) since $T$ is fixed in advance of testing. For non–informative sampling plans, the likelihood, up to a constant of proportionality, depends only on the life distribution model and the data, e.g., see (8.6c,d). This proportionality constant depends on the stopping rule, but not on the unknown parameters in $\boldsymbol{\theta}$. An informative stopping rule may come into existence when the withdrawal of non–failed units is random. In this case we have informative censoring when lifetime and withdrawal time have a common distribution which cannot be factored into the product of the two marginal distributions.

Before going into the details of how to discontinue a life test, we premise a general remark which is useful for a better understanding of censoring. Let $t$ be the time elapsed since the starting of the test and let $A(t)$ be the sum of the random number of failed test units up to and including $t$. $A(t)$ is a discrete function of a continuous variable. In a given life test we observe a realization (= trajectory) of $A(t)$, denoted $a(t)$, which may be depicted as a step function in the $(t, a)$–plane (see Fig. 8/2). The life test will be stopped at that very moment when $A(t)$ enters a special area $P$ of that plane. $P$ is specified by the chosen censoring criterion or criteria.

### 8.3.2   Singly censored tests

A life test is named **singly censored** (= **single–censored**) **from above** (= **on the right**) when all test units that have not failed up to a certain time $D$ are withdrawn from the test,

---

[3] For a historical account on the concepts see COHEN (1991).

which is thus terminated. **Censoring from below (= on the left)** means that the failure occurred prior to some designated left censoring time.[4] Censoring on the right is most common with life testing so that — in the sequel — censoring means censoring on the right unless stated otherwise. Left censoring might be employed to burn off "duds" or "freaks," i.e., early failing items which are atypical of the population targeted for study.[5] In either case a single censoring means that we know only the number of failed items outside a certain range, but their exact value remains unknown.

### 8.3.2.1 Type-I censoring

In **type–I censoring (= time censoring)**, testing is suspended when a preestablished and fixed testing time $T$ has been reached — the maximum duration is $D = T$. Presuming $S = 0$ (no stress) and $G = 0$ (no grouping)[6] in the following sections, we have two plans with respect to $R$:

$\{n, 0, G, S, T\}$ — no replacement of items failing prior to $T$,
$\{n, 1, G, S, T\}$ — replacement of items failing items.

A closer examination reveals that — ceteris paribus — these plans differ with respect to the area $P$ and may have differing effective test times. As failing items are replaced instantaneously with the $(R=1)$–plan, we will never run out of units on test and the effective and the maximum durations always coincide at $t = T$, which is non–random. The course of $A(t)$ and the number of failed items $A(t)$ are random. The area $P$ is given by the half–plane $t > T$ (see the left–hand side of Fig. 8/2).

The average number of failing items in $(0, T]$ is

$$\mathrm{E}\big(A(T)\big) = \left\{ \begin{array}{ll} n\,\mathrm{Pr}(X \leq T) & \text{for } R = 0 \\ n\,\mathrm{E}(N_T) & \text{for } R = 1, \end{array} \right\} \tag{8.4a}$$

where $\mathrm{E}(N_T) = \sum_{k=1}^{\infty} F_k(T)$ is the renewal function, because on each of the $n$ testing places we observe a renewal process. These processes are assumed to run independently. For $\mathrm{E}(N_T)$ in the WEIBULL case, see (4.47) ff.

Testing with the $(R = 0)$–plan may come to an end before $T$ when $T$ has been chosen too large with respect to the lifetime distribution so that all $n$ units fail before $T$, which happens with probability

$$\mathrm{Pr}(X_{n:n} < T) = \big[F(T)\big]^n, \text{ generally,} \tag{8.4b}$$

$$= \left[ 1 - \exp\left\{ - \left( \frac{T - a}{b} \right)^c \right\} \right]^n, \text{ for WEIBULL lifetime.} \tag{8.4c}$$

---

[4] For **doubly censoring (= censoring on both sides)**, see Sect. 8.3.4

[5] Left censoring occurs in other than life–testing contexts when the lower limit of resolution of a measuring device prevents the measurement of a fraction of the smallest ordered values in a sample.

[6] The case $G = 1$ will be discussed in Sect. 8.3.4.

This probability is lower the greater $n$ and/or the higher the reliability of the units. The effective test duration is

$$D = \min(T, X_{n:n}),\qquad(8.4\text{d})$$

which is random. Its average may be calculated as

$$E(D) = \int\limits_{0}^{T} x\, f_{n:n}(x)\,\mathrm{d}x + T\,\mathrm{Pr}\big(X_{n:n} > T\big).\qquad(8.4\text{e})$$

Now $P$ is a three–quarter plane

$$P = \{(t,a) : t > T \;\vee\; a \geq n\}.\qquad(8.4\text{f})$$

The right–hand side of Fig. 8/2 shows this area $P$ for $n = 10$ together with two trajectories $a_1(t)$ and $a_2(t)$ leading to differing test durations.

Figure 8/2:  Trajectory of a plan $\{n,\ 1,\ G,\ S,\ T\}$ on the left-hand side and two trajectories of a plan $\{10,\ 0,\ G,\ S,\ T\}$ on the right-hand side

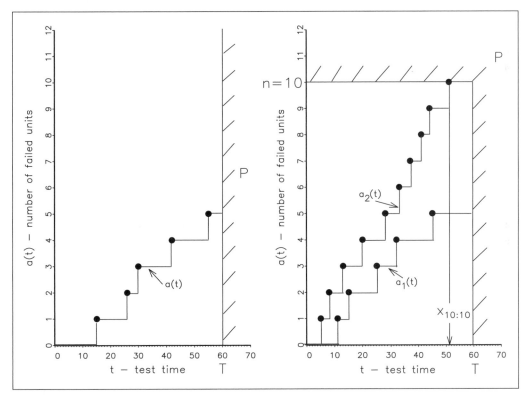

A singly censored type–I plan with $R = 0$ has an **expected time saving** $TS$ with respect to an uncensored test, i.e., to the lifetime $X_{n:n}$ of the last failing unit out of $n$ units, given by

$$TS = \mathrm{E}\big(X_{n:n}\big) - T.\qquad(8.4\text{g})$$

For $X \sim We(a, b, c)$, we use (5.34) to get

$$TS = a + b \left[ n\,\Gamma\!\left(1 + \frac{1}{c}\right) \sum_{i=0}^{n-1} \frac{(-1)^i \binom{n-1}{i}}{(i+1)^{1+(1/c)}} \right] - T. \tag{8.4h}$$

(8.4h) may be evaluated with presumptive values of $a$, $b$ and $c$ to get an idea of the cutting of testing time.

Considering the $(R=0)$–plan we define the **relative time saving** $w$ by

$$w \;\; := \;\; \frac{TS}{\mathrm{E}\!\left(X_{n:n}\right)} = 1 - \frac{T}{\mathrm{E}\!\left(X_{n:n}\right)}, \;\; \text{in general,}$$

$$= \;\; 1 - \frac{T}{b\,n\,\Gamma\!\left(1 + \dfrac{1}{c}\right) \displaystyle\sum_{i=0}^{n-1} \dfrac{(-1)^i \binom{n-1}{i}}{(i+1)^{1+(1/c)}}}, \;\; \text{for } X \sim We(0, b, c). \tag{8.4i}$$

Tab. 8/1 gives the scaled censoring time

$$\frac{T}{b} = (1 - w)\, n\,\Gamma\!\left(1 + \frac{1}{c}\right) \sum_{i=0}^{n-1} \frac{(-1)^i \binom{n-1}{i}}{(i+1)^{1+(1/c)}} \tag{8.4j}$$

for several combinations of $n$ and $c$ when $w$ is $0.1(0.1)0.5$. Tab. 8/1 thus helps to choose $n$ and $T$ when there is a reasonable guess of $b$ and $c$.

Let $n^*$ be the effective number of units tested

$$n^* \begin{cases} = n & \text{for } R = 0, \\[2mm] \geq n & \text{for } R = 1, \end{cases}$$

and let $f(\cdot)$ be the DF and $F(\cdot)$ the CDF of a unit's **true lifetime** $X_i$. By $Y_i$ we denote the unit's **observable lifetime**. The true lifetime $X_i$ of an individual will be observed only if $X_i \leq T$. The data from such a setup can be represented by the $n^*$ pairs of random variables $(Y_i, \delta_i)$,

$$Y_i = \min(X_i, T) \;\; \text{and} \;\; \delta_i = \begin{cases} 1 & \text{if } X_i \leq T, \\ 0 & \text{if } X_i > T. \end{cases} \tag{8.5a}$$

$\delta_i$ indicates whether the lifetime is complete or censored, and $Y_i$ is equal to $X_i$ if it is observed, and to $T$ if it is not. The joint density–probability of $Y_i$ and $\delta_i$ is

$$g(y_i, \delta_i) = f(y_i)^{\delta_i} \left[1 - F(T)\right]^{1 - \delta_i}. \tag{8.5b}$$

To see this, note that $Y_i$ is a mixed random variable with a continuous and a discrete component. For the discrete part we have

$$\Pr(Y_i = T) = \Pr(\delta_i = 0) = \Pr(X_i > T) = 1 - F(T). \tag{8.5c}$$

Table 8/1: Scaled censoring time $T/b$ to save $100\,w\%$ time against $X_{n.n}$ on the average

| | | | | | $w = 0.10$ | | | | | |
|---|---|---|---|---|---|---|---|---|---|---|
| $c$ / $n$ | 0.5 | 1.0 | 1.5 | 2.0 | 2.5 | 3.0 | 3.5 | 4.0 | 4.5 | 5.0 |
| 5 | 6.0095 | 2.0550 | 1.5158 | 1.3158 | 1.2129 | 1.1506 | 1.1088 | 1.0789 | 1.0565 | 1.0391 |
| 10 | 9.1158 | 2.6361 | 1.8083 | 1.5082 | 1.3556 | 1.2638 | 1.2025 | 1.1589 | 1.1262 | 1.1008 |
| 15 | 11.3320 | 2.9864 | 1.9727 | 1.6123 | 1.4310 | 1.3226 | 1.2507 | 1.1996 | 1.1614 | 1.1318 |
| 20 | 13.0859 | 3.2380 | 2.0863 | 1.6828 | 1.4814 | 1.3616 | 1.2824 | 1.2263 | 1.1844 | 1.1521 |
| 25 | 14.5505 | 3.4344 | 2.1728 | 1.7357 | 1.5190 | 1.3905 | 1.3058 | 1.2459 | 1.2013 | 1.1668 |
| 30 | 15.8149 | 3.5955 | 2.2424 | 1.7779 | 1.5487 | 1.4133 | 1.3242 | 1.2613 | 1.2145 | 1.1784 |
| 35 | 16.9313 | 3.7321 | 2.3005 | 1.8128 | 1.5732 | 1.4320 | 1.3393 | 1.2739 | 1.2253 | 1.1878 |
| 40 | 17.9335 | 3.8507 | 2.3503 | 1.8425 | 1.5939 | 1.4478 | 1.3520 | 1.2845 | 1.2344 | 1.1958 |
| 45 | 18.8448 | 3.9577 | 2.3995 | 1.8780 | 1.6251 | 1.4776 | 1.3820 | 1.3149 | 1.2657 | 1.2283 |
| 50 | 19.6848 | 4.0900 | 2.5318 | 2.0553 | 1.8467 | 1.7349 | 1.6747 | 1.6480 | 1.6245 | 1.6257 |

| | | | | | $w = 0.20$ | | | | | |
|---|---|---|---|---|---|---|---|---|---|---|
| $n$ | 0.5 | 1.0 | 1.5 | 2.0 | 2.5 | 3.0 | 3.5 | 4.0 | 4.5 | 5.0 |
| 5 | 5.3418 | 1.8267 | 1.3474 | 1.1696 | 1.0781 | 1.0227 | 0.9856 | 0.9590 | 0.9391 | 0.9236 |
| 10 | 8.1029 | 2.3432 | 1.6073 | 1.3406 | 1.2050 | 1.1233 | 1.0689 | 1.0301 | 1.0011 | 0.9785 |
| 15 | 10.0729 | 2.6546 | 1.7535 | 1.4331 | 1.2720 | 1.1756 | 1.1117 | 1.0663 | 1.0324 | 1.0061 |
| 20 | 11.6319 | 2.8782 | 1.8545 | 1.4958 | 1.3168 | 1.2103 | 1.1399 | 1.0900 | 1.0528 | 1.0240 |
| 25 | 12.9338 | 3.0528 | 1.9314 | 1.5428 | 1.3502 | 1.2360 | 1.1607 | 1.1075 | 1.0678 | 1.0372 |
| 30 | 14.0577 | 3.1960 | 1.9932 | 1.5803 | 1.3766 | 1.2562 | 1.1771 | 1.1211 | 1.0796 | 1.0475 |
| 35 | 15.0500 | 3.3174 | 2.0449 | 1.6114 | 1.3984 | 1.2729 | 1.1905 | 1.1323 | 1.0892 | 1.0559 |
| 40 | 15.9409 | 3.4228 | 2.0891 | 1.6378 | 1.4168 | 1.2869 | 1.2018 | 1.1418 | 1.0972 | 1.0629 |
| 45 | 16.7510 | 3.5179 | 2.1329 | 1.6693 | 1.4445 | 1.3135 | 1.2284 | 1.1688 | 1.1251 | 1.0919 |
| 50 | 17.4976 | 3.6355 | 2.2505 | 1.8270 | 1.6415 | 1.5422 | 1.4887 | 1.4649 | 1.4440 | 1.4450 |

| | | | | | $w = 0.30$ | | | | | |
|---|---|---|---|---|---|---|---|---|---|---|
| $n$ | 0.5 | 1.0 | 1.5 | 2.0 | 2.5 | 3.0 | 3.5 | 4.0 | 4.5 | 5.0 |
| 5 | 4.6741 | 1.5983 | 1.1789 | 1.0234 | 0.9434 | 0.8949 | 0.8624 | 0.8392 | 0.8217 | 0.8082 |
| 10 | 7.0900 | 2.0503 | 1.4064 | 1.1730 | 1.0544 | 0.9829 | 0.9353 | 0.9013 | 0.8759 | 0.8562 |
| 15 | 8.8138 | 2.3228 | 1.5343 | 1.2540 | 1.1130 | 1.0287 | 0.9728 | 0.9330 | 0.9033 | 0.8803 |
| 20 | 10.1779 | 2.5184 | 1.6227 | 1.3088 | 1.1522 | 1.0590 | 0.9974 | 0.9538 | 0.9212 | 0.8960 |
| 25 | 11.3171 | 2.6712 | 1.6899 | 1.3500 | 1.1814 | 1.0815 | 1.0156 | 0.9690 | 0.9343 | 0.9075 |
| 30 | 12.3004 | 2.7965 | 1.7441 | 1.3828 | 1.2045 | 1.0992 | 1.0299 | 0.9810 | 0.9446 | 0.9165 |
| 35 | 13.1688 | 2.9027 | 1.7893 | 1.4100 | 1.2236 | 1.1138 | 1.0417 | 0.9908 | 0.9530 | 0.9239 |
| 40 | 13.9483 | 2.9950 | 1.8280 | 1.4331 | 1.2397 | 1.1261 | 1.0516 | 0.9990 | 0.9601 | 0.9301 |
| 45 | 14.6571 | 3.0782 | 1.8663 | 1.4607 | 1.2640 | 1.1493 | 1.0749 | 1.0227 | 0.9845 | 0.9554 |
| 50 | 15.3104 | 3.1811 | 1.9692 | 1.5986 | 1.4363 | 1.3494 | 1.3026 | 1.2818 | 1.2635 | 1.2644 |

| | | | | | $w = 0.40$ | | | | | |
|---|---|---|---|---|---|---|---|---|---|---|
| $n$ | 0.5 | 1.0 | 1.5 | 2.0 | 2.5 | 3.0 | 3.5 | 4.0 | 4.5 | 5.0 |
| 5 | 4.0063 | 1.3700 | 1.0105 | 0.8772 | 0.8086 | 0.7670 | 0.7392 | 0.7193 | 0.7043 | 0.6927 |
| 10 | 6.0772 | 1.7574 | 1.2055 | 1.0054 | 0.9037 | 0.8425 | 0.8017 | 0.7726 | 0.7508 | 0.7339 |
| 15 | 7.5546 | 1.9909 | 1.3151 | 1.0749 | 0.9540 | 0.8817 | 0.8338 | 0.7997 | 0.7743 | 0.7546 |
| 20 | 8.7239 | 2.1586 | 1.3909 | 1.1219 | 0.9876 | 0.9077 | 0.8550 | 0.8175 | 0.7896 | 0.7680 |
| 25 | 9.7004 | 2.2896 | 1.4485 | 1.1571 | 1.0126 | 0.9270 | 0.8705 | 0.8306 | 0.8009 | 0.7779 |
| 30 | 10.5432 | 2.3970 | 1.4949 | 1.1852 | 1.0324 | 0.9422 | 0.8828 | 0.8408 | 0.8097 | 0.7856 |
| 35 | 11.2875 | 2.4881 | 1.5336 | 1.2085 | 1.0488 | 0.9547 | 0.8928 | 0.8492 | 0.8169 | 0.7919 |
| 40 | 11.9557 | 2.5671 | 1.5668 | 1.2284 | 1.0626 | 0.9652 | 0.9013 | 0.8563 | 0.8229 | 0.7972 |
| 45 | 12.5632 | 2.6384 | 1.5996 | 1.2520 | 1.0834 | 0.9851 | 0.9213 | 0.8766 | 0.8438 | 0.8189 |
| 50 | 13.1232 | 2.7267 | 1.6879 | 1.3702 | 1.2311 | 1.1566 | 1.1165 | 1.0987 | 1.0830 | 1.0838 |

| | | | | | $w = 0.50$ | | | | | |
|---|---|---|---|---|---|---|---|---|---|---|
| $n$ | 0.5 | 1.0 | 1.5 | 2.0 | 2.5 | 3.0 | 3.5 | 4.0 | 4.5 | 5.0 |
| 5 | 3.3386 | 1.1417 | 0.8421 | 0.7310 | 0.6738 | 0.6392 | 0.6160 | 0.5994 | 0.5869 | 0.5773 |
| 10 | 5.0643 | 1.4645 | 1.0046 | 0.8379 | 0.7531 | 0.7021 | 0.6681 | 0.6438 | 0.6257 | 0.6116 |
| 15 | 6.2955 | 1.6591 | 1.0959 | 0.8957 | 0.7950 | 0.7348 | 0.6948 | 0.6664 | 0.6452 | 0.6288 |
| 20 | 7.2699 | 1.7989 | 1.1591 | 0.9349 | 0.8230 | 0.7565 | 0.7125 | 0.6813 | 0.6580 | 0.6400 |
| 25 | 8.0836 | 1.9080 | 1.2071 | 0.9643 | 0.8439 | 0.7725 | 0.7255 | 0.6922 | 0.6674 | 0.6482 |
| 30 | 8.7860 | 1.9975 | 1.2458 | 0.9877 | 0.8604 | 0.7851 | 0.7357 | 0.7007 | 0.6747 | 0.6547 |
| 35 | 9.4063 | 2.0734 | 1.2780 | 1.0071 | 0.8740 | 0.7955 | 0.7440 | 0.7077 | 0.6807 | 0.6599 |
| 40 | 9.9631 | 2.1393 | 1.3057 | 1.0236 | 0.8855 | 0.8043 | 0.7511 | 0.7136 | 0.6858 | 0.6643 |
| 45 | 10.4694 | 2.1987 | 1.3330 | 1.0433 | 0.9028 | 0.8209 | 0.7678 | 0.7305 | 0.7032 | 0.6824 |
| 50 | 10.9360 | 2.2722 | 1.4066 | 1.1419 | 1.0259 | 0.9638 | 0.9304 | 0.9156 | 0.9025 | 0.9031 |

For values $y_i \leq T$ the continuous part is

$$\Pr(y_i \,|\, \delta_i = 1) = \Pr(y_i \,|\, X_i < T) = f(y_i)/F(T), \tag{8.5d}$$

where for convenience we have used the notation $\Pr(y_i \,|\, \delta_i = 1)$ to represent the probability density function of $Y_i$, given that $X_i < T$. The distribution of $(Y_i, \delta_i)$ thus has components

$$\Pr(Y_i = T, \delta_i = 0) = \Pr(\delta_i = 0) = 1 - F(T), \quad y_i = T,$$

$$\Pr(y_i, \delta_i = 1) = \Pr(y_i \,|\, \delta_i = 1)\Pr(\delta_i = 1) = f(y_i), \quad y_i < T.$$

These expressions can be combined into the single expression $\Pr(y_i, \delta_i) = f(y_i)^{\delta_i} \left[1 - F(T)\right]^{1-\delta_i}$. If pairs $(Y_i, \delta_i)$ are independent, the **likelihood function for type–I single censoring** is

$$L(\boldsymbol{\theta} \,|\, \mathrm{data}) = K \prod_{i=1}^{n^*} f(y_i \,|\, \boldsymbol{\theta})^{\delta_i} \left[1 - F(T \,|\, \boldsymbol{\theta})\right]^{1-\delta_i}, \tag{8.5e}$$

where $K$ denotes an ordering constant that does not depend on the parameters in $\boldsymbol{\theta}$. The form of $K$ depends on the underlying sampling and censoring mechanisms and is difficult to characterize in general. Because, for most models, $K$ is a constant not depending on the model parameters, it is common practice to take $K = 1$ and suppress $K$ from the likelihood expressions and computations.[7]

With respect to the $(R = 1)$–plan we remark that the observed lifetimes do not come in natural ascending order. In both cases, $R = 0$ as well as $R = 1$, the number of failing units is random.

### 8.3.2.2 Type-II censoring

Type–I censoring — especially in conjunction with replacement — is to be preferred when the testing facilities are limited and one or more other types of products are waiting to be tested with the same equipment. Type–I censoring is preferable too when the unit price of an item to be tested is low compared with the operating cost of the equipment for one unit of time. If instead the unit price is dominant and failing is destructive, one will be interested not in using too many items. In this case the test duration may be controlled by a given fixed number $r$ of failures leading to **type–II censoring (= failure censoring)** having a random test duration $D$. When $R = 0$, the test duration is given simply by

$$D = X_{r:n}, \tag{8.6a}$$

whereas for $R = 1$, the duration is more difficult to determine and is given by the time $X_{r:n}^*$ of the $r$–th failure (= renewal) of a superposed renewal process consisting of $n$ renewal processes running parallel. With the $(R=1)$–plan it is even possible to set $r > n$. It should

---

[7] ASHOUR/SHALABY (1983) show how to estimate $n$ in the case of type–I censoring.

be self–evident that — ceteris paribus — the expected duration of the $(R=1)$–plan is less than or equal to that of the $(R=0)$–plan given by

$$\mathrm{E}\big(X_{r:n}\big) = r \binom{n}{r} \Gamma\Big(1 + \frac{1}{c}\Big) \sum_{i=0}^{r-1} \frac{(-1)^i \binom{r-1}{i}}{(n - r + i + 1)^{1+(1/c)}}\,, \quad \text{for } X \sim We(a, b, c).$$

(8.6b)

Irrespective of the choice of $R$, type–II singly censored tests always have a fixed number of failures but a random duration. Thus the area $P$ is the half–plane $a \geq r$ (see Fig. 8/3).

Figure 8/3: Trajectory of a type-II censored test

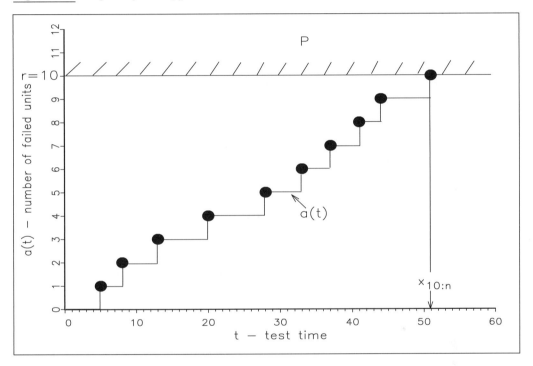

The **likelihood function for type–II singly censoring** has the general formula

$$L(\boldsymbol{\theta} \,|\, \mathrm{data}) = K \left[\prod_{i=1}^{r} f(x_i \,|\, \boldsymbol{\theta})\right] \big[1 - F(x_{r:n} \,|\, \boldsymbol{\theta})\big]^{n-r},$$

(8.6c)

where the ordering constant is given as

$$K = \frac{n!}{(n - r)!}\,.$$

(8.6d)

The form of the likelihood function (8.6c) is noteworthy: Each true lifetime contributes a term $f(x_i)$ to the likelihood, and each censoring time contributes a term $1 - F(x_{r:n})$. It can also be noted that although the genesis of (8.6c) is quite different from that of the likelihood function (8.5e) obtained for type–I censoring, the form of the observed likelihood function is the same in both cases when in (8.5e) $T$ is substituted by $x_{r:n}$ and the product is transformed using the number of failed and non–failed items in $(0, T]$. The presentation in (8.5e) is more general and turns out to have a wide applicability when censoring is random (see Sect. 8.3.4).

We will now present two approaches measuring the time saving under type–II censoring when $R = 0$. The first approach starts from the distribution of the random quotient

$$V := \frac{X_{r:n}}{X_{n:n}}, \ 0 \leq V \leq 1, \tag{8.7a}$$

which is the **relative waiting time** if failures $r + 1$ through to $n$ are not observed. The **relative time saving** is

$$W := 1 - V = 1 - \frac{X_{r:n}}{X_{n:n}}. \tag{8.7b}$$

$W$ thus tells what portion of time could be saved when in a given random sample of $n$ units they are only observed until the $r$–th failure instead of waiting up to the last failure. $W$ as well as $V$ vary at random with the sample chosen.

The second approach uses REET,[8] the **ratio of** of average type–II **censoring time to** the average **complete sampling time**:

$$REET := \frac{\mathrm{E}(X_{r:n})}{\mathrm{E}(X_{n:n})}; \tag{8.8}$$

thus, $REET$ tells what is to be saved by comparing the average duration of a great number of censored tests with the expected duration of a series of uncensored tests, each having a sample size of $n$. Because $X_{r:n}$ and $X_{n:n}$ are dependent variates, we have

$$\mathrm{E}(W) = 1 - \mathrm{E}\left(\frac{X_{r:n}}{X_{n:n}}\right) \neq 1 - REET = 1 - \frac{\mathrm{E}(X_{r:n})}{\mathrm{E}(X_{n:n})}. \tag{8.9}$$

In the WEIBULL case $\mathrm{E}(W)$ is infinite for all $c \in (0, 1]$. $W$ and its percentiles should be used when we have to judge only one test, whereas $REET$ is to be preferred when the test is to be executed a great number of times so that the long run behavior is what counts.

MUENZ/GRUEN (1977) give the following CDF for $V$:

$$\Pr(V \leq v) = n \sum_{i=r}^{n-1} \binom{n-1}{i} \int_{0}^{1} Q_v(s)^i \left[1 - Q_v(s)\right]^{n-1-i} s^{n-1} \, ds \tag{8.10a}$$

where

$$Q_v(s) = F\left[v \, F^{-1}(s)\right]\big/ s, \ 0 < s \leq 1, \tag{8.10b}$$

---

[8]  REET is the acronym for **r**atio of **e**xpected **e**xperimental **t**imes.

and $F(\cdot)$ is the CDF of the lifetime variate $X$. For large values of $n$ we can use the asymptotic formula, which is based on a normal approximation of the binomial distribution with continuity correction,

$$\lim_{n \to \infty} \Pr(V \le v) = n \int_0^1 \Phi\left[\frac{(n-1)Q_v(s) - r + 0.5}{\sqrt{(n-1)Q_v(s)\left[1 - Q_v(s)\right]}}\right] s^{n-1} \, ds, \qquad (8.10c)$$

where

$$\Phi(u) = \frac{1}{\sqrt{2\pi}} \int_{-\infty}^u \exp\{-u^2/2\} \, du$$

is the CDF of the standard normal distribution and $r$ goes to infinity such that $\lim_{n \to \infty}(r/n)$ is bounded away from 0 and 1.

When $X \sim We(0, b, c)$ the term $Q_v(s)$ turns into

$$Q_v(s) = \frac{1 - (1-s)^{v^c}}{s}, \quad 0 < s \le 1, \ 0 \le v \le 1, \ c \in \mathbb{R}^+. \qquad (8.10d)$$

$V$ is independent of the scale parameter $b$. Fig. 8/4 depicts (8.10a) for six different values of $c$ when $n = 10$ and $r = 7$. From Fig 8/4 we may read the 90% percentile of the relative waiting time $v_{0.9}$. Thus we have a 90% chance to save more than $100 \cdot (1 - v_{0.9})\%$ of the complete sampling time, where the saving goes from 15% for $c = 3$ to 58% for $c = 0.5$.

Figure 8/4: Distribution function of $V$ for $X \sim We(0, b, c)$

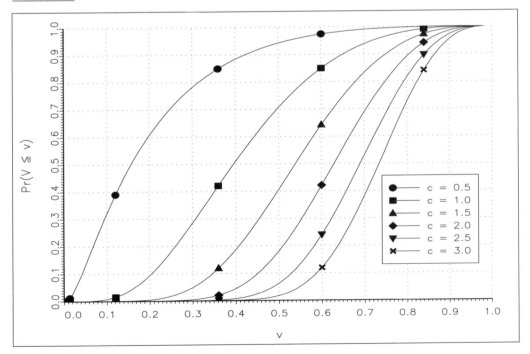

Table 8/2:   $REET$ when $X \sim We(0, b, c)$

|     |     | c |     |     |     |     |     |     |     |     |     |
| --- | --- | --- | --- | --- | --- | --- | --- | --- | --- | --- | --- |
| $n$ | $r$ | 0.5 | 1.0 | 1.5 | 2.0 | 2.5 | 3.0 | 3.5 | 4.0 | 4.5 | 5.0 |
| 10 | 5 | 0.0500 | 0.2207 | 0.3637 | 0.4675 | 0.5438 | 0.6016 | 0.6467 | 0.6828 | 0.7122 | 0.7367 |
|    | 6 | 0.0837 | 0.2891 | 0.4369 | 0.5371 | 0.6081 | 0.6605 | 0.7008 | 0.7326 | 0.7583 | 0.7796 |
|    | 7 | 0.1381 | 0.3746 | 0.5203 | 0.6129 | 0.6760 | 0.7217 | 0.7562 | 0.7831 | 0.8047 | 0.8223 |
|    | 8 | 0.2329 | 0.4885 | 0.6219 | 0.7009 | 0.7529 | 0.7895 | 0.8167 | 0.8377 | 0.8544 | 0.8680 |
|    | 9 | 0.4247 | 0.6595 | 0.7596 | 0.8144 | 0.8490 | 0.8727 | 0.8900 | 0.9031 | 0.9135 | 0.9218 |
| 20 | 10 | 0.0343 | 0.1863 | 0.3265 | 0.4321 | 0.5111 | 0.5716 | 0.6192 | 0.6574 | 0.6888 | 0.7150 |
|    | 12 | 0.0586 | 0.2451 | 0.3927 | 0.4965 | 0.5714 | 0.6274 | 0.6707 | 0.7051 | 0.7331 | 0.7562 |
|    | 14 | 0.0988 | 0.3197 | 0.4693 | 0.5677 | 0.6362 | 0.6862 | 0.7243 | 0.7542 | 0.7783 | 0.7981 |
|    | 16 | 0.1713 | 0.4218 | 0.5648 | 0.6525 | 0.7112 | 0.7531 | 0.7844 | 0.8087 | 0.8281 | 0.8439 |
|    | 18 | 0.3297 | 0.5843 | 0.7016 | 0.7677 | 0.8100 | 0.8392 | 0.8607 | 0.8771 | 0.8901 | 0.9006 |
| 30 | 15 | 0.0282 | 0.1699 | 0.3075 | 0.4133 | 0.4933 | 0.5551 | 0.6039 | 0.6432 | 0.6756 | 0.7027 |
|    | 18 | 0.0485 | 0.2239 | 0.3700 | 0.4750 | 0.5516 | 0.6093 | 0.6541 | 0.6898 | 0.7190 | 0.7431 |
|    | 21 | 0.0825 | 0.2927 | 0.4427 | 0.5435 | 0.6145 | 0.6667 | 0.7066 | 0.7381 | 0.7635 | 0.7845 |
|    | 24 | 0.1445 | 0.3879 | 0.5342 | 0.6259 | 0.6879 | 0.7325 | 0.7660 | 0.7921 | 0.8130 | 0.8301 |
|    | 27 | 0.2837 | 0.5427 | 0.6681 | 0.7401 | 0.7866 | 0.8190 | 0.8429 | 0.8613 | 0.8758 | 0.8876 |
| 40 | 20 | 0.0248 | 0.1597 | 0.2952 | 0.4009 | 0.4816 | 0.5441 | 0.5936 | 0.6337 | 0.6667 | 0.6943 |
|    | 24 | 0.0429 | 0.2106 | 0.3553 | 0.4608 | 0.5384 | 0.5972 | 0.6429 | 0.6795 | 0.7094 | 0.7342 |
|    | 28 | 0.0732 | 0.2757 | 0.4254 | 0.5276 | 0.6000 | 0.6536 | 0.6947 | 0.7272 | 0.7535 | 0.7752 |
|    | 32 | 0.1288 | 0.3661 | 0.5140 | 0.6081 | 0.6722 | 0.7185 | 0.7535 | 0.7808 | 0.8026 | 0.8205 |
|    | 36 | 0.2554 | 0.5149 | 0.6451 | 0.7210 | 0.7703 | 0.8049 | 0.8304 | 0.8501 | 0.8657 | 0.8783 |
| 50 | 25 | 0.0226 | 0.1525 | 0.2864 | 0.3919 | 0.4729 | 0.5359 | 0.5860 | 0.6266 | 0.6600 | 0.6881 |
|    | 30 | 0.0391 | 0.2012 | 0.3447 | 0.4505 | 0.5287 | 0.5882 | 0.6347 | 0.6719 | 0.7023 | 0.7276 |
|    | 35 | 0.0669 | 0.2636 | 0.4129 | 0.5159 | 0.5893 | 0.6439 | 0.6858 | 0.7191 | 0.7460 | 0.7682 |
|    | 40 | 0.1182 | 0.3505 | 0.4993 | 0.5949 | 0.6606 | 0.7081 | 0.7441 | 0.7723 | 0.7949 | 0.8134 |
|    | 45 | 0.2358 | 0.4946 | 0.6280 | 0.7066 | 0.7580 | 0.7941 | 0.8209 | 0.8415 | 0.8579 | 0.8712 |

The second approach gives the following formula when $X \sim We(0, b, c)$ and using (5.34):

$$REET = \binom{n-1}{r-1} \frac{\sum_{i=0}^{r-1} \frac{(-1)^i \binom{r-1}{i}}{(n-r+i+1)^{1+(1/c)}}}{\sum_{i=0}^{n-1} \frac{(-1)^i \binom{n-1}{i}}{(i+1)^{1+(1/c)}}}, \qquad (8.11a)$$

which — like $V$ above — is independent of $b$. Using (8.11a) to compute $REET$ is not recommended, especially when $r$ and/or $n$ is large. The binomial coefficient gets huge and its number of digits can exceed the number of precision digits of the computer software being used. For this reason Tab. 8/2 has been computed by evaluating the defining integrals in

$$E(X_{r:n}) = b\,n \int_0^1 F^{-1}(y) \binom{n-1}{r-1} y^{r-1} (1-y)^{n-r}\, \mathrm{d}y. \qquad (8.11b)$$

As $F^{-1}(y) = \left[ -\ln(1-y) \right]^{1/c}$ in the WEIBULL case, we get

$$\mathrm{E}(X_{r:n}) = b\,r \int_0^1 \binom{n}{r} \left[ -\ln(1-y) \right]^{1/c} y^{r-1} (1-y)^{n-r}\, dy \qquad (8.11c)$$

$$\mathrm{E}(X_{n:n}) = b\,n \int_0^1 \left[ -\ln(1-y) \right]^{1/c} y^{n-1}\, dy. \qquad (8.11d)$$

To avoid computing a huge binomial coefficient in (8.11c), take the logarithm of the integrand first and then recover its value by the exponent operation. Tab. 8/2 gives $REET$ for $n = 10(10)50$, $c = 0.5(0.5)5$ and $r = [0.5(0.1)0.9]\,n$, i.e., the censoring number $r$ is 50%, 60%, 70%, 80% and 90% of $n$.

HSIEH (1994) has studied a different version of $REET$, namely

$$REET^* = \frac{\mathrm{E}(X_{r:n})}{\mathrm{E}(X_{r:r})}. \qquad (8.11e)$$

The difference to (8.8) is found in the denominator. We think that $REET$ given by (8.8) is more apt to measure time saving with type–II single censoring. $REET^*$ uses different sets of information from differing sample sizes, $r$ complete lifetimes and $n - r$ censored lifetimes from a sample of size $n$ in the numerator and $r$ complete lifetimes from a sample of size $r$ in the denominator. $REET$, instead, has the same set of information in the numerator and denominator, namely $n$ observed lifetimes where in the numerator only $n - r$ life spans are known to be greater than $X_{r:n}$.

With respect to $R$ we have two type–II censored plans:

$$\{n,\ 0,\ G,\ S,\ X_{r:n}\} \quad - \quad \text{no replacement of items failing prior to } X_{r:n},$$
$$\{n,\ 1,\ G,\ S,\ X_{r:n}^*\} \quad - \quad \text{replacement of failing items.}$$

What plan to choose? It is easy to see that — for the same $n$ and the same $r$ — the $(R=1)$–plan has a shorter duration. But with respect to the cost depending on the number of used items, the $(R=1)$–plan is less favorable. $r$ items will be destroyed under both plans, but the $(R=1)$–plan uses $n + r$ items instead of only $n$ items with the $(R=0)$–plan.

### 8.3.2.3  Combined type-I and type-II censoring

If the product under life testing is highly reliable, the duration $D = X_{r:n}$ or $D = X_{r:n}^*$ of a type–II plan may become very long. To have a safeguard against those costs depending on $D$ as well as against those costs depending on the unit price of an item, one can combine type–I and type–II censoring and set, e.g., for $R = 0$,

$$D = \min(X_{r:n}, T)$$

resulting into the plan $\{n,\ 0,\ G,\ S,\ \min(X_{r:n}, T)\}$. The area $P$ in the $(t, a)$–plane, which when entered by the trajectory $a(t)$ determines the end of the test, is a three–quarter plane similar to that on the right–hand side of Fig. 8/2 where $n$ has to be changed against $r$. The duration as well as the number of failures are random but limited above by $T$ and $r$. The average duration $\mathrm{E}(D)$ is computed by (8.4e) with $X_{n:n}$ substituted by $X_{r:n}$ and $f_{n:n}(x)$ by $f_{r:n}(x)$. We mention that

$$\{n, \, 0, \, G, \, S, \, \min(X_{n:n}, T)\} = \{n, \, 0, \, G, \, S, \, T\},$$

but

$$\{n, \, 1, \, G, \, S, \, \min(X_{n:n}^*, T)\} \neq \{n, \, 1, \, G, \, S, \, T\}.$$

### 8.3.2.4 Indirect censoring

The inference of certain parametric distributions $F(x \,|\, \boldsymbol{\theta})$ depends on the sum of all observed lifetimes whether censored or not. This statistic is called **total time on test**, $TTT$ for short. With respect to an exponential distribution $TTT$ is a sufficient statistic, but for a WEIBULL distribution the only sufficient statistic is the entire dataset. When $TTT$ is to be used, the quality of the inferential process depends on the size of $TTT$, which — on its side — is determined by the test duration $D$, so that we write $TTT(D)$.

Besides the function $A(t)$ giving the number of failed items up to and including $t$, we introduce another random function $B(t)$ showing the number of items that have not failed or have not been removed otherwise up to and including $t$. A realization or trajectory of $B(t)$ is denoted by $b(t)$. $B(t)$, giving the number of units on test at time $t$, may also be named **stock function**. $B(t)$ is

- a constant function when failing items are replaced instantaneously

$$b(t) = n, \tag{8.12a}$$

- a descending step function which is complementary to $n$ when we run the $(R=0)$–plan

$$B(t) = n - A(t). \tag{8.12b}$$

$TTT$ is the integral of $B(t)$ (see Fig. 8/5).

So, when a special value $\omega$ for $TTT(D)$ has been chosen, which ensures the desired quality of the inferential statement, we have one more possibility to fix $D$ via $TTT(D) = \omega$, i.e.,

$$D = TTT^{-1}(\omega). \tag{8.13}$$

The test is suspended at that very moment when $TTT$ has reached $\omega$. We thus censor the long lifetimes not directly by $T$, $X_{r:n}$, $X_{r:n}^*$, $\min(T, X_{r:n})$ or $\min(T, X_{r:n}^*)$ but indirectly by the fulfilment of some other criterion.

Looking at $(R=1)$–plans with always $n$ items on test $TTT(D) = \omega$ is not an additional censoring criterion because

- the plan $\{n, \, 1, \, G, \, S, \, T\}$ with $D = T$ has

$$TTT(T) = n \cdot T, \tag{8.14a}$$

- the plan $\{n,\ 1,\ G,\ S,\ X^*_{r:n}\}$ with $D = X^*_{r:n}$ has

$$TTT(X^*_{r:n}) = n \cdot X^*_{r:n} \qquad (8.14b)$$

- and the plan $\{n,\ 1,\ G,\ S,\ \min(T, X^*_{r:n})\}$ has

$$TTT[\min(T, X^*_{r:n})] = n \cdot \min(T, X^*_{r:n}). \qquad (8.14c)$$

$(R=1)$–plans have a $TTT$ statistic which is unequivocally determined by either the deterministic duration $T$ or the time of the $r$–th failure $X^*_{r:n}$. $TTT$ always is a rectangle with height $n$ and width $T$ or $X^*_{r:n}$.

Figure 8/5:  Stock function of an $(R=0)$-plan and two approaches to calculate $TTT(D)$

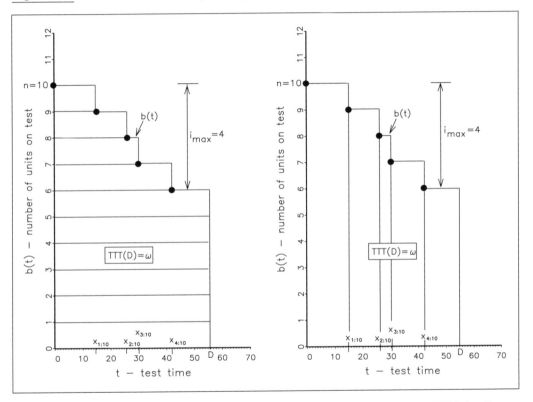

$(R=0)$–plans have a random course of $B(t)$ and thus a random area $TTT(t)$ leading to a non–trivial supplement of direct singly censoring. Looking at a given trajectory $b(t)$ of $B(t)$ (see Fig. 8/5), we have two possibilities to represent and calculate $TTT$. Let $i_{\max}$ be the number of the last failing item within $(0, D]$, i.e.,

$$i_{\max} = \max\{i : X_{i:n} \le D\}, \qquad (8.15a)$$

then:

1. $TTT(D)$ is the sum of $n$ **horizontal bars** representing **individual lifetimes** from which $i_{\max}$ are complete and $b(D) = n - i_{\max}$ are incomplete and of equal length $D$,

$$TTT(D) = \sum_{i=1}^{i_{\max}} X_{i:n} + (n - i_{\max})\, D \qquad (8.15b)$$

(see the left–hand side of Fig. 8/5).

2. $TTT(D)$ is the sum of $i_{\max} + 1$ **vertical bars**. The first $i_{\max}$ bars present integrals of **inter–failure times**.The last vertical bar is the integral of $n - i_{\max}$ durations expressing the accumulated time from the last failure $X_{i_{\max}:n}$ to the end of the test, $D$,

$$TTT(D) = \sum_{i=1}^{i_{\max}} (n - i + 1)\left(X_{i:n} - X_{i-1:n}\right) + (n - i_{\max})\left(D - X_{i_{\max}:n}\right) \quad (8.15c)$$

where $X_{0:n} = 0$.

(8.15b,c) together with Fig. 8/5 show

- that many failures lead to a heavy decline of $B(t)$ and thus prolong the test duration,

- that early failures also have a tendency to prolong test duration.

The area $P$ in the $(t, a)$–plane is not of simple form. The limiting line of $P$ depends on the random course of the stochastic process $A(t) = n - B(t)$.

Based on $TTT(t)$ we may introduce further types of life test plans.

1. A first type is denoted by

$$\{n,\ 0,\ G,\ S,\ TTT^{-1}(\omega)\},$$

where the test stops when $TTT = \omega$ has been reached. When we have fairly unreliable items, it might happen — in analogy to the plan $\{n,\ 0,\ G,\ S,\ T\}$ — that after the last item has failed, $TTT = \omega$ has not been reached, i.e., $TTT(X_{n:n}) < \omega$. The effective test duration of this first type is thus

$$D = \min\left(TTT^{-1}(\omega); X_{n:n}\right).$$

2. A second type is denoted by

$$\left\{n,\ 0,\ G,\ S,\ \min\left(TTT^{-1}(\omega); X_{r:n}\right)\right\}.$$

The test is terminated by the earlier of the following two events:
- the $r$–th failure, but $TTT(X_{r:n}) \leq \omega$,
- $TTT(D) = \omega$, but $A(D) < r$.

This second type — like the combined type–I and type–II censoring in Sect. 8.3.2.3 — is a safeguard against using up too many items.

3. A third type is

$$\left\{n,\ 0,\ G,\ S,\ \min\left(TTT^{-1}(\omega); T\right)\right\}.$$

The test stops when either
- a fixed time $T$ has elapsed, but $TTT(T) \leq \omega$, or
- $TTT(D) = \omega$, but $D < T$.

This type is a safeguard against a very long test duration.

4. A fourth type is

$$\left\{ n,\ 0,\ G,\ S,\ \min \left( TTT^{-1}(\omega); T; X_{r:n} \right) \right\},$$

safeguarding against both an unduly long test duration and using up too many items.

### 8.3.3 Multiply censored tests

In a typical case, the life test is terminated with a single stage of censoring, no further observations being recorded for any of the survivors. In many practical situations, the initial censoring results in withdrawal of only a portion of the survivors. Those which remain on test continue under observation until ultimate failure or until a subsequent stage of censoring is performed. For sufficiently large samples, censoring may be progressive through several stages, thus leading to a **progressively (multiply) censored test** or **hypercensored test**. Multiply censored tests will never have $R = 1$ because the reasons for hypercensoring contradict a replacement of failing items. Contrary to single censoring the censored lifetimes are not of equal length.

There are several reasons for multiple censoring:

1. Certain specimens must be withdrawn from a life test prior to failure for use as test objects in related experimentation or to be inspected more thoroughly.

2. In other instances progressively censored samples result from a compromise between the need for more rapid testing and the desire to include at least some extreme life spans in the sample data.

3. When test facilities are limited and when prolonged tests are expensive, the early censoring of a substantial number of sample specimens frees facilities for other tests while specimens that are allowed to continue on test until subsequent failure permit observation of extreme sample values.

If we look at a $k$–stage plan ($k \geq 2$), the stock function $B(t)$ not only does descend in a natural way by failing items but it is lowered deliberately by withdrawing a certain number $e_i$ ($e_i \geq 1$) of non–failed items at time $t_i$, $i = 1, 2, \ldots, k$ and $t_{i-1} < t_i$, $t_0 = 0$. $t_k$ represents the stopping time of testing. Generally, the numbers $e_i$ are fixed in advance — the simplest case being $e_1 = e_2 = \ldots = e_k = $ constant — and thus are non–random,[9] whereas the process of withdrawing is random in any case. There are situations (see the remarks on type–I progressive censoring further down), where the test stops before $t_k$. Then $e_k$ and even some of the $e_i$ preceding $e_k$ cannot be realized and the number of items withdrawn at the time of premature shopping is less than the scheduled number. We have the following ex–post representation of the sample size

$$n = a(t_k) + \sum_{i=1}^{k} e_i, \tag{8.16a}$$

---

[9] TSE/YUEN (1998) analyze a situation with random $e_i$ and a WEIBULL distributed lifetime.

$a(t_k)$ being the number of failures in $(0, t_k]$ and resulting in complete lifetimes. The functions $A(t)$ and $B(t)$ now are no longer complementary to $n$, instead we have

$$B(t) = \begin{cases} n - A(t) & \text{for} \quad 0 < t < t_1, \\ n - \left[ A(t) + \sum_{i=1}^{j} e_i \right] & \text{for} \quad t \geq t_j, \ j = 1, \ldots, k. \end{cases} \quad (8.16b)$$

There are different approaches to set the times $t_i$. In progressive censoring as in single censoring, a distinction is to be made between type–I censoring and type–II censoring. We will first discuss **type–I multiple censoring** where the withdrawal times are fixed in advance and denoted by $t_i = T_i$; the simplest case is

$$T_i = i \cdot \Delta t; \ \ i = 1, 2, \ldots, k,$$

i.e., the withdrawal times are equidistant.

We will denote a type–I progressively censored test by

$$\left\{ n; \ 0; \ G; \ S; \ (T_1, e_1), \ (T_1, e_2), \ \ldots, \ (T_k, e_k) \right\}.$$

But we cannot be sure that the test evolves in the predetermined way suggested by the notation and that the test is terminated at $t = T_k$ with a withdrawal of $e_k$ items. We may distinguish two cases of **premature termination**:

1. The test ends at $T_j$, $1 \leq j < k$, with either the withdrawal of the scheduled number $e_j$ or of fewer than $e_j$ items. The condition is

$$1 \leq n - \sum_{i=1}^{j-1} e_i - a(T_j - \varepsilon) \leq e_j, \ \varepsilon > 0, \ \text{and small.} \quad (8.17a)$$

   $a(T_j - \varepsilon)$ is the number of failures accumulated immediately before $T_j$.

2. The test ends at some time $t^*$ between two scheduled times, i.e., $T_{j-1} < t^* < T_j$, $1 \leq j < k$ and $T_0 = 0$, because the stock function $B(t)$ has come down to zero. The condition is

$$a(t^*) = n - \sum_{i=1}^{j-1} e_i. \quad (8.17b)$$

Summarizing the description of type–I progressive censoring, we can state that the number of items on test at the non–random points of withdrawal $T_i$ is a variate because the number of failing items in the fixed interval $(T_{i-1}, T_i]$ is stochastic. Denoting the number of failed items by $a(T_k)$ and the corresponding times to failure by $x_i$ the likelihood function for type–I multiple censoring is

$$L(\boldsymbol{\theta} \,|\, \text{data}) = \begin{cases} K_1 \prod_{i=1}^{k} \left[ 1 - F(T_i \,|\, \boldsymbol{\theta}) \right]^{e_i} & \text{for} \quad a(T_k) = 0, \\ K_2 \prod_{j=1}^{a(T_k)} f(x_j \,|\, \boldsymbol{\theta}) \prod_{i=1}^{k} \left[ 1 - F(T_i \,|\, \boldsymbol{\theta}) \right]^{e_i} & \text{for} \quad a(T_k) > 0, \end{cases} \quad (8.17c)$$

where $K_1$ and $K_2$ are constants not depending on $\boldsymbol{\theta}$. When there is a premature termination, the index $i$ (in the case of $a(T_k) > 0$) will also not reach $k$ with consequences for the last exponent $e_i$.

Looking to the literature it seems that type–I multiple censoring has not been as attractive as **type–II multiple censoring** where the points of withdrawal are given by failure times. Let $r_1 < r_2 < \ldots < r_k < n$ be $k$ failure numbers, $r_i \geq 1$ for all $i$, and $X_{r_i:n}$ the corresponding failure times. The test is thus scheduled to observe $r_k$ failures. Then the most general type–II hypercensored test is written as

$$\left\{ n;\ 0;\ G;\ S;\ (X_{r_1:n}, e_1),\ (X_{r_2:n}, e_2),\ \ldots,\ (X_{r_k:n}, e_k) \right\}.$$

The following relation with respect to failures and removals holds

$$n = r_k + \sum_{i=1}^{k} e_i. \tag{8.18a}$$

Here — contrary to type–I hypercensoring — there is no premature termination; the test always stops at $X_{r_k:n}$, which is random, and the last group of withdrawals always has the planned size

$$e_k = n - r_k - \sum_{i=1}^{k-1} e_i. \tag{8.18b}$$

The stock function always has the non–random value

$$b(X_{r_i:n}) = n - r_i - \sum_{j=1}^{i} e_j, \tag{8.18c}$$

at $t = X_{r_i:n}$, which is random. The likelihood function is

$$L(\boldsymbol{\theta}\,|\,\text{data}) = K \prod_{\ell=1}^{r_k} f(x_{\ell:n}\,|\,\boldsymbol{\theta}) \prod_{i=1}^{k} \left[ 1 - F(x_{r_i:n}\,|\,\boldsymbol{\theta}) \right]^{e_i}, \tag{8.18d}$$

with $K$ as a constant independent of $\boldsymbol{\theta}$.

A special, but more popular version of the plan just described has the **first $k$ natural numbers as its failure numbers**:

$$r_1 = 1,\ r_2 = 2, \ldots,\ r_k = k;$$

i.e., we have as many uncensored lifetimes (= failures) as there will be stages. We denote this plan as

$$\left\{ n;\ 0;\ G;\ S;\ (X_{1:n}, e_1),\ (X_{2:n}, e_2),\ \ldots,\ (X_{k:n}, e_k) \right\}.$$

Its likelihood function simply reads

$$L(\boldsymbol{\theta} \mid \text{data}) = \prod_{i=1}^{k} \widetilde{n}_i \, f(x_{i:n} \mid \boldsymbol{\theta}) \left[ 1 - F(x_{i:n} \mid \boldsymbol{\theta}) \right]^{e_i}, \tag{8.19a}$$

where

$$\widetilde{n}_i = n - i + 1 - \sum_{j=1}^{i-1} e_j.$$

The average test duration is not simply given by $\mathrm{E}(X_{k:n})$, using (8.6b) and substituting $k$ against $r$, because we successively reduce the initial sample size by $e_i$. We thus have to condition $X_{k:n}$ on the vector $\boldsymbol{e} = (e_1, \dots, e_k)'$. TSE /YUEN (1998) give the following rather complicated formula for the mean duration:

$$\mathrm{E}(D) = \mathrm{E}(X_{k:n} \mid \boldsymbol{e}) = K(\boldsymbol{e}) \sum_{\ell=0}^{e_1} \cdots \sum_{\ell_k=0}^{e_k} (-1)^A \frac{\binom{e_1}{\ell_1} \cdots \binom{e_k}{\ell_k}}{\prod\limits_{i=1}^{k-1} h(\ell_i)} \int_0^{\infty} x \, f(x \mid \boldsymbol{\theta}) \, F^{h(\ell_i)-1}(x \mid \boldsymbol{\theta}) \, \mathrm{d}x,$$

$$\tag{8.19b}$$

where

$$A = \sum_{i=1}^{k} \ell_i, \quad h(\ell_i) = \ell_1 + \dots + \ell_i + i,$$

$$K(\boldsymbol{e}) = n \, (n - 1 - e_1) \, (n - 2 - e_1 - e_2) \dots \left( n - k + 1 - \sum_{j=1}^{k-1} e_j \right).$$

In the WEIBULL case — $X \sim We(0, b, c)$ — (8.19b) turns into

$$\mathrm{E}(D) = K(\boldsymbol{e}) \, b \, \Gamma\left(1 + \frac{1}{c}\right) \sum_{\ell_1=0}^{e_1} \cdots \sum_{\ell_k=0}^{e_k} (-1)^A \frac{\binom{e_1}{\ell_1} \cdots \binom{e_k}{\ell_k}}{\prod\limits_{i=1}^{k-1} h(\ell_i)} \times$$

$$\sum_{j=0}^{h(\ell_i)-1} (-1)^j \binom{h(\ell_i) - 1}{j} \left( \frac{1}{j+1} \right)^{1+(1/c)}. \tag{8.19c}$$

One may combine type–II hypercensoring with type–I single censoring as a safeguard against a too long test duration:

$$\left\{ n; \, 0; \, G; \, S; \, (X_{1:n}, e_1), \, (X_{2:n}, e_2), \, \dots, \, (X_{k-1:n}, e_{k-1}), \, \min\left[ T, (X_{k:n}, e_k) \right] \right\}.$$

Now, the test stops at the $k$–th failure or at a fixed time $T$, whichever occurs first. The corresponding likelihood function is

$$
L(\boldsymbol{\theta}|\text{data}) = \begin{cases} \left[1 - F(T|\boldsymbol{\theta})\right]^n & \text{for } T < x_{1:n}, \\ K_1 \prod_{i=1}^{j} f(x_{i:n}|\boldsymbol{\theta}) \left[1 - F(x_{i:n}|\boldsymbol{\theta})\right]^{e_i} \left[1 - F(T|\boldsymbol{\theta})\right]^{e^*} & \text{for } \begin{cases} x_{j:n} \leq T < x_{j+1:n}, \\ j = 1, \ldots, k-1, \end{cases} \\ K_2 \prod_{i=1}^{k} f(x_{i:n}|\boldsymbol{\theta}) \left[1 - F(x_{i:n}|\boldsymbol{\theta})\right]^{e_i} & \text{for } T \leq x_{k:n}, \end{cases}
$$

$$(8.20)$$

with

$$
e^* = n - k - \sum_{i=1}^{j} e_j
$$

and $K_1, K_2$ as constants not involving $\boldsymbol{\theta}$.

JOHNSON (1964) introduced a special version of the type–II hypercensored test with $r_i = i; i = 1, \ldots, k$, which later on was called a **sudden death test**; see McCOOL (1970b, 1974a).[10] The design is as follows: The tester groups the $n$ test specimens randomly in $k$ sets of equal size $m$; i.e., we have $n = k\,m$. All $k$ sets are put on test at the same time. Upon the first failure within a set, the surviving $m - 1$ specimens of this set are removed and the test continues with $k - 1$ sets until the next failure in a set and so on until the first failure in the last remaining set. In practical applications, samples of this kind might arise from testing "throw away" units, each consisting of $m$ identical components, when the failure of any single component means failure of the unit; i.e., the unit is a series system of components with identical life distribution. Such a plan is usually feasible when test facilities are scarce but test material is relatively cheap. The likelihood function belonging to a sudden death test is given by (8.19a) after substitution of each $e_i$ by $m - 1$. To develop the formula for the average test duration of a sudden death test, we first have to notice that the time to the first failure within set $i$ is $X_{1:m}^{(i)}; i = 1, \ldots, k$. So the test stops after a duration given by the largest, the $k$–th order statistic of these $X_{1:m}^{(i)}$. Assuming $X \sim We(0, b, c)$, each $X_{1:m}^{(i)}$ is a sample minimum which — on its side — is again WEIBULL distributed:

$$
Z_i := X_{1:m}^{(i)} \sim We(0, b\,m^{-1/c}, c), \quad \text{see (5.32a,b)}. \tag{8.21a}
$$

Upon applying (5.34) to (8.21a) we get the average test duration $\mathrm{E}(D_1)$ of this **simultaneous sudden death test**

$$
\mathrm{E}(D_1) = \mathrm{E}(Z_{k:k}) = k\,b\,m^{-1/c}\,\Gamma\left(1 + \frac{1}{c}\right) \sum_{i=0}^{k-1}(-1)^i \binom{k-1}{i}\left(\frac{1}{i+1}\right)^{1+(1/c)}.
$$

$$(8.21b)$$

---

[10] These papers also contain point estimators and interval estimators for the percentiles of a WEIBULL distribution.

When the test facilities are extremely scarce, we have to run a **successive sudden death test**, first introduced by BALASOORIYA (1995) for exponentially distributed lifetime. This type of test has the following design: We have $k$ sets of equal size $m$, but we have only $m$ test facilities. A first set is put on test and upon the first failure the set is removed and the next set is put on test and so on. The $k$ sets are thus tested one after the other giving rise to $k$ complete lifetimes within an average test time[11]

$$E(D_2) = k\,E(X_{1:m}) = k\,b\,m^{-1/c}\,\Gamma\left(1 + \frac{1}{c}\right). \tag{8.22}$$

We will compare $E(D_1)$ and $E(D_2)$ to the average duration $E(D_3)$ of a type–II singly censored test when the test facilities are not limited at all, so that we may put all $n = k\,m$ units on test simultaneously and wait until the occurrence of the $k$–th failure. We get

$$E(D_3) = E(X_{k:k\,m})$$

$$= k\,m\,b\binom{k\,m-1}{k-1}\Gamma\left(1+\frac{1}{c}\right)\sum_{i=0}^{k-1}(-1)^i\binom{k-1}{i}$$

$$\times \left(\frac{1}{k\,m-k+i+1}\right)^{1+(1/c)}. \tag{8.23}$$

Fig. 8/6 shows $E(D_i)$; $i = 1, 2, 3$ for the reduced WEIBULL variate $U \sim We(0,1,c)$ when $k = 4$, $m = 5$ and $n = k \cdot m = 20$.[12] We generally have $E(D_3) < E(D_1) < E(D_2)$.

### 8.3.4 Further types of censoring

In (8.2) we have introduced the life test parameter $G$ encoding whether the data are grouped or not. For $G = 1$ (= grouping), the time axis is divided into $k + 1$ intervals, the first (last) being opened to the left (right):

$$(-\infty, t_1],\ (t_1, t_2],\ \ldots,\ (t_{k-1}, t_k],\ (t_k, \infty).$$

In the simplest case the inner intervals are of equal width:

$$t_{i+1} - t_i = \Delta;\ \ i = 1, 2, \ldots, k-1.$$

If we put $n$ units on test and we only register the number $d_i$ of units failing in interval $i$, we have **interval censoring**. $d_1$ units are censored on the left and $d_{k+1}$ are censored on the right. Let $(F(x \mid \boldsymbol{\theta})$ be the CDF of the lifetime; then the likelihood function has the general form

$$L(\boldsymbol{\theta} \mid \text{data}) = F(t_1 \mid \boldsymbol{\theta})]^{d_1}\left[1 - F(t_k \mid \boldsymbol{\theta})\right]^{d_{k+1}}\prod_{i=2}^{k}\left[F(t_i \mid \boldsymbol{\theta}) - F(t_{i-1} \mid \boldsymbol{\theta})\right]^{d_i}. \tag{8.24}$$

---

[11] When we need $\ell > k$ complete lifetimes the BALASOORIYA plan is inappropriate. We have to accumulate more than one failure in some test sets. WU et al. (2001) have analyzed this situation by a simulation approach to evaluate the rather complicated formula for the average test time. They also give parameter estimates and their confidence intervals.

[12] For numerical evaluation of (8.21b) and (8.30) one should use the integral representation according to (8.11c, d).

Figure 8/6: Average test duration for several variants of type-II hypercensoring for $k = 4$ and $m = 5$ and $X \sim We(0, 1, c)$

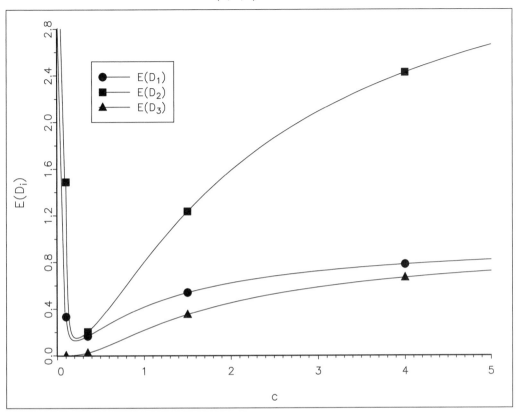

**Censoring times** are often effectively **random**, especially with field data. For example, in a medical trial patients may enter the study in a more or less random fashion, according to their time of diagnosis. Some of the patients are lost to follow–up because they move or die of an illness other than that under investigation or the study is terminated at some prearranged date. Censoring times, that is, the lengths of a patients's time under study, are random.

A very simple random censoring process that is often realistic is one in which each individual is assumed to have a lifetime $X$ and a **censoring time** $C$, with $X$ and $C$ independent continuous variates, having CDFs $F(x)$ and $G(c)$, respectively. $C$ may be the time of a competing risk. Let $(X_i, C_i)$; $i = 1, 2, \ldots, n$ be independent and, as in the case of type–I censoring (8.5a), define

$$Y_i = \min(X_i, C_i) \text{ and } \delta_i = \left\{ \begin{array}{ll} 1 & \text{if } X_i \leq C_i, \\ 0 & \text{if } X_i > C_i. \end{array} \right\} \tag{8.25a}$$

The data from observations on $n$ individuals consist of the pairs $(y_i, \delta_i)$. The joint density–probability of $(y_i, \delta_i)$ is easily obtained using $f(x)$ and $g(c)$, the DFs of $X$ and $C$. We have

$$
\begin{aligned}
\Pr(Y_i = y, \delta_i = 0) &= \Pr(X_i > y, C_i = y) \\
&= [1 - F(y)]\, g(y),
\end{aligned}
\tag{8.25b}
$$

$$
\begin{aligned}
\Pr(Y_i = y, \delta_i = 1) &= \Pr(C_i > y, X_i = y) \\
&= [1 - G(y)]\, f(y).
\end{aligned}
\tag{8.25c}
$$

These can be combined into the single expression

$$
\Pr(Y_i = y, \delta_i) = \left\{ f(y)\,[1 - G(y)] \right\}^{\delta_i} \left\{ g(y)\,[1 - F(y)] \right\}^{1 - \delta_i},
\tag{8.25d}
$$

and thus the joint density of the $n$ pairs $(y_i, \delta_i)$ is

$$
\left.
\begin{aligned}
\prod_{i=1}^{n} &\left\{ f(y_i)\,[1 - G(y_i)] \right\}^{\delta_i} \left\{ g(y_i)\,[1 - F(y_i)] \right\}^{1 - \delta_i} = \\
&= \left( \prod_{i=1}^{n} [1 - G(y_i)]^{\delta_i}\, g(y_i)^{1 - \delta_i} \right) \left( \prod_{i=1}^{n} f(y_i)^{\delta_i}\, [1 - F(y_i)]^{1 - \delta_i} \right).
\end{aligned}
\right\}
\tag{8.25e}
$$

If $G(y)$ and $g(y)$ do not involve any parameters in $\boldsymbol{\theta}$ of $F(y \,|\, \boldsymbol{\theta})$, then the first term on the right–hand side of (8.25e) can be neglected and the likelihood function taken to be

$$
L(\boldsymbol{\theta} \,|\, \text{data}) = \prod_{i=1}^{n} f(y_i \,|\, \boldsymbol{\theta})^{\delta_i}\, [1 - F(y_i \,|\, \boldsymbol{\theta})]^{1 - \delta_i},
\tag{8.25f}
$$

which is of the form (8.5e).

# 9 Parameter estimation — Graphical approaches

The great majority of papers on the WEIBULL distribution deal with the estimation of its parameters $a$ (location), $b$ (scale) and $c$ (shape). An initial approach, which is appropriate for selecting good estimators for the parameters of a given distribution, is to determine the minimal sufficient–statistic vector.[1] This is the vector of smallest dimension that includes functions of the observed data yielding all information useful in estimating the parameters and functions thereof. The minimal vector of sufficient statistics for complete samples of size $n$ from a two–parameter WEIBULL distribution and a Log–WEIBULL distribution consists of $(X_{1:n}, X_{2:n}, \ldots, X_{n:n})$ and $(X^*_{1:n}, X^*_{2:n}, \ldots, X^*_{n:n})$, where $X^*_{i:n} = \ln(X_{i:n})$. Also, when only the first $r$ of $n$ sample values are observable, then $(X_{1:n}, X_{2:n}, \ldots, X_{r:n})$ and $(X^*_{1:n}, X^*_{2:n}, \ldots, X^*_{r:n})$ are minimally sufficient. Thus, for $r > 2$, no joint complete sufficient statistics exist and hence no unique minimum–variance unbiased estimators of the parameters and functions. So, in the last fifty years a multitude of estimation procedures have come into existence which exploit the information of the $n$ sampled observations in different ways. In this and the following seven chapters we will present these techniques.

## 9.1 General remarks on parameter estimation

There are two approaches to estimation and several methods for each of them. The two approaches are **point estimation** and **interval estimation**. In point estimation, a numerical value for the the $k$–dimensional vector $\boldsymbol{\theta}$ of parameters is calculated. In interval estimation, a $k$–dimensional region is determined in such a way that this region covers the true vector $\boldsymbol{\theta}$ with a specified and predetermined probability $1 - \alpha$, $1 - \alpha \geq 0.90$ in general. $1 - \alpha$ is the **level of confidence**. If $k = 1$, this region is an interval, hence the name.

Let $X_i$; $i = 1, \ldots, n$ be the sample variates and $x_i$ the realization of $X_i$ in a given sample. The expression $\widehat{\Theta} = \widehat{\Theta}(X_1, \ldots, X_n)$ is called an **estimator**, and it is a random variable. $\widehat{\theta} = \widehat{\theta}(x_1, \ldots, x_n)$ is called an **estimate**; it is the numerical value obtained using the data available. In the case of censoring or grouping, the estimate depends on the observed data as well as on the censoring or grouping values.

Many methods are available for estimating the parameters, especially those of the WEIBULL distribution, each having its advantages and disadvantages, and there is no method which is best with respect to all characteristics that may be set up to measure its behavior. From the **viewpoint of statistical theory** we have the following list of desirable properties of estimators:[2]

---

[1] Sufficiency is explained in Sect. 9.1.

[2] The following listing deals with only a scalar–valued parameter. A generalization to a vector–valued parameter is straightforward and will be given in the following sections.

1. **Unbiasedness** — An estimator $\widehat{\Theta}_n$ of $\theta$ is said to be **unbiased** if

$$\mathrm{E}\big(\widehat{\Theta}_n\big) = \theta \;\; \forall\, n. \tag{9.1a}$$

Unbiasedness means that repeating the estimation process a great number of times — with the same sample size $n$ — will lead to an average of the estimates obtained which is (approximately) equal to $\theta$. An estimator with $\mathrm{E}(\widehat{\Theta}_n) \neq \theta$ is said to be **biased**. The **bias** is given by

$$\mathrm{B}\big(\widehat{\Theta}_n\big) := \mathrm{E}\big(\widehat{\Theta}_n\big) - \theta. \tag{9.1b}$$

$\mathrm{B}\big(\widehat{\Theta}_n\big) < (>)\, 0$ means systematic underestimation (overestimation) of $\theta$. If, in the case of $\mathrm{B}\big(\widehat{\Theta}_n\big) \neq 0$,

$$\lim_{n\to\infty} \mathrm{E}\big(\widehat{\Theta}_n\big) = \theta, \tag{9.1c}$$

$\widehat{\Theta}_n$ is called **asymptotically unbiased**.

2. **Median unbiasedness** — An estimator $\widehat{\Theta}_n$ of $\theta$ is said to be **median unbiased** if

$$\Pr\big(\widehat{\Theta}_n \leq \theta\big) = 0.5 \;\; \forall\, n, \tag{9.2}$$

i.e., $\theta$ is the median of the distribution of $\widehat{\Theta}_n$. Median unbiasedness means that the chances of overshooting or undershooting $\theta$ are equal.

3. **Normality** — An estimator $\widehat{\Theta}_n$ of $\theta$ is said to be

   - **normal**[3] if

$$\frac{\widehat{\Theta}_n - \mathrm{E}\big(\widehat{\Theta}_n\big)}{\sqrt{\mathrm{Var}\big(\widehat{\Theta}_n\big)}} \sim No(0,1) \;\; \forall\, n, \tag{9.3a}$$

   - and **asymptotically normal** if

$$\frac{\widehat{\Theta}_n - \mathrm{E}\big(\widehat{\Theta}_n\big)}{\sqrt{\mathrm{Var}\big(\widehat{\Theta}_n\big)}} \sim No(0,1) \;\; \text{for } n \to \infty. \tag{9.3b}$$

   Normality means that we can easily compute prediction intervals for $\widehat{\Theta}_n$ as well as confidence intervals for $\theta$ using the percentiles $u_P$ of the standard normal distribution.

4. **Linearity** — An estimator $\widehat{\Theta}_n$ of $\theta$ is said to be **linear** if it is of the form

$$\widehat{\Theta}_n = c_0 + \sum_{i}^{n} c_i X_i; \;\; c_0, c_i \in \mathbb{R}. \tag{9.4}$$

This estimator has the most easily computable linear form, but the weighting factors $c_i$ ensuring optimal estimators may be rather difficult to determine (see Sect. 10.2). In general, we have $c_0 = 0$.

---

[3]   $No(0,1)$ stands for the standard normal distribution with density $\varphi(u) = \exp(-u^2/2)/\sqrt{2\,\pi}$.

5. **Consistency** — An estimator $\widehat{\Theta}_n$ of $\theta$ is said to be **consistent** if some law of large numbers holds. We mostly use the **weak law of large numbers**:

$$\lim_{n \to \infty} \Pr(|\widehat{\Theta}_n - \theta| > \epsilon) = 0, \ \epsilon > 0 \text{ and small}, \tag{9.5a}$$

shortly written as,

$$\operatorname{plim} \widehat{\Theta}_n = \theta. \tag{9.5b}$$

6. **Sufficiency** — The estimator $\widehat{\Theta}_n$ of $\theta$ compresses the $n$ variates $X_1, \ldots, X_n$ into one variate. If there is no loss of information the estimator is called **sufficient**. One way to ascertain sufficiency is to use the **factorization criterion** of J. NEYMAN and R.A. FISHER:

$$f(x \mid \theta) = g_1(x) \, g_2\big[\widehat{\Theta}_n(x); \theta\big], \tag{9.6}$$

where $f(x \mid \theta)$ is the joint density of $x = (x_1, \ldots, x_n)$.

7. **Efficiency** — The estimate $\widehat{\theta}_n$ may deviate from the true parameter value $\theta$ for two reasons: bias and random sampling variation. The **mean squared error** (MSE) and the **root mean squared error** (RMSE $:= \sqrt{\text{MSE}}$) incorporate both types of deviation:

$$
\begin{aligned}
\operatorname{MSE}(\widehat{\Theta}_n) \ &:= \ \operatorname{E}\big[(\widehat{\Theta}_n - \theta)^2\big] \\
&= \ \underbrace{\operatorname{E}\big[\{\widehat{\Theta}_n - \operatorname{E}(\widehat{\Theta}_n)\}^2\big]}_{} \ + \ \underbrace{\{\operatorname{E}(\widehat{\Theta}_n) - \theta\}^2}_{} \\
&= \ \operatorname{Var}(\widehat{\Theta}_n) \ \ + \ \ \operatorname{B}(\widehat{\Theta}_n)^2.
\end{aligned} \tag{9.7a}
$$

Basically, efficiency of an estimator $\widehat{\Theta}_n$ is judged by means of $\operatorname{MSE}(\widehat{\Theta}_n)$,[4] either relative to another estimator (**relative efficiency**) or relative to an absolute standard (**absolute efficiency**). The most frequently used standard is the CAMÉR–RAO–**bound**:[5]

$$\operatorname{MSE}(\widehat{\Theta}_n) \geq \frac{1 + \dfrac{\partial \operatorname{B}(\widehat{\Theta}_n)}{\partial \theta}}{-\operatorname{E}\left[\dfrac{\partial^2 \mathcal{L}(\theta \mid x)}{\partial \theta^2}\right]}, \tag{9.7b}$$

$\mathcal{L}(\theta \mid x)$ being the **log–likelihood function**: $\mathcal{L}(\theta \mid x) := \ln[L(\theta \mid x)]$. The denominator of the term on the right–hand side is nothing but the FISHER **information**:

$$\operatorname{I}(\theta) := -\operatorname{E}\left\{\frac{\partial^2 \mathcal{L}(\theta \mid x)}{\partial \theta^2}\right\} = \operatorname{E}\left\{\left(\frac{\partial \mathcal{L}(\theta \mid x)}{\partial \theta}\right)^2\right\}. \tag{9.7c}$$

---

[4] In the case of unbiasedness $\operatorname{MSE}(\widehat{\Theta}_n)$ reduces to $\operatorname{Var}(\widehat{\Theta}_n)$

[5] (9.7b) holds only under the following regularity conditions: The limits of integration, i.e., the limits of variation of $x$, are finite and independent of $\theta$ and also, if these limits are infinite, provided that the integral resulting from the interchange of integration and differentiation is uniformly convergent for all $\theta$ and its integrand is a continuous function of $x$ and $\theta$. These are sufficient sets of conditions, but they are not necessary.

If an estimator $\widehat{\Theta}_n$ is unbiased (9.7b) reduces to

$$\mathrm{Var}\big(\widehat{\Theta}_n\big) \geq \frac{1}{-\mathrm{E}\left[\dfrac{\partial^2 \mathcal{L}(\theta \,|\, \boldsymbol{x})}{\partial \theta^2}\right]}, \tag{9.7d}$$

and an estimator with variance equal to this lower bound is called a **MVB estimator** (= minimum variance bound estimator). Estimators that are efficient in this sense are also termed **UMVUE** (= uniformly minimum variance unbiased estimator) and **BLUE** (= best linear unbiased estimator) if $\widehat{\Theta}_n$ is linear. The condition under which the MVB is attained reads

$$\frac{\partial \mathcal{L}(\theta \,|\, \boldsymbol{x})}{\partial \theta} = A(\theta)\,\big(\widehat{\theta}_n - \theta\big), \tag{9.7e}$$

where $A(\theta)$ is independent of the observations. It is easily shown that the right–hand side of (9.7d) reduces to $1/A(\theta)$ and

$$\min \mathrm{Var}(\widehat{\theta}_n) = 1/A(\theta). \tag{9.7f}$$

Besides the formal criteria listed above there are other, more **informal criteria** for the choice of a method of estimation:

1. **Applicability to censored samples** — This characteristic is very important for the WEIBULL distribution.

2. **Applicability to interval estimation** — This characteristic is relevant when judging the precision of estimation.

3. **Clarity of the approach** — This characteristic makes sense when the estimation results are to be communicated to statistical laymen.

4. **Simplicity of calculation** — This characteristic has lost its importance with the advent of electronic computers and has been replaced by the **availability of software** to implement the method on a computer.

There are a lot of ways to classify inferential procedures, especially estimation procedures. One criterion is according to the topic to be estimated and another is the basic philosophical or conceptual aspect of the approach.[6] The contents of Chapters 9 through 17 discuss these two aspects. Especially when estimating the distributional parameters of the WEIBULL distribution, we have stressed the philosophy behind the approach. We start with the classical approach (Chapters 9 through 13) of FISHER, NEYMAN, PEARSON and others. A classical approach rests only on sample data, takes the parameters as unknown constants, and works with the frequency concept of probability. The BAYES approach (Chapter 14) rests on sample data and prior information, takes the parameters as unknown, but variates with a given prior distribution and works with the frequency interpretation as well as with the subjective interpretation of probability. Finally, Chapter 15 reports on fiducial and structural inferences that are of minor importance in statistics.

---

[6] For an introduction into theories of statistical inference see BARNETT (1973).

## 9.2 Motivation for and types of graphs in statistics

Graphical approaches mainly serve four purposes:

- **data exploration** in the sense of finding a suitable model (see the works of TUKEY, MOSTELLER, HOAGLIN, and VELLEMAN written in the 1970s);
- **data analysis** in the sense of estimation and validation of a model (this will be the main topic of Section 9.3);
- **data presentation** as an appealing alternative to statistical tables;
- **tools in statistical work**, e.g., in the form of nomograms allowing quick reading of the values of complicated mathematical–statistical functions.

There are graphs that do not serve only one purpose.

In the following text, we will present graphs which play a dominant role in confirmatory data analysis, e.g., in statistical inference, of lifetime distributions, namely the following five types:

- the **QQ–plot** or quantile plot,
- the **PP–plot** or percent plot,
- plots on probability paper (= **probability plotting**),
- **hazard plotting** and
- the **TTT–plot**.

The first four techniques have a common basis: they start with the cumulative distribution function of a random variable. This CDF has to be a parametric function; more precisely, it should be a member of the location–scale family. The TTT–plot is mainly a technique in non–parametric statistics giving less information than hazard plotting and probability plotting.

### 9.2.1 PP-plots and QQ-plots

Both types of plots are suitable to compare

- two theoretical CDFs,
- two empirical CDFs or
- an empirical CDF to a theoretical CDF.

They do not help in estimation but they are the basis for probability plotting, and hazard plotting which help in estimating the distribution parameters.

Let $F_X(x)$ and $F_Y(y)$ be the CDFs of variates $X$ and $Y$, respectively, given in Fig. 9/1. From this display we may deduce two types of graphs, the QQ–plot or quantile plot and the PP–plot or percent plot.

For each value $P$ on the ordinate axis displaying the CDF, there are at most two values on the abscissa axis displaying the realizations of the variates, called quantiles:

$$x_P := Q_X(P) \text{ and } y_P := Q_Y(P).$$

A **QQ–plot** is a display where $y_P$ is plotted against $x_P$ with $P$ varying, $0 < P < 1$. Conversely, for each value $Q$ on the abscissa axis there are at most two values on the ordinate axis:

$$P_X(Q) := F_X(Q) \text{ and } P_Y(Q) := F_Y(Q).$$

Figure 9/1:  Explanation of the QQ-plot and the PP-plot

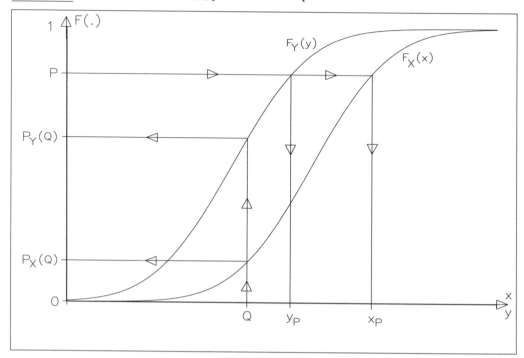

A **PP–plot** is a display where $F_Y(Q)$ is plotted against $F_X(Q)$ for $Q$ varying, $Q \in \mathbb{R}$. We have several modifications of these two basic displays. There is a special case where QQ–plot and PP–plot are identical: $X$ and $Y$ both are uniformly distributed in $[0, 1]$. This will be the case when the two variates are probability integral transforms.

We will first and briefly comment on the PP–plot which is less important. If $X$ and $Y$ are identically distributed, their CDFs will coincide in Fig. 9/1 and the resulting PP–plot will be a 45°–line running from $\big(P_X(Q), P_Y(Q)\big)] = (0, 0)$ to $\big(P_X(Q), P_Y(Q)\big) = (1, 1)$. Contrary to the QQ–plot the PP–plot will not be linear if one of the two variates is a linear transform of the other one; see Fig. 9/2 where $X \sim We(0, 1, 2)$ and $Y \sim We(-1, 2, 2)$.

Despite the missing clear sensitivity of a PP–plot against a linear transformation of the variates, it is of some importance. The PP–plot possesses a high discriminatory power in the region of high density because in that region the CDF, i.e., the value of $P$, is a more rapidly changing function of $Q$ than in the region of low density. Furthermore, the idea of the PP–plot is — contrary to that of a QQ–plot — transferable to a multivariate distribution.

For identically distributed variates $X$ and $Y$, the CDFs in Fig. 9/1 will coincide, we will get $x_P = y_P \,\forall\, P \in (0, 1)$, and the QQ–plot is a 45°–line running in the direction of the origin of the coordinate axes. If $Y$ is a positive linear transform of $X$, i.e., $Y = a + b\, X$, $b > 0$, then the QQ–plot will be a straight line, which easily shown:

$$F_Y(y) \;=\; F_X\!\left(\frac{y - a}{b}\right) \text{ and } b > 0$$

$$\Longrightarrow \; x_P \;=\; \frac{y_P - a}{b} \text{ or } y_P = a + b\, x_P.$$

This property of linear invariance renders the QQ–plot especially useful in statistical analysis because linearity is a form which is easily perceived by the human eye, as well.

Figure 9/2: PP–plot of $P_Y(Q) = 1 - \exp\{-[(y+1)/2]^2\}$ against $P_X(Q) = 1 - \exp\{-x^2\}$

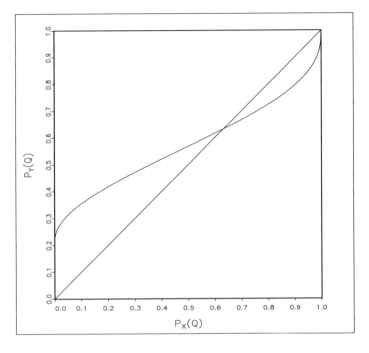

If one of the two distributions to be compared by a QQ–plot possesses very long tails with rather small DF–values, then the QQ–plot will emphasize this distributional difference whereas the difference in the "middle" region of the two distributions, where the density is relatively high, will be blurred somewhat. The reason for this kind of sensitivity of a QQ–plot is that a quantile changes rapidly with $P$ where the density is low and only changes slowly with $P$ where the density is high.

**Example 9/1:   QQ–plots when one of the two distributions is WEIBULL**
The two distributions to be compared in this example are always given in reduced form, i.e., their location parameters are set to 0 and their scale parameters are set to 1.

1. WEIBULL distribution and exponential distribution

$$
\begin{aligned}
x_P &= [-\ln(1-P)]^{1/c} &&- \quad \text{WEIBULL quantile} \\
y_P &= -\ln(1-P) &&- \quad \text{exponential quantile} \\
&= x_P^c
\end{aligned}
$$

The QQ–graph will be

- concave for $0 < c < 1$,
- linear for $c = 1$,
- convex for $c > 1$.

2. WEIBULL distribution and type–I distribution of the minimum

$$
\begin{aligned}
y_P &= \ln[-\ln(1-P)] \; - \; \text{type–I minimum quantile} \\
&= \ln\left(x_P^c\right)
\end{aligned}
$$

The QQ–graph will be concave.

3. WEIBULL distribution and uniform distribution in $[0,1]$

$$
\begin{aligned}
y_P &= P \; - \; \text{quantile of the uniform distribution} \\
&= 1 - \exp\left(-x_P^c\right)
\end{aligned}
$$

The QQ–graph is identical to the CDF of the reduced WEIBULL variate (see the left–hand side of Fig. 2/7).

4. WEIBULL distribution and standard normal distribution

$$
\begin{aligned}
y_P &= \Phi^{-1}(P) \; - \; \text{quantile of the standard normal distribution} \\
&= \Phi^{-1}\left(1 - \exp\left[-x_P^c\right]\right)
\end{aligned}
$$

The plot will be

- concave for $0 < c \lesssim 3.6$,

- linear for $c \approx 3.6$,

- convex for $c \gtrsim 3.6$.

It is also possible to use a QQ–plot for the comparison of two empirical CDFs. When both samples are of equal size $n$, the **empirical QQ–plot** consists of simply plotting $y_{i:n}$ over $x_{i:n}$ for $i = 1, 2, \ldots, n$. For samples of different sizes the procedure is as follows.

1. Let $n_1$ be the size of the smaller sample with observations $x_\nu$, $\nu = 1, 2, \ldots, n_1$ and $n_2$ be the size of the greater sample with observations $y_\kappa$, $\kappa = 1, 2, \ldots, n_2$.

2. The order of the empirical quantiles is chosen in such a way that the ordered $x$-observations are equal to the natural quantiles, i.e.,

$$
x_{p_\nu} = x_{\nu:n_1}; \quad p_\nu = \nu/n_1; \quad \nu = 1, 2, \ldots, n_1. \tag{9.8a}
$$

3. The $y$-quantile to be plotted over $x_{\nu:n_1}$ is an interpolated value:

$$
y_{p_\nu} = y_{\kappa:n_2} + \left(n_2\, p_\nu - \kappa\right)\left(y_{\kappa+1:n_2} - y_{\kappa:n_2}\right) \tag{9.8b}
$$

with

$$
\kappa < n\, p_\nu \leq \kappa + 1. \tag{9.8c}
$$

Fig. 9/3 shows four empirical QQ–plots for samples of equal size $n = 20$:

- upper row, left–hand side — The universes are the same:

$$X_\nu \sim We(0,1,2), \quad Y_\kappa \sim We(0,1,2).$$

The 20 points scatter around a 45°–line running through the origin.

- upper row, right–hand side — The universes are shifted against each other:

$$X_\nu \sim We(0,1,2), \quad Y_\kappa \sim We(3,1,2).$$

The 20 points scatter around a 45°–line shifted upwards.

- lower row, left–hand side — The universes are differently scaled:

$$X_\nu \sim We(0,1,2), \quad Y_\kappa \sim We(0,2,2).$$

Because $X$ has a smaller variance than $Y$ the 20 points scatter around a line which is steeper than a 45°–line

- lower row, right–hand side:

$$X_\nu \sim We(0,1,2), \quad Y_\kappa \text{ are reduced exponential variates.}$$

The 20 points scatter around a convex curve.

Figure 9/3: Four empirical QQ-plots, each involving at least one WEIBULL sample

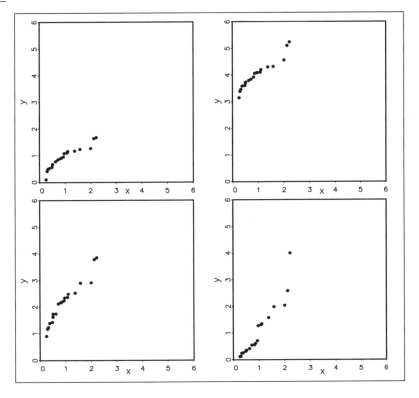

### 9.2.2   Probability plots

The two axes of a QQ–plot have a natural (= linear) scale. The most frequently used **probability papers** or **probability grids** result into a QQ–plot too, but

- on probability paper we compare empirical quantiles of a sample to the theoretical quantiles of a universe,

- the axes of a probability paper are generally non–linear and distorted in such a way that the data points will scatter around a straight line when the sample has been drawn from the given universe.

Probability paper is predominant in the location–scale family of distributions where it is easily implemented and has many advantages (see Sect. 9.2.2.3). We will first give the theoretical background and show how to construct a probability paper with emphasis on the WEIBULL–paper (Sect. 9.2.2.1). A special problem in application is the choice of the plotting position belonging to an ordered observation $x_{i:n}$. Sect. 9.2.2.2 treats this subject and shows the close connection to linear estimation procedures covered in Chapter 10.

#### 9.2.2.1   Theory and construction

When a variate $X$ has a two–parameter distribution, the parameters being $a$ ($a \in \mathbb{R}$) and $b$ ($b > 0$), which depends on only the **reduced variate**

$$U = \frac{X - a}{b}, \tag{9.9a}$$

i.e.,

$$F_X(x \mid a, b) = F_U\left(\frac{x - a}{b}\right), \tag{9.9b}$$

then $a$ is called **location parameter** and $b$ is the **scale parameter**. $a$ is not necessarily the mean and $b$ is not necessarily the standard deviation of $X$, an exception being the normal distribution. The distributions of $X$ obtained when $a$ and $b$ take on all values of their domains constitute a **location–scale family**. $U$ in (9.9a) has a distribution which is free of any other parameters, so we have

$$
\begin{aligned}
F_X(x \mid a, b) &= \Pr(X \leq x \mid a, b) \\
&= \Pr\left(U \leq \frac{x - a}{b}\right) \\
&= F_U(u) = P_u, \;\; u = (x - a)/b.
\end{aligned} \tag{9.9c}
$$

When $X$ is not of the location–scale type, a suitable transformation

$$X^* = g(X) \tag{9.9d}$$

may lead to the desired type of distribution, prominent examples being the lognormal and the WEIBULL distributions.

A probability paper for a location–scale distribution is constructed by taking the **vertical axis** (ordinate) of a rectangular system of coordinates to lay off the quantiles of the reduced variable,[7]

$$u_P = F^{-1}(P), \quad 0 < P < 1, \tag{9.9e}$$

but the labeling of this axis is according to the corresponding probability $P = F^{-1}(u_P)$ or $100\,P\%$. This procedure gives a scaling with respect to $P$ which — in general — is non–linear, an exception being the uniform distribution over $[0, 1]$. Despite this probability labeling, which is chosen for reasons of an easier application, the basis of this axis is a theoretical quantile function. Sometimes a second vertical axis is given with the quantile labeling (see Fig. 9/4) which will help to read the parameters $a$ and $b$. There are cases where the inverse $F^{-1}(P)$ cannot be given in closed, analytical form but has to be determined numerically, the normal distribution being the most prominent example of this case.

The second, **horizontal axis** of the system of coordinates is for the display of $X$, either in linear scaling or non–linear according to $g(X)$ when a transformation has been made. The quantiles of the distribution of $X$ or of $g(X)$ will lie on a straight line

$$x_P = a + b\,u_P. \tag{9.9f}$$

In applications to sample data, the ordered sample values — regarded as empirical quantiles — are laid off on the horizontal axis and the corresponding estimated probabilities, called plotting positions (Sect. 9.2.2.2), on the vertical axis.

Figure 9/4: Extreme value paper (upper part) and WEIBULL paper (lower part)

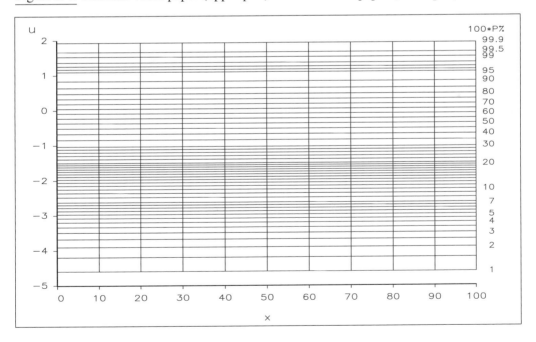

---

[7] The quantiles of some reduced variables bear special names: **probit** for the normal distribution, **logit** for the logistic distribution and **rankit** for the uniform distribution.

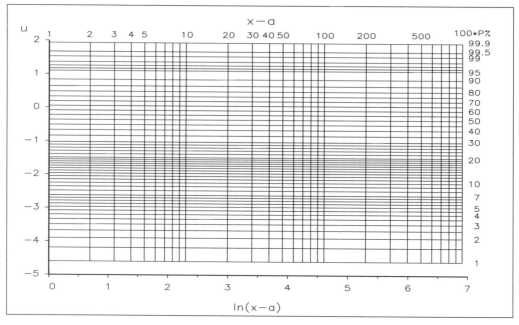

The type–I–minimum distribution, extreme value distribution for short, is one starting point for the construction of WEIBULL–probability–paper; a second approach is given further down. The extreme value distribution is a location–scale distribution. Looking at

$$P = F_X(x_P \mid a, b) = 1 - \exp\left[-\exp\left(\frac{x_P - a}{b}\right)\right]; \quad x \in \mathbb{R} \qquad (9.10a)$$

and — for the reduced variable — at

$$P = F_U(u_P) = 1 - \exp[-\exp(u_P)], \qquad (9.10b)$$

we get upon equating the two quantiles

$$u_P = \frac{x_P - a}{b} = \frac{1}{b} x_P - \frac{a}{b}, \qquad (9.10c)$$

where

$$u_P = \ln[-\ln(1 - P)], \quad 0 < P < 1 \qquad (9.10d)$$

is the reduced quantile. The **probability paper for the extreme value distribution** thus has a double log–scale on the ordinate and a linear scale on the abscissa (see the upper part of Fig. 9/4).

The three–parameter WEIBULL distribution with CDF

$$F_X(x \mid a, b, c) = 1 - \exp\left\{-\left(\frac{x - a}{b}\right)^c\right\}, \qquad (9.11a)$$

which is not a member of the location–scale family, may be transformed by taking

$$X^* = g(X) = \ln(X - a), \qquad (9.11b)$$

provided $a$ is known. We arrive at

$$F_{X^*}(x^* \mid a^*, b^*) = 1 - \exp\left[-\exp\left(\frac{x^* - a^*}{b^*}\right)\right]; \quad x^* \geq a^*, \ a^* \in \mathbb{R}, \ b^* > 0, \quad (9.11c)$$

where

$$a^* = \ln b, \qquad (9.11d)$$

$$b^* = 1/c. \qquad (9.11e)$$

Thus $X^* = \ln(X - a)$ has the extreme value distribution, and upon proceeding as above we get

$$\begin{aligned} u_P &= \frac{x_P^* - a^*}{b^*} = \frac{1}{b^*} \ln(x_P - a) - \frac{a^*}{b^*} \\ &= = c \ln(x_P - a) - c \ln b. \end{aligned} \qquad (9.11f)$$

The **probability paper for the WEIBULL distribution** has the same ordinate scaling as the extreme–value paper, but the abscissa is logarithmic according to (9.11b) (see the lower part of Fig. 9/4). Another starting point to construct the WEIBULL grid is to take the CDF in (9.11a) with the intention to its linearization:

$$\ln[-\ln(1 - P)] = c \ln(x_P - a) - c \ln b.$$

---

**Excursus: Further probability papers**

We list some popular location–scale distributions and the scaling of the axes of their corresponding probability papers.

1. Uniform distribution

$$F_X(x \mid a, b) = \frac{x - a}{b}, \quad a \leq x \leq a + b, \ a \in \mathbb{R}, \ b > 0$$

$$F_U(u) = u, \quad 0 \leq u \leq 1$$

$$u_P = P = \frac{x_P - a}{b}, \quad 0 \leq P \leq 1$$

Both axes have a linear scale.

2. Exponential distribution

$$F_X(x \mid a, b) = 1 - \exp\left\{\frac{x - a}{b}\right\}, \quad x \geq a$$

$$F_U(u) = 1 - \exp\{u\}, \quad u \geq 0$$

$$u_P = -\ln(1 - P) = \frac{x_P - a}{b}, \quad 0 \leq P < 1$$

The abscissa is linear; the ordinate is logarithmic.

3. Normal distribution

$$F_X(x \mid a, b) = \int_{-\infty}^{x} \frac{1}{b\sqrt{2\pi}} \exp\left\{-\frac{(y - a)^2}{2b^2}\right\} dy$$

$$F_U(u) = \int_{-\infty}^{u} \frac{1}{\sqrt{2\pi}} \exp\{-y^2/2\} dy =: \Phi(u)$$

$$u_P = \Phi^{-1}(P) = \frac{x_P - a}{b}, \quad 0 < P < 1$$

The abscissa is linear; the ordinate is scaled according to the numerically inverted CDF of the standard normal distribution.

4. Lognormal distribution

$$F_X(x \mid a, b, c) = \int_a^x \frac{1}{b\sqrt{2\pi}} \frac{1}{y-a} \exp\left\{-\frac{[\ln(y-a)-b]^2}{2c^2}\right\} dy, \quad x > a$$

$$X^* = \ln(X-a)$$

$$F_{X^*}(x^* \mid a^*, b^*) = \int_{-\infty}^{x^*} \frac{1}{b^*\sqrt{2\pi}} \exp\left\{-\frac{(y^*-a^*)^2}{2b^{*2}}\right\} dy^*$$

$$a^* = \mathrm{E}\big[\ln(X-a)\big] = b$$

$$b^* = \sqrt{\mathrm{Var}\big[\ln(X-a)\big]} = c$$

$$u_P = \Phi^{-1}(P) = \frac{\ln(x_P-a)-a^*}{b^*}, \quad 0 < P < 1$$

The ordinate is the same as that of the normal paper, but the abscissa is logarithmic.

5. Logistic distribution

$$F_X(x \mid a, b) = \left\{1 + \exp\left[-\frac{x-a}{b}\right]\right\}^{-1}$$

$$F_U(u) = \left\{1 + \exp(-u)\right\}^{-1}$$

$$u_P = \ln\left(\frac{P}{1-P}\right) = \frac{x_P-a}{b}, \quad 0 < P < 1$$

The abscissa is linear; the ordinate is scaled according to $\ln[P/(1-P)]$.

## 9.2.2.2 Plotting positions[8]

Let $x_{i:n}; \; i = 1, \ldots, n$ be the ordered observations of a sample from a location–scale distribution. The corresponding reduced observations would be

$$u_{i:n} = (x_{i:n} - a)/b, \tag{9.12a}$$

provided $a$ and $b$ are known. In the latter case we could even compute the corresponding $P$–value on the probability–labeled ordinate:

$$P_i = F_U\left(\frac{x_{i:n} - a}{b}\right) \tag{9.12b}$$

to be plotted over $x_{i:n}$. All these points would lie on a straight line. As $a$ and $b$ are unknown, we have to ask how to estimate $u_{i:n}$ or equivalently $P_i$. These estimates are called **plotting positions**. We have to bear in mind two aims:

1. achieving linearity when the distribution of $X$ has been chosen correctly,
2. efficient estimation of the parameters $a$ and $b$.

---

[8] Suggested reading for this section: BARNETT (1975), BLOM (1958), KIMBALL (1960).

This section treats only the first aim, whereas the second aim will be discussed in Chapter 10.

As two equivalent quantities can be laid off on the ordinate of a probability paper, the search for a plotting position can start at either $u_P$ or $P = F_U(u_P)$. The various plotting conventions are based wholly on the sample size $n$ and on the nature of $F_U(\cdot)$. The numerical values $x_{i:n}$ of the observations will not play a part. The conventions will indicate appropriate values $\widehat{P}_i$ on the $P$–scale or values $\widehat{u}_{i:n}$ on the axis for the reduced variate corresponding to the $\widehat{P}_i$. As in formulas (9.12a,b) the $\widehat{P}_i$ may then be expressed in terms of the $\widehat{u}_{i:n}$:

$$\widehat{P}_i = F_U(\widehat{u}_{i:n}) \tag{9.13a}$$

or conversely

$$\widehat{u}_{i:n} = F_U^{-1}(\widehat{P}_i). \tag{9.13b}$$

We first present the so–called "direct" method, see KIMBALL (1960, p. 549), since the rationale involved is based directly on the order number $i$ of $x_{i:n}$.

1. A naïve estimator — but simultaneously the maximum likelihood estimator (MLE) — of $F_X(x) = F_U([x - a]/b) = F_U(u)$ is the stair–case function

$$\widehat{P} = \widehat{F}_X(x) = \begin{cases} 0 & \text{for} \quad x < x_{1:n}, \\ \dfrac{i}{n} & \text{for} \quad x_{i:n} \leq x < x_{i+1:n}, \ i = 1, 2, \ldots, n-1, \\ 1 & \text{for} \quad x \geq x_{n:n}, \end{cases}$$

   leading to the plotting position

$$\widehat{P}_i = \frac{i}{n}. \tag{9.14a}$$

   A drawback of this proposal is that for all distributions with unlimited range to the right, $P = 1$ is not found on the probability scale, so that the largest sample value $x_{n:n}$ has to be omitted.

2. For this reason WEIBULL (1939b) has proposed

$$\widehat{P}_i = \frac{i}{n+1}. \tag{9.14b}$$

   Another rationale for this choice will be given below.

3. The **midpoint position** is

$$\widehat{P}_i = \frac{i - 0.5}{n}, \tag{9.14c}$$

   motivated by the fact that at $x_{i:n}$ the stair–case moves upward from $(i-1)/n$ to $i/n$. Thus we believe that $x_{i:n}$ is a quantile of order $P_i$ somewhere between $(i-1)/n$ and $i/n$, and the average of these two is the estimator in (9.14c).

4. BLOM (1958) has proposed

$$\widehat{P}_i = \frac{i - 0.375}{n + 0.25}. \tag{9.14d}$$

This position guarantees optimality of the linear fit on normal probability paper. Sometimes this plotting position is used for other than normal distributions.

There are plotting positions which rest on the theory of order statistics (see Chapter 5). A first approach along this line departs from the random portion $\Pi_i$ of sample values less than $X_{i:n}$ and tries to estimate $P_i$. A second approach — discussed further down — tries to estimate $u_{i:n}$ and departs from the distribution of $U_{i:n}$. The random portion $\Pi_i$ is defined as

$$\Pi_i = \Pr(X \leq X_{i:n}) = F_U\left(\frac{X_{i:n} - a}{b}\right) \tag{9.15a}$$

and has the following CDF:

$$
\begin{aligned}
F_{\Pi_i}(p) &= \Pr\left(\Pi_i \leq p\right) \\
&= \sum_{j=i}^{n} \binom{n}{j} p^j (1-p)^{n-j}.
\end{aligned}
\tag{9.15b}
$$

The binomial formula (9.15b) results from the fact that we have $n$ independent observations $X_i$, each of them having a probability $p$ to fall underneath the quantile $x_p = F_X^{-1}(p)$. Then $X_{i:n}$ will be smaller than $F_X^{-1}(p)$ if $i$ or more sample values will turn out to be smaller than $F_X^{-1}(p)$. (9.15b) is identical to the CDF of the beta distribution with parameters $i$ and $n - i + 1$, the DF being

$$f_{\Pi_i}(p) = \frac{n!}{(i-1)!\,(n-i)!}\, p^{i-1} (1-p)^{n-i}, \quad 0 \leq p \leq 1. \tag{9.15c}$$

Taking the mean, the median or the mode of $\Pi_i$ gives the following three plotting positions:

- $$\widehat{P}_i = \mathrm{E}(\Pi_i) = \frac{i}{n+1} - \quad \textbf{mean plotting position,}$$
  $$\tag{9.15d}$$
  which is equal to (9.14b).

- $$\widehat{P}_i \text{ such that } F_{\Pi_i}\left(\widehat{P}_i\right) = 0.5$$

This **median plotting position** cannot be given in closed form, but JOHNSON (1964) suggested the following approximation

$$\widehat{P}_i \approx \frac{i - 0.3}{n + 0.4}. \tag{9.15e}$$

- $$\widehat{P}_i = \frac{i-1}{n-1} - \quad \textbf{mode plotting position.}$$
  $$\tag{9.15f}$$

(9.15f) turns into $\widehat{P}_1 = 0$ for $i = 1$ and into $\widehat{P}_n = 1$ for $i = n$, and because most of the probability papers do not include the ordinate values $P = 0$ and $P = 1$, the mode plotting position is rarely used.

All plotting positions presented above are estimates of $P_i = F_X(X \leq x_{i:n})$, and all of them do not depend on the sampled distribution. Plotting positions on the scale of the reduced variable $U = (X - a)/b$ depend on the distribution of the ordered variates $U_{i:n}$, which in turn depend on the sampled distribution; see (5.1e). The **plotting position $\widehat{u}_{i:n}$** is chosen as one of the functional parameters of $U_{i:n}$, either the mean; see (5.18a)

$$\widehat{u}_{i:n} = \mathrm{E}(U_{i:n}) =: \alpha_{i:n} \tag{9.16a}$$

or the median

$$\widehat{u}_{i:n} = \widetilde{u}_{i:n} \tag{9.16b}$$

or the mode

$$\widehat{u}_{i:n} = u_{i:n}^*. \tag{9.16c}$$

All these plotting positions cannot be given in closed form and have to be computed numerically. That is the reason plotting positions, which are directly based on $U$, are not very popular.

Finally, there is a third possibility to determine plotting positions that is dependent on the sampled distribution too. The location–scale distributed variate $X$ and its reduced form $U$ are linked by a linear function, thus we have with respect to the ordered observations:

$$x_{i:n} = a + b\,u_{i:n}. \tag{9.17}$$

In (9.17) we only know $x_{i:n}$, and $a$, $b$ as well as the $u_{i:n}$ are unknown. We can choose values $\widehat{u}_{i:n}$ in such a manner that the estimators $\widehat{a}$ and $\widehat{b}$ are optimal in a given sense. We will revert to this approach in Chapter 10 because its emphasis is on directly estimating the parameters $a$ and $b$, and the calculation of $\widehat{u}_{i:n}$ is a by–product.

Tab. 9/1 summarizes all the plotting positions discussed above. With respect to the choice we can finally state that in most applications it does not matter much how $P_i$ or $u_{i:n}$ are estimated. One will only notice marked differences when the sample size is small. But even these differences are blurred when the straight line is fitted to the data point free–hand.

### 9.2.2.3   Advantages and limitations

Probability papers[9] are often preferred over the numerical analysis presented in later chapters because plots serve many purposes, which no single numerical method can. A plot has many advantages:

1. It is fast and simple to use. In contrast, numerical methods may be tedious to compute and may require analytic know–how or an expensive statistical consultant. Moreover, the added accuracy of numerical methods over plots often does not warrant the effort.

2. It presents data in an easily understandable form. This helps one to draw conclusions from data and also to present data to others. The method is easily understood, even by laymen.

---

[9] For a short remark on the history of probability plotting, see BARNETT (1975).

Table 9/1:  Plotting positions

| Name | Reduced variate $\widehat{u}_{i:n}$ | Probability $\widehat{P}_i$ |
|---|---|---|
| Naïve estimator | $\widehat{u}_{i:n} = F_U^{-1}(i/n)$ | $\widehat{P}_i = i/n$ |
| Midpoint position | $\widehat{u}_{i:n} = F_U^{-1}[(i - 0.5)/n]$ | $\widehat{P}_i = (i - 0.5)/n$ |
| BLOM position | $\widehat{u}_{i:n} = F_U^{-1}[(i - 0.375)/(n + 0.25)]$ | $\widehat{P}_i = (i - 0.375)/(n + 0.25)$ |
| Mean position | | |
| – with respect to $\Pi_i$ | $\widehat{u}_{i:n} = F_U^{-1}[i/(n + 1)]$ | $\widehat{P}_i = i/(n + 1)$ |
| – with respect to $U_{i:n}$ | $\widehat{u}_{i:n} = \alpha_{i:n}$ | $\widehat{P}_i = F_U(\alpha_{i:n})$ |
| Median position | | |
| – with respect to $\Pi_i$ | $\widehat{u}_{i:n} = F_U^{-1}[(i - 0.3)/(n + 0.4)]$ | $\widehat{P}_i = (i - 0.3)/(n + 0.4)$ |
| – with respect to $U_{i:n}$ | $\widehat{u}_{i:n} = \widetilde{u}_{i:n}$ | $\widehat{P}_i = F_U(\widetilde{u}_{i:n})$ |
| Mode position | | |
| – with respect to $\Pi_i$ | $\widehat{u}_{i:n} = F_U^{-1}[(i - 1)/(n - 1)]$ | $\widehat{P}_i = (i - 1)/(n - 1)$ |
| – with respect to $U_{i:n}$ | $\widehat{u}_{i:n} = u_{i:n}^*$ | $\widehat{P}_i = F_U(u_{i:n}^*)$ |
| Positions based on optimal estimation of $a$ and $b$ | no analytic formulas | |

3. It provides simple estimates for a distribution: its parameters, its percentiles, its per-centages failing and percentages surviving. When the paper is supplemented by aux-iliary scales one can even read the hazard rate, the cumulative hazard rate, the mean and the standard deviation.

4. It helps to assess how well a given theoretical distribution fits the data. Sometimes it is even possible to identify and estimate a mixture of two or at most three distribu-tions.

5. It applies to both complete and censored data. Graphical extrapolation into the cen-sored region is easily done.

6. It helps to spot unusual data. The peculiar appearance of a data plot or certain plotted points may reveal bad data or yield important insight when the cause is determined.

7. It lets one assess the assumptions of analytic methods which will be applied to the data in a later stage.

Some limitations of a data plot in comparison with analytic methods are the following:

1. It is not objective. Two people using the same plot may obtain somewhat different estimates. But they usually come to the same conclusion.

2. It does not provide confidence intervals or a statistical hypothesis test. However, a plot is often conclusive and leaves little need for such analytic results.

Usually a thorough statistical analysis combines graphical and analytical methods.

### 9.2.3 Hazard plot

The plotting of multiple and randomly censored data on probability paper causes some problems, and the plotting positions are not easy to compute (see Sect. 9.3.2). Plotting positions are comfortably determined as shown by NELSON (1972a) when **hazard paper** is used. We will first demonstrate how to construct a hazard paper with emphasis on the WEIBULL distribution and then shortly comment on the choice of the plotting position for this kind of paper.

We still analyze location–scale distributions where $F_X(x) = F_U(u)$ for $u = (x - a)/b$. With respect to the cumulative hazard rate (see (2.5) and Table 2/1), we thus have

$$H_X(x) = - \ln[1 - F_X(x)] = - \ln[1 - F_U(u)] = H_U(u), \quad u = \frac{x - a}{b}. \tag{9.18a}$$

Let $\Lambda$, $\Lambda > 0$ be a value of the CHR, then

$$u_\Lambda = H_U^{-1}(\Lambda) \tag{9.18b}$$

and consequently

$$x_\Lambda = a + b\, u_\Lambda \tag{9.18c}$$

$u_\Lambda$ and $x_\Lambda$ may be called **hazard quantile**, **h–quantile** for short. A hazard paper for a location–scale distribution is constructed by taking the vertical axis of a rectangular system of coordinates to lay off $u_\Lambda$, but the labeling of this axis is according to the corresponding CHR–value $\Lambda$. This procedure gives a scaling with respect to $\Lambda$ which — in general — is non–linear, an exception being the exponential distribution.

The probability grid and the hazard grid for one and the same distribution are related to one another because

$$\Lambda = - \ln(1 - P) \tag{9.18d}$$

or

$$P = 1 - \exp(-\Lambda), \tag{9.18e}$$

where $P$ is a given value of the CDF. Thus, a probability grid may be used for hazard plotting when the $P$–scaling of the ordinate is supplemented by a $\Lambda$–scaling (see Fig. 9/5). Conversely, a hazard paper may be used for probability plotting.

The reduced extreme value distribution has

$$F_U(u) = 1 - \exp[- \exp(u)],$$

so that the CHR–function is

$$\begin{aligned} H_U(u) &= - \ln[1 - F_U(u)] \\ &= - \ln\{\exp[- \exp(u)]\} \\ &= \exp(u) \end{aligned} \tag{9.19a}$$

and

$$u_\Lambda = \ln \Lambda \tag{9.19b}$$

and finally

$$x_\Lambda = a + b\, u_\Lambda = a + b \ln \Lambda. \tag{9.19c}$$

Figure 9/5: Hazard paper for the extreme value distribution (upper part) and the WEIBULL distribution (lower part)

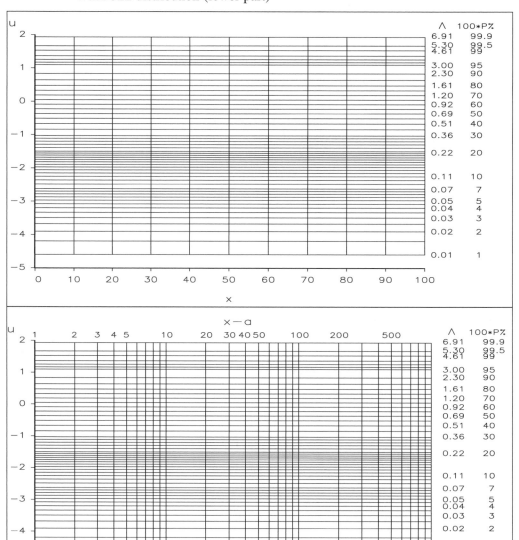

The **hazard paper of the extreme value distribution** has a log–scale on the ordinate and a linear scale on the abscissa (see the upper part of Fig. 9/5). When $X \sim We(a, b, c)$, we have an extreme value distribution for $\ln(X - a)$ with location parameter $a^* = \ln b$ and scale parameter $b^* = 1/c$. So the **hazard paper of the WEIBULL distribution** has the same ordinate as the extreme value hazard paper but a logarithmic scale on the abscissa (see the lower part of Fig. 9/5) and the linear relation, connecting the h–quantiles, reads

$$\ln(x_\Lambda - a) = a^* + b^* \ln \Lambda. \tag{9.20}$$

The cumulative hazard value $\Lambda_i$ for the $i$–th ordered observation $x_{i:n}$ has to be estimated.

For each uncensored lifetime the hazard rate $h(x \mid a, b)$ is estimated by the hazard value

$$\widehat{h}(x_{i:n} \mid a, b) = \frac{1}{n - i + 1}, \tag{9.21a}$$

where $n - i + 1$ is the number of sampled items that have not failed up to $x_{i:n}$. Thus, $n - i + 1$ is the size of the risk set exposed to fail at $x_{i:n}$. $n - i + 1$ is nothing but the **reverse rank**

$$r_i := n - i + 1, \tag{9.21b}$$

which results when all lifetime data — censored as well as uncensored — would be ordered in descending order. The **hazard plotting position** is estimated by

$$\widehat{\Lambda}_i = \widehat{H}_X(x_{i:n}) = \sum_{j=1}^{i} \frac{1}{r_j}, \tag{9.21c}$$

and the summation is only over those reverse ranks belonging to uncensored observations. NELSON (1972a) proved the unbiasedness of (9.21c) when the data are type–II multiply censored.

### 9.2.4 TTT-plot

The TTT–plot is a graph which mainly serves to discriminate between different types of aging, i.e., between constant, decreasing or increasing hazard rates. With respect to the WEIBULL distribution it thus indicates whether $c = 1$, $c < 1$ or $c > 1$. We will first present TTT–plots for uncensored life tests, make some remarks on the censored–data case and close by listing the advantages and limitations.

Let $0 = X_{0:n} \leq X_{1:n} \leq \ldots \leq X_{n:n}$ denote an ordered sample from a life distribution $F(x)$ with survival function $R(x) = 1 - F(x)$. The **total time on test (TTT) statistics**

$$TTT_i = \sum_{j=1}^{i} (n - j + 1)(X_{j:n} - X_{j-1:n}); \ i = 1, 2, \ldots, n \tag{9.22a}$$

have been introduced by EPSTEIN/SOBEL (1953) in connection with the inference of the exponential distribution. For a graphical illustration of $TTT_i$, see Fig. 8/4. The sample mean may be expressed as

$$\overline{X} = \frac{1}{n} TTT_n. \tag{9.22b}$$

The normalized quantity

$$TTT_i^* = \frac{TTT_i}{TTT_n} = \frac{TTT_i}{n\,\overline{x}}, \ 0 \leq TTT_i^* \leq 1 \tag{9.22c}$$

is called **scaled total time on test**. By plotting and connecting the points $(i/n, TTT_i^*)$; $i = 0, 1, \ldots, n$, where $TTT_0 = 0$, by straight line segments we obtain a curve called the **TTT–plot** (see Fig. 9/6). This plotting technique was first suggested by BARLOW/CAMPO (1975) and shows what portion of the total time on test has been accumulated by the portion $i/n$ of items failing first. The TTT–plot has some resemblance to the LORENZ–curve, the difference being that the latter is always strictly convex and lies beneath the 45°–line.

To see what is revealed by a TTT–plot we look at the exponential distribution with $F(x) = 1 - \exp(-x/b)$, $x \geq 0$. The theoretical counterpart to (9.22a) for this distribution is

$$
G_F^{-1}(P) := \int_0^{F^{-1}(P)} \exp(-x/b)\,\mathrm{d}x
$$

$$
= \int_0^{-b\,\ln(1-P)} \exp(-x/b)\,\mathrm{d}x
$$

$$
= b\,P. \tag{9.23a}
$$

This is called the **TTT–transform** of $F_X(x)$. The scale invariant transform being the theoretical counterpart to (9.22c) is

$$
\frac{G_F^{-1}(P)}{\mu} = \frac{b\,P}{b} = P,\ 0 \leq P \leq 1, \tag{9.23b}
$$

and the TTT–plot will be the $45°$–line. BARLOW/CAMPO (1975) have shown that the theoretical TTT–plot will be

- concave and lying above the $45°$–line when $F(x)$ has an increasing hazard rate,

- convex and lying below the $45°$–line when $F(x)$ has a decreasing hazard rate.

An empirical TTT–plot that

- takes its course randomly around the $45°$–line indicates a sample from an exponential distribution,

- is nearly concave (convex) and is mainly above (below) the $45°$–line indicates a sample from an IHR (DHR) distribution.

BARLOW/CAMPO (1975) formulated a test of $H_0$: "$F(x)$ is an exponential distribution" against $H_1$: "$F(x)$ is IHR (DHR)." If the TTT–plot is completely above (below) the $45°$–line $H_0$ is rejected in favor of IHR (DHR), the level of significance being $\alpha = 1/n$. Fig. 9/6 shows three empirical TTT–plots of samples from different WEIBULL distributions.

With respect to censored life tests the scaled total time on test is defined as

- 

$$
TTT_i^* = \frac{TTT_i}{TTT(T)} \tag{9.24a}
$$

for type–I singly censoring at $T$ and plotted against $i/k$; $i = 1, 2, \ldots, k$ and $k$ failures within $(0, T]$,

- 

$$
TTT_i^* = \frac{TTT_i}{TTT(x_{r:n})} \tag{9.24b}
$$

for type–II singly censoring at the $r$-th failure and plotted against $i/r$; $i = 1, 2, \ldots, r$.

Figure 9/6: TTT-plots for several WEIBULL samples of size $n = 20$

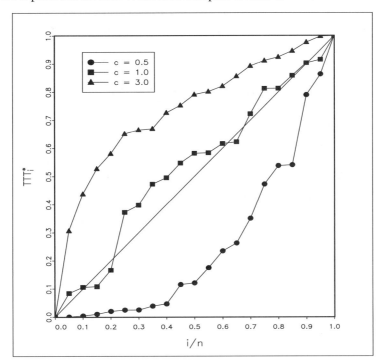

The plots generated in this way will generally lie above those of an uncensored sample of equal size. For the TTT–plot when the sample is multiply censored, see BERGMAN/KLEFSJÖ (1984) and WESTBERG/KLEFSJÖ (1994).

Compared to plots on probability paper or on hazard paper the TTT–plot has several advantages:

1. The TTT–plot is well motivated in theory as well as in practice.

2. The TTT–plot is scale invariant.

3. The TTT–plot does not need a special system of coordinates; it is simply displayed in the linearly scaled unit square.

4. Several distributions — even from different families — can be compared.

5. Its interpretation is plain.

The limitations are as follows:

1. It is only possible to give a rough classification into IHR, DHR or exponential or neither of them.

2. Parameter estimation is impossible as is the reading of percentages or of quantiles.

## 9.3  WEIBULL plotting techniques

WEIBULL probability plotting was introduced by KAO (1959) and was popularized by NELSON (1967). We will first present the procedure when the life test is either complete or singly censored (Sect. 9.3.1). The display of multiply censored data either on probability paper or on hazard paper is discussed in Sections 9.3.2.1 and 9.3.2.2. Two special problems (the existence of a location parameter $a \neq 0$ and the mixture of WEIBULL distributions) are treated in Sections 9.3.3.1 and 9.3.3.2.

### 9.3.1 Complete samples and singly censored samples[10]

The approach to graphical estimation involves two parts. The first part consists in plotting the data either on WEIBULL paper or on a paper with linear scales. This part depends on the data available, whether they are complete or censored, and on the type of censoring. The second part, which is the same for all types of data, comprises the fitting of a straight line to the scatter plot and the estimation, i.e., reading or computing the parameters from the fitted line.

**First Part**

1. Arrange the data in ascending order: $x_{1:n} \leq x_{2:n} \leq \ldots$.

2. Convert the data to logarithms: $x^*_{i:n} = \ln(x_{i:n})$.

3. Compute the plotting positions $\widehat{P}_i$ according to one of the methods presented in Tab. 9/1.

4. Compute the corresponding reduced variables $\widehat{u}_{i:n} = \ln[-\ln(1 - \widehat{P}_i)]$.

   Note to steps 3 and 4: If a direct determination of the plotting position $\widehat{u}_{i:n}$ via $\mathrm{E}(U_{i:n}) = \alpha_{i:n}$, the median $\widetilde{u}_{i:n}$ or the mode $u^*_{i:n}$, is preferred, steps 3 and 4 are skipped.

5. Plot $\widehat{u}_{i:n}$ on the ordinate versus $x^*_{i:n}$ on the abscissa.

6. Judge whether the points scatter randomly around a straight line. If not, there may exist a location parameter $a \neq 0$ or a mixture of WEIBULL distributions or no WEIBULL distribution at all.

With a WEIBULL–probability–paper at hand or with software, one can skip most of the steps above. For singly censored data, Steps 3 to 5 have to be applied to the complete data (see Examples 9/2 to 9/4), but for multiply censored data, Step 3 has to be modified (see Sect. 9.3.2). For grouped data $\widehat{P}_i$ has to be plotted versus the logarithm of the upper limit of class $i$.

**Second Part**

1. Determine the best straight–line fit either free–hand or by least squares with the following two options:

---

[10] Suggested reading for this section: BERGER/LAWRENCE (1974), DODSON (1994), GIBBONS/VANCE (1979), HOMAN (1989), MCCOOL (1974b), NELSON, L.S. (1967), NELSON, W. (1982), NELSON/THOMPSON (1971), PLAIT (1962), STEINECKE (1979).

1.1 Minimize the **departures in the horizontal direction**, i.e., along the axis for the $x_{i:n}^*$, which leads to the regression model

$$x_{i:n}^* = \frac{1}{c}\,\widehat{u}_{i:n} + \ln b + \epsilon_i. \tag{9.25a}$$

1.2 Minimize the **departures in the vertical direction**, i.e. along the axis for the $\widehat{u}_{i:n}$, giving the regression model

$$\widehat{u}_{i:n} = c\,x_{i:n}^* - c\,\ln b + \eta_i. \tag{9.25b}$$

$\epsilon_i$ and $\eta_i$ are random disturbances. A person choosing (9.25a) takes the plotting positions as fixed for a given sample size and argues that the life lengths are random and would change in another sample of the same size whereas the vertical positions will be unchanged. A person choosing (9.25b) takes the observed life lengths as fixed and wants to estimate the unknown value of the CDF at these points in an updated version, updated with respect to the initial estimates $\widehat{P}_i$ that do not incorporate the observed life lengths. Because (9.25b) is the inverse regression function of (9.25a) and not the inverse of a simple linear function, the two sets of parameter estimates resulting from (9.25a,b) will differ. The differences will be smaller the closer the plotted points lie to a straight line, i.e., the higher the correlation between $x_{i:n}^*$ and $\widehat{u}_{i:n}$.

2. Compute and/or read the estimates $\widehat{b}$ and $\widehat{c}$ for $b$ and $c$.

2.1 With respect to (9.25a) we find $\ln \widehat{b}$ as the abscissa value of the intersection of the regression line with a horizontal line running through $P \approx 0.6321$, as $u = \ln[-\ln(1 - 0.6321)] \approx 0.\widehat{c}$ is either the reciprocal slope of the estimated regression line or the reciprocal of the difference $(\ln x_{0.9340}^* - \ln \widehat{b})$. $\ln x_{0.9340}^*$ is the abscissa value of the intersection of the regression line with a horizontal line running through $P \approx 0.9340$, as $u = \ln[-(1 - 0.9340)] \approx 1$.

2.2 With respect to (9.25b) the slope of the regression line yields $\widehat{c}$. Compute $\widehat{u}(0)$, the ordinate intercept of the regression line, and find $\widehat{b} = \exp[-\widehat{u}(0)/\widehat{c}]$. We mention that there exist WEIBULL–probability–papers having auxiliary axes and scales to read directly off the estimates once the straight line has been fitted.

---

**Example 9/2: Graphical estimation for a complete life test (dataset #1)**
To demonstrate the diverse estimation approaches of this and the following chapters, we will depart from a basic set of complete lifetimes (dataset #1), which will be modified to give various models of censoring. Dataset #1 consists of $n = 20$ simulated observations from $We(0, 100, 2.5)$. The ordered observations $x_{i:20}$ and their corresponding plotting positions can be found in Tab. 9/2. Fig. 9/7 shows the data on WEIBULL–paper — without the $100 * P\%$–scale — using the plotting position $i/(n + 1)$ together with the OLS–fitted straight lines according to (9.25a) and (9.25b). The two lines nearly coincide because the correlation between $\widehat{u}$ and $x^*$ is nearly perfect ($r = 0.9865$). The estimated parameters $\widehat{b}$ and $\widehat{c}$, according to (9.25a) and (9.25b) based on this choice of the plotting position as well as on the other choices are displayed in Tab. 9/3.

Table 9/2: Dataset #1 $\left[X \sim We(0, 100, 2.5)\right]$ and diverse plotting positions

| $i$ | $x_{i:n}$ | Plotting Position | | | | | | |
|---|---|---|---|---|---|---|---|---|
| | | $\dfrac{i}{n}$ | $\dfrac{i-0.5}{n}$ | $\dfrac{i-0.375}{n+0.25}$ | $\dfrac{i}{n+1}$ | $\alpha_{i:n}$ | $\dfrac{i-0.3}{n+0.4}$ | $\dfrac{i-1}{n-1}$ |
| (1) | (2) | (3) | (4) | (5) | (6) | (7) | (8) | (9) |
| 1 | 35 | 0.05 | 0.025 | 0.0309 | 0.0476 | −3.5729 | 0.0343 | 0.0000 |
| 2 | 38 | 0.10 | 0.075 | 0.0802 | 0.0952 | −2.5471 | 0.0833 | 0.0526 |
| 3 | 42 | 0.15 | 0.125 | 0.1296 | 0.1429 | −2.0200 | 0.1324 | 0.1053 |
| 4 | 56 | 0.20 | 0.175 | 0.1790 | 0.1905 | −1.6584 | 0.1814 | 0.1579 |
| 5 | 58 | 0.25 | 0.225 | 0.2284 | 0.2381 | −1.3786 | 0.2304 | 0.2105 |
| 6 | 61 | 0.30 | 0.275 | 0.2778 | 0.2857 | −1.1471 | 0.2794 | 0.2632 |
| 7 | 63 | 0.35 | 0.325 | 0.3272 | 0.3333 | −0.9472 | 0.3284 | 0.3158 |
| 8 | 76 | 0.40 | 0.375 | 0.3765 | 0.3810 | −0.7691 | 0.3775 | 0.3684 |
| 9 | 81 | 0.45 | 0.425 | 0.4259 | 0.4286 | −0.6064 | 0.4265 | 0.4211 |
| 10 | 83 | 0.50 | 0.475 | 0.4753 | 0.4762 | −0.4548 | 0.4755 | 0.4737 |
| 11 | 86 | 0.55 | 0.525 | 0.5247 | 0.5238 | −0.3112 | 0.5245 | 0.5263 |
| 12 | 90 | 0.60 | 0.575 | 0.5741 | 0.5714 | −0.1727 | 0.5735 | 0.5789 |
| 13 | 99 | 0.65 | 0.625 | 0.6235 | 0.6190 | −0.0371 | 0.6225 | 0.6316 |
| 14 | 104 | 0.70 | 0.675 | 0.6728 | 0.6667 | 0.0979 | 0.6716 | 0.6842 |
| 15 | 113 | 0.75 | 0.725 | 0.7222 | 0.7143 | 0.2350 | 0.7206 | 0.7368 |
| 16 | 114 | 0.80 | 0.775 | 0.7716 | 0.7619 | 0.3776 | 0.7696 | 0.7895 |
| 17 | 117 | 0.85 | 0.825 | 0.8210 | 0.8095 | 0.5304 | 0.8186 | 0.8421 |
| 18 | 119 | 0.90 | 0.875 | 0.8704 | 0.8571 | 0.7022 | 0.8676 | 0.8947 |
| 19 | 141 | 0.95 | 0.925 | 0.9198 | 0.9048 | 0.9120 | 0.9167 | 0.9474 |
| 20 | 183 | 1.00 | 0.975 | 0.9691 | 0.9524 | 1.2232 | 0.9657 | 1.0000 |

Source of Column (7): WHITE (1967b)

Figure 9/7: Dataset #1 on WEIBULL-paper and OLS-fitted straight lines

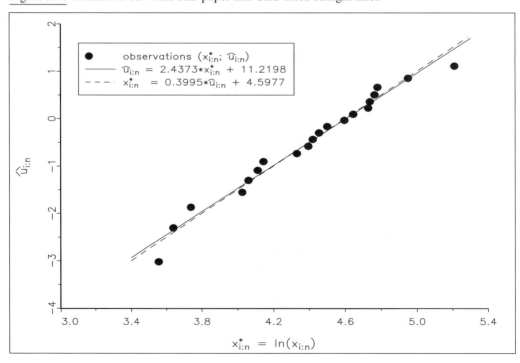

Table 9/3:  Estimates of $b$ and $c$ for dataset #1 depending on the choice of the plotting position and of the kind of regression

| Plotting position | Regression (9.25a) | | Regression (9.25b) | | Correlation $r\left(x^*, \widehat{u}\right)$ |
|---|---|---|---|---|---|
| | $\widehat{b}$ | $\widehat{c}$ | $\widehat{b}$ | $\widehat{c}$ | |
| $i/n$ | 93.6627 | 2.6874 | 94.0309 | 2.6316 | 0.9895 |
| $(i-0.5)/n$ | 97.9444 | 2.8760 | 98.7502 | 2.7599 | 0.9796 |
| $(i-0.375)/(n+0.25)$ | 98.3173 | 2.7586 | 99.0310 | 2.6626 | 0.9824 |
| $i/(n+1)$ | 99.2558 | 2.5029 | 99.8204 | 2.4373 | 0.9865 |
| $\alpha_{i:n}$ | 98.8892 | 2.8098 | 99.6950 | 2.7038 | 0.9809 |
| $(i-0.3)/(n+0.4)$ | 98.5240 | 2.6983 | 99.1898 | 2.6109 | 0.9837 |
| $(i-1)/(n-1)$ | 96.1792 | 2.9343 | 96.4875 | 2.8856 | 0.9917 |

With respect to the chosen plotting position, all estimates are close together, the only exception being the positions $(i-1)/(n-1)$ and $i/n$. Because the correlation between $x^*$ and $\widehat{u}$ is very high, the estimates resulting from using either (9.25a) or (9.25b) for each plotting position do not differ markedly. The best approximation to the true parameter values $b = 100$ and $c = 2.5$ is given by using (9.25a) in conjunction with the plotting position $i/(n+1)$.

When the censoring is simple type–I or type–II censoring, probability plotting can proceed as above except that the lifetimes associated with the non–failed elements are unknown and hence cannot be plotted. One can plot the failures that have occurred using their order numbers in the complete sample and perform the fitting operation using these points. Examples 9/3 and 9/4 will demonstrate the procedure by first singly censoring the dataset #1 at $T = 100$ (see Example 9/3) and afterwards by singly censoring the original set at the 15–th failure (see Example 9/4). Comparing the results of these two examples to those of Example 9/2 shows the effect of censoring on the parameter estimates.

**Example 9/3:  Type–I singly censored dataset #1 at $T = 100$**

Application of the censoring time $T = 100$ to the data in Tab. 9/2 gives 13 complete lifetimes, the longest being $x_{13:20} = 99$, and 7 censored lifetimes. We can thus use the entries of rows $i = 1$ to $i = 13$ to make a WEIBULL probability plot including the two straight line fits based on $i/(n+1)$ (see Fig. 9/8). Tab. 9/4 summarizes the estimation results for all plotting conventions.

The correlation between $x^*$ and $\widehat{u}$ has weakened a little bit compared with Tab. 9/3. All estimates have moved — but only by a small amount — from those of the complete sample so that their departure from the true values has grown somewhat.

<u>Figure 9/8</u>:  Dataset #1 type-I singly censored at $T = 100$ and OLS-fitted straight lines

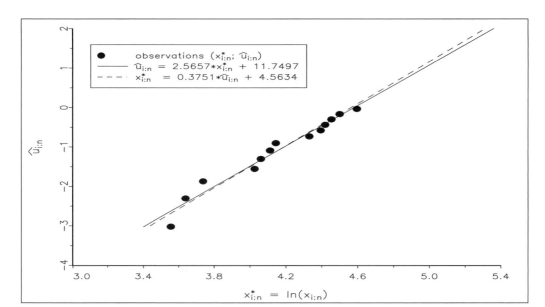

<u>Table 9/4</u>:  Estimates of $b$ and $c$ for dataset #1 singly censored at $T = 100$ depending on the choice of the plotting position and of the kind of regression

| Plotting position | Regression (9.25a) | | Regression (9.25b) | | Correlation $r(x^*, \widehat{u})$ |
|---|---|---|---|---|---|
| | $\widehat{b}$ | $\widehat{c}$ | $\widehat{b}$ | $\widehat{c}$ | |
| $i/n$ | 93.2981 | 2.6940 | 94.6823 | 2.5946 | 0.9814 |
| $(i - 0.5)/n$ | 92.2498 | 3.2092 | 94.6965 | 2.9994 | 0.9667 |
| $(i - 0.375)/(n + 0.25)$ | 93.3625 | 3.0331 | 95.4789 | 2.8656 | 0.9720 |
| $i/(n + 1)$ | 97.4686 | 2.5657 | 95.9090 | 2.6660 | 0.9810 |
| $\alpha_{i:n}$ | 93.3728 | 3.1172 | 95.6060 | 2.9368 | 0.9706 |
| $(i - 0.3)/(n + 0.4)$ | 93.9441 | 2.9438 | 95.9198 | 2.7955 | 0.9795 |
| $(i - 1)/(n - 1)$ | 95.3734 | 2.9718 | 96.3759 | 2.8869 | 0.9856 |

**Example 9/4:  Type-II singly censored dataset #1 at $r = 15$**

When the test would have been terminated with the 15th failure, Tab. 9/2 would give the longest complete lifetime $x_{15:20} = 113$. The WEIBULL probability plot and the OLS–fitted straight lines — based on $i/(n + 1)$ — in Fig. 9/9 now rest on the entries of rows $i = 1$ to $i = 15$ of Tab. 9/2. The set of estimation results is presented in Tab. 9/5.

The results from single type–II censoring lie in between those of single type–I censoring and those of the complete sample. The entries in Tab. 9/5 are a little bit closer to the true parameter values than those of Tab. 9/4. The close agreement of the results of both types of censoring is surely due to the fact that the censoring happened at nearly the same lifetime with nearly the same number of incomplete observations.

Figure 9/9: Dataset #1 type-II singly censored at $x_{15:20}$ and OLS-fitted straight lines

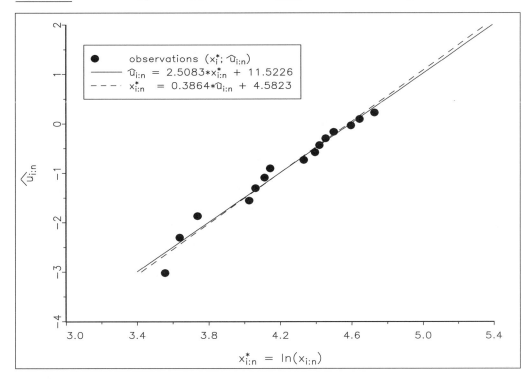

Table 9/5: Estimates of $b$ and $c$ for dataset #1 type-II singly censored at $x_{15:20}$ depending on the choice of the plotting position and of the kind of regression

| Plotting position | Regression (9.25a) | | Regression (9.25b) | | Correlation $r(x^*, \widehat{u})$ |
|---|---|---|---|---|---|
| | $\widehat{b}$ | $\widehat{c}$ | $\widehat{b}$ | $\widehat{c}$ | |
| $i/n$ | 94.7934 | 2.6233 | 95.7402 | 2.5463 | 0.9852 |
| $(i-0.5)/n$ | 94.9642 | 3.0618 | 96.8707 | 2.8892 | 0.9714 |
| $(i-0.375)/(n+0.25)$ | 95.8132 | 2.9087 | 97.4186 | 2.7728 | 0.9763 |
| $i/(n+1)$ | 97.7389 | 2.5880 | 98.8683 | 2.5083 | 0.9845 |
| $\alpha_{i:n}$ | 95.9666 | 3.0750 | 97.7036 | 2.8350 | 0.9748 |
| $(i-0.3)/(n+0.4)$ | 96.2549 | 2.8313 | 97.7372 | 2.7114 | 0.9786 |
| $(i-1)/(n-1)$ | 97.0572 | 2.8785 | 97.7805 | 2.8111 | 0.9882 |

The graphical approach essentially consists in a linearization of the WEIBULL distribution function $F(x_i) = 1 - \exp\{-(x_i/b)^c\}$ and substituting $F(x_i)$ by appropriate estimates. BERGER/LAWRENCE (1974) have compared the resulting linear regression estimators of $b$ and $c$ to those of a non–linear regression; i.e., $b$ and $c$ are directly estimated by a regression of $\widehat{F}(x_i)$ on $x_i$. Their Monte Carlo simulations showed "that non–linear regression does not represent an improvement in the estimation of WEIBULL distribution parameters, and that the mean squared error of both methods is significantly greater than the asymptotic variance of maximum likelihood estimators, and of best linear unbiased or best linear invariant estimators."

HOMAN (1989) compared the performance of six plotting conventions using least absolute deviation (L 1) and least squares (L 2) regression methods under conditions of small sample size, censoring and outlier contamination. His finding is: "A series of simulation experiments indicate that in small samples ($n = 10$), L 2 regression methods are superior in all ten situations of censoring and contamination." A similar result has been found by LAWRENCE/SHIER (1981) and SHIER/LAWRENCE (1984).

### 9.3.2 Multiply censored data

Two basic types of methods for the graphical analysis of multiply censored data have appeared in the literature. These are the probability plotting methods (Sect. 9.3.2.1) and the hazard plotting method (Sect. 9.3.2.2). The probability plotting methods employ a non–parametric estimate of the CDF that is plotted on the probability paper belonging to some parametric distributions. The hazard plotting method employs a non–parametric estimate of the CHR which is plotted on the hazard paper belonging to some parametric function. NELSON (1972a) gives a short review of the history of these two methods.

#### 9.3.2.1 Probability plotting[11]

Probability plotting involves calculating the sample reliability function. This function or its complement, the sample CDF, is plotted against age on probability paper. We will present the HERD–JOHNSON and the KAPLAN–MEIER estimators of $R(x)$, which are so–called **product–limit estimators**. These methods are better for small samples because they plot individual failure times. Another approach — not presented here — consists in applying one of the various **actuarial methods** for constructing life–tables.[12] The actuarial approach is used for large samples where the data are grouped into time intervals.

The two product–limit estimators have a common starting point. There are $n$ units in a multiply censored sample, ordered from the smallest to the largest observed times. Each censored observation time is marked by "+" (see Tab. 9/6). The observed times are numbered backwards with reverse ranks; the smallest is labeled $n$, the largest is labeled 1. If there should be a tie between a failure time and a censored time, the censored time is considered to be a little bit greater than the failure time, i.e., the failure will receive the higher reverse rank. The reliability function is estimated for only those times that are failures, but the censored times have an influence in the determination of the plotting positions of the failures. Let $i$ be the $i$–th failure with reverse rank $r_i = n - i + 1$, then the reliability is recursively calculated in both approaches. The **HERD–JOHNSON estimator**, HJ–estimator for short, is

$$\widehat{R}_i^* = \frac{r_i}{r_i + 1}\,\widehat{R}_{i-1}^*, \quad \widehat{R}_0^* = 1, \tag{9.26}$$

---

[11] Suggested reading for this section: DODSON (1994), HERD (1960), JOHNSON (1964), KAPLAN/MEIER (1958).

[12] See ELANDT–JOHNSON (1976).

and the **KAPLAN–MEIER estimator**, KM–estimator for short, is

$$\widehat{R}_i^+ = \frac{r_i - 1}{r_i}\, \widehat{R}_{i-1}^+, \quad \widehat{R}_0^+ = 1. \tag{9.27}$$

The corresponding failure probabilities

$$\widehat{F}_i^* = 1 - \widehat{R}_i^* \ \text{ and } \ \widehat{F}_i^+ = 1 - \widehat{R}_i^+$$

or their transforms

$$\widehat{u}_i^* = \ln\left[-\ln\left(\widehat{R}_i^*\right)\right] \ \text{ and } \ \widehat{u}_i^+ = \ln\left[-\ln\left(\widehat{R}_i^+\right)\right]$$

are the plotting positions. When the sample is complete,

- the HJ–estimator results in

$$\widehat{F}_i = \frac{i}{n+1}\,,$$

  which is the mean plotting position with respect to $\Pi_i$ (see Tab. 9/1), and

- the KM–estimator will be

$$\widehat{F}_i = \frac{i}{n}\,,$$

  which is the ML estimator of the CDF.

We further note that under KM–estimation we will get $\widehat{F}_n^+ = 1$ when the largest time in the sample is a failure time that cannot be plotted on the WEIBULL–paper. But despite this deficiency the KM–estimator is more popular, because it rests upon an intuitive idea, called "**redistribute–to–the–right**." The KM–estimator is defined by an algorithm that starts with an empirical distribution that puts mass $1/n$ at each observed time and then moves the mass of each censored observation by distributing it equally to all observed times to the right of it.

---

**Example 9/5: Randomly censored dataset #1.**

We have randomly censored dataset #1 by independently generating corresponding failure times $c_i$ of a competing risk, the $c_i$ coming from $We(0, 100, 3)$; i.e., the mean of the competing risk variable is greater than the mean of the original lifetime. The observed times $y_i = \min(x_i, c_i)$ are listed in column 2 of Tab. 9/6 together with an indicator "+" marking a censored observation.

<u>Table 9/6:</u>  HJ-estimate and KM-estimate of $R(x)$ for randomly censored dataset #1

|     |       | HERD–JOHNSON estimates | | | KAPLAN–MEIER estimates | | |
| --- | ----- | ----------------------- | --- | --- | ---------------------- | --- | --- |
| $i$ | $y_i$ | $\dfrac{r_i}{r_i+1}$ | $\widehat{R}^*_{i-1}$ | $\widehat{R}^*_i$ | $\dfrac{r_i-1}{r_i}$ | $\widehat{R}^+_{i-1}$ | $\widehat{R}^+_i$ |
| (1) | (2)   | (3)    | (4)    | (5)    | (6)    | (7)    | (8)    |
| 1   | 35    | 0.9524 | 1.0000 | 0.9524 | 0.9500 | 1.0000 | 0.9500 |
| 2   | 38    | 0.9500 | 0.9524 | 0.9048 | 0.9474 | 0.9500 | 0.9000 |
| 3   | 39+   | –      | –      | –      | –      | –      | –      |
| 4   | 40+   | –      | –      | –      | –      | –      | –      |
| 5   | 41+   | –      | –      | –      | –      | –      | –      |
| 6   | 42    | 0.9375 | 0.9048 | 0.8482 | 0.9333 | 0.9000 | 0.8400 |
| 7   | 51+   | –      | –      | –      | –      | –      | –      |
| 8   | 56    | 0.9231 | 0.8482 | 0.7830 | 0.9167 | 0.8400 | 0.7700 |
| 9   | 56+   | –      | –      | –      | –      | –      | –      |
| 10  | 61    | 0.9167 | 0.7830 | 0.7177 | 0.9091 | 0.7700 | 0.7000 |
| 11  | 67+   | –      | –      | –      | –      | –      | –      |
| 12  | 76    | 0.9000 | 0.7177 | 0.6459 | 0.8889 | 0.7000 | 0.6222 |
| 13  | 83+   | –      | –      | –      | –      | –      | –      |
| 14  | 86    | 0.8750 | 0.6459 | 0.5652 | 0.8571 | 0.6222 | 0.5333 |
| 15  | 94+   | –      | –      | –      | –      | –      | –      |
| 16  | 99    | 0.8333 | 0.5652 | 0.4710 | 0.8000 | 0.5333 | 0.4267 |
| 17  | 104   | 0.8000 | 0.4710 | 0.3768 | 0.7500 | 0.4267 | 0.3200 |
| 18  | 109+  | –      | –      | –      | –      | –      | –      |
| 19  | 117   | 0.6667 | 0.3768 | 0.2512 | 0.5000 | 0.3200 | 0.1600 |
| 20  | 143+  | –      | –      | –      | –      | –      | –      |

Columns 3 to 5 (6 to 8) display the calculation of the HJ–estimates (KM–estimates). Despite the differing estimators of $R(x)$, the OLS–estimates of $b$ and $c$ are not too far away from each other or from the true parameter values $b = 100$ and $c = 2.5$; see the following table:

| Method | Regression (9.25a) | | Regression (9.25b) | |
| ------ | --- | --- | --- | --- |
|        | $\widehat{b}$ | $\widehat{c}$ | $\widehat{b}$ | $\widehat{c}$ |
| HERD–JOHNSON | 103.1359 | 2.4432 | 105.0804 | 2.3460 |
| KAPLAN–MEIER | 96.5733  | 2.5733 | 98.3540  | 2.4736 |

The following excursus shows another approach to determine the plotting positions for a multiply censored dataset. The method has been suggested by DODSON (1994, pp. 21–25) and consists in recursively calculating a modified order of the failed items that afterwards is converted into one of those plotting positions in Tab. 9/1, which rest upon an order number.

**Excursus:** DODSON'S **method of calculating plotting positions for randomly censored datasets**

The order $o_j$ of the $j$–th failure (in ascending order) is given by

$$o_j = o_{j-1} + \Delta_j \qquad\qquad (9.28a)$$

with

$$\Delta_j = \frac{(n+1) - o_{j-1}}{1 + k_j}, \qquad\qquad (9.28b)$$

where

$$\begin{aligned}
o_{j-1} &= \text{the order of the previous failure,}\\
n &= \text{the total sample size of both censored and uncensored observations,}\\
\Delta_j &= \text{the increment for the } j\text{–th failure,}\\
k_j &= \text{the number of data points remaining in the dataset, including the current data point.}
\end{aligned}$$

Table 9/7:  DODSON'S estimation procedure for plotting positions belonging to the randomly censored dataset #1

| $i$ | $y_i$ | $j$ | $x_j$ | $o_{j-1}$ | $k_j$ | $\Delta_j$ | $o_j$ | $\widehat{F}(x_j)$ | $\widehat{u}_j$ |
|-----|-------|-----|-------|-----------|-------|------------|-------|--------------------|-----------------|
| (1) | (2) | (3) | (4) | (5) | (6) | (7) | (8) | (9) | (10) |
| 1 | 35 | 1 | 35 | 0 | 20 | 1 | 1 | 0.0343 | −3.3552 |
| 2 | 38 | 2 | 38 | 1 | 19 | 1 | 2 | 0.0833 | −2.4421 |
| 3 | 39+ | – | – | – | – | – | – | – | – |
| 4 | 40+ | – | – | – | – | – | – | – | – |
| 5 | 41+ | – | – | – | – | – | – | – | – |
| 6 | 42 | 3 | 42 | 2 | 15 | 1.1875 | 3.1875 | 0.1415 | −1.8801 |
| 7 | 51+ | – | – | – | – | – | – | – | – |
| 8 | 56 | 4 | 56 | 3.1875 | 13 | 1.2723 | 4.4598 | 0.2039 | −1.4783 |
| 9 | 56+ | – | – | – | – | – | – | – | – |
| 10 | 61 | 5 | 61 | 4.4598 | 11 | 1.3784 | 5.8382 | 0.2715 | −1.1496 |
| 11 | 67+ | – | – | – | – | – | – | – | – |
| 12 | 76 | 6 | 76 | 5.8382 | 9 | 1.5162 | 7.3544 | 0.3458 | −0.8572 |
| 13 | 83+ | – | – | – | – | – | – | – | – |
| 14 | 86 | 7 | 86 | 7.3544 | 7 | 1.7057 | 9.0601 | 0.4294 | −0.5779 |
| 15 | 94+ | – | – | – | – | – | – | – | – |
| 16 | 99 | 8 | 99 | 9.0601 | 5 | 1.9900 | 11.0501 | 0.5270 | −0.2895 |
| 17 | 104 | 9 | 104 | 11.0501 | 4 | 1.9900 | 13.0401 | 0.6245 | −0.0207 |
| 18 | 109+ | – | – | – | – | – | – | – | – |
| 19 | 117 | 10 | 117 | 13.0401 | 2 | 2.6533 | 15.6934 | 0.7546 | 0.3399 |
| 20 | 143+ | – | – | – | – | – | – | – | – |

The procedure is another way of realizing the idea of "redistribute–to–the–right." Tab. 9/7 demonstrates the calculation using the randomly censored dataset of Tab. 9/6 and using the median position $(o_j - 0.3)/(n + 0.4)$ to estimate $F(x_j)$ in Column 9 and the plotting position $\widehat{u}_j = \ln\{-\ln[1 - \widehat{F}(x_j)]\}$ in Column 10.

Taking the failure times and their corresponding $\widehat{u}_j$ from Tab. 9/7 the OLS–estimation of (9.25a) delivers $\widehat{b} = 99.4544$ and $\widehat{c} = 2.7685$, whereas (9.25b) gives $\widehat{b} = 102.0762$ and $\widehat{c} = 2.6066$. In both regressions the estimate for $c$ is higher than the HJ– and the KM–estimates. The estimate for $b$ is closer to the true value than the HJ– and KM–estimates.

### 9.3.2.2 Hazard Plotting[13]

It is much easier and simpler for multiply censored data to use hazard plotting as outlined in Sect. 9.2.3 than probability plotting based on the HJ–, KM– or DODSON–approaches. The probability and data scales on a hazard paper for a given location–scale distribution are exactly the same as those on the probability paper; compare Fig. 9/4 to Fig. 9/5 for the extreme value distribution and the WEIBULL distribution. Thus, a hazard plot is interpreted in just the same way as a probability plot, and the scales on hazard paper are used like those on probability paper. The additional cumulative hazard rate $\Lambda$ (see Fig. 9/5) is only a convenience for plotting multiply censored data. The hazard plotting position on the $\Lambda$–scale is estimated by (see (9.21c)):

$$\widehat{\Lambda}_j = \widehat{H}_X(x_j) = \sum_{\ell=1}^{j} \frac{1}{r_\ell},$$

where $x_1 < x_2 < \ldots$ are the complete lifetimes in ascending order, and the summation is thus over only those reverse ranks that belong to uncensored observations. The $\widehat{\Lambda}_j$ may be converted to the hazard quantiles $\widehat{u}_j$ of the extreme value distribution

$$\widehat{u}_j := u(\widehat{\Lambda}_j) = \ln\widehat{\Lambda}_j$$

for convenient plotting and to OLS–estimation of the WEIBULL parameters in (9.25a,b).

**Example 9/6: Hazard plotting of randomly censored dataset #1**

We use the randomly censored data in Tab. 9/6 to demonstrate the computation of the hazard plotting positions $\widehat{u}_j$ in Column 9 of Tab. 9/8. These plotting positions may be compared to those of the HERD–JOHNSON approach and the KAPLAN–MEIER approach in Columns 10 and 11. The hazard plotting positions deliver $\widehat{b} = 100.7052$ and $\widehat{c} = 2.4897$ from (9.25a) and $\widehat{b} = 102.4495$ and $\widehat{c} = 2.3934$ from (9.25b).

---

[13] Suggested reading for this section: NELSON (1970, 1972a, 1982).

Table 9/8: Hazard plotting for randomly censored dataset #1

| | | | | | Hazard plotting | | | | HJ | KM |
|---|---|---|---|---|---|---|---|---|---|---|
| $i$ | $y_i$ | $r_i$ | $j$ | $x_j$ | $r_j$ | $1/r_j$ | $\widehat{\Lambda}_j$ | $\widehat{u}_j$ | $\widehat{u}_j$ | $\widehat{u}_j$ |
| (1) | (2) | (3) | (4) | (5) | (6) | (7) | (8) | (9) | (10) | (11) |
| 1 | 35 | 20 | 1 | 35 | 20 | 0.0500 | 0.0500 | $-2.9957$ | $-3.0202$ | $-2.9702$ |
| 2 | 38 | 19 | 2 | 38 | 19 | 0.0526 | 0.1026 | $-2.2766$ | $-2.3018$ | $-2.2504$ |
| 3 | 39+ | 18 | – | – | – | – | – | – | – | – |
| 4 | 40+ | 17 | – | – | – | – | – | – | – | – |
| 5 | 41+ | 16 | – | – | – | – | – | – | – | – |
| 6 | 42 | 15 | 3 | 42 | 15 | 0.0667 | 0.1693 | $-1.7761$ | $-1.8041$ | $-1.7467$ |
| 7 | 51+ | 14 | – | – | – | – | – | – | – | – |
| 8 | 56 | 13 | 4 | 56 | 13 | 0.0769 | 0.2462 | $-1.4015$ | $-1.4079$ | $-1.3418$ |
| 9 | 56+ | 12 | – | – | – | – | – | – | – | – |
| 10 | 61 | 11 | 5 | 61 | 11 | 0.0909 | 0.3371 | $-1.0873$ | $-1.1036$ | $-1.0309$ |
| 11 | 67+ | 10 | – | – | – | – | – | – | – | – |
| 12 | 76 | 9 | 6 | 76 | 9 | 0.1111 | 0.4482 | $-0.8024$ | $-0.8277$ | $-0.7456$ |
| 13 | 83+ | 8 | – | – | – | – | – | – | – | – |
| 14 | 86 | 7 | 7 | 86 | 7 | 0.1429 | 0.5911 | $-0.5258$ | $-0.5611$ | $-0.4642$ |
| 15 | 94+ | 6 | – | – | – | – | – | – | – | – |
| 16 | 99 | 5 | 8 | 99 | 5 | 0.2000 | 0.7911 | $-0.2343$ | $-0.2838$ | $-0.1605$ |
| 17 | 104 | 4 | 9 | 104 | 4 | 0.2500 | 1.0411 | 0.0403 | $-0.0243$ | 0.1305 |
| 18 | 109+ | 3 | – | – | – | – | – | – | – | – |
| 19 | 117 | 2 | 10 | 117 | 2 | 0.5000 | 1.5411 | 0.4325 | 0.3232 | 0.6057 |
| 20 | 143+ | 1 | – | – | – | – | – | – | – | – |

### 9.3.3 Special problems

A scatter plot that is significantly different from a straight line on WEIBULL–paper may indicate the existence of either a location parameter $a \neq 0$ or a mixture of WEIBULL distributions.

#### 9.3.3.1 Three-parameter WEIBULL distribution[14]

Plotting $\widehat{u}_{i:n} = \ln\{-\ln[1 - \widehat{F}(x_{i:n})]\}$ against $x^*_{i:n} = \ln(x_{i:n})$ when the $x_{i:n}$ are sampled from $F(x) = 1 - \exp\{-[(x-a)/b]^c\}$ will result into an approximately convex (concave) curve if $a < 0$ $(a > 0)$ (see Fig. 2/8). The curved plot can be converted into a fairly straight line when

1. $c$ is known or

2. $a$ is known.

---

[14] Suggested reading for this section: DAVID (1975), JIANG/MURTHY (1997), KECECIOGLU (1991), LI (1994).

In the first case $F(x) = 1 - \exp\{-[(x - a)/b]^c\}$ is transformed to

$$z = \frac{1}{b}x - \frac{a}{b} \qquad (9.29a)$$

where

$$z = \left\{ -\ln\left[1 - F(x)\right]\right\}^{1/c}. \qquad (9.29b)$$

In the second case the linearizing transformation is

$$\ln(x - a) = \frac{1}{c}\ln\left\{ -\ln\left[1 - F(x)\right]\right\} + \ln b \qquad (9.30a)$$

or

$$\ln\left\{ -\ln\left[1 - F(x)\right]\right\} = c\ln(x - a) - c\ln b. \qquad (9.30b)$$

The second case is more relevant in practice than the first case. Graphical oriented methods for a three–parameter WEIBULL distribution mostly attempt to estimate $a$ first. Then, using this estimate, the problem is reduced to estimating the remaining parameters $b$ and $c$ as has been described in Sections 9.3.1 and 9.3.2. We will describe two such approaches, the "trial and error" approach and an approach suggested by DAVID (1975), working along this line.[15] Afterward, we present two approaches, which simultaneously estimate all three parameters, i.e., the approach of JIANG/MURTHY (1997) and a non–linear least squares regression of (9.30b). All four approaches will be demonstrated by using dataset #2 consisting of $n = 20$ observations from a WEIBULL distribution having $a = 15$, $b = 30$ and $c = 2.5$. The ordered dataset is given in Tab. 9/9. The plotting of these data is always done with $\widehat{P}_i = i/(n + 1)$, i.e., $\widehat{u}_{i:n} = \ln\left[ -\ln\left(1 - \widehat{P}_i\right)\right]$.

Table 9/9:   Ordered dataset #2 from $We(15, 30, 2.5)$

| $i$ | $x_{i:20}$ | $i$ | $x_{i:20}$ | $i$ | $x_{i:20}$ | $i$ | $x_{i:20}$ |
|---|---|---|---|---|---|---|---|
| 1 | 22.8 | 6 | 34.4 | 11 | 41.6 | 16 | 49.5 |
| 2 | 26.3 | 7 | 35.6 | 12 | 43.5 | 17 | 52.0 |
| 3 | 28.8 | 8 | 37.3 | 13 | 44.7 | 18 | 53.8 |
| 4 | 30.9 | 9 | 38.5 | 14 | 46.2 | 19 | 57.3 |
| 5 | 32.4 | 10 | 39.9 | 15 | 48.4 | 20 | 66.4 |

Trial and error is a rather simple but subjective method to estimate $a$ based on a graph. We iteratively assume a value $\widehat{a}$, construct a probability plot based on $\widehat{x}_{i:n}^* = \ln(x_{i:n} - \widehat{a})$ and $\widehat{u}_{i:n} = \ln\{ -\ln\left[1 - \widehat{F}(x_{i:n})\right]\}$, then revise $\widehat{a}$ and plot again until there is no more obvious curvature in the plot. When $\widehat{a}$ is too small (large), the WEIBULL plot is concave (convex)

---

[15] KECECIOGLU (1991, pp. 291–303.) gives more methods to estimate $a$.

and $\widehat{a}$ has to be raised (reduced). Adding a positive (negative) constant $con$ to variable $x$ has the effect that $\ln(x + con)$ moves further to the right (left) for $x$ being small than for $x$ being large and thus reduces a convex (concave) bending of a monotonically increasing function of $\ln x$.

**Example 9/7:   Estimating $a$ by trial and error in dataset #2**

Fig. 9/10 shows the working of the trial–and–error approach applied to dataset #2 of Tab. 9/9. The curve marked by solid circles displays $\widehat{u}_{i:n}$ as a function of $\ln x_{i:n}$ and is obviously concave; i.e., $a$ must be greater than zero. Introducing $\widehat{a} = 5$, $10$, $15$ and $20$ moves the curves to the left, reduces the concavity and finally leads to convexity when $\widehat{a} = 20$. As it is rather difficult to decide which of the five curves is closest to linearity, we have calculated the residual sum of squares $RSS(\widehat{a})$ for the linear fit (9.25b) to each curve: $RSS(0) = 0.273$, $RSS(5) = 0.173$, $RSS(10) = 0.080$, $RSS(15) = 0.059$ and $RSS(20) = 0.585$. Thus, $\widehat{a} = 15$ should be preferred.

Figure 9/10:  Graphical estimation of $a$ by trial and error

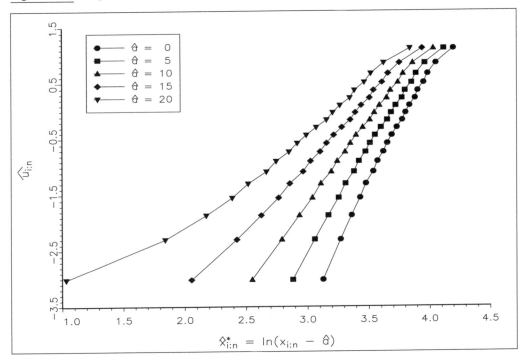

The graphical method proposed by DAVID (1975) requires the selection of three special points on the graph of $\widehat{u}_{i:n}$ against $x^*_{i:n} = \ln x_{i:n}$. The points have the coordinates

$$(x^*_1, u_1), \quad (x^*_2, u_2), \quad (x^*_3, u_3)$$

and have to satisfy the condition

$$u_3 - u_2 = u_2 - u_1 = con; \tag{9.31a}$$

i.e., they are equidistant on the $u$–axis. When $X \sim We(a,b,c)$, these points are supposed to lie on a straight line; therefore,

$$\frac{u_3 - u_2}{\ln(x_3 - a) - \ln(x_2 - a)} = \frac{u_2 - u_1}{\ln(x_2 - a) - \ln(x_1 - a)} \qquad (9.31b)$$

and

$$(x_3 - a)^{u_2 - u_1} (x_1 - a)^{u_3 - u_2} = (x_2 - a)^{u_3 - u_1}. \qquad (9.31c)$$

(9.31c) turns into

$$(x_3 - a)^{con} (x_1 - a)^{con} = (x_2 - a)^{2\,con} \qquad (9.31d)$$

observing (9.31a). The solution for $a$ is

$$\widehat{a} = \frac{x_2^2 - x_1\,x_3}{2\,x_2 - x_1 - x_3}. \qquad (9.31e)$$

**Example 9/8:  Estimating $a$ by DAVID's method**

Fig. 9/11 demonstrates how to apply DAVID's approach to dataset #2. The estimate $\widehat{a}$ will be subjective for two reasons. First, we have to approximate the data points $(x_{i:n}^*, \widehat{u}_{i:n})$ by a smooth curve. (Here, this step is unnecessary because the points lie on concavely banded smooth curve.) Second, we have to choose $con$ and a starting point $u_1$. Fig. 9/11 shows two choices:

- $con = 1$ and $u_1 = -1$ lead to $\widehat{a} \approx 19.2$,
- $con = 0.5$ and $u_1 = -2.5$ give $\widehat{a} \approx 15.6$

Figure 9/11: Graphical estimation of $a$ by the method of DAVID

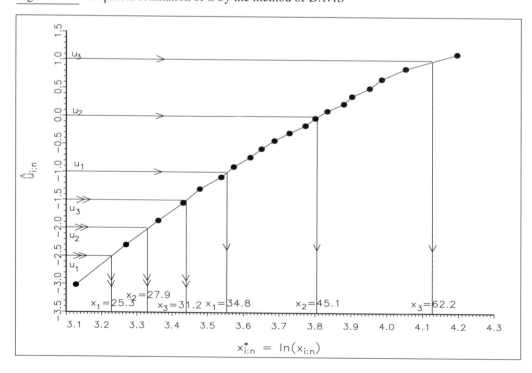

JIANG/MURTHY (1997) have proposed a method, which is based on a TAYLOR's series expansion and which simultaneously estimates all three WEIBULL parameters. A plot of $u = \ln\{-\ln[1 - F(x)]\} = \ln[-\ln R(x)]$ versus $x^* = \ln x$ is a curve that intersects the $x^*$–axis at $x_0^*$. Re–transforming $x_0^*$ gives

$$x_0 = \exp(x_0^*). \tag{9.32a}$$

Observing (9.30b) for $\ln[-\ln R(x)] = 0$ leads to

$$x_0 = a + b. \tag{9.32b}$$

Using (9.32b) the reliability function may be rewritten as

$$R(x) = \exp\left[-\left(\frac{x - x_0 + b}{b}\right)^c\right] \tag{9.32c}$$

or

$$-\ln R(x) = \left(\frac{z}{b} + 1\right)^c, \tag{9.32d}$$

where

$$z := x - x_0.$$

When $z/b < 1$ the TAYLOR's series expansion of (9.32d) is

$$-\ln R(x) \approx 1 + c\frac{z}{b} + \frac{c}{2}(c-1)\left(\frac{z}{b}\right)^2 \tag{9.33a}$$

or

$$y \approx \alpha + \beta z, \tag{9.33b}$$

where

$$y := -\frac{\ln R(x) + 1}{z}, \tag{9.33c}$$

$$\alpha := c/b, \tag{9.33d}$$

$$\beta := \frac{c(c-1)}{2b^2}. \tag{9.33e}$$

By plotting $y_i$ versus $z_i$ and fitting a straight line, e.g., by OLS regression, one can find estimates $\widehat{\alpha}$ and $\widehat{\beta}$ which may be re–transformed to estimates $\widehat{a}$, $\widehat{b}$ and $\widehat{c}$ using (9.32b) and (9.33d,e):

$$\widehat{c} = \frac{\widehat{\alpha}^2}{\widehat{\alpha}^2 - 2\widehat{\beta}}, \tag{9.34a}$$

$$\widehat{b} = \frac{\widehat{c}}{\widehat{\alpha}}, \tag{9.34b}$$

$$\widehat{a} = x_0 - \widehat{b}. \tag{9.34c}$$

**Example 9/9: Estimating $a$, $b$ and $c$ by the method of JIANG/MURTHY**
We apply the method described above to the data of Tab. 9/9. From Fig. 9/11 we find $x_0 = 45.1$. (9.34a–c) lead to

$$\widehat{c} = 2.03, \quad \widehat{b} = 24.90, \quad \widehat{a} = 20.20,$$

which may be compared with the true values $c = 2.5$, $b = 30$ and $a = 15$.

The function (9.30b) is non–linear. A non–linear least squares estimation using numerical derivatives produced the following estimates

$$\widehat{c} = 2.4978, \quad \widehat{b} = 31.8234, \quad \widehat{a} = 13.4868$$

within at most 15 iterations. The results are stable; i.e., starting with several initial values we always finished with the results above. Besides, the NLS–estimates are extremely close to the true values.

### 9.3.3.2   Mixed WEIBULL distributions[16]

Plotting $\widehat{u}_{i:n} = \ln\{ -\ln [1 - \widehat{F}(x_{i:n})]\}$ versus $x_{i:n}^* = \ln(x_{i:n})$ when the $x_{i:n}$ are sampled from a mixture of several WEIBULL distributions (see Sect. 3.3.6.4) will result in a curve which is not only curved in one direction, either convex or concave, but exhibits several types of curving, provided the sample size is sufficiently large. In the following text we will only comment upon the graphical estimation of a twofold mixture which has been explored in some detail by several authors.

The reliability function of a twofold WEIBULL mixture[17] is given by

$$R(x) = p \, \exp\left[-\left(\frac{x}{b_1}\right)^{c_1}\right] + (1 - p) \, \exp\left[-\left(\frac{x}{b_2}\right)^{c_2}\right], \quad 0 < p < 1 \qquad (9.35)$$

and thus has five parameters. When $c_1 \neq c_2$, then without loss of generality we take $c_2 > c_1$. Plotting

$$y(t) = \ln\{ -\ln [R(e^{\,t})]\} \quad \text{versus} \quad t = \ln x$$

gives the curve $\mathcal{C}$. This curve has to be fitted visually to the data points $\big(y(t_{i:n}), t_{i:n} = \ln(x_{i:n})\big)$.

The graphical estimation procedure depends on whether

1)  $c_1 = c_2$ or

2)  $c_2 > c_1$

and whether in the case $c_1 = c_2 = c$

---

[16]  Suggested reading for this section: BRIKC (1999), CRAN (1976), JENSEN/PETERSEN (1982), JIANG/KECECIOGLU (1992a), JIANG/MURTHY (1995), KAO (1959).

[17]  The mixed distributions are supposed to have $a_1 = a_2 = 0$.

1.1) $(b_2/b_1)^c \approx 1$ or

1.2) $(b_2/b_1)^c \gg 1$

and whether in the case $c_2 > c_1$

2.1) $b_1 \approx b_2$ or

2.2) $b_1 \gg b_2$ or

2.3) $b_1 \ll b_2$.

When $c_1 = c_2$, the curve $\mathcal{C}$ will have only one point of inflection called $T$, whereas curve $\mathcal{C}$ has three points of inflection for $c_2 > c_1$. The two subcases to case 1 are detected as follows: when the data points are scattered on either side of $T$, we have $(b_2/b_1)^c \approx 1$. When $(b_2/b_1)^c \gg 1$, the data are scattered either mainly on one side or on both sides of $T$. To discriminate between the three subclasses of case 2, we have to determine the point of intersection $I$ of the asymptote to the right hand side of $\mathcal{C}$ with the curve $\mathcal{C}$. When most of the data points are to the left (right) of $I$ we can assume $b_1 \ll b_2$ $(b_1 \gg b_2)$.

The reader is referred to the original paper of JIANG/MURTHY (1995) for a description of the procedure in each of the five cases. Besides the disadvantage of all graphical procedures in statistics, the graphical analysis of mixed WEIBULL distributions demands a large sample size to clearly perceive the underlying model. Fig. 9/12 demonstrates the latter assertion for the cases $c_1 = c_2$ (upper row) and $c_2 > c_1$ (lower row) with $p = 0.4$ in either case. The smooth curve in each graph represents the WEIBULL probability plot of (9.35).

Figure 9/12: WEIBULL probability plots of mixed distributions

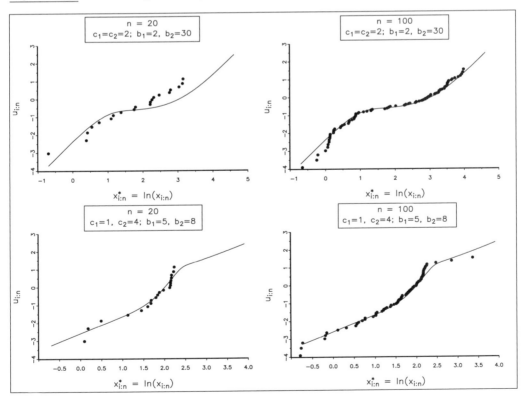

## 9.4   Nomograms and supporting graphs[18]

Several decades ago when electronic computers did not exist, mathematicians, statisticians, and engineers relied on nomograms and other graphical aids to find solutions to complicated mathematical functions. For example, the WEIBULL–probability–paper was supplemented by additional scales to find estimates of the WEIBULL parameters $b$ and $c$ after a straight line has been fitted to the data points; see e.g. LLOYD/LIPOW (1962, p. 164) or NELSON (1967). The WEIBULL–probability–paper of STEINECKE (1979) even has two extra scales to read off estimates of the mean and the standard deviation.

A nomogram designed by KOTELNIKOV (1964), which is reprinted and whose handling is described in JOHNSON/KOTZ/BALAKRISHNAN (1994, pp. 675–677), allows one to find an estimate of $c$ and of $F(x)$ when the mean and the standard deviation of the data are given.

A rather sophisticated nomogram of SEN/PRABHASHANKER (1980) permits one to find estimates of the three parameters $a$, $b$ and $c$, respectively, when we have estimates $\widehat{\mu}$, $\widehat{\sigma}$ and $\widehat{\alpha}_3$, $\alpha_3$ being the standardized third central moment; see (2.82a) and (2.93). Their nomogram is based on the method of moments (see Chapter 12).

A series of nomograms given by STONE/ROSEN (1984) facilitates the construction of confidence intervals for the parameters $b$ and $c$ and the percentiles of the two–parameter WEIBULL distribution. The point estimates are BLIEs (see Sect. 10.3). As STONE/ROSEN have stated, their graphs are reasonably accurate for sample sizes up to $n = 25$ and type–II singly censoring at about $r = 0.5\,n$ to $r = n$.

---

[18]   Suggested reading for this section:   KOTELNIKOV (1964), LIEBSCHER (1967), LLOYD/LIPOW (1962), NELSON (1967), NELSON/THOMPSON (1971), PLAIT (1962), SEN/PRABHASHANKER (1980), STONE/ROSEN (1984), TYURIN (1975).

# 10 Parameter estimation — Least squares and linear approaches

The topic of this chapter is linear estimation; i.e., the estimators are linear combinations of order statistics with suitably chosen coefficients. The parameters to be estimated are $a^*$ and $b^*$ of the type–I extreme value distribution of the minimum (GUMBEL's distribution or Log–WEIBULL distribution); see (9.11c). These parameters are linked to the WEIBULL parameters $b$ and $c$ by

$$a^* = \ln b \ \ \text{or} \ \ b = \exp(a^*) \tag{10.1a}$$

and

$$b^* = \frac{1}{c} \ \ \text{or} \ \ c = \frac{1}{b^*} . \tag{10.1b}$$

Not all of the favorable properties of linear estimators $\widehat{a^*}$ and $\widehat{b^*}$ are preserved under the non–linear transformation $\widehat{b}$ and $\widehat{c}$ according to (10.1a,b). Furthermore, the linear approach assumes that the location parameter $a$ of the WEIBULL distribution is zero or otherwise known, so that one can build upon the shifted variate $X - a$. With the assumption that the shape parameter $c$ is known, the linear estimators of the location and scale parameters $a$ and $b$ can be derived without using the Log–WEIBULL distribution, as will be shown in Sect. 10.6.2.1.

This chapter is arranged as follows: Sect. 10.1 shows how the linear estimators are linked to the graphical procedures of Chapter 9 and to least–squares regression. In Sections 10.2 and 10.3 we develop the best linear unbiased estimators (BLUEs) and the best linear invariant estimators (BLIEs) and show their relationship.

The BLUEs and BLIEs of $a^*$ and $b^*$ require sets of weights laid down in tables which are difficult to compute. Besides, the computations turn out to become unstable with $n$ increasing. So weights exist only for $n \leq 25$. Therefore, modifications and approximations are needed, which are presented in Sect. 10.4. Sometimes there are good reasons not to build upon all order statistics available but to work with only a few of them. The optimal choice of these order statistics is developed in Sect. 10.5. In Sect. 10.6 we show how to proceed when only a subset of the parameters is to be estimated. Finally, in Sect. 10.7 we treat miscellaneous problems of linear estimation.

## 10.1 From OLS to linear estimation[1]

In Sect. 9.3.1 we frequently applied the OLS principle of regression theory instead of an eye–fitting to arrive at a straight line[2] $\widehat{x^*} = \widehat{a^*} + \widehat{b^*}\,\widehat{u}$ without asking for the properties of $\widehat{a^*}$ and $\widehat{b^*}$. Now we will make up for what we have omitted to do.

The distribution of the Log–WEIBULL variate $X^* = \ln(X - a)$, with $X \sim We(a, b, c)$ and $a = 0$ or known, is a member of the location–scale family; i.e., the reduced variate $U = (X^* - a^*)/b^*$ has a distribution free of any parameters. The DF of $U$ is

$$f_U(u) = \exp(u - e^u), \ u \in \mathbb{R}.$$

The ordered reduced variates $U_{i:n} = (X^*_{i:n} - a^*)/b^*$ will also have parameter–free distributions with (see (5.18a–b)):

$$\mathrm{E}\big(U_{i:n}\big) \quad =: \quad \alpha_{i:n}; \ \ i = 1, 2, ..., n; \tag{10.2a}$$

$$\mathrm{Var}\big(U_{i:n}\big) \quad =: \quad \beta_{i,i:n}; \ \ i = 1, 2, ..., n; \tag{10.2b}$$

$$\mathrm{Cov}\big(U_{i:n}, U_{j:n}\big) \quad =: \quad \beta_{i,j:n}; \ \ i \neq j \text{ and } i, j = 1, 2, ..., n. \tag{10.2c}$$

These quantities have known values depending on the form of $f_U(u)$ but not on the parameters $a^*$ and $b^*$. Approaches to compute $\alpha_{i:n}$ and $\beta_{i,j:n}$ for the Log–WEIBULL distribution have been presented in Sect. 5.4.

Reverting to the original ordered variates

$$X^*_{i:n} = a^* + b^*\, U_{i:n}, \tag{10.3a}$$

we clearly have

$$\mathrm{E}\big(X^*_{i:n}\big) \quad = \quad a^* + b^*\, \alpha_{i:n}, \tag{10.3b}$$

$$\mathrm{Var}\big(X^*_{i:n}\big) \quad = \quad b^{*\,2}\, \beta_{i,i:n}, \tag{10.3c}$$

$$\mathrm{Cov}\big(X^*_{i:n}, X^*_{j:n}\big) \quad = \quad b^{*\,2}\, \beta_{i,j:n}. \tag{10.3d}$$

Introducing latent disturbance variates, $\epsilon_i$ (10.3a) may be written as

$$X^*_{i:n} = a^* + b^*\, \alpha_{i:n} + \epsilon_i; \ \ i = 1, 2, ..., n. \tag{10.3e}$$

$\epsilon_i$ is nothing but the difference between the variate $X^*_{i:n}$ and its mean, so $\mathrm{E}(\epsilon_i) = 0 \ \forall\ i$. We further have $\mathrm{Var}(\epsilon_i) = \mathrm{Var}\big(X^*_{i:n}\big)$ and $\mathrm{Cov}\big(\epsilon_i, \epsilon_j\big) = \mathrm{Cov}\big(X^*_{i:n}, X^*_{j:n}\big)$. The regression model (10.3e) does not fulfill all those conditions that are necessary for OLS estimators to have minimum variance within the class of unbiased linear estimators. The regressands

---

[1] Suggested reading for this section: BARNETT (1975), GANDER (1996), LLOYD (1952), NELSON/HAHN (1974), WHITE (1965, 1967b, 1969).

[2] The corresponding inverse regression leading to $\widehat{u} = \widehat{A^*} + \widehat{B^*}\,\widehat{x^*}$ is of minor interest. The resulting estimators are sometimes called "least squares estimators" (see GIBBONS/VANCE (1981)); e.g., LSMR when the median ranks $u_i = (i - 0.3)/(n + 0.4)$ are chosen as plotting positions. These estimators as well as those based on some other plotting positions have been examined by BALABAN/HASPERT (1972), HOSSAIN/HOWLADER (1996) and TSANG/JARDINE (1973).

$X_{i:n}^*$ are **heteroscedastic** (see (10.3c)) and **autocorrelated** (see (10.3d)). So, we have to apply the **general–least–squares** (GLS) principle of AITKEN (1935) to ensure the validity of the GAUSS–MARKOV theorem.[3] LLOYD (1952) was the first to apply the GLS method for estimating the parameters of a location–scale distribution; i.e., the results apply to each member of the location–scale family.

We write (10.3e) in matrix–vector form:

$$\boldsymbol{x}^* = \boldsymbol{P}\,\boldsymbol{\theta} + \boldsymbol{\epsilon} \tag{10.4a}$$

with

$$\boldsymbol{x}^* = \begin{pmatrix} X_{1:n}^* \\ \vdots \\ X_{n:n}^* \end{pmatrix}, \; \boldsymbol{P} = \begin{pmatrix} 1 & \alpha_{1:n} \\ \vdots & \vdots \\ 1 & \alpha_{n:n} \end{pmatrix}, \; \boldsymbol{\theta} = \begin{pmatrix} a^* \\ b^* \end{pmatrix}, \; \boldsymbol{\epsilon} = \begin{pmatrix} \epsilon_1 \\ \vdots \\ \epsilon_n \end{pmatrix}.$$

The variance–covariance matrix $\mathrm{Var}(\boldsymbol{x}^*)$ is given by

$$\boldsymbol{\Omega} := \mathrm{Var}(\boldsymbol{x}^*) = b^{*2}\,\boldsymbol{B} \tag{10.4b}$$

with

$$\boldsymbol{B} = \begin{pmatrix} \beta_{1,1:n} & \beta_{1,2:n} & \cdots & \beta_{1,n:n} \\ \beta_{1,2:n} & \beta_{2,2:n} & \cdots & \beta_{2,n:n} \\ \vdots & \vdots & \ddots & \vdots \\ \beta_{1,n:n} & \beta_{2,n:n} & \cdots & \beta_{n,n:n}, \end{pmatrix}$$

which is symmetric and positive–definite. The GLS estimator of the parameter vector $\boldsymbol{\theta}$ is

$$\widehat{\boldsymbol{\theta}} = \left(\boldsymbol{P}'\,\boldsymbol{B}^{-1}\,\boldsymbol{P}\right)^{-1}\boldsymbol{P}'\,\boldsymbol{B}^{-1}\,\boldsymbol{x}^*, \tag{10.4c}$$

which is BLUE. The $(2 \times n)$–matrix

$$\boldsymbol{C} := \left(\boldsymbol{P}'\,\boldsymbol{B}^{-1}\,\boldsymbol{P}\right)^{-1}\boldsymbol{P}'\,\boldsymbol{B}^{-1}, \tag{10.4d}$$

depending only on the known parameters $\alpha_{i:n}$ and $\beta_{i,j:n}$ of the ordered reduced Log–WEIBULL variates $U_{i:n}$, contains two rows of weights

$$\left(\boldsymbol{P}'\,\boldsymbol{B}^{-1}\,\boldsymbol{P}\right)^{-1}\boldsymbol{P}'\,\boldsymbol{B}^{-1} = \begin{pmatrix} a_1 & a_2 & \cdots & a_n \\ b_1 & b_2 & \cdots & b_n \end{pmatrix}, \tag{10.4e}$$

which linearly combine the $X_{i:n}^*$ into the estimators

$$\widehat{a}^* = \sum_{i=1}^n a_i\,X_{i:n}^* \quad \text{with} \quad \sum_{i=1}^n a_i = 1, \tag{10.4f}$$

$$\widehat{b}^* = \sum_{i=1}^n b_i\,X_{i:n}^* \quad \text{with} \quad \sum_{i=1}^n b_i = 0. \tag{10.4g}$$

---

[3] For a thorough representation of regression theory, the reader is referred to a modern textbook in econometrics, e.g., GREENE (2003), JUDGE/GRIFFITHS/HILL/LEE (1980), RINNE (2004) or STOCK/WATSON (2003).

The **theoretical variance–covariance matrix** of the GLS estimators is

$$\text{Var}\left(\widehat{\boldsymbol{\theta}}\right) = b^{*\,2}\,\boldsymbol{\Sigma} = b^{*\,2}\left(\boldsymbol{P}'\,\boldsymbol{B}^{-1}\,\boldsymbol{P}\right)^{-1} = b^{*\,2}\begin{pmatrix} A & C \\ C & B \end{pmatrix}. \tag{10.4h}$$

$\boldsymbol{\Sigma} := \left(\boldsymbol{P}'\,\boldsymbol{B}^{-1}\,\boldsymbol{P}\right)^{-1}$ may be called the **basic variance–covariance matrix**. Its elements $A$, $B$ and $C$ will play a prominent role in the following text. From (10.4h) we get

$$\text{Var}\left(\widehat{a^*}\right) = b^{*\,2}\,A, \quad \text{Var}\left(\widehat{b^*}\right) = b^{*\,2}\,B \quad \text{and} \quad \text{Cov}\left(\widehat{a^*},\widehat{b^*}\right) = b^{*\,2}\,C. \tag{10.4i}$$

In applications we have to use the **estimated variance–covariance matrix** where $b^*$ is substituted by its estimator $\widehat{b^*}$:

$$\widehat{\text{Var}}\left(\widehat{\boldsymbol{\theta}}\right) = \widehat{b^*}^{\,2}\,\boldsymbol{\Sigma}. \tag{10.4j}$$

---

**Excursus:  OLS estimators of $a^*$ and $b^*$**

The OLS estimator of $\boldsymbol{\theta}$ is

$$\widehat{\boldsymbol{\theta}}_{OLS} = (\boldsymbol{P}'\,\boldsymbol{P})^{-1}\,\boldsymbol{P}'\,\boldsymbol{x}^*. \tag{10.5a}$$

This estimator is linear and unbiased, but it has the variance–covariance matrix

$$\text{Var}\left(\widehat{\boldsymbol{\theta}}_{OLS}\right) = b^{*\,2}\left(\boldsymbol{P}'\,\boldsymbol{P}\right)^{-1}. \tag{10.5b}$$

The difference matrix $\text{Var}\left(\widehat{\boldsymbol{\theta}}_{OLS}\right) - \text{Var}\left(\widehat{\boldsymbol{\theta}}\right)$ is positive–definite so that $\text{Var}\left(\widehat{a^*}_{OLS}\right) > \text{Var}\left(\widehat{a^*}\right)$ and $\text{Var}\left(\widehat{b^*}_{OLS}\right) > \text{Var}\left(\widehat{b^*}\right)$.

---

**Excursus:  Choice of some other plotting positions (regressors) $\widehat{u}_{i:n} \neq \alpha_{i:n}$**

If instead of choosing the regressors $\alpha_{i:n}$ in the design matrix $\boldsymbol{P}$ we would have taken some other regressors (= plotting positions) $\widehat{u}_i$ and applied the OLS technique, the means, variances and co-variance of $\widehat{a^*}\left(\widehat{u}\right)_{OLS}$ and $\widehat{b^*}\left(\widehat{u}\right)_{OLS}$, will be

$$\text{E}\left[\widehat{a^*}(\widehat{u})_{OLS}\right] \;=\; b^*\,\widetilde{\boldsymbol{a}}'\,\boldsymbol{\alpha} + \widehat{a^*}, \tag{10.6a}$$

$$\text{E}\left[\widehat{b^*}(\widehat{u})_{OLS}\right] \;=\; b^*\,\widetilde{\boldsymbol{b}}'\,\boldsymbol{\alpha}, \tag{10.6b}$$

$$\text{Var}\left[\widehat{a^*}(\widehat{u})_{OLS}\right] \;=\; b^{*\,2}\,\widetilde{\boldsymbol{a}}'\,\boldsymbol{B}\,\widetilde{\boldsymbol{a}}, \tag{10.6c}$$

$$\text{Var}\left[\widehat{b^*}(\widehat{u})_{OLS}\right] \;=\; b^{*\,2}\,\widetilde{\boldsymbol{b}}'\,\boldsymbol{B}\,\widetilde{\boldsymbol{b}}, \tag{10.6d}$$

$$\text{Cov}\left[\widehat{a^*}(\widehat{u})_{OLS},\widehat{b^*}(\widehat{u})_{OLS}\right] \;=\; b^{*\,2}\,\widetilde{\boldsymbol{a}}'\,\boldsymbol{B}\,\widetilde{\boldsymbol{b}}, \tag{10.6e}$$

with $\boldsymbol{\alpha}' = (\alpha_{1:n}, \ldots, \alpha_{n:n})$. The elements of the OLS–weighting vectors $\widetilde{\boldsymbol{a}}$ and $\widetilde{\boldsymbol{b}}$, which lead to $\widehat{a^*}(\widehat{u})_{OLS} = \widetilde{\boldsymbol{a}}'\,\widehat{\boldsymbol{u}}$ and $\widehat{b^*}(\widehat{u})_{OLS} = \widetilde{\boldsymbol{b}}'\,\widehat{\boldsymbol{u}}$, are given by

$$\widetilde{a}_i = \frac{1}{n} - \frac{\overline{\widehat{u}}\left(\widehat{u}_i - \overline{\widehat{u}}\right)}{\sum\left(\widehat{u}_i - \overline{\widehat{u}}\right)^2}, \quad \overline{\widehat{u}} = \frac{1}{n}\sum \widehat{u}_i \quad \text{and} \quad \widetilde{b}_i = \frac{\widehat{u}_i - \overline{\widehat{u}}}{\sum\left(\widehat{u}_i - \overline{\widehat{u}}\right)^2}.$$

Unbiasedness of any estimator of $a^*$ thus requires

$$\widetilde{\boldsymbol{a}}'\,\boldsymbol{\alpha} = \sum_{i=1}^{n}\widetilde{a}_i\,\alpha_{i:n} = 0, \tag{10.7a}$$

whereas unbiasedness of any estimator of $b^*$ demands

$$\widetilde{b}' \, \alpha = \sum_{i=1}^{n} \widetilde{b}_i \, \alpha_{i:n} = 1. \tag{10.7b}$$

(10.7a,b) are satisfied by the weights given in (10.4e). The bias $B(\cdot)$ and the mean squared error $MSE(\cdot)$ associated with $\widehat{a}^*(\widehat{u})_{OLS}$ and $\widehat{b}^*(\widehat{u})_{OLS}$ are

$$B\big[\widehat{a}^*(\widehat{u})_{OLS}\big] = b^* \, \widetilde{a}' \, \alpha, \tag{10.8a}$$

$$B\big[\widehat{b}^*(\widehat{u})_{OLS}\big] = b^* \, \big(\widetilde{b}' \, \alpha - 1\big), \tag{10.8b}$$

$$MSE\big[\widehat{a}^*(\widehat{u})_{OLS}\big] = b^{*\,2} \, \big[\big(\widetilde{a}' \, \alpha\big)^2 + \widetilde{a}' \, B \, \widetilde{a}\big], \tag{10.8c}$$

$$MSE\big[\widehat{b}^*(\widehat{u})_{OLS}\big] = b^{*\,2} \, \big[\big(\widetilde{b}' \, \alpha - 1\big)^2 + \widetilde{b}' \, B \, \widetilde{b}\big]. \tag{10.8d}$$

Tab. 10/1 — compiled according to GIBBONS/VANCE (1979) — shows biases, variances and mean squared errors for OLS estimators, based on several plotting positions, as well as those for BLUEs, BLIEs and BLOM's estimators[4] when $n = 6$. The BLUEs and BLIEs incorporate $\alpha' = \big(\alpha_{1:n}, \ldots, \alpha_{n:n}\big)$ and $B$. BLOM's estimators require $\alpha$ but use an approximation of $B$ and are thus easier to compute than BLUEs and BLIEs. The BLIEs minimize the MSE as expected by their design, see Sect. 10.3. BLOM's estimators provide a good approximation for the BLUEs. The latter two estimators are better than all OLS estimators independent of the choice of their regressor. Despite their biases the BLIEs have an overall error that is smaller than that of the unbiased BLUEs.

Table 10/1:  Biases, variances and MSEs of estimators for $a^*$ and $b^*$ with $n = 6$

| Plotting position and estimation convention | $b^* = 1/c$ | | | $a^* = \ln b$ | | |
|---|---|---|---|---|---|---|
| | Bias$/b^*$ | Variance$/b^{*\,2}$ | MSE$/b^{*\,2}$ | Bias$/b^*$ | Variance$/b^{*\,2}$ | MSE$/b^{*\,2}$ |
| $\widehat{u}_{i:n} = \dfrac{i}{n+1}$ ; OLS | 0.228 | 0.249 | 0.301 | 0.001 | 0.194 | 0.194 |
| $\widehat{u}_{i:n} = \dfrac{i-0.375}{n+0.25}$ ; OLS | 0.22 | 0.175 | 0.176 | 0.054 | 0.197 | 0.200 |
| $\widehat{u}_{i:n} = \dfrac{i-0.5}{n}$ ; OLS | $-0.055$ | 0.151 | 0.154 | 0.073 | 0.199 | 0.204 |
| $\widehat{u}_{i:n} = \dfrac{i-0.3}{n+0.4}$ ; OLS | 0.066 | 0.190 | 0.194 | 0.044 | 0.197 | 0.198 |
| $\widehat{u}_{i:n} = \alpha_{i:n}$ ; OLS | 0.0 | 0.169 | 0.169 | 0.0 | 0.195 | 0.195 |
| BLOM | 0.0 | 0.137 | 0.137 | 0.0 | 0.192 | 0.192 |
| BLUE | 0.0 | 0.132 | 0.132 | 0.0 | 0.192 | 0.192 |
| BLIE | $-0.117$ | 0.103 | 0.117 | $-0.028$ | 0.190 | 0.190 |

## 10.2    BLUEs for the Log-WEIBULL distribution

Having presented the theory of linear parameter estimation, we will now elaborate the practical and numerical details. In the preceding section we have developed the linear estimation theory for uncensored samples. As has been proven by NELSON/HAHN (1974), the BLU–property is still valid under single type–II censoring. Let $r$, $2 \leq r \leq n$ be the

---

[4]  BLIEs will be explained in Sect. 10.3 and BLOM's estimators in Sect. 10.4.

censoring number, $X_{i:n}^*$ the first $r$ ordered and complete log–data, $b^{*2}B$ their variance–covariance matrix ($B$ of type $r \times r$) and $\alpha' = (\alpha_{1:n}, \ldots, \alpha_{r:n})$ the means of the first $r$ reduced order statistics. Then model (10.3e) with $i$ running from 1 to $r$ fulfills the pre–conditions of the generalized GAUSS–MARKOV theorem. So for type-II censored samples with failure number $r$, the BLUEs of $a^*$ and $b^*$ are given by formulas analogue to (10.4 f,g) when the first $r$ sample data, which are uncensored and thus represent failure times, are used for estimation. For this reason Sect. 10.2.1 jointly presents the complete sample case and single type–II censored sample case.

BLUEs of $a^*$ and $b^*$ still exist when samples are progressively type–II censored but their determination demands for some modifications and amendments, see Sect. 10.2.2. When samples are type–I censored the number of failures is random so that the usage of the formulas developed for type–II censoring — what usually is done in practice — is not correct, but it usually provides a satisfactory approximation for practical purposes. Exact BLUEs of $a^*$ and $b^*$ with type–I censoring can be deduced from their BLIEs. This approach is postponed to Sect. 10.3.3.

### 10.2.1   Complete and singly type-II censored samples

The evaluation of (10.4f–j) requires a huge amount of computation, which is why the co–efficients $a_i$ and $b_i$ and the matrix $\Sigma = (P'\,B^{-1}\,P)^{-1}$ have been computed once and for all and presented in tables. LIEBLEIN/ZELEN (1956) gave the first such tables with $n = 2(1)6$ and $r = 2(1)n$ with a precision of seven decimal places. Extended tables with $n = 2(1)20$ and $r = 2(1)n$ with seven significant decimal places are to be found in WHITE (1964b). Coefficients $a_i$, $b_i$ and the elements $A$, $B$ and $C$ of $\Sigma$ for $21 \le n, r \le 25$ may be derived[5] from the tables of the BLIEs given in MANN (1967b) or MANN/SCHAFER/SINGPURWALLA (1974). A procedure for linear estimation when $n > 25$ is described farther down.

The first and crucial step to arrive at the vectors $a' = (a_1, \ldots, a_r)$ and $b' = (b_1, \ldots, b_r)$ and the matrix $\Sigma$ is the computation of the mean vector $\alpha$ and the variance–covariance matrix $B$ of the ordered reduced Log–WEIBULL statistics. The adjacent computation of $a$, $b$ and $\Sigma$ according to (10.4e) and (10.4h) is a matter of simple linear algebra.[6] The algorithms to arrive at $\alpha$ and $B$ have been described in Sect. 5.4, and resulting tables for $2 \le r, n \le 20$ are found in WHITE (1964b), the precision being eight decimal places.

---

**Example 10/1:   Computation of $a, b, \Sigma$ for $n = 6$ when the sample is complete or censored at $r = 4$**

Tab. 10/2 displays $\alpha$ and $B$ as taken from WHITE (1964b). For $n = r = 6$ (complete sample) we have to take the full $(6 \times 6)$–matrix $B$ and the full $(6 \times 1)$–vector $\alpha$ and arrive at

$$\Sigma = (P'\,B^{-1}\,P')^{-1} = \begin{pmatrix} A & C \\ C & B \end{pmatrix} = \begin{pmatrix} 0.1912 & -0.0314 \\ -0.0314 & 0.1320 \end{pmatrix}$$

---

[5]   The conversion procedures is described in Sect. 10.3.1.

[6]   LIEBLEIN/ZELEN (1956) have used a different approach.

and at the $(2 \times 6)$–matrix $C$ of coefficients $\alpha_i$ and $b_i$

$$C = \left(P' \, B^{-1} \, P\right)^{-1} P' \, B^{-1} = \begin{pmatrix} a_1 & a_2 & \dots & a_6 \\ b_1 & b_2 & \dots & b_6 \end{pmatrix}$$

$$= \begin{pmatrix} 0.0489 & 0.0835 & 0.1211 & 0.1656 & 0.2255 & 0.3554 \\ -0.1458 & -0.1495 & -0.1267 & -0.0732 & 0.0360 & 0.4593 \end{pmatrix}.$$

These results and the results below have been rounded to four decimal places. For the censored case $(r = 4)$ we only have to use the upper left part of Tab. 10/2 as $B_{(4 \times 4)}$ and the first four $\alpha_{i:6}$ as $\alpha$. We get

$$\Sigma = \begin{pmatrix} A & C \\ C & B \end{pmatrix} = \begin{pmatrix} 0.3237 & 0.1020 \\ 0.1020 & 0.2697 \end{pmatrix}$$

and

$$C = \begin{pmatrix} a_1 \, a_2 \, a_3 \, a_4 \\ b_1 \, b_2 \, b_3 \, b_4 \end{pmatrix} = \begin{pmatrix} -0.0865 & -0.0281 & 0.0650 & 1.0496 \\ -0.2859 & -0.2655 & -0.1859 & 0.7372 \end{pmatrix}.$$

As expected the basic variances $A$ and $B$ are smaller for $r = n = 6$ than for $r = 4$.

Table 10/2:　First and second moments of the ordered and reduced Log-WEIBULL statistics for $n = 6$

| $i$ \ $j$ | $\beta_{i,j:6}$ | | | | | | $\alpha_{i:6}$ |
|---|---|---|---|---|---|---|---|
| | 1 | 2 | 3 | 4 | 5 | 6 | |
| 1 | 1.6449493406 | 0.59985669 | 0.33204512 | 0.20925462 | 0.13619097 | 0.08285414 | −2.3689751 |
| 2 | – | 0.64769956 | 0.36145546 | 0.22887905 | 0.14945321 | 0.09116185 | −1.2750458 |
| 3 | – | – | 0.40185510 | 0.25616501 | 0.16806468 | 0.10291529 | −0.6627159 |
| 4 | – | – | – | 0.29761598 | 0.19670624 | 0.12171630 | −0.1883853 |
| 5 | – | – | – | – | 0.24854556 | 0.15496732 | 0.2545345 |
| 6 | – | – | – | – | – | 0.24658202 | 0.7772937 |

For a sample of size $n > 20$ the tables in WHITE (1964b) can be used as follows. Randomly divide the sample into $m$ smaller subsamples that can be handled with available tables. The subsamples should be as large as possible and nearly equal in size. Obtain the linear estimates from each subsample. Suppose the BLUEs are $\widehat{a_j^*}$ and $\widehat{b_j^*}$ $(j = 1, \dots, m)$ with $A_j$, $B_j$ and $C_j$ as corresponding elements of the basic variance–covariance matrices $\Sigma_j$. Then the unbiased and minimum variance pooled linear estimators are

$$\widehat{a^*} = \left( \sum_{j=1}^{m} \widehat{a_j^*} / A_j \right) \Big/ \sum_{j=1}^{m} A_j^{-1}, \tag{10.9a}$$

$$\widehat{b^*} = \left( \sum_{j=1}^{m} \widehat{b_j^*} / B_j \right) \Big/ \sum_{j=1}^{m} B_j^{-1}. \tag{10.9b}$$

The variances and covariance of these pooled estimators are

$$\text{Var}\left(\widehat{a^*}\right) = b^{*2} \sum_{j=1}^{m} A_j^{-1}, \tag{10.9c}$$

$$\text{Var}\left(\widehat{b^*}\right) = b^{*2} \sum_{j=1}^{m} B_j^{-1}, \tag{10.9d}$$

$$\text{Cov}\left(\widehat{a^*}, \widehat{b^*}\right) = b^{*2} \left( \sum_{j=1}^{m} A_j^{-1} \sum_{j=1}^{m} B_j^{-1} \right) \sum_{j=1}^{m} C_j \left(A_j B_j\right)^{-1}. \tag{10.9e}$$

These formulas are also used to pool BLUEs from independent samples from different populations with common parameter values.

The estimators $\widehat{a^*}$ and $\widehat{b^*}$ may easily be transformed to estimators for the original WEIBULL parameters $b$ and $c$:

$$\widehat{b} = \exp(\widehat{a^*}), \tag{10.10a}$$

$$\widehat{c} = 1/\widehat{b^*}, \tag{10.10b}$$

thereby losing the BLU–property. For instance, we have the following first order approximations:[7]

$$\text{E}\left(\widehat{b}\right) \approx b \left[1 + 0.5 \sqrt{\text{Var}\left(\widehat{a^*}\right)}\right]^2, \tag{10.11a}$$

$$\text{E}\left(\widehat{c}\right) \approx c \left[1 + c^2 \text{Var}\left(\widehat{b^*}\right)\right]. \tag{10.11b}$$

$\widehat{a^*}$ and $\widehat{b^*}$ as well as $\widehat{b}$ and $\widehat{c}$ are point estimators. When $r$ becomes large, the linear estimators $\widehat{a^*}$ and $\widehat{b^*}$ tend to be normally distributed. Thus, we can construct the following crude two–sided $100\left(1 - \alpha\right)\%$ confidence intervals, where $\tau_{1-\alpha/2}$ is the $100\left(1 - \alpha/2\right)\%$ percentile of the standard normal distribution:

1. $a^*$ and $b$

$$\underline{\underline{a}}^* \leq a^* \leq \overline{\overline{a}}^* \tag{10.12a}$$

The lower and upper confidence limits are

$$\underline{\underline{a}}^* \approx \widehat{a^*} - \tau_{1-\alpha/2}\, \widehat{b^*}\, \sqrt{A}, \qquad \overline{\overline{a}}^* \approx \widehat{a^*} + \tau_{1-\alpha/2}\, \widehat{b^*}\, \sqrt{A}. \tag{10.12b}$$

For the WEIBULL scale parameter we get

$$\exp\left(\underline{\underline{a}}^*\right) \lesssim b \lesssim \exp\left(\overline{\overline{a}}^*\right). \tag{10.12c}$$

---

[7] FEI et al. (1995) have developed an approximate unbiased estimator for the shape parameter $c$:

$$\widehat{c} = \frac{1 - B}{\widehat{b^*}} \quad \text{with } \text{Var}\left(\widehat{c}\right) = \frac{c^2}{B^{-1} - 2},$$

where $B$ is defined in (10.4h). $\widehat{c}$ has a smaller MSE than $\widehat{b^*}$.

2. $b^*$ and $c$

$$\underline{\underline{b}}^* \leq b^* \leq \overline{\overline{b}}^* \tag{10.13a}$$

The confidence limits based on the WILSON–HILFERTY $\chi^2$–approximation as suggested by MANN/SCHAFER/SINGPURWALLA (1974) are

$$\begin{aligned}
\underline{\underline{b}}^* &\approx \widehat{b^*} \Big/ \big[1 - (B/9) + \tau_{1-\alpha/2}\sqrt{B/9}\,\big]^3, \\
\overline{\overline{b}}^* &\approx \widehat{b^*} \Big/ \big[1 - (B/9) - \tau_{1-\alpha/2}\sqrt{B/9}\,\big]^3.
\end{aligned} \tag{10.13b}$$

The shape parameter $c$ thus has the crude confidence interval

$$\frac{1}{\overline{\overline{b}}^*} \leq c \leq \frac{1}{\underline{\underline{b}}^*}. \tag{10.13c}$$

The limits given above yield one–sided $100\,(1-\alpha)\%$ confidence intervals when $\tau_{1-\alpha/2}$ is replaced by $\tau_{1-\alpha}$. When the sample size is small, exact confidence intervals can be built upon those confidence limits that have been constructed when the estimators are BLIEs. The necessary coefficients have been found by Monte Carlo simulation (see MANN/FERTIG/SCHEUER (1971) for $3 \leq r,\ n \leq 25$ and MANN/FERTIG (1973) for $3 \leq r,\ n \leq 16$). This approach will be described in Sect. 10.3.2.

---

**Example 10/2: BLUEs of the Log–WEIBULL parameters for the complete and type-II censored dataset #1**

Complete dataset #1[8]

From WHITE (1964b) we take the following coefficients:

$a' = (0.0139614,\ 0.0174774,\ 0.0207043,\ 0.0238575,\ 0.0268727,\ 0.0301410,$
$\qquad 0.0330898,\ 0.0360067,\ 0.0397778,\ 0.0434551,\ 0.0464597,\ 0.0503302,\ 0.0545472,$
$\qquad 0.0590297,\ 0.0637215,\ 0.0693656,\ 0.0757046,\ 0.0837306,\ 0.0945666,\ 0.1172105).$

$b' = (-0.0351831,\ -0.0376927,\ -0.0389337,\ -0.0393871,\ -0.0392059,\ -0.0383703,$
$\qquad -0.0372104,\ -0.0352635,\ -0.0327620,\ -0.0291613,\ -0.0255917,\ -0.0205968,\ -0.0142128,$
$\qquad -\ 0.0067182,\ 0.0029339,\ 0.0154276,\ 0.0321637,\ 0.0559418,\ 0.0950718,\ 0.2287506).$

The estimates are

$$\begin{aligned}
\widehat{a}^* &= 4.6068, & \widehat{b} &= 100.1608, \\
\widehat{b}^* &= 0.4035, & \widehat{c} &= 2.4782,
\end{aligned}$$

which are close to the true values $b = 100$ and $c = 2.5$. The estimated covariance is $\widehat{\mathrm{Cov}}\big(\widehat{a}^*, \widehat{b}^*\big) = \widehat{b}^{*\,2}\, C = 0.4035^2\,(-0.011984) = -0.00195$, and the estimated variances are

$$\begin{aligned}
\widehat{\mathrm{Var}}\big(\widehat{a}^*\big) &= \widehat{b}^{*\,2}\, A = 0.4035^2 \cdot 0.055929 = 0.00907, \\
\widehat{\mathrm{Var}}\big(\widehat{b}^*\big) &= \widehat{b}^{*\,2}\, B = 0.4035^2 \cdot 0.033133 = 0.00540,
\end{aligned}$$

---

[8] The observations are to be found in Example 9/2.

giving the crude 90% confidence intervals $(\tau_{0.95} = 1.645)$ :

$$4.4498 \leq a^* \leq 4.7638, \qquad 0.3064 \leq b^* \leq 0.5600,$$
$$85.61 \leq b \leq 117.19, \qquad 1.7857 \leq c \leq 3.2637.$$

The confidence intervals for the WEIBULL parameters cover both the true values.

**Dataset #1 singly censored at $r = 15$**

From WHITE (1964b) we take the following coefficients:

$a' = (-0.0034703, -0.0002371, 0.0033313, 0.0078548, 0.0103638, 0.0183819, 0.0214801,$
     $0.0248986, 0.0377571, 0.0348805, 0.0510985, 0.0540101, 0.0657545, 0.0753868, 0.5985093),$

$b' = (-0.0577508, -0.0606651, -0.0614806, -0.0603470, -0.0603206, -0.0542982, -0.0525228,$
     $-0.0494348, -0.0370986, -0.0386460, -0.0214342, -0.0158828, -0.0005047,$
     $0.0138450, 0.5565350).$

The estimates are

$$\widehat{a^*} = 4.6177, \qquad \widehat{b} = 101.2609,$$
$$\widehat{b^*} = 0.3967, \qquad \widehat{c} = 2.5402$$

with estimated covariance $\widehat{\mathrm{Cov}}(\widehat{a^*}, \widehat{b^*}) = \widehat{b^*}^2 \, C = 0.3967^2 \cdot 0.004873 = 0.000666$ and estimated variances

$$\widehat{\mathrm{Var}}(\widehat{a^*}) = \widehat{b^*}^2 \, A \;=\; 0.3967^2 \cdot 0.069542 = 0.01078,$$
$$\widehat{\mathrm{Var}}(\widehat{b^*}) = \widehat{b^*}^2 \, B \;=\; 0.3967^2 \cdot 0.054939 = 0.00851.$$

The crude 90% confidence intervals result as:

$$4.4669 \leq a^* \leq 4.7885, \qquad 0.2784 \leq b^* \leq 0.6075,$$
$$85.36 \leq b \leq 120.12, \qquad 1.6462 \leq c \leq 3.5920.$$

## 10.2.2  Progressively type-II censored samples[9]

The general scheme of progressive type–II censoring has been described in Sect. 8.3.3. Here, we will assume the most popular scheme with the removal of a prespecified number of non–failed units $e_j$, $e_j \leq 0$, at the random time $T_j$ of the $j$–th failure $(j = 1, 2, .., k)$, $k \geq 2$. The life test ends with the $k$–th failure at a random time $T_k$ with the withdrawal of the last $e_k$ surviving units. We have

$$n = k + \sum_{j=1}^{k} e_j \tag{10.14}$$

and we will observe $k$ complete lifetimes

$$T_1 < T_2 < \ldots < T_k.$$

---

[9]  Suggested reading for this section: FEI/KONG/TANG (1995), MANN (1971), THOMAN/WILSON (1972).

One has to be careful in interpreting $T_j$. Of course, we have $T_1 = X_{1:n}$, but $T_2, T_3, \ldots, T_k$ are not equal to $X_{2:n}, X_{3:n}, \ldots, X_{k:n}$ of a singly censored sample of equal size, which is censored at the $k$–th failure. This fact has been overlooked in the paper of FEI et al. (1995). For example, $T_2$ may be any of the failures times $X_{2:n}, \ldots, X_{2+e_1:n}$ of the $n$ sampled units. The following approach, which is based on THOMAN/WILSON (1972), derives the first two moments of the reduced order statistics

$$Y_j = \frac{T_j^* - a^*}{b^*}; \quad j = 1, \ldots, k \text{ and } T_j^* = \ln T_j, \tag{10.15a}$$

designated by

$$\mathrm{E}(\boldsymbol{y}) = \boldsymbol{\alpha} = (\alpha_1, \ldots, \alpha_k)' \text{ and } \boldsymbol{y}' = (Y_1, \ldots, Y_k), \tag{10.15b}$$

$$\mathrm{Var}(\boldsymbol{y}) = \boldsymbol{B} = (\beta_{i,j}); \quad i, j = 1, \ldots, k, \tag{10.15c}$$

and derived from the corresponding moments of the reduced order statistics in complete samples. Instead of (10.4e) we now have

$$(\boldsymbol{P}' \boldsymbol{B}^{-1} \boldsymbol{P})^{-1} \boldsymbol{P}' \boldsymbol{B}^{-1} = \begin{pmatrix} \alpha_1, \ldots, \alpha_k \\ \beta_1, \ldots, \beta_k \end{pmatrix}, \tag{10.15d}$$

and the estimators are

$$\widehat{a^*} = \sum_{j=1}^{k} \alpha_j \, T_j^*, \quad \widehat{b^*} = \sum_{j=1}^{k} \beta_j \, T_j^*, \tag{10.15e}$$

with variance–covariance matrix

$$\mathrm{Var}\begin{pmatrix} \widehat{a^*} \\ \widehat{b^*} \end{pmatrix} = b^{*2} \, (\boldsymbol{P}' \boldsymbol{B}^{-1} \boldsymbol{P})^{-1}. \tag{10.15f}$$

---

**Excursus: Derivation of the moments of the reduced order statistics in progressively type–II censored samples**

Consider the reduced order statistics $Y_1 < \ldots < Y_k$ from a type–II progressive censoring pattern $e_1, \ldots, e_k$. Let $U_{1:n} < \ldots < U_{n:n}$ represent the reduced order statistics from the complete sample, and $R_j$ the random rank of $Y_j$ in the complete sample; i.e., $Y_j = U_{R_j:n}$ for $j = 1, \ldots, k$. $r_j$ is a realization of $R_j$ in the random rank vector $(R_1, \ldots, R_k)$. $r_j$ can be described recursively for $j = 2, \ldots, k$ with $r_1 = 1$ by

$$r_j = \begin{cases} r_{j-1} + 1, \\ r_{j-1} + 2, \\ \vdots \\ j + r_1 + \ldots + r_{j-1}, \end{cases}, \quad j = 2, \ldots, k. \tag{10.16a}$$

The probability of a realization $(r_1, \ldots, r_k)$ is

---

[9] MANN (1971) has chosen a different approach.

$$P_\ell = \Pr(R_1 = r_1, \ldots, R_k = r_k) = \Pr(R_1 = r_1) \prod_{j=2}^{k} \Pr(R_j = r_j \mid R_1 = r_1, \ldots, R_{j-1} = r_{j-1}),$$

(10.16b)

with $\Pr(R_1 = 1) = 1$ and $\ell = 1, \ldots, m$; i.e., there are $m$ realizations of $(R_1, \ldots, R_k)$. For the ease of notation we have suppressed the second index $\ell$ in $r_{j\ell}$. THOMAN/WILSON (1972, p. 689) give the following solution for the conditional probabilities in (10.16 b):

$$\Pr\left(R_j = r_j \mid R_1 = 1, R_2 = r_2, \ldots, R_{j-1} = r_{j-1}\right) = c_j \frac{n - \sum_{i=1}^{j-1}(e_i + 1)}{n - r_j + 1},$$

(10.16c)

where

$$c_j = \begin{cases} 1 & \text{for} \quad r_j = r_{j-1} + 1 \\ \displaystyle\prod_{\nu=1}^{r_j - r_{j-1} - 1} \frac{\sum_{i=1}^{j-1}(e_i + 1) - r_{j-1} - \nu + 1}{n - r_{j-1} - \nu + 1} & \text{for} \quad r_j \geq r_{j-1} + 2. \end{cases}$$

(10.16d)

We define the following matrices $\boldsymbol{D}_\ell$ ($\ell = 1, \ldots, m$) of indicators $d_{\nu\kappa}^\ell$

$$d_{\nu\kappa}^\ell = \begin{cases} 1 & \text{for} \quad \kappa = r_\nu \\ 0 & \text{for} \quad \kappa \neq r_\nu \end{cases}; \quad \nu = 1, \ldots, k; \; \kappa = 1, \ldots, n,$$

(10.17a)

where $r_\nu$ are elements of the $\ell$–th realization of $(R_1, \ldots, R_k)$. Then, by writing $\boldsymbol{y} = \boldsymbol{D}_\ell \boldsymbol{u}$, $\boldsymbol{u}' = (U_{1:n}, \ldots, U_{n:n})$ and using $\boldsymbol{\alpha} = \mathrm{E}(\boldsymbol{u})$ and $\mathrm{Var}(\boldsymbol{u}) = \boldsymbol{B}$, we arrive at

$$\begin{aligned} \boldsymbol{\phi} = \mathrm{E}(\boldsymbol{y}) &= \mathrm{E}\,\mathrm{E}(\boldsymbol{y} \mid \boldsymbol{D}_\ell) \\ &= \mathrm{E}(\boldsymbol{D}_\ell \boldsymbol{\alpha}) \\ &= \left(\sum_{\ell=1}^{m} \boldsymbol{D}_\ell P_\ell\right)\boldsymbol{\alpha}, \end{aligned}$$

(10.17b)

$$\begin{aligned} \boldsymbol{B} &= \mathrm{E}(\boldsymbol{y}\,\boldsymbol{y}') - \boldsymbol{\phi}\,\boldsymbol{\phi}' \\ &= \mathrm{E}\,\mathrm{E}(\boldsymbol{D}_\ell \boldsymbol{u}\,\boldsymbol{u}'\,\boldsymbol{D}_\ell' \mid \boldsymbol{D}_\ell) - \boldsymbol{\phi}\,\boldsymbol{\phi}' \\ &= \mathrm{E}\left[\boldsymbol{D}_\ell(\boldsymbol{B} + \boldsymbol{\alpha}\,\boldsymbol{\alpha}')\boldsymbol{D}_\ell\right] - \boldsymbol{\phi}\,\boldsymbol{\phi}' \\ &= \sum_{\ell=1}^{m} \boldsymbol{D}_\ell(\boldsymbol{B} + \boldsymbol{\alpha}\,\boldsymbol{\alpha}')\boldsymbol{D}_\ell' P_\ell - \boldsymbol{\phi}\,\boldsymbol{\phi}'. \end{aligned}$$

(10.17c)

Evaluation of the formulas given in the preceding excursus is tedious[10] but can be accomplished by a suitable computer program. The following Example 10/3 shows how to proceed in a rather simple case.

**Example 10/3:  Computing the weights for the BLUEs of $a^*$ and $b^*$ and their variance–covariance matrix for $n = 6$ and the type–II censoring pattern $e_1 = e_2 = e_3 = 1$**

---

[10]  That is the reason THOMAN/WILSON (1972) have examined various and simpler approximations.

A sample of size $n = 6$ and a censoring scheme $e_1 = e_2 = e_3 = 1$ lead to $m = 5$ different realizations of the random vector $(R_1, R_2, R_3)$ which are listed together with their probabilities $P_\ell$ and indicator matrices $D_\ell$ in Table 10/3. The vector $\alpha$ and the matrix $B$ for the complete sample of $n = 6$ are taken from Table 10/2.

From (10.17b,c) we get

$$\alpha = (-2.3689751, -1.1525798, -0.28953358)',$$

$$B = \begin{pmatrix} 1.6449341 & 0.5462944 & 0.2437581 \\ - & 0.6585223 & 0.3013714 \\ - & - & 0.4484980 \end{pmatrix}.$$

If we had taken a singly censored sample of size $n = 6$, censored at $r = 3$, the corresponding vector and matrix would be

$$\alpha = (-2.3689751, -1.2750458, -0.6627159)',$$

$$B = \begin{pmatrix} 1.64499341 & 0.59985669 & 0.20925462 \\ - & 0.64769956 & 0.36145546 \\ - & - & 0.40185510 \end{pmatrix}.$$

The coefficients according to (10.15d) are

$$\begin{pmatrix} -0.171773 & 0.078395 & 1.093378 \\ -0.374725 & -0.255816 & 0.630541 \end{pmatrix},$$

and the variance–covariance matrix (10.15f) is

$$b^{*2} \begin{pmatrix} 0.534141 & 0.214164 \\ - & 0.344712 \end{pmatrix}.$$

A sample of $n = 6$ and singly censored at $r = 3$ would have the following coefficients to generate $\widehat{a^*}$ and $\widehat{b^*}$ :

$$\begin{pmatrix} -0.315397 & -0.203432 & 1.518829 \\ -0.446602 & -0.388649 & 0.835251 \end{pmatrix}$$

with corresponding variance–covariance matrix

$$b^{*2} \begin{pmatrix} 0.652941 & 0.333249 \\ - & 0.432116 \end{pmatrix}.$$

Thus, the single type–II censoring leads to greater variances of the estimators than progressive type–II censoring with the same $n$ and $k = r$.

Table 10/3:   Rank vectors, their probabilities and indicator matrices

| $\ell$ | $(r_1, r_2, r_3)_\ell$ | $P_\ell$ | $\boldsymbol{D}_\ell$ |
|:---:|:---:|:---:|:---:|
| 1 | $(1, 2, 3)$ | $6/15$ | $\begin{pmatrix} 100000 \\ 010000 \\ 001000 \end{pmatrix}$ |
| 2 | $(1, 2, 4)$ | $4/15$ | $\begin{pmatrix} 100000 \\ 010000 \\ 000100 \end{pmatrix}$ |
| 3 | $(1, 2, 5)$ | $2/15$ | $\begin{pmatrix} 100000 \\ 010000 \\ 000010 \end{pmatrix}$ |
| 4 | $(1, 3, 4)$ | $2/15$ | $\begin{pmatrix} 100000 \\ 001000 \\ 000100 \end{pmatrix}$ |
| 5 | $(1, 3, 5)$ | $1/15$ | $\begin{pmatrix} 100000 \\ 001000 \\ 000010 \end{pmatrix}$ |

## 10.3   BLIEs for Log-WEIBULL parameters

The subclass of the class of linear estimators for $a^*$ and $b^*$ considered so far, took into account only the unbiased linear estimators and searched for the minimum–variance (= best) estimator within this subclass. Sometimes it is appropriate to abandon unbiasedness, especially when there exist estimators within another subclass having a smaller overall–error than the minimum variance of the unbiased estimator. The overall–error is measured by the mean squared error (MSE) being the sum of the variance, giving the effect of random sampling errors, and the squared bias, giving the effect of systematic errors, which originate in the sampling process and/or in the processing of the sampled data. NANCY R. MANN (1972) has developed BLIEs as an alternative to the BLUEs.

### 10.3.1   BLUE versus BLIE[11]

Consider a variate whose distribution is a member of the location–scale family. A special case is $X^*$, the Log–WEIBULL variate. The distribution of the reduced variate

---

[11]   Suggested reading for this section: MANN (1965; 1967a; 1968a,b; 1969a; 1970b), MANN/SCHAFER/ SINGPURWALLA (1974).

$U = (X^* - a^*)/b^*$ is independent of the location parameters $a^*$ and the scale parameter $b^*$. For estimating the general parametric function

$$\psi = \ell_1 a^* + \ell_2 b^*, \tag{10.18}$$

the GAUSS–MARKOV theorem specifies the GLS estimator (see LLOYD (1952)), as the unique best among unbiased linear functions of $X^*_{1:n}, \ldots, X^*_{r:n}$ for all $n$ and all $r$, $2 \leq r \leq n$. The function $\psi$ in (10.18) includes $\psi = a^*$ and $\psi = b^*$ as special cases. Another special case is

$$x^*_P = a^* + u_P b^*, \ \ 0 < P < 1, \tag{10.19a}$$

the $100P\%$ percentile of $X^*$ where

$$u_P = \ln[-\ln(1 - P)] \tag{10.19b}$$

is the $100P\%$ percentile of the reduced Log–WEIBULL variate.

Let $\widehat{\psi}$, $\widehat{a^*}$ and $\widehat{b^*}$ be the BLUEs of $\psi$, $a^*$ and $b^*$, respectively, based on the first $r$ of $n$ ordered observations. It is well known that $\widehat{\psi}$ is equal to $\ell_1 \widehat{a^*} + \ell_2 \widehat{b^*}$ and that these estimators $\widehat{\psi}$, $\widehat{a^*}$ and $\widehat{b^*}$ enjoy all the large sample properties attributed to maximum–likelihood estimators, including asymptotic normality (see Sect. 11.2.1).

Suppose that $\widehat{a^*}$ with variance $b^{*2} A$ and $\widehat{b^*}^{2}$ with variance $b^{*2} B$ are the joint unique (with probability 1) uniformly minimum unbiased linear estimators of $a^*$ and $b^*$, respectively, and that $b^{*2} C$ is the covariance of $\widehat{a^*}$ and $\widehat{b^*}$; for $A$, $B$ and $C$, see (10.4h). MANN (1967a) stated and proved the following.

**Theorem:** In the class of linear estimators of $\psi$ based on the first $r$ of $n$ Log–WEIBULL order statistics and with mean squared–error loss independent of $a^*$, there is a unique best one given by

$$\widehat{\widehat{\psi}} = \ell_1 \left( \widehat{a^*} - \frac{C}{1 + B} \widehat{b^*} \right) + \ell_2 \frac{\widehat{b^*}}{1 + B} \tag{10.20a}$$

with mean squared error

$$
\begin{aligned}
\mathrm{MSE}\left( \widehat{\widehat{\psi}} \right) &= \mathrm{E}\left[ \left( \widehat{\widehat{\psi}} - \psi \right)^2 \right] \\
&= b^{*2} \left\{ \ell_1^2 A + 2\ell_1 \ell_2 C + \ell_2^2 B - \frac{(\ell_1 C + \ell_2 B)^2}{1 + B} \right\} \tag{10.20b} \\
&= b^{*2} \left\{ \ell_1^2 \left[ A - \frac{C^2}{1 + B} \right] + 2\ell_1 \ell_2 \frac{C}{1 + B} + \ell_2^2 \frac{B}{1 + B} \right\}
\end{aligned}
$$

for all $\ell_1$ and $\ell_2$. ∎

Let loss be defined as squared error divided by $b^{*2}$. Then $\widehat{\widehat{\psi}}$ is the best among linear estimators of $\psi$ invariant under location and scalar transformations, the best linear invariant estimator (BLIE). $\widehat{\widehat{\psi}}$ also is the unique admissible minimax linear estimator of $\psi$ based on $X^*_{1:n}, \ldots, X^*_{r:n}$. It also has all the asymptotic properties of the BLUE plus that of asymptotic unbiasedness.

From (10.20a,b) we extract the following BLIEs for $a^*$ and $b^*$ and their corresponding MSEs:

$$\widehat{\widehat{a^*}} \;=\; \widehat{a^*} - \frac{C}{1+B}\,\widehat{b^*}, \tag{10.21a}$$

$$\mathrm{MSE}\left(\widehat{\widehat{a^*}}\right) \;=\; \mathrm{E}\left[\left(\widehat{\widehat{a^*}} - a^*\right)^2\right]$$

$$=\; b^{*\,2}\left(A - \frac{C^2}{1+B}\right), \tag{10.21b}$$

$$\widehat{\widehat{b^*}} \;=\; \frac{\widehat{b^*}}{1+B}, \tag{10.22a}$$

$$\mathrm{MSE}\left(\widehat{\widehat{b^*}}\right) \;=\; \mathrm{E}\left[\left(\widehat{\widehat{b^*}} - b^*\right)^2\right],$$

$$=\; b^{*\,2}\,\frac{B}{1+B}. \tag{10.22b}$$

The expected cross–product of the estimation errors is

$$\mathrm{E}\left[\left(\widehat{\widehat{a^*}} - a^*\right)\left(\widehat{\widehat{b^*}} - b^*\right)\right] = b^{*\,2}\,\frac{C}{1+B}. \tag{10.23}$$

From (10.21a) and (10.22a) it is evident that coefficients $A_i$ and $B_i$ in

$$\widehat{\widehat{a^*}} \;=\; \sum_{i=1}^{r} A_i\, X_{i:n}, \;\; 2 \le r \le n, \tag{10.24a}$$

$$\widehat{\widehat{b^*}} \;=\; \sum_{i=1}^{r} B_i\, X_{i:n}, \;\; 2 \le r \le n, \tag{10.24b}$$

are related to the BLU–coefficients $a_i,\; b_i$ given by (10.4e), as follows

$$A_i \;=\; a_i - \frac{C}{1+B}\,b_i, \tag{10.25a}$$

$$B_i \;=\; \frac{b_i}{1+B}. \tag{10.25b}$$

These coefficients together with the basic MSE–terms in (10.21b), (10.22b) and (10.23)

$$\widetilde{A} \;:=\; A - \frac{C^2}{1+B}, \tag{10.25c}$$

$$\widetilde{B} \;:=\; \frac{1}{1+B}, \tag{10.25d}$$

$$\widetilde{C} \;:=\; \frac{C}{1+B} \tag{10.25e}$$

have been tabulated for $2 \le r, n \le 20$ by MANN (1967b) and are reprinted in DODSON (1994). An abridged table for $2 \le r, n \le 15$ is contained in MANN (1967a).

### 10.3.2 Type-II censored samples

For practical purposes the BLUEs and BLIEs are almost the same unless the observed number $r$ of order statistics is small. Moreover, for large $r$ and $n$ BLUEs and BLIEs and their mean squared errors are asymptotically equal. The choice of either estimator is mostly a matter of taste and tables available. Thus, there is no compelling reason to choose either minimum squared error or unbiasedness as essential for a good estimator.

In this section we will first give a tabular overview showing the conversion of coefficients, estimators and mean squared errors from the BLU approach to the BLI approach and vice versa (see Table 10/4). Then we will extend the aforementioned Example 10/2 to BLI estimation. Finally, we will comment upon exact confidence intervals for BLIEs.

**Example 10/4: BLIEs of the Log–WEIBULL parameters for the complete and type-II censored dataset #1**

We apply the conversion formulas from Table 10/4 to the results of BLU estimation given in Example 10/2.

Complete dataset #1

We have $\widehat{a^*} = 4.6068$ with $A = 0.055929$, $\widehat{b^*} = 0.4035$ with $B = 0.033153$ and $C = -0.011984$ from which we get the following

$$\widehat{\widetilde{b^*}} = \frac{\widehat{b^*}}{1+B} = \frac{0.4035}{1+0.033153} = 0.3906 \implies \widehat{\widetilde{c}} = \frac{1}{\widehat{\widetilde{b^*}}} = 2.5605,$$

$$\widehat{\widetilde{a^*}} = \widehat{a^*} - \frac{C}{1+B}\,\widehat{b^*} = 4.6068 + \frac{0.011984}{1+0.033153} = 4.6184 \implies \widehat{\widetilde{b}} = \exp\left(\widehat{\widetilde{a^*}}\right) = 101.3318,$$

$$\widehat{\mathrm{MSE}}\left(\widehat{\widetilde{b^*}}\right) = \widehat{\widetilde{b^*}}^2\,\widetilde{B} = \widehat{\widetilde{b^*}}^2\,\frac{B}{1+B} = 0.3906^2\,\frac{0.033153}{1+0.033153} = 0.00504,$$

$$\widehat{\mathrm{MSE}}\left(\widehat{\widetilde{a^*}}\right) = \widehat{\widetilde{b^*}}^2\,\widetilde{A} = \widehat{\widetilde{b^*}}^2\left(A - \frac{C^2}{1+B}\right) = 0.3906^2\left(0.055929 - \frac{(-0.011984)^2}{1+0.033153}\right) = 0.00851.$$

Comparing the estimation errors of the BLUEs and BLIEs, we have, as is to be expected,

$$\widehat{\mathrm{MSE}}\left(\widehat{\widetilde{a^*}}\right) = 0.00851 < \widehat{\mathrm{Var}}\left(\widehat{a^*}\right) = 0.00907,$$

$$\widehat{\mathrm{MSE}}\left(\widehat{\widetilde{b^*}}\right) = 0.00504 < \widehat{\mathrm{Var}}\left(\widehat{b^*}\right) = 0.00540.$$

Dataset #1 singly censored at $r = 15$

Starting with $\widehat{a^*} = 4.6177$, $A = 0.069542$, $\widehat{b^*} = 0.3967$, $B = 0.054939$ and $C = 0.004873$ from Example 10/2, we get the following results for the BLI approach:

$$\widehat{\widetilde{b^*}} = \frac{0.3967}{1.054939} = 0.3760 \implies \widehat{\widetilde{c}} = \frac{1}{0.3760} = 2.6593,$$

$$\widehat{\widetilde{a^*}} = 4.6177 - \frac{0.004873}{1.054939}\,0.3967 = 4.6159 \implies \widehat{\widetilde{b}} = \exp\left(\widehat{\widetilde{a^*}}\right) = 101.0788,$$

$$\widehat{\mathrm{MSE}}\left(\widehat{\widetilde{b^*}}\right) = 0.3760^2\,\frac{0.054939}{1.054939} = 0.00736,$$

$$\widehat{\mathrm{MSE}}\left(\widehat{\widetilde{a^*}}\right) = 0.3760^2\left(0.069542 - \frac{0.004873^2}{1.054939}\right) = 0.00983.$$

The estimation errors behave as is to be expected:

$$\widehat{\text{MSE}}\left(\widehat{\widehat{a^*}}\right) = 0.00983 < \widehat{\text{Var}}(\widehat{a^*}) = 0.01078,$$

$$\widehat{\text{MSE}}\left(\widehat{\widehat{b^*}}\right) = 0.00736 < \widehat{\text{Var}}\left(\widehat{b^*}\right) = 0.00851.$$

MANN/FERTIG(1973) have found by Monte Carlo simulation the percentiles of the following BLI–based variates:

$$W \quad := \quad \frac{\widehat{\widehat{b^*}}}{b^*}, \tag{10.26a}$$

$$T \quad := \quad \frac{\widehat{\widehat{a^*}} - a^*}{\widehat{\widehat{b^*}}} \tag{10.26b}$$

of order $P$ $(P = 0.02, \ 0.05, \ 0.10, \ 0.25, \ 0.50, \ 0.60, \ 0.75, \ 0.90, \ 0.95, \ 0.98)$ and for $3 \leq r, n \leq 16.$[12] let

$$w(P, n, r) \ \text{ and } \ t(P, n, r)$$

denote the percentiles of order $P$ for a sample of size $n$ singly censored at $r$, $r \leq n$. The exact two–sided $100\,(1 - \alpha)\%$ confidence intervals in terms of the BLIEs are

$$\frac{\widehat{\widehat{b^*}}}{w(1 - \alpha/2, n, r)} \ \leq \ b^* \ \leq \ \frac{\widehat{\widehat{b^*}}}{w(\alpha/2, n, r)}, \tag{10.27a}$$

$$\widehat{\widehat{a^*}} - \widehat{\widehat{b^*}}\, t(1 - \alpha/2, n, r) \ \leq \ a^* \ \leq \ \widehat{\widehat{a^*}} - \widehat{\widehat{b^*}}\, t(\alpha/2, n, r). \tag{10.27b}$$

Applying the conversion formulas of Table 10/4, we get these confidence intervals in terms of the BLUEs:

$$\frac{\widehat{b^*}}{(1 + B)\, w(1 - \alpha/2, n, r)} \ \leq \ b^* \ \leq \ \frac{\widehat{b^*}}{(1 + B)\, w(\alpha/2, n, r)}, \tag{10.28a}$$

$$\widehat{a^*} - \frac{\widehat{b^*}}{1 + B}\left[C + t(1 - \alpha/2, n, r)\right] \ \leq \ a^* \ \leq \ \widehat{a^*} - \frac{\widehat{b^*}}{1 + B}\left[C + t(\alpha/2, n, r)\right] \tag{10.28b}$$

---

[12]  In MANN/FERTIG/SCHEUER (1971) the tables are extended: $3 \leq r, n \leq 25$

Table 10/4: Relationships between the coefficients, MSEs and parameter estimators in the BLU and BLI approaches

| Item[*)] | Symbols for: | | From BLU to BLI | From BLI to BLU |
|---|---|---|---|---|
| | BLU | BLI | | |
| Coefficients to estimate | | | | |
| $a^*$ | $a_i$ | $A_i$ | $A_i = a_i - \dfrac{C}{1+B}\,b_i$ | $a_i = A_i + \dfrac{\widetilde{C}}{1-\widetilde{B}}\,B_i$ |
| $b^*$ | $b_i$ | $B_i$ | $B_i = \dfrac{b_i}{1+B}$ | $b_i = \dfrac{B_i}{1-\widetilde{B}}$ |
| $\left(\text{MSE}/b^{*\,2}\right)$ of the estimator for | | | | |
| $a^*$ | $A$ | $\widetilde{A}$ | $\widetilde{A} = A - \dfrac{C^2}{1+B}$ | $A = \widetilde{A} + \dfrac{\widetilde{C^2}}{1-\widetilde{B}}$ |
| $b^*$ | $B$ | $\widetilde{B}$ | $\widetilde{B} = \dfrac{B}{1+B}$ | $B = \dfrac{\widetilde{B}}{1-\widetilde{B}}$ |
| $\left(\text{Covariance}/b^{*\,2}\right)$ of the estimators for $a^*$ and $b^*$ | $C$ | $\widetilde{C}$ | $\widetilde{C} = \dfrac{C}{1+B}$ | $C = \dfrac{\widetilde{C}}{1-\widetilde{B}}$ |
| Estimators of | | | | |
| $a^*$ | $\widehat{a^*}$ | $\widehat{\widehat{a^*}}$ | $\widehat{\widehat{a^*}} = \widehat{a^*} - \dfrac{C}{1+B}\,\widehat{b^*}$ | $\widehat{a^*} = \widehat{\widehat{a^*}} + \dfrac{\widetilde{C}}{1-\widetilde{B}}\,\widehat{\widehat{b^*}}$ |
| $b^*$ | $\widehat{b^*}$ | $\widehat{\widehat{b^*}}$ | $\widehat{\widehat{b^*}} = \dfrac{\widehat{b^*}}{1-B}$ | $\widehat{b^*} = \dfrac{\widehat{\widehat{b^*}}}{1-\widetilde{B}}$ |

$*)$ In the case of BLU the variance and the MSE of an estimator are identical.

Denoting the lower (upper) confidence limit for $b^*$ by $\underline{b}^*$ $\left(\overline{\overline{b}}^*\right)$ and those for $a^*$ by $\underline{a}^*$ $\left(\overline{\overline{a}}^*\right)$, we find the following $100\,(1-\alpha)\%$ confidence intervals for the original WEIBULL parameters $b$ and $c$:

$$\exp\left(\underline{a}^*\right) \ \le \ b \ \le \ \exp\left(\overline{\overline{a}}^*\right)\,, \tag{10.29a}$$

$$\frac{1}{\overline{\overline{b}}^*} \ \le \ c \ \le \ \frac{1}{\underline{b}^*}\,. \tag{10.29b}$$

**Example 10/5: Exact confidence intervals for $b^*$ and $c$ using the complete dataset #1**

We set $1-\alpha = 0.90$. In terms of the BLIE in Example 10/4 and using (10.27a), we get

$$\frac{0.3906}{1.27} \ \le \ b^* \ \le \ \frac{0.3906}{0.70}$$

$$0.3076 \ \le \ b^* \ \le \ 0.5580$$

and for the WEIBULL shape parameter

$$\frac{1}{0.5580} \le c \le \frac{1}{0.3076}$$

$$1.7921 \le c \le 3.2510.$$

In terms of the BLUE in Example 10/2 and using (10.28a) we have

$$\frac{0.4035}{1.033153 \cdot 1.27} \le b^* \le \frac{0.4035}{1.033153 \cdot 0.70}$$

$$0.3075 \le b^* \le 0.5579$$

and for the WEIBULL shape parameter

$$\frac{1}{0.5579} \le c \le \frac{1}{0.3075}$$

$$1.7924 \le c \le 3.2520.$$

Compared with the approximate confidence intervals $0.3064 \le b^* \le 0.5600$ and $1.7857 \le c \le 3.2638$ the latter tends to be too wide.

### 10.3.3  Type-I censored samples

In a paper, which is not easy to read, MANN (1972b) showed how to construct BLIEs of the Log–WEIBULL parameters $a^*$ and $b^*$ when a sample of size $n$ is censored at a fixed and given time $T$. The BLIEs thus obtained may be converted to BLUEs with the formulas in Tab. 10/4.

Let $m$ be the realized number of failures within $(0, T]$, i.e., $X_{m:n} \le T$, and $X^*_{i:n} = \ln X_{i:n}$; $i = 1, \ldots, m$; and $T^* = \ln T$. The $m$ complete lifetimes are regarded as an ordered sample of size $m$ and using the BLI–coefficients $A_i$ and $B_i$ pertaining to $n = m$ the interim estimators

$$\widetilde{a}_m = \sum_{i=1}^{m} A_i \, X^*_{i:n} \quad \text{and} \quad \widetilde{b}_m = \sum_{i=1}^{m} B_i \, X^*_{i:n} \tag{10.30a}$$

are calculated. These estimators have to be adjusted by applying corrective coefficients $a(m, \lambda_m)$ and $b(m, \lambda_m)$ tabulated in MANN (1972b). These coefficients depend on $m$ and on

$$\lambda_m = \frac{\widetilde{b}_m}{T^* - \widetilde{a}_m}. \tag{10.30b}$$

The final BLIEs for type–I censored samples are

$$\widehat{\widetilde{a}}^*_I \;=\; \widetilde{a}_m - a(m, \lambda_m)\, \widetilde{b}_m, \tag{10.30c}$$

$$\widehat{\widetilde{b}}^*_I \;=\; b(m, \lambda_m)\, \widetilde{b}_m. \tag{10.30d}$$

# 10.4  Approximations to BLUEs and BLIEs

Many methods have been advanced for approximating the BLUEs and BLIEs of the Log–WEIBULL parameters $a^*$ and $b^*$. This is due in part to the fact that the weights for obtaining the estimators are available for only sample sizes and censoring numbers up to 25. Furthermore, the computation of the weights suffers from the extensive accumulation of rounding errors which result from the calculation of the extremely cumbersome functions involved in the determination of the order–statistics moments, especially of the covariances (see Section 5.4). There are several methods that have been derived for obtaining good approximations to the GLS–weights. For example, WHITE (1965) proposed **WLS–estimators** (weighted least squares) that depend on the means and variances, but not on the covariances of the order statistics. MCCOOL (1965) suggested estimators which are functions of weights for the BLUEs, available for an uncensored sample of size smaller than the sample size of interest. The weights of the proposed estimators are obtained as linear combinations of the latter ones with the aid of values of the hypergeometric probability function. In the following text we will present approximations which rest upon asymptotics (Sections 10.4.5 and 10.5.2) and some other methods (Sections 10.4.1 through 10.4.4). Approximations which rely upon a few optimally chosen order statistics will be presented in Sect. 10.5.

## 10.4.1  Least squares with various functions of the variable

The following approach is due to BAIN/ANTLE (1967) and applies to complete as well as censored samples. Suppose $f(x \,|\, \boldsymbol{\theta})$ denotes a density function with unknown parameters in the vector $\boldsymbol{\theta}$ and $X_1, \ldots, X_n$ denote a random sample of size $n$ from $f(x \,|\, \boldsymbol{\theta})$. Let $u(x, \boldsymbol{\theta})$ be some function of the variable and the parameters such that

1. the distribution of $u$ is independent of $\boldsymbol{\theta}$ and
2. the ordering of the $u(X_i, \boldsymbol{\theta})$ can be determined from the ordering of the $X_i$, so that $U_{i:n}$ denotes the $i$–th largest value of $U(X, \boldsymbol{\theta})$ for a given sample.

An estimate of $\boldsymbol{\theta}$ is the value $\widehat{\boldsymbol{\theta}}$ which maximizes — according to some criterion — the agreement between $U_{i:n}$ and $\mathrm{E}(U_{i:n})$. Clearly, the choice of the function $u(X, \boldsymbol{\theta})$ and of the criterion is rather arbitrary. One possible procedure is to choose a function which would provide a high correlation between $U_{i:n}$ and $\mathrm{E}(U_{i:n})$ and to choose the least–squares criterion.

BAIN/ANTLE (1967) have considered various functions of the variable and parameters when $f(x \,|\, \boldsymbol{\theta})$ is the two–parameter WEIBULL distribution: $f(x \,|\, b, c) = (c/b) \, (x/b)^{c-1} \, \exp\big[-(x/b)^c\big]$.

1. $u(x \,|\, b, c) = (x/b)^c$

    For this function we have

$$f(u) = \exp(-u), \;\; u \geq 0, \tag{10.31a}$$

    the reduced exponential distribution, where

$$\mathrm{E}(U_{i:n}) = \xi_i = \sum_{j=1}^{i} \frac{1}{n - j + 1}. \tag{10.31b}$$

Trying to minimize

$$\sum_{i=1}^{n} \left[ (X_{i:n}/b)^c - \xi_i \right]^2$$

would not result in a closed form of the estimator for $c$, so BAIN/ANTLE considered minimizing

$$\sum_{i=1}^{n} \left[ \ln(X_{i:n}/b)^c - \ln \xi_i \right]^2$$

which gives

$$\widehat{b}_1 \;=\; \prod_{i=1}^{n} X_{i:n}^{1/n} \Bigg/ \left( \prod_{i=1}^{n} \xi_i^{1/(n\,\widehat{c}_1)} \right), \tag{10.31c}$$

$$\widehat{c}_1 \;=\; s_{k_i\,Y_i} \big/ s_{Y_i\,Y_i}, \tag{10.31d}$$

where

$$s_{k_i\,Y_i} = \sum_{i=1}^{n} k_i\,Y_i - \frac{\sum k_i \sum Y_i}{n}, \quad k_i = \ln \xi_i, \quad Y_i = \ln X_{i:n}. \tag{10.31e}$$

The estimators given in MILLER/FREUND (1965) are the same as $\widehat{b}_1$ and $\widehat{c}_1$ with $\xi_i$ replaced by $-\ln\big[1 - (i-0.5)/n\big]$.

2. $u(x\,|\,b,c) = 1 - \exp\big[ -(x/b)^c \big]$

In this case we have

$$f(u) = 1, \;\; 0 \leq u < 1, \tag{10.32a}$$

the reduced uniform distribution where

$$E(U_{i:n}) = \eta_i = \frac{i}{n+1}. \tag{10.32b}$$

Simple estimators can be obtained by minimizing

$$\sum_{i=1}^{n} \Big\{ \ln\big[ \ln(1 - U_{i:n}) \big] - \ln\big[ \ln(1 - \eta_i) \big] \Big\}^2.$$

This gives

$$\widehat{b}_2 \;=\; \frac{\displaystyle\prod_{i=1}^{n} X_{i:n}^{1/n}}{\displaystyle\prod_{i=1}^{n} \Big\{ -\ln\big[1 - i/(n+1)\big] \Big\}^{1/(n\,\widehat{c}_2)}}, \tag{10.32c}$$

$$\widehat{c}_2 \;=\; s_{k_i\,Y_i} \big/ s_{Y_i\,Y_i}, \tag{10.32d}$$

where

$$k_i = \ln\Big\{ -\ln\big[1 - i/(n+1)\big] \Big\}, \quad Y_i = \ln X_{i:n}. \tag{10.32e}$$

These estimators have also been suggested by GUMBEL (1954). Note, that these are also the same as $\widehat{b}_1$ and $\widehat{c}_1$ in (10.31c,d) with $\xi_i$ replaced by $-\ln(1 - \eta_i)$.

3. $u(x \mid b, c) = -\ln\left[(x/b)^c\right]$

In this case we have

$$f(u) = \exp\left(-u - e^{-u}\right), \quad u \in \mathbb{R}, \tag{10.33a}$$

the reduced form of the type–I extreme value distribution of the maximum, where

$$\mathrm{E}\left(U_{i:n}\right) = \delta_i \tag{10.33b}$$

is not given in closed form. Estimators for $b$ and $c$ may be obtained by minimizing

$$\sum_{i=1}^{n} \left\{ -\ln\left[\left(X_{i:n}/b\right)^c\right] - \delta_{n-i+1} \right\}^2.$$

The estimators may be obtained by replacing $\ln \xi_i$ with $-\delta_{n-i+1}$ in the expression for $\widehat{b}_1$ and $\widehat{c}_1$ above.

The aforementioned estimators are still applicable if censored sampling is employed. The sums and products are over the set of the $r$ observed uncensored observations and $n$ — except in the subscripts — is replaced by $r$.

### 10.4.2 Linear estimation with linear and polynomial coefficients

The following approach, which is due to DOWNTON (1966), is only applicable when the sample is not censored. The observed fact — by DOWNTON and several other authors — that the efficiencies of linear estimators for $a^*$ and $b^*$ of the Log–WEIBULL distribution are not particulary sensitive to changes in the values of the coefficients suggests the possibility that efficient estimators might be found, where the coefficients have been chosen for convenience, rather than because they conform to some optimizing process. DOWNTON therefore suggested instead that estimators based on

$$V = \sum_{i=1}^{n} X_{i:n}^* \quad \text{and} \quad W = \sum_{i=1}^{n} i\, X_{i:n}^* \tag{10.34a}$$

could be used.[13] When $X_{i:n}^*$ are order statistics from the Log–WEIBULL distribution, we have

$$\mathrm{E}(V) = n\, a^* - n\, \gamma\, b^*, \tag{10.34b}$$

$$\mathrm{E}(W) = \frac{n(n+1)}{2}\, a^* + \left[\frac{n(n-1)}{2} \ln 2 - \frac{n(n+1)}{2}\gamma\right] b^*, \tag{10.34c}$$

where $\gamma \approx 0.57721$ is EULER's constant. Thus, unbiased estimators of $a^*$ and $b^*$ based on the linear coefficients defined by (10.34a) are obtained in the form

$$\widehat{a_1^*} = \frac{(n-1)\ln 2 - (n+1)\gamma}{n(n-1)\ln 2}\, V + \frac{2\gamma}{n(n-1)\ln 2}\, W, \tag{10.34d}$$

$$\widehat{b_1^*} = -\frac{n+1}{n(n-2)\ln 2}\, V + \frac{2}{n(n-1)\ln 2}\, W. \tag{10.34e}$$

---

[13] A generalization employing higher powers in $i$, $\sum i^s X_{i:n}$, $s \geq 2$, will be presented below.

One great advantage of these estimators is that their variances and covariance may be explicitly evaluated as follows:

$$\text{Var}\big(\widehat{a_1^*}\big) = \frac{b^{*\,2}}{n\,(n-1)}\,(1.112825\,n - 0.906557), \tag{10.34f}$$

$$\text{Var}\big(\widehat{b_1^*}\big) = \frac{b^{*\,2}}{n\,(n-1)}\,(0.804621\,n - 0.185527), \tag{10.34g}$$

$$\text{Cov}\big(\widehat{a^*},\widehat{b^*}\big) = -\frac{b^{*\,2}}{n\,(n-1)}(0.228707\,n - 0.586058). \tag{10.34h}$$

The CRAMÉR–RAO lower bounds for estimating $a^*$ and $b^*$ are given by

$$\text{Var}\big(\widehat{a^*}\big) \geq b^{*\,2}\big[1 + 6\,(1-\gamma)^2/\pi^2\big]/n, \tag{10.35a}$$

$$\text{Var}\big(\widehat{b^*}\big) \geq 6\,b^{*\,2}/\big(n\,\pi^2\big), \tag{10.35b}$$

when $\widehat{a^*}$ and $\widehat{b^*}$ are any unbiased estimators of $a^*$ and $b^*$. Dividing $\text{Var}\big(\widehat{a^*}\big)$ from (10.35a) by $\text{Var}\big(\widehat{a_1^*}\big)$ from (10.34f) and doing the same with the variances of the $b^*$–estimator, we get the relative efficiencies of the linear estimators as shown in Tab. 10/5. The asymptotic relative efficiencies $(n \to \infty)$ are

$$\text{ARE}\big(\widehat{a_1^*}\big) = \frac{1 + 6\,(1-\gamma^2)\,/\pi^2}{1.112825} = 0.9963, \tag{10.36a}$$

$$\text{ARE}\big(\widehat{b_1^*}\big) = \frac{6}{0.804621\,\pi^2} = 0.7556. \tag{10.36b}$$

From Tab. 10/5 we see that the estimator $\widehat{a_1^*}$ of $a^*$ is so efficient that there seems little point in trying to improve upon it, but this is not true of $b^*$.

Table 10/5: Efficiencies of linear unbiased estimators for the Log-WEIBULL distribution

| Parameter | $a^*$ | | | | | |
|---|---|---|---|---|---|---|
| Sample size $n$ | 2 | 3 | 4 | 5 | 6 | $\infty$ |
| BLUE | 0.8405 | 0.9173 | 0.9445 | 0.9582 | 0.9665 | 1 |
| BLOM's approximation[†] | 0.8405 | 0.9172 | 0.9437 | 0.9568 | 0.9645 | 1 |
| Linear coeff. approx. $\widehat{a_1^*}$ | 0.8405 | 0.9118 | 0.9383 | 0.9521 | 0.9607 | 0.9963 |
| Quadratic coeff. approx. $\widehat{a_2^*}$ | 0.8405 | 0.9173 | 0.9442 | 0.9579 | 0.9660 | 0.9987 |
| Parameter | $b^*$ | | | | | |
| Sample size $n$ | 2 | 3 | 4 | 5 | 6 | $\infty$ |
| BLUE | 0.4270 | 0.5879 | 0.6746 | 0.7296 | 0.7678 | 1 |
| BLOM's approximation[†] | 0.4270 | 0.5747 | 0.6539 | 0.7047 | 0.7407 | 1 |
| Linear coeff. approx. $\widehat{b_1^*}$ | 0.4270 | 0.5456 | 0.6013 | 0.6337 | 0.6548 | 0.7556 |
| Quadratic coeff. approx. $\widehat{b_2^*}$ | 0.4270 | 0.5878 | 0.6714 | 0.7226 | 0.7551 | 0.9364 |

† See Sect. 10.4.4

Suppose we need to obtain an unbiased estimator of

$$\psi = \ell_1\, a^* + \ell_2\, b^* \tag{10.37a}$$

by linear combination of order statistics,[14] and suppose that this estimator is of the form

$$\widehat{\psi} = \sum_{s=0}^{n} \theta_s \sum_{i=1}^{n} i^s\, X_{i:n}^*. \tag{10.37b}$$

DOWNTON (1966) explores only the case $p = 2$ leading to quadratic coefficients. The condition of unbiasedness gives the following restraints upon the coefficients $\theta_s$:

$$\sum_{s=0}^{n} \theta_s \sum_{i=1}^{n} i^s \;=\; \ell_1, \tag{10.38a}$$

$$\sum_{s=0}^{n} \theta_s \sum_{i=1}^{n} i^s\, \alpha_{i:n} \;=\; \ell_2, \tag{10.38b}$$

where $\alpha_{i:n} = \mathrm{E}(U_{i:n})$, $U_{i:n}$ being the reduced Log–WEIBULL order statistic. Introducing the sums of powers

$$p_{s,n} = \sum_{i=1}^{n} i^s, \tag{10.39a}$$

with

$$p_{0,n} \;=\; n, \tag{10.39b}$$

$$p_{1,n} \;=\; \frac{n\,(n+1)}{2}, \tag{10.39c}$$

$$p_{2,n} \;=\; \frac{n\,(n+1)\,(2\,n+1)}{6}, \tag{10.39d}$$

and letting $\boldsymbol{\theta}$ be the column vector of $\theta_s$ and $\boldsymbol{p}$ be the column vector with $p_{s,n}$, (10.38a) may be written as

$$\boldsymbol{\theta}'\, \boldsymbol{p} = \ell_1. \tag{10.40a}$$

Similarly, letting $\boldsymbol{q}$ be the column vector with elements

$$q_{s,n} = \sum_{i=1}^{n} i^s\, \alpha_{i:n},$$

(10.38b) becomes

$$\boldsymbol{\theta}'\, \boldsymbol{q} = \ell_2. \tag{10.40b}$$

---

[14] Special cases of $\psi$ are
- $\ell_1 = 1$ and $\ell_2 = 0 \;\Rightarrow\; a^*$,
- $\ell_1 = 0$ and $\ell_2 = 1 \;\Rightarrow\; b^*$,
- $\ell_1 = 1$ and $\ell_2 = u_P = \ln[-\ln(1 - P)] \;\Rightarrow\; x_P$, percentile of order $P$.

The first three elements of $q$ are

$$q_{0,n} = n\,\alpha_{1:1} \tag{10.41a}$$

$$q_{1,n} = \frac{n\,(n-1)}{2}\,\alpha_{2:2} + n\,\alpha_{1:1} \tag{10.41b}$$

$$q_{2,n} = \frac{n\,(n-1)\,(n-2)}{3}\,\alpha_{3:3} + \frac{3\,n\,(n-1)}{2}\,\alpha_{2:2} + n\,\alpha_{1:1}, \tag{10.41c}$$

based on a recurrence relationship for moments of order statistics, where

$$\alpha_{n:n} = \sum_{i=1}^{n}(-1)^i \binom{n}{i}\ln i - \gamma. \tag{10.41d}$$

Using

$$v_{i,j:n} = \mathrm{Cov}\big(U_{i:n}, U_{j:n}\big),$$

we define a matrix $W$ with elements

$$w_{ij} = w_{ji} = \sum_{r=1}^{n}\sum_{s=1}^{n} r^i\, s^j\, v_{r,s:n}; \quad i,j = 0,\dots,p. \tag{10.42}$$

Then, the most efficient estimator of $\psi$ in (10.37a) is obtained by choosing $\theta$ so as to minimize

$$\theta'\, W\, \theta + \lambda_1\, \theta'\, p + \lambda_2\, \theta'\, q,$$

where $\lambda_1$, $\lambda_2$ are LAGRANGE–multipliers, introduced to ensure the side conditions (10.40a,b). The solution is

$$\theta = W^{-1}\, \chi'\,\big[\chi\, W^{-1}\, \chi'\big]^{-1}\, \ell, \tag{10.43a}$$

where

$$\chi = \begin{pmatrix} p' \\ q' \end{pmatrix} \quad \text{and} \quad \ell = \begin{pmatrix} \ell_1 \\ \ell_2 \end{pmatrix}. \tag{10.43b}$$

With $\theta$ given by (10.43a) the variance of $\widehat{\psi}$ may be explicitly evaluated as

$$\mathrm{Var}\big(\widehat{\psi}\big) = b^{*\,2}\, \ell'\,\big[\chi\, W^{-1}\, \chi'\big]^{-1}\, \ell. \tag{10.44}$$

Based on another recurrence relationship for moments of order statistics, DOWNTON gives the elements $w_{ij}$ of $W$ in (10.42a) for $p = 2$ as

$$w_{00} = 1.644934\,n, \tag{10.45a}$$

$$w_{01} = w_{10} = 0.582241\,n^2 + 1.062694\,n, \tag{10.45b}$$

$$w_{02} = w_{20} = 0.331209\,n^3 + 0.753094\,n^2 + 0.560631\,n, \tag{10.45c}$$

$$w_{11} = 0.267653\,n^3 + 0.703537\,n^2 + 0.673744\,n, \tag{10.45d}$$

$$w_{12} = w_{21} = 0.169092\,n^4 + 0.519071\,n^3 + 0.614148\,n^2 + 0.342623\,n, \tag{10.45e}$$

$$w_{22} = 0.112793\,n^5 + 0.397472\,n^4 + 0.562378\,n^3 + 0.410243\,n^2 + 0.162049\,n \tag{10.45f}$$

The estimators based on this approach with $p = 2$ and quadratic coefficients are denoted by $\widehat{a_2^*}$ and $\widehat{b_2^*}$ with efficiencies given in Tab. 10/5 for $n = 2, \dots, 6$ and $n = \infty$.

The asymptotic estimators are

$$\widehat{a^*_{2,\infty}} \approx 0.308\,n^{-1}\sum_{i=1}^{n}X^*_{i:n} + 0.707\,n^{-2}\sum_{i=1}^{n}i\,X^*_{i:n} + 1.016\,n^{-3}\sum_{i=1}^{n}i^2\,X^*_{i:n}, \quad (10.46a)$$

$$\widehat{b^*_{2,\infty}} \approx -0.464\,n^{-1}\sum_{i=1}^{n}X^*_{i:n} - 3.785\,n^{-2}\sum_{i=n}^{n}i\,X^*_{i:n} + 7.071\,n^{-3}\sum_{i=1}^{n}i^2\,X^*_{i:n} \quad (10.46b)$$

with variance–covariance matrix

$$\operatorname{Var}\begin{pmatrix}\widehat{a^*_{2,\infty}} \\ \widehat{b^*_{2,\infty}}\end{pmatrix} \approx n^{-1}\begin{pmatrix} 1.110 & -0.251 \\ -0.251 & 0.649 \end{pmatrix}. \quad (10.46c)$$

A look at Tab. 10/5 reveals an efficiency of the quadratic version that is even higher than that of BLOM's "unbiased nearly best estimator," but the disadvantage of DOWNTON's method is to not be applicable to censored samples.

### Example 10/6: Linear estimation with linear and quadratic coefficients for uncensored dataset #1

From dataset #1 consisting of $n = 20$ uncensored observations of $X \sim We(0, 100, 2.5)$, we get

$$\sum_{i=1}^{20}\ln x_i = 87.7712; \quad \sum_{i=1}^{20}i\,\ln x_i = 970.14; \quad \sum_{i=1}^{20}i^2\,\ln x_i = 13588.62.$$

The estimates (10.34d,e) using linear coefficients are

$$\widehat{a^*_1} = 4.6013 \implies \widehat{b} = \exp(4.6013) = 99.6137,$$

$$\widehat{b^*_1} = 0.3686 \implies \widehat{c} = 1/0.3686 = 2.7130.$$

Using the approach with quadratic coefficients we first have

$$W = \begin{pmatrix} 32.8987 & 254.1503 & 2962.1222 \\ 254.1503 & 2436.1137 & 31459.8000 \\ 2962.1222 & 31459.8000 & 429199.4800 \end{pmatrix},$$

$$X = \begin{pmatrix} 20 & 210 & 2870 \\ -11.5443 & 10.4827 & 974.7756 \end{pmatrix},$$

the $\theta$–coefficients for $\widehat{a^*_2}$

$$\theta' = (0.014325,\ 0.001265,\ 0.000156),$$

the $\theta$–coefficients for $\widehat{b^*_2}$

$$\theta' = (-0.018442,\ -0.010878,\ 0.000924).$$

The estimates are

$$\widehat{a^*_2} = 4.6050 \implies \widehat{b} = \exp(4.6050) = 99.9830,$$

$$\widehat{b^*_2} = 0.3904 \implies \widehat{c} = 1/0.3904 = 2.5615,$$

which are closer to the true values than the estimates based on linear coefficients.

### 10.4.3  GLUEs of BAIN and ENGELHARDT

BAIN (1972) suggested simpler versions of the estimators in the Log–WEIBULL notation, which are linear and unbiased and apply to complete as well as censored samples. These estimators are identical to the BLUEs for $n = 2$ and for $n > 2$ when $r = 2$. Their form is similar but not identical for larger values of $r$. BAIN referred to these estimators as only **good linear unbiased estimators** (GLUEs). Further developments on the GLUEs are to be found in ENGELHARDT/BAIN (1973), ENGELHARDT (1975), BAIN/ENGELHARDT (1991a) and SMITH (1977). Tables are needed to apply the GLU–method, but these tables are less extensive than those needed for BLUEs.

The GLUEs for $\widehat{a^*}$ and $\widehat{b^*}$ of the Log–WEIBULL distribution are

$$\widehat{b^*} \;=\; \sum_{i=1}^{r} \frac{X_{r:n}^* - X_{i:n}}{n\,k_{r:n}} \;=\; \frac{r\,X_{r:n}^* - \sum\limits_{i=1}^{r} X_{i:n}^*}{n\,k_{r:n}}, \tag{10.47a}$$

$$\widehat{a^*} \;=\; X_{r:n}^* - c_{r:n}\,\widehat{b^*}. \tag{10.47b}$$

The unbiasing coefficients are defined as

$$k_{r:n} \;=\; \frac{1}{n}\,\mathrm{E}\!\left[\sum_{i=1}^{r}\left(U_{r:n}^* - U_{i:n}^*\right)\right], \tag{10.47c}$$

$$c_{r:n} \;=\; \mathrm{E}\!\left(U_{r:n}^*\right), \tag{10.47d}$$

where $U_{i:n}^* = (X_{i:n}^* - a^*)/b^*$. BAIN/ENGELHARDT (1991a, pp. 255/256) give tables for $k_{r:n}$ and $c_{r:n}$ for $r/n = 0.1(0.1)0.9$ and $n = 5, 10(10)100$ together with approximating formulas to quadratic interpolation for non–tabulated coefficients. For $r/n < 0.1$ interpolation is not possible and $k_{r:n}$ and $c_{r:n}$ have to be computed with the formulas for $\mathrm{E}\!\left(U_{i:n}\right) = \alpha_{i:n}$ given in Sect. 5.4.

ENGELHARD/BAIN (1973) have shown that the asymptotic relative efficiency of (10.47a,b) — relative to the BLUEs — is zero when $r = n$. Therefore, (10.47a,b) have to be modified as follows:

$$\widehat{b^*} \;=\; \frac{-\sum\limits_{i=1}^{s} X_{i:n}^* + \dfrac{s}{n-s}\sum\limits_{i=s+1}^{n} X_{i:n}^*}{n\,k_n}, \quad \text{for } r = n, \tag{10.48a}$$

$$\widehat{a^*} \;=\; \frac{1}{n}\sum_{i=1}^{n} X_{i:n}^* + \gamma\,\widehat{b^*}, \quad \text{for } r = n, \tag{10.48b}$$

$\gamma$ being EULER's constant. The unbiasing coefficients $k_n$ have been tabulated for $n = 2(1)60$. $s = [0.84\,n]$, the largest integer less than or equal to $0.84\,n$, leads to the smallest asymptotic variance of $\widehat{b^*}$.

BAIN/ENGELHARDT (1991a) provide tables showing the variances and covariance of $\widehat{a^*}$ and $\widehat{b^*}$ according to (10.47a,b) and (10.48a,b). SCHÜPBACH/HÜSLER (1983) have improved (10.47a,b) for $0.8 \leq r/n < 1$ by carrying over (10.48a,b) to censored samples.

### 10.4.4 BLOM's unbiased nearly best linear estimator

BLOM (1958, 1962) derived linear estimators for the location and scale parameters of a location–scale distribution, termed **unbiased nearly best linear**, depending upon the true expectations $\alpha_{i:n}$ of the reduced order statistics but not upon the exact variances and co-variances $\beta_{i,j:n}$. Instead he used the asymptotic variances and covariances. WHITE (1965) and ENGEMAN/KEEFE (1982) have carried over this approach to the Log–WEIBULL distribution.

Let $U_{i:n}$ be the ordered sample statistics of some reduced location–scale distributed variate, e.g., the Log–WEIBULL distribution with CDF

$$F(u) = 1 - \exp\{-e^u\}, \quad u \in \mathbb{R}. \tag{10.49a}$$

Then $Z_{i:n} := F(U_{i:n})$ are distributed as the order statistics of a random sample of size $n$ from a uniform $(0,1)$–distribution. Defining $G(z) = u$ as the inverse of $F(u) = z$ and expanding $u = G(z)$ in a TAYLOR series about $z = P$, we have

$$u = G(z) = G(P) + (z - P) G'(P) + \dots \tag{10.49b}$$

Thus we may expand the order statistics $U_{i:n}$ in terms of the order statistics $Z_{i:n}$.

Since the moments of $Z_{i:n}$ are

$$\mathrm{E}\big(Z_{i:n}\big) \;=\; \frac{i}{n+1} \;=\; P_i \tag{10.50a}$$

$$\mathrm{Cov}\big(Z_{i:n}, Z_{j:n}\big) \;=\; \frac{i\,(n-j+1)}{(n+1)^2\,(n+2)} \;=\; \frac{P_i\,(1-P_j)}{n+2}, \quad i \le j, \tag{10.50b}$$

we have expanding about $P_i = \mathrm{E}\big(Z_{i:n}\big)$

$$U_{i:n} \;=\; G(Z_{i:n}) \;=\; G(P_i) + (Z_{i:n} - P_i)\,G'(P_i) + \dots \tag{10.51a}$$

$$\mathrm{E}\big(U_{i:n}\big) \;=\; G(P_i) + 0 + \dots \tag{10.51b}$$

$$\mathrm{Cov}\big(U_{i:n}, U_{j:n}\big) \;=\; \frac{P_i\,(1-P_j)}{n+2}\,G'(P_i)\,G'(P_j) + \dots \tag{10.51c}$$

In the Log–WEIBULL case we have

$$G(z) \;=\; u \;=\; \ln[-\ln(1-P)] \tag{10.52a}$$

$$G'(z) \;=\; \frac{-1}{(1-z)\ln(1-z)}. \tag{10.52b}$$

We take as an approximation to the variance–covariance matrix $\boldsymbol{B}$ (see (10.4b)), the matrix $\boldsymbol{V}$, where

$$v_{ij} \;=\; \frac{P_i\,(1-P_j)}{n+2}\,G'(P_i)\,G'(P_j), \quad i \le j, \tag{10.53a}$$

$$v_{ji} \;=\; v_{ij}, \quad i \le j. \tag{10.53b}$$

The matrix $\boldsymbol{V}$ may be written as

$$\boldsymbol{V} = \boldsymbol{D}\,\boldsymbol{Z}\,\boldsymbol{D} \tag{10.53c}$$

where $D$ is a diagonal matrix with elements

$$G'(P_i) = -\frac{n+1}{(n+1-i)\ln\left(\frac{n+1-i}{n+1}\right)}, \qquad (10.53\text{d})$$

using (10.52b) and (10.50a). The elements of the symmetric $(n \times n)$ matrix $Z$ are

$$z_{ij} = \frac{P_i(1-P_j)}{(n+2)} = \frac{i(n+1-j)}{(n+1)^2(n+2)}, \quad \text{using (10.50a).} \qquad (10.53\text{e})$$

The inverse of $Z$ is

$$Z^{-1} = (n+1)(n+2)\begin{pmatrix} 2 & -1 & 0 & \cdots & 0 \\ -1 & 2 & -1 & \cdots & 0 \\ 0 & -1 & 2 & \cdots & 0 \\ & & & \ddots & \\ 0 & 0 & 0 & \cdots & 2 \end{pmatrix}. \qquad (10.53\text{f})$$

The elements of $D^{-1}$ are simply the reciprocal of $G'(P_i)$ given in (10.53d). Thus, $V^{-1}$ is simple to evaluate as follows:

$$V^{-1} = D^{-1} Z^{-1} D^{-1}. \qquad (10.53\text{g})$$

$D^{-1}$ is also diagonal having the reciprocal diagonal elements of $D$ on its diagonal.

Reverting to the original BLUEs or GLS estimators given in (10.4c) the **ABLUEs** (asymptatically best linear estimators) turn out to be

$$\widehat{\theta} = \begin{pmatrix} \widehat{a^*} \\ \widehat{b^*} \end{pmatrix} = (P'V^{-1}P)^{-1} P'V^{-1}x^*, \qquad (10.54)$$

where $x^*$ and $P$ are defined in (10.4a). $P$ contains the means $\alpha_{i:n} = \mathrm{E}(U_{i:n})$.

---

**Example 10/7: ABLUEs for singly type–II censored dataset #1**

We apply the ABLU–method to dataset #1 censored at $r = 15$. The results should be compared with those given in Example 10/2 for the original BLUEs. From (10.54) we have

$$\widehat{a^*} = 4.6297 \implies \widehat{b} = \exp(\widehat{a^*}) = 102.4833$$
$$\widehat{b^*} = 0.4184 \implies \widehat{c} = 1/\widehat{b^*} = 2.3901.$$

---

## 10.4.5 ABLIEs

**Asymptotically best linear estimators** (ABLIEs) of the parameters $a^*$ and $b^*$ of the Log–WEIBULL distribution have been found by JOHNS/LIEBERMAN (1966) on the basis of the material in CHERNOFF et al. (1967). They defined estimators in the form

$$\widehat{a^*} = \frac{1}{n} \sum_{i=1}^{r} J_1\left(\frac{i}{n+1}\right) X_{i:n}^*, \quad \widehat{b^*} = \frac{1}{n} \sum_{i=1}^{r} J_2\left(\frac{i}{n+1}\right) X_{i:n}^*, \quad r \le n, \qquad (10.55)$$

where $J_1(\cdot)$ and $J_2(\cdot)$ have been chosen so that the estimators are asymptotically jointly normal and efficient. $J_1(\cdot)$ and $J_2(\cdot)$ depend upon functions related to the incomplete digamma and trigamma functions. D'AGOSTINO (1971a,b) has simplified the approach of JOHNS/LIEBERMAN by showing that the weights in (10.55) which rely on extensive tables can be written in closed, simple to compute forms. The only requirements are knowledge of the proportion of available observations $P = r/n$ and a simple table look–up (Tab. 10/6) for four numbers dependent upon this proportion.

The estimators proposed by D'AGOSTINO are

$$\widehat{a^*} = \frac{1}{n}\left[H_1(P)\,L_1(P) - H_3(P)\,L_2(P)\right], \quad 0 < P = \frac{r}{n} \le 1, \qquad (10.56a)$$

$$\widehat{b^*} = \frac{1}{n}\left[H_2(P)\,L_2(P) - H_3(P)\,L_1(P)\right], \qquad (10.56b)$$

where

$$L_1(P) = \sum_{i=1}^{r-1} w_{i,n}\, X_{i:n}^* + w_{r,n,P}\, X_{r:n}^*,$$

$$w_{i,n} = \ln\left(\frac{n+1}{n+1-i}\right), \quad w_{r,n,P} = P - \sum_{i=1}^{r-1} w_{i,n},$$

and

$$L_2(P) = \sum_{i=1}^{r-1} v_{i,n}\, X_{i:n}^* + v_{r,n,P}\, X_{r:n}^*$$

$$v_{i,n} = \ln\left(\frac{n+1}{n+1-i}\right)\left[1 + \ln\ln\left(\frac{n+1}{n+1-i}\right)\right] - 1$$

$$= w_{i,n}\left(1 + \ln w_{i,n}\right) - 1$$

$$v_{r,n,P} = n\,H_4(P) - \sum_{i=1}^{r-1} v_{i,n}.$$

Tab. 10/6 has been compiled from D'AGOSTINO (1971a) and gives $H_i(P)$ for $i = 1, 2, 3, 4$ and for $P = 0.01(0.01)1$. The coefficient $H_4(P)$ is related to the first three coefficients by

$$H_4(P) = \frac{H_3(P)}{H_1(P)\,H_2(P) - H_3(P)^2}.$$

<u>Table 10/6:</u>  Components for the ABLIEs of $a^*$ and $b^*$

| $P$ | $H_1(P)$ | $H_2(P)$ | $H_3(P)$ | $H_4(P)$ | $P$ | $H_1(P)$ | $H_2(P)$ | $H_3(P)$ | $H_4(P)$ |
|---|---|---|---|---|---|---|---|---|---|
| 0.010 | 2213.1439 | 99.7495 | −459.1133 | −0.0460 | 0.510 | 2.4172 | 1.6760 | −0.8747 | −0.2662 |
| 0.020 | 809.3986 | 49.7490 | −194.3690 | −0.0781 | 0.520 | 2.3305 | 1.6374 | −0.8168 | −0.2594 |
| 0.030 | 438.3524 | 33.0818 | −115.7530 | −0.1050 | 0.530 | 2.2495 | 1.6001 | −0.7618 | −0.2523 |
| 0.040 | 279.8085 | 24.7479 | −79.4102 | −0.1284 | 0.540 | 2.1739 | 1.5642 | −0.7097 | −0.2450 |
| 0.050 | 195.7245 | 19.7474 | −58.9076 | −0.1492 | 0.550 | 2.1031 | 1.5296 | −0.6602 | −0.2374 |
| 0.060 | 145.1791 | 16.4135 | −45.9276 | −0.1679 | 0.560 | 2.0370 | 1.4961 | −0.6131 | −0.2295 |
| 0.070 | 112.1877 | 14.0320 | −37.0643 | −0.1849 | 0.570 | 1.9751 | 1.4637 | −0.5685 | −0.2213 |
| 0.080 | 89.3589 | 12.2458 | −30.6789 | −0.2004 | 0.580 | 1.9171 | 1.4324 | −0.5258 | −0.2129 |
| 0.090 | 72.8571 | 10.8563 | −25.8908 | −0.2146 | 0.590 | 1.8628 | 1.4021 | −0.4852 | −0.2042 |
| 0.100 | 60.5171 | 9.7447 | −22.1872 | −0.2277 | 0.600 | 1.8120 | 1.3728 | −0.4466 | −0.1952 |
| 0.110 | 51.0357 | 6.8350 | −19.2505 | −0.2397 | 0.610 | 1.7652 | 1.3443 | −0.4098 | −0.1859 |
| 0.120 | 43.5871 | 8.0769 | −16.8742 | −0.2507 | 0.620 | 1.7195 | 1.3168 | −0.3746 | −0.1764 |
| 0.130 | 37.6262 | 7.4353 | −14.9187 | −0.2608 | 0.630 | 1.6775 | 1.2900 | −0.3411 | −0.1666 |
| 0.140 | 32.7806 | 6.8852 | −13.2661 | −0.2702 | 0.640 | 1.6381 | 1.2640 | −0.3091 | −0.1565 |
| 0.150 | 28.7863 | 6.4085 | −11.9065 | −0.2787 | 0.650 | 1.6011 | 1.2388 | −0.2786 | −0.1462 |
| 0.160 | 25.4606 | 5.9912 | −10.7282 | −0.2865 | 0.660 | 1.5664 | 1.2142 | −0.2494 | −0.1355 |
| 0.170 | 22.6584 | 5.6230 | −9.7124 | −0.2936 | 0.670 | 1.5337 | 1.1904 | −0.2215 | −0.1246 |
| 0.180 | 20.2774 | 5.2955 | −8.8295 | −0.3001 | 0.680 | 1.5031 | 1.1671 | −0.1948 | −0.1135 |
| 0.190 | 18.2380 | 5.0025 | −8.0565 | −0.3060 | 0.690 | 1.4743 | 1.1445 | −0.1692 | −0.1020 |
| 0.200 | 16.4786 | 4.7388 | −7.3753 | −0.3113 | 0.700 | 1.4473 | 1.1224 | −0.1448 | −0.0903 |
| 0.210 | 14.9513 | 4.5000 | −6.7715 | −0.3160 | 0.710 | 1.4218 | 1.1009 | −0.1215 | −0.0783 |
| 0.220 | 13.6174 | 4.2830 | −6.2334 | −0.3202 | 0.720 | 1.3980 | 1.0800 | −0.0991 | −0.0661 |
| 0.230 | 12.4463 | 4.0847 | −5.7515 | −0.3239 | 0.730 | 1.3756 | 1.0595 | −0.0776 | −0.0535 |
| 0.240 | 11.4132 | 3.9029 | −5.3181 | −0.3270 | 0.740 | 1.3545 | 1.0395 | −0.0571 | −0.0407 |
| 0.250 | 10.4979 | 3.7356 | −4.9268 | −0.3297 | 0.750 | 1.3347 | 1.0159 | −0.0374 | −0.0275 |
| 0.260 | 9.6835 | 3.5811 | −4.5721 | −0.3320 | 0.760 | 1.3061 | 1.0008 | −0.0186 | −0.0141 |
| 0.270 | 8.9562 | 3.4379 | −4.2494 | −0.3337 | 0.770 | 1.2987 | 0.9821 | −0.0005 | −0.0004 |
| 0.280 | 8.3045 | 3.3050 | −3.9551 | −0.3351 | 0.780 | 1.2823 | 0.9638 | 0.0168 | 0.0176 |
| 0.290 | 7.7187 | 3.1812 | −3.6857 | −0.3360 | 0.790 | 1.2670 | 0.9458 | 0.0334 | 0.0279 |
| 0.300 | 7.1904 | 3.0655 | −3.4386 | −0.3365 | 0.800 | 1.2526 | 0.9282 | 0.0493 | 0.0425 |
| 0.310 | 6.7128 | 2.9573 | −3.2113 | −0.3366 | 0.810 | 1.2391 | 0.9109 | 0.0645 | 0.0575 |
| 0.320 | 6.2799 | 2.8558 | −3.0016 | −0.3363 | 0.820 | 1.2265 | 0.8939 | 0.0792 | 0.0726 |
| 0.330 | 5.8865 | 2.7604 | −2.8079 | −0.3357 | 0.830 | 1.2147 | 0.8773 | 0.0932 | 0.0082 |
| 0.340 | 5.5282 | 2.6705 | −2.6285 | −0.3346 | 0.840 | 1.2037 | 0.8609 | 0.1067 | 0.1041 |
| 0.350 | 5.2013 | 2.5857 | −2.4620 | −0.3332 | 0.850 | 1.1934 | 0.8447 | 0.1196 | 0.1203 |
| 0.360 | 4.9023 | 2.5056 | −2.3072 | −0.3315 | 0.860 | 1.1838 | 0.8288 | 0.1319 | 0.1369 |
| 0.370 | 4.6283 | 2.4298 | −2.1631 | −0.3294 | 0.870 | 1.1748 | 0.8131 | 0.1438 | 0.2538 |
| 0.380 | 4.3769 | 2.3579 | −2.0286 | −0.3269 | 0.880 | 1.1665 | 0.7976 | 0.1551 | 0.1711 |
| 0.390 | 4.1458 | 2.2896 | −1.9030 | −0.3241 | 0.890 | 1.1588 | 0.7822 | 0.1660 | 0.1889 |
| 0.400 | 3.9330 | 2.2247 | −1.7855 | −0.3210 | 0.900 | 1.1507 | 0.7670 | 0.1764 | 0.2070 |
| 0.410 | 3.7368 | 2.1630 | −1.6754 | −0.3176 | 0.910 | 1.1451 | 0.7520 | 0.1864 | 0.2255 |
| 0.420 | 3.5556 | 2.1041 | −1.5721 | −0.3136 | 0.920 | 1.1390 | 0.7370 | 0.1959 | 0.2446 |
| 0.430 | 3.3881 | 2.0479 | −1.4751 | −0.3097 | 0.930 | 1.1335 | 0.7220 | 0.2050 | 0.2661 |
| 0.440 | 3.2331 | 1.9942 | −1.3639 | −0.3053 | 0.940 | 1.1284 | 0.7070 | 0.2137 | 0.2842 |
| 0.450 | 3.0894 | 1.9428 | −1.2980 | −0.3006 | 0.950 | 1.1239 | 0.6920 | 0.2220 | 0.3048 |
| 0.460 | 2.9561 | 1.8937 | −1.2170 | −0.2956 | 0.960 | 1.1198 | 0.6768 | 0.2299 | 0.3261 |
| 0.470 | 2.8322 | 1.8465 | −1.1406 | −0.2903 | 0.970 | 1.1162 | 0.6613 | 0.2374 | 0.3482 |
| 0.480 | 2.7171 | 1.6013 | −1.0685 | −0.2847 | 0.980 | 1.1130 | 0.6452 | 0.2449 | 0.3723 |
| 0.490 | 2.6100 | 1.7579 | −1.0003 | −0.2788 | 0.990 | 1.1105 | 0.6282 | 0.2511 | 0.3957 |
| 0.500 | 2.5102 | 1.7162 | −0.9358 | −0.2726 | 1.000 | 1.1087 | 0.6079 | 0.2570 | 0.4228 |

D'AGOSTINO (1971a) shows how to update the estimators when the censoring number increases with $n$ fix. He further demonstrates that even for small samples the MSEs of (10.56a,b) compare favorably with other well-known estimating procedures. D'AGOSTINO (1971b) extends the foregoing approach to the case that either of the two parameters $a^*$, $b^*$ is known.

## 10.5 Linear estimation with a few optimally chosen order statistics[15]

In some experiments it is difficult, expensive or time–consuming to make measurements on all the $n$ individual objects with sufficient accuracy. It may be preferable to concentrate on the making of accurate measurements of a few selected items, which still provide high efficiency in estimation. The estimator we are interested in is LLOYD's BLUE, which is obtained by the GLS method. It will be found that in many cases the $n$–th observation must be used. Thus, the experimental time itself — the time until $X_{n:n}$ is available to be measured — will not necessarily be shortened by using only a few of the observations. However, time and money may be saved both in making accurate measurements on fewer observations and in the simpler calculation of the estimates. The chosen order statistics will be called **optimum–order statistics** and the corresponding BLUE will be called the **optimum BLUE**.

This section is arranged as follows:

1. In Section 10.5.1 we comment on choosing $k$ optimum–order statistics when the sample size is small.
2. Section 10.5.2 deals with the method of quantiles. The aim is to select $k$ sample quantiles when the sample size is very large and to build upon the asymptotic distribution of those sample quantiles.

### 10.5.1   Optimum-order statistics for small sample sizes

Assume, that we have $k$ ascending order statistics $X^*_{n_1:n}$, $X^*_{n_2:n}$, ..., $X^*_{n_k:n}$ from the Log–WEIBULL distribution (type–I extreme value distribution of the minimum), where $n_1$, $n_2$, ..., $n_k$ are the ranks such that

$$1 \le n_1 < n_2 < \ldots < n_k \le n.$$

Suppose that the sample size is small and we want to estimate $a^*$ and $b^*$ in (10.3e), but instead of processing all $n$ order statistics as described in Sect. 10.1, we now want to use only $k$ of the $n$ order statistics. Evidently this procedure is not sufficient and we will suffer a loss in efficiency.

There exist $\binom{n}{k}$ possible sets of $k$ order statistics in a sample of size $n$, so we have to choose an optimizing criterion to decide which set is to be taken. Let $\widehat{\boldsymbol{\theta}}_k = \left(\widehat{a^*_k}, \widehat{b^*_k}\right)'$ denote the

---

[15] Suggested reading for this section: CHAN/KABIR (1969), CHAN/MEAD (1971), HASSANEIN (1969, 1972), MANN (1970a), MANN/FERTIG (1977), OGAWA (1951).

estimator based on $k$ ($2 \leq k \leq n$) order statistics; i.e., $\boldsymbol{P}$ and $\boldsymbol{B}$ are of types $k \times 2$ and $k \times k$, respectively, with elements $\alpha_{n_i:n}$ and $\beta_{n_i,n_j:n}$ ($n_i$, $n_j = 1, 2, \ldots, k$). The basic variance–covariance matrix in (10.4h) is denoted by $\boldsymbol{\Sigma}_k$ and its elements are $A_k$, $B_k$ and $C_k$ where

$$A_k = \mathrm{Var}\left(\widehat{a_k^*}\right)\big/b^{*2}, \tag{10.57a}$$

$$B_k = \mathrm{Var}\left(\widehat{b_k^*}\right)\big/b^{*2}, \tag{10.57b}$$

$$C_k = \mathrm{Cov}\left(\widehat{a_k^*}, \widehat{b_k^*}\right)\big/b^{*2}. \tag{10.57c}$$

HASSANEIN (1969) minimizes the sum of the basic variances, i.e.,

$$A_k + B_k \stackrel{!}{=} \min_{(n_1,\ldots,n_k)}, \tag{10.58a}$$

whereas CHAN/MEAD (1971) minimize the generalized variance, i.e., the determinant of $\boldsymbol{\Sigma}_k$:

$$\left|\boldsymbol{\Sigma}_k\right| = A_k \cdot B_k - C_k^2 \stackrel{!}{=} \min_{(n_1,\ldots,n_k)}. \tag{10.58b}$$

Table 10/7: BLUES of $a^*$ and $b^*$ based on $k = 2$ order statistics with optimum ranks, coefficients, variances, covariances and relative efficiencies

| $n$ | $n_1$ | $n_2$ | $a_1$ | $a_2$ | $b_1$ | $b_2$ | $A_k$ | $B_k$ | $C_k$ | Eff$(\widehat{a_k^*})$ | Eff$(\widehat{b_k^*})$ | JEff$(\widehat{a_k^*},\widehat{b_k^*})$ |
|---|---|---|---|---|---|---|---|---|---|---|---|---|
| 3 | 1 | 3 | 0.19 | 0.81 | −0.48 | 0.48 | 0.43 | 0.37 | −0.05 | 0.94 | 0.93 | 0.88 |
| 4 | 1 | 4 | 0.23 | 0.77 | −0.39 | 0.39 | 0.34 | 0.26 | −0.07 | 0.85 | 0.86 | 0.77 |
| 5 | 2 | 5 | 0.39 | 0.61 | −0.57 | 0.57 | 0.26 | 0.22 | −0.06 | 0.88 | 0.75 | 0.68 |
| 6 | 2 | 6 | 0.38 | 0.62 | −0.49 | 0.49 | 0.23 | 0.17 | −0.06 | 0.83 | 0.78 | 0.67 |
| 7 | 2 | 7 | 0.37 | 0.63 | −0.44 | 0.44 | 0.21 | 0.14 | −0.05 | 0.78 | 0.79 | 0.65 |
| 8 | 2 | 8 | 0.36 | 0.64 | −0.40 | 0.40 | 0.19 | 0.12 | −0.05 | 0.74 | 0.79 | 0.62 |
| 9 | 3 | 9 | 0.45 | 0.55 | −0.48 | 0.48 | 0.16 | 0.11 | −004 | 0.77 | 0.75 | 0.60 |
| 10 | 3 | 10 | 0.44 | 0.56 | −0.44 | 0.44 | 0.15 | 0.09 | −0.04 | 0.74 | 0.76 | 0.59 |
| 11 | 3 | 11 | 0.43 | 0.57 | −0.42 | 0.42 | 0.14 | 0.08 | −0.03 | 0.71 | 0.77 | 0.57 |
| 12 | 4 | 12 | 0.49 | 0.51 | −0.47 | 0.47 | 0.13 | 0.08 | −0.03 | 0.74 | 0.73 | 0.55 |
| 13 | 4 | 13. | 0.48 | 0.52 | −0.44 | 0.44 | 0.12 | 0.07 | −0.03 | 0.71 | 0.74 | 0.54 |
| 14 | 4 | 14 | 0.47 | 0.53 | −0.42 | 0.42 | 0.12 | 0.07 | −0.03 | 0.69 | 0.75 | 0.53 |
| 15 | 4 | 15 | 0.46 | 0.54 | −0.40 | 0.40 | 0.11 | 0.06 | −0.02 | 0.67 | 0.75 | 0.52 |
| 16 | 5 | 16 | 0.51 | 0.49 | −0.44 | 0.44 | 0.10 | 0.06 | −0.02 | 0.70 | 0.72 | 0.51 |
| 17 | 5 | 17 | 0.50 | 0.50 | −0.42 | 0.42 | 0.10 | 0.05 | −0.02 | 0.68 | 0.73 | 0.50 |
| 18 | 5 | 18 | 0.49 | 0.51 | −0.41 | 0.41 | 0.09 | 0.05 | −0.02 | 0.66 | 0.74 | 0.49 |
| 19 | 5 | 18 | 0.40 | 0.60 | −0.45 | 0.45 | 0.08 | 0.05 | −0.02 | 0.74 | 0.66 | 0.49 |
| 20 | 5 | 19 | 0.40 | 0.60 | −0.44 | 0.44 | 0.08 | 0.05 | −0.02 | 0.72 | 0.66 | 0.48 |
| 21 | 6 | 20 | 0.44 | 0.56 | −0.47 | 0.47 | 0.07 | 0.05 | −0.02 | 0.74 | 0.65 | 0.48 |
| 22 | 6 | 21 | 0.43 | 0.57 | −0.45 | 0.45 | 0.07 | 0.05 | −0.02 | 0.73 | 0.66 | 0.48 |
| 23 | 6 | 22 | 0.43 | 0.57 | −0.44 | 0.44 | 0.07 | 0.04 | −0.02 | 0.71 | 0.67 | 0.48 |
| 24 | 6 | 23 | 0.42 | 0.58 | −0.43 | 0.43 | 0.07 | 0.04 | −0.02 | 0.70 | 0.68 | 0.47 |
| 25 | 7 | 24 | 0.45 | 0.55 | −0.45 | 0.45 | 0.06 | 0.04 | −0.01 | 0.72 | 0.66 | 0.47 |

Table 10/8:  BLUES of $a^*$ and $b^*$ based on $k = 3$ order statistics with optimum ranks, coefficients, variances, covariances and relative efficiencies

| $n$ | $n_1$ | $n_2$ | $n_3$ | $a_1$ | $a_2$ | $a_3$ | $b_1$ | $b_2$ | $b_3$ | $A_k$ | $B_k$ | $C_k$ | $\mathrm{Eff}(\widehat{a_k^*})$ | $\mathrm{Eff}(\widehat{b_k^*})$ | $\mathrm{JEff}(\widehat{a_k^*}, \widehat{b_k^*})$ |
|---|---|---|---|---|---|---|---|---|---|---|---|---|---|---|---|
| 4 | 1 | 2 | 4 | 0.08 | 0.27 | 0.65 | -0.25 | -0.26 | 0.51 | 0.31 | 0.23 | -0.04 | 0.96 | 0.99 | 0.95 |
| 5 | 1 | 3 | 5 | 0.09 | 0.38 | 0.53 | -0.24 | -0.28 | 0.52 | 0.24 | 0.18 | -0.04 | 0.95 | 0.95 | 0.91 |
| 6 | 1 | 3 | 6 | 0.08 | 0.38 | 0.54 | -0.19 | -0.28 | 0.47 | 0.21 | 0.14 | -0.04 | 0.92 | 0.95 | 0.88 |
| 7 | 1 | 3 | 7 | 0.06 | 0.37 | 0.57 | -0.15 | -0.28 | 0.43 | 0.19 | 0.11 | -0.03 | 0.88 | 0.95 | 0.84 |
| 8 | 1 | 4 | 8 | 0.07 | 0.44 | 0.49 | -0.16 | -0.30 | 0.46 | 0.16 | 0.10 | -0.03 | 0.89 | 0.91 | 0.81 |
| 9 | 1 | 4 | 9 | 0.06 | 0.43 | 0.51 | -0.13 | -0.30 | 0.43 | 0.15 | 0.09 | -0.03 | 0.85 | 0.92 | 0.79 |
| 10 | 2 | 6 | 10 | 0.14 | 0.48 | 0.38 | -0.27 | -0.20 | 0.47 | 0.13 | 0.08 | -0.03 | 0.90 | 0.85 | 0.77 |
| 11 | 2 | 7 | 11 | 0.14 | 0.51 | 0.35 | -0.26 | -0.19 | 0.46 | 0.11 | 0.08 | -0.03 | 0.90 | 0.83 | 0.75 |
| 12 | 2 | 7 | 12 | 0.12 | 0.51 | 0.37 | -0.23 | -0.21 | 0.44 | 0.11 | 0.07 | -0.02 | 0.88 | 0.84 | 0.74 |
| 13 | 2 | 7 | 13 | 0.11 | 0.50 | 0.39 | -0.21 | -0.22 | 0.43 | 0.10 | 0.06 | -0.02 | 0.86 | 0.85 | 0.73 |
| 14 | 2 | 8 | 14 | 0.11 | 0.53 | 0.36 | -0.21 | -0.22 | 0.43 | 0.09 | 0.06 | -0.02 | 0.86 | 0.83 | 0.72 |
| 15 | 3 | 10 | 15 | 0.18 | 0.54 | 0.28 | -0.30 | -0.13 | 0.43 | 0.08 | 0.06 | -0.02 | 0.88 | 0.79 | 0.71 |
| 16 | 3 | 11 | 16 | 0.18 | 0.56 | 0.26 | -0.29 | -0.12 | 0.41 | 0.08 | 0.05 | -0.02 | 0.88 | 0.79 | 0.70 |
| 17 | 3 | 11 | 17 | 0.16 | 0.56 | 0.28 | -0.27 | -0.14 | 0.41 | 0.06 | 0.05 | -0.02 | 0.87 | 0.79 | 0.69 |
| 18 | 4 | 13 | 18 | 0.23 | 0.55 | 0.23 | -0.34 | -0.06 | 0.41 | 0.07 | 0.04 | -0.02 | 0.88 | 0.77 | 0.689 |
| 19 | 4 | 14 | 19 | 0.23 | 0.56 | 0.21 | -0.34 | -0.06 | 0.40 | 0.07 | 0.05 | -0.02 | 0.87 | 0.76 | 0.68 |
| 20 | 4 | 15 | 20 | 0.23 | 0.57 | 0.20 | -0.33 | -0.05 | 0.38 | 0.05 | 0.04 | -0.02 | 0.86 | 0.76 | 0.68 |
| 21 | 4 | 15 | 21 | 0.20 | 0.58 | 0.22 | -0.31 | -0.07 | 0.39 | 0.06 | 0.04 | -0.02 | 0.86 | 0.76 | 0.67 |
| 22 | 5 | 17 | 22 | 0.27 | 0.56 | 0.17 | -0.36 | -0.01 | 0.37 | 0.06 | 0.04 | -0.02 | 0.86 | 0.75 | 0.66 |
| 23 | 5 | 18 | 23 | 0.26 | 0.57 | 0.17 | -0.36 | -0.00 | 0.36 | 0.06 | 0.04 | -0.02 | 0.85 | 0.75 | 0.66 |
| 24 | 5 | 18 | 24 | 0.24 | 0.58 | 0.18 | -0.34 | -0.02 | 0.37 | 0.05 | 0.04 | -0.01 | 0.86 | 0.75 | 0.65 |
| 25 | 5 | 19 | 25 | 0.24 | 0.58 | 0.18 | -0.34 | -0.02 | 0.36 | 0.05 | 0.04 | -0.01 | 0.85 | 0.75 | 0.65 |

Table 10/9:   BLUEs of $a^*$ and $b^*$ based on $k = 4$ order statistics with optimum ranks, coefficients, variances, covariances and relative efficiencies

| $n$ | $n_1$ | $n_2$ | $n_3$ | $n_4$ | $a_1$ | $a_2$ | $a_3$ | $a_4$ | $b_1$ | $b_2$ | $b_3$ | $b_4$ | $A_k$ | $B_k$ | $C_k$ | $\text{Eff}(\widehat{a_k^*})$ | $\text{Eff}(\widehat{b_k^*})$ | $\text{JEff}(\widehat{a_k^*}, \widehat{b_k^*})$ |
|---|---|---|---|---|---|---|---|---|---|---|---|---|---|---|---|---|---|---|
| 5 | 1 | 2 | 4 | 5 | 0.06 | 0.18 | 0.35 | 0.41 | −0.19 | −0.23 | −0.08 | 0.50 | 0.24 | 0.17 | −0.04 | 0.99 | 0.99 | 0.97 |
| 6 | 1 | 2 | 4 | 6 | 0.05 | 0.13 | 0.37 | 0.45 | −0.15 | −0.20 | −0.13 | 0.48 | 0.20 | 0.13 | −0.03 | 0.97 | 0.98 | 0.95 |
| 7 | 1 | 3 | 5 | 7 | 0.06 | 0.20 | 0.35 | 0.39 | −0.15 | −0.26 | −0.05 | 0.46 | 0.17 | 0.11 | −0.03 | 0.97 | 0.96 | 0.93 |
| 8 | 1 | 3 | 6 | 8 | 0.05 | 0.21 | 0.39 | 0.35 | −0.13 | −0.26 | −0.05 | 0.44 | 0.15 | 0.10 | −0.03 | 0.96 | 0.95 | 0.92 |
| 9 | 1 | 3 | 7 | 9 | 0.05 | 0.21 | 0.43 | 0.31 | −0.12 | −0.26 | −0.04 | 0.42 | 0.13 | 0.09 | −0.03 | 0.95 | 0.94 | 0.90 |
| 10 | 1 | 3 | 7 | 10 | 0.04 | 0.18 | 0.44 | 0.34 | −0.10 | −0.23 | −0.08 | 0.41 | 0.12 | 0.08 | −0.02 | 0.94 | 0.94 | 0.88 |
| 11 | 1 | 4 | 9 | 11 | 0.05 | 0.27 | 0.43 | 0.25 | −0.10 | −0.29 | 0.01 | 0.38 | 0.11 | 0.07 | −0.02 | 0.93 | 0.92 | 0.87 |
| 12 | 1 | 4 | 9 | 12 | 0.04 | 0.23 | 0.45 | 0.28 | −0.10 | −0.27 | −0.02 | 0.39 | 0.10 | 0.06 | −0.02 | 0.93 | 0.92 | 0.86 |
| 13 | 1 | 4 | 10 | 13 | 0.04 | 0.23 | 0.47 | 0.26 | −0.09 | −0.27 | −0.02 | 0.38 | 0.09 | 0.06 | −0.02 | 0.92 | 0.91 | 0.85 |
| 14 | 2 | 6 | 12 | 14 | 0.09 | 0.32 | 0.39 | 0.20 | −0.18 | −0.26 | 0.09 | 0.35 | 0.09 | 0.06 | −0.02 | 0.93 | 0.89 | 0.84 |
| 15 | 2 | 6 | 12 | 15 | 0.08 | 0.28 | 0.42 | 0.22 | −0.16 | −0.26 | 0.06 | 0.36 | 0.08 | 0.05 | −0.02 | 0.93 | 0.89 | 0.83 |
| 16 | 2 | 6 | 13 | 16 | 0.07 | 0.28 | 0.44 | 0.21 | −0.15 | −0.26 | 0.06 | 0.35 | 0.08 | 0.05 | −0.02 | 0.92 | 0.89 | 0.83 |
| 17 | 2 | 7 | 14 | 17 | 0.08 | 0.31 | 0.41 | 0.20 | −0.15 | −0.27 | 0.08 | 0.34 | 0.07 | 0.05 | −0.02 | 0.93 | 0.88 | 0.82 |
| 18 | 2 | 7 | 15 | 18 | 0.07 | 0.31 | 0.43 | 0.19 | −0.14 | −0.27 | 0.08 | 0.33 | 0.07 | 0.04 | −0.06 | 0.92 | 0.88 | 0.82 |
| 19 | 2 | 7 | 16 | 19 | 0.07 | 0.31 | 0.44 | 0.18 | −0.13 | −0.27 | 0.09 | 0.31 | 0.06 | 0.04 | −0.02 | 0.91 | 0.89 | 0.82 |
| 20 | 2 | 8 | 17 | 20 | 0.07 | 0.34 | 0.42 | 0.17 | −0.14 | −0.28 | 0.11 | 0.31 | 0.06 | 0.04 | −0.01 | 0.91 | 0.88 | 0.81 |
| 21 | 2 | 8 | 18 | 21 | 0.07 | 0.33 | 0.44 | 0.16 | −0.13 | −0.28 | 0.11 | 0.30 | 0.06 | 0.04 | −0.01 | 0.90 | 0.88 | 0.81 |
| 22 | 2 | 8 | 18 | 22 | 0.06 | 0.30 | 0.46 | 0.18 | −0.12 | −0.28 | 0.09 | 0.31 | 0.06 | 0.03 | −0.01 | 0.91 | 0.87 | 0.80 |
| 23 | 3 | 10 | 20 | 23 | 0.10 | 0.35 | 0.40 | 0.15 | −0.17 | −0.26 | 0.14 | 0.29 | 0.05 | 0.03 | −0.01 | 0.91 | 0.86 | 0.80 |
| 24 | 3 | 10 | 21 | 24 | 0.09 | 0.35 | 0.41 | 0.15 | −0.16 | −0.26 | 0.14 | 0.28 | 0.05 | 0.03 | −0.01 | 0.90 | 0.87 | 0.79 |
| 25 | 3 | 10 | 21 | 25 | 0.08 | 0.32 | 0.44 | 0.16 | −0.16 | −0.26 | 0.13 | 0.29 | 0.05 | 0.03 | −0.01 | 0.91 | 0.86 | 0.79 |

There are some sample sizes for which the optimum ranks resulting from (10.58a) and (10.58b) coincide. Tables 10/7 through 10/9 are based on the criterion given by (10.58b) and on the variances and covariances in WHITE (1964b). The tables show — for $k = 2, 3$ and 4 and $n = k + 1(1)25$ — the optimum ranks $n_1, \ldots, n_k$, as well as the weights $a_1, \ldots, a_k$ and $b_1, \ldots, b_k$ to construct the estimators

$$\widehat{a_k^*} = \sum_{j=1}^{k} a_j \, X_{n_j:n}^*, \quad \widehat{b_k^*} = \sum_{j=1}^{k} b_j \, X_{n_j:n}^*,$$

their variances $A_k$, $B_k$ and covariance $C_k$ and their efficiencies relative to the BLUE based upon the complete sample:

$$\begin{aligned}
\mathrm{Eff}\big(\widehat{a_k^*}\big) &= A_n/A_k, \\
\mathrm{Eff}\big(\widehat{b_k^*}\big) &= B_n/B_k, \\
\mathrm{JEff}\big(\widehat{a_k^*}, \widehat{b_k^*}\big) &= \frac{A_n \cdot B_n - C_n^2}{A_k \cdot B_k - C_k^2}.
\end{aligned}$$

For given $k$ the efficiencies decrease with $n$ growing.

## 10.5.2 Quantile estimators and ABLEs

When the sample size $n$ is large, the $k$ out of $n$ order statistics are selected by applying the asymptotic theory of sample quantiles. Consider $k$ real numbers such that

$$0 = \lambda_0 < \lambda_1 < \lambda_2 < \ldots < \lambda_k < \lambda_{k+1} = 1,$$

where the integer $k$ is less than $n$. Let the selected sample quantiles be $X_{n_1:n}^*, \ldots, X_{n_k:n}^*$, where $n_i = [n\,\lambda_i] + 1$ and $[n\,\lambda_i]$ is the greatest integer not exceeding $n\,\lambda_i$.[16] The unit interval is subdivided into $k + 1$ intervals by the set of points $\lambda_1, \ldots, \lambda_k$. Thus the estimator consists of the $k$ observations spaced at these points. The quantiles of a sample from a Log–WEIBULL distribution are asymptotically normal with

$$\mathrm{E}\big(X_{n_i:n}^*\big) = a^* + b^* u_i, \tag{10.59a}$$

$$\mathrm{Cov}\big(X_{n_i:n}^*, X_{n_j:n}^*\big) = \frac{b^{*2}}{n} \frac{\lambda_i\,(1 - \lambda_j)}{f_i\,f_j}; \ i, j = 1, \ldots, k, \tag{10.59b}$$

where

$$u_i = \ln[-\ln \lambda_i], \ f_i = f(u_i), \ f_j = f(u_j)$$

and

$$\lambda_i = \int_{-\infty}^{u_i} f(u)\,\mathrm{d}u \ \text{ with } \ f(u) = \exp\big[u - e^u\big].$$

A first approach to select an appropriate set of proportions giving linear estimators

$$\widehat{a^*} = \sum_{i=1}^{k} a_i \, X_{n_i:n}^*, \quad \widehat{b^*} = \sum_{i=1}^{k} b_i \, X_{n_i:n}^*$$

---

[16] Choosing $n_i = [n\,\lambda_i]$ results in estimators which are less efficient as shown by CHAN/MEAD (1971).

consists in minimizing $\text{Var}(\widehat{a^*}) + \text{Var}(\widehat{b^*})$ under the conditions

$$\sum_{i=1}^{k} a_i = 1, \quad \sum_{i=1}^{k} b_i = 0, \quad \sum_{i=1}^{k} a_i\, u_i = 0, \quad \sum_{i=1}^{k} b_i\, u_i = 0,$$

as proposed by HASSANEIN (1969). For $k = 2$, 3 and 4 he found results.

A second approach elaborated by HASSANEIN (1972) for the Log–WEIBULL parameters[17] results in the so–called ABLEs (asymptotically best linear estimators). This approach originates in a paper of OGAWA (1951) and applies the GLS–method to the large sample distribution of sample percentiles given in (10.59a,b). The optimum spacings $\lambda_1, \lambda_2, \ldots, \lambda_k$ are those which maximize the asymptotic relative efficiency of the estimators $\widehat{a^*}$ and $\widehat{b^*}$ given by

$$\text{ARE}(\widehat{a^*}, \widehat{b^*}) = \frac{\Delta + \dfrac{2\,k}{n} K_1}{|I|} \approx \frac{\Delta}{I}, \tag{10.60}$$

where

$$\Delta = K_1\, K_2 - K_3^2,$$

$$|I| = \text{E}\big[(f'/f)^2\big]\big\{\text{E}\big[(U\, f'/f)^2\big] - 1\big\} - \text{E}^2\big[U\, f'^2/f^2\big],$$

$$K_1 = \sum_{i=1}^{k+1} \frac{(f_i - f_{i-1})^2}{\lambda_1 - \lambda_{i-1}},$$

$$K_2 = \sum_{i=1}^{k+1} \frac{\big(f_i\, u_i - f_i\, u_{i-1}\big)^2}{\lambda_i - \lambda_{i-1}},$$

$$K_3 = \sum_{i=1}^{k+1} \frac{(f_i - f_{i-1})\big(f_i\, u_i - f_{i-1}\, u_{i-1}\big)}{\lambda_i - \lambda_{i-1}},$$

$$f_i = f(u_i); \quad i = 1, 2, \ldots, k,$$

$$f(u_0) = f(u_{k+1}) = 0,$$

$$f = f(U) = \exp\big(U - e^U\big).$$

For $k = 2(1)10$ Tab. 10/10 shows the optimum spacings $\lambda_i$, the percentiles $u_i$, the coefficients $a_i$, $b_i$ for the linear estimators $\widehat{a^*}$ and $\widehat{b^*}$ together with

$$\frac{K_2}{\Delta} = \frac{\text{Var}(\widehat{a^*})}{b^{*2}/n}, \quad \frac{K_1}{\Delta} = \frac{\text{Var}(\widehat{b^*})}{b^{*2}/n}, \quad \frac{K_3}{\Delta} = -\frac{\text{Cov}(\widehat{a^*}, \widehat{b^*})}{b^{*2}/n}$$

---

[17] CHAN/KABIR (1969) have used this approach to estimate the parameters of the type–I extreme value distribution for the largest value.

Table 10/10: Asymptotically optimum spacings $\lambda_i$, percentiles $u_i$, coefficients $a_i$ and $b_i$ for the estimators $\widehat{a^*}$ and $\widehat{b^*}$, their variances, covariances and asymptotic relative efficiencies $(x^*_{n_1:n} > x^*_{n_2:n} > \ldots > x^*_{n_k:n})$

| | $k=2$ | $k=3$ | $k=4$ | $k=5$ | $k=6$ | $k=7$ | $k=8$ | $k=9$ | $k=10$ |
|---|---|---|---|---|---|---|---|---|---|
| $\lambda_1$ | 0.0870 | 0.0550 | 0.0280 | 0.0180 | 0.0110 | 0.0080 | 0.0060 | 0.0040 | 0.0030 |
| $u_1$ | 0.8928 | 1.0648 | 1.2741 | 1.3906 | 1.5063 | 1.5745 | 1.6324 | 1.7086 | 1.7594 |
| $a_1$ | 0.5680 | 0.3386 | 0.1566 | 0.0994 | 0.0623 | 0.0439 | 0.0323 | 0.0228 | 0.0175 |
| $b_1$ | 0.4839 | 0.4372 | 0.2845 | 0.2047 | 0.1454 | 0.1112 | 0.0875 | 0.0663 | 0.0538 |
| $\lambda_2$ | 0.7340 | 0.4390 | 0.1930 | 0.1140 | 0.0710 | 0.0470 | 0.0330 | 0.0240 | 0.0180 |
| $u_2$ | −1.1736 | −0.1945 | 0.4978 | 0.7754 | 0.9727 | 1.1176 | 1.2271 | 1.3163 | 1.3906 |
| $a_2$ | 0.4320 | 0.5184 | 0.4316 | 0.3030 | 0.2027 | 0.1382 | 0.0981 | 0.0737 | 0.0557 |
| $b_2$ | −0.4839 | −0.1602 | 0.1526 | 0.2236 | 0.2189 | 0.1854 | 0.1531 | 0.1300 | 0.1079 |
| $\lambda_3$ | | 0.8500 | 0.6040 | 0.4040 | 0.2510 | 0.1630 | 0.1110 | 0.0790 | 0.0580 |
| $u_3$ | | −1.8170 | −0.6848 | −0.0983 | 0.3238 | 0.5955 | 0.7877 | 0.9315 | 1.0464 |
| $a_3$ | | 0.1430 | 0.3250 | 0.3673 | 0.3315 | 0.2649 | 0.2002 | 0.1512 | 0.1156 |
| $b_3$ | | −0.2770 | −0.2651 | −0.1012 | 0.0481 | 0.1254 | 0.1497 | 0.1481 | 0.1365 |
| $\lambda_4$ | | | 0.8960 | 0.7260 | 0.5470 | 0.3960 | 0.2790 | 0.1990 | 0.1450 |
| $u_4$ | | | −2.2090 | −1.1388 | −0.5053 | −0.0765 | 0.2442 | 0.4790 | 0.6580 |
| $a_4$ | | | 0.0868 | 0.1804 | 0.2564 | 0.2813 | 0.2652 | 0.2277 | 0.1865 |
| $b_4$ | | | −0.1720 | −0.2208 | −0.1733 | −0.0780 | 0.0121 | 0.0720 | 0.1026 |
| $\lambda_5$ | | | | 0.9310 | 0.7990 | 0.6520 | 0.5130 | 0.3930 | 0.2970 |
| $u_5$ | | | | −2.6381 | −1.4944 | −0.8493 | −0.4042 | −0.0683 | 0.1912 |
| $a_5$ | | | | 0.0499 | 0.1144 | 0.1727 | 0.2134 | 0.2278 | 0.2199 |
| $b_5$ | | | | −0.1063 | −0.1673 | −0.1680 | −0.1262 | −0.0639 | −0.0036 |
| $\lambda_6$ | | | | | 0.9510 | 0.8490 | 0.7270 | 0.6049 | 0.4910 |
| $u_6$ | | | | | −2.9909 | −1.8097 | −1.1431 | −0.6848 | −0.3406 |
| $a_6$ | | | | | 0.0327 | 0.0764 | 0.1208 | 0.1579 | 0.1824 |
| $b_6$ | | | | | −0.0718 | −0.1251 | −0.1439 | −0.1317 | −0.0981 |
| $\lambda_7$ | | | | | | 0.9640 | 0.8840 | 0.7820 | 0.6750 |
| $u_7$ | | | | | | −3.3060 | −2.0932 | −1.4028 | −0.9338 |
| $a_7$ | | | | | | 0.0226 | 0.0537 | 0.0874 | 0.1184 |
| $b_7$ | | | | | | −0.0508 | −0.0952 | −0.1191 | −0.1217 |
| $\lambda_8$ | | | | | | | 0.9730 | 0.9090 | 0.8240 |
| $u_8$ | | | | | | | −3.5983 | −2.3496 | −1.6420 |
| $a_8$ | | | | | | | 0.0163 | 0.0393 | 0.0652 |
| $b_8$ | | | | | | | −0.0371 | −0.0736 | −0.0980 |
| $\lambda_9$ | | | | | | | | 0.9790 | 0.9280 |
| $u_9$ | | | | | | | | −3.8576 | −2.5940 |
| $a_9$ | | | | | | | | 0.0122 | 0.0296 |
| $b_9$ | | | | | | | | −0.0281 | −0.0580 |
| $\lambda_{10}$ | | | | | | | | | 0.9840 |
| $u_{10}$ | | | | | | | | | −4.1271 |
| $a_{10}$ | | | | | | | | | 0.0092 |
| $b_{10}$ | | | | | | | | | −0.0214 |
| $\mathrm{Var}(\widehat{a^*})/(b^{*2}/n)$ | 1.5106 | 1.2971 | 1.2287 | 1.1924 | 1.1706 | 1.1567 | 1.1470 | 1.1401 | 1.1349 |
| $\mathrm{Var}(\widehat{b^*})/(b^{*2}/n)$ | 1.0749 | 0.9028 | 0.7933 | 0.7374 | 0.7043 | 0.6825 | 0.6676 | 0.6567 | 0.6486 |
| $\mathrm{Cov}(\widehat{a^*},\widehat{b^*})/(b^{*2}/n)$ | −0.3401 | −0.2579 | −0.2570 | −0.2674 | −0.2657 | −0.2628 | −0.2627 | −0.2616 | −0.2609 |
| $\mathrm{ARE}(\widehat{a^*})$ | 0.7339 | 0.8548 | 0.9023 | 0.9298 | 0.9471 | 0.9585 | 0.9666 | 0.9725 | 0.9769 |
| $\mathrm{ARE}(\widehat{b^*})$ | 0.5656 | 0.6734 | 0.7663 | 0.8244 | 0.8632 | 0.8907 | 0.9106 | 0.9257 | 0.9373 |
| $\mathrm{ARE}(\widehat{a^*},\widehat{b^*})$ | 0.4031 | 0.5504 | 0.6770 | 0.7526 | 0.8064 | 0.8445 | 0.8726 | 0.8936 | 0.9101 |

and their asymptotic relative efficiencies

$$\text{ARE}(\widehat{a^*}) = \frac{\{\text{E}[(U\,f'/f)^2] - 1\}/|I|}{K_2/\Delta},$$

$$\text{ARE}(\widehat{b^*}) = \frac{\text{E}[(f'/f)^2]/|I|}{K_1/\Delta}$$

and $\text{ARE}(\widehat{a^*}, b^*)$ according to (10.60) using

$$\text{E}[(f'/f)^2] = 1, \quad \text{E}[(U\,f'/f)^2] = 2.82371, \quad \text{E}[U\,f'^2/f^2] = -0.42279.$$

It occurs in practical situations that the results in Tab. 10/10 cannot be applied because of the censoring of the extreme ordered observations in the sample. To deal with this problem HASSANEIN (1972) has given special tables for the asymptotically optimum spacings when $\lambda_1$ is not less than 0.01 and $\lambda_k$ not more than 0.9. MANN/FERTIG (1977) give factors for correcting the small–sample bias in HASSANEIN's estimators.

When estimating a subset of the parameters, we sometimes will revert to the idea of using only a few order statistics, especially the first two of them or the last one.

## 10.6    Linear estimation of a subset of parameters

There are situations where we know the value of one or even two of the three WEIBULL parameters $a$, $b$ and $c$. In this section we will show how to linearly estimate the remaining unknown parameter(s). Sect. 10.6.1 deals with the problem of indirectly estimating $b = \exp(a^*)$ or $c = 1/b^*$ via the parameters $a^*$, $b^*$ of the Log–WEIBULL variate $X^* = \ln(X - a)$; i.e., the location parameter $a$ is assumed to be known as in all preceding sections of this chapter. In Sect. 10.6.2 we treat direct estimating a subset of $\{a, b, c\}$.

### 10.6.1    Estimation of one of the Log-WEIBULL parameters

We will first present the BLUEs of either $a^*$ or $b^*$ for $b^*$ or $a^*$ unknown.[18] In either case we have to reformulate the regression equation (10.4a) which has to be estimated by GLS.

First case:  BLUE of $a^*$ when $b^*$ is known

The regression equation now reads

$$\boldsymbol{x}^* - b^*\,\boldsymbol{\alpha} = a^*\,\boldsymbol{1} + \boldsymbol{\varepsilon}, \tag{10.61a}$$

where

$$\boldsymbol{\alpha}' = (\alpha_{1:n}, \dots, \alpha_{n:n}) \text{ and } \boldsymbol{1}' = (1, \dots, 1).$$

The GLS–estimator of $a^*$ is

$$\widehat{a^*} = (\boldsymbol{1}'\,\boldsymbol{B}^{-1}\,\boldsymbol{1})^{-1}\,\boldsymbol{1}'\,\boldsymbol{B}^{-1}\,(\boldsymbol{x}^* - b^*\,\boldsymbol{\alpha}) \tag{10.61b}$$

---

[18]  For a better understanding and for the definition of the symbols used here the reader should consult Sect. 10.1.

with variance

$$\mathrm{Var}\big(\widehat{a^*}\big) = b^{*2}\left(\mathbf{1}'\,\mathbf{B}^{-1}\,\mathbf{1}\right)^{-1}. \tag{10.61c}$$

The row vector $\left(\mathbf{1}'\,\mathbf{B}^{-1}\,\mathbf{1}\right)^{-1}\mathbf{1}'\,\mathbf{B}^{-1}$ contains the weights for linearly combining the modified observations $X^*_{i:n} - b^*\,\alpha_{i:n}$ in the column vector $\mathbf{x}^* - b^*\,\boldsymbol{\alpha}$. The elements of $\mathbf{B}$ and $\boldsymbol{\alpha}$ have to be taken from tables, e.g., from WHITE (1964b) for $n = 2(1)20$.

Second case: BLUE of $b^*$ when $a^*$ is known

The regression equation is

$$\mathbf{x}^* - a^*\,\mathbf{1} = b^*\,\boldsymbol{\alpha} + \boldsymbol{\varepsilon} \tag{10.62a}$$

giving the GLS–estimator

$$\widehat{b^*} = \left(\boldsymbol{\alpha}'\,\mathbf{B}^{-1}\,\boldsymbol{\alpha}\right)^{-1}\boldsymbol{\alpha}'\,\mathbf{B}^{-1}\left(\mathbf{x}^* - a^*\,\mathbf{1}\right) \tag{10.62b}$$

with variance

$$\mathrm{Var}\big(\widehat{b^*}\big) = b^{*2}\left(\boldsymbol{\alpha}'\,\mathbf{B}^{-1}\,\boldsymbol{\alpha}\right)^{-1}. \tag{10.62c}$$

We should mention that in both cases the results may be applied to singly type–II censored data where the index of the observations and of the elements in $\boldsymbol{\alpha}$ and in $\mathbf{B}$ runs from 1 to the censoring number $r$. If only $k$ of $n$ optimally chosen order statistics should be used, HASSANEIN (1969) shows how to proceed in both cases when $k = 2,\ 3$ and CHAN/MEAD (1971) when $k = 2,\ 3$ and 4. CHAN/KABIR (1969) should be consulted for the ABLEs. D'AGOSTINO (1971b) extends the ABLIE–approach of Section 10.4.5 to the two cases mentioned above as follows:

$$\widehat{a^*} = \frac{1}{P}\left[\frac{L_1(P)}{n} - b^*\,H_4(P)\right], \tag{10.63a}$$

$$\widehat{b^*} = H_5(P)\left[\frac{L_2(P)}{n} - a^*\,H_4(P)\right], \tag{10.63b}$$

where

$$H_5(P) = \frac{H_1(P)\,H_2(P) - H_3(P)^2}{H_1(P)}.$$

## 10.6.2   Estimation of one or two of the three WEIBULL parameters

In this section we will build estimators based on the original WEIBULL variate, either $X \sim We(a,b,c)$ or $U \sim We(0,1,c)$, the ordered variates being $X_{i:n}$ and $U_{i:n}$. For the special case $r = 2$, i.e., the sample is singly type–II censored by the second failure, the following estimators have been suggested by LEONE et al. (1960):

- $a = 0$, $b$ and $c$ unknown

  $\widehat{c}$ is found by solving the equation

$$\frac{X_{2:n}}{X_{1:n}} = n\left(\frac{n}{n-1}\right)^{1/\widehat{c}} - n + 1. \tag{10.64a}$$

- c known

$$\widehat{a} \;=\; X_{1:n} - \frac{X_{2:n} - X_{1:n}}{n} \left[ \left( \frac{n}{n-1} \right)^{1/c} - 1 \right]^{-1} \qquad (10.64\mathrm{b})$$

$$\widehat{b} \;=\; \frac{1}{n} \left( X_{2:n} - X_{1:n} \right) \frac{(n-1)^{1/c}}{1 - \left( \dfrac{n-1}{n} \right)^{1/c} \Gamma \left( 1 + \dfrac{1}{c} \right)} \qquad (10.64\mathrm{c})$$

- $b$ and $c$ known

$$\widehat{a} = X_{1:n} - b \, (n+1)^{-1/c} \qquad (10.64\mathrm{d})$$

- $a$ and $c$ known

$$\widehat{b} = (X_{\kappa+1:n} - a) \left\{ \ln \left[ \frac{n+1}{n-\kappa} \right] \right\}^{-1/c}, \qquad (10.64\mathrm{e})$$

where $\kappa$ is the largest integer not exceeding $n \left( 1 - e^{-1} \right)$.

The remainder of this section is organized as follows:

- Sect. 10.6.2.1 shows how to estimate $a$ and $b$ when $c$ is known by considering the two cases:
    1. usage of all available $n$ or $r$ order statistics,
    2. usage of $k$ out of $n$ order statistics when $n$ is either small (finite case) or great (infinite case).
- Section 10.6.2.2 enumerates approaches to estimate either the shape parameter $c$ or the scale parameter $b$.

### 10.6.2.1   Estimating $a$ and $b$ with $c$ known

The ordered reduced WEIBULL statistics

$$U_{i:n} = \frac{X_{i:n} - a}{b}$$

have a distribution and hence moments depending on $c$ (see (5.33) and (5.37) for the densities and (5.34) and (5.38a) for the moments). Thus, when $c$ is known we can compute the means, variances and covariances (see Sect. 5.2), or look up these values in one of the tables cited in Sect. 5.2. For given $c$ we denote

$$\mathrm{E}(\boldsymbol{u} \,|\, c) \;\; := \;\; \boldsymbol{\alpha}_{c}, \qquad (10.65\mathrm{a})$$
$$\mathrm{Var}(\boldsymbol{u} \,|\, c) \;\; := \;\; \boldsymbol{B}_{c}. \qquad (10.65\mathrm{b})$$

All the formulas developed in Sections 10.2 through 10.5 for estimating $a^{*}$ and $b^{*}$ of the Log–WEIBULL distribution can be carried over to estimate $a$ and $b$ by substituting $\boldsymbol{\alpha}$ and $\boldsymbol{B}$ with $\boldsymbol{\alpha}_{c}$ and $\boldsymbol{B}_{c}$.

The BLUEs of $a$ and $b$ based on a complete or a singly type–II censored sample are given by

$$\widehat{a} = -\boldsymbol{\alpha}'_c \, \boldsymbol{\Gamma} \, \boldsymbol{x}, \quad \widehat{b} = \mathbf{1}' \boldsymbol{\Gamma} \, \boldsymbol{x}, \tag{10.66a}$$

where

$$
\begin{aligned}
\boldsymbol{x}' &= \left( X_{1:n}, \dots, X_{n:n} \right) \ \text{ or } \ \boldsymbol{x}' = \left( X_{1:n}, \dots, X_{r:n} \right), \\
\boldsymbol{\Gamma} &= \boldsymbol{B}_c^{-1} \left( \mathbf{1} \, \boldsymbol{\alpha}'_c - \boldsymbol{\alpha}_c \, \mathbf{1}' \right) \boldsymbol{B}_c^{-1} / \Delta, \\
\Delta &= \left( \mathbf{1}' \boldsymbol{B}_c^{-1} \mathbf{1} \right) \left( \boldsymbol{\alpha}'_c \, \boldsymbol{B}_c^{-1} \boldsymbol{\alpha}_c \right) - \left( \mathbf{1}' \boldsymbol{B}_c^{-1} \boldsymbol{\alpha} \right)^2,
\end{aligned}
$$

with $\boldsymbol{\alpha}_c$ and $\mathbf{1}$ having $n$ or $r$ elements, respectively, and $\boldsymbol{B}_c$ being of type $(n \times n)$ or $(r \times r)$, respectively. The variances and covariance of $\widehat{a}$ and $\widehat{b}$ are

$$
\begin{aligned}
\text{Var}\big(\widehat{a}\big) &= b^2 \left( \boldsymbol{\alpha}'_c \, \boldsymbol{B}_c^{-1} \boldsymbol{\alpha}_c \right) / \Delta, \tag{10.66b} \\
\text{Var}\big(\widehat{b}\big) &= b^2 \left( \mathbf{1}' \boldsymbol{B}_c^{-1} \mathbf{1} \right) / \Delta, \tag{10.66c} \\
\text{Cov}\big(\widehat{a}, \widehat{b}\big) &= -b^2 \left( \mathbf{1}' \boldsymbol{B}_c^{-1} \boldsymbol{\alpha}_c \right) / \Delta. \tag{10.66d}
\end{aligned}
$$

GOVINDARAJULU/JOSHI (1968) have tabulated the coefficients $-\boldsymbol{\alpha}'_c \, \boldsymbol{\Gamma}$ and $\mathbf{1}' \boldsymbol{\Gamma}$ together with the basic covariance $-\left( \mathbf{1}' \boldsymbol{B}_c^{-1} \boldsymbol{\alpha}_c \right) / \Delta$ for $c = 3$, 5, 8 and 10 and $2 \leq r$, $n \leq 10$.

The reader intending to use only $k$ of $n$ order statistics to estimate $a$ and $b$ is referred to CHAN et al. (1974) and HASSANEIN (1971). The latter gives the optimum spacings $\lambda_1, \dots, \lambda_k$ and the weights for the ABLEs when $k = 2$, 4, 6 and $c = 3(1)10$, 15, 20. CHAN et al. have explored the finite sample case and tabulated the optimum ranks $n_1, \dots, n_k$, the weights and variances and covariances of $\widehat{a}$ and $\widehat{b}$ for $k = 3$ and $k = 4$ in combination with $n = 3(1)15$ and $c = 0.5$, 1.5, 2, 2.5, 3 and 4.

Further material on BLUEs of $a$ and $b$ is contained in ENGEMAN/KEEFE (1982), FRIEDMAN (1981), FRIEDMAN/GERTSBAKH (1981) and YILDIRIM (1990).

### 10.6.2.2 Estimating either $b$ or $c$

Estimation of the scale parameter $b$ when $a = 0$ and $c$ is known by the use of $k$ optimally selected order statistics has been discussed by CLARK (1964). The special case $k = 1$ has been explored by QUAYLE (1963) and MOORE/HARTER (1965, 1966). The latter developed a one–order–statistic estimator for the parameter $\lambda$ of the exponential variate $Y$ with density $f(y) = \lambda^{-1} \exp(y/\lambda)$ by

$$\widehat{\lambda} = d_{\ell,n} \, Y_{\ell:n}, \tag{10.67a}$$

where $\ell$ is the optimum order in a sample of size $n$, singly censored with the $r$–th ordered observation. If $Y$ has an exponential distribution with scale parameter $\lambda$, then $X = Y^{1/c}$ has a WEIBULL distribution with shape parameter $c$ and scale parameter $b = \lambda^{1/c}$. Thus an estimator of $b$ is

$$\widehat{b} = d_{\ell,n}^{1/c} \, X_{\ell:n}. \tag{10.67b}$$

The coefficients $d_{\ell,n}$ have been tabulated for $2 \leq n \leq 20$ in MOORE/HARTER (1965) and for $21 \leq n \leq 40$ in MOORE/HARTER (1966).

VOGT (1968) has constructed a median–unbiased estimator of $c$ when $a$ is known and $b$ does not matter. The estimator is based on the first and last order statistics in an uncensored sample. The estimator is given by

$$\widehat{c} = g_n \left[ \ln X_{n:n} - \ln X_{1:n} \right]^{-1}. \tag{10.68}$$

Some values of $g_n$ are the following:

| $n$ | 2 | 3 | 4 | 5 | 6 | 7 | 8 | 9 | 10 |
|-----|-----|-----|-----|-----|-----|-----|-----|-----|-----|
| $g_n$ | 0.477 | 0.802 | 1.009 | 1.159 | 1.278 | 1.375 | 1.458 | 1.530 | 1.594 |

There exist several estimators of $c$ which are based on sample quantiles. We first give estimators requiring $a = 0$ and where $b$ does not matter. Afterwards we show how to estimate $c$ independently of $a$ and $b$ using three sample quantiles.

For large samples, when $a$ is known to be zero, DUBBEY (1967e) has suggested to use

$$\widehat{c} = \frac{\ln\{ \ln(1 - p_k) / \ln(1 - p_i) \}}{\ln \left( X_{k:n} / X_{i:n} \right)}, \quad 0 < p_i < p_k < 1, \tag{10.69a}$$

with $k = [n\,p_k] + 1$, $i = [n\,p_i] + 1$. This estimator is asymptotically unbiased and normal. Its asymptotic variance is minimized when $p_i = 0.16731$ and $p_k = 0.97366$. Thus we have

$$\widehat{c}_{opt} = 2.9888 \left\{ \ln\left( \frac{X_{k:n}}{X_{i:n}} \right) \right\}^{-1}. \tag{10.69b}$$

MURTHY/SWARTZ (1975) derived unbiasing factors for $\widehat{c}$ in (10.69a).

The percentile of order $p$ of the three–parameter WEIBULL distribution is (see (2.57b)):

$$x_p = a + b\left[-\ln(1 - p)\right]^{1/c}.$$

Suppose that $\widehat{X}_{p_i}$, $\widehat{X}_{p_j}$, $\widehat{X}_{p_k}$ are estimators of $x_p$ corresponding to $F(x) = p_i$, $p_j$, $p_k$, respectively, e.g., $\widehat{X}_p = X_{[n\,p]+1:n}$. Given $0 < p_i < p_j < p_k < 1$, the three equations

$$\widehat{X}_{p_\ell} = \widehat{a} + \widehat{b}\left[ -\ln(1 - p_\ell) \right]^{1/\widehat{c}}; \quad \ell = i, j, k; \tag{10.70a}$$

can be solved, yielding the estimators $\widehat{a}$, $\widehat{b}$ and $\widehat{c}$. The equation for $\widehat{c}$ is

$$\widehat{c} = \frac{0.5 \ln\{ \ln(1 - p_k) / \ln(1 - p_j) \}}{\ln\{ (\widehat{X}_{p_k} - \widehat{X}_{p_j}) / (\widehat{X}_{p_j} - \widehat{X}_{p_i}) \}}. \tag{10.70b}$$

DUBEY (1967e) has discussed the optimal choice of $p_i$, $p_j$ and $p_k$. ZANAKIS/MANN (1982), departing from $0 < p_i < p_j < p_k < 1$ and

$$-\ln(1 - p_j) = \left\{ \ln(1 - p_i)\, \ln(1 - p_k) \right\}^{1/2}, \tag{10.70c}$$

showed that choosing $p_i = 0.16731$ and $p_k = 0.97366$, leading to $p_j = 0.20099$, gives

$$\widehat{c} = 1.4944 \left\{ \ln\left( \frac{\widehat{X}_{p_k} - \widehat{X}_{p_j}}{\widehat{X}_{p_j} - \widehat{X}_{p_i}} \right) \right\}^{-1}, \tag{10.70d}$$

which produces a rather poor behavior. The optimal values of $p_i$ and $p_k$ depend on $c$ but are not very sensitive with respect to $c$. According to ZANAKIS/MANN a minimum asymptotic variance of $\widehat{c}$ in (10.70b) is attained for $0.0086 < p_i < 0.0202$ and $0.9746 < p_k < 0.9960$.

## 10.7 Miscellaneous problems of linear estimation

The RAYLEIGH distribution is a special WEIBULL distribution having $c = 2$; i.e., its density function reads

$$f(x) = \frac{2\,x}{b^2} \exp\left\{ -\left( \frac{x}{b} \right)^2 \right\}, \quad x \geq 0, \; b > 0.$$

DYER/WHISENAND (1973a) presented the BLUE of $b$ for samples of size $n = 2(1)15$ where singly type–II censoring might occur. In a second paper, DYER/WHISENAND (1973b) developed the BLUE for $b$ based on $k$ out of $n$ ordered observations, $k = 2(1)4$ and $n = 2(1)22$. They also presented the ABLE (asymptotical BLUE) based on $k = 2(1)4$ order statistics. All the above estimators require table look–ups for the coefficients of the estimator, depending on $n$ and the censoring number $r$. Based on the technique of CHERNOFF et al. (1967), D'AGOSTINO/LEE (1975) suggested an ABLE for $b$ which does not rely on tabulated coefficients.

Let $X_{1:n}, \ldots, X_{n:n}$ represent the ordered observations from a RAYLEIGH distribution. Of these $X_{\ell:n}, \ldots, X_{u:n}$, $1 \leq \ell \leq u \leq n$, are available for estimation of $b$. Denoting

$$P := \frac{\ell - 1}{n}, \quad Q := \frac{u}{n},$$

$$\overline{P} := 1 - P, \quad \overline{Q} := 1 - Q,$$

the fractions $P$ and $\overline{Q}$ have been censored from the lower and upper tail of the sample, respectively. The ABLE of $b$ is

$$\widehat{b} = \left[ \sum_{i=\ell}^{u} b_{i,n} X_{i:n} / n + b_\ell X_{\ell:n} + b_u X_{u:n} \right] \bigg/ K_{\ell,u}, \tag{10.71a}$$

where

$$b_{i:n}^2 = -2 \ln\left( 1 - \frac{i}{n+1} \right), \tag{10.71b}$$

$$b_\ell^2 = c_\ell^2 \left( -2 \ln \overline{P} \right), \quad c_\ell = -\overline{P}\left[ 1 + \left( \ln \overline{P} \right)/P \right], \tag{10.71c}$$

$$b_u^2 = \overline{Q}^2 \left( -\ln \overline{Q} \right), \tag{10.71d}$$

$$K_{\ell,u} = \frac{1}{n} \sum_{i=\ell}^{u} b_{i,n}^2 + \frac{b_\ell^2}{c_\ell} + \frac{b_u^2}{\overline{Q}}. \tag{10.71e}$$

If $\ell = 1$ (no censoring from below), then $P = 0$ and $b_\ell = b_1 = 0$. If $u = n$ (no censoring from above), then $\overline{Q} = 0$ and $b_u = b_n = 0$. If $n \to \infty$, $K_{\ell,u}$ in (10.71e) can be approximated by a simple formula not involving the sum of terms from $\ell$ to $u$:

$$\lim_{n\to\infty} K_{\ell,u} = 2\left[(Q - P) + \frac{\overline{P}}{P}\left(\ln \overline{P}\right)^2\right]. \tag{10.71f}$$

When $\ell = 1$, we further get

$$\lim_{n\to\infty} K_{1,u} = 2\,Q. \tag{10.71g}$$

The efficiency of $\widehat{b}$ in (10.71a) relative to the BLUE is extremely good for $n \geq 5$.

DUBEY (1965) developed estimators, called pseudo least squares estimators, for the scale and shape parameters of a WEIBULL distribution[19] when the items to be sampled are inexpensive. The sampling plan is as follows: Draw random samples of a given inexpensive product from an infinite (large) lot of sizes $n(t_1)$, $n(t_2)$, ..., $n(t_k)$, not necessary of equal size, at the time points $t_1 < t_2 < \ldots < t_k$. Let $G(t_i)$ be the random number of good items observed in $n(t_i)$ at time $t_i$. The estimators of $b_2 = b^c$ and $c$ are found by minimizing

$$\sum_{i=1}^{k}(Y_i - c\,\ln t_i + \ln b_2)^2, \tag{10.72a}$$

where

$$Y_i := \ln\ln \widehat{R}^{-1}(t_i) = \ln\left\{\ln[n(t_i)] - \ln[G(t_i)]\right\} \tag{10.72b}$$

and reading

$$\widehat{c} = \sum_{i=1}^{k} w_i\,Y_i, \tag{10.72c}$$

$$\widehat{b}_2 = \exp\left[-\overline{Y} + \widehat{c}\,\overline{u}\right], \tag{10.72d}$$

where

$$w_i = \frac{\ln t_i - \overline{u}}{\sum_{i=1}^{k}(\ln t_i - \overline{u})^2}$$

$$\overline{Y} = \frac{1}{k}\sum_{i=1}^{k} y_i \quad \text{and} \quad \overline{u} = \frac{1}{k}\sum_{i=1}^{k}\ln t_i.$$

$\widehat{c}$ and $\widehat{b}_2$ are asymptotically normally distributed. The asymptotic variances and covariances

---

[19]  The location parameter is assumed to be zero.

of $\widehat{c}$ and $\widehat{b}_2$ are

$$\mathrm{Var}\big(\widehat{c}\,\big) \;\; = \;\; \sum_{i=1}^{k} w_i^2 \,\mathrm{Var}(Y_i), \tag{10.72e}$$

$$\mathrm{Var}\big(\widehat{b}_2\big) \;\; = \;\; b_2^2 \sum_{i=1}^{k} \left(\overline{u}\,w_i - \frac{1}{k}\right)^2 \mathrm{Var}(Y_i), \tag{10.72f}$$

$$\mathrm{Cov}\big(\widehat{c},\,\widehat{b}_2\big) \;\; = \;\; b_2 \sum_{i=1}^{k} w_i \left(\overline{u}\,w_i - \frac{1}{k}\right) \mathrm{Var}(Y_i), \tag{10.72g}$$

where the asymptotic variance of $Y_i$ is

$$\mathrm{Var}(Y_i) = b_2^2 \left\{ \exp\left(t_i^c/b_2\right) - 1 \right\} \Big/ n(t_i)\, t_i^{2c}. \tag{10.72h}$$

The DUBEY approach has some resemblance to that of HERBACH (1963) which is based on the number of failures. It is well known that if the time to failure $X$ has a WEIBULL distribution, i.e., $\Pr(X > x) = \exp\{[(x - a)/b]^c\}$, then the reduced variate $Z = (X - a)^c$ has an exponential distribution with mean $b^c$. It is equally well known that if the times between failures occur independently according to an exponential distribution, the number of failures in a fixed time interval has a POISSON distribution.

We finally mention a paper of SKINNER/HUMPHREYS (1999) that treats linear estimation when the WEIBULL variates are error–contaminated.

# 11  Parameter estimation — Maximum likelihood approaches[1]

Maximum likelihood (ML) is by far the most popular method of estimation. It is generally credited to R. A. FISHER (1890 – 1962), although its roots date back as far as J. H. LAMBERT (1728 – 1777), DANIEL BERNOULLI (1708 – 1782) and J. L. LAGRANGE (1736 – 1813) in the eighteenth century (see EDWARDS (1972) for a historical account). FISHER introduced this method as an alternative to the method of moments (see Chapter 12) and to the method of least squares (see the previous chapter). The former method FISHER criticized for being arbitrary in the choice of the moment equations and the latter for not being invariant under scale changes in the variables.

ML is considered to be one of the most versatile and reliable methods. Its properties will be presented and discussed in Sect. 11.2. ML copes with all types of samples, whether uncensored or censored in one way or the other and whether the data are grouped or not. One writes down the **likelihood function** (see Sect. 11.1) or its logarithm, the **log–likelihood function**. The likelihood function depends on the distribution of the sampled universe, its parameters and the data and the way these data have been collected. Sect. 11.1 shows how to set up a likelihood function. The maximum likelihood estimators are those parameter values that maximize the likelihood function, given the data available. The process of maximizing will be treated in Sect. 11.2 in conjunction with the statistical properties, which partly depend on the maximizing process and vice versa. The exact distribution of many maximum likelihood estimators (MLEs) and the confidence limits of the estimated parameters are not always known, as is the case with the WEIBULL distribution. However, they are given approximately by the large sample theory, which involves the **asymptotic variance–covariance matrix** and the **FISHER information matrix**. Sections 11.3 through 11.8 will go into the details of MLEs for different kinds of samples, including small–sample and large–sample theory.

## 11.1  Likelihood functions and likelihood equations

The likelihood is a tool of statistical inference used in testing (see Chapter 21) as well as in estimating. For a given, realized sample the likelihood numerically expresses the chance or plausibility of just having realized the sample at hand under any set of those parameters governing the sampled universe. The term *likelihood* is thus distinct from the term *inverse probability* used in BAYES inference (see Chapter 14). The likelihood is a function of the

---

[1]  Estimation by maximum likelihood is treated in almost every textbook on statistical inference. An easy to read introduction to the principles of likelihood is the monograph of EDWARDS (1972).

distributional parameters $\theta_j$ ($j = 1, \ldots, m$), collected in the vector $\boldsymbol{\theta}$, given the sampled data. Formally and depending on the nature of the sampled variate (continuous or discrete), the likelihood is a joint density, a joint probability or a combination thereof, but the roles of the variables and the parameters in this joint function have been interchanged.

As a rule, the data consist of a vector of $n$ independent observations so that the likelihood is a product of $n$ simple factors termed **likelihood elements** $L_i$, expressing the likelihood of an individual observation. Let $\boldsymbol{\theta}$ be the vector of the unknown parameters and $x_i$ an observation of a continuous variate $X$ with density function $f(x \,|\, \boldsymbol{\theta})$ and distribution function $F(x \,|\, \boldsymbol{\theta})$. Then we may have the following likelihood elements:

- **uncensored observation** $x_i$ of $X$

$$L_i(\boldsymbol{\theta}) := f(x_i \,|\, \boldsymbol{\theta}); \tag{11.1a}$$

- **interval censored observation**[2] of $X$

$$
\begin{aligned}
L_i(\boldsymbol{\theta}) &= \int_{t_{j-1}}^{t_j} f(x \,|\, \boldsymbol{\theta}) \, \mathrm{d}x, \quad x \in (t_{j-1}, t_j], \\
&= F(t_j \,|\, \boldsymbol{\theta}) - F(t_{j-1} \,|\, \boldsymbol{\theta}) \\
&= \Pr(t_{j-1} < X_i \le t_j \,|\, \boldsymbol{\theta});
\end{aligned} \right\} \tag{11.1b}
$$

- **observation singly censored on the left** at $T$

$$
\begin{aligned}
L_i &= \int_{-\infty}^{T} f(x \,|\, \boldsymbol{\theta}) \, \mathrm{d}x, \quad x_i \le T, \\
&= F(T \,|\, \boldsymbol{\theta}) \\
&= \Pr(X_i \le T);
\end{aligned} \right\} \tag{11.1c}
$$

- **observation singly censored on the right** at $T$

$$
\begin{aligned}
L_i &= \int_{T}^{\infty} f(x \,|\, \boldsymbol{\theta}) \, \mathrm{d}x, \quad x_i > T, \\
&= 1 - F(T \,|\, \boldsymbol{\theta}) = R(T \,|\, \boldsymbol{\theta}) \\
&= \Pr(X_i > T).
\end{aligned} \right\} \tag{11.1d}
$$

- **randomly censored observation**

  Let $C$, the censoring variate, be continuous with density $g(c \,|\, \boldsymbol{\psi})$ and distribution function $G(c \,|\, \boldsymbol{\psi})$. We will observe $Y_i = \min(X_i, C_i)$ together with the indicator

$$
\delta_i := I(X_i \le C_i) = \left\{ \begin{array}{ll} 1 & \text{for} \quad X_i \le C_i \\ 0 & \text{for} \quad X_i > C_i. \end{array} \right\}
$$

---

[2] Interval censored data are also termed **grouped data**.

When $X_i$ and $C_i$ are independent variates we have the following likelihood element of an observation $(y_i, \delta_i)$:

$$
L_i(\boldsymbol{\theta}, \boldsymbol{\psi}) = \left\{
\begin{array}{ll}
f(y_i \mid \boldsymbol{\theta})\,[1 - G(y_i \mid \boldsymbol{\psi})] & \text{for} \quad \delta_i = 1 \\[2mm]
g(y_i \mid \boldsymbol{\psi})\,[1 - F(y_i \mid \boldsymbol{\theta})] & \text{for} \quad \delta_i = 0.
\end{array}
\right\}
\tag{11.1e}
$$

Furthermore, if the parameter vectors $\boldsymbol{\theta}$ and $\boldsymbol{\psi}$ have no elements in common, (11.1e) reduces to

$$
\begin{aligned}
L_i(\boldsymbol{\theta}) \;&=\; con\; f(y_i \mid \boldsymbol{\theta})^{\delta_i}\, R(y_i \mid \boldsymbol{\theta})^{1-\delta_i} \\[2mm]
&=\; con \left\{
\begin{array}{ll}
f(y_i \mid \boldsymbol{\theta}) & \text{for} \quad \delta_i = 1 \\[2mm]
R(y_i \mid \boldsymbol{\theta}) & \text{for} \quad \delta_i = 0,
\end{array}
\right\}
\end{aligned}
\tag{11.1f}
$$

where $con$ is a constant not involving any parameters of $X$.

The **likelihood function** of an independent sample of size $n$ is

$$
L(\boldsymbol{\theta} \mid \text{data}) = K \prod_{i=1}^{n} L_i(\boldsymbol{\theta}),
\tag{11.2a}
$$

where $K$ is a combinatorial constant giving the number of ways the sample might have been observed and thus is free of any parameter in $\boldsymbol{\theta}$. To simplify the notation we will omit $K$ and the hint to the data, so that we write

$$
L(\boldsymbol{\theta}) = \prod_{i=1}^{n} L_i(\boldsymbol{\theta}).
\tag{11.2b}
$$

Strictly speaking we have $L(\boldsymbol{\theta}) \propto \prod_{i=1}^{n} L_i(\boldsymbol{\theta})$.

The idea of maximum likelihood estimation is to find that vector $\widehat{\boldsymbol{\theta}} := \widehat{\boldsymbol{\theta}}_{ML}$, which maximizes (11.2b), thus assigns the highest chance of having realized the data at hand. Maximization of (11.2b) is rarely done by some procedure of direct optimization, e.g., by the method of grid search, but usually by some gradient method built upon partial derivatives of (11.2b), assuming all parameters $\theta_1, \ldots, \theta_m$ to be continuous. The process of forming these derivatives is made easier by departing from the **log–likelihood function** which is a sum instead of a product:[3]

$$
\begin{aligned}
\mathcal{L}(\boldsymbol{\theta}) \;&:=\; \ln L(\boldsymbol{\theta}) \\[2mm]
&=\; \sum_{i=1}^{n} \mathcal{L}_i(\boldsymbol{\theta}), \quad \mathcal{L}_i(\boldsymbol{\theta}) := \ln L_i(\boldsymbol{\theta}).
\end{aligned}
\tag{11.2c}
$$

The partial derivative of $\mathcal{L}(\boldsymbol{\theta})$ with respect to parameter $\theta_i$ $(i = 1, \ldots, m)$ is denoted by $\partial \mathcal{L}(\boldsymbol{\theta})/\partial \theta_i$, and

$$
\frac{\partial \mathcal{L}(\boldsymbol{\theta})}{\partial \theta_i} = 0
\tag{11.3a}
$$

---

[3] As the logarithmic transformation is isotonic, the extremal points $\widehat{\theta}_\ell$ of $L(\boldsymbol{\theta})$ and $\mathcal{L}(\boldsymbol{\theta})$ will be the same.

is called a **likelihood equation**. The system of all $m$ likelihood equations,

$$\frac{\mathcal{L}(\boldsymbol{\theta})}{\partial \boldsymbol{\theta}} := \left( \frac{\mathcal{L}(\boldsymbol{\theta})}{\partial \theta_1}, \ldots, \frac{\mathcal{L}(\boldsymbol{\theta})}{\partial \theta_m} \right)' = \boldsymbol{o}, \tag{11.3b}$$

$\boldsymbol{o}$ being a vector of zeros, is often termed **system of normal equations**.

The solutions $\widehat{\theta}_1, \ldots, \widehat{\theta}_m$ are stationary points of the log–likelihood function and might include the MLE as the global maximum.

**Example 11/1:** **Normal equations of an uncensored sample from a three–parameter WEIBULL distribution**

For an uncensored sample of size $n$ from a three–parameter WEIBULL distribution, i.e., $X_i \overset{iid}{\sim} We(a, b, c)$, we get:

$$L_i(a, b, c) = \frac{c}{b} \left( \frac{x_i - a}{b} \right)^{c-1} \exp\left\{ -\left( \frac{x_i - a}{b} \right)^c \right\}, \tag{11.4a}$$

$$\mathcal{L}_i(a, b, c) = \ln c - c \ln b + (c - 1) \ln(x_i - a) - \left( \frac{x_i - a}{b} \right)^c, \tag{11.4b}$$

$$\mathcal{L}(a, b, c) = n \left[ \ln c - c \ln b \right] + (c - 1) \sum_{i=1}^{n} \ln(x_i - a) - \sum_{i=1}^{n} \left( \frac{x_i - a}{b} \right)^c, \tag{11.4c}$$

$$\frac{\partial \mathcal{L}(a, b, c)}{\partial a} = -(c - 1) \sum_{i=1}^{n} \frac{1}{x_i - a} + \frac{c}{b} \sum_{i=1}^{n} \left( \frac{x_i - a}{b} \right)^{c-1} = 0, \tag{11.4d}$$

$$\frac{\partial \mathcal{L}(a, b, c)}{\partial b} = -\frac{n\,c}{b} + \frac{c}{b} \sum_{i=1}^{n} \left( \frac{x_i - a}{b} \right)^c = 0, \tag{11.4e}$$

$$\frac{\partial \mathcal{L}(a, b, c)}{\partial c} = \frac{n}{c} - n \ln b + \sum_{i=1}^{n} \ln(x_i - a) - \sum_{i=1}^{n} \left( \frac{x_i - a}{b} \right)^c \ln\left( \frac{x_i - a}{b} \right) = 0. \tag{11.4f}$$

# 11.2 Statistical and computational aspects of MLEs

There are cases where the MLEs, i.e., the solution to (11.3b), are unique and can be written in closed form so that the exact distribution and the small–sample properties of $\widehat{\boldsymbol{\theta}}$ may be found. These properties are not always favorable and do not coincide with the most favorable properties in the asymptotic case ($n \to \infty$), which will be discussed in Sect. 11.2.1. When the MLEs cannot be given in closed form because (11.3b) is a system of interdependent non–linear equations, which will be the case for the WEIBULL distribution, we have to apply iterative methods, carried through to convergence or terminated after reaching some given stopping criterion, to calculate or approximate the MLE resulting in a so–called **iterated MLE**. Sect. 11.2.2 comments upon the iterative solution, but the details of those algorithms which are used in finding the roots of (11.3b) are postponed to Sections 11.3 – 11.8. The only way to judge an iterated MLE is by means of the asymptotic properties.

### 11.2.1 Asymptotic properties of MLEs

The following discussion motivates the ML theory for complete data, but it can be carried over to censored samples. It is a heuristic proof of theoretical results[4] and the key result is summarized in the following theorem using these notations:

$$\widehat{\boldsymbol{\theta}} \quad - \quad \text{MLE of } \boldsymbol{\theta},$$

$$\boldsymbol{\theta}_0 \quad - \quad \text{true value of the parameter vector } \boldsymbol{\theta},$$

$$\mathrm{E}_0(\cdot) \quad - \quad \text{expectation based on the true parameter values.}$$

**Theorem on MLE**

Under regularity conditions given below and if the likelihood function attains its maximum at $\widehat{\boldsymbol{\theta}}$ being the unique solution to (11.3b) and an interior point of the parameter space, the MLE has the following asymptotic properties:

1. **Consistency**

$$\mathrm{plim}\ \widehat{\boldsymbol{\theta}} = \boldsymbol{\theta}_0. \tag{11.5a}$$

2. **Asymptotic normality**

$$\widehat{\boldsymbol{\theta}} \overset{\mathrm{asym}}{\sim} No\big(\boldsymbol{\theta}_0, \big[\boldsymbol{I}(\boldsymbol{\theta}_0)\big]^{-1}\big), \tag{11.5b}$$

where

$$\boldsymbol{I}(\boldsymbol{\theta}_0) = -\mathrm{E}_0\big[\partial^2 \mathcal{L} / (\partial \boldsymbol{\theta}_0\, \partial \boldsymbol{\theta}_0')\big]. \tag{11.5c}$$

Thus the MLE $\widehat{\boldsymbol{\theta}}$ of $\boldsymbol{\theta}_0$ is asymptotically unbiased,

$$\lim_{n \to \infty} \mathrm{E}\big(\widehat{\boldsymbol{\theta}}_n\big) = \boldsymbol{\theta}_0 \tag{11.5d}$$

with asymptotic variance–covariance matrix

$$\lim_{n \to \infty} \mathrm{Var}\big(\widehat{\boldsymbol{\theta}}_n\big) = \big[\boldsymbol{I}(\boldsymbol{\theta}_0)\big]^{-1}. \tag{11.5e}$$

3. **Asymptotic efficiency**
   $\widehat{\boldsymbol{\theta}}$ is asymptotically efficient and achieves the CRAMÉR–RAO lower bound for a consistent estimator given in (11.5c).

4. **Invariance**
   The MLE of $\boldsymbol{\gamma}_0 = \varphi(\boldsymbol{\theta}_0)$ is $\varphi(\widehat{\boldsymbol{\theta}})$ if $\varphi(\cdot)$ is a continuous and continuously differentiable function. ∎

The following **regularity conditions**, under which the above theorem holds, are informal. A more rigorous treatment may be found in STUART/ORD (1991):

1. The first three derivatives of $\ln f(x_i \,|\, \boldsymbol{\theta})$ with respect to $\boldsymbol{\theta}$ are continuous for almost every $x_i$ and for all $\boldsymbol{\theta}$. This condition ensures the existence of a certain TAYLOR–series expansion and the finite variance of the derivatives of $\mathcal{L}(\boldsymbol{\theta})$.

---

[4] A thorough discussion of ML theory is to be found in the literature of advanced statistical theory, e.g., CRAMÉR (1971), LEHMANN (1983) or STUART/ORD (1991).

2. The conditions necessary to obtain the expectations of the first and second derivatives of $\ln f(x_i \mid \boldsymbol{\theta})$ are met.

3. For all values of $\boldsymbol{\theta}$, $\left| \partial^3 \ln f(x_i \mid \boldsymbol{\theta}) / (\partial \theta_j \, \partial \theta_k \, \partial \theta_\ell) \right|$ is less than a function that has a finite expectation. This condition will allow us to truncate the TAYLOR–series. ■

Densities that are "regular" in the sense above have three properties which are given without proof and which in turn are used in establishing the properties of MLEs.

**First property**
Because we are taking a random sample,

$$
\left.
\begin{aligned}
\mathcal{L}_i &= \ln\left[ f(X_i \mid \boldsymbol{\theta}) \right], \\[4pt]
\boldsymbol{g}_i &= \frac{\partial \mathcal{L}_i}{\partial \boldsymbol{\theta}}, \\[4pt]
\boldsymbol{H}_i &= \frac{\partial^2 \mathcal{L}_i}{\partial \boldsymbol{\theta} \, \partial \boldsymbol{\theta}'}
\end{aligned}
\right\} i = 1, \ldots, n
$$

are all random quantities. $\boldsymbol{H}_i$ is the **HESSIAN matrix** of $\mathcal{L}_i$. The notations $\boldsymbol{g}_i(\boldsymbol{\theta}_0)$ and $\boldsymbol{H}_i(\boldsymbol{\theta}_0)$ indicate the derivatives evaluated at $\boldsymbol{\theta}_0$.

**Second property**
The mean of the random vector $\boldsymbol{g}_i(\boldsymbol{\theta}_0)$ of first–order derivatives is

$$
\mathrm{E}_0\!\left[ \boldsymbol{g}_i(\boldsymbol{\theta}_0) \right] = \boldsymbol{o}. \tag{11.6a}
$$

**Third property**
The variance–covariance matrix of $\boldsymbol{g}_i(\boldsymbol{\theta}_0)$ is

$$
\mathrm{Var}\!\left[ \boldsymbol{g}_i(\boldsymbol{\theta}_0) \right] = \mathrm{E}_0\!\left[ \left( \frac{\partial \boldsymbol{g}_i(\boldsymbol{\theta}_0)}{\partial \boldsymbol{\theta}_0} \right) \left( \frac{\partial \boldsymbol{g}_i(\boldsymbol{\theta}_0)}{\partial \boldsymbol{\theta}_0'} \right) \right] \tag{11.6b}
$$

$$
= -\mathrm{E}_0\!\left[ \frac{\partial^2 \boldsymbol{g}_i(\boldsymbol{\theta}_0)}{\partial \boldsymbol{\theta}_0 \partial \boldsymbol{\theta}_0'} \right] = -\mathrm{E}_0\!\left[ \boldsymbol{H}_i(\boldsymbol{\theta}_0) \right]. \tag{11.6c}
$$

(11.6b) gives the variance–covariance matrix as the expected square (outer product) of the first derivative vector, which in general is easier to evaluate than (11.6c), being the negative of the expected second derivatives matrix.

The log–likelihood function is random too:

$$
\mathcal{L}(\boldsymbol{\theta} \mid \boldsymbol{x}) = \sum_{i=1}^{n} \mathcal{L}_i(\boldsymbol{\theta} \mid X_i).
$$

The first derivative vector of $\mathcal{L}(\boldsymbol{\theta} \mid \boldsymbol{x})$, termed **score vector**, is

$$
\boldsymbol{g} = \frac{\partial \mathcal{L}(\boldsymbol{\theta} \mid \boldsymbol{x})}{\partial \boldsymbol{\theta}} = \sum_{i=1}^{n} \frac{\partial \mathcal{L}_i}{\partial \boldsymbol{\theta}} = \sum_{i=1}^{n} \boldsymbol{g}_i. \tag{11.7a}
$$

Since we are just adding terms, it follows from (11.6a) that at $\boldsymbol{\theta}_0$ the **mean score** is

$$
\mathrm{E}_0\!\left[ \frac{\partial \mathcal{L}(\boldsymbol{\theta}_0 \mid \boldsymbol{x})}{\partial \boldsymbol{\theta}_0} \right] = \mathrm{E}_0\!\left[ \boldsymbol{g}_0 \right] = \boldsymbol{o}. \tag{11.7b}
$$

The HESSIAN of the log–likelihood is

$$H = \frac{\partial^2 \mathcal{L}(\boldsymbol{\theta} \mid \boldsymbol{x})}{\partial \boldsymbol{\theta} \, \partial \boldsymbol{\theta}'} = \sum_{i=1}^{n} \frac{\partial^2 \mathcal{L}_i}{\partial \boldsymbol{\theta} \, \partial \boldsymbol{\theta}'} = \sum_{i=1}^{n} \boldsymbol{H}_i. \tag{11.7c}$$

Evaluating once again at $\boldsymbol{\theta}_0$ by taking

$$\mathrm{E}_0[\boldsymbol{g}_0 \, \boldsymbol{g}_0'] = \mathrm{E}_0 \left[ \sum_{i=1}^{n} \sum_{j=1}^{n} \boldsymbol{g}_{0i} \, \boldsymbol{g}_{0j}' \right]$$

and, because of the first property above, dropping terms with unequal subscripts, we obtain

$$\begin{aligned} \mathrm{E}_0[\boldsymbol{g}_0 \, \boldsymbol{g}_0'] &= \mathrm{E}_0 \left[ \sum_{i=1}^{n} \boldsymbol{g}_{0i} \, \boldsymbol{g}_{0i}' \right] \\[2mm] &= \mathrm{E}_0 \left[ \sum_{i=1}^{n} \left( - \boldsymbol{H}_{0i} \right) \right] \\[2mm] &= -\mathrm{E}_0[\boldsymbol{H}_0], \end{aligned} \tag{11.7d}$$

so that the variance–covariance matrix of the score vector at $\boldsymbol{\theta}_0$ is

$$\begin{aligned} \mathrm{Var}_0 \left[ \frac{\partial \mathcal{L}(\boldsymbol{\theta}_0 \mid \boldsymbol{x})}{\partial \boldsymbol{\theta}_0} \right] &= \mathrm{E}_0 \left[ \left( \frac{\partial \mathcal{L}(\boldsymbol{\theta}_0 \mid \boldsymbol{x})}{\partial \boldsymbol{\theta}_0} \right) \left( \frac{\partial \mathcal{L}(\boldsymbol{\theta}_0 \mid \boldsymbol{x})}{\partial \boldsymbol{\theta}_0'} \right) \right] \tag{11.8a} \\[2mm] &= -\mathrm{E}_0 \left[ \frac{\partial^2 \mathcal{L}(\boldsymbol{\theta}_0 \mid \boldsymbol{x})}{\partial \boldsymbol{\theta}_0 \, \partial \boldsymbol{\theta}_0'} \right]. \tag{11.8b} \end{aligned}$$

This useful result is known as the **information matrix equality** since

$$\boldsymbol{I}(\boldsymbol{\theta}_0) := \mathrm{Var}_0 \left( \frac{\partial \mathcal{L}(\boldsymbol{\theta}_0 \mid \boldsymbol{x})}{\partial \boldsymbol{\theta}_0} \right) \tag{11.8c}$$

is the **FISHER information matrix**.

Based on the results (11.6a) through (11.8c), we now establish the asymptotic normality of $\widehat{\boldsymbol{\theta}}$, the MLE of $\boldsymbol{\theta}$. At the MLE, the gradient of the log–likelihood equals zero

$$\frac{\partial \mathcal{L}(\widehat{\boldsymbol{\theta}})}{\partial \widehat{\boldsymbol{\theta}}} = \boldsymbol{g}(\widehat{\boldsymbol{\theta}}) = \boldsymbol{o}.$$

(This is the sample statistic, not the expectation!) We now expand this set of equations in a second–order TAYLOR–series around the parameter $\boldsymbol{\theta}_0$. We will use the mean value theorem to truncate the TAYLOR–series at the second term:

$$\boldsymbol{g}(\widehat{\boldsymbol{\theta}}) = \boldsymbol{g}(\boldsymbol{\theta}_0) + \boldsymbol{H}(\widetilde{\boldsymbol{\theta}})(\widehat{\boldsymbol{\theta}} - \boldsymbol{\theta}_0) = \boldsymbol{o}. \tag{11.9a}$$

The HESSIAN is evaluated at a point $\widetilde{\boldsymbol{\theta}}$ that is between $\widehat{\boldsymbol{\theta}}$ and $\boldsymbol{\theta}_0$, i.e., $\widetilde{\boldsymbol{\theta}} = w \widehat{\boldsymbol{\theta}} + (1 - w) \boldsymbol{\theta}_0$ for some $0 < w < 1$. We than rearrange this function and multiply the result by $\sqrt{n}$ to obtain

$$\sqrt{n} \left( \widehat{\boldsymbol{\theta}} - \boldsymbol{\theta}_0 \right) = \left[ - \boldsymbol{H}(\widetilde{\boldsymbol{\theta}}) \right]^{-1} \left[ \sqrt{n} \, \boldsymbol{g}(\boldsymbol{\theta}_0) \right]. \tag{11.9b}$$

Because $\text{plim}\big(\widehat{\theta} - \theta_0\big) = o$, $\text{plim}\big(\widehat{\theta} - \widetilde{\theta}\big) = o$. The second derivatives are continuous functions. Therefore, if the limiting distribution exists, then

$$\sqrt{n}\,\big(\widehat{\theta} - \theta_0\big) \stackrel{\mathrm{d}}{\to} \big[-\boldsymbol{H}(\theta_0)\big]^{-1}\big[\sqrt{n}\,\boldsymbol{g}(\theta_0)\big]. \tag{11.9c}$$

Upon dividing $\boldsymbol{H}(\theta_0)$ and $\boldsymbol{g}(\theta_0)$ by $n$, we obtain

$$\sqrt{n}\,\big(\widehat{\theta} - \theta_0\big) \stackrel{\mathrm{d}}{\to} \left[-\frac{1}{n}\boldsymbol{H}(\theta_0)\right]^{-1}\big[\sqrt{n}\,\overline{\boldsymbol{g}}(\theta_0)\big]. \tag{11.9d}$$

We may apply the LINDEBERG–LEVY central limit theorem to $\big[\sqrt{n}\,\overline{\boldsymbol{g}}(\theta_0)\big]$ since it is $\sqrt{n}$–times the mean of a random sample. The limiting variance of $\big[\sqrt{n}\,\overline{\boldsymbol{g}}(\theta_0)\big]$ is $-\text{E}_0\left[\dfrac{1}{n}\boldsymbol{H}(\theta_0)\right]$, so

$$\sqrt{n}\,\overline{\boldsymbol{g}}(\theta_0) \stackrel{\mathrm{d}}{\to} No\left(o, -\text{E}_0\left[\frac{1}{n}\boldsymbol{H}(\theta_0)\right]\right). \tag{11.9e}$$

By virtue of CHEBYCHEV's inequality we have $\text{plim}\left[\dfrac{1}{n}\boldsymbol{H}(\theta_0)\right] = -\text{E}_0\left[\dfrac{1}{n}\boldsymbol{H}(\theta_0)\right]$. Since this result is a constant matrix, we can combine results to obtain

$$\left.\begin{array}{l}\left[-\dfrac{1}{n}\boldsymbol{H}(\theta_0)\right]^{-1}\big[\sqrt{n}\,\overline{\boldsymbol{g}}(\theta_0)\big] \stackrel{\mathrm{d}}{\to} \\[2mm] No\left(o, \left\{-\text{E}_0\left[\dfrac{1}{n}\boldsymbol{H}(\theta_0)\right]\right\}^{-1}\left\{-\text{E}_0\left[\dfrac{1}{n}\boldsymbol{H}(\theta_0)\right]\right\}\left\{-\text{E}_0\left[\dfrac{1}{n}\boldsymbol{H}(\theta_0)\right]\right\}^{-1}\right)\end{array}\right\} \tag{11.9f}$$

or

$$\sqrt{n}\,\big(\widehat{\theta} - \theta_0\big) \stackrel{\mathrm{d}}{\to} No\left(0, \left\{-\text{E}_0\left[\frac{1}{n}\boldsymbol{H}(\theta_0)\right]\right\}^{-1}\right), \tag{11.9g}$$

which gives the asymptotic distribution of the MLE:

$$\widehat{\theta} \stackrel{\text{asym}}{\sim} No\left(\theta_0, \big[\boldsymbol{I}(\theta_0)\big]^{-1}\right). \tag{11.9h}$$

---

**Example 11/2:** HESSIAN of $\mathcal{L}(\theta\,|\,x)$, FISHER information matrix $\boldsymbol{I}(\theta)$ and its inverse for an uncensored sample from a three–parameter WEIBULL distribution

Referring to (11.4d–f) and substituting the observed value $x_i$ by the variate $X_i$, we have the following elements of $\boldsymbol{H}(\boldsymbol{\theta}) = \partial^2 \mathcal{L}(\boldsymbol{\theta} \mid \boldsymbol{x}) / (\partial \boldsymbol{\theta} \, \partial \boldsymbol{\theta}')$ where $\boldsymbol{\theta}' = (a, b, c)$:[5]

$$\frac{\partial^2 \mathcal{L}}{\partial a^2} = -(c-1) \left[ \sum \left( \frac{1}{X_i - a} \right)^2 + \frac{c}{b^2} \sum \left( \frac{X_i - a}{b} \right)^{c-2} \right], \tag{11.10a}$$

$$\frac{\partial^2 \mathcal{L}}{\partial a \, \partial b} = \frac{\partial^2 \mathcal{L}}{\partial b \, \partial a} = -\left( \frac{c}{b} \right)^2 \sum \left( \frac{X_i - a}{b} \right)^{c-1}, \tag{11.10b}$$

$$\frac{\partial^2 \mathcal{L}}{\partial a \, \partial c} = \frac{\partial^2 \mathcal{L}}{\partial c \, \partial a} = -\sum \frac{1}{X_i - a} + \frac{c}{b} \sum \left( \frac{X_i - a}{b} \right)^{c-1} \ln \left( \frac{X_i - a}{b} \right) + \frac{1}{b} \sum \left( \frac{X_i - a}{b} \right)^{c-1}, \tag{11.10c}$$

$$\frac{\partial^2 \mathcal{L}}{\partial b^2} = \frac{c}{b^2} \left[ n - (c-1) \sum \left( \frac{X_i - a}{b} \right)^c \right], \tag{11.10d}$$

$$\frac{\partial^2 \mathcal{L}}{\partial b \, \partial c} = \frac{\partial^2 \mathcal{L}}{\partial c \, \partial b} = -\frac{1}{b} \left[ n - \sum \left( \frac{X_i - a}{b} \right)^c - c \sum \left( \frac{X_i - a}{b} \right)^c \ln \left( \frac{X_i - a}{b} \right) \right], \tag{11.10e}$$

$$\frac{\partial^2 \mathcal{L}}{\partial c^2} = -\frac{n}{c^2} - \sum \left( \frac{X_i - a}{b} \right)^c \left[ \ln \left( \frac{X_i - a}{b} \right) \right]^2. \tag{11.10f}$$

The expectations of (11.10a–f) with respect to each $X_i$ and multiplied by $(-1)$ are the elements of the FISHER information matrix $\boldsymbol{I}(\boldsymbol{\theta})$:[6]

$$-\mathrm{E} \left( \frac{\partial^2 \mathcal{L}}{\partial a^2} \right) = \frac{n \, (c-1)^2}{b^2} \Gamma \left( 1 - \frac{2}{c} \right), \tag{11.11a}$$

$$-\mathrm{E} \left( \frac{\partial^2 \mathcal{L}}{\partial a \, \partial b} \right) = -\mathrm{E} \left( \frac{\partial^2 \mathcal{L}}{\partial b \, \partial a} \right) = -\frac{n \, c^2}{b^2} \Gamma \left( 2 - \frac{1}{c} \right), \tag{11.11b}$$

$$-\mathrm{E} \left( \frac{\partial^2 \mathcal{L}}{\partial a \, \partial c} \right) = -\mathrm{E} \left( \frac{\partial^2 \mathcal{L}}{\partial c \, \partial a} \right) = -\frac{n}{b} \left( 1 - \frac{1}{c} \right) \Gamma \left( 1 - \frac{1}{c} \right) \left[ 1 + \psi \left( 1 - \frac{1}{c} \right) \right] \tag{11.11c}$$

$$-\mathrm{E} \left( \frac{\partial^2 \mathcal{L}}{\partial b^2} \right) = \frac{n \, c^2}{b^2}, \tag{11.11d}$$

$$-\mathrm{E} \left( \frac{\partial^2 \mathcal{L}}{\partial b \, \partial c} \right) = -\mathrm{E} \left( \frac{\partial^2 \mathcal{L}}{\partial c \, \partial b} \right) = -\frac{n}{b} \Gamma'(2) = -\frac{n \, (1 - \gamma)}{b} \approx -0.422784 \, \frac{n}{b}, \tag{11.11e}$$

$$-\mathrm{E} \left( \frac{\partial^2 \mathcal{L}}{\partial c^2} \right) = \frac{n}{c^2} \left[ 1 + \Gamma''(2) \right] = \frac{n}{c^2} \left[ \frac{\pi^2}{6} + (1 - \gamma^2) \right] \approx 1.823680 \, \frac{n}{c^2}. \tag{11.11f}$$

---

[5]  These results are also given in FUKUTA (1963) and HEO/SALAS/KIM (2001).

[6]  $\psi(\cdot)$ denotes the digamma function.

The symmetric variance–covariance matrix of $\widehat{\boldsymbol{\theta}}$ results as the inverse of $\boldsymbol{I}(\boldsymbol{\theta})$:

$$\text{Var}(\widehat{\boldsymbol{\theta}}) = \left[\boldsymbol{I}(\boldsymbol{\theta})\right]^{-1} = \frac{1}{n\,D} \begin{pmatrix} \dfrac{B\,b^2}{(c-1)^2} & -\dfrac{H\,b^2}{c\,(c-1)} & \dfrac{F\,b\,c}{c-1} \\[2mm] - & \dfrac{A\,b^2}{c^2} & b\,G \\[2mm] - & - & c^2\,C \end{pmatrix}, \qquad (11.12a)$$

where

$$A \;=\; \Gamma\!\left(1-\frac{2}{c}\right)\left[1+\Gamma''(2)\right] - \Gamma^2\!\left(1-\frac{1}{c}\right)\left[1+\psi\!\left(1-\frac{1}{c}\right)\right]^2, \qquad (11.12b)$$

$$B \;=\; 1+\Gamma''(2)-\left[\Gamma'(2)\right]^2 \;=\; \frac{\pi^2}{6}, \qquad (11.12c)$$

$$C \;=\; \Gamma\!\left(1-\frac{2}{c}\right) - \Gamma^2\!\left(1-\frac{1}{c}\right), \qquad (11.12d)$$

$$F \;=\; \Gamma\!\left(1-\frac{1}{c}\right)\left[1-\Gamma'(2)+\psi\!\left(1-\frac{1}{c}\right)\right], \qquad (11.12e)$$

$$G \;=\; \Gamma\!\left(1-\frac{2}{c}\right)\Gamma'(2) - \Gamma^2\!\left(1-\frac{1}{c}\right)\left[1+\psi\!\left(1-\frac{1}{c}\right)\right], \qquad (11.12f)$$

$$H \;=\; \Gamma\!\left(1-\frac{1}{c}\right)\left\{1+\Gamma''(2)-\Gamma'(2)\left[1+\psi\!\left(1-\frac{1}{c}\right)\right]\right\}, \qquad (11.12g)$$

$$D \;=\; B\,C+F^2. \qquad (11.12h)$$

The classical regularity conditions for the asymptotic properties of MLEs are not satisfied when $X \sim We(a,b,c)$ because the support of $X$ depends upon one of the parameters. It is easily checked from (11.11a–f) that, when $c > 2$, the FISHER information is finite, and it is widely known that the classical properties hold in this case. For $c \le 2$ the FISHER information is infinite (see (11.11a)), so the classical results are certainly not valid. SMITH (1985) confirms that the classical results hold when $c > 2$ and studies the case $c \le 2$ in detail. His surprising result is that, in the latter case, estimation of $a$ and the other parameters are asymptotically independent: each of the MLEs of $a$ and $b$, $c$ has the same asymptotic distribution when the other is unknown as when the other is known and these asymptotic distributions are independent. For $c = 2$ (RAYLEIGH distribution) the MLEs are asymptotically efficient and normally distributed but with different rates of convergence, $n^{-1/2}\,[\ln n]^{-1}$ for $b$ and $c$, $[n\,\ln n]^{-1/2}$ for $a$. For $1 < c < 2$ there exists a consistent sequence of MLEs as the sample size tends to infinity. For $c \le 1$ no consistent MLEs exist at all.

**Example 11/3:** $\mathcal{L}(\theta \,|\, x)$, **normal equations, HESSIAN, $I(\theta)$ and $\big[I(\theta)\big]^{-1}$ for an uncensored sample from a two–parameter WEIBULL distribution**

Let $x = (X_1, \ldots, X_n)$ with $X_i \overset{\text{iid}}{\sim} We(0, b, c)$ for $i = 1, \ldots, n$. The corresponding log–likelihood function is

$$\mathcal{L}(b, c) = n\left[\ln c - c \ln b\right] + (c-1)\sum_{i=1}^{n} \ln X_i - \sum_{i=1}^{n} \left(\frac{X_i}{b}\right)^c \tag{11.13}$$

with the following system of normal equations:

$$\frac{\partial \mathcal{L}(b, c)}{\partial b} = -\frac{n\,c}{b} + \frac{b}{c}\sum \left(\frac{X_i}{b}\right)^c = 0, \tag{11.14a}$$

$$\frac{\partial \mathcal{L}(b, c)}{\partial c} = \frac{n}{c} - n\ln b + \sum \ln X_i - \sum \left(\frac{X_i}{b}\right)^c \ln\left(\frac{X_i}{b}\right) = 0. \tag{11.14b}$$

The elements of $H(\theta)$ are as follows:[7]

$$\frac{\partial^2 \mathcal{L}}{\partial b^2} = \frac{c}{b^2}\left[n - (c-1)\sum \left(\frac{X_i}{b}\right)^c\right], \tag{11.15a}$$

$$\frac{\partial^2 \mathcal{L}}{\partial b\,\partial c} = \frac{\partial^2 \mathcal{L}}{\partial c\,\partial b} = -\frac{1}{b}\left[n - \sum \left(\frac{X_i}{b}\right)^c - c\sum \left(\frac{X_i}{b}\right)^c \ln\left(\frac{X_i}{b}\right)\right], \tag{11.15b}$$

$$\frac{\partial^2 \mathcal{L}}{\partial c^2} = \frac{n}{c^2} - \sum \left(\frac{X_i}{b}\right)^c \left[\ln\left(\frac{X_i}{b}\right)\right]^2. \tag{11.15c}$$

The FISHER matrix $I(\theta) = -\mathrm{E}\big[H(\theta)\big]$ has the following elements:

$$-\mathrm{E}\left(\frac{\partial^2 \mathcal{L}}{\partial b^2}\right) = \frac{n\,c^2}{b^2} \tag{11.16a}$$

$$-\mathrm{E}\left(\frac{\partial^2 \mathcal{L}}{\partial b\,\partial c}\right) = -\mathrm{E}\left(\frac{\partial^2 \mathcal{L}}{\partial c\,\partial b}\right) = -\frac{n}{b}\,\Gamma'(2) \approx -0.422784\,\frac{n}{b}, \tag{11.16b}$$

$$-\mathrm{E}\left(\frac{\partial^2 \mathcal{L}}{\partial c^2}\right) = \frac{n}{c^2}\left[1 + \Gamma''(2)\right] \approx 1.823680\,\frac{n}{c^2}. \tag{11.16c}$$

Finally, the asymptotic variance–covariance matrix of $\widehat{\theta} = \big(\widehat{b}, \widehat{c}\big)'$ is

$$\mathrm{Var}\big(\widehat{\theta}\big) = \big[I(\theta)\big]^{-1} = \frac{1}{n}\begin{pmatrix} 1.1087\,\dfrac{b^2}{c^2} & 0.2570\,b \\[2mm] 0.2570\,b & 0.6079\,c^2 \end{pmatrix}. \tag{11.17}$$

From (11.17) we see that in the limit, $\widehat{b}$ and $\widehat{c}$ are correlated independent of $b$ and $c$ with

$$\lim_{n\to\infty} \rho\big(\widehat{b}, \widehat{c}\big) = 0.3130.$$

Contrary to the three–parameter WEIBULL distribution, we have no problems concerning the asymptotic properties of the MLEs to hold in the two–parameter case.

---

[7] WATKINS (1998) also gives the third partial derivatives together with their expectations.

## 11.2.2 Iterated MLEs

The MLEs of the WEIBULL parameters cannot be given in closed form. Looking at the system of normal equations (11.4d–f) for the three–parameter case, one can simplify so that only two non–linear equations have to be solved simultaneously by iteration. From (11.4e) we get

$$b = \left[ \frac{1}{n} \sum (x_i - a)^c \right]^{1/c}. \tag{11.18a}$$

After substituting $b^c = \frac{1}{n} \sum (x_i - a)$ into (11.4d) and into (11.4f) and after some rearrangement, we arrive at

$$\frac{c-1}{c} \sum (x_i - a)^{-1} - n \frac{\sum (x_i - a)^{c-1}}{\sum (x_i - a)^c} = 0, \tag{11.18b}$$

$$\frac{1}{c} + \frac{1}{n} \sum \ln(x_i - a) - \frac{\sum (x_i - a)^c \ln(x_i - a)}{\sum (x_i - a)^c} = 0. \tag{11.18c}$$

(11.18b,c) do not involve $b$ and with the solution of (11.18b,c) for $\widehat{a}$ and $\widehat{c}$ we find $\widehat{b}$ from (11.18a).

With regard to the two–parameter WEIBULL distribution, we have from (11.14a):

$$b = \left[ \frac{1}{n} \sum x_i^c \right]^{1/c}. \tag{11.19a}$$

After substituting $b^c = \frac{1}{n} \sum x_i^c$ into (11.14b) and some rearrangement, we have

$$\frac{1}{c} + \frac{1}{n} \sum \ln x_i - \frac{\sum x_i^c \ln x_i}{\sum x_i^c} = 0. \tag{11.19b}$$

Thus only $\widehat{c}$ has to be found by iterating (11.19b). $\widehat{b}$ results from (11.19a) by inserting $\widehat{c}$ from (11.19b).

A lot of numerical techniques exist to solve a single non-linear equation such as (11.19b) iteratively or a system of interdependent non-linear equations such as (11.18b,c) (see JUDGE et al. (1980) for an introduction and overview). These techniques are either gradient methods using derivatives or direct search methods (derivative-free methods). Reviewing the great bulk of papers on ML estimation of WEIBULL parameters, practically none of these methods has been left out. This is especially true for the three–parameter case; see PAN-CHANG/GUPTA (1989), ZANAKIS/KYPARISIS (1986) and ZANAKIS (1977).

**Gradient methods** must be regarded as the most popular iterative technique for maximizing $\mathcal{L}(\boldsymbol{\theta})$. Generally, the gradient method is characterized as follows. A starting or initial point $\widehat{\boldsymbol{\theta}}_0$ is chosen and the method then proceeds to compute successive approximations to the estimate $\widehat{\boldsymbol{\theta}}$ according to

$$\widehat{\boldsymbol{\theta}}_{p+1} = \widehat{\boldsymbol{\theta}}_p - s_p \, \boldsymbol{Q}_p \, \boldsymbol{\gamma}_p, \quad p = 0, 1, 2, \ldots \tag{11.20a}$$

The scalar $s_p$ is the **step length**, often $s_p = 1 \, \forall \, p$. $\boldsymbol{Q}_p$, the **direction matrix**, is square and positive–definite and determines the direction of change from $\widehat{\boldsymbol{\theta}}_p$ to $\widehat{\boldsymbol{\theta}}_{p+1}$. $\boldsymbol{\gamma}_p$, the **gradient**, is the column vector of first partial derivatives of $\mathcal{L}(\boldsymbol{\theta})$ with respect to $\boldsymbol{\theta}$ and evaluated at $\widehat{\boldsymbol{\theta}}_p$, i.e.,

$$\boldsymbol{\gamma}_p := \left. \frac{\partial \mathcal{L}(\boldsymbol{\theta})}{\partial \boldsymbol{\theta}} \right|_{\widehat{\boldsymbol{\theta}}_p}. \tag{11.20b}$$

The gradient methods differ mainly with respect to the direction matrix. We mention only three choices; for more see JUDGE et al. (1980) or KENNEDY/GENTLE (1980).

1. **Method of steepest ascent**
   Here the choice is

   $$\boldsymbol{Q}_p = \boldsymbol{I}, \tag{11.21}$$

   where $\boldsymbol{I}$ is the **identity matrix**. Although this method is very simple, its use cannot be recommended in most cases because it may converge very slowly if the maximum is on a long and narrow ridge; i.e., if the objective function is ill–conditioned.

2. **NEWTON–RAPHSON method**
   Here the choice is

   $$\boldsymbol{Q}_p = \left[ \boldsymbol{H}(\widehat{\boldsymbol{\theta}}_p) \right]^{-1}; \tag{11.22a}$$

   i.e., we use the inverse of the HESSIAN matrix of $\mathcal{L}(\boldsymbol{\theta})$ evaluated at $\widehat{\boldsymbol{\theta}}_p$:

   $$\boldsymbol{H}(\widehat{\boldsymbol{\theta}}_p) = \left. \frac{\partial^2 \mathcal{L}(\boldsymbol{\theta})}{\partial \boldsymbol{\theta}\, \partial \boldsymbol{\theta}'} \right|_{\widehat{\boldsymbol{\theta}}_p}. \tag{11.22b}$$

   In this algorithm, the value for $\widehat{\boldsymbol{\theta}}$ at the $(p+1)$–th stage, with $s_p = 1$, can be interpreted as the solution to a linearization around $\widehat{\boldsymbol{\theta}}_p$ of the system of normal equations $\left( \partial \mathcal{L}(\boldsymbol{\theta})/\partial \boldsymbol{\theta} = \boldsymbol{o} \right)$:

   $$\boldsymbol{o} \approx \frac{\partial \mathcal{L}(\widehat{\boldsymbol{\theta}}_{p+1})}{\partial \boldsymbol{\theta}} \approx \frac{\partial \mathcal{L}(\widehat{\boldsymbol{\theta}}_p)}{\partial \boldsymbol{\theta}} + \frac{\partial^2 \mathcal{L}(\widehat{\boldsymbol{\theta}}_p)}{\partial \boldsymbol{\theta}\, \partial \boldsymbol{\theta}'} \left( \widehat{\boldsymbol{\theta}}_{p+1} - \widehat{\boldsymbol{\theta}}_p \right) \tag{11.22c}$$

   or using (11.20b) and (11.22b)

   $$\boldsymbol{o} \approx \boldsymbol{\gamma}_{p+1} \approx \boldsymbol{\gamma}_p + \boldsymbol{H}(\widehat{\boldsymbol{\theta}}_p) \left( \widehat{\boldsymbol{\theta}}_{p+1} - \widehat{\boldsymbol{\theta}}_p \right). \tag{11.22d}$$

3. **Method of scoring**
   Here the choice is

   $$\boldsymbol{Q}_p = \left\{ \mathrm{E}\!\left[ \boldsymbol{H}(\widehat{\boldsymbol{\theta}}_p) \right] \right\}^{-1}; \tag{11.23a}$$

   i.e., we use the inverse of the expected value of the HESSIAN matrix as the direction matrix. Remember — see (11.7) and (11.8–c) — that

   $$\mathrm{E}\!\left[ \boldsymbol{H}(\widehat{\boldsymbol{\theta}}) \right] = -\boldsymbol{I}(\widehat{\boldsymbol{\theta}}), \tag{11.23b}$$

   so that the elements of the true, theoretical FISHER information matrix $\boldsymbol{I}(\widehat{\boldsymbol{\theta}}_0)$ have to be multiplied by $(-1)$ and the true parameter values have to be replaced by their last estimates.

In the two–parameter case we will have no problems finding the MLEs using the NEWTON–RAPHSON algorithm when it is properly organized. We have to apply this algorithm to only one non–linear equation, namely to (11.19b), and $\widehat{b}$ then follows from (11.19a). We thus have to show that (11.19b) has only one admissible, i.e., a positive solution, and that the algorithm converges to that solution.[8]

Uniqueness of the solution to (11.19b)

We rewrite (11.19b) as

$$g(c) = \frac{1}{c} + \overline{z} - \frac{\sum_{i=1}^{n} z_i \exp(c\,z_i)}{\sum_{i=1}^{n} \exp(c\,z_i)} = 0 \qquad (11.24a)$$

with

$$z_i := \ln x_i \quad \text{and} \quad \overline{z} := \frac{1}{n} \sum_{i=1}^{n} z_i.$$

Figure 11/1: Likelihood equation of $c$ for dataset #1

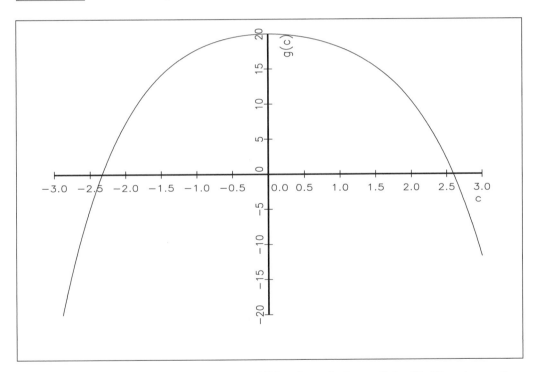

The roots of (11.24a) with respect to $c$ will be the solutions of the likelihood equation. Letting

$$w_i := z_i - \overline{z},$$

$g(c)$ becomes, after some rearrangement,

$$g(c) = \sum_{i=1}^{n} (1 - c\,w_i)\,\exp(c\,w_i) = 0. \qquad (11.24b)$$

---

[8] For further details see FARNUM/BOOTH (1997), GUPTA et al. (1998) and MCCOOL (1970a).

At $c = 0$ we have $g(0) = n \neq 0$. Hence $\hat{c} = 0$ cannot be a solution of the likelihood equation (11.19b). In addition, it can be easily seen that $g(c) \rightarrow -\infty$ as $c \rightarrow \infty$. Since at least one $w_i$ must be negative,[9] $g(c) \rightarrow -\infty$ as $c \rightarrow -\infty$. Differentiating (11.24b) with respect to $c$ gives

$$g'(c) = -\sum_{i=1}^{n} c\, w_i^2 \exp(c\, w_i). \qquad (11.24c)$$

Since $g'(0) = 0$, the function has a horizontal tangent at $c = 0$. For $0 < c < \infty$, $g'(c)$ is negative while for $-\infty < c < 0$, $g'(c)$ is positive. Thus the function $g(c)$ has a maximum (of $n$) at $c = 0$, a single negative root, which is non-admissible; and a single positive root $\hat{c}$ (see Fig. 11/1 for dataset #1 of Example 9/2).

Convergence of the NEWTON–RAPHSON algorithm to $\hat{c}$

The NEWTON–RAPHSON method for finding the single admissible root $\hat{c}$ is given by

$$\hat{c}_{p+1} = \hat{c}_p - \frac{g(\hat{c}_p)}{g'(\hat{c}_p)}\,; \quad p = 0, 1, 2, \ldots \qquad (11.25a)$$

We introduce two definitions:

1. $\hat{c}_p$ generated by (11.25a) **converges** to $\hat{c}$ **monotonically from above** if it converges to $\hat{c}$ and $\hat{c}_{p+1} < \hat{c}_p \; \forall \, p > 0$, $\hat{c}_0$ being the initial value.
2. $\hat{c}_p$ generated by (11.25a) **quadratically converges** to $\hat{c}$ if it converges to $\hat{c}$ and the limit $|\hat{c}_{p+1} - \hat{c}|/|\hat{c}_p - \hat{c}|^2$ exists.

Let $w_M$ be any positive $w_i$, $m$ be the number of $w_i$ such that $w_i > w_M$ and

$$C = \frac{1}{w_M} \max\{2,\, \ln(n - m) - \ln m\}. \qquad (11.25b)$$

Then as GUPTA et al. (1998) have proven, the following theorem holds.
**Theorem**:
For any positive $\hat{c}_0$, (11.25a) quadratically converges to the unique positive solution $\hat{c}$ of (11.24a). If $\hat{c}_0 \geq C$, $C$ given by (11.25b), then (11.25a) converges to $\hat{c}$ also monotonically from above. ∎

We remark that, although theoretically the NEWTON–RAPHSON method will converge monotonically for any $\hat{c}_0 \geq \hat{c}$, one would like to choose $\hat{c}_0$ close to $\hat{c}$ because $g(c)$ decreases exponentially. Usually the choice $\hat{c}_0 = C$ with $w_M = \max\{w_i\}$ is good, but for some examples, other $w_M$ with large $m$ bring a smaller value of $C$, which gives a better initial approximation $\hat{c}_0$.

The determination of MLEs for the three–parameter WEIBULL model is usually considered a non–trivial problem because of the complexity of the non–linear likelihood equations. Despite the availability of a large number of algorithms that tackle this problem (see the following sections), there is considerable dissatisfaction among practitioners, who report an inability to conveniently determine the desired MLEs. As PANCHANG/GUPTA (1989) have explored, the simple procedure outlined by LAWLESS (1982) (see Sect. 11.3.2.2) is

---

[9] We assume that not all observations $x_i$ are the same.

the only one that is guaranteed to yield parameter estimates that maximize the likelihood function for any sample.

The gradient methods essentially seek to maximize $\mathcal{L}(a, b, c)$ by obtaining a turning point which satisfies (11.4d–f). With regard to this approach, however, it is important to note the following features of the likelihood function:

1. It may have no turning point in the allowable parameter space defined by $b$, $c > 0$ and $a \leq \min\{x_i\}$.

2. It may have only one turning point, in which case ROCKETTE et al. (1974) have proven that it must be a saddle point.

3. It may have two turning points, in which case the one with the larger $\hat{a}$ is a saddle point; see ROCKETTE et al. (1974).

4. It also appears from empirical evidence that there cannot be more than two solutions though this has not been proven.

5. Even if there is a local maximum, the likelihood function may be greater at the corner point, as demonstrated by ROCKETTE et al. (1974).

## 11.3   Uncensored samples with non-grouped data

The log–likelihood function, the likelihood equations and the estimating procedure depend upon the type of sample, i.e., whether it is uncensored or in which way it is censored; furthermore on how the observations become available, i.e., whether they are grouped or not; and last but not least, which parameters of the WEIBULL distribution have to be estimated, especially whether we assume the two–parameter case ($b$ and $c$ unknown) or the three–parameter case ($a$, $b$ and $c$ unknown). Chapter 11 is structured according to these influential items.

Here, in Sect. 11.3 we assume an uncensored sample of size $n$ with all observations individually known. We thus have $n$ likelihood elements, each being a density; see (11.1a).

### 11.3.1   Two-parameter WEIBULL distribution

We will first show how to apply the ML–technique (Sect. 11.3.1.1). In Sect. 11.3.1.2, we want to find results pertaining to finite samples.

#### 11.3.1.1   Point and interval estimates for $b$ and $c$

The log–likelihood function of the two–parameter WEIBULL distribution $\mathcal{L}(b, c)$ is given in (11.13). BUNDAY/AL–MUTWALI (1981) have found ML estimates by directly maximizing the negative log–likelihood function. The formulas resulting from the more common methods based on derivatives have been developed in Sect. 11.2. The ML estimates in the following example have been found by applying the NEWTON–RAPHSON technique as laid down in (11.25a) and (11.19a).

**Example 11/4:   ML estimates for dataset #1**

The $n = 20$ observations of dataset #1, where $X_i \sim We(0, 100, 2.5) \; \forall \; i$, are listed in Tab. 9/2. Fig. 11/2 shows the corresponding log–likelihood function as a surface plot. Around the summit the mountain is rather flat.

The initial value $\widehat{c}_0$ for starting (11.25a) has been found from (11.25b) with $w_M = \max\{w_i\} = 0.8209$, $m = 1$ and $n = 20$ as $\widehat{c}_0 = 3.5867$. The iteration process giving

| $p$ | $\widehat{c}_p$ | $\widehat{b}_p$ |
|---|---|---|
| 0 | 3.5867 | 105.7048 |
| 1 | 2.9569 | 101.6209 |
| 2 | 2.6543 | 99.6031 |
| 3 | 2.5974 | 99.2195 |
| 4 | 2.5957 | 99.2079 |

came to a stop after $p = 4$ iterations giving an accuracy of $10^{-4}$.

Figure 11/2:  Surface plot of the log-likelihood function for dataset #1

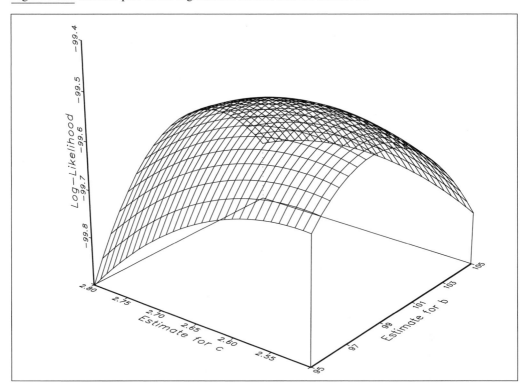

Fig 11/3 shows the path of the iterations together with the contour plot. The points for steps $p = 3$ and $p = 4$ practically coincide on the graph.

Applying $\widehat{c} = 2.5957$ and $\widehat{b} = 99.2079$ to (11.17), the estimated variance–covariance matrix results as

$$\widehat{\mathrm{Var}}(\boldsymbol{\theta}) = \begin{pmatrix} 80.9780 & 1.2748 \\ 1.2748 & 0.2048 \end{pmatrix}.$$

Figure 11/3: Contour plot of the log-likelihood function for dataset #1 and path of iterations

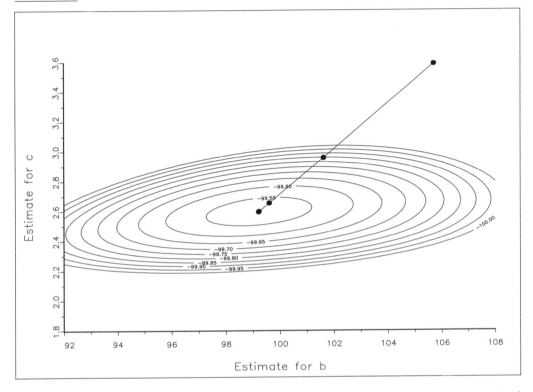

Based on the asymptotic normality we find the following two–sided 90% confidence intervals for $b$ and $c$:

$$84.4049 \leq b \leq 114.0109,$$

$$1.8513 \leq c \leq 3.3401,$$

both covering the true parameter values $b = 100$ and $c = 2.5$. The reader should compare these ML results with those of the BLUEs in Example 10/2.

We should mention a solution technique for $\widehat{b}$ and $\widehat{c}$ suggested by QUREISHI (1964). Upon substituting the value of $b$ given by (11.19a) into (11.13), the log–likelihood function, now depending on $c$ only, reads as follows:

$$\mathcal{L}(c) = n \left\{ \ln \left( \frac{n\,c}{\sum x_i^c} \right) - 1 \right\} + (c-1) \sum \ln x_i,$$

which can be evaluated over a finite set of values of $c$. We then can select that value of $c$ which maximizes $\mathcal{L}(c)$.

**Excursus:   MLEs for $b_2 = b^c$ and $c$**

Sometimes the WEIBULL distribution is written with the combined scale–shape parameter $b_2 = b^c$; see (2.27). With respect to this parametrization we have:

$$L(b_2, c) \;=\; \prod_{i=1}^{n} f(x_i \mid b_s, c) \;=\; \left(\frac{c}{b_2}\right)^n \prod_{i=1}^{n} x_i^{c-1} \exp\left\{-\frac{x_i^c}{b_2}\right\}, \quad (11.26a)$$

$$\mathcal{L}(b_2, c) \;=\; n\left[\ln c - \ln b_2\right] + (c-1)\sum_{i=1}^{n} \ln x_i - \frac{1}{b_2}\sum_{i=1}^{n} x_i^c, \quad (11.26b)$$

$$\frac{\partial \mathcal{L}(b_2, c)}{\partial b_2} \;=\; -\frac{n}{b_2} + \frac{1}{b_2^2}\sum_{i=1}^{n} x_i^c \;=\; 0, \quad (11.26c)$$

$$\frac{\partial \mathcal{L}(b_2, c)}{\partial c} \;=\; \frac{n}{c} + \sum_{i=1}^{n} \ln x_i - \frac{1}{b_2}\sum_{i=1}^{n} x_i^c \ln x_i \;=\; 0, \quad (11.26d)$$

$$\frac{\partial^2 \mathcal{L}(b_2, c)}{\partial b_2^2} \;=\; \frac{n}{b_2^2} - \frac{2}{b_2^3}\sum_{i=1}^{n} x_i^c, \quad (11.26e)$$

$$\frac{\partial^2 \mathcal{L}(b_2, c)}{\partial b_2\,\partial c} = \frac{\partial^2 \mathcal{L}(b_2, c)}{\partial c\,\partial b_2} \;=\; \frac{1}{b_2^2}\sum_{i=1}^{n} x_i^c \ln x_i, \quad (11.26f)$$

$$\frac{\partial^2 \mathcal{L}(b_2, c)}{\partial c^2} \;=\; -\frac{n}{c^2} - \frac{1}{b_2}\sum_{i=1}^{n} x_i^c \left(\ln x_i\right)^2, \quad (11.26g)$$

$$-\mathrm{E}\left(\frac{\partial^2 \mathcal{L}(b_2, c)}{\partial b_2^2}\right) \;=\; \frac{n}{b_2^2}, \quad (11.26h)$$

$$-\mathrm{E}\left(\frac{\partial^2 \mathcal{L}(b_2, c)}{\partial b_2\,\partial c}\right) = -\mathrm{E}\left(\frac{\partial^2 \mathcal{L}(b_2, c)}{\partial c\,\partial b_2}\right) \;=\; -\frac{n}{c\,b_2}\left[1 - \gamma + \ln b_2\right] \approx -\frac{n}{c\,b_2}\left[0.422784 + \ln b_2\right], \quad (11.26i)$$

$$-\mathrm{E}\left(\frac{\partial^2 \mathcal{L}(b_2, c)}{\partial c^2}\right) \;=\; \frac{n}{c^2}\left[1 + \gamma^2 - 2\,\gamma + \frac{\pi^2}{6} - 2\,(\gamma - 1)\,\ln b_2 + (\ln b_2)^2\right]$$
$$\approx \frac{n}{c^2}\left[1.82368 + 0.84557\,\ln b_2 + (\ln b_2)^2\right]. \quad (11.26j)$$

From (11.26h–j) we derive the following variance–covariance matrix of $\widehat{\boldsymbol{\theta}} = \left(\widehat{b}_2, \widehat{c}\right)'$:

$$\mathrm{Var}\left(\widehat{\boldsymbol{\theta}}\right) \approx \frac{1}{1.644934\,n} \begin{pmatrix} b_2^2\left[1.82368 + 0.84557\,\ln b_2 + (\ln b_2)^2\right] & c\,b_2\left[0.422784 + \ln b_2\right] \\ c\,b_2\left[0.422784 + \ln b_2\right] & c^2 \end{pmatrix}. \quad (11.27)$$

### 11.3.1.2   Finite sample results based on pivotal functions[10]

The distributional properties stated in the theorem on MLEs (see Sect. 11.2.1) hold only asymptotically, so it is quite natural to look for finite sample evidences. Finite sample results can be deduced from a well–known property of the MLEs of location and scale parameters; see ANTLE/BAIN (1969).

**Theorem on pivotal functions**
Let $a^*$ and $b^*$ be the location and scale parameters, respectively, of a location–scale distributed variate and $\widehat{a^*}$, $\widehat{b^*}$ the MLEs of $a^*$ and $b^*$. The **pivotal functions**

$$\frac{\widehat{b^*}}{b^*}, \quad \frac{\widehat{a^*} - a^*}{b^*} \quad \text{and} \quad \frac{\widehat{a^*} - a^*}{\widehat{b^*}}$$

are each distributed independently of $a^*$ and $b^*$.　■

Perhaps the best known pivotal function occurs when one samples from a normal distribution with unknown mean $\mu$ and unknown standard deviation $\sigma$ and forms — according to the third pivot of the theorem – the pivotal function

$$t := (\overline{X} - \mu)/(s/\sqrt{n})$$

for inference about $\mu$. The distribution of this pivotal function is the well-known $t$–distribution, depending only on the degrees of freedom $n - 1$, first given by W.S. GOSSET in 1908.

It is clear that the results of the above theorem will also apply to other types of parameters if they are appropriately related to location and scale parameters under some change of variable. This is the case with the two–parameter Weibull distribution. After a log–transformation of the WEIBULL variate,

$$a^* = \ln b \quad \text{and} \quad b^* = 1/c$$

are the location and scale parameters, respectively, of the type–I–extreme value distribution of the sample minimum. Thus by the theorem $(1/\widehat{c})/(1/c)$, $(\ln \widehat{b} - \ln b)\, c$ and $(\ln \widehat{b} - \ln b)\, \widehat{c}$ have distributions that are independent of the parameters. By the invariant property of the MLEs it follows that

$$\frac{c}{\widehat{c}}, \quad \left(\frac{\widehat{b}}{b}\right)^c \quad \text{and} \quad \left(\frac{\widehat{b}}{b}\right)^{\widehat{c}}$$

are distributed independently of all parameters.

The above results have been observed directly in THOMAN et al. (1969) also giving the pivotal distributions.[11] In what follows, $\widehat{c}_{11}$ and $\widehat{b}_{11}$ are used to denote the MLEs of $c$ and

---

[10]　Suggested reading for this section: ANTLE/BAIN (1969), BAIN/ENGELHARDT (1986), LAWLESS (1972, 1973b, 1974), LEMON (1975), THOMAN/BAIN/ANTLE (1969). See also Section 11.3.2.4 for additional pivotal functions.

[11]　Their results can be extended to the three–parameter distribution (see Sect. 11.3.2.4) and to right–censored samples (see Sect. 11.6).

$b$, respectively, when in fact the sampling is from a WEIBULL distribution with $b = 1$ and $c = 1$, i.e., from a reduced exponential distribution. THOMAN et al. then proved the following two theorems:

**Theorem on $\widehat{c}/c$**

$\widehat{c}/c$ is distributed independently of $b$ and $c$ and has the same distribution as $\widehat{c}_{11}$, denoted[12]

$$\widehat{c}_{11} \stackrel{\mathrm{d}}{=} \widehat{c}/c. \qquad \blacksquare$$

**Theorem on $\widehat{c}\ln(\widehat{b}/b)$**

$\widehat{c}\ln(\widehat{b}/b)$ is distributed independently of $b$ and $c$ and has the same distribution as $\widehat{c}_{11}\ln\widehat{b}_{11}$, denoted

$$\widehat{c}_{11}\ln\widehat{b}_{11} \stackrel{\mathrm{d}}{=} \widehat{c}\ln(\widehat{b}/b). \qquad \blacksquare$$

The distributions of $\widehat{c}_{11}$ and of $\widehat{c}_{11}\ln\widehat{b}_{11}$ dependent on $n$ but cannot be given in closed form. THOMAN et al. (1969) obtained the percentage points $\ell_1(n, P)$ of $\widehat{c}_{11} \stackrel{\mathrm{d}}{=} \widehat{c}/c$ and $\ell_2(n, P)$ of $\widehat{c}_{11}\ln\widehat{b}_{11} \stackrel{\mathrm{d}}{=} \widehat{c}\ln(\widehat{b}/b)$ by Monte Carlo methods for $P = 0.02, 0.05, 0.10, 0.25, 0.40(0.10), 0.80, 0.85(0.05), 0.95, 0.98$ and $n = 5(1)20(2)80(5)100, 110, 120$. Tables 11/1 and 11/2 are extracts of their tables.

Thus we have

$$\mathrm{Pr}\left[\frac{\widehat{c}}{c} \leq \ell_1(n, P)\right] = P \qquad (11.28\mathrm{a})$$

and

$$\mathrm{Pr}\left[\widehat{c}\ln(\widehat{b}/b) \leq \ell_2(n, P)\right] = P, \qquad (11.28\mathrm{b})$$

which serve to test statistical hypotheses on $b$ or $c$ and to set up confidence intervals. The two–sided confidence intervals of $b$ and $c$ with confidence level $1 - \alpha$ are

$$\frac{\widehat{c}}{\ell_1(n, 1 - \alpha/2)} \leq c \leq \frac{\widehat{c}}{\ell_1(n, \alpha/2)} \qquad (11.29\mathrm{a})$$

and

$$\widehat{b}\,\exp\left[-\ell_2(n, 1 - \alpha/2)/\widehat{c}\right] \leq b \leq \widehat{b}\,\exp\left[-\ell_2(n, \alpha/2)/\widehat{c}\right]. \qquad (11.29\mathrm{b})$$

Applying (11.29a,b) to $\widehat{c} = 2.5957$ and $\widehat{b} = 99.2079$ of Example 11/4, where $n = 20$, we have the following two–sided 90% confidence intervals:

$$1.7914 \leq c \leq 3.2815$$
$$84.3543 \leq b \leq 116.9920,$$

which are not very different from those based on the asymptotic normal distribution (see Example 11/4).

---

[12] The symbol $\stackrel{\mathrm{d}}{=}$ means that the quantities on each side have the same distribution (**distributional equivalence**).

Table 11/1:  Percentage points $\ell_1(n, P)$ of $\widehat{c}/c$

| $n$ \ $P$ | 0.02 | 0.05 | 0.10 | 0.90 | 0.95 | 0.98 |
|---|---|---|---|---|---|---|
| 5 | 0.604 | 0.683 | 0.766 | 2.277 | 2.779 | 3.518 |
| 6 | 0.623 | 0.697 | 0.778 | 2.030 | 2.436 | 3.067 |
| 7 | 0.639 | 0.709 | 0.785 | 1.861 | 2.183 | 2.640 |
| 8 | 0.653 | 0.720 | 0.792 | 1.747 | 2.015 | 2.377 |
| 9 | 0.665 | 0.729 | 0.797 | 1.665 | 1.896 | 2.199 |
| 10 | 0.676 | 0.738 | 0.802 | 1.602 | 1.807 | 2.070 |
| 11 | 0.686 | 0.745 | 0.807 | 1.553 | 1.738 | 1.972 |
| 12 | 0.695 | 0.752 | 0.811 | 1.513 | 1.682 | 1.894 |
| 13 | 0.703 | 0.759 | 0.815 | 1.480 | 1.636 | 1.830 |
| 14 | 0.710 | 0.764 | 0.819 | 1.452 | 1.597 | 1.777 |
| 15 | 0.716 | 0.770 | 0.823 | 1.427 | 1.564 | 1.732 |
| 16 | 0.723 | 0.775 | 0.826 | 1.406 | 1.535 | 1.693 |
| 17 | 0.728 | 0.729. | 0.829 | 1.388 | 1.510 | 1.660 |
| 18 | 0.734 | 0.784 | 0.832 | 1.371 | 1.487 | 1.630 |
| 19 | 0.739 | 0.788 | 0.835 | 1.356 | 1.467 | 1.603 |
| 20 | 0.743 | 0.791 | 0.838 | 1.343 | 1.449 | 1.579 |
| 22 | 0.752 | 0.798 | 0.843 | 1.320 | 1.418 | 1.538 |
| 24 | 0.759 | 0.805 | 0.848 | 1.301 | 1.392 | 1.504 |
| 26 | 0.766 | 0.810 | 0.852 | 1.284 | 1.370 | 1.475 |
| 28 | 0.772 | 0.815 | 0.856 | 1.269 | 1.351 | 1.450 |
| 30 | 0.778 | 0.820 | 0.860 | 1.257 | 1.334 | 1.429 |
| 32 | 0.783 | 0.824 | 0.863 | 1.246 | 1.319 | 1.409 |
| 34 | 0.788 | 0.828 | 0.866 | 1.236 | 1.306 | 1.392 |
| 36 | 0.793 | 0.832 | 0.869 | 1.227 | 1.294 | 1.377 |
| 38 | 0.797 | 0.835 | 0.872 | 1.219 | 1.283 | 1.363 |
| 40 | 0.801 | 0.839 | 0.875 | 1.211 | 1.273 | 1.351 |
| 42 | 0.804 | 0.842 | 0.877 | 1.204 | 1.265 | 1.339 |
| 44 | 0.808 | 0.845 | 0.880 | 1.198 | 1.256 | 1.329 |
| 46 | 0.811 | 0.847 | 0.882 | 1.192 | 1.249 | 1.319 |
| 48 | 0.814 | 0.850 | 0.884 | 1.187 | 1.242 | 1.310 |
| 50 | 0.817 | 0.852 | 0.886 | 1.182 | 1.235 | 1.301 |
| 52 | 0.820 | 0.854 | 0.888 | 1.177 | 1.229 | 1.294 |
| 54 | 0.822 | 0.857 | 0.890 | 1.173 | 1.224 | 1.286 |
| 56 | 0.825 | 0.859 | 0.891 | 1.169 | 1.218 | 1.280 |
| 58 | 0.827 | 0.861 | 0.893 | 1.165 | 1.213 | 1.273 |
| 60 | 0.830 | 0.863 | 0.894 | 1.162 | 1.208 | 1.267 |
| 62 | 0.832 | 0.864 | 0.896 | 1.158 | 1.204 | 1.262 |
| 64 | 0.834 | 0.866 | 0.897 | 1.155 | 1.200 | 1.256 |
| 66 | 0.836 | 0.868 | 0.899 | 1.152 | 1.196 | 1.251 |
| 68 | 0.838 | 0.869 | 0.900 | 1.149 | 1.192 | 1.246 |
| 70 | 0.840 | 0.871 | 0.901 | 1.146 | 1.188 | 1.242 |
| 72 | 0.841 | 0.872 | 0.903 | 1.144 | 1.185 | 1.237 |
| 74 | 0.843 | 0.874 | 0.904 | 1.141 | 1.182 | 1.233 |
| 76 | 0.845 | 0.875 | 0.905 | 1.139 | 1.179 | 1.229 |
| 78 | 0.846 | 0.876 | 0.906 | 1.136 | 1.176 | 1.225 |
| 80 | 0.848 | 0.878 | 0.907 | 1.134 | 1.173 | 1.222 |
| 85 | 0.852 | 0.881 | 0.910 | 1.129 | 1.166 | 1.213 |
| 90 | 0.855 | 0.883 | 0.912 | 1.124 | 1.160 | 1.206 |
| 95 | 0.858 | 0.886 | 0.914 | 1.120 | 1.155 | 1.199 |
| 100 | 0.861 | 0.888 | 0.916 | 1.116 | 1.150 | 1.192 |
| 110 | 0.866 | 0.893 | 0.920 | 1.110 | 1.141 | 1.181 |
| 120 | 0.871 | 0.897 | 0.923 | 1.104 | 1.133 | 1.171 |

<u>Table 11/2:</u> Percentage points $\ell_2(n, P)$ of $\widehat{c}\,\ln(\widehat{b}/b)$

| $n$ \ $P$ | 0.02 | 0.05 | 0.10 | 0.90 | 0.95 | 0.98 |
|---|---|---|---|---|---|---|
| 5 | −1.631 | −1.247 | −0.888 | 0.772 | 1.107 | 1.582 |
| 6 | −1.396 | −1.007 | −0.740 | 0.666 | 0.939 | 1.291 |
| 7 | −1.196 | −0.874 | −0.652 | 0.598 | 0.829 | 1.120 |
| 8 | −1.056 | −0.784 | −0.591 | 0.547 | 0.751 | 1.003 |
| 9 | −0.954 | −0.717 | −0.544 | 0.507 | 0.691 | 0.917 |
| 10 | −0.876 | −0.665 | −0.507 | 0.475 | 0.644 | 0.851 |
| 11 | −0.813 | −0.622 | −0.477 | 0.448 | 0.605 | 0.797 |
| 12 | −0.762 | −0.587 | −0.451 | 0.425 | 0.572 | 0.752 |
| 13 | −0.719 | −0.557 | −0.429 | 0.406 | 0.544 | 0.714 |
| 14 | −0.683 | −0.532 | −0.410 | 0.389 | 0.520 | 0.681 |
| 15 | −0.651 | −0.509 | −0.393 | 0.374 | 0.499 | 0.653 |
| 16 | −0.624 | −0.489 | −0.379 | 0.360 | 0.480 | 0.627 |
| 17 | −0.599 | −0.471 | −0.365 | 0.348 | 0.463 | 0.605 |
| 18 | −0.578 | −0.455 | −0.353 | 0.338 | 0.447 | 0.584 |
| 19 | −0.558 | −0.441 | −0.342 | 0.328 | 0.433 | 0.566 |
| 20 | −0.540 | −0.428 | −0.332 | 0.318 | 0.421 | 0.549 |
| 22 | −0.509 | −0.404 | −0.314 | 0.302 | 0.398 | 0.519 |
| 24 | −0.483 | −0.384 | −0.299 | 0.288 | 0.379 | 0.494 |
| 26 | −0.460 | −0.367 | −0.286 | 0.276 | 0.362 | 0.472 |
| 28 | −0.441 | −0.352 | −0.274 | 0.265 | 0.347 | 0.453 |
| 30 | −0.423 | −0.338 | −0.264 | 0.256 | 0.334 | 0.435 |
| 32 | −0.408 | −0.326 | −0.254 | 0.247 | 0.323 | 0.420 |
| 34 | −0.394 | −0.315 | −0.246 | 0.239 | 0.312 | 0.406 |
| 36 | −0.382 | −0.305 | −0.238 | 0.232 | 0.302 | 0 393 |
| 38 | −0.370 | −0.296 | −0.231 | 0.226 | 0.293 | 0.382 |
| 40 | −0.360 | −0.288 | −0.224 | 0.220 | 0.285 | 0.371 |
| 42 | −0.350 | −0.280 | −0.218 | 0.214 | 0.278 | 0.361 |
| 44 | −0.341 | −0.273 | −0.213 | 0.209 | 0.271 | 0.352 |
| 46 | −0.333 | −0.266 | −0.208 | 0.204 | 0.264 | 0.344 |
| 48 | −0.325 | −0.260 | −0.203 | 0.199 | 0.258 | 0.336 |
| 50 | −0.318 | −0.254 | −0.198 | 0.195 | 0.253 | 0.328 |
| 52 | −0.312 | −0.249 | −0.194 | 0.191 | 0.247 | 0.321 |
| 54 | −0.305 | −0.244 | −0.190 | 0.187 | 0.243 | 0.315 |
| 56 | −0.299 | −0.239 | −0.186 | 0.184 | 0.238 | 0.309 |
| 58 | −0.294 | −0.234 | −0.183 | 0.181 | 0.233 | 0.303 |
| 60 | −0.289 | −0.230 | −0.179 | 0.177 | 0.229 | 0.297 |
| 62 | −0.284 | −0.226 | −0.176 | 0.174 | 0.225 | 0.292 |
| 64 | −0.279 | −0.222 | −0.173 | 0.171 | 0.221 | 0.287 |
| 66 | −0.274 | −0.218 | −0.170 | 0.169 | 0.218 | 0.282 |
| 68 | −0.270 | −0.215 | −0.167 | 0.166 | 0.214 | 0.278 |
| 70 | −0.266 | −0.211 | −0.165 | 0.164 | 0.211 | 0.274 |
| 72 | −0.262 | −0.208 | −0.162 | 0.161 | 0.208 | 0.269 |
| 74 | −0.259 | −0.205 | −0.160 | 0.159 | 0.205 | 0.266 |
| 76 | −0.255 | −0.202 | −0.158 | 0.157 | 0.202 | 0.262 |
| 78 | −0.252 | −0.199 | −0.155 | 0.155 | 0.199 | 0.258 |
| 80 | −0.248 | −0.197 | −0.153 | 0.153 | 0.197 | 0.255 |
| 85 | −0.241 | −0.190 | −0.148 | 0.148 | 0.190 | 0.246 |
| 90 | −0.234 | −0.184 | −0.144 | 0.143 | 0.185 | 0.239 |
| 95 | −0.227 | −0.179 | −0.139 | 0.139 | 0.179 | 0.232 |
| 100 | −0.221 | −0.174 | −0.136 | 0.136 | 0.175 | 0.226 |
| 110 | −0.211 | −0.165 | −0.129 | 0.129 | 0.166 | 0.215 |
| 120 | −0.202 | −0.158 | −0.123 | 0.123 | 0.159 | 0.205 |

The theorem on $\widehat{c}/c$ confirms the feeling expressed by JOHNSON et al. (1994, p. 657) that the bias in $\widehat{c}$ is independent of the true values of $b$ and $c$, but depends only on $n$. THOMAN et al. (1969) have found (see Tab. 11/3) the unbiasing factors $B(n)$ such that $\mathrm{E}[B(n)\widehat{c}] = c$, based on the generated distribution of $\widehat{c}_{11}$. Tab. 11/3 shows that $\widehat{c}$ is overestimated, in the case of $n = 20$ by 7.41% of its mean.[13]

Table 11/3: Unbiasing factors $B(n)$ for the MLE of $c$

| $n$ | 5 | 6 | 7 | 8 | 9 | 10 | 11 | 12 | 13 | 14 | 15 | 16 |
|---|---|---|---|---|---|---|---|---|---|---|---|---|
| $B(n)$ | 0.669 | 0.752 | 0.792 | 0.820 | 0.842 | 0.859 | 0.872 | 0.883 | 0.893 | 0.901 | 0.908 | 0.914 |
| $n$ | 18 | 20 | 22 | 24 | 26 | 28 | 30 | 32 | 34 | 36 | 38 | 40 |
| $B(n)$ | 0.923 | 0.931 | 0.938 | 0.943 | 0.947 | 0.951 | 0.955 | 0.958 | 0.960 | 0.962 | 0.964 | 0.966 |
| $n$ | 42 | 44 | 46 | 48 | 50 | 52 | 54 | 56 | 58 | 60 | 62 | 64 |
| $B(n)$ | 0.968 | 0.970 | 0.971 | 0.972 | 0.973 | 0.974 | 0.975 | 0.976 | 0.977 | 0.978 | 0.979 | 0.980 |
| $n$ | 66 | 68 | 70 | 72 | 74 | 76 | 78 | 80 | 85 | 90 | 100 | 120 |
| $B(n)$ | 0.980 | 0.981 | 0.981 | 0.982 | 0.982 | 0.983 | 0.983 | 0.984 | 0.985 | 0.986 | 0.987 | 0.990 |

Source: THOMAN et al. (1969, p. 449) — Reprinted with permission from *Technometrics*. Copyright 1969 by the American Statistical Association. All rights reserved.

There are two approaches to **approximate the distributions of the pivotal functions**. BAIN/ENGELHARDT (1981, 1986) have suggested a $\chi^2$–approximation to $c/\widehat{c}$ and a $t$–approximation to $\widehat{c} \ln(\widehat{b}/b)$. These approximations are based on an association between the normal and the extreme–value distributions, established via the generalized gamma distribution. The one–sided confidence limits of level $P$ are

$$\widehat{c}_P = \widehat{c} \sqrt{\chi^2_{0.822\,(n-1),1-P}/(0.822\,n)}, \qquad (11.30a)$$

$$\widehat{b}_P = \widehat{b} \exp\left\{ -1.053\,t_{n-1,1-P}/\left(\widehat{c}\sqrt{n-1}\right) \right\}, \qquad (11.30b)$$

where $\chi^2_{\nu,\alpha}$ is the $\chi^2$–percentile of order $\alpha$ with $\nu$ degrees of freedom and $t_{\nu,\alpha}$ the $t$–percentile. Given the data of Example 11/4 ($\widehat{c} = 2.5957$, $\widehat{b} = 99.2079$ and $n = 20$) and a confidence level of 90%, we find from (11.30a,b) the following two–sided confidence intervals:

$$1.7756 \leq c \leq 3.2518,$$

$$84.4631 \leq b \leq 116.5294,$$

which should be compared with those based on the pivots and those given in Example 11/4.

The second approach is due to LAWLESS (1972, 1973b, 1974). He advocates the use of **conditional confidence intervals**[14] which are derived from the conditional distributions of $\widehat{b}^*/b^*$ and $(\widehat{a}^* - a^*)/\widehat{b}^*$, each conditioned on the **ancillary statistics** $(\ln x_i - \widehat{a}^*)/b^*$. This approach involves numerical integration. The conditional and the unconditional confidence

---

[13] Further methods of unbiasing $\widehat{c}$ and $\widehat{b}$ are found with HIROSE (1999), MAZZANTI et al. (1997), MONTANARI et al. (1997), ROSS (1996) and WATKINS (1996). MCCOOL (1970a) gives median unbiased estimators of $c$.

[14] The confidence intervals based on pivotal distributions are unconditional.

intervals are nearly equivalent. We thus recommend using the approach based on the tables of THOMAN et al. (1969) given above.

## 11.3.2   Three-parameter WEIBULL distribution

Finding the maximum of the likelihood function (11.4c) is no trivial task. We will first review on how this process has been managed in the past (Sect. 11.3.2.1). The complicated non–linearity of (11.4c) has caused many algorithms to fail. In Sect. 11.3.2.2 we propose a simple procedure that guarantees to yield parameter estimates maximizing the likelihood function for any sample. Finally, in Sect. 11.3.2.3 we will describe a procedure that is hybrid insofar as the likelihood equation for $a$ is replaced with alternative functional relationships. Sect. 11.3.2.4 reports on finite sample results.

### 11.3.2.1   History of optimizing the WEIBULL log-likelihood [15]

The log–likelihood function

$$
\mathcal{L}(a,b,c) = n\left[\ln c - c\ln b\right] + (c-1)\sum_{i=1}^{n}\ln(x_i - a) - \sum_{i=1}^{n}\left(\frac{x_i - a}{b}\right)^c \tag{11.31a}
$$

is to be maximized with respect to $a$, $b$ and $c$ and satisfying the constraints[16]

$$
0 \le a \le \min_{1\le i\le n}\{x_i\}, \tag{11.31b}
$$

$$
b,\ c > 0. \tag{11.31c}
$$

When $a$, $b$ and $c$ are all unknown, the log–likelihood function clearly is not bounded.[17] Sometimes an interior relative maximum will exist and sometimes not. ROCKETTE et al. (1974) conjectured and partly proved that when there is an interior relative maximum, there will also be a saddle point, both given by solutions of the likelihood equations (11.4d–f). If $a$, $b$ and $c$ are unknown and $c$ is restricted to $c \ge 1$, then, when no solution exists for the likelihood equations, the maximum value for the log–likelihood function occurs at the **corner point**:

$$
\widehat{a} = \min(x_i),\quad \widehat{b} = \frac{1}{n}\sum_{i=1}^{n}\left(x_i - \widehat{a}\right),\quad \widehat{c} = 1. \tag{11.32}
$$

---

[15]   Suggested reading for this section: PANCHANG/GUPTA (1989), ROCKETTE/ANTLE/KLIMKO (1974), ZANAKIS (1977), ZANAKIS/KYPARISI (1986).

[16]   Sometimes the constraints read $0 \le a \le \min\{x_i\} - \epsilon$, $b \ge \epsilon$ and $c \ge \epsilon$, where a small positive constant $\epsilon$, say $10^{-8}$, is recommended instead of 0 for most iteration procedures, in order to avoid any pathological situation in the logarithmic and exponential terms.

[17]   The following results also hold under censoring on the right. When $0 < c \le 1$, the smallest observation is hyper–efficient for $a$ (see DUBEY (1966c) and Sect. 13.5.1) but no true MLE exist for the other two parameters.

Also, even when an interior relative maximum exists, it may or may not exceed the value of the log–likelihood at this corner point. ROCKETTE et al. (1974) believe (on the basis of hundreds of numerical investigations) that there will either be no solution or two solutions and in very special cases exactly one solution.

Computational difficulties encountered by some optimizing techniques are caused by the geometry of $\mathcal{L}(a, b, c)$ and its pathology when $a$ approaches its upper bound $x_{1:n}$. In summary, the location parameter causes computational difficulties because of the logarithmic terms. Parameter $b$ creates scaling problems (large values of $b$) or logarithmic term problems (low values of $b$). Low values of $c$, i.e., $c \leq 2$, may favor non–concavity of $\mathcal{L}(a, b, c)$, whereas high values of $c$ may cause slow convergence.

There are two principal approaches to find the MLEs: directly maximizing (11.31a) or solving the system of likelihood equations (11.4d–f). The first approach has been taken by only a few researchers. PIKE (1966) substituted $b$ of (11.18a) into (11.31a) yielding

$$\mathcal{L}(a, c) = n\left[ \ln c + \ln n - 1 \right] - n \ln \left\{ \sum_{i=1}^{n} (x_i - a)^c \right\} + (c - 1) \sum_{i=1}^{n} \ln(x_i - a) \quad (11.33)$$

and then used the derivative–free search procedure of HOOKE/JEEVES (1961) to find the bivariate maximum $(\widehat{a}, \widehat{c})$, which in turn produces $\widehat{b}$ according to (11.18a). ZANAKIS (1977) used the adaptive pattern search of BUFFA/TAUBERT (1972), a later version of the HOOKE/JEEVES algorithm. SEYMORE/MAKI (1969) used the ZOUTENDIJK method of feasible directions to obtain the joint MLEs of (11.31a). These methods do not guarantee a global optimum. Moreover, the pattern search method does not utilize second order information and therefore will tend to be slower than the NEWTON–RAPHSON technique.

We now review some of the gradient techniques. HARTER/MOORE (1965a) applied the **rule of false position** (iterative linear interpolation on first derivative equations by using second derivatives; see BARNETT (1966)) to simultaneous equations (11.4d–f) for estimating the three unknown parameters, one at a time in the cyclic order $b$, $c$, $a$. The method always yields positive estimates of $b$ and $c$. However, an iteration may lead to an estimate of $a$ equal to $\min\{x_i\}$, thus causing logarithmic difficulty. This is avoided each time by censoring the smallest sample observation(s), which subsequently is used only as an upper bound for the location parameter. However, this results in a partial loss of information, which may reduce the accuracy of the solution. HAAN/BEER (1967) used, in a trial- and-error scheme, a combination of the rule of false position, the **secant method** and the **golden section** univariate maximization search. It is well known, however, that a one–at–a–time search may fail if sharp ridges exist. WINGO (1972, 1973) employed a **modified quasi–linearization algorithm**. As pointed out by BROWN/WINGO (1975), this algorithm is a version of a modified NEWTON–RAPHSON method tailored to the case of bounded variables. Even though the method of WINGO is very fast, it sometimes converges to a saddle point instead of an interior maximum point. ARCHER (1980) proposed a **hybrid technique** combining the value of false position for $a$ and WINGO's method for $b$ and $c$, but it may converge to a saddle point also.

### 11.3.2.2   A non-failing algorithm [18]

The iterative procedures discussed in the preceding section do not make explicit use of the constraints given in (11.31b,c). Their role is usually limited to altering the parameter values to the specified limit if, during an iteration, the resulting value violates the constraints. The approach presented now is rather simple and guarantees to find the maximizing parameter set. We will call it **profile likelihood method**. Its idea goes back to LAWLESS, (1982) and it has been further explored and elaborated by PANCHANG/GUPTA (1989), LOCK-HART/STEPHENS (1994) and QIAO/TSOKOS (1995). The following representation rests mainly upon the version of PANCHANG/GUPTA.

The approach is based on the philosophy that the constraint (11.31b) $0 \leq a \leq x_{min} :=$ $\min\{x_i\}$ defines an allowable parameter space in which to look for a solution.[19] The domain $[0, x_{min} - \epsilon]$ is divided into a suitable number $J$ of intervals of size $\Delta a$, and $a$ is then assigned values $a_j = (j - 1)\Delta a$, $j = 1, 2, \ldots, J + 1$. $\epsilon$ is a small positive number introduced to prevent any pathological situation in the likelihood equations. For a given value of $a = a_j$, equations (11.4e,f) are solved to get $b = b_j$ and $c = c_j$. This is equivalent to solving the two–parameter MLE problem since the $x_i - a_j$ may be considered as another sample $y_i$:

$$y_i := x_i - a_j.$$

$c_j$ is obtained by solving (from 11.19b)

$$f(c \mid a_j) = \frac{n}{c} + \sum_i \ln y_i - \frac{n \sum_i y_i^c \ln y_i}{\sum_i y_i^c} = 0 \tag{11.34a}$$

by NEWTON–RAPHSON and then $b_j$ is found by inserting $c_j$ into (11.19a):

$$b_j = \left[\frac{1}{n} \sum_i y_i^{c_j}\right]^{1/c_j}. \tag{11.34b}$$

Finally, we calculate the log–likelihood at $(a_j, b_j, c_j)$

$$\mathcal{L}(a_j, b_j, c_j) = n\left\{\ln\left(\frac{c_j}{b_j^{c_j}}\right) - 1\right\} + (c_j - 1)\sum_i \ln(x_i - a_j). \tag{11.34c}$$

The above process is repeated for all $a_j$ and thus $\mathcal{L}(a, b, c)$ is obtained as a function of $a$:

$$\mathcal{L}^*(a) := \max_{b,c} \mathcal{L}(a, b, c). \tag{11.34d}$$

$\mathcal{L}^*(a)$ is the **profile log–likelihood function** that is to be scanned to determine the maximum.

We make the following remarks to the above algorithm:

---

[18]  Suggested reading for this section: LAWLESS (1982, pp. 192–194), LOCKHART/STEPHENS (1994), PANCHANG/GUPTA (1989) and QIAO/TSOKOS (1995).

[19]  If there is strong evidence for shelf–ageing, i.e., $a < 0$, the constraint has to be extended.

1. For $j = 1$, i.e., $a = 0$, PANCHANG/GUPTA (1989) have found a good initial estimate to be

$$c_{0,1} = \frac{1}{n \ln x_{max} - \sum \ln x_i} \, . \tag{11.34e}$$

2. An initial estimate $c_{0,j}$ $(j = 2, 3, ...)$ for solving (11.34.a), suggested by PANCHANG/GUPTA and based on the observation that $\widehat{c}$ decreases as $\widehat{a}$ increases, is

$$c_{0,j} = c_{j-1}; \; j = 2, 3, \dots \tag{11.34f}$$

3. The number of intervals $J$ depends on the accuracy that is desired in the estimators. To reduce the computational burden one may start with a broad grid for $a$ to locate the area containing the maximum and then zoom into its vicinity using a finer grid and so on.

4. Even if the iterated solution $(\widehat{a}, \widehat{b}, \widehat{c})$ is a local maximum, it may not be the maximum likelihood estimate. This could be the corner point (11.32). So, one should check for this by comparing the log–likelihood at the iterated solution to the corner–point solution.

**Example 11/5: Applications of the non–failing algorithm**

We have applied the PANCHANG–GUPTA algorithm to different datasets. Fig. 11/4 shows different courses of the profile log–likelihood function $\mathcal{L}^*(a)$.

Figure 11/4: Profile log-likelihood functions

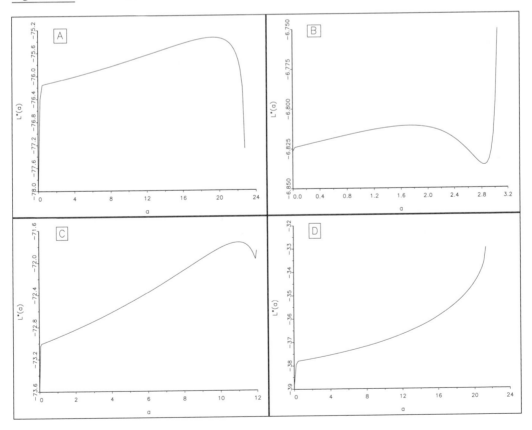

The graph in part A belongs to dataset #2 (see Tab. 9/9) which was simulated using $a = 15$, $b = 30$ and $c = 2.5$. The maximum of $\mathcal{L}^*(a)$ occurs at $\hat{a} = 19.4$ accompanied by $\hat{b} = 25.0$ and $\hat{c} = 2.2$. The corresponding estimated variance–covariance matrix according to (11.12a) reads

$$\widehat{\mathrm{Var}}(\hat{\theta}) = \begin{pmatrix} 2.8279 & 2.2981 & -0.0983 \\ 2.2981 & 9.0257 & 0.2414 \\ -0.0983 & 0.2414 & 0.1505 \end{pmatrix}.$$

The 95% confidence intervals are

$$\begin{array}{ccccl} 16.10 & \leq & a & \leq & 22.70, \quad \text{not covering the true parameter,} \\ 19.11 & \leq & b & \leq & 30.89, \\ 1.44 & \leq & c & \leq & 2.96. \end{array}$$

The graph in part B rests upon the sample data $x' = (3.1,\ 4.6,\ 5.6,\ 6.8)$ taken from ROCKETTE et al. (1974). The local maximum at 1.73 has $(\hat{a},\ \hat{b},\ \hat{c}) = (1.73,\ 3.72,\ 2.71)$ with likelihood 0.0011. The local minimum at 2.83 has $(\hat{a}, \hat{b},\ \hat{c}) = (2.83,\ 2.39,\ 1.44)$ and is a saddle point of the likelihood function. However the likelihood is maximum with 0.0013 at the corner point $(\hat{a},\ \hat{b},\ \hat{c}) = (3.1,\ 1.925,\ 1)$. The graph in part C belongs to a dataset on ocean wave heights (PETRUASKAS/AAGARD (1971)) and looks similar to that of ROCKETTE's data, but here the local maximum is not at the corner, we have $(\hat{a},\ \hat{b},\ \hat{c}) = (11,\ 11.8,\ 1.51)$. The graph in part D belongs to a simulated dataset $x' = (21.3,\ 22.5,\ 23.3,\ 24.9,\ 29.3,\ 29.8,\ 31.3,\ 35.5,\ 42.5,\ 57.8)$, the true parameters being $a = 20, b = 10$, $c = 1.0$. The maximum of the graph belongs to the corner point $(\hat{a},\ \hat{b},\ \hat{c}) = (21.3,\ 10.4,\ 1)$.

Fig. 11/4 does not include another possible course of profile log–likelihood. $\mathcal{L}^*(a)$ may decline monotonically, from $a = 0$ to $a = x_{min}$. In this case the three–parameter WEIBULL model should be dropped in favor of the two–parameter version.

### 11.3.2.3   Modified ML estimation

COHEN/WHITTEN (1982) proposed **modified maximum likelihood estimators** (MMLEs) for $a$, $b$ and $c$ as an alternative for use when the MLEs are likely to be unsatisfactory, i.e., when $c < 2.0$. However, some of these MMLEs appear to offer advantages with respect to ease of computation, bias and/or variance for all values of $c$. The various modified estimators are obtained by replacing the first of the three likelihood equations (11.4d–f), i.e., $\partial \mathcal{L}(a, b, c)/\partial a = 0$, with alternative functional relationships, enumerated MMLE–I through MMLE-V.

MMLE–I
$\partial \mathcal{L}(a, b, c)/\partial a = 0$ is replaced with

$$\mathrm{E}\big[F(X_{r:n})\big] = F(x_{r:n}). \tag{11.35a}$$

Since $\mathrm{E}\big[F(X_{r:n})\big] = r/(n+1)$, we get

$$\frac{r}{n+1} = 1 - \exp\left\{-\left(\frac{x_{r:n} - a}{b}\right)^c\right\}, \tag{11.35b}$$

which reduces to

$$\left(\frac{x_{r:n} - a}{b}\right)^c = -\ln\left[\frac{n+1-r}{n+1}\right]. \tag{11.35c}$$

COHEN/WHITTEN (1982) — although $1 \leq r \leq n$ — restrict consideration to $r = 1$, arguing that the first order statistic $x_{1:n}$ contains more information about $a$ than any of the other order statistics and often more information than all the other order statistics combined.[20] With $r = 1$ we have

$$\left(\frac{x_{1:n} - a}{b}\right)^c = -\ln\left(\frac{n}{n+1}\right). \tag{11.35d}$$

The system of equations to be solved now consists of (11.35d) and — from (11.18c) — of

$$\sum_{i=1}^{n}(1 - c\,w_i)\,\exp\{c\,w_i\} = 0, \tag{11.35e}$$

where $w_i = \ln(x_i - a) - \frac{1}{n}\sum\ln(x_i - a)$, and of

$$b = \left[\frac{1}{n}\sum_{i=1}^{n}(x_i - a)^c\right]^{1/c}. \tag{11.35f}$$

A first approximation $a_1$ ($a_1 < x_{min}$) is selected and (11.35e) is solved for $c_1$, and $b_1$ follows from (11.35f). When (11.35d) is solved by $a_1$, $b_1$ and $c_1$, the task is completed. Otherwise the cycle of computations is repeated with a new approximation $a_2$ and continues until we find a pair of values $(a_i, a_j)$ such that $|a_i - a_j|$ is sufficiently small and such that

$$\left(\frac{x_{1:n} - a_i}{b_i}\right)^{c_i} \gtrless -\ln\left(\frac{n}{n+1}\right) \lessgtr \left(\frac{x_{1:n} - a_j}{b_j}\right)^{c_j}.$$

We then interpolate linearly for the required estimates $\widehat{a}$, $\widehat{b}$, $\widehat{c}$.

MMLE–II
$\partial\mathcal{L}(a, b, c)/\partial a = 0$ is replaced with

$$E(X_{1:n}) = x_{1:n}, \tag{11.36a}$$

where

$$E(X_{1:n}) = a + \frac{b}{n^{1/c}}\Gamma\left(1 + \frac{1}{c}\right). \tag{11.36b}$$

We then have to solve (11.35e,f) and

$$a + \frac{b}{n^{1/c}}\Gamma\left(1 + \frac{1}{c}\right) = x_{1:n} \tag{11.36c}$$

by the same iteration as described above.

---

[20] When outliers are suspected of being present, values of $r > 1$ might be justified.

MMLE–III

Now $\partial \mathcal{L}(a, b, c)/\partial a = 0$ is replaced with

$$E(X) = \bar{x}, \tag{11.37a}$$

where

$$E(X) = a + b\, \Gamma\left(1 + \frac{1}{c}\right). \tag{11.37b}$$

Then (11.35e,f) and

$$a + b\, \Gamma\left(1 + \frac{1}{c}\right) = \bar{x} \tag{11.37c}$$

have to be solved iteratively.

MMLE–IV

In this case the replacing equation is

$$\text{Var}(X) = s^2 \tag{11.38a}$$

or

$$b^2 \left[\Gamma\left(1 + \frac{2}{c}\right) - \Gamma^2\left(1 + \frac{1}{c}\right)\right] = \frac{1}{n-1}\sum_{i=1}^{n}(x_i - \bar{x})^2. \tag{11.38b}$$

MMLE–V

This time, the population median is equated to the sample median $\tilde{x}$:

$$a + b\,(\ln 2)^{1/c} = \tilde{x}. \tag{11.39}$$

COHEN/WHITTEN (1982) recommend that MMLE–I or MMLE–II be employed regardless of the value of $c$. They have further shown by simulation studies that the entries in the asymptotic variance–covariance matrix (11.12) provide a reasonable approximation to the variances and covariances of the MLEs when $c > 2$.

**Example 11/6:  Applications of MMLEs**

We have applied MMLE–I and MMLE–II to the four datasets used in Example 11/5 and which led to Fig. 11/4. The MMLEs failed to give any estimator for ROCKETTE's dataset. Tab. 11/4 compares the results of the non–failing algorithm to those of the MMLE–approaches.

Table 11/4:   Comparison of the results of ML approaches

| Dataset | Non–failing algorithm | | | MMLE–I | | | MMLE–II | | |
|---|---|---|---|---|---|---|---|---|---|
| leading to: | $\hat{a}$ | $\hat{b}$ | $\hat{c}$ | $\hat{a}$ | $\hat{b}$ | $\hat{c}$ | $\hat{a}$ | $\hat{b}$ | $\hat{c}$ |
| Figure 11/4A | 19.4 | 25.0 | 2.2 | 5.2 | 40.2 | 3.64 | 12.1 | 33.3 | 2.96 |
| Figure 11/4B | 3.1 | 1.925 | 1 | – | – | – | – | – | – |
| Figure 11/4C | 11 | 11.8 | 1.51 | 9.2 | 14.0 | 1.93 | 9.95 | 13.13 | 1.77 |
| Figure 11/4D | 21.3 | 10.4 | 1 | 19.3 | 13.4 | 1.24 | 19.4 | 13.3 | 1.22 |

#### 11.3.2.4 Finite sample results

Finite sample results used to construct confidence intervals, which are more correct than those based on the asymptotic normal distribution, can be found from suitably chosen pivotal functions. Based on the approach of THOMAN et al. (1969) the results in Tab. 11/5 have been found by LEMON (1975). In what follows, $\widehat{a}_{abc}$, $\widehat{b}_{abc}$ and $\widehat{c}_{abc}$ are used to denote the MLEs of the parameters $a$, $b$ and $c$ when in fact sampling is from the three–parameter WEIBULL distribution with the parameter values given in the suffix. So $\widehat{c}_{01c}$ is the MLE of the $c$ when sampling from a WEIBULL distribution with parameters $a = 0$, $b = 1$ and $c$.

Table 11/5: Pivotal functions for WEIBULL distributions

| Parameters | | Pivotal functions |
|---|---|---|
| known | unknown | |
| none | $a,\ b,\ c$ | $\widehat{c}_{abc} \stackrel{\mathrm{d}}{=} \widehat{c}_{01c}; \quad \dfrac{\widehat{b}_{abc}}{b} \stackrel{\mathrm{d}}{=} \widehat{b}_{01c}; \quad \dfrac{\widehat{a}_{abc} - a}{\widehat{b}_{abc}} \stackrel{\mathrm{d}}{=} \dfrac{\widehat{a}_{01c}}{\widehat{b}_{01c}}$ |
| $c$ | $a,\ b$ | $\dfrac{\widehat{b}_{abc}}{b} \stackrel{\mathrm{d}}{=} \widehat{b}_{01c}; \quad \dfrac{\widehat{a}_{abc} - a}{\widehat{b}_{abc}} \stackrel{\mathrm{d}}{=} \dfrac{\widehat{a}_{01c}}{\widehat{b}_{01c}}$ |
| $b$ | $a,\ c$ | $\widehat{c}_{abc} \stackrel{\mathrm{d}}{=} \widehat{c}_{01c}; \quad \dfrac{\widehat{a}_{abc} - a}{b} \stackrel{\mathrm{d}}{=} \widehat{a}_{01c}$ |
| $a$ | $b,\ c$ | $\dfrac{\widehat{c}_{abc}}{c} \stackrel{\mathrm{d}}{=} \widehat{c}_{a11}; \quad \widehat{c}_{abc} \ln\left(\dfrac{\widehat{b}_{abc}}{b}\right) \stackrel{\mathrm{d}}{=} \widehat{c}_{a11} \ln b_{a11}$ |
| $a,\ b$ | $c$ | $\dfrac{\widehat{c}_{abc}}{c} \stackrel{\mathrm{d}}{=} \widehat{c}_{a11}$ |
| $a,\ c$ | $b$ | $c \ln\left(\dfrac{\widehat{b}_{abc}}{b}\right) \stackrel{\mathrm{d}}{=} \widehat{b}_{a11}$ |
| $b,\ c$ | $a$ | $\dfrac{\widehat{a}_{abc} - a}{b} \stackrel{\mathrm{d}}{=} \widehat{a}_{01c}$ |

Source: LEMON (1975, p. 252) — Reprinted with permission from *Technometrics*. Copyright 1975 by the American Statistical Association. All rights reserved.

The principal advantage of working with the pivotal function is that their distributions are independent of one or more of the unknown parameters and can be obtained by a Monte Carlo simulation of the "reduced" distribution. Tab. 11/5 contains pivotal functions equated to their equivalent "reduced" forms for the MLEs of the three–parameter WEIBULL distribution. For example, the pivotal function $\widehat{c}_{abc} \stackrel{d}{=} \widehat{c}_{01c}$ is distributed independently of the parameters $a$ and $b$ and the reduced WEIBULL distribution has parameters $a = 0$, $b = 1$ and $c$. The pivotal functions given in Tab. 11/5 are valid for all types of censoring. But it must be noted that the distributions of these pivotal functions are directly dependent on the sample size and the type of censoring used. So these distributions have to be hand–tailored by simulation in each individual case.

Looking into Tab. 11/5 we recognize that when both $a$ and $c$ are unknown, the pivotal distributions depends on the unknown value of $c$. In these cases LEMON (1975) proposes to approximate the distribution by using $c = \widehat{c}_{abc}$, and hence, approximate confidence intervals for $b$ and $c$ are obtained by Monte Carlo methods; e.g., $\widehat{b}_{01\widehat{c}}$ is used for $\widehat{b}_{01c}$. It is not possible to find a confidence interval for $c$ via a pivotal distribution when $c$ and $a$ are unknown; instead, one has to consider bootstrapping.

---

**Example 11/7: Working with pivotal distributions when all three WEIBULL parameters are unknown**

In Example 11/5 we have estimated the parameters belonging to dataset #2 of size $n = 20$ and found $\widehat{a} = 19.4$, $\widehat{b} = 25.0$ and $\widehat{c} = 2.2$. Here we generated 20,000 samples of size 20, WEIBULL distributed with $a = 0$, $b = 1$ and $c = \widehat{c} = 2.2$. In each sample we determined $\widehat{a}_{0,1,2.2}$ by (11.18b), the pivotal function $\widehat{b}_{0,1,2.2}$ by (11.18a) and the pivotal function $\widehat{a}_{0,1,2.2}/\widehat{b}_{0,1,2.2}$. From the distributional equivalences and the two resulting empirical distributions, we computed

$$\Pr\left(0.671 \leq \widehat{b}_{0,1,2.2} = \frac{\widehat{b}_{abc}}{b} \leq 1.184\right) = 0.95$$

$$\Pr\left(-0.504 \leq \frac{\widehat{a}_{0,1,2.2}}{\widehat{b}_{0,1,2.2}} = \frac{\widehat{a}_{abc} - a}{\widehat{b}_{abc}} \leq 0.406\right) = 0.95.$$

Inserting the estimates $\widehat{a}_{abc} = \widehat{a} = 19.4$ and $\widehat{b}_{abc} = \widehat{b} = 25.0$ from dataset #2, we found the following approximate 95% confidence intervals:

$$21.11 \leq b \leq 36.98 \quad \text{and} \quad 9.25 \leq a \leq 26.75$$

---

# 11.4 Uncensored samples with grouped data [21]

Grouped data arise in a natural way when $n$ items on test are not monitored but are inspected for the number of surviving items at given points in time $t_1 < t_2 < \ldots < t_{k-1}$ (see Sect. 8.3 on life test plans). Introducing $t_0 := 0$ and $t_k := \infty$, we thus have $k$ time–intervals $[t_0, t_1), [t_1, t_2), \ldots, [t_{k-1}, t_k)$ with $n_j$ failed items in the $j$–th interval $[t_{j-1}, t_j)$; $j = 1, 2, \ldots, k$; $n = \sum_{j=1}^{k} n_j$. In the following we will show only how to proceed in the two–parameter case with $b$ and $c$ unknown.[22]

The likelihood element $L_j(b, c)$ of an item which failed in the $j$–th interval is

$$L_j(b, c) = \int_{t_{j-1}}^{t_j} \frac{c}{b}\left(\frac{x}{b}\right)^{c-1} \exp\left\{-\left(\frac{x}{b}\right)^c\right\} dx$$

$$= F(t_j \mid b, c) - F(t_{j-1} \mid b, c)$$

$$= \exp\left\{-\left(\frac{t_{j-1}}{b}\right)^c\right\} - \exp\left\{-\left(\frac{t_j}{b}\right)^c\right\}; \quad j = 1, \ldots, k \qquad (11.40a)$$

with $L_1(b, c) = 1 - \exp\left\{-\left(\frac{t_1}{b}\right)^c\right\}$ and $L_k(b, c) = \exp\left\{-\left(\frac{t_{k-1}}{b}\right)^c\right\}$. The likelihood

---

[21] Suggested reading for this section: ARCHER (1980, 1982), CHENG/CHEN (1988), FLYGARE/AUSTIN/BUCKWALTER (1985), HIROSE/LAI (1977), OSTROUCHOV/MEEKER (1988), RAO/RAO/NARASIMHAM (1994).

[22] For the three–parameter case, see ARCHER (1980).

function results as

$$L(b,c) = \frac{n!}{\prod\limits_{j=1}^{k} n_j!} \prod_{j=1}^{k} L_j(b,c)^{n_j} \qquad (11.40\text{b})$$

with $\sum\limits_{j=1}^{k} n_j = n$. Neglecting the pre–factor not involving the parameters, the log–likelihood function is

$$\mathcal{L}(b,c) = \sum_{j=1}^{k} n_j \ln\left[L_j(b,c)\right] = \sum_{j=1}^{k} n_j \ln\left[\exp\left\{-\left(\frac{t_{j-1}}{b}\right)^c\right\} - \exp\left\{-\left(\frac{t_j}{b}\right)^c\right\}\right].$$

$$(11.40\text{c})$$

Introducing

$$u_j := t_j/b,$$

the system of likelihood equations on simplification is given by

$$\sum_{j=1}^{k-1} \frac{n_j\left(t_j^c - t_{j-1}^c\right)}{\exp\left\{u_j^c - u_{j-1}^c\right\} - 1} - \sum_{j=2}^{k} n_j\, t_{j-1}^c = 0, \qquad (11.41\text{a})$$

$$n_1\, t_1^c \frac{\ln(u_1)}{\exp\{u_1^c\} - 1} + \sum_{j=2}^{k-1} n_j \frac{t_j^c \ln(u_j) - t_{j-1}^c \ln(u_{j-1})}{\exp\{u_j^c - u_{j-1}^c\} - 1} - \sum_{j=2}^{k} n_j\, t_{j-1}^c \ln(u_{j-1}) = 0.$$

$$(11.41\text{b})$$

With respect to (11.40c) we remark that empty intervals, i.e., $n_j = 0$, do not contribute to the log–likelihood.

Let $\ell \le k$ be the number of non–zeros $n_j'$s, denoted by $n_\iota^*$, and $[v_\iota, w_\iota]$ be the redefined interval corresponding to $n_\iota^*$. CHENG/CHEN (1988) proved the following.

Theorem:

The MLE $(\widehat{b}, \widehat{c})$ of $(b, c)$ exists and is unique if and only if

- $\ell \ge 3$ or

- $\ell = 2$, $w_1 \ne v_2$ and $v_1 \ne 0$ or

- $\ell = 2$, $w_1 \ne v_2$ and $w_2 \ne \infty$. ∎

Let

$$\begin{aligned}
y_j &:= (t_j/b)^c, \\
A_j &:= y_j - y_{j-1}, \\
B_j &:= y_j \ln(y_j) - y_{j-1} \ln(y_{j-1}), \\
C_j &:= \exp\{y_j\} - \exp\{y_{j-1}\}.
\end{aligned}$$

The elements $I_{\nu\kappa}$ of the information matrix $\mathbf{I}(\theta)$ are given by

$$I_{bb} = E\left[-\frac{\partial^2 \mathcal{L}}{\partial b^2}\right] = n\left(\frac{c}{b}\right)^2 \sum_{j=1}^{k-1} \frac{A_j^2}{C_j}, \tag{11.42a}$$

$$I_{bc} = E\left[-\frac{\partial^2 \mathcal{L}}{\partial b\,\partial c}\right] = \frac{n}{b}\left[\frac{y_1^2 \ln(y_1)}{\exp(y_1) - 1} + \sum_{j=2}^{k-1} \frac{A_j B_j}{C_j}\right], \tag{11.42b}$$

$$I_{cc} = E\left[-\frac{\partial^2 \mathcal{L}}{\partial c^2}\right] = \frac{n}{c^2}\left[\frac{[y_1 \ln(y_1)]^2}{\exp(y_1) - 1} + \sum_{j=2}^{k-1} \frac{B_j^2}{c_j}\right]. \tag{11.42c}$$

RAO et al. (1994) derived the optimal interval length $t_j - t_{j-1} = \Delta$; $j = 1, \ldots, k-1$ for an equi–class grouped sample by minimizing the generalized asymptotic variance of the estimators, i.e., the determinant $I_{bb}\,I_{cc} - I_{bc}^2$ of the variance–covariance matrix. FLYGARE et al. (1985) presented the MLEs for the two–parameter WEIBULL distribution when the data are given as intervals; i.e., two known data $\ell_j$ and $u_j$ surround the unknown value $x_j$ of the response variable. Their approach is somewhat similar to the grouped–data approach, the only exception being $j = 1, 2, \ldots, n$ and $n_j = 1 \,\forall\, j$.

## 11.5   Samples censored on both sides[23]

Before going into the details of the prevailing right–censoring mechanisms in Sections 11.6 and 11.7, we will briefly comment on the more general doubly censoring scheme. Let $n_\ell$ ($n_r$) be the number of observations censored on the left at $t_\ell$ (on the right at $t_r$). Assuming the three–parameter WEIBULL distribution, we further introduce the order statistics

$$x_i := x_{i:n}; \quad i = n_\ell + 1, \ldots, n - n_r,$$

and

$$u_i := \frac{x_i - a}{b}, \quad v_\ell = \frac{t_\ell - a}{b}, \quad v_r = \frac{t_r - a}{b}.$$

The log–likelihood function is

$$\mathcal{L}(a, b, c) =$$

$$n_\ell \ln\left[1 - \exp\left\{-v_\ell^c\right\}\right] - n_r\, v_r^c + (n - n_r - n_\ell) \ln\left(\frac{c}{b}\right) + (c-1) \sum_{i=n_\ell+1}^{n-n_r} \left[\ln(u_i) - u_i^c\right]. \tag{11.43a}$$

The resulting system of likelihood equations is given by

$$\frac{\partial \mathcal{L}(a, b, c)}{\partial a} =$$

$$-n_\ell \left(\frac{c}{b}\right) v_\ell^{c-1} \frac{\exp\left\{-v_\ell^c\right\}}{1 - \exp\left\{-v_\ell^c\right\}} + n_r \left(\frac{c}{b}\right) v_r^c - \frac{1}{b} \sum_{i=n_\ell+1}^{n-n_r} \left[\frac{c-1}{u_i} - c\, u_i^{c-1}\right] = 0, \tag{11.43b}$$

---

[23] Suggested reading for this section: ARCHER (1980), HARTER/MOORE (1967a), LEMON (1975) and ZANAKIS/KYPARISIS (1986).

$$\frac{\partial \mathcal{L}(a,b,c)}{\partial b} =$$

$$-n_\ell \left(\frac{c}{b}\right) v_\ell^c \frac{\exp\{-v_\ell^c\}}{1 - \exp\{-v_\ell^c\}} + n_r \left(\frac{c}{b}\right) v_r^c - (n - n_r - n_\ell)\frac{c}{b} + \frac{c}{b}\sum_{i=n_\ell+1}^{n-n_r} u_i^c = 0, \quad (11.43c)$$

$$\frac{\partial \mathcal{L}(a,b,c)}{\partial c} =$$

$$n_\ell\, v_\ell^c \ln(v_\ell) \frac{\exp\{-v_\ell^c\}}{1 - \exp\{-v_\ell^c\}} - n_r\, v_r^c \ln(v_r) + \frac{n - n_r - n_\ell}{c} + \sum_{i=n_\ell+1}^{n-n_r} (1 - u_i^c)\ln(u_i) = 0.$$

$$(11.43d)$$

HARTER/MOORE (1967b) have derived the asymptotic variance–covariance matrix for the solutions $\hat{a}$, $\hat{b}$ and $\hat{c}$ of (11.43b–d). As ANTLE/BAIN (1969) have noted the pivotal functions considered above are valid for type–II censoring from both sides. We notice that $n_\ell = n_r = 0$ is the complete–sample case (see (11.4c–f)) $n_r = 0$ and $n_\ell > 0$ is left–censoring only and $n_\ell = 0$, $n_r > 0$ is singly right-censoring (see Section 11.6).

Suppose the uncensored observations, instead of being observed individually, are divided into $k$ groups with $t_0$ the boundary of the first group coinciding with $t_\ell$ and $t_k$, the upper boundary of the last group coinciding with $t_r$. Using the normalized boundaries

$$w_j := \frac{t_j - a}{b}; \quad j = 0, 1, \ldots, k$$

together with

$$g_j := \exp(-w_j^c),$$

the log–likelihood function and the system of likelihood equations are given by

$$\mathcal{L}(a,b,c) = n_\ell \ln\left[1 - \exp\{w_0^c\}\right] - n_r\, w_k^c + \sum_{j=1}^{k} n_j \ln\left[g_{j-1} - g_j\right], \quad (11.44a)$$

$$\frac{\partial \mathcal{L}(a,b,c)}{\partial a} =$$

$$-n_\ell\left(\frac{c}{b}\right) w_0^{c-1} \frac{g_0}{1 - g_0} + n_r\left(\frac{c}{b}\right) w_k^{c-1} + \frac{c}{b}\sum_{j=1}^{k} n_j \frac{w_{j-1}^{c-1} g_{j-1} - w_j^{c-1} g_j}{g_{j-1} - g_j} = 0, \quad (11.44b)$$

$$\frac{\partial \mathcal{L}(a,b,c)}{\partial b} =$$

$$-n_\ell\left(\frac{c}{b}\right) w_0^c \frac{g_0}{1 - g_0} + n_r\left(\frac{c}{b}\right) w_k^c + \frac{c}{b}\sum_{j=1}^{k} n_j \frac{w_{j-1}^c g_{j-1} - w_j^c g_j}{g_{j-1} - g_j} = 0, \quad (11.44c)$$

$$\frac{\partial \mathcal{L}(a,b,c)}{\partial c} =$$

$$n_\ell\, w_0^c \ln(w_0)\frac{g_0}{1 - g_0} - n_r\, w_k^c \ln(w_k) - \sum_{j=1}^{k} n_j \frac{w_{j-1}^c \ln(w_{j-1})\, g_{j-1} - w_j^c \ln(w_j)\, g_j}{g_{j-1} - g_j} = 0.$$

$$(11.44d)$$

$n_j$; $j = 1, \ldots, k$ is the number of observations in the $j$–th group. Computational techniques for solving (11.43b–d) or (11.44b–d) are reviewed in ARCHER (1980).

## 11.6    Samples singly censored on the right

Censoring on the right and especially its single version is the most popular way to collect lifetime data. That is the reason it has attracted the attention of many researchers, the focus being on type II where the number $n - r$ of unobserved highest lifetimes is fixed. Type–I censoring with a fixed censoring time $T$ should be treated differently because the number $n - k$ of surviving items is random, but in most papers both types are treated alike,[24] an exception being the papers of SIRVANCI/YANG (1984), SIRVANCI (1984)[25] and MENDENHALL/LEHMAN (1960) who derived the mean and the variance of the scale parameter when the shape parameter is assumed known.

This section is organized as follows: We first treat the two–parameter distribution (Sect. 11.6.1) with respect to finding and statistically evaluating the estimators; finally (Sect. 11.6.2) we comment upon the three–parameter distribution. The results of single censoring on the right have a lot in common with the complete–sampling case (Sect. 11.3).

### 11.6.1    Two-parameter WEIBULL distribution

The following two subsections are devoted to the case of non–grouped data only. CHACE (1976) has analyzed censored and grouped life test data, but only with respect to calculating the estimates. BHATTACHARYA (1962) has studied a similar life–testing problem. $N_i$ items ($i = 1, 2, \ldots, k$) are put on test, e. g., in $k$ different plants or departments of an enterprise. The testing intervals $(0, T_i]$ are different and we know only the numbers $n_i$ of failed items up to $t_i$. Let $P_i = \int_0^{t_i} \frac{c}{b} \left(\frac{x}{b}\right)^{c-1} \exp\left\{-\left(\frac{x}{b}\right)^c\right\} \, dx$ be the probability of failure in $(0, t_i]$. The MLEs of $b$ and $c$ are derived from the likelihood function $\prod_{i=1}^{k} \binom{N_i}{n_i} P_i^{n_i} (1 - P_i)^{N_i - n_i}$.

#### 11.6.1.1    Solving the likelihood equations[26]

We note — for short — the first $r$ ordered variates from a sample of size $n$ by $X_1 \leq X_2 \leq \ldots \leq X_r$, $r < n$. The likelihood elements are

$$L_i(b, c) = \begin{cases} f(X_i \mid b, c) = \dfrac{c}{b} \left(\dfrac{X_i}{b}\right)^{c-1} \exp\left\{-\left(\dfrac{X_i}{b}\right)^c\right\} & \text{for } i = 1, \ldots, r, \\[2em] 1 - F(X_r \mid b, c) = \exp\left\{-\left(\dfrac{X_r}{b}\right)^c\right\} & \text{for } i = r+1, \ldots, n, \end{cases} \tag{11.45a}$$

---

[24]  This procedure is all right as long as one interprets the type–I results as conditional, the condition being the number $k$ of realized failures within $(0, T]$.

[25]  See Sect. 13.5. 2 for details of this approach.

[26]  Suggested reading for this section: BAIN/ENGELHARDT (1991a, pp. 211 ff.), COHEN (1965, 1966), FARNUM/BOOTH (1997), KEATS/LAWRENCE/WANG (1997), MCCOOL (1970a).

from which the likelihood function results as

$$L(b, c) = \frac{n!}{(n-r)!} \left(\frac{c}{b}\right)^r \prod_{i=1}^{r} \left(\frac{X_i}{b}\right)^{c-1} \exp\left\{-\left(\frac{X_i}{b}\right)^c\right\} \prod_{j=r+1}^{n} \exp\left\{-\left(\frac{X_r}{b}\right)^c\right\}$$

$$= \frac{n!}{(n-r)!} \left(\frac{c}{b}\right)^r \left[\prod_{i=1}^{r} \left(\frac{X_i}{b}\right)\right]^{c-1} \exp\left\{-\sum_{i=1}^{r} \left(\frac{X_i}{b}\right)^c - (n-r)\left(\frac{X_r}{b}\right)^c\right\}.$$

$$(11.45b)$$

Neglecting the combinatorial pre–factor, which does not involve any of the unknown parameters, we get the log–likelihood function

$$\mathcal{L}(b, c) = r\left[\ln c - c \ln b\right] + (c-1)\sum_{i=1}^{r} \ln X_i - \frac{1}{b^c}\left[\sum_{i=1}^{r} X_i^c + (n-r)X_r^c\right], \quad (11.45c)$$

resulting in the likelihood equations

$$\frac{\partial \mathcal{L}(b, c)}{\partial b} = -\frac{r\,c}{b} + \frac{c}{b}\left[\sum_{i=1}^{r} \left(\frac{X_i}{b}\right)^c + (n-r)\left(\frac{X_r}{b}\right)^c\right] = 0, \quad (11.45d)$$

$$\frac{\partial \mathcal{L}(b, c)}{\partial c} = \frac{r}{c} - r\ln b + \sum_{i=1}^{r} \ln X_i - \frac{1}{b^c}\left[\sum_{i=1}^{r} X_i^c \ln\left(\frac{X_i}{b}\right) + (n-r)X_r^c \ln\left(\frac{X_r}{b}\right)\right] = 0. \quad (11.45e)$$

(11.45c–e) and (11.13) through (11.14a–b) coincide when $r = n$. (11.45d) is reduced to

$$b = \left[\frac{\sum_{i=1}^{r} X_i^c + (n-r)X_r^c}{r}\right]^{1/c} \quad (11.46a)$$

and (11.45c) — upon substituting (11.46a) and some rearrangement — reduces to

$$\frac{1}{c} + \frac{1}{r}\sum_{i=1}^{r} \ln X_i - \frac{\sum_{i=1}^{r} X_i^c \ln X_i + (n-r)X_r^c \ln X_r}{\sum_{i=1}^{r} X_i^c + (n-r)X_r^c} = 0. \quad (11.46b)$$

(11.46a,b) and (11.19a,b) coincide for $r = n$. The solution $\widehat{c}$ to (11.46b) may be found by the NEWTON–RAPHSON method — see below — and gives $\widehat{b}$ when inserted into (11.46a).

FARNUM/BOOTH (1997) state the condition for the existence and uniqueness of the solution to (11.46a,b). Introducing the realizations $x_i$ and $x_r$ of the variates we rewrite (11.46b) into

$$h(c) = \frac{1}{c}, \quad (11.47a)$$

where

$$h(c) = \frac{\sum\limits_{i=1}^{r} x_i^c \ln x_i + (n-r) x_r^c \ln x_r}{\sum\limits_{i=1}^{r} x_i^c + (n-r) x_r^c} - \frac{1}{r} \sum\limits_{i=1}^{r} \ln x_i. \qquad (11.47b)$$

The auxiliary function $h(c)$ has the following properties (for a proof see FARNUM/BOOTH):

- $h(c)$ is increasing for $c \geq 0$,
- $\lim_{c \to \infty} h(c) = \ln x_r - \frac{1}{r} \sum\limits_{i=1}^{r} \ln x_i =: V$,
- $h(0) = \left(1 - \frac{r}{n}\right) V$.

Fig. 11/5 shows the graph of $h(c)$ for dataset #1 singly censored at $r = 15$ (see Tab. 9/2) along with the graph of $1/c$.

Figure 11/5: Graphs of $h(c)$ and $c^{-1}$ for dataset #1 singly censored at $r = 15$

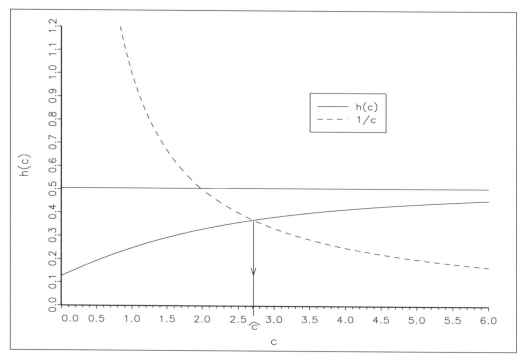

With the aid of Fig. 11/5 the following propositions — also proved by FARNUM/BOOTH — are easily established:

1. If $V > 0$, the estimates $\widehat{b}$ and $\widehat{c}$ are unique.
2. If $V > 0$, then the following bounds on $\widehat{b}$ and $\widehat{c}$ hold:

$$\widehat{c} > 1/V,$$

$$\widehat{b} > \left\{ \frac{1}{n} \left[ \sum_{i=1}^{r} x_i^{1/V} + (n-r) x_r^{1/V} \right] \right\}^V.$$

3. $V = 0$ iff $x_i = x_r \; \forall \; i$; i.e., all items fail simultaneously. In this case we will have no solution to (11.47a).

$V$ measures the variation in the failure data. It is always non–negative and is equal to zero only when all observations are the same.

We may apply the NEWTON–RAPHSON formula (11.25a) developed for the complete–sample case

$$\widehat{c}_{p+1} = \widehat{c}_p - \frac{g(\widehat{c}_p)}{g'(\widehat{c}_p)}, \quad p = 0, 1, 2, \ldots, \tag{11.48a}$$

where

$$g(c) = \sum_{j=1}^{n} (1 - c\,w_j) \exp(c\,w_j) \tag{11.48b}$$

$$g'(c) = -\sum_{j=1}^{n} c\,w_j^2 \exp(c\,w_j) \tag{11.48c}$$

with

$$w_j = \ln x_j - \frac{1}{r} \sum_{i=1}^{r} \ln x_i; \quad j = 1, 2, \ldots, n$$

and

$$x_j = \begin{cases} x_i & \text{for} \quad j = i = 1, \ldots, r, \\ x_r & \text{for} \quad j = r + 1, \ldots, n. \end{cases}$$

Looking at Fig. 11/5 the bound $1/V$ provides a good value to start (11.48a). But this starting value can be improved by choosing a value between $\left(1 - \dfrac{r}{n}\right) V$ and $V$, and then converting that value into an initial value for $\widehat{c}$ by using the $c^{-1}$–curve. FARNUM/BOOTH (1997) have suggested using

$$\widehat{c}_0 = \left[\left(1 - \frac{r}{2\,n}\right) V\right]^{-1}. \tag{11.48d}$$

---

**Example 11/8:  MLEs for dataset #1 singly censored at $r = 15$**

Dataset #1 consists of $n = 20$ observations generated from $We(0, 100, 2.5)$ and censored at $r = 15$ (see Tab. 9/2). The quantity $V$ is

$$V = \ln x_{15} + \frac{1}{15} \sum_{i=1}^{15} \ln x_i = 0.50502210.$$

We further get

$$\widehat{c} > 1/V = 1.9801,$$

$$\widehat{b} > \left\{\frac{1}{20}\left[\sum_{i=1}^{15} x_i^{1/V} + (20 - 15)\, x_{15}^{1/V}\right]\right\}^{V} = 86.6923,$$

$$\widehat{c}_0 = 3.1682.$$

(11.48a–c) produced $\widehat{c} = 2.7114$ and (11.46a) gave $\widehat{b} = 99.2851$. Both estimates come very near to the true values.

### 11.6.1.2 Statistical properties of the estimators[27]

We will first give results based on the asymptotic normal distribution of the MLEs and we will conclude with finite–sample results based on pivotal functions.

We start with the construction of the asymptotic variance–covariance matrix of $\boldsymbol{\theta} = (\widehat{b}, \widehat{c})'$ based on the FISHER information matrix $I(\boldsymbol{\theta})$. The means of $-\partial^2 \mathcal{L}/\partial b^2, -\partial^2 \mathcal{L}/\partial b\,\partial c$ and $-\partial^2 \mathcal{L}/\partial c^2$, which are the elements of $I(\boldsymbol{\theta})$, cannot be easily evaluated in small samples and hence large sample approximations have to be applied when $n \to \infty$ and $P$ is being fixed in $r = n\,P$, i.e.,

$$\lim_{n \to \infty} \frac{r}{n} = P. \tag{11.49a}$$

As $n \to \infty$, $x_r$ tends to $x_P$ :

$$x_P = b\left[-\ln(1-P)\right]^{1/c}. \tag{11.49b}$$

We further introduce

$$s := (x_P/b)^c. \tag{11.49c}$$

FUKUTA (1966) gives the following FISHER information matrix:

$$I(\boldsymbol{\theta}) = \begin{pmatrix} \dfrac{c^2}{b^2}\,nP & -\dfrac{n}{b}\left\{P + \dfrac{\partial}{\partial\lambda}\gamma(\lambda\,|\,s)\Big|_{\lambda=1}\right\} \\[2ex] -\dfrac{n}{b}\left\{P + \dfrac{\partial}{\partial\lambda}\gamma(\lambda\,|\,s)\Big|_{\lambda=1}\right\} & \dfrac{n}{c^2}\left\{P + 2\dfrac{\partial}{\partial\lambda}\gamma(\lambda\,|\,s)\Big|_{\lambda=1} + \dfrac{\partial^2}{\partial\lambda^2}\gamma(\lambda\,|\,s)\Big|_{\lambda=1}\right\} \end{pmatrix}, \tag{11.50a}$$

where

$$\gamma(\lambda\,|\,s) = \int_0^s u^{\lambda-1}\,e^{-u}\,\mathrm{d}u \tag{11.50b}$$

is the incomplete gamma function. From (11.50a) the variances of $\widehat{b}$ and $\widehat{c}$ follow as

$$\mathrm{Var}(\widehat{b}) = \frac{P + 2\dfrac{\partial}{\partial\lambda}\gamma(\lambda\,|\,s)\Big|_{\lambda=1} + \dfrac{\partial^2}{\partial\lambda^2}\gamma(\lambda\,|\,s)\Big|_{\lambda=1}}{n\,c^2\,\Delta}, \tag{11.51a}$$

$$\mathrm{Var}(\widehat{c}) = \frac{P\,c^2}{n\,b^2\,\Delta}, \tag{11.51b}$$

where

$$\Delta = \frac{\dfrac{\partial^2}{\partial\lambda^2}\gamma(\lambda\,|\,s)\Big|_{\lambda=1} - \left[\dfrac{\partial}{\partial\lambda}\gamma(\lambda\,|\,s)\Big|_{\lambda=1}\right]^2}{b^2}. \tag{11.51c}$$

---

[27] Suggested reading for this section: BAIN/ENGELHARDT (1986, 1991a pp. 218 ff.), BILLMAN/ANTLE/BAIN (1972), COHEN (1965, 1966), FUKUTA (1963), HARTER/MOORE (1965b), KAHLE (1996), MEEKER/NELSON (1977), SCHULZ (1983).

It is interesting to see how the variances (11.51a,b) in a censored sample compare with those of an uncensored sample of equal sample size given in (11.17). Let $\text{Var}_c(\widehat{b})$ and $\text{Var}_c(\widehat{c})$ be the variances of the censored sample — cf. (11.51a,b) — and $\text{Var}_u(\widehat{b})$ and $\text{Var}_u(\widehat{c})$ those of the uncensored sample, then the **efficiencies** are defined by

$$\varepsilon(\widehat{b}) = \frac{\text{Var}_u(\widehat{b})}{\text{Var}_c(\widehat{b})} \quad \text{and} \quad \varepsilon(\widehat{c}) = \frac{\text{Var}_u(\widehat{c})}{\text{Var}_c(\widehat{c})}.$$

Tab. 11/6 gives these efficiencies for some values of $P$ as reported by FUKUTA 1963.

Table 11/6:   Efficiencies of $\widehat{b}$ and $\widehat{c}$ (uncensored and censored samples)

| $P$ | $\varepsilon(\widehat{b})$ in % | $\varepsilon(\widehat{c})$ in % |
|------|------|------|
| 0.98 | 99.6 | 94.2 |
| 0.96 | 99.1 | 90.5 |
| 0.94 | 98.2 | 85.8 |
| 0.92 | 97.2 | 82.1 |
| 0.90 | 96.3 | 79.2 |
| 0.80 | 88.3 | 65.4 |
| 0.70 | 76.4 | 54.0 |
| 0.60 | 60.1 | 43.7 |
| 0.50 | 44.8 | 35.7 |
| 0.40 | 27.3 | 26.8 |

**Excursus:**    **Simultaneous confidence region for $b$ and $c$ based on the observed information**

Some authors replace the expected information with the observed information to ease the computation. KAHLE (1996) used the following observed information matrix instead of (11.50a):

$$I^*(\boldsymbol{\theta}) = \begin{pmatrix} I_{11}^* & I_{12}^* \\ I_{21}^* & I_{22}^* \end{pmatrix} \tag{11.52}$$

with elements

$$I_{11}^* = \frac{\widehat{c}^2}{\widehat{b}^2} \frac{r}{n},$$

$$I_{12}^* = I_{21}^* = -\frac{1}{b} \left\{ \frac{1}{n} \sum_{i=1}^{r} \left(\frac{x_i}{\widehat{b}}\right)^{\widehat{c}} \ln\left[\left(\frac{x_i}{\widehat{b}}\right)^{\widehat{c}}\right] + \left(1 - \frac{r}{n}\right) \left(\frac{x_r}{\widehat{b}}\right)^{\widehat{c}} \ln\left[\left(\frac{x_r}{\widehat{b}}\right)^{\widehat{c}}\right] \right\},$$

$$I_{22}^* = \frac{1}{\widehat{b}^2} \left\{ \frac{r}{n} + \frac{1}{n} \sum_{i=1}^{r} \left(\frac{x_i}{\widehat{b}}\right)^{\widehat{c}} \ln^2\left[\left(\frac{x_i}{\widehat{b}}\right)^{\widehat{c}}\right] + \left(1 - \frac{r}{n}\right) \left(\frac{x_r}{\widehat{b}}\right)^{\widehat{c}} \ln^2\left[\left(\frac{x_r}{\widehat{b}}\right)^{\widehat{c}}\right] \right\}.$$

A **simultaneous confidence region** for $\boldsymbol{\theta} = (b, c)'$ based on the asymptotically normal distributed vector $\widehat{\boldsymbol{\theta}} = (\widehat{b}, \widehat{c})'$ is given by

$$(\widehat{\boldsymbol{\theta}} - \boldsymbol{\theta})' \, I^*(\boldsymbol{\theta})^{-1} \, (\widehat{\boldsymbol{\theta}} - \boldsymbol{\theta}) < g, \tag{11.53a}$$

where

$$g = -2 \ln \alpha \tag{11.53b}$$

is the quantile of the $\chi^2$–distribution with two degrees of freedom and $1 - \alpha$ is the level of confidence. Introducing

$$C_1 \;:=\; \frac{r}{n} + \frac{1}{n} \sum_{i=1}^{r} \left(\frac{x_i}{\widehat{b}}\right)^{\widehat{c}} \ln^2\left[\left(\frac{x_i}{\widehat{b}}\right)^{\widehat{c}}\right] + \left(1 - \frac{r}{n}\right) \left(\frac{x_r}{\widehat{b}}\right)^{\widehat{c}} \ln^2\left[\left(\frac{x_r}{\widehat{b}}\right)^{\widehat{c}}\right],$$

$$C_2 \;:=\; \frac{r}{n},$$

$$C_3 \;:=\; \frac{1}{n} \sum_{i=1}^{r} \left(\frac{x_i}{\widehat{b}}\right)^{\widehat{c}} \ln\left[\left(\frac{x_i}{\widehat{b}}\right)^{\widehat{c}}\right] + \left(1 - \frac{r}{n}\right) \left(\frac{x_r}{\widehat{b}}\right)^{\widehat{c}} \ln\left[\left(\frac{x_r}{\widehat{b}}\right)^{\widehat{c}}\right],$$

$$\widetilde{c} \;:=\; \frac{c - \widehat{c}}{c},$$

KAHLE (1996b, p. 34) suggested the following method of constructing the confidence region:

1. Upper and lower bounds for c are

$$\frac{\widehat{c}}{1 + B} \le c \le \frac{\widehat{c}}{1 - B} \tag{11.53c}$$

   with

$$B = \sqrt{\frac{C_2\, g}{n\, (C_1\, C_2 - C_3^2)}}.$$

2. For every $c$ in $\left[\dfrac{\widehat{c}}{1 + B}, \dfrac{\widehat{c}}{1 - B}\right]$, the parameter $b$ is varying in

$$\frac{\widehat{b}\, c}{c + \dfrac{C_3}{C_2}\, \widetilde{c} + A} \le b \le \frac{\widehat{b}\, c}{c + \dfrac{C_3}{C_2}\, \widetilde{c} - A} \tag{11.53d}$$

   with

$$A = \sqrt{\frac{g}{n\, C_2} - \widetilde{c}^2\, \frac{C_1\, C_2 - C_3^2}{C_2^2}}.$$

Fig. 11/6 shows the 90%–confidence region for $b$ and $c$ belonging to dataset #1 censored at $r = 15$.

Finite sample results for testing hypotheses concerning $b$ or $c$ or for constructing confidence intervals for $b$ and $c$ rest upon the distribution of pivotal functions (cf. Sect. 11.3.2.4). BILLMAN et al. (1972) obtained the percentage points in Tables 11/7 and 11/8 by simulation.

Figure 11/6:  Simultaneous 90%-confidence region for $b$ and $c$ (dataset #1 singly censored at $r = 15$)

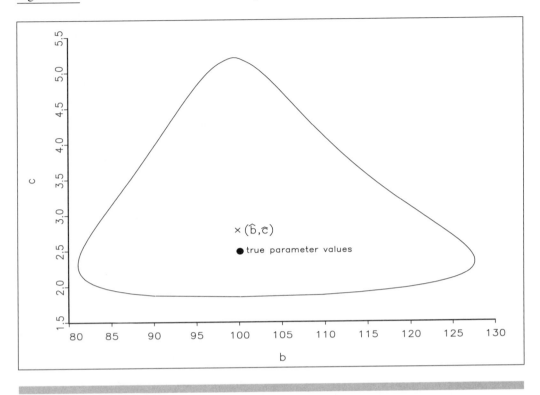

Table 11/7:  Percentage points $v_P$ such that $\Pr\left\{\sqrt{n}\left[\dfrac{\widehat{c}}{c} - \mathrm{E}\left(\dfrac{\widehat{c}}{c}\right)\right] < v_P\right\} = P$

| $n$ | $r/n$ | $P$ 0.01 | 0.05 | 0.10 | 0.90 | 0.95 | 0.99 | $\mathrm{E}(\widehat{c}/c)$ |
|---|---|---|---|---|---|---|---|---|
| 40 | 1.00 | −1.60 | −1.20 | −0.98 | 1.10 | 1.60 | 2.41 | 1.036 |
|    | 0.75 | −2.15 | −1.56 | −1.30 | 1.59 | 2.09 | 3.30 | 1.060 |
|    | 0.50 | −2.62 | −2.09 | −1.74 | 2.09 | 2.95 | 4.90 | 1.098 |
| 60 | 1.00 | −1.59 | −1.21 | −0.97 | 1.05 | 1.43 | 2.18 | 1.024 |
|    | 0.75 | −2.03 | −1.50 | −1.19 | 1.48 | 2.03 | 3.11 | 1.036 |
|    | 0.50 | −2.62 | −2.02 | −1.62 | 1.96 | 2.70 | 4.28 | 1.060 |
| 80 | 1.00 | −1.62 | −1.20 | −0.97 | 1.04 | 1.40 | 2.10 | 1.019 |
|    | 0.75 | −2.02 | −1.51 | −1.21 | 1.49 | 1.99 | 3.06 | 1.027 |
|    | 0.50 | −2.60 | −1.98 | −1.61 | 1.99 | 2.74 | 4.22 | 1.047 |
| 100 | 1.00 | −1.63 | −1.20 | −0.95 | 1.04 | 1.36 | 2.06 | 1.016 |
|     | 0.75 | −1.99 | −1.46 | −1.16 | 1.46 | 1.92 | 2.90 | 1.022 |
|     | 0.50 | −2.54 | −1.93 | −1.54 | 1.96 | 2.59 | 4.00 | 1.035 |
| 120 | 1.00 | −1.64 | −1.26 | −0.97 | 0.99 | 1.33 | 2.06 | 1.012 |
|     | 0.75 | −2.05 | −1.48 | −1.15 | 1.45 | 1.97 | 2.88 | 1.018 |
|     | 0.50 | −2.61 | −1.93 | −1.55 | 1.89 | 2.51 | 3.81 | 1.030 |
| ∞ | 1.00 | −1.81 | −1.28 | −0.99 | 0.99 | 1.28 | 1.81 | 1.000 |
|   | 0.75 | −2.35 | −1.66 | −1.29 | 1.29 | 1.66 | 2.35 | 1.000 |
|   | 0.50 | −3.05 | −2.15 | −1.68 | 1.68 | 2.15 | 3.05 | 1.000 |

Source:  BILLMAN et al. (1972, p. 833) — Reprinted with permission from *Technometrics*. Copyright 1972 by the American Statistical Association. All rights reserved.

<u>Table 11/8:</u>  Percentage points $w_P$ such that $\Pr\{\sqrt{n}\,\widehat{c}\,\ln(\widehat{b}/b) < w_P\} = P$

| $n$ | $r/n$ \ $P$ | 0.01 | 0.05 | 0.10 | 0.90 | 0.95 | 0.99 |
|---|---|---|---|---|---|---|---|
| 40 | 1.00 | −2.58 | −1.82 | −1.41 | 1.39 | 1.80 | 2.62 |
|    | 0.75 | −3.29 | −2.25 | −1.69 | 1.40 | 1.85 | 2.71 |
|    | 0.50 | −6.21 | −3.77 | −2.91 | 1.63 | 2.16 | 2.96 |
| 60 | 1.00 | −2.48 | −1.78 | −1.38 | 1.37 | 1.77 | 2.56 |
|    | 0.75 | −3.22 | −2.16 | −1.68 | 1.42 | 1.84 | 2.66 |
|    | 0.50 | −5.37 | −3.56 | −2.69 | 1.67 | 2.18 | 3.01 |
| 80 | 1.00 | −2.51 | −1.76 | −1.37 | 1.37 | 1.76 | 2.49 |
|    | 0.75 | −3.11 | −2.10 | −1.61 | 1.43 | 1.85 | 2.65 |
|    | 0.50 | −5.14 | −3.45 | −2.62 | 1.71 | 2.16 | 3.08 |
| 100 | 1.00 | −2.45 | −1.74 | −1.37 | 1.35 | 1.73 | 2.50 |
|    | 0.75 | −3.12 | −2.09 | −1.60 | 1.44 | 1.85 | 2.61 |
|    | 0.50 | −4.92 | −3.34 | −2.49 | 1.78 | 2.26 | 3.19 |
| 120 | 1.00 | −2.44 | −1.73 | −1.35 | 1.35 | 1.74 | 2.48 |
|    | 0.75 | −3.01 | −2.01 | −1.58 | 1.45 | 1.86 | 2.63 |
|    | 0.50 | −4.50 | −3.17 | −2.44 | 1.75 | 2.27 | 3.13 |
| $\infty$ | 1.00 | −2.45 | −1.73 | −1.35 | 1.35 | 1.73 | 2.45 |
|    | 0.75 | −2.69 | −1.90 | −1.48 | 1.48 | 1.90 | 2.69 |
|    | 0.50 | −3.69 | −2.61 | −2.03 | 2.03 | 2.61 | 3.69 |

<u>Source:</u> BILLMAN et al. (1972, p. 834) — Reprinted with permission from *Technometrics*. Copyright 1972 by the American Statistical Association. All rights reserved.

From Tab. 11/7 we construct the two–sided $(1 - \alpha)$–confidence interval for $c$ as

$$\frac{\widehat{c}}{\mathrm{E}\left(\dfrac{\widehat{c}}{c}\right) + v_{1-\alpha/2}/\sqrt{n}} \le c \le \frac{\widehat{c}}{\mathrm{E}\left(\dfrac{\widehat{c}}{c}\right) + v_{\alpha/2}/\sqrt{n}} , \tag{11.54a}$$

and from Tab. 11/8 we find the two–sided $(1 - \alpha)$–confidence interval for $b$ as

$$\frac{\widehat{b}}{\exp\left\{\dfrac{w_{1-\alpha/2}}{\widehat{c}\sqrt{n}}\right\}} \le b \le \frac{\widehat{b}}{\exp\left\{\dfrac{w_{\alpha/2}}{\widehat{c}\sqrt{n}}\right\}} . \tag{11.54b}$$

## 11.6.2   Three-parameter WEIBULL distribution

The likelihood function in the three–parameter case with censoring on the right at the $r$–th failure is

$$L(a,b,c) = \frac{n!}{(n-r)!}\left(\frac{c}{b}\right)^r \left[\prod_{i=1}^{r}\left(\frac{X_i - a}{b}\right)\right]^{c-1} \times$$
$$\exp\left\{-\sum_{i=1}^{r}\left(\frac{X_i - a}{b}\right)^c - (n-r)\left(\frac{X_r - a}{b}\right)^c\right\}. \tag{11.55a}$$

Omitting the factorial constants we have the following log–likelihood function

$$\mathcal{L}(a,b,c) = r\left(\ln c - c\ln b\right) + (c-1)\sum_{i=1}^{r}\ln(X_i - a) -$$
$$\frac{1}{b^c}\left\{\sum_{i=1}^{r}(X_i - a)^c + (n-r)(X_r - a)^c\right\}, \tag{11.55b}$$

which coincides with (11.4c) when $n = r$. (11.55b) leads to the following likelihood equations:

$$\frac{\partial\mathcal{L}(a,b,c)}{\partial a} = -(c-1)\sum_{i=1}^{r}\frac{1}{X_i - a} +$$
$$\frac{c}{b}\left\{\sum_{i=1}^{r}\left(\frac{X_i - a}{b}\right)^{c-1} + (n-r)\left(\frac{X_r - a}{b}\right)^{c-1}\right\} = 0, \tag{11.55c}$$

$$\frac{\partial\mathcal{L}(a,b,c)}{\partial b} = -\frac{rc}{b} + \frac{c}{b}\left\{\sum_{i=1}^{r}\left(\frac{X_i - a}{b}\right)^{c} + (n-r)\left(\frac{X_r - a}{b}\right)^{c}\right\} = 0, \tag{11.55d}$$

$$\frac{\partial\mathcal{L}(a,b,c)}{\partial c} = \frac{r}{c} - r\ln b + \sum_{i=1}^{r}\ln(X_i - a) - \sum_{i=1}^{r}\left(\frac{X_i - a}{b}\right)^{c}\ln\left(\frac{X_i - a}{b}\right) -$$
$$(n-r)\left(\frac{X_r - a}{b}\right)^{c}\ln\left(\frac{X_r - a}{b}\right) = 0, \tag{11.55e}$$

which coincide with (11.4d–f) when $n = r$. A solution to the system (11.55c–e) may be found by applying the profile likelihood method which is described in Sect. 11.3.2.2. FUKUTA (1963, p. 5) gives the following asymptotic variances of the MLE's:

$$\text{Var}(\hat{a}) = \frac{\Delta_{33}}{\Delta}\frac{b^2}{n(c-1)^2}, \tag{11.56a}$$

$$\text{Var}(\hat{b}) = \frac{\Delta_{22}}{\Delta}\frac{b^2}{nc^2}, \tag{11.56b}$$

$$\text{Var}(\hat{c}) = \frac{\Delta_{11}}{\Delta}\frac{c^2}{n}, \tag{11.56c}$$

where

$$\Delta = \begin{vmatrix} P+2\left.\frac{\partial\gamma(\lambda|s)}{\partial\lambda}\right|_{\lambda=1}+\left.\frac{\partial^2\gamma(\lambda|s)}{\partial\lambda^2}\right|_{\lambda=1} & -\left\{P+\left.\frac{\partial\gamma(\lambda|s)}{\partial\lambda}\right|_{\lambda=1}\right\} & -\left\{\gamma(1-\tfrac{1}{c}|s)+\left.\frac{\partial\gamma(\lambda|s)}{\partial\lambda}\right|_{\lambda=1-\frac{1}{c}}\right\} \\ * & P & \gamma(1-\tfrac{1}{c}|s) \\ * & * & \gamma(1-\tfrac{2}{c}|s) \end{vmatrix},$$

and $\Delta_{ij}$ being the cofactor of $a_{ij}$ in $\Delta = |a_{ij}|$. $P$, $s$ and $\gamma(.|.)$ are defined in Sect. 11.6.1.2.

## 11.7 Samples progressively censored on the right[28]

The motivation for progressively (or multiply) censored samples has been given in Sect. 8.3.3. ML estimation of the parameters $b$ and $c$ or $a$, $b$ and $c$ of the two–parameter or three–parameter WEIBULL distributions, respectively, will be discussed here.

Let $n$ designate the **total sample size**, i.e., the number of items simultaneously put on test, and $m$ the **number which fail** and therefore result in completely determined life spans. Suppose that censoring occurs in $k$ **stages** $(k > 1)$ at **times** $t_j > t_{j-1}$; $j = 2, \ldots, k$ and that $e_j$ **surviving items** are **removed** (censored) from further observation at the $j$–th stage. Thus we have

$$n = m + \sum_{j=1}^{k} e_j. \tag{11.57a}$$

The numbers $e_j$ with

$$0 \leq r_j \leq n - m - \sum_{\ell=1}^{j-1} e_\ell \tag{11.57b}$$

are fixed in advance together with $n$ and $m$,[29] but the $e_j$ items are chosen randomly from the items still on test at $t_j$.

Two types of censoring are generally recognized.[30] In **type–I progressive censoring** the $t_j$ **are fixed** at $T_j$:

$$t_j = T_j; \; j = 1, \ldots, k,$$

and the number of survivors immediately before these times are random variables with realizations $n - \sum_{\ell=1}^{j-1} e_\ell - \nu$, where $\nu$ is the realized number of failures before $T_j$ and $e_0 := 0$. In **type–II progressive censoring** the $t_j$ **coincide with times of failure** $X_j$ and are random:

$$t_j = X_j; \; j = 1, \ldots, k.$$

Thus the number $k$ of stages and the number $m$ of complete life spans coincide and the number of surviving items immediately before $X_j$ are fixed and amount to $n - (j - 1) - \sum_{\ell=1}^{j-1} e_\ell$. We will first (Sect. 11.7.1) discuss type–I censoring where the formulas are a little bit more cumbersome than in the type–II case (Sect. 11.7.2).

---

[28] Suggested reading for this section: BALAKRISHNAN/AGGARWALA (2000), BALASOORIYA/LOW (2004), BALASOORIYA/SAW/GADAG (2000), CACCIARI/MONTANARI (1987), COHEN (1965, 1975), FEI/KONG/TANG (1995), LEMON (1975), NG/CHAN/BALAKRISHNAN (2004), RINGER/SPRINKLE (1972), TSE/YANG/YUEN (2000), VIVEROS/BALAKRISHNAN (1994), WINGO (1973), WU (2002), YUEN/TSE (1996).

[29] YUEN/TSE (1996) consider the case where $e_j$ is random and the realization of a uniform discrete distribution with probability $1/[n - m - (e_1 + \ldots + e_{j-1}) + 1]$. TSE et al. (2000) take a binomial distribution instead.

[30] LEMON (1975) developed MLE based on various left and right progressively censored situations.

### 11.7.1 Type-I censoring

Let $f(x \mid \boldsymbol{\theta})$ and $F(x \mid \boldsymbol{\theta})$ be the density and distribution functions, respectively. For a type–I progressively censored sample the likelihood function is

$$L(\boldsymbol{\theta} \mid \boldsymbol{x}) = C \prod_{i=1}^{m} f(x_i \mid \boldsymbol{\theta}) \prod_{j=1}^{k} \left[ 1 - F(T_j \mid \boldsymbol{\theta}) \right]^{e_j}, \qquad (11.58a)$$

where $C$ is a combinatorial constant and generally $m \neq k$. In the first product the index $i$ is running over failures whereas in the second product we have a different index $j$ running over the fixed censoring times $T_j$. Inserting the two–parameter WEIBULL density and distribution functions into (11.58a), neglecting the constant $C$ and taking logarithm, we have the log–likelihood function

$$\mathcal{L}(b, c) = m \left[ \ln c - c \ln b \right] + (c - 1) \sum_{i=1}^{m} \ln x_i - \sum_{i=1}^{m} \left( \frac{x_i}{b} \right)^c - \sum_{j=1}^{k} e_j \left( \frac{T_j}{b} \right)^c. \quad (11.58b)$$

The system of equations to be solved for $b$ and $c$ results as

$$\frac{\partial \mathcal{L}(b,c)}{\partial b} = -\frac{mc}{b} + \frac{c}{b} \left[ \sum_{i=1}^{m} \left( \frac{x_i}{b} \right)^c + \sum_{j=1}^{k} e_j \left( \frac{T_j}{b} \right)^c \right] = 0, \qquad (11.58c)$$

$$\frac{\partial \mathcal{L}(b,c)}{\partial c} = \frac{m}{c} - m \ln b + \sum_{i=1}^{m} \ln x_i - \left[ \sum_{i=1}^{m} \left( \frac{x_i}{b} \right)^c \ln \left( \frac{x_i}{b} \right) + \sum_{j=i}^{k} e_j \left( \frac{T_j}{b} \right)^c \ln \left( \frac{T_j}{b} \right) \right] = 0. \tag{11.58d}$$

Without censoring $(e_j = 0 \ \forall \ j)$, (11.58b–d) will coincide with (11.13) through (11.14 a,b). On eliminating $b$ between (11.58c) and (11.58b), we have

$$\frac{1}{c} + \frac{1}{m} \sum_{i=1}^{m} \ln x_i - \frac{\sum\limits_{i=1}^{m} x_i^c \ln x_i + \sum\limits_{j=1}^{k} e_j T_j^c \ln T_j}{\sum\limits_{i=1}^{m} x_i^c + \sum\limits_{j=1}^{k} e_j T_j^c} = 0, \qquad (11.58e)$$

to be solved by the NEWTON–RAPHSON method. With $\widehat{c}$ thus determined, it follows from (11.58c) that

$$\widehat{b} = \left\{ \frac{1}{m} \left[ \sum_{i=1}^{m} x_i^{\widehat{c}} + \sum_{j=1}^{k} e_j T_j^c \right] \right\}^{1/\widehat{c}}. \qquad (11.58f)$$

The asymptotic variance–covariance matrix is approximated by:

$$\widehat{\mathrm{Var}}(\widehat{b}, \widehat{c}) \approx \begin{pmatrix} -\dfrac{\partial^2 \mathcal{L}}{\partial b^2} \Big|_{\widehat{b}, \widehat{c}} & -\dfrac{\partial^2 \mathcal{L}}{\partial b \partial c} \Big|_{\widehat{b}, \widehat{c}} \\[2ex] -\dfrac{\partial^2 \mathcal{L}}{\partial c \partial b} \Big|_{\widehat{b}, \widehat{c}} & -\dfrac{\partial^2 \mathcal{L}}{\partial c^2} \Big|_{\widehat{b}, \widehat{c}} \end{pmatrix}^{-1} \qquad (11.59a)$$

with

$$-\frac{\partial^2 \mathcal{L}}{\partial b^2}\bigg|_{\widehat{b},\widehat{c}} = -\frac{\widehat{c}}{\widehat{b}^2}\left\{m - (\widehat{c}+1)\left[\sum_{i=1}^{m}\left(\frac{x_i}{\widehat{b}}\right)^{\widehat{c}} + \sum_{j=1}^{k}\left(\frac{T_j}{\widehat{b}}\right)^{\widehat{c}}\right]\right\}, \qquad (11.59b)$$

$$-\frac{\partial^2 \mathcal{L}}{\partial b\partial c}\bigg|_{\widehat{b},\widehat{c}} = -\frac{\partial^2 \mathcal{L}}{\partial c\partial b}\bigg|_{\widehat{b},\widehat{c}} = \frac{1}{\widehat{b}}\left\{m - \widehat{c}\left[\sum_{i=1}^{m}\left(\frac{x_i}{\widehat{b}}\right)^{\widehat{c}}\ln\left(\frac{x_i}{\widehat{b}}\right) + \right.\right.$$
$$\left.\left.\sum_{j=1}^{k}e_j\left(\frac{T_j}{\widehat{b}}\right)^{\widehat{c}}\ln\left(\frac{T_j}{\widehat{b}}\right) + \sum_{i=1}^{m}\left(\frac{x_i}{\widehat{b}}\right)^{\widehat{c}} + \sum_{j=1}^{k}e_j\left(\frac{T_j}{\widehat{b}}\right)^{\widehat{c}}\right]\right\}, \qquad (11.59c)$$

$$-\frac{\partial^2 \mathcal{L}}{\partial c^2}\bigg|_{\widehat{b},\widehat{c}} = \frac{m}{\widehat{c}^2} + \sum_{i=1}^{m}\left(\frac{x_i}{\widehat{b}}\right)^{\widehat{c}}\left[\ln\left(\frac{x_i}{\widehat{b}}\right)\right]^2 + \sum_{j=1}^{k}\left(\frac{T_j}{\widehat{b}}\right)^{\widehat{c}}\left[\ln\left(\frac{T_j}{\widehat{b}}\right)\right]^2. \qquad (11.59d)$$

## 11.7.2  Type-II censoring

For a type–II progressively censored sample the likelihood function is

$$L(\boldsymbol{\theta}\,|\,\boldsymbol{x}) = C\prod_{i=1}^{m}f(x_i\,|\,\boldsymbol{\theta})\left[1 - F(x_i\,|\,\boldsymbol{\theta})\right]^{e_i} \qquad (11.60a)$$

with

$$C = n\,(n - 1 - e_1)\,(n - 2 - e_1 - e_2)\cdot\ldots\cdot(n - m + 1 - e_1 - \ldots - e_{m-1}).$$

The log–likelihood for the two–parameter WEIBULL distribution may then be written as

$$\mathcal{L}(b,c) = m\left[\ln c - c\,\ln b\right] + (c-1)\sum_{i=1}^{m}\ln x_i - \sum_{i=1}^{m}(e_i+1)\left(\frac{x_i}{b}\right)^c, \qquad (11.60b)$$

and hence we have the likelihood equations for $b$ and $c$ to be

$$\frac{\partial \mathcal{L}(b,c)}{\partial b} = -\frac{mc}{b} + \frac{c}{b}\sum_{i=1}^{m}(e_i+1)\left(\frac{x_i}{b}\right)^c = 0, \qquad (11.60c)$$

$$\frac{\partial \mathcal{L}(b,c)}{\partial c} = \frac{m}{c} - m\ln b + \sum_{i=1}^{m}\ln x_i - \sum_{i=1}^{m}(e_i+1)\left(\frac{x_i}{b}\right)^c\ln\left(\frac{x_i}{b}\right) = 0. \; (11.60d)$$

Equation (11.60c) yields the MLE of $b$ to be

$$\widehat{b} = \left\{\frac{1}{m}\sum_{i=1}^{m}(e_i+1)\,x_i^{\widehat{c}}\right\}^{1/\widehat{c}}. \qquad (11.61a)$$

Equation (11.60d), in conjunction with the MLE of $b$ in (11.61a), reduces to

$$\frac{1}{\widehat{c}} + \frac{1}{m}\sum_{i=1}^{m}\ln x_i - \frac{\displaystyle\sum_{i=1}^{m}(e_i+1)\,x_i^{\widehat{c}}\ln x_i}{\displaystyle\sum_{i=1}^{m}(e_i+1)\,x_i^{\widehat{c}}} = 0, \qquad (11.61b)$$

which may be solved by NEWTON–RAPHSON method.

The asymptotic variance–covariance matrix for $\widehat{b}$ and $\widehat{c}$ can be approximated by inverting the matrix composed of the negatives of the second partial derivatives of the logarithm of log-likelihood evaluated at the estimates, see (11.59a), but with elements

$$-\frac{\partial^2 \mathcal{L}}{\partial b^2} = -\frac{\widehat{c}}{\widehat{b}^2}\left[m + (\widehat{c}+1)\sum_{i=1}^{m}\left(\frac{x_i}{b}\right)^c\right], \tag{11.62a}$$

$$-\frac{\partial^2 \mathcal{L}}{\partial b \partial c} = -\frac{\partial^2 \mathcal{L}}{\partial c \partial b} = \frac{1}{\widehat{b}}\left\{m - \sum_{i=1}^{m}(e_i+1)\left(\frac{x_i}{\widehat{b}}\right)^{\widehat{c}} - \widehat{c}\sum_{i=1}^{m}(e_i+1)\left(\frac{x_i}{\widehat{b}}\right)^{\widehat{c}}\ln\left(\frac{x_i}{\widehat{b}}\right)\right\}, \tag{11.62b}$$

$$-\frac{\partial \mathcal{L}}{\partial c^2} = \frac{m}{\widehat{c}^2} + \sum_{i=1}^{m}(e_i+1)\left(\frac{x_i}{\widehat{b}}\right)^{\widehat{c}}\left[\ln\left(\frac{x_i}{\widehat{b}}\right)\right]^2. \tag{11.62c}$$

An exact confidence interval for $c$ and an exact joint confidence region for $b$ and $c$ have been given by CHEN (1998) and WU (2002). A two–sided $100\,(1-\alpha)\%$ **confidence interval for $c$** is

$$\varphi\big(x_1,\ldots,x_m; F_{2\,m-2,2,\alpha/2}\big) \le c \le \varphi\big(x_1,\ldots,x_m; F_{2\,m-2,2,1-\alpha/2}\big), \tag{11.63a}$$

where $0 < \alpha < 1$. $F_{\nu_1,\nu_2,P}$ is the percentile of order $P$ of the $F$–distribution and $\varphi(x_1,\ldots,x_m;t)$ is the solution for $\widehat{c}$ of the equation

$$\frac{\sum\limits_{i=1}^{m}(e_i+1)\,x_i^{\widehat{c}} - n\,x_1^{\widehat{c}}}{n\,(m-1)\,x_1^{\widehat{c}}} = t. \tag{11.63b}$$

A $100\,(1-\alpha)\%$ **joint confidence region** for $b$ and $c$ is determined by the following two inequalities:

$$\varphi\big(x_1,\ldots,x_m; F_{2\,m-2,2,(1-\sqrt{1-a})/2}\big) \le c \le \varphi\big(x_1,\ldots,x_m; F_{2\,m-2,2,(1+\sqrt{1-a})/2}\big) \tag{11.64a}$$

$$\left\{\frac{2\sum\limits_{i=1}^{m}(e_i+1)\,x_i^c}{\chi^2_{2\,m,(1+\sqrt{1-\alpha})/2}}\right\}^{1/c} \le b \le \left\{\frac{2\sum\limits_{i=1}^{m}(e_i+1)\,x_i^c}{\chi^2_{2\,m,(1-\sqrt{1-\alpha})/2}}\right\}^{1/c}. \tag{11.64b}$$

The results above rest upon the following facts:

1. $Y_i := (X_i/b)^c;\ i = 1,\ldots,m$ is a progressively type–II censored sample from an exponential distribution with mean 1.

2. The transformations

$$S_1 = n\,Y_1$$
$$S_2 = (n - e_1 - 1)\,(Y_2 - Y_1)$$
$$\vdots$$
$$S_m = (n - e_1 - \ldots - e_{m-1} - m + 1)\,(Y_m - Y_{m-1})$$

are generalized spacings which are iid as an exponential distribution with mean 1.

3. Then

$$V = 2\,S_1 = 2\,n\,Y_1 \sim \chi^2(2),$$

$$U = 2\sum_{i=2}^{m} S_i = 2\left\{\sum_{i=1}^{m}(e_i+1)Y_i - n\,Y_1\right\} \sim \chi^2(2\,m - 2).$$

$U$ and $V$ are independent variables.

4. It is easy to show that

$$T_1 = \frac{U}{(m-1)\,V} = \frac{\sum_{i=1}^{m}(e_i+1)\,Y_i - n\,Y_1}{n\,(m-1)\,Y_1} = \frac{\sum_{i=1}^{m}(e_i+1)\,X_i^c - n\,X_1^c}{n\,(m-1)\,X_1^c} \sim F(2\,m-2,2),$$

$$T_2 = U + V = 2\sum_{i=1}^{m}(e_i+1)\,Y_i \sim \chi^2(2\,m)$$

and $T_1$ and $T_2$ are independent.

---

**Example 11/9:   Confidence interval for $c$ and joint confidence region for $b$ and $c$ based on dataset #1 progressively type–II censored**

We have taken dataset #1 (see Tab. 9/2), consisting of $X_i \overset{\text{iid}}{\sim} We(0, 100, 2.5)$; $i = 1, 2, \ldots, 20$ and generated a progressively type–II censored sample with the following observations and censoring scheme:

| $i$   | 1  | 2  | 3  | 4  |
|-------|----|----|----|----|
| $x_i$ | 35 | 38 | 81 | 99 |
| $r_i$ | 3  | 3  | 5  | 5  |

The ML estimates according to (11.61a,b) and with $m = 4$ are

$$\widehat{c} = 2.4347, \quad \widehat{b} = 146.5101.$$

$\widehat{b}$ is a rather bad estimate, compared with the true value $b = 100$, whereas $\widehat{c}$ compares rather well with to the true value $c = 2.5$.

To find a 95% confidence interval for $c$ according to (11.63a), we need the percentiles

$$F_{6,2,0.025} = 1.3774 \text{ and } F_{6,2,0.975} = 39.3$$

and get

$$2.1646 \leq c \leq 5.4725.$$

To obtain a 95% joint confidence region for $b$ and $c$, we need the following percentiles:

$$F_{6,2,0.0127} = 0.1015, \quad F_{6,2,0.9873} = 78.1,$$

$$\chi^2_{8,0.0127} = 1.7687, \quad \chi^2_{8,0.9873} = 19.4347.$$

The 95% joint confidence region according to (11.64a,b) is determined by the following two inequalities:

$$0.4275 \leq c \leq 6.1626,$$

$$\left( \frac{2 \sum\limits_{i=1}^{4} (e_i + 1)\, x_i^c}{19.4347} \right)^{1/c} \leq b \leq \left( \frac{2 \sum\limits_{i=1}^{4} (e_i + 1)\, x_i^c}{1.7687} \right)^{1/c}.$$

In order to clearly see the shape of this confidence region, logarithmic scaling has been used for the scale parameter $b$ in Fig. 11/7. The region is wide when $c$ is small and gets smaller with $c$ increasing. It is also wide when $b$ is small and becomes smaller the greater $b$.

Figure 11/7: 95% joint confidence region for $c$ and $\log b$

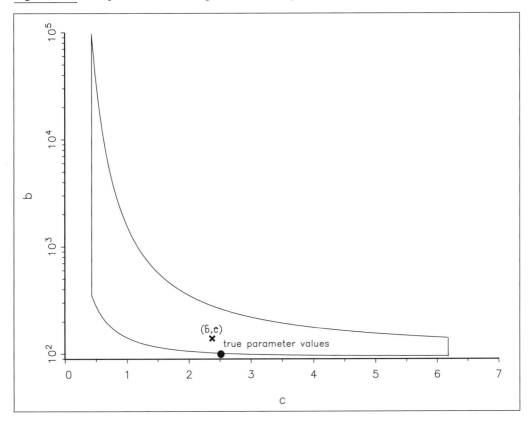

## 11.8 Randomly censored samples

When the censoring variate and the lifetime variate are independent and have no parameters in common, we have a rather simple likelihood function; see (8.25f). We can simplify this likelihood function even more by separately indexing the data. The failure times are denoted by $x_i$ with $i$ running from 1 to $m$; the censoring times are denoted by $x_j$ with $j$

running from 1 to $k$. Then the likelihood function turns out to be

$$L(\boldsymbol{\theta} \,|\, \boldsymbol{x}) = C \prod_{i=1}^{m} f(x_i \,|\, \boldsymbol{\theta}) \prod_{j=1}^{k} \left[1 - F(x_j \,|\, \boldsymbol{\theta})\right], \tag{11.65}$$

which is the same as (11.58a), the likelihood function of a progressively type–I censored sample with $T_j$ replaced by the censoring time $x_j$ and $e_j = 1 \ \forall \ j$. Thus the formulas of Sect. 11.7.1 apply to randomly censored samples as well.

# 12 Parameter estimation — Methods of moments

Estimation of the parameters of a parametric distribution using moments is perhaps the oldest method. The concept of statistical moments was introduced by KARL PEARSON (1857 – 1936), and he advocated its use in parameter estimation. Its drawback compared with other techniques, e.g., maximum likelihood or least squares, is that it is applicable to only uncensored samples. On the other hand, the asymptotic variance–covariance matrix of ML estimators in the three–parameter case exists only if $c > 2$, whereas this matrix can be given for all values of $c$ when estimation is done by moments. The moment estimators of the WEIBULL parameters cannot be given in closed form as is the case with nearly all competing estimation techniques.

This chapter is organized as follows: Sect. 12.1 reports on the traditional method of moments. The moments are of integral order, and we use as many moments as there are parameters to be estimated. In the three–parameter case the equation involving the third sample moment may be replaced by some other functional relationship giving rise to modified methods of moments (Sect. 12.2). Sections 12.3 through 12.5 deal with methods based on alternative kinds of moments. All moments used in this chapter are those of the original WEIBULL variate. We may also find estimators of the WEIBULL parameters which are based on the moments of the log–transformed WEIBULL variate, but these are discussed in Sect. 13.4.1.

## 12.1 Traditional method of moments

When the distribution function of a variate depends on $m$ parameters $\theta_1, \ldots, \theta_m$, collected in the vector $\boldsymbol{\theta}$, the moments of this variate will also depend on these parameters. Taking $\mu'_r = \mathrm{E}(X^r)$, $r$ a non–negative integer, these moments about zero will be functions of $\boldsymbol{\theta}$:

$$\mu'_r = g_r(\boldsymbol{\theta}); \quad r = 1, 2, \ldots \tag{12.1a}$$

Now we write down — starting at $r = 1$ — as many different moments as there are parameters to be estimated.

$$\left. \begin{aligned} \mu'_1 &= g_1(\boldsymbol{\theta}) \\ &\vdots \\ \mu'_m &= g_m(\boldsymbol{\theta}). \end{aligned} \right\} \tag{12.1b}$$

Then we optimally estimate the population moments $\mu'_r$ by their corresponding sample moments:

$$\widehat{\mu'_r} = \frac{1}{n} \sum_{i=1}^{n} X_i^r. \tag{12.1c}$$

By equating $\mu'_r$ to $\widehat{\mu'_r}$ the moment estimator $\widehat{\boldsymbol{\theta}}$ follows as the solution of the system

$$\left.\begin{aligned}\widehat{\mu'_1} &= g_1(\widehat{\boldsymbol{\theta}}) \\ &\vdots \\ \widehat{\mu'_m} &= g_m(\widehat{\boldsymbol{\theta}}).\end{aligned}\right\} \tag{12.1d}$$

Unlike ML estimation, linear estimation or minimum distance estimation (see Chapter 13), the method of moments is not an optimizing technique. Moment estimators are always consistent, asymptotically normal, but sometimes not efficient and not sufficient. It is even possible to get non–plausible estimates, e.g., $\widehat{a} > x_{1:n}$ in the three–parameter WEIBULL distribution.

### 12.1.1   Two-parameter WEIBULL distribution[1]

We will present three different ways to estimate $b$ and $c$ when $X \sim We(0, b, c)$. The first approach, due to MIHRAM (1977), is built upon

$$\mu = \mu'_1 = \mathrm{E}(X) = b\,\Gamma(1 + 1/c) \tag{12.2a}$$

and

$$\mu'_2 = \mathrm{E}(X^2) = b^2\,\Gamma(1 + 2/c). \tag{12.2b}$$

The other two approaches also use (12.2a), but instead of (12.2b) BLISCHKE/SCHEUER (1986) take the squared coefficient of variation and NEWBY (1980) takes the simple coefficient of variation.

MIHRAM's approach

The following moment ratio

$$R(c) = \frac{\left[\mathrm{E}(X)\right]^2}{\mathrm{E}(X^2)} = \frac{\Gamma^2(1 + 1/c)}{\Gamma(1 + 2/c)} \tag{12.3a}$$

is independent of the scale parameter $b$. The sample statistic

$$\widehat{R} = \frac{\left(\sum\limits_{i=1}^{n} x_i/n\right)^2}{\sum\limits_{i=1}^{n} x_i^2/n} \tag{12.3b}$$

equated to $R(c)$ gives a unique[2] estimate $\widehat{c}$ of $c$.

$$\varphi(c) = R(c) - \widehat{R} = 0 \tag{12.3c}$$

can be solved for $c \equiv \widehat{c}$ by using

---

[1]   Suggested reading for this section: BLISCHKE/SCHEUER (1986), MIHRAM (1977), NEWBY (1980, 1982).

[2]   $R(c)$ is a monotone and decreasing function of $c$, see Fig. 12/1

- a table (see Tab. 12/1) or

- a graph of $R(c)$ versus $c$ (see Fig. 12/1) or

- the NEWTON–RAPHSON method

$$\widehat{c}_{p+1} = \widehat{c}_p - \frac{\varphi_M(\widehat{c}_p)}{\varphi'_M(\widehat{c}_p)}; \quad p = 0, 1, 2, \ldots$$

with

$$\varphi'_M(c) = -\frac{2\,\Gamma^2(1+1/c)\,\psi(1+1/c)}{c^2\,\Gamma(1+2/c)}. \tag{12.3d}$$

With $\widehat{c}$ we find from (12.2a)

$$\widehat{b} = \frac{\overline{x}}{\Gamma(1+1/\widehat{c})}. \tag{12.3e}$$

Monte Carlo analysis shows that $c$ is underestimated by $\widehat{c}$ and there is a tendency of the bias and the standard deviation of $\widehat{c}$ to diminish with $n$ increasing. A comparison of the distribution of $\widehat{c}$ with that of the MLE of $c$ shows remarkable similarities whenever $c > 1$.

Figure 12/1: $R(c) = \Gamma^2(1+1/c)/\Gamma(1+2/c)$ versus $c$

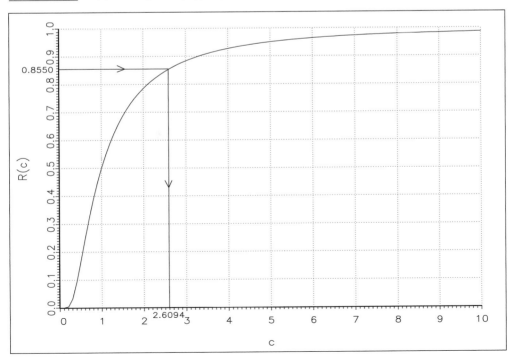

Table 12/1: $R(c) = \Gamma^2(1 + c/1)/\Gamma(1 + 2/c)$ versus $c$

| $c$ | $R(c)$ | $c$ | $R(c)$ | $c$ | $R(c)$ | $c$ | $R(c)$ |
|------|--------|------|--------|------|--------|------|--------|
| 0.1 | 0.0000 | 2.2 | 0.8129 | 5.2 | 0.9534 | 8.2 | 0.9794 |
| 0.2 | 0.0040 | 2.4 | 0.8354 | 5.4 | 0.9564 | 8.4 | 0.9803 |
| 0.3 | 0.0331 | 2.6 | 0.8542 | 5.6 | 0.9591 | 8.6 | 0.9811 |
| 0.4 | 0.0920 | 2.8 | 0.8700 | 5.8 | 0.9616 | 8.8 | 0.9819 |
| 0.5 | 0.1667 | 3.0 | 0.8833 | 6.0 | 0.9638 | 9.0 | 0.9827 |
| 0.6 | 0.2445 | 3.2 | 0.8947 | 6.2 | 0.9659 | 9.2 | 0.9833 |
| 0.7 | 0.3186 | 3.4 | 0.9046 | 6.4 | 0.9677 | 9.4 | 0.9840 |
| 0.8 | 0.3863 | 3.6 | 0.9131 | 6.6 | 0.9695 | 9.6 | 0.9846 |
| 0.9 | 0.4467 | 3.8 | 0.9205 | 6.8 | 0.9711 | 9.8 | 0.9852 |
| 1.0 | 0.5000 | 4.0 | 0.9270 | 7.0 | 0.9725 | 10.0 | 0.9857 |
| 1.2 | 0.5881 | 4.2 | 0.9328 | 7.2 | 0.9739 | 10.2 | 0.9862 |
| 1.4 | 0.6562 | 4.4 | 0.9379 | 7.4 | 0.9752 | 10.4 | 0.9867 |
| 1.6 | 0.7095 | 4.6 | 0.9424 | 7.6 | 0.9763 | 10.6 | 0.9872 |
| 1.8 | 0.7516 | 4.8 | 0.9465 | 7.8 | 0.9774 | 10.8 | 0.9876 |
| 2.0 | 0.7854 | 5.0 | 0.9502 | 8.0 | 0.9785 | 11.0 | 0.9881 |

### BLISCHKE/SCHEUER's approach

The squared coefficient of variation of the two–parameter WEIBULL distribution depends only on $c$ and is

$$\frac{\sigma^2}{\mu^2} = \frac{\Gamma(1 + 2/c)}{\Gamma^2(1 + 1/c)} - 1. \tag{12.4a}$$

$\widehat{c}$ is found by solving

$$\phi_{BS}(c) = \frac{\sigma^2}{\mu^2} - 1 - \frac{s^2}{\overline{x}^2} = 0, \tag{12.4b}$$

by using either a table (see Tab. 12/2) or a graphical aid (see Fig. 12/2) or iteratively by means of the NEWTON–RAPHSON formula

$$\widehat{c}_{p+1} = \widehat{c}_p - \frac{\varphi_{BS}(\widehat{c}_p)}{\varphi'_{BS}(\widehat{c}_p)}; \quad p = 0, 1, 2, \ldots;$$

where

$$\varphi'_{BS}(c) = -\frac{2\,\Gamma(1 + 2/c)\,\psi(1 + 2/c)}{c^2\,\Gamma^2(1 + 1/c)} - 1. \tag{12.4c}$$

$\widehat{b}$ is found with $\widehat{c}$ from (12.4b) using (12.3e).

Figure 12/2: $\sigma^2/\mu^2 = \Gamma(1 + 2/c)/\Gamma^2(1 + 1/c) - 1$ versus $c$

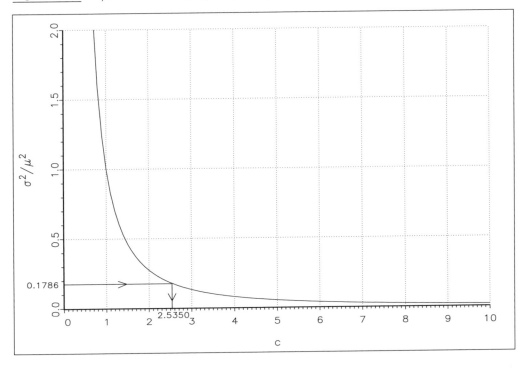

Table 12/2: $\sigma^2/\mu^2 = \Gamma(1 + 2/c)/\Gamma^2(1 + 1/c) - 1$ versus $c$

| $c$ | $\sigma^2/\mu^2$ | $c$ | $\sigma^2/\mu^2$ | $c$ | $\sigma^2/\mu^2$ | $c$ | $\sigma^2/\mu^2$ |
|---|---|---|---|---|---|---|---|
| 0.1 | 184.8 $10^3$ | 2.2 | 0.2302 | 5.2 | 0.0488 | 8.2 | 0.0210 |
| 0.2 | 251.0 | 2.4 | 0.1970 | 5.4 | 0.0456 | 8.4 | 0.0201 |
| 0.3 | 29.24 | 2.6 | 0.1707 | 5.6 | 0.0426 | 8.6 | 0.0192 |
| 0.4 | 9.865 | 2.8 | 0.1495 | 5.8 | 0.0400 | 8.8 | 0.0184 |
| 0.5 | 5.000 | 3.0 | 0.1321 | 6.0 | 0.0376 | 9.0 | 0.0177 |
| 0.6 | 3.091 | 3.2 | 0.1177 | 6.2 | 0.0353 | 9.2 | 0.0169 |
| 0.7 | 2.139 | 3.4 | 0.1055 | 6.4 | 0.0333 | 9.4 | 0.0163 |
| 0.8 | 1.589 | 3.6 | 0.0952 | 6.6 | 0.0315 | 9.6 | 0.0156 |
| 0.9 | 1.239 | 3.8 | 0.0864 | 6.8 | 0.0298 | 9.8 | 0.0150 |
| 1.0 | 1.000 | 4.0 | 0.0787 | 7.0 | 0.0282 | 10.0 | 0.0145 |
| 1.2 | 0.7004 | 4.2 | 0.0721 | 7.2 | 0.0268 | 10.2 | 0.0139 |
| 1.4 | 0.5238 | 4.4 | 0.0662 | 7.4 | 0.0255 | 10.4 | 0.0134 |
| 1.6 | 0.4095 | 4.6 | 0.0611 | 7.6 | 0.0242 | 10.6 | 0.0130 |
| 1.8 | 0.3305 | 4.8 | 0.0565 | 7.8 | 0.0231 | 10.8 | 0.0125 |
| 2.0 | 0.2732 | 5.0 | 0.0525 | 8.0 | 0.0220 | 11.0 | 0.0121 |

$\widehat{\theta} = (\widehat{b}, \widehat{c})$ is asymptotically normal and the asymptotic variance–covariance matrix follows from standard asymptotic theory and from a result concerning the JACOBIAN of the inverse of a transformation. Let

$$\hat{c} = h_1(\hat{\mu}, \hat{\sigma}^2),$$
$$\hat{b} = h_2(\hat{\mu}, \hat{\sigma}^2)$$

denote the implicit solutions of (12.4b) and of $\hat{b} = \hat{\mu}/\Gamma(1 + 1/\hat{c})$. We get

$$\mathrm{Var}\begin{pmatrix} \hat{c} \\ \hat{b} \end{pmatrix} = D\,\Sigma\,D', \tag{12.5a}$$

where

$$D = \begin{pmatrix} \dfrac{\partial h_1}{\partial \hat{\mu}} & \dfrac{\partial h_1}{\partial \hat{\sigma}^2} \\[2mm] \dfrac{\partial h_2}{\partial \hat{\mu}} & \dfrac{\partial h_2}{\partial \hat{\sigma}^2} \end{pmatrix},$$

and

$$\Sigma = \mathrm{Var}\begin{pmatrix} \hat{\mu} \\ \hat{\sigma}^2 \end{pmatrix} = \begin{pmatrix} \dfrac{\sigma^2}{n} & \dfrac{\mu_3}{n} \\[3mm] \dfrac{\mu_3}{n} & \dfrac{1}{n}\left[\mu_4 - \dfrac{n-3}{n-1}\sigma^4\right] \end{pmatrix} \tag{12.5b}$$

with[3]

$$\sigma^2 = \mu_2 = b^2\left[\Gamma_2 - \Gamma_1^2\right],$$
$$\mu_3 = b^3\left[\Gamma_3 - 3\,\Gamma_2\,\Gamma_1 + 2\,\Gamma_1^3\right],$$
$$\mu_4 = b^4\left[\Gamma_4 - 4\,\Gamma_3\,\Gamma_1 + 6\,\Gamma_2\,\Gamma_1^2 - 3\,\Gamma_1^4\right].$$

The matrix $D$ is obtained as the inverse of derivatives of $\hat{\mu}$ and $\hat{\sigma}^2$ with respect to $\hat{b}$ and $\hat{c}$ :

$$D = \begin{pmatrix} \dfrac{\partial \hat{\mu}}{\partial \hat{c}} & \dfrac{\partial \hat{\mu}}{\partial \hat{b}} \\[3mm] \dfrac{\partial \hat{\sigma}^2}{\partial \hat{c}} & \dfrac{\partial \hat{\sigma}^2}{\partial \hat{b}} \end{pmatrix}^{-1}. \tag{12.5c}$$

The derivatives in (12.5c) are as follows:[4]

$$\frac{\partial \hat{\mu}}{\partial \hat{c}} = -\frac{\hat{b}}{\hat{c}^2}\,\hat{\Gamma}_1\,\hat{\psi}_1; \quad \frac{\partial \hat{\mu}}{\partial \hat{b}} = \hat{\Gamma}_1;$$

$$\frac{\partial \hat{\sigma}^2}{\partial \hat{c}} = \frac{2\,\hat{b}^2}{\hat{c}^2}\left(\hat{\Gamma}_1^2\,\hat{\psi}_1 - \hat{\Gamma}_2\,\hat{\psi}_2\right); \quad \frac{\partial \hat{\sigma}^2}{\partial \hat{b}} = 2\,\hat{b}\left(\hat{\Gamma}_2 - \hat{\Gamma}_1^2\right).$$

---

[3]  We write $\Gamma_r = \Gamma(1 + r/c)$ for short.

[4]  We write $\hat{\Gamma}_r = (1 + r/\hat{c})$ and $\hat{\psi}_r = (1 + r/\hat{c})$ for short.

NEWBY's approach

NEWBY (1980) takes the coefficient of variation, which is independent of $b$, to estimate $c$ by solving

$$\varphi_N(c) = \frac{\sigma}{\mu} = \frac{\sqrt{\Gamma(1+2/c) - \Gamma^2(1+1/c)}}{\Gamma(1+1/c)} - \frac{s}{\bar{x}} = 0 \qquad (12.6a)$$

using either a graph (see Fig. 12/3) or a table (see Tab. 12/3) or by applying the NEWTON–RAPHSON algorithm

$$\widehat{c}_{p+1} = \widehat{c}_p - \frac{\varphi_N(\widehat{c}_p)}{\varphi'_N(\widehat{c}_p)}; \quad p = 0, 1, 2, \ldots$$

with

$$\varphi'_N(c) = \frac{\Gamma^2(1+1/c)\,\psi(1+1/c) - \Gamma(1+2/c)\,\psi(1+2/c)}{c^2\,\Gamma(1+1/c)\,\sqrt{\Gamma(1+2/c) - \Gamma^2(1+1/c)}}. \qquad (12.6b)$$

$\widehat{b}$ is found to be

$$\widehat{b} = \frac{\bar{x}}{\Gamma(1+1/\widehat{c})}.$$

Figure 12/3: $\sigma/\mu = \sqrt{\Gamma(1+2/c) - \Gamma^2(1+1/c)}\Big/\Gamma(1+1/c)$ versus $c$

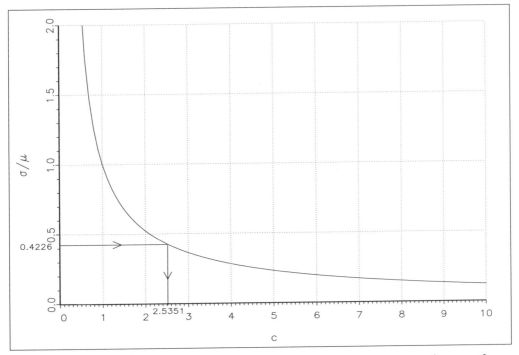

The variance–covariance matrix of the asymptotically normal estimator $\widehat{\boldsymbol{\theta}} = (\widehat{c}, \widehat{b})'$ — based on the same ideas as in BLISCHKE/SCHEUER's approach — is found by NEWBY to be

$$C = \frac{1}{n}\,T\,H\,T, \qquad (12.7a)$$

Table 12/3:  $\sigma/\mu = \sqrt{\Gamma(1 + 2/c) - \Gamma^2(1 + 1/c)}\Big/\Gamma(1 + 1/c)$ versus $c$

| $c$ | $\sigma/\mu$ | $c$ | $\sigma/\mu$ | $c$ | $\sigma/\mu$ | $c$ | $\sigma/\mu$ |
|-----|-----|-----|-----|-----|-----|-----|-----|
| 0.1 | 429.8  | 2.2 | 0.4798 | 5.2 | 0.2210 | 8.2 | 0.1450 |
| 0.2 | 15.84  | 2.4 | 0.4438 | 5.4 | 0.2135 | 8.4 | 0.1417 |
| 0.3 | 5.408  | 2.6 | 0.4131 | 5.6 | 0.2065 | 8.6 | 0.1387 |
| 0.4 | 3.141  | 2.8 | 0.3866 | 5.8 | 0.1999 | 8.8 | 0.1357 |
| 0.5 | 2.236  | 3.0 | 0.3634 | 6.0 | 0.1938 | 9.0 | 0.1329 |
| 0.6 | 1.758  | 3.2 | 0.3430 | 6.2 | 0.1880 | 9.2 | 0.1301 |
| 0.7 | 1.462  | 3.4 | 0.3248 | 6.4 | 0.1826 | 9.4 | 0.1275 |
| 0.8 | 1.261  | 3.6 | 0.3085 | 6.6 | 0.1774 | 9.6 | 0.1250 |
| 0.9 | 1.113  | 3.8 | 0.2939 | 6.8 | 0.1726 | 9.8 | 0.1226 |
| 1.0 | 1.000  | 4.0 | 0.2805 | 7.0 | 0.1680 | 10.0 | 0.1203 |
| 1.2 | 0.8369 | 4.2 | 0.2684 | 7.2 | 0.1637 | 10.2 | 0.1181 |
| 1.4 | 0.7238 | 4.4 | 0.2573 | 7.4 | 0.1596 | 10.4 | 0.1159 |
| 1.6 | 0.6399 | 4.6 | 0.2471 | 7.6 | 0.1556 | 10.6 | 0.1139 |
| 1.8 | 0.5749 | 4.8 | 0.2377 | 7.8 | 0.1519 | 10.8 | 0.1119 |
| 2.0 | 0.5227 | 5.0 | 0.2291 | 8.0 | 0.1484 | 11.0 | 0.1099 |

where

$$T = \begin{pmatrix} 1 & 0 \\ 0 & b \end{pmatrix}, \quad H = \begin{pmatrix} h_{11} & h_{12} \\ h_{12} & h_{22} \end{pmatrix}.$$

The standardized asymptotic variance–covariance matrix $H$ depends only on $c$ and is given in Tab. 12/4 for selected values of $c$.

Thus we have with respect to NEWBY's estimators:

$$\mathrm{Var}\big(\widehat{c}\big) \;=\; \frac{1}{n}\, h_{11}(c), \qquad\qquad (12.7\mathrm{b})$$

$$\mathrm{Var}\big(\widehat{b}\big) \;=\; \frac{b^2}{n}\, h_{22}(c), \qquad\qquad (12.7\mathrm{c})$$

$$\mathrm{Cov}\big(\widehat{c},\widehat{b}\big) \;=\; \frac{b}{n}\, h_{12}(c). \qquad\qquad (12.7\mathrm{d})$$

NEWBY (1980) has shown that the joint asymptotic efficiency[5] of his estimator compared with that of the MLEs is above 69% for all values of $c$ greater than unity. A Monte Carlo comparison of the method of moments to the ML method has been done by SAYLOR (1977).

---

[5]  This efficiency is the ratio of the determinants of the variance–covariance matrices.

Table 12/4: Elements of the standardized asymptotic variance–covariance matrix of
NEWBY's moment estimators

| $c$ | $h_{11}$ | $h_{22}$ | $h_{12}$ |
|---|---|---|---|
| 0.4 | 2.7056 | 112.0224 | 16.7355 |
| 0.6 | 1.1332 | 6.4539 | 2.0048 |
| 0.8 | 0.9459 | 2.1008 | 0.7092 |
| 1.0 | 1.0000 | 1.1787 | 0.4228 |
| 1.2 | 1.1603 | 0.7884 | 0.3294 |
| 1.4 | 1.3957 | 0.5718 | 0.2922 |
| 1.6 | 1.6973 | 0.4326 | 0.2755 |
| 1.8 | 2.0626 | 0.3434 | 0.2675 |
| 2.0 | 2.4918 | 0.2778 | 0.2634 |
| 2.2 | 2.9858 | 0.2295 | 0.2611 |
| 2.4 | 3.5459 | 0.1928 | 0.2598 |
| 2.6 | 4.1736 | 0.1642 | 0.2588 |
| 2.8 | 4.8703 | 0.1416 | 0.2579 |
| 3.0 | 5.6371 | 0.1233 | 0.2571 |
| 3.2 | 6.4756 | 0.1084 | 0.2563 |
| 3.4 | 7.3867 | 0.0960 | 0.2553 |
| 3.6 | 8.3715 | 0.0857 | 0.2543 |
| 3.8 | 9.4311 | 0.0769 | 0.2532 |
| 4.0 | 10.5663 | 0.0694 | 0.2520 |
| 4.2 | 11.7780 | 0.0630 | 0.2508 |
| 4.4 | 13.0668 | 0.0574 | 0.2495 |

Source: NEWBY (1982, p. 90) — Reprinted with permission from *Technometrics*.

**Example 12/1: Moment estimates for dataset #1**

Dataset #1 consists of $n = 20$ uncensored observation from $We(0, 100, 2.5)$ (see Tab. 9/2). Using
MIHRAM's estimators (see (12.3c,e)), we have $\hat{c} = 2.6094$ and $\hat{b} = 99.0086$. The formulas of
BLISCHKE/SCHEUER give $\hat{c} = 2.5350$ and $\hat{b} = 99.0898$. NEWBY's formulas lead to $\hat{c} = 2.5351$
and $\hat{b} = 99.0897$ and interpolating for $\hat{c}$ in Tab. 12/4 to an estimated variance–covariance matrix

$$
\widehat{C} = \frac{1}{20} \begin{pmatrix} 1 & 0 \\ 0 & 99.0897 \end{pmatrix} \begin{pmatrix} 3.9699 & 0.1735 \\ 0.1735 & 0.2591 \end{pmatrix} \begin{pmatrix} 1 & 0 \\ 0 & 99.0897 \end{pmatrix}
$$

$$
= \begin{pmatrix} 0.1985 & 0.8596 \\ 0.8596 & 127.2021 \end{pmatrix} \begin{pmatrix} \widehat{\mathrm{Var}}(\hat{c}) & \widehat{\mathrm{Cov}}(\hat{c},\hat{b}) \\ \widehat{\mathrm{Cov}}(\hat{c},\hat{b}) & \widehat{\mathrm{Var}}(\hat{b}) \end{pmatrix}.
$$

## 12.1.2   Three-parameter WEIBULL distribution[6]

We need three moments to estimate the parameters $a$, $b$ and $c$ of $We(a, b, c)$. These moments are as follows:

$$\mu = \mathrm{E}(X) = a + b\,\Gamma_1, \tag{12.8a}$$

to estimate $a$ as

$$\widehat{a} = \overline{x} - \widehat{b}\,\widehat{\Gamma}_1, \quad \overline{x} = \frac{1}{n}\sum_{i=1}^{n} x_i\,, \tag{12.8b}$$

using $\widehat{b}$ and $\widehat{c}$ from (12.9b) and (12.10b);

$$\sigma = \sqrt{\mathrm{E}\big[(X - \mu)^2\big]} = b\,\sqrt{\Gamma_2 - \Gamma_1^2} \tag{12.9a}$$

to estimate $b$ as

$$\widehat{b} = \frac{s}{\sqrt{\widehat{\Gamma}_2 - \widehat{\Gamma}_1^2}}\,, \quad s^2 = \frac{1}{n-1}\sum_{i=1}^{n}(x_i - \overline{x})^2; \tag{12.9b}$$

and the **skewness coefficient**

$$\alpha_3 = \frac{\mu_3}{\mu_2^{3/2}} = \frac{\mathrm{E}\big[(X - \mu)^3\big]}{\{\mathrm{E}\big[(X - \mu)^2\big]\}^{3/2}} = \frac{\Gamma_3 - 3\,\Gamma_2\,\Gamma_1 + 2\,\Gamma_1^3}{\big(\Gamma_2 - \Gamma_1^2\big)^{3/2}} \tag{12.10a}$$

to estimate $\widehat{c}$ by solving[7]

$$\varphi(c) = \alpha_3 - \widehat{\alpha}_3 = \alpha_3 - \frac{\dfrac{1}{n}\sum_{i=1}^{n}(x_i - \overline{x})^3}{\left[\dfrac{1}{n}\sum_{i=1}^{n}(x_i - \overline{x})^2\right]^{3/2}} = 0. \tag{12.10b}$$

The solution of (12.10b) is found by using a graph of $\alpha_3(c)$ versus $c$ (see Fig.2/23), or a table (see Tab. 12/5), or by applying the NEWTON–RAPHSON method with

$$\varphi'(c) = \frac{1}{c^2(\Gamma_2 - \Gamma_1^2)^{5/2}}\left\{\big[-3\,\Gamma_3' + 6\,\Gamma_2'\,\Gamma_1 + 3\,\Gamma_1'\,\Gamma_2 - 6\,\Gamma_1'\,\Gamma_1^2\big]\big[\Gamma_2 - \Gamma_1^2\big] + \right.$$
$$\left.\big[3\,\Gamma_2' - 3\,\Gamma_1'\,\Gamma_1\big]\big[\Gamma_3 - 3\,\Gamma_2\,\Gamma_1 + 2\,\Gamma_1^3\big]\right\}. \tag{12.10c}$$

---

[6]   Suggested reading for this section: BOWMAN/SHENTON (1987, 2001), DUBEY (1966b,d,g; 1967b–d), HEO/BOES/SALAS (2001), NEWBY (1984), RAVENIS (1964), SEN/PRABHASHANKER (1980).

[7]   The standardized moments

$$\alpha_k = \frac{\mu_k}{\mu_2^{k/2}}\,;\quad k = 0, 1, \ldots$$

are independent of $a$ and $b$ for $k \geq 3$ and each of them can be used to estimate $c$ (see RAVENIS (1964)) for $k = 3$ and 4 and DUBEY (1966g) for $k = 3$, 4 and 5)). It is not recommended to take $k \geq 4$ because the sampling variability of $\widehat{\alpha}_k$ increases with $k$ and because $\alpha_k$ is not a monotone function of $c$ for $k \geq 4$ (see Fig. 2/23).

HEO/SALAS/XIM (2001) have found an approximating equation to solve (12.10b):

$$\widehat{c} = -0.729268 - 0.338679\,\widehat{\alpha}_3 + 4.96077\,(\widehat{\alpha}_3 + 1.14)^{-1.0422} + 0.683609\big[\ln(\widehat{\alpha}_3 + 1.14)\big]^2,$$
(12.10d)

which is valid for $-1.08 \le \widehat{\alpha}_3 \le 6.0$ $(0.52 \le \widehat{c} \le 100)$ with a determination coefficient $R^2 = 0.999999$. It is even possible to find $\widehat{a}$, $\widehat{b}$ and $\widehat{c}$ based on $\overline{x}$, $s$ and $\widehat{\alpha}_3$ by using the nomogram of SEN/PRABHASHANKER (1980).

Table 12/5: $\alpha_3(c)$ versus $c$

| $c$ | $\alpha_3(c)$ | $c$ | $\alpha_3(c)$ | $c$ | $\alpha_3(c)$ | $c$ | $\alpha_3(c)$ |
|------|---------------|------|---------------|------|----------------|------|----------------|
| 0.1 | $6.990 \cdot 10^4$ | 2.2 | 0.5087 | 5.2 | $-0.2810$ | 8.2 | $-0.5461$ |
| 0.2 | 190.1 | 2.4 | 0.4049 | 5.4 | $-0.3062$ | 8.4 | $-0.5579$ |
| 0.3 | 28.33 | 2.6 | 0.3155 | 5.6 | $-0.3299$ | 8.6 | $-0.5692$ |
| 0.4 | 11.35 | 2.8 | 0.2373 | 5.8 | $-0.3522$ | 8.8 | $-0.5802$ |
| 0.5 | 6.619 | 3.0 | 0.1681 | 6.0 | $-0.3733$ | 9.0 | $-0.5907$ |
| 0.6 | 4.593 | 3.2 | 0.1064 | 6.2 | $-0.3932$ | 9.2 | $-0.6008$ |
| 0.7 | 3.498 | 3.4 | 0.0509 | 6.4 | $-0.4121$ | 9.4 | $-0.6105$ |
| 0.8 | 2.815 | 3.6 | 0.0006 | 6.6 | $-0.4300$ | 9.6 | $-0.6199$ |
| 0.9 | 2.345 | 3.8 | $-0.0453$ | 6.8 | $-0.4470$ | 9.8 | $-0.6289$ |
| 1.0 | 2.000 | 4.0 | $-0.0872$ | 7.0 | $-0.4632$ | 10.0 | $-0.6376$ |
| 1.2 | 1.521 | 4.2 | $-0.1259$ | 7.2 | $-0.4786$ | 10.2 | $-0.6461$ |
| 1.4 | 1.198 | 4.4 | $-0.1615$ | 7.4 | $-0.4933$ | 10.4 | $-0.6542$ |
| 1.6 | 0.9620 | 4.6 | $-0.1946$ | 7.6 | $-0.5074$ | 10.6 | $-0.6621$ |
| 1.8 | 0.7787 | 4.8 | $-0.2254$ | 7.8 | $-0.5209$ | 10.8 | $-0.6697$ |
| 2.0 | 0.6311 | 5.0 | $-0.2541$ | 8.0 | $-0.5337$ | 11.0 | $-0.6771$ |

NEWBY(1984) gives the asymptotic variance–covariance matrix:

$$\mathrm{Var}\begin{pmatrix} \widehat{c} \\ \widehat{b} \\ \widehat{a} \end{pmatrix} = \frac{1}{n}\,T H T$$
(12.11)

where

$$T = \begin{pmatrix} 1 & 0 & 0 \\ 0 & b & 0 \\ 0 & 0 & b \end{pmatrix}$$

The elements $h_{ij}$ of the symmetric standardized variance–covariance matrix $H$ are listed in Tab. 12/6. BOWMAN/SHENTON (1987, 2001) have given series expressions for the moments of these moment estimators.

<u>Table 12/6</u>:  Elements of the standardized variance–covariance matrix $H$ in the three-parameter case

| $c$ | $h_{11}$ | $h_{12}$ | $h_{13}$ | $h_{22}$ | $h_{23}$ | $h_{33}$ |
|-----|----------|----------|----------|----------|----------|----------|
| 0.4 | 104.3561 | 1098.2257 | $-1271.4367$ | 11601.5710 | $-13474.1443$ | 15735.0050 |
| 0.6 | 17.3525 | 66.0829 | $-43.8139$ | 258.1763 | $-171.0589$ | 115.4578 |
| 0.8 | 9.5787 | 18.3163 | $-11.2638$ | 37.6456 | $-22.5719$ | 14.2566 |
| 1.0 | 8.0000 | 9.3823 | $-6.0000$ | 12.5034 | $-7.5367$ | 5.0000 |
| 1.2 | 8.0002 | 6.5716 | $-4.4690$ | 6.4185 | $-4.0137$ | 2.8570 |
| 1.4 | 8.7500 | 5.5130 | $-3.9775$ | 4.2454 | $-2.7919$ | 2.1203 |
| 1.6 | 10.0459 | 5.1722 | $-3.9274$ | 3.2923 | $-2.2884$ | 1.8326 |
| 1.8 | 11.8587 | 5.2177 | $-4.1353$ | 2.8389 | $-2.0836$ | 1.7394 |
| 2.0 | 14.2302 | 5.5251 | $-4.5372$ | 2.6350 | $-2.0322$ | 1.7520 |
| 2.2 | 17.2420 | 6.0460 | $-5.1130$ | 2.5768 | $-2.0747$ | 1.8337 |
| 2.4 | 21.0050 | 6.7649 | $-5.8617$ | 2.6156 | $-2.1834$ | 1.9676 |
| 2.6 | 25.6553 | 7.6828 | $-6.7925$ | 2.7265 | $-2.3446$ | 2.1455 |
| 2.8 | 31.3530 | 8.8096 | $-7.9202$ | 2.8957 | $-2.5513$ | 2.3633 |
| 3.0 | 38.2821 | 10.1609 | $-9.2634$ | 3.1158 | $-2.7997$ | 2.6193 |
| 3.2 | 46.6510 | 11.7557 | $-10.8433$ | 3.3825 | $-3.0881$ | 2.9126 |
| 3.4 | 56.6926 | 13.6161 | $-12.6831$ | 3.6932 | $-3.4158$ | 3.2435 |
| 3.6 | 68.6653 | 15.7658 | $-14.8077$ | 4.0469 | $-3.7828$ | 3.6123 |
| 3.8 | 82.8535 | 18.2305 | $-17.2434$ | 4.4431 | $-4.1894$ | 4.0197 |
| 4.0 | 99.5681 | 21.0370 | $-20.0176$ | 4.8817 | $-4.6363$ | 4.4667 |
| 4.2 | 119.1471 | 24.2134 | $-23.1589$ | 5.3630 | $-5.1242$ | 4.9541 |
| 4.4 | 141.9564 | 27.7888 | $-26.6966$ | 5.8876 | $-5.6539$ | 5.4828 |
| 4.6 | 168.3898 | 31.7933 | $-30.6610$ | 6.4560 | $-6.2264$ | 6.0540 |
| 4.8 | 198.8700 | 36.2576 | $-35.0833$ | 7.0690 | $-6.8425$ | 6.6685 |
| 5.0 | 233.8485 | 41.2135 | $-39.9952$ | 7.7274 | $-7.5031$ | 7.3274 |

**Example 12/2:   Moment estimates for dataset #2**

We take dataset #2 from Tab. 9/9 which is a sample of $n = 20$ from $We(15, 30, 2.5)$. The empirical moments are

$$\overline{x} = 41.5250, \quad s = 11.0908, \quad \widehat{\alpha}_3 = 0.3424.$$

Using (12.10b) we find[8] $\widehat{c} = 2.5368$, which — inserted into (12.9b) — gives $\widehat{b} = 29.5855$, and $\widehat{b}$, $\widehat{c}$ together with (12.8b) give $\widehat{a} = 15.2651$. All three estimates are very close to their true values.

---

[8]  (12.10d) gives a rather bad solution: $\widehat{c} = 2.1171$.

## 12.2 Modified method of moments[9]

In Sect. 11.3.2.3 we presented several modifications of the maximum likelihood procedure for the three–parameter case. There we substituted the likelihood equation
$\partial \mathcal{L}(a, b, c)/\partial a = 0$ by one of five alternative relationships connecting $a$, $b$ and $c$. Here
we will substitute the moment equation (12.10b) by one out of three alternatives.[10]

<u>MME–I</u>

A first modification suggested by COHEN/WHITTEN (1982) consists in equating the population median $x_{0.5}$ to the sample median $\widetilde{x}$:

$$x_{0.5} - \widetilde{x} = 0$$

resulting in (see (2.57b))

$$a + b \, (\ln 2)^{1/c} = \widetilde{x}. \tag{12.12a}$$

Elimination of $a$ and $b$ from (12.12a) by means of (12.8b) and (12.9b) yields

$$\frac{s^2}{\left(\overline{x} - \widetilde{x}\right)^2} = \frac{\Gamma_2 - \Gamma_1^2}{\left[\Gamma_1 - (\ln 2)^{1/c}\right]^2}. \tag{12.12b}$$

The right–hand side of (12.12b) is no monotone function of $c$ (see Fig. 12/4) and consequently the solution of (12.12b) is not unique. Thus, this modification is not be recommended.

Figure 12/4: $\left[\Gamma_2 - \Gamma_1^2\right] \big/ \left[\Gamma_1 - (\ln 2)^{1/c}\right]^2$ versus $c$

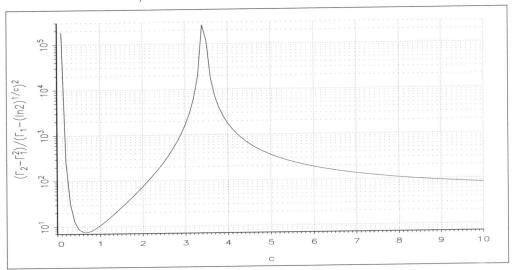

---

[9]  Suggested reading for this section: BALAKRISHNAN/COHEN (1991, pp. 275–278), COHEN/WHITTEN
(1982), COHEN/WHITTEN/DING (1984), WHITTEN/COHEN(1996).

[10]  Two of the five alternatives of Sect. 11.3.2.3 are the moment equations for the mean (12.8b) and the
standard deviation (12.9b) and thus are not eligible here.

MME–II[11]

In this case we replace (12.10b) by

$$E(X_{1:n}) - x_{1:n} = 0;$$

i.e., the expectation of the first order statistic is equated to the sample minimum. Because of (5.32d) we get

$$a + \frac{b}{n^{1/c}} \Gamma_1 = x_{1:n}. \tag{12.13a}$$

Upon elimination of $a$ and $b$ from (12.13a) by means of (12.8b) and (12.9b), we get

$$\frac{s^2}{(\bar{x} - x_{1:n})^2} = \frac{\Gamma_2 - \Gamma_1^2}{\left[(1 - n^{-1/c})\Gamma_1\right]^2}, \tag{12.13b}$$

which can be solved for $\hat{c}$ either by iteration or by using a table or a graph of $\left[\Gamma_2 - \Gamma_1^2\right]/\left[(1 - n^{-1/c})\Gamma_1\right]^2$ versus $c$ (see Fig. 12/5). Then $\hat{b}$ and $\hat{a}$ are found from (12.9b) and (12.8b) or from

$$\hat{b} = \frac{(\bar{x} - x_{1:n})\, n^{1/\hat{c}}}{(n^{1/\hat{c}} - 1)\, \hat{\Gamma}_1} \tag{12.13c}$$

Figure 12/5: $\left[\Gamma_2 - \Gamma_1^2\right]/\left[(1 - n^{-1/c})\Gamma_1\right]^2$ versus $c$

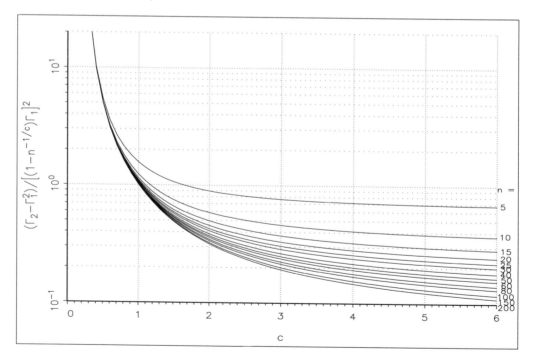

[11] For more details on this modification, see COHEN/WHITTEN/DING (1984).

and

$$\widehat{a} = \frac{x_{1:n}\, n^{1/\widehat{c}} - \overline{x}}{n^{1/\widehat{c}} - 1}.$$ (12.13d)

MME–III[12]

This modification consists in replacement of (12.10b) by

$$\mathrm{E}\big[F(X_{1:n})\big] = F(x_{1:n}).$$

Since $\mathrm{E}\big[F(X_{1:n})\big] = 1/(n+1)$ and $F(x_{1:n}) = 1 - \exp\left\{-\left(\frac{x_{1:n} - a}{b}\right)^{c}\right\}$, we get

$$\left(\frac{x_{1:n} - a}{b}\right)^{c} = -\ln\left(\frac{n}{n+1}\right).$$ (12.14a)

Elimination of $a$ and $b$ from (12.14a) by means of (12.8b) and (12.9b) leads to

$$\frac{s^2}{\left(\overline{x} - x_{1:n}\right)^2} = \frac{\Gamma_2 - \Gamma_1^2}{\left\{\Gamma_1 - \left[-\ln\left(\frac{n}{n+1}\right)\right]^{1/c}\right\}^2},$$ (12.14b)

which can be solved for $\widehat{c}$ using techniques similar to those employed in the solution of (12.12b) or (12.13b). A graph of $\left[\Gamma_2 - \Gamma_1^2\right]\Big/\left\{\Gamma_1 - \left[-\ln(n/(n+1))\right]^{1/c}\right\}^2$ versus $c$ is given in Fig. 12/6. $\widehat{c}$ from (12.14b) gives $\widehat{b}$ and $\widehat{a}$ by inserting into (12.9b) and (12.8b) or from

$$\widehat{b} = \frac{\overline{x} - x_{1:n}}{\widehat{\Gamma}_1 - \left[-\ln\left(\frac{n}{n+1}\right)\right]^{1/\widehat{c}}}$$ (12.14c)

and

$$\widehat{a} = \frac{x_{1:n}\,\widehat{\Gamma}_1 - \overline{x}\left[-\ln\left(\frac{n}{n+1}\right)\right]^{1/\widehat{c}}}{\widehat{\Gamma}_1 - \left[-\ln\left(\frac{n}{n+1}\right)\right]^{1/\widehat{c}}}.$$ (12.14d)

---

[12] For more detail on this modification see, WHITTEN/COHEN (1996).

Figure 12/6: $\left[\Gamma_2 - \Gamma_1^2\right] \Big/ \left\{\Gamma_1 - \left[-\ln\left(\dfrac{n}{n+1}\right)\right]^{1/c}\right\}^2$ versus $c$

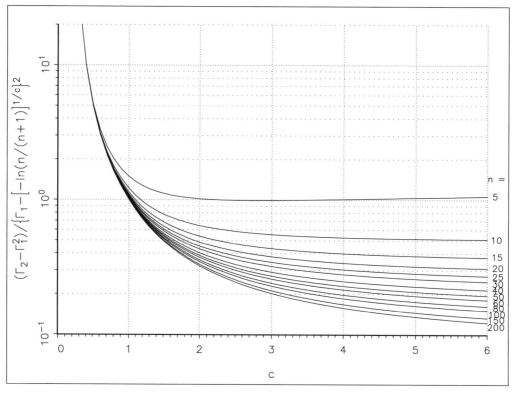

The asymptotic variance–covariance matrix of the MLEs (see (11.12a–h)) is not strictly applicable for the modified moment estimators, but simulation studies carried out by CO-HEN/WHITTEN (1982) disclosed close agreement between simulated variances and corresponding asymptotic variances when $c > 2$. In order to avoid possible computational difficulties in applications, $c$ should be greater than 2.2.

## 12.3   W. WEIBULL's approaches to estimation by moments[13]

WEIBULL (1961, p. 226) introduced the following class of moments of a continuous variable $X$ having the distribution function $F(x)$:

$$\overline{\mu}_r := \int_{x_a}^{x_b} \left[1 - F(x)\right]^r \, dx; \quad r = 1, 2, 3, \ldots \tag{12.15}$$

with limits $x_a$, $x_b$ chosen according to whether the distribution is complete or truncated. These moments are named **WEIBULL moments** or **vertical moments**. For a three–

---

[13]   Suggested reading for this section: CRAN (1988), SCHREIBER (1963), WEIBULL (1961, 1967b).

parameter WEIBULL distribution, (12.15) turns into

$$\overline{\mu}_r = \int_a^\infty \left[ \exp\left\{ -\left( \frac{x-a}{b} \right)^c \right\} \right]^r \mathrm{d}x$$

$$= a + \frac{b}{r^{1/c}} \Gamma_1 . \tag{12.16}$$

The parameters $a$, $b$ and $c$ are explicit functions of these moments:

$$a = \frac{\overline{\mu}_1 \overline{\mu}_4 - \overline{\mu}_2^2}{\overline{\mu}_1 + \overline{\mu}_4 - 2\overline{\mu}_2} , \tag{12.17a}$$

$$b = \frac{\overline{\mu}_1 - a}{\Gamma_1} , \tag{12.17b}$$

$$c = \frac{\ln 2}{\ln\left(\overline{\mu}_1 - \overline{\mu}_2\right) - \ln(\overline{\mu}_2 - \overline{\mu}_4)} . \tag{12.17c}$$

Given the ordered sample values $x_{1:n} \le x_{2:n} \le \dots \le x_{n:n}$, we estimate the distribution function as

$$\widehat{F}(x) = \begin{cases} 0 & \text{for } x < x_{1:n} , \\[2mm] \dfrac{i}{n} & \text{for } x_{i:n} \le x < x_{i+1:n}; \ i = 1, 2, \dots, n-1, \\[2mm] 1 & \text{for } x \ge x_{n:n}, \end{cases} \tag{12.18a}$$

and the WEIBULL moments as

$$\overline{m}_r := \widehat{\overline{\mu}}_r = \int_a^\infty \left[ 1 - \widehat{F}(x) \right]^r \mathrm{d}x$$

$$= \sum_{i=0}^{n-1} \left( 1 - \frac{i}{n} \right)^r \left( x_{i+1:n} - x_{i:n} \right), \quad x_{0:n} := 0. \tag{12.18b}$$

It is easily seen that $\overline{m}_1 = \overline{x}$. The estimators of $a$, $b$ and $c$ are obtained from (12.17a–c) by substituting $\overline{m}_r$ for $\overline{\mu}_r$. From (12.17c) and (12.17b) it can be shown that the estimators $\widehat{c}$ and $\widehat{b}$ are non–positive and hence inadmissible when $\overline{m}_2 \ge (\overline{m}_1 + \overline{m}_n)/2$. In addition it is possible that the estimator $\widehat{a}$ is inadmissible by exceeding $x_{1:n}$. In these cases alternative methods of estimation have to be used.

CRAN (1988) has found the following properties of the estimators based on $\overline{m}_r$ by Monte Carlo methods.

- For all $(c, n)$–combinations studied, $\widehat{a}$ appears to be negatively biased whereas $\widehat{b}$ and $\widehat{c}$ are positively biased.

- All three estimators show considerable variation. For $\widehat{a}$ and $\widehat{c}$ the variation increases dramatically as $c$ increases, the pattern of variation of $\widehat{b}$ is a concave–upwards function of $c$ with a minimum lying between $c = 1$ and $c = 2$.

- $\widehat{a}$ and $\widehat{b}$ as well as $\widehat{a}$ and $\widehat{c}$ are negatively correlated and $\widehat{b}$, $\widehat{c}$ have a positive correlation.

The procedure described above has the following advantages over the conventional method of moments.

- The sample WEIBULL moments (12.18b) are functions of differences of the observations, rather than powers and thus are less sensitive to sample variation.

- The parameter estimates can be given in closed form.

- It is easy to decide whether $a = 0$. For this purpose the estimators of the three–parameter model are compared with those of the two–parameter model where $a = 0$:

$$\widehat{c}^* = \frac{\ln 2}{\ln \overline{m}_1 - \ln \overline{m}_2} \quad \text{and} \quad \widehat{b}^* = \frac{\overline{m}_1}{\Gamma\left(1 + \dfrac{1}{\widehat{c}^*}\right)}.$$

If they are reasonably close then it can be assumed that $a = 0$.

WEIBULL proposed still another type of moments useful to estimate the parameters (see WEIBULL (1967b)) the **moments about the sample minimum**:

$$R_j := \frac{1}{n} \sum_{i=1}^{n} \left(X_{i:n} - X_{1:n}\right)^j; \quad j = 1, 2. \tag{12.19}$$

As $n \to \infty$, the expected value of $R_2 / R_1^2$ tends to a function $\phi(c)$ of the shape parameter $c$. Some values of this function are to be found in Tab. 12/7 compiled from WEIBULL (1967b).

Table 12/7: Limit of the ratio $R_2 / R_1^2$ of moments about the smallest sample value

| $c$ | $\phi(c)$ | $c$ | $\phi(c)$ |
|------|-----------|------|-----------|
| 0.00 | 1.0000 | 0.60 | 1.3801 |
| 0.01 | 1.0002 | 0.70 | 1.5045 |
| 0.10 | 1.0145 | 0.80 | 1.6480 |
| 0.20 | 1.0524 | 0.90 | 1.8124 |
| 0.30 | 1.1093 | 1.00 | 2.0000 |
| 0.40 | 1.1831 | 1.50 | 3.3953 |
| 0.50 | 1.2732 | 2.00 | 6.0000 |

By solving $\phi(c) = R_2 / R_1^2$, an asymptotic estimator of $c$ can be derived which is markedly more accurate than the estimator based on central moments when $c \geq 0.5$. The asymptotic efficiency (compared with MLEs) decreases from $97.59\%$ (when $c = 0.5$) to $60.79\%$ (when $c = 1.00$), while that of a central moment estimator decreases from $17.09\%$ to $7.60\%$ over the same range of values of $c$; see WEIBULL (1967b, p. 11).

## 12.4   Method of probability weighted moments

The probability weighted moment (PWM) of order $r$ of the three–parameter WEIBULL distribution is given in GREENWOOD et al. (1979) as

$$
\begin{aligned}
A_r &= \mathrm{E}\{X\,[1 - F(x)]^r\} \\
&= \frac{1}{r+1}\left[a + b\,(r+1)^{-1/c}\,\Gamma\!\left(1 + \frac{1}{c}\right)\right]; \quad r = 0, 1, 2, \ldots, \quad (12.20)
\end{aligned}
$$

where $A_0 = \mu_1' = \mu$. Likewise, the sample PWMs are

$$
\left.
\begin{aligned}
\widehat{A}_0 &= \frac{1}{n}\sum_{i=1}^{n} x_i \\
\widehat{A}_r &= \frac{1}{n}\sum_{i=1}^{n} x_i\,\frac{(n-i)\,(n-i-1)\cdots(n-i-r+1)}{(n-1)\,(n-2)\cdots(n-r)}; \quad r = 1, 2, \ldots
\end{aligned}
\right\} \quad (12.21)
$$

From (12.20) the first three population PWMs are

$$
\begin{aligned}
A_0 &= a + b\,\Gamma_1, & (12.22\text{a}) \\
A_1 &= \left[a + b\,2^{-1/c}\,\Gamma_1\right]/2, & (12.22\text{b}) \\
A_2 &= \left[a + b\,3^{-1/c}\,\Gamma_1\right]/3. & (12.22\text{c})
\end{aligned}
$$

By substituting these three PWMs by the corresponding sample PWMs $\widehat{A}_0$, $\widehat{A}_1$ and $\widehat{A}_2$, the PWM estimator of the shape parameter $c$ is the solution of

$$
\frac{3^{-1/c} - 1}{2^{-1/c} - 1} = \frac{3\,\widehat{A}_2 - \widehat{A}_0}{2\,\widehat{A}_1 - \widehat{A}_0}. \quad (12.23\text{a})
$$

(12.23a) can be solved numerically for $c$ by using the NEWTON–RAPHSON algorithm with

$$
g(c) = \frac{3^{-1/c} - 1}{2^{-1/c} - 1} - \frac{3\,\widehat{A}_2 - \widehat{A}_0}{2\,\widehat{A}_1 - \widehat{A}_0} \quad (12.23\text{b})
$$

and

$$
g'(c) = \frac{1}{c^2\,(2^{-1/c} - 1)^2}\left[3^{-1/c}\,(2^{-1/c} - 1)\,\ln 3 + 2^{-1/c}\,(3^{-1/c} - 1)\,\ln 2\right]. \quad (12.23\text{c})
$$

The PWM estimators of $b$ and $a$ follow with $\widehat{c}$ from (12.23a) as

$$
\widehat{b} = \frac{\widehat{A}_0 - 2\,A_1}{\left(1 - 2^{-1/\widehat{c}}\right)\widehat{\Gamma}_1} \quad (12.23\text{d})
$$

and

$$
\widehat{a} = \widehat{A}_0 - \widehat{b}\,\widehat{\Gamma}_1. \quad (12.23\text{e})
$$

For a two–parameter WEIBULL distribution $(a = 0)$, the PWM estimators may be obtained from (12.22a) as

$$
\widehat{c} = \frac{\ln 2}{\ln\left(\widehat{A}_0/\widehat{A}_1\right)} \quad (12.24\text{a})
$$

and

$$\widehat{b} = \frac{\widehat{A}_0}{\widehat{\Gamma}_1}.$$  (12.24b)

HEO/BOES/SALAS (2001) have derived the rather difficult variance–covariance matrix of the PWM estimators.

## 12.5  Method of fractional moments

The moments used in the previous sections have had an integer order $r$. There exist some papers[14] which allow for fractional order $(r \in \mathbb{R}^+)$ of moments to estimate the WEIBULL parameters $b$ and $c$. In these papers the orders $r_1$ and $r_2$ of the two moments are chosen by applying some optimizing criterion so that the orders found depend on that criterion.

RAFIQ/AHMAD (1999) decided to minimize the determinant $D$ of the asymptotic variance–covariance matrix of the moment estimators. Let $r_1$ and $r_2$ be the order of the two moments; then

$$D = \frac{k_{r_1} k_{r_2} - k_{r_1} - k_{r_2} - k_{r_{12}}^2 + 2 k_{r_{12}}}{b^2 n^2 (a_1 r_2 - b_1 r_1)^2},$$  (12.25)

where

$$a_1 \quad = \quad -\frac{r_1}{c^2} \psi(1 + r_1/c), \qquad b_1 \quad = \quad -\frac{r_2}{c^2} \psi(1 + r_2/c),$$

$$k_{r_1} \quad = \quad \frac{\Gamma(1 + 2 r_1/c)}{\Gamma^2(1 + r_1/c)}, \qquad k_{r_2} \quad = \quad \frac{\Gamma(1 + 2 r_2/c)}{\Gamma^2(1 + r_2/c)},$$

$$k_{r_{12}} \quad = \quad \frac{\Gamma(1 + [r_1 + r_2]/c)}{\Gamma(1 + r_1/c) \Gamma(1 + r_2/c)}.$$

For the values of $r_1$ and $r_2$, which minimize $D$, the ratios $r_1/c$ and $r_2/c$ are constant, regardless of the values of $r_1$, $r_2$ and $c$ and hence the ratios $r_1/r_2$ or $r_2/r_1$: $r_1/c \approx 0.193$, $r_2/c \approx 1.186$ and $r_2/r_1 \approx 6.15$. The value of the scale parameter $b$ is of no importance. The overall asymptotic relative efficiency (compared to ML) in this approach is higher for the method of fractional moments than for the traditional method of moments using $r_1 = 1$ and $r_2 = 2$.

MUKHERJEE/SASMAL (1984) have found the optimal orders $r_1$ and $r_2$ so that the overall relative efficiency of the moment estimators compared with the MLEs is maximized. This efficiency turns out to be a function of $c$ alone. Tab. 12/8 shows that it is recommendable to use fractional moments of orders less than 1 when $c < 1$. The results, valid for the traditional method of moments, are found in the row with $r_1 = 1$ and $r_2 = 2$.

---

[14]  See MUKHERJEE/SASMAL (1984) and RAFIG/AHMAD (1999)

Table 12/8: Overall relative efficiency of moment estimators compared with MLEs

| $r_1$ | $r_2$ | $c = 0.25$ | $c = 0.50$ | $c = 0.75$ | $c = 1.00$ | $c = 2.00$ | $c = 4.00$ |
|-------|-------|-----------|-----------|-----------|-----------|-----------|-----------|
| 0.10 | 0.20 | 0.9852 | 0.8519 | 0.2518 | 0.6926 | 0.5946 | 0.5562 |
| 0.25 | 0.50 | 0.6079 | 0.9759 | 0.9750 | 0.9075 | 0.7381 | 0.6211 |
|      | 0.75 | 0.3800 | 0.9059 | 0.9937 | 0.9650 | 0.1030 | 0.6572 |
|      | 1.00 | 0.2154 | 0.7951 | 0.9743 | 0.9913 | 0.8570 | 0.6974 |
|      | 1.25 | 0.6108 | 0.6672 | 0.9248 | 0.9906 | 0.9007 | 0.7301 |
| 0.50 | 0.75 | 0.0067 | 0.7400 | 0.9810 | 0.9845 | 0.8635 | 0.7047 |
|      | 1.00 | 0.0889 | 0.6079 | 0.8829 | 0.9759 | 0.9076 | 0.7382 |
|      | 1.25 | 0.0436 | 0.4865 | 0.8146 | 0.9479 | 0.9412 | 0.7723 |
|      | 1.50 | 0.0206 | 0.3800 | 0.7333 | 0.9059 | 0.9658 | 0.8032 |
|      | 1.75 | 0.0145 | 0.5674 | 0.6463 | 0.8539 | 0.9821 | 0.8314 |
|      | 2.00 | 0.0037 | 0.2154 | 0.5644 | 0.7951 | 0.9923 | 0.1570 |
| 0.75 | 1.00 | 0.0158 | 0.4504 | 0.7679 | 0.9311 | 0.9411 | 0.7778 |
|      | 1.25 | 0.0057 | 0.3434 | 0.7073 | 0.8816 | 0.9662 | 0.8008 |
|      | 1.50 | 0.0070 | 0.2584 | 0.6380 | 0.8243 | 0.9808 | 0.8356 |
|      | 1.75 | 0.0030 | 0.1916 | 0.5225 | 0.7631 | 0.9881 | 0.8614 |
|      | 2.00 | 0.0013 | 0.1398 | 0.4511 | 0.6998 | 0.9888 | 0.8046 |
| 1.00 | 1.25 | 0.0052 | 0.2401 | 0.5885 | 0.8072 | 0.9793 | 0.8380 |
|      | 1.50 | 0.0024 | 0.1737 | 0.1506 | 0.7400 | 0.9845 | 0.8635 |
|      | 1.75 | 0.0010 | 0.1248 | 0.0624 | 0.6731 | 0.9830 | 0.8860 |
|      | 2.00 | 0.0004 | 0.0889 | 0.1308 | 0.6079 | 0.9759 | 0.9075 |
|      | 2.50 | 0.0546 | 0.0436 | 0.2426 | 0.4865 | 0.9479 | 0.9693 |
| 1.50 | 2.00 | 0.0000 | 0.0353 | 0.2215 | 0.4504 | 0.9311 | 0.9428 |
|      | 2.50 | 0.0000 | 0.0159 | 0.1383 | 0.3434 | 0.8816 | 0.9662 |
| 2.00 | 2.50 | 0.0003 | 0.0067 | 0.0315 | 0.2401 | 0.8072 | 0.9793 |
|      | 3.00 | 0.0000 | 0.0023 | 0.0095 | 0.1737 | 0.7400 | 0.9845 |
| 3.00 | 4.00 | 0.0000 | 0.0000 | 0.0016 | 0.0353 | 0.4604 | 0.9311 |

# 13 Parameter estimation — More classical approaches and comparisons

The estimation procedures contained in this chapter are generally much easier to apply than those of the previous chapters because they are mostly given by analytic formulas. The estimates can be used as quick rough estimates or as starting points for an iterative procedure employed in deriving "better" estimates such as maximum likelihood. Sections 13.1 through 13.4 deal with methods which give estimators for all the parameters of either a two–parameter or a three–parameter WEIBULL distribution. Sect. 13.5 shows how to estimate a single parameter. We finish this chapter by comparing classical approaches to parameter estimation (Sect. 13.6).

## 13.1 Method of percentiles[1]

The design of the **method of percentiles** or **method of quantiles** is similar to the method of moments (see Sect. 12.1), the difference being that here the role of the moments is taken by percentiles. For any given cumulative probability $P$, $0 < P < 1$, the $100\,P\%$ percentile $x_P$ of a WEIBULL population is

$$x_P = a + b\left[-\ln(1-P)\right]^{1/c}. \tag{13.1a}$$

For a sample of size $n$ we declare the $100\,P\%$ percentile $\widehat{x}_P$ to be

$$\widehat{x}_P = \left\{\begin{array}{ll} x_{n\,P:n}, & \text{if } n\,P \text{ is an integer,} \\[2mm] x_{[n\,P]+1:n}, & \text{if } n\,P \text{ is not an integer.} \end{array}\right\} \tag{13.1b}$$

In order to derive estimators we have to set up as many population percentiles — of different order, of course — as there are parameters to be estimated. These population percentiles are functions of the parameters. After equating them to the sample percentiles, the percentile (or quantile) estimators follow as the solutions of these equations.

### 13.1.1 Two-parameter WEIBULL distribution

In the two parameter case $(a = 0)$ we find from (13.1a)

$$\ln\left[-\ln(1-P)\right] = c\left(\ln x_P - \ln b\right). \tag{13.2a}$$

---

[1] Suggested reading for this section: DUBEY (1966g, 1967c,e,f), SCHMID (1997), ZANAKIS (1979), ZANAKIS/MANN (1982).

For two real numbers $P_1$ and $P_2$ such that $0 < P_1 < P_2 < 1$, we have

$$\ln\left[-\ln(1-P_i)\right] = c\left(\ln x_{P_i} - \ln b\right); \quad i = 1, 2. \tag{13.2b}$$

Solving (13.2b) for $c$ gives

$$c = \frac{\ln\left[-\ln(1-P_1)\right] - \ln\left[-\ln(1-P_2)\right]}{\ln x_{P_1} - \ln x_{P_2}}. \tag{13.2c}$$

Therefore, a percentile estimator of $c$, based on two ordered sample values, is:

$$\widehat{c} = \frac{\ln\left[-\ln(1-P_1)\right] - \ln\left[-\ln(1-P_2)\right]}{\ln \widehat{x}_{P_1} - \ln \widehat{x}_{P_2}}. \tag{13.2d}$$

To simplify the notation we rewrite (13.2d) as

$$\widehat{c} = \frac{k}{\ln y_1 - \ln y_2}, \tag{13.2e}$$

where

$$k \quad := \quad \ln\left[-\ln(1-P_1)\right] - \ln\left[-\ln(1-P_2)\right], \tag{13.2f}$$

$$y_i \quad := \quad \widehat{x}_{P_i}; \quad i = 1, 2. \tag{13.2g}$$

The crucial point is the selection of $P_1$ and $P_2$. We shall determine $P_1$ and $P_2$ such that the variance of $\widehat{c}$ is minimum. Now $\widehat{c}$ is asymptotically normal with mean $c$ and variance

$$\mathrm{Var}\left(\widehat{c}\right) = \frac{c^2}{n\,k^2}\left[\frac{q_1}{k_1^2} + \frac{q_2}{k_2^2} - 2\frac{q_1}{k_1\,k_2}\right], \tag{13.2h}$$

where

$$q_i \quad := \quad \frac{P_i}{1-P_i}; \quad i = 1, 2; \tag{13.2i}$$

$$k_i \quad := \quad -\ln(1-P_i); \quad i = 1, 2. \tag{13.2j}$$

The minimizing values of $P_1$ and $P_2$ are found by iteration and are

$$P_1^*(c) = 0.167307, \qquad P_2^*(c) = 0.973664, \tag{13.2k}$$

and the minimum variance, which is independent of $b$, is

$$\min_{P_1, P_2} \mathrm{Var}\left(\widehat{c}\right) = \frac{0.916275\,c^2}{n}. \tag{13.2l}$$

We have shown that the 17th and the 97th sample percentiles asymptotically yield the percentile estimator of the shape parameter $c$ in a class of two–observation percentile estimators, where $b$ is unknown. Because the MLE of $c$ has a variance of $c^2/\left[n\,\psi'(1)\right] \approx 0.6079\,c^2/n$ (see (11.17)), we have an efficiency of about 66%.

From (13.2b) we may form a percentile estimator for the scale parameter in three obvious ways:

$$\widehat{b}_1 = \exp\left\{ \ln y_1 - \frac{\ln\left[ -\ln(1 - P_1)\right]}{\widehat{c}} \right\}, \tag{13.3a}$$

$$\widehat{b}_2 = \exp\left\{ \ln y_2 - \frac{\ln\left[ -\ln(1 - P_2)\right]}{\widehat{c}} \right\}, \tag{13.3b}$$

$$\widehat{b}_3 = \exp\left\{ \frac{1}{2} \sum_{i=1}^{2} \left[ \ln y_i - \frac{\ln\left[ -\ln(1 - P_i)\right]}{\widehat{c}} \right] \right\}, \tag{13.3c}$$

where $\widehat{c}$ is given by (13.2d). It is easily seen that (13.3a–c) are identical and that they can be expressed more conveniently as

$$\widehat{b} = \exp\{w \ln y_1 + (1 - w) \ln y_2\}, \tag{13.3d}$$

where

$$w := 1 - \frac{\ln k_1}{k}. \tag{13.3e}$$

We select $P_1$ and $P_2$ so as to minimize the variance of (13.3d). $\widehat{b}$ is asymptotically normal with mean $b$ and variance

$$
\begin{aligned}
\mathrm{Var}\big(\widehat{b}\big) &= \frac{b^2}{n c^2} \left\{ \frac{P_1}{1 - P_1} \frac{w}{k_1} \left[ \frac{w}{k_1} + \frac{2(1 - w)}{k_2} \right] + \frac{P_2}{1 - P_2} \frac{(1 - w)^2}{k_2^2} \right\} \\
&= \frac{b^2}{n c^2 k^2} \left\{ q_1 \frac{k - \ln k_1}{k_1} \left[ \frac{k - \ln k_1}{k_1} + \frac{2 \ln k_1}{k_2} \right] + q_2 \frac{\ln^2 k_1}{k_2^2} \right\}.
\end{aligned} \tag{13.3f}
$$

The minimizing values of $P_1$ and $P_2$ again have to be found by iteration:

$$P_1^*(b) = 0.3977778, \quad P_2^*(b) = 0.821111. \tag{13.4a}$$

The corresponding minimum variance of $\widehat{b}$ is

$$\min_{P_1, P_2} \mathrm{Var}\big(\widehat{b}\big) = \frac{1.359275 \, b^2}{n c^2}. \tag{13.4b}$$

We have found that 40th and the 82nd sample percentiles asymptotically yield the best percentile estimator of the scale parameter $b$, when the shape parameter is unknown, in a class of two–observation percentile estimators. The efficiency of $\widehat{b}$ with (13.4a) is about 82% when compared with the MLE of $b$, which has a variance of $b^2 \left(1 + \frac{\psi^2(2)}{\psi'(1)}\right) \big/ n c^2 \approx 1.1087 \, b^2 / (n c^2)$.

Notice, that the pairs $\{P_1^*(c), \, P_2^*(c)\}$ for optimum $\widehat{c}$ and $\{P_1^*(b), \, P_2^*(b)\}$ for optimum $\widehat{b}$ are not identical, i.e. we have to use four sample percentiles. If we want to use only the same two sample percentiles for estimating both $b$ and $c$, the orders $P_1^*(b, c)$ and $P_2^*(b, c)$ are determined such that the generalized variance (the determinant of the asymptotic variance–covariance matrix) is minimum. This matrix has the elements $\mathrm{Var}\big(\widehat{b}\,\big)$ and $\mathrm{Var}\big(\widehat{c}\,\big)$ given by (13.3f) and (13.2h) and

$$\mathrm{Cov}\big(\widehat{b},\widehat{c}\,\big) = \frac{b}{n\,k^2}\left\{\frac{q_1}{k_1}\left(\frac{k-\ln k_1}{k_2} - \frac{\ln k_1}{k_2} - \frac{k-\ln k_1}{k_1}\right) + \frac{q_2\,\ln k_1}{k_2^2}\right\}. \tag{13.5a}$$

DUBEY (1967e, p. 125) gives the minimizing orders as

$$P_1^*(b,c) = 0.238759, \quad P_2^*(b,c) = 0.926561 \tag{13.5b}$$

and a corresponding minimum generalized variance of $1.490259\,b^2/n^2$. The joint asymptotic efficiency of the percentile estimators based on the 24th and 93rd percentiles is about 41% when compared with their MLEs whose asymptotic generalized variance, i.e., the determinant of the matrix given in (11.17), is $0.6079\,b^2/n^2$.

---

**Excursus:   Percentile estimator of $c$ when $b$ is known and of $b$ when $c$ is known**

When only one of the two parameters is unknown, we need one percentile–equation for estimation.

$c$ to be estimated assuming $b = b_0$ to be known

A single–observation percentile estimator of $c$ is

$$\widehat{c}(b_0) = \frac{\ln\big[-\ln(1-P)\big]}{\ln\widehat{x}_P - \ln b_0}. \tag{13.6a}$$

The variance of $\widehat{c}(b_0)$ is

$$\mathrm{Var}\big(\widehat{c}\,|\,b_0\big) = c^2\,P/n\,(1-P)\,\ln^2(1-P)\,\ln^2[-\ln(1-P)]. \tag{13.6b}$$

The minimum variance of $1.9681\,c^2/n$ is attained with $P^*(c\,|\,b_0) = 0.1121$. Because the MLE of $c$, when $b$ is known, is given by $c^2/n\,\big[\psi'(1) + \psi^2(2)\big] \approx c^2/1.8237\,n$, we have an asymptotic efficiency of about 28%.

$b$ to be estimated assuming $c = c_0$ to be known

A single–observation percentile estimator of $b$ is

$$\widehat{b}(c_0) = \frac{\widehat{x}_P}{[-\ln(1-P)]^{1/c_0}} \tag{13.7a}$$

with variance

$$\mathrm{Var}\big(\widehat{b}\,|\,c_0\big) = \frac{b^2}{n\,c_0^2}\,\frac{P}{1-P}\,\ln^{-2}(1-P). \tag{13.7b}$$

The minimum of (13.7b) is $1.5446\,b^2/n\,c_0^2$ when $P^*(b\,|\,c_0) = 0.797$. The MLE of $b$ for $c$ known is $b^2/n\,c_0^2$ so that the asymptotic efficiency compared with the MLE is about 65%.

**Example 13/1:   Percentile estimates of $b$ and $c$ for dataset #1**

We will estimate $b$ and $c$ for dataset #1 given in Tab. 9/2 by using only one pair of sample percentiles whose orders are fixed in (13.5b). We have $\widehat{x}_{0.24} = x_{[20\cdot0.24]+1:20} = x_{5:20} = y_1 = 58$ and $\widehat{x}_{0.93} = x_{[20\cdot0.93]+1:20} = x_{19:20} = y_2 = 141$. From (13.2d) we get

$$\widehat{c} = \frac{\ln[-\ln 0.76] - \ln[-\ln 0.07]}{\ln 58 - \ln 141} \approx 2.5566,$$

which is rather near to the true value $c = 2.5$. To compute $\widehat{b}$ according to (13.3d), we first determine $w$, given $P_1 = 0.24$ and $P_2 = 0.93$:

$$w = 1 - \frac{\ln k_1}{k} = 1 - \frac{\ln[-\ln(1 - P_1)]}{\ln[-\ln(1 - P_1)] - \ln[-\ln(1 - P_2)]} \approx 0.4307.$$

Then we find

$$\widehat{b} = \exp\{0.4307 \ln 58 + 0.5693 \ln 141\} = 96.1743,$$

which is also near to its true value $b = 100$.

We conclude this section by giving percentile estimators that utilize all $n$ observations of the sample assuming $n$ is even, i.e., $n = 2\,j$:

$$\widehat{c} = \frac{\sum_{i=1}^{j}\{\ln[-\ln(1 - P_i)] - \ln[-\ln(1 - P_{i+1})]\}}{\sum_{i=1}^{j}(\ln y_i - \ln y_{i+1})}, \tag{13.8a}$$

where $y_i$ are sample percentiles corresponding to $0 < P_1 < P_2 < \ldots < P_n < 1$. In practice the $P_i$ are chosen in such a manner that $P_i = i/(n+1)$, corresponding to the ordered observations $y_1 < y_2 < \ldots < y_n$. The scale parameter is estimated by

$$\widehat{b} = \exp\left\{\frac{1}{n}\left[\sum_{i=1}^{n}\ln y_i - \frac{1}{\widehat{c}}\sum_{i=1}^{n}\ln[-\ln(1 - P_i)]\right]\right\}, \tag{13.8b}$$

where $\widehat{c}$ is given by (13.8a).

### 13.1.2   Three-parameter WEIBULL distribution

In order to derive percentile estimators of the three WEIBULL parameters, we need three percentiles. Suppose that $y_1$, $y_2$, $y_3$ are three sample percentiles according to (13.1b) and (13.2g) corresponding to three cumulative probabilities $0 < P_1 < P_2 < P_3 < 1$. Then the three estimators for $a$, $b$ and $c$ are derived by solving the system

$$y_s = \widehat{a} + \widehat{b}\,[-\ln(1 - P_s)]^{1/\widehat{c}}; \quad s = 1, 2, 3. \tag{13.9a}$$

It follows that

$$\frac{y_3 - y_2}{y_2 - y_1} = \frac{\left[-\ln(1 - P_3)\right]^{1/\widehat{c}} - \left[-\ln(1 - P_2)\right]^{1/\widehat{c}}}{\left[-\ln(1 - P_2)\right]^{1/\widehat{c}} - \left[-\ln(1 - P_1)\right]^{1/\widehat{c}}} \cdot \tag{13.9b}$$

$\widehat{c}$ is found from (13.9b) by iteration. Using this solution $\widehat{c}$ we further have

$$\widehat{b} = \frac{y_s - y_r}{\left[-\ln(1 - P_s)\right]^{1/\widehat{c}} - \left[-\ln(1 - P_r)\right]^{1/\widehat{c}}}; \quad (r < s) \in \{1, 2, 3\}, \tag{13.9c}$$

and with $\widehat{b}$ and $\widehat{c}$ we finally arrive at

$$\widehat{a} = y_s - \widehat{b}\left[-\ln(1 - P_s)\right]^{1/\widehat{c}}, \quad s \in \{1, 2, 3\}. \tag{13.9d}$$

We will not adhere to the estimators (13.9b–d), but recommend the approaches of DUBEY (1967f) and SCHMID (1997). This approach needs only the choice of $P_1$ and $P_3$ whereas $P_2$ follows from

$$-\ln(1 - P_2) = \sqrt{\left[-\ln(1 - P_1)\right]\left[-\ln(1 - P_3)\right]} \tag{13.10a}$$

as

$$P_2 = 1 - \exp\left\{-\sqrt{\left[-\ln(1 - P_1)\right]\left[-\ln(1 - P_3)\right]}\right\}. \tag{13.10b}$$

Further down we will show how to find optimal values of $P_1$ and $P_3$.

As $-\ln(1 - P_2)$ is the geometric mean of $-\ln(1 - P_1)$ and $-\ln(1 - P_3)$, it follows that $P_1 < P_2 < P_3$ and hence

$$a \leq x_{P_1} < x_{P_2} < x_{P_3} < \infty. \tag{13.10c}$$

To obtain a percentile estimator of the location parameter $a$, we note that (13.1a) and (13.10a) yield

$$\left(\frac{x_{P_2} - a}{b}\right)^c = \frac{\sqrt{(x_{P_1} - a)^c (x_{P_3} - a)^c}}{b^c}. \tag{13.11a}$$

Solving (13.11a) for $x_{P_2}$ and using the inequality between the geometric and arithmetic means gives

$$x_{P_2} < \frac{x_{P_1} + x_{P_3}}{2}, \tag{13.11b}$$

whereas solving (13.11a) for $a$ implies

$$a = \frac{x_{P_1} x_{P_3} - x_{P_2}^2}{x_{P_1} + x_{P_3} - 2 x_{P_2}}. \tag{13.11c}$$

The denominator of (13.11c) is not zero because of (13.11b). Replacing the percentiles by the sample percentiles $y_i = \widehat{x}_{P_i}$ $(i = 1, 2, 3)$, we obtain the percentile estimator of the location parameter proposed by DUBEY (1967f):

$$\widehat{a} = \frac{y_1 y_3 - y_2^2}{y_1 + y_3 - 2 y_2}, \quad \text{if } y_1 + y_3 \neq 2 y_2. \tag{13.11d}$$

Next we derive a percentile estimator of the shape parameter. With the help of (13.10a), equation (13.9b) turns into

$$\ln\left(\frac{y_3-y_2}{y_2-y_1}\right)=\ln\left\{\frac{\left[-\ln(1-P_3)\right]^{1/(2\widehat{c})}\left\{\left[-\ln(1-P_3)\right]^{1/(2\widehat{c})}-\left[-\ln(1-P_1)\right]^{1/(2\widehat{c})}\right\}}{\left[-\ln(1-P_1)\right]^{1/(2\widehat{c})}\left\{\left[-\ln(1-P_3)\right]^{1/(2\widehat{c})}-\left[-\ln(1-P_3)\right]^{1/(2\widehat{c})}\right\}}\right\}$$

(13.12a)

or equivalently

$$\widehat{c}=\frac{1}{2}\frac{\ln\left\{\dfrac{-\ln(1-P_3)}{-\ln(1-P_1)}\right\}}{\ln\left\{\dfrac{y_3-y_2}{y_2-y_1}\right\}}\,,\ \text{if}\ y_1\neq y_2,\ y_2\neq y_3\ \text{and}\ y_1+y_3\neq 2\,y_2.$$

(13.12b)

Contrary to (13.9b) we now have an explicit estimator of $c$.

There also exists an estimator of the scale parameter which is based on (13.10a). This estimator is given by SCHMID (1997) as

$$\widehat{b}=\frac{(y_2-y_1)^{2\,v}\,(y_3-y_2)^{2\,(1-v)}}{y_1+y_3-2\,y_2}\,,\ \text{if}\ y_1+y_3\neq 2\,y_2,$$

(13.13a)

where

$$v:=1-\frac{\ln\left[-\ln(1-P_1)\right]}{\ln\left[-\ln(1-P_1)\right]-\ln\left[-\ln(1-P_3)\right]}.$$

(13.13b)

Table 13/1: Optimal percentile orders for estimating $a$

| $c$ | Optimal values | | |
|---|---|---|---|
| | $P_1$ | $P_2$ | $(n/b^2)\,\text{Var}\,(\widehat{a})$ |
| 2.5 | 0.0011 | 0.9994 | 1.374 |
| 3.0 | 0.0037 | 0.9986 | 2.978 |
| 3.5 | 0.0062 | 0.9982 | 5.090 |
| 4.0 | 0.0084 | 0.9979 | 7.696 |
| 4.5 | 0.0102 | 0.9977 | 10.790 |
| 5.0 | 0.0118 | 0.9975 | 14.367 |
| 7.5 | 0.0169 | 0.9971 | 39.407 |
| 10.0 | 0.0196 | 0.9970 | 76.292 |

Source: SCHMID (1997, p. 781) — Reprinted with permission from
*Communications in Statistics — Theory and Methods.*
Copyright 1997 by Marcel Dekker, Inc. All rights reserved.

The estimators (13.11d), (13.12b) and (13.13a) are consistent and their joint asymptotic distribution is three–variate normal with mean vector $(a,b,c)'$ and a rather complicated looking variance–covariance matrix $V$ which is given in SCHMID (1997). This author has searched the minima of the variances directly. These minima and the corresponding optimal $P_1$ and $P_3$ depend on the shape parameter and are given in Tables 13/1 through 13/3. The dependence is not very sensitive (see also ZANAKIS/MANN (1982)), and for most sample sizes we get $y_1 = x_{1:n}$ and $y_3 = x_{n:n}$. The optimal pairs $(P_1, P_3)$ are different for each parameter. Table 13/4 gives an optimal pair for estimating all three parameters with the same set of percentiles. This table is based on minimizing the generalized variance of three estimators, i.e., the determinant of the variance–covariance matrix $V$.

Table 13/2:  Optimal percentile orders for estimating $b$

| $c$ | Optimal values | | |
|---|---|---|---|
| | $P_1$ | $P_2$ | $(n/b^2)\,\mathrm{Var}\big(\hat{b}\big)$ |
| 0.5 | 0.0797 | 0.8086 | 5.619 |
| 1.0 | 0.0103 | 0.7373 | 1.573 |
| 1.5 | $5 \cdot 10^{-6}$ | 0.6475 | 0.771 |
| 2.0 | 0.0002 | 0.4944 | 0.797 |
| 2.5 | 0.0007 | 0.3244 | 1.487 |
| 3.0 | 0.0063 | 0.99996 | 3.886 |
| 3.5 | 0.0082 | 0.9995 | 6.102 |
| 4.0 | 0.0103 | 0.9990 | 8.742 |
| 4.5 | 0.0119 | 0.9985 | 11.857 |
| 5.0 | 0.0132 | 0.9982 | 15.449 |
| 7.5 | 0.0175 | 0.9974 | 40.538 |
| 10.0 | 0.0199 | 0.9971 | 77.450 |

Table 13/3:  Optimal percentile orders for estimating $c$

| $c$ | Optimal values | | |
|---|---|---|---|
| | $P_1$ | $P_3$ | $n\,\mathrm{Var}\big(\hat{c}\big)$ |
| 0.5 | 0.0086 | 0.9746 | 0.230 |
| 1.0 | 0.0048 | 0.9817 | 1.028 |
| 1.5 | 0.0028 | 0.9887 | 3.155 |
| 2.0 | 0.0033 | 0.9920 | 9.096 |
| 2.5 | 0.0051 | 0.9932 | 23.215 |
| 3.0 | 0.0072 | 0.9939 | 51.070 |
| 3.5 | 0.0092 | 0.9944 | 99.544 |
| 4.0 | 0.0109 | 0.9947 | 176.936 |
| 4.5 | 0.0124 | 0.9949 | 292.964 |
| 5.0 | 0.0136 | 0.9951 | 458.760 |
| 7.5 | 0.0179 | 0.9957 | 2522.943 |
| 10.0 | 0.0202 | 0.9960 | 8314.388 |

<u>Table 13/4:</u>  Optimal percentile orders for estimating all three WEIBULL parameters

| $c$ | Optimal values | | |
|---|---|---|---|
| | $P_1$ | $P_3$ | $(n^3/b^4)\det \boldsymbol{V}$ |
| 1.5 | $5 \cdot 10^{-c}$ | 0.9487 | 0.09819 |
| 2.0 | 0.0025 | 0.9366 | 1.029 |
| 2.5 | 0.0076 | 0.9467 | 2.885 |
| 3.0 | 0.0126 | 0.9524 | 5.607 |
| 3.5 | 0.0167 | 0.9560 | 9.157 |
| 4.0 | 0.0202 | 0.9585 | 13.514 |
| 4.5 | 0.0231 | 0.9603 | 18.666 |
| 5.0 | 0.0255 | 0.9617 | 24.605 |
| 7.5 | 0.0333 | 0.9656 | 65.978 |
| 10.00 | 0.0372 | 0.9674 | 126.686 |

**Example 13/2:  Percentiles estimates of $a$, $b$ and $c$ for dataset #2**

The $n = 20$ observations of dataset #2 (Tab. 9/9) have been generated as $X_i \overset{\text{iid}}{\sim} We(15, 30, 2.5)$.
From Tab. 13/4 we choose $P_1 = 0.0076$ and $P_3 = 0.9467$ (for $c = 2.5$) and find from (13.10b):

$$P_2 = 1 - \exp\left\{-\sqrt{(-\ln 0.9924)(-\ln 0.0533)}\right\} \approx 0.1389.$$

The three sample percentiles are

$$
\begin{aligned}
y_1 &= \widehat{x}_{[20 \cdot 0.0076]+1:20} = \widehat{x}_{1:20} = 22.8, \\
y_2 &= \widehat{x}_{[20 \cdot 0.1389]+1:20} = \widehat{x}_{3:20} = 28.8, \\
y_3 &= \widehat{x}_{[20 \cdot 0.9467]+1:20} = \widehat{x}_{19:20} = 57.3.
\end{aligned}
$$

From (13.11d) we then find

$$\widehat{a} = \frac{y_1\, y_3 - y_2^2}{y_1 + y_3 - 2\, y_2} = \frac{22.8 \cdot 57.3 - 28.8^2}{22.8 + 57.3 - 2 \cdot 28.8} = 21.2.$$

As $\widehat{a} = 21.2 < x_{1:20}$ we have an admissible estimate, but it is not near to the true value $a = 15$.
From (13.12b) we get

$$\widehat{c} = \frac{1}{2}\,\frac{\ln\left\{\dfrac{-\ln 0.0533}{-\ln 0.9924}\right\}}{\ln\left\{\dfrac{57.3 - 28.8}{28.8 - 22.8}\right\}} \approx \frac{1}{2}\,\frac{\ln 384.2979}{\ln 4.75} \approx 1.9098.$$

This estimate differs considerably from the true value $c = 2.5$.

From (13a,b) we get

$$v = 1 - \frac{\ln[-\ln 0.9924]}{\ln[-\ln 0.9924] - \ln[-\ln 0.0533]} \approx 0.1807$$

and

$$\widehat{b} = \frac{(28.8 - 22.8)^{2 \cdot 0.1807} \, (57.3 - 28.8)^{2 \cdot 0.8193}}{22.8 + 57.3 - 2 \cdot 28.8} \approx 20.5565.$$

This estimate also differs considerably from the true value $b = 30$.

## 13.2 Minimum distance estimators[2]

The minimum distance method of parameter estimation is best explained by consideration of one of the simplest cases. Let $X_1, X_2, \ldots, X_n$ be iid variates with cumulative distribution $G$, thought to be an element of $\Gamma = \{F_{\boldsymbol{\theta}}, \boldsymbol{\theta} \in \Omega\}$, a parameterized set of continuous distribution functions, and let $G_n$ denote the usual empirical distribution function. Let $\delta(G_n, F_{\boldsymbol{\theta}})$ be some measure of the "distance" between $G_n$ and $F_{\boldsymbol{\theta}}$, such as

$$\delta_K(G_n, F_{\boldsymbol{\theta}}) = \sup_{x \in \mathbb{R}} \left| G_n(x) - F_{\boldsymbol{\theta}}(x) \right|, \tag{13.14a}$$

$$\delta_C(G_n, F_{\boldsymbol{\theta}}) = \int_{-\infty}^{\infty} \left[ G_n(x) - F_{\boldsymbol{\theta}}(x) \right]^2 \omega_{\boldsymbol{\theta}}(x) \, \mathrm{d}F_{\boldsymbol{\theta}}(x), \tag{13.14b}$$

$$\delta_A(G_n, F_{\boldsymbol{\theta}}) = \int_{-\infty}^{\infty} \left[ G_n(x) - F_{\boldsymbol{\theta}}(x) \right]^2 \left\{ F_{\boldsymbol{\theta}}(x) \left[ 1 - F_{\boldsymbol{\theta}}(x) \right] \right\}^{-1} \mathrm{d}F_{\boldsymbol{\theta}}(x), \tag{13.14c}$$

the KOLMOGOROV, the weighted CRAMÉR–VON MISES and the ANDERSON–DARLING discrepancies, respectively.[3]  When the weight function $\omega_{\boldsymbol{\theta}}(x)$ in (13.14b) is chosen as $\omega_{\boldsymbol{\theta}}(x) \equiv 1$ we have the (ordinary) CRAMÉR–VON MISES statistic. The ANDERSON–DARLING statistic is another special case of (13.14b), the weight function being $\omega_{\boldsymbol{\theta}}(x) \equiv \left\{ F_{\boldsymbol{\theta}}(x) \left[ 1 - F_{\boldsymbol{\theta}}(x) \right] \right\}^{-1}$.

The **minimum distance estimator** (MDE) of $\boldsymbol{\theta}$ is chosen to be any vector $\widehat{\boldsymbol{\theta}}_n$ in $\Omega$ such that

$$\delta\left(G_n, F_{\widehat{\boldsymbol{\theta}}_n}(x)\right) = \inf_{\boldsymbol{\theta} \in \Omega} \delta\left(G_n, F_{\boldsymbol{\theta}}\right), \tag{13.15}$$

---

[2]  Suggested reading for this section: CARMODY/EUBANK/LARICCIA (1984), GALLAGHER/MOORE (1990), GANDER (1996), HOBBS/MOORE/MILLER (1985), KAO (1964), LARICCIA (1982), PARR/SCHUCANY (1980).

[3]  For further measures of distance, see PARR/SCHUCANY (1980). These measures also serve as test statistics in a goodness–of–fit test (see Chapter 22).

a value $\widehat{\boldsymbol{\theta}}_n$ minimizing the distance between $G_n$ and $F_{\boldsymbol{\theta}}$. Discrepancies can be chosen to measure the distance between empirical and theoretical distribution functions, characteristic functions, hazard functions, density functions, quantile functions, or other such quantities.

Computation of MDEs based on the discrepancy between empirical and theoretical distribution functions and using the ordinary CRAMÉR–VON MISES discrepancy is quite simple. Applying the computational formula for this statistic, we observe that[4]

$$\delta_C\big(G_n, F_{\boldsymbol{\theta}}\big) = \frac{1}{12\,n} + \sum_{i=1}^{n} \left[\frac{2\,i - 1}{2\,n} - F_{\boldsymbol{\theta}}(X_{i:n})\right]^2. \tag{13.16}$$

Thus, computation of the MDE becomes a non–linear least–squares problem. Most simply, the data can be fed into a non–linear regression program with $(2\,i - 1)/(2\,n)$ playing the role of the dependent variable[5] and $F_{\boldsymbol{\theta}}(X_{i:n})$ the role of the model.

**Example 13/3:  MDE of $b$ and $c$ for dataset #1**

We have applied (13.16) to dataset #1 of Tab. 9/2, using the GAUSS–NEWTON method with analytical derivatives of $F(x\,|\,b, c) = 1 - \exp\big\{-(x/b)^c\big\}$ :

$$\frac{\partial F(x\,|\,b, c)}{\partial b} = -\frac{c}{b}\left(\frac{x}{b}\right)^c \exp\left\{-\left(\frac{x}{b}\right)^c\right\},$$

$$\frac{\partial F(x\,|\,b, c)}{\partial c} = \left(\frac{x}{b}\right)^2 \ln\left(\frac{x}{b}\right) \exp\left\{-\left(\frac{x}{b}\right)^c\right\}.$$

The estimates turned out as follows: $\widehat{b} = 97.6428$, $\widehat{c} = 2.6428$, both being close to the true values $b = 100$ and $c = 2.5$. Fig. 13/1 shows $F(x\,|\,97.6428, 2.6428)$, marked by $\blacklozenge$, and the staircase function $(2\,i - 1)/(2\,n)$, marked by $\bullet$. The CRAMÉR-VON MISES statistic is the sum of the squared vertical distances between $\blacklozenge$ and $\bullet$ plus $1/(12\,n)$ and amounts to 0.0208.

MDEs are consistent and robust, i.e., insensitive with respect to outliers and to the choice of an incorrect distribution; see GALLAGHER/MOORE (1990). When the sample size is sufficiently great so that grouping makes sense the $\chi^2$–minimum method (see KAO (1964)) may be applied as another technique of MDE.

---

[4] When the ANDERSON–DARLING discrepancy is used, the computational formula is

$$\delta_A\big(G_n, F_{\boldsymbol{\theta}}\big) = -n - \sum_{i=1}^{n} \frac{2\,i - 1}{n}\left\{\ln F_{\boldsymbol{\theta}}(X_{i:n}) + \ln\big[1 - F_{\boldsymbol{\theta}}(X_{n+1-i:n})\big]\right\}.$$

[5] $(2\,i - 1)/(2\,n)$ is the midpoint plotting position (see Tab. 9/1).

Figure 13/1: Empirical distribution function and MD-estimated WEIBULL CDF function of dataset #1

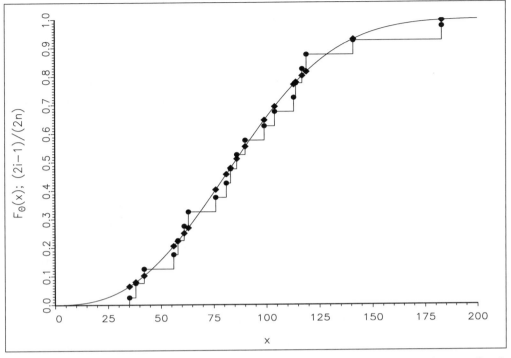

CARMODY et al. (1984) have suggested a **minimum quantile distance estimator** for the three–parameter WEIBULL distribution. Let

$$Q(P) = -\ln(1-P), \ \ 0 < P < 1 \tag{13.17a}$$

denote the quantile function of the reduced exponential distribution. Then the quantile function corresponding to the three–parameter WEIBULL distribution with $\boldsymbol{\theta} = (a, b, c)'$ is

$$Q(P, \boldsymbol{\theta}) = a + b\, Q(p)^{1/c}. \tag{13.17b}$$

Define the sample quantile function by

$$\widehat{Q}(P) = X_{i:n} \ \ \text{for} \ \ \frac{i-1}{n} < P \le \frac{i}{n}; \ \ i = 1, 2, \ldots, n. \tag{13.17c}$$

For a given set of $k < n$ percentile points $\boldsymbol{P} = (P_1, \ldots, P_k)'$ with $0 < P_1 < \ldots < P_k < 1$, let

$$\widehat{\boldsymbol{Q}}_{\boldsymbol{P}} = \left( \widehat{Q}(P_1), \ldots, \widehat{Q}(P_k) \right)'$$

and

$$\boldsymbol{Q}_{\boldsymbol{P}}(\boldsymbol{\theta}) = \left( Q(P_1, \boldsymbol{\theta}), \ldots, Q(P_k, \boldsymbol{\theta}) \right)'.$$

Then the minimum quantile distance estimator of $\boldsymbol{\theta}$ is the vector that minimizes the quadratic form

$$D(\boldsymbol{\theta}) \equiv \left( \widehat{\boldsymbol{Q}}_{\boldsymbol{P}} - \boldsymbol{Q}_{\boldsymbol{P}}(\boldsymbol{\theta}) \right)' \boldsymbol{W}(\boldsymbol{\theta}) \left( \widehat{\boldsymbol{Q}}_{\boldsymbol{P}} - \boldsymbol{Q}_{\boldsymbol{P}}(\boldsymbol{\theta}) \right) \tag{13.17d}$$

as a function of $\boldsymbol{\theta}$. A good choice of the user defined matrix of weights $\boldsymbol{W}(\boldsymbol{\theta})$ suggested by CARMODY et al. (1984) is

$$\boldsymbol{W}^*(g) = \boldsymbol{H}_P(g)\,\boldsymbol{R}_P^{-1}\,\boldsymbol{H}_P(g) \tag{13.17e}$$

where

$$\boldsymbol{R}_P = \min(P_i, P_j) - P_i\,P_j$$

and $\boldsymbol{H}_P(g)$ the $k \times k$ diagonal matrix having its $i$–th diagonal element be $g(1 - P_i)\,Q(P_i)^{(c-1)/c}$. Since $\boldsymbol{R}_P$ is well known to have a tridiagonal inverse, whose typical row has non–zero entries

$$-(P_i - P_{i-1})^{-1}, \quad \frac{P_{i+1} - P_i}{(P_{i+1} - P_i)\,(P_i - P_{i-1})}, \quad -(P_{i+1} - P_i)^{-1},$$

the elements of (13.17e) can easily be evaluated.

The estimator $\widehat{\boldsymbol{\theta}}$ has no closed form but can be computed without difficulty by using a standard minimization routine. The estimator is unique, consistent and asymptotically normal. It should also be noted that, provided the $\widehat{Q}(P_i)$'s are selected from the uncensored portions of the data, this estimation technique requires no modification for use with type–II right, left or doubly censored samples.

## 13.3    Some hybrid estimation methods

Hybrid methods are characterized by the fact that the estimators for the two or three WEIBULL parameters, do not originate in the same approach. We have already encountered two such hybrid methods in Sections 11.3.2.3 and 12.2 where the MLEs and MMEs for $b$ and $c$ were combined with another kind of estimating approach for the location parameter $a$.

Perhaps the oldest hybrid technique was proposed by WEIBULL himself; see WEIBULL (1967a). He combined the maximum likelihood method and the linear estimation method based on order statistics. The method starts with an unbiased estimation of the scale and location parameters for a properly chosen set of shape parameter values by use of the best linear method. For each such set of three parameter values, the corresponding likelihood of the sample is computed. Finally, that set which yields the maximum likelihood is determined by interpolation and accepted as the best estimate.

Most of the hybrid methods which are used today combine simple estimators of the three parameters as given in Sect. 13.5. ZANAKIS (1979) has run a simulation study for seventeen simple estimators. Among the estimators considered the best in a MSE–sense are the following:

- for the location parameter $a$:

$$\widehat{a} = \frac{X_{1:n}\,X_{n:n} - X_{2:n}^2}{X_{1:n} + X_{n:n} - 2\,X_{2:n}} \,. \tag{13.18}$$

$\widehat{a}$ is the nearest neighbor of $X_{1:n}$. It is almost always permissable; i.e., $\widehat{a} \le X_{1:n}$ because for all practical purposes, $X_{2:n}$ is closer to $X_{1:n}$ than $X_{n:n}$. Although unlikely, the opposite might occur if the sample size is extremely small. If that happens, then use $\widehat{a} = X_{1:n}$.

- for the scale parameter $b$:

$$\widehat{b} = X_{[0.63\,n]+1:n} - \widehat{a} \qquad (13.19a)$$

with $\widehat{a}$ according to (13.18). For smaller values of $c$ HASSANEIN's (1972) ABLE combined with the MANN/FERTIG (1977) small–sample correction (see Sect. 10.5.2) is preferred:

$$\widehat{b} = \exp\left\{ \sum_{i=1}^{10} a_i \ln(X_{n_i:n} - \widehat{a}) + 0.000595\,\widehat{c} \right\} \qquad (13.19b)$$

with $\widehat{a}$ according to (13.18) and $\widehat{c}$ according to

$$\widehat{c} = 0.979811 \Big/ \sum_{i=1}^{10} b_i \ln(X_{n_i:n} - \widehat{a}). \qquad (13.19c)$$

The weights $a_i$, $b_i$ are given in Tab. 10/10 together with the spacings $\lambda_i$ leading to $n_i = [n\,\lambda_i]$. For smaller $c$ and smaller sample size one should take the simple linear unbiased estimator for the two parameters of the extreme–value (and two–parameter WEIBULL) distributions, proposed by ENGELHARDT/BAIN (1977) and extended to the three–parameter WEIBULL as follows:

$$\widehat{b} = \exp\left\{ \frac{0.5772}{\widehat{c}} + \frac{1}{n} \sum_{i=1}^{n} \ln\left(X_{i:n} - \widehat{a}\right) \right\} \qquad (13.19d)$$

with $\widehat{a}$ according to (13.18) and $\widehat{c}$ according to

$$\widehat{c} = \frac{n\,k_n}{-\sum_{i=1}^{s} \ln\left(X_{i:n} - \widehat{a}\right) + \frac{s}{n-s} \sum_{i=s+1}^{n} \ln X_{i:n}} \qquad (13.19e)$$

where $s = [0.84\,n]$ and $k_n$ as unbiasing factor (see also Sect. 10.4.3 for this approach).

- for the shape parameter $c$:

$$\widehat{c} = \frac{\ln\left[-\ln(1-P_2)\right] - \ln\left[-\ln(1-P_1)\right]}{\ln\left(\widehat{x}_{P_2} - \widehat{a}\right) - \ln\left(\widehat{x}_{P_1} - \widehat{a}\right)}. \qquad (13.20)$$

This is the percentile estimator (13.2d) modified by $\widehat{a}$ according to (13.18) and with $P_1 = 0.16731$ and $P_2 = 0.97366$. However, (13.19c) and (13.19e) have a slight edge if $c \leq 1.0$.

WYCKOFF et al. (1980) proposed the following procedure for the complete–sample case:[6] The first order statistic $X_{1:n}$ is used as an initial estimate of $a$. An initial estimate of $c$,

---

[6] They also give formulas for a type–II censored sample, but these formulas need tabulated coefficients depending on $n$ and the censoring number $r$.

based on $\widehat{a}_1 = X_{1:n}$, is derived from DUBEY's (1967e) formula

$$\widehat{c}_1 = \frac{2.9888}{\ln\left(\widehat{x}_{P_2} - X_{1:n}\right) - \ln\left(\widehat{x}_{P_1} - X_{1:n}\right)}$$

which is identical to (13.20). In order to re–estimate $a$, $X_{1:n}$ is set equal to its expected value

$$\begin{aligned} E(X_{1:n}) &= a + b\,\Gamma\left(1 + \frac{1}{c}\right)\Big/ n^{1/c} \\ &= a + \left[E(X) - a\right]\Big/ n^{1/c}. \end{aligned}$$

Estimating $E(X)$ by $\overline{X} = \dfrac{1}{n}\sum\limits_{i=1}^{n} X_i$ and $c$ by $\widehat{c}_1$ leads to

$$\widehat{a}_2 = \frac{X_{1:n} - \overline{X}\, n^{-1/\widehat{c}_1}}{1 - n^{-1/\widehat{c}_1}}. \tag{13.21}$$

The parameters $b$ and $c$ are then estimated by assuming $a = \widehat{a}_2$ and using the estimators (3.19d,e) of ENGELHARDT/BAIN (1977). KAPPENMAN (1985b) compared the WYCKOFF et al. (1980) estimators with the modified maximum likelihood estimators MMLE–I and MMLE–II of COHEN/WHITTEN (1982) (see Sect. 11.3.2.3) by running a Monte Carlo experiment for $c = 0.5(0.5)2.5, 3.5$ with 500 repetitions each. The MSE's for $\widehat{a}$ and $\widehat{c}$ using the ENGELHARDT/BAIN–procedure were always considerably smaller than those for the MMLEs. The same was true for the $b$–estimators, except for $c = 0.5$. These conclusions are also essentially true for bias comparisons, except for the case where $c = 3.5$.

KAPPENMAN (1981) further compared the estimation procedure recommended by ZANAKIS (1979) and consisting of (13.18) for $\widehat{a}$, (3.19a) for $\widehat{b}$ and (13.20a) for $\widehat{c}$ to a procedure where he substituted (13.19a) by the following estimator of $a$:

$$\widehat{a} = 2\,X_{1:n} - (e-1)\sum_{i=1}^{n} \frac{X_{i:n}}{e^i}. \tag{13.22}$$

This alternative estimator was essentially derived by COOKE (1979) although he was not working on the WEIBULL estimation problem per se. It is derived by expressing the expectation of the first order statistic, $X_{1:n}$, in terms of an integral, the integrand being $X_{1:n}$ times its density. Then one uses integration by parts, replaces $E(X_{1:n})$ by $X_{1:n}$, replaces the population distribution function by the empirical distribution function and solves for $a$.[7] The results of KAPPENMAN's Monte Carlo study indicate that the estimator (13.22) is better in the sense of MSE than the one given by (13.18) when the true value of $c$ is at least one. When the shape parameter's value is less than one, the two estimators appear to perform equally well. In addition, this simulation study indicated that the performance of the estimators given by (13.19a) and (13.20) are improved when $\widehat{a}$ is substituted by (13.22). Thus we recommend the following hybrid procedure as apparently the best one:

---

[7] For more details of this estimator see, Sect. 13. 5.1

- Estimate $a$ by (13.22),

- Insert this estimate into (13.19a) to estimate $b$,

- Insert the estimator of $a$ into (13.20) to estimate $c$,

We successfully applied this procedure to establish capability indices for the tensile strength of ultrasonically welded splices; see RINNE et al. (2004).

## 13.4 Miscellaneous approaches

There have appeared several more procedures to estimate the WEIBULL parameters. We do not persist in the completeness of the following enumeration.

### 13.4.1 MENON's estimators

A very popular method has been proposed by MENON (1963). **MENON's estimators** are used as initial values in most computer programs to estimate $b$ and $c$ by ML. These estimators are nothing but the re–transformed moment estimators of the parameters of the Log–WEIBULL distribution (see Sect. 3.3.4). If $X \sim We(0, b, c)$, then $X^* = \ln X$ has the type–I–minimum distribution (or Log–WEIBULL distribution) with density

$$f(x^*) = \frac{1}{b^*} \exp\left\{ \frac{1}{b^*} \left( x^* - a^* \right) - \exp\left[ \frac{1}{b^*} \left( x^* - a^* \right) \right] \right\}, \quad x^* \geq 0, \qquad (13.23a)$$

distribution function

$$F(x^*) = 1 - \exp\left\{ \exp\left[ \frac{1}{b^*} \left( x^* - a^* \right) \right] \right\}, \qquad (13.23b)$$

mean

$$E(X^*) = a^* - \gamma\, b^*, \quad \gamma \approx 0.577216, \qquad (13.23c)$$

and variance

$$\mathrm{Var}(X^*) = \frac{b^{*^2} \pi^2}{6} \approx 1.644934\, b^{*^2}. \qquad (13.23d)$$

The parameters $b$ and $c$ are linked to $a^*$ and $b^*$ as

$$a^* = \ln b \quad \text{and} \quad b = \exp(a^*), \qquad (13.24a)$$

$$b^* = \frac{1}{c} \quad \text{and} \quad c = \frac{1}{b^*} . \qquad (13.24b)$$

We estimate $E(X^*)$ by

$$\widehat{E(X^*)} = \overline{X^*} = \frac{1}{n} \sum_{i=1}^{n} \ln X_i \qquad (13.25a)$$

and $\mathrm{Var}(X^*)$ by

$$\widehat{\mathrm{Var}(X^*)} = S_{X^*}^2 = \frac{1}{n-1} \sum_{i=1}^{n} \left( \ln X_i - \overline{X^*} \right)^2. \qquad (13.25b)$$

Thus the moment estimators of $a^*$ and $b^*$ are

$$\widehat{b^*} = \frac{\sqrt{6}}{\pi} S_{X^*} \approx 0.779697 \, S_{X^*}, \tag{13.26a}$$

$$\widehat{a^*} = \overline{X^*} + \gamma \, \widehat{b^*} \approx \overline{X^*} + 0.450054 \, S_{X^*}. \tag{13.26b}$$

MENON (1963) has shown that

- $\widehat{b^*}$ is asymptotically normal with mean

$$\mathrm{E}\left(\widehat{b^*}\right) = b^* + b^* \, O(1/n) \tag{13.27a}$$

and variance

$$\mathrm{Var}\left(\widehat{b^*}\right) = \frac{1.1 \, b^{*2}}{n} + b^{*2} \, O(1/n), \tag{13.27b}$$

and has asymptotic efficiency of 55% as the CRAMÉR–RAO lower bound of a regular unbiased estimate of $b^*$ is $0.61 \, b^{*2}/n$;

- $\widehat{a^*}$ is asymptotically normal with mean

$$\mathrm{E}\left(\widehat{a^*}\right) = a^* + b^* \, O(1/n) \tag{13.28a}$$

and variance

$$\mathrm{Var}\left(\widehat{a^*}\right) = \frac{1.2 \, b^{*2}}{n} + b^{*2} \, O(1/n^2), \tag{13.28b}$$

and has asymptotic efficiency of 92% as the CRAMÉR–RAO lower bound of a regular unbiased estimate of $a^*$ is $1.1 \, b^{*2}/n$.

Referring to (13.23a,b) and (13.25a) through (13.27b), we can state that MENON's estimators of the original WEIBULL parameters are

$$\widehat{c} = \frac{1}{\widehat{b^*}} \tag{13.29a}$$

with

$$\widehat{c} \overset{\mathrm{asym}}{\sim} No\left(c, \frac{1.1 \, c^2}{n}\right), \tag{13.29b}$$

and

$$\widehat{b} = \exp\left(\widehat{a^*}\right) \tag{13.30a}$$

with

$$\widehat{b} \overset{\mathrm{asym}}{\sim} No\left(b, \frac{1.2 \, b^2}{n \, c^2}\right). \tag{13.30b}$$

**Example 13/4:** MENON's estimates of $b$ and $c$ for dataset #1

MENON's approach applied to dataset #1 ($X_i \overset{iid}{\sim} We(0, 100, 2.5)$ for $i = 1, 2, \ldots, n$; see Tab. 9/2) yields,

$$\overline{x^*} = 4.3886, \quad s^2_{X^*} = 0.1949,$$

$$\widehat{b^*} = 0.3442, \quad \widehat{a^*} = 4.7328,$$

$$\widehat{c} = 2.9050, \quad \widehat{b} = 113.6125,$$

$$\widehat{\mathrm{Var}}(\widehat{b^*}) = 0.0065, \quad \widehat{\mathrm{Var}}(\widehat{a^*}) = 0.0071,$$

$$\widehat{\mathrm{Var}}(\widehat{c}) = 0.4642, \quad \widehat{\mathrm{Var}}(\widehat{b}) = 91.7710.$$

Both estimates $\widehat{b}$ and $\widehat{c}$ are rather different from the true values $b = 100$ and $c = 2.5$.

## 13.4.2 Block estimators of HÜSLER/SCHÜPBACH

The block estimators proposed by HÜSLER/SCHÜPBACH (1986) and SCHÜPBACH/HÜSLER (1983) are a further development of BAIN's (1972) GLUEs (see Sect. 10.4.3). They consist of linear combinations of the partial means of two or three blocks of the ordered Log–WEIBULL observations $X^*_{i:n} = \ln X_{i:n}$.

When two blocks are chosen, these are

$$X^*_{1:n}, \ldots, X^*_{s:n} \quad \text{and} \quad X^*_{s+1:n}, \ldots, X^*_{r:n}$$

with $r < n$. The estimators of $b^* = 1/c$ and $a^* = \ln b$, given in SCHÜPBACH/HÜSLER(1983), are

$$\widehat{b^*} = -\frac{s}{n\, b_{r,n}} \left( \frac{1}{s} \sum_{i=1}^{s} X^*_{i:n} \right) + \frac{s}{n\, b_{r,n}} \left( \frac{1}{r-s} \sum_{i=s+1}^{r} X^*_{i:n} \right) \quad (13.31\text{a})$$

$$\widehat{a^*} = \frac{1}{r} \sum_{i=1}^{r} X^*_{i:n} - c_{r,n}\, \widehat{b^*}, \quad (13.31\text{b})$$

where $b_{r,n}$ and $c_{r,n}$ are unbiasing constants and $s$ is chosen to minimize the variance of $\widehat{b^*}$. The estimators have good small sample and asymptotic efficiencies.

In HÜSLER/SCHÜPBACH (1986) the idea is extended to three blocks:

$$X^*_{1:n}, \ldots, X^*_{s:n} \quad \text{with mean} \quad \overline{X^*_1} = \frac{1}{s} \sum_{i=1}^{s} X^*_{i:n},$$

$$X^*_{s+1:n}, \ldots, X^*_{t:n} \quad \text{with mean} \quad \overline{X^*_2} = \frac{1}{t-s} \sum_{i=s+1}^{t} X^*_{i:n},$$

$$X^*_{t+1:n}, \ldots, X^*_{r:n} \quad \text{with mean} \quad \overline{X^*_3} = \frac{1}{r-t} \sum_{i=t+1}^{r} X^*_{i:n}.$$

The estimators are

$$\widehat{b^*} \;=\; d_1\,\overline{X_1^*} + d_2\,\overline{X_2^*} + d_3\,\overline{X_3^*}, \tag{13.32a}$$

$$\widehat{a^*} \;=\; e_1\,\overline{X_1^*} + e_2\,\overline{X_2^*} + e_3\,\overline{X_3^*}. \tag{13.32b}$$

The factors $\{d_i\}$ and $\{e_i\}$ are chosen such that $\mathrm{E}(\widehat{b^*}) = b^*$ and $\mathrm{E}(\widehat{a^*}) = a^*$. Furthermore, if $s$ and $t$ are also optimally chosen (possible criteria are given in HÜSLER/SCHÜPBACH (1986)), $\widehat{b^*}$ and $\widehat{a^*}$ according to (13.32a,b) are more efficient than the two–block estimators (13.31a,b).

### 13.4.3   KAPPENMAN's estimators based on the likelihood ratio

Let $X_1, \ldots, X_n$ represent a sample of a continuous type variate whose density function is $f(x \mid b, c)$, where $b$ and $c$ are unknown parameters and $b$ a scale parameter and $c$ not a location parameter. The most powerful scale invariant test of $H_1 : c = c_1$ versus $H_2 : c = c_2$ rejects $H_1$ whenever $L_1/L_2 < k$ (This is the idea of a likelihood–ratio test.), where

$$L_j = \int_0^\infty \left[ \prod_{i=1}^n f(v\,x_i \mid 1, c_j) \right] v^{n-1} \mathrm{d}v; \quad j = 1, 2. \tag{13.33}$$

This test suggests a procedure for selecting one from among several plausible values of $c$, say $c_1, \ldots, c_m$. If $L_p = \max\{L_1, \ldots, L_m\}$, where $L_j$ $(j = 1, \ldots, m)$ is given by (13.33), then the selected value is $c_p$. Based on this idea KAPPENMAN (1985a) suggests a method for estimating $c$. The estimate of $c$ is the value of $c$ which maximizes

$$L(c) = \int_0^\infty \left[ \prod_{i=1}^n f(v\,x_i \mid 1, c) \right] v^{n-1} \mathrm{d}v. \tag{13.34a}$$

(13.34a) turns into

$$L(c) = c^{n-1} \left( \prod_{i=1}^n x_i^{c-1} \right) \Gamma(n) \left/ \left( \sum_{i=1}^n x_i^c \right)^n \right. \tag{13.34b}$$

when $f(x \mid b, c)$ is the two–parameter WEIBULL density. The value of $c$ which maximizes $L(c)$ is the solution to the equation

$$\frac{n-1}{c} + \sum_{i=1}^n \ln x_i - \frac{n \sum_{i=1}^n x_i^c \ln x_i}{\sum x_i^c} = 0, \tag{13.34c}$$

which is the partial derivative of the logarithm of $L(c)$, set equal to zero. (13.34c) is slightly different from (11.19b) in finding the MLE of $c$. (11.19b) is obtained from (13.34c) by replacing $n-1$ in the first terms on the left–hand side by $n$. Both estimates must be worked out by an iterative procedure, e.g., by the NEWTON–RAPHSON method; see (11.25a).

KAPPENMAN has done a Monte Carlo investigation to compare the performance of the MLE of $c$ with that of $\widehat{c}$ according to (13.34c). The bias and the MSE for the solution to (13.34c) are smaller than the bias and MSE, for the MLE for sample sizes up to $n = 100$.

As far as estimation of the scale parameter $b$ is concerned, there are at least three possibilities. The MLE for $b$ is given by (11.19a). One might also use (11.19a), replacing the MLE of $c$ by the solution to (13.34c). The third possibility is $\exp\{\widehat{a}\}$, where $\widehat{a}$ is the value of $a$ which maximizes

$$\int_0^\infty \left[ \prod_{i=1}^n \exp\{v\,(x_i^* - a) - \exp[v\,(x_i^* - a)]\} \right] v^{n-1}\,dv,$$

where $x_i^* = \ln x_i$. Simulation studies of KAPPENMAN indicate that all three estimators perform equally well, as far as bias and MSE are concerned.

### 13.4.4  KAPPENMAN's estimators based on sample reuse

Let $X_1, \ldots, X_n$ be a random sample of $n$ observations of a variate with density function $f(x \mid b, c)$, where $b$ and $c$ represent unknown parameters. Given that $X_j = x_j$, for $j = 1, 2, \ldots, i-1, i+1, \ldots, n$, an estimated density function for $X_i$ is, for any given value of $c$, $f(x_i \mid \widehat{b}_i, c)$. Here $\widehat{b}_i = g(x_1, \ldots, x_{i-1}, x_{i+1}, \ldots, n; c)$ is the ML estimate of $b$, based upon the given value of $c$ and the observed values of $X_1, \ldots, X_{i-1}, X_{i+1}, \ldots, X_n$. Set

$$L(c \mid x_1, \ldots, x_n) = \prod_{i=1}^n f(x_i \mid \widehat{b}_i, c), \qquad (13.35)$$

where $x_i$ is the observed value of $X_i$, for $i = 1, \ldots, n$. Then the **sample reuse estimate** (SRE) of $c$ is the value of $c$ which maximizes (13.35) or, equivalently, the logarithm of (13.35). If this value is $\widehat{c}$, then $g(x_1, \ldots, x_n; \widehat{c})$ is the sample reuse estimate of $b$.

Note that SR estimation of $c$ is similar to, but different from, ML estimation of $c$.[8] To obtain the SRE of $c$, one essentially reuses each sample observation $n - 1$ times for estimation of $b$ whereas for ML estimation of $c$, each observation is used only once for estimation of $c$. That is, if $\widehat{b}_i$ in (13.35), is based upon all $n$ sample observations, instead of just $n - 1$ of them (and is then the same for all $i$), then the value of $c$ maximizing (13.35) is the MLE of $c$.

For the two–parameter WEIBULL density $f(x \mid b, c) = (c/b)\,(x/b)^{c-1}\,\exp\{-(x/b)^c\}$ we get — see (11.19a) — the following MLE

$$\widehat{b}_i = \left[ \frac{1}{n-1} \sum_{j \neq i} x_j^c \right]^{1/c}. \qquad (13.36a)$$

If we find the partial derivative of the logarithm of (13.35) and set it equal to zero, we obtain — compare that to (11.19b) — the equation

$$\frac{n}{c} + \sum_{i=1}^n \frac{\left[ (n-1) - \sum_{j \neq i} q_{ij}^c \right] \sum_{j \neq i} q_{ij}^c \ln q_{ij}}{\left( \sum_{j \neq i} q_{ij}^c \right)^2} = 0, \qquad (13.36b)$$

---

[8]  When there is only one unknown parameter to be estimated, SR estimation is equivalent to ML estimation.

where $q_{ij} = x_j/x_i$. The value of $c$ satisfying (13.36b) and to be found by an application of the NEWTON–RAPHSON method is the SR estimate of $c$.

A Monte Carlo study of KAPPENMAN revealed that the bias and MSE of the SRE of $c$ are substantially smaller than the bias and MSE, of the MLE for sample sizes as large as $n = 75$.

### 13.4.5   Confidence intervals for $b$ and $c$ based on the quantiles of beta distributions

The following approach, which is due to BRIKC (1990), gives only interval estimates of $b$ and $c$ and no point estimates. The distribution function $F(x) = 1 - \exp\{-(x/b)^c\}$ may be written as

$$c\left[\ln x - \ln b\right] = \ln \ln \left(\frac{1}{1 - F(x)}\right) =: k(F). \tag{13.37}$$

Define the events $A$, $B$ in terms of two values $x_1$, $x_2$; $x_1 < x_2$ as follows:

$$\Pr(A) = \Pr(X \le x_1) = F(x_1), \tag{13.38a}$$

$$\Pr(B) = \Pr(X \le x_2) = F(x_2). \tag{13.38b}$$

If $\widehat{F}_{11}$, $\widehat{F}_{12}$ are the lower and upper confidence limits for $F(x_1)$ and $\widehat{F}_{21}$, $\widehat{F}_{22}$ are the corresponding values for $F(x_2)$, and if these limits are known, then confidence limits $\widehat{c}_1$, $\widehat{c}_2$ for $c$ and $\widehat{b}_1$, $\widehat{b}_2$ for $b$ follow by

$$
\left.
\begin{aligned}
\widehat{c}_2 \left(\ln x_1 - \ln \widehat{b}_2\right) &= \ln \ln \frac{1}{1 - \widehat{F}_{11}} =: k_{11}, \\[2mm]
\widehat{c}_1 \left(\ln x_1 - \ln \widehat{b}_1\right) &= \ln \ln \frac{1}{1 - \widehat{F}_{12}} =: k_{12}, \\[2mm]
\widehat{c}_1 \left(\ln x_2 - \ln \widehat{b}_2\right) &= \ln \ln \frac{1}{1 - \widehat{F}_{21}} =: k_{21}, \\[2mm]
\widehat{c}_2 \left(\ln x_2 - \ln \widehat{b}_1\right) &= \ln \ln \frac{1}{1 - \widehat{F}_{22}} =: k_{22};
\end{aligned}
\right\} \tag{13.39}
$$

and

$$
\left.
\begin{aligned}
\widehat{c}_1 &= \frac{k_{22}\,k_{21} - k_{11}\,k_{12}}{(k_{11} + k_{22})\,(\ln x_2 - \ln x_1)}, \\[3mm]
\widehat{c}_2 &= \frac{k_{22}\,k_{21} - k_{11}\,k_{12}}{(k_{12} + k_{21})\,(\ln x_2 - \ln x_1)}, \\[3mm]
\widehat{b}_1 &= \exp\left\{\frac{k_{12}\,(k_{11} + k_{22})\,\ln x_2 - k_{22}\,(k_{12} + k_{21})\,\ln x_1}{k_{11}\,k_{12} - k_{21}\,k_{22}}\right\}, \\[3mm]
\widehat{b}_2 &= \exp\left\{\frac{k_{11}\,(k_{12} + k_{21})\,\ln x_2 - k_{21}\,(k_{11} + k_{22})\,\ln x_1}{k_{11}\,k_{12} - k_{21}\,k_{22}}\right\}.
\end{aligned}
\right\} \tag{13.40}
$$

$(1 - \alpha_1 - \alpha_2)\, 100\%$ confidence limits for $P = \Pr(A) = F(x_1)$ are given by

$$\widehat{P}_1 \;=\; 1 - \beta_{n-r,r+1,\alpha_1}, \tag{13.41a}$$

$$\widehat{P}_2 \;=\; \beta_{r+1,n-r,\alpha_2}, \tag{13.41b}$$

where $\beta_{.,\,.,\,\alpha_1}$ and $\beta_{.,\,.,\,\alpha_2}$ are the upper $(\alpha_1, \alpha_2)$ quantiles of a beta distribution with parameters equal to the first two subscripts. $n$ is the sample size, $r$ is the number of $x$–values which are less than or equal to $x_1$. The procedure may also be carried out for $P = \Pr(B) = F(x_2)$.

It is only left to determine the values of $x_1$ and $x_2$. They are obtained by ordering the sample $x_{1:n} \le x_{2:n} \le \ldots \le x_{n:n}$ and obtaining two numbers $r_1$, $r_2$ by

$$r_1 = 1 + [0.2\,n], \quad r_2 = [0.8\,n], \tag{13.42}$$

which are stated to be appropriate values. $x_1, x_2$ are then defined to be

$$\left.\begin{aligned}
x_1 &= \frac{x_{r_1:n} + x_{r_1+1:n}}{2}, \\[2mm]
x_2 &= \frac{x_{r_2:n} + x_{r_2+1:n}}{2}.
\end{aligned}\right\} \tag{13.43}$$

Thus the algorithm may be summarized as follows, given $\alpha_1$ and $\alpha_2$:

1. Determine $r_1$ and $r_2$ by (13.42) and find $x_1$ and $x_2$ by (13.43).

2. Use a table of the inverse beta distribution to find confidence limits for $\Pr(A) = F(x_1)$ and $\Pr(B) = F(x_2)$ by (13.41a,b).

3. Determine $k_{11}$, $k_{12}$, $k_{21}$ and $k_{22}$ by (13.39).

4. Solve (13.40) to obtain the confidence intervals to the two parameters.

### 13.4.6   Robust estimation [9]

ADATIA/CHAN (1982) propose two estimators of the scale parameter of the two–parameter WEIBULL distribution: the **maximin estimator** (based on a mixture of WEIBULL distributions and the procedure of constructing the BLIE) and the **adaptive estimator** (based on an adaptive procedure which chooses between the maximin estimator and the BLIE from individual WEIBULL distributions). These two estimators are robust in the sense that their efficiencies are high regardless of the true value of the shape parameter, provided that it is in a prescribed interval. These two estimators are also more robust than many commonly used estimators including the MLE, BLIE and BLUE. ADATIA/CHAN (1985) extended their 1982 approach to all three parameters of the three–parameter WEIBULL distribution.

The MLEs are very sensitive to the occurrence of upper and lower outliers, especially when $c > 1$. HE/FUNG (1999) have considered the **method of medians estimator** for the two–parameter WEIBULL model. The estimator is obtained by equating the sample median of the likelihood score function — see (11.7a) — to its population counter part. As an

---

[9]  Suggested reading for this section: HUBER (1981).

M–estimator, it has a bounded influence function and is highly robust against outliers. It is easy to compute as it requires solving only one equation instead of a pair of equations as for most other M–estimators. Furthermore, no assumptions or adjustments are needed for the estimator when there are some possibly censored observations at either end of the sample. About 16% of the largest observations and 34% of the smallest observations may be censored without affecting the calculations.

### 13.4.7 Bootstrapping [10]

With complete or singly type–II censored samples, confidence intervals for the WEIBULL parameters can be computed based on published tables of pivotal percentage points (see Sect. 11.3.1.2). These intervals are exact, apart from the Monte Carlo sampling error inherent in the tables. When the censoring is progressive, however, the exact methods generally do not carry through. ROBINSON (1983) has proposed a simple approximation to handle the progressive censoring case. The algorithm for computing confidence limits is a type of bootstrapping. The method is applicable to any model that is transformable to location–scale form and has invariant estimators.

A paper of SEKI/YOKOYAMA (1996) proposes bootstrap robust estimator methods for the WEIBULL parameters. It applies bootstrap estimators of order statistics to the parametric estimation procedure. Estimates of the WEIBULL parameters are equivalent to the estimates using the extreme–value distribution. Therefore, the bootstrap estimators of order statistics for the parameters of the extreme–value distribution are examined. Accuracy and robustness for outliers are studied using Monte Carlo experiments which indicate adequate efficiency of the proposed estimators for data with some outliers.

## 13.5 Further estimators for only one of the WEIBULL parameters

Many estimators exist for each of the WEIBULL parameters which are mostly heuristic and do not presuppose a preceding estimation of the other parameter(s). Most of these estimators are easily and quickly evaluated from the observations available and thus serve as starting points for more sophisticated estimation procedures.

### 13.5.1 Location parameter[11]

We will present three estimators of the location parameter which do not rest upon estimators of the scale and the shape parameters and which have easy to evaluate analytic formulas. These estimators are the following:

- the sample minimum,
- COOKE's (1979) estimator and
- ZANAKIS' (1979) estimator

---

[10]  Suggested reading for this section: EFRON (1979, 1982).

[11]  Suggested reading for this section: COOKE (1979), DUBEY (1966c, 1967f), KAPPENMAN (1981), WYCKOFF/BAIN/ENGELHARDT (1980), ZANAKIS (1979).

The location parameter $a$ is a threshold and lower bound to the variate. Thus is quite natural to take the **sample minimum** $X_{1:n}$ as an estimator $\hat{a}$:

$$\hat{a} = \min_{1 \le i \le n} \{X_i\} = X_{1:n} . \tag{13.44}$$

Before exploring the statistical properties of $\hat{a} = X_{1:n}$, we remark that (13.44) is the MLE of $a$ for $0 < c \le 1$. Looking at the normal equations (11.4d–f), we see from (11.4d)

$$\frac{\partial \mathcal{L}(a,b,c)}{\partial a} = -(c-1) \sum_{i=1}^{n} \frac{1}{x_i - a} + \frac{c}{b} \sum_{i=1}^{n} \left( \frac{x_i - a}{b} \right)^{c-1}$$

that $\partial \mathcal{L}(a,b,c)/\partial a > 0$ for $0 < c \le 1$, which implies that the log–likelihood function (11.4a) is monotone increasing in $a$. Thus the maximum value of $\mathcal{L}(a,b,c)$ occurs at the maximum admissible value of $a$ which is clearly the first ordered sample observation.

When $X_i \overset{iid}{\sim} We(a,b,c)$, we have $Y := \min_{1 \le i \le n} \{X_i\} \sim We(a, b\, n^{-1/c}, c)$, as has been shown in Sect. 3.1.4. So the estimator (13.44) has a WEIBULL distribution with the same location and scale parameters as the sampled distribution but with a scale parameter

$$\tilde{b} = b\, n^{-1/c} < b \quad \text{for } n > 1. \tag{13.45a}$$

The density of $Y := \hat{a} = X_{1:n}$ is

$$f(y) = \frac{c}{\tilde{b}} \left( \frac{y-a}{\tilde{b}} \right)^{c-1} \exp\left\{ - \left( \frac{y-a}{\tilde{b}} \right)^{c} \right\} . \tag{13.45b}$$

The moments about zero are

$$E(Y^r) = \sum_{j=0}^{r} \binom{r}{j} a^j \, \tilde{b}^{r-j} \, \Gamma\left( \frac{r-j}{c} + 1 \right), \tag{13.45c}$$

leading to

$$E(Y) = a + \tilde{b}\, \Gamma\left( 1 + \frac{1}{c} \right) = a + \frac{b}{\sqrt[c]{n}} \Gamma\left( 1 + \frac{1}{c} \right) \tag{13.45d}$$

and

$$\begin{aligned}
\text{Var}(Y) &= \tilde{b}^2 \left[ \Gamma\left( 1 + \frac{2}{c} \right) - \Gamma^2\left( 1 + \frac{1}{c} \right) \right] \\
&= \frac{b^2}{\sqrt[c]{n^2}} \left[ \Gamma\left( 1 + \frac{2}{c} \right) - \Gamma^2\left( 1 + \frac{1}{c} \right) \right] . \tag{13.45e}
\end{aligned}$$

We see from (13.45d) that $\hat{a}$ is not an unbiased estimator of $a$; however, it approaches $a$ extremely rapidly if $c$ is not too near unity. The bias can be removed leading to

$$\hat{a}_u = \hat{a} - \frac{b}{\sqrt[c]{n}} \Gamma\left( 1 + \frac{1}{c} \right) \tag{13.45f}$$

when $b$ and $c$ are known. Applying estimators $\hat{b}$ and $\hat{c}$ to (11.45f) does not guarantee unbiasedness. From (13.45e) we see that the variance of $\hat{a}$ (or of $\hat{a}_u$) approaches zero extremely rapidly for $0 < c < 1$. Even for $c = 1$ (exponential distribution), it reduces to

$$\text{Var}(\widehat{a}) = \text{Var}(\widehat{a}_u) = \frac{b^2}{n^2}, \tag{13.45g}$$

which is of order $O(n^{-2})$. An estimator possessing such property, is called **hyper–efficient**.[12] It is further seen from (13.45a,b) that $\widehat{a} = \min\{X_i\}$ does not tend to a normal distribution with $n \to \infty$ for $0 < c \leq 1$. In other words the common asymptotic properties of the MLE do not hold good here.

We now turn to the estimator of $a$ which was proposed by KAPPENMAN (1981) and is based upon COOKE (1979). The approach is not restricted to the three–parameter WEIBULL distribution but is valid for each variate $X$ with support $(a, \theta)$, where $a$ is known to be finite. The estimator of $a$ to be constructed applies whether $\theta$ is unknown or known; in particular, when $\theta = \infty$, the only assumption required is $\int_a^\infty x^2 \, \mathrm{d}F(x) < \infty$. We start by considering $\widehat{a} = Y = X_{1:n}$ in (13.44) which overestimates $a$; see (13.45d). The CDF of $Y$, when sampling from a population with distribution $F_X(x)$ is

$$F_Y(y) = \big[1 - F_X(y)\big]^n = \big[R_X(y)\big]^n \tag{13.46a}$$

so

$$\mathrm{E}(Y) = -\int_a^\theta y \, \mathrm{d}\big[R_X(y)\big]^n. \tag{13.46b}$$

When we integrate (11.46b) by parts, we get

$$\mathrm{E}(Y) = a + \int_a^\theta \big[R_X(y)\big]^n \mathrm{d}y \tag{13.46c}$$

or

$$a = \mathrm{E}(Y) - \int_a^\theta \big[1 - F_X(y)\big]^n \mathrm{d}y. \tag{13.46d}$$

(13.46d) suggest, the estimator

$$\widehat{a}_C = X_{1:n} - \int_{x_{n:n}}^{x_{n:n}} \big[1 - \widehat{F}_{(}y)\big]^n \mathrm{d}y, \tag{13.46e}$$

where $\widehat{F}_X(y)$ is the empirical distribution function based on the order statistics $X_{i:n}$; i.e.,

$$\widehat{F}_X(y) = \begin{cases} 0 & \text{for } y < X_{i:n}; \\ \dfrac{i}{n} & \text{for } X_{i:n} \leq y < X_{i+1:n}; \; i = 1, \ldots, n-1; \\ 1 & \text{for } y \geq X_{n:n}. \end{cases} \tag{13.46f}$$

---

[12] An efficient estimator has a variance of order $O(n^{-1})$.

Now

$$\int_{x_{1:n}}^{x_{n:n}} \left[1 - \widehat{F}_X(y)\right]^n \, dy = \sum_{i=1}^{n-1} (1 - \frac{i}{n})^n \, (x_{i+1:n} - x_{i:n}) \tag{13.46g}$$

and (14.46e) turns into

$$\widehat{a}_C = 2 \, X_{1:n} - \sum_{i=1}^{n} \left[\left\{1 - \frac{i-1}{n}\right\}^n - \left\{1 - \frac{i}{n}\right\}^n\right] X_{i:n}. \tag{13.46h}$$

For large $n$ we can approximate (11.46h):

$$\widehat{a}_C = 2 \, X_{1:n} - (e - 1) \sum_{i=1}^{n} e^{-i} \, X_{i:n}. \tag{13.46i}$$

ZANAKIS (1979) recommends — based on Monte Carlo studies — an estimator which uses three order statistics: $X_{1:n}$; $X_{2:n}$ and $X_{n:n}$; see Sect. 13.3 and (13.18):

$$\widehat{a}_Z = \frac{X_{1:n} \, X_{n:n} - X_{2:n}^2}{X_{1:n} + X_{n:n} - 2 \, X_{2:n}} .$$

This estimator is almost always permissible ($\widehat{a}_Z \leq X_{1:n}$) and might be regarded as a special percentile estimator; see (13.11d). A similar estimator introduced by DUBEY (1966g) is

$$\widehat{a}_D = \frac{X_{1:n} \, X_{3:n} - X_{2:n}^2}{X_{1:n} + X_{3:n} - 2 \, X_{2:n}} .$$

Confidence bounds for the threshold $a$ based on special goodness–of–fit statistics are presented by MANN/FERTIG (1975b) and SOMERVILLE (1977).

### 13.5.2 Scale parameter

Simple estimators of the scale parameter always presuppose the knowledge of the shape parameter or of its estimate; e.g., see ZANAKIS (1979) and Sect. 13.3. We will present two estimators of $b$:

- the one–order estimator of MOORE/HARTER (1965, 1966) with known shape parameter and

- the estimator of SIRVANCI (1984) with estimated shape parameter under singly type–I censoring.

Looking at the two–parameter WEIBULL density function

$$f(x) = \frac{c}{b} \left(\frac{x}{b}\right)^{c-1} \exp\left\{-\left(\frac{x}{b}\right)^c\right\}$$

and making the change of variable

$$Y = X^c, \tag{13.47a}$$

we obtain

$$g(y) = \frac{1}{b^c} \exp\left\{-\frac{y}{b^c}\right\}, \tag{13.47b}$$

which is the familiar exponential density function with parameter

$$\lambda = b^c. \tag{13.47c}$$

Now if we take the time to the $r$–th failure $X_{r:n}$ for the WEIBULL population and calculate $Y_{r:n} = X_{r:n}^c$, then the $Y_{r:n}$'s are exponentially distributed where $\lambda = b^c$. Therefore, if we want a one–order statistic estimator for $b$, we can use the best unbiased estimator of HARTER (1961):

$$\widehat{\lambda} = w_r\, Y_{r:n} = w_r\, X_{r:n}^c, \tag{13.47d}$$

where

$$w_r = \sum_{j=1}^{r} \frac{1}{n-j+1}. \tag{13.47e}$$

Now taking the $c$–th root of both sides of (13.47c) and using (13.47d), we get

$$\widehat{b} = \widehat{\lambda}^{1/c} = w_r^{1/c}\, X_{r:n}. \tag{13.47f}$$

A feature of type–I censoring, which distinguishes it from type–II censoring, is that — in addition to the failure times, $X_{1:n} \leq X_{2:n} \leq \ldots \leq X_{m:n} \leq T$ — the failure count $m$ is also observed as part of the data. This provides the opportunity of using the failure count as a basis for statistical inference. For this, the **censoring level** $P$ is expressed in terms of the parameters $b$ and $c$ as

$$P = 1 - \exp\left\{-\left(\frac{T}{b}\right)^c\right\}. \tag{13.48a}$$

Assuming that estimators $\widehat{P}$ and $\widehat{c}$ are available; (13.48a) can be solved for $b$ obtaining the estimator, see SIRVANCI (1984).

$$\widehat{b} = \frac{T}{\left[-\ln(1-\widehat{P})\right]^{1/\widehat{c}}}, \quad 0 < m < n. \tag{13.48b}$$

An estimator of $P$ is

$$\widehat{P} = \frac{m}{n}, \tag{13.48c}$$

the proportion of failed items in $(0, T]$. For the reciprocal shape parameter

$$c_1 := 1/c,$$

SIRVANCI/YANG (1984) have derived the estimator

$$\widehat{c_1} = \sum_{i=1}^{m} (\ln T - \ln X_{i:n})/m\, h(P); \quad 0 < m < n, \tag{13.48d}$$

where

$$h(P) = \ln\ln\left(\frac{1}{1-P}\right) - \frac{1}{P} \int_{0}^{P} \ln\ln\left(\frac{1}{1-t}\right) dt \tag{13.48e}$$

is a bias correction factor. SIRVANCI (1984) has derived the exact moments of

$$\ln \widehat{b} = \ln T - \widehat{c}_1 \, \ln\ln\left(\frac{1}{1-P}\right),$$  (13.48f)

conditioned on the observed number of failures $m$. He also demonstrated that $\ln \widehat{b}$ is asymptotically normal.

### 13.5.3 Shape parameter

The shape parameter $c$ as being responsible for the type of aging has attracted the greatest interest of statisticians, so it is quite natural that there exists a lot of estimators for this parameter. We will present the following estimation approaches:

- tabular estimators based on descriptive statistics measuring skewness and kurtosis (= peakedness), proposed by DUBEY (1966g, 1967c);

- estimators for the special case that there are two failures in one or more samples; see JAECH (1964) and BAIN (1973);

- estimators of the reciprocal shape parameter, but not derived from the Log–WEIBULL distribution; see MIHRAM (1969) and MURTHY/SWARTZ (1975);

- estimators based on the ratio of the sample arithmetic and geometric means; see MIHRAM (1973) and CHEN (1997);

- shrunken estimators as proposed by SINGH/BHATKULIKAR (1978), PANDEY (1983), PANDEY/SINGH (1984), PANDEY et al. (1989) and PANDEY/SINGH (1993).

In Sect. 12.1.2 we have seen that the skewness coefficient based on moments [13]

$$\alpha_3 = \frac{\mu_3}{\mu_2^{3/2}} = \frac{\Gamma_3 - 3\,\Gamma_2\,\Gamma_1 + 2\,\Gamma_1^3}{(\Gamma_2 - \Gamma_1^2)^{3/2}}$$

depends only on $c$ and may be used to estimate $c$ by equating the sample skewness

$$\frac{\frac{1}{n}\sum(X_i - \overline{X})^3}{\left[\frac{1}{n}\sum(X_i - \overline{X})^2\right]^{3/2}}$$

to $\alpha_3$ and solving for $c$. There exist different statistics to measure skewness (see Sect. 2.9.4). DUBEY (1967c) has proposed the following **measures of skewness** which depend only on $c$ and which may be tabulated to find an estimate of $c$ when the population measure is

---

[13]   Remember: $\Gamma_i := \Gamma(1 + i/c)$ and $\widehat{\Gamma}_i := \Gamma(1 + i/\widehat{c})$.

equated to the sample measure:

$$S_1 = \frac{\text{mean} - \text{median}}{\text{standard deviation}} = \frac{\mu - x_{0.5}}{\sigma} = \frac{\Gamma_1 - (\ln 2)^{1/c}}{(\Gamma_2 - \Gamma_1^2)^{0.5}}, \tag{13.49a}$$

$$S_2 = \frac{\text{mean} - \text{mode}}{\text{standard deviation}} = \frac{\mu - x^*}{\sigma}$$

$$= \begin{cases} \dfrac{\Gamma_1 - (1 - \frac{1}{c})^{1/c}}{(\Gamma_2 - \Gamma_1^2)^{0.5}} & \text{for} \quad c > 1 \\[3mm] \dfrac{\Gamma_1}{(\Gamma_2 - \Gamma_1^2)^{0.5}} & \text{for} \quad 0 < c \le 1 \end{cases}, \tag{13.49b}$$

$$S_3 = \frac{\text{median} - \text{mode}}{\text{standard deviation}} = \frac{x_{0.5} - x^*}{\sigma}$$

$$= \begin{cases} \dfrac{(\ln 2)^{1/c} - (1 - \frac{1}{c})^{1/c}}{(\Gamma_2 - \Gamma_1^2)^{0.5}} & \text{for} \quad c > 1 \\[3mm] \dfrac{(\ln 2)^{1/c}}{(\Gamma_2 - \Gamma_1^2)^{0.5}} & \text{for} \quad 0 < c \le 1 \end{cases}, \tag{13.49c}$$

$$S_4 = \frac{(x_{1-\alpha} - x_{0.5}) - (x_{0.5} - x_\alpha)}{x_{1-\alpha} - x_\alpha}$$

$$= \frac{\left[ -\ln \alpha \right]^{1/c} - 2 \left[ \ln 2 \right]^{1/c} - \left[ -\ln(1-\alpha) \right]^{1/c}}{\left[ -\ln \alpha \right]^{1/c} - \left[ -\ln(1-\alpha) \right]^{1/c}}, \quad 0 < \alpha < 0.5. \tag{13.49d}$$

We add the following remarks to these measures:

- For a continuous distribution function, we have $-1 \le S_1 \le +1$.
- The three measures are not independent, e.g.,

$$S_3 = S_2 - S_1.$$

- When $S_2$ or $S_3$ is to be used, we should have a large sample size to find the sample mode from the grouped data.
- $S_4$ is a percentile estimator and thus a generalization of the estimators found in Sect. 13.1.2.

The **kurtosis coefficient** $\alpha_4$ based on moments (see (2.101)) depends only on $c$:

$$\alpha_4 = \frac{\mu_4}{\mu_2^2} = \frac{\Gamma_4 - 4\,\Gamma_3\,\Gamma_1 + 6\,\Gamma_2\,\Gamma_1^2 - 3\,\Gamma_1^4}{\left(\Gamma_2 - \Gamma_1^2\right)^2}$$

and may be used to estimate $c$ by equating the sample kurtosis

$$\frac{\frac{1}{n} \sum (X_i - \overline{X})^4}{\left[ \frac{1}{n} \sum (X_i - \overline{X})^2 \right]^2}$$

to $\alpha_4$ and solving for $c$. A percentile oriented measure of kurtosis is; see (2.102c):

$$
\begin{aligned}
L &= \frac{x_{0.975} - x_{0.025}}{x_{0.75} - x_{0.25}} \\
&= \frac{\left[ -\ln 0.025 \right]^{1/c} - \left[ -\ln 0.975 \right]^{1/c}}{\left[ -\ln 0.25 \right]^{1/c} - \left[ -\ln 0.75 \right]^{1/c}} .
\end{aligned}
\tag{13.50}
$$

When there are two failures in a sample of size $n$ occurring at times $X_{1:n}$ and $X_{2:n}$, JAECH (1964) gives the following estimator of $c$ :

$$
\widehat{c} = \frac{1}{\ln Z} ,
\tag{13.51a}
$$

based on the ratio of time at failure for the second failure item to time at failure for first failed item

$$
Z = \frac{X_{2:n}}{X_{1:n}} .
\tag{13.51b}
$$

This estimator is a special case $(r = 2)$ of the GLUE of BAIN/ENGELHARDT (see Sect. 10.4.3), as shown by BAIN (1973). When there are $k$ samples of size $n$ each with two failures, JAECH (1964) gives the estimator

$$
\widehat{c}_k = \frac{k}{\sum\limits_{i=1}^{k} \ln Z_i} ,
\tag{13.51c}
$$

where $Z_i$ is the ratio of $X_{2:n}$ to $X_{1:n}$ in the $i$–th sample. A generalization of (13.51c) for samples of unequal sizes $n_i$ is found in BAIN (1973).

The moment estimator of $c$ in the two–parameter case, derived from the ratio $\left[ \mathrm{E}(X) \right]^2 / \mathrm{E}(X^2) = \Gamma_1^2 / \Gamma_2$; see (Sect. 12.1.1) and proposed by MIHRAM (1977) has its origin in the estimator of the reciprocal $c_1 := 1/c$; see MIHRAM (1969). Another estimator of $c_1$ has been set up by MURTHY/SWARTZ (1975). Motivated by the percentile representation (see (13.2c)):

$$
c = \frac{\ln\ln \left[ 1/(1 - P_1) \right] - \ln\ln \left[ 1/(1 - P_2) \right]}{\ln x_{P_1} - \ln x_{P_2}} ,
$$

they proposed for the reciprocal shape parameter $c_1$:

$$
\widehat{c}_1 = \left( \ln X_{k:n} - \ln X_{\ell:n} \right) B(n, k, \ell),
\tag{13.52}
$$

where $B(n, k, \ell)$ is an unbiasing factor determined by $\mathrm{E}\big(\widehat{c}_1\big) = c_1$. The choice of the orders $k$ and $\ell$ has been made to minimize the variance of the estimator. The efficiency $\mathrm{Eff}\big(\widehat{c}_1\big)$ relative to the CRAMÉR–RAO lower limit approaches 70% with $n \to \infty$.

Let

$$\overline{X}_a = \frac{1}{n} \sum_{i=1}^{n} X_i,$$

$$\overline{X}_g = \left\{ \prod_{i=1}^{n} X_i \right\}^{1/n} = \prod_{i=1}^{n} X_i^{1/n}$$

be the arithmetic and geometric means, respectively. When $X_i \stackrel{iid}{\sim} We(0, b, c)$, we notice that

$$\mathrm{E}(\overline{X}_a) = \mathrm{E}(X) = b\Gamma_1, \quad \mathrm{Var}(\overline{X}_a) = \frac{b^2}{n} \left[ \Gamma_2 - \Gamma_1^2 \right],$$

$$\mathrm{E}(\overline{X}_g) = b\,\Gamma^n\left(1 + \frac{1}{n\,c}\right), \quad \mathrm{Var}(\overline{X}_g) = b^2 \left\{ \Gamma^n\big(1 + 2/(n\,c)\big) - \Gamma^{2n}\big(1 + 1/(n\,c)\big) \right\}.$$

The **ratio of the arithmetic mean to the geometric mean**,

$$R = \frac{\overline{X}_a}{\overline{X}_g}, \tag{13.53a}$$

could be expected to have a distribution independent of the scale parameter $b$. MIHRAM (1973) gives the mean of $R$ for finite sample size $n$ as

$$\mathrm{E}(R) = \rho_n(c) = \Gamma\left(1 + \frac{n-1}{n\,c}\right)\, \Gamma^{n-1}\big(1 - 1/(n\,c)\big), \tag{13.53b}$$

provided that $c > 1/n$, and for $n \to \infty$ as

$$\lim_{n \to \infty} \rho_n(c) = \rho_\infty(c) = \Gamma\left(1 + \frac{1}{c}\right)\, \exp\{-\psi(1)/c\}, \tag{13.53c}$$

where $\psi(1) = -\gamma \approx -0.57722$. MIHRAM suggests as a WEIBULL shape estimator

$$\widehat{c} = \rho_\infty^{-1}(R), \tag{13.53d}$$

which can be read from Fig. 13/2.

The estimator given in (13.53d) is consistent for $c$, but as shown by MIHRAM the asymptotic efficiency is higher than 0.5 as long as $c \le 1$, but declines rapidly when $c > 1$.

CHEN (1997) has built a confidence interval for an estimator that uses a sample statistic analogous to $R$ in (13.53a) and is valid for type–II censored samples. His estimator starts from the reduced exponentially distributed order statistic

$$Y_{i:n} = \left(\frac{X_{i:n}}{b}\right)^c \tag{13.54a}$$

Figure 13/2: $\rho_\infty(c) = \Gamma(1 + 1/c) \, \exp\{-\psi(1)/c\}$

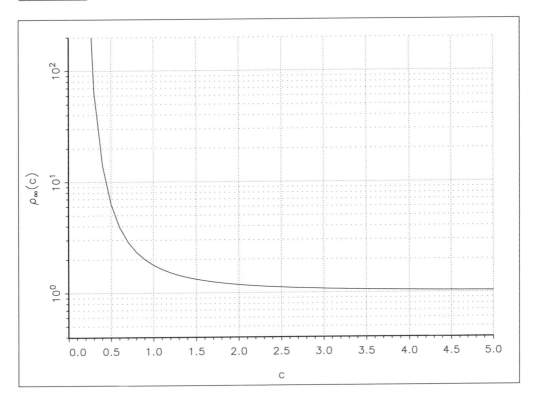

and forms the pseudo arithmetic mean

$$\overline{Y}_a^* = \frac{1}{n} \left[ \sum_{i=1}^{r-1} Y_{i:n} + (n - r + 1) \, Y_{r:n} \right] \tag{13.54b}$$

and the pseudo geometric mean

$$\overline{Y}_g^* = \left( \prod_{i=1}^{r-1} Y_{i:n} \, Y_{r:n}^{n-r+1} \right)^{1/n}. \tag{13.54c}$$

When $r = n$ (no censoring), (13.54b,c) will turn into proper means. The ratio

$$\xi(c; n, r) := \frac{\overline{Y}_a^*}{\overline{Y}_g^*} = \frac{\frac{1}{n} \left[ \sum_{i=1}^{r-1} Y_{i:n}^c + (n - r + 1) \, Y_{r:n}^c \right]}{\left( \prod_{i=1}^{r-1} Y_{i:n} \, Y_{r:n}^{n-r+1} \right)^{c/n}} \tag{13.54d}$$

is seen not to depend on any parameter and its distribution may be found by Monte Carlo simulation. Let $\xi_{1-\alpha}(n, r)$ be the upper $1 - \alpha$ critical value, then

$$\Pr\left[ \xi_{\alpha/2}(n, r) \le \xi(c; n, r) \le \xi_{1-\alpha/2}(n, r) \right] = 1 - \alpha \tag{13.54e}$$

offers a $(1 - \alpha) \, 100\%$ confidence interval for $\xi(c; n, r)$. The corresponding confidence interval $(c_L, c_U)$ for the shape parameter may be found with $c_L$ and $c_U$ as solutions of $c$ for the equations

$$\xi(c; n, r) = \xi_{\alpha/2}(n, r); \quad \xi(c; n, r) = \xi_{1-\alpha/2}(n, r).$$

CHEN (1997) gives a table of percentiles $\xi_P(n, r)$.

**Shrunken estimation** is on the borderline between classical statistical inference and BAYESIAN inference. The idea of shrunken estimation of the shape parameter $c$ is to combine a guessed or a priori given value $c_o$ with an estimate $\widehat{c}$, which is coming from observed sample data.[14] The papers which have been written on this topic (in chronological order of a publishing: SINGH/BHATKULILAR (1978), PANDEY (1983), PANDEY/SINGH (1984), PANDEY et al. (1989) and PANDEY/SINGH (1993)) all have in common that the data supported estimate $\widehat{c}$ is derived from BAIN's (1972) GLUE of $b^* = 1/c$; see (10.47). Sometimes the result of a test on the validity of $c_0$ is incorporated into the estimation procedure. We will give no details on shrunken estimation technique as we believe that it is more fruitful to apply BAYES estimation (see Chapter 14) when there exists prior knowledge on the parameters to be estimated.

## 13.6    Comparisons of classical estimators

The existence of the great number of estimators for the WEIBULL parameters raises a double question: What estimator is the best one, and how to measure its quality? — The answers depend on the sample size.

In the **asymptotic case** ($n \rightarrow \infty$ or $n$ very large), the benchmark is given by the CRAMÉR–RAO lower bound of the variance of unbiased estimators and this bound is reached by the MLEs; see (11.17) for the two–parameter case and (11.12) for the three–parameter case when the sample is uncensored. Other estimators do not reach this lower bound and their asymptotic relative efficiency is lower than $100\%$, as has been shown when presenting these estimators in the preceding sections and chapters.

However, the superiority of the MLEs is not guaranteed when the **sample size** is finite and **small**. The size of the systematic error of an estimator (the bias) and the size of its random or sample error (measured by the variance or the standard deviation) or the size of the combined systematic and random errors (measured by the mean square error, MSE, or its root, RMSE) cannot be given by closed and analytical formulas when estimating WEIBULL parameters. So a great number of papers have been written on Monte Carlo experiments to gain insight into the comparative behavior of several estimators. Most of the Monte Carlo investigations refer to the two–parameter WEIBULL distribution and to complete or single type–II censored samples and we will discuss these studies in some detail[15] and in

---

[14]   The difference between shrunken estimation and BAYESIAN estimation is that the former one works without a prior distribution of $c$ which expresses the uncertainty about $c$.

[15]   A comparison of the ML and moment estimation techniques for the three–parameter and complete sample case has been made by SAYLOR (1977). His results are as follows: The MM location parameter estimators are more efficient than the ML location parameter estimators in all tests. The MMEs appear to be more efficient than the MLEs for sample sizes less than or equal to 50. But SAYLOR's spectrum of tested parameter values and sample sizes is not wide and the number of Monte Carlo repetitions is rather

chronological order.

BAIN/ANTLE (1967) compared their estimators (10.31c,d) with those given by GUMBEL (see (10.32c,d)), by MILLER/FREUND (see Sect. 10.4.1), and by MENON (see Sect. 13.4.1). Comparisons were made using the variances and the biases of the estimators for complete samples of size $n = 5(5)30$. All of these estimators were fairly good. MENON's estimators are good if the bias is removed from the estimator of $c$ for small $n$. MENON's estimators seem better for larger sample sizes although they could not be used for censored data.

Monte Carlo investigations of HARTER/MOORE (1968) indicate that MLEs and BLUEs for $b^* = 1/c$ have nearly equal MSEs. For estimating $a^* = \ln b$ the two methods are also comparable except when $n$ is very small, in which case the linear estimator is slightly preferable. HARTER/MOORE state that since the MSEs of MLEs have been shown to differ but little from those of the BLIEs and to be considerably smaller than those of the BLUEs in some cases, especially those involving strongly asymmetric censoring, the use of the MLEs has much to recommend it.

A combination of Monte Carlo experiments and exact small–sample and asymptotic results has been used by MANN (1968a) to compare the expected loss (with loss equal to squared error) of a great number of point estimators. Comparisons of MSEs for MLEs, BLIEs and BLUEs are also given in MANN et al. (1974).

THOMAN et al. (1969) compared the MLEs and MENON's estimators for complete samples of sizes ranging from $n = 5$ to $n = 100$. The biases for the two estimators of $c$ are nearly equal. Both are highly biased for small $n$, i.e., for $n \leq 10$. However, multiplicative unbiasing factors were given so that both estimators of $c$ could be rendered unbiased. The variances of the unbiased estimators were obtained and the unbiased MLE was clearly superior even for small values of $n$, i.e., $5 \leq n \leq 10$.

McCOOL(1970c) evaluated the median unbiased MLE and an LSE with median rank plotting position proposed by JOHNSON (1964). The tenth percentile and the shape parameter $c$ were the parameters considered. A simulation study with observations from two sets was conducted. One set consisted of $n = 10$ with the five largest values censored. The second set was a complete sample of size $n = 10$. The precision of the estimates, as measured by the ratio of the upper to lower 5% points, was uniformly better for ML than for the least squares procedure.

GROSS/LURIE (1977) compared the estimators of BAIN/ANTLE (see (10.31c,d)) and of GUMBEL (see (10.32c,d)) to the MLEs for samples of sizes $n = 10, 25, 50, 100$ and

---

small. HOBBS et al. (1985) and GALLAGHER/MOORE (1990) compared the MLEs with the MDEs of $a$, $b$ and $c$. Their main finding is as follows: "Whether the data were WEIBULL or generated from other distributions, minimum distance estimation using the ANDERSON–DARLING goodness–of–fit statistic on the location parameter and maximum likelihood on the shape and scale parameters was the best or close to the best estimation technique" (GALLAGHER/MOORE, 1990, p. 575). This statement was also affirmed by HOBBS et al. (1985).

Progressively type–II censored samples of a two–parameter WEIBULL distribution with four types of linear estimators (BLUE, approximated BLUE, unweighed regression and linearized MLE) have been studied by THOMAN/WILSON (1972). The biases and the variances and covariances turned out to be smaller with the approximate BLUEs and the linearized MLEs. RINGER/SPRINKLE (1972), comparing MMEs, MLEs and MENON's estimators in the progressive censoring case, concluded that MENON's estimators are best for $c < 1$ and that for $c \geq 1$ MLEs and MENON's estimators are best.

several values of $c$ and $b$. Comparisons were made with respect to bias and MSE. Their conclusions were that for samples of size $n = 10$, the estimators of BAIN/ANTLE and GUMBEL were superior. For $n = 25$, the MLEs were comparable. For $n \geq 50$, MLEs were best on the basis of overall bias and MSE for the two parameters.

KUCHII et al. (1979) as well as KAIO/OSAKI (1980) compared the MLE, MME and a method of WEIBULL probability paper (WPP) for $n = 5$, 10, 15 combined with $c = 0.5$, 1, 5 ($b = 1$ in each case). Their finding was that WPP is relatively better with respect to bias, standard deviation and RMSE, but the number of replications being only 100 gives no representative results.

A Monte Carlo simulation study on the performance of several simple estimators by ZANAKIS (1979) has already been reviewed in Sect. 13.3.

GIBBONS/VANCE (1979, 1981) compared seven estimators including MLE, linear estimators, least squares estimators and moment estimators. The performance of these estimators with respect to MSE was studied in complete and type–II censored samples of $n = 10$ and $n = 25$. No estimator outperformed all the others in all situations. One estimator, the moment estimator, was uniformly worse than the MLE.

ENGEMAN/KEEFE (1982) compared four estimators (GLS, OLS, MLE and MENON's estimator) for $n = 5(5)25$ and 16 combinations of the parameters ($b = 0.1$, 1, 10, 100 and $c = 0.5$, 1, 2, 4). The comparison is based on observed relative efficiencies, i.e., the ratio of CRAMÉR–RAO lower bound to the observed MSE. The GLS estimator was found to be best for estimating $c$ and a close competitor to the MLE of $b$.

CHAO/HWANG (1986) studied the mean length of three exact confidence intervals, based on conditional MLE (see LAWLESS (1973b, 1978)), MLE and BLIE for $n = 4$, 5, 8, 10, 15, 20 and several censoring levels, but for only one combination of $b$ and $c$. Their finding was that in very small sample sizes or in heavily censored samples the three results are fairly close, but generally the MLE produced the shortest mean lengths for both parameters.

AL–BAIDHANI/SINCLAIR (1987) compared GLS, OLS, MLE, hazard plotting and two mixed methods (GLS of $c$ and $b$ with hazard plotting, GLS of $c$ and MLE of $b$) for $n = 10$ and 25 and the same 16 combinations of $b$ and $c$ as used by ENGEMANN/KEEFE (1982). Their finding was that GLS is best for $c$ and the best method for $b$ depends on the size of $c$, the parameter estimates being assessed by their relative efficiency.

HOSSAIN/HOWLADER (1996) considered several unweighed least squares estimators, differing with respect to the plotting position, and the MLE. The spectrum of sample sizes was very wide ($n = 5$, 8, 10, 12, 15, 18, 22, 25, 30, 40, 50) for $c = 1.5$, 2.5, 5, 100 and $b = 1$. For $n > 20$ MLE has a slight edge over the OLS in terms of MSE whereas for $n \leq 20$ OLS with plotting position $i/(n + 1)$ is recommended.

The findings of the cited Monte Carlo studies do not point in the same direction. Thus we can say that the choice of an estimator is mainly a matter of taste and availability of computer programs. But we strongly recommend not to rely on only one approach but to evaluate at least two estimation procedures and pool their results.

# 14 Parameter estimation — BAYESIAN approaches

The framework for the results in the preceding chapters has been random sample of $n$ observations that are either complete or censored in one way or the other. We have regarded these observations as realizations of independent and identically distributed random variables with cumulative distribution function $F(x \mid \boldsymbol{\theta})$ and density function $f(x \mid \boldsymbol{\theta}) = \mathrm{d}F(x \mid \boldsymbol{\theta})/\mathrm{d}x$, where $\boldsymbol{\theta}$ is a labeling vector of parameters. We also assumed that these parameters are constants with unknown values. However, we agreed upon the parameter space, i.e., the set of all possible values of the parameters, which we denoted by $\Omega$. Mostly we have assumed that $\Omega$ is the natural parameter space; i.e., $\Omega$ contains those values of $\boldsymbol{\theta}$ for which the density function $f(x \mid \boldsymbol{\theta})$ is well defined, i.e., $f(x \mid \boldsymbol{\theta}) \geq 0$ and $\int_{-\infty}^{\infty} f(x \mid \boldsymbol{\theta}) \, \mathrm{d}x = 1$.

But there are cases in which one can assume a little more about the unknown parameters. For example, we could assume that $\boldsymbol{\theta}$ is itself a realization of a random vector, denoted by $\boldsymbol{\Theta}$, with density function $g(\boldsymbol{\theta})$. In the WEIBULL model the scale parameter $b$ may be regarded as varying from batch to batch over time, and this variation is represented by a probability distribution over $\Omega$. Thus, the set–up is now as described in Sect. 14.1.

## 14.1 Foundations of BAYESIAN inference[1]

### 14.1.1 Types of distributions encountered

There are $n$ items put on test and it is assumed that their recorded lifetimes form a random sample of size $n$ from a population with density function $f(x \mid \boldsymbol{\theta})$, called the **sampling model**. This density is conditioned on $\boldsymbol{\theta}$. The **joint conditional density of the sampling vector** $\boldsymbol{X} = (X_1, \ldots, X_n)'$ is

$$f(\boldsymbol{x} \mid \boldsymbol{\theta}) = \prod_{i=1}^{n} f(x_i \mid \boldsymbol{\theta}), \tag{14.1a}$$

assuming an independent sample. Given the sample data $\boldsymbol{x} = (x_1, \ldots, x_n)'$, $f(\boldsymbol{x} \mid \boldsymbol{\theta})$ may be regarded as a function, not of $\boldsymbol{x}$, but of $\boldsymbol{\theta}$. When so regarded, (14.1a) is referred to as the **likelihood function** of $\boldsymbol{\theta}$ given $\boldsymbol{x}$, which is usually written as $L(\boldsymbol{\theta} \mid \boldsymbol{x})$ to ensure its distinct interpretation apart from $f(\boldsymbol{x} \mid \boldsymbol{\theta})$, but formally we have

$$f(\boldsymbol{x}, \mid \boldsymbol{\theta}) \equiv L(\boldsymbol{\theta} \mid \boldsymbol{x}). \tag{14.1b}$$

---

[1] Suggested reading for this section: BARNETT (1973), BOX/TIAO (1973), JEFFREYS (1961), LINDLEY (1965) and SAVAGE (1962) for the basic theory and foundations, MARTZ/WALLER (1982) for applications in life–testing and reliability.

We agree to regard $\theta$ as realization of a random vector $\Theta$ having the **prior density** $g(\theta)$, called the **prior model**. This prior density is crucial to BAYESIAN inference and has always been a point of criticism to this approach. We will revert to the possibilities of choosing $g(\theta)$ by the end of this section. The **joint density of $X$ and $\Theta$** is found by applying the multiplication theorem of probabilities to be

$$f(x, \theta) = g(\theta)\, f(x \mid \theta). \tag{14.1c}$$

The **marginal density of $X$** may be obtained according to

$$f(x) = \int_{\Omega} f(x \mid \theta)\, g(\theta)\, \mathrm{d}\theta, \tag{14.1d}$$

where the integration is taken over the admissible range $\Omega$ of $\theta$. The **conditional density of** $\Theta$, given the date $x$, is found by using **BAYES' theorem** (THOMAS BAYES, 1701 – 1761):

$$g(\theta \mid x) = \frac{f(x, \theta)}{f(x)} = \frac{g(\theta)\, f(x \mid \theta)}{f(x)}\,. \tag{14.1e}$$

$g(\theta \mid x)$ is called the **posterior density of** $\Theta$. This **posterior model** is the main object of study and the basis of estimating $\theta$ and testing hypotheses on $\theta$. In the remainder of this chapter we sometimes refer to the prior distribution and posterior distributions simply as the "prior" and "posterior."

The prior model is assumed to represent the totality of subjective information available concerning the parameter vector $\Theta$ prior to the observation $x$ of the sample vector $X$. Thus, it is not functionally dependent upon $x$. On the other hand, the sampling model depends on the parameters in $\theta$ and is thus a conditional distribution. The posterior model tells us what is known about $\theta$, given knowledge of the data $x$. It is essentially an updated version of our prior knowledge about $\theta$ in light of the sample data — hence, the name *posterior model*. It is intuitive that the posterior model should represent a modification of the subjective knowledge about $\theta$ expressed by the prior model in light of the observed sample data. If the sample data support our subjective opinion about $\theta$, then the posterior model should reflect increased confidence in the subjective notions embodied in the prior model. On the other hand, if the sample data do not support the subjective information, the posterior model should reflect a weighted consideration of both assessments.

If we regard $f(x \mid \theta)$ in BAYES' theorem as the likelihood function $L(\theta \mid x)$, then we may write BAYES' theorem (14.1e) as

$$g(\theta \mid x) \propto g(\theta)\, L(\theta \mid x), \tag{14.1f}$$

which says that the posterior distribution is proportional to the product of the prior distribution and the likelihood function. The constant of proportionality necessary to ensure that the posterior density integrates to one is the integral of the product, which is the marginal density of $X$; see (14.1d). Thus, in words we have the **fundamental relationship** given by

$$
\begin{aligned}
\text{posterior distribution} \quad &\propto \quad \text{prior distribution} \times \text{likelihood} \\
&= \frac{\text{prior distribution} \times \text{likelihood}}{\text{marginal distribution}}.
\end{aligned}
$$

BAYES' theorem provides a mechanism for continually **updating our knowledge** about $\theta$ as more sample data become available. Suppose that we have an initial set of sample data $x_1$, as well as a prior distribution for $\Theta$. BAYES' theorem says that

$$g(\theta \mid x_1) \propto g(\theta) \, L(\theta \mid x_1). \tag{14.2a}$$

Suppose now that we later obtain a second set of sample data $x_2$ that is statistically independent of the first sample. Then

$$\begin{aligned} g(\theta \mid x_1, x_2) &\propto g(\theta) \, L(\theta \mid x_1) \, L(\theta \mid x_2) \\ &\propto g(\theta \mid x_1) \, L(\theta \mid x_2), \end{aligned} \tag{14.2b}$$

as the combined likelihood may be factored as a product. It is observed that (14.2b) is of the same form as (14.2a) except that the posterior distribution of $\Theta$ given $x_1$ assumes the role of the prior distribution $g(\theta)$. This process may be repeated. In fact, each sample observation can be processed separately if desired.

### 14.1.2 BAYESIAN estimation theory

Once the posterior has been obtained, it becomes the main object of further inquiries into the parameters incorporated in $\Theta$. We will first comment upon estimation.

In general, BAYESIAN point estimation is attached to a **loss function** indicating the loss coming up when the estimate $\widehat{\theta}$ deviates from the true value $\theta$. The loss should be zero if and only if $\widehat{\theta} = \theta$. For this reason the loss function $\ell(\widehat{\theta}, \theta)$ is often assumed to be of the form

$$\ell(\widehat{\theta}, \theta) = h(\theta) \, \varphi(\theta - \widehat{\theta}), \tag{14.3a}$$

where $\varphi(\cdot)$ is a non–negative function of the error $\theta - \widehat{\theta}$ such that $\varphi(0) = 0$ and $h(\cdot)$ is a non–negative weighting function that reflects the relative seriousness of a given error for different values of $\theta$. In determining BAYES' estimators based on this loss function, the function $h(\cdot)$ can be considered as a component of the prior $g(\theta)$. For this reason, it is frequently assumed that the function $h(\cdot)$ in (14.3a) is a constant.

When the parameter $\theta$ is one–dimensional, the loss function can often be expressed as

$$\ell(\widehat{\theta}, \theta) = A \, |\theta - \widehat{\theta}|^B, \tag{14.3b}$$

where $A > 0$ and $B > 0$. If $B = 2$, the loss function is quadratic and is called a **squared–error loss function**. If $B = 1$, the loss function is piecewise linear and the loss is proportional to the absolute value of the estimation error and is called an **absolute–error loss function**.

When the loss function in the one–dimensional case is specified as

$$\ell(\widehat{\theta}, \theta) = A \left( \theta - \widehat{\theta} \right)^2, \tag{14.4a}$$

the **BAYES estimator**, for any specified prior $g(\theta)$, will be the estimator that minimizes the posterior risk given by

$$E\left[A \, (\Theta - \widehat{\theta})^2 \mid x \right] = \int_\Omega A \, (\theta - \widehat{\theta})^2 \, g(\theta \mid x) \, d\theta, \tag{14.4b}$$

provided this expectation exists.[2] After adding and subtracting $E(\Theta \mid x)$ and simplifying, we have

$$E\left[A\left(\Theta - \widehat{\theta}\right)^2 \mid x\right] = A\left[\widehat{\theta} - E(\Theta \mid x)\right]^2 + A\,\mathrm{Var}(\Theta \mid x), \tag{14.4c}$$

which is clearly minimized when

$$\widehat{\theta} = E(\Theta \mid x) = \int_\Omega \theta\, g(\theta \mid x)\, d\theta. \tag{14.4d}$$

Thus, for a squared–error loss function the BAYES estimator is simply the **posterior mean** of $\Theta$ given $x$.

A generalization of the squared–error loss function when we have to estimate a vector $\boldsymbol{\theta}$ is the **quadratic loss function**

$$\ell(\widehat{\boldsymbol{\theta}}, \boldsymbol{\theta}) = (\boldsymbol{\theta} - \widehat{\boldsymbol{\theta}})'\, A\,(\boldsymbol{\theta} - \widehat{\boldsymbol{\theta}}), \tag{14.5a}$$

where $A$ is a symmetric non–negative definite matrix. We shall suppose that the mean vector $E(\Theta \mid x)$ and variance–covariance matrix $\mathrm{Cov}(\Theta \mid x)$ of the posterior distribution of $\Theta$ exist. The BAYES estimator is the estimator that minimizes the posterior risk given by

$$
\begin{aligned}
E\left[(\Theta - \widehat{\boldsymbol{\theta}})'\, A\,(\Theta - \widehat{\boldsymbol{\theta}}) \mid x\right] &= E\left\{[\Theta - E(\Theta \mid x)]'\, A\,[\Theta - E(\Theta \mid x)] \mid x\right\} \\
&\quad + \left[E(\Theta \mid x) - \widehat{\boldsymbol{\theta}}\right]'\, A\,\left[E(\Theta \mid x) - \widehat{\boldsymbol{\theta}}\right] \\
&= \mathrm{tr}[A\,\mathrm{Cov}(\Theta \mid x)] + \left[E(\Theta \mid x) - \widehat{\boldsymbol{\theta}}\right]'\, A\,\left[E(\Theta \mid x) - \widehat{\boldsymbol{\theta}}\right] \tag{14.5b}
\end{aligned}
$$

This is clearly minimized when

$$\widehat{\boldsymbol{\theta}} = E(\Theta \mid x) = \int_\Omega \boldsymbol{\theta}\, g(\boldsymbol{\theta} \mid x)\, d\boldsymbol{\theta} \tag{14.5c}$$

and thus the posterior mean $E(\Theta \mid x)$ is a BAYES estimator for $\boldsymbol{\theta}$.

Assuming the absolute–error loss function the BAYES estimator will be the estimator that minimizes $E[A \mid \Theta - \widehat{\theta} \mid \mid x]$. The value of $\widehat{\theta}$ minimizing this posterior risk is the **median** of the posterior distribution.

Another BAYES estimator for either a univariate or vector parameter is the value of the parameter that maximizes $g(\boldsymbol{\theta} \mid x)$, i.e., that satisfies the relation

$$g(\widehat{\boldsymbol{\theta}} \mid x) = \max_{\boldsymbol{\theta}} g(\boldsymbol{\theta} \mid x). \tag{14.6}$$

This estimator is the **mode** of $g(\boldsymbol{\theta} \mid x)$. Although such an estimator may not be a BAYES estimator for any standard loss function, it is a reasonable estimator because it measures the location of the posterior distribution analogous to the posterior mean and posterior median. But we can state that the most popular BAYES estimator is given by the posterior mean (14.4d) or (14.5c).

Having obtained $g(\theta \mid x)$, one may ask, "How likely is it that $\theta$ lies within a specified interval $[\theta_\ell, \theta_u]$?" This is not the same as the classical confidence interval interpretation for $\theta$ because there $\theta$ is a constant and it is meaningless to make a probability statement about

---

[2] The posterior risk for $A = 1$ resembles the MSE of classical inference.

a constant.[3] BAYESIANS call this interval based on the posterior a **credible interval**, an interval which contains a certain fraction of the degree of belief. The interval[4] $[\theta_\ell, \theta_u]$ is said to be a $(1 - \alpha)$–credible region for $\theta$ if

$$\int_{\theta_\ell}^{\theta_u} g(\theta \mid x)\, d\theta = 1 - \alpha. \tag{14.7a}$$

Now we have to distinguish between three types of credible intervals.

1. For the **shortest credible interval** we have to minimize $I = \theta_u - \theta_\ell$ subject to condition (14.7a) which requires

$$g(\theta_\ell \mid x) = g(\theta_u \mid x). \tag{14.7b}$$

2. A **highest posterior density interval** (**HPD–interval**) satisfies the following two conditions:

   (a) for a given probability $1 - \alpha$, the interval should be as short as possible,

   (b) the posterior density at every point inside the interval must be greater than for every point outside so that the interval includes more probable values of the parameter and excludes less probable values.

3. An **equal–tail** $(1 - \alpha)$**–credible interval** for $\theta$ is given by

$$\int_{-\infty}^{\theta_\ell} g(\theta \mid x)\, d\theta = \int_{\theta_u}^{\infty} g(\theta \mid x)\, d\theta = \alpha/2. \tag{14.7c}$$

For a unimodal but not necessarily symmetrical posterior density, shortest credible and HPD–intervals are one and the same. All these types of credible regions are the same when $g(\theta \mid x)$ is unimodal and symmetric. The most popular type is the equal–tail interval because it is the easiest one to determine.

### 14.1.3 Prior distributions

As stated previously, the prior $g(\theta)$ represents all that is known or assumed about the parameter $\theta$ (either scalar or vector) prior to the observation of sample data. Thus, the information summarized by the prior may be either objective or subjective or both. Two examples of objective input to the prior distribution are operational data or observational data from a previous comparable experiment. Subjective information may include an engineer's quantification of personal experience and judgments, a statement of one's degree of belief regarding the parameter, design information and personal opinions. The latter type of information is extremely valuable in life–testing and reliability estimation where sample data are difficult to obtain when the items have a long time to failure or are expensive to obtain as the items have a high price. In some cases we may evaluate the informational content of a prior by the number of sampled items saved and leading to the same estimate.

---

[3] In classical statistics the confidence limits are random and the probability statement refers to the interval and is the probability that it covers over the fixed but unknown value of $\theta$.

[4] The following ideas can be carried over to a vector parameter leading to regions in $\Omega$.

Prior distributions may be categorized in different ways. One common classification is a dichotomy that separates "proper" and "improper" priors. A **proper prior** is one that allocates positive weights that total one to the possible values of the parameter. Thus a proper prior is a weight function that satisfies the definition of a probability mass function or a density function. An **improper prior** is any weight function that sums or integrates over the possible values of the parameter to a value other than one, say $K$. If $K$ is finite, then an improper prior can induce a proper prior by normalizing the function. Other classifications of priors, either by properties, e.g., non–informative, or by distributional forms, e.g., beta, gamma or uniform distributions, will be met in the following sections.

One general class of prior distribution is called **non–informative priors** or **priors of ignorance**. Rather than a state of complete ignorance, the non–informative prior refers to the case when relatively little or very limited information is available a priori. In other words a priori information about the parameter is not considered substantial relative to the information expected to be provided by the sample of empirical data. Further, it frequently means that there exists a set of parameter values that the statistician believes to be equally likely choices for the parameter. One way of expressing indifference is to select a prior distribution that is locally uniform, i.e., a prior that is approximately uniformly distributed over the interval of interest. BOX/TIAO (1973) give the following definition:

If $\phi(\theta)$ is a one–to–one transformation of $\theta$, we shall say that a prior of $\Theta$ that is locally proportional to $|d\phi(\theta)/d\theta|$ is **non–informative** for the parameter $\theta$ if, in terms of $\phi$, the likelihood curve is **data translated**; that is, the data $x$ only serve to change the location of the likelihood $L(\theta \,|\, x)$.

A general rule to find a non–informative prior has been proposed by JEFFREYS (1961), known as **JEFFREYS' rule**:

$$g(\theta) = \text{constant } \sqrt{I(\theta)} \tag{14.8a}$$

for a one–dimensional prior, where $I(\theta)$ is the FISHER information and

$$g(\boldsymbol{\theta}) = \text{constant } \sqrt{|\boldsymbol{I}(\boldsymbol{\theta})|} \tag{14.8b}$$

for a multi–dimensional prior, where $|\boldsymbol{I}(\boldsymbol{\theta})|$ is the determinant of the information matrix, see (11.8).

Another type of prior is the **conjugate prior distribution** introduced by RAIFFA/SCHLAIFER (1961). A conjugate prior distribution, say $g(\theta)$, for a given sampling distribution, say $f(x \,|\, \theta)$, is such that the posterior distribution $g(\theta \,|\, x)$ and the prior $g(\theta)$ are members of the same family of distributions.

When the chosen prior $g(\theta)$ depends on one or more parameters, we have to find their values unless they are given together with the function $g(\cdot)$. Sometimes the statistician has some subjective information about characteristics of the distribution of $\Theta$; e.g., he has an impression of moments or quantiles. Then he can set up moment equations or percentile equations which connect these characteristics with the unknown parameters of $g(\boldsymbol{\theta})$ and solve for these parameters.

We will encounter the types of priors mentioned above in Sect. 14.2, but the assignment of a prior contains an element of risk for the investigator. It is appropriate to ask if there is any alternative to the use of an assumed prior or of having to ignore altogether that the prior distribution exists. An alternative that assumes the existence of a prior but of unknown distributional form is the **empirical BAYES procedure** (see Sect. 14.3). Past data is used to bypass the necessity for identification of the correct prior distribution.

## 14.2 Two-parameter WEIBULL distribution[5]

We will present three classes of problems:

- the shape parameter $c$ is known and the scale parameter $b$ is random (Sect. 14.2.1),

- the scale parameter is known and the shape parameter is random (Sect. 14.2.2),

- both parameters are random variables (Sect. 14.2.3).

BAYESIAN analysis is applied to a reparameterized version of the WEIBULL distribution, either to

$$f(x \mid \lambda, c) = \lambda\, c\, x^{c-1} \exp\left(-\lambda x^c\right), \qquad (14.9a)$$

where

$$\lambda = b^{-c} \qquad (14.9b)$$

is a scale factor, or to

$$f(x \mid \theta, c) = \frac{c}{\theta}\, x^{c-1} \exp\left(-x^c/\theta\right), \qquad (14.10a)$$

where

$$\theta = \lambda^{-1} = b^c. \qquad (14.10b)$$

These versions separate the two original parameters $b$ and $c$ and thus simplify the algebra in the subsequent BAYESIAN manipulations.

### 14.2.1 Random scale parameter and known shape parameter[6]

In WEIBULL reliability analysis it is frequently the case that the value of the shape parameter is known.[7] In this case the variate $Y = X^c$ has an exponential distribution and the BAYESIAN estimators follow directly from this transformation and the results for the exponential distribution, see MARTZ/WALLER (1982, Chapter 8). We will present the BAYESIAN estimators of $\lambda = b^{-c}$ for a non–informative prior, a uniform prior and a gamma prior.

When $c$ is assumed known, we can make the following transformation

$$Y = X^c, \qquad (14.11a)$$

and the parent WEIBULL densities (14.9a) and (14.10a) turn into the exponential densities

$$f(y \mid \lambda) = \lambda \exp(-\lambda y) \qquad (14.11b)$$

---

[5] BAYES estimation of the parameters of the three–parameter WEIBULL distribution is discussed in SINHA/ SLOAN (1988), SMITH/NAYLOR (1987) and TSIONAS (2000).

[6] Suggested reading for this section: CANAVOS (1983), CANAVOS/TSOKOS (1973), HARRIS/ SINGPURWALLA (1968, 1969), MARTZ/WALLER (1982), PAPADOPOULOS/TSOKOS (1975), SINHA (1986b), SOLAND (1968b) and TSOKOS (1972).

[7] SOLAND (1968b) gives a justification for this situation.

and

$$f(y \mid \theta) = \frac{1}{\theta} \exp(-y/\theta). \tag{14.11c}$$

We further have the following likelihood function of $\lambda$ when a sample of size $n$ is simply type–II censored on the right, $r \leq n$ being the censoring number:

$$
\begin{aligned}
L(\lambda \mid \boldsymbol{y}, n, r) \quad &\propto \quad \prod_{i=1}^{r} \left[ \lambda \exp(-\lambda\, y_i) \right] \prod_{j=r+1}^{n} \exp(-\lambda\, y_r), \quad y_i = x_{i:n}^{c} \\
&= \quad \lambda^r \exp\left\{ -\lambda \left[ \sum_{i=1}^{r} y_i + (n-r)\, y_r \right] \right\} \\
&= \quad \lambda^r \exp(-\lambda\, T), \tag{14.12a}
\end{aligned}
$$

where

$$T := \sum_{i=1}^{r} x_{i:n}^{c} + (n-r)\, x_{r:n}^{c} \tag{14.12b}$$

is the **rescaled total time on test** in which the lifetimes $x_i$ and survival time $x_r$ are "rescaled" by raising it to a power $c$. We remark that (14.12a,b) include the complete–sample case where $r = n$.

We first want to study BAYESIAN estimation of $\lambda$ when the **prior** distribution is **non–informative** according to JEFFREYS' rule. The log–likelihood function belonging to (14.12a) is

$$\mathcal{L}(\lambda \mid \boldsymbol{y}, n, r) = r \ln \lambda - \lambda\, T + \text{constant}. \tag{14.13a}$$

Consequently

$$\frac{\partial \mathcal{L}}{\partial \lambda} = \frac{r}{\lambda} - T \tag{14.13b}$$

and

$$\frac{\partial^2 \mathcal{L}}{\partial \lambda^2} = -\frac{r}{\lambda^2}, \tag{14.13c}$$

and the FISHER information is

$$I(\lambda) = \frac{r}{\lambda^2}, \tag{14.13d}$$

so that the non–informative prior according to (14.8a) is given by

$$g(\lambda) \propto \sqrt{I(\lambda)} \propto \frac{1}{\lambda} \quad \text{for} \quad \lambda > 0, \tag{14.13e}$$

and $\phi(\lambda) = \ln \lambda$ is the transformation for which the approximately non–informative prior is locally uniform.

The prior density given in (14.13e) is improper, but the posterior density — according to (14.1e) — will be proper:

$$
\begin{aligned}
g(\lambda \mid T, r) \quad &= \quad \frac{\lambda^{r-1} \exp(-\lambda\, T)}{\displaystyle\int_{0}^{\infty} \lambda^{r-1} \exp(-\lambda\, T)\, \mathrm{d}\lambda} \\
&= \quad \frac{T^r}{\Gamma(r)} \lambda^{r-1} \exp(-\lambda\, T). \tag{14.14a}
\end{aligned}
$$

This posterior density is recognized as the density of the gamma distribution $Ga(r, T)$, so that the posterior mean and posterior variance are

$$E(\Lambda \mid T, r) = r/T, \tag{14.14b}$$

$$\text{Var}(\Lambda \mid T, r) = r/T^2. \tag{14.14c}$$

(14.14b) is nothing but the BAYESIAN estimator $\widehat{\lambda}_B$ of $\lambda$ with respect to squared–error loss. We further remark that this BAYESIAN estimator is equal to the classical MLE resulting from (see (14.13b))

$$\frac{\partial \mathcal{L}}{\partial \lambda} = \frac{r}{\lambda} - T = 0.$$

After re–substitution we write

$$\widehat{\lambda}_B = r \left\{ \sum_{i=1}^{r} x_{i:n}^c + (n - r) \, x_{r:n}^c \right\}^{-1}. \tag{14.14d}$$

---

**Excursus:   A generalization of the non–informative prior $1/\lambda$**

When we introduce a power $d$, $d > 0$ to (14.13e), we arrive at the following generalization of the non–informative prior:

$$g(\lambda) = \frac{1}{\lambda^d} ; \quad \lambda > 0, \ d > 0. \tag{14.15a}$$

(14.15a) and (14.12a) give the following:

$$g(\lambda \mid T, r, d) = \frac{T^{r+1-d}}{\Gamma(r + 1 - d)} \, \lambda^{r-d} \exp(-\lambda T), \ d < r + 1, \tag{14.15b}$$

$$E(\Lambda \mid T, r, d) = \frac{r - d + 1}{T}, \ d < r + 1, \tag{14.15c}$$

$$\text{Var}(\Lambda \mid T, r, d) = \frac{r - d + 1}{T^2}, \ d < r + 1. \tag{14.15d}$$

---

Since the conditional distribution of $(\Lambda \mid T, r)$ is a gamma distribution (see (14.14a)), the transformed variable $2\,T\,\Lambda$ follows the $\chi^2(2\,r)$–distribution. This fact can be used to obtain credible intervals for $\Lambda$. The **equal–tail $100\,(1 - \alpha)\%$ credible interval** is

$$\frac{\chi^2_{2r,\alpha/2}}{2\,T} \leq \Lambda \leq \frac{\chi^2_{2r,1-\alpha/2}}{2\,T}, \tag{14.16}$$

where $\chi^2_{2r,P}$ is the percentile of order $P$ of the $\chi^2(2\,r)$–distribution.

---

**Example 14/1:   Bayesian estimation of $\lambda = b^{-c}$ for dataset #1 censored at $r = 15$ using**
$\qquad\qquad\quad$ **$g(\lambda) \propto 1/\lambda$**

Dataset #1 (see Tab. 9/2) is a sample of $n = 20$ from $We(a, b, c) = We(0, 100, 2.5)$. We will use these data censored at $r = 15$, i.e., $x_{r:20} = 113$, and assume $c = 2.5$. The rescaled total time on test results as

$$T = \sum_{i=1}^{15} x_{i:20}^{2.5} + 5 \, x_{15:20}^{2.5} = 1{,}477{,}067.2,$$

so that the BAYESIAN estimator is

$$\widehat{\lambda}_B = \frac{r}{T} = \frac{15}{1{,}477{,}067.2} = 1.015526 \cdot 10^{-5}.$$

The true value is $\lambda = 100^{-2.5} = 1 \cdot 10^{-5}$. The equal–tail 95% credible interval is

$$\frac{16.791}{2 \cdot 1{,}477{,}067.2} \leq \Lambda \leq \frac{46.979}{2 \cdot 1{,}477{,}067.2} \implies 5.683899 \cdot 10^{-6} \leq \Lambda \leq 1.590280 \cdot 10^{-5}$$

and includes the true value $10^{-5}$. Fig. 14/1 shows $g(\lambda) = 1/\lambda$ and $g(\lambda \,|\, T, r)$ together with the lower and upper limits of the 95%–interval.

Figure 14/1: Non-informative prior and posterior density for dataset #1 censored at $r = 15$

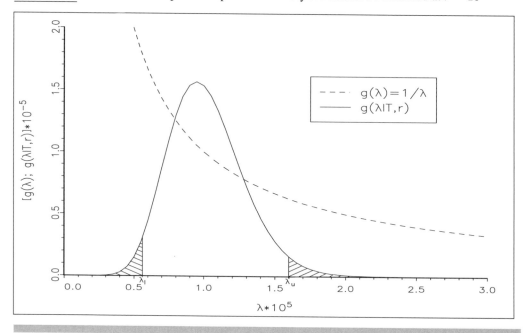

We will now consider the **uniform prior**[8]

$$g(\lambda \,|\, A, B) = \left\{ \begin{array}{ll} \dfrac{1}{B - A} & \text{for} \quad A \leq \lambda \leq B, \\[2mm] 0 & \text{elsewhere.} \end{array} \right\} \tag{14.17a}$$

When we combine (14.17a) and the likelihood function (14.12a) according to (14.1e), we first get the following posterior density

$$g(\lambda \,|\, T, r, A, B) = \frac{\lambda^r \exp(-\lambda T)}{\displaystyle\int_A^B \lambda^r \exp(-\lambda T)\, d\lambda}. \tag{14.17b}$$

---

[8] The BAYESIAN estimator based on a uniform prior for $\theta$ in (14.10a,b) is to be found in CANAVOS (1983), CANAVOS/TSOKOS (1973) and PAPADOPOULOS/TSOKOS (1975).

Introducing $y = \lambda T$, the denominator of (14.17b) becomes

$$
\int\limits_A^B \lambda^r \exp(-\lambda T)\, d\lambda \;=\; \int\limits_{AT}^{BT} \frac{y^r \exp(-y)}{T^{r+1}}\, dy
$$

$$
=\; \frac{1}{T^{r+1}} \left\{ \gamma(r+1 \,|\, B\,T) - \gamma(r+1 \,|\, A\,T) \right\}, \quad (14.17c)
$$

where $\gamma(\cdot \,|\, \cdot)$ represents the incomplete gamma function (see the excursus on the gamma function in Sect. 2.9). Substituting (14.17c) into (14.17b) we finally have

$$
g(\lambda \,|\, T, r, A, B) = \frac{T^{r+1}\, \lambda^r \exp(-\lambda T)}{\gamma(r+1 \,|\, B\,T) - \gamma(r+1 \,|\, A\,T)}. \quad (14.17d)
$$

The posterior mean and posterior variance are

$$
\mathrm{E}(\Lambda \,|\, T, r, A, B) \;=\; \frac{\gamma(r+2 \,|\, B\,T) - \gamma(r+2 \,|\, A\,T)}{T\left\{ \gamma(r+1 \,|\, B\,T) - \gamma(r+1 \,|\, A\,T) \right\}}, \quad (14.17e)
$$

$$
\mathrm{Var}(\Lambda \,|\, T, r, A, B) \;=\; \frac{\gamma(r+3 \,|\, B\,T) - \gamma(r+3 \,|\, A\,T)}{T^2\left\{ \gamma(r+1 \,|\, B\,T) - \gamma(r+1 \,|\, A\,T) \right\}}
$$

$$
-\left[ \mathrm{E}(\Lambda \,|\, T, r, A, B) \right]^2. \quad (14.17f)
$$

(14.17e) is the BAYESIAN estimator with respect to squared–error loss. The estimator $\widehat{\lambda} = r/T$ is the mode of (14.17d).

The limits $\lambda_\ell$ and $\lambda_u$ of an equal–tail $100\,(1-\alpha)\%$ credible interval

$$
\lambda_\ell \leq \Lambda \leq \lambda_u
$$

are found from solving

$$
\mathrm{Pr}(\Lambda < \lambda_\ell \,|\, T, r, A, B) \;=\; \int\limits_A^{\lambda_\ell} \frac{T^{r+1}\, \lambda^r \exp(-\lambda T)}{\gamma(r+1 \,|\, B\,T) - \gamma(r+1 \,|\, A\,T)}\, d\lambda
$$

$$
=\; \frac{\gamma(r+1 \,|\, \lambda_\ell\, T) - \gamma(r+1 \,|\, A\,T)}{\gamma(r+1 \,|\, B\,T) - \gamma(r+1 \,|\, A\,T)} = \frac{\alpha}{2} \quad (14.18a)
$$

and

$$
\mathrm{Pr}(\Lambda > \lambda_u \,|\, T, r, A, B) \;=\; \int\limits_{\lambda_u}^{B} \frac{T^{r+1}\, \lambda^r \exp(-\lambda T)}{\gamma(r+1 \,|\, B\,T) - \gamma(r+1 \,|\, A\,T)}\, d\lambda
$$

$$
=\; \frac{\gamma(r+1 \,|\, B\,T) - \gamma(r+1 \,|\, \lambda_u\, T)}{\gamma(r+1 \,|\, B\,T) - \gamma(r+1 \,|\, A\,T)} = \frac{\alpha}{2} \quad (14.18b)
$$

using an incomplete gamma subroutine and a search program.

**Example 14/2:  BAYESIAN estimation of $\lambda = b^{-c}$ for dataset #1 censored at $r = 15$ using a uniform prior**

We alter Example 14/1 by introducing the uniform prior

$$g(\lambda \,|\, 0.5 \cdot 10^{-5}, \; 1.5 \cdot 10^{-5}) = 10^5 \quad \text{for} \quad 0.5 \cdot 10^{-5} \le \lambda \le 1.5 \cdot 10^{-5}.$$

Fig. 14/2 shows this prior density and the resulting posterior density. The prior mean is $10^{-5}$ and the posterior mean (= BAYESIAN estimate) is $1.038118 \cdot 10^{-5}$. The 95% credible interval according to (14.18a,b) is

$$0.6256 \cdot 10^{-5} \le \Lambda \le 1.4500 \cdot 10^{-5}.$$

Figure 14/2: Uniform prior and posterior density for dataset #1 censored at $r = 15$

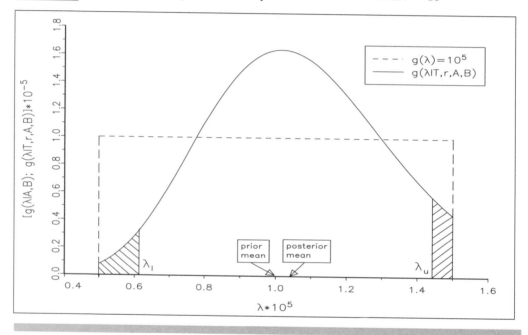

**Excursus:  A general uniform prior density**

TSOKOS (1972) has used the following general uniform prior for the parameter $\theta = \lambda^{-1}$ in (14.10a):

$$g(\theta \,|\, \alpha, \beta) = \left\{ \begin{array}{ll} \dfrac{(\alpha - 1)(\alpha\,\beta)^{\alpha-1}}{\beta^{\alpha-1} - \alpha^{\alpha-1}} \dfrac{1}{\theta^\alpha} & \text{for} \quad \alpha \le \theta \le \beta \\[2mm] 0 & \text{elsewhere,} \end{array} \right\} \tag{14.19a}$$

which for $\alpha = 0$ reduces to the uniform density on $[0, \beta]$. Now the posterior density is

$$g(\theta \,|\, T, r, \alpha, \beta) = \frac{T^{r+\alpha-1} \, \exp(-T/\theta)}{\theta^{r+\alpha} \left\{ \gamma(r + \alpha - 1 \,|\, \frac{T}{\alpha}) - \gamma(r + \alpha - 1 \,|\, \frac{T}{\beta}) \right\}} \tag{14.19b}$$

with

$$E(\Theta \mid T, r, \alpha, \beta) = T \, \frac{\gamma(r + \alpha - 2 \mid \frac{T}{\alpha}) - \gamma(r + \alpha - 2 \mid \frac{T}{\beta})}{\gamma(r + \alpha - 1 \mid \frac{T}{\alpha}) - \gamma(r + \alpha - 1 \mid \frac{T}{\beta})} \qquad (14.19c)$$

and

$$E(\Theta^2 \mid T, r, \alpha, \beta) = T^2 \, \frac{\gamma(r + \alpha - 3 \mid \frac{T}{\alpha}) - \gamma(r + \alpha - 3 \mid \frac{T}{\beta})}{\left\{ \gamma(r + \alpha - 1 \mid \frac{T}{\alpha}) - \gamma(r + \alpha - 1 \mid \frac{T}{\beta}) \right\}^2} \qquad (14.19d)$$

and

$$\text{Var}(\Theta \mid T, r, \alpha, \beta) = E(\Theta^2 \mid T, r, \alpha, \beta) - \left[ E(\Theta \mid T, r, \alpha, \beta) \right]^2. \qquad (14.19e)$$

We now take a **gamma prior** for $\lambda$ using the following parametrization; see (3.24b):

$$g(\lambda \mid b, d) = \frac{\lambda^{d-1} \exp(-\lambda/b)}{b^d \, \Gamma(d)} \, ; \quad \lambda \geq 0; \; b, d > 0 \, . \qquad (14.20a)$$

Combining the likelihood function (14.12a) and (14.20a), the posterior density results as

$$g(\lambda \mid T, r, b, d) = \frac{\lambda^{d+r-1} \exp[-\lambda \, (T + 1/b)]}{\Gamma(d + r) \left( \dfrac{b}{bT + 1} \right)^{d+r}} \, , \qquad (14.20b)$$

which is recognized as a gamma density where the parameters have changed: $b \implies b/(bT + 1)$ and $d \implies d + r$. Thus, the gamma prior is a **conjugate prior**.

The gamma prior has mean

$$E(\Lambda \mid b, d) = d \, b \qquad (14.20c)$$

and variance

$$\text{Var}(\Lambda \mid b, d) = b^2 \, d, \qquad (14.20d)$$

whereas the mean and variance of the posterior distribution are

$$E(\Lambda \mid T, r, b, d) \;\; = \;\; \frac{b \, (d + r)}{bT + 1} \, , \qquad (14.20e)$$

$$\text{Var}(\Lambda \mid T, r, b, d) \;\; = \;\; \frac{b^2 \, (d + r)}{(bT + 1)^2} \, . \qquad (14.20f)$$

The BAYESIAN estimator (14.20e) converges almost surely to the MLE given by $r/T$ as $r$ approaches infinity. Thus the likelihood tends to dominate the prior as the number $r$ of failures specified (and corresponding rescaled total time on test) increases. The limits $\lambda_\ell$ and $\lambda_u$ of an equal–tailed $100\,(1 - \alpha)\%$ credible interval are found — using the familiar relationship between gamma and $\chi^2$–distributions — to be

$$\lambda_\ell = \frac{b}{2\,(b\,T + 1)}\,\chi^2_{2r+2d,\alpha/2}; \quad \lambda_u = \frac{b}{2\,(b\,T + 1)}\,\chi^2_{2r+2d,1-\alpha/2}. \tag{14.20g}$$

**Example 14/3:  Bayesian estimation of $\lambda = b^{-c}$ for dataset #1 censored at $r = 15$ using a gamma prior**

Assuming the gamma prior density (14.20a), we first have to determine the values of the parameters $b$ and $d$. Past experience with the $We(0, \lambda, 2.5)$–distributed items showed a mean value of $1.5 \cdot 10^{-5}$

Figure 14/3: Prior gamma density and posterior gamma density for dataset #1 censored at $r = 15$

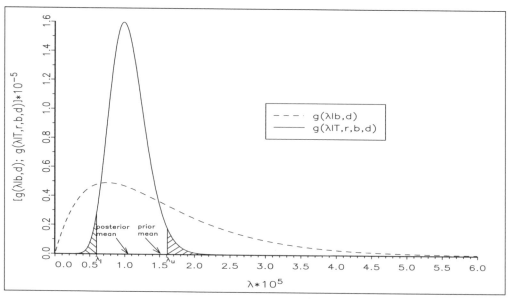

for $\lambda$ and a variance of $1.125 \cdot 10^{-10}$. By equating these values to the prior mean (14.20c) and prior variance (14.20d), we find $b = 0.75 \cdot 10^{-5}$ and $d = 2$. From the data of set #1 we have $r = 15$ and $T = 1,477,067.2$ so the BAYESIAN estimator (14.20e) results as $\widehat{\lambda}_B = 1.055638 \cdot 10^{-5}$. Fig. 14/3 shows the prior and posterior densities together with their means and the equal–tailed 95% credible interval according to (14.20g,h):

$$0.614948 \cdot 10^{-5} \leq \Lambda \leq 1.613449 \cdot 10^{-5}.$$

If one decides to work with the parameter $\theta = \lambda^{-1} = b^c$, the conjugate prior of $\Theta$ is given by the **inverted gamma distribution**.[9] The prior density of this distribution is

$$g(\theta \mid \mu, \nu) = \left(\frac{\mu}{\theta}\right)^{\nu+1} \frac{\exp\left(-\dfrac{\mu}{\theta}\right)}{\mu\,\Gamma(\nu)}\,; \quad \theta > 0;\; \mu, \nu > 0 \tag{14.21a}$$

---

[9] See CANAVOS (1983), CANAVOS/TSOKOS (1973), PAPADOPOULUS/TSOKOS (1975) and TSOKOS (1972).

with mean and variance

$$E(\Theta \mid \mu, \nu) = \frac{\mu}{\nu - 1}, \quad \nu > 1, \tag{14.21b}$$

$$\mathrm{Var}(\Theta \mid \mu, \nu) = \frac{\mu^2}{(\nu - 1)^2 \, (\nu - 2)}, \quad \nu > 2. \tag{14.21c}$$

The posterior density is

$$g(\theta \mid T, r, \mu, \nu) = \left(\frac{T + \mu}{\theta}\right)^{r + \nu - 1} \frac{\exp\left(-\dfrac{T + \mu}{\theta}\right)}{(T + \mu)\,\Gamma(r + \nu)}, \tag{14.21d}$$

and the parameters have changed as follows:

$$\nu \implies r + \nu, \qquad \mu \implies T + \mu.$$

The posterior mean (= BAYESIAN estimator under squared loss) and posterior variance are

$$E(\Theta \mid T, r, \mu, \nu) = \widehat{\theta}_B = \frac{T + \mu}{r + \nu - 1}, \quad r + \nu > 1, \tag{14.21e}$$

$$\mathrm{Var}(\Theta \mid T, r, \mu, \nu) = \frac{(T + \mu)^2}{(r + \nu - 1)^2 \, (r + \mu - 2)}, \quad r + \nu > 2. \tag{14.21f}$$

## 14.2.2 Random shape parameter and known scale parameter[10]

There are only two papers on BAYESIAN estimation when the WEIBULL shape parameter is random. The first paper of HARRIS/SINGPURWALLA (1968) assumes that the shape parameter $c$ in the parametrization of (14.9a) has a **two–point prior**:

$$\Pr(C = c) = \left\{ \begin{array}{ll} p & \text{for} \quad c = c_1 \\ 1 - p & \text{for} \quad c = c_2 \end{array} \right\}, \quad 0 < p < 1, \tag{14.22a}$$

and that the loss function might be defined by the following **loss table**, where $l_1$, $l_2 > 0$:

| chosen value \ true value | $c_1$ | $c_2$ |
|---|---|---|
| $c_1$ | 0 | $l_2$ |
| $c_2$ | $l_1$ | 0 |

The $c$ that minimizes expected risk would be that $c$ associated with

$$\min_{c_1, c_2}\{l_2\,(1 - p)\,f(\boldsymbol{x} \mid c_2),\ l_1\,p\,f(\boldsymbol{x} \mid c_1)\},$$

and thus

$$\widehat{c} = c_1$$

---

[10] Suggested reading for this section: HARRIS/SINGPURWALLA (1968), TSOKOS (1972).

if

$$l_1 \, p_1 \, f(\boldsymbol{x} \,|\, c_1) < l_2 \, (1 - p_2) \, f(\boldsymbol{x} \,|\, c_2)$$

or — for a non–censored sample — if

$$\frac{l_1 \, p_1}{l_2 \, (1 - p_2)} \left(\frac{c_1}{c_2}\right)^n \prod_{i=1}^{n} x_i^{c_1 - c_2} \, \exp\left\{-\lambda \left[\sum_{i=1}^{n} x_i^{c_1} - \sum_{i=1}^{n} x_i^{c_2}\right]\right\} < 1, \qquad (14.22\text{b})$$

and $\widehat{c} = c_2$ when the inequality (14.22b) is reversed.

Assuming a **uniform prior** for the shape parameter $c$ as done by TSOKOS (1972):

$$g(c \,|\, A, B) = \left\{ \begin{array}{cc} \dfrac{1}{B - A} & \text{for } A \le c \le B; \ A, B > 0 \\[2mm] 0 & \text{elsewhere} \end{array} \right\}, \qquad (14.23\text{a})$$

we arrive at the following posterior density upon combining (14.23a) with the likelihood

$$L(c \,|\, \boldsymbol{x}, n, r) = \frac{n!}{(n - r)!} \left\{ \left(\frac{c}{\theta}\right)^r \prod_{i=1}^{r} x_{i:n}^{c-1} \, e^{-1/\theta} \left[\sum_{i=1}^{r} x_{i:n}^{c} + (n - r) \, x_{r:n}^{c}\right] \right\} \qquad (14.23\text{b})$$

of a type–II censored sample:

$$g(c \,|\, \boldsymbol{x}, n, r, A, B) = \frac{c^r \prod_{i=1}^{r} x_{i:n}^{c} \, e^{-1/\theta} \left[\sum_{i=1}^{r} x_{i:n}^{c} + (n - r) \, x_{r:n}^{c}\right]}{\int_{A}^{B} c^r \prod_{i=1}^{r} x_{i:n}^{c} \, e^{-1/\theta} \left[\sum_{i=1}^{r} x_{i:n}^{c} + (n - r) \, x_{r:n}^{c}\right] \mathrm{d}c}. \qquad (14.23\text{c})$$

There exists no closed form solution of the integral which appears in the denominator of (14.23c). So we have to utilize numerical techniques to obtain the BAYESIAN estimator $\widehat{c}_B$:

$$\widehat{c}_B = \int_{A}^{B} c \, g(c \,|\, \boldsymbol{x}, n, r, A, B) \, \mathrm{d}c. \qquad (14.23\text{d})$$

### 14.2.3 Random scale and random shape parameters[11]

The first paper on BAYESIAN estimation of both randomly varying WEIBULL parameters is that of SOLAND (1969a). As it is not possible to find a family of continuous joint prior distributions on the two parameters that is closed under sampling, SOLAND used a family of prior distributions that places continuous distributions on the scale parameter and discrete distributions on the shape parameter.

---

[11] Suggested reading for this section: CANAVOS/TSOKOS (1973), SINHA/GUTTMAN (1988), SOLAND (1969a), SUN (1997), TSOKOS/RAO (1976). With respect to BAYESIAN shrinkage estimators, the reader is referred to PANDEY/UPADHYAY (1985).

SINHA/GUTTMAN (1988) suppose that little is known a–priori about $\theta$ and $c$, so that JEF-FREY's **vague prior** is appropriate for this situation,[12] i.e.

$$g(\theta, c) \propto \frac{1}{\theta\, c}\,. \tag{14.24a}$$

(14.24a) combined with the likelihood of a non–censored sample

$$L(\theta, c \mid \boldsymbol{x}) \propto \left(\frac{c}{\theta}\right)^n \prod_{i=1}^{n} x_i^{c-1} \exp\left[-\frac{1}{\theta} \sum_{i=1}^{n} x_i^c\right] \tag{14.24b}$$

gives the following joint posterior of $(\theta,\ c)$:

$$g(\theta, c \mid \boldsymbol{x}) \propto \frac{c^{n-1}}{\theta^{n+1}} \prod_{i=1}^{n} x_i^{c-1} \exp\left[-\frac{1}{\theta} \sum_{i=1}^{n} x_i^c\right]. \tag{14.24c}$$

From (14.24c), after integrating out $\theta$, we have the **marginal posterior of** $c$

$$g(c \mid \boldsymbol{x}) = K_{c,n}\, c^{n-1} \prod_{i=1}^{n} x_i^{c-1} \bigg/ \left(\sum_{i=1}^{n} x_i^c\right)^n, \tag{14.25a}$$

where

$$K_{c,n}^{-1} = \int_0^\infty c^{n-1} \prod_{i=1}^{n} x_i^{c-1} \bigg/ \left(\sum_{i=1}^{n} x_i^c\right)^n \mathrm{d}c. \tag{14.25b}$$

Assuming the squared–error–loss function, the BAYESIAN estimator follows as

$$\widehat{c}_B = \mathrm{E}(C \mid \boldsymbol{x}) = K_{c,n} \int_0^\infty c^n \prod_{i=1}^{n} x_i^{c-1} \bigg/ \left(\sum_{i=1}^{n} x_i^c\right)^n \mathrm{d}c. \tag{14.25c}$$

At this point we again have to resort to numerical integration for (14.25b,c) because the integrals involved do not exist in closed form. Similarly, the **marginal posterior of** $\theta$ is found after integrating out $c$ in (14.24c):

$$g(\theta \mid \boldsymbol{x}) = K_{\theta,n}\, \theta^{-(n+1)} \int_0^\infty c^{n-1} \prod_{i=1}^{n} x^{c-1} \exp\left[-\frac{1}{\theta} \sum_{i=1}^{n} x_i^c\right] \mathrm{d}c, \tag{14.26a}$$

where

$$K_{\theta,n}^{-1} = \Gamma(n) \int_0^\infty c^{n-1} \prod_{i=1}^{n} x_i^c \bigg/ \left(\sum_{i=1}^{n} x_i^c\right)^n \mathrm{d}c. \tag{14.26b}$$

The BAYESIAN estimator of $\theta$ is given by numerical integration:

$$\widehat{\theta}_B = \mathrm{E}(\Theta \mid \boldsymbol{x}) = K_{\theta,n}\, \Gamma(n-1) \int_0^\infty c^{n-1} \prod_{i=1}^{n} x_i^{c-1} \exp\left[-\frac{1}{\theta} \sum_{i=1}^{n} x_i^c\right] \mathrm{d}c. \tag{14.26c}$$

---

[12] A thorough discussion of non–informative priors for WEIBULL parameters is found in SUN (1997).

A BAYESIAN solution by assuming independent distributions of $\Theta$ and $C$ has been found by CANAVOS/TSOKOS (1973). Specifically, they consider the inverted gamma prior distribution (14.21a) and the uniform prior distribution $g(\theta) = 1/\delta$ $(0 < \theta < \delta)$ for the scale parameter while for the shape parameter the uniform prior $g(c) = 1/(B - A)$ with $A \leq c \leq B$ is chosen. The resulting BAYESIAN estimators under quadratic loss function have complicated formulas which have to be evaluated by numerical integration.

## 14.3    Empirical BAYES estimation[13]

The concept of empirical BAYES, EB for short, introduced by ROBBINS (1955), is in some sense a hybrid combination of both classical methods and BAYESIAN methods, due to the fact that the parameter is treated as a random variable having an unknown prior distribution with a frequency interpretation. There are two main classes of EB estimators:

1. The first class consists of those methods that attempt to approximate the BAYESIAN estimator without explicitly estimating the unknown prior distribution. This is the traditional EB procedure as introduced by ROBBINS (1955).

2. The second class of EB estimators consists of those methods in which the unknown prior distribution is explicitly estimated. The so–called **smooth EB methods** developed by BENNETT/MARTZ (1973), LEMON/KRUTCHKOFF (1969) and MARITZ (1967, 1970) are based on this approach. These methods have the appeal of being universally applicable regardless of the specific distributional family to which the conditional density function $f(x \mid \theta)$ belongs. So we will present this type of EB estimator.

Assume that the problem of estimating a realization $\theta$ of a stochastic parameter $\Theta$, when a random sample of measurements $\boldsymbol{x} = (x_1, \ldots, x_n)'$ has been observed, occurs periodically and independently with likelihood function $L(\theta \mid \boldsymbol{x})$ and a fixed but unknown prior distribution $G(\theta)$. Let

$$\left(\boldsymbol{x}_1; \widehat{\theta}_1\right), \ \left(\boldsymbol{x}_2; \widehat{\theta}_2\right), \ \ldots, \ \left(\boldsymbol{x}_k; \widehat{\theta}_k\right); \ \ k \geq 2 \tag{14.27}$$

denote the sequence of experiences (or samples) where $\boldsymbol{x}_j$ is the vector of measurements of length $n_j$ from the $j$–th experiment, and $\widehat{\theta}_j$ is the corresponding MVU estimator of $\theta_j$. It is possible to estimate the current realization $\theta_k$, when $\boldsymbol{x}_k$ has been observed, by constructing an EB decision function of the form

$$\delta_k\left(\boldsymbol{x}_k\right) = \frac{\displaystyle\sum_{j=1}^{k} \theta_j \, h\left(\widehat{\theta}_k \mid \theta_j\right) \mathrm{d}G_k(\theta_j)}{\displaystyle\sum_{j=1}^{k} h\left(\widehat{\theta}_k \mid \theta_j\right) \mathrm{d}G_k(\theta_j)}, \tag{14.28}$$

where $h\left(\widehat{\theta}_k \mid \theta_j\right)$ is the density function of the MVU estimator $\widehat{\theta}_k$.

---

[13]   Suggested reading for this section: BENNETT (1977), BENNETT/MARTZ (1973), CANAVOS (1983), COUTURE/MARTZ (1972), DEY/KUO (1991), MARTZ/WALLER (1982, Chapter 13).

MARITZ (1967) and LEMON/KRUTCHKOFF (1969) suggested replacing the unknown prior distribution in (14.28) by a step function having steps of equal height $1/k$ at each of the $k \geq 2$ classical estimates $\widehat{\theta}_1, \ldots, \widehat{\theta}_k$ yielding a smooth EB estimator of the form

$$D_k(1) = \frac{\sum\limits_{j=1}^{k} \widehat{\theta}_j \, h\big(\widehat{\theta}_k \,|\, \widehat{\theta}_j\big)}{\sum\limits_{j=1}^{k} h\big(\widehat{\theta}_k \,|\, \widehat{\theta}_j\big)} \, , \tag{14.29}$$

where $\widehat{\theta}_j$ is the MVU estimate of $\theta_j$ determined as a result of the $j$–th experiment.

The rationale for this procedure is easily understood. If the decision process was such that the true value of $\theta$ was revealed immediately after each decision, then one could use an estimate of $dG(\theta)$, the frequency with which each value of $\theta$ occurred in $k$ repetitions of the decision-making process. Thus, $dG(\theta)$ could be approximated by

$$d\widehat{G}(\theta) = \frac{1}{k} \sum_{j=1}^{k} \delta(\theta_j, \theta),$$

where

$$\delta(\theta_j, \theta) = \begin{cases} 1 & \text{if } \theta_j = \theta \\ 0 & \text{if } \theta_j \neq \theta, \end{cases}$$

and

$$\mathrm{E}\big(\Theta \,|\, \widehat{\theta}_k\big) \approx \frac{\sum\limits_{\theta} \theta \, h\big(\widehat{\theta}_k \,|\, \theta\big) \, d\widehat{G}(\theta)}{\sum\limits_{\theta} h\big(\widehat{\theta}_k \,|\, \theta\big) \, d\widehat{G}(\theta)}$$

$$= \frac{\sum\limits_{j=1}^{k} \theta_j \, h\big(\widehat{\theta}_k \,|\, \theta_j\big)}{\sum\limits_{j=1}^{k} h\big(\widehat{\theta}_k \,|\, \theta_j\big)} \, . \tag{14.30}$$

Since the values $\theta_1, \ldots, \theta_k$ generally remain unknown, we simply replace each $\theta_j$ in (14.30) by its corresponding estimate $\widehat{\theta}_j$ and obtain the EB estimator as defined by (14.29). Also suggested by LEMON/KRUTCHKOFF (1969), as a second possible EB estimator, is the estimator obtained by a second iteration of (14.29) with each classical estimate $\widehat{\theta}_j$ replaced by the corresponding EB estimate $D_j(1)$ from the first iteration. This estimator may be denoted by $D_k(2)$.

Along the lines of the previous paragraph we will now give a smooth EB estimator for the WEIBULL scale parameter $\theta$ (see (14.10a,b)), when the shape parameter $c$ is assumed known.[14] Suppose the existence of a sequence of experiences such as (14.27), where each

---

[14] Estimation of the shape parameter is treated by COUTURE/MARTZ (1972) and MARTZ/WALLER (1982, pp. 639–640).

measurement is a value of a variate having a density function given by (14.10a). It is known that for the $j$–th experiment the MVU estimator of $\theta_j$ is given by

$$\widehat{\theta}_j = \frac{1}{n_j} \sum_{i=1}^{n_j} x_{ij}^{\alpha}; \quad j = 1, 2, \ldots, k \tag{14.31a}$$

while the distribution of the current MVU estimator $\widehat{\theta}_k$ is gamma with density function

$$h\big(\widehat{\theta}_k \mid \theta\big) = \frac{n_k^{n_k}}{\Gamma(n_k)\,\theta^{n_k}}\, \widehat{\theta}_k^{\,n_k-1}\, \exp\big\{-n_k\,\widehat{\theta}_k/\theta\big\}, \quad \widehat{\theta}_k > 0. \tag{14.31b}$$

Upon substitution of (14.31b) into (14.29) and after simplification, one obtains the smooth EB estimator

$$\widehat{\theta}_k = \frac{\displaystyle\sum_{j=1}^{k} \left[ \exp\big\{-n_k\,\widehat{\theta}_k/\widehat{\theta}_j\big\} \Big/ \widehat{\theta}_j^{\,n_k-1} \right]}{\displaystyle\sum_{j=1}^{k} \left[ \exp\big\{-n_k\,\widehat{\theta}_k/\widehat{\theta}_j\big\} \Big/ \widehat{\theta}_j^{\,n_k} \right]} \tag{14.31c}$$

for the current realization $\theta_k$ as a function of the $k$ experiences.

# 15 Parameter estimation — Further approaches

The prevailing statistical philosophy underlying statistical inference is either the classical approach (see Chapters 10 – 13) or the BAYESIAN approach (see Chapter 14). But there are some more approaches, e.g., fiducial inference, likelihood inference and structural inference,[1] which are of minor importance, especially in inference of the WEIBULL distribution. This chapter will briefly present the fiducial and the structural approaches.

## 15.1 Fiducial inference[2]

It has been said that fiducial inference as put forward by R.A. FISHER is not so much a theory as a collection of examples. The subject has been one of confusion and controversy since its introduction in 1930.

### 15.1.1 The key ideas

FISHER considered inference about a real parameter $\theta$ from a maximum likelihood estimate $T$, having a continuous cumulative distribution $F(T, \theta)$. If $F(T, \theta) = 1 - P$ has a unique solution $\theta_P(T)$, this is the fiducial $100\,P$ percent point of $\theta$. For the case when $\theta_P(T)$ increases with $T$, $\Pr\left[\theta < \theta_P(T)\right] = P$ and $P$ is the **fiducial probability**, equal to the confidence level for the $\theta$–intervals $\{\theta < \theta_P(T)\}$. If, for fixed $T$, the definable fiducial probabilities take all values in $(0, 1)$, the pairs $[P,\ \theta_P(T)]$, $0 < P < 1$ formally constitute a cumulative distribution for what FISHER called the **fiducial distribution**. When $F(T, \theta)$ is differentiable with respect to $\theta$, the fiducial distribution has a formal density $\mathrm{d}f = -\partial F(T, \theta)/\partial\theta\,\mathrm{d}\theta$, while the distribution of the statistic for a given value of the parameter is $\mathrm{d}f = \partial F(T, \theta)/\partial T\,\mathrm{d}T$. FISHER was keen to promote fiducial probability against posterior probability.

Several key ideas can be illustrated by the case of a single observation $X$ from a normal distribution with mean $\mu$ and unit variance, i.e., $X \sim No(\mu, 1)$. If we make $Z = X - \mu$, then $Z \sim No(0, 1)$. A quantity like $Z$, which depends on the observation $X$ and the parameter $\mu$ and whose distribution is free of the parameter, is called a **pivotal quantity** or **pivot**. The fiducial argument consists of writing $\mu = X - Z$ and asserting that when we have no knowledge about $\mu$ except the value $x$, our uncertainty about $\mu$ is summarized by saying that $\mu$ equals $x$ minus an unknown value of a standard normal variate. In short, $\mu \sim No(x, 1)$, which is the fiducial distribution of $\mu$. The values $x \pm 1.96$ include all but $5\%$ of the distribution and so would be called the $95\%$ fiducial limits of $\mu$.

---

[1] A comparative description of these inference approaches is given by BARNETT (1973).

[2] Suggested reading for this section: PEDERSEN (1978).

### 15.1.2 Application to the WEIBULL parameters

The general results of HORA/BUEHLER (1966) can be applied directly to the two–parameter WEIBULL distribution, but the concepts involved and the relationship to BAYESIAN methods are more transparent in the setting of the location and scale parameter extreme value distribution of the logarithm of the WEIBULL data.

If $X \sim We(0, b, c,)$, then $Y = \ln X$ has the extreme value distribution with location and scale parameters that can be taken as $a^* = \ln b$ and $b^* = 1/c$, respectively. For simplicity in presentation consider the case of uncensored data, and write $\widehat{a^*}$ and $\widehat{b^*}$ for the maximum likelihood estimators based on a sample $y_i = \ln x_i$, $i = 1, 2, \ldots, n$.

Fiducial inferences (see BOGDANOFF/PIERCE (1973)) are obtained from the known distributions of the pivotal quantities $(\widehat{a^*} - a^*)/\widehat{b^*}$ and $\widehat{b^*}/b^*$,[3] except that the distributions used are those conditional on the ancillary statistic

$$z' = (z_3, \ldots, z_n)' = \left( \frac{y_3 - y_1}{y_2 - y_1}, \ldots, \frac{y_n - y_1}{y_2 - y_1} \right)'. \tag{15.31a}$$

For example, let $\zeta$ be the function of $z$ such that

$$\Pr\left( \frac{\widehat{a^*} - a^*}{\widehat{b^*}} \geq -\zeta \,\Big|\, z \right) = \gamma. \tag{15.31b}$$

For the data $(\widehat{a^*}, \widehat{b^*}, z)$ the $\gamma$–level upper fiducial limit for $a^*$ is given by $\widehat{a^*} + \zeta \widehat{b^*}$. Similarly the conditional distribution of $\widehat{b^*}/b^*$ given $z$ yields fiducial limits for $b^*$. Fiducial limits for certain types of functions $\theta(a^*, b^*)$, e.g., for percentiles of $Y$, can be obtained in similar ways.

It has been shown by HORA/BUEHLER (1960) that for any location and scale parameter problem fiducial limits obtained in this manner are exactly the same as BAYESIAN limits using the improper prior distribution

$$g(a^*, b^*) \propto \frac{1}{b^*}; \quad a^* \in \mathbb{R}, \, b^* > 0. \tag{15.31c}$$

Thus in the preceding case the posterior distribution of $a^*$ satisfies

$$\Pr(a^* \leq \widehat{a^*} + \zeta \widehat{b^*} \,|\, z) = \gamma. \tag{15.31d}$$

These results hold not only for the case of uncensored data but for any censoring procedure such that the problem remains one of location and scale. Type–II censoring clearly satisfies this condition, whereas type–I censoring does not.

## 15.2 Structural inference

The method of structural inference was developed by D.A.S. FRASER, first announced in FRASER (1966). The monograph FRASER (1968) details the theory.

---

[3] For these pivotal distributions see Sect. 11.3.1.2.

### 15.2.1 The key ideas

The BAYESIAN method constructs probability density functions for the parameters of a statistical model. BAYESIAN conclusions are therefore strong in the sense that probability statements on parameter values can be constructed. On the other hand, the BAYESIAN method necessitates the specification of a prior parameter density function on the basis of non–data information. Because this specification carries an element of subjectivity, the resulting conclusions are correspondingly subjective. An alternative approach to inference, that of **structural inference**, retains the power of the BAYESIAN method in that it also constructs probability densities of model parameters; on the other hand, the specification of a "prior" density is made on the basis of the **structure** inherent in the model underlying the analysis.

The structural approach is also firmly rooted in the classical tradition in many aspects. There are, however, two absolutely crucial differences of attitude as compared with the classical approach. These concern

1. the formulation of the classical model and

2. the manner in which inferences are expressed about unknown quantities ("physical constants, relationships"), broadly interpretable as what we have previously called parameters.

With respect to the **statistical model** a distinction is drawn at the outset between the "exterior" nature of the usual classical model, and the concern of the "measurement model" (or its generalization as a "structural model") of the structural approach with the internal mechanism governing the observable behavior of the situation under study. In the classical model we typically observe some "response variable" whose observed value $x$, as a reflection of an unknown quantity $\theta$, is governed by a probability distribution $p_\theta(x)$ on the sample space. FRASER refers to this as a "black box" approach, which ignores the fact that we often know (directly or indirectly) much more about the internal relationship of $x$ and $\theta$ than is represented through such a model. He argues that situations commonly contain identifiable sources of variation, such as errors of measurement, variation in the quality of products, and effect of randomization in designed experiments. He calls these **error variables**, though in the wider applications of the method this term needs to be liberally interpreted. In many cases they truly determine the probabilistic structure of the situation, their variational behavior is well understood, and they constitute the natural basis for expressing a statistical model.

These considerations provide the motivation for formulating the statistical model in a particular way. Instead of regarding the data $x$ as being generated by the distribution $p_\theta(x)$, a quantity $e$ (the error variable) is introduced which is expressed in terms of $x$ and $\theta$, and which is assumed to have a known distribution which does not depend on $\theta$. Thus the statistical model consists of two parts:

- a statement of the probability distribution of the error variable (independent of $\theta$) and

- a statement of the relationship between the observational data $x$ (which are known) and the unknown $\theta$, on the one hand, and the unknown but realized value $e$ of the error variable on the other.

With respect to **structural inference** the structural model tells us how $x$ is related to $\theta$ and $e$ and also the probability mechanism under which $e$ is generated. This relationship frequently expresses the observed response (the data $x$) in terms of a simple kind of transformation of the error $e$, governed by the value of $\theta$. The basis for statistical inference in the structural approach is to essentially reverse this relationship and to interpret $\theta$ as being obtained as an associated transformation of $e$ by the operation of the unknown data $x$. Thus $\theta$ is expressed in terms of $x$ and $e$, the probability mechanism by which $e$ is obtained is assumed known (in principle), so that through this inverse relationship we have a probability statement concerning $\theta$.

Let us consider a simple example referring to an unknown location parameter $\theta$. The random error variable $e$ has two realizations (tossing of a coin):

$$\Pr(e = +1) = P \quad \text{and} \quad \Pr(e = -1) = 1 - P.$$

The coin is tossed, but the result is kept secret and only the sum $x = \theta + e$ is announced. The statistician now says about the parameter $\theta$ that with probability $P$ we have the parameter value $x - 1$ and with probability $1 - P$ the parameter value will be $x + 1$ because $e$ takes on the values $+1$ and $-1$ with these probabilities and we have $\theta = x - e$. The probability distribution of $e$ which depends neither on $x$ nor on $\theta$ is transferred to $\theta$ based upon the relationship $\theta = x - e$.

### 15.2.2  Application to the WEIBULL parameters[4]

We take an uncensored sample of size $n$ from a WEIBULL population with density function

$$f(x \mid b, c) = \frac{c}{b} \left(\frac{x}{b}\right)^{c-1} \exp\left\{-\left(\frac{x}{b}\right)^c\right\} \quad \text{for } x \geq 0. \tag{15.5a}$$

Consider a logarithmic transformation of the measurement variable $x$

$$y = \ln x, \quad \mathrm{d}x/\mathrm{d}y = \exp(x), \tag{15.5b}$$

which maps the measurement density (15.5a) onto the location–scale model

$$g(y \mid a^*, b^*) = \frac{1}{b^*} \exp\left\{\frac{x - a^*}{b^*} - \exp\left(\frac{x - a^*}{b^*}\right)\right\}, \quad y \in \mathbb{R}, \tag{15.5c}$$

where $a^* = \ln b$ and $b^* = 1/c$. The joint structural probability density for the parameters of a general location–scale model is derived by FRASER (1968, p. 64), and is given as

$$\varphi^+\left(a^*, b^* \mid y_1, \ldots, y_n\right) = k \left(\frac{s_y}{b^*}\right)^n \prod_{i=1}^{n} g^+\left(y_i \mid a^*, b^*\right) \frac{1}{s_y \, b^*}. \tag{15.5d}$$

Here, the measurement model $g^+(\cdot)$ is given by (15.5c), $k$ is a normalizing constant, and $s_y$ is the sample standard deviation in terms of the transformed data $y_i$.

---

[4]  Suggested reading for this section: BURY (1973), BURY/BERNHOLTZ (1971), SHERIF/TAN (1978), TAN/SHERIF (1974).

In the density (15.5d), the parameters $a^*$ and $b^*$ are treated as random variables, whereas the sample $x_1, \ldots, x_n$ is taken as a known set of fixed numbers which condition the density (15.5d). Substituting for $g^+(\cdot)$ in (15.5d), transforming variables from $(a^*, b^*)$ to $(b, c)$, and changing constants from $y_i$ to $x_i$, the **joint structural density** is obtained **for the WEIBULL parameters** $b$ and $c$:

$$\varphi(b, c \mid x_1, \ldots, x_n) = k_1 \, b^{-nc-1} \, c^{n-1} \prod_{i=1}^{n} x_i^c \, \exp\left\{ -b^{-c} \sum_{i=1}^{n} x_i^c \right\}. \tag{15.6a}$$

The normalizing constant $k_1$ is given by

$$k_1^{-1} = \Gamma(n) \int_0^\infty c^{n-2} \prod_{i=1}^{n} x_i^c \left( \sum_{i=1}^{n} x_i^c \right)^{-n} dc. \tag{15.6b}$$

The joint density (15.6a) is conditional on the the the sample $x_1, \ldots, x_n$ only; no prior information is introduced to the analysis.

The **marginal structural density of the WEIBULL shape parameter** $c$ is obtained from (15.6a) as

$$\varphi(c \mid x_1, \ldots, x_n) = k_1 \, \Gamma(n) \, c^{n-2} \prod_{i=1}^{n} x_i^c \left( \sum_{i=1}^{n} x_i^c \right)^{-n}, \tag{15.7a}$$

and the **marginal density of the scale parameter** $b$ is similarly found as

$$\varphi(b \mid x_1, \ldots, x_n) = k_1 \, b^{-1} \int_0^\infty c^{n-1} \prod_{i=1}^{n} x_i^c \, b^{-nc} \, \exp\left\{ -b^{-c} \sum_{i=1}^{n} x_i^c \right\} dc. \tag{15.7b}$$

Given a sample of $n$ observations $x_i$ from a two–parameter WEIBULL population, the densities (15.7a,b), their moments and corresponding distribution functions can be computed. Probability statements concerning the possible parameter values for the population sampled follow directly. It is to be emphasized that such statements are probabilities in the classical sense; they are not "confidence" statements with a sample frequency interpretation, nor are they BAYESIAN probabilities which require the assumption of a prior parameter density

Occasionally the nature of the statistical analysis is such that conditional parameter densities are derived from the joint density (15.6a) as

$$\varphi(c \mid b_0; x_1, \ldots, x_n) = k_2 \, c^{n-1} \prod_{i=1}^{n} x_i^c \, \exp\left\{ -b_0^{-c} \sum_{i=1}^{n} x_i^c \right\} \tag{15.8a}$$

and

$$\varphi(b \mid c_0; x_1, \ldots, x_n) = c_0 \, b^{-nc_0-1} \, [\Gamma(n)]^{-1} \left( \sum_{i=1}^{n} x_i^{c_0} \right)^n \exp\left\{ -b^{c_0} \sum_{i=1}^{n} x_i^{c_0} \right\}, \tag{15.8b}$$

where $k_2$ is a normalizing constant.

# 16 Parameter estimation in accelerated life testing[1]

There are two methods to shorten the test duration of a planned life test, namely acceleration and censoring. Of course, both instruments may be combined. Engineers reduce the testing time by applying some sort of stress on the items on test and thus shorten the individual lifetimes, whereas statisticians limit the test duration by fixing it more or less directly via some sort of censoring.

We will first (Sect. 16.1) review the most common relationships connecting a stress variable to some quantifiable life measure. The following sections deal with parameter estimation when accelerated life testing (ALT) is done in one of three different ways,[2] each using the aforementioned parametric life–stress functions. The main assumption in ALT is that the acceleration factor is known or that there is a known mathematical model to justify the relation between lifetime and stress. In some situations such models do not exist or are very hard to assume. Therefore, the partially acceleration life test (PALT) is a good candidate to perform the life test in such cases. Also, PALT is used for problems where it is desired to test only at a specified acceleration condition and then the data are extrapolated to normal use condition. Models for PALT are presented in Sect. 16.5.

## 16.1 Life–stress relationships

Sometimes we have good reasons based on physical and/or chemical considerations to use a special mathematical function relating some quantifiable life measure to one or more stress variables. This function contains unknown parameters which have to be estimated together with the WEIBULL parameters. The estimation procedure to be used in the following sections will be maximum likelihood. It offers a very powerful method in estimating all the parameters appearing in accelerated testing models, making possible the analysis of very complex models.[3]

The quantifiable life measure mostly used in WEIBULL–ALT is the **characteristic life** (see (2.57b) and (2.58d)), which is the percentile of order $P = (e-1)/e \approx 0.6321$:

$$x_{0.6321} = a + b \left(\ln e\right)^{1/c} = a + b. \tag{16.1a}$$

---

[1] An excellent monograph on accelerated life testing, including different lifetime distributions, is NELSON (1990).

[2] See Sect. 8.2 for a description of the common types of acceleration of tests.

[3] There are some papers using the linear estimation approach, e.g. ESCOBAR/MEEKER (1986b), MANN (1972a, 1978b), MAZZUCHI/SOYER/NOPETEK (1997) or NELSON (1990, chapter 4). BUGAIGHIS (1988) has compared MLE and BLUE for parameters of an ALT model and stated that the two kinds of estimators are of comparable efficiencies, measured by MSE. However, where there are two or fewer uncensored observations, at any level(s) of the acceleration variable, the MLE is clearly favored.

When $X \sim We(0, b, c)$, which is always assumed to hold in ALT, we have

$$x_{0.6321} = b. \tag{16.1b}$$

Thus, the **scale parameter**[4] is seen to **depend on** some non–random **stress variable** $s$ or variables $s_1, s_2, \ldots, s_\ell$. The single stress relationships are

- the ARRHENIUS **relationship**,
- the EYRING **relationship**,
- the **inverse power relationship** sometimes called the **inverse power law** or simply the **power law** and
- other single stress relationships.

Multivariate relationships are

- a **general log–linear function**,
- **temperature–humidity dependence** and
- **temperature–non thermal dependance**.

### ARRHENIUS life–stress model

The ARRHENIUS relationship is widely used to model product life as a function of temperature. It is derived from the reaction rate equation proposed by the Swedish physical chemist SVANTE AUGUST ARRHENIUS (1859 – 1927) in 1887:

$$r = A^+ \exp\{-E/(kT)\}, \tag{16.2}$$

where

- $r$ is the speed of a certain chemical reaction;
- $A^+$ is an unknown non–thermal constant, that is, a characteristic of the product failure mechanism and test conditions;
- $E$ is the activity energy of the reaction, usually in electron–volts;
- $k$ is the BOLTZMANN's constant ($8.6171 \times 10^{-5}$ electron–volts per CELSIUS degree),
- $T$ is the absolute temperature expressed in KELVIN grades (see the following excursus on temperature scales).

Product failure is a result of degradation due to chemical reactions or metal diffusion. The product is assumed to fail when some critical amount of the chemical has reacted (or diffused):

$$\text{critical amount} = \text{rate} \times \text{time to failure}$$

or

$$\text{time to failure} = \frac{\text{critical amount}}{\text{rate}},$$

saying that **nominal life** $\tau$ to failure ("life") is inversely proportional to the rate given in (16.2):

$$\tau = A \exp\{E/(kT)\}, \tag{16.3a}$$

---

[4] GLASER (1984, 1995) supposes both the shape and the scale parameters are expressible as functions of the testing environment.

where $A$ is a constant depending on the the size and geometry of the product and its fabrication and testing method. The natural logarithm of (16.3a) is

$$\ln \tau = \gamma_0 + \frac{\gamma_1}{T}, \tag{16.3b}$$

where $\gamma_0 := \ln A$ and $\gamma_1 := E/k$. Thus, the log of nominal life $\tau$ is a linear function of the inverse absolute temperature $s := 1/T$. Raising the temperature (equivalent to lessening $s$) should reduce $\tau$, so $\gamma_1$ must be positive. Nominal life $\tau$ is usually taken to be the WEIBULL scale parameter $b$ which also is the percentile of order $P \approx 0.6321$. Thus, an increasing temperature causes the proportion $0.6321$ of failing items to occur within a shorter span of time; i.e., raising the temperature causes a displacement of the distribution to the left. The **ARRHENIUS acceleration factor** $AF_A$ refers to the ratio of the nominal life between the use level $T_u$ and a higher stress level $T_a$:

$$AF_A = \frac{\tau(T_u)}{\tau(T_a)} = \exp\left\{\gamma_1 \left(\frac{1}{T_u} - \frac{1}{T_a}\right)\right\}, \tag{16.3c}$$

saying that specimens run $AF_A$ times longer at temperature $T_u$ than at $T_a$.

The ARRHENIUS–WEIBULL MODEL combines a WEIBULL life distribution with an ARRHENIUS dependence of life on temperature with the following assumptions:

- At absolute temperature $T$, the product life $X \sim We(0, b(T), c)$ or $Y = \ln X \sim Ex_I^m\left(a^*(T), b^*\right)$, i.e., a type–I–minimum extreme value distribution with location parameter $a^*(T) = \ln b(T)$ and scale parameter $b^* = 1/c$; see (3.21a,b).

- The WEIBULL shape parameter $c$ is independent of the temperature as is the extreme value parameter $b^*$.

- The WEIBULL scale parameter is

$$b(T) = \exp\left\{\gamma_0 + \gamma_1/T\right\}, \tag{16.3d}$$

and the extreme value location parameter

$$a^*(T) = \ln b(T) = \gamma_0 + \gamma_1/T \tag{16.3e}$$

is a linear function of the inverse of $T$.

The parameters $c, \gamma_0$ and $\gamma_1$ are characteristic of the product and have to be estimated from data (see Sect. 16.2 to 16.4).

---

**Excursus: Temperature scales**

There exist several scales for measuring temperature. Most of them were established in the 18th century. Of major importance are the following:

- the **CELSIUS scale** (°C) of ANDERS CELSIUS (1701 – 1744) from 1742,

- the **KELVIN scale** (°K) of Lord KELVIN (WILLIAM THOMSON, 1824 – 1907) from 1848,

- the **FAHRENHEIT scale** (°F) of GABRIEL FAHRENHEIT (1686 – 1736) from 1714,

- the **RANKINE scale** (°Ra) of WILLIAM RANKINE (1820 – 1872) from 1859,

- the **RÉAUMUR scale** (°Re) of RÉNÉ RÉAUMUR (1683 – 1757) from 1730.

Other scales go back to JOSEPH–NICOLAS DELISLE (1688 – 1768) from 1732, Sir ISAAC NEWTON (1643 – 1727) about 1700 and OLE RØMER (1644 – 1710) from 1701.

At three prominent points we have the following correspondences:

- **Absolute zero**

$$0°K \equiv 0°Ra \equiv -273.15°C \equiv -459.67°F \equiv -218.52°Re,$$

- **Freezing point of water** (with normal atmospheric pressure)

$$0°C \equiv 273.15°K \equiv 32°F \equiv 491.67°Ra \equiv 0°Re,$$

- **Boiling point of water** (with normal atmospheric pressure)

$$100°C \equiv 373.15°K \equiv 212°F \equiv 671.67°Ra \equiv 80°Re.$$

Tab. 16/1 shows how to convert one scale another one.

Table 16/1:  Conversion formulas for temperature scales

| | $T_K$ | $T_C$ | $T_{Re}$ | $T_F$ | $T_{Ra}$ |
|---|---|---|---|---|---|
| $T_K =$ | – | $T_C + 273.15$ | $1.25\,T_{Re} + 273.15$ | $(T_F + 459.67)/1.8$ | $T_{Ra}/1.8$ |
| $T_C =$ | $T_K - 273.15$ | – | $1.25\,T_{Re}$ | $(T_F - 32)/1.8$ | $T_{Ra}/1.8 - 273.15$ |
| $T_{Re} =$ | $0.8\,(T_K - 273.15)$ | $0.8\,T_C$ | – | $(T_F - 32)/2.25$ | $T_{Ra}/2.25 - 218.52$ |
| $T_F =$ | $1.8\,T_K - 459.67$ | $1.8\,T_C + 32$ | $2.25\,T_{Re} + 32$ | – | $T_{Ra} - 459.67$ |
| $T_{Ra} =$ | $1.8\,T_K$ | $1.8\,T_C + 491.67$ | $2.25\,T_{Re} + 491.67$ | $T_F + 459.67$ | – |

## EYRING life–stress model

An alternative to the ARRHENIUS relationship for temperature acceleration is the EYRING relationship (HENRY EYRING, 1901 – 1981) based on quantum mechanics. However, the EYRING relationship is also often used for stress variables other than temperature, such as humidity. The relationship for nominal life $\tau$ (corresponding to the scale parameter $b$ in the WEIBULL case) as a function of absolute Temperature $T$ is

$$\tau = \frac{A}{T}\,\exp\{B/(k\,T)\}. \tag{16.4a}$$

Here $A$ and $B$ are constants characteristic of the product and its test method; $k$ is BOLTZ-MANN's constant. The **EYRING acceleration factor** $AF_E$ is given by

$$AF_E = \frac{\tau(T_u)}{\tau(T_a)} = \frac{T_a}{T_u}\,\exp\left\{\frac{B}{k}\left(\frac{1}{T_u} - \frac{1}{T_a}\right)\right\}. \tag{16.4b}$$

Comparing (16.4a) with the ARRHENIUS relationship (16.3a), it can be seen that the only difference between the two relationships is the factor $1/T$ in (16.4a). For the small range of absolute temperature in most applications, $A/T$ is essentially constant and (16.4a) is close to (16.3a). In general, both relationships yield very similar results. The **EYRING–WEIBULL model** is similar to the ARRHENIUS–WEIBULL model with the exception that

$b(T)$ is given by (16.4a) and that there is not such a simple relationship between $b(T)$ and $a^* = \ln b(T)$ of the Log–WEIBULL variate.

**Inverse power law (IPL)**

The inverse power law is commonly used for non–thermal accelerated stress, e.g., higher voltage. The inverse power relationship between nominal life $\tau$ of a product and a (positive) stress variable $V$ is

$$\tau(V) = a/V^{\gamma_1}, \tag{16.5a}$$

where $A$ (positive) and $\gamma_1$ are parameters to be estimated. Equivalent forms are

$$\tau(V) = (A^*/V)^{\gamma_1} \quad \text{and} \quad \tau(V) = A^{**}(V_0/V)^{\gamma_1},$$

where $V_0$ is a specified (standard) level of stress. The **acceleration factor** $AF_I$ **of the inverse power law** is

$$AF_I = \frac{\tau(V_u)}{\tau(V_a)} = \left(\frac{V_a}{V_u}\right)^{\gamma_1}. \tag{16.5b}$$

The natural logarithm of (16.5a) is

$$\ln \tau(V) = \gamma_0 + \gamma_1(-\ln V), \quad \gamma_0 = \ln A, \tag{16.5c}$$

showing that the log of nominal life, $\ln \tau$, is a linear function of the transformed stress $s = -\ln V$. The **IPL–WEIBULL model** is similar to the ARRHENIUS–WEIBULL model when (16.5a) holds for the scale parameter. When looking at the Log–WEIBULL variate (= type–I–minimum extreme value distribution), the location parameter $a^*(s) = \ln b(s)$ is of the same linear form in both models:

$$a^*(s) = \ln b(s) = \gamma_0 + \gamma_1 s, \tag{16.5d}$$

but the transformed stress $s$ is different:

$$s = \frac{1}{T} \quad \text{for the ARRHENIUS–WEIBULL model}$$

and

$$s = -\ln V \quad \text{for the IPL–WEIBULL model.}$$

**Other single stress relationships**

In the following formulas $s$ may be a transformed stress or an original stress variable:

- the **exponential relationship**

$$\tau = \exp(\gamma_0 + \gamma_1 s), \tag{16.6a}$$

- the **exponential–power relationship**

$$\tau = \exp\left(\gamma_0 + \gamma_1 s^{\gamma_2}\right), \tag{16.6b}$$

- the **polynomial relationship**

$$\ln \tau = \sum_{i=1}^{k} \gamma_i s^i. \tag{16.6c}$$

**Multivariate relationships**

A general and simple relationship for nominal life $\tau$ with two or more stress variables $s_j$, $j = 1, \ldots, J$ is the **log–linear form**

$$\ln \tau = \gamma_0 + \sum_{j=1}^{J} \gamma_j \, s_j. \tag{16.7}$$

The **temperature–humidity relationship**, a variant of the EYRING relationship, has been proposed when temperature $T$ and humidity $U$ are the accelerated stresses in a test. The combination model is given by

$$\tau(T, U) = A \, \exp\left\{ \frac{B}{T} + \frac{C}{U} \right\}. \tag{16.8}$$

When temperature $T$ and a second non–thermal variable $V$ are the accelerated stresses of a test, then the ARRHENIUS and the inverse power law models can be combined to yield

$$\tau(T, V) = A \, \exp(B/T)/V^C. \tag{16.9}$$

## 16.2 ALT using constant stress models

The most common stress loading is **constant stress**. Each specimen is run at a constant stress level and there must be at least two different levels in order to estimate all the parameters of the life–stress relationship. Fig. 16/1 depicts a constant–stress model with three stress levels. There the history of a specimen is depicted as moving along a horizontal line until it fails at a time shown by $\times$. A non–failed specimen has its age shown by an arrow.[5]

When in use, most products run at constant stress, thus a constant–stress test mimics actual use. Moreover, such testing is simple and has a number of advantages:

1. In most tests it is easier to maintain a constant stress level than changing the stress either in steps (see Sect. 16.3) or continuously (see Sect. 16.4).

2. Accelerated tests for constant stress are better developed and empirically verified for some materials and products.

3. Data analysis for reliability estimation is well developed and computerized.

Before turning to parameter estimation for different types of parameterized life–stress relationships in Sections 16.2.1 to 16.2.3, we present a model where the WEIBULL shape

---

[5] BUGAIGHIS (1995) examined by simulation whether failure times, actually derived under type–I censorship but treated as though type–II censorship were used, lead to an important reduction in efficiency of the MLEs. He showed that the exchange of censorship results in only minor reduction in the efficiency suggesting that results under type–II censorship apply to corresponding (in terms of observed failures) type–I situations. The only exception to this general rule occurs when the number of test items at each level of stress drops below 5.

Figure 16/1:  A constant-stress model

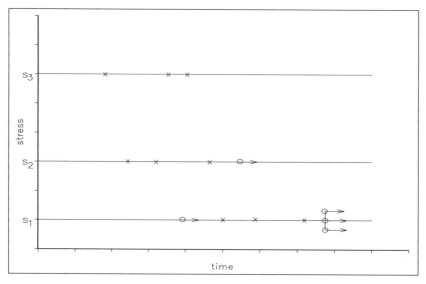

parameter is assumed to not depend on the stress level[6] but the scale parameter depends on stress either in a non–systematic way or according to an unknown function. The results are not suitable to extrapolate the scale parameter under normal design stress.

The definitions and assumptions of this model are as follows:

1. There are $m$ $(m \geq 2)$ stress levels $s_i$; $i = 1, \dots, m$.

2. The number of specimens tested at each level without censoring is $n$.

3. The observed life of specimen $j$ $(j = 1, \dots, n)$ at stress level $i$ is denoted by $X_{ij}$.

4. The lives $X_{i1}, \dots, X_{in}$ (at stress level $i$) are independently identically distributed as:

$$X_{ij} \overset{\text{iid}}{\sim} We(0, b_i, c) \; \forall \; i;$$

   i.e., the scale parameter $b_i$ depends on $i$, but the shape parameter does not vary with $i$.

From Sect. 11.2.2 we find the ML equation for a **pooled estimation** of $c$ by solving

$$\sum_{i=1}^{m} \left[ \frac{\sum\limits_{j=1}^{n} X_{ij}^c \ln X_{ij}}{\sum\limits_{j=1}^{n} X_{ij}^c} \right] - \frac{m}{c} - \frac{\sum\limits_{i=1}^{m} \sum\limits_{i=1}^{n} \ln X_{ij}}{n} = 0. \qquad (16.10a)$$

The MLE $\widehat{c}$ is found by some iterative procedure, e.g., by the NEWTON–RAPHSON algorithm; see (11.24) – (11.25). The MLE of $b_i$, corresponding to the pooled MLE $\widehat{c}$ of $c$ is; see (11.18a)

$$\widehat{b}_i = \left[ \frac{1}{n} \sum_{j=1}^{n} X_{ij}^{\widehat{c}} \right]^{1/\widehat{c}} \quad \forall \; i. \qquad (16.10b)$$

---

[6]  In Sect. 21.1.1 we will show how to test the hypothesis of a constant shape parameter.

### 16.2.1 Direct ML procedure of the IPL–WEIBULL model

Most approaches to parameter estimation in accelerated life testing use a rather universal log–linear life–stress function in combination with the Log–WEIBULL distribution (see Sect. 16.2.3). Here we will show how to directly estimate an IPL–WEIBULL without linearizing the power law and without forming the natural logarithm of the WEIBULL variate. The work described in this section applies to the analysis of **censored ALT** in which

- a two–parameter WEIBULL distribution governs at each stress level,

- the WEIBULL scale parameter varies inversely with a power of the stress level $s_i$

$$b_i = \gamma_0 \, s_i^{-\gamma_1}; \quad i = 1, \ldots, m, \qquad (16.11a)$$

- the shape parameter $c$ is invariant with stress,

thus,

$$X_{ij} \overset{iid}{\sim} We(0, b_i, c) \; \forall \, i. \qquad (16.11b)$$

At stress level $s_i$ the sample size is $n_i$, the censoring number being $r_i$, $r_i \leq n_i$. The total numbers are

$$n = \sum_{i=1}^{m} n_i, \quad r = \sum_{i=1}^{m} r_i.$$

For $i = 1, \ldots, m$ the specimens are subjected to stress $s_i$ and run until the occurrence of the first $r_i$ failures. Both $n_i$ and $r_i$ are specified in advance of testing. Let $X_{ij}$ denote the $j$–th order statistic in a sample of $n_i$ WEIBULL variates. Under the assumption (16.11b) the log–likelihood function for stress level $s_i$ is, to within an additive constant; see (11.45c)

$$\mathcal{L}_i(b_i, c) = r_i \left[ \ln c - c \ln b_i \right] + (c - 1) \sum_{j=1}^{r_i} \ln X_{ij} - \sum_{j=1}^{n_i} \left( \frac{X_{ij}}{b_i} \right)^c ; \; i = 1, \ldots, m, \quad (16.12a)$$

where in the last sum we have $X_{ij} = X_{ir_i}$ for $j = r_i + 1, \ldots, n_i$. Summing over all levels $i$ we arrive at the log–likelihood function for the complete ALT model:

$$\mathcal{L}(b_1, \ldots, b_m, c) = r \ln c + c \sum_{i=1}^{m} r_i \ln b_i + (c - 1) Y_0 - \sum_{i=1}^{m} \sum_{j=1}^{n_i} \left( \frac{X_{ij}}{b_i} \right)^c, \quad (16.12b)$$

where

$$Y_0 := \sum_{i=1}^{m} \sum_{j=1}^{n_i} \ln X_{ij} \; \text{ with } \; X_{ij} = X_{ir_i} \; \text{ for } \; j = r_i + 1, \ldots, n_i.$$

Inserting (16.11a) into (16.12b) and differentiating, the MLE of $\gamma_0$ is

$$\widehat{\gamma}_0 = \left\{ \frac{\sum_{i=1}^{m} s_i^{\widehat{\nu}} Z_i}{r} \right\}^{1/\widehat{c}}, \qquad (16.12c)$$

where

$$Z_i := \sum_{j=1}^{n_i} X_{ij}^{\widehat{c}} \quad \text{with} \quad X_{ij} = X_{ir_i} \quad \text{for} \quad j = r_i + 1, \ldots, n_i;$$

$$\widehat{\nu} := \widehat{\gamma}_1 \widehat{c}.$$

The estimates $\widehat{\gamma}_1$ and $\widehat{c}$ are found by the simultaneous solution of

$$\frac{1}{\widehat{c}} + \frac{Y_0}{r} - \frac{\displaystyle\sum_{i=1}^{m} s_i^{\widehat{\nu}} \sum_{j=1}^{n_i} X_{ij}^{\widehat{c}} \ln X_{ij}}{\displaystyle\sum_{i=1}^{m} s_i^{\widehat{\nu}} Z_i} = 0, \tag{16.12d}$$

$$\sum_{i=1}^{m} r_i \ln s_i - \frac{r \displaystyle\sum_{i=1}^{m} s_i^{\widehat{\nu}} Z_i \ln s_i}{\displaystyle\sum_{i=1}^{m} s_i^{\widehat{\nu}} Z_i} = 0. \tag{16.12e}$$

The MLE of $b(s)$ at any stress $s$ is

$$\widehat{b}(s) = \widehat{\gamma}_0 \, s^{-\widehat{\gamma}_1}. \tag{16.12f}$$

It is readily shown that when all testing is at a single stress, $(s_i = s \ \forall \ i)$ $\gamma_1$ becomes non–estimable as (16.12e) reduces to the trivial identity $0 = 0$, and the estimator of $c$ reduces to the single sample expression (11.46b). Furthermore, the estimators of $\gamma_1$, $c$ and $b(s)$ are invariant with respect to the scale used to measure stress.

## 16.2.2   Direct ML estimation of an exponential life–stress relationship

At stress level $s_i$ $(i = 1, \ldots, m)$, $n_i$ specimens are put on test until they fail, $X_{ij}$ being the times to failure and $r_i$ being the number of failed specimens $(j = 1, \ldots, r_i)$, or they are withdrawn from test without failure, $c_{i\ell}$ being the censoring times[7] and $k_i := n_i - r_i$ $(r_i \leq n_i)$ being the number of censored life times $(\ell = 1, \ldots, k_i)$. The data at each level of stress are modeled as statistically independent observations from a WEIBULL distribution with a common shape parameter $c$ and a scale parameter which is an exponential function of the stress level:

$$b_i = \exp(\gamma_0 + \gamma_1 s_i). \tag{16.13a}$$

(16.13a) is a variant of the ARRHENIUS life–temperature relationship (16.3d).

The likelihood and the log–likelihood associated with the data at stress $s_i$ are

$$L_i(b_i, c) = \prod_{j=1}^{r_i} \frac{c}{b_i} \left(\frac{X_{ij}}{b_i}\right)^{c-1} \exp\left\{-\left(\frac{X_{ij}}{b_i}\right)^c\right\} \prod_{\ell=1}^{k_i} \exp\left\{-\left(\frac{c_{i\ell}}{b_i}\right)^c\right\} \quad and \tag{16.13b}$$

$$\mathcal{L}_i(b_i, c) = r_i \ln c + (c-1) \sum_{j=1}^{r_i} \ln X_{ij} - r_i c \ln b_i - b_i^{-c} \left[\sum_{j=1}^{r_i} X_{ij}^c + \sum_{\ell=1}^{k_i} c_{i\ell}^c\right]. \tag{16.13c}$$

---

[7]  This notation allows for different ways of censoring, type–I, type–II, progressive or even at random.

Use (16.13a) to introduce the model parameters $\gamma_0$ and $\gamma_1$ into (16.13c). The log–likelihood of the entire dataset is

$$\mathcal{L}(\gamma_0, \gamma_1, c) = \sum_{i=1}^{m} \mathcal{L}_i(b_i, c)$$

$$= r \ln c + (c - 1) S_x - c\left(\gamma_0\, r + \gamma_1 S_s\right) - \exp(-\gamma_0\, c)\, S_{0,0}(c, \gamma_1) \quad (16.13\text{d})$$

where

$$r := \sum_{i=1}^{m} r_i,$$

$$S_x := \sum_{i=1}^{m} \sum_{j=1}^{r_i} \ln X_{ij},$$

$$S_s := \sum_{i=1}^{m} r_i\, s_i,$$

$$S_{\alpha,\beta}(c, \gamma_1) := \sum_{i=1}^{m} s_i^{\alpha} \exp(-\gamma_1\, c\, s_i) \left[ \sum_{j=1}^{r_i} X_{ij}^c\, (\ln X_{ij})^{\beta} + \sum_{\ell=1}^{k_i} c_{ij}^c\, (\ln c_{ij})^{\beta} \right]$$

for non–negative integers $\alpha$, $\beta$ with $0^0 = 1$.

This notation has been introduced by WATKINS (1994) and lends itself to efficient differentiation and computer programming. For $c$, $\gamma_1$ fixed, the $\gamma_0$ which maximizes (16.13d) is

$$\gamma_0 = \frac{1}{c} \ln\left( \frac{S_{0,0}(c, \gamma_1)}{r} \right). \quad (16.14\text{a})$$

Substituting (16.14a) into (16.13d) and ignoring any terms which are independent of model parameters yields the reduced log–likelihood

$$\mathcal{L}^*(c, \gamma_1) = r \ln c + (c - 1)\, S_x - r \ln S_{0,0}(c, \gamma_1) - c\, \gamma_1\, S_s. \quad (16.14\text{b})$$

The aim is to maximize (16.13d) with respect to $\gamma_0$, $\gamma_1$ and $c$. The approach is to maximize (16.14b) — yielding MLEs of $c$ and $\gamma_1$ — and then calculate the MLE of $\gamma_0$ using these maximized values of $c$ and $\gamma_1$ in (16.14a). We find the MLEs of $c$ and $\gamma_1$ as the solutions of the likelihood equations:

$$\frac{\partial \mathcal{L}^*(c, \gamma_1)}{\partial c} = 0 \quad \text{and} \quad \frac{\partial \mathcal{L}^*(c, \gamma_1)}{\partial \gamma_1} = 0.$$

The key formulas to find the first and second partial derivatives of $\mathcal{L}^*(\gamma_1, c)$ are

$$\frac{\partial S_{\alpha,\beta}(c, \gamma_1)}{\partial c} = S_{\alpha,\beta+1}(c, \gamma_1) - \gamma_1\, S_{\alpha+1,\beta}(c, \gamma_1), \quad (16.15\text{a})$$

$$\frac{\partial S_{\alpha,\beta}(c, \gamma_1)}{\partial \gamma_1} = -c\, S_{\alpha+1,\beta}(c, \gamma_1). \quad (16.15\text{b})$$

The likelihood equations, which have to be solved by some numeric algorithm, result as

$$\frac{\partial \mathcal{L}^*(c, \gamma_1)}{\partial c} = \frac{r}{c} + S_x - r \frac{S_{0,1}(c, \gamma_1) - c S_{1,0}(c, \gamma_1)}{S_{0,0}(c, \gamma_1)} - c S_s = 0, \quad (16.16a)$$

$$\frac{\partial \mathcal{L}^*(c, \gamma_1)}{\partial \gamma_1} = \frac{r c S_{1,0}(c, \gamma_1)}{S_{0,0}(c, \gamma_1)} - c S_s = 0. \quad (16.16b)$$

### 16.2.3   MLE of a log–linear life–stress relationship[8]

Lifetimes of test items follow a two–parameter WEIBULL distribution with CDF

$$F(x) = 1 - \exp\left\{ -\left( \frac{x}{b(s)} \right)^c \right\}, \quad (16.17a)$$

where the scale parameter is now given by

$$b(s) = \exp(\gamma_0 + \gamma_1 s). \quad (16.17b)$$

$s$ is either a transformed stress or the original stress variable. Generally, it is more convenient to work with

$$Y = \ln X$$

since the parameters in the distribution of $Y$ appear as location and scale parameters. If $X$ has a two–parameter WEIBULL distribution, then $Y$ has the smallest extreme value distribution with CDF

$$F(y) = 1 - \exp\left\{ -\exp\left( \frac{y - a^*(s)}{b^*} \right) \right\}; \quad y \in \mathbb{R}, \ a^* \in \mathbb{R}, \ b^* > 0, \quad (16.18a)$$

where

$$a^*(s) = \ln b(s) = \gamma_0 + \gamma_1 s, \quad (16.18b)$$

$$b^* = 1/c. \quad (16.18c)$$

In order to find an optimum accelerated censored life test,[9] NELSON/MEEKER (1978) and NELSON (1990) used a reparameterized model with the **stress factor**

$$\xi_1 = \frac{s - s_h}{s_d - s_h}, \quad (16.19a)$$

where $s_h$ is the (probably transformed) **highest stress** and is specified. $s_d$ is the **design stress**. For $s = s_h$ we get $\xi_1 = 0$, and for the design stress $s = s_d$ we have $\xi_1 = 1$. (16.18b) may be written in terms of $\xi_1$:

$$a^*(\xi_1) = \beta_0 \xi_0 + \beta_1 \xi_1. \quad (16.19b)$$

---

[8]  Suggested reading for this section: BARBOSA/LOUZADA–NETO (1994), BUGAIGHIS (1993), GLASER (1995), HAGWOOD/CLOUGH/FIELDS (1999), KLEIN/BASU (1981, 1982), MEEKER (1984), MEEKER/NELSON (1975), NELSON/MEEKER (1978), ODELL/ANDERSON/D'AGOSTINO (1992), SEO/YUM (1991), VAN DER WIEL/MEEKER (1990).

[9]  The optimization criterion is to minimize the large sample variance of the MLE of the $100\,P$–th percentile of this smallest extreme value distribution at a specified — usually design — stress.

Here the new coefficients $\beta_0$ and $\beta_1$ are related to the previous $\gamma_0$ and $\gamma_1$ by

$$\beta_0 = \gamma_0 + \gamma_1\, s_h \quad \text{and} \quad \beta_1 = \gamma_1\, (s_d - s_h).$$

To simplify the derivations in the ML procedure we have introduced the pseudo–variable

$$\xi_0 = 1.$$

We now provide the log–likelihood of an observation at a transformed stress $\xi_1$ and with type–I censoring. The indicator function in terms of the log–censoring time $\eta$ is

$$\delta = \begin{cases} 1 & \text{if} \quad Y \leq \eta \text{ (uncensored observation)}, \\ 0 & \text{if} \quad Y > \eta \text{ (censored observation)}. \end{cases}$$

Let

$$Z = \frac{Y - a^*(\xi_1)}{b^*} = \frac{Y - \beta_0\,\xi_0 - \beta_1\,\xi_1}{b^*} \tag{16.20a}$$

be the reduced log–failure time with DF

$$\psi(z) = \exp\{z - \exp(z)\}, \tag{16.20b}$$

so that the DF of $Y$ is

$$f(y) = \frac{1}{b^*}\, \psi\left(\frac{y - \beta_0\,\xi_0 - \beta_1\,\xi_1}{b^*}\right). \tag{16.20c}$$

The CDF at the the reduced censoring time $\zeta = [\eta - a^*(\xi_1)]/b^*$ is

$$\Psi := \Psi(\zeta) = 1 - \exp\{-\exp(\zeta)\}. \tag{16.20d}$$

Now, the log–likelihood $\mathcal{L}$ of a possibly type–I censored observation at a stress factor $\xi_1$ is

$$\mathcal{L} = \delta\left[-\ln b^* - e^z + z\right] + (1 - \delta)\ln(1 - \Psi). \tag{16.20e}$$

Suppose, the $i$-th observation $y_i = \ln x_i$ corresponds to a value $\xi_{1i}$ with the corresponding log–likelihood

$$\mathcal{L}_i := \mathcal{L}(\beta_0, \beta_1, b^* \,|\, y_i, \xi_{1i}, \eta), \tag{16.20f}$$

then the sample log-likelihood for $n$ independent observations is

$$\mathcal{L}(\beta_0, \beta_1, b^* \,|\, \mathbf{y}, \boldsymbol{\xi_1}, \eta) = \sum_{i=1}^{n} \mathcal{L}_i. \tag{16.20g}$$

For a single observation the three first partial derivatives are:

$$\frac{\partial \mathcal{L}}{\partial \beta_j} = \frac{\xi_1}{b^*}\left[\delta\,(e^z - 1) + (1 - \delta)\,e^\zeta\right]; \quad j = 0, 1; \tag{16.20h}$$

$$\frac{\partial \mathcal{L}}{\partial b^*} = \frac{1}{b^*}\left[\delta\,(z\,e^z - z - 1) + (1 - \delta)\,\zeta\,e^\zeta\right]. \tag{16.20i}$$

These three expressions, when summed over all test units and set equal to zero, are the likelihood equations. They have to be solved by some numerical root-finding procedure. NELSON/MEEKER (1978) also give the FISHER information matrix, which is a prerequisite to finding a solution for their optimization problem.

## 16.3 ALT using step–stress models[10]

Tests at constant, high stresses can run too long because there is usually a great scatter in failure times. Step–stress testing like progressive–stress testing (see Sect. 16.4) is intended to reduce test time and to assure that failures occur quickly enough. A step–stress test runs through a pattern of specified stresses, each for a specified time, until the specimen fails or the test is stopped with specimens unfailed when a certain censoring time has been reached (see the upper part of Fig. 16/2). The step–stress test requires the stress setting of a unit to be changed at specified times (**time–step**) or upon the occurrence of a fixed number of failures (**failure–step**). Often different step–stress patterns are used on different specimens.

Figure 16/2: Cumulative exposure model of NELSON (1980)

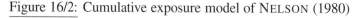

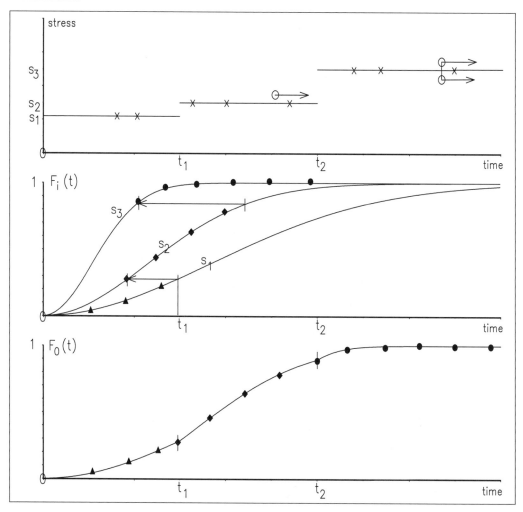

For a step–stress pattern, there is a CDF for time to failure under test, i.e., $F_0(t)$ in Fig. 16/2. Data from this CDF are observed in the test. Not interested in life under a step–stress pattern, one usually wants the life distribution under constant stress, which units see in use.

---

[10] Suggested reading for this section: BAI/KIM (1993), MILLER/NELSON (1983), NELSON (1980, 1990).

To analyze data, one needs a model that relates the distribution (or cumulative exposure) under step–stressing to the distribution under constant stress. The widely used **cumulative exposure model** (**CE model**) goes back to NELSON (1980): "The model assumes that the remaining life of specimens depends only on the current cumulative fraction failed and current stress — regardless how the fraction accumulated. Moreover, if held at the current stress, survivors will fail according to the CDF of that stress but starting at the previously accumulated fraction failed."

Fig. 16/2 depicts this CE model. The upper part shows a step–stress pattern with three steps. The figure gives the failure times ($\times$) and censoring times ($\circ\!\!\rightarrow$) of test specimens. The middle part depicts the three CDF's for the constant stresses $s_1, s_2$ and $s_3$. The arrows show that the specimens first follow the CDF or $s_1$ up to the first hold time $t_1$. When the stress increases from $s_1$ to $s_2$, the non–failed specimens continue along the CDF for $s_2$, starting at the accumulated fraction failed. Similarly, when the stress increases from $s_2$ to $s_3$, the non–failed specimens continue along the next CDF, starting at the accumulated fraction failed. The CDF $F_0(t)$ for life under the step–stress pattern is shown in the lower part of Fig. 16/2 and consists of the segments of the CDF's for the constant stresses.

We now describe how to find $F_0(t)$. Suppose that, for a particular pattern, step $i$ runs at stress $s_i$, starts at time $t_{i-1}$ ($t_0 = 0$) and runs to time $t_i$. The CDF of time to failure for units running at a a constant stress $s_i$ is denoted by $F_i(t)$.

The population cumulative fraction of units failing in **step 1** is

$$F_0(t) = F_1(t), \ \ 0 \leq t \leq t_1. \tag{16.21}$$

**Step 2** has an **equivalent start time** $\tau_1$, which would have produced the same population cumulative fraction failing, as shown in the middle part of Fig. 16/2. So $\tau_1$ is the solution of

$$F_2(\tau_1) = F_1(t_1) \tag{16.22a}$$

and the population cumulative fraction failing in step 2 by total time $t$ is

$$F_0(t) = F_2\big[(t - t_1) + \tau_1\big], \ \ t_1 \leq t \leq t_2. \tag{16.22b}$$

**Step 3** has an equivalent start time $\tau_2$ that is the solution of

$$F_3(\tau_2) = F_2(t_2 - t_1 + \tau_1), \tag{16.23a}$$

and the segment of $F_0(t)$ in step 3 is

$$F_0(t) = F_3\big[(t - t_2) + \tau_2\big], \ \ t_2 \leq t. \tag{16.23b}$$

In general, **step $i$** has the equivalent time $\tau_{i-1}$ that is the solution of

$$F_i(\tau_{i-1}) = F_{i-1}(t_{i-1} - t_{i-2} + \tau_{i-2}) \tag{16.24a}$$

and

$$F_0(t) = F_i\big[(t - t_{i-1}) + \tau_{i-1}\big], \ \ t_{i-1} \leq t \leq t_i. \tag{16.24b}$$

A different step–stress pattern would have a different $F_0(t)$.

**Example 16/1:   CDF under step–stress for the inverse power law**

The preceding specializes to the IPL–WEIBULL model where

$$b = \gamma_0 \, s^{-\gamma_1}, \quad \gamma_0 \text{ and } \gamma_1 \text{ are positive parameters,} \tag{16.25a}$$

as follows. By (16.25a), the CDF for the fraction of specimens failing by time $t$ for the constant stress $s_i$ is

$$F_i(t) = 1 - \exp\left\{ -\left( \frac{t \, s_i^{\gamma_1}}{\gamma_0} \right)^c \right\}. \tag{16.25b}$$

Then, for step 1 (16.21) becomes

$$F_0(t) = 1 - \exp\left\{ -\left( \frac{t \, s_1^{\gamma_1}}{\gamma_0} \right)^c \right\}, \quad 0 \le t \le t_1. \tag{16.25c}$$

The equivalent time $\tau_1$ at $s_2$ is given by (16.22a) as

$$\tau_1 = t_1 \left( \frac{s_1}{s_2} \right)^{\gamma_1}. \tag{16.25d}$$

For step 2 the segment of $F_0(t)$ is

$$F_0(t) = 1 - \exp\left\{ -\left[ \frac{(t - t_1 + \tau_1) \, s_2^{\gamma_1}}{\gamma_0} \right]^c \right\}, \quad t_1 \le t \le t_2. \tag{16.25e}$$

Similarly, for step 3,

$$\tau_2 = (t_2 - t_1 + \tau_1) \left( \frac{s_2}{s_3} \right)^{\gamma_1}, \tag{16.25f}$$

and the corresponding segment of $F_0(t)$ is

$$F_0(t) = 1 - \exp\left\{ -\left[ \frac{(t - t_2 + \tau_2) \, s_3^{\gamma_1}}{\gamma_0} \right]^c \right\}, \quad t_2 \le t \le t_3. \tag{16.25g}$$

In general, for step $i$, with $t_0 = \tau_0 = 0$:

$$\tau_{i-1} = (t_{i-1} - t_{i-2} + \tau_{i-2}) \left( \frac{s_{i-1}}{s_i} \right)^{\gamma_1}, \quad i \ge 2, \tag{16.25h}$$

$$F_0(t) = 1 - \exp\left\{ -\left[ \frac{(t - t_{i-1} + \tau_{i-1}) \, s_i^{\gamma_1}}{\gamma_0} \right]^c \right\}, \quad t_{i-1} \le t \le t_i. \tag{16.25i}$$

Once one has estimates of $c, \gamma_0$ and $\gamma_1$, one can then use the procedure above to estimate $F_0(t)$ under any varying stress patterns that might occur in actual use.

---

BAI/KIM (1973) have carried over formulas (16.21) to (16.24b) to the Log–WEIBULL distribution with a log–linear life–stress relationship (see Sect. 16.2.3) and under type–I censoring. They give the log–likelihood together with its first and second partial derivatives. For the special case of two stress levels $s_1$ and $s_2$, they also give the FISHER information which is needed in finding the optimum design of this simple step–stress test. The optimum plan — low stress and stress change time — is obtained, which minimizes the asymptotic variance of the MLE of the median life at design stress.

# 16.4   ALT using progressive stress models[11]

In a progressive–stress test, the stress applied to a test unit is continuously increasing in time. A widely used progressive–stress test is the **ramp–test** where the stress is linearly increasing (see Fig. 16/3). In particular, a ramp–test with two different linearly increasing stresses is a **simple ramp–test**. There should be at least two different stress rates in order to estimate the parameters of the life–stress relationship.

Figure 16/3: A simple ramp-test situation

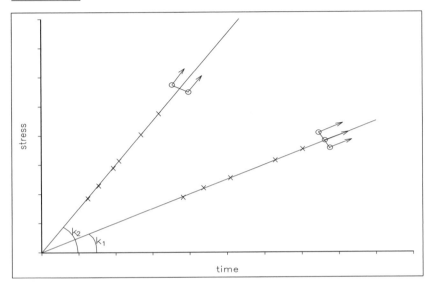

In the following we will analyze a simple ramp–test, the assumptions being

1. $k_1$ and $k_2$ $(k_1 < k_2)$ are the stress rates.

2. At any constant stress $s$, the lifetimes of a unit follow a WEIBULL distribution with scale parameter $b(s)$ and shape parameter $c$, and the inverse power law holds for $b(s)$:

$$b(s) = 1e^{\gamma_0} \left( \frac{s_0}{s} \right)^{\gamma_1}, \qquad (16.26)$$

where $\gamma_0$ and $\gamma_1$ are unknown positive parameters and $s_0$ is the known design stress.

3. For the effect of changing stress, the CE model of NELSON (1980) holds (see Sect. 16.3).

4. The stress applied to test units is continuously increased with constant rate $k_1$ (or $k_2$) from zero.

5. $n_1$ units $(n_1 < n)$ randomly chosen among $n$ are allocated to stress rate $k_1$ and the remaining $n_2 = n - n_1$ units to stress rate $k_2$.

6. The test is continued until all test units fail or a prescribed censoring time $T$ is reached.

---

[11]   Suggested reading for this section: BAI/CHA/CHUNG (1992), BAI/CHUN/CHA (1997).

The stress at time $t$ (from assumption 4) is:

$$s(t) = k\,t. \tag{16.27}$$

From the linear CE model and the inverse power law, the CDF of the lifetime $X$ of a unit tested under stress rate $k$ is

$$
\begin{aligned}
\Pr(X \le x) &= 1 - \exp\left\{ - \left[ \int\limits_0^x \frac{1}{b\big(s(u)\big)} \, du \right]^c \right\} \\
&= 1 - \exp\left\{ - \left( \frac{x}{\theta} \right)^{(\gamma_1 + 1)\,c} \right\},
\end{aligned}
\tag{16.28a}
$$

where

$$\theta^{\gamma_1 + 1} = e^{\gamma_0} \, (\gamma_1 + 1) \left( \frac{s_0}{k} \right)^{\gamma_1}; \tag{16.28b}$$

i.e., we have a WEIBULL distribution with scale parameter $\theta$ and shape parameter $(\gamma_1 + 1)\,c$. Then $Y = \ln X$ follows the smallest extreme value distribution with location parameter

$$a^* = \ln \theta = \left[ \gamma_0 + \ln(\gamma_1 + 1) + \gamma_1 \ln \left( \frac{s_0}{k} \right) \right] \Big/ (\gamma_1 + 1) \tag{16.29a}$$

and scale parameter

$$b^* = 1/c = \left[ (\gamma_1 + 1)\, c \right]^{-1}. \tag{16.29b}$$

We introduce the **stress–rate factor** as

$$\xi = \frac{k}{k_2}. \tag{16.29c}$$

For the high stress–rate, $k = k_2$, we have $\xi = 1$. The low stress–rate factor is $\xi = k_1/k_2 < 1$. The location parameter of the log–lifetime distribution of a unit tested under stress-rate $k$ is

$$a^*(\xi) = \left[ \gamma_0 + \ln(\gamma_1 + 1) + \gamma_1 \ln \left( \frac{s_0}{k_2} \right) - \gamma_1 \ln \xi \right] \Big/ (\gamma_1 + 1). \tag{16.29d}$$

Introducing the reduced log–failure time

$$Z = \frac{Y - a^*}{b^*} \tag{16.30a}$$

and the reduced log–censoring time

$$\zeta = \frac{\ln T - a^*}{b^*}, \tag{16.30b}$$

we have the following indicator function:

$$\delta(Z) = \begin{cases} 1 & \text{if} \quad Y \le \ln T, \\ 0 & \text{if} \quad Y > \ln T. \end{cases} \tag{16.30c}$$

The log–likelihood of a single observation $y$ is

$$\mathcal{L}(\gamma_0, \gamma_1, b^* \,|\, y) = \delta(z) \left[ - \ln b^* - e^z + z \right] + \left[ 1 - \delta(z) \right] \ln \left[ 1 - \Psi(\zeta) \right], \tag{16.31a}$$

where $\Psi(\cdot)$ is the CDF of the reduced smallest extreme value. Let the log–likelihood of unit $i\,(i = 1, \ldots, n)$ be $\mathcal{L}_i$. The log–likelihood $\mathcal{L}_0$ for $n$ statistically independent observations is

$$\mathcal{L}_0 = \sum_{i=1}^{n} \mathcal{L}_i. \tag{16.31b}$$

For a single observation we have

$$\frac{\partial \mathcal{L}_i}{\partial \gamma_j} = \frac{u_j}{b^*} \left\{ \delta(z) \left( e^z - 1 \right) + \left[ 1 - \delta(z) \right] e^\zeta \right\}; \quad j = 0, 1, \tag{16.32a}$$

$$\frac{\partial \mathcal{L}_i}{\partial b^*} = \frac{1}{b^*} \left\{ \delta(z) \left( z \, e^z - z - 1 \right) + \left[ 1 - \delta(z) \right] \zeta \, e^\zeta \right\}, \tag{16.32b}$$

where

$$u_0 = (\gamma_1 + 1)^{-1},$$

$$u_1 = \left[ 1 - \ln \left( \frac{s_0}{k_2} \right) - \ln \xi - \gamma_0 - \ln(\gamma_1 + 1) \right] \Big/ (\gamma_1 + 1)^2.$$

The parameter values that solve these equations summed over all test units are the MLEs.

## 16.5   Models for PALT[12]

ALT models in the preceding sections assumed a life–stress relationship of known mathematical form but with unknown parameters. Partially accelerated life tests (PALT) do not have such a life–stress relationship. In step PALT, the test unit is first run at use condition, and if the unit does not fail by the end of the specified time $\tau$, the test is switched to a higher level and continued until the unit fails or the test is censored. Thus, the total lifetime $Y$ of the unit in step PALT is given as follows:

$$Y = \begin{cases} X & \text{for} \quad X \leq \tau, \\ \tau + \beta^{-1} \left( X - \tau \right) & \text{for} \quad X < \tau, \end{cases} \tag{16.33a}$$

where $X$ is the lifetime of an item at use condition, $\tau$ is the stress change time and $\beta$ is the acceleration factor, usually $\beta > 1$. Because the switching to the higher stress level can be regarded as tampering with the ordinary life test, $Y$ is called a **tampered random variable**, $\tau$ is called the tampering point and $\beta^{-1}$ is called the tampering coefficient. If the observed value of $Y$ is less than the tampering point, it is called non–tampered observation; otherwise, it is called a tampered observation. The model (16.33a) is referred to as a **TRV model** (= tampered random variable model). Another approach in step PALT is the **TFR model** (= tampered failure rate model) defined as

$$h^*(x) = \begin{cases} h(x) & \text{for} \quad X \leq \tau, \\ \alpha \, h(x) & \text{for} \quad X > \tau. \end{cases} \tag{16.33b}$$

---

[12] Suggested reading for this section: ABDEL–GHALY/ATTIA/ABDEL–GHANI (2002), BHATTACHARYYA/SOEJOETI (1989).

In the following we will analyze the TRV model (16.33a), whereas BHAT-TACHARYYA/SOEJOETI (1989) apply the ML procedure the TFR model.

Assuming the lifetime of the test item to follow the two–parameter WEIBULL distribution with scale parameter $b$ and shape parameter $c$, the CDF of the total lifetime $Y$ as defined by (16.33a) is given by

$$
f(y) = \left\{
\begin{array}{ll}
0, & y \leq 0, \\[2mm]
\left(\dfrac{c}{b}\right)\left(\dfrac{y}{b}\right)^{c-1} \exp\left\{-\left(\dfrac{y}{b}\right)^{c}\right\}, & 0 < y \leq \tau, \\[4mm]
\left(\dfrac{\beta c}{b}\right)\left[\dfrac{\beta(y-\tau)+\tau}{b}\right]^{c-1} \exp\left\{-\left[\dfrac{\beta(y-\tau)+\tau}{b}\right]^{c}\right\}, & y > \tau.
\end{array}
\right\}
$$

$$(16.34)$$

We will obtain the MLEs of the acceleration factor $\beta$, the scale parameter $b$ and the shape parameter $c$ using both the type–I and the type–II censoring.

**The case of type–I censoring**

All the $n$ test units are tested first under the use condition. If the test unit does not fail by the tampering point $\tau$, the test is continued at a higher stress level until either all units have failed or a censoring time $T$ has been reached. The observed values of $Y$ are denoted by

$$
y_{1:n} \leq y_{2:n} \leq y_{n_u:n} \leq \tau \leq y_{n_u+1:n} \leq \cdots y_{n_u+n_a:n} \leq T.
$$

$n_u$ is the number of items failed at use condition, and $n_a$ is the number of items failed at accelerated conditions. We introduce two indicator functions:

$$
\delta_{1i} := \left\{
\begin{array}{ll}
1 & \text{if} \quad Y_i \leq \tau, \\[2mm]
0 & \text{if} \quad Y_i > \tau,
\end{array}
\right.
$$

$$(16.35a)$$

$$
\delta_{2i} := \left\{
\begin{array}{ll}
1 & \text{if} \quad \tau < Y_i \leq T, \\[2mm]
0 & \text{if} \quad Y_i > T,
\end{array}
\right.
$$

$$(16.35b)$$

where $\bar{\delta}_{1i}$ and $\bar{\delta}_{2i}$ are their complements and $i = 1, 2, \ldots, n$. The numbers of failing items can be written as

$$
n_u = \sum_{i=1}^{n} \delta_{1i} \quad \text{and} \quad n_a = \sum_{i=1}^{n} \delta_{2i}.
$$

$$(16.35c)$$

The likelihood function results as

$$
L(\beta, b, c \,|\, \boldsymbol{y}) = \prod_{i=1}^{n} \left\{ \left(\frac{c}{b}\right)\left(\frac{y_i}{b}\right)^{c-1} \exp\left[-\left(\frac{y_i}{b}\right)^{c}\right] \right\}^{\delta_{1i}} \times
$$

$$
\left\{ \left(\frac{\beta c}{b}\right)\left[\frac{\beta(y_i-\tau)+\tau}{b}\right]^{c-1} \exp\left[-\left(\frac{\beta(y_i-\tau)+\tau}{b}\right)^{c}\right] \right\}^{\delta_{2i}} \times \quad (16.36a)
$$

$$
\exp\left[-\left(\frac{\beta(T-\tau)+\tau}{b}\right)^{c}\right]^{\bar{\delta}_{1i}\bar{\delta}_{2i}},
$$

and the corresponding log–likelihood function is

$$\mathcal{L}(\beta, b, c \,|\, \boldsymbol{y}) \;=\; (n_u + n_a)\,(\ln c - c \ln b) + n_a \ln \beta \;+$$

$$(c-1)\left\{ \sum_{i=1}^{n} \delta_{1i} \ln y_i + \sum_{i=1}^{n} \delta_{2i} \ln\left[\beta\,(y_i - \tau) + \tau\right]\right\} - \frac{1}{b^c}\,Q, \quad (16.36b)$$

where

$$Q = \sum_{i=1}^{n} \delta_{1i}\, y_i^c + \sum_{i=1}^{n} \delta_{2i}\left[\beta\,(y_i - \tau) + \tau\right]^c + (n - n_u - n_a)\left[\beta\,(T - \tau) + \tau\right]^c. \quad (16.36c)$$

The reader should compare (16.36a–b) with the formulas (11.45b–c) when there is no acceleration.

The first derivatives of (16.36b) with respect to $\beta, b$ and $c$ are given by

$$\frac{\partial \mathcal{L}}{\partial \beta} = \frac{n_a}{\beta} + (c-1) \sum_{i=1}^{n} \delta_{2i}\, \frac{y_i - \tau}{\beta\,(y_i - \tau) + \tau} - \frac{1}{b^c}\, Q_1, \quad (16.37a)$$

$$\frac{\partial \mathcal{L}}{\partial b} = -(n_u + n_a)\,\frac{c}{b} + \frac{c}{b^{c+1}}\, Q, \quad (16.37b)$$

$$\frac{\partial \mathcal{L}}{\partial c} = (n_u + n_a)\left(\frac{1}{c} - \ln b\right) + \sum_{i=1}^{n} \delta_{1i} \ln y_i + \sum_{i=1}^{n} \delta_{2i} \ln\left[\beta\,(y_i - \tau) + \tau\right] - (16.37c)$$
$$\frac{1}{b^c}\,(Q_2 - Q \ln b),$$

where

$$Q_1 = \frac{\partial Q}{\partial \beta} = c\left\{ \sum_{i=1}^{n} \delta_{2i}\,(y_i - \tau)\left[\beta\,(y_i - \tau) + \tau\right]^{c-1} + \right.$$
$$\left. (n - n_u - n_a)\,(T - \tau)\left[\beta\,(T - \tau) + \tau\right]^{c-1}\right\} \quad (16.37d)$$

and

$$Q_2 = \frac{\partial Q}{\partial c} = \sum_{i=1}^{n} \delta_{1i}\, y_i^c \ln y_i + \sum_{i=1}^{n} \delta_{2i}\left[\beta\,(y_i - \tau) + \tau\right]^c \ln\left[\beta\,(y_i - \tau) + \tau\right] +$$
$$(n - n_u - n_a)\left[\beta\,(T - \tau) + \tau\right]^c \ln\left[\beta\,(T - \tau) + \tau\right]. \quad (16.37e)$$

Setting (16.37a–c) equal to zero, we can reduce the three non–linear equations to two:

$$\frac{n_a}{\widehat{\beta}} + (\widehat{c} - 1) \sum_{i=1}^{n} \delta_{2i}\, \frac{y_i - \tau}{\widehat{\beta}\,(y_i - \tau) + \tau} - (n_u + n_a)\,\frac{Q_1}{Q} = 0, \quad (16.37f)$$

$$(n_u + n_a)\,\frac{1}{\widehat{c}} + \sum_{i=1}^{n} \delta_{1i} \ln y_i + \sum_{i=1}^{n} \delta_{2i} \ln\left[\widehat{\beta}\,(y_i - \tau) + \tau\right] - (n_u + n_a)\,\frac{Q_2}{Q} = 0. \quad (16.37g)$$

and the MLE of $b$ is expressed by

$$\widehat{b} = \left( \frac{Q}{n_u + n_a} \right)^{1/\widehat{c}}. \tag{16.37h}$$

**The case of type–II censoring**

All of the $n$ units are tested first under use condition, and if the test unit does not fail by $\tau$, the test is continued at an accelerated condition until the predetermined number of failures $r$ is reached. The observed values of the total lifetime $Y$ are denoted by

$$y_{1:n} \leq y_{2:n} \leq y_{n_u:n} \leq \tau \leq y_{n_u+1:n} \leq \cdots y_{r:n},$$

where $r = n_u + n_a$, The indicator function $\delta_{1i}$ is the same as (16.35a), but $\delta_{2i}$ is now defined as

$$\delta_{2i} := \begin{cases} 1 & \text{if} \quad \tau < Y_i \leq Y_{r:n}, \\ 0 & \text{if} \quad Y_i > Y_{r:n}. \end{cases} \tag{16.38}$$

The ML procedure is the same as above, but $T$ has to be substituted by $y_{r:n}$ in (16.36a) through (16.36h).

# 17 Parameter estimation for mixed WEIBULL models

Finite mixture distributions[1] have been widely used in nearly all fields of applied statistical science to model heterogeneous populations. Such a population is a composition of $m$ ($m \geq 2$) different subpopulations. When dealing with lifetime distributions the reason for the existence of several subpopulations is that a unit can have more than one failure mode. For example, if some units in a batch are freaks (e.g., have gross defects due to the material processed or to improper manufacturing), then such units will have short lifetimes or will fail at lower stress levels. The remaining units will have longer lifetimes and will endure higher stress. In this case, the DF of the time to failure can have a bimodal or multimodal shape.[2] Consequently, the finite mixed WEIBULL distribution is a good candidate to represent these times to failure.

The mixed WEIBULL model requires some statistical analysis such as estimation of parameters. The graphical estimation approach has been laid out in Sect. 9.3.3.2. This chapter deals with analytical and numerical estimation procedures: classical estimation in Sect. 17.1 and BAYESIAN methods in Sect. 17.2. In both approaches the number $m$ of subpopulations is assumed known, but two different types of situations commonly arise which are of importance when doing statistical inference. In one case it is possible to assign each unit to the appropriate subpopulation, while in the other case such information is not available. JIANG/KECECIOGLU (1992b) refer to these cases as **postmortem data** (first case) and **non–postmortem data** (second case). Therefore, in the first case the data would consist of $n$ failure times grouped according to the subpopulations

$$\{(x_{11}, \ldots, x_{1n_1}),\ (x_{21}, \ldots, x_{2n_2}), \ldots, (x_{m1}, \ldots, x_{mn_m})\},$$

where it is assumed that $n_1, \ldots, n_m$ are the observed frequencies in the sample belonging to subpopulations $\pi_1, \ldots, \pi_m$, respectively. Assuming that $f(x \mid \boldsymbol{\theta}_j)$ is the DF corresponding to $\pi_j$, the likelihood of the sample is given by

$$L(\boldsymbol{\theta}, \boldsymbol{p} \mid \boldsymbol{x}) = \frac{n!}{n_1! \cdots n_m!} \, p_1^{n_1} \cdots p_m^{n_m} \prod_{j=1}^{m} \left\{ \prod_{i=1}^{n_j} f(x_{ji} \mid \boldsymbol{\theta}_j) \right\}, \qquad (17.1a)$$

where the vector $\boldsymbol{p} = (p_1, \ldots, p_m)'$ contains the proportions $p_j$ (= relative sizes of the subpopulations). In the second case the failure time distribution of a unit is given by

$$F(x \mid \boldsymbol{\theta}, \boldsymbol{p}) = \sum_{j=1}^{m} p_j \, F(x \mid \boldsymbol{\theta}_j)$$

---

[1]  The theory and the description of a mixture of WEIBULL distributions has been given in Sect. 3.3.6.4.

[2]  An example for this phenomenon is the life table, and we will present the decomposition of German life tables into WEIBULL subpopulations in Sect. 17.1.2.2.

with corresponding DF

$$f(x \mid \boldsymbol{\theta}, \boldsymbol{p}) = \sum_{j=1}^{m} p_j \, f(x \mid \boldsymbol{\theta}_j),$$

and the likelihood function results as

$$L(\boldsymbol{\theta}, \boldsymbol{p} \mid \boldsymbol{x}) = \prod_{i=1}^{n} f(x_i \mid \boldsymbol{\theta}, \boldsymbol{p}) = \prod_{i=1}^{n} \left\{ \sum_{j=1}^{m} p_j \, f(x_i \mid \boldsymbol{\theta}_j) \right\}. \qquad (17.1b)$$

## 17.1  Classical estimation approaches

We will first describe how to estimate the unknown parameters by the method of moments (Sect. 17.1.1), and afterwards we will turn to the more versatile ML procedures (Sect. 17.1.2).

### 17.1.1  Estimation by the method of moments[3]

This estimation method requires uncensored data and a mixture of only two subpopulations. The latter requirement is necessary to arrive at a manageable set of equations. In a mixture of two two–parameter WEIBULL distributions there are five unknown parameters:

$p$ — the mixing proportion,
$b_1$, $b_2$ — the scale parameters, and
$c_1$, $c_2$ — the shape parameters.

The first five moments about the origin of the mixed WEIBULL distribution with CDF

$$
\begin{aligned}
F(x) &= p\,F_1(x) + (1-p)\,F_2(x) \\
&= p\left[1 - \exp\left\{-\left(\frac{x}{b_1}\right)^{c_1}\right\}\right] + (1-p)\left[1 - \exp\left\{-\left(\frac{x}{b_2}\right)^{c_2}\right\}\right] \quad (17.2a)
\end{aligned}
$$

are

$$\mu'_r = p\,b_1^r\,\Gamma\left(1 + \frac{r}{c_1}\right) + (1-p)\,b_2^r\,\Gamma\left(1 + \frac{r}{c_2}\right); \; r = 1, \ldots, 5. \qquad (17.2b)$$

When $c_1 = c_2 = c$ and $c$ is assumed known, we can apply the approach of RIDER (1961) and equate the first three moments $\mu'_1$, $\mu'_2$, $\mu'_3$ of (17.2b) to the sample moments

$$m'_r = \frac{1}{n} \sum_{i=1}^{n} X_i^r; \; r = 1, 2, 3:$$

$$\widehat{p}\,\widehat{b_1}\,\Gamma\left(1 + \frac{1}{c}\right) + (1 - \widehat{p})\,\widehat{b_2}\,\Gamma\left(1 + \frac{1}{c}\right) = m'_1, \qquad (17.3a)$$

$$\widehat{p}\,\widehat{b_1^2}\,\Gamma\left(1 + \frac{2}{c}\right) + (1 - \widehat{p})\,\widehat{b_2^2}\,\Gamma\left(1 + \frac{2}{c}\right) = m'_2, \qquad (17.3b)$$

$$\widehat{p}\,\widehat{b_1^3}\,\Gamma\left(1 + \frac{3}{c}\right) + (1 - \widehat{p})\,\widehat{b_2^3}\,\Gamma\left(1 + \frac{3}{c}\right) = m'_3. \qquad (17.3c)$$

---

[3] Suggesting reading for this section: FALLS (1970), RIDER (1961).

RIDER (1961) gave the following solution of the system (17.3a–c):

$$\widehat{b}_j = \frac{d_1\,d_2 - d_3 \pm \sqrt{d_3^2 - 6\,d_1\,d_2\,d_3 - 4\,d_1^2\,d_2^2 + 4\,d_1^3\,d_3 + 4\,d_2^3}}{d_1^2 - d_2}\,, \qquad (17.4a)$$

where for simplification he has set

$$d_j = m'_j\,\Gamma\left(1 + \frac{j}{c}\right);\; j = 1, 2, 3.$$

The estimator of the mixing proportion is

$$\widehat{p} = \frac{d_1 - \widehat{b}_2}{\widehat{b}_1 - \widehat{b}_2}\,. \qquad (17.4b)$$

We now turn to the general case where all five parameters $p, b_1, b_2, c_1, c_2$ are unknown. In order to avoid the solution of a system of five dependent non–linear equations

$$m'_r = p\,b_1^r\,\Gamma\left(1 + \frac{r}{c_1}\right) + (1 - p)\,b_2^r\,\Gamma\left(1 + \frac{r}{c_2}\right);\; r = 1, \dots, 5.$$

FALLS (1970) suggested to estimate $p$ by the graphical approach of KAO (1959):

1. Plot the sample CDF for the mixed data on WEIBULL–probability–paper (see Fig. 9/4) and visually fit a curve among these points (= WEIBULL plot).

2. Starting at each end of the WEIBULL plot, draw two tangent lines and denote them by $\widehat{p\,F_1}(x)$ and $\widehat{(1 - p)\,F_2}(x)$, which are estimates of $p\,F_1(x)$ and $(1 - p)\,F_2(x)$, respectively.

3. At the intersection of both tangent lines drop a vertical line on the percent scale which gives the estimate $\widehat{p}$ of $p$.

The solution of $m'_r = \mu'_r$ $(r = 1, \dots, 4)$ with $p$ substituted by $\widehat{p}$ will not be unique. When confronted with more than one set of acceptable estimates $\widehat{b}_1, \widehat{b}_2, \widehat{c}_2, \widehat{c}_2$, FALLS (1970) suggested to adopt PEARSON's procedure and chose the set which produces the closest agreement between $m'_5$ and the theoretical moment $\mu'_5$ when evaluated by the estimates.

## 17.1.2 Estimation by maximum likelihood

We will first present the solution of the special case of a two–fold mixture (Sect. 17.1.2.1) and then turn to the general case with more than two subpopulations (Sect. 17.1.2.2).

### 17.1.2.1 The case of two subpopulations

When we have **postmortem data** that are type–I censored, $T$ being the censoring time, there are $r_1$ failures in subpopulation 1 and $r_2$ in subpopulation 2, found in a sample of $n$

test units. The — unknown — relative sizes of these subpopulations are $p$ and $q = 1 - p$, respectively. The log–likelihood is

$$
\begin{aligned}
\mathcal{L}(\theta_1, \theta_2, c_1, c_2, p \,|\, \boldsymbol{x}) = {} & (n - r) \ln\left\{ p \exp\left[ -\frac{T^{c_1}}{\theta_1} \right] + (1 - p) \exp\left[ -\frac{T^{c_2}}{\theta_2} \right] \right\} + \\
& r_1 \ln p + r_2 \ln(1 - p) + \sum_{j=1}^{r_1} \left[ \ln c_1 - \ln \theta_1 + (c_1 - 1) \ln x_{1j} + \ln T - \frac{x_{1j}^{c_1}}{\theta_1} \right] + \\
& \sum_{j=1}^{r_2} \left[ \ln c_2 - \ln \theta_2 + (c_2 - 1) \ln x_{2j} - \frac{x_{2j}^{c_2}}{\theta_2} \right],
\end{aligned}
\tag{17.5}
$$

where $\theta_j = b_j^{c_j}$; $j = 1, 2$; and $r = r_1 + r_2$. Differentiating with respect to the five unknown parameters in turn, we have

$$
\frac{\partial \mathcal{L}}{\partial \theta_1} = \frac{k \, (n - r) \, T^{c_1}}{\theta_1^2} - \frac{r_1}{\theta_1} + \frac{1}{\theta_1^2} \sum_{j=1}^{r_1} x_{1j}^{c_1},
\tag{17.6a}
$$

$$
\frac{\partial \mathcal{L}}{\partial \theta_2} = \frac{(1 - k) \, (n - r) \, T^{c_2}}{\theta_2^2} - \frac{r_2}{\theta_2} + \frac{1}{\theta_2^2} \sum_{j=1}^{r_2} x_{2j}^{c_2},
\tag{17.6b}
$$

$$
\frac{\partial \mathcal{L}}{\partial c_1} = -\frac{k \, (n - r) \, T^{c_1} \ln T}{\theta_1} + \frac{r_1}{c_1} + \sum_{j=1}^{r_1} \ln x_{1j} \left[ 1 - \frac{x_{1j}^{c_1}}{\theta_1} \right],
\tag{17.6c}
$$

$$
\frac{\partial \mathcal{L}}{\partial c_2} = -\frac{(1 - k) \, (n - r) \, T^{c_2} \ln T}{\theta_2} + \frac{r_2}{c_2} + \sum_{j=1}^{r_2} \ln x_{2j} \left[ 1 - \frac{x_{2j}^{c_2}}{\theta_2} \right],
\tag{17.6d}
$$

$$
\frac{\partial \mathcal{L}}{\partial p} = \frac{k \, (n - r) + r_1}{p} - \frac{(1 - k) \, (n - r) + r_2}{1 - p},
\tag{17.6e}
$$

where

$$
k = \left\{ 1 + \frac{1 - p}{p} \exp\left[ \frac{T^{c_1}}{\theta_1} + \frac{T^{c_2}}{\theta_2} \right] \right\}^{-1}.
$$

The resulting estimators are

$$
\widehat{p} = \frac{r_1}{n} + \widehat{k} \, \frac{n - r}{n},
\tag{17.7a}
$$

$$
\widehat{\theta}_1 = \frac{\widehat{k} \, (n - r) \, T^{\widehat{c}_1} + \displaystyle\sum_{j=1}^{r_1} x_{1j}^{\widehat{c}_1}}{r_1},
\tag{17.7b}
$$

$$
\widehat{\theta}_2 = \frac{\left(1 - \widehat{k}\right) (n - r) \, T^{\widehat{c}_2} + \displaystyle\sum_{j=1}^{r_2} x_{2j}^{\widehat{c}_2}}{r_2},
\tag{17.7c}
$$

where $\widehat{c}_1$ and $\widehat{c}_2$ are the solutions of (17.6c–d) when set equal to zero. SINHA (1986b, pp. 111 ff.) gives an algorithm to solve these two simultaneous equations.

Turning to **non–postmortem data** and assuming a three–parameter WEIBULL distribution in each subpopulation, the likelihood equations resulting for an uncensored sample are

$(j = 1, 2)$:

$$\widehat{p} = \frac{1}{n} \sum_{i=1}^{n} f(1 \mid x_i), \tag{17.8a}$$

$$0 = (c_j - 1) \sum_{i=1}^{n} f(j \mid x_i)(x_i - a_j)^{-1} - \frac{c_j}{b_j^{c_j}} \sum_{i=1}^{n} f(j \mid x_i)(x_i - a_j)^{c_j - 1}, \tag{17.8b}$$

$$b_j = \left\{ \left[ \sum_{i=1}^{n} f(j \mid x_i)(x_i - a_j)^{c_j} \right] \Big/ \sum_{i=1}^{n} f(j \mid x_i) \right\}^{1/c_j}, \tag{17.8c}$$

$$c_j = \left\{ \sum_{i=1}^{n} \left[ \left( \frac{x_i - a_j}{b_j} \right)^{c_j} - 1 \right] \ln\left( \frac{x_i - a_j}{b_j} \right) f(j \mid x_i) \Big/ \sum_{i=1}^{n} f(j \mid x_i) \right\}^{-1}, \tag{17.8d}$$

where

$$f(j \mid x) = p_j \frac{f_j(x)}{f(x)}, \ p_1 = p, \ p_2 = 1 - p,$$

$$f(x) = p f_1(x) + (1 - p) f_2(x),$$

$$f_j(x) = \left( \frac{c_j}{b_j} \right) \left( \frac{x - a_j}{b_j} \right)^{c_j - 1} \exp\left\{ - \left( \frac{x - a_j}{b_j} \right)^{c_j} \right\}.$$

Solving the set of seven equations given by (17.8a–d) is computationally difficult due largely to equation (17.8b) which is not in fixed–point form. Because of the computational difficulties of directly solving the likelihood equations and possible numerical instabilities, WOODWARD/GUNST (1987) have considered the use of minimum distance estimators (MDE) (see Sect. 13.2), using the CRAMÉR–VON MISES distance.

### 17.1.2.2 The case of $m$ subpopulations ($m \geq 2$)

JIANG/KECECIOGLU (1992b) have suggested an ML procedure to estimate all the parameters of a mixture of $m$ two–parameter WEIBULL distributions for non–postmortem data as well as for postmortem data coming from time–censored samples. The parameter vector $\theta$ contains the parameters of an $m$–mixed–WEIBULL distribution:

$$\theta = (p_1, \ldots, p_m, b_1, \ldots, b_m, c_1, \ldots, c_m)'.$$

Let

$$f_i(c \mid b_i, c_i) = \frac{c_i}{b_i} \left( \frac{x}{b_i} \right)^{c_i - 1} \exp\left\{ - \left( \frac{x}{b_i} \right)^{c_i} \right\}; \ i = 1, \ldots, m \tag{17.9a}$$

and

$$R_i(x \mid b_i, c_i) = \exp\left\{ - \left( \frac{x}{b_i} \right)^{c_i} \right\}; i = 1, \ldots, m \tag{17.9b}$$

be the DF and CCDF of the $i$–th subpopulation, respectively; then

$$f(x \mid \boldsymbol{\theta}) = \sum_{i=1}^{m} p_i\, f_i(c \mid b_i, c_i), \tag{17.9c}$$

$$R(x \mid \boldsymbol{\theta}) = \sum_{i=1}^{m} p_i\, R_i(x \mid b_i, c_i) \tag{17.9d}$$

are the DF and the CCDF of the mixed WEIBULL distribution, respectively.

We first consider the **case of non–postmortem data** coming from a time–terminated sample of size $n$, $T$ being the censoring time. The vector of the ordered times to failure is

$$\boldsymbol{x} = (x_1, \ldots, x_r)',\ r \leq n.$$

The algorithm for finding the MLEs of the parameters is given by JIANG/KECECIOGLU (1992b) as follows:

1. Begin with an initial guess $p_i^{(0)}$, $b_i^{(0)}$ and $c^{(0)}$; $i = 1, \ldots, m$.

2. In iteration $h$ $(h \geq 1)$, calculate the probabilities

$$\mathrm{Pr}^{(h)}(i \mid x_j) = p_i^{(h-1)} \frac{f_i\big(x_j \mid b_i^{(h-1)}, c_i^{(h-1)}\big)}{f\big(x_j \mid \boldsymbol{\theta}^{(h-1)}\big)}\ \forall\, i, j, \tag{17.10a}$$

$$\mathrm{Pr}^{(h)}(i \mid T) = p_i^{(h-1)} \frac{R_i\big(T \mid b_i^{(h-1)}, c_i^{(h-1)}\big)}{R\big(T \mid \boldsymbol{\theta}^{(h-1)}\big)}\ \forall\, i. \tag{17.10b}$$

(17.10a) is the probability that the unit comes from subpopulation $i$, knowing that it failed at time $x_j$. Similarly, for the $n - r$ surviving units, the conditional probability of a unit's belonging to subpopulation $i$, given that it survived until $T$, is given by (17.10b).

3. In iteration $h$, find the MLEs of $p_i^{(h)}, b_i^{(h)}, c_i^{(h)}$, given the $\mathrm{Pr}^{(h)}(i \mid x_j)$ and $\mathrm{Pr}^{(h)}(i \mid T)$ in the following substeps:

   a) Use the NEWTON–RAPHSON iteration to find the MLE for $c_i^{(h)}$ from

$$\left. g\big(c_i^{(h)}\big) = \sum_{j=1}^{r} \mathrm{Pr}^{(h)}(i \mid x_j) \ln x_j + \frac{1}{c_i^{(h)}} \sum_{j=1}^{r} \mathrm{Pr}^{(h)}(i \mid x_j) - \right.$$
$$\frac{\left[\sum_{j=1}^{r} \mathrm{Pr}^{(h)}(i \mid x_j)\, x_j^{c_i^{(h)}} \ln x_j + (n-r)\,\mathrm{Pr}^{(h)}(i \mid T)\, T^{c_i^{(h)}} \ln T\right]\left[\sum_{j=1}^{r} \mathrm{Pr}^{(h)}(i \mid x_j)\right]}{\sum_{j=1}^{r} \mathrm{Pr}^{(h)}(i \mid x_j)\, x_j^{c_i^{(h)}} + (n-r)\,\mathrm{Pr}^{(h)}(i \mid T)\, T^{c_i^{(h)}}}$$

$$\left. = 0. \right\}$$

$$\tag{17.10c}$$

b) Calculate $b_i^{(h)}$ as

$$b_i^{(h)} = \left[ \frac{\sum_{j=1}^{r} \Pr^{(h)}(i \,|\, x_j)\, x_j^{c_i^{(h)}} + (n-r)\, \Pr^{(h)}(i \,|\, T)\, T^{c_i^{(h)}}}{\sum_{j=1}^{r} \Pr^{(h)}(i \,|\, x_j)} \right]^{1/c_i^{(h)}} . \qquad (17.10d)$$

c) Calculate $p_i^{(h)}$ as

$$p_i^{(h)} = \frac{1}{n} \left[ \sum_{j=1}^{r} \Pr^{(h)}(i \,|\, x_j) + (n-r)\, \Pr^{(h)}(i \,|\, T) \right] . \qquad (17.10e)$$

4. Repeat steps 2 and 3 until the desired accuracy has been reached.

**Example 17/1: German life table as a mixture distribution**

In Germany we have had twelve sets of life tables since the foundation of the German Empire in 1871. With the exception of the last table for 2000/02 all life tables have been set up around the year of a population census. From these tables we have extracted the conditional probabilities of death (= the annual rates of mortality) $q_x$, being the discrete version of the hazard rate, and we have depicted them in Fig. 17/1. This figure pertains to females, but the same course of $q_x$ is to be found with males, their curves lying on a somewhat higher level.

Figure 17/1: Annual mortality rates $q_x$ for females in German life tables from 1871 to 2002

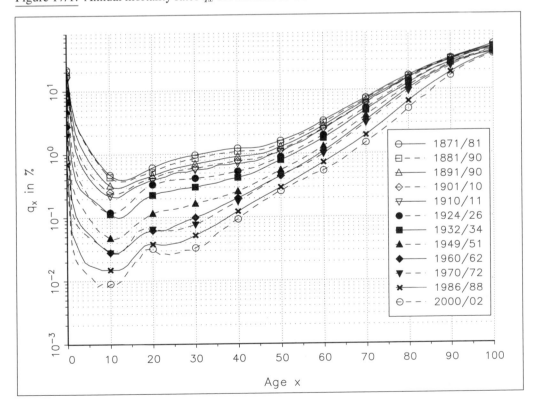

We clearly may distinguish three different types of behavior of the $q_x$:

1. For young age (up to $x \approx 10$) $q_x$ is steadily decreasing with $x$, expressing high infant mortality.

2. For the youth ($10 \lesssim x \lesssim 25$) $q_x$ is first increasing and then slightly decreasing, a phenomenon mainly caused by traffic accidents.

3. Finally, starting at $x \approx 25$ the $q_x$ are steadily increasing, expressing death by "wear and tear."

So we may look at the life table population as consisting of three subpopulations having different courses of the rate of mortality.

Subpopulations 1 and 3 each have a monotone hazard rate and thus may be modeled by a WEIBULL distribution. To model subpopulation 2 with a non–monotone hazard rate we have decided to take an inverse WEIBULL distribution (see Sect. 3.3.2).[4] We have to adapt the JIANG/KECECIOGLU algorithm with respect to the second subpopulation. As shown by the following results the modified algorithm also works.

Fig. 17/2 depicts the estimated densities of the three subpopulations and the total population for females of the life table 1891/1900. In Fig. 17/3 we show the fit of the empirical frequencies of deaths by the estimated mixture density, pertaining to females of the life table 1891/1900. The fit is rather good.

Figure 17/2: Estimated densities of the subpopulations and the total population (females, life table 1891/1900)

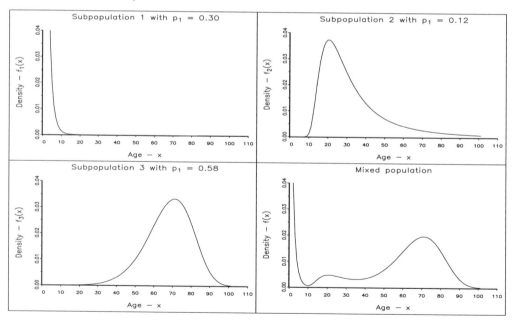

---

[4] For a mixture of inverse WEIBULL distributions see SULTAN et al. (2007).

Figure 17/3: Mixture density and empirical frequencies of death (females, life table 1891/1900)

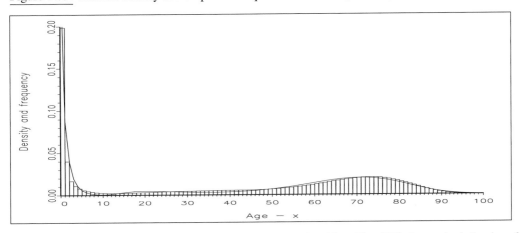

The following figures exhibit results for all the $2 \times 12$ life tables. Fig. 17/4 shows the behavior of the three proportions $p_1$, $p_2$, $p_3$ with time. As to be expected $p_1$ decreases and $p_3$ increases over the last 130 years, mainly due to medical progress.

Figure 17/4: Proportions of the three subpopulations

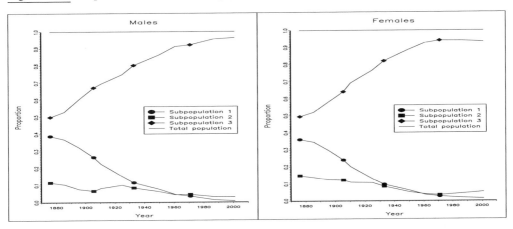

Fig. 17/5 and Fig. 17/6 show the estimated scale parameters and shape parameters, respectively.

Figure 17/5: Estimated scale parameters of the three subpopulations

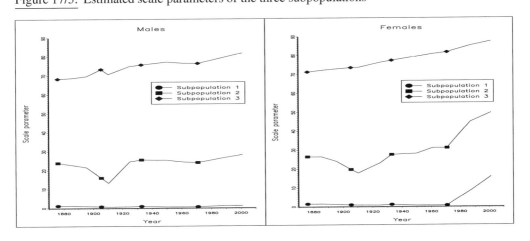

Figure 17/6: Estimated shape parameters of the three subpopulations

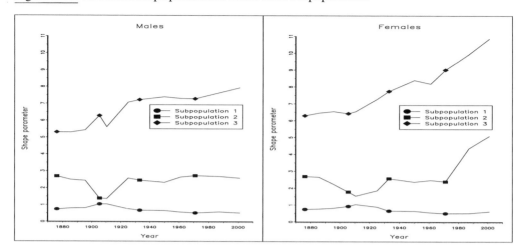

Finally, Fig. 17/7 gives the mean ages in the three subpopulations as well as in the mixed total population. They all behave as to be expected. We have also depicted the life expectation of a newborn, $e_0$, as taken from the life table which should be compared with the mean in the mixed population. Both are in close agreement demonstrating the plausibility of the approach.

Figure 17/7: Mean ages and life expectation

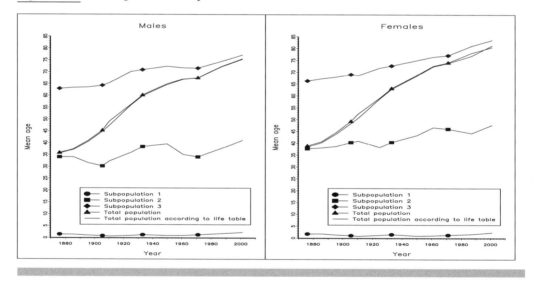

We finally turn to the **case of postmortem date**. $n_i$ test units belong to subpopulation $i$, $n = \sum_{i=1}^{m} n_i$. For time–termination, the life test data consist of $r_i$ $(r_i \leq n_i)$ times to failure $x_{ji}$ $(j = 1, \ldots, r_i)$ from subpopulation $i$, and $n - r$ survivors at time $T$, where $r = \sum_{i=1}^{m} r_i$. The algorithm for postmortem data is just a special case of the algorithm for non–postmortem data, where in (17.10a) $\mathrm{Pr}^{(h)}(i \mid x_j)$ is replaced by 0, if $x_j$ does not belong to subpopulation $i$, and by 1, otherwise. Instead of (17.10e) we have

$$p_i^{(h)} = \frac{n_i}{n} \ \forall \ h;$$

i.e., step 3c can be skipped.

# 17.2   BAYESIAN estimation approaches[5]

We will first show how to proceed when there are only two subpopulations (Sect. 17.2.1) and then turn to the general case of $m \geq 2$ subpopulations (Sect. 17.2.2).

## 17.2.1   The case of two subpopulations

When we have **non–postmortem data,** we can consider two cases:

1. the two shape parameters $c_1$ and $c_2$ are known, and $\lambda_1$, $\lambda_2$, $p$ are to be estimated;

2. all five parameters are unknown.

The mixed WEIBULL DF and CDF are

$$f(x) \;=\; p\,\lambda_1\,c_1\,x^{c_1-1}\exp\big(-\lambda_1\,x^{c_1}\big) + (1-p)\,\lambda_2\,c_2\,x^{c_2-1}\exp\big(-\lambda_2\,x^{c_2}\big)\text{and} \quad (17.11a)$$
$$R(x) \;=\; p\,\exp\big(-\lambda_1\,x^{c_1}\big) + (1-p)\,\exp\big(-\lambda_2\,x^{c_2}\big), \qquad\qquad (17.11b)$$

respectively, where
$$\lambda_i = b_i^{-c_i};\; i = 1, 2.$$

Case 1:  $c_1$ and $c_2$ known, type-II censoring

A very flexible prior for $p$ is the beta distribution with DF

$$g(p \mid \alpha, \beta) = \frac{\Gamma(\alpha+\beta)}{\Gamma(\alpha)\,\Gamma(\beta)}\,p^{\alpha-1}\,(1-p)^{\beta-1};\; 0 < p < 1,\; \alpha > 0,\; \beta > 0. \qquad (17.12a)$$

The prior for $\lambda_i$ is the gamma distribution (see (14.20a)) where the DFs are now

$$g(\lambda_i \mid \gamma_i, \delta_i) = \frac{\lambda_i^{\delta_i-1}\,\exp(-\gamma_i\,\lambda_i)\,\gamma_i^{\delta_i}}{\Gamma(\delta_i)};\; \lambda_i > 0,\; \gamma_i > 0;\; i = 1, 2. \qquad (17.12b)$$

The joint posterior DF of $\phi = (p,\,\lambda_1,\,\lambda_2)'$ takes the form

$$g(\phi \mid c_1, c_2, \boldsymbol{x}) \propto p^{\alpha-1}\,(1-p)^{\beta-1}\prod_{i=1}^{2}\lambda_i^{\delta_i-1}\,\exp(-\gamma_i\,\lambda_i)\,[R(x_{r:n})]^{n-r}\prod_{j=1}^{r}f(x_{j:n}).$$
$$(17.12c)$$

AHMAD et al. (1997) used LINDLEY's (1980) approximation to find the BAYES estimators.

Case 2:  All parameters unknown, type–II censoring

We keep the priors for $p$, $\lambda_1$ and $\lambda_2$ of case 1. Priors for $c_1$ and $c_2$ are assumed to be discrete probability mass functions and are given by[6]

$$\Pr(c_i = c_{ik}) = q_{ik} \geq 0;\; k = 1, 2, \ldots, d_i;\; \sum_{k=1}^{d_i} q_{ik};\; i = 1, 2. \qquad (17.13a)$$

---

[5] Suggested reading for this section: AHMAD/MOUSTAFA/ABD–ELRAHMAN (1997), SINHA (1987), TSIONAS (2002).

[6] See Sect. 14.2.2, where we used a two–point prior.

Here, the joint posterior probability function of $\psi = (\phi', c_1, c_2)'$, where $\psi = (p, \lambda_1, \lambda_2)'$, can be written as

$$g(\psi \mid x) \propto q_{1k}\, q_{1\ell}\, g(\phi \mid c_{1k}, c_{2\ell}, x), \tag{17.13b}$$

where $g(\phi \mid c_{1k}, c_{2\ell}, x)$ is as given by (17.12c) with $c_1$ and $c_2$ replaced by $c_{1k}$ and $c_{2\ell}$, respectively. The BAYES estimators may be found — using LINDLEY's approximation — as described in AHMAD et al. (1997).

Turning to a situation with **postmortem data** and a type-I censoring at time $T$, SINHA (1987) proposed taking the following priors, when the WEIBULL DF is given by

$$f_i(x) = \frac{c_i}{\theta_i}\, x^{c_i - 1}\, \exp\left\{ -\frac{x^{c_i}}{\theta_i} \right\}; \; i = 1, 2,$$

where

$$\theta_i = \lambda_i^{-1} = b_i^{c_i}.$$

1. a uniform distribution for $p$

$$g(p) = 1; \; 0 < p < 1; \tag{17.14a}$$

2. a uniform distribution for $c_i$

$$g(c_i) = \frac{1}{5}; \; 0 < c_i < 5; \; i = 1, 2; \tag{17.14b}$$

3. a vague prior for $(\theta_1, \theta_2)$

$$g(\theta_1, \theta_2) \propto \frac{1}{\theta_1\, \theta_2}; \; \theta_i > 0; \; i = 1, 2. \tag{17.14c}$$

The joint posterior DF of $\psi = (p, \theta_1, \theta_2, c_1, c_2)'$ is given by

$$g(\psi \mid x) \propto c_1^{r_1}\, c_2^{r_2}\, \kappa_1^{c_1}\, \kappa_2^{c_2} \sum_{k=0}^{n-r} \binom{n-r}{k} p^{n-k-r_2}\, (1-p)^{r_2-k} \times$$
$$\left[ \exp\left\{ -\frac{\sum_{j=1}^{r_1} x_{1j}^{c_1} + (n-r-k)\, T^{c_1}}{\theta_1} \right\} \Bigg/ \theta_1^{r_1+1} \right] \left[ \exp\left\{ -\frac{\sum_{j=1}^{r_2} x_{2j}^{c_2} + k\, T^{c_2}}{\theta_2} \right\} \Bigg/ \theta_2^{r_2+1} \right], \tag{17.14d}$$

where

- $r_1$ ($r_2$) units from subpopulation 1 (2) have failed during the interval $(0, T]$,

- $r = r_1 + r_2$, $r \leq n$,

- $x_{1j}$ ($x_{2j}$) is the failure time of the $j$–th unit belonging to subpopulation 1 (2),

- $\kappa_1 = \prod_{j=1}^{r_1} x_{1j}$, $\kappa_2 = \prod_{j=1}^{r_2} x_{2j}$.

For this scenario Sinha (1987) has found the following Bayes estimators using

$$
\eta_1 = \prod_{j=1}^{r_1} t_{1j};
$$

$$
\eta_2 = \prod_{j=1}^{r-2} t_{2j};
$$

$$
t_{ij} = x_{ij}/T;\ i = 1, 2;
$$

$$
C^{-1} = \sum_{k=0}^{n-r} \binom{n-r}{k} B(n - k - r_2 + 1, r_2 + k + 1) \times
$$

$$
\int_0^5 \frac{c_1^{r_1}\, \eta_1^{c_1}}{\left[\sum\limits_{j=1}^{r_1} t_{1j}^{c_1} + (n - r + k)\right]^{r_1}}\, dc_1 \int_0^5 \frac{c_2^{r_2}\, \eta_2^{c_2}}{\left[\sum\limits_{j=1}^{r_2} t_{2j}^{c_2} + k\right]^{r_2}}\, dc_2\ :
$$

$$
\widehat{p} = C \sum_{k=0}^{n-r} \binom{n-r}{k} B(n - k - r_2 + 2, r_2 + k + 1) \times
$$
$$
\int_0^5 \frac{c_1^{r_1}\, \eta_1^{c_1}}{\left[\sum\limits_{j=1}^{r_1} t_{1j}^{c_1} + (n - r + k)\right]^{r_1}}\, dc_1 \int_0^5 \frac{c_2^{r_2}\, \eta_2^{c_2}}{\left[\sum\limits_{j=1}^{r_2} t_{2j}^{c_2} + k\right]^{r_2}}\, dc_2,
$$
(17.15a)

$$
\widehat{c}_1 = C \sum_{k=0}^{n-r} \binom{n-r}{k} B(n - k - r_2 + 1, r_2 + k + 1) \times
$$
$$
\int_0^5 \frac{c_1^{r_1+1}\, \eta_1^{c_1}}{\left[\sum\limits_{j=1}^{r_1} t_{1j}^{c_1} + (n - r + k)\right]^{r_1}}\, dc_1 \int_0^5 \frac{c_2^{r_2}\, \eta_2^{c_2}}{\left[\sum\limits_{j=1}^{r_2} t_{2j}^{c_2} + k\right]^{r_2}}\, dc_2,
$$
(17.15b)

$$
\widehat{c}_2 = C \sum_{k=0}^{n-r} \binom{n-r}{k} B(n - k - r_2 + 1, r_2 + k + 1) \times
$$
$$
\int_0^5 \frac{c_1^{r_1}\, \eta_1^{c_1}}{\left[\sum\limits_{j=1}^{r_1} t_{1j}^{c_1} + (n - r + k)\right]^{r_1}}\, dc_1 \int_0^5 \frac{c_2^{r_2+1}\, \eta_2^{c_2}}{\left[\sum\limits_{j=1}^{r_2} t_{2j}^{c_2} + k\right]^{r_2}}\, dc_2,
$$
(17.15c)

$$
\widehat{\theta}_1 = \frac{C}{r_1 - 1} \sum_{k=0}^{n-r} \binom{n-r}{k} B(n - k - r_2 + 1, r_2 + k + 1) \times
$$
$$
\int_0^5 \frac{c_1^{r_1}\, (T\, \eta_1)^{c_1}}{\left[\sum\limits_{j=1}^{r_1} t_{1j}^{c_1} + (n - r + k)\right]^{r_1-1}}\, dc_1 \int_0^5 \frac{c_2^{r_2}\, \eta_2^{c_2}}{\left[\sum\limits_{j=1}^{r_2} t_{2j}^{c_2} + k\right]^{r_2}}\, dc_2,
$$
(17.15d)

$$
\widehat{\theta}_2 = \frac{C}{r_2 - 1} \sum_{k=0}^{n-r} \binom{n-r}{k} B(n - k - r_2 + 1, r_2 + k + 1) \times
$$

$$
\int_0^5 \frac{c_1^{r_1}\, \eta_1^{c_1}}{\left[\sum_{j=1}^{r_1} t_{1j}^{c_1} + (n - r + k)\right]^{r_1}}\, dc_1 \int_0^5 \frac{c_2^{r_2}\, (T\,\eta_2)^{c_2}}{\left[\sum_{j=1}^{r_2} t_{2j}^{c_2} + k\right]^{r_2 - 1}}\, dc_2. \qquad (17.15e)
$$

## 17.2.2   The case of $m$ subpopulations ($m \geq 2$)

Little research has been done so far for Bayesian analysis of finite mixtures of more than two Weibull distributions. Tsionas (2002) gave an approach for an uncensored sample of non–postmortem data assuming that the $m$ Weibull distributions have a common but unknown location parameter $a$. Thus, the DF of the $i$–th Weibull distribution is ($i = 1, 2, \ldots, n$):

$$
f(x) = \lambda_i\, c_i\, (x - a)^{c_i - 1} \exp\left[ -\lambda_i\, (x - a)^{c_i} \right], \; x > a.
$$

He chose independent priors:

$$
g(\lambda_i) \;\propto\; \lambda_i^{n_i - 1} \exp(-s_i\, \lambda_i), \qquad\qquad\qquad (17.16a)
$$

$$
g(c_i) \;\propto\; c_i^{n_i - 1} \exp(-q_i\, c_i), \qquad\qquad\qquad (17.16b)
$$

$$
g(\boldsymbol{p}) \;\propto\; \prod_{i=1}^{m} p_i^{n_i - 1}, \; \boldsymbol{p} = (p_1, \ldots, p_m)'. \qquad\qquad (17.16c)
$$

The first two priors are gamma distributions, and the third is a Dirichlet distribution defined on the unit simplex. From the form of the conditioned distributions, given by Tsionas, these priors are conditionally conjugate. This property may help in prior elicitation if we consider a fictitious sample of size $n_i$ in each mixing component so that the parameters $n_i$ reflect the prior importance of each component; see (17.16c). Within each component the parameters $\lambda_i$ and $c_i$ are gamma distributed a priori as shown in (17.16a–b), so the hyperparameters of these equations may be elicited using standard results for the moments of the gamma distribution. Further results of this approach may be found in the paper of Tsionas.

# 18 Inference of WEIBULL processes

The WEIBULL process, also called power law process, RASCH–WEIBULL process, WEIBULL–POISSON process or WEIBULL intensity function, has been introduced and described in Sect. 4.3. It is a special non–homogeneous POISSON process with **mean value function**

$$\Lambda(t) = \left(\frac{t}{b}\right)^c, \quad t \geq 0; \ b, c > 0, \tag{18.1a}$$

giving the expected number of events in $(0, t]$, and **intensity function**

$$\lambda(t) = \frac{d\Lambda(t)}{dt} = \frac{c}{b}\left(\frac{t}{b}\right)^{c-1}. \tag{18.1b}$$

Some authors prefer a different parametrization: $\Lambda(t) = \theta\, t^c$, $\lambda(t) = \theta\, c\, t^{c-1}$, where $\theta = b^{-c}$. Essentially, all of the research on WEIBULL processes has been motivated by applications in which events are failures of a repairable system. Improvement of a system, or **reliability improvement**, may occur if the system is in a development program. During testing, deficiencies are identified and subsequent redesign efforts are initiated to develop corrections for the problem areas. If the modifications introduced into the system during the test are effective, then the system reliability should increase over the testing phase. On the other hand, if only **minimal repairs** are made each time a failure occurs, the system will be deteriorating with time.

A characterization of a WEIBULL process is given by the sequence of successive failure times $T_1$, $T_2$, ..., where $T_n$ represents the time until the $n$–th failure. The time to first failure $T_1$ has the WEIBULL distribution with hazard rate function equal to (18.1b) and CDF

$$F(t_1) = 1 - \exp\left\{-\left(\frac{t_1}{b}\right)^c\right\}, \quad 0 \leq t_1 < \infty. \tag{18.2a}$$

It is also true that the conditional failure times $T_n$, $n \geq 2$, given $T_1 = t_1, \ldots, T_{n-1} = t_{n-1}$, follow a WEIBULL distribution which is truncated below the point $t_{n-1}$:

$$F(t_n | T_1 = t_1, \ldots, T_{n-1} = t_{n-1}) = 1 - \exp\left\{-\left(\frac{t_n}{b}\right)^c + \left(\frac{t_{n-1}}{b}\right)^c\right\}, \quad t_{n-1} \leq t_n < \infty. \tag{18.2b}$$

The joint density of the first $n$ times of failure $T_1, \ldots, T_n$, therefore, has a particularly simple form:

$$f(t_1, \ldots, t_n) = \left(\frac{c}{b}\right)^n \prod_{i=1}^{n} \left(\frac{t_i}{b}\right)^{c-1} \exp\left\{-\left(\frac{t_n}{b}\right)^c\right\}, \quad 0 < t_1 < t_1 < \ldots < t_n < \infty. \tag{18.2c}$$

In order to do inference of a WEIBULL process, it is necessary to adopt a method of collecting data. A very common way to obtain data is **failure truncation** (see Sect. 18.1) whereby the process is observed for a fixed number $n$ of failures, leading to an ordered set of data $t_1 < t_2 < \ldots < t_n$. Another way to obtain data is **time truncation** (see Sect. 18.2) whereby the process is observed for a fixed length of time $t^*$. In this case, the data have one of the following forms, $N(t^*)$ being the random number of failures:

1. $N(t^*) = 0$ or

2. $N(t^*) = n > 0$ and $0 < t_1 < t_2 < \ldots < t_n < t^*$.

Notice that, with time truncation, the observed number of failures is part of the dataset. There are some other ways of observing the process, e.g., **distance sampling**, where the process is observed at given points $\tau_1, \tau_2, \ldots$ in time. Inference in this and other cases will be presented in Sect. 18.3. A graphical estimation of $b$ and $c$ based on DUANE's plot is given in Sect. 18.4.

## 18.1 Failure truncation[1]

We will first present results for the case that there has only been one WEIBULL process under observation from $t = 0$ onwards (Sect. 18.1.1) and then turn to the more general case of several processes where the observation not necessarily started at $t = 0$ (Sect. 18.1.2).

### 18.1.1 The case of one observed process

When $n$, the number of failures to be observed, is fixed the first $n$ successive failure times $T_1, \ldots, T_n$ of a WEIBULL process are random variables having the joint DF given by (18.2c). Looking at (18.2c) as the likelihood function the MLEs of $b$ and $c$ are easily found to be

$$\widehat{c} = n \Big/ \sum_{i=1}^{n} \ln \left( \frac{T_n}{T_i} \right), \tag{18.3a}$$

$$\widehat{b} = T_n / n^{1/\widehat{c}}. \tag{18.3b}$$

In order to construct confidence intervals for $b$ and $c$ or to test hypotheses on these parameters, we need the distributions of $\widehat{b}$ and $\widehat{c}$.

The variable

$$Z = \frac{2nc}{\widehat{c}}, \tag{18.4a}$$

---

[1] Suggested reading for this section: CHAN/RUEDA (1992), CROW (1982), ENGELHARDT (1988), ENGELHARDT/BAIN (1978), FINKELSTEIN (1976), LEE/LEE (1978), RIGDON/BASU (1988, 1989).

as noted by FINKELSTEIN (1976), is a pivotal quantity, i.e., its distribution is free of unknown parameters, and it is has the $\chi^2$–distribution with $\nu = 2\,(n-1)$ degrees of freedom:

$$Z = \frac{2\,n\,c}{\widehat{c}} \sim \chi^2(2\,(n-1)). \tag{18.4b}$$

Thus, an upper $1 - \alpha$ level confidence interval for $c$ is given by

$$c \geq \frac{\widehat{c}}{2\,n}\,\chi^2_{2\,(n-1),\alpha}. \tag{18.4c}$$

The variable

$$W = \left(\frac{\widehat{b}}{b}\right)^{\widehat{c}} \tag{18.5a}$$

also is a pivotal quantity — for a proof see FINKELSTEIN (1976) — with CDF given by LEE/LEE (1978) as

$$F(w) = \int_0^\infty H\left[(n\,w)^{z/2\,n}; 2\,n\right]\,h\left[z; 2\,(n-1)\right]\,\mathrm{d}z, \tag{18.5b}$$

where $H[\cdot\,;\nu]$ and $h[\cdot\,;\nu]$ represent the CDF and DF of $\chi^2(\nu)$, respectively. Tabulated percentiles of $W$, obtained by Monte Carlo simulation, are given in FINKELSTEIN (1976) and may be used to construct confidence intervals for $b$. Tab. 18/1 is an extraction of the FINKELSTEIN table.

BAIN/ENGELHARDT (1980a) have derived the following asymptotic normal approximation:

$$\sqrt{n}\,\widehat{c}\,\ln\left(\frac{\widehat{b}}{b}\right)\bigg/\ln n \sim No(0,1). \tag{18.5c}$$

The unusual standardizing factor $\sqrt{n}/\ln n$, rather than $\sqrt{n}$, suggests that larger sample sizes may be needed to obtain high precision for inferences on $b$.

If the parameter $c$ is known, then

$$S = 2\left(\frac{T_n}{b}\right)^c \tag{18.6a}$$

would be used for inference on the parameter $b$. LEE/LEE (1978) show that

$$2\left(\frac{T_n}{b}\right)^c \sim \chi^2(2\,n). \tag{18.6b}$$

If testing of the system is planned to end at the $(n+k)$–th failure, then it may be desired to **predict the time** $T_{n+k}$ when testing will be completed, based on $T_1, \ldots, T_n$. The function

$$U = 2\left[\left(\frac{T_{n+k}}{b}\right)^c - \left(\frac{T_n}{b}\right)^c\right] \tag{18.7a}$$

<u>Table 18/1:</u> Percentage points $w_P$, such that $\Pr\left[\left(\dfrac{\hat{b}}{b}\right)^{\hat{c}} \leq w_P\right] = P$

| P \ n | 0.02 | 0.05 | 0.10 | 0.20 | 0.30 | 0.50 | 0.70 | 0.80 | 0.90 | 0.95 | 0.98 |
|---|---|---|---|---|---|---|---|---|---|---|---|
| 2 | — | 0.002 | 0.06 | 0.33 | 0.60 | 1.5 | 8.0 | 61.0 | 2380.0 | * | * |
| 3 | 0.09 | 0.23 | 0.40 | 0.58 | 0.79 | 1.6 | 4.7 | 13.0 | 113.0 | 2730.0 | * |
| 4 | 0.26 | 0.36 | 0.47 | 0.65 | 0.86 | 1.6 | 3.7 | 8.0 | 35.0 | 210.0 | 5140 |
| 5 | 0.31 | 0.40 | 0.49 | 0.67 | 0.87 | 1.5 | 3.2 | 6.1 | 19.0 | 75.0 | 629. |
| 6 | 0.33 | 0.42 | 0.51 | 0.68 | 0.88 | 1.5 | 3.0 | 5.2 | 14.0 | 43.0 | 220. |
| 7 | 0.34 | 0.42 | 0.51 | 0.68 | 0.87 | 1.4 | 2.8 | 4.6 | 11.0 | 28.0 | 126. |
| 8 | 0.35 | 0.42 | 0.51 | 0.68 | 0.86 | 1.4 | 2.6 | 4.1 | 9.2 | 21.0 | 69. |
| 9 | 0.35 | 0.42 | 0.51 | 0.68 | 0.85 | 1.3 | 2.4 | 3.8 | 7.8 | 17.0 | 49. |
| 10 | 0.35 | 0.42 | 0.51 | 0.67 | 0.85 | 1.3 | 2.3 | 3.6 | 7.1 | 15.0 | 42. |
| 12 | 0.35 | 0.43 | 0.52 | 0.67 | 0.84 | 1.3 | 2.2 | 3.2 | 5.9 | 11.0 | 27. |
| 14 | 0.36 | 0.43 | 0.52 | 0.67 | 0.83 | 1.3 | 2.0 | 2.9 | 5.1 | 9.0 | 19. |
| 16 | 0.36 | 0.43 | 0.52 | 0.68 | 0.83 | 1.2 | 2.0 | 2.7 | 4.8 | 8.0 | 16. |
| 18 | 0.36 | 0.43 | 0.53 | 0.68 | 0.83 | 1.2 | 1.9 | 2.6 | 4.3 | 6.9 | 13. |
| 20 | 0.36 | 0.44 | 0.53 | 0.68 | 0.83 | 1.2 | 1.9 | 2.5 | 4.1 | 6.4 | 11. |
| 22 | 0.36 | 0.44 | 0.53 | 0.68 | 0.82 | 1.2 | 1.8 | 2.4 | 3.9 | 5.9 | 9.8 |
| 24 | 0.37 | 0.44 | 0.53 | 0.67 | 0.81 | 1.2 | 1.8 | 2.3 | 3.6 | 5.5 | 9.2 |
| 26 | 0.37 | 0.45 | 0.53 | 0.67 | 0.81 | 1.2 | 1.7 | 2.3 | 3.4 | 5.1 | 8.2 |
| 28 | 0.37 | 0.45 | 0.53 | 0.67 | 0.81 | 1.2 | 1.7 | 2.2 | 3.3 | 4.9 | 7.9 |
| 30 | 0.37 | 0.45 | 0.53 | 0.67 | 0.81 | 1.1 | 1.7 | 2.2 | 3.2 | 4.7 | 7.1 |
| 35 | 0.38 | 0.46 | 0.54 | 0.68 | 0.82 | 1.1 | 1.6 | 2.1 | 3.0 | 4.3 | 6.5 |
| 40 | 0.39 | 0.46 | 0.55 | 0.69 | 0.82 | 1.1 | 1.6 | 2.0 | 2.8 | 3.9 | 5.9 |
| 45 | 0.39 | 0.47 | 0.55 | 0.70 | 0.83 | 1.1 | 1.6 | 1.9 | 2.7 | 3.7 | 5.4 |
| 50 | 0.39 | 0.47 | 0.56 | 0.70 | 0.83 | 1.1 | 1.5 | 1.9 | 2.6 | 3.5 | 5.0 |
| 60 | 0.40 | 0.49 | 0.57 | 0.71 | 0.83 | 1.1 | 1.5 | 1.8 | 2.4 | 3.2 | 4.4 |
| 70 | 0.42 | 0.49 | 0.58 | 0.71 | 0.82 | 1.1 | 1.4 | 1.7 | 2.3 | 3.0 | 4.0 |
| 80 | 0.43 | 0.51 | 0.59 | 0.71 | 0.82 | 1.1 | 1.4 | 1.7 | 2.2 | 2.8 | 3.8 |
| 90 | 0.44 | 0.51 | 0.60 | 0.71 | 0.83 | 1.1 | 1.4 | 1.7 | 2.1 | 2.6 | 3.5 |
| 100 | 0.45 | 0.52 | 0.60 | 0.72 | 0.83 | 1.1 | 1.4 | 1.6 | 2.1 | 2.6 | 3.3 |

\* — greater than $10^{10}$

is needed to solve the problem of predicting $T_{n+k}$. $U$ is a pivotal quantity and

$$U \sim \chi^2(2\,k), \tag{18.7b}$$

as shown by LEE/LEE (1978).[2] The random variables $Z$ in (18.4a), $S$ in (18.6a) and $U$ are independent. One approach to find an upper prediction limit is the following. The

---

[2] The prediction of $T_{n+k}$ based on the pivotal quantity
$$Y = (n-1)\,\hat{c}\,\ln(T_{n+k}/T_n)$$
is considered by ENGELHARDT (1988). The distribution of $Y$ is rather complicated, but for $k = 1$ an explicit form of the lower $1 - \alpha$ prediction limit $T_L$ for $T_{n+1}$ can be obtained as follows:
$$T_L = T_n \exp\left\{\left[(1-\alpha)^{-1/(n-1)} - 1\right]\Big/\hat{c}\right\}.$$

distribution of

$$V = \left(\frac{T_{n+k}}{T_n}\right)^{\widehat{c}} \tag{18.8a}$$

is found, and then to obtain, say, an upper prediction limits, the percentage point $v_\alpha$ is chosen so that

$$\Pr(V \le v_\alpha) = 1 - \alpha. \tag{18.8b}$$

It follows that

$$1 - \alpha = \Pr\left[\left(\frac{T_{n-k}}{T_n}\right)^{\widehat{c}} \le v_\alpha\right] = \Pr\left[T_{n+k} \le T_n \, v_\alpha^{1/\widehat{c}}\right], \tag{18.8c}$$

and $t_n \, v_\alpha^{1/\widehat{c}}$ is an upper $1 - \alpha$ level prediction limits for $T_{n+k}$. The distribution of $V$ is found by showing that it can be represented as a function of $Z$, $S$, $U$. We have

$$\frac{n \, U}{k \, S} = \frac{n}{k}\left[\left(\frac{T_{n+k}}{T_n}\right)^c - 1\right] \sim F(2\,k, 2\,n). \tag{18.8d}$$

If $c$ were known, (18.8d) could be used to construct the predictive limits for $T_{n+k}$. When $c$ is unknown, let us write, using (18.8d):

$$\begin{aligned}
V &= \left[\left(\frac{T_{n+k}}{T_n}\right)^c\right]^{\widehat{c}/c} \\
&= \left[\frac{k}{n}F(2\,k, 2\,n) + 1\right]^{2\,n/Z}, \tag{18.8e}
\end{aligned}$$

where $F(2\,k, 2\,n)$ and $Z$ of (18.4a) are independent. The CDF of $V$ is therefore given by

$$\Pr(V \le v) = \int_0^\infty G\left[\frac{n}{k}\left(v^{z/2\,n} - 1\right); 2\,k, 2\,n\right] h\left[z; 2\,(n-1)\right] \mathrm{d}z, \tag{18.8f}$$

where $G[\,\cdot\,; \nu_1, \nu_2]$ is the CDF of $F(\nu_1, \nu_2)$ and $h[\,\cdot\,; \nu]$ is the DF of $\chi^2(\nu)$.

If at time $t_n$ no further improvements are planned, then it may be assumed that the system has a constant failure rate which takes the current value of the intensity function $\lambda(t_n)$. Consequently, the system's life length has an exponential distribution with mean ($= $ MTBF) $\lambda^{-1}(t_n)$, and the current system reliability is represented by $R(t_0) = \exp[-\lambda(t_n)\,t_0]$ for some time interval $[0, t_0)$ of its useful life.

The current value of the **intensity function** $\lambda(t_n)$ has the MLE

$$\widehat{\lambda}(t_n) = n\,\widehat{c}/t_n, \tag{18.9a}$$

which does not involve the scale parameter. Confidence limits for $\lambda(t_n)$, $\lambda^{-1}(t_n)$ and $R(t_o)$ can be based on the pivotal quantity

$$Q = \frac{\lambda(t_n)}{\widehat{\lambda}(t_n)} = \frac{Z\,S}{4\,n^2}, \tag{18.9b}$$

which has the CDF

$$F(q) = \int\limits_0^\infty H\big[4\,n^2\,q/z; 2\,n\big]\,h\big[z; 2\,(n-1)\big]\,\mathrm{d}z, \tag{18.9c}$$

where $H[\cdot; \cdot]$ and $h[\cdot; \cdot]$ are defined in the context of (18.5b). A lower $1-\alpha$ confidence limit for $\lambda(t_n)$ is

$$\lambda_L = \widehat{\lambda}(t_n)\,q_\alpha, \tag{18.9d}$$

where $q_\alpha$ is the $100\,\alpha\%$ percentile of $Q$, and a lower $1-\alpha$ confidence limit for $R(t_0)$ is

$$R_L = \exp\big\{-\widehat{\lambda}(t_n)\,t_0\,q_{1-\alpha}\big\}. \tag{18.9e}$$

This is also related to the mean time between failure (= MTBF) $1/\lambda(t_n)$). The corresponding lower $1-\alpha$ confidence limit is

$$M_L = 1\big/\big\{\widehat{\lambda}(t_n)\,q_{1-\alpha}\big\}. \tag{18.9f}$$

Tabulated values $\rho_1$ and $\rho_2$ are given in CROW (1982) such that

$$\Pr\big[\rho_1\,\widehat{M}(t_n) < M(t_n) < \rho_2\,\widehat{M}(t_n)\big] = 1-\alpha, \tag{18.9g}$$

where $\widehat{M}(t_n) = t_n/(n\,\widehat{c})$. Tab. 18/2 is an extraction of the CROW–table.
It is straightforward to show also that

$$\sqrt{n}\left[\frac{\widehat{M}(t_n)}{M(t_n)} - 1\right] \overset{\text{asym}}{\sim} No(0, 2). \tag{18.10a}$$

Thus, for large $n$ approximate $1-\alpha$ two–sided confidence intervals are of the form given in (18.9g), where

$$\left.\begin{aligned}
\rho_1 &\approx \big[1 + u_{1-\alpha/2}\,\sqrt{2/n}\big]^{-1}, \\
\rho_2 &\approx \big[1 - u_{1-\alpha/2}\,\sqrt{2/n}\big]^{-1},
\end{aligned}\right\} \tag{18.10b}$$

$u_{1-\alpha/2}$ being the $100\,(1-\alpha/2)\%$ percentile of the standard normal distribution.

## 18.1.2  The case of more than one observed process

We consider the parametrization

$$\lambda(t) = \theta\, c\, t^{c-1}, \ \theta = b^{-c}, \tag{18.11}$$

<u>Table 18/2:</u>  Values $\rho_1$, $\rho_2$, such that $\rho_1\widehat{M}(t_n)$, $\rho_2\widehat{M}(t_n)$ are $1-\alpha$ confidence limits of $M(t_n)$

| $1-\alpha$ $n$ | 0.80 | | 0.90 | | 0.95 | | 0.98 | |
|---|---|---|---|---|---|---|---|---|
| | $\rho_1$ | $\rho_2$ | $\rho_1$ | $\rho_2$ | $\rho_1$ | $\rho_2$ | $\rho_1$ | $\rho_2$ |
| 2 | .8065 | 33.76 | .5552 | 72.67 | .4099 | 151.5 | .2944 | 389.9 |
| 3 | .6840 | 8.927 | .5137 | 14.24 | .4054 | 21.96 | .3119 | 37.60 |
| 4 | .6601 | 5.328 | .5174 | 7.651 | .4225 | 10.65 | .3368 | 15.96 |
| 5 | .6568 | 4.000 | .5290 | 5.424 | .4415 | 7.147 | .3603 | 9.995 |
| 6 | .6600 | 3.321 | .5421 | 4.339 | .4595 | 5.521 | .3815 | 7.388 |
| 7 | .6656 | 2.910 | .5548 | 3.702 | .4760 | 4.595 | .4003 | 5.963 |
| 8 | .6720 | 2.634 | .5668 | 3.284 | .4910 | 4.002 | .4173 | 5.074 |
| 9 | .6787 | 2.436 | .5780 | 2.989 | .5046 | 3.589 | .4327 | 4.469 |
| 10 | .6852 | 2.287 | .5883 | 2.770 | .5171 | 3.286 | .4467 | 4.032 |
| 11 | .6915 | 2.170 | .5979 | 2.600 | .5285 | 3.054 | .4595 | 3.702 |
| 12 | .6975 | 2.076 | .6067 | 2.464 | .5391 | 2.870 | .4712 | 3.443 |
| 13 | .7033 | .1.998 | .6150 | 2.353 | .5488 | 2.721 | .4821 | 3.235 |
| 14 | .7087 | 1.933 | .6227 | 2.260 | .5579 | 2.597 | .4923 | 3.064 |
| 15 | .7139 | 1.877 | .6299 | 2.182 | .5664 | 2.493 | .5017 | 2.921 |
| 16 | .7188 | 1.829 | .6367 | 2.144 | .5743 | 2.404 | .5106 | 2.800 |
| 17 | .7234 | 1.788 | .6431 | 2.056 | .5818 | 2.327 | .5189 | 2.695 |
| 18 | .7278 | 1.751 | .6491 | 2.004 | .5888 | 2.259 | .5267 | 2.604 |
| 19 | .7320 | 1.718 | .6547 | 1.959 | .5954 | 2.200 | .5341 | 2.524 |
| 20 | .7360 | 1.688 | .6601 | 1.918 | .6016 | 2.147 | .5411 | 2.453 |
| 21 | .7398 | 1.662 | .6652 | 1.881 | .6076 | 2.099 | .5478 | 2.390 |
| 22 | .7434 | 1.638 | .6701 | 1.848 | .6132 | 2.056 | .5541 | 2.333 |
| 23 | .7469 | 1.616 | .6747 | 1.818 | .6186 | 2.017 | .5601 | 2.281 |
| 24 | .7502 | 1.596 | .6791 | 1.790 | .6237 | 1.982 | .5659 | 2.235 |
| 25 | .7534 | 1.578 | .6833 | 1.765 | .6286 | 1.949 | .5714 | 2.192 |
| 26 | .7565 | 1.561 | .6873 | 1.742 | .6333 | 1.919 | .5766 | 2.153 |
| 27 | .7594 | 1.545 | .6912 | 1.720 | .6378 | 1.892 | .5817 | 2.116 |
| 28 | .7622 | 1.530 | .6949 | 1.700 | .6421 | 1.866 | .5865 | 2.083 |
| 29 | .7649 | 1.516 | .6985 | 1.682 | .6462 | 1.842 | .5912 | 2.052 |
| 30 | .7676 | 1.504 | .7019 | 1.664 | .6502 | 1.820 | .5957 | 2.023 |
| 35 | .7794 | 1.450 | .7173 | 1.592 | .6681 | 1.729 | .6158 | 1.905 |
| 40 | .7894 | 1.410 | .7303 | 1.538 | .6832 | 1.660 | .6328 | 1.816 |
| 45 | .7981 | 1.378 | .7415 | 1.495 | .6962 | 1.606 | .6476 | 1.747 |
| 50 | .8057 | 1.352 | .7513 | 1.460 | .7076 | 1.562 | .6605 | 1.692 |
| 60 | .8184 | 1.312 | .7678 | 1.407 | .7267 | 1.496 | .6823 | 1.607 |
| 70 | .8288 | 1.282 | .7811 | 1.367 | .7423 | 1.447 | .7000 | 1.546 |
| 80 | .8375 | 1.259 | .7922 | 1.337 | .7553 | 1.409 | .7148 | 1.499 |
| 100 | .8514 | 1.225 | .8100 | 1.293 | .7759 | 1.355 | .7384 | 1.431 |

of the WEIBULL intensity function. The number of systems (= processes) which are all assumed to have the same intensity function as given in (18.11a)[3] is $k$ and the $\ell$-th system is observed from $S_\ell$ ($S_\ell \geq 0$) until the occurrence of the $n_\ell$-th failure. The total number of failures is

$$n = \sum_{\ell=1}^{k} n_\ell,$$

and the times of failure are denoted as $T_{\ell i}$, $i = 1, \ldots, n_\ell$, assembled in vector $\boldsymbol{t}_\ell$, $\ell = 1, \ldots, k$. Remember, that $T_{\ell i} \geq S_\ell$. The likelihood function is

$$L(\theta, c \,|\, \boldsymbol{t}_1, \ldots, \boldsymbol{t}_k) = \theta^n \, c^n \left( \prod_{\ell=1}^{k} \prod_{i=1}^{n_\ell} T_{\ell i}^{c-1} \right) \exp\left\{ -\theta \sum_{\ell=1}^{k} \left( T_{\ell i}^c - S_\ell^c \right) \right\}, \qquad (18.12a)$$

and the MLEs are the solutions of the equations

$$\frac{1}{\widehat{c}} - \frac{\sum_{\ell=1}^{k} \left( T_{\ell n_\ell}^{\widehat{c}} \ln T_{\ell n_\ell} - S_\ell^{\widehat{c}} \ln S_\ell \right)}{\sum_{\ell=1}^{k} \left( T_{\ell n_\ell}^{\widehat{c}} - S_\ell^{\widehat{c}} \right)} = -\frac{1}{n} \sum_{\ell=1}^{k} \sum_{i=1}^{n_\ell} \ln T_{\ell i}, \qquad (18.12b)$$

$$\widehat{\theta} = n \bigg/ \sum_{\ell=1}^{k} \left( T_{\ell n_\ell}^{\widehat{c}} - S_\ell^{\widehat{c}} \right). \qquad (18.12c)$$

For $k = 1$ and $S_1 = 0$ and observing $0 \ln 0 = 0$, the solutions of (18.12b,c) can be given in closed form

$$\widehat{c} = n \bigg/ \sum_{i=1}^{n} \ln\left( \frac{T_n}{T_i} \right), \qquad (18.13a)$$

$$\widehat{\theta} = n \bigg/ T_n^{\widehat{c}}, \qquad (18.13b)$$

which coincide with (18.3a,b) because $\theta = b^{-c}$.

CHAN/RUEDA (1992) have shown that the MLEs do not exist when $S_\ell > 0$ and

$$-\frac{1}{n} \sum_{\ell=1}^{k} \sum_{i=1}^{n_\ell} \ln T_i \geq -\frac{\sum_{\ell=1}^{k} \left[ \left( \ln T_{\ell n_\ell} \right)^2 - \left( \ln S_\ell \right)^2 \right]}{2 \sum_{\ell=1}^{k} \left( \ln T_{\ell n_\ell} - \ln S_\ell \right)}. \qquad (18.14)$$

They also give a formula for the probability of this non-existence.

---

[3] A test for the equality of several intensity functions is given by LEE (1980).

## 18.2 Time truncation[4]

In this section we will proceed as in Sect. 18.1 and first look at the case of only one system under observation and afterwards turn to the case of more than one system to be observed.

### 18.2.1 The case of one observed process

The process is observed from $t = 0$ to a preselected time $t^*$. The dataset consists of the random number of failures $N(t^*)$ and the random times to failure $0 < T_1 < T_2 < \ldots < T_{N(t^*)} \le t^*$. Suppose $N(t^*) = n$ is the number of failures in $(0, t^*]$. If $n = 0$, then a limited amount of statistical analysis is possible on

$$\Lambda(t^*) = \mathrm{E}\big[N(t^*)\big] = \left(\frac{t^*}{b}\right)^c. \tag{18.15a}$$

In particular the MLE is

$$\widehat{\Lambda}(t^*) = 0 \tag{18.15b}$$

and a lower $1 - \alpha$ confidence limit for $\Lambda(t^*)$ is

$$\Lambda_L = \frac{1}{2}\chi^2_{2,\alpha}. \tag{18.15c}$$

The following results pertain to the case $N(t^*) = n > 0$.

The joint density function of the successive failure times $T_1, \ldots < T_n$ and $N(t^*)$ is

$$f(t_1, \ldots, t_n, n) = \left(\frac{c}{b}\right)^n \prod_{i=1}^n \left(\frac{t_i}{b}\right)^{c-1} \exp\left\{-\left(\frac{t^*}{b}\right)^c\right\}, \quad 0 < t_1 < \ldots < t_n \le t^*. \tag{18.16a}$$

The joint MLEs follow as

$$\widehat{c} = n \Big/ \sum_{i=1}^n \ln(T_i/t^*), \tag{18.16b}$$

$$\widehat{b} = t^*/n^{1/\widehat{c}}. \tag{18.16c}$$

(18.6a–c) are similar to (18.2c) and (18.3a,b) with $T_n$ substituted by $t^*$ and the summation in (18.16b) is over $i$ from 1 to $n$ instead of $n - 1$.

Because $N(t^*)$ is sufficient for $b$ when $c$ is fixed, the conditional distribution of $\widehat{c}$, given $N(t^*) = n$, is free of $b$. Furthermore, since the conditional distribution of $Z$ as defined by (18.4a), given $N(t^*) = n$, is $\chi^2(2n)$, a conditional upper $1 - \alpha$ confidence interval for $c$ is found to be

$$c \ge \frac{\widehat{c}}{2n}\chi^2_{2n,\alpha}. \tag{18.17}$$

Due to the time truncation, $b$ is no longer a scale parameter, and the pivotal property of $W$ as given in (18.5a) does not longer hold in this case. BAIN/ENGELHARDT (1980a) give the

---

[4] Suggested reading for this section: BAIN/ENGELHARDT (1980a), CHAN/RUEDA (1992), CROW (1982), ENGELHARDT (1988), RIGDON/BASU (1989, 1990a).

following approximate upper $1 - \alpha$ confidence interval for $b$:

$$b \geq \hat{b} \left\{ n \left[ (n + 1) \, w_{1-\alpha,n+1} \right]^{1/(2\,n+2)} \right\}^{1/\hat{c}}, \tag{18.18}$$

where $w_{1-\alpha,n+1}$ is the $100\,(1 - \alpha)$ percentage point of $W$ in Tab. 18.1.

Conservative confidence intervals for $\lambda(t^*)$, the MTBF $\lambda(t^*)^{-1}$ and $R(t_0)$ can be found using the conditional probability density of $N(t^*)$ and $\lambda(t^*)$ given $y = n/\hat{c}$; see ENGEL-HARDT (1988):

$$f(n, \lambda(t^*)\,|\,y) = \frac{\left[ t^* \, \lambda(t*) \, y \right]^n}{n!\,(n-1)!} \left\{ \sum_{k=1}^{\infty} \frac{\left[ t^* \, \lambda(t^*) \, y \right]^k}{k!\,(k-1)!} \right\}^{-1}, \quad n = 1, 2, \ldots \tag{18.19}$$

A conservative $1 - \alpha$ confidence limit for $\lambda(t^*)$ is given by the largest solution $\lambda_L = \lambda_1$ of the inequality

$$\sum_{k=1}^{\infty} f(k, \lambda_1\,|\,y) \leq \alpha. \tag{18.20}$$

A conservative lower $1 - \alpha$ confidence limit for $R(t_0)$ follows as

$$R_L = \exp\left( - \lambda_2 \, t_0 \right), \tag{18.21a}$$

where $\lambda_2$ is the smallest solution of

$$\sum_{k=1}^{n} f(k, \lambda\,|\,y) \leq \alpha. \tag{18.21b}$$

The corresponding lower $1 - \alpha$ confidence limit $M_L$ for the instantaneous MTBF $M(t^*) = \lambda(t^*)^{-1}$ is

$$M_L = 1/\lambda_2. \tag{18.22a}$$

CROW (1982) derives a normal approximation of $\widehat{M}(t^*)$ leading to the following lower $1 - \alpha$ confidence limit

$$M_L = n^2 \big/ \left( n + u_\alpha \sqrt{n/2} \right)^2, \tag{18.22b}$$

$u_\alpha$ being the percentile of order $\alpha$ of the standard normal distribution.

### 18.2.2 The case of more than one observed process

As in Sect. 18.1.2 we take the parametrization

$$\lambda(t) = \theta \, c \, t^{c-1}, \ \theta = b^{-c}.$$

The number of systems or processes under study is $k$ and the $\ell$–th system is observed from $S_\ell$ to $t_\ell^*$, $\ell = 1, \ldots, k$. Suppose $N_\ell = N(t_\ell^*) = n_\ell$ failures are observed on the $\ell$–th system at times

$$T_{\ell 1} < T_{\ell 2} < \ldots < T_{\ell n_\ell} \leq t_\ell^*.$$

The random variable $N_\ell$ has a POISSON distribution with mean

$$\Lambda(t_\ell^*) = \int_{S_\ell}^{t_\ell^*} \theta \, c \, t^{c-1} \, \mathrm{d}t = \theta \left(t_\ell^{*c} - S_\ell^c\right). \tag{18.23a}$$

Given $N_\ell = n_\ell$, the failure times $T_{\ell 1}, \ldots T_{\ell n_\ell}$ have the same distribution as the order statistics corresponding to $n_\ell$ independent random variables having distribution function

$$F(t) = \begin{cases} 0 & \text{for} \quad t < S_\ell, \\ \left(t^c - S_\ell^c\right)\big/\left(t_\ell^{*c} - S_\ell^c\right) & \text{for} \quad S_\ell \le t \le t_\ell^*, \\ 1 & \text{for} \quad t > t_\ell^*. \end{cases} \tag{18.23b}$$

The likelihood function based only on the $\ell$–th system is

$$\begin{aligned} &n! \, \frac{c^{n_\ell} \prod_{i=1}^{n_\ell} T_{\ell i}^{c-1}}{\left(t_\ell^* - S_\ell^c\right)^{n_\ell}} \, \frac{\exp\left\{-\theta \left(t_\ell^{*c} - S_\ell^c\right)\right\} \theta^{n_\ell} \left(t_\ell^{*c} - S_\ell^*\right)^{n_\ell}}{n_\ell!} \\ &= \theta^{n_\ell} \, c^{n_\ell} \left(\prod_{i=1}^{n_\ell} T_{\ell i}^{c-1}\right) \exp\left\{-\theta \left(t_\ell^{*c} - S_\ell^c\right)\right\}. \end{aligned} \tag{18.24a}$$

Let $n = \sum_{\ell=1}^k n_\ell$. The likelihood function based on all $k$ systems is

$$L(\theta, c \,|\, t_1, \ldots, t_k, n) = \theta^n \, c^n \left(\prod_{\ell=1}^k \prod_{i=1}^{n_\ell} T_{\ell i}^{c-1}\right) \exp\left\{-\theta \sum_{\ell=1}^k \left(t_\ell^{*c} - S_\ell^c\right)\right\}. \tag{18.24b}$$

If $n > 0$, then the MLEs of $\theta$ and $c$ are the solutions of the equations

$$\frac{1}{\widehat{c}} - \frac{\sum_{\ell=1}^k \left(t_\ell^{*\widehat{c}} \ln t_\ell^* - S_\ell^{\widehat{c}} \ln S_\ell\right)}{\sum_{\ell=1}^k \left(t_\ell^{*\widehat{c}} - S_\ell^{\widehat{c}}\right)} = -\frac{1}{n} \sum_{\ell=1}^k \sum_{i=1}^{n_\ell} \ln T_{\ell i}, \tag{18.25a}$$

$$\widehat{\theta} = n \bigg/ \sum_{\ell=1}^k \left(t_\ell^{*\widehat{c}} - S_\ell^{\widehat{c}}\right). \tag{18.25b}$$

CHAN/RUEDA (1992) show that the MLE of $c$ may not exist if $S_\ell > 0$ for each of the $k$ systems. A unique solution exists for (18.25a) if $S_\ell = 0$ for some $\ell$ and $0 < T_{\ell i} < \max\left(t_1^*, \ldots, t_k^*\right)$ for each $\ell$ and $i$.

## 18.3 Other methods of collecting data

Besides time truncation and failure truncation there are other sampling designs to observe for a WEIBULL process. LEE et al. (1988) discuss two such approaches:

- **same–shape sampling** and

- **equal–distance sampling**.

The latter approach is a special case of what is called "data form C" by MØLLER (1976).

The same–shape sampling works as follows: The WEIBULL process of interest $\{N(t)\}$ with mean function $\Lambda(t) = \theta\, t^c$ is observed at the waiting times $\{T_n^*\}$ of an independent WEIBULL process $\{N^*(t)\}$ with mean function $\Lambda^*(t) = t^c$, i.e., with the same known shape parameter. The data are of the form $\boldsymbol{N} = \big(N(T_1^*), \ldots, N(T_r^*)\big)'$ or $\boldsymbol{Y}^* = (Y_1^*, \ldots, Y_r^*)'$, where $Y_i^* = N(T_i^*) - N(T_{i-1}^*)$ with $T_0^* = 0$. Then $Y_1^*, \ldots, Y_r^*$ are iid with a negative binomial distribution having parameters 1 and $(1 + \theta)^{-1}$, and $N\big(T_r^*\big) = \sum_{k=1}^r Y_k^*$ is also a negative binomial but with parameters $r$ and $(1 + \theta)^{-1}$; see LEE et al. (1988). The MLE of $\theta$, given $c$, is

$$\widehat{\theta} = \frac{N\big(T_r^*\big)}{r}. \tag{18.26}$$

In equal–distance sampling $\{N(t)\}$ is observed at times $0 < t_1 < \ldots < t_r$, where $\Lambda(t_\ell) = \ell\,\Lambda(t_1)$ which means that[5]

$$t_\ell = \ell^{1/c}\, t_1; \ \ell = 1, 2, \ldots, r. \tag{18.27a}$$

Here, the data are $\boldsymbol{N} = \big(N(t_1), \ldots, N(t_r)\big)'$ or $\boldsymbol{Y} = (Y_1, \ldots, Y_r)'$, where $Y_\ell = N(t_\ell) - N(t_{\ell-1})$ with $t_0 = 0$. It can be shown that $Y_1, \ldots, Y_r$ are iid with a POISSON distribution having the parameter $\Lambda(t_1) = \theta\, t_1^c$. $N(t_r) = \sum_{k=1}^r Y_k$ is POISSON distributed with parameter $r\,\Lambda(t_1) = r\,\theta\, t_1^c$. The MLE of $\theta$, given $c$, is

$$\widehat{\theta} = \frac{N(t_r)}{t_r^c} = \frac{N\big(t_1\, \ell^{1/c}\big)}{r\, t_1^c}. \tag{18.27b}$$

## 18.4  Estimation based on DUANE's plot

We assume a system undergoing reliability improvement as described in the beginning of this chapter. The standard estimator of the MTBF for a system with a constant repair or failure rate — an HPP system — is $T/r$, with $T$ denoting the total time the system was observed and $r$, $r > 0$, being the number of failures in $(0, T]$. If we calculate successive MTBF estimators every time $T_i$ $(i = 1, 2, \ldots)$ a failure occurs for a system undergoing reliability improvement testing, we typically see a sequence of mostly increasing numbers:

$$\text{MTBF}_i = T_i/i; \ i = 1, 2, \ldots; \tag{18.28}$$

i.e., the MTBF increases when we really have an improvement of system reliability.

DUANE's (1964) and other reliability engineers very often observed that when $\log \text{MTBF}_i$ is plotted versus $\log T_i$, the points tended to lie on an increasing straight line. This type of plot is called a **DUANE's plot** and the slope $\beta$ of the fitted line through the points is called a **reliability growth slope**. This slope should as a rule of thumb lie between 0.3 and 0.6. The lower end expresses minimally effective testing.

---

[5] MØLLER (1976) assumes in what he called "data form C," that the times of observation $t_\ell$ are linked by a function such as (18.27a) or any other type.

What is the rationale of this procedure? If the failure process is a WEIBULL process with mean value function (see (4.14a)),

$$\Lambda(t) = \mathrm{E}(N_t) = \left(\frac{t}{b}\right)^c ; \ t \geq 0; \ b, \ c > 0,$$

we are plotting estimators of

$$\frac{t}{\Lambda(t)} = \frac{t}{a\,t^c} = \frac{1}{a}t^{1-c}, \tag{18.29a}$$

where

$$a = 1/b^c \tag{18.29b}$$

versus the time of failure $t$. This is the same as plotting

$$\frac{1}{a}t^{\beta} \quad \text{versus} \quad t$$

with the reliability growth slope

$$\beta = 1 - c. \tag{18.29c}$$

On log–log paper this will be a straight line[6] with slope $\beta$ and intercept of $-\log a$ (when $t = 1 \implies \log t = 0$). The above mentioned rule of thumb for $\beta$ : $0.3 \leq \beta \leq 0.6$ thus means for the shape parameter $c$ : $0.4 \leq c \leq 0.7$, i.e., the existence of a decreasing hazard rate.

Putting forward the regression model for $r$ failures with failure times $T_i$ $(i = 1, \ldots, r)$ :

$$\mathrm{MTBF}_i^* = \alpha + \beta\, T_i^* + \epsilon \tag{18.30a}$$

with

$$\mathrm{MTBF}_i^* = \log \mathrm{MTBF}_i \quad \text{and} \quad T_i^* = \log T_i,$$

we find the OLS estimators of $\alpha$ and $\beta$ to be

$$\widehat{\alpha} = \frac{\sum \mathrm{MTBF}_i^* \sum \left(T_i^*\right)^2 - \sum \left(\mathrm{MTBF}_i^*\, T_i^*\right) \sum T_i^*}{r \sum \left(T_i^*\right)^2 - \left(\sum T_i^*\right)^2}, \tag{18.30b}$$

$$\widehat{\beta} = \frac{r \sum \left(\mathrm{MTBF}_i^*\, T_i^*\right) - \sum T_i^* \sum \mathrm{MTBF}_i}{r \sum \left(T_i^*\right)^2 - \left(\sum T_i^*\right)^2}. \tag{18.30c}$$

The original WEIBULL parameters will be estimated as

$$\widehat{c} = 1 - \widehat{\beta}, \tag{18.30d}$$

$$\widehat{b} = 10^{\widehat{\alpha}/\widehat{\beta}}. \tag{18.30e}$$

---

[6] A significant departure from a straight line would be evidence for the underlying process not to be WEIBULL.

**Example 18/1:   Parameter estimation using DUANE's plot**

A reliability growth test lasted 150 hours. Eight failures at times 3, 8, 15, 34, 56, 80, 121 and
150 were recorded. The input to Fig. 18/1 and to the OLS estimation formulas are as follows:

| Failure No. $i$ | 1 | 2 | 3 | 4 | 5 | 6 | 7 | 8 |
|---|---|---|---|---|---|---|---|---|
| System age $t_i$ | 3 | 8 | 15 | 34 | 56 | 80 | 121 | 150 |
| Succ. MTBF$_i$ | 3 | 4 | 5 | 8.5 | 11.2 | 13.33 | 17.29 | 18.75 |

Fig. 18/1 shows the corresponding DUANE plot. The estimated parameters of the straight line are
as follows:
$$\widehat{\alpha} = 0.1810, \quad \widehat{\beta} = 0.4955,$$
resulting in the WEIBULL parameter estimates:
$$\widehat{c} = 0.5045, \quad \widehat{b} = 2.2841.$$

Figure 18/1:  DUANE's plot

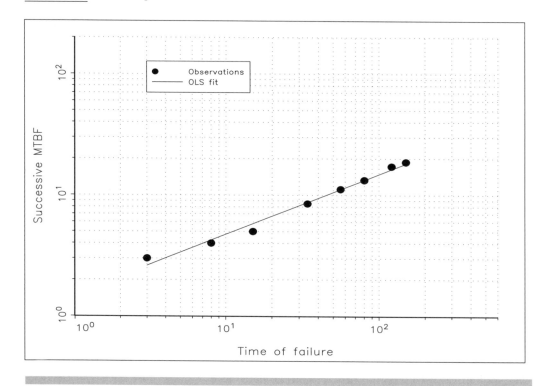

# 19 Estimation of percentiles and reliability including tolerance limits

People working with the WEIBULL distribution in practice are generally more interested in estimating a percentile of given order $P$ or the reliability at a specified mission time $x_0$ than in estimating the parameters $a$, $b$ and $c$ of this distribution. Of course, most estimation procedures for these two quantities rest upon estimators of the WEIBULL parameters.

We will first demonstrate how percentiles and reliability and their confidence limits are related to each other and where emphasis is on tolerance limits and tolerance intervals (Sect. 19.1). Classical inference approaches to reliability are presented in Sect. 19.2 and to percentiles in Sect. 19.3. The construction of tolerance intervals is described in Sect. 19.4. This chapter is completed by BAYESIAN inference approaches to percentiles and reliability in Sect. 19.5.

## 19.1 Percentiles, reliability and tolerance intervals

There are two related problems which are among the most important associated with the life–testing situation.

- First, one may suppose that a required life $x_0$, e.g., a guaranteed life, is specified and an estimate of the survival proportion or reliability $R(x_0) = \Pr(X > x_0)$ is sought together with a confidence interval for $R(x_0)$.

- For the second and alternative problem a survival probability $\gamma$ is given[1] — as a rule $\gamma$ will be a high value near unity — and an estimate of the corresponding lifetime, the **reliable life**, is sought which is nothing but the percentile $x_P$, of order $P = 1 - \gamma$. Besides finding an estimate of $x_P$, one is also interested in having a confidence interval for $x_P$.

Reliable life and reliability are linked as follows:

$$\Pr(X \le x_{1-\gamma}) = F(x_{1-\gamma}) = 1 - R(x_{1-\gamma}) = 1 - \gamma \implies x_{1-\gamma} = F^{-1}(1-\gamma) \quad (19.1a)$$

$$\Pr(X > x_{1-\gamma}) = R(x_{1-\gamma}) = 1 - F(x_{1-\gamma}) = \gamma \implies x_{1-\gamma} = R^{-1}(\gamma). \quad (19.1b)$$

---

[1] In the ball–bearing industry, the tradition is to specify $\gamma = 0.9$. Another choice is $\gamma = 0.5$ leading to the median or half–time period $x_{0.5}$.

When $X \sim We(a, b, c)$, these relationships are as follows:

$$F(x_{1-\gamma}) = 1 - \exp\left\{-\left(\frac{x_{1-\gamma} - a}{b}\right)^c\right\} = 1 - \gamma \implies x_{1-\gamma} = 1 + b\,[-\ln\gamma]^{1/c},$$

$$R(x_{1-\gamma}) = \exp\left\{-\left(\frac{x_{1-\gamma} - a}{b}\right)^c\right\} = \gamma \implies x_{1-\gamma} = 1 + b\,[-\ln\gamma]^{1/c}.$$

Let $R_L$ be a lower $(1 - \alpha)$ confidence limit for $R(x_0)$, i.e.,

$$\Pr\left[R(x_0) \geq R_L\right] = 1 - \alpha, \tag{19.2a}$$

then $F_U = 1 - R_L$ will be an upper $(1 - \alpha)$ confidence limit for $F(x_0)$:

$$\Pr\left[1 - R(x_0) \leq 1 - R_L\right] = \Pr\left[F(x_0) \leq F_U\right] = 1 - \alpha. \tag{19.2b}$$

A **two–sided** interval $\left[L_t(\boldsymbol{x}, 1 - \alpha, \gamma),\ U_t(\boldsymbol{x}, 1 - \alpha, \gamma)\right]$ depending on the sample data $\boldsymbol{x}$ is said to be a $(1 - \alpha)$ probability **tolerance interval** for proportion $\gamma$ if

$$\Pr\left\{\int_{L_t(\boldsymbol{x},1-\alpha,\gamma)}^{U_t(\boldsymbol{x},1-\alpha,\gamma)} f(x \mid \boldsymbol{\theta})\,\mathrm{d}x \geq \gamma\right\} = \Pr\left\{F[U_t(\boldsymbol{x}, 1-\alpha, \gamma)|\boldsymbol{\theta}] - F[L_t(\boldsymbol{x}, 1-\alpha, \gamma)|\boldsymbol{\theta}] \geq \gamma\right\} = 1-\alpha,$$

$$\tag{19.3}$$

where $f(x \mid \boldsymbol{\theta})$ and $F(x \mid \boldsymbol{\theta})$ are the DF and CDF of $X$, respectively. Thus, with probability $1 - \alpha$ a two–sided $(1 - \alpha, \gamma)$ tolerance interval includes at least a central portion $\gamma$ of the underlying distribution, and at most a portion $1 - \gamma$ of the smallest and largest realizations of $X$ is outside of $\left[L_t(\boldsymbol{x}, 1-\alpha, \gamma),\ U_t(\boldsymbol{x}, 1-\alpha, \gamma)\right]$. Because the limits $L_t(\boldsymbol{x}, 1-\alpha, \gamma)$ and $U_t(\boldsymbol{x}, 1 - \alpha, \gamma)$ depend upon the sampled data, they are random and the tolerance interval is a variate too.

**One–sided tolerance intervals** are defined similarly:

- $L_o(\boldsymbol{x}, 1 - \alpha, \gamma)$ is a **lower** $(1 - \alpha, \gamma)$ tolerance limit if

$$\Pr\left\{\int_{L_o(\boldsymbol{x},1-\alpha,\gamma)}^{\infty} f(x \mid \boldsymbol{\theta})\,\mathrm{d}x \geq \gamma\right\} = \Pr\left\{R[L_o(\boldsymbol{x}, 1 - \alpha, \gamma) \mid \boldsymbol{\theta}] \geq \gamma\right\} = 1 - \alpha \tag{19.4a}$$

or

$$\Pr\left\{\int_{-\infty}^{L_o(\boldsymbol{x},1-\alpha,\gamma)} f(x \mid \boldsymbol{\theta})\,\mathrm{d}x \leq 1 - \gamma\right\} = \Pr\left\{F[L_o(\boldsymbol{x}, 1 - \alpha, \gamma) \mid \boldsymbol{\theta}] \leq 1 - \gamma\right\} = 1 - \alpha. \tag{19.4b}$$

Thus, the lower $(1 - \alpha, \gamma)$ tolerance interval $\left[L_o(\boldsymbol{x}, 1 - \alpha, \gamma), \infty\right)$ includes at least the portion $\gamma$ of largest values of $X$ or excludes at most the portion $1 - \gamma$ of smallest values of $X$, both statements having probability $1 - \alpha$. Note that a lower $(1 - \alpha, \gamma)$ tolerance limit is also an $(\alpha, 1 - \gamma)$ upper tolerance limit.

- $U_o(\boldsymbol{x}, 1 - \alpha, \gamma)$ is an **upper** $(1 - \alpha, \gamma)$ tolerance limits if

$$\Pr\left\{\int_{-\infty}^{U_o(\boldsymbol{x},1-\alpha,\gamma)} f(x \,|\, \boldsymbol{\theta})\,\mathrm{d}x \geq \gamma\right\} = \Pr\{F[U_o(\boldsymbol{x}, 1 - \alpha, \gamma) \,|\, \boldsymbol{\theta}] \geq \gamma\} = 1 - \alpha \quad (19.5\mathrm{a})$$

or

$$\Pr\left\{\int_{U_o(\boldsymbol{x},1-\alpha,\gamma)}^{\infty} f(x \,|\, \boldsymbol{\theta})\,\mathrm{d}x \leq 1 - \gamma\right\} = \Pr\{R[U_o(\boldsymbol{x}, 1 - \alpha, \gamma) \,|\, \boldsymbol{\theta}] \leq 1 - \gamma\} = 1 - \alpha.$$
$$(19.5\mathrm{b})$$

Note that an upper $(1 - \alpha, \gamma)$ tolerance limit is also an $(\alpha, 1 - \gamma)$ lower tolerance limit.

Suppose $x_{1-\gamma}$ denotes the $(1 - \gamma)$ percentile such that $F(x_{1-\gamma}) = 1 - \gamma$, then $L_o(\boldsymbol{x}, 1 - \alpha, \gamma)$ will be a lower $(1 - \alpha, \gamma)$ tolerance limit if it is a $(1 - \alpha)$ level lower confidence limit for $x_{1-\gamma}$, because

$$\Pr\{L_o(\boldsymbol{x}, 1 - \alpha, \gamma) \leq x_{1-\gamma}\} = \Pr\{F[L_o(\boldsymbol{x}, 1 - \alpha, \gamma)] \leq 1 - \gamma\} = 1 - \alpha. \quad (19.6)$$

Thus one–sided tolerance limits are directly related to one–sided confidence limits on percentiles. They are also indirectly related to confidence limits on reliability. For tolerance limits, the fraction $1 - \gamma$ is fixed and the limit $L_o(\boldsymbol{x}, 1 - \alpha, \gamma)$ is a random variable. For reliability, the time $x_0$ is fixed and the fraction $R_L$, indicating the lower confidence limit for $R(x_0)$, is a random variable. Thus, the interpretation and application of the two methods are slightly different and would depend on which factor was kept fixed. However, computationally there is a direct relationship between the two methods and one can be obtained from the other. This is helpful because in some cases one procedure may be easier to develop than the other. For example, suppose the tolerance limit problem has been solved, so that $L_o(\boldsymbol{x}, 1 - \alpha, \gamma)$ can be calculated for a specified $\gamma$ and $1 - \alpha$. Then for a given set of sample data, determine what value of $\gamma$ would have made $L_o(\boldsymbol{x}, 1 - \alpha, \gamma) = x_0$. That value of $\gamma$ then is the lower confidence limit $R_L$ for $R(x_0)$.

**Example 19/1: Confidence limit for $R(x_0)$ and tolerance limit of an exponential distribution**

For an exponential distribution with DF

$$f(x \,|\, b) = \frac{1}{b}\,\exp\left\{-\frac{x}{b}\right\};\ x \geq 0,\ b > 0, \quad (19.7\mathrm{a})$$

the percentile of order $\gamma$ is

$$x_\gamma = -b\,\ln(1 - \gamma). \quad (19.7\mathrm{b})$$

The MLE of $b$ is

$$\widehat{b} = \frac{1}{n} \sum_{i=1}^{n} X_i, \qquad (19.7c)$$

and the MLE of $x_\gamma$ results as

$$\widehat{X}_\gamma = -\widehat{b} \ln(1 - \gamma) = -\ln(1 - \gamma) \frac{1}{n} \sum_{i=1}^{n} X_i. \qquad (19.7d)$$

Because

$$2 \sum_{i=1}^{n} \frac{X_i}{b} \sim \chi^2(2\,n), \qquad (19.7e)$$

the lower $(1 - \alpha)$ confidence limit $\widehat{b}_L$ of $b$ is

$$\widehat{b}_L = \frac{2\,n\,\widehat{b}}{\chi^2_{2n,1-\alpha}} \qquad (19.7f)$$

leading to the lower $(1 - \alpha)$ confidence limit $\widehat{X}_{\gamma,L}$ of $x_\gamma$

$$\widehat{X}_{\gamma,L} = -\widehat{b}_L \ln(1 - \gamma) = -\frac{2\,n\,\widehat{b} \ln(1 - \gamma)}{\chi^2_{2n,1-\alpha}}, \qquad (19.7g)$$

which also is the lower $(1 - \alpha, \gamma)$ tolerance limit of the exponential distribution:

$$L_o(\boldsymbol{x}, 1 - \alpha, \gamma) = \widehat{X}_{\gamma,L} = -\frac{2\,n\,\widehat{b} \ln(1 - \gamma)}{\chi^2_{2n,1-\alpha}}. \qquad (19.7h)$$

We now determine what value of $\gamma$ would have made $L_o(\boldsymbol{x}, 1 - \alpha, \gamma) = x_0$. That value of $\gamma$ is then the lower confidence limits $\widehat{R}_L$ of $R(x_0)$. Setting

$$L_o(\boldsymbol{x}, 1 - \alpha, \gamma) = -\widehat{b}_L \ln(1 - \gamma) = x_0 \qquad (19.7i)$$

gives

$$\widehat{R}_L = 1 - \gamma = \exp\left\{ -\frac{x_0}{\widehat{b}_L} \right\} = \exp\left\{ -\frac{x_0 \chi^2_{2n,1-\alpha}}{2\,n\,\widehat{b}} \right\}. \qquad (19.7j)$$

## 19.2 Classical methods of estimating reliability $R(x)$

Most of the classical approaches to reliability estimation center around ML, but there are other methods too. We mention the following:

- the approach of BASU (1964) to derive the minimum variance unbiased estimator of $R(x)$ using the theorems of RAO–BLACKWELL and LEHMANN–SCHEFFÉ when the shape parameter $c$ is assumed known,

- the procedure of JOHNS/LIEBERMAN (1966) and HERBACH (1970) to derive exact asymptotically efficient confidence bounds for $R(x)$,

- the linear estimation procedure (least squares method) of ENGELHARDT/BAIN (1977) and ERTO/GUIDA (1985b),

- the simplified linear estimators proposed by BAIN (1978, Sect. 4.2), BAIN/ ENGELHARDT (1991a, Sect. 4.2) or KINGSTON/PATEL (1981).

A general drawback to these methods is that they rely on a lot of tables and that they are usually worse than the ML based estimators. Besides, the ML procedure to estimate $R(x)$ has been extensively elaborated. So we will present ML–oriented procedures (Sect. 19.2.2), but we start with an easily applicable non–parametric estimator which fits all continuous distributions (Sect. 19.2.1).

## 19.2.1 Non-parametric approaches[2]

A non–parametric or distribution–free estimator of the reliability $R(x)$ at age $x$ is the sample fraction that survives an age $x$, $n$ being the sample size. Let $Z$ be the number of items failing beyond age $x$, then the estimator is

$$\widehat{R}(x) = \frac{Z}{n}. \tag{19.8a}$$

$\widehat{R}(x)$ is unbiased with variance

$$\text{Var}\left[\widehat{R}(x)\right] = \frac{R(x)\left[1 - R(x)\right]}{n}, \tag{19.8b}$$

and $Z$ is binomial with parameters $n$ and $P = R(x)$.

The entire sample reliability function is obtained by estimating $R(x)$ for all $x$–values observed. The sample reliability function $\widehat{R}(x)$ is a decreasing staircase function which decreases by $1/n$ at each data value and is constant between data values. So it needs to be calculated only at each data value. Let $x_{1:n} \leq x_{2:n} \leq \ldots \leq x_{n:n}$ denote the ordered observations; then,

$$\widehat{R}(x) = \begin{cases} 1 & \text{for} \quad x < x_{1:n}, \\ \dfrac{n-i}{n} & \text{for} \quad x_{i:n} \leq x < x_{i-1:n}; \ i = 1, \ldots, n-1, \\ 0 & \text{for} \quad x \geq x_{n:n}. \end{cases} \tag{19.9}$$

Binomial limits apply to reliability $P = R(x)$ at age $x$. Based on a familiar relationship between the CDF of the binomial distribution and the CDF of the $F$–distribution,[3] we have the following two–sided $(1 - \alpha)$ confidence interval for $R(x)$ when $x_{i:n} \leq x < x_{i+1:n}$:

$$\frac{n-i}{(n-i) + (i+1)F_{2(i+1),2(n-i),1-\alpha/2}} \leq R(x) \leq \frac{(n-i+1)F_{2(n-i+1),2i,1-\alpha/2}}{i + (n-i+1)\,F_{2(n-i+1),2i,1-\alpha/2}}. \tag{19.10a}$$

For $\widehat{R}(x) = 1$, i.e., for $i = 0$, we have

$$\sqrt[n]{\alpha/2} \leq R(x) \leq 1, \tag{19.10b}$$

---

[2] Suggested reading for this section: DODSON (1994, Chapter 3), KAO (1959), NELSON (1982, Section 6.7).

[3] There is also a relationship between the binomial distribution and the beta distribution which may be used to construct confidence intervals for $R(x)$; see KAO (1959).

and when $\widehat{R}(x) = 0$, i.e., for $i = n$, we have

$$0 \leq R(x) \leq 1 - \sqrt[n]{\alpha/2}. \tag{19.10c}$$

One–sided $(1 - \alpha)$ confidence intervals are

$$\left.\begin{aligned} R(x) &\geq \frac{n - i}{(n - i) + (i + 1)\, F_{2(i+1),2(n-i),1-\alpha}}\,, \quad 1 \leq i \leq n - 1, \\ R(x) &\geq \sqrt[n]{\alpha}, \; i = 0, \end{aligned}\right\} \tag{19.10d}$$

and

$$\left.\begin{aligned} R(x) &\leq \frac{(n - i + 1)\, F_{2(n-i+1),2i,1-\alpha}}{i + (n - i + 1)\, F_{2(n-i+1),2i,1-\alpha}}\,, \quad 1 \leq i \leq n - 1, \\ R(x) &\leq 1 - \sqrt[n]{\alpha}, \; i = n. \end{aligned}\right\} \tag{19.10e}$$

For $i$ and $n - i$ large, a normal approximation may be used.

$$\widehat{R}(x) - u_{1-\alpha/2}\sqrt{\frac{\widehat{R}(x)\left[1 - \widehat{R}(x)\right]}{n}} \leq R(x) \leq \widehat{R}(x) + u_{1-\alpha/2}\sqrt{\frac{\widehat{R}(x)\left[1 - \widehat{R}(x)\right]}{n}}. \tag{19.10f}$$

Fig. 19/1 shows the application of (19.10d) to the $n = 20$ observations of dataset #1, generated from $We(0, 100, 2.5)$. This figure gives the true reliability function $R(x) = \exp\left\{-(x/100)^{2.5}\right\}$ together with the staircase function $\widehat{R}(x)$ given by (19.9) and the lower confidence limits of a one–sided 95% confidence interval.

Figure 19/1: Distribution-free estimate of $R(x)$ for dataset #1

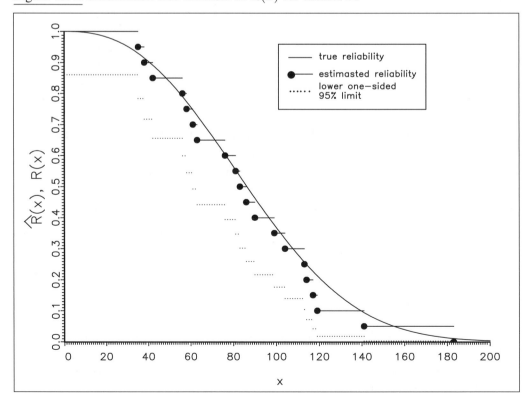

The distribution–free approach has two drawbacks:

1. It gives distinct estimates only at those values of $X$ which have been realized in the sample.

2. Because the procedure has to be valid for all continuous distributions, the confidence interval tends to be too wide compared with what would be expected if a method tailored for the distribution at hand were used.

## 19.2.2 Maximum likelihood approaches[4]

We will first present results for the two–parameter WEIBULL distribution and conclude this section by some hints to the three–parameter WEIBULL distribution.

When we insert the MLEs $\widehat{b}$ and $\widehat{c}$, as given in Chapter 11, into the reliability function:

$$\widehat{R}(x) = \exp\left\{-\left(\frac{x}{\widehat{b}}\right)^{\widehat{c}}\right\},\qquad (19.11)$$

we will get the MLE of $R(x)$ because the MLEs are functional invariant; i.e., when $\widehat{\boldsymbol{\theta}}$ is the MLE of $\boldsymbol{\theta}$, $g(\widehat{\boldsymbol{\theta}})$ will also be the MLE of $g(\boldsymbol{\theta})$, provided $g(\cdot)$ is strictly monotone. Although the MLEs $\widehat{b}$ and $\widehat{c}$ are computationally tedious to calculate, it has been shown by several authors that they are usually better than more convenient estimators of $b$ and $c$. Thus the MLE of $R(x)$ in (19.11) might be expected to have good properties.

In Sect. 11.3.1.2 it has been shown for the complete–sample case and in Sect. 11.6.1.2 for the singly censored–sample case that there exist exact distributions for $\widehat{b}$ and $\widehat{c}$ based on suitably chosen pivots to construct confidence intervals for $b$ and $c$ and to test hypotheses. A pivotal quantity for $\widehat{R}(x)$ is not available, but it can be shown that the distribution of $\widehat{R}(x)$ depends only on $R(x)$ and not on $x$, $b$ and $c$ individually. This follows since

$$
\begin{aligned}
-\ln \widehat{R}(x) \;&=\; \left(x \big/ \widehat{b}\,\right)^{\widehat{c}} \\[2mm]
&=\; \left\{\frac{(x/b)^c}{(\widehat{b}/b)^c}\right\}^{\widehat{c}/c} \\[2mm]
&=\; \left\{\frac{-\ln R(x)}{(\widehat{b}/b)^c}\right\}^{\widehat{c}/c},\qquad (19.12a)
\end{aligned}
$$

which is a function only of $R(x)$ and the pivots given in Sect. 11.3.1.2. Another form, useful in tolerance and percentile problems is

$$-\ln\left\{-\ln \widehat{R}(x)\right\} = \frac{\widehat{c}}{c}\left[\ln\left(\frac{\widehat{b}}{b}\right)^c - \ln\left\{-\ln R(x)\right\}\right].\qquad (19.12b)$$

---

[4] Suggested reading for this section: ACHCAR/FOGO (1997), BAIN (1978), BAIN/ENGELHARDT (1981, 1986, 1991a), BILLMAN/ANTLE/BAIN (1972), DODSON (1994), HENTZSCHEL (1989), KLEYLE (1978), MOORE/HARTER/ANTOON (1981), SINHA (1986b), SRINIVASAN/WHARTON (1975), THOMAN/BAIN/ANTLE (1970).

(19.12a) makes it feasible to study the distribution of $\widehat{R}(x)$ empirically as has been done by THOMAN et al. (1970) for complete samples and by BILLMAN et al. (1972) for censored samples.

The properties of $\widehat{R}(x)$ as a point estimator are as follows:

- The bias is quite small, especially for high values of $R(x)$ which normally are of interest, and it does not seem worthwhile to eliminate the bias.

- The variance is approximately equal to the CRAMÉR–RAO lower bound (CRLB) for a regular unbiased estimator of $R(x)$ (see (19.13a–c)) especially for the reliability values of interest.

Thus the MLE $\widehat{R}(x)$ is a very good point estimator of $R(x)$.

**Exact confidence limits** for $R(x)$ based on $\widehat{R}(x)$ have been determined using Monte Carlo methods by THOMAN et al. (1970) for a complete sample and by BILLMAN et al. (1972) for censored samples when the censoring is 25% and 50%. Tables 19/1 and 19/2 give the 95% lower confidence limits for $R(x)$ without and with censoring.

---

**Example 19/2:   Exact confidence limits for $R(x)$ using dataset #1**

The MLEs from dataset #1 are (see Example 11/4):

$$\widehat{b} = 99.2079 \quad \text{and} \quad \widehat{c} = 2.5957$$

leading to

$$\widehat{R}(x) = \exp\left\{-\left(\frac{x}{99.2079}\right)^{2.5957}\right\}$$

whereas the true reliability function in this case is

$$R(x) = \exp\left\{-\left(\frac{x}{100}\right)^{2.5}\right\}.$$

Fig. 19/2 shows $\widehat{R}(x)$ and $R(x)$ together with the lower 95% confidence limit as given in Tab. 19/1.

---

The standard procedure for obtaining **approximate confidence intervals** for $R(x)$ when $n$ is large is to assume that $\widehat{R}(x)$ is normally distributed with mean $R(x)$ and variance $\mathrm{Var}(R)$ equal to the CRLB, which is found by evaluating

$$\mathrm{AVar}(R) = \mathrm{AVar}(\widehat{b})\left(\frac{\partial R(x)}{\partial b}\right)^2 + 2\,\mathrm{ACov}(\widehat{b}, \widehat{c})\,\frac{\partial R(x)}{\partial b}\frac{\partial R(x)}{\partial c} + \mathrm{AVar}(\widehat{c})\left(\frac{\partial R(x)}{\partial c}\right)^2.$$

$$(19.13a)$$

Figure 19/2: $\widehat{R}(x)$ and $R(x)$ for dataset #1 with lower 95% confidence limit $\widehat{R}_L(x)$

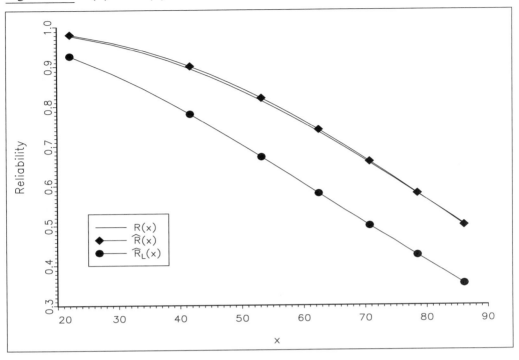

Table 19/1:  95% lower confidence limit for $R(x)$ — No censoring

| $\widehat{R}(x)$ \ $n$ | 8 | 10 | 12 | 15 | 18 | 20 | 25 | 30 | 40 | 50 | 75 | 100 |
|---|---|---|---|---|---|---|---|---|---|---|---|---|
| .50 | – | – | .308 | .329 | .343 | .353 | .356 | .379 | .394 | .404 | .420 | .432 |
| .52 | – | .308 | .325 | .346 | .361 | .371 | .384 | .398 | .413 | .423 | .439 | .452 |
| .54 | .300 | .323 | .341 | .363 | .378 | .389 | .402 | .416 | .432 | .442 | .459 | .471 |
| .56 | .316 | .339 | .358 | .381 | .396 | .407 | .421 | .435 | .451 | .461 | .478 | .491 |
| .58 | .331 | .355 | .376 | .398 | .414 | .425 | .440 | .454 | .471 | .481 | .498 | .510 |
| .60 | .347 | .372 | .393 | .416 | .432 | .443 | .459 | .473 | .490 | .500 | .517 | .530 |
| .62 | .363 | .389 | .411 | .434 | .450 | .462 | .478 | .493 | .510 | .519 | .537 | .551 |
| .64 | .380 | .406 | .428 | .452 | .469 | .480 | .497 | .512 | .530 | .539 | .558 | .571 |
| .66 | .396 | .424 | .445 | .471 | .488 | .499 | .517 | .532 | .550 | .559 | .579 | .592 |
| .68 | .414 | .443 | .464 | .490 | .507 | .519 | .536 | .552 | .570 | .580 | .599 | .612 |
| .70 | .432 | .461 | .483 | .510 | .527 | .538 | .557 | .573 | .591 | .601 | .620 | .633 |
| .72 | .450 | .481 | .502 | .530 | .547 | .559 | .577 | .594 | .612 | .622 | .642 | .654 |
| .74 | .469 | .500 | .523 | .550 | .568 | .580 | .598 | .616 | .633 | .644 | .663 | .675 |
| .76 | .489 | .520 | .544 | .572 | .590 | .602 | .620 | .638 | .654 | .666 | .684 | .697 |
| .78 | .509 | .542 | .567 | .594 | .612 | .625 | .643 | .661 | .676 | .688 | .707 | .719 |
| .80 | .529 | .564 | .590 | .617 | .636 | .648 | .666 | .683 | .700 | .711 | .729 | .741 |
| .82 | .552 | .587 | .614 | .641 | .660 | .672 | .689 | .706 | .724 | .734 | .752 | .763 |
| .84 | .576 | .611 | .638 | .667 | .685 | .697 | .714 | .730 | .748 | .758 | .775 | .786 |
| .86 | .602 | .638 | .664 | .693 | .710 | .723 | .740 | .755 | .772 | .783 | .799 | .809 |
| .88 | .629 | .666 | .692 | .721 | .737 | .750 | .767 | .781 | .798 | .808 | .823 | .833 |
| .90 | .661 | .696 | .722 | .751 | .766 | .780 | .795 | .809 | .824 | .834 | .848 | .857 |
| .92 | .695 | .729 | .755 | .782 | .798 | .811 | .825 | .838 | .853 | .862 | .874 | .882 |
| .94 | .735 | .767 | .792 | .817 | .832 | .845 | .858 | .869 | .882 | .890 | .901 | .908 |
| .96 | .782 | .812 | .835 | .857 | .872 | .882 | .893 | .903 | .915 | .921 | .930 | .935 |
| .98 | .844 | .869 | .890 | .907 | .918 | .926 | .935 | .943 | .950 | .955 | .962 | .965 |

<u>Table 19/2:</u> 95% lower confidence limit for $R(x)$ — Censoring

| $\widehat{R}(x)$ \ $n$ | 25% Censoring | | | | | 50% Censoring | | | | |
|---|---|---|---|---|---|---|---|---|---|---|
| | 40 | 60 | 80 | 100 | 120 | 40 | 60 | 80 | 100 | 120 |
| .70 | .594 | .626 | .624 | .625 | .643 | .600 | .623 | .628 | .639 | .646 |
| .72 | .613 | .644 | .643 | .647 | .662 | .614 | .641 | .647 | .659 | .664 |
| .74 | .632 | .662 | .664 | .669 | .681 | .632 | .660 | .667 | .678 | .683 |
| .76 | .651 | .680 | .684 | .691 | .701 | .651 | .679 | .686 | .698 | .702 |
| .78 | .671 | .699 | .705 | .713 | .722 | .671 | .699 | .707 | .719 | .722 |
| .80 | .692 | .719 | .726 | .736 | .743 | .691 | .719 | .727 | .741 | .742 |
| .82 | .714 | .740 | .748 | .759 | .764 | .712 | .740 | .749 | .761 | .762 |
| .84 | .737 | .761 | .771 | .782 | .786 | .734 | .761 | .771 | .782 | .784 |
| .86 | .760 | .784 | .795 | .806 | .809 | .757 | .784 | .793 | .805 | .806 |
| .88 | .785 | .808 | .819 | .830 | .832 | .781 | .807 | .817 | .827 | .829 |
| .90 | .811 | .833 | .844 | .854 | .856 | .807 | .831 | .841 | .851 | .852 |
| .92 | .839 | .860 | .870 | .879 | .881 | .834 | .857 | .866 | .876 | .877 |
| .94 | .869 | .888 | .897 | .904 | .906 | .863 | .883 | .892 | .902 | .903 |
| .95 | .885 | .903 | .911 | .917 | .920 | .878 | .879 | .906 | .915 | .917 |
| .96 | .902 | .919 | .926 | .932 | .933 | .894 | .913 | .920 | .929 | .931 |
| .97 | .920 | .935 | .941 | .946 | .948 | .913 | .929 | .936 | .946 | .947 |
| .98 | .940 | .953 | .957 | .962 | .963 | .933 | .947 | .952 | .960 | .961 |
| .99 | .964 | .972 | .976 | .978 | .979 | .957 | .968 | .971 | .975 | .977 |
| .9925 | .970 | .978 | .981 | .983 | .984 | .965 | .974 | .977 | .980 | .981 |
| .995 | .978 | .984 | .986 | .988 | .988 | .973 | .980 | .983 | .985 | .986 |
| .996 | .981 | .986 | .988 | .990 | .990 | .976 | .983 | .985 | .988 | .989 |
| .997 | .985 | .989 | .991 | .992 | .992 | .980 | .986 | .988 | .990 | .991 |
| .998 | .988 | .992 | .993 | .994 | .995 | .985 | .990 | .992 | .993 | .994 |
| .9985 | .991 | .994 | .995 | .996 | .996 | .987 | .992 | .993 | .994 | .995 |
| .999 | .993 | .995 | .996 | .997 | .997 | .990 | .994 | .995 | .996 | .996 |

<u>Source:</u> BILLMAN/ANTLE/BAIN (1972, p. 837) — Reprinted with permission from *Technometrics*. Copyright 1972 by the American Statistical Association. All rights reserved.

The asymptotic variances and covariance have to be taken from (11.17) when the sample is complete and from the inverse of (11.50a) when the sample is singly censored on the right. We thus get

- for the complete–sample case

$$\text{AVar}_u(R) = \frac{R^2 (\ln R)^2}{n} \left\{ 1.1087 - 0.5140 \ln(-\ln R) + 0.6079 [\ln(-\ln R)]^2 \right\},$$
(19.13b)

- and for the censored–sample case

$$\text{AVar}_c(R) = \frac{R^2 (\ln R)^2}{n} \left\{ c_{11} - 2 c_{12} \ln(-\ln R) + c_{22} [\ln(-\ln R)]^2 \right\}, \quad (19.13c)$$

where the factors $c_{11}$, $c_{12}$ and $c_{22}$ depend on the amount of censoring (see Tab. 19/3) and $r$ is the censoring number.

Table 19/3: Coefficients of the asymptotic variance of $\widehat{R}$ for uncensored samples

| $r/n$ | $c_{11}$ | $c_{22}$ | $c_{12}$ |
|-------|----------|----------|----------|
| 1.0 | 1.108665 | 0.607927 | 0.257022 |
| 0.9 | 1.151684 | 0.767044 | 0.176413 |
| 0.8 | 1.252617 | 0.928191 | 0.049288 |
| 0.7 | 1.447258 | 1.122447 | −0.144825 |
| 0.6 | 1.811959 | 1.372781 | −0.446603 |
| 0.5 | 2.510236 | 1.716182 | −0.935766 |
| 0.4 | 3.933022 | 2.224740 | −1.785525 |
| 0.3 | 7.190427 | 3.065515 | −3.438601 |
| 0.2 | 16.478771 | 4.738764 | −7.375310 |
| 0.1 | 60.517110 | 9.744662 | −22.187207 |

Source: BAIN (1978, p. 218) — Reprinted with permission from *Statistical Analysis of Reliability and Life–Testing Models.* Copyright 1978 by Marcel Dekker, Inc. All rights reserved.

The true reliability $R$ could be replaced by $\widehat{R}$ in the expressions (19.13b,c) for the asymptotic variances, and an approximate lower $(1 - \alpha)$ confidence limit would be

$$\widehat{R}_{L,1} = \widehat{R} - u_{1-\alpha} \left[\mathrm{AVar}_i(\widehat{R})\right]^{1/2}; \; i = u, c. \tag{19.14a}$$

$\widehat{R}_{L,1}$ is usually too large. But, as THOMAN et al. (1970) report, this direct approximation can be improved considerably by using an iterative procedure. This procedure calls for replacing $\mathrm{AVar}_i(\widehat{R})$ by $\mathrm{AVar}_i(\widehat{R}_{L,1})$ to obtain $\widehat{R}_{L,2}$. Then $\widehat{R}_{L,3}$ is obtained in the same way, replacing $\mathrm{AVar}_i(\widehat{R})$ by $\mathrm{AVar}(\widehat{R}_{L,2})$, etc. Thus we obtain the sequence

$$\widehat{R}_{L,j} = \widehat{R} - u_{1-\alpha} \left[\mathrm{AVar}_i(\widehat{R}_{L,j-1})\right]^{1/2}; \; i = u, c; \; j = 2, 3, \ldots \tag{19.14b}$$

THOMAN et al. (1970) observed that after four or five iterations the changes in $\widehat{R}_{L,j}$ were less than 0.00005, and the values of $\widehat{R}_{L,j}$ were very close to the values given in Tab. 19/1 when the sample is uncensored. The maximum difference between the values in Tab. 19/1 and the iterated lower limits was 0.005 for $n \geq 40$. We thus only need better approximations to the confidence limits for smaller $n$ when there are no tables of the true confidence limits available. The following papers contain such approximations:

- ACHCAR/FOGO (1997) suggested the re–parametrization

$$T(R) = \left[\left(\frac{R}{1 - R}\right)^{\lambda} - 1\right] \Big/ \lambda$$

  to improve the normality of the likelihood for $R$.

- BAIN/ENGELHARDT (1981, 1986) introduced approximations of the pivotal distributions.

- KLEYLE (1978) modified the $F$–approximation for obtaining approximate lower confidence limits on WEIBULL percentiles, as had been proposed by LAWLESS (1975).

- MOORE et al. (1981) found Monte Carlo generated estimates of $\mathrm{AVar}(\widehat{R})$ for smaller sample sizes before applying a normal approximation to $R_L$.

- SRINIVASAN/WHARTON (1975) constructed confidence bounds on the entire CDF using KOLMOGOROV–SMIRNOV type statistics.

When we have a **three–parameter WEIBULL distribution**, the MLE of $R(x)$ is given by

$$\widehat{R}(x) = \exp\left\{ - \left( \frac{x - \widehat{a}}{\widehat{b}} \right)^{\widehat{c}} \right\}, \tag{19.15a}$$

where $\widehat{a}$, $\widehat{b}$ and $\widehat{c}$ are the MLEs of $a$, $b$ and $c$ as described in Sect. 11.3.2.2. An approximate lower $(1 - \alpha)$ confidence limit, based on the normal distribution, is given by

$$\widehat{R}_L = \widehat{R} - u_{1-\alpha} \left[ \mathrm{AVar}(\widehat{R}) \right]^{1/2}, \tag{19.15b}$$

where

$$
\begin{aligned}
\mathrm{AVar}(\widehat{R}) \;=\; & \mathrm{AVar}(\widehat{a}) \left( \frac{\partial R(x)}{\partial a} \right)^2 + \mathrm{AVar}(\widehat{b}) \left( \frac{\partial R(x)}{\partial b} \right)^2 + \mathrm{AVar}(\widehat{c}) \left( \frac{\partial R(x)}{\partial c} \right)^2 + \\
& 2\,\mathrm{ACov}(\widehat{a},\widehat{b}) \frac{\partial R(x)}{\partial a} \frac{\partial R(x)}{\partial b} + 2\,\mathrm{ACov}(\widehat{a},\widehat{c}) \frac{\partial R(x)}{\partial a} \frac{\partial R(x)}{\partial c} + \\
& 2\,\mathrm{ACov}(\widehat{b},\widehat{c}) \frac{\partial R(x)}{\partial b} \frac{\partial R(x)}{\partial c} \,.
\end{aligned}
\tag{19.15c}
$$

The asymptotic variances and covariances can be taken from (11.12a).

## 19.3 Classical methods of estimating percentiles $x_P$

We will proceed in this section as has been done in Sect. 19.2; i.e., we first give a non–parametric estimator (Sect. 19.3.1) and then discuss ML based techniques (Sect. 19.3.2). The following papers rely on other estimation approaches:

- DODSON (1994) gives a normal approximation to the confidence limits of percentiles of the Log–WEIBULL distribution which afterwards is transformed to WEIBULL.

- Linear estimation is applied to find estimators and confidence limits of Log–WEIBULL percentiles by HASSANEIN et al. (1984), and MANN (1969b, 1969c, 1970a).

- LAWLESS (1974, 1975) relies on the pivotal distribution of the MLEs of the Log–WEIBULL distribution.

- SCHNEIDER/WEISSFELD (1989) improve the $t$–like statistic for Log–WEIBULL percentiles.

- WHITE (1966) gives confidence intervals for Log–WEIBULL percentiles based on the method of moments.

### 19.3.1 A non-parametric approach

Suppose the ordered observations from a continuous distributions are $X_{1:n} \leq X_{2:n} \leq \dots \leq X_{n:n}$. Also, suppose one wishes to estimate the percentile of order $P$ of that distribution where $i \leq P(n+1) \leq i+1$ for some $i = 1, \dots, n-1$. Then a point estimator of $x_P$ is

$$\widehat{X}_P = \left\{1 - [(n+1)P - i]\right\} X_{i:n} + \left\{(n+1)P - i\right\} X_{i+1:n}, \qquad (19.16)$$

which is the weighted average of the two ordered observations $X_{i:n}$ and $X_{i+1:n}$. If $P = 0.5$ and $n$ is odd, the estimator is taken to be $X_{(n+1)/2:n}$. An upper one–sided $(1-\alpha)$ confidence interval for $x_P$ is

$$x_P \geq X_{k:n}, \qquad (19.17a)$$

where $k$ is chosen such that

$$\sum_{i=0}^{k-1} \binom{n}{i} P^i (1-P)^{n-i} \leq \alpha < \sum_{i=0}^{k} \binom{n}{i} P^i (1-P)^{n-i}. \qquad (19.17b)$$

Similarly, a lower one–sided $(1-\alpha)$ confidence interval is

$$x_P \leq X_{m:n} \qquad (19.18a)$$

with $m$ so that

$$\sum_{i=m}^{n} \binom{n}{i} P^i (1-P)^{n-i} \leq \alpha < \sum_{i=m-1}^{n} \binom{n}{i} P^i (1-P)^{n-i}. \qquad (19.18b)$$

A two–sided $(1-\alpha)$ confidence interval is

$$X_{k:n} \leq x_P \leq X_{m:n},$$

where $k$ and $m$ follow from (19.17b) and (19.18b) with $\alpha$ replaced by $\alpha/2$. For large $n$, approximate $k$ and/or $m$ follow from a normal approximation of the binomial distribution with continuity correction.

### 19.3.2 Maximum likelihood approaches[5]

We start estimating $x_P$ when the shape parameter $c$ is assumed known, then we turn to the two–parameter WEIBULL distribution with both parameters assumed unknown and finally to the three–parameter case.

For **c known** the usual MLE of the scale parameter $b$ is

$$\widehat{b} = \left\{ \frac{\sum\limits_{i=1}^{r} X_{i:n}^c + (n-r) X_{r:n}^c}{r} \right\}^{1/c}, \qquad (19.19a)$$

---

[5] Suggested reading for this section: BAIN (1978), BAIN/ENGELHARDT (1991a), HEO/BOES/SALAS (2001), HIROSE (1991), McCOOL (1969, 1970a, 1970b, 1974a), MEEKER (1986), MEEKER/NELSON (1976), NELSON (1985), SCHAFER/ANGUS (1979).

(see (11.46a)) where $r$ is the number of uncensored observations. If there are no failures, then we have $r = 0$ and $\widehat{b} = \infty$, indicating that the true $b$ is likely to be much greater than the total running time of the sample.[6] According to NELSON (1985) the lower limit $\widehat{b}_L$ of a one–sided $(1 - \alpha)$ confidence interval $b \geq \widehat{b}_L$ is given by

$$\widehat{b}_L = \widehat{b} \left\{ \frac{2\,r}{\chi^2_{2(r+1),1-\alpha}} \right\}^{1/c} \quad \text{for } r \geq 1, \tag{19.19b}$$

or

$$\widehat{b}_L = \widehat{b} \left\{ \frac{2 \sum\limits_{i=1}^{n} X^c_{i:n}}{\chi^2_{2(r+2),1-\alpha}} \right\}^{1/c} \quad \text{for } r \geq 0. \tag{19.19c}$$

Note that (19.19c) also applies for $r = 0$, but (19.19b) is limited to $r \geq 1$. Inserting $\widehat{b}$ as given by (19.19a) into $x_P = b \left[ -\ln(1 - P) \right]^{1/c}$ gives the usual MLE of $x_P$:

$$\widehat{x}_P = \widehat{b} \left[ -\ln(1 - P) \right]^{1/c}. \tag{19.20a}$$

The lower limit $\widehat{x}_{P,L}$ of a one–sided $(1 - \alpha)$ confidence interval $x_P \geq \widehat{x}_{P,L}$ is

$$\widehat{x}_{P,L} = \widehat{b}_L \left[ -\ln(1 - P) \right]^{1/c}. \tag{19.20b}$$

We will give exact confidence limits for $x_P$ when $X \sim We(0, b, c)$ with both **b and c unknown** in the context of estimating tolerance limits (see Sect. 19.4.1). Here we will provide two approximate large–sample confidence limits for $x_P$.

The first method, proposed by MEEKER/NELSON (1976), starts with the MLE of the Log–WEIBULL percentiles which are corrected to approximate normality by applying a variance factor $V$. This factor depends on the order $P$ of the percentile and on the censoring time (type–II censoring) and can be read from a nomogram. This nomogram includes a curve pertaining to the case of no–censoring, i.e., a censoring of infinity. The limits of the approximate two–sided $(1 - \alpha)$ confidence interval for the WEIBULL percentile of order $P$ are

$$\widehat{x}_{P,L} = \widehat{b} \left[ -\ln(1 - p) \right]^{1/\widehat{c}} \Big/ \Psi, \tag{19.21a}$$

$$\widehat{x}_{P,U} = \widehat{b} \left[ -\ln(1 - p) \right]^{1/\widehat{c}} \Psi, \tag{19.21b}$$

where $\widehat{b}$ and $\widehat{c}$ are the MLEs and

$$\Psi = \exp\left\{ \frac{u_{1-\alpha/2}}{\widehat{c}} \left( \frac{V}{n} \right)^{1/2} \right\}. \tag{19.21c}$$

A one–sided limit is obtained by replacing the percentile $u_{1-\alpha/2}$ of the standard normal distribution by $u_{1-\alpha}$.

---

[6]  When $r = 0$, some people use a 50% confidence limit resulting from (19.19c) as an estimator. This estimator tends to be low (= conservative).

A second method, proposed by NELSON (1982, p. 232), which is valid only for uncensored samples, starts with a simple estimator of the Log–WEIBULL percentile $x_P$ and its approximate confidence limits which afterwards are transformed to limits of the WEIBULL percentile $x_P$. The estimator of $y_P$ is

$$\widehat{y}_P = \widehat{a}^* + z_P\, \widehat{b}^* \tag{19.22a}$$

with

$$z_P = \ln[-\ln(1-P)] \;-\; \begin{cases} P\text{--th percentile of the reduced} \\ \text{Log–WEIBULL distribution,} \end{cases} \tag{19.22b}$$

$$\widehat{a}^* = \overline{Y} + 0.5772\,\widehat{b}^* \;-\; \text{simple estimator of } a^* = \ln b, \tag{19.22c}$$

$$\widehat{b}^* = 0.7797\,S \;-\; \text{simple estimator of } b^* = 1/c, \tag{19.22d}$$

$$\overline{Y} = \frac{1}{n}\sum_{i=1}^{n} Y_i \;-\; \text{sample mean of the Log–WEIBULL data } Y_i = \ln X_i, \tag{19.22e}$$

$$S^2 = \frac{1}{n-1}\sum_{i=1}^{n}\left(Y_i - \overline{Y}\right)^2 \;-\; \text{sample variance of the Log–WEIBULL data.} \tag{19.22f}$$

Two–sided approximate $(1-\alpha)$ confidence limits for $y_P$ are

$$\widehat{y}_{P,L} = \widehat{y}_P - u_{1-\alpha/2}\,\widehat{b}^*\left[\left(1.1680 + 1.1000\,z_P^2 - 0.1913\,z_P\right)/n\right]^{1/2}, \tag{19.23a}$$

$$\widehat{y}_{P,U} = \widehat{y}_P + u_{1-\alpha/2}\,\widehat{b}^*\left[\left(1.1680 + 1.1000\,z_P^2 - 0.1913\,z_P\right)/n\right]^{1/2}, \tag{19.23b}$$

leading to the $(1-\alpha)$ confidence limits for the WEIBULL percentile $x_P$:

$$\widehat{x}_{P,L} = \exp\left(\widehat{y}_{P,L}\right), \quad \widehat{x}_{P,U} = \exp\left(\widehat{y}_{P,U}\right). \tag{19.24}$$

---

### Example 19/3: Approximate confidence intervals for $x_{0.1}$

For the $n = 20$ observations of dataset #1 we have found the MLEs (see Example 11/4):

$$\widehat{b} = 99.2079 \quad \text{and} \quad \widehat{c} = 2.5957.$$

The point estimate of $x_{0.1}$ is

$$\widehat{x}_{0.1} = \widehat{b}\left[-\ln 0.9\right]^{1/\widehat{c}} = 41.6899.$$

The nomogram in MEEKER/NELSON (1976) gives $V = 5$, so the two–sided 95% confidence interval for $x_{0.1}$ according to (19.21a–c) is

$$41.6899 \Big/ \exp\left\{\frac{1.96}{2.5958}\sqrt{\frac{5}{20}}\right\} \;\leq\; x_{0.1} \;\leq\; 41.6899\,\exp\left\{\frac{1.96}{2.5958}\sqrt{\frac{5}{20}}\right\}$$

$$28.5801 \;\leq\; x_{0.1} \;\leq\; 60.8131.$$

The second method of NELSON (1982) first gives:

$$\widehat{y}_{0.1} = 3.7306, \quad \widehat{y}_{0.1,L} = 3.3263, \quad \widehat{y}_{0.1,U} = 4.1342.$$

The resulting confidence interval for the WEIBULL percentile $x_{0.1}$ according to (19.24) is

$$27.8354 \leq x_{0.1} \leq 62.4400,$$

which is wider than that of the first approximation.

Approximate confidence limits for

$$x_P = a + b \left[ -\ln(1-P) \right]^{1/c} \tag{19.25}$$

of the **three–parameter WEIBULL distribution** are given by

$$\left. \begin{array}{rcl} \widehat{x}_{P,L} &=& \widehat{x}_P - u_{1-\alpha/2}\sqrt{\mathrm{AVar}\left(\widehat{x}_P\right)}, \\[2mm] \widehat{x}_{P,U} &=& \widehat{x}_P + u_{1-\alpha/2}\sqrt{\mathrm{AVar}\left(\widehat{x}_P\right)}. \end{array} \right\} \tag{19.26a}$$

$\widehat{x}_P$ is found by substituting $a$, $b$ and $c$ in (19.25) by their MLEs (see Sections 11.3.2.2 and 11.3.2.3). The asymptotic variance of $\widehat{x}_P$ is

$$\left. \begin{array}{rcl} \mathrm{AVar}\left(\widehat{x}_P\right) &=& \left(\dfrac{\partial x_P}{\partial a}\right)^2 \mathrm{AVar}\left(\widehat{a}\right) + \left(\dfrac{\partial x_P}{\partial b}\right)^2 \mathrm{AVar}\left(\widehat{b}\right) + \left(\dfrac{\partial x_P}{\partial c}\right)^2 \mathrm{AVar}\left(\widehat{c}\right) + \\[3mm] && 2\dfrac{\partial x_P}{\partial a}\dfrac{\partial x_P}{\partial b}\mathrm{ACov}\left(\widehat{a},\widehat{b}\right) + 2\dfrac{\partial x_P}{\partial a}\dfrac{\partial x_P}{\partial c}\mathrm{ACov}\left(\widehat{a},\widehat{c}\right) + \\[3mm] && 2\dfrac{\partial x_P}{\partial b}\dfrac{\partial x_P}{\partial c}\mathrm{ACov}\left(\widehat{b},\widehat{c}\right). \end{array} \right\} \tag{19.26b}$$

The asymptotic variances and covariances in (19.26b) can be taken from (11.12a) in conjunction with (11.12b–h) yielding

$$\mathrm{AVar}\left(\widehat{x}_P\right) = \frac{b^2}{nD} \left\{ \frac{B}{(c-1)^2} - \frac{2\,\beta^{1/c}}{c\,(c-1)} \left(H + F \ln \beta\right) + \frac{\beta^{1/c}}{c^2} \left[A - 2\,G \ln \beta + (\ln \beta)^2\right] \right\} \tag{19.26c}$$

with $\beta = -\ln(1-P)$ and $A$, $B$, $D$, $F$, $G$, $H$ defined in (11.12b–h). Of course, $b$ and $c$ have to be replaced by their MLEs $\widehat{b}$ and $\widehat{c}$ before applying (19.26c).

## 19.4    Tolerance intervals

We will first give distribution–free tolerance intervals (Sect. 19.4.1) before showing how to obtain exact and small-sample tolerance intervals for the WEIBULL distribution based on the MLEs of the distribution parameters (Sect. 19.4.2). The tolerance limits will also be the confidence limits of the percentiles as shown in (19.6).

### 19.4.1    A non-parametric approach

Distribution–free tolerance intervals are based on the extreme sample values, provided $X$ is continuous. The **one–sided tolerance intervals**

$$\left[X_{1:n}, \infty\right) \quad \text{and} \quad \left(-\infty, X_{n:n}\right]$$

each include at least the portion $\gamma$ of the distribution of $X$ with probability $1 - \alpha$, where $\alpha$, $\gamma$ and $n$ are linked by

$$\gamma^n = \alpha. \tag{19.27}$$

When two of the three quantities are given, (19.27) may be easily solved for the third quantity.

The **two–sided tolerance interval**

$$\left[ X_{1:n}, X_{n:n} \right]$$

includes at least the portion $\gamma$ of the distribution of $X$ with probability $1 - \alpha$, where $\alpha$, $\gamma$ and $n$ are linked by

$$n\,\gamma^{n-1} - (n-1)\,\gamma^n = \alpha. \tag{19.28}$$

When $n$ and $\gamma$ are given, (19.28) directly gives $\alpha$. For the other two possibilities ($\alpha$ and $\gamma$ given or $\alpha$ and $n$ given), (19.28) must be solved iteratively for the third quantity.

### 19.4.2  Maximum likelihood approaches[7]

The case of an **uncensored sample** is considered first. Starting from (19.12b) we introduce the generalized quantity

$$
\begin{aligned}
U_R &= \sqrt{n}\,\frac{\widehat{c}}{c}\left[ \ln\left( \frac{\widehat{b}}{b} \right)^c - \ln\left\{ -\ln R(x) \right\} \right] + \sqrt{n}\,\ln\left\{ -\ln R(x) \right\} \\
&= \sqrt{n}\left[ -\ln\left\{ -\widehat{R}(x) \right\} + \ln\left\{ -\ln R(x) \right\} \right].
\end{aligned} \tag{19.29}
$$

The distribution of $U_R$ depends only on $R := R(x)$ and has been tabulated in BAIN (1978, pp. 228 ff.). It is interesting to note that for the tolerance limit problem, $U_R$ may be expressed in the following form (letting $R = \gamma$ in this case):

$$U_\gamma = \widehat{c}\,\sqrt{n}\,\ln\left[ \frac{\widehat{x}_{1-\gamma}}{x_{1-\gamma}} \right], \tag{19.30}$$

providing a very convenient pivotal quantity for determining confidence limits for $x_{1-\gamma}$.

As defined in (19.4a) a lower $(1 - \alpha)$ probability tolerance limit for proportion $\gamma$ is a function $L(x) := L_o(x, 1 - \alpha, \gamma)$ of the sample such that

$$\Pr\left\{ \int_{L(x)}^{\infty} f(x \mid \theta)\,dx \geq \gamma \right\} = 1 - \alpha.$$

It has been shown in (19.6) that $L_o(x, 1 - \alpha, \gamma)$ may also be interpreted as a lower $(1 - \alpha)$ confidence limit for the $(1 - \gamma)$ percentile $x_{1-\gamma}$ of the distribution. Using (19.30) a **lower** $(1 - \alpha, \gamma)$ **tolerance limit** for the WEIBULL distribution based on the MLEs $\widehat{b}$ and $\widehat{c}$ is

---

[7]  Suggested reading for this section: BAIN (1978), BAIN/ENGELHARDT (1981, 1986, 1991a), EN-
GELHARDT/BAIN (1977), ISAIC–MANIU/VODA (1980), LAWLESS (1975), MANN (1977, 1978a),
MANN/FERTIG (1973).

given by

$$L_o(\boldsymbol{x}, 1 - \alpha, \gamma) = \widehat{x}_{1-\gamma} \exp\left\{ \frac{-u_{1-\alpha}(\gamma)}{\widehat{c}\sqrt{n}} \right\}$$

$$= \widehat{b}\left(-\ln\gamma\right)^{1/\widehat{c}} \exp\left\{ \frac{-u_{1-\alpha}(\gamma)}{\widehat{c}\sqrt{n}} \right\}, \qquad (19.31a)$$

where $u_{1-\alpha}(\gamma)$ is tabulated in Tab. 19/4a for $1-\alpha = 0.95$ and in Tab. 19/4b for $1-\alpha = 0.05$ with each table covering several values of $\gamma$. An **upper $(1 - \alpha, \gamma)$ tolerance limit** is given by

$$U_o(\boldsymbol{x}, 1 - \alpha, \gamma) = L_o(\boldsymbol{x}, \alpha, 1 - \gamma). \qquad (19.31b)$$

The one–sided tolerance limits may be combined to obtain a conservative **two–sided tolerance interval**. A two–sided $(1 - \alpha)$ probability tolerance interval for proportion $\gamma$ is given by

$$\left[ L_o(\boldsymbol{x}, 1 - \alpha_1, \gamma_1), U_o(\boldsymbol{x}, 1 - \alpha_2, \gamma_2) \right], \qquad (19.32)$$

where $\gamma = \gamma_1 + \gamma_2 - 1$ and $1 - \alpha = 1 - \alpha_1 - \alpha_2$. Exact two–sided intervals would require the generation of additional tables. As BAIN (1978, p. 244) has stated, the above method provides a good solution which is nearly exact.

As mentioned in Sect. 19.1 there is a **direct connection between tolerance limits and confidence limits on reliability**. Consequently, Tables 19/4a and 19/4b may be used to compute confidence limits on $R(x)$. For specified $x$, $1 - \alpha$ and $\widehat{R}(x)$, the lower $(1 - \alpha)$ level confidence limit $R_L$ is the value of $\gamma$ which satisfies

$$x = L_o(\boldsymbol{x}, 1 - \alpha, \gamma) = \widehat{b}\left(-\ln\gamma\right)^{1/\widehat{c}} \exp\left\{ \frac{-u_{1-\alpha}(\gamma)}{\widehat{c}\sqrt{n}} \right\}. \qquad (19.33a)$$

Expressing this in terms of $\widehat{R}(x)$ gives $R_L$ as the value of $\gamma$ for which

$$-\sqrt{n}\,\ln\left[ -\ln\widehat{R}(x) \right] = -\sqrt{n}\,\ln(-\ln\gamma) + u_{1-\alpha}(\gamma). \qquad (19.33b)$$

This requires trying a few values of $\gamma$ until the right–hand side of (19.33b) becomes equal to the left–hand side, which is fixed.

Of course, the tables which allow a direct reading of the confidence limits on $R(x)$ — see Tables 19/1 and 19/2 — can be used inversely to obtain tolerance limits. This is of value in the **censored–sample case** since there are no tables provided for the direct calculation of tolerance limits in that case. To determine $L_o(\boldsymbol{x}, 1 - \alpha, \gamma)$, set $R_L = \gamma$ in the body of Tab. 19/1 (uncensored–sample case) or Tab. 19/2 (censored–sample case) and then search for the value of $R(x)$, say $R_c$, from the left–hand side, which would have resulted in $R_L = \gamma$ for that $n$ and $1 - \alpha$. Then set $\widehat{R}_c = \exp\left\{ -(x/\widehat{b})^{1/\widehat{c}} \right.$, and solve for

$$x = L_o(\boldsymbol{x}, 1 - \alpha, \gamma) = \widehat{b}\left(-\ln\widehat{R}_c\right)^{1/\widehat{c}}. \qquad (19.34)$$

Table 19/4a:   Percentage points $u_{0.95}(\gamma)$ for lower tolerance limits and for confidence limits of percentiles

| $n$ \ $\gamma$ | 0.50 | 0.60 | 0.70 | 0.75 | 0.80 | 0.85 | 0.90 | 0.95 | 0.98 |
|---|---|---|---|---|---|---|---|---|---|
| 10 | 2.579 | 3.090 | 3.792 | 4.257 | 4.842 | 5.613 | 6.712 | 8.468 | 10.059 |
| 11 | 2.526 | 3.016 | 3.685 | 4.130 | 4.690 | 5.433 | 6.489 | 8.191 | 9.741 |
| 12 | 2.482 | 2.956 | 3.603 | 4.033 | 4.573 | 5.293 | 6.314 | 7.949 | 9.412 |
| 13 | 2.446 | 2.903 | 3.534 | 3.952 | 4.478 | 5.174 | 6.168 | 7.744 | 9.111 |
| 14 | 2.415 | 2.863 | 3.473 | 3.880 | 4.393 | 5.074 | 6.038 | 7.565 | 8.853 |
| 15 | 2.388 | 2.824 | 3.424 | 3.823 | 4.326 | 4.992 | 5.932 | 7.404 | 8.594 |
| 16 | 2.362 | 2.789 | 3.380 | 3.772 | 4.264 | 4.916 | 5.835 | 7.267 | 8.388 |
| 18 | 2.323 | 2.733 | 3.305 | 3.683 | 4.162 | 4.794 | 5.679 | 7.038 | 8.015 |
| 20 | 2.292 | 2.689 | 3.247 | 3.634 | 4.079 | 4.691 | 5.548 | 6.846 | 7.724 |
| 22 | 2.263 | 2.656 | 3.194 | 3.556 | 4.011 | 4.606 | 5.444 | 6.688 | 7.463 |
| 24 | 2.241 | 2.622 | 3.150 | 3.508 | 3.949 | 4.537 | 5.353 | 6.554 | 7.246 |
| 26 | 2.221 | 2.597 | 3.116 | 3.463 | 3.901 | 4.472 | 5.271 | 6.444 | 7.078 |
| 28 | 2.204 | 2.568 | 3.080 | 3.424 | 3.853 | 4.419 | 5.205 | 6.343 | 6.911 |
| 30 | 2.188 | 2.548 | 3.051 | 3.391 | 3.812 | 4.371 | 5.141 | 6.254 | 6.781 |
| 32 | 2.175 | 2.530 | 3.027 | 3.361 | 3.779 | 4.328 | 5.089 | 6.176 | 6.641 |
| 34 | 2.160 | 2.509 | 3.003 | 3.330 | 3.744 | 4.286 | 5.036 | 6.110 | 6.554 |
| 36 | 2.151 | 2.498 | 2.982 | 3.307 | 3.714 | 4.248 | 4.990 | 6.041 | 6.438 |
| 38 | 2.136 | 2.480 | 2.959 | 3.280 | 3.681 | 4.217 | 4.948 | 5.984 | 6.362 |
| 40 | 2.128 | 2.468 | 2.941 | 3.258 | 3.662 | 4.187 | 4.912 | 5.931 | 6.263 |
| 42 | 2.122 | 2.451 | 2.923 | 3.235 | 3.636 | 4.154 | 4.871 | 5.883 | 6.196 |
| 44 | 2.113 | 2.443 | 2.906 | 3.218 | 3.616 | 4.133 | 4.840 | 5.836 | 6.123 |
| 46 | 2.106 | 2.430 | 2.890 | 3.202 | 3.588 | 4.104 | 4.806 | 4.798 | 6.064 |
| 48 | 2.096 | 2.420 | 2.876 | 3.181 | 3.568 | 4.074 | 4.778 | 5.756 | 6.021 |
| 50 | 2.089 | 2.406 | 2.864 | 3.161 | 3.550 | 4.052 | 4.749 | 5.719 | 5.968 |
| 52 | 2.080 | 2.403 | 2.849 | 3.152 | 3.534 | 4.031 | 4.721 | 5.688 | 5.921 |
| 54 | 2.076 | 2.390 | 2.837 | 3.139 | 3.513 | 4.013 | 4.700 | 5.650 | 5.857 |
| 56 | 2.069 | 2.382 | 2.822 | 3.121 | 3.495 | 3.989 | 4.674 | 5.626 | 5.837 |
| 58 | 2.068 | 2.371 | 3.811 | 3.108 | 3.481 | 3.976 | 4.650 | 5.596 | 5.781 |
| 60 | 2.056 | 2.365 | 2.805 | 3.099 | 3.471 | 3.958 | 4.637 | 5.560 | 6.725 |
| 64 | 2.052 | 2.354 | 2.777 | 3.073 | 3.440 | 3.920 | 4.589 | 5.518 | 5.688 |
| 68 | 2.041 | 2.336 | 2.763 | 3.044 | 3.414 | 3.893 | 4.549 | 5.474 | 5.641 |
| 72 | 2.032 | 2.327 | 2.741 | 3.030 | 3.386 | 3.861 | 4.520 | 5.429 | 5.575 |
| 76 | 2.027 | 2.313 | 2.729 | 3.009 | 3.366 | 3.836 | 4.486 | 5.395 | 5.536 |
| 80 | 2.017 | 2.301 | 2.711 | 2.988 | 3.346 | 3.811 | 4.460 | 5.356 | 5.483 |
| $\infty$ | 1.932 | 2.163 | 2.487 | 2.700 | 2.965 | 3.311 | 3.803 | 4.653 | 5.786 |

<u>Table 19/4b:</u> Percentage points $u_{0.05}(\gamma)$ for lower tolerance limits and for confidence limits of percentiles

| $n$ \ $\gamma$ | 0.02 | 0.05 | 0.10 | 0.15 | 0.20 | 0.25 | 0.30 | 0.40 | 0.50 |
|---|---|---|---|---|---|---|---|---|---|
| 10 | −3.725 | −3.317 | −2.966 | −2.741 | −2.571 | −2.436 | −2.322 | −2.145 | −2.022 |
| 11 | −3.569 | −3.180 | −2.852 | −2.646 | −2.488 | −2.366 | −2.263 | −2.107 | −2.002 |
| 12 | −3.443 | −3.075 | −2.764 | −2.569 | −2.422 | −2.308 | −2.215 | −2.076 | −1.987 |
| 13 | −3.349 | −2.992 | −2.693 | −2.508 | −2.369 | −2.262 | −2.175 | −2.049 | −1.974 |
| 14 | −3.270 | −2.922 | −2.634 | −2.453 | −2.324 | −2.224 | −2.142 | −2.026 | −1.963 |
| 15 | −3.203 | −2.865 | −2.583 | −2.412 | −2.286 | −2.190 | −2.112 | −2.008 | −1.950 |
| 16 | −3.148 | −2.819 | −2.544 | −2.375 | −2.252 | −2.161 | −2.089 | −1.990 | −1.942 |
| 18 | −3.063 | −2.740 | −2.473 | −2.315 | −2.198 | −2.114 | −2.047 | −1.962 | −1.928 |
| 20 | −2.996 | −2.678 | −2.419 | −2.266 | −2.156 | −2.077 | −2.014 | −1.938 | −1.916 |
| 22 | −2.941 | −2.630 | −2.378 | −2.226 | −2.121 | −2.042 | −1.986 | −1.920 | −1.907 |
| 24 | −2.900 | −2.586 | −2.342 | −2.193 | −2.092 | −2.015 | −1.961 | −1.903 | −1.898 |
| 26 | −2.860 | −2.554 | −2.310 | −2.165 | −2.066 | −1.996 | −1.945 | −1.889 | −1.889 |
| 28 | −2.836 | −2.528 | −2.286 | −2.141 | −2.043 | −1.976 | −1.928 | −1.875 | −1.886 |
| 30 | −2.810 | −2.502 | −2.262 | −2.118 | −2.027 | −1.957 | −1.914 | −1.865 | −1.881 |
| 32 | −2.789 | −2.482 | −2.240 | −2.102 | −2.009 | −1.942 | −1.897 | −1.858 | −1.875 |
| 34 | −2.752 | −2.454 | −2.216 | −2.086 | −1.995 | −1.932 | −1.891 | −1.851 | −1.875 |
| 36 | −2.718 | −2.423 | −2.196 | −2.068 | −1.981 | −1.922 | −1.880 | −1.851 | −1.875 |
| 38 | −2.694 | −2.403 | −2.182 | −2.051 | −1.967 | −1.913 | −1.870 | −1.846 | −1.871 |
| 40 | −2.669 | −2.383 | −2.163 | −2.041 | −1.955 | −1.900 | −1.868 | −1.837 | −1.869 |
| 42 | −2.637 | −2.364 | −2.145 | −2.026 | −1.945 | −1.895 | −1.862 | −1.837 | −1.870 |
| 44 | −2.620 | −2.347 | −2.136 | −2.014 | −1.938 | −1.886 | −1.853 | −1.834 | −1.867 |
| 46 | −2.597 | −2.332 | −2.123 | −2.005 | −1.927 | −1.881 | −1.847 | −1.827 | −1.869 |
| 48 | −2.577 | −2.313 | −2.106 | −1.993 | −1.920 | −1.873 | −1.845 | −1.825 | −1.867 |
| 50 | −2.566 | −2.304 | −2.100 | −1.985 | −1.910 | −1.862 | −1.834 | −1.827 | −1.870 |
| 52 | −2.545 | −2.285 | −2.084 | −1.973 | −1.905 | −1.863 | −1.834 | −1.820 | −1.871 |
| 54 | −2.535 | −2.277 | −2.079 | −1.967 | −1.897 | −1.855 | −1.825 | −1.818 | −1.870 |
| 56 | −2.507 | −2.259 | −2.065 | −1.958 | −1.894 | −1.851 | −1.829 | −1.822 | −1.867 |
| 58 | −2.498 | −2.245 | −2.056 | −1.955 | −1.890 | −1.846 | −1.823 | −1.816 | −1.870 |
| 60 | −2.478 | −2.237 | −2.052 | −1.949 | −1.883 | −1.839 | −1.823 | −1.816 | −1.871 |
| 64 | −2.456 | −2.214 | −2.032 | −1.933 | −1.873 | −1.835 | −1.811 | −1.811 | −1.868 |
| 68 | −2.440 | −2.200 | −2.020 | −1.919 | −1.865 | −1.825 | −1.809 | −1.809 | −1.868 |
| 72 | −2.409 | −2.179 | −2.002 | −1.915 | −1.851 | −1.819 | −1.802 | −1.811 | −1.871 |
| 76 | −2.388 | −2.160 | −1.966 | −1.906 | −1.849 | −1.816 | −1.799 | −1.808 | −1.870 |
| 80 | −2.379 | −2.154 | −1.985 | −1.893 | −1.844 | −1.810 | −1.792 | −1.802 | −1.874 |
| ∞ | −2.041 | −1.859 | −1.728 | −1.669 | −1.647 | −1.650 | −1.673 | −1.771 | −1.932 |

<u>Source:</u> BAIN (1978, p. 240) — Reprinted with permission from *Statistical Analysis of Reliability and Life–Testing Models*. Copyright 1978 by Marcel Dekker, Inc. All rights reserved.

There exist several **approximate methods** to find tolerance limits; see BAIN/ENGELHARDT (1981, 1986, 1991a) and ENGELHARDT/BAIN (1977), LAWLESS (1975) and MANN (1978a). The following simple method is due to BAIN/ENGELHARDT (1991a). Suppose $X \sim We(0, b, c)$, then a lower $(1 - \alpha)$ tolerance limit for the proportion $\gamma$ (or a lower $(1 - \alpha)$ level confidence limit for the percentile $x_{1-\gamma}$) based on the MLEs $\widehat{b}$ and $\widehat{c}$ is given by

$$L_o(\boldsymbol{x}, 1 - \alpha, \gamma) = \widehat{x}_{1-\gamma} \, \exp\left\{-\frac{d + \lambda_{1-\gamma}}{\widehat{c}}\right\}, \tag{19.35a}$$

where

$$\lambda_{1-\gamma} = \ln(-\ln\gamma), \tag{19.35b}$$

$$d = \frac{c_{12}u_{1-\alpha}^2 - n\lambda_{1-\gamma} + u_{1-\alpha}\left[(c_{12}^2 - c_{11}c_{22})u_{1-\alpha}^2 + nc_{11} - 2nc_{12}\lambda_{1-\gamma} + nc_{22}\lambda_{1-\gamma}^2\right]^{1/2}}{n - c_{22}\,u_{1-\alpha}}. \tag{19.35c}$$

$c_{11}$, $c_{12}$ and $c_{22}$ are the coefficients of the asymptotic variance of $\widehat{R}$ as given in Tab. 19/3 and $u_{1-\alpha}$ is the $(1-\alpha)$ percentile of the standard normal distribution. Substituting $\widehat{x}_{1-\gamma}$ by $\widehat{b}\,(-\ln\gamma)^{1/\widehat{c}}$ in (19.35a) gives the equivalent form

$$L_o(\boldsymbol{x}, 1 - \alpha, \gamma) = \widehat{b}\,\exp(-d/\widehat{c}). \tag{19.35d}$$

---

**Example 19/4: Confidence intervals for percentiles and tolerance intervals pertaining to dataset #1**

The MLEs of $b$ and $c$ resulting from the $n = 20$ uncensored observations in Tab. 9/2 are $\widehat{b} = 99.2079$ and $\widehat{c} = 2.5957$. The point estimate of the percentile $x_{0.10}$ is

$$\widehat{x}_{0.10} = \widehat{b}\,(-\ln 0.90)^{1/\widehat{c}} = 99.2079\,(-\ln 0.90)^{1/2.5957} = 41.69.$$

From (19.31a) and Tab. 19/4a the lower $(1 - \alpha = 0.95,\ \gamma = 0.90)$ tolerance limit is

$$L_o(\boldsymbol{x}, 0.95, 0.90) = 41.69 \exp\left\{\frac{-5.548}{2.5957\,\sqrt{20}}\right\} = 25.85,$$

which is also the lower limit of the one–sided 0.95 level confidence interval for $x_{0.10}$, i.e., $x_{0.10} \geq 25.85$.

For an upper $(0.95,\ 0.90)$ tolerance limit $U_o(\boldsymbol{x}, 0.95, 0.90)$, we use (19.31b), here:

$$U_o(\boldsymbol{x}, 0.95, 0.90) = L_o(\boldsymbol{x}, 0.05, 0.10),$$

together with Tab. 19.4b and find

$$U_o(\boldsymbol{x}, 0.95, 0.90) = \widehat{x}_{0.90} \exp\left\{\frac{2.419}{2.5957\,\sqrt{20}}\right\} = 136.8015 \cdot 1.2317 = 168.50.$$

This is also the upper limit of the one–sided 0.95 level confidence interval for $x_{0.90}$, i.e., $x_{0.9} \leq 168.50$.

When we bring together $L_o(\boldsymbol{x}, 0.95, 0.90) = 25.85$ and $U_o(\boldsymbol{x}, 0.95, 0.90) = 168.50$ the resulting two–sided tolerance interval $[25.85, 168.50]$ has — see (19.32) — approximate probability $1 - \alpha = 0.90$ for proportion $\gamma = 0.80$.

If one had wished to determine the $(0.95,\ 0.90)$ lower tolerance limits from the reliability table the procedure would have been to find 0.90 in the $(n = 20)$–column of Tab. 19/1, then read $\widehat{R}_c \approx 0.9682$ from the left–hand side. Then from (19.34)

$$L_o(\boldsymbol{x}, 0.95, 0.90) \approx 99.2079\,(-\ln 0.9682)^{1/2.5957} = 26.44.$$

Due to rounding errors or other discrepancies, there is a slight difference between this result and the result above.

On the other hand suppose one wished to determine the lower $0.95$ confidence limit for $R(25.85)$ by using the tolerance limit tables. This requires finding the value of $\gamma$ which satisfies (19.33b). Now $-\sqrt{n}\,\ln[-\ln\widehat{R}(25.85)] = -\sqrt{20}\,\ln[-\ln 0.97] = 15.61$, and trying various values of $\gamma$, we find from Tab. 19/4a that for $\gamma = \widehat{R}_L = 0.90$:[8]

$$-\sqrt{n}\,\ln[-\ln\gamma] + u_{1-\alpha}(\gamma) = -\sqrt{20}\,\ln[-\ln 0.90] + 5.548 = 15.61.$$

If we had used Tab. 19/1 with $\widehat{R}(25.85) = 0.97$ to directly read off $\widehat{R}_L$, we would have found $\widehat{R}_L = 0.904$. The slight difference between $0.904$ and $0.90$ again is due to round–off errors and linear interpolation.

We finally mention some special papers on tolerance limit calculation:

- ISAIC–MANIU/VODA (1980) suppose that the shape parameter $c$ is known.

- MANN (1978a) shows how to proceed when the data come from accelerated life testing.

- MANN/FERTIG (1973) build upon the BLIEs of $b$ and $c$.

## 19.5 BAYESIAN approaches[9]

We will first address BAYESIAN estimation of a percentile (or of reliable life) and then BAYESIAN estimation of reliability. The latter topic has been treated rather intensively in the literature.

The **reliable life** is defined as the time $x_R$ for which $100\,R\%$ of the population will survive. Taking the form

$$f(x) = \lambda\,c\,x^{c-1}\,\exp\left(-\lambda\,x^c\right), \quad \lambda = b^{-c}, \tag{19.36a}$$

of the two–parameter WEIBULL density, the reliable life $x_R = x_{1-P}$ is defined as

$$x_R = \lambda^{-1/c}\,(-\ln R)^{1/c}, \tag{19.36b}$$

where $R$ is a specified proportion. We restrict our consideration to the case where $c$ is assumed known, and consider point and interval estimators for $x_R$ when $\lambda$ is the realization of a variate $\Lambda$ having one of the following prior distributions:

- uniform distribution over $[A, B]$; see (14.17a);

- non–informative prior; see (14.13e);

- gamma distribution; see (14.20a).

---

[8]  $\widehat{R}_L = 0.90$ agrees with result above, where $L_o(\boldsymbol{x}, 0.95, 0.90)$ was found to be $25.85$.

[9]  Suggested reading for this section: ABDEL–WAHID/WINTERBOTTOM (1987), CANAVOS/TSOKOS (1973), ERTO/GUIDA (1985a), MARTZ/WALLER (1982), PADGETT/TSOKOS (1979), PANDEY (1987), PAPADOPOULOS/TSOKOS (1975), SINGPURWALLA/SONG (1988), SINHA (1986a), TSOKOS (1972), TSOKOS/RAO (1976).

Assuming a sample of size $n$ which is type–II censored on the right ($r \leq n$), we observe the following re–scaled total time on test (see (14.12b)):

$$T := \sum_{i=1}^{r} X_{i:n}^c + (n-r)\, X_{r:n}^c. \tag{19.36c}$$

The posterior density of $\Lambda$ when the prior is uniform, non–informative or gamma is given in (14.17a), (14.14a) and (14.20b), respectively. Results of BAYESIAN estimation of $x_R$ can be traced back to the results of BAYESIAN estimation of **mean time to failure** (MTTF) as has been proven by MARTZ/WALLER (1982).

---

**Excursus: MTTF estimation**

The WEIBULL mean to failure is

$$\mu = \lambda^{-1/c}\, \Gamma\!\left(1 + \frac{1}{c}\right), \tag{19.37a}$$

which is considered to be random when $\lambda$ is a realization of $\Lambda$:

$$M = \Lambda^{-1/c}\, \Gamma\!\left(1 + \frac{1}{c}\right). \tag{19.37b}$$

MARTZ/WALLER (1982, pp. 413–418) give the following results for $M$:

1. Uniform prior on $\Lambda$; see (14.17a)

   BAYESIAN point estimator using squared–error loss:

$$\mathrm{E}(M\,|\,T,A,B) = \frac{T^{1/c}\, \Gamma\!\left(1 + \frac{1}{c}\right)\left[\gamma\!\left(r + 1 - \frac{1}{c}\,\middle|\, B\,T\right) - \gamma\!\left(r + 1 - \frac{1}{c}\,\middle|\, A\,T\right)\right]}{\gamma(r + 1\,|\,B\,T) - \gamma(r + 1\,|\,A\,T)} \tag{19.38a}$$

   Posterior variance of $M$:

$$\left.\begin{aligned}\mathrm{Var}(M\,|\,T,A,b) \;=\; & \frac{T^{2/c}\, \Gamma^2\!\left(1 + \frac{1}{c}\right)\, \gamma\!\left(r + 1 - \frac{2}{c}\,\middle|\, b\,T\right) - \gamma\!\left(r + 1 - \frac{2}{c}\,\middle|\, A\,T\right)}{\gamma(r + 1\,|\,B\,T) - \gamma(r + 1\,|\,A\,T)} - \\[2mm] & \mathrm{E}^2(M\,|\,T,A,B)\end{aligned}\right\} \tag{19.38b}$$

   An equal–tail $100\,(1-\alpha)\%$ credible interval for $M$ is easily obtained from the corresponding interval $[\lambda_\ell, \lambda_u]$ given in (14.18a,b) as $[m_\ell, m_u]$, where

$$m_\ell = \Gamma\!\left(1 + \frac{1}{c}\right)\lambda_\ell^{-1/c}, \quad m_u = \Gamma\!\left(1 + \frac{1}{c}\right)\lambda_u^{-1/c}. \tag{19.38c}$$

2. Non–informative prior on $\Lambda$; see (14.13e)

$$\mathrm{E}(M\,|\,T) \;=\; \frac{T^{1/c}\, \Gamma\!\left(1 + \frac{1}{c}\right)\Gamma\!\left(r - \frac{1}{c}\right)}{\Gamma(r)} \tag{19.39a}$$

$$\mathrm{Var}(M\,|\,T) \;=\; \frac{T^{2/c}\, \Gamma^2\!\left(1 + \frac{1}{c}\right)\Gamma\!\left(r - \frac{2}{c}\right)}{\Gamma(r)} - \mathrm{E}^2(M\,|\,T) \tag{19.39b}$$

An equal–tail $100\,(1-\alpha)\%$ credible interval for $M$ is given by

$$\left[\Gamma\left(1+\frac{1}{c}\right)\left\{\chi^2_{2r,\alpha/2}\big/(2\,T)\right\}^{-1/c},\ \Gamma\left(1+\frac{1}{c}\right)\left\{\chi^2_{2r,1-\alpha/2}\big/(2\,T)\right\}^{-1/c}\right]. \qquad (19.39c)$$

3. Gamma prior on $\Lambda$; see (14.20a)

$$E(M\,|\,T,b,d) \;=\; \frac{\Gamma\left(d+r-\dfrac{1}{c}\right)\Gamma\left(1+\dfrac{1}{c}\right)}{\Gamma(d+r)\,[b/(b\,T+1)]^{1/c}} \qquad (19.40a)$$

$$\text{Var}(M\,|\,T,b,d) \;=\; \frac{\Gamma\left(d+r-\dfrac{2}{c}\right)\Gamma\left(1+\dfrac{1}{c}\right)}{\Gamma(d+r)\,[b/(b\,T+1)]^{2/c}} - E^2(M\,|\,T,b,d) \qquad (19.40b)$$

An equal–tail $100\,(1-\alpha)\%$ credible region for $M$ is given by

$$\left[\Gamma\left(1+\frac{1}{c}\right)\left\{\frac{b\,\chi^2_{2(r+d),1-\alpha/2}}{2\,b\,T+2}\right\}^{-1/c},\ \Gamma\left(1+\frac{1}{c}\right)\left\{\frac{b\,\chi^2_{2(r+d),\alpha/2}}{2\,b\,T+2}\right\}^{-1/c}\right]. \qquad (19.40c)$$

It is observed that $x_R$ in (19.36b) is of the same functional form in $\lambda$ as $\mu$ in (19.37a). Thus, all of the results for MTTF given in the excursus above apply here as well, the only change being that the term $\Gamma(1+1/c)$ must be replaced by the term $(-\ln R)^{1/c}$ in all the point and interval estimation equations.

We now turn to BAYESIAN estimation of **WEIBULL reliability** given by

$$r(x) = \exp\left\{-\lambda\,x^c\right\}. \qquad (19.41)$$

We first assume that $c$ **is known** and $\boldsymbol{\lambda}$ **is a realization of the variate** $\boldsymbol{\Lambda}$. The prior (posterior) distribution of $R = R(x)$ may be obtained directly from the prior (posterior) distribution of $\Lambda$ by means of a transformation. Since $r = r(x)$ is a monotonic function of $\lambda$, the unique inverse exists which may be represented as

$$\lambda = -\frac{\ln r}{x^c}. \qquad (19.42a)$$

Letting $g_\lambda(\cdot)$ represent either the prior or posterior distribution of $\Lambda$, the corresponding prior or posterior distribution of $R$, denoted by $g_r(\cdot)$, may be obtained as

$$g_r(r) = g_\lambda\left(-\frac{\ln r}{x^c}\right)\left(\frac{1}{x^c}\right). \qquad (19.42b)$$

The BAYESIAN point estimator of $r$ is the mean of the posterior distribution of $R$ under squared–error loss. There are two techniques that can be used to determine this mean. As the rescaled total time on test $T$ (see (19.36c)) is sufficient for estimating $\lambda$, we can find the posterior mean of $R$ given $T$ either by directly calculating the mean of the posterior distribution of $R$ or by taking the expectation of $R$ with respect to the posterior distribution of $\Lambda$ given $T$.

When **scale and shape parameters are random variables** the BAYESIAN estimator of $r(x)$ turns out to be of complicated form. CANAVOS/TSOKOS (1973) studied two cases, both assuming a type–II censored sample and independent prior distributions:

- the shape and the scale parameters both have a uniform prior,
- the shape parameter has a uniform prior and the scale parameter has an inverted gamma prior.

PANDEY (1987) also assumes independent priors, a uniform prior for the shape parameter and a non–informative prior for the scale parameter, the observations being randomly censored.

# 20 Prediction of future random quantities

Statistical **prediction** is the process by which values for unknown observables (potential observations yet to be made or past ones which are no longer available) are inferred based on current observations and other information at hand. Whereas statistical estimation is concerned about how to get information on a distribution, usually its parameters, percentiles, DF, CDF or CCDF, the aim of prediction is to make inference on the value of some statistic of a sample and to enclose it within **prediction limits**.

We will comment on both **new–sample prediction** (= **two–sample prediction**) and **within–sample prediction** (= **one–sample prediction**). For new–sample prediction, data from a past sample of size $n_1$ is used to make predictions on one or more future units in a second sample of size $n_2$ from the same process or population. For example, based on previous (possibly censored) life test data, one could be interested in predicting the following:

- time to failure of a new item $(n_2 = 1)$;
- time until the $r$–th failure in a future sample of $n_2$ units, $n_2 \geq r$;
- number of failures by time $t^*$ in a future sample of $n_2$ units.

For within–sample prediction, the problem is to predict future events in a sample or process based on early date from that sample or process. For example, if $n$ units are followed until censoring time $t_c$ and there are $k$ observed failure times, $x_{1:n}, \ldots, x_{k:n}$, one could be interested in predicting the following:

- time of next failure;
- time until $\ell$ additional failures, $\ell \leq n - k$;
- number of additional failures in a future interval $(t_c, t_w)$.

We will present classical prediction methods in Sect. 20.1 and BAYESIAN prediction methods in Sect. 20.2.

## 20.1 Classical prediction

### 20.1.1 Prediction for a WEIBULL process[1]

We consider a non–homogeneous POISSON process with intensity function

$$\lambda(t) = \frac{c}{b} \left( \frac{t}{b} \right)^{c-1}, \tag{20.1a}$$

---

[1] Suggested reading for this section: BAIN (1978, pp. 317 ff.), ENGELHARDT/BAIN (1978) and Chapter 18, this book.

called a WEIBULL process. Suppose, $T_1, \ldots, T_n$ denote the first $n$ successive occurrence times of such a process. The time to first occurrence, $T_1$, follows a WEIBULL distribution with hazard function given by (20.1a). The conditional occurrence time $T_n$, given $T_{n-1} = t_{n-1}, \ldots, T_1 = t_1$, follows a truncated WEIBULL distribution with truncation point $t = t_{n-1}$. The joint density of $T_1, \ldots, T_n$ is given by

$$f(t_1, \ldots, t_n) = \left(\frac{c}{b}\right)^n \prod_{i=1}^{n} \left(\frac{t_i}{b}\right)^{c-1} \exp\left\{ -\left(\frac{t_n}{b}\right)^c \right\}; \quad 0 < t_1 \ldots t_n < \infty; \qquad (20.1b)$$

and the MLEs of $b$ and $c$ are

$$\hat{c} = n \left/ \sum_{1=1}^{n-1} \ln\left(T_n/T_i\right), \right. \qquad (20.1c)$$

$$\hat{b} = T_n / n^{1/\hat{c}}. \qquad (20.1d)$$

Consider a repairable system, and suppose $n$ breakdowns have occurred and that the times of occurrence follow a WEIBULL process. Perhaps the most natural question concerns when the next failure will occur. This suggests that a prediction interval for $T_{n+1}$, or more generally for $T_{n+m}$, would be quite useful. A $(1 - \alpha)$ level **lower prediction limit** for $T_{n+m}$ is a statistic $T_L(n, m, 1 - \alpha)$ such that

$$\Pr\big(T_L(n, m, 1 - \alpha) \leq T_{n+m}\big] = 1 - \alpha; \qquad (20.2)$$

i.e., the $(n + m)$-th failure after having observed the $n$-th failure will not occur before $T_L(n, m, 1 - \alpha)$ with probability $1 - \alpha$, or we will have to wait at least another $T_L(n, m, 1 - \alpha) - t_n$ time units for $m$ additional failures to occur.

Consider first the case $m = 1$. The limit $T_L$ should be a function of the sufficient statistics (20.1c,d) and the probability must be free of parameters. From Sect. 18.1.1 we have the following distributional results, see BAIN (1978, p. 317/318):

1) $\quad U_1 = 2\left(\dfrac{T_{n+1}}{b}\right)^c \sim \chi^2(2(n+1)),$ $\qquad (20.3a)$

2) $\quad U = 2\left(\dfrac{T_n}{b}\right)^c \sim \chi^2(2n),$ $\qquad (20.3b)$

3) $\quad V = 2n\dfrac{c}{\hat{c}} = 2c\displaystyle\sum_{i=1}^{n-1} \ln\left(\dfrac{T_n}{T_i}\right) \sim \chi^2(2(n-1)),$ $\qquad (20.3c)$

4) $\quad 2nW = 2n\ln\left(\dfrac{U_1}{U}\right) = 2nc\ln\left(\dfrac{T_{n+1}}{T_n}\right) \sim \chi^2(2),$ $\qquad (20.3d)$

5) $\qquad\qquad U_1, U$ and $W$ are independent of $V,$ $\qquad (20.3e)$

6) $\quad Y = 2n(n-1)\dfrac{W}{V} = (n-1)\hat{c}\ln\left(\dfrac{T_{n+1}}{T_n}\right) \sim F(2, 2(n-1)).$ $\quad (20.3f)$

From (20.3f) we have

$$\Pr\left[(n-1)\,\widehat{c}\,\ln\left(\frac{T_{n+1}}{T_n}\right) \geq F_{2,2(n-1),\alpha}\right] = 1 - \alpha, \tag{20.4a}$$

and as the $100\,\alpha$ percentage point of $F(2, 2\,(n-1))$ is

$$F_{2,2(n-1),\alpha} = (n-1)\left[(1-\alpha)^{-1/(n-1)} - 1\right], \tag{20.4b}$$

we arrive at

$$\Pr\left[T_{n+1} \geq T_n\,\exp\left\{\frac{(1-\alpha)^{-1/(n-1)} - 1}{\widehat{c}}\right\}\right] = 1 - \alpha \tag{20.4c}$$

so that

$$T_L(n, 1, 1 - \alpha) = T_n\,\exp\left\{\frac{(1-\alpha)^{-1/(n-1)} - 1}{\widehat{c}}\right\}. \tag{20.4d}$$

We now turn to the more general case $m \geq 1$. ENGELHARDT/BAIN (1978) obtained

$$T_L(n, m, 1 - \alpha) = T_n\,\exp\left\{\frac{y_\alpha}{(n-1)\,\widehat{c}}\right\}, \tag{20.5a}$$

where $T_n$ and $\widehat{c}$ are determined from the first $n$ failure times and $y_\alpha$ is the solution of

$$\alpha = Q(y; n, m) \tag{20.5b}$$

with

$$Q(y; n, m) = \sum_{j=1}^{m} k_j \left\{1 - \left[\frac{1 + (n+j-1)\,y}{n\,(n-1)}\right]^{-(n-1)}\right\} \tag{20.5c}$$

and

$$k_j = \frac{(-1)^{j-1}\,(n+m-1)!}{(n-1)!\,(m-j)!\,(j-1)!\,(n+j-1)!}. \tag{20.5d}$$

For the case $m = 1$ we have the explicit form given in (20.4d). A convenient approximation to (20.5a) is

$$T_L(n, m, 1 - \alpha) \approx T_n\,\exp\left\{\frac{\nu\,F_{\nu,2(n-1),\alpha}}{2\,(n-1)\,d\,\widehat{c}}\right\} \tag{20.5e}$$

with

$$\nu = 2\left(\sum_{i=n}^{n+m-1}\frac{1}{i}\right)^2 \bigg/ \sum_{i=n}^{n+m-1}\frac{1}{i^2}, \tag{20.5f}$$

$$d = \sum_{i=n}^{n+m-1}\frac{1}{i} \bigg/ \left(n\sum_{i=n}^{n+m-1}\frac{1}{i^2}\right). \tag{20.5g}$$

The sums in (20.5f,g) can be further approximated as:

$$\sum_{i=n}^{n+m-1} \frac{1}{i} \approx \ln(n+m-0.5) - \ln(n-0.5), \tag{20.5h}$$

$$\sum_{i=n}^{n+m-1} \frac{1}{i^2} \approx \frac{1}{n-0.5} - \frac{1}{n+m-0.5}. \tag{20.5i}$$

## 20.1.2 One-sample prediction[2]

Let $X_{1:n}, \ldots, X_{n:n}$ be the order statistics of a two–parameter WEIBULL distribution. Having observed the first $k$ failure times $X_{1:n}, \ldots, X_{k:n}$, we want to predict the $r$–th ordered observation time $X_{r:n}$, $k < r \le n$. $X_{r:n}$ will be the time to complete the test under type–II censoring with censoring number $r$. Setting $r = n$ leads to a prediction of the largest observation time or of the test duration when there is no censoring. A solution to this prediction problem is given for three cases:

- The shape parameter $c$ is assumed known.
- There is partial knowledge on $c$; i.e., $c$ is some distinct value in the interval $[c_\ell, c_u]$.
- $c$ is completely unknown.

First case: $c$ known

When $c$ is known the transformed WEIBULL variate

$$Y := X^c$$

has an exponential distribution with DF

$$f(y) = \lambda \exp(-\lambda x), \ \ \lambda = b^{-c},$$

and the prediction problem is solved by applying the well–known solutions for this distribution as found in LAWLESS (1971).

Let

$$S_k = \sum_{i=1}^{n} Y_{i:n} + (n-k)\, Y_{k:n}; \ \ Y_{i:n} := X_{i:n}^c; \tag{20.6a}$$

($S_k$ is the transformed total time on test up to the $k$–th failure), and consider (for given $k < r \le n$) the non–negative variate

$$U := U_{k,r,n} = \frac{Y_{r:n} - Y_{k:n}}{S_k}. \tag{20.6b}$$

LAWLESS (1971) gives the following DF of $U$:

$$f(u) = \frac{k}{B(r-k, n-r+1)} \sum_{i=0}^{r-k-1} \binom{r-k-1}{i} (-1)^i \big[1 + (n-r+i+1)\, u\big]^{-k-1}, \ u > 0,$$

$$\tag{20.6c}$$

---

[2] Suggested reading for this section: ADATIA/CHAN (1982), BALASOORIYA/CHAN (1983), HSIEH (1996), KAMINSKY/NELSON (1975), KAMINSKY/MANN/NELSON (1975), LAWLESS (1971), WRIGHT/SINGH(1981).

where $B(\cdot,\cdot)$ is the beta function. Integration yields the CCDF

$$
\left.
\begin{aligned}
\Pr(U > t) &= \frac{k}{B(r-k,n-r+1)} \sum_{i=0}^{r-k-1} \frac{\binom{r-k-1}{i}(-1)^i}{n-r+i+1} \left[1+(n-r+i+1)\,t\right]^{-k} \\
&= P(t;k,r,n).
\end{aligned}
\right\}
$$
(20.6d)

The distribution of $U$ does not involve $\lambda$.

Probability statements about $U$ give prediction statements about $Y_{r:n}$ on the basis of observed $S_k$ and $Y_{k:n}$. For example, the statement $\Pr(U > t_\alpha) = 1 - \alpha$ yields the prediction statement:

$$
\Pr\left[Y_{r:n} > Y_{k:n} + t_\alpha\,S_k\right] = 1 - \alpha,
$$
(20.7a)

giving the one–sided $100\,(1-\alpha)\%$ prediction interval for $Y_{r:n}$. From the transformation $Y = X^c$ we find the $(1-\alpha)$ level **lower prediction limit for the $r$–th WEIBULL order statistic** as

$$
X_L(n,r,1-\alpha) = \left(Y_{k:n} + t_\alpha\,S_k\right)^{1/c}.
$$
(20.7b)

For given values of $n$, $r$, $k$ and $t$, the probabilities $P(t;k,r,n)$ defined in (20.6d) are easily evaluated on a computer. If, for specified $n$, $r$ and $k$ values of $t$ are desired which make $\Pr(U > t)$ equal to some specified value, such as $\alpha = 0.05$, these values can be found using a simple iteration scheme.

There are two special cases concerning $r$: $r = n$ and $r = k + 1$. If we wish to predict the largest observation $Y_{n:n}$ on the basis of the $k$ smallest, (20.6d) simplifies to

$$
\begin{aligned}
\Pr(U > t) &= (n-k) \sum_{i=0}^{n-k-1} \frac{\binom{n-k-1}{i}(-1)^i}{(i-1)\left[1+(i+1)\,t\right]^k} \\
&= 1 - \sum_{i=0}^{n-k} \binom{n-k}{i}(-1)^i\,(1+i\,t)^{-k}
\end{aligned}
$$
(20.8a)

or

$$
\Pr(U \le t) = \sum_{i=0}^{n-k} \binom{n-k}{i}(-1)^i\,(1+i\,t)^{-k}.
$$
(20.8b)

In the case where $k = r - 1$ (prediction of the next failure time after $Y_{k:n}$), we have

$$
k\,(n-k)\,\frac{Y_{k+1:n} - Y_{k:n}}{S_k} \sim F(2,2\,k)
$$
(20.9a)

so that

$$
\Pr\left[Y_{k+1:n} > Y_{k:n} + \frac{F_{2,2k,\alpha}}{k\,(n-k)}\,S_k\right] = 1 - \alpha.
$$
(20.9b)

**Example 20/1: 95% level lower prediction intervals using dataset #1**

The $n = 20$ observations of dataset #1 in Tab. 9/2 have been generated with $b = 100$ and $c = 2.5$. We set $1 - \alpha = 0.95$, $k = 10$ and want to give lower prediction limits for $X_{11:20}$ and $X_{20:20}$. We further set $c = 2.5$ and transform $x_{i:20}$ $(i = 1, \ldots, 10)$ to $y_{i:20} = x_{i:20}^{2.5}$ and find $y_{10:20} = 62,761.78$, $s_{10} = 937,015.22$.

The 95% lower prediction limit for $Y_{20:20}$ is found with $t = 0.11997$ from (20.8a) yielding

$$Y_L(20, 20, 0.95) = 175,175.5 \text{ and } X_L(20, 20, 0.95) = 125.13.$$

The realization of $X_{20:20}$ in Tab. 9/2 is $x_{20:20} = 183$ and thus lies within this prediction interval.

The 95% lower prediction limit for $Y_{11:20}$ is found with $F_{2,20,0.05} = 0.0514$ from (20.9b) yielding

$$Y_L(20, 11, 0.95) = 63,243.40 \text{ and } X_L(20, 11, 0.95) = 83.25.$$

The realization $x_{11:20} = 86$ in Tab. 9/2 lies within this prediction interval.

WRIGHT/SINGH (1981) also assumed $c$ known and considered the first $k$ order statistics in a sample of size $n$ when

- $n$ is a fixed number or
- $n$ is the realization of a binomially distributed random variable.

Second case: $c$ within $[c_\ell, c_u]$

ADATIA/CHAN (1982) used the maximin estimator and the adaptive estimator of the scale parameter $b$, both of which are robust (see Sect. 13.4.6), to construct prediction intervals for $X_{r:n}$, assuming $c$ is within $[c_\ell, c_u]$. These predictors are taken to be linear functions of the first $k$ order statistics with weights determined from the guaranteed efficiency required. Necessary tables are provided.

Under the same setup BALASOORIYA/CHAN (1983) carried out a robustness study of four types of predictors for $X_{r:n}$ assuming that it is known only that $c$ has one of the four values 1, 1.5, 2, 2.5. They considered the following four predictors:

- BLUP (best linear unbiased predictor),
- BLIP (best linear invariant predictor),
- FLUP (final linear unbiased predictor),
- CROSS (cross–validatory predictive function).

They showed that CROSS is the best in the sense of guaranteed efficiency; in addition, this method does not require the knowledge of the covariances of WEIBULL order statistics.

Third case: $c$ is completely unknown[3]

A simple linear unbiased predictor for $Y_{r:n} = \ln X_{r:n}$ based on $Y_{k:n}$ $(k < r)$ is

$$\widehat{Y}_{r:n} = Y_{k:n} + \widehat{b}^* \left[ E(U_{r:n}) - E(U_{k:n}) \right], \tag{20.10a}$$

---

[3] General results on BLUP and BLIP of order statistics in location and scale families are to be found in KAMINSKY/NELSON (1975) and KAMINSKY et al. (1975). HSIEH (1996) gives predictor intervals based on quantiles of relevant pivotal statistics.

where $\mathrm{E}(U_{k:n})$ and $\mathrm{E}(U_{r:n})$ are the expectations of the $k$–th and the $r$–th reduced extreme value order statistics in a sample of size $n$ (For tables of these expectations, see WHITE (1967b) or DAVID (1981).) and $\widehat{b}^*$ is the BLUE of $b^* = 1/c$. The predictor of the $r$–th WEIBULL order statistic follows as

$$\widehat{X}_{r:n} = \exp\left(\widehat{Y}_{r:n}\right). \tag{20.10b}$$

Two–sided $(1 - \alpha)$ prediction limits for $Y_{r:n}$ are

$$Y_L(n, r, 1 - \alpha) = Y_{k:n} + \frac{\widehat{b}^*}{F_{\nu_2, \nu_1, 1-\alpha/2}} \left[\mathrm{E}(U_{r:n}) - \mathrm{E}(U_{k:n})\right], \tag{20.11a}$$

$$Y_U(n, r, 1 - \alpha) = Y_{k:n} + \widehat{b}^* F_{\nu_1, \nu_2, 1-\alpha/2} \left[\mathrm{E}(U_{r:n}) - \mathrm{E}(U_{k:n})\right], \tag{20.11b}$$

where the degrees of freedom in the $F$–percentiles are

$$\nu_1 = \frac{2\left[\mathrm{E}(U_{r:n}) - \mathrm{E}(U_{k:n})\right]^2}{\mathrm{Var}(U_{r:n}) + \mathrm{Var}(U_{k:n}) - 2\,\mathrm{Cov}(u_{r:n}, U_{k:n})},$$

$$\nu_2 = 2/B_{n,r}, \quad \mathrm{Var}(\widehat{b}^*) = b^{*2} B_{n,r}.$$

The variance factors $B_{n,r}$ to give the variance of the BLUE of $b^*$ are found in WHITE (1964b). Prediction limits for the $r$–th WEIBULL order statistic $X_{r:n}$ follow from (20.11a,b) as

$$X_L(n, r, 1 - \alpha) = \exp\left[Y_L(n, r, 1 - \alpha)\right], \tag{20.11c}$$

$$X_U(n, r, 1 - \alpha) = \exp\left[Y_U(n, r, 1 - \alpha)\right]. \tag{20.11d}$$

We mention that an upper prediction limit for $X_{n:n}$ is related to an **outlier test**. If the $n$–th observation is above a $(1 - \alpha)$ upper prediction limit for $X_{n:n}$, it is a statistically significant outlier at level $\alpha$.

### 20.1.3   Two-sample prediction[4]

We have a first sample of size $n_1$ with observations $X_{1:n_1}, \ldots, X_{n_1:n_1}$. This sample may be censored and we want to predict $Y_{r:n_2}$ in a future sample of size $n_2$ from the same distribution. We will first give results valid for any continuous distribution. Then we will turn to the WEIBULL distribution showing how to predict the first order statistic $Y_{1:n_2}$ or any order statistic $Y_{r:n_2}$.

Suppose $X_{i:n_1}$ $(i = 1, \ldots, n)$ are the ordered observations in a sample from **any continuous distribution** and the second sample of size $n_2$ is from the same distribution. To predict the $r$–th largest future observation $Y_{r:n_2}$, suppose

$$i \leq \frac{(n_1 + 1)\, r}{n_2 + 1} \leq i + 1$$

---

[4]  Suggested reading for this section: ENGELHARDT/BAIN (1979, 1982), FERTIG/MEYER/MANN (1980), HEWETT/MOESCHBERGER (1976), LAWLESS (1973a), MANN (1970d), MANN/SAUNDERS (1969), MEE/KUSHARY (1994), PANDEY/UPADHYAY (1986), SHERIF/TAN (1978), YANG/SEE/XIE (2003).

for some $i = 1, \ldots, n_1$. Then a predictor for $Y_{r:n_2}$ is

$$\widehat{Y}_{r:n_2} = \left[ i + 1 - \frac{(n_1 + 1)\, r}{n_2 + 1} \right] X_{i:n_1} + \left[ \frac{(n_1 + 1)\, r}{n_2 + 1} - i \right] X_{i+1:n_1}; \qquad (20.12)$$

i.e., the predictor is the percentile of order $r/n_1$ of the past sample, perhaps, interpolating between two past observations. If we use $\widehat{Y}_{1:n_2} = X_{1:n_1}$ (smallest observation in the first sample), the probability of $Y_{1:n_2}$ being above $X_{1:n_1}$ is

$$1 - \alpha = \frac{n_1}{n_1 + n_2}; \qquad (20.13)$$

i.e., $X_{1:n_1}$ is a $(1 - \alpha)$ lower prediction limit for $Y_{1:n_2}$. Similarly, $X_{n_1:n_1}$ is an upper prediction limit for $Y_{n_2:n_2}$ and the probability of it being below $X_{n_1:n_1}$ is also given by (20.13).

We now assume that both samples come from a **two–parameter WEIBULL distribution** and first give methods to predict the smallest observation $Y_{1:n_2}$ in a future sample of size $n_2$. An interesting application of $Y_{1:n_2}$ is the following: A value $1 - \alpha$ is specified for the probability that no failures in a lot of identically distributed items yet to be manufactured will occur before the expiration of the **warranty period**. The warranty period, which must satisfy an assurance criterion at the prescribed probability level regardless of the true parameters within the distribution, is to be calculated from a small preliminary sample of size $n_1$.

MANN/SAUNDERS (1969) and MANN (1970d) — setting $1 - \alpha = 0.95$ — derived an expression for the lower limit of $Y_{1:n_2}$ (the warranty period) as a function of three suitably chosen order statistics $X_{r:n_1}$, $X_{p:n_1}$ and $X_{q:n_1}$. MANN presents tables giving $\nu$, $r$, $p$ and $q$ for $n_1 = 10(1)25$ and $n_2 = 2(1)n_1 - 3$ such that

$$\Pr\bigl(Y_{1:n_2} > X_{r:n_1} + \nu\,(X_{p:n_1} - X_{q:n_1})\bigr) = 0.95. \qquad (20.14)$$

HEWETT/MOESCHBERGER (1976) derived an inequality which is useful in constructing conservative simultaneous prediction intervals for functions of $\ell$ future samples of equal sizes $n_2 = \ldots = n_\ell = n^*$, using results from one previous sample. Let the function be the minimum $Y_{1:n_i}$ $(i = 2, \ldots, \ell)$, each predicted by (20.14), then

$$\Pr\left[ \bigcap_{i=2}^{\ell} \{ Y_{1:n_i} > X_{r:n_1} + \nu\,(X_{p:n_1} - X_{q:n_1}) \} \right] \geq (1 - \alpha)^{\ell-1}. \qquad (20.15)$$

We now turn to predict $Y_{1:n_2}$ based not only on three selected order statistics from the first sample but on all the information contained in this — possibly type–II censored — sample. ENGELHARDT/BAIN (1979) used the logarithmic transformation to change to the type–I smallest extreme value distribution and then discussed the construction of prediction limits for the minimum and, in general, the $r$-th smallest observation in a future sample.[5] These intervals are based on some closed–form estimators of the Log–WEIBULL parameters $a^* = \ln b$ and $b^* = 1/c$ and $F$–approximations to the resulting pivotal quantities. ENGELHARDT/BAIN (1982) subsequently presented a simplified approximation for the lower prediction limit of $Y^*_{1:n_2} = \ln Y_{1:n_2}$:

$$\Pr\!\left[Y^*_{1:n_2} > \widehat{a}^* - t_{1-\alpha}\,\widehat{b}^*\right] = 1 - \alpha, \qquad (20.16a)$$

where $\widehat{a}^*$ and $\widehat{b}^*$ are the GLUEs given in (10.47a,b) with censoring number $k$ and

$$t_{1-\alpha} \approx (A + B) - \left\{(A + B)^2 - C + 2\,A\,\ln\!\left[-(1/n_2)\ln(1-\alpha)\right]\right\}^{1/2}. \qquad (20.16b)$$

The quantities appearing in (20.16b) are defined as:

$$A = g\operatorname{Var}\!\left(\frac{\widehat{b}^*}{b^*}\right), \quad g = 1 + \left[5 + \ln(n_2)/2\right]/k,$$

$$B = \operatorname{Cov}\!\left(\frac{\widehat{a}^*}{b^*}, \frac{\widehat{b}^*}{b^*}\right)\Big/ \operatorname{Var}\!\left(\frac{\widehat{b}^*}{b^*}\right),$$

$$C = \operatorname{Var}\!\left(\frac{\widehat{a}^*}{b^*}\right)\Big/ \operatorname{Var}\!\left(\frac{\widehat{b}^*}{b^*}\right),$$

with tables of the variances and covariances given in ENGELHARDT/BAIN (1991a). The limiting approximation of $t_{1-\alpha}$ is

$$t_{1-\alpha} \approx -\ln\!\left[-(1/n_2)\ln(1-\alpha)\right]. \qquad (20.16c)$$

The $(1 - \alpha)$ lower prediction for the WEIBULL order statistic $Y_{1:n_2}$ is

$$\Pr\!\left[Y_{1:n_2} > \widehat{b}\,\exp\!\left(-t_{1-\alpha}/\widehat{c}\right)\right] = 1 - \alpha; \quad \widehat{b} = \exp(\widehat{a}^*), \ \widehat{c} = 1/\widehat{b}^*. \qquad (20.16d)$$

It may also be of interest to consider an **interval estimate for the expectation of $Y^*_{r:n}$**. A lower $(1 - \alpha)$ confidence limit for $\operatorname{E}\!\left(Y^*_{r:n_2}\right)$ is simply a lower $(1 - \alpha)$ level confidence limit for proportion $\gamma$, where $\gamma = \exp\!\left[-\exp(c_{r,n_2})\right]$ as shown by BAIN/ENGELHARDT (1991a) who also give a table for the coefficients $c_{r,n_2}$. Thus, any of the tolerance limits results in

---

[5]  We mention some other papers on prediction in the WEIBULL case:

- LAWLESS (1973a) gives conditional confidence intervals for $Y_{1:n_2}$, conditional on ancillary statistics, requiring numerical integration.

- SHERIF/TAN (1978) have discussed the prediction of $Y_{1:n_2}$ when the available sample is type–II progressively censored.

- PANDEY/UPADHYAY (1986) assumed that the scale parameter is known and then developed prediction limits by using the shrunken estimator of the unknown shape parameter $c$ proposed by PANDEY (1983).

- MEE/KUSHARY (1994) used a simulation–based procedure.

- There are two papers concerning the prediction of a single future observation ($n_2 = 1$): FERTIG et al. (1980), along the lines of ENGELHARDT/BAIN (1979), used pivotal quantities based on the BLIEs of $a^*$ and $b^*$ of the Log–WEIBULL distribution. YANG et al. (2003) built upon transformations (BOX–COX and KULLBACK–LEIBLER) of the WEIBULL data to near normality.

Sect. 19.4.2 can be applied here. The lower $(1 - \alpha)$ level confidence limit for $E(Y^*_{r:n_2})$ is given by

$$L(1 - \alpha, \gamma) = \widehat{a}^* - d\,\widehat{b}^*, \tag{20.17}$$

where $\gamma = \exp\big[-\exp(c_{r,n_2})\big]$ and $d$ is given in (19.35c), but evaluated with $\lambda_{1-\gamma} = c_{r,n_2}$.

## 20.1.4 Prediction of failure numbers[6]

Suppose $n$ sample units are put on test at time 0, and by some censoring time $t_c$, the cumulative number of failures is $N_1$. We would like to have a prediction (point and interval) for the future added number $N_2$ of units that fail by time $t_w$, which for example may be the end of a warranty period; i.e., $N_2$ is the number of failures in the interval $(t_c, t_w)$. Denote by $N_3$ the remaining number of non–failed units at time $t_w$. $(N_1,\ N_2,\ N_3)$ is a random vector having a trinomial distribution with corresponding probabilities $(p_1,\ p_2,\ p_3)$, where

$$N_1 + N_2 + N_3 = n \quad \text{and} \quad p_1 + p_2 + p_3 = 1.$$

Assuming a two–parameter WEIBULL distribution these probabilities are

$$p_1 \;=\; 1 - \exp\left\{-\left(\frac{t_c}{b}\right)^c\right\}, \tag{20.18a}$$

$$p_2 \;=\; \exp\left\{-\left(\frac{t_c}{b}\right)^c\right\} - \exp\left\{-\left(\frac{t_w}{b}\right)^c\right\}, \tag{20.18b}$$

$$p_3 \;=\; \exp\left\{-\left(\frac{t_w}{b}\right)^c\right\}. \tag{20.18c}$$

We will assume that the shape parameter $c$ is known while $b$ is unknown.

Conditional on $N_1$, the number of additional failures $N_2$ has a binomial distribution with parameters $n - N_1$ and $\rho$, where

$$\rho = \frac{p_1}{p_1 + p_2} \tag{20.19a}$$

is the conditional probability that a sample unit fails in $(t_c, t_w)$ given it survived until $t_c$. For a given $N_1$ and interval $[\,\overline{N}_2, \underline{N}_2\,]$ the conditional coverage probability $CP(\cdot)$ of the specific prediction interval with nominal confidence level $1 - \alpha$ is

$$
\begin{aligned}
CP\big[PI(1-\alpha)\,\big|\,N_1, b\big] \;=\;\; & \Pr\big[\underline{N}_2 \le N_2 \le \overline{N}_2\big|N_1, b\big] \\[2mm]
=\;\; & \left\{
\begin{aligned}
& \sum_{i=0}^{\overline{N}_2} \binom{n - N_1}{i} \rho^i\,(1-\rho)^{n-N_1-i} - \\[2mm]
& \sum_{i=0}^{N_2-1} \binom{n - N_1}{i} \rho^i\,(1-\rho)^{n-N_1-i}.
\end{aligned}
\right\}
\end{aligned}
\tag{20.19b}
$$

---

[6] Suggested reading for this section: MEEKER/ESCOBAR (1998, Section 12.7), NELSON (2000), NORDMAN/ MEEKER (2002).

The actual conditional coverage probability in (20.19b) is random because $[\underline{N}_2, \overline{N}_2]$ depends on $N_1$, which varies from sample to sample. This probability is also unknown here because $\rho$ depends on the unknown scale parameter $b$ through (20.19a) and (20.18a,b).

Given the observed (non–zero) failures $N_1$ by $t_c$, the MLE of $b$ is

$$\widehat{b} = \frac{t_c}{\left\{ -\ln\left[1 - N_1/n\right] \right\}^{1/c}} .$$

(20.20a)

NELSON (2000) suggested the following **point estimator** for $N_2$:

$$\widehat{N}_2 = n\,\widehat{p}_2,$$

(20.20b)

where the estimator $\widehat{p}_2$ is obtained by substituting $\widehat{b}$ of (20.20a) into (20.18b), yielding

$$\widehat{p}_2 = \left(1 - \frac{N_1}{n}\right) - \left(1 - \frac{N_1}{n}\right)^{(t_w/t_c)^c} .$$

(20.20c)

We now derive **prediction bounds** for $N_2$, considering two cases.

First case: $p_1$ and $p_2$ are small

$p_1$ and $p_2$ being small, we have

$$
\begin{aligned}
\frac{p_2}{p_1} &= \frac{\left[1 - \exp\left\{-\left(\frac{t_w}{b}\right)^c\right\}\right] - \left[1 - \exp\left\{-\left(\frac{t_c}{b}\right)^c\right\}\right]}{1 - \exp\left\{-\left(\frac{t_c}{b}\right)^c\right\}} \\[2mm]
&\approx \frac{\left(\frac{t_w}{b}\right)^c - \left(\frac{t_c}{b}\right)^c}{\left(\frac{t_c}{b}\right)^c} = \left(\frac{t_w}{t_c}\right)^c - 1.
\end{aligned}
$$

(20.21)

The approximation does not depend on the unknown scale parameter $b$. NELSON (1972b) has given a conservative $(1-\alpha)$ confidence interval for the trinomial probability ratio $p_1/p_2$ as $[g_L(N_2, N_1, \alpha_1), g_U(N_2, N_1, \alpha_2)]$ with $\alpha_1 + \alpha_2 = \alpha$ where

$$
g_L(N_2, N_!, \alpha_1) = 
\begin{cases}
\dfrac{N_1}{(N_2 + 1)\, F_{2(N_2+1),2N_1,1-\alpha_1}} & \text{for} \quad N_1 \neq 0 \\[4mm]
0 & \text{for} \quad N_1 = 0
\end{cases}
$$

(20.22a)

and

$$g_U(N_2, N_1, \alpha_2) = \frac{(N_1 + 1)\, F_{2(N_1+1),2N_2,1-\alpha_2}}{N_2} .$$

(20.22b)

Then, using the approximation (20.21) for small $p_1$ and $p_2$, the preceding limits will provide an approximate $(1 - \alpha)$ confidence interval for $\left[(t_w/t_c)^c - 1\right]^{-1}$, i.e.,

$$\Pr\left[g_L(N_2, N_1, \alpha_1) \leq \frac{1}{(t_w/t_c)^c - 1} \leq g_U(N_2, N_1, \alpha_2)\right] \approx 1 - \alpha_1 - \alpha_2.$$

(20.22c)

The $FLOOR$[7] of the smallest, positive real $n_2$ satisfying the left–hand inequality in (20.22c) is a one–sided **lower** approximate $(1 - \alpha_1)$ **prediction bound** for $N_2$, denoted by $\underline{N}_2$. If $N_1 = 0$, we define the trivial lower bound to be 0. The $CEILING$[8] of the largest, positive real $n_2$ satisfying the right–hand inequality in (20.22c) is a one–sided **upper** approximate $(1 - \alpha_2)$ **prediction bound** for $N_2$, denoted by $\overline{N}_2$. In certain circumstances the prediction bounds may fall outside the sample space of $N_2$, namely bounds greater than $n - N_1$. If the computed value of $\overline{N}_2$ is greater than $n - N_1$, we reset the upper bound to $n - N_1$. Likewise, if the procedure produces a lower prediction bound greater than $n - N_1$, we redefine the bound to be $n - N_1 - 1$. We remark that the prediction bounds do not depend on the sample size $n$.

NELSON (2000) gave simpler bound for large $N_2$:

$$\underline{N}_2 = FLOOR\left\{0.5\left[(t_w/t_c)^c - 1\right]\chi^2_{2N_1,\alpha_1} - 1\right\}, \tag{20.23a}$$

$$\overline{N}_2 = CEILING\left\{0.5\left[(t_w/t_c)^c - 1\right]\chi^2_{2(N_1+1),1-\alpha_2}\right\}. \tag{20.23b}$$

If $N_1$ or if $0.5\left[(t_w/t_c)^c - 1\right]\chi^2_{2N_1,\alpha_1} < 1$, we define $\underline{N}_2$ to be 0.

Second case: More general values of $p_1$ and $p_2$

For this case NELSON (2000) suggested approximate prediction bounds for $N_2$ based on a likelihood ratio statistic. The maximum of the two–parameter multinomial likelihood for the observed set $(n_1, n_2, n_2)$ is

$$\mathcal{L}^*(n_1, n_2) = K \left(\frac{n_1}{n}\right)^{n_1} \left(\frac{n_2}{n}\right)^{n_2} \left(1 - \frac{n_1}{n} - \frac{n_2}{n}\right)^{n-n_1-n_2} \tag{20.24a}$$

where

$$K = \frac{n!}{n_1!\,n_2!\,(n - n_1 - n_2)!}.$$

Under the WEIBULL distribution model with probabilities $(p_1, p_2, p_3)$ given by (20.18a–c), the one–parameter constrained sample likelihood is

$$\mathcal{K}(b; n_1, n_2) = K \left[1 - \exp\left\{-\left(\frac{t_c}{b}\right)^c\right\}\right]^{n_1} \left[\exp\left\{-\left(\frac{t_c}{b}\right)^c\right\} - \exp\left\{-\left(\frac{t_w}{b}\right)^c\right\}\right]^{n_2} \times$$

$$\left[\exp\left\{-\left(\frac{t_w}{b}\right)^c\right\}\right]^{n-n_1-n_2}. \tag{20.24b}$$

The MLE $\widehat{b}(n_1, n_2)$ of $b$ must be found numerically by maximizing (20.24b). We denote the maximum of (20.24b) by $\mathcal{K}^*(n_1, n_2)$.

The log–likelihood ratio statistic comparing the constrained WEIBULL likelihood with the unconstrained multinomial likelihood is

$$Q(n_1, n_2) = -2\left\{\ln\left[\mathcal{K}^*(n_1, n_2)\right] - \ln\left[\mathcal{L}^*(n_1, n_2)\right]\right\}. \tag{20.24c}$$

---

[7] FLOOR means rounding "down" to the nearest integer.

[8] CEILING means rounding "up" to the nearest integer.

If the true distribution is WEIBULL, then approximately

$$Q(n_1, n_2) \sim \chi^2(1). \tag{20.24d}$$

Hence, for the random variables $N_1$ and $N_2$,

$$\Pr\left[Q(N_1, N_2) \le \chi^2_{1, 1-\alpha}\right] \approx 1 - \alpha. \tag{20.24e}$$

Given the cumulative number of failures $N_1$ by $t_c$, the set of $n_2$–values for which

$$Q(N_1, n_2) \le \chi^2_{1, 1-\alpha} \tag{20.24f}$$

is fulfilled, provides an approximate $(1 - \alpha)$ prediction region for $N_2$. In particular, the $FLOOR$ and $CEILING$ of the respective smallest and largest positive real values that satisfy (20.24f) yield the approximate two–sided $(1 - \alpha)$ prediction interval for $N_2$. A one–sided (lower or upper) $(1 - \alpha)$ prediction bound can be obtained from the appropriate end point of a two–sided $(1 - 2\,\alpha)$ prediction interval.

## 20.2 BAYESIAN prediction[9]

**One–sample prediction** of $X_{r:n}$ based on $k$ preceding observations $X_{1:n}, \ldots, X_{k:n}$, $k < r$, is considered by EVANS/NIGM (1980b) using the non–informative prior

$$g(b, c) \propto \frac{1}{b\,c}; \quad b > 0, \; c > 0. \tag{20.25a}$$

They derive the predictive survival function of

$$Z_r := X_{r:n} - X_{k:n}$$

as

$$\Pr(Z_r > z \,|\, \text{data}) = \frac{I(z)}{I(0)}, \tag{20.25b}$$

where

$$I(z) = \sum_{i=0}^{r-k-1} \frac{(-1)^i}{n-r+1+1} \binom{r-k-1}{i}$$

$$\times, \int_0^\infty c^{k-2}\, u^{c-1} \left[t^c + (n - r + 1 + i)\, z^c\right]^{-k} \mathrm{d}c, \tag{20.25c}$$

$$I(0) = B(r - k, n - k + 1) \int_0^\infty c^{k-2}\, u^{c-1}\, t^{-kc}\, \mathrm{d}c \tag{20.25d}$$

with

$$u := \prod_{i=1}^{k} X_{i:n}, \quad t^c := \sum_{i=1}^{k} X_{i:n}^c + (n - k)\, X_{k:n}^c.$$

A lower $(1 - \alpha)$ prediction bound for any $X_{r:n}$, $k < r \le n$, is

$$X_L(n, r, 1 - \alpha) = X_{k:n} + d_r, \tag{20.25e}$$

[9] Suggested reading for this section: ASHOUR/RASHWAN (1981), CALABRIA/PULCINI (1994, 1995), DELLAPORTAS/WRIGHT (1991), DIXIT (1994), EVANS/NIGM (1980a,b), LINGAPPAIAH (1977, 1990), NIGM (1989, 1990), TZIAFETAS (1987).

where $I(d_r)/I(0) = 1 - \alpha$. The solution $d_r$ can be derived iteratively. The simplest form occurs when $r = k + 1$. In this case $d_r$ is the solution $d$ of

$$\frac{\int\limits_0^\infty c^{k-2}\, u^c\, [t^c + (n-k)\, d]^{-k}\, dc}{\int\limits_0^\infty c^{k-2}\, u^c\, t^{-k\,c}\, dc} = 1 - \alpha. \qquad (20.25f)$$

EVANS/NIGM (1980b) also give a simplified approximation based on SOLAND (1969a).

NIGM (1989) has discussed the derivation of BAYESIAN prediction bounds for $X_{r:n}$ using an informative prior for the parameters $\lambda = b^{-c}$ and $c$ of the bivariate form

$$g(\lambda, c) = c^{2a}\, \lambda^{a + h/\phi(c)}\, \exp\{-c\, g\}\, \exp\{-d\, \lambda\, \psi(c)\}, \left\{\begin{array}{c} a > -1 \\ g, h, d > 0 \\ \lambda, c > 0 \end{array}\right\}, \qquad (20.26)$$

where $\phi(c)$ and $\psi(c)$ are increasing functions of $c$. Another paper — NIGM (1990) — assumes type–I censoring together with a gamma prior for $\lambda$ when $c$ is known and, when both parameters are unknown, a gamma prior for $\lambda$ with discrete probabilities $P_1, \ldots, P_\ell$ for a restricted set $\{c_1, \ldots, c_\ell\}$ of $c$. LINGAPPAIAH (1990) studies one–order BAYESIAN prediction when an outlier is present.

**Two–sample BAYESIAN prediction**[10] is discussed by EVANS/NIGM (1980a). The papers by NIGM (1989, 1990) also contain results on two–sample prediction. DELLAPORTAS/WRIGHT (1991) have described a numerical approach and the method of evaluating the posterior expectations. LINGAPPAIAH (1990) and DIXIT (1994) assume that the WEIBULL distribution is contaminated by one or more outliers. LINGAPPAIAH (1997) describes how to proceed when the prediction of a future order statistic can be based on more than one preceding samples. BAYESIAN prediction of a single future observation, i.e., $n_2 = 1$, is presented in TZIAFETAS (1987) after assigning WEIBULL and uniform priors for the scale parameter $b$ and shape parameter $c$, respectively.

BAYESIAN prediction of the **future number of failures** is treated in a paper of CALABRIA/PULCINI (1995). Both, the one–sample and two–sample prediction problems are dealt with, and some choices of the prior densities for the WEIBULL parameters are discussed which are relatively easy to work with and allow different degrees of knowledge on the failure mechanism to be incorporated in the prediction procedure. The authors also derive useful relations between the predictive distribution of the number of future failures and the predictive distribution of the future failure times.

---

[10] CALABRIA/PULCINI (1994) give results when both samples come from an inverse WEIBULL distribution and ASHOUR/RASHWAN (1981) when the common distribution is a compound WEIBULL model.

# 21 WEIBULL parameter testing

This chapter deals with statistical testing of hypotheses on parameters, either the parameters $a$, $b$ and $c$ appearing in the WEIBULL distribution function (= **function parameters**) (see Sect. 21.1) or parameters which depend on $a$, $b$ and/or $c$ such as the mean, the variance, a reliability or a percentile (= **functional parameters**) (see Sect. 21.2). In some cases a test of a function parameter is equivalent to a test of a functional parameter, for example,

- when $b$ and $c$ are known a hypothesis on $a$ also is a hypothesis on the mean $\mu = a + b$ $\Gamma(1 + 1/c)$,
- when $c$ is known a hypothesis on $b$ also is a hypothesis on the variance $\sigma^2 = b^2$ $\left[\Gamma(1 + 2/c) - \Gamma^2(1 + 1/c)\right]$,

or a test of a functional parameter is equivalent to a test on a function parameter, e.g., a test on the percentile $x_P = b\left[-\ln(1 - P)\right]^{1/c}$ is also a test on the scale parameter $b$ when $P = 1 - 1/e \approx 0.6321$.

Testing a hypothesis on only one parameter using data of only one sample (= one–sample problem) is intimately related to setting up a confidence interval for that parameter. A $(1 - \alpha)$ confidence interval for a parameter $\theta$ contains all those values of $\theta$ which — when put under the null hypothesis — are not significant, i.e., cannot be rejected, at level $\alpha$. Thus,

- a two–sided $(1 - \alpha)$ level confidence interval for $\theta$, $\widehat{\theta}_\ell \le \theta \le \widehat{\theta}_u$, will not reject $H_0 : \theta = \theta_0$ in favor of $H_A : \theta \ne \theta_0$ for all $\theta_0 \in \left[\widehat{\theta}_\ell, \widehat{\theta}_u\right]$ when $\alpha$ is chosen as level of significance or
- a one–sided $(1 - \alpha)$ confidence interval $\theta \le \widehat{\theta}_u$ $(\theta \ge \widehat{\theta}_\ell)$ will not reject $H_0 : \theta \le \theta_0$ $(H_0 : \theta \ge \theta_0)$ for all $\theta_0 \le \widehat{\theta}_u$ $(\theta_0 \ge \widehat{\theta}_\ell)$ when the level of significance is $\alpha$.

So, when in the preceding chapters we have given confidence intervals, these may be used to test hypotheses in the one–sample case. For most hypotheses it is also possible to find a **likelihood–ratio test**. But this test requires greater sample sizes to hold properly.

## 21.1 Testing hypotheses on function parameters

Perhaps the most important parameter of a WEIBULL distribution is the shape parameter $c$ which is responsible for the behavior of its hazard rate. Thus, we start in Sect. 21.1.1 with hypotheses concerning the shape parameter. The following two sections 21.1.2 and 21.1.3 are devoted to the scale parameter $b$ and the location parameter $a$, respectively. The last section 21.1.4 presents tests for joint hypotheses on any two or all three types of WEIBULL function parameters.

### 21.1.1 Hypotheses concerning the shape parameter $c$

We first show in Sect. 21.1.1.1 how to test hypotheses on one parameter (one–sample problem) followed by testing hypotheses on two or more shape parameters in Sect. 21.1.1.2

### 21.1.1.1 Tests for one parameter[1]

We will first provide **test procedures based on the MLE** $\widehat{c}$ of $c$. The quantity $\widehat{c}/c$ is a pivotal function (see Sect. 11.3.1.2 for complete samples and Sect. 11.6.1.2 for singly censored samples), and its percentage points have been found by Monte Carlo simulation.

Tab. 11/1 gives the percentage points $\ell_1(n, P)$ for **uncensored samples** such that

$$\Pr\left[\frac{\widehat{c}}{c} \leq \ell_1(n, P)\right] = P.$$

Introducing

$$\ell_{n,P} := \ell_1(n, P)$$

for convenience we can state the following:

1. A size $\alpha$ test of

$$H_0: c \leq c_0 \quad \text{against} \quad H_A: c > c_0 \tag{21.1a}$$

is to reject $H_0$ if

$$\widehat{c} > c_0 \, \ell_{n,1-\alpha}. \tag{21.1b}$$

The power of this test, i.e., the probability of rejecting $H_0$ when $c = c^*$, is

$$\Pr\left[\frac{\widehat{c}}{c_0} > \ell_{n,1-\alpha} \,\middle|\, c^*\right] = \Pr\left[\frac{\widehat{c}}{c^*} > \frac{c_0}{c^*} \ell_{n,1-\alpha}\right]$$

$$= 1 - \Pr\left[\frac{\widehat{c}}{c^*} \leq \frac{c_0}{c^*} \ell_{n,1-\alpha}\right]. \tag{21.1c}$$

The probabilities in (21.1c) may be obtained by interpolation for $P$ from Tab. 11/1.

2. A size $\alpha$ test of

$$H_0: c \geq c_0 \quad \text{against} \quad H_A: c < c_0 \tag{21.2a}$$

is to reject $H_0$ if

$$\widehat{c} < c_0 \, \ell_{n,\alpha}. \tag{21.2b}$$

The power of this test is

$$\Pr\left[\frac{\widehat{c}}{c_0} < \ell_{n,\alpha} \,\middle|\, c^*\right] = \Pr\left[\frac{\widehat{c}}{c^*} < \frac{c_0}{c^*} \ell_{n,\alpha}\right]. \tag{21.2c}$$

3. A size $\alpha$ test of

$$H_0: c = c_0 \quad \text{against} \quad H_A: c \neq c_0 \tag{21.3a}$$

is to reject $H_0$ if

$$\widehat{c} < c_0 \, \ell_{n,\alpha/2} \quad \text{or} \quad \widehat{c} > c_0 \, \ell_{n,1-\alpha/2} \tag{21.3b}$$

---

[1] Suggested reading for this section: BAIN (1978), BAIN/ENGELHARDT (1991a), BILL-MAN/ANTLE/BAIN (1972), CHANDRA/CHAUDHURI (1990a,b), CHEN (1997), McCOOL (1970a), THOMAN/BAIN/ANTLE (1969).

with power

$$\Pr\left[\frac{\widehat{c}}{c_0} < \ell_{n,\alpha/2}\Big|c^*\right] + \Pr\left[\frac{\widehat{c}}{c_0} > \ell_{n,1-\alpha/2}\Big|c^*\right]$$

$$= \Pr\left[\frac{\widehat{c}}{c^*} < \frac{c_0}{c^*}\,\ell_{n,\alpha/2}\right] + \Pr\left[\frac{\widehat{c}}{c^*} > \frac{c_0}{c^*}\,\ell_{n,1-\alpha/2}\right]$$

$$= \left\{\begin{array}{l} 1 + \Pr\left[\dfrac{\widehat{c}}{c^*} < \dfrac{c_0}{c^*}\,\ell_{n,\alpha/2}\right] - \\[3mm] \Pr\left[\dfrac{\widehat{c}}{c^*} \le \dfrac{c_0}{c^*}\,\ell_{n,1-\alpha/2}\right] \end{array}\right\}. \tag{21.3c}$$

For $c_0 = 1$ this test also is for the hypothesis that $X$ exponentially distributed.

For large $n$ we may use the normal approximation of $\widehat{c}$:

$$\widehat{c} \overset{\text{asym}}{\sim} No\left(c, \frac{0.6079\,c^2}{n}\right) \quad \text{or} \quad U = \frac{\widehat{c} - c}{c\sqrt{0.6079}}\sqrt{n} \overset{\text{asym}}{\sim} No(0,1). \tag{21.4a}$$

Then, for example, testing $H_0: c \le c_0$ against $H_A: c > c_0$ with size $\alpha$, the approximate critical region for $H_0$ is

$$\widehat{c} > c_0\left(1 + u_{1-\alpha}\sqrt{\frac{0.6079}{n}}\right) \tag{21.4b}$$

with approximate power

$$\Pr\left[\frac{\widehat{c} - c_0}{c_0\sqrt{0.6079}}\sqrt{n} > u_{1-\alpha}\Big|c^*\right] = \Pr\left[U > \frac{c_0}{c^*}u_{1-\alpha} + \frac{c_0 - c^*}{c^*\sqrt{0.6079}}\sqrt{n}\right]. \tag{21.4c}$$

$u_{1-\alpha}$ is the percentile of order $1 - \alpha$ of the standard normal distribution. The reader will easily find the approximating results for the other two tests.

---

**Example 21/1: Testing hypotheses on $c$ using dataset #1**

Dataset #1 as given in Tab. 9/2 is an uncensored sample of size $n = 20$. The MLE of $c$ is $\widehat{c} = 2.5957$. The test statistic using $c_0 = 2$ is

$$\frac{\widehat{c}}{c_0} = 1.2979.$$

We first want to test $H_0: c \le 2$ against $H_A: c > 2$ with $\alpha = 0.05$. From Tab. 11/1 we take the critical value $\ell_{20,0.95} = 1.449$, and as $1.2979 \not> 1.449$, we cannot reject $H_0: c \le 2$. (We would have rejected $H_0: c \le c_0$ for any $c_0 \le 1.79$.) In order to find the power function, we have used the original and more detailed table of $\ell_{n,P}$ in THOMAN et al. (1969) and found the following:

| $\Pr\left[\dfrac{\widehat{c}}{c^*} > \dfrac{c_0}{c^*}\ell_{20,0.95}\right]$ | 0.98 | 0.95 | 0.90 | 0.75 | 0.60 | 0.50 | 0.40 | 0.30 | 0.25 | 0.20 | 0.15 | 0.10 | 0.05 | 0.02 |
|---|---|---|---|---|---|---|---|---|---|---|---|---|---|---|
| $c^*$ | 3.90 | 3.66 | 3.45 | 3.12 | 2.90 | 2.77 | 2.64 | 2.51 | 2.44 | 2.36 | 2.26 | 2.16 | 2.00 | 1.84 |

Applying the normal approximation (21.4b), we find the critical region

$$\hat{c} > 2 \left( 1 + 1.645 \sqrt{\frac{0.6079}{20}} \right) = 2.62,$$

and as $\hat{c} = 2.5957 \not> 2.61$, we cannot reject $H_0: c \leq 2$, too.

Fig. 21/1 shows the exact power function as given in the table above and the approximate power function (21.4c). We see that for $2 < c^* < 3.6$, the approximation leads to an enlarged power or, stated otherwise in this region, the approximated type–II error is too small. We also have included another pair of power functions, assuming that the sample size is $n = 120$. We see that both power functions become steeper (i.e., we have a better discrimination between $H_0$ and $H_A$) and that the approximation is nearly perfect.

We now test whether the sample comes from an exponential distribution; i.e., we test $H_0: c = 1$ against $H_A: c \neq 1$. We take $\alpha = 0.10$ and find the following acceptance region for $H_0$:

$$\ell_{20,0.05} = 0.0791 \leq \frac{\hat{c}}{c_0} \leq \ell_{20,0.95} = 1.449.$$

Because $\hat{c}/c = 2.5951$ is not within this region, we have to reject $H_0$ which is in accordance with the fact that the data have been simulated using a WEIBULL distribution with $c = 2.5$.

Figure 21/1: Exact and approximate power functions of testing $H_0: c \leq 2$ against $H_A: c > 2$ with $\alpha = 0.05$ for an uncensored sample

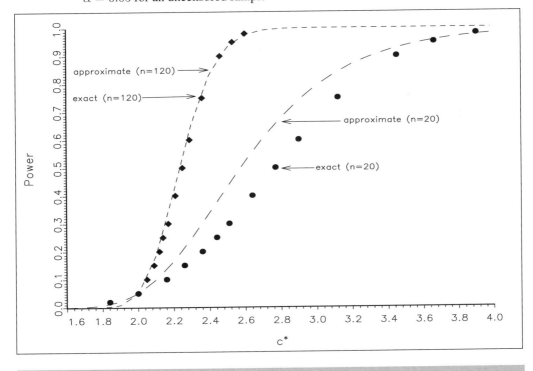

When we have a **singly censored sample**, $r < n$ being the the censoring number,[2] we could work with the percentage points of $\sqrt{n}\left[\widehat{c}/c - E(\widehat{c}/c)\right]$ in Tab. 11/7, but we prefer to use Tab. 21/1 which gives the percentage points $\kappa_P$ such that

$$\Pr\left[\sqrt{n}\left(\frac{\widehat{c}}{c} - 1\right) \leq \kappa_P\right] = P,$$

taken from BAIN/ENGELHARDT (1991a), because this table contains values for sample sizes as small as $n = 5$.

<u>Table 21/1:</u> Percentage points $\kappa_P$ such that $\Pr\left[\sqrt{n}\left(\widehat{c}/c - 1\right) \leq \kappa_P\right] = P$ for censored samples

| $n$ | $r/n$ | $P$ 0.01 | 0.05 | 0.10 | 0.90 | 0.95 | 0.99 |
|---|---|---|---|---|---|---|---|
| 5 | 1.00 | −0.99 | −0.72 | −0.52 | 2.87 | 4.06 | 7.71 |
| | 0.60 | −1.18 | −0.82 | −0.56 | 8.38 | 12.95 | 31.40 |
| 10 | 1.00 | −1.14 | −0.83 | −0.63 | 1.91 | 2.64 | 4.29 |
| | 0.50 | −1.53 | −1.11 | −0.78 | 5.21 | 7.21 | 13.46 |
| 20 | 1.00 | −1.28 | −0.92 | −0.72 | 1.54 | 1.99 | 3.08 |
| | 0.75 | −1.53 | −1.14 | −0.88 | 2.33 | 2.83 | 4.34 |
| | 0.50 | −1.79 | −1.32 | −1.00 | 3.27 | 4.43 | 7.20 |
| 40 | 1.00 | −1.37 | −0.97 | −0.75 | 1.33 | 1.83 | 2.64 |
| | 0.75 | −1.77 | −1.18 | −0.92 | 1.97 | 2.47 | 3.68 |
| | 0.50 | −2.00 | −1.47 | −1.12 | 2.71 | 3.57 | 5.52 |
| 80 | 1.00 | −1.45 | −1.03 | −0.80 | 1.21 | 1.57 | 2.27 |
| | 0.75 | −1.78 | −1.27 | −0.97 | 1.73 | 2.23 | 3.30 |
| | 0.50 | −2.18 | −1.56 | −1.19 | 2.41 | 3.16 | 4.64 |
| 120 | 1.00 | −1.51 | −1.13 | −0.84 | 1.12 | 1.46 | 2.19 |
| | 0.75 | −1.89 | −1.32 | −0.99 | 1.61 | 2.13 | 3.04 |
| | 0.50 | −2.34 | −1.66 | −1.28 | 2.16 | 2.78 | 4.08 |
| $\infty$ | 1.00 | −1.81 | −1.28 | −0.99 | 0.99 | 1.28 | 1.81 |
| | 0.75 | −2.35 | −1.66 | −1.29 | 1.29 | 1.66 | 2.35 |
| | 0.50 | −3.05 | −2.15 | −1.68 | 1.68 | 2.15 | 3.05 |

Introducing

$$T_n := \sqrt{n}\left(\frac{\widehat{c}}{c} - 1\right), \tag{21.5}$$

---

[2] Strictly speaking the following procedure only holds for type-II censoring, but it also is all right as long as one interprets the type–I results as conditional, the condition being the number of realized failures in $(0, T]$.

where $\widehat{c}$ is the MLE of $c$, we can state the following results concerning tests of hypotheses on $c$ with size $\alpha$:

1. $H_0: c \leq c_0$ against $H_A: c > c_0$

   Critical region for $H_0$:

   $$\widehat{c} > c_0 \left( \frac{\kappa_{1-\alpha}}{\sqrt{n}} + 1 \right) \quad \text{or} \quad \sqrt{n} \left( \frac{\widehat{c}}{c_0} - 1 \right) > \kappa_{1-\alpha} \tag{21.6a}$$

   Power function for an alternative value $c^*$ of $c$:

   $$\Pr\left[ \sqrt{n} \left( \frac{\widehat{c}}{c_0} - 1 \right) > \kappa_{1-\alpha} \,\Big|\, c^* \right] = \Pr\left[ T_n > \frac{c_0}{c^*} \kappa_{1-\alpha} + \sqrt{n} \left( \frac{c_0}{c^*} - 1 \right) \right] \tag{21.6b}$$

   These probabilities may be obtained from Tab. 21/1 by interpolation for $P$.

2. $H_0: c \geq c_0$ against $H_A: c < c_0$

   Critical region for $H_0$:

   $$\widehat{c} < c_0 \left( \frac{\kappa_{\alpha}}{\sqrt{n}} + 1 \right) \quad \text{or} \quad \sqrt{n} \left( \frac{\widehat{c}}{c_0} - 1 \right) < \kappa_{\alpha} \tag{21.7a}$$

   Power function:

   $$\Pr\left[ \sqrt{n} \left( \frac{\widehat{c}}{c_0} - 1 \right) < \kappa_{\alpha} \,\Big|\, c^* \right] = \Pr\left[ T_n < \frac{c_0}{c^*} \kappa_{\alpha} + \sqrt{n} \left( \frac{c_0}{c^*} - 1 \right) \right] \tag{21.7b}$$

3. $H_0: c = c_0$ against $H_A: c \neq c_0$

   Critical region for $H_0$:

   $$\left. \begin{array}{c} \widehat{c} < c_0 \left( \dfrac{\kappa_{\alpha/2}}{\sqrt{n}} + 1 \right) \quad \cup \quad \widehat{c} > c_0 \left( \dfrac{\kappa_{1-\alpha/2}}{\sqrt{n}} + 1 \right) \\[2ex] \text{or} \\[2ex] \sqrt{n} \left( \dfrac{\widehat{c}}{c_0} - 1 \right) < \kappa_{\alpha/2} \quad \cup \quad \sqrt{n} \left( \dfrac{\widehat{c}}{c_0} - 1 \right) > \kappa_{1-\alpha/2} \end{array} \right\} \tag{21.8a}$$

   Power function:

   $$\left. \begin{array}{c} \Pr\left[ \sqrt{n} \left( \dfrac{\widehat{c}}{c_0} - 1 \right) < \kappa_{\alpha/2} \cup \sqrt{n} \left( \dfrac{\widehat{c}}{c_0} - 1 \right) > \kappa_{1-\alpha/2} \,\Big|\, c^* \right] = \\[2ex] \Pr\left[ T_n < \dfrac{c_0}{c^*} \kappa_{\alpha/2} + \sqrt{n} \left( \dfrac{c_0}{c^*} - 1 \right) \right] + \Pr\left[ T_n > \dfrac{c_0}{c^*} \kappa_{1-\alpha/2} + \sqrt{n} \left( \dfrac{c_0}{c^*} - 1 \right) \right] \end{array} \right\} \tag{21.8b}$$

Asymptotic results suggest that for heavy censoring, approximately

$$\frac{2\,r\,c}{\widehat{c}} \sim \chi^2(2\,(r-1)), \tag{21.9a}$$

while for complete samples the degrees of freedom in (21.9a) should be $\nu = 2\,(n-1)$. BAIN/ENGELHARDT (1991a, p. 223) suggest the following approximation

$$d\,r \left( \frac{\widehat{c}}{c} \right)^{1+p^2} \sim \chi^2(d\,(r+1)), \tag{21.9b}$$

where

$$p = \frac{r}{n} \quad \text{and} \quad d = \frac{2}{(1+p^2)^2 \, p \, c_{22}},$$

with $c_{22}$ given in Tab. 19/3. To demonstrate the working of this approximation, we take the test of $H_0$ : $c \leq c_0$ against $H_A$ : $c > c_0$. The critical region of a size $\alpha$ test results as

$$d \, r \left( \frac{c_0}{\widehat{c}} \right)^{1+p^2} < \chi^2_{d\,(r-1),\alpha} \tag{21.10a}$$

and the corresponding power function is

$$\Pr \left[ d \, r \left( \frac{c_0}{\widehat{c}} \right)^{1+p^2} < \chi^2_{d\,(r-1),\alpha} \, \middle| \, c^* \right] = \Pr \left[ d \, r \left( \frac{c^*}{\widehat{c}} \right)^{1+p^2} < \left( \frac{c^*}{c_0} \right)^{1+p^2} \chi^2_{d\,(r-1),\alpha} \right]$$

$$= \Pr \left[ \chi^2_{d\,(r-1)} < \left( \frac{c^*}{c_0} \right)^{1+p^2} \chi^2_{d\,(r-1),\alpha} \right]. \tag{21.10b}$$

---

**Example 21/2:** **Testing $H_0$ : $c \leq 2$ against $H_A$ : $c > 2$ with $\alpha = 0.05$ using dataset #1 censored at $r = 15$**

When dataset #1 in Tab. 9/2 is censored at $r = 15$, we find the MLE of $c$ to be $\widehat{c} = 2.7114$ (see Example 11/8). From Tab. 21/1 (with $n = 20$, $r/n = 0.75$ and $P = 1 - \alpha = 0.95$), we take $\kappa_{0.95} = 2.83$,. As

$$\sqrt{n} \left( \frac{\widehat{c}}{c_0} - 1 \right) = 1.5907 \not> \kappa_{0.95} = 2.83,$$

we cannot reject $H_0$. We further find from Tab. 21/1 that we have a power of 0.90 at $c^* = 4.96$.

Working with the approximating formulas (21.10a,b), we first interpolate for $c_{22}$ in Tab. 19/3: $c_{22} \approx 1.025$. With $p = 0.75$ we find $d = 1.066$. The test statistic is

$$d \, r \left( \frac{c_0}{\widehat{c}} \right)^{1+p^2} = 1.066 \cdot 15 \cdot \left( \frac{2}{2.7114} \right)^{1.5625} = 9.939$$

and the $\chi^2$-percentile is

$$\chi^2_{1.066 \cdot 14, 0.05} = \chi^2_{14.924, 0.05} = 7.2081.$$

Because $d \, r (c_0/\widehat{c})^{1+p^2} = 9.939 \not< \chi^2_{14.924, 0.05} = 7.2981$, we cannot reject $H_0$ as above. The approximate power at $c^* = 4.96$ according to (21.10b) is

$$\Pr \left[ \chi^2_{14.924} < \left( \frac{4.96}{2} \right)^{1.5625} \cdot 7.2081 \right] = \Pr[\chi^2_{14.924} < 29.80] = 0.9877,$$

which is somewhat higher than 0.90 from the exact procedure. However, it should be remembered that $n = 20$ is not great enough for the approximation to hold properly.

Testing for $c$ may also be based on any other type of estimator. For example, the GLUE of Sect. 10.4.3 has been applied by BAIN (1978, p. 275) and by CHANDRA/CHAUDHURI (1990a,b). There are even tests that do not require an estimate of $c$:

- WONG/WONG (1982) took the extremal quotient $X_{r:n}/X_{1:n}$ as test statistic.

- CHEN (1997) used the following ratio of the arithmetic mean to the geometric mean as a test statistic

$$Q = \frac{\frac{1}{n}\sum_{i=1}^{r-1} X_{i:n}^{c_0} + (n-r+1)\, X_{r:n}^{c_0}}{\left[\prod_{i=1}^{r-1} X_{i:n}\, X_{r:n}^{n-r+1}\right]^{c_0/n}}. \tag{21.11}$$

### 21.1.1.2   Tests for $k \geq 2$ parameters[3]

We start with the **case $k = 2$ using MLEs**. For convenience of reference the two populations about which statements are to be made will be called populations 1 and 2. The parameters, sample sizes, numbers of failures and MLEs are subscripted with the population number from which they derive. MCCOOL (1970a) has shown that the following random variable follows a distribution which depends upon the sample sizes and the numbers of failures in the two samples but not upon the population parameters:

$$W\left(n_1, n_2, r_1, r_2\right) = \frac{\widehat{c}_1/c_1}{\widehat{c}_2/c_2}. \tag{21.12a}$$

If the subscripts 1 and 2 are interchanged, $W \Rightarrow 1/W$ and $1/W$ has the same distribution as $W$. This implies for the percentile of order $P$:

$$w_P = \frac{1}{w_{1-P}} \ \forall\, P \tag{21.12b}$$

and that $w_{0.5} = 1$. Using the percentiles $w_P(n_1, n_2, r_1, r_2)$ of the relevant $W$–distribution hypotheses on the ratio of the two population parameters may be tested with size $\alpha$:

1. $H_0: c_1 = c_2$ against $H_A: c_1 > c_2$

   Reject $H_0$ when
   $$\widehat{c}_1/\widehat{c}_2 > w_{1-\alpha}(n_1, n_2, r_1, r_2), \tag{21.13a}$$

2. $H_0: c_1 = c_2$ against $H_A: c_1 < c_2$

   Reject $H_0$ when
   $$\widehat{c}_1/\widehat{c}_2 < w_{\alpha}(n_1, n_2, r_1, r_2), \tag{21.13b}$$

3. $H_0: c_1 = c_2$ against $H_A: c_1 \neq c_2$

   Accept $H_0$ when
   $$w_{\alpha/2}(n_1, n_2, r_1, r_2) < \widehat{c}_1/\widehat{c}_2 < w_{1-\alpha/2}(n_1, n_2, r_1, r_2). \tag{21.13c}$$

---

[3] Suggested reading for this section: BAIN (1978, pp. 286 ff.), BAIN/ENGELHARDT (1991a) BILIKAM/ MOORE/PETRICK (1979), ENGELHARDT/BAIN (1979), MCCOOL (1970a), THOMAN/BAIN (1969).

The percentiles $w_P(n_1, n_2, r_2, r_2)$ have to be found by Monte Carlo techniques. For the case

$$n_1 = n_2 = n \quad \text{and} \quad r_1 = r_2 = n,$$

i.e., for uncensored samples of equal size, a table of

$$w_P^* := w_P(n, n, n, n)$$

is provided by THOMAN/BAIN (1969) which is reproduced here as Tab. 21/2. Because this table gives percentiles only of order $P > 0.5$, points $w_P^*$ for $P \leq 0.5$ can be found by using the fact that $w_P^* = 1/w_{1-P}^*$; see (21.12b).

The power of a size $\alpha$ test of $H_0: c_1 = c_2$ against $H_A: c_1 = k c_2$, $k > 1$, having critical region $\widehat{c}_1/\widehat{c}_2 > w_{1-\alpha}^*$, is

$$\Pr\left[\frac{\widehat{c}_1}{\widehat{c}_2} > w_{1-\alpha}^* \middle| c_1 = k c_2\right] = \Pr\left[\frac{\widehat{c}_1/c_1}{\widehat{c}_2/c_2} > \frac{1}{k} w_{1-\alpha}^*\right], \tag{21.14}$$

which can be obtained from Tab. 21/2 by interpolating for $P$. THOMAN/BAIN (1969) give power curves for certain values of $n$ and for $\alpha = 0.05$ and $0.10$. The procedure can, of course, be generalized to a test of

$$H_0: c_1 = k c_2 \quad \text{against} \quad H_A: c_1 = k' c_2. \tag{21.15a}$$

For the case when $k < k'$ the rejection interval becomes

$$\widehat{c}_1/\widehat{c}_2 > k w_{1-\alpha}^* \tag{21.15b}$$

with power function

$$\Pr\left[\frac{\widehat{c}_1}{\widehat{c}_2} > k w_{1-\alpha}^* \middle| c_1 = k' c_2\right] = \Pr\left[\frac{\widehat{c}_1/c_1}{\widehat{c}_2/c_2} > \frac{k}{k'} w_{1-\alpha}^*\right]. \tag{21.15c}$$

The **case $k > 2$ using MLEs** is discussed by BILIKAM et al. (1979) applying a maximum likelihood ratio test. Unfortunately, the critical values of a size $\alpha$ test for $H_0: c_1 = c_2 = \ldots = c_k$ against $H_A: c_i \neq c_j$ for at least one pair $(i \neq j)$ have to be found by Monte Carlo techniques depending on the individual sample sizes $n_1, \ldots, n_k$ and the individual failure numbers $r_1, \ldots, r_k$ (type–II censoring). BILIKAM et al. give critical points for $k = 2$ and certain combinations $(n_1, n_2)$, both samples being uncensored.

Tests using the GLUEs of BAIN/ENGELHARDT (see Sect. 10.4.3) are presented in

- ENGELHARDT/BAIN (1979) for the case $k = 2$ and

- BAIN/ENGELHARDT (1991a) for the case $k \geq 2$.

These tests require a lot of tables for calculating both the test statistics and the critical values.

Table 21/2: Percentage points $w_P^*$ such that $\Pr\left[\left(\widehat{c}_1/c_1\right)\Big/\left(\widehat{c}_2/c_2\right) \leq w_P^*\right] = P$

| $P$ / $n$ | 0.60 | 0.70 | 0.75 | 0.80 | 0.85 | 0.90 | 0.95 | 0.98 |
|---|---|---|---|---|---|---|---|---|
| 5 | 1.158 | 1.351 | 1.478 | 1.636 | 1.848 | 2.152 | 2.725 | 3.550 |
| 6 | 1.135 | 1.318 | 1.418 | 1.573 | 1.727 | 1.987 | 2.465 | 3.146 |
| 7 | 1.127 | 1.283 | 1.370 | 1.502 | 1.638 | 1.869 | 2.246 | 2.755 |
| 8 | 1.119 | 1.256 | 1.338 | 1.450 | 1.573 | 1.780 | 2.093 | 2.509 |
| 9 | 1.111 | 1.236 | 1.311 | 1.410 | 1.524 | 1.711 | 1.982 | 2.339 |
| 10 | 1.104 | 1.220 | 1.290 | 1.380 | 1.486 | 1.655 | 1.897 | 2.213 |
| 11 | 1.098 | 1.206 | 1.273 | 1.355 | 1.454 | 1.609 | 1.829 | 2.115 |
| 12 | 1.093 | 1.195 | 1.258 | 1.334 | 1.428 | 1.571 | 1.774 | 2.036 |
| 13 | 1.088 | 1.186 | 1.245 | 1.317 | 1.406 | 1.538 | 1.727 | 1.922 |
| 14 | 1.084 | 1.177 | 1.233 | 1.301 | 1.386 | 1.509 | 1.688 | 1.917 |
| 15 | 1.081 | 1.170 | 1.224 | 1.288 | 1.369 | 1.485 | 1.654 | 1.870 |
| 16 | 1.077 | 1.164 | 1.215 | 1.277 | 1.355 | 1.463 | 1.624 | 1.829 |
| 17 | 1.075 | 1.158 | 1.207 | 1.266 | 1.341 | 1.444 | 1.598 | 1.793 |
| 18 | 1.072 | 1.153 | 1.200 | 1.257 | 1.329 | 1.426 | 1.574 | 1.762 |
| 19 | 1.070 | 1.148 | 1.194 | 1.249 | 1.318 | 1.411 | 1.553 | 1.733 |
| 20 | 1.068 | 1.144 | 1.188 | 1.241 | 1.308 | 1.396 | 1.534 | 1.708 |
| 22 | 1.064 | 1.136 | 1.178 | 1.227 | 1.291 | 1.372 | 1.501 | 1.663 |
| 24 | 1.061 | 1.129 | 1.169 | 1.216 | 1.276 | 1.351 | 1.473 | 1.625 |
| 26 | 1.058 | 1.124 | 1.162 | 1.206 | 1.263 | 1.333 | 1.449 | 1.593 |
| 28 | 1.055 | 1.119 | 1.155 | 1.197 | 1.252 | 1.318 | 1.428 | 1.566 |
| 30 | 1.053 | 1.114 | 1.149 | 1.190 | 1.242 | 1.304 | 1.409 | 1.541 |
| 32 | 1.051 | 1.110 | 1.144 | 1.183 | 1.233 | 1.292 | 1.393 | 1.520 |
| 34 | 1.049 | 1.107 | 1.139 | 1.176 | 1.224 | 1.281 | 1.378 | 1.500 |
| 36 | 1.047 | 1.103 | 1.135 | 1.171 | 1.217 | 1.272 | 1.365 | 1.483 |
| 38 | 1.046 | 1.100 | 1.131 | 1.166 | 1.210 | 1.263 | 1.353 | 1.467 |
| 40 | 1.045 | 1.098 | 1.127 | 1.161 | 1.204 | 1.255 | 1.342 | 1.453 |
| 42 | 1.043 | 1.095 | 1.124 | 1.156 | 1.198 | 1.248 | 1.332 | 1.439 |
| 44 | 1.042 | 1.093 | 1.121 | 1.152 | 1.193 | 1.241 | 1.323 | 1.427 |
| 46 | 1.041 | 1.091 | 1.118 | 1.149 | 1.188 | 1.235 | 1.314 | 1.416 |
| 48 | 1.040 | 1.088 | 1.115 | 1.145 | 1.184 | 1.229 | 1.306 | 1.405 |
| 50 | 1.039 | 1.087 | 1.113 | 1.142 | 1.179 | 1.224 | 1.299 | 1.396 |
| 52 | 1.038 | 1.085 | 1.111 | 1.139 | 1.175 | 1.219 | 1.292 | 1.387 |
| 54 | 1.037 | 1.083 | 1.108 | 1.136 | 1.172 | 1.215 | 1.285 | 1.378 |
| 56 | 1.036 | 1.081 | 1.106 | 1.133 | 1.168 | 1.210 | 1.279 | 1.370 |
| 58 | 1.036 | 1.080 | 1.104 | 1.131 | 1.165 | 1.206 | 1.274 | 1.363 |
| 60 | 1.035 | 1.078 | 1.102 | 1.128 | 1.162 | 1.203 | 1.268 | 1.355 |
| 62 | 1.034 | 1.077 | 1.101 | 1.126 | 1.159 | 1.199 | 1.263 | 1.349 |
| 64 | 1.034 | 1.076 | 1.099 | 1.124 | 1.156 | 1.196 | 1.258 | 1.342 |
| 66 | 1.033 | 1.075 | 1.097 | 1.122 | 1.153 | 1.192 | 1.253 | 1.336 |
| 68 | 1.032 | 1.073 | 1.096 | 1.120 | 1.151 | 1.189 | 1.249 | 1.331 |
| 70 | 1.032 | 1.072 | 1.094 | 1.118 | 1.148 | 1.186 | 1.245 | 1.325 |
| 72 | 1.031 | 1.071 | 1.093 | 1.116 | 1.146 | 1.184 | 1.241 | 1.320 |
| 74 | 1.031 | 1.070 | 1.091 | 1.114 | 1.143 | 1.181 | 1.237 | 1.315 |
| 76 | 1.030 | 1.069 | 1.090 | 1.112 | 1.141 | 1.179 | 1.233 | 1.310 |
| 78 | 1.030 | 1.068 | 1.089 | 1.111 | 1.139 | 1.176 | 1.230 | 1.306 |
| 80 | 1.030 | 1.067 | 1.088 | 1.109 | 1.137 | 1.174 | 1.227 | 1.301 |
| 90 | 1.028 | 1.063 | 1.082 | 1.102 | 1.128 | 1.164 | 1.212 | 1.282 |
| 100 | 1.026 | 1.060 | 1.078 | 1.097 | 1.121 | 1.155 | 1.199 | 1.266 |
| 120 | 1.023 | 1.054 | 1.071 | 1.087 | 1.109 | 1.142 | 1.180 | 1.240 |

## 21.1.2 Hypotheses concerning the scale parameter $b$[4]

We start with the **one–sample test** of a hypothesis on $b$ **using the MLE** $\widehat{b}$ of $b$. It has been shown in Chapter 11 that $\widehat{c}\ln\left(\widehat{c}/b\right)$ is a pivotal quantity whose distribution solely depends on $n$ when the sample is uncensored or on $n$ and $r$ when the sample is singly type–II censored on the right.[5] Percentage points $u_P$ of $U = \widehat{c}\ln\left(\widehat{b}/b\right)$ are to be found in Tab. 11/2, called $\ell_2(n, P)$, for complete samples. For censored sample with censoring fraction $r/n = 1$, 0.75 and 0.5, we find percentage points $w_P$ of $W = \sqrt{n}\,\widehat{c}\ln\left(\widehat{b}/b\right)$ in Tab. 11/8. Here we give another Tab. 21/3 of $W = \sqrt{n}\,\widehat{c}\ln\left(\widehat{b}/b\right)$, which contains percentage points for $n$ as small as 5.

Table 21/3: Percentage points of $w_P$ such that $\Pr\left[\sqrt{n}\,\widehat{c}\ln\left(\widehat{b}/b\right) \leq w_P\right] = P$

| $n$ | $r/n$ | $P$ 0.01 | 0.05 | 0.10 | 0.90 | 0.95 | 0.99 |
|---|---|---|---|---|---|---|---|
| 5 | 1.00 | −5.17 | −2.85 | −2.00 | 1.72 | 2.47 | 4.29 |
|  | 0.60 | −29.71 | −11.82 | −7.54 | 1.44 | 2.26 | 5.12 |
| 10 | 1.00 | −3.18 | −2.08 | −1.60 | 1.51 | 2.01 | 3.16 |
|  | 0.50 | −14.76 | −7.98 | −5.49 | 1.40 | 1.90 | 3.02 |
| 20 | 1.00 | −2.77 | −1.94 | −1.48 | 1.42 | 1.87 | 2.77 |
|  | 0.75 | −4.02 | −2.59 | −1.99 | 1.39 | 1.83 | 2.75 |
|  | 0.50 | −7.85 | −4.98 | −3.63 | 1.49 | 1.97 | 2.84 |
| 40 | 1.00 | −2.58 | −1.82 | −1.41 | 1.39 | 1.80 | 2.62 |
|  | 0.75 | −3.29 | −2.25 | −1.69 | 1.39 | 1.85 | 2.61 |
|  | 0.50 | −6.21 | −3.77 | −2.91 | 1.63 | 2.16 | 2.96 |
| 80 | 1.00 | −2.51 | −1.76 | −1.37 | 1.37 | 1.76 | 2.49 |
|  | 0.75 | −3.11 | −2.10 | −1.61 | 1.43 | 1.85 | 2.65 |
|  | 0.50 | −5.14 | −3.45 | −2.62 | 1.71 | 2.16 | 3.08 |
| 120 | 1.00 | −2.44 | −1.73 | −1.35 | 1.35 | 1.74 | 2.48 |
|  | 0.75 | −3.01 | −2.01 | −1.58 | 1.45 | 1.86 | 2.63 |
|  | 0.50 | −4.50 | −3.17 | −2.44 | 1.75 | 2.27 | 3.13 |
| ∞ | 1.00 | −2.45 | −1.73 | −1.35 | 1.35 | 1.73 | 2.45 |
|  | 0.75 | −2.69 | −1.90 | −1.48 | 1.48 | 1.90 | 2.69 |
|  | 0.50 | −3.69 | −2.61 | −2.03 | 2.03 | 2.61 | 3.69 |

Source: BAIN/ENGELHARDT (1991a, p. 231) — Reprinted with permission from *Statistical Analysis of Reliability and Life–Testing Models*. Copyright 1991 by Marcel Dekker, Inc. All rights reserved.

Assuming an uncensored sample[6] we can state the following:[7]

---

[4] Suggested reading for this section: BAIN (1978), BILLMAN/ANTLE/BAIN (1972), CHAUDHURI/CHANDRA (1989), ENGELHARDT/BAIN (1979), McCOOL (1970a, 1977), PAUL/THIAGARAJAH (1992), SCHAFER/ SHEFFIELD (1976), THOMAN/BAIN (1969), THOMAN/BAIN/ANTLE (1969).

[5] The results are also valid for type–I censoring when interpreted as conditional on the number of realized failures in $(0, T]$.

[6] For a right–censored sample we have to use $w_P$ instead of $u_P$, either from Tab. 11/8 or from Tab. 21/3.

[7] Notice that we also have to calculate the MLE $\widehat{c}$ of $c$.

1. A size $\alpha$ test of

$$H_0: b \le b_0 \text{ against } H_A: b > b_0 \qquad (21.16a)$$

is to reject $H_0$ if

$$\widehat{b} > b_0 \exp\left(\frac{u_{1-\alpha}}{\widehat{c}}\right). \qquad (21.16b)$$

The power of this test at $b = b^*$ is

$$\Pr\left[\widehat{b} > b_0 \exp\left(\frac{u_{1-\alpha}}{\widehat{c}}\right) \Big| b^*\right] = \Pr\left[U > u_{1-\alpha} + \widehat{c} \ln\left(\frac{b_0}{b^*}\right)\right]. \qquad (21.16c)$$

The probabilities in (21.16c) may be obtained by interpolating for $P$ in Tab. 11/2.

2. A size $\alpha$ test of

$$H_0: b \ge b_0 \text{ against } H_A: b < b_0 \qquad (21.17a)$$

is to reject $H_0$ if

$$\widehat{b} < b_0 \exp\left(\frac{u_\alpha}{\widehat{c}}\right) \qquad (21.17b)$$

with power

$$\Pr\left[\widehat{b} < b_0 \exp\left(\frac{u_\alpha}{\widehat{c}}\right) \Big| b^*\right] = \Pr\left[U < u_\alpha + \widehat{c} \ln\left(\frac{b_0}{b^*}\right)\right]. \qquad (21.17c)$$

3. A size $\alpha$ test of

$$H_0: b = b_0 \text{ against } H_A: b \ne b_0 \qquad (21.18a)$$

is to reject $H_0$ if

$$\widehat{b} < b_0 \exp\left(\frac{u_{\alpha/2}}{\widehat{c}}\right) \text{ or } \widehat{b} > b_0 \exp\left(\frac{u_{1-\alpha/2}}{\widehat{c}}\right) \qquad (21.18b)$$

with power

$$\left. \begin{array}{l} \Pr\left[\widehat{b} < b_0 \exp\left(\frac{u_{\alpha/2}}{\widehat{c}}\right) \Big| b^*\right] + \Pr\left[\widehat{b} > b_0 \exp\left(\frac{u_{1-\alpha/2}}{\widehat{c}}\right) \Big| b^*\right] = \\[2mm] \Pr\left[U < u_{\alpha/2} + \widehat{c} \ln\left(\frac{b_0}{b^*}\right)\right] + \Pr\left[U > u_{1-\alpha/2} + \widehat{c} \ln\left(\frac{b_0}{b^*}\right)\right]. \end{array} \right\} \qquad (21.18c)$$

For $n \to \infty$ and $\lim_{n \to \infty} r/n = p$, we may use the following **normal approximation**; see ENGELHARDT/BAIN (1991a, p. 220):

$$T := \widehat{c} \sqrt{\frac{n}{c_{11}}} \ln\left(\frac{\widehat{b}}{b}\right) \overset{\text{asym}}{\sim} No(0, 1), \qquad (21.19a)$$

where $c_{11}$ depends on $p$ and is given in Tab. 19/3. Taking the size $\alpha$ test of $H_0:\ b \le b_0$ against $H_A:\ b > b_0$ as an example, the critical region is $\tau_{1-\alpha}$, being the $(1-\alpha)$ percentile of the standard normal distribution:

$$\ln \widehat{b} > \ln b_0 + \frac{\tau_{1-\alpha}}{\widehat{c}} \sqrt{\frac{c_{11}}{n}} \tag{21.19b}$$

with power

$$\Pr\left[\widehat{c}\sqrt{\frac{n}{c_{11}}} \ln\left(\frac{\widehat{b}}{b_0}\right) > \tau_{1-\alpha}\Big|b^*\right] = \Pr\left[T > \tau_{1-\alpha} + \widehat{c}\sqrt{\frac{n}{c_{11}}} \ln\left(\frac{b_0}{b^*}\right)\right]. \tag{21.19c}$$

Approximating results for the other two types of tests follow easily.

---

**Example 21/3:   Testing a hypothesis on $b$ using dataset #1**

Dataset #1 in Tab. 9/2 is an uncensored sample of size $n = 20$ from a two–parameter WEIBULL distribution with $b = 100$ and $c = 2.5$. The MLEs are $\widehat{b} = 99.2079$ and $\widehat{c} = 2.5957$. We want to test $H_0:\ b \le 80$ against $H_A:\ b > 80$ with $\alpha = 0.05$. The critical region according to (21.16b) is

$$\widehat{b} > b_0 \exp\left(\frac{u_{0.95}}{\widehat{c}}\right) \quad \Longrightarrow \quad \widehat{b} > 80 \exp\left(\frac{0.421}{2.5957}\right) = 94.09.$$

As $\widehat{b} = 99.2079 > 94.09$, we have to reject $H_0$.

Applying the normal approximation to this test, we have — according to (21.19b) — the critical region

$$\ln \widehat{b} > \ln b_0 + \frac{\tau_{1-\alpha}}{\widehat{c}} \sqrt{\frac{c_{11}}{n}} \quad \Longrightarrow \quad \ln \widehat{b} > \ln 80 + \frac{1.645}{2.5957} \sqrt{\frac{1.1087}{20}} \approx 4.53.$$

As $\ln \widehat{b} \approx 4.60 > 4.53$ the approximate procedure also leads to a rejection of $H_0$.

The power function of the exact test (21.16c) as well as that of the approximate test (21.19c) are depicted in Fig. 21/2. The approximation overestimates the true power. This figure also shows another pair of power functions that result if we assume a sample of size $n = 120$, the MLEs $\widehat{b}$ and $\widehat{c}$ being the same as above. The power increases and the approximating is nearly perfect.

---

**Two–sample tests** of $b_1 = b_2$ from two WEIBULL distributions based on MLEs have been proposed by THOMAN/BAIN (1969) and SCHAFER/SHEFFIELD (1976). Both approaches require the equality of the two shape parameters, i.e., $c_1 = c_2$, so that before applying the following results one should execute a pretest of $H_0:\ c_1 = c_2$ against $H_A:\ c_1 \ne c_2$. The test of $H_0:\ b_1 = b_2,\ c_1 = c_2$ versus $H_A:\ b_1 = k\,b_2,\ c_1 = c_2,\ k > 1$ is equivalent to the more general problem of testing $H_0:\ b_1 = k'\,b_2,\ c = c_2$ versus $H_A:\ b_1 = k''\,b_2,\ c_1 = c_2$. For example, under the null hypothesis the random variable $W = k'\,X_2$ has a WEIBULL distribution with shape parameter $c_W = c_2$ and scale parameter $b_W = k'\,b_2$. Thus, the hypothesis $b_1 = k'\,b_2,\ c_1 = c_2$ is equivalent to $b_1 = b_W,\ c_1 = c_W$. The data $X_{2i}$ must be multiplied by $k'$ and the test carried out as described below.

Figure 21/2: Exact and approximate power functions of testing $H_0 : b \le 80$ against $H_A :$ $b > 80$ with $\alpha = 0.05$ (uncensored sample)

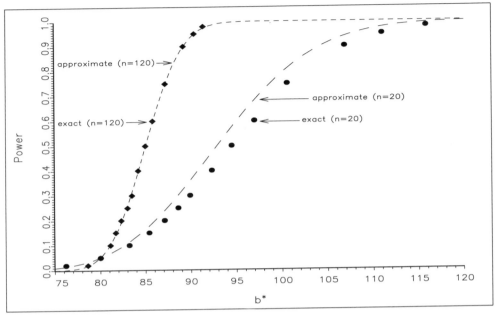

The tests presented in THOMAN/BAIN (1969) and SCHAFER/SHEFFIELD (1976) require equal sample sizes $n_1 = n_2 = n$ and no censoring.[8] In Chapter 11 it has been shown that $c \ln \left( \widehat{b}/b \right)$ has the same distribution as $\ln b^*$ and that $\widehat{c} \ln \left( \widehat{b}/b \right)$ has the same distribution as $\widehat{c}^* \ln b^*$, $\widehat{c}^*$ and $\widehat{b}^*$ being the MLEs when sampling from the special WEIBULL distribution having shape and scale parameters both equal to unity, i.e., from the reduced exponential distribution. For the case of two independent samples, it follows similarly that

$$c \left[ \ln \left( \widehat{b}_1/b_1 \right) - \ln \left( \widehat{b}_2/b_2 \right) - (1/c) \ln M \right]$$

has the same distribution as

$$\ln \widehat{b}_1^* - \ln \widehat{b}_2^* - \ln M$$

if $c_1 = c_2 = c$, and also that

$$Z(M) = \left[ (\widehat{c}_1 + \widehat{c}_2)/2 \right] \left[ \ln \left( \widehat{b}_1/b_1 \right) - \ln \left( \widehat{b}_2/b_2 \right) - (1/c) \ln M \right] \qquad (21.20a)$$

has the same distribution as

$$Z^*(M) = \left[ (\widehat{c}_1^* + \widehat{c}_2^*)/2 \right] \left[ \ln \widehat{b}_1^* - \ln \widehat{b}_2^* - \ln M \right] \qquad (21.20b)$$

if $c_1 = c_2 = c$. Let $G_M(\cdot)$ be the common CDF of $Z(M)$ and $Z^*(M)$. We will denote $Z(M)$ by $Z$ when $M = 1$.

---

[8] It is no restriction that the two samples are uncensored and assumed equal so far as methodology is concerned, but up to now there exist only tabulated percentage points of the test statistic for this special case.

A test of $H_0 : b_1 = b_2,\ c_1 = c_2$ against $H_A : b_1 = k b_2,\ c_1 = c_2,\ k > 1$ can now be carried out by using the fact that

$$\Pr\left[\frac{\widehat{c}_1 + \widehat{c}_2}{2}\left(\ln \widehat{b}_1 - \ln \widehat{b}_2\right) < z \big| H_0\right] = G_1(z). \tag{21.21a}$$

$H_0$ is rejected with size $\alpha$ when

$$\frac{\widehat{c}_1 + \widehat{c}_2}{2}\left(\ln \widehat{b}_1 - \ln \widehat{b}_2\right) > z_{1-\alpha}, \tag{21.21b}$$

where $z_{1-\alpha}$ is such that $G_1(z_{1-\alpha}) = 1 - \alpha$. Percentage points $z_{1-\alpha}$ have been found using Monte Carlo techniques by THOMAN/BAIN (1969) and are reproduced in Tab. 21/4. The power of this test can also be expressed in terms of $G_M$:

$$\Pr\left[\frac{\widehat{c}_1 + \widehat{c}_2}{2}\left(\ln \widehat{b}_1 - \ln \widehat{b}_2\right) > z_{1-\alpha} \big| H_A\right] = 1 - G_{k^c}(z_{1-\alpha}). \tag{21.21c}$$

The probability on the left–hand side of (21.21c) clearly is

$$\Pr\left[\frac{\widehat{c}_1 + \widehat{c}_2}{2}\left\{\ln\left(\frac{\widehat{b}_1}{b_1}\right) - \ln\left(\frac{\widehat{b}_2}{b_2}\right) + \ln k\right\} > z_{1-\alpha}\right]$$

and

$$\frac{\widehat{c}_1 + \widehat{c}_2}{2}\left\{\ln\left(\frac{\widehat{b}_1}{b_1}\right) - \ln\left(\frac{\widehat{b}_2}{b_2}\right) + \ln k\right\} = \frac{\widehat{c}_1 + \widehat{c}_2}{2}\left\{\ln\left(\frac{\widehat{b}_1}{b_1}\right) - \ln\left(\frac{\widehat{b}_2}{b_2}\right) + \frac{1}{c}\ln k^c\right\}.$$

The distribution of the latter random variable does not depend on $b_1$ or $b_2$, but only depends on $c$ through $k^c$ and its CDF is $G_{k^c}(z)$, which must be found by Monte Carlo techniques.

Up to now we have assumed that $c_1 = c_2$. However, as THOMAN/BAIN remarked, one may apply the percentage points in Tab. 21/4 to give conservative tests when $c_1 \neq c_2$. For example, consider the test of $H_0 : b_1 \geq b_2$ against $H_A : b_1 < b_2$ and we have $c_1 \leq c_2$. The test procedure to reject $H_0$ if

$$\frac{\widehat{c}_1 + \widehat{c}_2}{2}\ln\left(\frac{\widehat{b}_1}{\widehat{b}_2}\right) < z_\alpha \tag{21.22a}$$

is conservative; i.e., its probability of a type–I error will not exceed $\alpha$. This follows since, under $H_0$:

$$\Pr\left[\frac{\widehat{c}_1 + \widehat{c}_2}{2}\ln\left(\frac{\widehat{b}_1}{\widehat{b}_2}\right) < z_\alpha\right] \leq \Pr\left[\frac{\widehat{c}_1 + \widehat{c}_2}{2}\left\{\ln\left(\frac{\widehat{b}_1}{b_1}\right) - \ln\left(\frac{\widehat{b}_2}{b_2}\right)\right\} < z_\alpha\right]$$

$$\leq \Pr\left[\frac{\widehat{c}_1/c_1 + \widehat{c}_2/c_2}{2}\left\{c_1\ln\left(\frac{\widehat{b}_1}{b_1}\right) - c_2\ln\left(\frac{\widehat{b}_2}{b_2}\right)\right\} < z_\alpha\right]$$

$$= G_1(z_\alpha) = \alpha. \tag{21.22b}$$

In a similar manner it can be seen that if $c_1 \geq c_2$ in the above test then the power of the test of $H_0 : b_1 \geq b_2$ against $H_A : b_1 = k b_2$ with $k < 1$ will be at least the power of the corresponding tests with $c_1 = c_2$, i.e. $G_{(1/k)c_1}(z_\alpha)$.

Table 21/4: Percentage points $z_P$ such that $\Pr\left[\dfrac{\widehat{c}_1 + \widehat{c}_2}{2}\left(\ln \widehat{b}_1 - \ln \widehat{b}_2\right) \leq z_P\right] = P$

| $n$ \ $P$ | 0.60 | 0.70 | 0.75 | 0.80 | 0.85 | 0.90 | 0.95 | 0.98 |
|---|---|---|---|---|---|---|---|---|
| 5 | 0.228 | 0.476 | 0.608 | 0.777 | 0.960 | 1.226 | 1.670 | 2.242 |
| 6 | 0.190 | 0.397 | 0.522 | 0.642 | 0.821 | 1.050 | 1.404 | 1.840 |
| 7 | 0.164 | 0.351 | 0.461 | 0.573 | 0.726 | 0.918 | 1.215 | 1.592 |
| 8 | 0.148 | 0.320 | 0.415 | 0.521 | 0.658 | 0.825 | 1.086 | 1.421 |
| 9 | 0.136 | 0.296 | 0.383 | 0.481 | 0.605 | 0.757 | 0.992 | 1.294 |
| 10 | 0.127 | 0.277 | 0.356 | 0.449 | 0.563 | 0.704 | 0.918 | 1.195 |
| 11 | 0.120 | 0.261 | 0.336 | 0.423 | 0.528 | 0.661 | 0.860 | 1.115 |
| 12 | 0.115 | 0.248 | 0.318 | 0.401 | 0.499 | 0.625 | 0.811 | 1.049 |
| 13 | 0.110 | 0.237 | 0.303 | 0.383 | 0.474 | 0.594 | 0.770 | 0.993 |
| 14 | 0.106 | 0.227 | 0.290 | 0.366 | 0.453 | 0.567 | 0.734 | 0.945 |
| 15 | 0.103 | 0.218 | 0.279 | 0.352 | 0.434 | 0.544 | 0.704 | 0.904 |
| 16 | 0.099 | 0.210 | 0.269 | 0.339 | 0.417 | 0.523 | 0.676 | 0.867 |
| 17 | 0.096 | 0.203 | 0.260 | 0.328 | 0.403 | 0.505 | 0.654 | 0.834 |
| 18 | 0.094 | 0.197 | 0.251 | 0.317 | 0.389 | 0.488 | 0.631 | 0.805 |
| 19 | 0.091 | 0.191 | 0.244 | 0.308 | 0.377 | 0.473 | 0.611 | 0.779 |
| 20 | 0.089 | 0.186 | 0.237 | 0.299 | 0.366 | 0.459 | 0.593 | 0.755 |
| 22 | 0.085 | 0.176 | 0.225 | 0.284 | 0.347 | 0.435 | 0.561 | 0.712 |
| 24 | 0.082 | 0.168 | 0.215 | 0.271 | 0.330 | 0.414 | 0.534 | 0.677 |
| 26 | 0.079 | 0.161 | 0.206 | 0.259 | 0.316 | 0.396 | 0.510 | 0.646 |
| 28 | 0.076 | 0.154 | 0.198 | 0.249 | 0.303 | 0.380 | 0.490 | 0.619 |
| 30 | 0.073 | 0.149 | 0.191 | 0.240 | 0.292 | 0.366 | 0.472 | 0.595 |
| 32 | 0.071 | 0.144 | 0.185 | 0.232 | 0.282 | 0.354 | 0.455 | 0.574 |
| 34 | 0.069 | 0.139 | 0.179 | 0.225 | 0.273 | 0.342 | 0.441 | 0.555 |
| 36 | 0.067 | 0.135 | 0.174 | 0.218 | 0.265 | 0.332 | 0.427 | 0.537 |
| 38 | 0.065 | 0.131 | 0.169 | 0.212 | 0.258 | 0.323 | 0.415 | 0.522 |
| 40 | 0.064 | 0.127 | 0.165 | 0.206 | 0.251 | 0.314 | 0.404 | 0.507 |
| 42 | 0.062 | 0.124 | 0.160 | 0.201 | 0.245 | 0.306 | 0.394 | 0.494 |
| 44 | 0.061 | 0.121 | 0.157 | 0.196 | 0.239 | 0.298 | 0.384 | 0.482 |
| 46 | 0.059 | 0.118 | 0.153 | 0.192 | 0.234 | 0.292 | 0.376 | 0.470 |
| 48 | 0.058 | 0.115 | 0.150 | 0.188 | 0.229 | 0.285 | 0.367 | 0.460 |
| 50 | 0.057 | 0.113 | 0.147 | 0.184 | 0.224 | 0.279 | 0.360 | 0.450 |
| 52 | 0.056 | 0.110 | 0.144 | 0.180 | 0.220 | 0.273 | 0.353 | 0.440 |
| 54 | 0.055 | 0.108 | 0.141 | 0.176 | 0.215 | 0.268 | 0.346 | 0.432 |
| 56 | 0.054 | 0.106 | 0.138 | 0.173 | 0.212 | 0.263 | 0.340 | 0.423 |
| 58 | 0.053 | 0.104 | 0.136 | 0.170 | 0.208 | 0.258 | 0.334 | 0.416 |
| 60 | 0.052 | 0.102 | 0.134 | 0.167 | 0.204 | 0.254 | 0.328 | 0.408 |
| 62 | 0.051 | 0.100 | 0.131 | 0.164 | 0.201 | 0.250 | 0.323 | 0.402 |
| 64 | 0.050 | 0.099 | 0.129 | 0.162 | 0.198 | 0.246 | 0.317 | 0.395 |
| 66 | 0.049 | 0.097 | 0.127 | 0.159 | 0.195 | 0.242 | 0.313 | 0.389 |
| 68 | 0.049 | 0.095 | 0.125 | 0.157 | 0.192 | 0.238 | 0.308 | 0.383 |
| 70 | 0.048 | 0.094 | 0.123 | 0.154 | 0.190 | 0.235 | 0.304 | 0.377 |
| 72 | 0.047 | 0.092 | 0.122 | 0.152 | 0.187 | 0.231 | 0.299 | 0.372 |
| 74 | 0.046 | 0.091 | 0.120 | 0.150 | 0.184 | 0.228 | 0.295 | 0.366 |
| 76 | 0.046 | 0.090 | 0.118 | 0.148 | 0.182 | 0.225 | 0.291 | 0.361 |
| 78 | 0.045 | 0.089 | 0.117 | 0.146 | 0.180 | 0.222 | 0.288 | 0.357 |
| 80 | 0.045 | 0.087 | 0.115 | 0.144 | 0.178 | 0.219 | 0.284 | 0.352 |
| 90 | 0.042 | 0.082 | 0.109 | 0.136 | 0.168 | 0.207 | 0.268 | 0.332 |
| 100 | 0.040 | 0.077 | 0.103 | 0.128 | 0.160 | 0.196 | 0.255 | 0.315 |
| 120 | 0.036 | 0.070 | 0.094 | 0.117 | 0.147 | 0.179 | 0.233 | 0.287 |

Source: THOMAN/BAIN (1969, p. 810) — Reprinted with permission from *Technometrics*. Copyright 1969 by

SCHAFER/SHEFFIELD (1976) did not use the MLEs $\widehat{c}_1$ and $\widehat{c}_2$ in the first and second sample, respectively, but they used a pooled estimator $\widehat{\widehat{c}}$ which results as the solution of

$$\frac{2\,n}{\widehat{\widehat{c}}} - n \left\{ \frac{\sum\limits_{i=1}^{n} X_{1i}^{\widehat{\widehat{c}}} \ln X_{1i}}{\sum\limits_{i=1}^{n} X_{1i}^{\widehat{\widehat{c}}}} + \frac{\sum\limits_{i=1}^{n} X_{2i}^{\widehat{\widehat{c}}} \ln X_{2i}}{\sum\limits_{i=1}^{n} X_{2i}^{\widehat{\widehat{c}}}} \right\} + \sum_{i=1}^{n} \ln X_{1i} + \sum_{i=1}^{n} \ln X_{2i} = 0. \quad (21.23a)$$

$\widehat{\widehat{c}}$ is taken instead of $(\widehat{c}_1 + \widehat{c}_2)/2$ in the above formulas where we also have to substitute $\widehat{b}_1$ and $\widehat{b}_2$ by

$$\widehat{\widehat{b}}_1 = \left( \sum_{i=1}^{n} X_{1i}^{\widehat{\widehat{c}}} \Big/ n \right)^{1/\widehat{\widehat{c}}} \quad \text{and} \quad \widehat{\widehat{b}}_2 = \left( \sum_{i=1}^{n} X_{2i}^{\widehat{\widehat{c}}} \Big/ n \right)^{1/\widehat{\widehat{c}}}, \quad (21.23b)$$

respectively. Of course, the percentage points in Tab. 21/4 do not apply to the test statistic

$$\widehat{\widehat{c}} \left( \ln \widehat{\widehat{b}}_1 + \ln \widehat{\widehat{b}}_2 \right).$$

SCHAFER/SHEFFIELD (1976) provided a table of the matching percentiles. The SCHAFER/SHEFFIELD approach dominates the THOMAN/BAIN approach insofar as it has a little more power.

Readers interested in a **$k$–sample test** of $H_0: b_1 = b_2 = \ldots = b_k$ against $H_A: b_i \neq b_j$ for at least one pair of indices $(i, j)$, $i \neq j$, are referred to

- McCOOL (1977) who develops a procedure in analogy to the one–way analysis of variance using certain ratios of MLEs of the common but unknown shape parameters;

- CHAUDHURI/CHANDRA (1989) who also present an ANOVA test, but based on sample quantiles;

- PAUL/THIAGARAJAH (1992) who compare the performance of several test statistics.

In all three papers the $k$ WEIBULL populations are assumed to have a common but unknown shape parameter.

Readers who prefer GLUEs over MLEs are referred to BAIN (1978), ENGELHARDT/BAIN (1979) and BAIN/ENGELHARDT (1991a) where the one–sample test and the $k$–sample test $(k \geq 2)$ are treated.

### 21.1.3   Hypotheses concerning the location parameter $a$[9]

Perhaps the most interesting hypothesis concerning the location parameter $a$ is whether $a = 0$. McCOOL (1998) has suggested an elegant procedure to test $H_0 : a = 0$ versus

---

[9]  Suggested reading for this section: DUBEY (1966a), McCOOL (1998), SCHAFER (1975).

$H_A$: $a > 0$. This procedure uses two MLEs of the shape parameter:

- one is based on all $r$, $r \leq n$, order statistics and
- one is based just on the first $r_1 < r$ early order statistics.

The idea behind this approach is readily grasped in the context of graphical estimation (see Sect. 9.3).

For the two–parameter WEIBULL distribution it is readily shown that

$$
\begin{aligned}
y(x) &:= \ln\{\ln[1/(1 - F(x))]\} \\
&= c \ln x - c \ln b.
\end{aligned}
\tag{21.24a}
$$

Thus, in this case $y(x)$ is a linear function of $\ln x$ having slope $c$ and intercept $-c \ln b$. For the three–parameter WEIBULL distribution we have

$$
\begin{aligned}
y(x) &:= \ln\{\ln[1/(1 - F(x))]\} \\
&= c \ln(x - a) - c \ln b, \quad x \geq a.
\end{aligned}
\tag{21.24b}
$$

The slope of a plot of $y(x)$ against $\ln x$ in the three–parameter case is

$$
\frac{dy(x)}{d \ln x} = \frac{c\,x}{x - a}, \quad x \geq a.
\tag{21.24c}
$$

This slope is infinite at $x = a$ and decreases monotonically thereafter with $x$ to an asymptote of $c$.

Let $x_{1:n} \leq x_{2:n} \leq \ldots \leq x_{n:n}$ denote the ordered observations in a sample drawn from a two– or three–parameter WEIBULL distribution. An estimate $\widehat{y}(x_{i:n})$ may be computed by substituting $F(x)$ in (21.24a,b) with one of the plotting positions of Tab. 9/2. If the sample is drawn from a two–parameter distribution, $\widehat{y}(x_{i:n})$ will tend to plot against $\ln x_{i:n}$ as a straight line with slope $c$. If $a > 0$, i.e., if the population is a three–parameter distribution, then $\widehat{y}(x_{i:n})$ will tend to be a concave function of $\ln x_{i:n}$ approaching a constant slope $c$ for large $x_{i:n}$–values.

If the data were regarded as coming from a two–parameter WEIBULL distribution, a graphical estimate of the shape parameter, $\widehat{c}_A$, could be found as the slope of the straight line that best fits **all** the sampled data. If only a subset comprising some number $r_1$ of the smaller ordered values were used in graphically estimating the shape parameter, the estimate $\widehat{c}_L$ would be obtained. For three-parameter WEIBULL data, $\widehat{c}_L$ will tend to exceed $\widehat{c}_A$. On the other hand, when the sample is drawn from a two–parameter WEIBULL distribution ($a = 0$), $\widehat{c}_L$ and $\widehat{c}_A$ will be comparable.

The MLE of $c$ is the solution of

$$
\frac{1}{\widehat{c}} + \frac{1}{r} \sum_{i=1}^{r} \ln X_{i:n} - \frac{\sum\limits_{i=1}^{r} X_{i:n}^{\widehat{c}} \ln X_{i:n} + (n - r)\, X_{r:n}^{\widehat{c}} \ln X_{r:n}}{\sum\limits_{i=1}^{r} X_{i:n}^{\widehat{c}} + (n - r)\, X_{r:n}^{\widehat{c}}} = 0.
\tag{21.25}
$$

It is well known that $\widehat{c}/c$ is a pivotal function having a distribution which depends on only $n$ and $r$.

Denoting the solution of (21.25) as $\widehat{c}_A$ and the solution of (21.25) with censoring at $r_1 < r$ as $\widehat{c}_L$, the distribution of

$$W = \frac{\widehat{c}_L}{\widehat{c}_A} \tag{21.26a}$$

will depend only on $n$, $r_1$ and $r$, when the underlying distribution is WEIBULL with $a = 0$. When the underlying distribution is WEIBULL with $a > 0$, the mean value of $\widehat{c}_L$ will exceed the mean value of $\widehat{c}_A$. MCCOOL (1998) determined the percentiles of $W$ by Monte Carlo sampling for specified $r_1$, $r$ and $n$ (see Tab. 21/5). One may reject $H_0 : a = 0$ against $H_A : a > 0$ at level $\alpha$,[10] if

$$\frac{\widehat{c}_L}{\widehat{c}_A} > w_{1-\alpha}. \tag{21.26b}$$

In Tab. 21/5 the value of $r_1$ has been selected to give maximum power in detecting a non–zero location parameter for a specified $n$ and $r$. MCCOOL (1998) also gave the power of this test as again found by Monte Carlo sampling.

A more general test concerning the contents of $H_0$ and $H_A$ is given by SCHAFER (1975), but it is restrictive in the sense that the shape parameter $c$ must be known. The hypotheses are $H_0 : a = a_0$ and $H_A : a = a_1$, $a_1 \neq a_0$. The test statistic is

$$S = \frac{X_{a:n} - a_0}{X_{b:n} - X_{a:n}}, \quad a < b. \tag{21.27a}$$

$H_0$ cannot be rejected for $k_1 < S < k_2$, and the critical values $k_1$ and $k_2$ are chosen so that

$$1 - \alpha = \int_{k_1}^{k_2} f(s)\,\mathrm{d}s. \tag{21.27b}$$

The DF of $S$, $f(s)$, is given by SCHAFER (1975) as

$$f(s) = C \sum_{j=0}^{b-a-1} \sum_{i=0}^{a-1} \binom{b-a-1}{j}\binom{a-1}{i}(-1)^{i+j}$$

$$\times \frac{c\left(s^{-1}+1\right)^{c-1}}{s^2\left[(b-a-j+i) + (s^{-1}+1)^c\,(n-b+j+1)\right]^2} \tag{21.27c}$$

with

$$C = \frac{n!}{(a-1)!\,(b-a-1)!\,(n-b)!}.$$

More tests for the location parameter can be found in the following section and in Sect. 21.2.1.

---

[10] For a test of $H_0 : a = 0$, against $H_A : a < 0$, the procedure has to be modified because under $H_A$ $\widehat{c}_L$ will fall short of $\widehat{c}_A$.

Table 21/5:   Percentage points $w_P$ such that $\Pr\left(\widehat{c}_L/\widehat{c}_A \le w_P\right) = P$

| $n$ | $r_1$ | $r$ | $w_{0.50}$ | $w_{0.90}$ | $w_{0.95}$ |
|---|---|---|---|---|---|
| 10 | 5 | 6 | 0.988 | 1.488 | 1.789 |
| 10 | 5 | 7 | 1.035 | 1.730 | 2.132 |
| 10 | 5 | 8 | 1.077 | 1.902 | 2.352 |
| 10 | 5 | 9 | 1.116 | 2.022 | 2.517 |
| 10 | 5 | 10 | 1.138 | 2.126 | 2.683 |
| 15 | 5 | 10 | 1.141 | 2.094 | 2.579 |
| 15 | 5 | 15 | 1.223 | 2.408 | 3.055 |
| 20 | 5 | 6 | 0.990 | 1.498 | 1.759 |
| 20 | 5 | 10 | 1.137 | 2.073 | 2.582 |
| 20 | 5 | 12 | 1.172 | 2.198 | 2.784 |
| 20 | 5 | 15 | 1.210 | 2.345 | 2.974 |
| 20 | 5 | 18 | 1.238 | 2.417 | 3.135 |
| 20 | 5 | 20 | 1.254 | 2.491 | 3.188 |
| 25 | 5 | 10 | 1.146 | 2.118 | 2.622 |
| 25 | 5 | 14 | 1.199 | 2.321 | 2.924 |
| 25 | 5 | 15 | 1.211 | 2.238 | 2.850 |
| 25 | 5 | 18 | 1.237 | 2.460 | 3.087 |
| 25 | 5 | 20 | 1.250 | 2.515 | 3.148 |
| 25 | 5 | 25 | 1.278 | 2.540 | 3.278 |
| 30 | 5 | 6 | 0.990 | 1.467 | 1.734 |
| 30 | 5 | 10 | 1.139 | 2.079 | 2.602 |
| 30 | 5 | 15 | 1.213 | 2.340 | 2.915 |
| 30 | 5 | 20 | 1.256 | 2.457 | 3.119 |
| 30 | 5 | 25 | 1.278 | 2.544 | 3.224 |
| 30 | 5 | 30 | 1.294 | 2.600 | 3.279 |
| 40 | 7 | 15 | 1.098 | 1.755 | 2.074 |
| 40 | 7 | 20 | 1.136 | 1.888 | 2.240 |
| 40 | 7 | 25 | 1.157 | 1.937 | 2.299 |
| 40 | 7 | 30 | 1.172 | 1.984 | 2.364 |
| 40 | 7 | 40 | 1.198 | 2.039 | 2.430 |
| 50 | 7 | 25 | 1.152 | 1.941 | 2.292 |
| 50 | 7 | 30 | 1.165 | 1.995 | 2.342 |
| 50 | 7 | 40 | 1.182 | 2.049 | 2.420 |
| 50 | 7 | 50 | 1.191 | 2.070 | 2.466 |
| 60 | 7 | 30 | 1.167 | 2.014 | 2.406 |
| 60 | 7 | 40 | 1.183 | 2.062 | 2.470 |
| 60 | 7 | 50 | 1.199 | 2.097 | 2.527 |
| 60 | 7 | 60 | 1.203 | 2.128 | 2.564 |
| 80 | 9 | 40 | 1.122 | 1.771 | 2.054 |
| 80 | 9 | 50 | 1.133 | 1.793 | 2.091 |
| 80 | 9 | 60 | 1.140 | 1.824 | 2.108 |
| 80 | 9 | 70 | 1.146 | 1.844 | 2.121 |
| 80 | 9 | 80 | 1.150 | 1.850 | 2.143 |
| 100 | 9 | 50 | 1.132 | 1.787 | 2.087 |
| 100 | 9 | 60 | 1.140 | 1.809 | 2.101 |
| 100 | 9 | 70 | 1.147 | 1.827 | 2.129 |
| 100 | 9 | 80 | 1.151 | 1.837 | 2.152 |
| 100 | 9 | 90 | 1.152 | 1.840 | 2.146 |
| 100 | 9 | 100 | 1.155 | 1.846 | 2.173 |

Source: McCool (1998, p. 121) — Reprinted with permission from *Journal of Quality Technology*. Copyright 1998 by the American Society for Quality Control. All rights reserved.

## 21.1.4   Hypotheses concerning two or more parameters

DUBEY (1966a) has compiled a number of hypotheses concerning the parameters $a$, $\theta$ $(= b^c)$ and $c$ and several statistics which are suited for their testing. The null hypotheses are as follows:

1.  $H_0^{(1)}$: $a = a_0$, $\theta = \theta_0$, $c = c_0$;

2.  $H_0^{(2)}$: $a = a_0$, $\theta = \theta_0$ and $c$ known;

3.  $H_0^{(3)}$: $a = a_0$, $c = c_0$ and $\theta$ known;

4.  $H_0^{(4)}$: $\theta = \theta_0$, $c = c_0$ and $a$ known;

5.  $H_0^{(5)}$: $a = a_0$, $\theta$ and $c$ known;

6.  $H_0^{(6)}$: $\theta = \theta_0$, $a$ and $c$ known;

7.  $H_0^{(7)}$: $c = c_0$, $a$ and $\theta$ known;

8.  $H_0^{(8)}$: $a = a_0$, $c = c_0$ and $\theta$ unknown;

9.  $H_0^{(9)}$: $a = a_0$, $c$ known and $\theta$ unknown;

10. $H_0^{(10)}$: equality of $k$ $(k \geq 2)$ WEIBULL populations.

The proposed statistics and their distributions are, where $r \leq n$ and $1 \leq a < b \leq n$:

$$S_1 = \frac{2}{\theta_0} \left[ \sum_{i=1}^{r} \left(X_{i:n} - a_0\right)^{c_0} + (n - r) \left(X_{r:n} - a_0\right)^{c_0} \right] \sim \chi^2(2\,r); \quad (21.28a)$$

$$S_1^* = \frac{2}{\theta_0} \sum_{i=1}^{n} \left(X_{i:n} - a_0\right)^{c_0} \sim \chi^2(2\,n); \quad (21.28b)$$

$$S_2 = \left\{ \begin{array}{l} \frac{2}{\theta_0} \left\{ \sum\limits_{i=a+1}^{b-1} \left[ \left(X_{i:n} - a_0\right)^{c_0} - \left(X_{a:n} - a_0\right)^{c_0} \right] + \right. \\ \left. (n - b + 1) \left[ \left(X_{b:n} - a_0\right)^{c_0} - \left(X_{a:n} - a_0\right)^{c_0} \right] \right\} \quad \text{with } S_1 \sim \chi^2(2\,(b - a)); \end{array} \right\}$$
$$(21.29a)$$

$$S_2^* = \frac{2}{\theta_0} \left\{ \sum_{i=2}^{n} \left[ \left(X_{i:n} - a_0\right)^{c_0} + \left(X_{1:n} - a_0\right)^{c_0} \right] \right\} \sim \chi^2(2\,(n - 1)); \quad (21.29b)$$

$$S_3 = \frac{n - b + 1}{b - a} \left[ \exp\left\{ \frac{1}{\theta_0} \left[ \left(X_{b:n} - a_0\right)^{c_0} - \left(X_{a:n} - a_0\right)^{c_0} \right] \right\} - 1 \right] \sim F(2(b - a), 2(n - b + 1));$$
$$(21.30)$$

$$S_4 = \frac{n - a + 1}{a} \left[ \exp\left\{ \frac{1}{\theta_0} \left(X_{a:n} - a_0\right)^{c_0} \right\} - 1 \right] \sim F(2\,a, 2\,(n - a + 1)); \quad (21.31)$$

$$S_5 = \frac{n(b-1)\left(X_{1:n}-a_0\right)^{c_0}}{\sum\limits_{i=2}^{b-1}\left[\left(X_{i:n}-a_0\right)^{c_0}-\left(X_{1:n}-a_0\right)^{c_0}\right]+(n-b+1)\left[\left(X_{b:n}-a_0\right)^{c_0}-\left(X_{1:n}-a_0\right)^{c_0}\right]}$$

$$\sim F(2,2(b-1)); \tag{21.32}$$

$$S_6 = \frac{n(b-a)\left(X_{a:n}-a_0\right)^{c_0}}{\sum\limits_{i=a+1}^{b-1}\left[\left(X_{i:n}-a_0\right)^{c_0}-\left(X_{a:n}-a_0\right)^{c_0}\right]+(n-b+1)\left[\left(X_{b:n}-a_0\right)^{c_0}-\left(X_{a:n}-a_0\right)^{c_0}\right]} : \tag{21.33a}$$

$$f(s_6) = \frac{(n-1)!}{(a-1)!\,(n-a)!}\sum_{k=0}^{a-1}\binom{a-1}{k}(-1)^k\left[1+\frac{n+k-a+1}{n(b-a)}s_6\right]^{a-b-1}, \quad s_6 > 0; \tag{21.33b}$$

$$S_7 = \frac{\left(X_{a:n}-a_0\right)^{c_0}}{\left(X_{b:n}-a_0\right)^{c_0}-\left(X_{a:n}-a_0\right)^{c_0}}, \tag{21.34a}$$

$$f(s_7) = \begin{cases} \dfrac{n!}{(a-1)!(b-a-1)!(n-b)!}\sum\limits_{k=0}^{a-1}\sum\limits_{m=0}^{b-a-1}\binom{a-1}{k}\times \\[2mm] \binom{b-a-1}{m}\dfrac{(-1)^{k+m}}{\left[(n-a+k+1)s_7+(n-b+m+1)\right]^2}, \quad s_7 > 0; \end{cases} \tag{21.34b}$$

$$S_8 = \sum_{j=1}^{k}\frac{2}{\theta_{0,j}}\left\{\sum_{i=1}^{r_j}\left(X_{i:n_j}-a_{0,j}\right)^{c_{0,j}}+(n_j-r_j)\left(X_{r_j:n_j}-a_{0,j}\right)^{c_{0,j}}\right\}\sim\chi^2\left(2\sum_{j=1}^{k}r_j\right), \tag{21.35}$$

where $X_{1:n_j},\ldots,X_{r_j:n_j}$ are the first $r_j$, $(r_j \le n_j)$ ordered observations from the $j-$th $(j=1,\ldots,k)$ WEIBULL distribution;

$$S_9 = \begin{cases} \sum\limits_{j=1}^{k}\left\{\dfrac{2}{\theta_{0,j}}\sum\limits_{i=a_j+1}^{b_j-1}\left[\left(X_{i:n_j}-a_{0,j}\right)^{c_{0,j}}-\left(X_{a_j:n_j}-a_{0,j}\right)^{c_{0,j}}\right]+\right. \\[3mm] \left. (n_j-b_j+1)\left[\left(X_{b_j:n_j}-a_{0,j}\right)^{c_{0,j}}-\left(X_{a_j:n_j}-a_{0,j}\right)^{c_{0,j}}\right]\right\}\sim\chi^2\left(2\sum\limits_{j=1}^{k}(b_j-a_j)\right) \\[3mm] \text{where } 1\le a_j < b_j \le n_j;\; j=1,\ldots,k. \end{cases} \tag{21.36}$$

The statistics $S_1$, $S_1^*$, $S_2$, $S_2^*$, $S_3$ and $S_4$ can be used to test the hypotheses $\mathrm{H}_0^{(1)}$ through $\mathrm{H}_0^{(7)}$. The proper choice of test functions based on these statistics depends on the type of data available. Because these statistics are non–negative, it seems reasonable to construct a critical region for the rejection of $H_0^{(j)}$; $j = 1, \ldots, 7$, in the upper tail of the distribution unless other considerations suggest otherwise. Similarly, the statistics $S_5$, $S_6$ and $S_7$ can be used to test $H_0^{(8)}$ and $H_0^{(9)}$. The statistics $S_8$ and $S_9$ can be used to test $H_0^{(10)}$ by assuming $a_{0,j}$, $\theta_{0,j}$ and $c_{0,j}$ to be $a_0$, $\theta_0$ and $c_0$, respectively, for $k$ WEIBULL distributions.

## 21.2 Testing hypotheses on functional parameters

Sometimes we are not interested in knowledge and inference of location, scale and shape parameters per se, but in some other quantities which depend on these parameters. Most important among these quantities are the mean, the variance, the reliability and the percentile.

### 21.2.1 Hypotheses concerning the mean $\mu$

The mean of the three–parameter WEIBULL distribution is

$$\mu = a + b\,\Gamma\left(1 + \frac{1}{c}\right). \tag{21.37}$$

A test on $\mu$ when **$b$ and $c$ are known** is equivalent to a test on the location parameter $a$. For definiteness consider the one–sided test of $H_0 : a = a_0$ against $H_A : a > a_0$. Because $X_{1:n}$ is the MLE of $a$ and because $X_{1:n}$ is a sufficient statistic for $a$ when $c = 1$ we decide to base the test on $X_{1:n}$. This is a very easy test to perform since the statistic

$$Z := n\,\left(\frac{X_{1:n} - a_0}{b}\right)^c \tag{21.38a}$$

is a reduced exponential variable with DF $f(z) = z\,\exp(-z)$. Thus, a size $\alpha$ test is to reject $H_0 : a = a_0$ in favor of $H_A : a > a_0$ when

$$Z > -\ln\alpha. \tag{21.38b}$$

The power of this one–sided test is

$$\Pr\big(Z > -\ln\alpha \,\big|\, a\big) = \exp\left\{\left[\frac{a_0 - a}{b} + \left(\frac{-\ln\alpha}{n}\right)^{1/c}\right]^c\right\}. \tag{21.38c}$$

Critical regions and power functions for other hypotheses based on $Z$ follow easily. As BAIN/THOMAN (1968) state that a test using $Z$ when $n$ is large performs better than a text based $\overline{X}_n$ in the sense of having higher relative efficiency when $c < 2$.

A method based on an appropriate single order statistic can also be used performing tests concerning $\mu$ when **$c$ is known**; see BAIN/THOMAN (1968). The mean of the WEIBULL distribution is a known percentile when $c$ is known, not depending on $a$ and $b$:

$$F(\mu) = 1 - \exp\big\{-\big[\Gamma(1 + 1/c)\big]^c\big\}. \tag{21.39}$$

Thus, the usual method of obtaining confidence intervals or tests for percentiles (see Sect. 21.2.4) can be applied. We have

$$\Pr\left(X_{r:n} < \mu\right) = B\left[F(\mu); r, n - r + 1\right], \tag{21.40a}$$

where $B[x; p, q]$ is the CDF of the beta distribution given by

$$\begin{aligned}
B[x; p, q] &= \frac{\Gamma(p + q)}{\Gamma(p)\,\Gamma(q)} \int_0^x u^{p-1}\,(1 - u)^{q-1} \mathrm{d}u \\
&= 1 - \sum_{i=0}^{p-1} \binom{p + q - 1}{i} x^i\,(1 - x)^{p+q-1-i}, \quad \text{for integer } p \text{ and } q. \tag{21.40b}
\end{aligned}$$

Let $B_P(p, q)$ denote the percentile of order $P$, then $X_{r:n}$ is a $100\,P\%$ lower confidence limit for $\mu$ if $r$ and $n$ are chosen so that $B_P(r, n - r + 1) = F(\mu)$. In terms of a binomial CDF, $r$ must be chosen so that

$$\sum_{i=1}^{r-1} \binom{n}{i} \left[F(\mu)\right]^i \left[1 - F(\mu)\right]^{n-i} = 1 - P. \tag{21.40c}$$

Because $r$ only assumes integer values, the exact level will usually not be obtained. In order to have a confidence level of at least $P$, one chooses $r$ as the smallest integer so that

$$\sum_{i=1}^{r-1} \binom{n}{i} \left[F(\mu)\right]^i \left[1 - F(\mu)\right]^{n-i} \geq 1 - P. \tag{21.40d}$$

In terms of hypothesis testing suppose the null hypothesis $H_0 : \mu = \mu_0$ is to be tested, at significance level $\alpha$, against $H_A : \mu < \mu_0$. $H_0$ should be rejected if $X_{r^*:n} < \mu_0$, where $r^*$ is chosen so that $X_{r^*:n}$ is a lower confidence limit for $\mu$ with confidence at least $1 - \alpha$.

When $n$ is large and $b$ **and** $c$ **are unknown**, hypotheses on $\mu$ can be tested using the conventional $t$–test.

## 21.2.2   Hypotheses concerning the variance $\sigma^2$

The variance of the three–parameter as well as of the two–parameter WEIBULL distribution is

$$\sigma^2 = b^2 \left[\Gamma\left(1 + \frac{2}{c}\right) - \Gamma^2\left(\frac{1}{c}\right)\right]. \tag{21.41}$$

So tests concerning $\sigma^2$ can be done using the procedures described in Sect. 21.1.2.

We consider the test of $H_0 : b = b_0$ against $H_A : b > b_0$ where $a$ is unknown. This is comparable to a test on $\sigma$ with $\mu$ unknown. We introduce the following two test statistics, assuming $c$ is known:

$$W = 2 \sum_{i=1}^{n} \left(\frac{X_i - X_{1:n}}{b_0}\right)^c, \tag{21.42a}$$

$$U = 2 \sum_{i=1}^{n} \left(\frac{X_i - a}{b_0}\right)^c. \tag{21.42b}$$

If $a$ were known, the best test of $H_0$ against $H_A$ would be to reject $H_0$ for large values of $U$. In particular, a size $\alpha$ test in this case would be to reject $H_0$ if

$$U > \chi^2_{2n,\alpha}. \tag{21.42c}$$

Since $W \leq U$, a conservative test is given by rejecting $H_0$ if $W > \chi^2_{2n,\alpha}$ when $a$ is unknown. As an example of conservativeness, the proper value when $c = 1$ is $\chi^2_{2(n-1),\alpha}$. BAIN/THOMAN (1968) state that the conservativeness will increase as $c$ increases.

### 21.2.3 Hypotheses on the reliability $R(x)$[11]

In Sect. 19.2.2 we have shown how to construct confidence intervals for the reliability $R(x)$ when it has been estimated as

$$\widehat{R}(x) = \exp\left\{-\left(\frac{x}{\widehat{b}}\right)^{\widehat{c}}\right\},$$

where $\widehat{b}$ and $\widehat{c}$ are the MLEs of $b$ and $c$, respectively, so $\widehat{R}(x)$ is MLE too. Using the following $(1 - \alpha)$ level confidence intervals belonging to $\widehat{R}(x)$

$$\Pr\left[R(x) \geq \widehat{R}_{\ell,\alpha}(x)\right] = 1 - \alpha, \tag{21.43a}$$

$$\Pr\left[R(x) \leq \widehat{R}_{u,1-\alpha}(x)\right] = 1 - \alpha, \tag{21.43b}$$

$$\Pr\left[\widehat{R}_{\ell,\alpha/2}(x) \leq R(x) \leq \widehat{R}_{u,1-\alpha/2}(x)\right] = 1 - \alpha, \tag{21.43c}$$

we have to reject

- $H_0$: $R(x) \geq R_0(x)$ against $H_A$: $R(x) < R_0(x)$ when $R_0(x) < \widehat{R}_{\ell,\alpha}(x)$,

- $H_0$: $R(x) \leq R_0(x)$ against $H_A$: $R(x) > R_0(x)$ when $R_0(x) > \widehat{R}_{u,1-\alpha}(x)$,

- $H_0$: $R(x) = R_0(x)$ against $H_A$: $R(x) \neq R_0(x)$ when $R_0(x) < \widehat{R}_{\ell,\alpha/2}(x)$ or $R_0(x) > \widehat{R}_{u,1-\alpha/2}(x)$,

the level of significance being $\alpha$.

**Example 21/4: Testing $H_0$: $R(25) \geq R_0(25) = 0.90$ against $H_A$ : $R(25) < R_0(25) = 0.90$ using dataset #1 ($\alpha = 0.05$)**

The MLEs of $b$ and $c$ belonging to dataset #1 (Tab. 9/2) are

$$\widehat{b} = 99.2079 \quad \text{and} \quad \widehat{c} = 2.5957$$

giving the point estimate of $R(25)$ as

$$\widehat{R}(25) = \exp\left\{-\left(\frac{25}{99.2079}\right)^{2.5957}\right\} = 0.9724.$$

---

[11] Suggested reading for this section: BAIN/ENGELHARDT (1991a, pp. 232 ff.), MOORE/HARTER/ANTOON (1981).

From Tab. 19/1 we find by linear interpolation ($n = 20$): $\widehat{R}_{\ell,0.05}(25) = 0.9093$, Thus

$$R_0(25) = 0.90 < \widehat{R}_{\ell,0.05}(25) = 0.9093,$$

and we have to reject $H_0$: $R(25) \geq 0.90$.

## 21.2.4   Hypotheses concerning the percentile $x_P$[12]

For testing hypotheses on $x_P$ we may either use the confidence interval given in Sect. 19.3.2 or construct critical regions. Here, we will follow the second way, assuming an uncensored sample and MLEs.

In (19.29) we have introduced the random quantity

$$U_R = \sqrt{n}\left[ -\ln\{ -\ln \widehat{R}(x)\} + \ln\{ -\ln R(x)\}\right]$$

which may be expressed in the following form to be used for constructing confidence intervals or testing hypotheses on $x_P$:

$$U_\gamma = \widehat{c}\,\sqrt{n}\,\ln\left\{\frac{\widehat{x}_{1-\gamma}}{x_{1-\gamma}}\right\}, \quad 1 - \gamma = P. \tag{21.44a}$$

Let $F_\gamma(\cdot)$ denote the CDF of $U_\gamma$, so that

$$F_\gamma\big[u_{1-\alpha}(\gamma)\big] = 1 - \alpha. \tag{21.44b}$$

Tab. 19/4a and Tab. 19/4b give percentage points $u_{0.95}(\gamma)$ and $u_{0.05}(\gamma)$, respectively. The power of the following tests can also be expressed in terms of this family of distributions.

A size $\alpha$ test of

- $H_0$: $x_{1-\gamma} \leq x_{1-\gamma}^{(0)}$ against $H_A$: $x_{1-\gamma} > x_{1-\gamma}^{(0)}$ is to reject $H_0$ if

$$\sqrt{n}\,\widehat{c}\,\ln\left\{\frac{\widehat{x}_{1-\gamma}}{x_{1-\gamma}^{(0)}}\right\} > u_{1-\alpha}(\gamma), \tag{21.45a}$$

- $H_0$: $x_{1-\gamma} \geq x_{1-\gamma}^{(0)}$ against $H_A$: $x_{1-\gamma} < x_{1-\gamma}^{(0)}$ is to reject $H_0$ if

$$\sqrt{n}\,\widehat{c}\,\ln\left\{\frac{\widehat{x}_{1-\gamma}}{x_{1-\gamma}^{(0)}}\right\} < u_\alpha(\gamma), \tag{21.45b}$$

- $H_0$: $x_{1-\gamma} = x_{1-\gamma}^{(0)}$ against $H_A$: $x_{1-\gamma} \neq x_{1-\gamma}^{(0)}$ is to reject $H_0$ if

$$\sqrt{n}\,\widehat{c}\,\ln\left\{\frac{\widehat{x}_{1-\gamma}}{x_{1-\gamma}^{(0)}}\right\} < u_{\alpha/2}(\gamma) \quad \text{or} \quad \sqrt{n}\,\widehat{c}\,\ln\left\{\frac{\widehat{x}_{1-\gamma}}{x_{1-\gamma}^{(0)}}\right\} > u_{1-\alpha/2}(\gamma). \tag{21.45c}$$

---

[12] Suggested reading for this section: BAIN/ENGELHARDT (1991a, pp. 229 ff.), McCOOL (1970a).

For example, the power of the test (21.45a) when an alternative value $x^*_{1-\gamma}$ is assumed true may be expressed as a function of $\left(x^{(0)}_{1-\gamma}\big/ x^*_{1-\gamma}\right)^c$ and is given in BAIN (1978, p. 245) as

$$\Pr\left[\text{reject } H_0:\ x_{1-\gamma} \le x^{(0)}_{1-\gamma}\ \big|\ x_{1-\gamma} = x^*_{1-\gamma}\right] = 1 - F_Q\left[c\,\ln\left\{\frac{x^{(0)}_{1-\gamma}}{x^*_{1-\gamma}}\right\} + u_{1-\alpha}(\gamma)\right],$$

(21.46)

where

$$Q = \exp\left\{-\left(\frac{x^{(0)}_{1-\gamma}}{x^*_{1-\gamma}}\right)^c(-\ln\gamma)\right\}.$$

The power function depends on the unknown shape parameter $c$.

---

**Example 21/5: Testing** $H_0:\ x_{10} \le 30$ **against** $H_A:\ x_{0.10} > 30$ **using dataset #1** $(\alpha = 0.05)$

Using the MLEs $\widehat{b} = 99.2079$ and $\widehat{c} = 2.5957$, we find

$$\widehat{x}_{0.10} = \widehat{b}\,(-\ln 0.90)^{1/\widehat{c}} = 41.69,$$

so that the test statistic in (21.45a) is

$$\sqrt{n}\,\widehat{c}\,\ln\left\{\frac{\widehat{x}_{0.10}}{x^{(0)}_{0.10}}\right\} = \sqrt{20}\cdot 2.5927\cdot\ln\left(\frac{41.69}{30}\right) = 3.82.$$

From Tab. 19/4a we take $u_{0.05}(0.90) = 5.548$, so we cannot reject $H_0$ because $3.82 \not> u_{0.95}(0.90) = 5.548$.

# 22 WEIBULL goodness-of-fit testing and related problems

Before applying the WEIBULL distribution to solve any real problem one should ask whether this distribution is the "right model," i.e., whether it adequately fits the sampled data. The question here is: "WEIBULL or not?" The answer is given by carrying out a **goodness–of–fit test** (see Sect. 22.1).

Sometimes we have competing distributions which for some reasons seem equally appropriate to describe the sampled data. Thus the question is: "WEIBULL or what other distribution?" Here we have a **discrimination problem** (see Sect. 21.2).

A special discriminating problem arises when we are sure that all competing distributions are WEIBULL. Then we want to select the one which is best according to a given criterion, or we want to select those which are better than a given control distribution. **Selection** will be treated in Sect. 22.3. We should mention that some authors use *selection* when the choice is between different families of distributions, but we have restricted this term for choosing among the members of one family, namely the WEIBULL family.

## 22.1 Goodness-of-fit testing[1]

There are several approaches to see whether a given dataset can be adequately described by some distributional model. On the one hand, we have graphical procedures such as the hazard plot or the probability plot for location–scale type distributions (see Chapter 9), and on the other hand, there exist many numerical procedures which are proper tests insofar as they are characterized by a level of significance indicating the probability of erroneously rejecting the distribution asserted under $H_0$. The latter class of procedures comprises the following:

- tests based on $\chi^2$ as a measure of fit;
- tests based on the empirical distribution function (**EDF test**);
- tests using higher moments like skewness and kurtosis;
- tests based on order statistics like correlation tests, regression tests and gap–ratio tests;
- tests based on the sample entropy;
- tests based on the empirical characteristic function.

Most goodness–of–fit testing in the WEIBULL case relies upon the EDF statistic or on certain functions of order statistics.

---

[1] Suggested reading for this section is the monograph of D'AGOSTINO/STEPHENS (1986).

### 22.1.1 Tests of $\chi^2$-type[2]

The $\chi^2$ goodness–of–fit test asks for great sample sizes which are seldom met in life testing. On the other hand, this type of test does not cause trouble when the hypothetical distribution is not fully specified, i.e., when some or even all of its parameters have to be estimated and/or when the sample is censored in one way or another. Let $F(x \mid \boldsymbol{\theta})$ be the CDF of a variate $X$ depending on one or more parameters collected in a vector $\boldsymbol{\theta}$. The null hypothesis is

$$\mathrm{H}_0: \; F_X(x) = F_0(x \mid \boldsymbol{\theta}),$$

where $\boldsymbol{\theta}$ is either known or partially or even completely unknown. The $\chi^2$ goodness–of–fit test consists of the following steps:

1. The $n$ sampled data are classified or grouped, $n_i$ being the empirical frequency in class $i$; $i = 1, \ldots, k$. We ask for $n_i \geq 10 \; \forall \, i$ because the test only holds asymptotically. When there are $n_i < 10$, we have to combine adjacent classes.

2. Let there be $m$ unknown (unspecified) parameters in $\boldsymbol{\theta}$. These have to be estimated; usually the MLEs are used.

3. For class $i$ with limits $x_{i-1}$ and $x_i$, we have to calculate

$$P_i = F_0(x_i \mid \boldsymbol{\theta}) - F_0(x_{i-1} \mid \boldsymbol{\theta}) \tag{22.1a}$$

or

$$\widehat{P}_i = F_0(x_i \mid \widehat{\boldsymbol{\theta}}) - F_0(x_{i-1} \mid \widehat{\boldsymbol{\theta}}). \tag{22.1b}$$

4. Then we calculate the expected frequency for each class $i$ either as

$$E_i = n \, P_i \;\; \text{or as} \;\; E_i = n \, \widehat{P}_i, \tag{22.1c}$$

demanding no $E_i$ less than 1 and not more than 20% of the $E_i$ being less than 5, otherwise we again have to combine adjacent classes.

5. We calculate the test statistic

$$\chi^2 = \sum_{i=1}^{k} \frac{(n_i - E_i)^2}{E_i}. \tag{22.1d}$$

$k \geq 2$ is the number of classes after probable combinations.

6. Reject $\mathrm{H}_0: \; F_X(x) = F_0(x \mid \boldsymbol{\theta})$ with level $\alpha$ when

$$\chi^2 > \chi^2_{k-m-1, 1-\alpha}, \tag{22.1e}$$

$m$ being the number of estimated parameters.

$\chi^2$–tests are generally less powerful than EDF tests and special purpose tests of fit.

### 22.1.2 Tests based on EDF statistics

In the following we consider tests of fit which start with the empirical distribution function (EDF). The EDF is a step function, calculated from the sample observing any mode of censoring, which estimates the population distribution function. The EDF statistics are measures of the discrepancy between the EDF and a given distribution function, and are used for testing the fit of the sample to the distribution which may be completely specified

---

[2] Suggested reading for this section: FREEMAN/FREEMAN/KOCH (1978).

(**simple hypothesis**) or may contain parameters which have to be estimated from the sample (**composite hypothesis**).

### 22.1.2.1 Introduction

Given an uncensored sample, the EDF is defined by

$$
F_n(x) = \begin{cases} 0 & \text{for} \quad x < X_{1:n}; \\ \dfrac{i}{n} & \text{for} \quad X_{i:n} \leq x < X_{i+1:n}; \ i = 1, \dots, n-1; \\ 1 & \text{for} \quad x \geq X_{n:n}. \end{cases} \tag{22.2}
$$

A statistic measuring the discrepancy between $F_n(x)$ and a hypothetical distribution with CDF $F_0(x \mid \boldsymbol{\theta})$ will be called an **EDF statistic**. EDF statistics have already been used in Sect. 13.2 to define minimum distance estimators. The EDF statistics most often used are based on the vertical differences between $F_n(x)$ and $F_0(x \mid \boldsymbol{\theta})$ (see Fig. 22/1 further down) and are divided into two classes, the **supremum class** and the **quadratic class**.

The following formulas give the definitions of the supreme statistics.

- $D^+$ is the largest vertical difference when $F_n(x)$ is greater than $F_0(x \mid \boldsymbol{\theta})$:

$$
D^+ := \sup_x \left\{ F_n(x) - F_0(x \mid \boldsymbol{\theta}) \right\}. \tag{22.3a}
$$

- $D^-$ is the largest vertical difference when $F_n(x)$ is less than $F_0(x \mid \boldsymbol{\theta})$:

$$
D^- := \sup_x \left\{ F_0(x \mid \boldsymbol{\theta}) - F_n(x) \right\}. \tag{22.3b}
$$

- The most well–known EDF statistic is the **Kolmogorov–Smirnov statistic**

$$
D := \sup_x \left\{ \left| F_n(x) - F_0(x \mid \boldsymbol{\theta}) \right| \right\} = \max \left\{ D^+, D^- \right\}. \tag{22.3c}
$$

- A closely related statistic is the **Kuiper statistic**

$$
V := D^+ + D^-. \tag{22.3d}
$$

$D^+$ can be used to give a one–sided test, i.e., to test that the distribution is $F_0(X \mid \boldsymbol{\theta})$ against alternatives $F_A(x \mid \boldsymbol{\theta})$ for which, ideally, $F_A(x \mid \boldsymbol{\theta}) \geq F_0(X \mid \boldsymbol{\theta}) \ \forall \ x$. $D^+$ will give good power if $F_A(x \mid \boldsymbol{\theta})$ is greater than $F_0(X \mid \boldsymbol{\theta})$ over most of the range, implying that the true mean is less than that hypothesized. Similarly, $D^-$ will be used to guard against alternatives with a larger mean. $D$ gives a test against shifts in mean in either direction and thus leads to a two–sided test. Statistic $V$, besides its use on the circle, will detect a shift in variance, represented by clusters of $Z_i = F_0(X_{i:n} \mid \boldsymbol{\theta})$, or values of $Z_i$ in two groups near 0 and 1.

The quadratic class is given by the CRAMÉR–VON MISES family

$$
Q := n \int_{-\infty}^{\infty} \left[ F_n(x) - F_0(x \mid \boldsymbol{\theta}) \right]^2 w(x) \, \mathrm{d}F_0(x \mid \boldsymbol{\theta}), \tag{22.4a}
$$

where $w(x)$ is a weight function. Most important in this family are the following:

- the **CRAMÉR–VON MISES** statistic for $\omega(x) = 1$

$$W^2 := n \int_{-\infty}^{\infty} \left[ F_n(x) - F_0(x \,|\, \boldsymbol{\theta}) \right]^2 \mathrm{d}F_0(x \,|\, \boldsymbol{\theta}), \qquad (22.4\mathrm{b})$$

- the **ANDERSON–DARLING** statistic for $\omega(x) = \left\{ F_0(x \,|\, \boldsymbol{\theta}) \left[ 1 - F_0(x \,|\, \boldsymbol{\theta}) \right] \right\}^{-1}$

$$A^2 := n \int_{-\infty}^{\infty} \frac{\left[ F_n(x) - F_0(x \,|\, \boldsymbol{\theta}) \right]^2}{F_0(x \,|\, \boldsymbol{\theta}) \left[ 1 - F_0(x \,|\, \boldsymbol{\theta}) \right]} \, \mathrm{d}F_0(x \,|\, \boldsymbol{\theta}), \qquad (22.4\mathrm{c})$$

- the **WATSON** statistic

$$U^2 := n \int_{-\infty}^{\infty} \left\{ F_n(x) - F_0(x \,|\, \boldsymbol{\theta}) - \int_{-\infty}^{\infty} \left[ F_n(x) - F_0(x \,|\, \boldsymbol{\theta}) \right] \mathrm{d}F_0(x \,|\, \boldsymbol{\theta}) \right\}^2 \mathrm{d}F_0(x \,|\, \boldsymbol{\theta}),$$

$$(22.4\mathrm{d})$$

as a modification of $W^2$ and devised originally for the circle like the KUIPER statistic $V$. $U^2$ and $V$ are invariant with respect to the origin of $x$.

A goodness–of–fit test using a member of the quadratic class is usually more powerful than a test based on a member of the supreme class, because $Q$ is explicitly defined on all discrepancies. $A^2$ gives weight to observations in the tails and so tends to detect alternatives where more such observations will arise.

The foregoing definitions of EDF statistics can be turned into straightforward computing formulas when $F_n(x)$ is continuous. Let

$$Z_i = F_0(X_{i:n} \,|\, \boldsymbol{\theta}), \qquad (22.5\mathrm{a})$$

then

$$D^+ = \max_{1 \le i \le n} \left\{ \frac{i}{n} - Z_i \right\}, \qquad (22.5\mathrm{b})$$

$$D^- = \max_{1 \le i \le n} \left\{ Z_i - \frac{i}{n} \right\}, \qquad (22.5\mathrm{c})$$

$$D = \max \left\{ D^+, D^- \right\}, \qquad (22.5\mathrm{d})$$

$$V = D^+ + D^-, \qquad (22.5\mathrm{e})$$

$$W^2 = \frac{1}{12\,n} + \sum_{i=1}^{n} \left[ Z_i - \frac{2\,i - 1}{2\,n} \right]^2, \qquad (22.5\mathrm{f})$$

$$U^2 = W^2 - n \left( \overline{Z} - 0.5 \right)^2, \quad \overline{Z} = \frac{1}{n} \sum_{i=1}^{n} Z_i, \qquad (22.5\mathrm{g})$$

$$A^2 = -n - \frac{1}{n} \sum_{i=1}^{n} (2\,i - 1) \left[ \ln Z_i + \ln(1 - Z_{n+1-i}) \right] \qquad (22.5\mathrm{h})$$

$$= -n - \frac{1}{n} \sum_{i=1}^{n} \left\{ (2\,i - 1) \ln Z_i + (2\,n + 1 - 2\,i) \ln(1 - Z_i) \right\}. \qquad (22.5\mathrm{i})$$

The KOLMOGOROV–SMIRNOV statistic has also been used to construct so–called **consonance sets**, see SALVIA (1979), and **confidence bands** for the CDF, see SRINIVASAN/WHARTON (1975). $D$ has also been applied to the plotting coordinates of the **stabilized probability plot**; see COLES (1989) and KIMBER (1985). To assess whether an ordered sample $X_{1:n} \leq \ldots \leq X_{n:n}$ has been drawn from a parent population of location–scale type with distribution function $F_0\big[(x-a)/b\big]$, the plotting coordinates, $(R_i, S_i)$, of the stabilized plot are calculated as

$$R_i = \frac{2}{\pi} \sin^{-1} \left\{ \sqrt{F_0\big[(X_{i:n} - a)/b\big]} \right\}, \tag{22.6a}$$

$$S_i = \frac{2}{\pi} \sin^{-1} \left\{ \sqrt{(i - 0.5)/n} \right\}. \tag{22.6b}$$

The properties of this plot are as follows:

- Deviations from the line joining the points $(0,0)$ and $(1,1)$ indicate departures from the proposed distribution.
- The plotted points have approximately equal variances due to the asymptotic properties of $S_i$. (As $n \to \infty$ and $i/n \to p$, the asymptotic variance of $S_i \sqrt{n}$ is $\pi^{-2}$ and thus independent of $p$.)

The KOLMOGOROV–SMIRNOV statistic based on $S_i$ and $R_i$ is

$$D_{sp} = \max_{1 \leq i \leq n} \left| R_i - S_i \right|. \tag{22.6c}$$

COLES (1989) shows that $D_{sp}$ has a high power when the parameters are estimated using a procedure due to BLOM (1958) and gives critical values for $D_{sp}$ when $F_0(\cdot)$ is the extreme–value distribution of the minimum which after a log–transformation is also applicable to the WEIBULL distribution.

### 22.1.2.2 Fully specified distribution and uncensored sample

When the null hypothesis is

$$H_0: \ F(x) = F_0(x \,|\, \boldsymbol{\theta} = \boldsymbol{\theta_0}), \tag{22.7}$$

i.e., we declare a certain distribution with fixed parameter vector $\boldsymbol{\theta_0}$ to be the parent population of the sample, the test is given by the following steps:

1. Put the complete sample data $X_i$ in ascending order: $X_{1:n} \leq \ldots \leq X_{n:n}$.
2. Calculate $Z_i = F_0\big(X_{i:n} \,|\, \boldsymbol{\theta_0}\big)$, $i = 1, \ldots, n$.
3. Calculate the statistic desired from formulas (22.5b–i).
4. Modify the test statistic as in Tab. 22/1 using the modification[3] for the upper tail, and compare with the appropriate table of upper tail percentage points. If the statistic

---

[3] The modifications are made in order to have a good approximation to the asymptotic distribution whose percentage points are given in Tab. 22/1. The quadratic statistics converge more rapidly to their asymptotic distributions, and the modifications are relatively minor. The supremum statistics converge more slowly.

exceeds[4] the value in the upper tail given at level $\alpha$, $H_0$ is rejected at significance level $\alpha$.

Table 22/1: Modifications and percentage points of EDF statistics based on an uncensored sample for testing a hypothesis on a completely specified distribution

| Statistic | | Significance level $\alpha$ | | | | | | | |
|---|---|---|---|---|---|---|---|---|---|
| $T$ | Modified form $T^*$ | 0.25 | 0.15 | 0.10 | 0.05 | 0.025 | 0.01 | 0.005 | 0.001 |
| | | upper tail percentage points | | | | | | | |
| $D^+(D^-)$ | $D^+\left(\sqrt{n}+0.12+0.11/\sqrt{n}\right)$ | 0.828 | 0.973 | 1.073 | 1.224 | 1.358 | 1.518 | 1.628 | 1.859 |
| $D$ | $D\left(\sqrt{n}+0.12+0.11/\sqrt{n}\right)$ | 1.019 | 1.138 | 1.224 | 1.358 | 1.480 | 1.628 | 1.731 | 1.950 |
| $V$ | $V\left(\sqrt{n}+0.155+0.24/\sqrt{n}\right)$ | 1.420 | 1.537 | 1.620 | 1.747 | 1.862 | 2.001 | 2.098 | 2.303 |
| $W^2$ | $\left(W^2-0.4/n+0.6/n^2\right)\left(1.0+1.0/n\right)$ | 0.209 | 0.284 | 0.347 | 0.461 | 0.581 | 0.743 | 0.869 | 1.167 |
| $U^2$ | $\left(U^2-0.1/n+0.1/n^2\right)\left(1.0+0.8/n\right)$ | 0.105 | 0.131 | 0.152 | 0.187 | 0.222 | 0.268 | 0.304 | 0.385 |
| $A^2$ | for all $n\geq 5$ | 1.248 | 1.610 | 1.933 | 2.492 | 3.070 | 3.880 | 4.500 | 6.000 |
| | | lower tail percentage points | | | | | | | |
| $D$ | $D\left(\sqrt{n}+0.275-0.04/\sqrt{n}\right)$ | – | 0.610 | 0.571 | 0.520 | 0.481 | 0.441 | – | – |
| $V$ | $V\left(\sqrt{n}+0.41-0.26/\sqrt{n}\right)$ | – | 0.976 | 0.928 | 0.861 | 0.810 | 0.755 | – | – |
| $W^2$ | $\left(W^2-0.03/n\right)\left(1.0+0.05/n\right)$ | – | 0.054 | 0.046 | 0.037 | 0.030 | 0.025 | – | – |
| $U^2$ | $\left(U^2-0.02/n\right)\left(1.0+0.35/n\right)$ | – | 0.038 | 0.033 | 0.028 | 0.024 | 0.020 | – | – |
| $A^2$ | for all $n\geq 5$ | – | 0.399 | 0.346 | 0.283 | 0.240 | 0.201 | – | – |

Source: D'AGOSTINO/STEPHENS (1986, p. 105) — Reprinted with permission from *Goodness–of–fit Techniques*. Copyright 1986 by Marcel Dekker, Inc. All rights reserved.

**Example 22/1:  Testing whether dataset #1 comes from $We(a=0, b=80, c=1.3)$**

The sample of size $n=20$ (dataset #1 in Tab. 9/2) has been generated from $We(a=0, b=100, c=2.5)$, but here we maintain that it comes from $We(a=0, b=80, c=1.3)$; i.e., we are going to test

$$H_0:\ F(x)=F_0(x\,|\,\boldsymbol{\theta}=\boldsymbol{\theta_0})=1-\exp\left\{-\left(\frac{x}{80}\right)^{1.3}\right\}.$$

Fig. 22/1 shows $F_n(x)$ together with $F_0(x\,|\,\boldsymbol{\theta}=\boldsymbol{\theta_0})$, $\boldsymbol{\theta_0}=(0,80,1.3)'$. The test statistics (22.5b–i) and their modifications according to Tab. 22/1 are given in the following table.

| Statistic | $D^+$ | $D^-$ | $D$ | $V$ | $W^2$ | $U^2$ | $A^2$ |
|---|---|---|---|---|---|---|---|
| original | 0.0872 | 0.3169 | 0.3169 | 0.4040 | 0.6185 | 0.2551 | 3.0445 |
| modified | 0.4025 | 1.4629 | 1.4629 | 1.8912 | 0.6300 | 0.2604 | 3.0445 |

---

[4]  A small value of this statistic will indicate that the $Z_i$ are **superuniform**, i.e., more regular than expected for an ordered uniform sample. To detect superuniformity the lower tail of, say, $D$ would be used.

Reference to the percentage points on the same line as the modification in Tab. 22/1 shows that $D$ is significant at the 5% level and $V$, $W^2$, $U^2$ as well as $A^2$ are all significant at the 2.5% level.

<u>Figure 22/1</u>: EDF for dataset #1 and $F_0(x \mid \boldsymbol{\theta}) = 1 - \exp\left\{ - \left( \dfrac{x}{80} \right)^{1.3} \right\}$

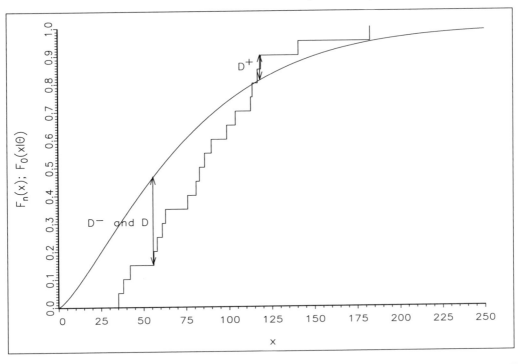

With respect to the power of the above statistics, we can state the following:

1. $D$ is often much less powerful than $W^2$ and $A^2$.

2. $D^+$, $D^-$, $A^2$, $W^2$ and $D$ will detect an error in the mean of $F_0(x \mid \boldsymbol{\theta})$ as specified, and $U^2$ and $V$ will detect an error in the variance.

3. $A^2$ often behaves similarly to $W^2$, but is on the whole more powerful for tests when $F_0(x \mid \boldsymbol{\theta})$ departs from the true distribution in the tails, especially when there appears to be too many outliers for $F_0(x \mid \boldsymbol{\theta})$ as specified. In goodness–of–fit testing, departures in the tails are often important to detect, and $A^2$ is the recommended statistic.

### 22.1.2.3    Fully specified distribution and censored sample

When the sample is censored, we have to adapt the EDF statistic of the previous sections, and we also have to use other percentage points than those given in Tab. 22/1. We will discuss only singly–censoring on the right as the most important case in life testing; other forms of censoring are discussed in D'AGOSTINO/STEPHENS (1986).

The probability integral transformation may be made for the observations available, giving a set of $Z_i = F_0(X_{i:n} \mid \boldsymbol{\theta})$ which itself is censored. Suppose, the $X$–set is right–censored of type–I, then the values of $X$ are known to be less than the fixed value $x^*$, and the available

$Z_i$ are $Z_1 < Z_2 < \ldots < Z_r < t$, where $t = F(x^* \mid \boldsymbol{\theta})$. If the censoring is of type–II, there are again $r$ values of $Z_i$, with $Z_r$ the largest and $r$ fixed.

<u>KOLMOGOROV–SMIRNOV statistic</u>

The modification for type–I censored date is

$$_1D_{t,n} = \max_{1 \leq i \leq r} \left\{ \frac{i}{n} - Z_i, \; Z_i - \frac{i-1}{n}, t - \frac{r}{n} \right\}, \qquad (22.8a)$$

and for type–II censored data, we have

$$_2D_{t,r} = \max_{1 \leq i \leq r} \left\{ \frac{i}{n} - Z_i, \; Z_i - \frac{i-1}{n} \right\}. \qquad (22.8b)$$

D'AGOSTINO/STEPHENS (1986) have suggested the following procedure to test $H_0$ : $F(x) = F_0(x \mid \boldsymbol{\theta} = \boldsymbol{\theta_0})$, when the values of $Z_i = F(X_{i:n} \mid \boldsymbol{\theta}); \; i = 1, \ldots, r$ have been calculated.

<u>Type–I censoring</u>

1. Calculate $_1D_{t,n}$ from (22.8a).

2. Modify $_1D_{t,n}$ to $D_t^*$ according to [5]

$$D_t^* = \; _1D_{t,n} \sqrt{n} + 0.19/\sqrt{n} \; \text{ for } \; n \geq 25, \, t = p \geq 0.25. \qquad (22.9a)$$

3. Refer to Tab. 22/2 and reject $H_0$ at significance level $\alpha$ if $D_t^*$ exceeds the tabulated value for $\alpha$.

<u>Type–II censoring</u>

1. Calculate $_2D_{r,n}$ from (22.8b).

2. Modify $_2D_{r,n}$ to $D_r^*$ according to [5]

$$D_r^* = \; _2D_{r,n} \sqrt{n} + 0.24/\sqrt{n} \; \text{ for } \; n \geq 25, \, t = p \geq 0.40. \qquad (22.9b)$$

3. Refer to Tab. 22/2 and reject $H_0$ at significance level $\alpha$ if $D_r^*$ exceeds the tabulated value for $\alpha$.

---

[5] For values of $n < 25$ or for censoring more extreme than the range given above, refer to tables of BARR/ DAVIDSON (1973) or DUFOUR/MAAG (1978).

Table 22/2: Upper tail asymptotic percentage points of $\sqrt{n}\,D$, $W^2$ and $A^2$ for type-I or type-II censored data

| $p$ \ $\alpha$ | 0.50 | 0.25 | 0.15 | 0.10 | 0.05 | 0.025 | 0.01 | 0.005 |
|---|---|---|---|---|---|---|---|---|
| | | | | $\sqrt{n}\,D$ | | | | |
| 0.2 | 0.4923 | 0.6465 | 0.7443 | 0.8155 | 0.9268 | 1.0282 | 1.1505 | 1.2361 |
| 0.3 | 0.5889 | 0.7663 | 0.8784 | 0.9597 | 1.0868 | 1.2024 | 1.3419 | 1.4394 |
| 0.4 | 0.6627 | 0.8544 | 0.9746 | 1.0616 | 1.1975 | 1.3209 | 1.4696 | 1.5735 |
| 0.5 | 0.7204 | 0.9196 | 1.0438 | 1.1334 | 1.2731 | 1.3997 | 1.5520 | 1.6583 |
| 0.6 | 0.7649 | 0.9666 | 1.0914 | 1.1813 | 1.3211 | 1.4476 | 1.5996 | 1.7056 |
| 0.7 | 0.7975 | 0.9976 | 1.1208 | 1.2094 | 1.3471 | 1.4717 | 1.6214 | 1.7258 |
| 0.8 | 0.8183 | 1.0142 | 1.1348 | 1.2216 | 1.3568 | 1.4794 | 1.6272 | 1.7306 |
| 0.9 | 0.8270 | 1.0190 | 1.1379 | 1.2238 | 1.3581 | 1.4802 | 1.6276 | 1.7308 |
| 1.0 | 0.8276 | 1.0192 | 1.1379 | 1.2238 | 1.3581 | 1.4802 | 1.6276 | 1.7308 |
| | | | | $W^2$ | | | | |
| 0.2 | 0.010 | 0.025 | 0.033 | 0.041 | 0.057 | 0.074 | 0.094 | .110 |
| 0.3 | 0.022 | 0.046 | 0.066 | 0.083 | 0.115 | 0.147 | 0.194 | .227 |
| 0.4 | 0.037 | 0.076 | 0.105 | 0.136 | 0.184 | 0.231 | 0.295 | .353 |
| 0.5 | 0.054 | 0.105 | 0.153 | 0.186 | 0.258 | 0.330 | 0.427 | .488 |
| 0.6 | 0.070 | 0.136 | 0.192 | 0.241 | 0.327 | 0.417 | 0.543 | .621 |
| 0.7 | 0.088 | 0.165 | 0.231 | 0.286 | 0.386 | 0.491 | 0.633 | .742 |
| 0.8 | 0.103 | 0.188 | 0.259 | 0.321 | 0.430 | 0.544 | 0.696 | .816 |
| 0.9 | 0.115 | 0.204 | 0.278 | 0.341 | 0.455 | 0.573 | 0.735 | .865 |
| 1.0 | 0.119 | 0.209 | 0.284 | 0.347 | 0.461 | 0.581 | 0.743 | .869 |
| | | | | $A^2$ | | | | |
| 0.2 | 0.135 | 0.252 | 0.333 | 0.436 | 0.588 | 0.747 | 0.962 | 1.129 |
| 0.3 | 0.204 | 0.378 | 0.528 | 0.649 | 0.872 | 1.106 | 1.425 | 1.731 |
| 0.4 | 0.275 | 0.504 | 0.700 | 0.857 | 1.150 | 1.455 | 1.872 | 2.194 |
| 0.5 | 0.349 | 0.630 | 0.875 | 1.062 | 1.419 | 1.792 | 2.301 | – |
| 0.6 | 0.425 | 0.756 | 1.028 | 1.260 | 1.676 | 2.112 | 2.707 | – |
| 0.7 | 0.504 | 0.882 | 1.184 | 1.451 | 1.920 | 2.421 | 3.083 | – |
| 0.8 | 0.588 | 1.007 | 1.322 | 1.623 | 2.146 | 2.684 | 3.419 | – |
| 0.9 | 0.676 | 1.131 | 1.467 | 1.798 | 2.344 | 2.915 | 3.698 | – |
| 1.0 | 0.779 | 1.248 | 1.610 | 1.933 | 2.492 | 3.070 | 3.880 | 4.500 |

<u>Table 22/3:</u> Upper tail percentage points of $_2W_{r,n}^2$ and for type-II censored data

| $p$ | $n$ | $\alpha$ 0.50 | 0.25 | 0.15 | 0.10 | 0.05 | 0.025 | 0.01 |
|---|---|---|---|---|---|---|---|---|
| 0.2 | 20 | 0.006 | 0.018 | 0.038 | 0.058 | 0.099 | 0.152 | 0.243 |
| | 40 | 0.008 | 0.018 | 0.032 | 0.046 | 0.084 | 0.128 | 0.198 |
| | 60 | 0.009 | 0.020 | 0.031 | 0.044 | 0.074 | 0.107 | 0.154 |
| | 80 | 0.009 | 0.021 | 0.031 | 0.043 | 0.069 | 0.097 | 0.136 |
| | 100 | 0.009 | 0.022 | 0.031 | 0.043 | 0.066 | 0.092 | 0.127 |
| | $\infty$ | 0.010 | 0.025 | 0.031 | 0.041 | 0.057 | 0.074 | 0.094 |
| 0.4 | 10 | 0.022 | 0.056 | 0.101 | 0,144 | 0.229 | 0.313 | 0.458 |
| | 20 | 0.029 | 0.062 | 0.095 | 0.132 | 0.209 | 0.297 | 0.419 |
| | 40 | 0.033 | 0.067 | 0.100 | 0.128 | 0.191 | 0.267 | 0.381 |
| | 60 | 0.034 | 0.070 | 0.102 | 0.130 | 0.189 | 0.256 | 0.354 |
| | 80 | 0.035 | 0.071 | 0.103 | 0.132 | 0.187 | 0.251 | 0.342 |
| | 100 | 0.035 | 0.072 | 0.103 | 0.132 | 0.187 | 0.248 | 0.335 |
| | $\infty$ | 0.037 | 0.076 | 0.105 | 0.135 | 0.184 | 0.236 | 0.307 |
| 0.6 | 10 | 0.053 | 0.107 | 0.159 | 0.205 | 0.297 | 0.408 | 0.547 |
| | 20 | 0.062 | 0.122 | 0.172 | 0.216 | 0.302 | 0.408 | 0.538 |
| | 40 | 0.067 | 0.128 | 0.180 | 0.226 | 0.306 | 0.398 | 0.522 |
| | 60 | 0.068 | 0.131 | 0.184 | 0.231 | 0.313 | 0.404 | 0.528 |
| | 80 | 0.068 | 0.132 | 0.186 | 0.233 | 0.316 | 0.407 | 0.531 |
| | 100 | 0.069 | 0.133 | 0.187 | 0.235 | 0.318 | 0.409 | 0.532 |
| | $\infty$ | 0.070 | 0.136 | 0.192 | 0.241 | 0.327 | 0.417 | 0.539 |
| 0.8 | 10 | 0.085 | 0.158 | 0.217 | 0.266 | 0.354 | 0.453 | 0.593 |
| | 20 | 0.094 | 0.172 | 0.235 | 0.289 | 0.389 | 0.489 | 0.623 |
| | 40 | 0.099 | 0.180 | 0.247 | 0.303 | 0.401 | 0.508 | 0.651 |
| | 60 | 0.100 | 0.183 | 0.251 | 0.308 | 0.410 | 0.520 | 0.667 |
| | 80 | 0.101 | 0.184 | 0.253 | 0.311 | 0.415 | 0.526 | 0.675 |
| | 100 | 0.101 | 0.185 | 0.254 | 0.313 | 0.418 | 0.529 | 0.680 |
| | $\infty$ | 0.103 | 0.188 | 0.259 | 0.320 | 0.430 | 0.544 | 0.700 |
| 0.9 | 10 | 0.094 | 0.183 | 0.246 | 0.301 | 0.410 | 0.502 | 0.645 |
| | 20 | 0.109 | 0.194 | 0.263 | 0.322 | 0.431 | 0.536 | 0.675 |
| | 40 | 0.112 | 0.199 | 0.271 | 0.330 | 0.437 | 0.546 | 0.701 |
| | 60 | 0.113 | 0.201 | 0.273 | 0.333 | 0.442 | 0.553 | 0.713 |
| | 80 | 0.114 | 0.202 | 0.274 | 0.335 | 0.445 | 0.558 | 0.718 |
| | 100 | 0.114 | 0.202 | 0.275 | 0.336 | 0.447 | 0.561 | 0.722 |
| | $\infty$ | 0.115 | 0.204 | 0.278 | 0.341 | 0.455 | 0.573 | 0.735 |
| 0.95 | 10 | 0.103 | 0.198 | 0.266 | 0.324 | 0.430 | 0.534 | 0.676 |
| | 20 | 0.115 | 0.201 | 0.275 | 0.322 | 0.444 | 0.551 | 0.692 |
| | 40 | 0.115 | 0.205 | 0.280 | 0.329 | 0.448 | 0.557 | 0.715 |
| | 60 | 0.116 | 0.207 | 0.280 | 0.338 | 0.451 | 0.562 | 0.724 |
| | 80 | 0.117 | 0.208 | 0.281 | 0.340 | 0.453 | 0.566 | 0.729 |
| | 100 | 0.117 | 0.208 | 0.282 | 0.341 | 0.454 | 0.569 | 0.735 |
| | $\infty$ | 0.118 | 0.208 | 0.283 | 0.346 | 0.460 | 0.579 | 0.742 |
| 1.0 | 10 | 0.117 | 0.212 | 0.288 | 0.349 | 0.456 | 0.564 | 0.709 |
| | 20 | 0.116 | 0.212 | 0.288 | 0.350 | 0.459 | 0.572 | 0.724 |
| | 40 | 0.115 | 0.211 | 0.288 | 0.350 | 0.461 | 0.576 | 0.731 |
| | 100 | 0.115 | 0.211 | 0.288 | 0,351 | 0.462 | 0.578 | 0.736 |
| | $\infty$ | 0.119 | 0.209 | 0.284 | 0.347 | 0.461 | 0.581 | 0.743 |

<u>Source:</u> D'AGOSTINO/STEPHENS (1986, pp. 116/117) — Reprinted with permission from *Goodness–of–fit Techniques.* Copyright 1986 by Marcel Dekker, Inc. All rights reserved.

**Table 22/4:** Upper tail percentage points of $_2A^2_{r,n}$ and for type-II censored data

| $p$ | $n$ | $\alpha$ 0.50 | 0.25 | 0.15 | 0.10 | 0.05 | 0.025 | 0.01 |
|-----|-----|------|------|------|------|------|-------|------|
| | 20 | 0.107 | 0.218 | 0.337 | 0.435 | 0.626 | 0.887 | 1.278 |
| | 40 | 0.119 | 0.235 | 0.337 | 0.430 | 0.607 | 0.804 | 1.111 |
| | 60 | 0.124 | 0.241 | 0.341 | 0.432 | 0.601 | 0.785 | 1.059 |
| 0.2 | 80 | 0.127 | 0.243 | 0.344 | 0.433 | 0.598 | 0.775 | 1.034 |
| | 100 | 0.128 | 0.245 | 0.345 | 0.434 | 0.596 | 0.769 | 1.019 |
| | $\infty$ | 0.135 | 0.252 | 0.351 | 0.436 | 0.588 | 0.747 | 0.962 |
| | 10 | 0.214 | 0.431 | 0.627 | 0.803 | 1.127 | 1.483 | 2.080 |
| | 20 | 0.241 | 0.462 | 0.653 | 0.824 | 1.133 | 1.513 | 2.011 |
| | 40 | 0.261 | 0.487 | 0.681 | 0.843 | 1.138 | 1.460 | 1.903 |
| 0.4 | 60 | 0.265 | 0.493 | 0.686 | 0.848 | 1.142 | 1.458 | 1.892 |
| | 80 | 0.268 | 0.496 | 0.688 | 0.850 | 1.144 | 1.457 | 1.887 |
| | 100 | 0.269 | 0.497 | 0.689 | 0.851 | 1.145 | 1.457 | 1.884 |
| | $\infty$ | 0.275 | 0.504 | 0.695 | 0.857 | 1.150 | 1.455 | 1.872 |
| | 10 | 0.354 | 0.673 | 0.944 | 1.174 | 1.577 | 2.055 | 2.774 |
| | 20 | 0.390 | 0.713 | 0.984 | 1.207 | 1.650 | 2.098 | 2.688 |
| | 40 | 0.408 | 0.730 | 1.001 | 1.229 | 1.635 | 2.071 | 2.671 |
| 0.6 | 60 | 0.413 | 0.739 | 1.011 | 1.239 | 1.649 | 2.084 | 2.683 |
| | 80 | 0.416 | 0.743 | 1.017 | 1.244 | 1.655 | 2.091 | 2.689 |
| | 100 | 0.418 | 0.746 | 1.020 | 1.248 | 1.659 | 2.095 | 2.693 |
| | $\infty$ | 0.425 | 0.756 | 1.033 | 1.260 | 1.676 | 2.112 | 2.707 |
| | 10 | 0.503 | 0.913 | 1.237 | 1.498 | 2.021 | 2.587 | 3.254 |
| | 20 | 0.547 | 0.952 | 1.280 | 1.558 | 2.068 | 2.570 | 3.420 |
| | 40 | 0.568 | 0.983 | 1.321 | 1.583 | 2.088 | 2.574 | 3.270 |
| 0.8 | 60 | 0.574 | 0.991 | 1.330 | 1.596 | 2.107 | 2.610 | 3.319 |
| | 80 | 0.578 | 0.995 | 1.335 | 1.603 | 2.117 | 2.629 | 3.344 |
| | 100 | 0.580 | 0.997 | 1.338 | 1.607 | 2.123 | 2.640 | 3.359 |
| | $\infty$ | 0.588 | 1.007 | 1.350 | 1.623 | 2.146 | 2.684 | 3.419 |
| | 10 | 0.639 | 1.089 | 1.435 | 1.721 | 2.281 | 2.867 | 3.614 |
| | 20 | 0.656 | 1.109 | 1.457 | 1.765 | 2.295 | 2.858 | 3.650 |
| | 40 | 0.666 | 1.124 | 1.478 | 1.778 | 2.315 | 2.860 | 3.628 |
| 0.9 | 60 | 0.670 | 1.128 | 1.482 | 1.784 | 2.325 | 2.878 | 3.648 |
| | 80 | 0.671 | 1.130 | 1.485 | 1.788 | 2.330 | 2.888 | 3.661 |
| | 100 | 0.673 | 1.131 | 1.486 | 1.790 | 2.332 | 2.893 | 3.668 |
| | $\infty$ | 0.676 | 1.136 | 1.492 | 1.798 | 2.344 | 2.915 | 3.698 |
| | 10 | 0.707 | 1.170 | 1.525 | 1.842 | 2.390 | 2.961 | 3.745 |
| | 20 | 0.710 | 1.177 | 1.533 | 1.853 | 2.406 | 2.965 | 3.750 |
| | 40 | 0.715 | 1.184 | 1.543 | 1.860 | 2.416 | 2.968 | 3.743 |
| 0.95 | 60 | 0.717 | 1.186 | 1.545 | 1.263 | 2.421 | 2.977 | 3.753 |
| | 80 | 0.718 | 1.187 | 1.546 | 1.865 | 2.423 | 2.982 | 3.760 |
| | 100 | 0.719 | 1.188 | 1.547 | 1.866 | 2.424 | 2.984 | 3.763 |
| | $\infty$ | 0.720 | 1.190 | 1.550 | 1.870 | 2.430 | 2.995 | 3.778 |
| 1.0 | all $n$ | 0.775 | 1.248 | 1.610 | 1.933 | 2.492 | 3.070 | 3.880 |

Quadratic statistics

The computing formulas differ slightly for the two types of censoring. For type–II censoring we have

$$_2W^2_{r,n} = \sum_{i=1}^{r} \left( Z_i - \frac{2i-1}{2n} \right)^2 + \frac{r}{12n} + \frac{n}{3} \left( Z_r - \frac{r}{n} \right)^3, \tag{22.10a}$$

$$_2U^2_{r,n} = {}_2W^2_{r,n} - nZ_r \left( \frac{r}{n} - \frac{Z_r}{2} - \frac{r\overline{Z}}{nZ_r} \right), \quad \overline{Z} = \frac{1}{n}\sum_{i=1}^{r} Z_i, \tag{22.10b}$$

$$_2A^2_{r,n} = \left\{ \begin{array}{l} -\dfrac{1}{n}\sum_{i=1}^{r}(2i-1)\left[\ln Z_i - \ln(1-Z_i)\right] - 2\sum_{i=1}^{r}\ln(1-Z_i) - \\[2mm] \dfrac{1}{n}\left[(r-n)^2\ln(1-Z_r) - r^2\ln Z_r + n^2 Z_r\right]. \end{array} \right\} \tag{22.10c}$$

For type–I censoring, suppose $t$ $(t < 1)$ is the fixed censoring value of $Z$, i.e., $t = F(x^* \mid \boldsymbol{\theta})$. This value is added to the sample set $\{Z_1, \ldots, Z_r\}$ and the statistics are now calculated by (22.9a,b) with $r$ replaced by $r+1$, and with $Z_{r+1} = t$. They will be called $_1W^2_{t,n}$, $_1U^2_{t,n}$ and $_1A^2_{t,n}$. It is possible to have $r = n$ observations less than $t$, so that when the value $t$ is added, the new sample size is $n+1$. $_1W^2_{t,n}$ and $_2W^2_{r,n}$ have the same asymptotic distribution, similar to the other statistics. Asymptotic percentage points of $W^2$ are given in Tab. 22/3 and for $A^2$ in Tab. 22/4.

The steps for testing $H_0\colon F(x) = F_0(x \mid \boldsymbol{\theta} = \boldsymbol{\theta_0})$ with right–censored data are as follows:

1. Calculate the statistic as required.

2. Refer to Tab. 22/2 for type–I data and to Tab. 22/3 or Tab. 22/4 for type–II data, entering at $n$ and at $p = r/n$.

**Example 22/2:**  **Testing whether dataset #1 singly right–censored at $r = 15$ comes from $We(a = 0, b = 80, c = 1.3)$**

The EDF corresponding to the censored data is shown in Fig. 22/2 together with $F_0(x \mid \boldsymbol{\theta_0})$, $\boldsymbol{\theta_0} = (0, 80, 1.3)'$. The test statistics result as

| Statistic | $_2D_{15,20}$ | $_2W^2_{15,20}$ | $_2A^2_{15,20}$ |
|-----------|---------------|-----------------|-----------------|
| original  | 0.3169        | 0.6695          | 2.9523          |
| modified  | 1.4707        | –               | –               |

When we use the modified value of $_2D^2_{15,20}$, i.e., $_2D^*_{15,20} = 1.4707$, and go to Tab. 22/2 — despite $n = 20$ not being greater than 25 — we find for $p = r/n = 15/20 = 0.75$ by interpolation that we can reject $H_0$ at level 0.05. The table of DUFOUR/MAAG (1978), which has to be used with $_2D_{15,20} = 0.3169$, gives a level of 0.025 for rejecting $H_0$. For $_2W^2_{15,20} = 0.6695$, we find from Tab. 22/3 — by interpolation — a significance level less than 0.01, and for $_2A^2_{15,20} = 2.9523$ from Tab. 22/4 — again by interpolation — a significance level of approximately 0.01.

Figure 22/2: EDF and $F_0(x \mid \boldsymbol{\theta}) = 1 - \exp\left\{ - (x/80)^{1.3} \right\}$ for dataset #1 censored at $r = 15$

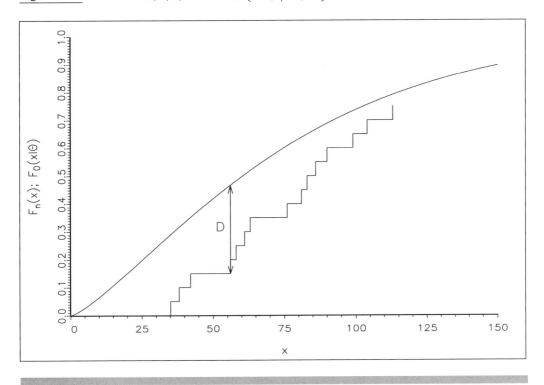

#### 22.1.2.4 Testing a composite hypothesis[6]

A composite hypothesis is characterized by the fact that some or all of the parameters in $\boldsymbol{\theta}$ are unknown and have to be estimated before calculating $F_0(X_{i:n} \mid \boldsymbol{\theta})$. When the unknown components of $\boldsymbol{\theta}$ are location and scale parameters, and if these are estimated by appropriate methods, e.g., by ML, the distributions of EDF statistics will not depend on the true values of these unknown parameters, but they depend on the family tested and on the sample size $n$. When there is censoring, the distributions of the EDF statistics will also depend on the mode of censoring and the censoring fraction; see AHO et al. (1983, 1985). When unknown parameters are not location or scale parameters, for example, when the shape parameter of a three–parameter WEIBULL distribution is unknown, the null distribution of the EDF statistic will depend on the true values of these parameters. However, if this dependence is very slight, then a set of tables, to be used with the estimated value of the shape parameter, can still be valuable; see, for example, LOCKHART/STEPHENS (1994) concerning the goodness–of–fit test of the three–parameter WEIBULL distribution. Nevertheless, the exact distributions of EDF statistics when testing a composite hypothesis are very difficult to find and Monte Carlo studies have to be used to find percentage points.

---

[6] Suggested reading for this section: AHO/BAIN/ENGELHARDT (1983, 1985), CHANDRA/ SINPUR-WALLA/ STEPHENS (1981), KHAMIS (1997), LIAO/SHIMOKAWA (1999), LITTELL/ MCCLAVE/OFFEN (1979), LOCKHART/STEPHENS (1994), SHIMOKAWA/LIAO (1999), STEPHENS (1977), WOZNIAK (1994).

In the following we will present only tests using one of the EDF statistics $D^+$, $D^-$, $D$, $W^2$, $U^2$ and $A^2$ for an uncensored sample when the two–parameter WEIBULL distribution has been transformed to an extreme–value distribution which is of the location–scale type.[7] Goodness–of–fit testing of a composite hypothesis with censored data will be presented in Sect. 22.1.3.

The null hypothesis in this section is

$$\text{H}_0: \text{ “}X_1, \ldots, X_n \text{ come from } We(a, b, c).\text{”}$$

We consider the case where $a$ is known. When $a = 0$, $\text{H}_0$ becomes

$$\text{H}_0: \text{ “The set } \{X_1, \ldots, X_n\} \text{ comes from } We(0, b, c).\text{”}$$

If $a \neq 0$, but we have $a = a_0$, $a_0$ known, the transformed set $\{X_1 - a_0, \ldots, X_n - a_0\}$ equally comes from $We(0, b, c)$ under $\text{H}_0$. In considering $\text{H}_0$ we distinguish three cases:

- Case 1: $c$ is known and $b$ is unknown.

- Case 2: $b$ is known and $c$ is unknown.

- Case 3: Both $b$ and $c$ are unknown.

For the test of $\text{H}_0$, the tables of percentage points for the **type–I extreme value distribution of the maximum** (see Tab. 22/5 and Tab. 22/6) may be used. Let

$$Y = -\ln X, \tag{22.11a}$$

where $X \sim We(0, b, c)$. The CDF of $Y$ becomes

$$F(y \mid \eta, \phi) = \exp\left\{-\exp\left[-\frac{y - \eta}{\phi}\right]\right\}, \quad y \in \mathbb{R}, \tag{22.11b}$$

with

$$\eta = -\ln b \quad \text{and} \quad \phi = 1/c.$$

A test for $\text{H}_0$: “The set $\{X_1, \ldots, X_n\}$ comes from $We(0, b, c)$” is made by testing that $Y$ has the above extreme–value distribution, with one or both of $\eta$ and $\phi$ unknown. The test procedure is as follows:

1. Make the transformation $Y_i = -\ln X_i$, $i = 1, \ldots, n$.

2. Arrange the $Y_i$ in ascending order. (If the $X_i$ were given in ascending order, the $Y_i$ will be in descending order!)

3. Test that the $Y$–sample is from (22.11b) as follows:

---

[7] KHAMIS (1997) used a $\delta$–corrected KOLMOGOROV–SMIRNOV test. LIAO/SHIMOKAWA (1999) introduced a new goodness–of–fit test with a test statistic being a combination of $D$, $W^2$ and $A^2$. The performance of $D$, $A^2$ and $W^2$ is studied by LITTELL et al. (1979), SHIMOAWA/LIAO (1999) and WOZNIAK (1994).

3.1 Estimate the unknown parameters

$$\widehat{\phi} = \frac{1}{n}\sum_{j=1}^{n}Y_j - \frac{\sum\limits_{j=1}^{n}Y_j \exp\left(-Y_j/\widehat{\phi}\right)}{\sum\limits_{j=1}^{n}\exp\left(-/\widehat{\phi}\right)} \qquad (22.11c)$$

by iteration and then

$$\widehat{\eta} = -\widehat{\phi}\,\ln\left\{\frac{1}{n}\sum_{j=1}^{n}\exp\left(-Y_j/\widehat{\phi}\right)\right\}. \qquad (22.11d)$$

In case 1 ($\phi = 1/c$ is known.) $\widehat{\eta}$ is given by (22.11d) with $\phi$ replacing $\widehat{\phi}$. In case 2 ($\eta = -\ln b$ is known.) $\widehat{\phi}$ is given by solving

$$\widehat{\phi} = \frac{1}{n}\left[\sum_{j=1}^{n}(Y_j - \eta) - \sum_{j=1}^{n}(Y_j - \eta)\exp\left\{-\frac{Y_j - \eta}{\widehat{\phi}}\right\}\right]. \qquad (22.11e)$$

3.2 Calculate

$$Z_i = F(Y_{i:n} \mid \eta, \phi),$$

where $F(\cdot)$ is given by (22.11b), using estimated parameters when necessary.

3.3 Use (22.5b–i) to calculate the EDF statistics.

3.4 Modify the test statistics as shown in Tab. 22/5 or use Tab. 22/6 and compare with the the upper tail percentage points given.

Table 22/5: Modifications and upper tail percentage points of $W^2$, $U^2$ and $A^2$ for the type-I extreme value distribution of the maximum

| Statistic | | Modification | Significance level $\alpha$ | | | | |
|---|---|---|---|---|---|---|---|
| | | | 0.25 | 0.10 | 0.05 | 0.025 | 0.01 |
| $W^2$ | Case 1 | $W^2\left(1+0.16/n\right)$ | 0.116 | 0.175 | 0.222 | 0.271 | 0.338 |
| | Case 2 | None | 0.186 | 0.320 | 0.431 | 0.547 | 0.705 |
| | Case 3 | $W^2\left(1+0.2/\sqrt{n}\right)$ | 0.073 | 0.102 | 0.124 | 0.146 | 0.175 |
| $U^2$ | Case 1 | $U^2\left(1+0.16/n\right)$ | 0.090 | 0.129 | 0.159 | 0.189 | 0.230 |
| | Case 2 | $U^2\left(1+0.15/\sqrt{n}\right)$ | 0.086 | 0.123 | 0.152 | 0.181 | 0.220 |
| | Case 3 | $U^2\left(1+0.2/\sqrt{n}\right)$ | 0.070 | 0.097 | 0.117 | 0.138 | 0.165 |
| $A^2$ | Case 1 | $A^2\left(1+0.3/n\right)$ | 0.736 | 1.062 | 1.321 | 1.591 | 1.959 |
| | Case 2 | None | 1.060 | 1.725 | 2.277 | 2.854 | 3.640 |
| | Case 3 | $A^2\left(1+0.2/\sqrt{n}\right)$ | 0.474 | 0.637 | 0.757 | 0.877 | 1.038 |

<u>Table 22/6:</u> Upper tail percentage points of the supreme statistics for testing
the type-I extreme value distribution of the maximum

| Statistic | $n$ | \multicolumn{4}{c}{Significance level $\alpha$} | Statistic | $n$ | \multicolumn{4}{c}{Significance level $\alpha$} |
|---|---|---|---|---|---|---|---|---|---|---|---|
| | | 0.10 | 0.05 | 0.025 | 0.01 | | | 0.10 | 0.05 | 0.025 | 0.01 |
| $\sqrt{n}\,D^+$ | 10 | 0.872 | 0.969 | 1.061 | 1.152 | $\sqrt{n}\,D^+$ | 10 | 0.99 | 1.14 | 1.27 | 1.42 |
| | 20 | 0.878 | 0.979 | 1.068 | 1.176 | | 20 | 1.00 | 1.15 | 1.28 | 1.43 |
| Case 1 | 50 | 0.882 | 0.987 | 1.070 | 1.193 | Case 2 | 50 | 1.01 | 1.17 | 1.29 | 1.44 |
| | $\infty$ | 0.886 | 0.996 | 1.094 | 1.211 | | $\infty$ | 1.02 | 1.17 | 1.30 | 1.46 |
| $\sqrt{n}\,D^-$ | 10 | 0.773 | 0.883 | 0.987 | 1.103 | $\sqrt{n}\,D^-$ | 10 | 1.01 | 1.16 | 1.28 | 1.41 |
| | 20 | 0.810 | 0.921 | 1.013 | 1.142 | | 20 | 1.01 | 1.15 | 1.28 | 1.43 |
| Case 1 | 50 | 0.840 | 0.950 | 1.031 | 1.171 | Case 2 | 50 | 1.00 | 1.14 | 1.29 | 1.45 |
| | $\infty$ | 0.886 | 0.996 | 1.094 | 1.211 | | $\infty$ | 1.02 | 1.17 | 1.30 | 1.46 |
| $\sqrt{n}\,D$ | 10 | 0.934 | 1.026 | 1.113 | 1.206 | $\sqrt{n}\,D$ | 10 | 1.14 | 1.27 | 1.39 | 1.52 |
| | 20 | 0.954 | 1.049 | 1.134 | 1.239 | | 20 | 1.15 | 1.28 | 1.40 | 1.53 |
| Case 1 | 50 | 0.970 | 1.067 | 1.148 | 1.263 | Case 2 | 50 | 1.16 | 1.29 | 1.41 | 1.53 |
| | $\infty$ | 0.995 | 1.094 | 1.184 | 1.298 | | $\infty$ | 1.16 | 1.29 | 1.42 | 1.53 |
| $\sqrt{n}\,V$ | 10 | 1.43 | 1.55 | 1.65 | 1.77 | $\sqrt{n}\,V$ | 10 | 1.39 | 1.49 | 1.60 | 1.72 |
| | 20 | 1.46 | 1.58 | 1.69 | 1.81 | | 20 | 1.42 | 1.54 | 1.64 | 1.76 |
| Case 1 | 50 | 1.48 | 1.59 | 1.72 | 1.84 | Case 2 | 50 | 1.45 | 1.56 | 1.67 | 1.79 |
| | $\infty$ | 1.53 | 1.65 | 1.77 | 1.91 | | $\infty$ | 1.46 | 1.58 | 1.69 | 1.81 |

| Statistic | $n$ | \multicolumn{4}{c}{Significance level $\alpha$} |
|---|---|---|---|---|---|
| | | 0.10 | 0.05 | 0.025 | 0.01 |
| $\sqrt{n}\,D^+$ | 10 | 0.685 | 0.755 | 0.842 | 0.897 |
| | 20 | 0.710 | 0.780 | 0.859 | 0.926 |
| Case 3 | 50 | 0.727 | 0.796 | 0.870 | 0.940 |
| | $\infty$ | 0.734 | 0.808 | 0.877 | 0.957 |
| $\sqrt{n}\,D^-$ | 10 | 0.700 | 0.766 | 0.814 | 0.892 |
| | 20 | 0.715 | 0.785 | 0.843 | 0.926 |
| Case 3 | 50 | 0.724 | 0.796 | 0.860 | 0.944 |
| | $\infty$ | 0.733 | 0.808 | 0.877 | 0.957 |
| $\sqrt{n}\,D$ | 10 | 0.760 | 0.819 | 0.880 | 0.944 |
| | 20 | 0.779 | 0.843 | 0.907 | 0.973 |
| Case 3 | 50 | 0.790 | 0.856 | 0.922 | 0.988 |
| | $\infty$ | 0.803 | 0.874 | 0.939 | 1.007 |
| $\sqrt{n}\,V$ | 10 | 1.287 | 1.381 | 1.459 | 1.535 |
| | 20 | 1.323 | 1.428 | 1.509 | 1.600 |
| Case 3 | 50 | 1.344 | 1.453 | 1538 | 1.639 |
| | $\infty$ | 1.372 | 1.477 | 1.557 | 1.671 |

**Example 22/3: Testing whether dataset #1 comes from $We(a = 0, b, c)$**

After the $n = 20$ observations in dataset #1 have been transformed according to (22.11a) the MLEs of the extreme value distribution result as

$$\widehat{\eta} = -\ln\widehat{b} = -4.5972, \quad \widehat{\phi} = 1/\widehat{c} = 0.3853.$$

Fig. 22/3 shows the EDF together with the estimated CDF $F(y \mid -4.5972, 0.3853)$. The fit is extremely good. The test statistics are as follows:

$$W^2\left(1 + 0.2/\sqrt{n}\right) = 0.0242, \quad U^2\left(1 + 0.2/\sqrt{n}\right) = 0.0229, \quad A^2\left(1 + 0.2/\sqrt{n}\right) = 0.2138$$
$$\sqrt{n}\,D^+ = 0.2896, \quad \sqrt{n}\,D^- = 0.4526, \quad \sqrt{n}\,D = 0.4523, \quad \sqrt{n}\,V = 0.7421.$$

Comparing these values with the percentage points in Tab. 22/5 and Tab. 22/6 for case 3, we find that $H_0$ : "Dataset #1 comes from a WEIBULL distribution" cannot be rejected at any reasonable level of significance.

Figure 22/3: EDF for transformed dataset #1 and CDF of the extreme value distribution of the maximum with $\widehat{\eta} = -4.5973$ and $\widehat{\phi} = 0.3853$

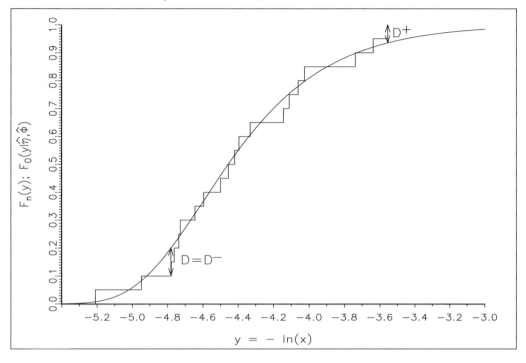

### 22.1.3 Tests using other than EDF statistics

The tests presented in Sect. 22.1.3 are especially appropriate for testing composite hypotheses. Some of these tests do not even require estimates of the unknown parameters.

#### 22.1.3.1 Tests based on the ratio of two estimates of scale[8]

The test statistic which will be used in the following is a modification of the well–known $W$ statistic introduced by SHAPIRO/WILK to see whether a given dataset can be fitted by a normal distribution. The test statistic is obtained as the ratio of two linear estimators of the scale parameter $b^* = 1/c$ of the Log–WEIBULL distribution. The procedure requires an uncensored sample.

The modified $W$ statistic suggested by SHAPIRO/BRAIN (1987) has been further modified by ÖZTÜRK/KORUKOGLU (1988) to give a higher power. Let $Y_{i:n} = \ln X_{i:n}$ be the transformed WEIBULL order statistics. The first estimator of $b^* = 1/c$ is the linear estimator

---

[8] Suggested reading for this section: ÖZTÜRK/KORUKOGLU (1988), SHAPIRO/BRAIN (1987).

suggested by D'AGOSTINO (1971a)

$$\widehat{b^*} = \frac{1}{n} \left[ 0.6079 \sum_{i=1}^{n} w_{n+i}\, Y_{i:n} - 0.2570 \sum_{i=1}^{n} w_i\, Y_{i:n} \right] \qquad (22.12a)$$

with weights

$$w_i \;=\; \ln\left( \frac{n+1}{n+1-i} \right); \quad i = 1, \ldots, n-1; \qquad (22.12b)$$

$$w_n \;=\; n - \sum_{i=1}^{n-1} w_i; \qquad (22.12c)$$

$$w_{n+i} \;=\; w_i \left( 1 + \ln w_i \right) - 1; \quad i = 1, \ldots, n-1; \qquad (22.12d)$$

$$w_{2n} \;=\; 0.4228\, n - \sum_{i=1}^{n-1} w_{n+i}. \qquad (22.12e)$$

The second linear estimator is the probability–weighted estimator [9]

$$\widehat{\sigma} = \frac{\sum_{i=1}^{n} (2\,i - n - 1)\, Y_{i:n}}{\left[ 0.693147\, n\, (n-1) \right]}. \qquad (22.13)$$

The test statistic then results as

$$W^* = \widehat{b^*} / \widehat{\sigma}. \qquad (22.14a)$$

In a typical sample from the Log–WEIBULL distribution, both of the statistics $\widehat{b^*}$ and $\widehat{\sigma}$ estimate the same parameter $b^*$. Since $\widehat{b^*}$ and $\widehat{\sigma}$ are linear unbiased estimators, the value of $W^*$ in such case must be close to 1. Deviations from the null distribution will result in $W^*$ values that are expected to shift away from 1 in both directions. Hence, the proposed test is two–sided.

Usually, it is more convenient to use the standardized version of the statistic rather than the statistic itself. The standardized statistic suggested by ÖZTÜRK/KORUKOGLU (1988) is

$$\widetilde{W^*} = \frac{W^* - 1.0 - 0.13/\sqrt{n} + 1.18/n}{0.49/\sqrt{n} - 0.36\, n}. \qquad (22.14b)$$

Tab. 22/7 gives percentage points $\widetilde{W}^*_{n,P}$ that have been found by performing Monte Carlo experiments.

---

**Example 22/4: Modified $W$–test applied to dataset #1**

We have applied the test procedure given above to dataset #1 in Tab. 9/2 ($n = 20$). The two estimates of $b^*$ are

$$\widehat{b^*} = 0.3950 \quad \text{and} \quad \widehat{\sigma} = 0.3686,$$

thus $W^* = 1.0716$ and $\widetilde{W^*} = 1.1084$. Assuming a significance level $\alpha = 0.05$, we find from Tab. 22/7 with $n = 20$:

$$-1.819 \leq \widetilde{W^*} = 1.1084 \leq 2.334.$$

Thus, $H_0$ : "Dataset #1 comes from a WEIBULL distribution" cannot be rejected.

---

[9] SHAPIRO/BRAIN (1987) take the usual estimator $\widehat{\sigma} = \left[ (1/n) \sum_{i=1}^{n} (Y_{i:n} - \overline{Y})^2 \right]^{1/2}$.

Table 22/7: Percentage points $\widetilde{W}^*_{n,P}$

| $P$ / $n$ | 0.005 | 0.025 | 0.05 | 0.10 | 0.50 | 0.90 | 0.95 | 0.975 | 0.995 |
|---|---|---|---|---|---|---|---|---|---|
| 3 | −1.106 | −1.065 | −1.019 | −0.933 | −0.181 | 0.791 | 0.941 | 1.024 | 1.055 |
| 4 | −1.486 | −1.355 | −1.203 | −1.036 | −0.202 | 1.038 | 1.437 | 1.740 | 2.298 |
| 5 | −1.728 | −1.494 | −1.323 | −1.113 | −0.165 | 1.212 | 1.684 | 2.079 | 2.853 |
| 6 | −1.877 | −1.587 | −1.395 | −1.159 | −0.136 | 1.307 | 1.804 | 2.240 | 3.110 |
| 8 | −2.043 | −1.687 | −1.470 | −1.206 | −0.100 | 1.390 | 1.898 | 2.362 | 3.300 |
| 7 | −1.974 | −1.646 | −1.440 | −1.188 | −0.115 | 1.360 | 1.866 | 2.230 | 3.237 |
| 9 | −2.094 | −1.716 | −1.491 | −1.219 | −0.089 | 1.409 | 1.915 | 2.382 | 3.330 |
| 10 | −2.132 | −1.737 | −1.507 | −1.228 | −0.080 | 1.419 | 1.923 | 3.391 | 3.342 |
| 15 | −2.241 | −1.795 | −1.547 | −1.249 | −0.057 | 1.429 | 1.915 | 2.373 | 3.305 |
| 20 | −2.292 | −1.819 | −1.562 | −1.256 | −0.047 | 1.418 | 1.889 | 2.334 | 3.238 |
| 25 | −2.322 | −1.833 | −1.570 | −1.258 | −0.041 | 1.405 | 1.864 | 2.298 | 3.176 |
| 30 | −2.342 | −1.841 | −1.574 | −1.259 | −0.038 | 1.393 | 1.843 | 2.267 | 3.124 |
| 40 | −2.367 | −1.851 | −1.578 | −1.259 | −0.034 | 1.372 | 1.808 | 2.217 | 3.041 |
| 50 | −2.382 | −1.857 | −1.580 | −1.258 | −0.032 | 1.355 | 1.781 | 2.179 | 2.978 |
| 60 | −2.393 | −1.861 | −1.581 | −1.256 | −0.030 | 1.342 | 1.759 | 2.149 | 2.928 |
| 70 | −2.401 | −1.863 | −1.581 | −1.255 | −0.030 | 1.331 | 1.742 | 2.124 | 2.887 |
| 80 | −2.407 | −1.865 | −1.582 | −1.254 | −0.029 | 1.321 | 1.727 | 2.103 | 2.853 |
| 90 | −2.412 | −1.866 | −1.582 | −1.253 | −0.029 | 1.313 | 1.714 | 2.085 | 2.824 |
| 100 | −2.417 | −1.868 | −1.581 | −1.252 | −0.028 | 1.306 | 1.703 | 2.070 | 2.798 |
| 110 | −2.420 | −1.868 | −1.581 | −1.252 | −0.028 | 1.299 | 1.693 | 2.056 | 2.776 |
| 120 | −2.423 | −1.869 | −1.581 | −1.251 | −0.028 | 1.293 | 1.685 | 2.044 | 2.755 |
| 130 | −2.426 | −1.870 | −1.581 | −1.250 | −0.028 | 1.288 | 1.677 | 2.033 | 2.737 |
| 140 | −2.428 | −1.870 | −1.581 | −1.249 | −0.027 | 1.283 | 2.670 | 2.023 | 2.721 |
| 150 | −2.430 | −1.871 | −1.580 | −1.249 | −0.027 | 1.279 | 1.663 | 2.013 | 2.706 |
| 200 | −2.438 | −1.872 | −1.579 | −1.246 | −0.027 | 1.262 | 1.637 | 1.977 | 2.645 |
| 300 | −2.447 | −1.874 | −1.578 | −1.243 | −0.02V | 1.239 | 1.603 | 1.929 | 2.568 |
| 400 | −2.452 | −1.874 | −1.576 | −1.240 | −0.027 | 1.224 | 1.581 | 1.899 | 2.518 |
| 500 | −2.455 | −1.874 | −1.575 | −1.238 | −0.027 | 1.213 | 1.565 | 1.876 | 2.481 |
| 750 | −2.461 | −1.875 | −1.573 | −1.235 | −0.028 | 1.194 | 1.538 | 1.839 | 2.420 |
| 1000 | −2.464 | −1.875 | −1.572 | −1.233 | −0.028 | 1.182 | 1.520 | 1.814 | 2.380 |

Source: ÖZTÜRK/KORUKOGLU (1988, pp. 1386) — Reprinted with permission from *Communications in Statistics — Simulation and Computation*. Copyright 1988 by Marcel Dekker, Inc. All rights reserved.

## 22.1.3.2 Tests based on spacings and leaps[10]

Differences in ordered variates, $X_{i+1:n} - X_{i:n}$, from any distribution having a density are asymptotically exponential and asymptotically independent. Thus the quantities

$$\ell_i = \frac{X_{i+1} - X_{i:n}}{\mathrm{E}\left(X_{i+1} - X_{i:n}\right)} \; ; i = 1, \ldots, n-1 \qquad (22.15a)$$

---

[10] Suggested reading for this section: GIBSON/HIGGINS (2000), LITTELL/MCCLAVE/OFFEN (1979), LOCKHART/ O'REILLY/STEPHENS (1986), MANN/FERTIG (1975a), MANN/SCHEUER/FERTIG (1973), SOMERVILLE (1977), THIAGARAJAN/HARRIS (1976) TIKU/SINGH (1981).

are asymptotically exponentially distributed with mean 1, and they are asymptotically independent. Tukey called them **leaps**, whereas the numerators, $X_{i+1:n} - X_{i:n}$, are called **gaps**. Thus, $2\,\ell_i$ is asymptotically $\chi^2(2)$, and for a sample censored at the $m$–th observation ($m = n$ for an uncensored sample) and for $r + s + 1 \leq m \leq n$:

$$L(r, s, m, n) := \frac{\dfrac{1}{r} \sum\limits_{j=m-r}^{m-1} \ell_j}{\dfrac{1}{s} \sum\limits_{j=1}^{s} \ell_j} \overset{\text{asym}}{\sim} F(2\,r, 2\,s), \tag{22.15b}$$

but this approximation is not equally good for all percentiles, so special tables have to be constructed.

Goodness–of–fit tests for a Weibull distribution with variate $X$ are performed via the extreme value distribution of the minimum, i.e., using the transformation $Y = \ln X$, where $a^* = \ln b$ and $b^* = 1/c$. In calculating the expected values appearing in the denominator of $\ell_i$ Mann/Scheuer/Fertig (1973) used the expected values of the reduced extreme value order statistics

$$Y_{i:n}^* = (Y_{i:n} - a^*)/b*,$$

which, for example, are tabulated in White (1967b) for $i = 1(1)50(5)100$. Of course, the value of $L$ is not affected if one uses

$$\ell_i^* = \frac{Y_{i+1:n} - Y_{i:n}}{\mathrm{E}\big(Y_{i+1:n}^*\big) - \mathrm{E}\big(Y_{i:n}^*\big)} \tag{22.15c}$$

instead of (22.15a). The differences $\mathrm{E}\big(Y_{i+1:n}^*\big) - \mathrm{E}\big(Y_{i:n}^*\big)$ for $n = 3(1)25$, $i = 1(1)n - 1$, are tabulated in Mann/Scheuer/Fertig (1973). Sampling studies by these authors reveal that one should form $L$ with the average of the first $m/2$ or $(m - 1)/2$ (whichever is an integer) leaps in the denominator and the remaining leaps in the numerator. For convenience in calculating percentage points these authors introduced a transformation of $L$ which takes on only values in the unit interval. This new statistic is

$$S = \frac{r}{s} L \bigg/ \left(1 + \frac{r}{s} L\right) \tag{22.16a}$$

$$= \sum_{j=m-r}^{m-1} \ell_j^* \bigg/ \sum_{j=1}^{m-1} \ell_j^*. \tag{22.16b}$$

Taking $r = \big[(m - 1)/2\big]$, where $[x]$ denotes the greatest integer contained in $x$, the final test statistic results as

$$S = \frac{\sum\limits_{i=[m/2]+1}^{m-1} \dfrac{Y_{i+1:n} - Y_{i:n}}{\mathrm{E}\big(Y_{i+1:n}^*\big) - \mathrm{E}\big(Y_{i:n}^*\big)}}{\sum\limits_{i=1}^{m-1} \dfrac{Y_{i+1:n} - Y_{i:n}}{\mathrm{E}\big(Y_{i+1:n}^*\big) - \mathrm{E}\big(Y_{i:n}^*\big)}}. \tag{22.16c}$$

The test is one–sided and values of $S$ greater than $S_{n,m,1-\alpha}$ are evidence against $H_0$ : "$X$ has a WEIBULL distribution." The percentiles $S_{n,m,1-\alpha}$ are given in MANN/SCHEUER/FERTIG (1973) for $n = 3(1)25$, $i = 1(1)m - 1$, $m = 3(1)n$.

Note that $S$ is much simpler to calculate than the classical EDF statistics, because the parameters $a^*$ and $b^*$ do not need to be estimated and $F_0(X_{i:n} \mid \widehat{\boldsymbol{\theta}})$ need not to calculated. The test is also valid for complete samples ($m = n$) as well as type–II right–censored samples at $m < n$. The performance of $S$ compared with several other test statistics has been studied by LITTELL et al. (1979) and LOCKHART et al. (1986) showing that, generally, $S$ has good power. A modification of $S$ has been suggested by MANN/FERTIG (1975a) to test for the two–parameter versus three–parameter WEIBULL distribution. Their methodology for testing for a zero threshold parameter can be used to obtain a confidence interval for this parameter; also see SOMERVILLE (1977) for this topic. A similar ratio–type goodness–of–fit test has been suggested by THIAGARAJAN/HARRIS (1976). This test can be applied directly to the WEIBULL distribution because the shape parameter is assumed known.

---

**Example 22/5: Application of the MANN–SCHEUER–FERTIG test to dataset #1 censored at $m = 15$**

We first form $\ell_i^* = \left(Y_{i+1:20} - Y_{i:20}\right) \big/ \left[\mathrm{E}\left(Y_{i+1:20}^*\right) - \mathrm{E}\left(Y_{i:20}^*\right)\right]$ for $i = 1, \ldots, 14$, where $Y_{i:20} = \ln X_{i:n}$. The differences of the expected values are taken from the table in MANN/SCHEUER/FERTIG (1973). The sum in the numerator runs from $[m/2] + 1 = 8$ to $m = 14$ and in the denominator from 1 to $m - 1 = 14$.

| $i$ | $X_{i:20}$ | $y_{i:20}$ | $y_{i+1:20} - y_{i:20}$ | $\mathrm{E}\left(Y_{i+1:20}^*\right) - \mathrm{E}\left(Y_{i:20}^*\right)$ | $\ell_i^*$ |
|-----|------------|------------|--------------------------|---------------------------------------------------------------------------|------------|
| 1   | 35         | 3.5553     | 0.0823                   | 1.0259                                                                     | 0.0802     |
| 2   | 38         | 3.6376     | 0.1001                   | 0.5270                                                                     | 0.1899     |
| 3   | 42         | 3.7377     | 0.2877                   | 0.3617                                                                     | 0.7954     |
| 4   | 56         | 4.0254     | 0.0350                   | 0.2798                                                                     | 0.1254     |
| 5   | 58         | 4.0604     | 0.0505                   | 0.2314                                                                     | 0.2179     |
| 6   | 61         | 4.1109     | 0.0322                   | 0.1999                                                                     | 0.1614     |
| 7   | 63         | 4.1431     | 0.1876                   | 0.1782                                                                     | 1.0529     |
| 8   | 76         | 4.3307     | 0.0637                   | 0.1627                                                                     | 0.3917     |
| 9   | 81         | 4.3944     | 0.0244                   | 0.1515                                                                     | 0.1609     |
| 10  | 83         | 4.4188     | 0.0355                   | 0.1437                                                                     | 0.2471     |
| 11  | 86         | 4.4543     | 0.0455                   | 0.1384                                                                     | 0.3283     |
| 12  | 90         | 4.4998     | 0.0953                   | 0.1356                                                                     | 0.7030     |
| 13  | 99         | 4.5951     | 0.0492                   | 0.1350                                                                     | 0.3648     |
| 14  | 104        | 4.6443     | 0.0830                   | 0.1371                                                                     | 0.6053     |
| 15  | 113        | 4.7273     | —                        | —                                                                         | —          |

We calculate

$$S = \frac{2.8012}{5.4243} = 0.5164.$$

$H_0$ : "Dataset #1 censored at $m = 15$ comes from a WEIBULL distribution" cannot be rejected at any reasonable level of significance because $S$ is less than $S_{20,15,0.75} = 0.59$.

### 22.1.3.3 Correlation tests[11]

Correlation tests and the closely related regression tests arise most naturally when the unknown parameters of the tested distribution are location and scale parameters. Suppose $F_0(x)$ is $F(u)$ with $u = (x - \alpha)/\beta$, so that $\alpha$ is a location parameter and $\beta$ a scale parameter, and suppose any other parameters in $F(u)$ are known. If a sample of values $U_i$ were taken from $F(u)$ with $\alpha = 0$ and $\beta = 1$, we could construct a sample $X_i$ from $F_0(x)$ by calculating

$$X_i = \alpha + \beta U_i, \quad i = 1, \ldots, n. \tag{22.17a}$$

Let $m_i = \mathrm{E}(U_{i:n})$, then

$$\mathrm{E}(X_{i:n}) = \alpha + \beta m_i, \tag{22.17b}$$

and a plot of $X_{i:n}$ against $m_i$ should be an approximate straight line with intercept $\alpha$ on the vertical axis and slope $\beta$. The values $m_i$ are the most natural to plot along the horizontal axis, but for most distributions they are difficult to calculate. So most authors propose alternatives $T_i$, which are convenient functions of $i$. Then (22.17b) can be replaced by the regression model

$$X_{i:n} = \alpha + \beta T_i + \varepsilon_i, \tag{22.17c}$$

where $\varepsilon_i$ is an error variable, which for $T_i = m_i$ will have mean 0. A frequent choice for $T_i$ is

$$T_i = F^{-1}\left(\frac{i}{n+1}\right); \quad i = 1, \ldots, n. \tag{22.17d}$$

Now, three main approaches to testing how well the data fit to (22.17c) can be found in the literature:

1. A test is based on the correlation coefficient $R(X, T)$ between the paired sets $\{X_i\}$ and $\{T_i\}$.

2. After estimates $\widehat{\alpha}$ and $\widehat{\beta}$ are found by a suitable method, a test is based on the sum of squared residuals $(X_{i:n} - \widehat{X}_{i:n})^2$, where $\widehat{X}_{i:n} = \widehat{\alpha} + \widehat{\beta} T_i$.

3. The scale parameter is estimated as $\widehat{\beta}$, and the squared value is compared with another estimate of $\beta^2$, for example, the one, that is obtained from the sample variance (see Sect. 22.1.3.1).

These three methods are closely connected, and we will apply only the first method, because $R^2(X, T)$ is consistent against all alternatives, whereas methods 2 and 3 can yield statistics that are not consistent against certain classes of alternatives.

---

[11] Suggested reading for this section: COLES (1989), D'AGOSTINO/STEPHENS (1986), GIBSON/HIGGINS (2000), LITTELL/MCCLAVE/OFFEN (1979), SMITH/BAIN (1976).

Table 22/8: Upper percentage points of $Z = n \left[ 1 - R^2(X, T) \right]$

| $p$ | $n$ | $\alpha$ 0.50 | 0.25 | 0.15 | 0.10 | 0.05 | 0.025 | 0.01 |
|-----|-----|------|------|------|------|------|-------|------|
| 0.2 | 20 | 1.62 | 2.81 | 3.52 | 4.07 | 4.90 | 5.69 | 7.19 |
|     | 40 | 2.58 | 4.12 | 5.18 | 6.09 | 7.48 | 8.92 | 10.47 |
|     | 60 | 3.01 | 4.68 | 5.89 | 6.90 | 8.75 | 10.45 | 13.02 |
|     | 80 | 3.33 | 5.11 | 6.45 | 7.59 | 9.71 | 11.79 | 14.50 |
|     | 100 | 3.56 | 5.42 | 6.88 | 8.12 | 10.42 | 12.84 | 15.47 |
| 0.4 | 10 | 0.81 | 1.38 | 1.75 | 2.04 | 2.52 | 2.93 | 3.39 |
|     | 20 | 1.27 | 1.98 | 2.51 | 2.94 | 3.76 | 4.56 | 5.46 |
|     | 40 | 1.62 | 2.47 | 3.09 | 3.66 | 4.66 | 5.59 | 6.85 |
|     | 60 | 1.77 | 2.71 | 3.39 | 3.93 | 4.92 | 5.88 | 7.35 |
|     | 80 | 1.88 | 2.86 | 3.58 | 4.16 | 5.19 | 6.25 | 7.72 |
|     | 100 | 1.95 | 2.95 | 3.72 | 4.33 | 5.41 | 6.55 | 7.99 |
| 0.6 | 10 | 0.74 | 1.17 | 1.49 | 1.75 | 2.16 | 2.61 | 3.18 |
|     | 20 | 0.98 | 1.49 | 1.88 | 2.20 | 2.72 | 3.28 | 4.03 |
|     | 40 | 1.15 | 1.73 | 2.15 | 2.49 | 3.08 | 3.77 | 4.66 |
|     | 60 | 1.23 | 1.82 | 2.25 | 2.61 | 3.26 | 3.92 | 4.77 |
|     | 80 | 1.28 | 1.89 | 2.34 | 2.71 | 3.35 | 4.04 | 4.91 |
|     | 100 | 1.32 | 1.94 | 2.41 | 2.78 | 3.41 | 4.12 | 5.03 |
| 0.8 | 10 | 0.64 | 0.99 | 1.25 | 1.46 | 1.79 | 2.14 | 2.58 |
|     | 20 | 0.79 | 1.19 | 1.48 | 1.71 | 2.14 | 2.58 | 3.29 |
|     | 40 | 0.90 | 1.32 | 1.63 | 1.85 | 2.27 | 2.70 | 3.32 |
|     | 60 | 0.94 | 1.37 | 1.68 | 1.94 | 2.38 | 2.79 | 3.37 |
|     | 80 | 0.97 | 1.40 | 1.72 | 1.98 | 2.41 | 2.82 | 3.37 |
|     | 100 | 0.99 | 1.42 | 1.74 | 1.99 | 2.41 | 2.84 | 3.35 |
| 0.9 | 10 | 0.61 | 0.93 | 1.24 | 1.37 | 1.71 | 2.08 | 2.51 |
|     | 20 | 0.74 | 1.13 | 1.42 | 1.64 | 2.03 | 2.44 | 3.05 |
|     | 40 | 0.84 | 1.23 | 1.53 | 1.77 | 2.17 | 2.59 | 3.14 |
|     | 60 | 0.88 | 1.28 | 1.57 | 1.80 | 2.19 | 2.59 | 3.17 |
|     | 80 | 0.91 | 1.31 | 1.59 | 1.81 | 2.20 | 2.60 | 3.18 |
|     | 100 | 0.92 | 1.32 | 1.60 | 1.82 | 2.20 | 2.60 | 3.18 |
| 0.95 | 10 | 0.61 | 0.94 | 1.23 | 1.41 | 1.76 | 2.13 | 2.60 |
|     | 20 | 0.75 | 1.14 | 1.44 | 1.68 | 2.11 | 2.57 | 3.20 |
|     | 40 | 0.85 | 1.28 | 1.60 | 1.84 | 2.28 | 2.73 | 3.33 |
|     | 60 | 0.90 | 1.33 | 1.63 | 1.88 | 2.30 | 2.74 | 3.39 |
|     | 80 | 0.93 | 1.35 | 1.65 | 1.89 | 2.31 | 2.75 | 3.43 |
|     | 100 | 0.95 | 1.36 | 1.66 | 1.90 | 2.32 | 2.75 | 3.45 |
| 1.0 | 10 | 0.61 | 0.95 | 1.23 | 1.45 | 1.81 | 2.18 | 2.69 |
|     | 20 | 0.82 | 1.30 | 1.69 | 2.03 | 2.65 | 3.36 | 4.15 |
|     | 40 | 1.04 | 1.67 | 2.23 | 2.66 | 3.63 | 4.78 | 6.42 |
|     | 60 | 1.20 | 1.93 | 2.57 | 3.18 | 4.33 | 5.69 | 7.79 |
|     | 80 | 1.32 | 2.14 | 2.87 | 3.55 | 4.92 | 6.54 | 8.86 |
|     | 100 | 1.41 | 2.30 | 3.09 | 3.82 | 5.38 | 7.22 | 9.67 |

Source: D'AGOSTINO/STEPHENS (1986, pp. 226) — Reprinted with permission from *Goodness–of–fit Techniques*. Copyright 1986 by Marcel Dekker, Inc. All rights reserved.

The statistic used here is

$$Z = n \left[ 1 - R^2(X, T) \right], \tag{22.17e}$$

which is a transformation of the estimated **coefficient of non–determination** $1 - R^2(X, T)$. High values of $Z$ indicate a poor fit and thus lead to a rejection of the hypothesis that the sample comes from the hypothetical distribution. The test is one–sided. $Z$ often has an asymptotic distribution or the distribution can easily be simulated by Monte Carlo techniques.

For testing $H_0$ : "The dataset comes from a WEIBULL distribution" using $Z$ we first have to transform the WEIBULL data to extreme value data by calculating $Y = \ln X$. For the extreme value distribution we have

$$T_i = -\ln\left\{-\ln\left[\frac{i}{n+1}\right]\right\}. \tag{22.17f}$$

The test statistic does not require estimates of the WEIBULL or the extreme value parameters and can be calculated for complete as well as for type–II censored samples, $r$ being the censoring number. Tab. 22/8 gives the upper tail percentage points of $Z$, and the table is entered at $p = r/n$ and $n$.

A comparison of several goodness–of–fit techniques by LITTELL et al. (1979) reveals that $Z$ is not, relatively, worse than the EDF tests using $D$, $W^2$ and $A^2$ and the statistics of MANN/ SCHEUER/FERTIG (see Sect. 22.1.3.2) for the alternative distributions chosen. COLES (1989) has applied a correlation test to the plotting coordinates of the stabilized plot, given by (22.6a,b), which exhibited a higher power than $D_{sp}$ as given in (22.6c).

---

**Example 22/6:  Application of the correlation test to dataset #1 censored at $r = 15$**

For dataset #1 in Tab. 9/2 ($n = 20$) censored at $r = 15$, we calculate — based on $y_{i:n} = \ln x_{i:n}$ :
$$R^2(X, T) = 0.3639 \quad \text{and} \quad Z = n\left[1 - R^2(X, T)\right] = 0.7210.$$

With $\alpha = 0.05$, $n = 20$ and $p = 15/20 = 0.75$, we interpolate $z_{0.95}(n = 20, p = 0.75) = 2.285$. Because $Z = 0.7210 \not> z_{0.95}(n = 20, p = 0.75) = 2.285$, we cannot reject $H_0$ that the sample comes from a WEIBULL distribution.

---

# 22.2   Discrimination between WEIBULL and other distributions

There are situations where we have to choose among two or more distinct distributions, all being apt to describe lifetime. We start by discriminating between the two–parameter and the three–parameter WEIBULL distributions (Sect. 22.2.1). Then we look at pairs of distributions where one member of the pair is a WEIBULL distribution (Sect. 22.2.2). Finally, we will discuss how to proceed when there are two or more other distributions besides the WEIBULL (Sect. 22.2.3).

## 22.2.1   Discrimination between the two-parameter and three-parameter WEIBULL distributions

When we have to decide whether a sample comes from $We(0, b, c)$ or $We(a, b, c)$, $a \neq 0$, it seems quite natural to calculate an estimate $\hat{a}$ of $a$, e.g., the MLE, and use $\hat{a}$ as a test statistic. But as this approach is rather cumbersome, due to the impossibility of directly giving the distribution of $\hat{a}$ (see Sect. 11.3.2.4), we suggest an indirect procedure to test

$$H_0: \ a = 0 \quad \text{versus} \quad H_A: \ a > 0.$$

This approach is a modification of the goodness–of–fit test based on spacings and leaps (Sect. 21.1.3.2). The approach is due to MANN/FERTIG (1975). Its test statistic is

$$F = \frac{k \sum_{i=k+1}^{m-1} \ell_i^*}{(m - k - 1) \sum_{i=1}^{k} \ell_i^*} , \qquad (22.18a)$$

with $\ell_i^*$ defined in (22.15c) and $m$ the number of uncensored observations. ($m = n$ means that the sample is uncensored.) The recommended values for $k$ can be taken from Tab. 22/9 for $m$ small. For $m \geq 15$, take $k = [m/3]$.

Table 22/9: Recommended values for $k$

| Size of test | $m$ | | | | |
|---|---|---|---|---|---|
| $\alpha$ | 3 | 4 | 5 | $6 - 8$ | $9 - 14$ |
| 0.25 | 1 | 1 | 1 | 2 | 2 |
| 0.20 | 1 | 1 | 2 | 2 | 2 |
| 0.15 | 1 | 2 | 2 | 3 | 4 |
| 0.10 | 1 | 2 | 2 | 3 | 4 |
| 0.05 | 1 | 2 | 3 | 3 | 4 |
| 0.01 | 1 | 2 | 3 | 3 | 4 |

Source: MANN/FERTIG (1975a, p. 241) — Reprinted with permission from *Technometrics*.
Copyright 1975 by the American Statistical Association. All rights reserved.

As the notation in (22.18a) suggests, the test statistic — under $H_0$ — is $F$–distributed:

$$F \overset{\text{approx}}{\sim} F(2(m - k - 1), 2k), \qquad (22.18b)$$

and $H_0$: $a = 0$ is rejected in favor of $H_A$: $a > 0$ at level $\alpha$ if $F > F_{2(m-k-1),2k,1-\alpha}$.

This test can also be stated in terms of the beta distribution, where

$$P_{k,m} = \sum_{i=k+1}^{m-1} \ell_i^* \Big/ \sum_{i=1}^{m-1} \ell_i^* \overset{\text{approx}}{\sim} Be(m - k - 1, k), \qquad (22.18c)$$

and $H_0$ is rejected if $P_{k,m} > b_{1-\alpha}(m - k - 1, k)$, where $b_{1-\alpha}(m - k - 1, k)$ is the $(1 - \alpha)$ percentile of $Be(m - k - 1, k)$.

**Example 22/7: Testing $H_0$: $a = 0$ versus $H_A$: $a > 0$ using the modified dataset #1 censored at $m = 15$**

Dataset #1 with $n = 20$ in Tab. 9/2 comes from $We(0, 100, 25)$. We add 150 to each observation so that we have a sample from $We(150, 100, 2.5)$. We censor this dataset at $m = 15$ so that we can use $E(Y_{i+1:20}^*) - E(Y_{i:20}^*)$ given in the work table of Example 22/5.

| $i$ | $x_{i:20}$ | $y_{i:20}$ | $y_{i+1:20} - y_{i:20}$ | $E\big(Y^*_{i+1:20}\big) - E\big(Y^*_{i:20}\big)$ | $\ell^*_i$ |
|---|---|---|---|---|---|
| 1  | 185 | 5.2204 | 0.0160 | 1.0259 | 0.0157 |
| 2  | 188 | 5.2364 | 0.0211 | 0.5270 | 0.0399 |
| 3  | 192 | 5.2575 | 0.0704 | 0.3617 | 0.1946 |
| 4  | 206 | 5.3279 | 0.0096 | 0.2798 | 0.0345 |
| 5  | 208 | 5.3375 | 0.0144 | 0.2314 | 0.0619 |
| 6  | 211 | 5.3519 | 0.0094 | 0.1999 | 0.0472 |
| 7  | 213 | 5.3613 | 0.0592 | 0.1782 | 0.3325 |
| 8  | 226 | 5.4205 | 0.0219 | 0.1627 | 0.1345 |
| 9  | 231 | 5.4424 | 0.0086 | 0.1515 | 0.0569 |
| 10 | 233 | 5.4510 | 0.0128 | 0.1437 | 0.0890 |
| 11 | 236 | 5.4638 | 0.0168 | 0.1384 | 0.1214 |
| 12 | 240 | 5.4806 | 0.0369 | 0.1356 | 0.2715 |
| 13 | 249 | 5.5175 | 0.0198 | 0.1350 | 0.1472 |
| 14 | 254 | 5.5373 | 0.0349 | 0.1371 | 0.2539 |
| 15 | 263 | 5.5722 | — | — | — |

We have to use $k = m/3 = 15/3 = 5$ and the $F$–statistic is calculated as $F = 2.3307$ and the $P$–statistic as $P_{5,15} = 0.8075$. Choosing $\alpha = 0.05$, we can reject $H_0 : a = 0$ with both statistics:

$$F = 2.3307 > F_{18,10,0.95} = 2.2153 \quad \text{and} \quad P_{5,15} = 0.8075 > b_{0.95}(9,5) = 0.7995.$$

## 22.2.2 Discrimination between WEIBULL and one other distribution

There exist several other distributions besides the WEIBULL which may be used to model lifetime as a random variable. We will show how to discriminate between the WEIBULL and one of its closest competitors.

### 22.2.2.1 WEIBULL versus exponential distribution[12]

Two approaches can be found in the literature to discriminate between WEIBULL and exponential distributions, i.e., a parameter test of $H_0 : c = 1$ versus $H_A : c \neq 1$ and a likelihood–ratio–type test.

We first turn to the **parameter test**. When we want to discriminate between a two–parameter WEIBULL distribution $We(0, b, c)$ with CDF

$$F_W(x \mid b, c) = 1 - \exp\big\{ - (x/b)^c \big\}$$

and a one–parameter exponential distribution $Ex(b)$ with CDF

$$F_E(x \mid b) = 1 - \exp\big( - x/b \big)$$

based on the MLE $\hat{c}$ of $c$, we refer to the tests given in Sect. 21.1.1.1. In the same way it would be useful to discriminate between the three–parameter WEIBULL distribution

---

[12] Suggested reading for this section: ANTLE/KLIMKO/RADEMAKER/ROCKETTE (1975), BAIN/ ENGELHARDT (1991a), CHEN (1997), ENGELHARDT/BAIN (1975), GUPTA/KUNDU (2003), HAGER/BAIN/ANTLE (1971), ZELEN/DANNEMILLER (1961).

$We(a, b, c)$ with CDF

$$F_W(x \mid a, b, c) = 1 - \exp\left\{-\left[(x - a)/b\right]^c\right\}$$

and the two–parameter exponential distribution $Ex(a, b)$ with CDF

$$F_E(x \mid a, b) = 1 - \exp\left\{-(x - a)/b\right\}.$$

It is possible to derive such a test and relate it back to the two–parameter WEIBULL case by using the idea of subtracting the sample minimum from each observation to eliminate the unknown location parameter. Suppose $X_{1:n}, \ldots, X_{r:n}$ denote the $r$ smallest ordered observations from $We(a, b, c)$ and let $X^*_{i:n-1} = X_{i+1:n} - X_{1:n}$; $i = 1, \ldots, r - 1$. Under $H_0:\ c = 1$, the starred observations represent the first $r - 1$ ordered observations from a sample of size $n - 2$ from $We(0, b, 1) = Ex(b)$. Thus, any test procedure previously available for testing $H_0:\ c = 1$ when $a = 0$ or $a$ known still holds when $a$ is unknown if it is applied to the starred observations.[13]

A **likelihood–ratio–type test statistic** can also be developed for this problem.[14] That is, consider a test of $H_0$ : "The sample follows $Ex(b_E)$." against $H_A$ : "The sample follows $We(0, b_W, c_W)$." Consider the statistic

$$
\lambda = \frac{\displaystyle\max_{b_E} \prod_{i=1}^{n} \frac{1}{b_E} \exp\left(-\frac{X_i}{b_E}\right)}{\displaystyle\max_{b_W, c_W} \prod_{i=1}^{n} \frac{c_W}{b_W} \left(\frac{X_i}{b_W}\right)^{c_W - 1} \exp\left\{-\left(\frac{X_i}{b_W}\right)^{c_W}\right\}}
$$

$$
= \frac{\widehat{b}_E^{-n} \exp\left\{-\displaystyle\sum_{i=1}^{n} \frac{X_i}{\widehat{b}_E}\right\}}{\left(\dfrac{\widehat{c}_W}{\widehat{b}_W^{\widehat{c}_W}}\right)^{n} \displaystyle\prod_{i=1}^{n} X_i^{\widehat{c}_W - 1} \exp\left\{-\displaystyle\sum_{i=1}^{n} \left(\frac{X_i}{\widehat{b}_W}\right)^{\widehat{c}_W}\right\}}, \tag{22.19a}
$$

where $\widehat{b}_E$, $\widehat{b}_W$ and $\widehat{c}_W$ are the respective MLEs. $\lambda$ is called the **ratio of the maximized likelihoods**. The MLE of $b_E$ is

$$\widehat{b}_E = \overline{X} = \frac{1}{n} \sum_{i=1}^{n} X_i,$$

and for the MLEs of $b_W$ and $c_W$, we have

$$\widehat{b}_W^{\widehat{c}_W} = \frac{1}{n} \sum X_i^{\widehat{c}_W}.$$

---

[13] For more general tests of $H_0:\ c = c_0$, $c_0 \neq 1$, the starred observations can still be used to eliminate the unknown parameter $a$; however, they will no longer be distributed as ordered exponential variables under $H_0$ and new percentage points would need to be determined for the test statistics.

[14] See DUMONCEAUX/ANTLE/HAAS (1973a) for a general representation of likelihood–ratio tests for discriminating between two models with unknown location and scale parameters.

Thus $\lambda$ reduces to

$$\lambda = \frac{\widehat{b}_W^{\,n\,\widehat{c}_W}}{\left(\overline{X}\,\widehat{c}_W\right)^n \prod\limits_{i=1}^{n} X_i^{\widehat{c}_W}}.\tag{22.19b}$$

It can be shown that $\lambda$ is independent of $b_E$ under $H_0$, so $\lambda$ may be used as a test statistic. HAGER et al. (1971) have calculated percentage points of $\lambda$ by Monte Carlo experiments (see Tab. 22/10). $H_0$ : "The sample comes from $Ex(b_E)$" is rejected in favor of $H_A$ : " The sample comes from $We(0, b_W, c_W)$" at level $\alpha$ when $\lambda < \lambda_\alpha$.

Table 22/10: Percentage points $\lambda_\alpha$ such that $\Pr(\lambda < \lambda_\alpha) = \alpha$

| $\alpha$ $\diagdown$ $n$ | 0.01 | 0.02 | 0.05 | 0.10 |
|---|---|---|---|---|
| 10 | 0.0198 | 0.0379 | 0.0941 | 0.193 |
| 20 | 0.0248 | 0.0462 | 0.1160 | 0.214 |
| 30 | 0.0289 | 0.0584 | 0.1335 | 0.239 |
| 50 | 0.0399 | 0.0657 | 0.1435 | 0.252 |

There is a motivation, or even a justification, for this procedure. Suppose, one is interested in testing one hypothesis, $H_1$, against another, $H_2$, where $H_1$ asserts that the DF of a variate is a certain, completely specified, density function and $H_2$ asserts that the DF is a second, completely specified, density function. If either $H_1$ or $H_2$ is true, then the test which minimizes the sum of the probability of a type–I error and the probability of a type–II error is the one which rejects $H_1$ if, and only if, the joint density function of the sample under $H_2$ exceeds the joint density function of the sample under $H_2$.

---

**Example 22/8: Testing exponential distribution against WEIBULL distribution using dataset #1**

Dataset #1 of Tab. 9/2 has $n = 20$ and $\widehat{b}_W = 99.2079$, $\widehat{c}_W = 2.5957$ and $\overline{x} = 87.95$. The test statistic is $\lambda = 3.42 \cdot 10^{-43}$ which is far below $\lambda_{0.05} = 0.0248$, so we can reject the hypothesis of an exponential distribution with great confidence.

---

GUPTA/KUNDU (2003) have applied the ratio of the maximized likelihoods to discriminate between the WEIBULL and the **generalized exponential distribution** with CDF

$$F(x \mid b, d) = \left(1 - \exp\left\{-x/b\right\}\right)^d; \quad b, d > 0.$$

### 22.2.2.2 WEIBULL versus gamma distribution[15]

Another competitor to the WEIBULL distribution with respect to modeling lifetime is the gamma distribution. We want to discriminate between these distributions by applying a test using the ratio of the maximized likelihoods. Suppose

$$f_G(x \mid b_g, c_G) = \frac{1}{b_G^{c_G} \, \Gamma(c_G)} \, x^{c_G - 1} \, \exp\left(-\frac{x}{b_G}\right) \tag{22.20a}$$

denotes the DF of the gamma distribution and

$$f_W(x \mid b_W, c_W) = \frac{c_W}{b_W} \left(\frac{x}{b_W}\right)^{c_W - 1} x^{c_W - 1} \exp\left\{-\left(\frac{x}{b_W}\right)^{c_W}\right\} \tag{22.20b}$$

denotes the DF of the WEIBULL distribution. The natural logarithm of the ratio of the maximized likelihoods, where the nominator is given by the gamma likelihood, follows as

$$T = n \left\{ \widehat{c}_G \ln\left(\frac{\widetilde{X} \, \widehat{c}_G}{\overline{X}}\right) + \widehat{c}_W \ln\left(\frac{\widehat{b}_W}{\widetilde{X}}\right) - \widehat{c}_G + 1 - \ln\Gamma\left(\widehat{c}_G\right) - \ln\widehat{c}_W \right\} \tag{22.21}$$

with

$$\overline{X} = \frac{1}{n} \sum_{i=1}^{n} X_i \quad \text{and} \quad \widetilde{X} = \left(\prod_{i=1}^{n} X_i\right)^{1/n}.$$

The MLEs of the WEIBULL parameters are given by (11.19a) and (11.19b) and the MLEs of the gamma parameters follow from

$$\psi(\widehat{c}_G) - \ln\widehat{c}_G = \ln\left\{\left(\prod_{i=1}^{n} X_i\right)^{1/n} \Big/ \overline{X}\right\} \quad \text{and} \quad \widehat{b}_G = \overline{X}/\widehat{c}_G.$$

BAIN/ENGELHARDT (1980) believe the decision should be based on the rule: "Choose the gamma distribution as the appropriate model for the $n$ data points only if $T > 0$, otherwise choose the WEIBULL model" (p. ?). This rule is equivalent to choosing the distribution that gives a greater maximized likelihood.

To measure the performance of this rule the quantity PCS (**probability of correct selection**) has been introduced. PCS is given by $\Pr(T > 0)$ under a gamma distribution and by $\Pr(T \leq 0)$ under a WEIBULL distribution. The first of these probabilities is free of $b_G$, but depends on $c_G$, so we denote

$$\text{PCS}(c_G) = \Pr(T > 0 \mid c_G). \tag{22.22a}$$

Similarly, if the sample is from a WEIBULL population, the PCS depends only on $c_W$:

$$\text{PCS}(c_W) = \Pr(T \leq 0 \mid c_W). \tag{22.22b}$$

---

[15] Suggested reading to this section: BAIN/ENGELHARDT (1980b), BALASOORIYA/ABEYSINGHE (1994), CHEN (1987), FEARN/NEBENZAHL (1991), KAPPENMAN (1982), PARR/WEBSTER (1965), VOLODIN (1974).

BAIN/ENGELHARDT (1980b) have found the PCSs in Tab. 22/11 by Monte Carlo experiments. As seen in this table, if $c_G = c_W = 1$, then PCS is near 0.5 as to be expected because in this case both models coincide to the exponential distribution and both are actually correct, so there is no reason to prefer one over the other.

<u>Table 22/11:</u> Probability of correct selection (Gamma versus WEIBULL)

| Gamma distribution: $\Pr(T > 0 \mid c_G)$ | | | | | | |
|---|---|---|---|---|---|---|
| $c_G$ \ $n$ | 0.5 | 1 | 2 | 4 | 8 | 16 |
| 10 | 0.60 | 0.51 | 0.53 | 0.57 | 0.61 | 0.64 |
| 20 | 0.66 | 0.50 | 0.58 | 0.64 | 0.68 | 0.71 |
| 40 | 0.70 | 0.49 | 0.62 | 0.72 | 0.77 | 0.81 |
| 80 | 0.76 | 0.50 | 0.67 | 0.81 | 0.87 | 0.89 |
| 160 | 0.83 | 0.48 | 0.75 | 0.89 | 0.95 | 0.97 |
| WEIBULL distribution: $\Pr(T \leq 0 \mid c_W)$ | | | | | | |
| $c_W$ \ $n$ | 0.5 | 1 | 2 | 4 | 8 | 16 |
| 10 | 0.53 | 0.49 | 0.59 | 0.63 | 0.65 | 0.66 |
| 20 | 0.59 | 0.50 | 0.65 | 0.70 | 0.73 | 0.74 |
| 40 | 0.70 | 0.51 | 0.70 | 0.81 | 0.84 | 0.87 |
| 80 | 0.80 | 0.50 | 0.80 | 0.89 | 0.93 | 0.95 |
| 160 | 0.90 | 0.52 | 0.86 | 0.96 | 0.99 | 0.99 |

<u>Source:</u> BAIN/ENGELHARDT (1980b, p. 377) — Reprinted with permission from *Communications in Statistics — Theory and Methods*. Copyright 1980 by Marcel Dekker, Inc. All rights reserved.

**Example 22/9: Discrimination between WEIBULL and gamma distributions using dataset #1**

The parameters of the fitted WEIBULL distribution are

$$\widehat{b}_W = 99.2079 \quad \text{and} \quad \widehat{c}_W = 2.5957,$$

and the parameters of the fitted gamma distribution are

$$\widehat{b}_G = 15.0855 \quad \text{and} \quad \widehat{c}_G = 5.8301.$$

Fig. 22/4 shows both fitted distributions where the $n = 20$ data points have been marked on the density curves. The value of the discriminating quantity $T$ is $T = 0.4359$, so that we have to choose the gamma distribution, but we know that dataset #1 in Tab. 9/2 has been generated from $We(80, 100, 2.5)$. According to Tab. 22/11 the PCS for the WEIBULL distribution with $c_W = 2.5$ and for $n = 20$ is about 66% or 2 out of 3, so we have had bad luck with this sample.

<u>Figure 22/4:</u> WEIBULL and gamma distributions fitted to dataset #1

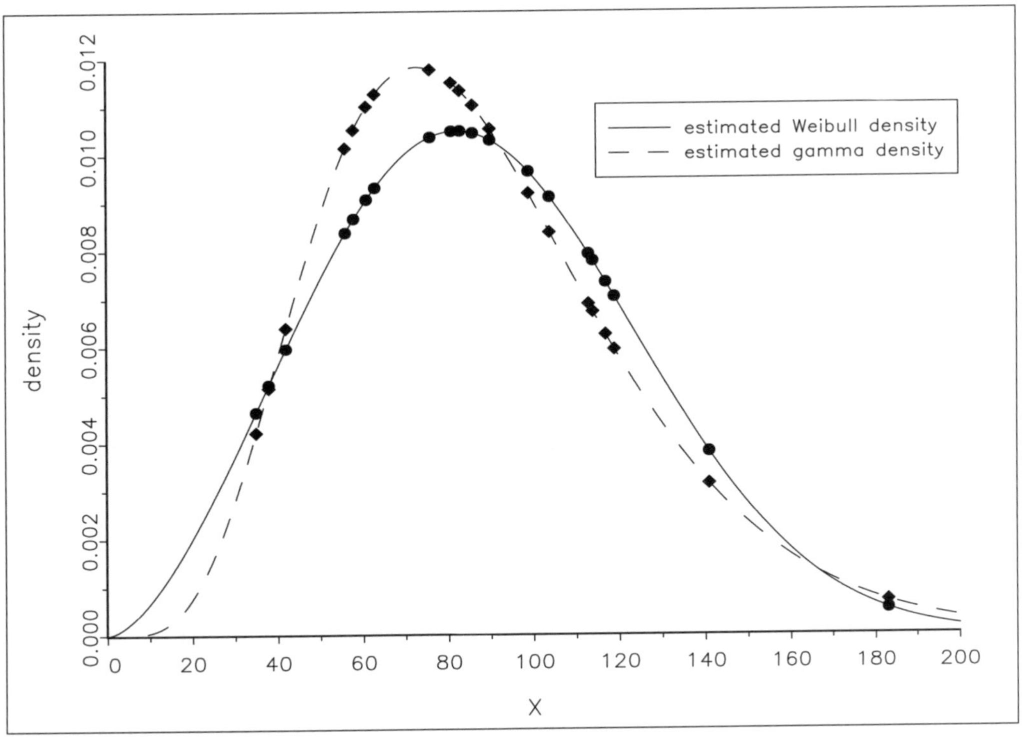

FEARN/NEBENZAHL (1991) have shown that $T$ is approximately normally distributed so that it is possible to find the sample size needed for a user specified value of PCS. Discrimination between WEIBULL and generalized gamma distributions is treated by PARR/WEBSTER (1965) and VOLODIN (1974). BALASOORIYA/ABEYSINGHE (1994) have chosen a discrimination procedure which is based on the deviation between the observed and the predicted order statistics at the far end of the sample, where the prediction rests upon the observations in the first part of the sample.

### 22.2.2.3 WEIBULL versus lognormal distribution[16]

The lognormal distribution with DF

$$f_L(x \mid b_L, c_L) = \frac{1}{\sqrt{2\pi} \, c_L \, x} \, \exp\left\{ -\frac{(\ln x - \ln b_L)^2}{2 \, c_L^2} \right\} \qquad (22.23)$$

is another competitor to the the WEIBULL distribution. In order to discriminate between these two distributions we use $T$, the natural logarithm of the ratio of the maximized like-

---

[16] Suggested reading for this section: DUMONCEAUX/ANTLE/HAAS (1973b), KAPPENMAN (1982, 1988), KUNDU/ MANGLICK (no year).

lihoods, where the nominator is given by the maximized lognormal likelihood:

$$T = n \left\{ 0.5 - \ln \left[ \widehat{c}_L \, \widehat{c}_W \left( \frac{\widehat{b}_L}{\widehat{b}_W} \right)^{\widehat{c}_W} \sqrt{2\pi} \right] \right\} \tag{22.24}$$

with the MLEs

$$\widehat{b}_L = \left( \prod_{i=1}^n X_i \right)^{1/n} \quad \text{and} \quad \widehat{c}_L^2 = \frac{1}{n} \sum_{i=1}^n \left( \ln X_i - \ln \widehat{b}_L \right)^2.$$

The guiding rule is: "Choose the lognormal distribution if $T > 0$; otherwise, choose the WEIBULL distribution."

It can be shown that the PCSs do not depend on the value of the scale parameters $b_L$ or $b_W$, but they depend on the shape parameters $c_L$ or $c_W$, respectively. Tab. 22/12 gives the PCSs $\Pr(T > 0 \,|\, c_L)$ and $\Pr(T \leq 0 \,|\, c_w)$ generated by Monte Carlo techniques.

It can be further shown that the distribution of $T$ in (22.24) is independent of all four parameters $b_L$, $c_L$, $b_W$ and $c_W$ and that the distribution of $T$ can be approximated by a normal distribution. KUNDU/MANGLICK (no year) give the following approximations:

$$\text{PCS(lognormal)} = \text{PCS}(T > 0 \,|\, \text{lognormal}) \approx \Phi\left( \frac{0.0810614\, n}{\sqrt{0.2182818\, n}} \right), \tag{22.25a}$$

$$\text{PCS(WEIBULL)} = \text{PCS}(T \leq 0 \,|\, \text{WEIBULL}) \approx \Phi\left( \frac{-0.0905730\, n}{\sqrt{0.2834081\, n}} \right), \tag{22.25b}$$

where $\Phi(\cdot)$ is the CDF of the standard normal distribution. Now, to determine the **minimum sample size** required to achieve at least PCS(lognormal) = $P$, we equate $P$ to (22.25a) and obtain

$$n = \frac{0.2182818\, u_P^2}{0.0810614^2}, \tag{22.26a}$$

where $u_P$ is the percentile of order $P$ of the standard normal distribution. When the data follow a WEIBULL distribution, we equate $P$ to (22.25b) and obtain

$$n = \frac{0.2834081\, u_P^2}{0.0905730^2}. \tag{22.26b}$$

Therefore, to achieve an overall protection level $P$, we need at least

$$\begin{aligned} n &= u_P^2 \, \max\left( \frac{0.2834081}{0.090530^2}, \frac{0.2182818}{0.0810614^2} \right) \\ &= u_P^2 \, \max(34.5, 33.2) = 34.5 \, u_P^2; \end{aligned} \tag{22.26c}$$

e.g., for $P = 0.95$ and $u_{0.95} = 1.6449$, we find $n = 57$.

Table 22/12: Probability of correct selection (Lognormal versus WEIBULL)

| | Lognormal distribution: $\Pr(T > 0 \mid c_L)$ | | | | |
|---|---|---|---|---|---|
| $c_L$ <br> $n$ | 3 | 1.5 | 1 | 0.5 | 0.25 |
| 10 | 0.67 | 0.67 | 0.69 | 0.66 | 0.66 |
| 30 | 0.85 | 0.84 | 0.82 | 0.82 | 0.85 |
| 50 | 0.90 | 0.90 | 0.90 | 0.91 | 0.90 |
| 100 | 0.97 | 0.97 | 0.96 | 0.97 | 0.98 |
| 200 | 1.00 | 1.00 | 1.00 | 1.00 | 1.00 |
| | WEIBULL distribution: $\Pr(T \leq 0 \mid c_W)$ | | | | |
| $c_W$ <br> $n$ | 0.5 | 1 | 1.5 | 2 | 3 |
| 10 | 0.68 | 0.69 | 0.68 | 0.67 | 0.67 |
| 30 | 0.85 | 0.83 | 0.84 | 0.84 | 0.84 |
| 50 | 0.91 | 0.92 | 0.90 | 0.90 | 0.91 |
| 100 | 0.98 | 0.97 | 0.97 | 0.97 | 0.98 |
| 200 | 1.00 | 1.00 | 1.00 | 1.00 | 0.99 |

Source: KAPPENMAN (1982, p. 668) — Reprinted with permission from *Communications in Statistics — Theory and Methods*. Copyright 1982 by Marcel Dekker, Inc. All rights reserved.

DUMONCEAUX/ANTLE/HAAS (1973b) considered the discrimination problem as a testing of a hypothesis problem, namely:

Problem 1 — $H_0$: "lognormal" versus $H_A$: "WEIBULL,"

Problem 2 — $H_0$: "WEIBULL" versus $H_A$: "lognormal."

They provide the exact critical regions and the power of the chosen likelihood–ratio test based on Monte Carlo simulations. The asymptotic results of KUNDU/MANGLICK given above can be used for testing the above two pairs of hypotheses as follows:

1. For Problem 1, reject $H_0$ at level $\alpha$ if

$$T < 0.0810614\, n - u_\alpha \sqrt{0.2182818\, n}. \tag{22.27a}$$

2. For Problem 2, reject $H_0$ at level $\alpha$ if

$$T > -0.090570\, n + u_\alpha \sqrt{0.2834081\, n}. \tag{22.27b}$$

KAPPENMAN (1988) has suggested a simple method for choosing between the lognormal and the WEIBULL models that does not require estimates of the distribution parameters and that performs — in terms of PCS — as well as the ratio-of–maximized–likelihood method. The method is summarized as follows. Let $X_{i:n}, \ldots, X_{n:n}$ represent the order statistics from a random sample. Set $Y_i = \ln X_{i:n}$. Compute

$$r = \frac{A_3 - A_2}{A_2 - A_1}, \tag{22.28}$$

where $A_1$ is the average of the first $0.05\, n$ $Y_i$'s, $A_3$ is the average of the last $0.05\, n$ $Y_i$'s, and $A_2$ is the average of the $Y_i$'s that remain after the first and last $0.2\, n$ $Y_i$'s are discarded.

For the order statistics averages, fractional observations are to be used if $0.05\,n$ and $0.2\,n$ are not integers. For example, if $n = 50$,

$$A_1 = \frac{Y_1 + Y_2 + 0.5\,Y_3}{2.5}, \quad A_2 = \frac{1}{30}\sum_{i=11}^{40} Y_i, \quad A_3 = \frac{0.5\,Y_{48} + Y_{49} + Y_{50}}{2.5}.$$

If $r > 0.7437$, select the lognormal distribution; otherwise, select the WEIBULL distribution.

---

**Example 22/10: Discrimination between WEIBULL and lognormal distributions using dataset #1**

The parameters of the fitted WEIBULL distribution are

$$\widehat{b}_W = 99.2079 \quad \text{and} \quad \widehat{c}_W = 2.5957,$$

and the parameters of the fitted lognormal distribution are

$$\widehat{b}_L = 80.5245 \quad \text{and} \quad \widehat{c}_L = 0.4303.$$

Fig. 22/5 shows both fitted distributions with the $n = 20$ data points marked on the density curves. The value of of the discriminating quantity $T$ is $T = 0.2410$, so that we have to choose the lognormal distribution, but we know that dataset #1 has been generated from $We(0, 100, 2.5)$. According to Tab. 22/12 the PCS for the WEIBULL distribution with $c_W = 2.5$ and $n = 20$ is about 75% or 3 out of 4, so — as in Example 22/9 — we again have bad luck with this sample.

Figure 22/5: WEIBULL and lognormal densities fitted to dataset #1

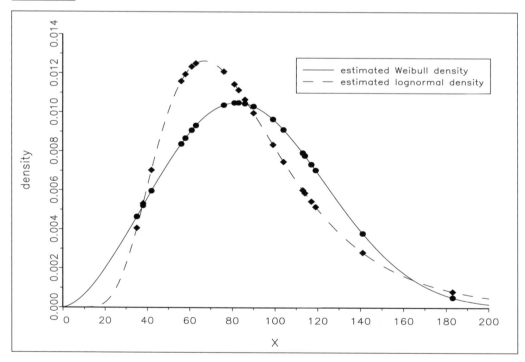

Applying the KAPPENMAN method to this problem, we first have

$$A_1 = \ln x_{1:20} = 3.5553, \quad A_2 = \frac{1}{12}\sum_{i=5}^{16}\ln x_{i:20} = 4.4263, \quad A_3 = \ln x_{20:20} = 5.2095$$

and then
$$r = \frac{A_3 - A_2}{A_2 - A_1} = \frac{5.2095 - 4.4263}{4.4263 - 3.5553} = 0.8991.$$

Because $r = 0.8991 > 0.7477$ we again have to choose the lognormal distribution.

### 22.2.3 Discrimination between WEIBULL and more than one other distribution[17]

It seems quite natural to extend the idea of selecting between two distributions using the ratio of the maximized likelihoods to more than two distributions and to choose that distribution that yields the largest value of the likelihood or the log–likelihood when applied to the sample at hand. KAPPENMAN (1982) has studied this approach for choosing between the gamma, lognormal and WEIBULL distributions and estimated PCSs by Monte Carlo techniques. The PCSs of these distributions do not depend on the shape parameter. Tab. 22/13 shows the results which indicate that for each "sample size – shape parameter value" combination it is much less likely to appropriately select the lognormal model when three models are to be considered instead just two (see Tab. 22/12). The same thing can be said about the WEIBULL model except for the cases where the sample size is small and the value of the shape parameter is less than one. This statement cannot be made about the gamma model, however. When the value of $c_G$ exceeds one, rather large sample sizes are needed to appropriately select the gamma model with any substantial probability. All of this appears to indicate that the gamma distribution is more WEIBULL–like than the lognormal distribution and is more lognormal–like than is the WEIBULL distribution.

KENT/QUESENBERRY (1982) propose a selection statistic that is essentially the value of the **density function of a scale transformation maximal invariant** and select that family with the largest value. The families considered include the exponential, gamma, lognormal and WEIBULL. QUESENBERRY/QUESENBERRY (1982) extend the above idea to type–I censored samples. In both papers the PCSs turn out to be smaller than when selection is based on the maximum likelihood.

TAYLOR/JAKEMAN (1985) propose a procedure which selects the distribution on the basis of the ratio of the KOLMOGOROV–SMIRNOV statistic $\sup_x |F_n(x) - F_0(x)|$ to a given $(1 - \alpha)$ point of the KS–statistic, here: $1 - \alpha = 0.95$. Using such a ratio, the model with the lowest value of this ratio is selected. The simulation results comparing their procedure with the ratio–of–maximized–likelihood approach shows that no one method is likely to provide the "best" selection criterion.

Still another selection procedure has been proposed by PANDEY et al. (1991). It rests upon the $F$–statistic that measures the closeness of a regression line for the cumulative hazard rate or for its logarithm to the data points. With respect to PCS their method is not always superior to the other methods when the choice is between the exponential, WEIBULL and finite–range PARETO distributions.

---

[17] Suggested reading for this section: KAPPENMAN (1982), KENT/QUESENBERRY (1982), PANDEY/FERDOUS/UDDIN (1991), QUESENBERRY/QUESENBERRY (1982), TAYLOR/JAKEMAN (1985).

Table 22/13: Probability of correct selection between gamma, lognormal and WEIBULL
models

| WEIBULL distribution | | | | | |
|---|---|---|---|---|---|
| $c_W$ $n$ | 0.5 | 1 | 1.5 | 2 | 3 |
| 10 | 0.20 | 0.43 | 0.55 | 0.59 | 0.62 |
| 30 | 0.50 | 0.42 | 0.64 | 0.69 | 0.74 |
| 50 | 0.62 | 0.50 | 0.68 | 0.73 | 0.81 |
| 100 | 0.82 | 0.48 | 0.73 | 0.80 | 0.90 |
| 200 | 0.93 | 0.49 | 0.78 | 0.90 | 0.95 |
| Gamma distribution | | | | | |
| $c_G$ $n$ | 0.5 | 1 | 2 | 3 | 4 |
| 10 | 0.58 | 0.30 | 0.15 | 0.19 | 0.18 |
| 30 | 0.70 | 0.41 | 0.36 | 0.37 | 0.37 |
| 50 | 0.72 | 0.42 | 0.51 | 0.45 | 0.51 |
| 100 | 0.77 | 0.50 | 0.61 | 0.64 | 0.66 |
| 200 | 0.84 | 0.51 | 0.76 | 0.84 | 0.82 |
| Lognormal distribution | | | | | |
| $c_L$ $n$ | 3 | 1.5 | 1 | 0.5 | 0.25 |
| 10 | 0.67 | 0.66 | 0.66 | 0.57 | 0.55 |
| 30 | 0.85 | 0.84 | 0.79 | 0.67 | 0.60 |
| 50 | 0.90 | 0.90 | 0.87 | 0.75 | 0.66 |
| 100 | 0.97 | 0.97 | 0.94 | 0.84 | 0.68 |
| 200 | 1.00 | 1.00 | 0.99 | 0.90 | 0.74 |

Source: KAPPENMAN (1982, p. 670) — Reprinted with permission from *Communications in Statistics
— Theory and Methods*. Copyright 1982 by Marcel Dekker, Inc. All rights reserved.

## 22.3   Selecting the better of several WEIBULL distributions[18]

There are three quantities that are commonly used to measure the goodness of a lifetime
distribution, here: the WEIBULL distribution:

- the mean or MTTF (mean time to failure)

$$\mathrm{E}(X) = a + b\,\Gamma\left(1 + \frac{1}{c}\right), \tag{22.29a}$$

- the percentile of a given order $P^*$

$$x_{P^*} = a + b\left[-\ln(1 - P^*)\right]^{1/c}, \tag{22.29b}$$

---

[18] Suggested reading for this section: BHANDARI/MUKHERJEE (1992), GILL/MEHTA (1994), HILL/
PATEL (2000), HSU (1982), KINGSTON/PATEL (1980, 1982), QUREISHI (1964), QUREISHI/
NABARVIAN/ALANEN (1965), RADEMAKER/ANTLE (1975), SCHAFER/SHEFFIELD (1976),
THOMAN/BAIN (1969), TSENG (1988), TSENG/WU (1990).

- the reliability at a given time $t^*$

$$R(t^*) = \exp\left\{-\left(\frac{t^* - a}{b}\right)^c\right\}. \tag{22.29c}$$

For all three quantities the following statement holds: "The higher the value of the quantity, the better the corresponding distribution." When we transfer this statement to the estimates of these quantities, it does not necessarily hold due to sampling errors in the estimated parameters $\widehat{a}$, $\widehat{b}$ and $\widehat{c}$.

To ease the selection or ordering process of several WEIBULL distributions, most authors make some assumptions:

- The threshold or location parameter $a$ is known and/or is equal for all distributions considered.

- The shape parameter $c$ is known and/or is equal for all distributions considered.

- The samples taken from each population and the possible censoring numbers or censoring times are of equal size.

We will present only approaches for selecting the better of two populations and shortly comment on papers treating the case of $k > 2$ populations.

If both WEIBULL populations have the same location parameter $a$ and have the same (though perhaps unknown) value of their shape parameters, then the one with the larger scale parameter would be the better in terms of $R(t^*)$ for all values of $t^*$. THOMAN/BAIN (1969) make use of this assumption and give the probability of correct selection for the simple selection rule: "Select the population with the larger MLE of $b$" disregarding the MLE of $c$.[19] This procedure can be recommended when one assuredly assumes that the shape parameters are equal, but when the actual ratio of the shape parameters is greater than 1.2 their method ought not be used.

QUREISHI (1964) present decision rules in a quality control context for selecting the WEIBULL population with the larger mean life. Since he also assumes equal shape parameters, his procedure would be selecting the WEIBULL with the larger reliability at all times. However, his procedure is quite sensitive to possible differences in the shape parameters.

RADEMAKER/ANTLE (1975) propose a simple decision rule which does not require that the shape parameters be equal. If $R_i(t^* \mid a, b_i, c_i)$ is the reliability at time $t^*$ for population $i(i = 1, 2)$, then we simply select the first population if

$$R_1\big(t^* \mid a, \widehat{b}_1, \widehat{c}_1\big) > R_2\big(t^* \mid a, \widehat{b}_2, \widehat{c}_2\big),$$

and otherwise select the second population. As $\widehat{b}_1$, $\widehat{c}_1$, $\widehat{b}_2$ and $\widehat{c}_2$ are the MLEs, the estimated reliabilities are also MLEs. In order to evaluate this appealing rule, RADE-MAKER/ANTLE presented the PCSs for equal sample sizes $n_1 = n_2 = n$ and equal censoring numbers $r_1 = r_2 = r$ based on Monte Carlo techniques (see Tab. 22/14). The

---

[19]  SCHAFER/SHEFFIELD (1976) have slightly modified the procedure of THOMAN/BAIN by introducing a pooled estimator of the unknown shape parameter (see Sect. 21.1.2).

simulated PCSs produce the answer for all arrangements of the parameters $b_1$, $c_1$, $b_2$, $c_2$ and $t^*$ for which $R_1(t^* \mid a, b_1, c_1) = R_1$ and $R_2(t^* \mid a, b_2, c_2) = R_2$.

Table 22/14: Probability of correct selection of the better of two WEIBULL distributions

| $n$ | $r$ | $(R_1, R_2) =$ $(0.80, 0.90)$ | $(R_1, R_2) =$ $(0.90, 0.95)$ | $(R_1, R2) =$ $(.095, 0.98)$ |
|---|---|---|---|---|
| 20 | 20 | 0.869 | 0.807 | 0.860 |
| 20 | 15 | 0.847 | 0.781 | 0.784 |
| 20 | 10 | 0.845 | 0.765 | 0.762 |
| 40 | 40 | 0.948 | 0.892 | 0.902 |
| 40 | 30 | 0.933 | 0.861 | 0.863 |
| 40 | 20 | 0.930 | 0.844 | 0.836 |
| 60 | 60 | 0.978 | 0.939 | 0.947 |
| 60 | 45 | 0.965 | 0.910 | 0.910 |
| 60 | 30 | 0.961 | 0.897 | 0.891 |
| 80 | 80 | 0.991 | 0.964 | 0.970 |
| 80 | 60 | 0.980 | 0.940 | 0.940 |
| 80 | 40 | 0.980 | 0.927 | 0.920 |
| 100 | 100 | 0.996 | 0.976 | 0.982 |
| 100 | 75 | 0.993 | 0.960 | 0.965 |
| 100 | 50 | 0.992 | 0.955 | 0.949 |
| 120 | 120 | 0.998 | 0.984 | 0.990 |
| 120 | 90 | 0.997 | 0.974 | 0.973 |
| 120 | 60 | 0.995 | 0.964 | 0.960 |
| 160 | 160 | 0.998 | 0.993 | 0.993 |
| 160 | 120 | 0.998 | 0.986 | 0.986 |
| 160 | 90 | 0.997 | 0.984 | 0.980 |

Source: RADEMAKER/ANTLE (1975, p. 18) — Reprinted with permission from
*Transactions on Reliability*. Copyright 1975 by IEEE. All rights reserved.

For planning an experiment it seems worthwhile to have another table giving the sample size needed to attain a given PCS for given pairs $(R_1, R_2)$. Such table has also been calculated by RADEMAKER/ANTLE and is reproduced as Tab. 22/15. The reader may object to Tables 22/14 and 22/15, to enter them one must assume values of $R_1(t^* \mid a, b_1, c_1)$ and $R_2(t^* \mid a, b_2, c_2)$, the unknown quantities of interest. However, the approach is necessary because one cannot have a PCS unless one specifies something about the difference or some lower bound on the difference in the actual probabilities.

The papers on selecting between $k > 2$ populations are rather difficult to read and still more difficult to implement. BHANDARI/MUKERJEE (1992) give two– and multi–stage procedures for selecting the most reliable of several WEIBULL populations with a common shape parameter. HSU (1982) derives a method of constructing optimal procedures to select a subset of the $k$ populations containing the best population which controls the size of the selected subset and which maximizes the minimum probability of making a correct selection; the sample sizes may be different. KINGSTON/PATEL (1980) give a ranking procedure when the unknown parameters are either equal or different. KINGSTON/PATEL

(1982) and TSENG (1988) classify $k$ populations by their reliabilities with respect to a control population. TSENG/WU (1990) propose a locally optimal selection rule when the shape parameters are either known or are unknown but have some prior distributions. We finally mention two papers — GILL/MEHTA (1994) and HILL/PATEL (2000) — treating the related problem of selecting the best of several WEIBULL populations with respect to the shape parameter.

Table 22/15: Sample size $n$ required for given PCS to select the better of two WEIBULL distributions

| $R_1$ | $R_2$ | PCS = 0.90 | | | PCS = 0.95 | | |
|---|---|---|---|---|---|---|---|
| | | Censoring levels | | | Censoring levels | | |
| | | None | $0.25\,n$ | $0.50\,n$ | None | $0.25\,n$ | $0.50\,n$ |
| 0.75 | 0.80 | > 160 | > 160 | > 160 | > 160 | > 160 | > 160 |
| 0.75 | 0.85 | 35 | 38 | 39 | 54 | 64 | 65 |
| 0.75 | 0.90 | 13 | 15 | 16 | 22 | 27 | 28 |
| 0.80 | 0.85 | 122 | 135 | > 160 | > 160 | > 160 | > 160 |
| 0.80 | 0.90 | 25 | 29 | 30 | 41 | 49 | 51 |
| 0.80 | 0.95 | < 10 | < 10 | 11 | 15 | 16 | 17 |
| 0.85 | 0.90 | 81 | 94 | 97 | 156 | > 160 | > 160 |
| 0.85 | 0.95 | 14 | 19 | 20 | 25 | 30 | 33 |
| 0.85 | 0.98 | < 10 | < 10 | 10 | 10 | 13 | 14 |
| 0.90 | 0.95 | 42 | 55 | 63 | 68 | 90 | 96 |
| 0.90 | 0.98 | 11 | 15 | 17 | 18 | 21 | 29 |
| 0.90 | 0.99 | < 10 | < 10 | 11 | 11 | 12 | 16 |
| 0.95 | 0.98 | 40 | 56 | 66 | 62 | 87 | 103 |
| 0.95 | 0.99 | 12 | 21 | 25 | 26 | 34 | 44 |
| 0.98 | 0.99 | 97 | 118 | > 160 | > 160 | > 160 | > 160 |

Source: RADEMAKER/ANTLE (1975, p. 19) — Reprinted with permission from *Transactions on Reliability*. Copyright 1975 by IEEE. All rights reserved.

# III

# Appendices

## Table of the gamma, digamma and trigamma functions

| $z$ | $\Gamma(z)$ | $\psi(z)$ | $\psi'(z)$ | $z$ | $\Gamma(z)$ | $\psi(z)$ | $\psi'(z)$ |
|---|---|---|---|---|---|---|---|
| 1.000 | 1.00000 | $-0.57722$ | 1.64494 | 1.250 | 0.90640 | $-0.22745$ | 1.19733 |
| 1.005 | 0.99714 | $-0.56902$ | 1.63299 | 1.255 | 0.90539 | $-0.22148$ | 1.19073 |
| 1.010 | 0.99433 | $-0.56089$ | 1.62121 | 1.260 | 0.90440 | $-0.21555$ | 1.18419 |
| 1.015 | 0.99156 | $-0.55281$ | 1.60959 | 1.265 | 0.90344 | $-0.20964$ | 1.17772 |
| 1.020 | 0.98884 | $-0.54479$ | 1.59812 | 1.270 | 0.90250 | $-0.20377$ | 1.17132 |
| 1.025 | 0.98617 | $-0.53683$ | 1.58680 | 1.275 | 0.90160 | $-0.19793$ | 1.16498 |
| 1.030 | 0.98355 | $-0.52892$ | 1.57562 | 1.280 | 0.90072 | $-0.19212$ | 1.15871 |
| 1.035 | 0.98097 | $-0.52107$ | 1.56460 | 1.285 | 0.89987 | $-0.18634$ | 1.15250 |
| 1.040 | 0.97844 | $-0.51327$ | 1.55371 | 1.290 | 0.89904 | $-0.18059$ | 1.14636 |
| 1.045 | 0.97595 | $-0.50553$ | 1.54297 | 1.295 | 0.89824 | $-0.17488$ | 1.14028 |
| 1.050 | 0.97350 | $-0.49784$ | 1.53236 | 1.300 | 0.89747 | $-0.16919$ | 1.13425 |
| 1.055 | 0.97110 | $-0.49021$ | 1.52188 | 1.305 | 0.89672 | $-0.16353$ | 1.12829 |
| 1.060 | 0.96874 | $-0.48263$ | 1.51154 | 1.310 | 0.89600 | $-0.15791$ | 1.12239 |
| 1.065 | 0.96643 | $-0.47509$ | 1.50133 | 1.315 | 0.89531 | $-0.15231$ | 1.11654 |
| 1.070 | 0.96415 | $-0.46761$ | 1.49125 | 1.320 | 0.89464 | $-0.14674$ | 1.11076 |
| 1.075 | 0.96192 | $-0.46018$ | 1.48129 | 1.325 | 0.89400 | $-0.14120$ | 1.10502 |
| 1.080 | 0.95973 | $-0.45280$ | 1.47145 | 1.330 | 0.89338 | $-0.13569$ | 1.09935 |
| 1.085 | 0.95757 | $-0.44547$ | 1.46174 | 1.335 | 0.89278 | $-0.13021$ | 1.09373 |
| 1.090 | 0.95546 | $-0.43818$ | 1.45214 | 1.340 | 0.89222 | $-0.12475$ | 1.08816 |
| 1.095 | 0.95339 | $-0.43094$ | 1.44266 | 1.345 | 0.89167 | $-0.11933$ | 1.08265 |
| 1.100 | 0.95135 | $-0.42375$ | 1.43330 | 1.350 | 0.89115 | $-0.11393$ | 1.07719 |
| 1.105 | 0.94935 | $-0.41661$ | 1.42405 | 1.355 | 0.89066 | $-0.10856$ | 1.07179 |
| 1.110 | 0.94740 | $-0.40951$ | 1.41491 | 1.360 | 0.89018 | $-0.10321$ | 1.06643 |
| 1.115 | 0.94547 | $-0.40246$ | 1.40588 | 1.365 | 0.88974 | $-0.09789$ | 1.06113 |
| 1.120 | 0.94359 | $-0.39546$ | 1.39695 | 1.370 | 0.88931 | $-0.09260$ | 1.05587 |
| 1.125 | 0.94174 | $-0.38849$ | 1.38813 | 1.375 | 0.88891 | $-0.08733$ | 1.05067 |
| 1.130 | 0.93993 | $-0.38157$ | 1.37942 | 1.380 | 0.88854 | $-0.08209$ | 1.04551 |
| 1.135 | 0.93816 | $-0.37470$ | 1.37080 | 1.385 | 0.88818 | $-0.07688$ | 1.04040 |
| 1.140 | 0.93642 | $-0.36787$ | 1.36229 | 1.390 | 0.88785 | $-0.07169$ | 1.03534 |
| 1.145 | 0.93471 | $-0.36108$ | 1.35388 | 1.395 | 0.88755 | $-0.06652$ | 1.03033 |
| 1.150 | 0.93304 | $-0.35433$ | 1.34556 | 1.400 | 0.88726 | $-0.06138$ | 1.02536 |
| 1.155 | 0.93141 | $-0.34762$ | 1.33734 | 1.405 | 0.88700 | $-0.05627$ | 1.02043 |
| 1.160 | 0.92980 | $-0.34095$ | 1.32921 | 1.410 | 0.88676 | $-0.05118$ | 1.01556 |
| 1.165 | 0.92823 | $-0.33433$ | 1.32117 | 1.415 | 0.88655 | $-0.04611$ | 1.01072 |
| 1.170 | 0.92670 | $-0.32774$ | 1.31323 | 1.420 | 0.88636 | $-0.04107$ | 1.00593 |
| 1.175 | 0.92520 | $-0.32119$ | 1.30537 | 1.425 | 0.88618 | $-0.03606$ | 1.00118 |
| 1.180 | 0.92373 | $-0.31469$ | 1.29760 | 1.430 | 0.88604 | $-0.03106$ | 0.99648 |
| 1.185 | 0.92229 | $-0.30822$ | 1.28992 | 1.435 | 0.88591 | $-0.02609$ | 0.99182 |
| 1.190 | 0.92089 | $-0.30179$ | 1.28232 | 1.440 | 0.88581 | $-0.02114$ | 0.98720 |
| 1.195 | 0.91951 | $-0.29540$ | 1.27481 | 1.445 | 0.88572 | $-0.01622$ | 0.98262 |
| 1.200 | 0.91817 | $-0.28904$ | 1.26738 | 1.450 | 0.88566 | $-0.01132$ | 0.97808 |
| 1.205 | 0.91686 | $-0.28272$ | 1.26003 | 1.455 | 0.88562 | $-0.00644$ | 0.97358 |
| 1.210 | 0.91558 | $-0.27644$ | 1.25276 | 1.460 | 0.88560 | $-0.00158$ | 0.96912 |
| 1.215 | 0.91433 | $-0.27019$ | 1.24557 | 1.465 | 0.88561 | 0.00325 | 0.96470 |
| 1.220 | 0.91311 | $-0.26398$ | 1.23845 | 1.470 | 0.88563 | 0.00807 | 0.96032 |
| 1.225 | 0.91192 | $-0.25781$ | 1.23141 | 1.475 | 0.88568 | 0.01286 | 0.95597 |
| 1.230 | 0.91075 | $-0.25167$ | 1.22445 | 1.480 | 0.88575 | 0.01763 | 0.95166 |
| 1.235 | 0.90962 | $-0.24556$ | 1.21756 | 1.485 | 0.88584 | 0.02237 | 0.94739 |
| 1.240 | 0.90852 | $-0.23949$ | 1.21075 | 1.490 | 0.88595 | 0.02710 | 0.94316 |
| 1.245 | 0.90745 | $-0.23346$ | 1.20400 | 1.495 | 0.88608 | 0.03181 | 0.93896 |

## Table of the gamma, digamma and trigamma functions (Continuation)

| $z$ | $\Gamma(z)$ | $\psi(z)$ | $\psi'(z)$ | $z$ | $\Gamma(z)$ | $\psi(z)$ | $\psi'(z)$ |
|---|---|---|---|---|---|---|---|
| 1.500 | 0.88623 | 0.03649 | 0.93480 | 1.750 | 0.91906 | 0.24747 | 0.76410 |
| 1.505 | 0.88640 | 0.04115 | 0.93068 | 1.755 | 0.92021 | 0.25129 | 0.76130 |
| 1.510 | 0.88659 | 0.04580 | 0.92658 | 1.760 | 0.92137 | 0.25509 | 0.75852 |
| 1.515 | 0.88680 | 0.05042 | 0.92253 | 1.765 | 0.92256 | 0.25887 | 0.75576 |
| 1.520 | 0.88704 | 0.05502 | 0.91850 | 1.770 | 0.92376 | 0.26264 | 0.75302 |
| 1.525 | 0.88729 | 0.05960 | 0.91451 | 1.775 | 0.92499 | 0.26640 | 0.75030 |
| 1.530 | 0.88757 | 0.06417 | 0.91056 | 1.780 | 0.92623 | 0.27015 | 0.74760 |
| 1.535 | 0.88786 | 0.06871 | 0.90663 | 1.785 | 0.92749 | 0.27388 | 0.74491 |
| 1.540 | 0.88818 | 0.07323 | 0.90274 | 1.790 | 0.92877 | 0.27760 | 0.74225 |
| 1.545 | 0.88851 | 0.07774 | 0.89888 | 1.795 | 0.93007 | 0.28130 | 0.73960 |
| 1.550 | 0.88887 | 0.08222 | 0.89505 | 1.800 | 0.93138 | 0.28499 | 0.73697 |
| 1.555 | 0.88924 | 0.08669 | 0.89126 | 1.805 | 0.93272 | 0.28867 | 0.73436 |
| 1.560 | 0.88964 | 0.09114 | 0.88749 | 1.810 | 0.93408 | 0.29234 | 0.73177 |
| 1.565 | 0.89005 | 0.09556 | 0.88376 | 1.815 | 0.93545 | 0.29599 | 0.72920 |
| 1.570 | 0.89049 | 0.09997 | 0.88005 | 1.820 | 0.93685 | 0.29963 | 0.72664 |
| 1.575 | 0.89094 | 0.10436 | 0.87638 | 1.825 | 0.93826 | 0.30325 | 0.72410 |
| 1.580 | 0.89142 | 0.10874 | 0.87273 | 1.830 | 0.93969 | 0.30687 | 0.72158 |
| 1.585 | 0.89191 | 0.11309 | 0.86911 | 1.835 | 0.94114 | 0.31047 | 0.71907 |
| 1.590 | 0.89243 | 0.11743 | 0.86552 | 1.840 | 0.94261 | 0.31406 | 0.71658 |
| 1.595 | 0.89296 | 0.12175 | 0.86196 | 1.845 | 0.94410 | 0.31764 | 0.71411 |
| 1.600 | 0.89352 | 0.12605 | 0.85843 | 1.850 | 0.94561 | 0.32120 | 0.71165 |
| 1.605 | 0.89409 | 0.13033 | 0.85493 | 1.855 | 0.94714 | 0.32475 | 0.70922 |
| 1.610 | 0.89468 | 0.13460 | 0.85145 | 1.860 | 0.94869 | 0.32829 | 0.70679 |
| 1.615 | 0.89529 | 0.13885 | 0.84800 | 1.865 | 0.95025 | 0.33182 | 0.70439 |
| 1.620 | 0.89592 | 0.14308 | 0.84458 | 1.870 | 0.95184 | 0.33534 | 0.70199 |
| 1.625 | 0.89657 | 0.14729 | 0.84118 | 1.875 | 0.95345 | 0.33884 | 0.69962 |
| 1.630 | 0.89724 | 0.15149 | 0.83781 | 1.880 | 0.95507 | 0.34233 | 0.69726 |
| 1.635 | 0.89793 | 0.15567 | 0.83447 | 1.885 | 0.95672 | 0.34581 | 0.69491 |
| 1.640 | 0.89864 | 0.15983 | 0.83115 | 1.890 | 0.95838 | 0.34928 | 0.69259 |
| 1.645 | 0.89937 | 0.16398 | 0.82786 | 1.895 | 0.96006 | 0.35274 | 0.69027 |
| 1.650 | 0.90012 | 0.16811 | 0.82459 | 1.900 | 0.96177 | 0.35618 | 0.68797 |
| 1.655 | 0.90088 | 0.17223 | 0.82135 | 1.905 | 0.96349 | 0.35962 | 0.68569 |
| 1.660 | 0.90167 | 0.17633 | 0.81813 | 1.910 | 0.96523 | 0.36304 | 0.68342 |
| 1.665 | 0.90247 | 0.18041 | 0.81493 | 1.915 | 0.96699 | 0.36645 | 0.68116 |
| 1.670 | 0.90330 | 0.18447 | 0.81176 | 1.920 | 0.96877 | 0.36985 | 0.67892 |
| 1.675 | 0.90414 | 0.18853 | 0.80862 | 1.925 | 0.97058 | 0.37324 | 0.67670 |
| 1.680 | 0.90500 | 0.19256 | 0.80550 | 1.930 | 0.97240 | 0.37662 | 0.67449 |
| 1.685 | 0.90588 | 0.19658 | 0.80240 | 1.935 | 0.97424 | 0.37999 | 0.67229 |
| 1.690 | 0.90678 | 0.20059 | 0.79932 | 1.940 | 0.97610 | 0.38334 | 0.67010 |
| 1.695 | 0.90770 | 0.20457 | 0.79626 | 1.945 | 0.97798 | 0.38669 | 0.66793 |
| 1.700 | 0.90864 | 0.20855 | 0.79323 | 1.950 | 0.97988 | 0.39002 | 0.66578 |
| 1.705 | 0.90960 | 0.21251 | 0.79022 | 1.955 | 0.98180 | 0.39335 | 0.66363 |
| 1.710 | 0.91057 | 0.21645 | 0.78724 | 1.960 | 0.98374 | 0.39666 | 0.66150 |
| 1.715 | 0.91157 | 0.22038 | 0.78427 | 1.965 | 0.98570 | 0.39996 | 0.65939 |
| 1.720 | 0.91258 | 0.22429 | 0.78133 | 1.970 | 0.98768 | 0.40325 | 0.65728 |
| 1.725 | 0.91361 | 0.22819 | 0.77840 | 1.975 | 0.98969 | 0.40653 | 0.65519 |
| 1.730 | 0.91467 | 0.23208 | 0.77550 | 1.980 | 0.99171 | 0.40980 | 0.65312 |
| 1.735 | 0.91574 | 0.23595 | 0.77262 | 1.985 | 0.99375 | 0.41306 | 0.65105 |
| 1.740 | 0.91683 | 0.23980 | 0.76976 | 1.990 | 0.99581 | 0.41631 | 0.64900 |
| 1.745 | 0.91793 | 0.24364 | 0.76692 | 1.995 | 0.99790 | 0.41955 | 0.64696 |
|  |  |  |  | 2.000 | 1.00000 | 0.42278 | 0.64493 |

# Abbreviations

| | |
|---|---|
| ABLE | asymptotically best linear estimator or estimate |
| ABLIE | asymptotically best linear invariant estimator or estimate |
| ABLUE | asymptotically best linear unbiased estimator or estimate |
| ALT | accelerated life test |
| ARE | asymptotic relative efficiency |
| BED | bivariate exponential distribution |
| BLIE | best linear invariant estimator or estimate |
| BLIP | best linear invariant predictor |
| BLUE | best linear unbiased estimator or estimate |
| BLUP | best linear unbiased predictor |
| BWD | bivariate WEIBULL distribution |
| CCDF | complementary cumulative distribution function or reliability function |
| CDF | cumulative distribution function |
| CE | cumulative exposure |
| CHR | cumulative hazard rate |
| CRLB | CRAMÉR–RAO lower bound |
| DAF | decreasing aging factor |
| DF | density function |
| DFR | decreasing failure (= hazard) rate |
| DFRA | decreasing failure (= hazard) rate average |
| DHR | decreasing hazard rate |
| DHRA | decreasing hazard rate average |
| DIHR | decreasing interval hazard rate average |
| DMRL | decreasing mean residual life |
| EB | empirical BAYES |
| EDF | empirical distribution function |
| FLUP | final linear unbiased predictor |
| GLS | general least squares |
| GLUE | good linear unbiased estimator or estimate |
| HNBUE | harmonic new better than used in expectation |
| HNWUE | harmonic new worse than used in expectation |
| HPD | highest posterior density |
| HPP | homogeneous POISSON process |
| HR | hazard rate |

| | |
|---|---|
| IAF | increasing aging factor |
| IFR | increasing failure (= hazard) rate |
| IFRA | increasing failure (= hazard) rate average |
| IHR | increasing hazard rate |
| IHRA | increasing hazard rate average |
| iid | identically and independently distributed |
| IDD | infinitely divisible distributed |
| IIHR | increasing interval hazard rate average |
| IMRL | increasing mean residual life |
| IPL | inverse power law |
| MDE | minimum distance estimator or estimate |
| ML | maximum likelihood |
| MLE | maximum likelihood estimator or estimate |
| MME | modified method of moments estimator or estimate |
| MMLE | modified maximum likelihood estimator or estimate |
| MRL | mean residual life |
| MSE | mean squared error |
| MTBF | mean time between failures |
| MTTT | mean time to failure |
| MVBE | minimum variance bound estimator or estimate |
| MVE | multivariate exponential distribution |
| MVU | minimum variance unbiased |
| MWD | multivariate WEIBULL distribution |
| NBU | new better than used |
| NBUE | new better than used in expectation |
| NHPP | non–homogeneous POISSON process |
| NLS | non–linear least squares |
| NWU | new worse than used |
| NWUE | new worse than used in expectation |
| OLS | ordinary least squares |
| PALT | partially accelerated life test |
| PCS | probability of correct selection |
| PWM | probability weighted moment |
| REET | ratio of expected experimental times |
| RMSE | root mean squared error |

| | |
|---|---|
| RP | renewal process |
| RSS | residual sum of squares |
| SRE | sample reuse estimator or estimate |
| TFR | tampered failure rate |
| TRV | tampered random variable |
| TTT | total time on test |
| UMVUE | uniformly minimum variance unbiased estimator or estimate |
| WLS | weighted least squares |
| WPP | WEIBULL–probability–paper |
| WRP | WEIBULL renewal process |

# Mathematical and statistical notations

| | |
|---|---|
| $\sim$ | distributed as |
| $\overset{iid}{\sim}$ | independently and identically distributed |
| $\overset{asym}{\sim}$ | asymptotically distributed |
| $\overset{d}{=}$ | equality in distribution |
| $:=$ | equal by definition |
| $\overset{d}{\to}$ | convergent in distribution |
| $\forall$ | for all |
| $\circ$ | inner product operator |
| $*$ | convolution operator |
| $\wedge$ | compounding operator |
| $\propto$ | proportional to |
| $\alpha_r$ | $\alpha$–coefficient of order $r$ |
| $B(a, b)$ | complete beta function |
| $Be(a, b)$ | beta distribution with parameters $a$, $b$ |
| $\beta_r$ | $\beta$–coefficient of order $r$ |
| $CEILING$ | rounding "up" to nearest integer |
| $\mathrm{CF}(\cdot)$ | coefficient of variation of |
| $\chi^2(\nu)$ | $\chi^2$– distribution with $\nu$ degrees of freedom |
| $\mathrm{Cov}(\cdot)$ | covariance of |
| $\mathrm{d}$ | differential operator |
| $\mathrm{E}(\cdot)$ | expectation of |
| $\mathrm{erf}(\cdot)$ | Gaussian error function of |
| $Ev_i(a, b)$ | type–I extreme value distribution of the minimum with parameters $a$, $b$ |
| $Ev_{ii}(a, b, c)$ | type–II extreme value distribution of the minimum with parameters $a$, $b$, $c$ |
| $Ev_{iii}(a, b, c)$ | type–III extreme value distribution of the minimum with parameters $a$, $b$, $c$ |
| $Ev_I(a, b)$ | type–I extreme value distribution of the maximum with parameters $a$, $b$ |
| $Ev_{II}(a, b, c)$ | type–II extreme value distribution of the maximum with parameters $a$, $b$, $c$ |
| $Ev_{III}(a, b, c)$ | type–III extreme value distribution of the maximum with parameters $a$, $b$, $c$ |
| $Ex(\lambda)$ | exponential distribution with parameter $\lambda$ |

| | |
|---|---|
| $f(\cdot)$ | density function of |
| $F(\cdot)$ | cumulative distribution of |
| $F(\nu_1, \nu_2)$ | $F$–distribution with $\nu_1$, $\nu_2$ degrees of freedom |
| $FLOOR$ | rounding "down" to nearest integer |
| $Ga(\cdots)$ | gamma distribution with parameters $\cdots$ |
| $\gamma$ | EULER's constant |
| $\gamma(\cdot\|\cdot)$ | incomplete gamma function |
| $\Gamma(\cdot)$ | complete gamma function |
| $\Gamma(\cdot\|\cdot)$ | complementary incomplete gamma function |
| $\Gamma_r$ | $\Gamma(1 + r/c)$ |
| $I(\cdot)$ | entropy of |
| $I(\boldsymbol{\theta})$ | FISHER information for parameter vector $\boldsymbol{\theta}$ |
| $I_p(a, b)$ | PEARSON's incomplete beta function |
| $K_X(\cdot)$ | cumulant generating function |
| $\kappa_r(\cdot)$ | cumulant of order $r$ of |
| $\ell(\cdot, \cdot)$ | loss function |
| $L(\boldsymbol{\theta} \| \mathrm{data})$ | likelihood function for parameter vector $\boldsymbol{\theta}$ |
| $\mathcal{L}(\boldsymbol{\theta} \| \mathrm{data})$ | log–likelihood function for parameter vector $\boldsymbol{\theta}$ |
| lim | limit |
| $Lw(a, b)$ | $= Ev_i(a, b)$ — Log–WEIBULL distribution with parameters $a$, $b$ |
| $M$ | sample mid–range |
| $M_X(\cdot)$ | raw moment generating function of $X$ |
| $\mu_{i:n}$ | expectation of $X_{i:n}$ |
| $\mu_r(\cdot)$ | central moment (= moment about the mean) of order $r$ of |
| $\mu_{[r]}(\cdot)$ | ascending factorial central moment of order $r$ of |
| $\mu_{(r)}(\cdot)$ | descending factorial central moment of order $r$ of |
| $\mu_{i:n}^{(k)}$ | $k$–th moment about zero of $X_{i:n}$ |
| $\mu_r'(\cdot)$ | raw moment (= moment about zero) of order $r$ of |
| $\mu_{[r]}'(\cdot)$ | ascending factorial raw moment of order $r$ of |
| $\mu_{(r)}'(\cdot)$ | descending factorial raw moment of order $r$ of |

| | |
|---|---|
| $\mu_r^*(\cdot)$ | standardized moment of order $r$ of |
| $No(\mu, \sigma^2)$ | normal distribution with parameters $\mu$, $\sigma^2$ |
| $\nu_r(\cdot)$ | absolute central moment of order $r$ of |
| $\nu_r'(\cdot)$ | absolute raw moment of order $r$ of |
| $o(g(x)) = f(x)$ | $\lim_{x \to x_0} \dfrac{f(x)}{g(x)} = 0$ — LANDAU symbol |
| $O(g(x)) = f(x)$ | $\lim_{x \to x_0} \dfrac{f(x)}{g(x)} < \text{constant}$ — LANDAU symbol |
| $\Omega$ | parameter space |
| $\varphi_X(\cdot)$ | characteristic function of $X$ |
| $\Phi(\cdot)$ | cumulative distribution function of $No(0, 1)$ |
| plim | probability limit |
| $Po(\lambda)$ | POISSON distribution |
| $\psi(\cdot)$ | digamma function |
| $\psi'(\cdot)$ | trigamma function |
| $\mathbb{R}$ | set of real numbers |
| $R(\cdot)$ | CCDF (= or reliability function) |
| $Re(a, b)$ | uniform (= rectangular) distribution with parameters $a$, $b$ |
| $\sigma(\cdot)$ | standard deviation of |
| $\text{sign}(\cdot)$ | sign function |
| $\mathfrak{s}_{(k,j)}$ | STIRLING number of the first kind |
| $\mathfrak{S}_{(k,j)}$ | STIRLING number of the second kind |
| $T_X(\cdot)$ | information generating function of |
| $\text{tr}(\cdot)$ | trace of |
| $\text{Var}(\cdot)$ | variance |
| $We(a, b, c)$ | $= Ev_{iii}(a, b, c)$ — WEIBULL distribution with parameters $a$, $b$, $c$ |
| $\overline{X}$ | sample mean of $X$ |
| $\widetilde{X}$ | sample median of $X$ |
| $X_{i:n}$ | $i$–th order statistic in a sample of size $n$ |
| $Z_X(\cdot)$ | central moment generating function of $X$ |
| $\mathbb{Z}$ | set of non–negative integers |

# Bibliography

**A-A-A-A-A**

ABDEL–GHALY, A.A. / ATTIA, A.F. / ABDEL–GHANI, M.M. (2002): The maximum likelihood estimates in step partially accelerated life tests for the Weibull parameters in censored data; *Communications in Statistics — Theory and Methods* **31**, 551–573

ABDEL–WAHID, A.A. / WINTERBOTTOM, A. (1987): Approximate Bayesian estimates for the Weibull reliability and hazard rate from censored data; *Journal of Statistical Planning and Inference* **16**, 277–283

ABDELHAFEZ, M.E. / THOMAN, D.R. (1991): Bootstrap confidence bounds for the Weibull and extreme value regression model with randomly censored data; *The Egyptian Statistical Journal* **35**, 95–109

ABERNETHY, R.B. (2006): *The New Weibull Handbook*, 5th ed.

ABRAMOWITZ, M. / STEGUN, I.A. (eds.) (1965): *Handbook of Mathematical Functions;* Dover, New York

ABRAMS, K. / ASHBY, D. / ERRINGTON, D. (1996): A Bayesian approach to Weibull survival model: Application to a cancer clinical trial; *Lifetime Data Analysis* **2**, 159–174

ACHCAR, J.A. (1991): A Bayesian approach in the detection of outliers for normal and Weibull distributions; *Revista de Matemática e Estatística* **9**, 91–108

ACHCAR, J.A. / BROOKMEYER, R. / HUNTER, W.G (1985): An application of Bayesian analysis to medical following–up data; *Statistics in Medicine* **4**, 509–520

ACHCAR, J.A. / FOGO, J.C. (1997): A useful reparameterization for the reliability in the Weibull case; *Computational Statistics & Data Analysis* **24**, 387–400

ADATIA, A. / CHAN, L.K. (1982): Robust procedures for estimating the scale parameter and predicting future order statistics of Weibull distributions; *IEEE Transactions on Reliability* **31**, 491–498

ADATIA, A. / CHAN, L.K. (1985): Robust estimates of the three–parameter Weibull distribution; *IEEE Transactions on Reliability* **34**, 347–351

AHMAD, K.E. / MOUSTAFA, H.M. / ABD–ELRAHMAN, A.M. (1997): Approximate Bayes estimation for mixtures of two Weibull distributions under Type–2 censoring; *Journal of Statistical Computation & Simulation* **58**, 269–285

AHMAD, S. (1972): The hazard rate and mean time to failure of a bivariate Weibull distribution; *University of Karachi Journal of Science* **1**, 51–53

AHO, M. / BAIN, L.J. / ENGELHARDT, M. (1983): Goodness–of–fit tests for the Weibull distribution with unknown parameters and censored sampling; *J. Statist. Comp. Simul.* **18**, 59–69

AHO, M. / BAIN, L.J. / ENGELHARDT, M. (1985): Goodness–of–fit tests for the Weibull distribution with unknown parameters and heavy censoring; *J. Statist. Comp. Simul.* **21**, 213–225

AITKEN, A.C. (1935): On least squares and linear combinations of observations; *Proc. Roy. Soc. Edinb.* **55**, 42–48

AITKIN, M. / LAIRD, N. / FRANCIS, B. (1983): A reanalysis of the Stanford heart transplant data (with discussion); *Journal of the American Statistical Association* **78**, 264–292

AL–BAIDHANI, F.A. / SINCLAIR, C.D. (1987): Comparison of methods of estimation of parameters of the Weibull distribution; *Communications in Statistics — Simulation and Computation* **16**, 373–384

AL–HUSSAINI, E.K. / ABD–EL–HAKIM, N.S. (1989): Failure rate of the inverse Gaussian–Weibull mixture model; *Annals of the Institute of Statistical Mathematics* **41**, 617–622

AL–HUSSAINI, E.K. / ABD–EL–HAKIM, N.S. (1990): Estimation of the parameters of the inverse Gaussian–Weibull mixture model; *Communications in Statistics — Theory and Methods* **19**, 1607–1622

AL–HUSSAINI, E.K. / ABD–EL–HAKIM, N.S. (1992): Efficiency of schemes of sampling from the inverse Gaussian–Weibull mixture model; *Communications in Statistics — Theory and Methods* **21**, 3147–3169

ALI, M. / KHAN, A.H. (1996): Ratio and inverse moments of order statistics from Weibull and exponential distributions; *Journal of Applied Statistical Science* **4**, 1–8

ALI, M.A. (1997): Characterization of Weibull and exponential distributions; *Journal of Statistical Studies* **17**, 103–106

ALI KHAN, M.S. / KHALIQUE, A. / ABOUAMMOH, A.M. (1989): On estimating parameters in a discrete Weibull distribution; *IEEE Transactions on Reliability* **38**, 348–350

ALMEIDA, J.B. (1999): Application of Weibull statistics to the failure of coatings; *Journal of Material Processing Technology* **93**, 257–263)

AL–MUTAIRI, D.K. / AGARWAL, S.K. (1999): Two types of generalized Weibull distributions and their applications under different environmental conditions; *Statistica* **59**, 269–227

ALPINI, R. / FATTORINI, L. (1993): Empirical performance of some goodness–of–fit tests for the Weibull and type I extreme value distributions; *Statistica Applicata* **5**, 129–148

ANDERSON, K.M. (1991): A nonproportional hazards Weibull accelerated failure time regression model; *Biometrics* **47**, 281–288

ANTLE, C.E. / BAIN, L.J. (1969): A property of maximum likelihood estimators of location and scale parameters; SIAM REVIEW **11**, 251–253

ANTLE, C.E. / BAIN, L.J. (1988): Weibull distribution; in KOTZ, S. / JOHNSON, N.L. / READ, C.B. (eds.): *Encyclopedia of Statistical Sciences* **9**, Wiley, New York, 549–556

ANTLE, C.E. / KLIMKO, L.A. / RADEMAKER, A.W. / ROCKETTE, H.E. (1975): Upper bounds for the power of invariant tests for the exponential distribution with Weibull alternatives; *Technometrics* **17**, 357–360

APT, K.E. (1976): Applicability of the Weibull distribution function to atmospheric radioactivity data; *Atmos. Environ.* **10**, 771–781

ARCHER, N.P. (1980): A computational technique for maximum likelihood estimation with Weibull models; *IEEE Transactions on Reliability* **29**, 57–62

ARCHER, N.P. (1982): Maximum likelihood estimation with Weibull models when the data are grouped; *Communications in Statistics — Theory and Methods* **11**, 199–207

ARNOLD, B.C. / BALAKRISHNAN, N. (1989): *Relations, Bounds and Approximations for Order Statistics*; Lecture Notes in Statistics **53**, Springer, New York

AROIAN, L.A. (1965): Some properties of the conditional Weibull distribution; *Transactions of the 19th Technical Conference of the American Society for Quality Control*, 361–368

AROIAN, L.A./ ROBINSON, D.E. (1966): Sequential life tests for the exponential distribution with changing parameters; *Technometrics* **8**, 217–227

ARORA, M.S. (1974): The four parameter generalized Weibull family and its derivatives; *Abstract 711–62–6, Notices of the Institute of Mathematical Statistics*, A–240

ASCHER, H. / FEINGOLD, H. (1984): *Repairable Systems Reliability*; Marcel Dekker, New York

ASHOUR, S.K. (1985): Estimation of the parameters of mixed Weibull–exponential models from censored data; *Tamkang Journal of Mathematics* **16**, 103–111

ASHOUR, S.K. (1987a): Multi–censored sampling in mixed Weibull distributions; *Journal of the Indian Association for Production, Quality and Reliability* **12**, 51–56

ASHOUR, S.K. (1987b): Bayesian estimation of mixed Weibull exponential in life testing; *Applied Stochastic Models and Data Analysis* **3**, 51–57

ASHOUR, S.K. / JONES, P.W. (1977): Shortest confidence intervals for the shape parameter of a Weibull distribution based on two adjacent failure times; *Technometrics* **19**, 201–204

ASHOUR, S.K. / RASHWAN, D.R. (1981): Bayesian prediction for compound Weibull model; *Communications in Statistics — Theory and Methods* **10**, 1613–1624

ASHOUR, S.K. / SHALABY, O.A. (1983): Estimating sample size with Weibull failure; *Mathematische Operationsforschung und Statistik, Series Statistics* **14**, 263–268

ATTIA, A.F. (1998): Empirical Bayes approach for accelerated life tests considering the Weibull distribution; *The Egyptian Statistical Journal* **42**, 22–31

**B-B-B-B-B**

BAGANHA, M.P. / FERRER, G. / PYKE, D.F. (1999): The residual life of the renewal process — A simple algorithm; *Naval Research Logistics Quarterly* **46**, 435–443

BAHLER, C. / HILL, R.R. JR. / BYERS, R.A. (1989): Comparison of logistic and Weibull functions: The effect of temperature on cumulative germination of alfalfa; *Crop Science* **29**, 142–146

BAI, D.S. / CHA, M.S. / CHUNG, S.W. (1992): Optimum simple ramp–tests for the Weibull distribution and type–I censoring; *IEEE Transactions on Reliability* **41**, 407–413

BAI, D.S. / CHUN, Y.R. / CHA, M.S. (1997): Time censored ramp tests with stress bound for Weibull life distribution; *IEEE Transactions on Reliability* **46**, 99–107

BAI, D.S. / KIM, M.S. (1993): Optimum simple step–stress accelerated life tests for Weibull distributions and type II censoring; *Naval Research Logistics Quarterly* **40**, 193–210

BAIN, L J. (1972): Inferences based on censored sampling from the Weibull or extreme–value distribution; *Technometrics* **14**, 693–702

BAIN, L.J. (1973): Results for one or more independent censored samples from the Weibull or extreme–value distribution; *Technometrics* **15**, 279–288

BAIN, L.J. (1978): *Statistical Analysis of Reliability and Life–Testing Models — Theory and Methods*; Marcel Dekker, New York

BAIN, L.J. / ANTLE, C.E. (1967): Estimation of parameters of the Weibull distribution; *Technometrics* **9**, 621–627

BAIN, L.J. / ANTLE, C.E. (1970): Inferential procedures for the Weibull and generalized gamma distribution; *ARL Technical Report No. 70–0266*

BAIN, L.J. / ANTLE, C.E. / BILLMAN, B.R. (1971): Statistical analyses for the Weibull distribution with emphasis on censored sampling; *ARL Technical Report No. 71–0242*

BAIN, L.J. / ENGELHARDT, M. (1977): Simplified statistical procedures for the Weibull or extreme–value distribution; *Technometrics* **19**, 323–331

BAIN, L.J. / ENGELHARDT, M. (1980a): Inferences on the parameters and current system reliability for a time truncated Weibull process; *Technometrics* **22**, 421–426

BAIN, L.J. / ENGELHARDT, M. (1980b): Probability of correct selection of Weibull versus gamma based on likelihood ratio; *Communications in Statistics — Theory and Methods* **9**, 375–381

BAIN, L.J. / ENGELHARDT, M. (1981) Simple approximate distributional results for confidence and tolerance limits for the Weibull distribution based on maximum likelihood estimators; *Technometrics* **23**, 15–20

BAIN, L.J. / ENGELHARDT, M.(1986): Approximate distributional results based on the maximum likelihood estimators for the Weibull distribution; *Journal of Quality Technology* **18**, 174–181

BAIN, L.J. / ENGELHARDT, M. (1991a): *Statistical Analysis of Reliability and Life–Testing Models — Theory and Methods,* 2nd ed.; Marcel Dekker, New York

BAIN, L.J. / ENGELHARDT, M. (1991b): Reliability test plans for one–shot devices based on repeated samples; *Journal of Quality Control* **23**, 304–311

BAIN, L.J. / THOMAN, D.R. (1968): Some tests of hypotheses concerning the three–parameter Weibull distribution; *Journal of the American Statistical Association* **63**, 853–860

BALABAN, H.S. / HASPERT, K. (1972): Estimating Weibull parameters for a general class of devices from limited failure data; *IEEE Transactions on Reliability* **21**, 111–117

BALAKRISHNAN, N. / AGGARWALA, R. (2000): *Progressive Censoring: Theory, Methods and Applications*; Birkhäuser, Boston

BALAKRISHNAN, N. / CHAN, P.S. (1993a): Extended tables of means, variances and covariances of Weibull order statistics; *Report*, McMaster University, Hamilton, Ontario

BALAKRISHNAN, N. / CHAN, P.S. (1993b): Extended tables of the best linear unbiased estimators of the location and scale parameters of the Weibull distribution based on complete and Type–II censored data; *Report*, McMaster University, Hamilton, Ontario

BALAKRISHNAN, N. / COHEN, A.C. (1991): *Order Statistics and Inference: Estimation Methods;* Academic Press, San Diego, California

BALAKRISHNAN, N. / JOSHI, P.C. (1981): A note on order statistics from Weibull distribution; *Scandinavian Actuarial Journal*, 121–122

BALAKRISHNAN, N. / KOCHERLAKOTA, S. (1985): On the double Weibull distribution: Order statistics and estimation; *Sankhyā B* **47**, 161–178

BALAKRISHNAN, N. / RAO, C.R. (eds.) (1998): *Order Statistics: Theory and Methods*; Handbook of Statistics, Vol. 16, Elsevier, Amsterdam

BALASOORIYA, U. (1995): Failure–censored reliability sampling plans for the exponential distribution; *Journal of Statistical Computer Simulation* **52**, 337–349

BALASOORIYA, U. / ABEYSINGHE, T. (1994): Selecting between gamma and Weibull distribution: An approach based on predictions of order statistics; *Journal of Applied Statistics* **21**, 17–27

BALASOORIYA, U. / CHAN, K.L. (1983): The prediction of future order statistics in the two–parameter Weibull distributions — A robust study; *Sankhyā B* **45**, 320–329

BALASOORIYA, U. / LOW, C.-K. (2004): Competing causes of failure and reliability tests for Weibull lifetimes under type I progressive censoring; *IEEE Transactions on Reliability* **53**, 29–36

BALASOORIYA, U. / SAW, S.L.C. / GADAG, V. (2000): Progressively censored reliability sampling plans for the Weibull distribution; *Technometrics* **42**, 160–167

BALOGH, A. / DUKATI, F. (1973): Sampling procedures and plans based on the Weibull distribution for reliability testing (in Hungarian); *Hiradastechnika* **24**, 1–8

BARBOSA, E.P. / LOUZADA-NETO, F. (1994): Analysis of accelerated life tests with Weibull failure distribution via generalized linear models; *Communications in Statistics — Simulation and Computation* **23**, 455–465

BARLOW, R.E. (1979): Geometry of the total time on test transform; *Naval Research Logistics Quarterly* **26**, 393–402

BARLOW, R.E. / CAMPO, R. (1975): Time on test processes and applications to failure data analysis; in BARLOW / FUSSSELL / SINGPURWALLA (eds.): *Reliability and Fault Tree Analysis*, SIAM, Philadelphia, 451–481

BARLOW, R.E. / HSIUNG, J.H. (1983): Expected information from a life test experiment; *The Statistician* **32**, 35–45

BARLOW, R.E. / MARSHALL, A.W. / PROSCHAN, F. (1963): Properties of probability distributions with monotone hazard rates; *Annals of Mathematical Statistics* **34**, 375–389

BARLOW, R.E. / PROSCHAN, F. (1965): *Mathematical Theory of Reliability;* Wiley, New York

BARLOW, R.E. / PROSCHAN, F. (1975): *Statistical Theory of Reliability and Life Testing;* Holt, Rinehart and Winston, New York

BARLOW, R.E. / PROSCHAN, F. (1988): Life distribution models and incomplete data; in KRISH-NAIAH, P.R. / RAO, C.R. (eds.): *Quality Control and Reliability*, Handbook of Statistics, Vol. 7, North–Holland, Amsterdam, 225–250

BARNDORFF-NIELSEN, O. (1978): *Information and Exponential Families;* Wiley, Chichester

BARNETT, V. (1966): Evaluation of the maximum likelihood estimator where the likelihood equation has multiple roots; *Biometrika* **53**, 151–165

BARNETT, V. (1973): *Comparative Statistical Inference;* Wiley, New York

BARNETT, V. (1975): Probability plotting methods and order statistics; *Applied Statistics* **24**, 95–108

BARR, D.R. / DAVIDSON, T. (1973): A Kolmogorov–Smirnov test for censored samples; *Technometrics* **15**, 739–757

BARROS, V.R. / ESTEVAN, E.A. (1983): On the evaluation of wind power from short wind records; *Journal of Climate and Applied Meteorology* **22**, 1116–1123

BARTHOLOMEW, D.J. (1963): An approximate solution of the integral equation of renewal theory; *Journal of the Royal Statistical Society B* **25**, 432–441

BARTLETT, M.S. (1962): *Stochastic Processes*; Cambridge University Press, Cambridge

BARTOLUCCI, A.A. / SINGH, K.P. / BARTOLUCCI, A.D. / BAE, S. (1999): Applying medical survival data to estimate the three–parameter Weibull distribution by the method of weighted moments; *Mathematics and Computers in Simulation* **48**, 385–392

BASU, A.P. (1964): Estimates of reliability for some distributions useful in life testing; *Technometrics* **6**, 215–219

BASU, S. / BASU, A.P. / MUKHOPADHYAY, C. (1999): Bayesian analysis for masked system failure data using non–identical Weibull models; *Journal of Statistical Planning and Inference* **78**, 255–275

BAXTER, L.R. / SCHEUER, E.M. / BLISCHKE, W.R. / McCONALOGUE, D.J. (1981): Renewal tables (Tables of functions arising in renewal theory); *Research Paper, Graduate School of Business Administration and School of Business, University of Southern California, Los Angeles*

BAXTER, L.R. / SCHEUER, E.M. / BLISCHKE, W.R. / McCONALOGUE, D.J. (1982): On the tabulation of the renewal function; *Technometrics* **24**, 151–156

BEG, M.I. / ALI, M.M. (1989): Characterization of the exponential and Weibull distributions; *Pakistan Journal of Statistics B* **5**, 287–291

BEGUM, A.A. / KHAN, A.H. (1997): Concomitants of order statistics from Gumbel's bivariate Weibull distribution; *Calcutta Statistical Association Bulletin* **47**, 133–140

BENJAMINI, Y. / KRIEGER, A.M. (1999): Skewness — Concepts and Measures; in KOTZ, S. / READ, C.B. / BANKS, D.L. (eds.): *Encyclopedia of Statistical Sciences,* Update Volume 3, Wiley, New York, 663–670

BENNETT, G.K. (1977): Basic concepts of empirical Bayes methods with some results for the Weibull distribution; in TSOKOS, C.P. / SHIMI, I.N. (eds.): *The Theory and Applications of Reliability,* Vol. II, Academic Press, New York, 181–202

BENNETT, G.K. / MARTZ, H.F. JR. (1955): A continuous empirical Bayes smoothing technique; *Biometrika* **59**, 361–368

BENNETT, G.K. / MARTZ, H.F. JR. (1973): An empirical Bayes estimator for the scale parameter of the two–parameter Weibull distribution; *Naval Research Logistics Quarterly* **20**, 387–393

BENNETT, J.G. (1936): Broken Coal; *Journal of the Institute of Fuels* **15**

BERGER, J.O. / SUN, D. (1993): Bayesian analysis for the poly–Weibull distribution; *Journal of the American Statistical Association* **88**, 1412–1418

BERGER, R.W. / LAWRENCE, K.D. (1974): Estimating Weibull parameters by linear and nonlinear regression; *Technometrics* **16**, 617–619

BERGMAN, B. / KLEFSJÖ, B. (1984): The total time on test concept and its use in reliability theory; *Operations Research* **31**, 506–606

BERRETTONI, J.N. (1964): Practical applications of the Weibull distribution; *Industrial Quality Control* **21**, 71–79

BERRY, G.L. (1975): Design of carcinogenesis experiments using the Weibull distribution; *Biometrika* **62**, 321–328

BERRY, G.L. (1981): The Weibull distribution as a human performance descriptor; *IEEE Transactions on Systems, Man, Cybernetics* **11**, 501–504

BEYER, R. / LAUSTER, E. (1990): Life testing plans in view of prior knowledge (in German); *Qualität und Zuverlässigkeit* **35**, 93–98

BHANDARI, S.K. / MUKHERJEE, R. (1992): Two– and multi–stage selection procedures for Weibull populations; *Journal of Statistical Planning and Inference* **30**, 73–81

BHATTACHARYA, S.K. (1962): On a probit analogue used in a life–test based on the Weibull distribution; *Australian Journal of Statistics* **4**, 101–105

BHATTACHARYA, S.K. (1967): Bayesian approach to life testing and reliability estimation; *Journal of the American Statistical Association* **62**, 48–62

BHATTACHARYYA, G.K. / SOEJOETI, Z. (1989): A tampered failure rate model for step–stress accelerated life test; *Communications in Statistics — Theory and Methods* **18**, 1627–1643

BILIKAM, J.E. / MOORE, A.H. / PETRICK, G. (1979): $k$–sample maximum likelihood ratio test for change of Weibull shape parameter; *IEEE Transactions on Reliability* **28**, 47–50

BILLMAN, B.R. / ANTLE, C.E. / BAIN, L.J. (1972): Statistical inference from censored Weibull samples; *Technometrics* **14**, 831–840

BJARNASON, H. / HOUGAARD, P. (2000): Fisher information for two gamma frailty bivariate Weibull models; *Lifetime Data Analysis* **5**, 59–71

BLISCHKE, W.R. / JOHNS, M.V. / TRUELOVE, A.J. / MURDLE, P.B. (1966): Estimation of the location parameters of Pearson type III and Weibull distributions in the non–regular case and other results in non–regular estimation; *ARL Technical Report No. 66–0233*

BLISCHKE, W.R. / SCHEUER, E.M. (1986): Tabular aids for fitting Weibull moment estimators; *Naval Research Logistics Quarterly* **33**, 145–153

BLOM, G. (1958): *Statistical Estimates and Transformed Beta Variables;* Stockholm

BLOM, G. (1962): Nearly best linear estimates of location and scale parameters; in SARHAN/GREENBERG (eds.): *Contributions to Order Statistics*; Wiley, New York, 34–46

BLUMENTHAL, S. (1967): Proportional sampling in life length studies; *Technometrics* **9**, 205–218

BOES, D.C. (1989): Regional flood quantile estimation for a Weibull model; *Water Resources Research* **25**, 979–990

BOGDANOFF, D.A. / PIERCE, D.A. (1973): Bayes–fiducial inference for the Weibull distribution; *Journal of the American Statistical Association* **68**, 659–664

BONDESSON, L. (1983): On preservation of classes of life distributions under reliability operations — Some complementary results; *Naval Research Logistics Quarterly* **30**, 443–447

BOORLA, R. / ROTENBERGER, K. (1997): Load variability of two–bladed helicopters; *Journal of American Helicopters* **42**, 15–26

BORTKIEWICZ, L. VON (1922): Range and mean error (in German); *Sitzungsberichte der Berliner Mathematischen Gesellschaft* **27**, 3–33

BOWMAN, K.O. / SHENTON, L.R. (1987): Moment series for moment estimators of the parameters of a Weibull density; in KEINER, K.W. / SACHER, R.S. / WILKINSON, J.W. (eds.): *Proceedings of the Fourteenth Symposium on the Interface;* Springer, Berlin, 174–186

BOWMAN K.O. / SHENTON, L.R. (2000): Maximum likelihood and the Weibull distribution; *Far East Journal of Theoretical Statistics* **4**, 391–422

BOWMAN K.O. / SHENTON, L.R. (2001): Weibull distributions when the shape parameter is defined; *Computational Statistics & Data Analysis* **36**, 299–310

BOX, G.E.P. / TIAO, G.C. (1973): *Bayesian Inference in Statistical Analysis*; Addison–Wesley

BRÄNNÄS, K. 1986): Prediction in a duration model; *Journal of Forecasting* **5**, 97–103

BRIKC, D.M. (1990): Interval estimation of the parameters $\beta$ and $\eta$ of the two parameter Weibull distribution; *Microelectronics and Reliability* **30**, 39–42

BRIKC, D.M. (1999): Grapho–numerical method for estimating the parameters of a 3–parameter Weibull distribution; *Journal of the Indian Association for Production, Quality and Reliability* **24**, 73–78

BRONIATOWSKI, M. (1993): On the estimation of the Weibull tail coefficient; *Journal of Statistical Planning and Inference* **35**, 349–365

BROWN, G.C. / RUTEMILLER, H.C. (1973): Evaluation of $Pr(x \geq y)$ when both $X$ and $Y$ are from three–parameter Weibull distributions; *IEEE Transactions on Reliability* **22**, 78–82

BROWN, G.C. / WINGO, D.R. (1975): Comments on "MLE of Weibull parameters by quasilinearization"; *IEEE Transactions on Reliability* **24**, 158–159

BRYSON, M.C. / SIDDIQUI, M.M. (1969): Some criteria for ageing; *Journal of the American Statistical Association* **64**, 1472–1483

BUCKLAND, W.R. (1964): *Statistical Assessment of the Life Characteristic;* Griffin, London

BUFFA, E.S. / TAUBERT, W.H. (1972): *Production–Inventory Systems — Planning and Control*; R.D. Irwin, Homewood, IL.

BUGAIGHIS, M.M. (1988): Efficiencies of MLE and BLUE for parameters of an accelerated life–test model; *IEEE Transaction on Reliability* **37**, 230–233

BUGAIGHIS, M.M. (1993): Percentiles of pivotal ratios for the MLE of the parameters of a Weibull regression model; *IEEE Transactions on Reliability* **42**, 97–99

BUGAIGHIS, M.M. (1995): Exchange of censorship types and its impact on the estimation of parameters of a Weibull regression model; *IEEE Transactions on Reliability* **44**, 496–499

BUNDAY, B.D. / AL–MUTWALI, J. (1981): Direct optimization for calculating maximum likelihood estimates of parameters of the Weibull distribution; *IEEE Transactions on Reliability* **30**, 367–369

BURR, I.W. (1942): Cumulative frequency functions; *Ann. Math. Statist.* **13**, 215–232

BURRIDGE, J. (1981): Empirical Bayes analysis of survival time data; *Journal of the Royal Statistical Society B* **43**, 65–75

BURY, K.V. (1972): Bayesian decision analysis of the hazard rate for a two–parameter Weibull process; *IEEE Transactions on Reliability* **21**, 159–169

BURY, K.V. (1973): Structural inference on the type I extreme value distribution; *Statistische Hefte* **14**, 111–122

BURY, K.V. / BERNHOLTZ, B. (1971): On structural inferences applied to the Weibull distribution; *Statistische Hefte* **12**, 177–192

## C-C-C-C-C

CACCIARI, M. (1995): Estimation of parameters in like–Weibull distribution (in Italian); *Statistica* **55**, 361–373

CACCIARI, M. / CONTIN, A. / MONTANARI, G.C. (1995): Use of mixed–Weibull distribution for the identification of PD phenomena; *IEEE Transactions on Dielectrics and Electrical Insulation* **2**, 1166–1179

CACCIARI, M. / MONTANARI, G.C. (1985): An approximate method for calculating parameters and tolerance limits for percentiles of a Weibull distribution valid for progressive censored tests (in Italian); *Metron* **43**, 67–84

CACCIARI, M. / MONTANARI, G.C. (1987): A method to estimate the Weibull parameters for progressively censored tests; *IEEE Transactions on Reliability* **36**, 87–93

CALABRIA, R. / PULCINI, G. (1989): Confidence limits for reliability and tolerance limits in the inverse Weibull distribution; *Reliability Engineering and System Safety* **24**, 77–85

CALABRIA, R. / PULCINI, G. (1990): On the maximum likelihood and least squares estimation in the inverse Weibull distribution; *Statistica Applicata* **2**, 53–66

CALABRIA, R. / PULCINI, G. (1994): Bayes two–sample prediction for the inverse Weibull distribution; *Communications in Statistics — Theory and Methods* **23**, 1811–1824

CALABRIA, R. / PULCINI, G. (1995): Bayes prediction of number of failures in Weibull samples; *Communications in Statistics — Theory and Methods* **24**, 487–499

CAMPOS, J.L. (1975): Application of the Weibull distribution to some instances of cell survival, of neoplastic survival, and of ageing; *British Journal of Radiology* **48**, 913–917

CANAVOS, G.C. (1983): A performance comparison of empirical Bayes and Bayes estimators of the Weibull and gamma scale parameters; *Naval Research Logistics Quarterly* **30**, 465–470

CANAVOS, G.C. / TSOKOS, C.P. (1973): Bayesian estimation of life parameters in the Weibull distribution; *Operations Research* **21**, 755–763

CANCHO, V.G. / BOLFARINE, H. (2001): Modelling the presence of immunes by using the exponentiated Weibull model; *Journal of Applied Statistics* **28**, 659–671

CANCHO, V.G. / BOLFARINE, H. / ACHCAR, J.A. (1999): A Bayesian analysis for the exponentiated Weibull distribution; *Journal of Applied Statistical Science* **8**, 227–242

CANFIELD, R.V. / BORGMANN, L.E. (1975): Some distributions of time to failure; *Technometrics* **17**, 263–268

CARLIN, J. / HASLETT, J. (1982): The probability distribution of wind power from a dispersed area of wind turbine generators; *Journal of Climate and Applied Meteorology* **21**, 303–313

CARMODY, T.J. / EUBANK, R.L. / LaRICCIA, V.N. (1984): A family of minimum quantile distance estimators for the three–parameter Weibull distribution; *Statistische Hefte* **25**, 69–82

CASTILLO, E. / GALAMBOS, J. (1990): Bivariate distributions with Weibull conditionals; *Analysis Mathematica* **16**, 3–9

CASTILLO, E. / MANTALBÁN, A. / FERNANDEZ–CANTELLI, A. / ARROYO, V. (1991): Two–step estimation method for a five parameter fatigue Weibull regression model; *Proceedings of the ICOSCO–I Conference,* Vol. II, 293–308

CHACE, E.F. (1976): Right–censored grouped life test data analysis assuming a two–parameter Weibull distribution function; *Microelectronics and Reliability* **15**, 497–499

CHAN, L.K. / CHENG, S.W. / MEAD, E.R. (1974): Simultaneous estimation of location and scale parameters of the Weibull distribution; *IEEE Transactions on Reliability* **23**, 335–342

CHAN, L.K. / KABIR, A.B.M.L. (1969): Optimum quantiles for the linear estimation of the extreme–value distribution in complete and censored samples; *Naval Research Logistics Quarterly* **16**, 381–404

CHAN, L.K. / MEAD, E.R. (1971): Linear estimation of the parameters of the extreme–value distribution based on suitably chosen order statistics; *IEEE Transactions on Reliability* **20**, 74–83

CHAN, W. / RUEDA, N.G. (1992): Nonexistence of the maximum likelihood estimates in the Weibull process; *Naval Research Logistics Quarterly* **39**, 359–368

CHANDRA, M. / SINGPURWALLA, N.D. / STEPHENS, M.A. (1981): Kolmogorov statistics for tests of fit for the extreme–value and Weibull distributions; *Journal of the American Statistical Association* **76**, 729–731

CHANDRA, N.K. / CHAUDHURI, A. (1990a): On testimating the Weibull shape parameter; *Communications in Statistics — Simulation and Computation* **19**, 637–648

CHANDRA, N.K. / CHAUDHURI, A. (1990b): On the efficiency of a testimator for the Weibull shape parameter, *Communications in Statistics — Theory and Methods* **19**, 1247–1259

CHANG, S.-C. (1998): Using parametric statistical models to estimate mortality structure: The case of Taiwan; *Journal of Actuarial Practice* **8**, No. 1 & 2

CHAO, A. / HWANG, S.J. (1986): Comparison of confidence intervals for the parameters of the Weibull and extreme–value distributions; *IEEE Transactions on Reliability* **35**, 111–113

CHARERNKAVANICH, D. / COHEN, A.C. (1984): Estimation in the singly truncated Weibull distribution with an unknown truncation point; *Communications in Statistics — Theory and Methods* **13**, 843–857

CHAUDHURI, A. / CHANDRA, N.K. (1989): A test for Weibull populations; *Statistics & Probability Letters* **7**, 377–380

CHAUDHURI, A. / CHANDRA, N.K. (1990): A characterization of the Weibull distribution; *Journal of the Indian Association for Productivity, Quality and Reliability* **15**, 69–72

CHAUDURY, M.L. (1995): On computations of the mean and variance of the number of renewals: A unified approach; *Journal of the Operational Research Society* **46**, 1352–1364

CHEN, J.M. / LIN, C.S. (2003): Optimal replenishment scheduling for inventory items with Weibull distributed deterioration and time–varying demand; *Journal of Information & Optimization Sciences* **24**, 1–21

CHEN, W.W.S. (1987): Testing gamma and Weibull distributions: A comparative study; *Estadística* **39**, 1–26

CHEN, Z. (1997): Statistical inference about the shape parameter of the Weibull distribution; *Statistics & Probability Letters* **36**, 85–90

CHEN, Z. (1998): Joint estimation for the parameters of Weibull distribution; *Journal of Statistical Planning and Inference* **66**, 113–120

CHEN, Z. (2000): A new two–parameter lifetime distribution with bathtub shape or increasing failure rate function; *Statistics & Probability Letters* **49**, 155–161

CHENG, K.F. / CHEN, C.H. (1988): Estimation of the Weibull parameters with grouped data; *Communications in Statistics — Theory and Methods* **17**, 325–341

CHENG, S.W. / FU, J.C. (1982): Estimation of mixed Weibull parameters in life testing; *IEEE Transactions on Reliability* **31**, 377–381

CHERNOFF, H. / GASTWIRTH, J.L. / JONES, M.V. JR. (1967): Asymptotic distribution of linear combinations of functions of order statistics to estimation; *Annals of Mathematical Statistics* **38**, 52–72

CHIN, A.C. / QUEK, S.T. / CHEU, R.L. (1991) Traffic conflicts in expressway merging; *Journal of Transport Engineering* **117**, 633–643

CHIOU, K.-C. / TONG, L.-I. (2001): A novel means of estimating quantiles for 2–parameter Weibull distribution under the right random censoring model; *Advances and Applications in Statistics* **1**, 1–26

CHRISTENSEN, E.R. / CHEN, C.-Y. (1985): A general non–interactive multiple toxicity model including probit, logit and Weibull transformations; *Biometrics* **41**, 711–725

CHRISTOFFERSON, R.D. / GILLETTE, D.A. (1987): A simple estimator of the shape factor of the two–parameter Weibull distribution; *Journal of Climate and Applied Meteorology* **26**, 323–325

CHRISTOPEIT, N. (1994): Estimating parameters of an extreme value distribution by the method of moments; *Journal of Statistical Planning and Inference* **41**, 173–186

CLARK, L.J. (1964): Estimation of the scale parameter of the Weibull probability function by the use of one order and of $m$ order statistics; *Unpublished Thesis*, Air Force Institute of Technology, Dayton Air Force Base, OH, USA

COBB, E.B. (1989): Estimation of the Weibull shape parameter in small–sample bioassay; *Journal of Statistical Computation & Simulation* **31**, 93–101

COHEN, A.C. (1963): Progressively censored samples in life testing; *Technometrics* **5**, 327–339

COHEN, A.C. (1965): Maximum likelihood estimation in the Weibull distribution based on complete and on censored samples; *Technometrics* **7**, 579–588

COHEN, A.C. (1966): Life testing and early failure; *Technometrics* **8**, 539–545

COHEN, A.C. (1969): A four–parameter generalized Weibull distribution and its inference; *NASA Contractor Report No. 61293, Contract NAS 8-11175*, Marshall Space Flight Center, Huntsville, AL

COHEN, A.C. (1973): The reflected Weibull distribution; *Technometrics* **15**, 867–873

COHEN, A.C. (1975): Multi–censored sampling in the three–parameter Weibull distribution; *Technometrics* **17**, 347–351

COHEN, A.C. (1991): *Truncated and Censored Sample;* Marcel Dekker, New York

COHEN, A.C. / WHITTEN, B.J. (1982): Modified maximum likelihood and modified moment estimators for the three–parameter Weibull distribution; *Communication in Statistics — Theory and Methods* **11**, 2631–2656

COHEN, A.C. / WHITTEN, B.J. (1988): *Parameter estimation in Reliability and Life Span Models;* Marcel Dekker, New York

COHEN, A.C. / WHITTEN, B.J. / DING, Y. (1984): Modified moment estimator for the three–parameter Weibull distribution; *Journal of Quality Technology* **16**, 159–167

COLEMAN, R. (1976): The superposition of the backward and forward processes of a renewal process; *Stochastic Processes and Their Applications* **4**, 135–148

COLEMAN, R. (1981): The moments of forward recurrence time; *Imperial College*, Department of Mathematics, London

COLES, S.G. (1989): On goodness–of–fit–tests for the two–parameter Weibull distribution derived from the stabilized probability plot; *Biometrika* **76**, 593–598

COLVERT, R.E. / BOARDMAN, T.J. (1976): Estimation in the piecewise constant hazard rate model; *Communications in Statistics – Theory and Methods* **1**, 1013–1029

CONRADSEN, K. / NIELSEN, L.B. / PRAHM, L.P. (1984): Review of the Weibull statistics for estimation of wind speed distribution; *Journal of Climate and Applied Meteorology* **23**, 1173–1183

CONSTANTINE, A.G. / ROBINSON, N.I. (1997): The Weibull renewal function for moderate to large arguments; *Computational Statistics & Data Analysis* **24**, 9–27

CONTIN, A. / CACCIARI, M. / MONTANARI, G.C. (1994) Estimation of Weibull distribution parameters for partial discharge inference; *IEEE 1994 Annual Report*, 71–78

COOK, L. (1978): Characterization of exponential and Weibull distributions through truncated statistics; *Scandinavian Actuarial Journal*, 165–168

COOKE, P. (1979): Statistical inference for bounds of random variables; *Biometrika* **66**, 367–374

COUTURE, D.J. / MARTZ, H.F. JR. (1972): Empirical Bayes estimation in the Weibull distribution; *IEEE Transactions on Reliability* **21**, 75–83

COX, D.R. (1972): Regression models and life tables; *Journal of the Royal Statistical Society B* **34**, 4187–4220

COX, D.R. (1979): A note on the graphical analysis of survival data; *Biometrika* **66**, 188–190

COX, D.R. / ISHAM, V. (1980): *Point Processes;* Chapman and Hall, London

CRAMER, E. / KAMPS, U. (1998): Sequential $k$–out–of–$n$ systems with Weibull components; *Economic Quality Control* **13**, 227–239

CRAMÉR, H. (1971): *Mathematical Methods of Statistics*; 12th ed., Princeton University Press, Princeton, NJ

CRAN, G.W. (1976): Graphical estimation methods for Weibull distributions; *Microelectronics and Reliability* **15**, 47–52

CRAN, G.W. (1988): Moment estimators for the three–parameter Weibull distribution; *IEEE Transactions on Reliability* **37**, 360–363

CROW, L.H. (1974): Reliability analysis for complex, repairable systems; in PROSCHAN/SERFLING (eds.): *Reliability and Biometry: Statistical Analysis of Life Lengths,* SIAM, Philadelphia, 379–410

CROW, L.H. (1982): Confidence interval procedures for the Weibull process with applications to reliability growth; *Technometrics* **24**, 67–72

CROWDER, M. (1985): A distributional model for repeated failure time measurements; *Journal of the Royal Statistical Society B* **47**, 447–452

CROWDER, M. (1989): A multivariate distribution with Weibull connections; *Journal of the Royal Statistical Society B* **51**, 93–100

CROWDER, M. (1990): On some nonregular tests for a modified Weibull model; *Biometrika* **77**, 499–506

CROWDER, M. / KIMBER, A. (1997): A score test for the multivariate Burr and other Weibull mixture distributions; *Scandinavian Journal of Statistics* **24**, 419–432

CULLEN, D.E. / HENDRICKSON, A. / MAXIM, L.D. (1977): Experimental design for sensitivity testing: The Weibull model; *Technometrics* **19**, 405–412

**D-D-D-D-D**

D'AGOSTINO, R.B. (1971a): Linear estimation of the Weibull parameters; *Technometrics* **13**, 171–182

D'AGOSTINO, R.B. (1971b): Addendum to "Linear estimation of the Weibull parameters"; *Technometrics* **13**, 428–429

D'AGOSTINO, R.B. / LEE, A.F.S. (1975): Asymptotically best linear unbiased estimation of the Rayleigh parameter for complete and tail–censored samples; *IEEE Transactions on Reliability* **24**, 156–157

D'AGOSTINO, R.B. / STEPHENS, M.A. (eds.) (1986): *Goodness–of–Fit Techniques;* Marcel Dekker, New York

DALLAS, A.C. (1982): Some results on record values from the exponential and Weibull law; *Acta Mathematica of Academy of Sciences of Hungary* **40**, 307–311

DANZIGER, L. (1970): Planning censored life tests for estimation the hazard rate of a Weibull distribution with prescribed precision; *Technometrics* **12**, 408–412

DASGUPTA, N. / XIE, P. / CHENEY, M.O. / BROEMELING, L. / MIELKE, C.K. JR. (2000): The Spokane Heart Study: Weibull regression and coronary artery disease; *Communications in Statistics — Simulation and Computation* **29**, 747–761

DAVID, H.A. (1981): *Order Statistics,* 2nd ed.; Wiley, New York

DAVID, H.A. / GROENEVELD, R.A. (1982): Measures of local variation in a distribution: Expected length of spacings and variances of order statistics; *Biometrika* **69**, 227–232

DAVID, H.A. / MOESCHBERGER, M.L. (1978): *The Theory of Competing Risks;* Macmillan, New York

DAVID, J. (1975): Détermination sans tatonnement du coefficient gamma de la loi de Weibull; *Revue de Statistique Appliquée* **23**, 81–85

DAVISON, A.C. / LOUZADA–NETO, F. (2000): Inference for the poly–Weibull model; *The Statistician* **49**, 189–196

DE BRAGANCA PEREIRA, B. (1984): On the choice of a Weibull model: *Estadística* **36**, 157–163

DEKKERS, A.L. / EINMAHL, J.H.J. / DE HAAN, L. (1989): A moment estimator for the index of an extreme–value distribution; *Annals of Statistics* **17**, 1833–1855

DELLAPORTAS, P. / WRIGHT, D.E. (1991): Numerical prediction for the two–parameter Weibull distribution; *The Statistician* **40**, 365–372

DEUTSCHER NORMENAUSSCHUSS (ed.) (1967): *DIN 40041 — Reliability of Electrical Components (Terminology)* (in German); Beuth, Berlin

DEUTSCHER NORMENAUSSCHUSS (ed.) (1970): *DIN 40042 — Reliability of Electrical Appliances, Installations and Systems (Terminology)* (in German); Beuth, Berlin

DEWANJI, A. / KREWSKI, D. / GODDARD, M.J. (1993): A Weibull model for the estimation of tumorigenic potency; *Biometrics* **49**, 367–377

DEY, D.K. / KUO, L. (1991): A new empirical Bayes estimator with type II censored data; *Computational Statistics & Data Analysis* **12**, 271–279

DIXIT, U.J. (1994): Bayesian approach to prediction in the presence of outliers for Weibull distribution; *Metrika* **41**, 127–136

DIXON, J.C. / SWIFT, R.H. (1984): The directional variation of wind speed and Weibull probability parameters; *Atmospheric Environment* **18**, 2041–2047

DODD, E.L. (1923): The greatest and the least variate under general laws of error; *Trans. Amer. Math. Soc.* **25**, 525–539

DODSON, B. (1994): *Weibull Analysis;* ASQC Quality Press, Milwaukee

DOOB, J.L. (1953): *Stochastic Processes;* Wiley, New York

DOURGNON, F. / REYROLLE, J. (1966): Tables de la fonction de répartition de la loi de Weibull; *Revue de Statistique Appliquée* **14**, 83–116

DOWNTON, F. (1966): Linear estimates of parameters in the extreme value distribution; *Technometrics* **8**, 3–17

DRAPELLA, A. (1993): Complementary Weibull distribution: Unknown or just forgotten; *Quality and Reliability Engineering International* **9**, 383–385

DUAN, J. / SELKER, J.S. / GRANT, G.E. (1998): Evaluation of probability density functions in precipitation models for the Pacific Northwest; *Journal of the American Water Ressource Association* **34**, 617–627

DUANE, J.T. (1964): Learning curve approach to reliability; *IEEE Transactions on Aerospace* **2**, 563–566

DUBEY, S.D. (1960): Contributions to statistical theory of life testing and reliability; *Ph. D. Dissertation*, Michigan State University, East Lansing

DUBEY, S.D. (1965): Asymptotic properties of several estimators of Weibull parameters; *Technometrics* **7**, 423–434

DUBEY, S.D. (1966a) Some test functions for the parameters of the Weibull distributions; *Naval Research Logistics Quarterly* **13**, 113–128

DUBEY, S.D. (1966b): On some statistical inferences for Weibull laws; *Naval Research Logistics Quarterly* **13**, 227–251

DUBEY, S.D. (1966c): Hyper–efficient estimator of the location parameter of the Weibull laws; *Naval Research Logistics Quarterly* **13**, 253–264

DUBEY, S.D. (1966d): Asymptotic efficiencies of the moment estimators for the parameters of the Weibull laws; *Naval Research Logistics Quarterly* **13**, 265–288

DUBEY, S.D. (1966e): Characterization theorems for several distributions and their applications; *Journal of the Industrial Mathematics Society* **16**, 1–22

DUBEY, S.D. (1966f): Transformations for estimation of parameters; *Journal of the Indian Statistical Association* **4**, 109–124

DUBEY, S.D. (1966g): Comparative performance of several estimators of the Weibull parameters; *Proceedings of the 20th Technical Conference of the American Society for Quality Control*, 723–735

DUBEY, S.D. (1967a): Normal and Weibull distributions; *Naval Research Logistics Quarterly* **14**, 67–79

DUBEY, S.D. (1967b): Revised tables for asymptotic efficiencies of the moment estimators for the parameters of the Weibull laws; *Naval Research Logistics Quarterly* **14**, 261–267

DUBEY, S.D. (1967c): Some simple estimators for the shape parameter of the Weibull laws; *Naval Research Logistics Quarterly* **14**, 489–512

DUBEY, S.D. (1967d): Monte Carlo study of the moment and maximum likelihood estimators of Weibull parameters; *Trabajos de Estadística* **18**, 131–141

DUBEY, S.D. (1967e): Some percentile estimators for Weibull parameters; *Technometrics* **9**, 119–129

DUBEY, S.D. (1967f): On some permissible estimators of the location parameter of the Weibull distribution and certain other distributions; *Technometrics* **9**, 293–307

DUBEY, S.D. (1968): A compound Weibull distribution; *Naval Research Logistics Quarterly* **15**, 179–188

DUBEY, S.D. (1972): Statistical contributions to reliability engineering; *ARL Technical Report No. 72–0120*

DUBEY, S.D. (1973): Statistical treatment of certain life testing and reliability problems; *ARL Technical Report No. 73–0155*

DUFOUR, R. / MAAG, U.R. (1978): Distribution results for modified Kolmogorov–Smirnov statistics for truncated and censored data; *Technometrics* **20**, 29–32

DUMONCEAUX, R. / ANTLE, C.E. / HAAS, G. (1973a): Likelihood ratio test for discrimination between two models with unknown location and scale parameters; *Technometrics* **15**, 19–27

DUMONCEAUX, R. / ANTLE, C.E. / HAAS, G. (1973b): Discrimination between the log–normal and the Weibull distributions; *Technometrics* **15**, 923–926

DURHAM, S.D. / PADGETT, W.J. (1997): Cumulative damage model for system failure with application to carbon fibers and composites; *Technometrics* **39**, 34–44

DYER, A.R. (1975): An analysis of the relationship of systolic blood pressure, serum cholesterol and smoking to 14–year mortality in the Chicago Gas Company Study, Part I – Total mortality in exponential–Weibull model, Part II – Coronary and cardiovascular–venal mortality in two competing risk models; *Journal of Chronic Diseases* **28**, 565–578

DYER, D.D. / WHISENAND, C.W. (1973a): Best linear unbiased estimator of the parameter of the Rayleigh distribution — Part I: Small sample theory for censored order statistics; *IEEE Transactions on Reliability* **22**, 27–34

DYER, D.D. / WHISENAND, C.W. (1973b): Best linear unbiased estimator of the parameter of the Rayleigh distribution — Part II: Optimum theory for selected order statistics; *IEEE Transactions on Reliability* **22**, 229–231

**E-E-E-E-E**

EDWARDS, A.W.F. (1972): *Likelihood*; Cambridge University Press, Cambridge

EFRON, B. (1979): Bootstrap methods: Another look at the jackknife; *The Annals of Statistics* **7**, 1–26

EFRON, B. (1982): *The Jackknife, the Bootstrap and Other Resampling Plans*; SIAM No. 38, Philadelphia

ELANDT–JOHNSON, R.C. (1976): A class of distributions generated from distributions of exponential type; *Naval Research Logistics Quarterly* **23**, 131–138

ELANDT–JOHNSON, R.C. / JOHNSON, N.L. (1980): *Survival Models and Data Analysis;* Wiley, New York

EL–ARISHY, S.M. (1993): Characterizations of the Weibull, Rayleigh and Maxwell distributions using truncated moments of order statistics; *The Egyptian Statistical Journal* **37**, 198–212

ELDERTON, W.P. / JOHNSON, N.L. (1969): *System of Frequency Curves*; Cambridge University Press, Cambridge

EL–DIN, M.M.M. / MAHMOUD, M.A.W. / YOUSSEF, S.E.A. (1991): Moments of order statistics from parabolic and skewed distributions and a characterization of Weibull distribution; *Communications in Statistics — Simulation and Computation* **20**, 639–645

ELLIS, W.C. / TUMMALA, V.M.R. (1983): Bayesian approach to maximum likelihood estimation of parameters of Weibull distribution; *ASA Proceedings of Business & Economic Statistics Section,* 574–575

ELLIS, W. C. / TUMMALA, V.M.R. (1986): Minimum expected loss estimators of the shape and scale parameters of the Weibull distribution; *IEEE Transactions on Reliability* **35**, 212–215

EMOTO, S.E. / MATTHEWS, P.C. (1990): A Weibull model for dependent censoring; *The Annals of Statistics* **18**, 1556–1577

ENGELHARDT, M. (1975): On simple estimation of the parameters of the Weibull or extreme–value distribution; *Technometrics* **17**, 369–374

ENGELHARDT, M. (1988): Weibull processes; in KOTZ, S. / JOHNSON, N.L. / READ, C.B. (eds.): *Encyclopedia of Statistical Sciences* **9**, Wiley, New York, 557–561

ENGELHARDT, M. / BAIN, L.J. (1973): Some complete and censored sampling results for the Weibull or extreme–value distribution; *Technometrics* **15**, 541–549

ENGELHARDT, M. / BAIN, L.J. (1974): Some results on point estimation for the two–parameter Weibull or extreme–value distribution; *Technometrics* **16**, 49–56

ENGELHARDT, M. / BAIN, L.J.(1975): Tests of two–parameter exponentiality against three–parameter Weibull alternatives; *Technometrics* **17**, 353–356

ENGELHARDT, M. / BAIN, L.J. (1977): Simplified statistical procedures for the Weibull or extreme–value distribution; *Technometrics* **19**, 323–331

ENGELHARDT, M. / BAIN, L.J. (1978): Prediction intervals for the Weibull process; *Technometrics* **20**, 167–169

ENGELHARDT, M. / BAIN, L.J. (1979): Prediction limits and two–sample problems with complete and censored Weibull data; *Technometrics* **21**, 233–237

ENGELHARDT, M. / BAIN, L.J. (1982): On prediction limits for samples from a Weibull or extreme–value distribution; *Technometrics* **24**, 147–150

ENGEMAN, R.M. / KEEFE, T.J. (1982): On generalized least squares estimation of the Weibull distribution; *Communications in Statistics — Theory and Methods* **11**, 2181–2193

ENGEMAN, R.M. / KEEFE, T.J. (1985): Two–step estimators of the scale parameter of the Weibull distribution; *Computers and Biomedical Research* **18**, 391–396

EPSTEIN, B. (1958): The exponential distribution and its role in life testing; *Industrial Quality Control* **15**, 4–9

EPSTEIN, B. (1960): Elements of the theory of extreme values; *Technometrics* **2**, 27–41

EPSTEIN, B. / SOBEL, M. (1953): Life testing; *Journal of the American Statistical Association* **48**, 486–502

ERTO, P. (1982): New practical Bayes estimators for the two–parameter Weibull distribution; *IEEE Transactions on Reliability* **31**, 194–197

ERTO, P. (1989): Genesis, properties and identification of the inverse Weibull lifetime model (in Italian); *Statistica Applicata* **1**, 117–128

ERTO, P. / GUIDA, M. (1985a): Estimation of Weibull reliability from few life tests; *Quality and Reliability Engineering International* **1**, 161–164

ERTO, P. / GUIDA, M. (1985b): Tables for exact lower confidence limits for reliability and quantiles, based on least–squares of Weibull parameters; *IEEE Transactions on Reliability* **34**, 219–223

ERTO, P. / PALLOTTA, G. (2007): A new control chart for Weibull technological processes; *Quality Technology & Quantitative Management* **4**, 553–567

ERTO, P. / RAPONE, M. (1984): Non–informative and practical Bayesian confidence bounds for reliable life in the Weibull model; *Reliability Engineering* **7**, 181–191

ESARY, J.D. / PROSCHAN, F. (1963): Relationship between system failure rate and component failure rates; *Technometrics* **5**, 183–189

ESCOBAR, L.A. / MEEKER, W.Q. JR. (1986a): Elements of the Fisher information matrix for the smallest extreme value distribution and censored data; *Applied Statistics* **35**, 80–86

ESCOBAR, L.A. / MEEKER, W.Q. JR. (1986b): Planning accelerated life tests with type II censored data; *Journal of Statistical Computation & Simulation* **23**, 273–297

EVANS, I.G. / NIGM, A.M. (1980a): Bayesian prediction for two–parameter Weibull lifetime models; *Communications in Statistics — Theory and Methods* **9**, 649–658

EVANS, I.G. / NIGM, A.M. (1980b) Bayesian one sample prediction for the two–parameter Weibull distribution; *IEEE Transactions on Reliability* **29**, 410–413

**F-F-F-F-F**

FALLS, W. (1970): Estimation of parameters in compound Weibull distributions; *Technometrics* **12**, 399–407

FANG, Z. / PATTERSON, B.R. / TURNER, M.E. (1993): Modeling particle size distribution by Weibull distribution function; *Materials Characterization* **31**, 177–182

FAREWELL, V.T. (1986): Mixture models in survival analysis: Are they worth the risk? *The Canadian Journal of Statistics* **14**, 257–262

FAREWELL, V.T. / PRENTICE, R.L. (1977): A study of distributional shape in life testing; *Technometrics* **19**, 69–75

FARNUM, N.R. / BOOTH, P. (1997): Uniqueness of maximum likelihood estimation of the two–parameter Weibull distribution; *IEEE Transactions on Reliability* **46**, 523–525

FAUCHON, J. / MARTINEAU, G. / HERBIN, G. / BAHUAUD, J. (1976): Étude d'une loi de probabilité a cinq paramètres généralisant la loi de Weibull; *Sciences et Techniques de l'Armement* **50**, 683–700

FEARN, D.H. / NEBENZAHL, E. (1991): On the maximum likelihood ratio method of deciding between the Weibull and gamma distributions; *Communications in Statistics — Theory and Methods* **20**, 579–593

FEI, H. / KONG, F. / TANG, Y. (1995): Estimation for the two–parameter Weibull distribution and extreme–value distribution under multiple type–II censoring; *Communications in Statistics — Theory and Methods* **24**, 2087–2104

FELLER, W. (1966): *An Introduction to Probability Theory and Its Applications*, Vols. 1 and 2; Wiley, New York

FERTIG, K.W. / MANN, N.R. (1980): Life–test sampling plans for two–parameter Weibull populations; *Technometrics* **22**, 165–177

FERTIG, K.W. / MEYER, M.E. / MANN, N.R. (1980): On constructing prediction intervals for samples from a Weibull or extreme value distribution; *Technometrics* **22**, 567–573

FINKELSTEIN, J.M. (1976): Confidence bounds on the parameters of the Weibull process; *Technometrics* **18**, 115–117

FISHER, R.A. / TIPPETT, L.H.C. (1928): Limiting forms of the frequency distribution of the largest and smallest member of a sample; *Proceedings of the Cambridge Philosophical Society* **24**, 180–90

FLEHINGER, B.J. / LEWIS, P.A. (1959): Two parameter lifetime distributions for reliability studies of renewal processes; *IBM Journal for Research and Development* **1**, 58–73

FLYGARE, M.E. / AUSTIN, J.A. / BUCKWALTER, R.M. (1985): Maximum likelihood estimation for the two–parameter Weibull distribution based on interval data; *IEEE Transactions on Reliability* **34**, 57–59

FOK, S.L. / MITCHELL, B.G. / SMART, J. / MARSDEN, B.J. (2001): A numerical study on the application of the Weibull theory to brittle materials; *Engineering Fracture Mechanics* **68**, 1171–1179

FOWLKES, E.B. (1987): *Folio of Distributions: A Collection of Theoretical Quantile–Quantile Plots;* Marcel Dekker, New York

FRAHM, P. (1995): Combining economic design of $\overline{X}$–charts and block replacement under Weibull shock model; *Economic Quality Control* **10**, 77–115

FRANCK, J.R. (1988): A simple explanation of the Weibull distribution and its applications; *Reliability Review* **8**, 6–9

FRASER, D.A.S. (1966): Structural probability and a generalization; *Biometrika* **53**, 1–8

FRASER, D.A.S. (1968): *The Structure of Inference;* Wiley, New York

FRASER, R.P. / EISENKLAM, P. (1956): Liquid atomization and the drop size of sprays; *Transactions of the Institute of Chemical Engineers* **34**, 294–319

FRÉCHET, M.R. (1927): Sur la loi de probabilité de l'écart maximum; *Annales de la Société Polonaise de Mathématique,* Cracovie, **6**, 93–116

FREEMAN, D.H. / FREEMAN, J.L. / KOCH, G.G. (1978): A modified chi–squared approach for fitting Weibull models to synthetic life tables; *Biometrical Journal* **20**, 29–40

FREIMER, M. / MUDHOLKAR, G.S. / LIN, C.T. (1989): Extremes, extreme spacings and outliers in the Tukey and Weibull families; *Communications in Statistics — Theory and Methods* **18**, 4261–4274

FREUDENTHAL, A.M. / GUMBEL, E.J. (1954): Minimum life in fatigue; *Journal of the American Statistical Association* **49**, 575–597

FREUND, J.E. (1961): A bivariate extension of the exponential distribution; *Journal of the American Statistical Association* **56**, 971–977

FRIEDMAN, L. (1981): Parameter estimation in a minimum–type model by least squares methods — A Monte Carlo study; *Communications in Statistics — Theory and Methods* **10**, 463–487

FRIEDMAN, L. / GERTSBAKH, I.B. (1980): Maximum likelihood estimation in a minimum–type model with exponential and Weibull failure modes; *Journal of the American Statistical Association* **75**, 460–465

FRIEDMAN, L. / GERTSBAKH, I.B. (1981): Parameter estimation in a minimum–type scheme; *Communications in Statistics — Theory and Methods* **10**, 439–462

FROM, S.G. (2001): Some new approximations for the renewal function; *Communications in Statistics — Simulation and Computation* **30**, 113–128

FUKUTA, J. (1963): Estimation of parameters in the Weibull distribution and its efficiency; *Research Report No. 13,* Faculty of Engineering, Gifu University, Japan

FUNG, K.Y. / PAUL, S.R. (1985): Comparison of outlier detection procedures in Weibull and extreme–value distribution; *Communications in Statistics — Simulation and Computation* **14**, 895–917

**G-G-G-G-G**

GAEDE, K. W. (1977): *Reliability — Mathematical Models* (in German); Hanser, München/Wien

GALAMBOS, J. (1975): Characterization of probability distributions by properties of order statistics I; in PATIL et al. (eds.): *Statistical Distributions in Scientific Work* **3**, Reidel, Dordrecht, 71–88

GALAMBOS, J. (1978): *The Asymptotic Theory of Extreme Order Statistics;* Wiley, New York

GALAMBOS, J. (1981): Failure time distributions: Estimates and asymptotic results; in TAILLIE, C. / PATIL, G.P. / BALDESSARI, B.A. (eds.): *Statistical Distributions in Scientific Work* **5**, Reidel, Dordrecht, 309–317

GALAMBOS, J. / KOTZ, S. (1978): *Characterizations of Probability Distributions;* Lecture Notes in Mathematics **675**, Springer, Heidelberg

GALETTO, F. (1988): Comment on "New practical Bayes estimators for the two–parameter Weibull distribution"; *IEEE Transactions on Reliability* **37**, 562–565

GALLACE, L. (1973/4): Practical applications of the Weibull distribution to power–hybrid burn–in; *RCA Eng.* **19**, 58–61

GALLAGHER, M.A. / MOORE, A.H. (1990): Robust minimum–distance estimation using the three–parameter Weibull distribution; *IEEE Transactions on Reliability* **39**, 575–580

GANDER, F. (1996): Point and interval estimation of Weibull parameters from censored samples (in German); *Ph.D. Dissertation*, Fachbereich Wirtschaftswissenschaften der Freien Universität Berlin

GANDOMI, A. / NOORBALOOCHI, S. / BOZORGNIA, A. (2001): Linex Bayes estimation of Weibull distribution; *Advances and Applications in Statistics* **1**, 107–118

GARG, A. / KALAGNANAM, J.R. (1998): Approximations for the renewal function; *IEEE Transactions on Reliability* **47**, 66–72

GEHAN, E.A. / SIDDIQUI, M.M. (1973): Simple regression methods for survival methods; *Journal of the American Statistical Association* **68**, 848–856

GERONTIDIS, I. / SMITH, R.L. (1982): Monte Carlo generation of order statistics from general distributions; *Applied Statistics* **31**, 238–243

GERTSBAKH, I.B. / KAGAN, A.M. (1999): Characterization of the Weibull distribution by properties of the Fisher information and Type–I censoring; *Statistics & Probability Letters* **42**, 99–105

GEURTS, J.H.J. (1983): Optimal age replacement versus condition based replacement: Some theoretical and practical considerations; *Journal of Quality Technology* **15**, 171–179

GHARE, P.M. (1981): Sequential tests under Weibull distribution; *Proceedings of the Annual Reliability and Maintenance Symposium,* 375–380

GHOSH, A. (1999): A FORTRAN program for fitting Weibull distribution and generating samples; *Computers and Geosciences* **25**, 729–738

GIBBONS, D.I. / VANCE, L.C. (1979): A simulation study of estimators for the parameters and percentiles in the two–parameter Weibull distribution; *General Motors Research Laboratories, GMR–3041,* Warren/Michigan

GIBBONS, D.I. / VANCE, L.C. (1981): A simulation study of estimators for the two–parameter Weibull distribution; *IEEE Transactions on Reliability* **30**, 61–66

GIBSON, E.W.B. / HIGGINS, J.J. (2000): Gap–ratio goodness of fit tests for Weibull or extreme value distribution assumptions with left or right censored data; *Communications in Statistics — Simulation and Computation* **29**, 541–557

GILL, A.N. / MEHTA, G.P. (1994): A class of subset selection procedures for Weibull population; *IEEE Transactions on Reliability* **43**, 65–70

GIMENEZ, P. / BOLFARINE, H. / COLOSIMO, E.A. (1999): Estimation in Weibull regression model with measurement error; *Communications in Statistics — Theory and Methods* **28**, 495–510

GITTUS, J.H. (1967): On a class of distribution functions; *Applied Statistics* **16**, 45–50

GLASER, R.E. (1984): Estimation for a Weibull accelerated life testing model; *Naval Research Logistics Quarterly* **31**, 559–570

GLASER, R.E. (1995): Weibull accelerated life testing with unreported early failures; *IEEE Transactions on Reliability* **44**; 31–36

GLASSER, G.J. (1967): The age replacement problem; *Technometrics* **9**, 83–91

GNEDENKO, B.V. (1943): Sur la distribution limite du terme maximum d'une série aléatoire; *Annals of Mathematics* **44**, 423–453

GNEDENKO, B.V. / BELJAJEW, J.K. / SOLOWJEW, A.D. (1968): *Mathematische Methoden der Zuverlässigkeitstheorie*, 2 Bände; Akademie–Verlag, Berlin (English Translation: *Mathematical Methods in Reliability Theory;* Academic Press, New York, 1968)

GOLDMAN, A.I. (1984): Survivorship analysis when cure is possible: A Monte Carlo study; *Statistics in Medicine* **3**, 153–163

GOODE, H.P. / KAO, J.H.K. (1961a): An adaption of the MIL-STD-105B plans to reliability and life testing applications; *Trans. 15th Ann. Conf.*, 245–259

GOODE, H.P. / KAO, J.H.K. (1961b): Sampling plans based on the Weibull distribution; *Proceedings of the 7th Nat. Symp. Rel. Qual. Cont.*, 24–40

GOODE, H.P. / KAO, J.H.K. (1962): Sampling procedures and tables for life and reliability testing based on the Weibull distribution (Hazard rate criterion); *Proceedings of the 8th Nat. Symp. Rel. Qual. Cont.*, 37–58

GOODE, H.P. / KAO, J.H.K. (1963): Weibull tables for bio–assaying and fatigue testing; *Proceedings of the 9th Nat. Symp. Rel. Qual. Cont.*, 270–286

GOODE, H.P. / KAO, J.H.K. (1964): Hazard rate sampling plans for the Weibull distribution; *Industrial Quality Control* **20**, 30–39

GORSKI, A.C. (1968); Beware of the Weibull euphoria; *IEEE Transactions on Reliability* **17**, 202-203

GOTTFRIED, P. / ROBERTS, H.R. (1963): Some pitfalls of the Weibull distribution; *Proceedings of the 9th Nat. Symp. Rel. Qual. Cont.*, 372–379

GOVINDARAJULU, Z. (1964): Best linear unbiased estimation of location and scale parameters of Weibull distribution using ordered observations (Abstract); *Technometrics* **6**, 117–118

GOVINDARAJULU, Z. / JOSHI, M. (1968): Best linear unbiased estimation of location and scale parameters of Weibull distribution using ordered observations; *Reports of Statistical Application Research, Japanese Union of Scientists and Engineers* **15**, 1–14

GREEN, E.J. / ROESCH, F.A. JR. / SMITH, A.F.M. / STRAWDERMAN, W.E. (1993): Bayesian estimating for the three–parameter Weibull distribution with tree diameter data; *Biometrics* **50**, 254–269

GREENE, W.H. (2003): *Econometric Analysis,* 5th ed.; Prentice–Hall, Upper Saddle River, NJ

GREENWOOD, J.A. / LANDWEHR, J.M. / MATALAS, N.C. / WALLIS, J.R. (1979): Probability weighted moments: Definition and relation to several distributions expressible in inverse form; *Water Resourc. Res.* **15**, 1040–1054

GROENEVELD, R.A. (1986): Skewness for the Weibull family; *Statistica Neerlandica* **40**, 135–140

GROENEVELD, R.A. (1998): Skewness, Bowley's measure of; in KOTZ, S. / READ, C.B. / BANKS, D.L. (eds.) *Encyclopedia of Statistical Sciences,* Update Volume 2, Wiley, New York, 619–621

GROSS, A.J. (1971): Monotonicity properties of the moments of truncated gamma and Weibull density functions; *Technometrics* **13**, 851–857

GROSS, A.J. / CLARK, V.A. (1975): *Survival Distributions: Reliability Applications in the Biomedical Sciences;* Wiley, New York

GROSS, A.J. / LURIE, D. (1977): Monte Carlo comparisons of parameter estimators of the two–parameter Weibull distribution; *IEEE Transactions on Reliability* **26**, 356–358

GUMBEL, E.J. (1954): *Statistical Theory of Extreme Values and Some Practical Applications;* U.S. Government Printing Office, Washington, D.C.

GUMBEL, E.J. (1958): *Statistics of Extremes;* Columbia University Press, New York

GUPTA, P.L. / GUPTA, R.C. (1996): Aging characteristics of the Weibull mixtures; *Probability in the Engineering and Informational Sciences* **10**, 591–600

GUPTA, P.L. / GUPTA, R.C. / LVIN, S.J. (1998): Numerical methods for the maximum likelihood estimation of Weibull parameters; *Journal of Statistical Computing & Simulation* **62**, 1–7

GUPTA, R.D. / KUNDU, D. (2001): Exponentiated exponential family: An alternative to gamma and Weibull distributions; *Biometrical Journal* **43**, 117–130

GUPTA, R.D. / KUNDU, D. (2003): Discriminating between Weibull and generalized exponential distributions: *Computational Statistics & Data Analysis* **43**, 179–196

GUPTA, R.D. / KUNDU, D. / MAGLICK, A. (no year): Probability of correct selection of gamma versus GE or Weibull versus GE based on likelihood ratio statistic; submitted to publication

GURKER, W. (1995): Some remarks on the skewness and excess of the Weibull distribution (in German); *Allgemeines Statistisches Archiv* **79**, 107–114

GURVICH, M.R. / DIBENEDETTO, A.T. / RANDE, S.V. (1997): A new statistical distribution for characterizing the random strength of brittle materials; *Journal of Material Science* **52**, 33–37

## H-H-H-H-H

HAAN, C.T. / BEER, C.E. (1967): Determination of maximum likelihood estimators for the three parameter Weibull distribution; *Iowa State Journal of Science* **42**, 37–42

HACKL, P. (1978): Nonparametric estimation in life testing with reduced information (in German); *Metrika* **25**, 179–182

HÄRTLER, G. (1967): On some special features when the mean is used to characterize Weibull distributed random variables (in German); *Abhandlungen der Deutschen Akademie der Wissenschaften Berlin* **IV**, 73–75

HAGER, H.W. / BAIN, L.J. / ANTLE, C.E. (1971): Reliability estimation for the generalized gamma distribution and robustness of the Weibull model; *Technometrics* **13**, 547–557

HAGIWARA, Y. (1974): Probability of earthquake occurrence as obtained from a Weibull distribution analysis of crustal strain; *Tectonophysics* **23**, 313–318

HAGWOOD, C. / CLOUGH, R. / FIELDS, R. (1999): Estimation of the stress–threshold for the Weibull inverse power law; *IEEE Transactions on Reliability* **48**, 176–181

HAHN, G.J. / GODFREY, J.T. (1964): Estimation of Weibull distribution parameters with differing test times for unfailed units (Abstract); *Technometrics* **6**, 118

HAHN, G.J. / GODFREY, J.T. / RENZI, N.A. (1960): Computer programs to estimate parameters of Weibull density from life test data and to calculate survival probabilities and percentiles; *General Electric Company Report No. 60GL235*

HALLINAN, A.J. JR. (1993): A review of the Weibull distribution; *Journal of Quality Technology* **25**, 85–93

HAMDAN, M.A. (1972): On a characterization by conditional expectations; *Technometrics* **14**, 497–499

HANAGAL, D.D. (1996): A multivariate Weibull distribution; *Economic Quality Control* **11**, 193–200 (Correction, *Ibid.*, **12**, 59)

HARRIS, C.M. / SINGPURWALLA, N.D. (1968): Life distributions derived from stochastic hazard functions; *IEEE Transactions on Reliability* **17**, 70–79

HARRIS, C.M. / SINGPURWALLA, N.D. (1969): On estimation in Weibull distributions with random scale parameters; *Naval Research Logistics Quarterly* **16**, 405–410

HARTER, H.L. (1960): Expected values of exponential, Weibull, and gamma order statistics; *ARL Technical Report No. 60–292*

HARTER, H.L. (1961): Estimating the parameters of negative exponential populations from one or two order statistics; *Annals of Mathematical Statistics* **32**, 1078–1090

HARTER, H.L. (1967): Maximum–likelihood estimation of the parameters of a four–parameter generalized gamma population for complete and censored samples; *Technometrics* **9**, 159–165

HARTER, H.L. (1970): *Order Statistics and Their Use in Testing and Estimation;* Volumes 1 and 2, U.S. Government Printing Office, Washington, D.C.

HARTER, H.L. (1986): Realistic models for system reliability; in BASU, A.P. (ed.): *Reliability and Quality Control;* North–Holland, Amsterdam, 201–207

HARTER, H.L. (1988): Weibull, log–Weibull and gamma order statistics; in KRISHNAIAH, P.R./ RAO, C.R. (eds.): *Handbook of Statistics Vol. 7: Quality Control and Reliability,* North–Holland, Amsterdam, 433–466

HARTER, H.L. / DUBEY, S.D. (1967): Theory and tables for tests of hypotheses concerning the mean and the variance of a Weibull population; *ARL Technical Report No. 67–0059*

HARTER, H.L. / MOORE, A.H. (1965a): Maximum–likelihood estimation of the parameters of gamma and Weibull populations from complete and from censored samples; *Technometrics* **7**, 639–643 (Correction, *Ibid.*, **9**, 195

HARTER, H.L. / MOORE, A.H. (1965b): Point and interval estimators, based on $m$ order statistics, for the scale parameter of a Weibull population with known shape parameter; *Technometrics* **7**, 405–422

HARTER, H.L. / MOORE, A.H. (1967a): A note on estimation from type I extreme–value distribution; *Technometrics* **9**, 325–331

HARTER, H.L. / MOORE, A.H. (1967b): Asymptotic variances and covariances of maximum likelihood estimators, from censored samples, of the parameters of Weibull and gamma populations; *Annals of Mathematical Statistics* **38**, 557–570

HARTER, H.L. / MOORE, A.H. (1968): Maximum likelihood estimation from doubly censored samples of the parameters of the first asymptotic distribution of extreme values; *Journal of the American Statistical Association* **63**, 889–901

HARTER, H.L. / MOORE, A.H. (1976): An evaluation of exponential and Weibull test plans; *IEEE Transactions on Reliability* **25**, 100–104

HARTER, H.L. / MOORE, A.H. / WIEGAND, R.P. (1985): Sequential tests of hypotheses for system reliability modeled by a two–parameter Weibull distribution; *IEEE Transactions on Reliability* **34**, 352–355

HASSANEIN, K.M. (1969): Estimation of the parameters of the extreme-value distribution by use of two or three order statistics; *Biometrika* **56**, 429–436

HASSANEIN, K.M. (1971): Percentile estimators for the parameters of the Weibull distribution; *Biometrika* **58**, 673–676

HASSANEIN, K.M. (1972): Simultaneous estimation of the parameters of the extreme value distributions by sample quantiles; *Technometrics* **14**, 63–70

HASSANEIN, K.M. / LEGLER, W.K. (1975): On minimum variance stratification for estimating the mean of a population; *Scandinavian Actuarial Journal*, 207–214

HASSANEIN, K.M. / SALEH, A.K.M.E. / BROWN, E.F. (1984): Quantile estimates in complete and censored samples from extreme–value and Weibull distributions; *IEEE Transactions on Reliability* **33**, 370–373

HAUGH, L.D. / DELANO, J. (1984): The effect of successive censoring and truncation on inference for the Weibull distribution: A Monte Carlo study; *ASA Proceedings of Statistical Computing Section*, 233–238

HE, X. / FUNG, W.K. (1999): Method of medians for lifetime data with Weibull models; *Statistics in Medicine* **18**, 1993–2009

HEIKKILÄ, H.J. (1999): New models for pharmacokinetic data based on a generalized Weibull distribution; *Journal of Biopharmaceutical Statistics* **9**, 89–107

HELLER, R.A. (1985): The Weibull distribution did not apply to its founder; in EGGWERTZ, S. / LIND, N.C. (eds.): *Probabilistic Methods in the Mechanics of Solids and Structures*, Springer–Verlag, Berlin etc., XIII–XVI

HENDERSON, G. / WEBBER, N. (1978): Waves in severe storms in the Central English Channel; *Coastal Engineering* **2**, 95–110

HENTZSCHEL, J. (1989): Parametric evaluation of censored samples: The Weibull distribution (in German); *Mathematische Operationsforschung und Statistik, Series Statistics* **20**, 383–395

HEO, J.–H. / BOES, D.C. / SALAS, J.D. (2001): Regional flood frequency analysis based on a Weibull model, Part I: Estimation and asymptotic variances; *Journal of Hydrology* **242**, 157–170

HEO, J.–H. / SALAS, J.D. / KIM, K.-D. (2001): Estimation of confidence intervals of quantiles for the Weibull distribution; *Stochastic Environmental Research and Risk Assessment* **15**, 284–309

HERBACH, L.H. (1963): Estimation of the three Weibull Parameters, using an overlooked consequence of a well–known property of the Weibull distribution (Abstract); *Annals of Mathematical Statistics* **34**, 681–682

HERBACH, L.H. 1970): An exact asymptotically efficient confidence bound for reliability in the case of a Gumbel distribution; *Technometrics* **12**, 700–701

HERD, G.R. (1960): Estimation of reliability from incomplete data; *Proceedings of the Sixth National Symposium on Reliability and Quality Control*, 202–217

HEWETT, J.E. / MOESCHBERGER, M.L. (1976): Some approximate simultaneous prediction intervals for reliability analysis; *Technometrics* **18**, 227–230

HILL, J.S. / PATEL, J.K. (2000): On selecting populations better than a control: Weibull populations case; *Communications in Statistics — Simulation and Computation* **29**, 337–360

HINKLEY, D.V. (1975): On power transformations to symmetry; *Biometrika* **62**, 101–111

HIROSE, H. (1991): Percentile point estimation in the three–parameter Weibull distribution by the extended maximum likelihood estimate; *Computational Statistics & Data Analysis* **11**, 309–331

HIROSE, H. (1997): Lifetime assessment by intermittent inspection under the mixture Weibull power law model with application to XLPE cables; *Lifetime Data Analysis* **3**, 179–189

HIROSE H. (1999): Bias correction for the maximum likelihood estimates in the two parameter Weibull distribution; *IEEE Transactions on Dielectric and Electric Insulation* **3**, 66–68

HIROSE, H. / LAI, T.L. (1997): Inference from grouped data in three–parameter Weibull models with application to breakdown-voltage experiments; *Technometrics* **39**, 199–210

HJORTH, U. (1980): A reliability distribution with increasing, decreasing, constant, and bath–tub–shaped failure rates; *Technometrics* **22**, 99–107

HOBBS, J.P. / MOORE, A.H. / MILLER, R.M. (1985): Minimum–distance estimation of the parameters of the three–parameter Weibull distribution; *IEEE Transactions on Reliability* **34**, 495–496

HOINKES, L.A. / PADGETT, W.J. (1994): Maximum likelihood estimation from record breaking data for the Weibull distribution; *Quality and Reliability Engineering International* **10** 5–13

HOMAN, S.M. (1989): A comparison of plotting rules under $L_1$ and $L_2$ estimation of the Weibull scale and shape parameters in situations of small samples with possible censoring and outliers; *Communications in Statistics — Simulation and Computation* **18**, 121–143

HOMAN, S.M. / PARDO, E.S. / TRUDEU, M. (1987): An application of Weibull modelling and discriminant analysis of alcoholism relapse among veterans; *ASA Proceedings of Social Statistics Section*, 222-224

HOOKE, R. / JEEVES, T.A. (1961): "Direct search" solution of numerical and statistical problems; *Journal of the Association for Computing Machinery* **8**, 212–229

HORA, R.B. / BUEHLER, R.J. (1960): Fiducial theory and invariant estimation; *Annals of Mathematical Statistics* **35**, 643–656

HOSONO, J. / OHTA, H. / KASE, S. (1980): Design of single sampling plans for doubly exponential characteristics; Paper delivered at the Workshop on Statistical Quality Control, Berlin

HOSSAIN, A. / HOWLADER, H.A. (1996): Unweighted least squares estimation of Weibull parameters; *Journal of Statistical Computation & Simulation* **54**, 265–271

HOUGAARD, P. (1986): Survival models for heterogeneous populations derived from stable distributions; *Biometrika* **73**, 387–396

HSIEH, H.K. (1994): Average type–II censoring times for two–parameter Weibull distributions; *IEEE Transactions on Reliability* **43**, 91–96

HSIEH, H.K. (1996): Prediction intervals for Weibull observations, based on early–failure data; *IEEE Transactions on Reliability* **45**, 666–676

HSIEH, P.I. / LU, M.W. / FROHMAN, T.F. (1987): Simulation study on the Weibull distribution; *Reliability Review* **7**, 53–56

HSU, T.–A. (1982): On some optimal selection procedures for Weibull populations; *Communications in Statistics — Theory and Methods* **11**, 2657–2668

HUBER, P.J. (1981): *Robust Statistics*; Wiley, New York

HÜRLIMANNN, W. (2000): General location transform of the order statistics from the exponential, Pareto and Weibull, with application to maximum likelihood estimation; *Communications in Statistics — Theory and Methods* **29**, 2535–2545

HÜSLER, J. / SCHÜPBACH, M. (1986): On simple block estimators for the parameters of the extreme–value distribution; *Communications in Statistics — Simulation and Computation* **15**, 61–76

HUGHES, D.W. (1978): Module mass distribution and the Rosin and Weibull statistical functions; *Earth and Planet. Sci. Lett.* **39**, 371–376

HUILLET, T. / RAYNAUD, H.F. (1999): Rare events in a log–Weibull scenario — Application to earth quake data; *European Physical Journal* **B 12**, 457–469

HWANG, H.S. (1996): A reliability prediction model for missile systems based on truncated Weibull distribution; *Computers and Industrial Engineering* **31**, 245–248

## I-I-I-I-I

IDA, M. (1980): The application of the Weibull distribution to the analysis of the reaction time; *Japanese Psychological Research* **22**, 207–212

ISAIC–MANIU, A. / VODA, V.G. (1980): Tolerance limits for the Weibull distribution using the maximum likelihood estimator for the scale parameter; *Bulletin Mathématique de la Société de Science Mathématique de la R.S. Roumanie,* **24**, 273–276

ISHIOKA, T. (1990): Generating the Weibull random numbers using the ratio of uniform deviates; *Journal of the Japanese Society for Quality Control* **20**, 127–129

ISHIOKA, T. / NONAKA, Y. (1991): Maximum likelihood estimation of Weibull parameters for two competing risks; *IEEE Transactions on Reliability* **40**, 71–74

## J-J-J-J-J

JAECH, J.L. (1964): Estimation of Weibull distribution shape parameter when no more than two failures occur per lot; *Technometrics* **6**, 415–422

JAECH, J.L. (1968): Estimation of Weibull parameters from grouped failure data; Paper presented at the American Statistical Association Annual Meeting at Pittsburgh, August

JAISINGH, L.R. / DEY, D.K. / GRIFFITH, W.S. (1993): Properties of a multivariate survival distribution generated by a Weibullian–Gaussian mixture; *IEEE Transactions on Reliability* **42**, 618–622

JANARDAN, K.G. (1978): A new functional equation analogous to Cauchy–Pexider functional equation and its application; *Biometrical Journal* **20**, 323–329

JANARDAN, K.G. / SCHAEFFER, D.J. (1978): Another characterization of the Weibull distribution; *The Canadian Journal of Statistics* **6**, 77-78

JANARDAN, K.G. / TANEJA, V.S. (1979a): Characterization of the Weibull distribution by properties of order statistics; *Biometrical Journal* **21**, 3–9

JANARDAN, K.G. / TANEJA, V.S. (1979b): Some theorems concerning characterization of the Weibull distribution; *Biometrical Journal* **21**, 139–144

JANDHYALA, V.K. / FOTOPOOULOS, S.B. / EVAGGELOPOULOS, N. (1999): Change–point methods for Weibull models with applications to detection of trends in extreme temperatures; *Environmetrics* **10**, 547–564

JAYARAM, Y.G. (1974): A Bayesian estimate of reliability in the Weibull distribution; *Microelectronics and Reliability* **13**, 29–32

JEFFREYS, H. (1961): *Theory of Probability*; Oxford University Press, Oxford.

JENSEN, F. / PETERSEN, N.E. (1982): *Burn–In — An Engineering Approach to the Design and Analysis of Burn–In Procedures*; Wiley, New York

JEWELL, N.P. (1982): Mixtures of exponential distributions; *The Annals of Statistics* **10**, 479–484

JIANG, H. / SANO, M. / SEKINE, M. (1997): Weibull raindrop–size distribution and its application to rain attenuation; *IEE Proceedings — Microwaves Antennas and Propagation* **144**, 197–200

JIANG, R. / MURTHY, D.N.P. (1995): Modelling failure–data by mixture of two Weibull distributions: A graphical approach; *IEEE Transactions on Reliability* **44**, 477–488

JIANG, R. / MURTHY, D.N.P. (1997): Comment on a general linear–regression analysis applied to the three–parameter Weibull distribution; *IEEE Transactions on Reliability* **46**, 389–393

JIANG, R. / MURTHY, D.N.P. (1998): Mixture of Weibull distributions: Parametric characterization of failure rate function; *Applied Stochastic Models and Data Analysis* **14**, 47–65

JIANG, R. / MURTHY, D.N.P. (1999): The exponentiated Weibull family: A graphical approach; *IEEE Transactions on Reliability* **48**; 68–72

JIANG, R. / MURTHY, D.N.P. / JI, P. (2001): Models involving two inverse Weibull distributions; *Reliability Engineering and System Safety* **73**, 73–81

JIANG, S. / KECECIOGLU, D. (1992a): Graphical representation of two mixed Weibull distributions; *IEEE Transactions on Reliability* **41**, 241–247

JIANG, S. / KECECIOGLU, D. (1992b): Maximum likelihood estimates, from censored data, for mixed–Weibull distributions; *IEEE Transactions on Reliability* **41**, 248–255

JOHNS, M.V. / LIEBERMAN, G.J. (1966): An exact asymptotically efficient confidence bound for reliability in the case of the Weibull distribution; *Technometrics* **8**, 135–175

JOHNSON, L.G. (1964): *The Statistical Treatment of Fatigue Experiments;* Amsterdam

JOHNSON, L.G. (1968): The probabilistic basis of cumulative damage; *Transactions of the 22nd Technical Conference of the American Society of Quality Control,* 133–140

JOHNSON, N.L. (1949): System of frequency curves generated by methods of translation; *Biometrika* **64**, 149–176

JOHNSON, N.L. (1966): Cumulative sum control charts and the Weibull distribution; *Technometrics* **8**, 481–491

JOHNSON, N.L. / KOTZ, S. (1994): Comment on: "A general linear regression analysis applied to the three–parameter Weibull distribution"; *IEEE Transactions on Reliability* **43**, 603

JOHNSON, N.L. / KOTZ, S. / BALAKRISHNAN, N. (1994): *Continuous Univariate Distributions,* Vol. I, 2nd ed.; Wiley, New York

JOHNSON, N.L. / KOTZ, S. / BALAKRISHNAN, N. (1995): *Continuous Univariate Distributions,* Vol. II, 2nd ed.; Wiley, New York

JOHNSON, N.L. / KOTZ, S. / KEMP, A.W. (1992): *Univariate Discrete Distributions,* 2nd ed.; Wiley, New York

JOHNSON, R.A. / HASKELL, J.H. (1983): Sampling properties of estimators of a Weibull distribution of use in the lumber industry; *The Canadian Journal of Statistics* **11**, 155–169

JOHNSON, R.A. / HASKELL, J.H. (1984): An approximate lower tolerance bound for the three parameter Weibull applied to lumber characterization; *Statistics & Probability Letters* **2**, 67–76

JUDGE, G.E. / GRIFFITHS, W.E. / HILL, R.C. / LEE, T.–C. (1980): *The Theory and Practice of Econometrics;* Wiley, New York

JUN, C.–H. / BALAMURALI, S. / LEE, S.-H. (2006): Variables sampling plans for Weibull distributed lifetimes under sudden death testing; *IEEE Transactions on Reliability* **55**, 53–58

**K-K-K-K-K**

KABIR, A.B.M.Z. (1998): Estimation of Weibull parameters for irregular interval group failure data with unknown failure times; *Journal of Applied Statistics* **25**, 207–219

KAGAN, A.M. / LINNIK, Y.U. / RAO, C.R. (1973): *Characterization Problems in Mathematical Statistics*; Wiley, New York

KAHLE, W. (1996a): Bartlett corrections for the Weibull distribution; *Metrika* **43**, 257–263

KAHLE, W. (1996b): Estimation of the parameters of the Weibull distribution for censored samples; *Metrika* **44**, 27–40

KAIO, N. / OSAKI, S. (1980): Comparisons of point estimation methods in the two–parameter Weibull distribution; *IEEE Transactions on Reliability* **29**, 21

KALBFLEISCH, J.D. / LAWLESS, J.F. (1988): Estimation of reliability in field–performance studies (with discussion); *Technometrics* **30**, 365–378

KALBFLEISCH, J.D. / PRENTICE, R.L. (1980): *The Statistical Analysis of Failure Time Data;* Wiley, New York

KAMAT, S.J. (1977): Bayesian estimation of system reliability for Weibull distribution using Monte Carlo simulation, in TSOKOS, C.P. / SHIMI, I.N. (eds.): *The Theory and Applications of Reliability,* Vol. II; Academic Press, New York, 123–132

KAMINSKY, K.S. / MANN, N.R. / NELSON, P.I. (1975): Best and simplified linear invariant prediction of order statistics in location and scale families; *Biometrika* **62**, 525–527

KAMINSKY, K.S. / NELSON, P.I. (1975): Best linear unbiased prediction of order statistics in location and scale families; *Journal of the American Statistical Association* **70**, 145–150

KAMPS, U. (1988): Distance measures in a one–parameter class of density functions; *Communications in Statistics — Theory and Methods* **17**, 2013–2019

KAMPS, U. (1991): Inequalities for moments of order statistics and characterizations of distributions; *Journal of Statistical Planning and Inference* **27**, 397–404

KANG, S.-B. / CHO, Y.-S. / CHOI, S.-H. (2001): Estimation of the shape parameter for the three–parameter Weibull distribution; *Journal of Information & Optimization Sciences* **22**, 35–42

KANJI, G.K. / ARIF, O.H. (2001): Median rankit control charts for Weibull distribution; *Total Quality Management* **12**, 629–642

KAO, J.H.K. (1956): The Weibull distribution in life–testing of electron tubes (Abstract); *Journal of the American Statistical Association* **51**, 514

KAO, J.H.K. (1958): Computer methods for estimating Weibull parameters in reliability studies; *Transactions of IRE — Reliability and Quality Control* **13**, 15–22

KAO, J.H.K. (1959): A graphical estimation of mixed Weibull parameters in life–testing electronic tubes; *Technometrics* **1**, 389–407

KAO, J.H.K. (1964): Savings in test time when comparing Weibull scale parameters; *Technometrics* **6**, 471

KAPLAN, E.L. / MEIER, P. (1958): Nonparametric estimation from incomplete data; *Journal of the American Statistical Association* **53**, 457–481

KAPPENMAN, R.F. (1981): Estimation of Weibull location; *Journal of Statistical Computing & Simulation* **13**, 245–254

KAPPENMAN, R.F. (1982): On a method for selecting a distributional model; *Communication in Statistics — Theory and Methods* **11**, 663–672

KAPPENMAN, R.F. (1983): Parameter estimation via a sample reuse; *Journal of Statistical Computing & Simulation* **16**, 213–222

KAPPENMAN, R.F. (1985a): A testing approach to estimation; *Communications in Statistics —
Theory and Methods* **14**, 2365–2377

KAPPENMAN, R.F. (1985b): Estimation for the three–parameter Weibull, lognormal and gamma
distributions; *Computational Statistics & Data Analysis* **3**, 11–23

KAPPENMAN, R.F. (1988): A simple method for choosing between the lognormal and Weibull
models; *Statistics & Probability Letters* **7**, 123–126

KAPUR, K.C. / LAMBERSON, L.R. (1977): *Reliability in Engineering Design;* Wiley, New York

KAR, T.R. / NACHLOS, J.A. (1997): Coordinated warranty and burn–in strategies; *IEEE Transactions on Reliability* **46**, 512–518

KARLIN, S. (1969): *A First Course in Stochastic Processes*; Academic Press, London / New York

KAUFMANN, N. / LIPOW, M. (1964): Reliability life analysis using the Weibull distribution; *Proc.
11th West. Reg. Conf. ASQC,* 175–187

KEATING, J.P. (1984): A note on estimation of percentiles and reliability in the extreme–value
distribution; *Statistics & Probability Letters* **2**, 143–146

KEATS, J.B. / LAWRENCE, F.P. / WANG, F.K. (1997): Weibull maximum likelihood parameter
estimates with censored data; *Journal of Quality Technology* **29**, 105–110

KECECIOGLU, D. (1991): *Reliability Engineering Handbook;* Prentice–Hall, Englewood Cliffs

KELLER, A.Z. / GIBLIN, M.T. / FARNWORTH, N.R. (1982): Reliability analysis of commercial
vehicle engines; *Reliability Engineering* **10**, 15–25

KENNEDY, W.J. JR. / GENTLE, J.E. (1980): *Statistical Computing*; Marcel Dekker, New York

KENT, J. / QUESENBERRY, C.P. (1982): Selecting among probability distributions used in reliability theory; *Technometrics* **24**, 59–65

KESHRAN, K. / SARGENT, G. / CONRAD, H. (1980): Statistical analysis of the Hertzian fracture
of Pyrex glass using the Weibull distribution function; *Journal of Material Science* **15**, 839–844

KHAMIS, H.J. (1997): The $\delta$–corrected Kolmogorov–Smirnov test for two–parameter Weibull distribution; *Journal of Applied Statistics* **24**, 301–317

KHAN, A.H. / ALI, M.M. (1987): Characterization of probability distributions through higher
order gap; *Communications in Statistics — Theory and Methods* **16**, 1281–1287

KHAN, A.H. / BEG, M.I. (1987): Characterization of the Weibull distribution by conditional variance; *Sankhyā A* **49**, 268–271

KHAN, A.H. / KHAN, R.U. / PARVEZ, S. (1984): Inverse moments of order statistics from Weibull
distributions; *Scandinavian Actuarial Journal*, 91–94

KHAN, A.H. / PARVEZ, S. / YAQUB, M. (1983): Recurrence relations between product moments
of order statistics; *Journal of Statistical Planning and Inference* **8**, 175–183

KHAN, A.H. / YAQUB, M. / PARVEZ, S. (1983): Recurrence relations between moments of order
statistics; *Naval Research Logistics Quarterly* **30**, 419–441

KHIROSHI, S. / MIEKO, N. (1963): On the graphical estimation of the parameter of the Weibull
distribution from small samples; *Bulletin of the Electrotechnical Laboratory* **27**, 655–663

KHIROSHI, S. / SIDERU, T. / MINORU, K. (1966): On the accuracy of estimation of the parameters
of the Weibull distribution from small samples; *Bulletin of the Electrotechnical Laboratory* **30**, 753-
765

KIES, J.A. (1958): The strength of glass; *Naval Research Lab. Report No. 5093;* Washington, D.C.

KIM, B.H. / CHANG, I.H. / KANG, C.K. (2001): Bayesian estimation for the reliability in Weibull stress–strength systems using noninformative priors; *Far East Journal of Theoretical Statistics* **5**, 299–315

KIMBALL, B.F. (1956): The bias in certain estimates of the parameters of the extreme–value distribution; *Annals of Mathematical Statistics* **27**, 758–767

KIMBALL, B.F. (1960): On the choice of plotting positions on probability paper; *Journal of the American Statistical Association* **55**, 546–560

KIMBER, A. (1996): A Weibull based score test for heterogeneity; *Lifetime Data Analysis* **2**, 63–71

KIMBER, A.C. (1985): Tests for the exponential, Weibull and Gumbel distributions based on the stabilized probability plot; *Biometrika* **72**, 661–663

KINGSTON, J.V. / PATEL, J.K. (1980): Selecting the best one of several Weibull populations; *Communications in Statistics — Theory and Methods* **9**, 383–398

KINGSTON, J.V. / PATEL, J.K. (1981): Interval estimation of the largest reliability of $k$ Weibull populations; *Communications in Statistics — Theory and Methods* **10**, 2279–2298

KINGSTON, J.V. / PATEL, J.K. (1982): Classifying Weibull populations with respect to control; *Communications in Statistics — Theory and Methods* **11**, 899–909

KLEIN, J.P. / BASU, A.P. (1981): Weibull accelerated life tests when there are competing causes of failure; *Communications in Statistics — Theory and Methods* **10**, 2073–2100

KLEIN, J.P. / BASU, A.P. (1982): Accelerated life tests under competing Weibull causes of failure; *Communications in Statistics — Theory and Methods* **11**, 2271–2286

KLEYLE, R. (1978): Approximate lower confidence limits for the Weibull reliability function; *IEEE Transactions on Reliability* **27**, 153–160

KOGELSCHATZ, H. (1993): Emil Julius Gumbel — Appreciation of his scientific work (in German); *Allgemeines Statistisches Archiv* **77**, 433–440

KOLAR–ANIC, L. / VELKOVIC, S. / KAPOR, S. / DUBLJEVIC, B. (1975): Weibull distribution and kinetics of heterogeneous processes; *Journal of Chem. Phys.* **63**, 633–668

KOTANI, K. / ISHIKAWA, T. / TAMIYA, T. (1997): Simultaneous point and interval predictions in the Weibull distributions; *Statistica* **57**, 221–235

KOTELNIKOV, V.P. (1964): A nomogram connecting the parameters of Weibull's distribution with probabilities (in Russian, English translation); *Teoriya Veroyatnostei i ee Primeneniya,* **9**, 670–673

KOTZ, S. / BALAKRISHNAN, N. / JOHNSON, N.L. (2000): *Continuous Multivariate Distributions,* Vol. I; Wiley, New York

KOTZ, S. / NADARAJAH, S. (2000): *Extreme Value Distributions: Theory and Applications;* Imperial College Press, River Edge, NJ

KRISHNAIAH, P.R. / RAO, C.R. (eds.) (1988) *Quality Control and Reliability; Handbook of Statistics,* Vol. 7, North–Holland, Amsterdam

KROHN, C.A. (1969): Hazard versus renewal rate of electronic items; *IEEE Transactions on Reliability* **18**, 64–73

KÜBLER, H. (1979): On the fitting of the three–parameter distributions lognormal, gamma and Weibull; *Statistische Hefte* **20**, 68–125

KUCHII, S. / KAIO, N. / OSAKI, S. (1979): Simulation comparisons of point estimation methods in the two–parameter Weibull distribution; *Microelectronics and Reliability* **19**, 333–336

KULASEKERA, K.B. (1994): Approximate MLE's of the parameters of a discrete Weibull distribution with type I censored data; *Microelectronics and Reliability* **34**, 1185–1188

KULKARNI, A. / SATYANARAYAN, K. / ROHTAGI, P. (1973): Weibull analysis of strength of coir fibre; *Fibre Science Technology* **19**, 56–76

KULLDORF, G. (1973): A note on the optimum spacing of sample quantiles from the six extreme value distributions; *The Annals of Statistics* **1**, 562–567

KUNDU, D. / GUPTA, R.D. / MANGLICK, A. (no year): Discriminating between the log–normal and generalized exponential distributions; to appear in *Journal of Statistical Planning and Inference*

KUNDU, D. / MANGLICK, A. (no year): Discriminating between the Weibull and log–normal distributions; submitted to *Naval Research Logistics*

KWON, Y.W. / BERNER, J. (1994): Analysis of matrix damage evaluation in laminated composites; *Engineering Fracture Mechanics* **48**, 811–817

**L-L-L-L-L**

LAGAKOS, S.W. (1979): General right censoring and its impact on the analysis of failure data; *Biometrika* **66**, 139–156

LAI, C.D. / XIE, M. / MURTHY, D.N.P. (2003): Modified Weibull model; *IEEE Transactions on Reliability* **52**, 33–37

LAI, C.D. / XIE, M. / MURTHY, D.N.P. (2005): Reply to "On some recent modifications of Weibull distribution"; *IEEE Transactions on Reliability* **54**, 562

LAMBERSON, L.R. / DESOUZA, D.I. JR. (1987): Bayesian Weibull estimation; *ASQC Technical Conference Transactions*, 497–506

LANDES, J.D. (1993): Two criteria statistical model for transition fracture toughness; *Fatigue and Fracture of Engineering Materials & Structure* **16**, 1161–1174

LARICCIA, V.N. (1982): Asymptotic properties of $L^2$ quantile distance–estimators; *The Annals of Statistics* **10**, 621–524

LAWLESS, J.F. (1971): A prediction problem concerning samples from the exponential distribution, with application in life testing; *Technometrics* **13**, 725–730

LAWLESS, J.F. (1972): Confidence intervals for the parameters of the Weibull distribution; *Utilitas Mathematica* **2**, 71–87

LAWLESS, J.F. (1973a): On the estimation of safe life when the underlying life distribution is Weibull; *Technometrics* **15**, 857–865

LAWLESS, J.F. (1973b): Conditional versus unconditional confidence intervals for the parameters of the Weibull distribution; *Journal of the American Statistic Association* **68**, 665–669

LAWLESS, J.F. (1974): Approximations to confidence intervals for parameters in the extreme–value and Weibull distribution; *Biometrika* **61**, 123–129

LAWLESS, J.F. (1975): Construction of tolerance bounds for the extreme–value and Weibull distribution; *Technometrics* **17**, 255–261

LAWLESS, J.F. (1978): Confidence interval estimation for the Weibull and extreme–value distributions (with discussion); *Technometrics* **20**, pp 355–364

LAWLESS, J.F. (1982): *Statistical Models and Methods for Lifetime Data;* Wiley, New York

LAWRENCE, K.D. / SHIER, D.R. (1981): A comparison of least squares and least absolute deviation regression models for estimating Weibull parameters; *Communications in Statistics — Simula-*

*tion and Computation* **10**, 315–326

LEADBETTER, M.R. (1974): On extreme values in stationary sequences; *Zeitschrift für Wahrscheinlichkeitstheorie* **28**, 289–303

LECOUTRE, B. / MABIKA, B. / DERZKO, G. (2002): Assessment and monitoring in clinical trials when survival curves have distinct shapes: A Bayesian approach with Weibull modelling; *Statistics in Medicine* **21**, 663–674

LEE, E.T. (1980): *Statistical Methods for Survival Data Analysis;* Lifetime Learning Publication, Belmont, CA

LEE, L. (1979): Multivariate distributions having Weibull properties; *Journal of Multivariate Analysis* **9**, 267–277

LEE, L. (1980): Comparing rates of several independent Weibull processes; *Technometrics* **22**, 427–430

LEE, L. / LEE, S.K. (1978): Some results on inference for the Weibull process; *Technometrics* **20**, 41–46

LEE, S.M. / BELL, C.B. / MASON, A.L. (1988): Discrimination, signal detection, and estimation for Weibull–type Poisson processes; *IEEE Transactions on Information Theory* **34**, 576–580

LEEMIS, L.M. (1986): Lifetime distribution identities; *IEEE Transactions on Reliability* **35**, 170–174

LEHMAN, E.H. JR. (1963): Shapes, moments and estimators of the Weibull distribution; *IEEE Transactions on Reliability* **12**, 32–38

LEHMANN, E.L. (1983): *Theory of Point Estimation*; Wiley, New York

LEMOINE, A.J. / WENOCUR, M.L. (1985): On failure modelling; *Naval Research Logistics Quarterly* **32**, 497–508

LEMON, G.H. (1975): Maximum likelihood estimation for the three–parameter Weibull distribution based on censored samples; *Technometrics* **17**, 247–254

LEMON, G.H. / KRUTCHKOFF, R.C. (1969): An empirical Bayes smoothing technique; *Biometrika* **56**, 361–365

LEMON, G.H. / WATTIER, J.B. (1976): Confidence and "A" and "B" allowable factors for the Weibull distribution; *IEEE Transactions on Reliability* **25**, 16–19

LEONE, F.D. / RUTENBERG, Y.H. / TOPP, C.W. (1960): Order statistics and estimators for the Weibull distribution; *Case Statistical Laboratory Publication, No. 1026,* Case Institute of Technology, Cleveland OH

LI, Y.–M. (1994): A general linear–regression analysis applied to the three–parameter Weibull distribution; *IEEE Transactions on Reliability* **43**, 255–263

LIAO, M. / SHIMOKAWA, T. (1999): A new goodness–of–fit test for type–1 extreme–value and 2–parameter Weibull distributions with estimated parameters; *Journal of Statistical Computation & Simulation* **64**, 23–48

LIEBERMAN, G.J. (1962): Estimates of the parameters of the Weibull distribution (Abstract); *Annals of Mathematical Statistics* **33**, 1483

LIEBLEIN, J. (1953): On the exact evaluation of the variances and covariances of order statistics in samples from the extreme–value distribution; *Annals of Mathematical Statistics* **24**, 282–287

LIEBLEIN, J. (1955): On moments of order statistics from the Weibull distribution; *Annals of Mathematical Statistics* **26**, 330–333

LIEBLEIN, J. / ZELEN, M. (1956): Statistical investigation of the fatigue life of deep–groove ball bearings; *Journal of Research of the National Bureau of Standards* **57**, 273–316

LIEBSCHER, U. (1967): Projectively distorted probability paper for time–dependent observations (in German); *Qualitätskontrolle* **12**, 141–144

LIERTZ, H. / OESTREICH, U.H.P. (1976): Application of Weibull distribution to mechanical reliability of optical waveguides for cables; *Siemens Forschungs- und Entwicklungsber.* **5**, 129–135

LIN, C.–L. / HU, J. (1998): A modified checking procedure for the Weibull failure distributions; *Journal of the Chinese Statistical Association* **36**, 227–245

LINDLEY, D.V. (1965): *Introduction to Probability and Statistics from a Bayesian Viewpoint, Part 2 — Inference*; Cambridge University Press, City?

LINDLEY, D.V. (1980): Approximate Bayesian method; *Trabajos de Estadistica* **32**, 223–237

LINGAPPAIAH, G.S. (1977): Bayesian approach to prediction in complete and censored samples from Weibull population; *Metron* **35**, 167–179

LINGAPPAIAH, G.S. (1983): A study of shift in parameter of a Weibull life test model; *Industrial Mathematics* **33**, 139–148

LINGAPPAIAH, G.S. (1990): Inference in life tests based on a Weibull model when an outlier is present; *Journal of the Indian Association for Productivity, Quality and Reliability* **15**, 1–10

LISEK, B. (1978): Comparability of special distributions; *Mathematische Operationsforschung und Statistik, Series Statistics* **9**, 587–598

LITTELL, R.C. / MCCLAVE, J.T. / OFFEN, W.W. (1979): Goodness–of–fit tests for the two–parameter Weibull distribution; *Communications in Statistics — Simulation and Computation* **8**, 257–269

LLOYD, E.H. (1952): Least–squares estimation of location and scale parameters using order statistics; *Biometrika* **39**, 88–95

LLOYD, D.K. / LIPOW, M. (1962): *Reliability: Management, Methods and Mathematics;* Prentice-Hall, Englewood Cliffs, NJ

LOCHNER, R.H. / BASU, A.P. / DIPONZIO, M. (1974): Covariances of Weibull order statistics; *Naval Research Logistics Quarterly* **21**, 543–547

LOCKHART, R.A. / O'REILLY, F. / STEPHENS, M.A. (1986): Tests for the extreme–value and Weibull distributions based on normalized spacings; *Naval Research Logistics Quarterly* **33**, 413–421

LOCKHART, R.A. / STEPHENS, M.A. (1994): Estimation and tests of fit for the three–parameter Weibull distribution; *Journal of the Royal Statistical Society B* **56**, 491–501

LOMAX, K.S. (1954): Business failures: Another example of the analysis of failure data; *Journal of the American Statistical Association* **49**, 847–852

LOMICKI, Z.A. (1966): A note on the Weibull renewal process; *Biometrika* **53**, 375–381

LOMICKI, Z.A. (1973): Some aspects of the statistical approach to reliability; *Journal of the Royal Statistical Society A* **136**, 395–419

LOONEY, S.W. (1983): The use of the Weibull distribution in bioassay; *ASA Proceedings of Statistical Computing Section,* 272–277

LOUZADA–NETO, F. / ANDRADE, C.S. (2001): Detecting non–identifiability in the poly–Weibull model; *Brazilian Journal of Probability and Statistics* **15**, 143–154

LOVE, G.E. / GUO, R. (1996): Utilizing Weibull failure rates in repair limit analysis for equipment replacement / Preventative maintenance decisions; *Journal of the Operational Research Society* **47**, 1366–1376

LU, J.–C. (1989): Weibull extensions on the Freund and Marshall–Olkin bivariate exponential models; *IEEE Transactions on Reliability* **38**, 615–619

LU, J.–C. (1990): Least squares estimation for the multivariate Weibull model of Hougaard based on accelerated life test of system and component; *Communications in Statistics — Theory and Methods* **19**, 3725–3739

LU, J.–C. (1992a): Bayes parameter estimation from bivariate Weibull model of Marshall–Olkin for censored data; *IEEE Transactions on Reliability* **41**, 608–615

LU, J.–C. (1992b): Effects of dependence on modelling system reliability and mean life via a multivariate Weibull distribution; *Journal of the Indian Association for Productivity, Quality and Reliability* **17**, 1–22

LU, J.–C. / BHATTACHARYYA, G.K. (1990): Some new constructions of bivariate Weibull models; *Annals of the Institute of Statistical Mathematics* **42**, 543–559

LUN, I.Y.F. / LAM, J.C. (2000): A study of Weibull parameters using long–term wind observations; *Renewable Energy* **20**, 145–153

**M-M-M-M-M**

MACARTHUR, R.H. (1957): On the relative abundance of bird species; *Proceedings of the National Academy of Sciences USA* **43**, 293–295

MACARTHUR, R.H. (1960): On the relative abundance of species; *American Naturalist* **94**, 25–36

MACGILLIVRAY, H.L. (1992): Shape properties of the g– and h– and Johnson families; *Communications in Statistics — Theory and Methods* **21**, 1233–1250

MACKISACK, M.S. / STILLMAN, R.H. (1996): A cautionary tale about Weibull analysis; *IEEE Transactions on Reliability* **45**, 244–248

MAHMOUD, M.A.W. (1996): On a stress–strength model in Weibull case; *The Egyptian Statistical Journal* **40**, 119–126

MAJESKE, K.D. / HERRIN, G.D. (1995): Assessing mixture model goodness of fit with application to automobile warranty data; *Proceedings of Annual Reliability and Maintainability Symposium*, 378–383

MAKINO, M. (1984): Mean hazard rate and its application to the normal approximation of the Weibull distribution; *Naval Research Logistics Quarterly* **31**, 1–8

MALIK, M.A.K. (1975): A note on the physical meaning of Weibull distribution; *IEEE Transactions on Reliability* **24**, 95

MALIK, M.A.K. / TRUDEL, R. (1982): Probability density function of quotient of order statistics from the Pareto, power and Weibull distributions; *Communications in Statistics — Theory and Methods* **11**, 801–814

MANDELBAUM, J. / HARRIS, C.M. (1982): Parameter estimation under progressive censoring conditions for a finite mixture of Weibull distributions; in ZANAKIS, S.H. / RUSTAGI, J.S. (eds.): *Optimization in Statistics,* North–Holland, Amsterdam, 239–260

MANN, N.R. (1965): Point and interval estimates for reliability parameters when failure times have the two–parameter Weibull distribution; *Ph.D. Thesis*, University of California at Los Angeles

MANN, N.R. (1966): Exact three–order–statistics confidence bounds on reliability parameters under Weibull assumptions; *ARL Technical Report* 67–0023

MANN, N.R. (1967a): Tables for obtaining the best linear invariant estimates of parameters of the Weibull distribution; *Technometrics* **9**, 629–645

MANN, N.R. (1967b): Results on location and scale parameters estimation with application to the extreme–value distribution; *ARL Technical Report 67–0023*

MANN, N.R. (1968a): Point and interval estimation procedures for the two–parameter Weibull and extreme–value distributions; *Technometrics* **10**, 231–256

MANN, N.R. (1968b): Results on statistical estimation and hypothesis testing with application to the Weibull and extreme–value distributions; *ARL Technical Report No. 68–0068*

MANN, N.R. (1969a): Optimum estimators for linear functions of location and scale parameters; *Annals of Mathematical Statistics* **40**, 2149–2155

MANN, N.R. (1969b): Cramér–Rao efficiencies of best linear invariant estimators of parameters of the extreme–value distribution under type II censoring from above; *SIAM Journal on Applied Mathematics* **17**, 1150–1162

MANN, N.R. (1969c): Exact three-order-statistics confidence bounds on reliable life for a Weibull model with progressive censoring; *Journal of the American Statistical Association* **64**, 306–315

MANN, N.R. (1970a): Estimators and exact confidence bounds for Weibull parameters based on a few ordered observations; *Technometrics* **12**, 345–361

MANN, N.R. (1970b): Estimation of location and scale parameters under various models of censoring and truncation; *ARL Technical Report No. 70–0026*

MANN, N.R. (1970c): Extension of results concerning tolerance bounds and warranty periods for Weibull Modells; *ARL Technical Report No. 70–0010*

MANN, N.R. (1970d): Warranty periods based on three ordered sample observations from a Weibull population; *IEEE Transactions on Reliability* **19**, 167–171

MANN, N.R. (1971): Best linear invariant estimation for Weibull parameters under progressive censoring; *Technometrics* **13**, 521–533

MANN, N.R. (1972a): Design of over–stress life–test experiments when failure times have the two–parameter Weibull distribution; *Technometrics* **14**, 437–451

MANN, N.R. (1972b): Best linear invariant estimation of Weibull parameters: Samples censored by time and truncated distributions; *International Conference on Structural Safety and Reliability (Proceedings)*, Washington, D.C., 107–114

MANN, N.R. (1977): An *F* approximation for two–parameter Weibull and lognormal tolerance bounds based on possibly censored data; *Naval Research Logistics Quarterly* **24**, 187–196

MANN, N.R. (1978a): Weibull tolerance intervals associated with moderate to small proportions for use in a new formulation of Lanchaster combat theory; *Naval Research Logistics Quarterly* **25**, 121–128

MANN, N.R. (1978b): Calculation of small–sample Weibull tolerance bounds for accelerated testing; *Communications in Statistics — Theory and Methods* **7**, 97–112

MANN, N.R. / FERTIG, K.W. (1973): Tables for obtaining Weibull confidence bounds and tolerance bounds based on best linear invariant estimates of parameters of the extreme–value distribution; *Technometrics* **15**, 87–101

MANN, N.R. / FERTIG, K.W. (1975a): A goodness–of–fit test for the two–parameter versus three–parameter Weibull; confidence bounds for threshold; *Technometrics* **17**, 237–245

MANN, N.R. / FERTIG, K.W. (1975b): Simplified efficient point and interval estimators for Weibull parameters; *Technometrics* **17**, 361–368

MANN, N.R. / FERTIG, K.W. (1977): Efficient unbiased quantile estimators for moderate–size complete samples from extreme–value and Weibull distributions; confidence bounds and tolerance and prediction intervals; *Technometrics* **19**, 87–94

MANN, N.R. / FERTIG, K.W. / SCHEUER, E.M. (1971): Confidence and tolerance bounds and a new goodness–of–fit test for the two–parameter Weibull or extreme–value distribution; *ARL Technical Report No. 71–0077*

MANN, N.R. / SAUNDERS, S.C. (1969): On evaluation of warranty assurance when life has a Weibull distribution; *Biometrika* **56**, 615–625

MANN, N.R. / SCHAFER, R.E. / SINGPURWALLA, N.D. (1974): *Methods for Statistical Analysis of Reliability and Life Data;* Wiley, New York

MANN, N.R. / SCHEUER, E.M. / FERTIG, K.W. (1973): A new goodness–of–fit test for the two–parameter Weibull or extreme–value distribution with unknown parameters; *Communications in Statistics* **2**, 383–400

MARITZ, J.S. (1967): Smooth empirical Bayes estimation for continuous distributions; *Biometrika* **54**, 435–450

MARITZ, J.S. (1970): *Empirical Bayes Methods;* Methuen, London

MAROHN, F. (1991): Global sufficiency of extreme order statistics in location models of Weibull type; *Probability Theory and Related Fields* **88**, 261–268

MARQUINA, N. (1979): A simple approach to estimating Weibull parameters; *ASA Proceedings of Statistical Computing Section,* 365–367

MARSHALL, A.W. / OLKIN, I. (1967): A multivariate exponential distribution; *Journal of the American Statistical Association* **63**, 30–44

MARSHALL, A.W. / OLKIN, I. (1997): A new method for adding a parameter to a family of distributions with application to the exponential and Weibull families; *Biometrika* **84**, 641–652

MARTINEZ, S. / QUINTANA, F. (1991): On a test for generalized upper truncated Weibull distributions; *Statistics & Probability Letters* **12**, 273–279

MARTZ, H.F. JR. / LIAN, M.G. (1977): Bayes and empirical Bayes point and interval estimation of reliability for the Weibull model; in TSOKOS, C.P. / SHIMI, I.N. (eds.): *The Theory and Applications of Reliability,* Vol. II, Academic Press, New York, 203–233

MARTZ, H.F. JR. / WALLER, R.A. (1982): *Bayesian Reliability Analysis;* Wiley, New York

MAZZANTI, G. / MONTANARI, G.C. / CACCIARI, M. / CONTIN, A. (1996): A method for the improvement of maximum–likelihood estimate accuracy of the Weibull parameters; *Metron* **54**, 65–82

MAZZUCHI, T.A. / SOYER, R. / VOPATEK, A.L. (1997): Linear Bayesian inference for accelerated Weibull model; *Lifetime Data Analysis* **3**, 63–75

MCCOOL, J.I. (1965): The construction of good linear unbiased estimates from the best linear estimates for smaller sample size; *Technometrics* **7**, 543–552

MCCOOL, J.I. (1966) Censored Sample Size Selection for Life Tests; *Report of the Engineering and Research Center of SKF Industries Inc.,* King of Prussia, Pennsylvania

MCCOOL, J.I. (1969): Unbiased maximum likelihood estimation of a Weibull percentile when the shape parameter is known; *IEEE Transactions on Reliability* **18**, 78–79

McCool, J.I. (1970a): Inference on Weibull percentiles and shape parameter from maximum likelihood estimates; *IEEE Transactions on Reliability* **19**, 2–10

McCool, J.I. (1970b): Inference on Weibull percentiles from sudden death tests using maximum likelihood; *IEEE Transactions on Reliability* **19**, 177–179

McCool, J.I. (1970c): Evaluating Weibull endurance data by the method of maximum likelihood; *ASLE (American Society of Lubrication Engineers) Transactions* **13**, 189–202

McCool, J.I. (1974a): Analysis of sudden death tests of bearing endurance; *ASLE Transactions* **17**, 8–13

McCool, J.I. (1974b): Inferential techniques for Weibull populations; *ARL Technical Report No. 74–0180*

McCool, J.I. (1974c): Using the Weibull distribution; *Paper presented at 42nd Annual Meeting of the National Lubricating Grease Institute,* Chicago, 20–23

McCool, J.I. (1975a): Inferential techniques for Weibull populations II; *ARL Technical Report No. 75–0233*

McCool, J.I. (1975b): Multiple comparison of Weibull parameters; *IEEE Transactions on Reliability* **24**, 186–192

McCool, J.I. (1976): Estimation of Weibull Parameters with competing–mode censoring; *IEEE Transactions on Reliability* **25**, 25–31

McCool, J.I. (1977): Analysis of variance for Weibull populations; in Tsokos, C.P. / Shimi, I.N. (eds.): *The Theory and Applications of Reliability,* Vol. II; Academic Press, New York, 335–379

McCool, J.I. (1979a): Analysis of single classification experiments based on censored samples from the two–parameter Weibull distribution; *Journal of Statistical Planning and Inference* **3**, 39–68

McCool, J.I. (1979b): Estimation of Weibull shape parameter for samples of size 2; *IEEE Transactions on Reliability* **28**, 158–160

McCool, J.I. (1980): Confidence limits for Weibull regression with censored data; *IEEE Transactions on Reliability* **29**, 145–150

McCool, J.I. (1991): Inference on $P(Y < X)$ in the Weibull case; *Communications in Statistics — Simulation and Computation* **20**, 129–148

McCool, J.I. (1996): The analysis of a two way factorial design with Weibull response; *Communications in Statistics — Simulation and Computation* **25**, 263–286

McCool, J.I. (1998): Inference on the Weibull location parameter; *Journal of Quality Technology* **30**, 119–126

McElhone, D.H. / Larsen, N.C. (1969): Expected values, variances and covariances of order statistics drawn from a Weibull density with integer shape parameter; *U.S. Atomic Energy Commission R&D Report IS–2156*, Ames Laboratory, Iowa State University

McEwen, R.P. / Parresol, B.R. (1991): Moment expressions and summary statistics for the complete and truncated Weibull distribution; *Communications in Statistics — Theory and Methods* **20**, 1361–1372

McWilliams, T.P. (1989): Economic control chart designs and the in–control time distributions; *Journal of Quality Technology* **21**, 103–110

Mee, R.W. / Kushary, D. (1994): Prediction limits for the Weibull distribution utilizing simulation; *Computational Statistics & Data Analysis* **17**, 327–336

MEEKER, W.Q. JR. (1984): A comparison of accelerated life test plans for Weibull and lognormal distributions under type I censoring; *Technometrics* **26**, 157–171

MEEKER, W.Q. JR. (1986): Planning life tests in which units are inspected for failure; *IEEE Transactions on Reliability* **35**, 571–578

MEEKER, W.Q. JR. (1987): Limited failure population life tests: Application to integrated circuit reliability; *Technometrics* **29**, 51–65

MEEKER, W.Q. JR. / ESCOBAR, L.A. (1998): *Statistical Methods for Reliability Data;* Wiley, New York

MEEKER, W.Q. JR. / ESCOBAR, L.A. / HILL, D.A. (1992): Sample sizes for estimating Weibull hazard function from censored samples; *IEEE Transactions on Reliability* **41**, 133–138

MEEKER, W.Q. JR. / NELSON, W. (1975): Optimum accelerated life–tests for the Weibull and Extreme Value Distributions; *IEEE Transactions on Reliability* **24**, 321–332

MEEKER, W.Q. JR. / NELSON, W. (1976): Weibull percentile estimates and confidence limits from singly censored data by maximum likelihood; *IEEE Transactions on Reliability* **25**, 20–24

MEEKER, W.Q. JR. / NELSON, W. (1977): Weibull variances and confidence limits by maximum likelihood for singly censored data; *Technometrics* **19**, 473–476

MELNYK, M. / SCHWITTER, J.P. (1965): Weibull distribution and Monte–Carlo–Simulation to assess the life length of machinery (in German); *Die Unternehmung* **19**, 206–212

MENDENHALL, W. (1958): A bibliography on life testing and related topics; *Biometrika* **45**, 521–543

MENDENHALL, W. / LEHMAN E.H. JR. (1960): An approximation to the negative moments of the positive binomial useful in life testing; *Technometrics* **2**, 227–242

MENON, M.W. (1963): Estimation of the shape and the scale parameters of the Weibull distribution; *Technometrics* **5**, 175–182

MEYNA, A. (1982): Calculating the probability of failure of appliances without collecting times to failure for arbitrarily distributed times of delivery (in German); *Qualität und Zuverlässigkeit* **28**, 36–39

MIHRAM, G.A. (1969): Complete sample estimation techniques for reparameterized Weibull distributions; *IEEE Transactions on Reliability* **18**, 190–195

MIHRAM, G.A. (1973): A consistent shape parameter estimator for the Weibull distribution; *Proceedings 1973 Annual Reliability and Maintainability Symposium,* Philadelphia, 1–9

MIHRAM, G.A. (1975): A generalized extreme–value density; *South African Statistical Journal* **9**, 153–162

MIHRAM, G.A. (1977): Weibull shape parameter: Estimation by moments; *American Society for Quality Control, 31st Annual Technical Conference Proceedings,* 315–322

MILLER, D.W. (1966): Degree of normality of maximum likelihood estimates of the shape parameters of the Weibull failure distribution; *Unpublished Thesis*, Washington University, St. Louis, MO

MILLER, G. (1989): Inference on a future reliability parameter with the Weibull process model; *Naval Research Logistics Quarterly* **31**, 91–96

MILLER, I. / FREUND, J.E. (1965): *Probability and Statistics for Engineers;* Prentice–Hall, Englewood Cliffs, NJ

MILLER, R.G. JR. (1981): *Survival analysis;* Wiley, New York

MILLER, R.W. / NELSON, W. (1983): Optimum simple step–stress plans for accelerated life testing; *IEEE Transaction on Reliability* **32**, 59–65

MISES, R. VON (1923): On the range in a series of observations (in German); *Sitzungsberichte der Berliner Mathematischen Gesellschaft* **222**, 3–8

MISES, R. VON (1936): La distribution de la plus grande de $n$ valeurs; *Revue Math. de l'Union Interbalcanique* **1**, 1–20

MITTAL, M.M. / DAHIYA, R.C. (1989): Estimating the parameters of a truncated Weibull distribution; *Communications in Statistics — Theory and Methods* **18**, 2027–2042

MOESCHBERGER, M.L. (1974): Life tests under dependent competing causes of failure; *Technometrics* **16**, 39–47

MØLLER, S.K. (1976): The Rasch–Weibull process; *Scandinavian Journal of Statistics* **3**, 107–115

MONTANARI, G.C. / CAVALLINI, A. / TOMMASINI, L. / CACCIARI, M. / CONTIN, A. (1995): Comparison of random number generators for Monte Carlo estimates of Weibull parameters; *Metron* **53**, 55–77

MONTANARI, G.C. / MAZZANTI, G. / CACCIARI, M. / FOTHERGILL, J.C. (1997): In search of convenient techniques for reducing bias in the estimates of Weibull parameters for uncensored tests; *IEEE Transactions of Dielectric and Electric Insulation* **4**, 306–313

MOORE, A.H. / BILIKAM, J.E. (1980): Admissible, minimax and equivariant estimates of life distributions from type II censored samples; *IEEE Transaction on Reliability* **29**, 401–405

MOORE, A.H. / HARTER, H.L. (1965): One–order–statistic estimation of the scale parameters of Weibull populations; *IEEE Transactions on Reliability* **14**, 100–106

MOORE, A.H. / HARTER, H.L. (1966): Point and interval estimation, from one order statistic, of the location parameter of an extreme–value distribution with known scale parameter, and of the scale parameter of a Weibull distribution with known shape parameter; *IEEE Transactions on Reliability* **15**, 120–126

MOORE, A.H. / HARTER, H.L. / ANTOON, D.F. (1981): Confidence intervals and tests of hypotheses for the reliability of a two–parameter Weibull system; *IEEE Transactions on Reliability* **30**, 468–470

MOOTHATHU, T.S.K. (1990): A characterization property of Weibull, exponential and Pareto distributions; *Journal of the Indian Statistical Association* **28**, 69–74

MORRIS, C. / CHRISTIANSEN, C. (1995): Fitting Weibull duration models with random effects; *Lifetime Data Analysis* **1**; 347–359

MOSLER, K. / SCHEICHER, C. (2005): Homogeneity testing in a Weibull mixture model; to be published

MU, F.C. / TAN, C.H. / XU, M.Z. (2000): Proportional difference estimate method of determining the characteristic parameters of monomodal and multimodal Weibull distributions of time–dependent dielectric breakdown; *Solid–State Electronics* **44**, 1419–1424

MUDHOLKAR, G.S. / HUTSON, A.D. (1996): The exponentiated Weibull family: Some properties and a flood application; *Communications in Statistics — Theory and Methods* **25**, 3059–3083

MUDHOLKAR, G.S. / KOLLIA, G.D. (1994): Generalized Weibull family: A structural analysis; *Communications in Statistics — Theory and Methods* **23**, 1149–1171

MUDHOLKAR, G.S. / SRIVASTAVA, D.K. (1993): Exponentiated Weibull family for analyzing bathtub failure–rate data; *IEEE Transactions on Reliability* **42**, 299–302

MUDHOLKAR, G.S. / SRIVASTAVA, D.K. / FREIMER, M. (1995): The exponentiated Weibull–family: A Reanalysis of bus–motor–failure data; *Technometrics* **37**, 436–445

MUDHOLKAR, G.S. / SRIVASTAVA, D.K. / KOLLIA, G.D. (1996): A generalization of the Weibull distribution with application to the analysis of failure data; *Journal of the American Statistical Association* **91**, 1575–1583

MUENZ, L.R. / GREEN, S.B. (1977): Time savings in censored life testing; *Journal of the Royal Statistical Society B* **39**, 269–275

MUKHERJEE, S.P. / ROY, D. (1987): Failure rate transform of the Weibull variate and its properties; *Communication in Statistics — Theory and Methods* **16**, 281–291

MUKHERJEE, S.P. / SASMAL, B.C. (1984): Estimation of Weibull parameters using fractional moments; *Calcutta Statistical Association Bulletin* **33**, 179–186

MUKHERJEE, S.P. / SINGH, N.K. (1998): Sampling properties of an estimator of a new process capability index for Weibull distributed quality characteristics; *Quality Engineering* **10**, 291–294

MUNFORD, A.G. / SHAHANI, A.K. (1973): An inspection policy for the Weibull case; *Operations Research Quarterly* **24**, 453–458

MURALIDHARAN, K. (1999): Testing for the shape parameter in Weibull processes: A conditional approach; *Statistical Methods* **1**, 36–40

MURALIDHARAN, K. / SHANUBHOGUE, A. / JANI, P.N. (2001): Conditional tests for the shape parameter in the mixture of a degenerate and a Weibull distribution; *Journal of Applied Statistical Science* **10**, 299–313

MURTHY, D.N.P. / XIE, M. / JIANG, R. (2004): *Weibull Models*; Wiley, New York

MURTHY, V.K. (1968): A new method of estimating the Weibull shape parameter; *ARL Tech. Rep. No. 68–0076*

MURTHY, V.K. / SWARTZ, G.B. (1971): Contributions to the reliability; *ARL Tech. Rep. No. 71–0060*

MURTHY, V.K. / SWARTZ, G.B. (1975): Estimation of Weibull parameters from two–order statistics; *Journal of the Royal Statistical Society B* **37**, 96–102

MUSSON, T.A. (1965): Linear estimation of the location and scale parameters of the Weibull and gamma distribution by the use of order statistics; *Unpublished Thesis*, Air Force Institute of Technology, Dayton Air Force Base, OH

**N-N-N-N-N**

NADARAJAH, S. / KOTZ, S. (2005): On some recent modifications of Weibull distribution; *IEEE Transactions on Reliability* **54**, 561–562

NADLER, P. / ERFURT, F. (1971): Statistical evaluation of measurements by the Weibull distribution function; *Hermsdorfer Techn. Mitt.* **11**, 966–973

NAIR, V.K.R. / GEETHA, K.G. (1997): Characterizations of bivariate Weibull and Gumbels bivariate exponential distributions; *Journal of the Indian Association for Production, Quality and Reliability* **22**, 135–139

NAKAGAWA, T. (1978): Approximate calculation of block replacement with Weibull failure times; *IEEE Transactions on Reliability* **27**, 268–269

NAKAGAWA, T. (1979): Approximate calculation of inspection policy with Weibull failure times; *IEEE Transactions on Reliability* **28**, 403–404

NAKAGAWA, T. (1986): Periodic and sequential preventative maintenance policies; *Journal of Applied Probability* **23**, 536–542

NAKAGAWA, T. / OSAKI, S. (1975): The discrete Weibull distribution; *IEEE Transactions on Reliability* **24**, 300–301

NAKAGAWA, T. / YASUI, K. (1978): Approximate calculation of block replacement with Weibull failure times; *IEEE Transactions on Reliability* **27**, 268–296

NAKAGAWA, T. / YASUI, K. (1981): Calculation of age–replacement with Weibull failure times; *IEEE Transactions on Reliability* **30**, 163–164

NAPIERLA, M. / POTRZ, J. (1976): A simulation program to optimize maintenance for complicated production (in German); *Angewandte Informatik* **18**, 284–289

NATHAN, R.J. / MCMAHON, T.A. (1990): Practical aspects of low–flow frequency analysis; *Water Resources Research* **25**, 2135–2141

NELSON, L.S. (1967): Weibull probability paper; *Industrial Quality Control* **23**, 452–453

NELSON, P.R. (1979): Control charts for Weibull process with standards given; *IEEE Transactions on Reliability* **28**, 283–288

NELSON, W. (1970): Hazard plotting methods for analysis of life data with different failure modes; *Journal of Quality Technology* **2**, 126–149

NELSON, W. (1972a): Theory and applications of hazard plotting for censored failure data; *Technometrics* **14**, 945–966

NELSON, W. (1972b): Statistical methods for the ratio of two multinomial proportions; *The American Statistician* **26**, 22–27

NELSON, W. (1980): Accelerated life testing: Step–stress models and data analysis; *IEEE Transactions on Reliability* **29**, 103–108

NELSON, W. (1982): *Applied Life Data Analysis;* Wiley, New York

NELSON, W. (1985): Weibull analysis of reliability data with few or no failures; *Journal of Quality Technology* **17**, 140–146

NELSON, W. (1990): *Accelerated Testing — Statistical Models, Test Plans, and Data Analysis;* Wiley, New York

NELSON, W. (2000): Weibull prediction of a future number of failures; *Quality and Reliability Engineering International* **16**, 23–26

NELSON, W. / HAHN, G.J. (1974): Linear estimation of a regression relationship from censored data — Part II: Best linear unbiased estimation and theory; *Technometrics* **15**, 133–150

NELSON, W. / MEEKER, W.Q. JR. (1978): Theory for optimum accelerated censored life tests for Weibull and extreme value distributions; *Technometrics* **20**, 171–177

NELSON, W. / THOMPSON, V.C. (1971): Weibull probability papers; *Journal of Quality Technology* **3**, 45–50

NESHYBA, S. (1980): On the size distribution of Antarctic icebergs; *Cold Region Science and Technology* **1**, 241–248

NEWBY, M.J. (1979): The simulation of order statistics from life distributions; *Applied Statistics* **28**, 298–301

NEWBY, M.J. (1980): The properties of moment estimators for the Weibull distribution based on the sample coefficient of variation; *Technometrics* **22**, 187–194

NEWBY, M.J. (1982): Corrigendum to "The properties of moment estimators ...";  *Technometrics* **24**, 90

NEWBY, M.J. (1984): Properties of moment estimators for the three–parameter Weibull distribution; *IEEE Transactions on Reliability* **33**, 192–195

NEWBY, M.J. / WINTERTON, J. (1983): The duration of industrial stoppages; *Journal of the Royal Statistical Society A* **146**, 62–70

NG, H.K.T. (2005): Parameter estimation for a modified Weibull distribution, for progressively type–II censored samples; *IEEE Transactions on Reliability* **54**, 374–380

NG, H.K.T. / CHAN, P.S. / BALAKRISHNAN, N. (2004): Optimal progressive censoring plans for the Weibull distribution; *Technometrics* **46**, 470–481

NIGM, A.M. (1989): An informative Bayesian prediction for the Weibull lifetime distribution; *Communications in Statistics — Theory and Methods* **18**, 897–911

NIGM, A.M. (1990): Prediction bounds using type I censoring for the Weibull distribution; *Mathematische Operationsforschung und Statistik, Series Statistics,* **21**, 227-237

NIGM, E.M. / EL–HAWARY, H.M. (1996): On order statistics from the Weibull distribution; *The Egyptian Statistical Journal* **40**, 80–92

NILSSON, A.A. / GLISSON, T.H. (1980): On the derivation and numerical evaluation of the Weibull–Rician distribution; *IEEE Transactions on Aerospace and Electronic Systems* **16**, 864–867

NIRUPAMA DEVI, K. / SRINIVASA RAO, K. / LAKSHMINARAYANA, J. (2001): Perishable inventory models with mixture of Weibull distributions having demand as power function of time; *Assam Statistical Review* **15**, 70–80

NORDEN, R.H. (1972): A survey of maximum likelihood estimation I; *International Statistical Review* **40**, 329–354

NORDEN, R.H. (1973): A survey of maximum likelihood estimation II; *International Statistical Review* **41**, 39–58

NORDMAN, D.J. / MEEKER, W.Q. JR. (2002): Weibull prediction intervals for a failure number of failures; *Technometrics* **44**, 15–23

NOSSIER, A. / EL–DEBEIKY, S. / HASHAD, I. / A–BARY, H. (1980): Effects of temperature on the breakdown property of dielectrics; *IEEE Transactions on Electrical Insulation;* **EI–15**, 502–505

NOUAKA, Y. / KANIE, H. (1985): Estimation of Weibull shape–parameters for two independent competing risks; *IEEE Transactions on Reliability* **34**, 53–56

**O-O-O-O-O**

OAKES, D. (1979): On log–linear hazards; *Journal of the Royal Statistical Society B* **41**, 274–275

OAKES, D. (1981): Survival times: Aspects of partial likelihood (with discussion); *International Statistical Review* **49**, 235–264

ODELL, P.M. / ANDERSON, K.M. / D'AGOSTINO, R.B. (1992): Maximum likelihood estimation for interval–censored data using a Weibull–based accelerated failure time model; *Biometrics* **48**, 951–959

OESTREICH, U.H.P. (1975): The application of the Weibull–distribution to the mechanical reliability of optical fibers for cables; *First European Conference on Optical Fibre Communication* **28**, 73–75

OFFICE OF THE ASSISTANT SECRETARY OF DEFENSE (ed.) (1961): Sampling procedures and tables for life and reliability testing based on the Weibull distribution (Mean life criterion); *Technical*

*Report 3*, Washington, D.C.

OFFICE OF THE ASSISTANT SECRETARY OF DEFENSE (ed.) (1962): Sampling procedures and tables for life and reliability testing based on the Weibull distribution (Hazard rate criterion); *Technical Report 4*, Washington, D.C.

OFFICE OF THE ASSISTANT SECRETARY OF DEFENSE (ed.) (1963): Sampling procedures and tables for life and reliability testing based on the Weibull distribution (Reliable life criterion); *Technical Report 6*, Washington, D.C.

OFFICE OF THE ASSISTANT SECRETARY OF DEFENSE (ed.) (1965): Factors and procedures for applying MIL-STD-105D sampling plans to life and reliability testing; *Technical Report 7*, Washington, D.C.

OFUSU, J.B. (1990): Design of experiments for selection from gamma and Weibull populations; *Pakistan Journal of Statistics A* **6**, 71–84

OGAWA, J. (1951): Contribution to the theory of systematic statistics I; *Osaka Mathematical Journal* **3**, 175–213

OGDEN, J.E. (1978): A Weibull shelf–life model for pharmaceuticals; *ASQC Technical Conference Transactions,* 574–580

OHL, H.L. (1974): *Weibull–Analysis* (in German); Schriftenreihe der Adam Opel AG No. 9, 2nd ed.; Rüsselsheim

ORD, J.K. (1972): *Families of Frequency Curves;* Charles Griffin, London

OSTROUCHOV, G. / MEEKER, W.Q. JR. (1988): Accuracy of approximate confidence bounds computed from interval censored Weibull and lognormal data; *Journal of Statistical Computation & Simulation* **29**, 43–76

OUYANG, L.Y. (1987): On characterizations of probability distributions based on conditional expected values; *Tamkang Journal of Mathematics* **18**, 113–122

ÖZTÜRK, A. / KORUKOGLU, S. (1988): A new test for the extreme value distribution; *Communications in Statistics — Simulation and Computation* **17**, 1375–1393

**P-P-P-P-P**

PACE, L. / SALVAN, A. (1987): A class of tests for exponentiality against monotone failure rate alternatives with incomplete data; *Journal of Statistical Planning and Inference* **16**, 249–256

PADGETT, W.J. / DURHAM, S.D. / MASON, M. (1995): Weibull analysis of the strength of carbon fibres using linear and power law models for the length effect; *Journal of Composite Materials* **29**, 1873–1884

PADGETT, W.J. / SPURRIER, J.D. (1985): On discrete failure models; *IEEE Transactions on Reliability* **34**, 253–256

PADGETT, W.J. / SPURRIER, J.D. (1990): Shewhart–type charts for percentiles of strength distributions; *Journal of Quality Technology* **22**, 283–288

PADGETT, W.J. / TSOKOS, C.P. (1979): Bayes estimation of reliability using an estimated prior distribution; *Operations Research* **27**, 1142–1157

PAGONIS, V. / MIAN, S.M. / KITIS, G. (2001): Fit of first order thermoluminescence glow peaks using the Weibull distribution function; *Radiation Protection Dosimetry* **93**, 11–17

PALISSON, F. (1989): Determination of the parameters of the Weibull model using an actuarial method (in French); *Revue de Statistique Appliquée* **37**, 5–39

PANCHANG, V.G. / GUPTA, R.C. (1989): On the determination of three–parameter Weibull MLE's; *Communications in Statistics — Simulation and Computation* **18**, 1037–1057

PANDEY, B.N. / MALIK, H.J. / SRIVASTVA, R. (1989): Shrinkage testimators for the shape parameter of Weibull distribution and type II censoring; *Communications in Statistics — Theory and Methods* **18**, 1175–1199

PANDEY, B.N. / SINGH, K.N. (1984): Estimating the shape parameter of the Weibull distribution by shrinkage towards an interval; *South African Statistical Journal* **18**, 1–11

PANDEY, M. (1983): Shrunken estimators of Weibull shape parameter in censored samples; *IEEE Transactions on Reliability* **32**, 200–203

PANDEY, M. (1987): Bayes estimator of survival probability from randomly censored observations with Weibull distribution of time to death; *Biometrical Journal* **29**,491–496

PANDEY, M. (1988): A Bayesian approach to shrinkage estimation of the scale parameter of a Weibull distribution; *South African Statistical Journal* **22**, 1–13

PANDEY, M. / FERDOUS, J. / UDDIN, M.B. (1991): Selection of probability distribution for life testing data; *Communications in Statistics — Theory and Methods* **20**, 1373-1388

PANDEY, M. / SINGH, U.S. (1993): Shrunken estimators of Weibull shape parameter from type II censored samples; *IEEE Transactions on Reliability* **42**, 81–86

PANDEY, M. / UPADHYAY, S.K. (1985): Bayes shrinkage estimators of Weibull Parameters; *IEEE Transactions on Reliability* **34**, 491–494

PANDEY, M. / UPADHYAY, S.K. (1986): Approximate prediction limit for Weibull failure based on preliminary test estimator; *Communications in Statistics — Theory and Methods* **15**, 241–250

PAPADOPOULOS, A.S. / TSOKOS, C.P. (1975): Bayesian confidence bounds for the Weibull failure model; *IEEE Transactions on Reliability* **24**, 21–26

PAPADOPOULOS, A.S. / TSOKOS, C.P. (1976): Bayesian analysis of the Weibull failure model with unknown scale and shape parameters; *Statistica Neerlandica* **36**, 547–560

PARK, K.S. / YEE, S.R. (1984): Present worth of service cost for consumer product warranty; *IEEE Transactions on Reliability* **33**, 424–426

PARK, W.J. (1983): Percentiles of pooled estimates of Weibull parameters; *IEEE Transactions on Reliability* **32**, 91–94

PARKHIDEH, B. / CASE, K.E. (1989): The economic design of a dynamic $\bar{x}$-control chart; *IEEE Transactions on Reliability* **21**, 313–323

PARR, V.B. / WEBSTER, J.T. (1965): A method for discriminating between failure density functions used in reliability predictions; *Technometrics* **7**, 1–10

PARR, W.C. / SCHUCANY, W.R. (1980): Minimum distance and robust estimation; *Journal of the American Statistical Association* **75**, 616–624

PARZEN, E. (1962): *Stochastic Processes*; Holden Day, San Francisco

PATEL, J.K. (1975): Bounds on moments of linear functions of order statistics from Weibull and other restricted families; *Journal of the American Statistic Association* **70**, 670–673

PATEL, J.K. / KAPADIA, C.H. / OWEN, D.B. (1976): *Handbook of Statistical Distributions;* Marcel Dekker, New York

PATEL, J.K. / READ, C.B. (1975): Bounds on conditional moments of Weibull and other monotone failure rate families; *Journal of the American Statistical Association* **70**, 238–244

PATEL, S.R. / PATEL, N.P. (1989): On shrinkage type Bayes estimators of the scale parameter of Weibull distribution; *Metron* **47**, 351–359

PATRA, K. / DEY, D.K. (1999): A multivariate mixture of Weibull distributions in reliability modelling; *Statistics & Probability Letters* **45**, 225–235

PAUL, S.R. / THIAGARAJAH, K. (1992): Hypothesis tests for the one–way layout of type II censored data from Weibull populations; *Journal of Statistical Planning and Inference* **33**, 367–380

PAVIA, E.J. / O'BRIEN, J.J. (1986): Weibull statistics of wind speed over the ocean; *Journal of Climate and Applied Meteorology* **25**, 1324–1332

PAWLAS, P. / SZYNAL, D. (2000a): Characterizations of the inverse Weibull distribution and generalized extreme value distributions by moments of $k$–th record values; *Applicationes Mathematicae* **27**, 197–202

PAWLAS, P. / SZYNAL, D. (2000b): Recurrence relations for single and product moments of the $k$–th record values from Weibull distributions, and a characterization; *Journal of Applied Statistical Science* **10**; 17–26

PAWLAS, P. / SZYNAL, D. (2001): Recurrence relations for single and product moments of lower generalized order statistics from the inverse Weibull distribution; *Demonstratio Mathematica* **34**, 353–358

PEARSON, E.S. / HARTLEY, H.O. (1966, 1972): *Biometrika Tables for Statisticians,* Vol. 1, 2; Cambridge University Press, Cambridge

PEDERSEN, J.G. (1978): Fiducial inference; *International Statistical Review* **46**, 147–170

PERRY, J.N. (1962): Semiconductor burn–in and Weibull statistics; *Semiconductor Reliability* **2**, 80–90

PETO, R. / LEE, P. (1973): Weibull distributions for continuous-carcinogenesis experiments; *Biometrics* **29**, 457–470

PETRUASKAS, A. / AAGARD, P.M. (1971): Extrapolation of historical storm data for estimating design wave heights; *Soc. of Pet. Engr.* **11**

PHANI, K.K. (1987): A new modified Weibull distribution function; *Communications of the American Ceramic Industry,***70**, 182–184

PIERRAT, L. (1992): Estimation of the failure probability by the convolution of two Weibull distributions (in French); *Revue de Statistique Appliquée* **40**, 5–13

PIKE, M.C. (1966): A new method of analysis of a certain class of experiments in carcinogenesis; *Biometrics* **22**, 142–161

PINDER, J.E. III. / WIENER, J.G. / SMITH, M.H. (1978): The Weibull distribution: A new method of summarizing survivorship data; *Ecology* **59**, 175–179

PLAIT, A. (1962): The Weibull distribution — With tables; *Industrial Quality Control* **19**, 17–26

PORTIER, C.J. / DINSE, G.E. (1987): Semiparametric analysis of tumor incidence rates in survival/sacrifice experiments; *Biometrics* **43**, 107–114

PRESCOTT, P. / WALDEN, A.T. (1980): Maximum likelihood estimation of the parameter of the generalized extreme–value distribution; *Biometrika* **67**, 723–724

PROCASSINI, A.A. / ROMANO, A. (1961): Transistor reliability estimates improved with Weibull distribution function; *Motorola Engineering Bulletin* **9**, 16–18

PROCASSINI, A.A. / ROMANO, A. (1962): Weibull distribution function in reliability analysis; *Semiconductor Reliability* **2**, 29–34

PROSCHAN, F. (1963): Theoretical explanation of observed decreasing failure rate; *Technometrics* **5**, 375–383

## Q-Q-Q-Q-Q

QIAO, H. / TSOKOS, C.P. (1994): Parameter estimation of the Weibull probability distribution; *Mathematics and Computers in Simulation* **37**, 47–55

QIAO, H. / TSOKOS, C.P. (1995): Estimation of the three–parameter Weibull probability distribution; *Mathematics and Computers in Simulation* **39**, 173–185

QUAYLE, R.J. (1963): Estimation of the scale parameter of the Weibull probability density function by use of one-order statistic; *Unpublished Thesis*, Air Force Institute of Technology, Wright–Patterson Air Force Base, Dayton, Ohio

QUESENBERRY, C.P. / QUESENBERRY, S. (1982): Selecting among Weibull, lognormal and gamma distribution using complete and censored samples; *Naval Research Logistics Quarterly* **29**, 557–569

QUREISHI, F.S. (1964): The discrimination between two Weibull processes; *Technometrics* **6**, 57–75

QUREISHI, F.S. / NABARVIAN, K.J. / ALANEN, J.D. (1965): Sampling inspection plans for discriminating between two Weibull processes; *Technometrics* **7**, 589–601

QUREISHI, F.S. / SHEIKH, A.K. (1997): Probabilistic characterization of adhesive wear in metals; *IEEE Transactions on Reliability* **46**, 38–44

## R-R-R-R-R

RADEMAKER, A.W. / ANTLE, C.E. (1975): Sample size for selection the better of two Weibull populations; *IEEE Transactions on Reliability* **24**, 17–20

RAFIG, M. / AHMAD, M. (1999): Estimation of the parameters of the two parameter Weibull distribution by the method of fractional moments; *Pakistan Journal of Statistics* **15**, 91–96

RAHIM, M.A. (1998): Economic design of $X$–control charts under truncated Weibull shock models; *Economic Quality Control* **13**, 79–105

RAIFFA, H. / SCHLAIFER, R. (1961): *Applied Statistical Decision Theory*; Harvard University Press,

RAJA RAO, B. / TALWALKER, S. (1989): Bounds on life expectancy for the Rayleigh and Weibull distributions; *Mathematical Bioscience* **96**, 95–115

RAMALHOTO, M.F. / MORAIS, M. (1998): EWMA control charts for the scale parameter of a Weibull control variable with fixed and variable sampling intervals; *Economic Quality Control* **13**, 23–46

RAMALHOTO, M.F. / MORAIS, M. (1999): Shewhart control charts for the scale parameters of a Weibull control variable with fixed and variable sampling intervals; *Journal of Applied Statistics* **26**, 129–160

RAMBERG, J.S. / TADIKAMALLA, P.R. (1974): An algorithm for generating gamma variates based on the Weibull distribution; *American Institute of Industrial Engineers Transactions* **6**, 257–260

RAMMLER, E. (1937): Laws governing the size–distribution of crushed materials (in German); *VDI–Beiheft Verfahrenstechnik*, 161–168

RAO, A.N.V. / TSOKOS, C.P. (1976): Bayesian analysis of the Weibull failure model under stochastic variation of the shape and scale parameters; *Metron* **34**, 201–217

RAO, A.V.D. / NARASIMHAM, V.L. (1989): Linear estimation in double Weibull distribution;

*Sankhyā B* **51**, 24–64

RAO, A.V. / RAO, A.V.D. / NARASIMHAM, V.L. (1991): Optimum linear unbiased estimation of the scale parameter by absolute values of order statistics in the double exponentiated and the double Weibull distributions; *Communications in Statistics — Simulation and Computation* **20**, 1139–1158

RAO, A.V. / RAO, A.V.D. / NARASIMHAM, V.L. (1994): Asymptotically optimal grouping for maximum likelihood estimation of Weibull parameters; *Communications in Statistics — Simulation and Computation* **23**, 1077–1096

RAO, B.R. / TALWALKER, S. / KUNDU, D. (1991 Confidence intervals for the relative relapse rate of placebo vs. Mercaptopurine group of acute leukemic patients; *Biometrical Journal* **33**, 579–598

RAO, C.R. / SHANBHAG, D.N. (1998): Recent approaches to characterizations based on order statistics and record values; in BALAKRISHNAN / RAO: *Order Statistics — Theory and Methods* (Handbook of Statistics, **16**), North–Holland, Amsterdam, 231–256

RAO, P. / OLSON, R. (1974): Tunnelling machine research: Size distribution of rock fragments produced by rolling disk cutter; *Bureau of Mines Report, No. BUMINES–RI–78–82*, Washington, D.C.

RAO TUMMALA, V.M. (1980): Minimum expected loss estimators of reliability and shape parameter of Weibull distributions; *Industrial Mathematics* **30**, 61–67

RATNAPARKHI, M.V. / PARK, W.J. (1986): Lognormal distribution: Model for life and residual strength of composite materials; *IEEE Transactions on Reliability* **35**, 312–315

RAUHUT, J. (1987): Estimating the lifetime distribution — especially the Weibull distribution — from observation in workshops (in German); *Schriften für Operations Research und Datenverarbeitung im Bergbau* **9**, Essen

RAVENIS, J.V.J. II. (1963): Life testing: Estimating the parameters of the Weibull distribution; *IEEE International Convention Record* **6**, 18–33

RAVENIS, J.V.J. II. (1964): Estimating Weibull–distribution parameters; *Electro–Technology,* 46–54

RAWLINGS, J.O. / CURE, W.W. (1985): The Weibull function as a dose–response model to describe ozone effects on crop yields; *Crop Science* **25**, 807–814

REICHELT, C. (1978): Estimating the parameters of the Weibull distribution (in German); *Fortschrittsberichte der VDI–Zeitschriften*, Reihe 1 **56**, Düsseldorf

REISER, B. / BAR LEV, S. (1979): Likelihood inference for life test data; *IEEE Transaction on Reliability* **28**, 38–42

REVFEIM, K.J.A. (1983): On the analysis of extreme rainfalls; *Journal of Hydrology* **22**, 107–117

REVFEIM, K.J.A. (1984): Generating mechanisms of, and parameters for, the extreme value distribution; *Australian Journal of Statistics* **26**, 151–159

REYROLLE, J. (1979): Viellissement et maintenance d'un système fortement modulaire: Estimation et factorisation des paramètres d'un modèle de Weibull; *Revue de Statistique Appliquée* **27**, 5–14

RIDER, P.R. (1961): Estimating the parameters of mixed Poisson, binomial, and Weibull distributions by the method of moments; *Bulletin de l'Institut International de Statistique* **39**, 225–232

RIGDON, S.E. / BASU, A.P. (1988): Estimating the intensity function of a Weibull process at the current time: Failure truncated case; *Journal of Statistical Computing & Simulation* **30**, 17–38

RIGDON, S.E. / BASU, A.P. (1989): The power law process: A model for the reliability of repairable systems; *Journal of Quality Technology* **21**, 251–260

RIGDON, S.E. / BASU, A.P. (1990a): Estimating the intensity function of a power law process at the current time: Time truncated case; *Communications in Statistics — Simulation and Computation* **19**, 1079–1104

RIGDON, S.E. / BASU, A.P. (1990b): The effect of assuming a homogeneous Poisson process when the true process is a power law process; *Journal of Quality Technology* **22**, 111–117

RINGER, L.J. / SPRINKLE, E.E. (1972): Estimation of the parameters of the Weibull distribution from multicensored samples; *IEEE Transaction on Reliability* **21**, 46–51

RINK, G. / DELL, T.R. / SWITZER, G. / BONNER, F.T. (1979): Use of the [three–parameter] Weibull distribution to quantify sweetgum germinating data; *Sivae Genetica* **28**, 9–12

RINNE, H. (1972a): Reliability parameters (in German); *Zeitschrift für wirtschaftliche Fertigung* **67**, 193–198

RINNE, H. (1972b): Stand–by coefficients (in German); *Zeitschrift für wirtschaftliche Fertigung* **67**, 431–435

RINNE, H. (1973): Optional preventative strategies for maintenance (in German); *Zeitschrift für Operations Research* **17**, 13–24

RINNE, H. (1974): Control and renewal of stocks being in reserve (in German); *Statistische Hefte* **16**, 256–275

RINNE, H. (1975): *Maintenance Strategies: A Contribution to the Theory of Maintenance* (in German); Anton Hain, Meisenheim

RINNE, H. (1978): Classification of life testing plans (in German); *Zeitschrift für wirtschaftliche Fertigung* **73**, 55–58

RINNE, H. (1979): Maintenance and renewal theory (in German); in BECKMANN, M.J. (ed.): *Handwörterbuch der Mathematischen Wirtschaftswissenschaften*, Vol. 3; Gabler, Wiesbaden, 79–83

RINNE, H. (1981a): Optimizing censored life testing of Type II for the one–parameter exponential distribution (in German); *Qualität und Zuverlässigkeit* **26**, 370–374

RINNE, H. (1981b): Parameter estimation for the Weibull distribution (in German); in FANDEL, H. et al. (eds.): *Operations Proceedings 1980;* Springer, Heidelberg, 171–180

RINNE, H. (1985): Estimating the lifetime distribution of private motor–cars using prices of used cars: The Teissier model (in German); in BUTTLER, G. et al. (eds.): *Statistik zwischen Theorie und Praxis (Festschrift f[u]r K.-A. Schäffer);* Vandenhoeck & Ruprecht, Göttingen, 172–184

RINNE, H. (1988): On the relationship between some notions in measuring economic concentration and life testing (in German); *Allgemeines Statistisches Archiv* **72**, 378–198

RINNE, H. (1995): Genesis of the Weibull distribution (in German); in RINNE, H. / RÜGER, B. / STRECKER, H. (eds.): *Grundlagen der Statistik und ihre Anwendungen (Festschrift für Kurt Weichselberger);* Physica, Heidelberg, 76–86

RINNE, H. (2004): *Ökonometrie*; Verlag Vahlen, München

RINNE, H. / DIETERLE, H. / STROH, G. (2004): Capability indices for asymmetrical distributions (in German); *Qualität und Zuverlässigkeit* **49, No. 11**, 27–31

ROBBINS, H. (1955): An empirical Bayes approach to statistics; *Proceedings of the 3rd Berkeley Symposium on Mathematical Statistics and Probability*, 157–163

ROBERTS, H.R. (1975): Life tests experiments with hypercensored samples; *Proceedings of the 18th Annual Quality Control Conference,* Rochester Society for Quality Control

ROBINSON, J.A. (1983): Bootstrap confidence intervals in location–scale models with progressive censoring; *Technometrics* **25**, 179–187

ROCA, J.L. (1979a): Using the Weibull distribution function for the analysis of data obtained by accelerated tests; *Rev. Telegr. Electron.* **67**, 45–48

ROCA, J.L. (1978b): Weibull distribution function I, Introduction; *Rev. Telegr. Electron.* **66**, 775–779

ROCA, J.L. (1978c): The Weibull distribution function II; *Rev. Telegr. Electron.* **67**, 920–922

ROCKETTE, H.E. / ANTLE, C.E. / KLIMKO, L.A. (1974): Maximum likelihood estimation with the Weibull model; *Journal of the American Statistical Association* **69**, 246–249

RODRIGUEZ, R.N. (1977): A guide to the Burr type XII distribution; *Biometrika* **64**, 129–134

RÖNNEBECK, H. / TITTES, E. (1994): Construction of confidence intervals (in German); *Qualität und Zuverlässigkeit* **39**, 666–670

ROSE, M.S. / GILLIS, A.M. / SHELDON, R.S. (1999): Evaluation of the bias in using the time to the first event when the inter–event intervals have a Weibull distribution; *Statistics in Medicine* **18**, 139–154

ROSIN, P. / RAMMLER, E. (1933): The laws governing the fineness of powdered coal; *Journal of the Institute of Fuels* **6**, 29–36

ROSIN, P. / RAMMLER, E. / SPERLING, K. (1933): Size of powdered coal and its meaning to grinding (in German); *Bericht C 52 des Reichskohlenrats;* Berlin

ROSS, R. (1996): Bias and standard deviation due to Weibull parameter estimation for small data sets; *IEEE Transactions on Dielectric and Electric Insulation* **3**, 18–42

ROSS, S.M. (1970): *Applied Probability Models With Optimization Applications*; Holden–Day, San Francisco

ROSS, S.M. (1980): *Introduction to Probability Models*, 2nd ed.; Academic Press, New York

ROUSU, D.D. (1973): Weibull skewness and kurtosis as a function of the shape parameter; *Technometrics* **15**, 927–930

ROY, D. (1984): A characterization of the generalized gamma distribution; *Calcutta Statistical Association Bulletin* **33**, 137–141 (Correction, *Ibid.*, **34**, 125)

ROY, D. / GUPTA, R.P. (1992): Classifications of discrete lives; *Microelectronics and Reliability* **32**, 1459–1473

ROY, D. / MUKHERJEE, S.P. (1986): A note on the characterizations of the Weibull distribution; *Sankhyā A* **48**, 250–253

ROY, D. / MUKHERJEE, S.P. (1988): Multivariate extensions of univariate life distributions; *Journal of Multivariate Analysis* **67**, 72–79

ROY, S.D. / KAKOTY, S. (1997): An economic design of CUSUM control chart for controlling normal means with Weibull in–control time distribution; *Journal of the Indian Association for Production, Quality and Reliability* **22**, 23–39

S-S-S-S-S

SAHU, S.K. / DEY, D.K. / ASLANIDOU, H. / SINHA, D. (1997): A Weibull regression model with gamma frailties for multivariate survival data; *Lifetime Data Analysis* **3**, 123–137

SALVIA, A.A. (1979): Consonance sets for 2–parameter Weibull and exponential distributions; *IEEE Transactions on Reliability* **28**, 300–302

SALVIA, A.A. (1996): Some results on discrete mean residual life; *IEEE Transactions on Reliability* **45**, 359–361

SALVIA, A.A. / BOLLINGER, R.C. (1982): On discrete hazard functions; *IEEE Transactions on Reliability* **31**, 458–459

SANDHYA, E. / SANTEESH, S. (1997): On exponential mixtures, mixed Poisson processes and generalized Weibull and Pareto models; *Journal of the Indian Statistical Association* **35**, 45–50

SANDOH, H. / FUJII, S. (1991): Designing an optimal life test with Type I censoring; *Naval Research Logistics Quarterly* **38**, 23–31

SARHAN, A.E. / GREENBERG, B.G. (1962): *Contributions to Order Statistics;* Wiley, New York

SARKAR, S.K. (1987): A continuous bivariate exponential distribution; *Journal of the American Statistical Association* **82**, 667–675

SAUNDERS, S.C. (1968): On the determination of a safe life for distributions classified by failure rate; *Technometrics* **10**, 361–377

SAVAGE, L.J. (1962): *The Foundations of Statistical Inference: A Discussion*; Methuen, London

SAYLOR, F. (1977): A Monte Carlo comparison of the method–of–moments to the maximum–likelihood estimates of Weibull parameters for CAS data; in TSOKOS, C.P. / SHIMI, I.N., (eds.): *The Theory and Applications of Reliability,* Vol. II, Academic Press, New York, 517–529

SCALLAN, A.J. (1999): Regression modelling of interval–censored failure time data using the Weibull distribution; *Journal of Applied Statistics* **26**, 613–618

SCHAFER, D.B. (1975): A note on a simple test function for the Weibull distribution location parameter; *Scandinavian Actuarial Journal,* 1–5

SCHAFER, R.B. (1974): Confidence bands for the minimum fatigue life; *Technometrics* **16**, 113–123

SCHAFER, R.E. (1966): Availability of the standardized Weibull distribution; *IEEE Transaction on Reliability* **15**, 132–133

SCHAFER, R.E. / ANGUS, J.E. (1979): Estimation of Weibull quantiles with minimum error in the distribution function; *Technometrics* **21**, 367–370

SCHAFER, R.E. / SHEFFIELD, T.S. (1976): On procedures for comparing two Weibull populations; *Technometrics* **18**, 231–235

SCHMID, F. / TREDE, M. (2003): Simple tests for peakedness, fat tails and leptokurtosis based on quantiles; *Computational Statistics & Data Analysis* **43**, 1–12

SCHMID, U. (1997): Percentile estimators for the three–parameter Weibull distribution for use when all parameters are unknown; *Communications in Statistics — Theory and Methods* **26**, 765–785

SCHMITTLEIN, D.C. / MORRISON, D.G. (1981): The median residual lifetime: A characterization theorem and an application; *Operation Research* **29**, 392–399

SCHNEIDER, H. (1989): Failure–censored variables–sampling plans for lognormal and Weibull distributions; *Technometrics* **31**, 199–206

SCHNEIDER, H. / WEISSFELD, L.A. (1989): Interval estimation based on censored samples from the Weibull distribution; *Journal of Quality Technology* **21**, 179–186

SCHOLZ, F.–W. (1990): Characterization of the Weibull distribution; *Computational Statistics & Data Analysis* **10**, 289–292

SCHREIBER, H.H. (1963): Mathematical–statistical evaluation of roller bearing lifetime experiments (in German); *Qualitätskontrolle* **8**, 59–66, 71–79

SCHULZ, W. (1983): The planning of sample sizes for maximum likelihood estimation of the parameters of the Weibull distributions; *Biometrical Journal* **25**, 783–789

SCHÜPBACH, M. / HÜSLER, J. (1983): Simple estimators for the parameters of the extreme–value distribution based on censored data; *Technometrics* **25**, 189–192

SCHÜTTE, T. / SALKA, O. / ISRAELSSON, S. (1987): The use of the Weibull distribution for thunderstorm parameters; *Journal of Climate and Applied Meteorology* **26**, 457–463

SCHWENKE, J.R. (1987): Using the Weibull distribution in modelling pharmacokinetic data; *ASA Proceedings of the Biopharmaceutical Section,* 104–109

SEGURO, J.V. / LAMBERT, T.W. (2000): Modern estimation of the parameters of the Weibull wind speed distribution; *Journal of Wind Engineering and Industrial Aerodynamics* **85**, 75–84

SEKI, T. / YOKOYAMA, S.–I. (1996): Robust parameter estimation using the bootstrap method for two–parameter Weibull distribution; *IEEE Transactions on Reliability* **45**, 34–41

SEKINE, M. / MUSHA, T. / TOMITA, Y. / HAGISAWA, T. / IRABU, T. / KIUCHI, E. (1979): On Weibull–distributed weather clutter; *IEEE Trans. Aerosp. and Electron. Syst.* **15**, 824–830

SELKER, J.S. / HAITH, D.A. (1990): Development and testing of single–parameter precipitation distribution; *Water Resources Research* **26**, 2733–2740

SEN, I. / PRABHASHANKER, V. (1980): A nomogram for estimating the three parameters of the Weibull distribution; *Journal of Quality Technology* **12**, 138–143

SEO, S.–K. / YUM, B.–J. (1991) Accelerated life test plans under intermittent inspection and type–I censoring: The case of Weibull failure distribution; *Naval Research Logistics Quarterly* **38**, 1–22

SEYMORE, G.E. / MAKI, W.R. (1969). Maximum likelihood estimation of the Weibull distribution using nonlinear programming; Paper presented at the 36th ORSA National Meeting

SHAHANI, A.K. / NEWBOLD, S.B. (1972): An inspection policy for the detection of failure in the Weibull case; *Quality Engineering* **36**, 8–10

SHAKED, M. / LI, M. (1997): Aging first–passage times; in KOTZ, S. / READ, C.B. / BANKS, D.L. (eds.): *Encyclopedia of Statistical Sciences*, Update Volume 1, Wiley, New York, 11–20

SHALABY, O.A. (1993): The Bayes risk for the parameters of doubly truncated Weibull distributions; *Microelectronics and Reliability* **33**, 2189–2192

SHALABY, O.A. / AL–YOUSSEF, M.H. (1992): Bayesian analysis of the parameters of a doubly truncated Weibull distribution; *The Egyptian Statistical Journal* **36**, 39–56

SHANBHAG, D.N. (1970): The characterizations for exponential and geometric distributions; *Journal of the American Statistical Association* **65**, 1256–1259

SHAPIRO, S.S. / BRAIN, C.W. (1987): W–test for the Weibull distribution; *Communications in Statistics — Simulation and Computation* **16**, 209–219

SHARIF, H. / ISLAM, M. (1980): Weibull distribution as a general model for forecasting technological change; *Technological Forecasting and Social Change* **18**, 247–256

SHEIKH, A.K. / BOAH, J.K. / HANSEN, D.A. (1990): Statistical modelling of pitting corrosion and pipeline reliability; *Corrosion* **46**, 190–196

SHERIF, A. / TAN, P. (1978): On structural predictive distribution with type II progressively censored Weibull data; *Statistische Hefte* **19**, 247–255

SHESHADRI, V. (1968): A characterization of the normal and Weibull distributions; *Canadian Math. Bull.* **12**, 257–260

SHIER, D.R. / LAWRENCE, K.D. (1984): A comparison of robust regression techniques for the estimation of Weibull parameters; *Communications in Statistics — Simulation and Computation* **13**, 743–750

SHIMIZU, R. / DAVIES, L. (1981): General characterization theorems for the Weibull and the stable distributions; *Sankhyā A* **43**, 282–310

SHIMOKAWA, T. / LIAO, M. (1999): Goodness–of–fit tests for type-I extreme–value and 2–parameter Weibull distribution; *IEEE Transactions on Reliability* **48**, 79–86

SHIOMI, H. / NAKAHARA, M. (1963): On confidence of Weibull parameter estimated by sight from small number of samples I (in Japanese); *Bulletin of the Electrotechnical Laboratory* **30**, 655–663

SHIOMI, H. / TAKAGI, S. / HARADA, M. (1963): On confidence of Weibull parameter by sight from small number of samples II (in Japanese); *Bulletin of the Electrotechnical Laboratory* **30**, 753–765

SHOR, M. / LEVANON, N. (1991): Performance of order statistics CFAR; *IEEE Transactions on Aerospace and Electronic Systems* **27**, 214–224

SHORACK, G.R. (1972): The best test of exponentiality against gamma alternatives; *Journal of the American Statistical Association* **67**, 213–214

SIEGEL, G. / WÜNSCHE, S. (1979): Approximating the renewal function (in German); *Mathematische Operationsforschung und Statistik, Series Optimization* **10**, 265–275

SILCOCK, H. (1954): The phenomenon of labour turnover; *Journal of the Royal Statistical Society A* **117**, 429–440

SINGH, H. / SHUKLA, S.K. (2000): Estimation in the two–parameter Weibull distribution with prior information; *Journal of the Indian Association for Production, Quality and Reliability* **25**, 107–118

SINGH, J. / BHATKULIKAR, S.G. (1978): Shrunken estimation in Weibull distribution; *Sankhyā B* **39**, 382–393

SINGH, V.P. / CRUISE, J.F. / MA, M. (1990): A comparative evaluation of the estimators of the Weibull distribution by Monte Carlo simulation; *Journal of Statistical Computation & Simulation* **36**, 229–241

SINGPURWALLA, N.D. (1971): Statistical fatigue models: A survey; *IEEE Transactions on Reliability* **20**, 185–189

SINGPURWALLA, N.D. (1988): An interactive PC–based procedure for reliability assessment incorporating expert opinion and survival data; *Journal of the American Statistical Association* **83**, 43–51

SINGPURWALLA, N.D. / AL–KHAYYAL, F.A. (1977): Accelerated life tests using the power law model for the Weibull model; in TSOKOS, C.P. / SHIMI, I.N. (eds.): *The Theory and Applications of Reliability,* Vol. II, Academic Press New York, 381–399

SINGPURWALLA, N.D. / CASTELLINO, V.C. / GOLDSCHEN, D.Y. (1975): Inference from accelerated life tests using Eyring type re–parameterizations; *Naval Research Logistics Quarterly* **22**, 289–296

SINGPURWALLA, N.D. / SONG, M.S. (1988): Reliability analysis using Weibull lifetime data and expert opinion; *IEEE Transactions on Reliability* **37**, 340–347

SINHA, S.K. (1982): Reliability estimation with type I censoring of a mixture of Weibull distributions; *Publications of the Institute of Statistics, University Paris IV,* **27**, 75–92

SINHA, S.K. (1986a): Bayesian estimation of the reliability function and hazard rate of a Weibull failure time distribution; *Trabajos de Estadística* **37**, 47–56

SINHA, S.K. (1986b): *Reliability and Life Testing*; Wiley, New York

SINHA, S.K. (1987): Bayesian estimation of the parameters and reliability function of a mixture of Weibull life distributions; *Journal of Statistical Planning and Inference* **16**, 377–387

SINHA, S.K. / GUTTMAN, I. (1988): Bayesian analysis of life–testing problems involving the Weibull distribution; *Communications in Statistics — Theory and Methods* **17**, 343–356

SINHA, S.K. / SLOAN, J.A. (1988): Bayesian estimation of the parameters and reliability function of the 3–parameter Weibull distribution; *IEEE Transactions on Reliability* **37**, 364–369

SINHA, S.K. / SLOAN, J.A. (1989): Prediction interval for a mixture of Weibull failure time distributions: A Bayesian approach; *South African Statistical Journal* **23**, 119–130

SIRVANCI, M. (1984): An estimator for the scale parameter of the two parameter Weibull distribution for type I singly right–censored data; *Communications in Statistics — Theory and Methods* **13**, 1759–1768

SIRVANCI, M. / YANG, G. (1984): Estimation of the Weibull parameters under type I censoring; *Journal of the American Statistical Association* **79**, 183–187

SKINNER, C.J. / HUMPHREYS, K. (1999): Weibull regression for lifetimes measured with errors; *Lifetime Data Analysis* **5**, 23–37

SLYMEN, D.J. / LACHENBRUCH, P.A. (1984): Survival distributions arising from two families and generated by transformations; *Communications in Statistics — Theory and Methods* **13**, 1179–1201

SMITH, R.L. (1985): Maximum likelihood estimation in a class of nonregular cases; *Biometrika* **72**, 67–90

SMITH, R.L. / NAYLOR, J.C. (1987): A comparison of maximum likelihood and Bayesian estimators for the three–parameter Weibull distribution; *Applied Statistics* **36**, 358–369

SMITH, R.M. (1977): Some results on interval estimation for the two parameter Weibull or extreme–value distribution; *Communications in Statistics — Theory and Methods* **6**, 1311–1321

SMITH, R.M. / BAIN, L.J. (1975): An exponential power life–testing distribution; *Communications in Statistics — Theory and Methods* **4**, 469–481

SMITH, R.M. / BAIN, L.J. (1976): Correlation type goodness–of–fit statistics with censored sampling; *Communications in Statistics — Theory and Methods* **5**, 119–132

SMITH, W.L. (1958): Renewal theory and its ramifications; *Journal Royal Statistical Society B* **20**, 243–302

SMITH, W.L. (1959): On the cumulants of renewal processes; *Biometrika* **46**

SMITH, W.L. / LEADBETTER, M.R. (1963): On the renewal function for the Weibull distribution; *Technometrics* **5**, 393–396

SOLAND, R.M. (1966): Use of the Weibull distribution in Bayesian decision theory; *Technical Paper RAC–TP 225,* Research Analysis Corporation, McLean, VA

SOLAND, R.M. (1968a): Renewal functions for gamma and Weibull distribution with increasing hazard rate; *Technical Paper RAC–TP 329,* Research Analysis Corporation, McLean, VA

SOLAND, R.M. (1968b): Bayesian analysis of the Weibull process with unknown scale parameter and its application to acceptance sampling; *IEEE Transactions on Reliability* **17**, 84–90

SOLAND, R.M. (1968c): A renewal theoretic approach to the estimation of future demand for replacement parts; *Operations Research* **16**, 36–51

SOLAND, R.M. (1969a): Bayesian analysis of the Weibull process with unknown scale and shape parameters; *IEEE Transactions on Reliability* **18**, 181–184

SOLAND, R.M. (1969b): Availability of renewal functions for gamma and Weibull distributions with increasing hazard rate; *Operations Research* **17**, 536–543

SOMERVILLE, P.N. (1977): Some aspects of the use of the Mann–Fertig statistic to obtain confidence interval estimates for the treshold parameter of the Weibull; in TSOKOS, C.P. / SHIMI, I.N., (eds.): *The Theory and Applications of Reliability,* Vol. I, Academic Press, New York, 423–432

SPLITSTONE, D. (1967): Estimation of the Weibull shape and scale parameters; *Master thesis*, Iowa State University

SPURRIER, J.D. / WEIER, D.R. (1981): Bivariate survival model derived from a Weibull distribution; *IEEE Transactions on Reliability* **30**, 194–197

SRINIVASAN, R. / WHARTON, R.M. (1975): Confidence bands for the Weibull distribution; *Technometrics* **17**, 375–380

SRIVASTAVA, J. (1987): More efficient and less time–consuming designs for life testing; *Journal of Statistical Planning and Inference* **16**, 389–413

SRIVASTAVA, J. (1989): Advances in the statistical theory of comparison of lifetimes of machines under the generalized Weibull distribution; *Communications in Statistics — Theory and Methods* **18**, 1031–1045

SRIVASTAVA, M.S. (1967): A characterization of the exponential distribution; *American Mathematical Monthly* **74**, 414–416

SRIVASTAVA, T.N. (1974): Life tests with periodic change in the scale parameter of a Weibull distribution; *IEEE Transactions on Reliability* **23**, 115–118

STACY, E.W. (1962): A generalization of the gamma distribution; *Annals of Mathemagtical Stastistics* **33**, 1187–1192

STACY, E.W. / MIHRAM, G.A. (1965): Parameter estimation for a generalized gamma distribution; *Technometrics* **7**, 349–357

STALLARD, N. / WHITEHEAD, A. (2000): Modified Weibull multi–state models for the analysis of animal carciogenicity; *Environmental and Ecological Statistics* **7**, 117–133

STANGE, K. (1953a): A generalization of the Rosin–Rammler–Sperling size distribution for finely ground material (in German); *Mitteilungsblatt für Mathematische Statistik* **5**, 143–158

STANGE, K. (1953b): On the laws of size distribution in grinding processes (in German); *Ingenieur–Archiv* **21**, 368–380

STAUFFER, H.B. (1979): A derivation of the Weibull distribution; *Journal of Theoretical Biology* **81**, 55–63

STEIN, W.E. / DATTERO, R. (1984): A new discrete Weibull distribution; *IEEE Transactions on Reliability* **33**, 196–197

STEINECKE, V. (1979): *Weibull Probability Paper — Explanation and Handling* (in German); 2nd ed.; DGQ–Schrift Nr. 17–25, Beuth, Berlin

STEPHENS, M.A. (1977): Goodness–of–fit for the extreme value distribution; *Biometrika* **64**, 583–588

STOCK, J.H. / WATSON, M.W. (2003): *Introduction to Econometrics*; Addison–Wesley, Boston

STONE, G.C. / ROSEN, H. (1984): Some graphical techniques for estimating Weibull confidence intervals; *IEEE Transactions on Reliability* **33**, 362–369

STONE, G.C. / VAN HEESWIJK, R.G. (1977): Parameter estimation for the Weibull distribution; *IEEE Trans. Electr. Insul.* **12**, 253–261

STRUTHERS, C.A. / FAREWELL, V.T. (1989): A mixture model for time to AIDS data with left truncation and uncertain origin; *Biometrika* **76**, 814–817

STUART, A. / ORD, J.K. (1991): *Kendall's Advance Theory of Statistics*; Edward Arnold, London

STUMP, F.B. (1968): Nearly best unbiased estimation of the location and scale parameters of the Weibull distribution by use of order statistics; *Master thesis*, Air Force Institute of Technology, Wright–Patterson Air Force Base, Dayton, OH

SUGIURA, N. / GOMI, A. (1985): Pearson diagrams for truncated normal and truncated Weibull distributions; *Biometrika* **72**, 219–222

SULTAN, K.S. / BALAKRISHNAN, N. (1999/2000): Higher order moments of record values from Rayleigh and Weibull distributions and Edgeworth approximate inference; *Journal of Applied Statistical Science* **9**, 193–209

SULTAN, K.S. et al. (2007): Mixture of two inverse Weibull distributions: Properties and estimation; *Computational Statistics & Data Analysis* **51**, 5377–5387

SUN, D. (1997): A note on noninformative priors for Weibull distributions; *Journal of Statistical Planning and Inference* **61**, 319–338

SUN, D. / BERGER, J.O. (1994): Bayesian sequential reliability for Weibull related distributions; *Annals of the Institute of Statistical Mathematics* **46**, 221–249

SUZUKI, Y. (1988): On selection of statistical models; in MATUSITA, K. (ed.): *Statistical Theory and Data Analysis II*, North–Holland, Amsterdam. 309–325

SWARTZ, G.B. (1973): The mean residual lifetime function; *IEEE Transactions on Reliability* **22**, 108–109

SWARTZ, G.B. / MURTHY, V.K. (1971): Further contributions to reliability; *ARL Technical Report No. 71–0261*

**T-T-T-T-T**

TADIKAMALLA, P.R. (1978): Application of the Weibull distribution in inventory control; *Journal of Operational Research Society* **29**, 77–83

TADIKAMALLA, P.R. (1980a): A look at the Burr and related distributions; *International Statistical Review* **48**, 337–344

TADIKAMALLA, P.R. (1980b): Age replacement policies for Weibull failure times; *IEEE Transactions on Reliability* **29**, 88–90 (Correction, *Ibid.*, **35**, 174)

TADIKAMALLA, P.R. / SCHRIBER, T.J. (1977): Sampling from the Weibull and gamma distributions in GPSS; *Simuletter* **9**, 39–45

TAILLIE, C. / PATIL, G.P. / BALDEROSSI, B. (eds.): (1981): *Statistical Distributions in Scientific Work* (Proceeding of the NATO Advanced Study Institute, Trieste, Italy, July–Aug. 1980)
Vol. 4 – Models, Structure and Characterizations
Vol. 5 – Inferential Problems and Properties
Vol. 6 – Applications in Physical, Social and Life Sciences
Reidel, Dordrecht

TALKNER, P. / WEBER, R.O. (2000): Power spectrum and detrended fluctuation analysis: Application to daily temperatures; *Physical Review* **E 62**, 150–160

TALWALKER, S. (1977): A note on characterization by conditional expectation; *Metrika* **23**, 129–136

TAN, P. / SHERIF, A. (1974): Some remarks on structural inference applied to Weibull distribution; *Statistische Hefte* **15**, 335–341

TANAKA, H. (1998): On reliability estimation in the Weibull case; *Journal of the Japanese Statistical Society* **28**, 193–203

TANG, D. (1989): Confidence interval estimation for the coefficient of variation of the Weibull distribution (in Chinese); *Chinese Journal of Applied Probability and Statistics* **5**, 276–282

TANG, L.C. / LU, Y. / CHEW, E.P. (1999): Mean residual life of lifetime distributions; *IEEE Transactions on Reliability* **48**, 73–78

TARUMOTO, M.H. / WADA, C.Y. (2001): A bivariate Weibull and its competing risk models; *Brazilian Journal of Probability and Statistics* **15**, 221–239

TATE, R.F. (1959): Unbiased estimation: Functions of location and scale parameters; *Annals of Mathematical Statistics* **30**, 341–366

TAYLOR, J.A. / JAKEMAN, A.J. (1985): Identification of a distributional model; *Communications in Statistics — Simulation and Computation* **14**, 363–371

THEILER, G. / TÖVISSI, L. (1976): Limit laws for extreme values, Weibull distribution and some statistical problems of control global analysis; *Economic Computation and Economic Cybernetics Studies and Research* **4**, 39–61

THIAGARAJAN, T.R. / HARRIS, C.M. (1976): A ratio–type goodness–of–fit test for 2–parameter Weibull distributions; *IEEE Transactions on Reliability* **25**, 340–343

THOMAN, D.R. / BAIN, L.J. (1969): Two samples tests in the Weibull distribution; *Technometrics* **11**, 805–815

THOMAN, D.R. / BAIN, L.J. / ANTLE, C.E. (1969): Inferences on the parameters of the Weibull distribution; *Technometrics* **11**, 445–460

THOMAN, D.R. / BAIN, L.J. / ANTLE, C.E. (1970): Maximum likelihood estimation, exact confidence intervals for reliability and tolerance limits in the Weibull distribution; *Technometrics* **12**, 363–371

THOMAN, D.R. / WILSON, W.M. (1972): Linear order statistic estimation for the two–parameter Weibull and extreme–value distributions from Type II progressively censored samples; *Technometrics* **14**, 679–691

THOMPSON, J.R. (1968): Some shrinkage techniques for estimating the mean; *Journal of the American Statistical Association* **63**, 113–123

THORELLI, H.B. / HIMMELBAUER, W.G. (1967): Executive salaries: Analysis of dispersion pattern; *Metron* **26**, 114–149

TIAGO DE OLIVEIRA, J. (1981): Statistical choice of univariate extreme models; in TAILLIE, C. / PATIL, G.P. / BALDESSARI, B.A. (eds.): *Statistical Distributions in Scientific Work,* Vol. 6, Reidel, Dordrecht, 367–387

TIAGO DE OLIVEIRA, J. / LITTAUER, S.B. (1976): Mean square invariant forecasters for the Weibull distribution; *Naval Research Logistics Quarterly* **23**, 487–511

TIKU, M.L. / SINGH, M. (1981): Testing the two parameter Weibull distribution; *Communications in Statistics A* **10**, 907–918

TIPPETT, L.H.C. (1925): On the extreme individuals and the range of samples taken from a normal population; *Biometrika* **17**, 364–387

TITTERINGTON, M. / SMITH, A.F.M. / MAKOV, V.E. (1985); *Statistical Analysis of Finite Mixture Distributions;* Wiley, New York

TITTES, E. (1973): Exploitation of life tests by means of the Weibull distribution (in German); *Qualität und Zuverlässigkeit* **18**, 108–113, 163–165

TITTES, E. (1976): Statistical methods as a means to analyze results of experiments (in German); *Bosch Technische Berichte* **5**, 208–215

TSANG, A.H.C. / JARDINE, A.K.S. (1993): Estimation of 2–parameter Weibull distribution from incomplete data with residual lifetimes; *IEEE Transactions on Reliability* **42**, 291–298

TSE, S.-K. / YANG, C. / YUEN, H.-K. (2000): Statistical analysis of Weibull distributed lifetime data and type I progressive censoring with binomial removals; *Journal of Applied Statistics* **27**, 1033–1043

TSE, S.-K. / YUEN, H.-K. (1998): Expected experiment times for the Weibull distribution under progressive censoring with random removals; *Journal of Applied Statistics* **25**, 75–83

TSENG, S.-T. (1988): Selecting more reliable Weibull populations than a control; *Communications in Statistics — Theory and Methods* **17**, 169–181

TSENG, S.-T. / WU, H.-J. (1990): Selecting, under type–II censoring, Weibull populations that are more reliable; *IEEE Transactions on Reliability* **39**, 193–198

TSIONAS, E.G. (2000): Posterior analysis, prediction and reliability in three–parameter Weibull distributions; *Communications in Statistics — Theory and Methods* **29**, 1435–1449

TSIONAS, E.G. (2002): Bayesian analysis of finite mixtures of Weibull distributions; *Communications in Statistics — Theory and Methods* **31**, 37–118

TSOKOS, C.P. (1972): Bayesian approach to reliability using the Weibull distribution with unknown parameters and its computer simulation; *Reports of Statistical Application Research (Union of Japanese Scientists and Engineers)* **19**, 123–134

TSOKOS, C.P. / RAO, A.N.V. (1976): Bayesian analysis of the Weibull failure model under stochastic variation of the shape and scale parameters; *Metron* **34**, 201–217

TSOKOS, C.P. / SHIMI, I.N. (eds.) (1977): *The Theory and Applications of Reliability (With Emphasis on Bayesian and Nonparametric Methods),* Vol. I and II; Academic Press, New York

TSUMOTO, M. / IWATA, M. (1975): An application of Weibull distribution to impulse breakdown of cross–linked polyethylene power cable; *Fujikura Technical Review* **7**, 19–22

TSUMOTO, M. / OKIAI, R. (1974): A new application of Weibull distribution to impulse breakdown of oil–filled cable; *IEEE Transactions Power Appar. and Syst.* **93**, 360–366

TULLER, S.E. / BRETT, A.C. (1984): The characteristics of wind velocity that favor the fitting of a Weibull distribution in wind speed analysis; *Journal of Climate and Applied Meteorology* **23**, 124–134

TYURIN, V.P. (1975): Nomograms for determining the reliability indices for a Weibull distribution; *Zavodstaya Laboratoriya* **41**, 585–589

TZIAFETAS, G.N. (1987): On the construction of Bayesian prediction limits for the Weibull distribution; *Mathematische Operationsforschung und Statistik, Series Statistics* **18**, 623–628

## U-U-U-U-U

UMBACH, D. / ALI, M. (1996): Conservative spacings for the extreme value and Weibull distributions; *Calcutta Statistical Association Bulletin* **46**, 169–180

UNGERER, A. (1980): Time series for the capital stock in different economic sectors of Baden–Württemberg (in German); *Jahrbuch für Statistik und Landeskunde von Baden-Württemberg* **25**, 23–24

USHER, J.S. (1996): Weibull component reliability–prediction in the presence of masked data; *IEEE Transactions on Reliability* **45**, 229–232

## V-V-V-V-V

VANCE, L.C. (1979): Estimating parameters of a Weibull distribution; *American Society for Quality Control, 33rd Annual Technical Conference Transactions,* 113–118

VAN DER AUWERA, L. / DE MEYER, F. / MALET, L.M. (1980): The use of the Weibull three-parameter model for estimating mean wind power densities; *Journal of Applied Meteorology* **19**, 819–825

VAN DER WIEL, S.A. / MEEKER, W.Q. JR. (1990): Accuracy of approx confidence bounds using censored Weibull regression data from accelerated life tests; *IEEE Transactions on Reliability* **39**, 346–351

VAN MONTFORT, M.A.J. (1970): On testing that the distribution of extremes is of type I when type II is the alternative; *Journal of Hydrology* **11**, 421–427

VAN WAGNER, F.R. (1966): Optimum Weibull sequential sampling plans; *Annual Technical Conference Transactions of the American Society of Quality Control,* 704–72

VAN ZWET, W.R. (1979): Mean, median and mode II; *Statistica Neerlandica* **33**, 1–33

VASUDEVA RAO, A. / DATTATRYA RAO, A.V. / NARASIMHAM, V.L. (1991a): Optimum linear unbiased estimation of the scale parameter by absolute values of order statistics in double exponential and the double Weibull distribution; *Communications in Statistics — Simulation and Computation* **20**, 1139–1158

VASUDEVA RAO, A. / DATTATRYA RAO, A.V. / NARASIMHAM, V.L. (1991b): Asymptotic relative efficiencies of the ML estimates of the Weibull parameters in grouped type I right censored samples; *Reports of Statistical Application Research, Union of Japanese Scientists and Engineers* **38**, 1–12

VENKATA RAMANAIAH, M. / RAMARAGHAVA REDDY, A. / BALASIDDAMUM, P. (2002): A maximum likelihood estimation in Weibull distribution based on censored and truncated samples; *Assam Statistical Review* **16**, 97–106

VIERTL, R. (1981): On the dependence of Polyoxymethylen roller bearing life time on pressure (in German); *Qualität und Zuverlässigkeit* **26**, 65–68

VIVEROS, R. / BALAKRISHNAN, N. (1994): Interval estimation of parameters of life from progressively censored data; *Technometrics* **36**, 84–91

VODA, V.G. (1978): Concerning an application of the Weibull distribution to reliability aspects in meteorology; *Metrologia Applicata* **25**, 123–124

VODA, V.G. (1989): New models in durability tool–testing: Pseudo–Weibull distribution; *Kybernetika* **25**, 209–215

VOGT, H. (1968): Estimating parameters and percentiles of life time distributions from small sample sizes (in German); *Metrika* **14**, 117–131

VOLODIN, I.N. (1974): On the discrimination of gamma and Weibull distributions; *SIAM – Theory of Probability and its Applications* **19**, 383–390

**W-W-W-W-W**

WALLENIUS, K.T. / KORKOTSIDES, A.S. (1990): Exploratory model analysis using CDF knotting with applications to distinguishability, limiting forms, and moment ratios to the three parameter Weibull family; *Journal of Statistical Computation & Simulation* **35**, 121–133

WANG, Y.H. (1976): A functional equation and its application to the characterization of the Weibull and stable distributions; *Journal of Applied Probability* **13**, 385–391

WANG, Y. / CHAN, Y.C. / GUI, Z.L. / WEBB, D.P. / LI, L.T. (1997): Application of Weibull distribution analysis to the dielectric failure of multilayer ceramic capacitors; *Material Science and Engineering* **B 47**, 197–203

WATKINS, A.J. (1994): Review: Likelihood method for fitting Weibull log–linear models to accelerated life test data; *IEEE Transactions on Reliability* **43**, 361–385

WATKINS, A.J. (1996): On maximum likelihood estimation for the two parameter Weibull distribution; *Microelectronics and Reliability* **43**, 361–365

WATKINS, A.J. (1998): On expectations associated with maximum likelihood estimation in the Weibull distribution; *Journal of the Italian Statistical Society* **7**, 15–26

WATSON, G.S. / WELLS, W.T. (1961): On the possibility of improving the mean useful life of items by eliminating those with short lives; *Technometrics* **3**, 281–298

WEIBULL, W. (1939a): A statistical theory of the strength of material; *Ingeniörs Vetenskaps Akademiens Handligar Report No. 151,* Stockholm

WEIBULL, W. (1939b): The phenomenon of rupture in solids; *Ingeniörs Vetenskaps Akademiens Handligar Report No. 153,* Stockholm

WEIBULL, W. (1949): A statistical representation of fatigue failures in solids; *Kungliga Tekniska Hogskolans Handligar (Royal Institute of Technology), No. 27*

WEIBULL, W. (1951): A statistical distribution function of wide applicability; *Journal of Applied Mechanics* **18**, 293–297

WEIBULL, W. (1952): Statistical design of fatigue experiments; *Journal of Applied Mechanics* **19**, 109–113

WEIBULL, W. (1959): Statistical evaluation of data from fatigue and creep–rupture tests; Part I, Fundamental concepts an general methods; *WADC TR59–400, Part I, Wright Air Development Center*, Wright–Patterson Air Force Base, Ohio

WEIBULL, W. (1961): *Fatigue Testing and Analysis of Results;* Pergamon Press, New York

WEIBULL, W. (1967a): Estimation of distribution parameters by a combination of the best linear order statistic method and maximum likelihood; *Air Force Materials Laboratory Technical Report AFML-TR-67-105*, Wright–Patterson Air Force Base, Dayton, OH.

WEIBULL, W. (1967b): Moments about smallest sample value; *Air Force Materials Laboratory Technical Report AFML-TR-67-375*, Wright–Patterson Air Force Base, Dayton, OH.

WEIBULL, W. / ODQUIST, F.K.G. (eds.) (1956): *Colloquium on Fatigue,* Stockholm May 25–27, 1955; Proceedings, Springer, Berlin

WESTBERG, U. / KLEFSJÖ, B. (1994): TTT–plotting for censored data based on piecewise exponential estimator; *Int. Jour. Reliability, Quality and Safety Engineering* **1**, 1–13

WHALEN, A.D. (1971): *Detection of Signals in Noise;* Academic Press, New York

WHITE, J.S. (1964a): Weibull renewal analysis; *Aerospace Reliability and Maintainability Conference Washington D.C.*, Society of Automotive Engineers Inc., New York, 639–649

WHITE, J.S. (1964b): Least–squares unbiased censored linear estimation for the log Weibull (extreme value) distribution; *Journal of Industrial Mathematics* **14**, 21–60

WHITE, J.S. (1965): Linear estimation for the log Weibull distribution; *General Motors Research Publication GMR–481*

WHITE, J.S. (1966): A technique for estimating Weibull percentage points; *General Motors Research Publication GMR–572*

WHITE, J.S. (1967a): Estimating reliability from the first two failures; *General Motors Research Publication GMR–669*

WHITE, J.S. (1967b): The moments of log Weibull order statistics; *General Motors Research Publication GMR–717*

WHITE, J.S. (1969): The moments of the log Weibull order statistics; *Technometrics* **11**, 373–386

WHITEHEAD, J. (1980): Fitting Cox's regression model to survival data using GLIM; *Applied Statistics* **29**, 268–275

WHITTEN, B.J. / COHEN, A.C. (1981): Percentiles and other characteristics of the four–parameter generalized gamma distribution; *Communications in Statistics — Simulation and Computation* **10**, 175–219 (Correction, *Ibid.*, **10**, 543)

WHITTEN, B.J. / COHEN, A.C. (1996): Further considerations of modified moment estimators for Weibull and lognormal parameters; *Communications in Statistics — Simulation and Computation* **25**, 749–784

WHITTENMORE, A. / ALTSCHULER, B. (1976): Lung cancer incidence in cigarette smokers: Further analysis of DOLL and HILL's data for British physicians; *Biometrica* **32**, 805–816

WILKS, D.S. (1989): Rainfall intensity, the Weibull distribution, and estimation of daily surface runoff; *Journal of Applied Meteorology* **28**, 52–58

WILLIAMS, J.S. (1978): Efficient analysis of Weibull survival data from experiments on heterogeneous patient populations; *Biometrics* **34**, 209–222

WILSON, R.B. (1965): Two notes on estimating shape parameters; *RAND Corporation Memorandum RM-4459-PR*

WINGO, D.R. (1972): Maximum likelihood estimation of the parameters of the Weibull distribution by modified quasilinearization (progressively censored samples); *IEEE Transactions on Reliability* **21**, 89–93

WINGO, D.R. (1973): Solution of the three–parameter Weibull equations by constrained quasilinearization (progressively censored samples); *IEEE Transactions on Reliability* **22**, 96–102

WINGO, D.R. (1988): Methods for fitting the right–truncated Weibull distribution to life–test and survival data; *Biometrical Journal* **30**, 545–551

WINGO, D.R. (1989): The left–truncated Weibull distribution: Theory and computation; *Statistische Hefte* **30**, 39–48

WINGO, D.R. (1998): Parameter estimation for a doubly truncated Weibull distribution; *Microelectronics and Reliability* **38**, 613–617

WOLSTENHOLME, L.C. (1996): An alternative to the Weibull distribution; *Communications in Statistics — Simulation and Computation* **25**, 119–137

WONG, P.G. / WONG, S.P. (1982): A curtailed test for the shape parameters of the Weibull distribution; *Metrika* **29**, 203–209

WONG, R.K.W. (1977): Weibull distribution, iterative likelihood techniques and hydrometeorological data; *Journal of Applied Meteorology* **16**, 1360–1364

WOO, J. / LEE, K.H. (1983): Estimation of the scale parameter in the Weibull distribution based on quasi–range; *Journal of the Korean Statistical Society* **12**, 69–80

WOODWARD, W.A. / FRAWLEY, W.H. (1980): One–sided tolerance limits for a broad class of lifetime distributions; *Journal of Quality Technology* **12**, 130–137

WOODWARD, W.A. / GUNST, R.F. (1987): Using mixtures of Weibull distributions to estimate mixing proportions; *Computational Statistics & Data Analysis* **5**, 163–176

WOZNIAK, P.J. (1994): Power of goodness of fit tests for the two–parameter Weibull distribution with estimated parameters; *Journal of Statistical Computation & Simulation* **50**, 153–161

WRIGHT, W.P. / SINGH, N. (1981): A prediction interval in life testing: Weibull distribution; *IEEE Transactions on Reliability* **30**, 466–467

WU, J.W. / LEE, W.C. (2003): An EOQ inventory model for items with Weibull deterioration shortages and time varying demand; *Journal of Information & Optimization Sciences* **24**, 103–122

WU, J.W. / TSAI, T.R. / OUYANG, L.Y. (2001): Limited failure–censored life test for the Weibull distribution; *IEEE Transactions on Reliability* **50**, 107–111

WU, S.J. (2002): Estimation of the parameters of the Weibull distribution with progressively censored data; *Journal of the Japan Statistical Society* **32**, 155–163

WYCKOFF, J. / BAIN, L.J. / ENGELHARDT, M. (1980): Some complete and censored sampling results for the three–parameter Weibull distribution; *Journal of Statistical Computation & Simulation* **11**, 139–151

## X-X-X-X-X

XIE, M. / GOH, T.N. / TANG, Y. (2002): A modified Weibull extension with bathtub–shaped failure rate function; *Reliability Engineering and Systems Safety* **76**, 279–285

XU, S. / BARR, S. (1995): Probabilistic model of fracture in concrete and some effects on fracture toughness; *Magazine of Concrete Research* **47**, 311–320

## Y-Y-Y-Y-Y

YAMADA, S. / HISHITANI, J. / OSAKI, S. (1993): Software–reliability growth with a Weibull test–effort: A model and application; *IEEE Transactions on Reliability* **42**, 100–106

YANG, L. / ENGLISH, J.R. / LANDERS, T.L. (1995): Modelling latent and patent failures of electronic products; *Microelectronics and Reliability* **35**, 1501–1510

YANG, Z. / SEE, S.P. / XIE, M. S.P. (2003): Transformation approaches for the construction of Weibull prediction interval; *Computational Statistics & Data Analysis* **43**, 357–368

YANNAROS, N. (1994): Weibull renewal processes; *Annals of the Institute of Statistical Mathematics* **46**, 641–648

YAZHOU, J. / WANG, M. / ZHIXIN, J. (1995): Probability distribution of machining center failures; *Engineering and System Safety* **50**, 121–125

YE, C. (1996): Estimation of parameters for a dependent bivariate Weibull distribution; *Chinese Journal of Applied Probability and Statistics* **12**, 195–199

YILDIRIM, F. (1990): Least squares estimation of the Weibull parameters; *Pakistan Journal of Statistics, Series A* **6**, 93–104

YILDIRIM, F. (1996): Monte Carlo comparison of parameter estimates of the three–parameter Weibull distribution; *Journal of Statistical Research* **30**, 121–126

YUEN, H.-K. / TSE, S.-K. (1996): Parameters estimation for Weibull distributed lifetimes under progressive censoring with random removals; *Journal of Statistical Computation & Simulation* **55**, 57–71

YUEN FUNG, K. / PAUL, S.R. (1985): Comparisons of outlier detection procedures in Weibull or extreme–value distribution; *Communications in Statistics — Simulation and Computations* **14**, 895–919

**Z-Z-Z-Z-Z**

ZACKS, S. (1984): Estimating the shift to wear–out of systems having exponential–Weibull life distributions; *Operations Research* **32**, 741–749

ZACKS, S. / FENSKE, W.J. (1973): Sequential determination of inspection epochs for reliability systems with general lifetime distributions; *Naval Research Logistics Quarterly* **20**, 377–386

ZAHER, A.M. / ISMAIL, M.A. / BAHAA, M.S. (1996): Bayesian type I censored designs for the Weibull lifetime model: Information based criterion; *The Egyptian Statistical Journal* **40**, 127–150

ZANAKIS, S.H. (1976): Least square estimation of the three-parameter Weibull distribution via pattern search with transformations; *TIMS/Orsa Bull.*, **1**, 164

ZANAKIS, S.H. (1977): Computational experience with some nonlinear optimization algorithms in deriving maximum likelihood estimate for the three–parameter Weibull distribution; *TIMS Studies in Management Sciences* **7**, 63–77

ZANAKIS, S.H. (1979): A simulation study of some simple estimators for the three–parameter Weibull distribution; *Journal of Statistical Computation & Simulation* **9**, 101–116

ZANAKIS, S.H. / KYPARISIS, J. (1986): A review of maximum likelihood estimation methods for the three–parameter Weibull distribution; *Journal of Statistical Computation & Simulation* **25**, 53–73

ZANAKIS, S.H. / MANN, N.R. (1982): A good simple percentile estimator of the Weibull shape parameter for use when all three parameters are unknown; *Naval Research Logistics Quarterly* **29**, 419–428

ZELEN, M. (1963): *Statistical Theory of Reliability;* University of Wisconsin Press, Madison

ZELEN, M. / DANNEMILLER, M.C. (1961): The robustness of life testing procedures derived from the exponential distribution; *Technometrics* **3**, 29–49

ZHANG, L. / CHEN, G. (2004): EWMA charts for monitoring the mean of censored Weibull life-times; *Journal of Quality Technology* **36**, 321–328

ZHANG, Y. (1982): Plotting positions of annual flood extremes considering extraordinary values; *Water Resources Research* **18**, 859–864

ZHONG, G. (2001): A characterization of the factorization of hazard function by the Fisher information under type II censoring with application to the Weibull family; *Statistics & Probability Letters* **52**, 249–253

ZUO, M.J. / JIANG, R. / YAM, R.C.M. (1999): Approaches for reliability modelling of continuous–state devices; *IEEE Transactions on Reliability* **48**, 9–18

ZUR NIEDEN, A. / STEINERT, M. (1983): Steps to balance reliability and durability (in German); *Qualität und Zuverlässigkeit* **28**, 2–6

# Author index

# Subject index